The Chemistry and Technology of Coal
Third Edition

CHEMICAL INDUSTRIES
A Series of Reference Books and Textbooks

Founding Editor

HEINZ HEINEMANN
Berkeley, California

Series Editor

JAMES G. SPEIGHT
CD & W, Inc.
Laramie, Wyoming

MOST RECENTLY PUBLISHED

The Chemistry and Technology of Coal, Third Edition, James G. Speight

Practical Handbook on Biodiesel Production and Properties, Mushtaq Ahmad, Mir Ajab Khan, Muhammad Zafar, and Shazia Sultana

Introduction to Process Control, Second Edition, Jose A. Romagnoli and Ahmet Palazoglu

Fundamentals of Petroleum and Petrochemical Engineering, Uttam Ray Chaudhuri

Advances in Fluid Catalytic Cracking: Testing, Characterization, and Environmental Regulations, edited by Mario L. Occelli

Advances in Fischer-Tropsch Synthesis, Catalysts, and Catalysis, edited by Burton H. Davis and Mario L. Occelli

Transport Phenomena Fundamentals, Second Edition, Joel Plawsky

Asphaltenes: Chemical Transformation during Hydroprocessing of Heavy Oils, Jorge Ancheyta, Fernando Trejo, and Mohan Singh Rana

Chemical Reaction Engineering and Reactor Technology, Tapio O. Salmi, Jyri-Pekka Mikkola, and Johan P. Warna

Lubricant Additives: Chemistry and Applications, Second Edition, edited by Leslie R. Rudnick

Catalysis of Organic Reactions, edited by Michael L. Prunier

The Scientist or Engineer as an Expert Witness, James G. Speight

Process Chemistry of Petroleum Macromolecules, Irwin A. Wiehe

Interfacial Properties of Petroleum Products, Lilianna Z. Pillon

Clathrate Hydrates of Natural Gases, Third Edition, E. Dendy Sloan and Carolyn Koh

Chemical Process Performance Evaluation, Ali Cinar, Ahmet Palazoglu, and Ferhan Kayihan

Hydroprocessing of Heavy Oils and Residua, edited by James G. Speight and Jorge Ancheyta

Process Chemistry of Lubricant Base Stocks, Thomas R. Lynch

The Chemistry and Technology of Coal

Third Edition

James G. Speight

CRC Press
Taylor & Francis Group
Boca Raton London New York

CRC Press is an imprint of the
Taylor & Francis Group, an **informa** business

CRC Press
Taylor & Francis Group
6000 Broken Sound Parkway NW, Suite 300
Boca Raton, FL 33487-2742

First issued in paperback 2016

© 2013 by Taylor & Francis Group, LLC
CRC Press is an imprint of Taylor & Francis Group, an Informa business

No claim to original U.S. Government works

ISBN 13: 978-1-138-19922-4 (pbk)
ISBN 13: 978-1-4398-3646-0 (hbk)

Visit the Taylor & Francis Web site at
http://www.taylorandfrancis.com

and the CRC Press Web site at
http://www.crcpress.com

Contents

PART I *Character and Properties*

PART II   Technology and Utilization

Preface to the Third Edition

The success of the first and second editions of this book has been the primary factor in the decision to publish a third edition.

In addition, the demand for coal products, particularly liquid fuels (gasoline and diesel fuel) and chemical feedstocks (such as aromatics and olefins), is increasing throughout the world. Traditional markets, such as North America and Europe, are experiencing a steady increase in demand, whereas emerging Asian markets, such as India and China, are witnessing a rapid surge in demand for liquid fuels. This has resulted in a tendency for existing refineries to seek fresh refining approaches to optimize efficiency and throughput. In addition, the increasing use of the heavier feedstocks for refineries is forcing technology suppliers/licensors to revamp their refining technologies in an effort to cater to the growing customer base.

Coal is again being seriously considered as an alternate source of fuel to petroleum, not so much for the production of liquids but more for the generation of power and as a source of chemicals. The reaffirmation by the United States to a clean environment through the passages of the various environmental laws as well as serious consideration to the effects of combustion products (carbon dioxide) on the atmosphere are all given consideration in this new book.

As in the second edition, this edition has references cited throughout the book so that the reader might use them for more detail. However, it should be noted that only selected references could be employed, lest the reference lists outweigh the actual book. It would have been impossible to include all of the relevant references. Thus, where possible, references such as review articles, other books, and those technical articles with substantial introductory material have been used in order to pass on the most information to the reader.

The reader may also be surprised at the number of older references that are included. The purpose of this is to remind the reader that there is much valuable work cited in the older literature—work that is still of value, and even though in some cases there has been similar work performed with advanced equipment, the older work has stood the test of time. This is particularly true of some of the older concepts of the chemical and physical structure of coal. Many of the ideas are still pertinent, and we should not forget these valuable contributions to coal science and technology.

It is the purpose of this book to provide the reader with a detailed overview of the chemistry and technology of coal as it evolves into the twenty-first century. With this in mind, many of the chapters that appear in the third edition have been rewritten to include the latest developments in the refining industry. Updates on the evolving processes and new processes as well as the various environmental regulations are presented. As part of this update, many chapters contain a section listing the relevant patents that have been issued as the industry evolves. However, the book still maintains its initial premise, that is, to introduce the reader to the science of coal, beginning with its formation in the ground and eventually leading to the production of a wide variety of products and petrochemical intermediates. However, the book will also prove useful for those scientists and engineers already engaged in the coal industry as well as in the catalyst manufacturing industry who wish to gain a general overview or update of the science of coal.

As always, I am indebted to my colleagues in many different countries who have continued to engage me in lively discussions and who have offered many thought-provoking comments. Thanks are also due to those colleagues who have made constructive comments on the previous editions, which were of great assistance in writing this edition. For such discussions and commentary, I continue to be grateful.

I am particularly indebted to those colleagues who have contacted me from time to time to ask whether I would change anything fundamental in the still-popular second edition of this book.

Preparing this updated and revised third edition gave me that chance. Since the first publication of this book, researchers have made advances in areas relating to the use of coal and the environmental aspects of coal use. However—and there are those who will sorely disagree with me—very little progress has been made on the so-called average structure of coal because coal does *not* have an average structure. As a result, the sections relating to the use of coal have been expanded in this book, but the sections relating to the average structure of coal, while being acknowledged as an area of study, have not been expanded to any great extent.

The book has been adjusted, polished, and improved for the benefit of new readers as well as for the benefit of readers of the two previous editions.

Finally, my sincerest thanks are due to Sharida Hassanali, who provided valuable assistance in typing and formatting the original manuscript.

Dr. James G. Speight
Laramie, Wyoming

Preface to the First Edition

There are no documented records of when or how mankind first discovered that a certain "black rock" would burn. However, it is known that coal was employed as a fuel in China circa 1100 BC and that Welsh Bronze Age cultures had used coal for funeral pyres. There are many other instances where coal receives some mention in the historical literature, but the consistent use of coal is believed to have evolved in England in the Middle Ages, later becoming the prime force behind the Industrial Revolution. From that point until the early decades of the twentieth century, coal has emerged as a major source of energy. However, the subsequent emergence of petroleum as a plentiful and cheap source of energy led to the "demotion" of coal to a "mere" source of combustible energy (into which the use of petroleum was steadily making inroads). Nevertheless, recent energy crises have served to emphasize that petroleum would no longer remain the cheap commodity to which mankind has grown accustomed. Indeed, assessments of the availability of petroleum have indicated that supplies of the more conventional crude oils could be virtually exhausted within the foreseeable future. This, coupled with the drastic increases in the price of the available petroleum, has caused a major shift in the emphasis of energy policies. As a result, there has been an "outburst" of serious attempts to produce liquid fuels from the so-called unconventional sources such as coal, oil sands (often referred to as tar sands or bituminous sands), and oil shale. Indeed, this re-emphasis of the value of these unconventional liquid fuel sources has helped reinstate coal to its once-enjoyed popularity and, perhaps, to a leading (even unique) position of being a major source of energy. In fact, power generation—once the sole domain of coal but since intruded upon by the shift to oil-fired generating plants—may also be returned to coal as the predominant source of combustible energy. Thus, it came about during the winter of 1976–1977 that the author was instrumental in initiating (with a colleague, John F. Fryer) a teaching course relating to the chemistry and technology of coal that ran parallel with the course relating to the chemistry and technology of petroleum, both of which were offered through the Faculty of Extension at the University of Alberta. The courses ran for several years and were also offered in the shortened multiday format through the Faculty of Continuing Education at the University of Calgary. This book is the result of the copious notes collected and employed for the course and is intended to be a companion volume to *The Chemistry and Technology of Petroleum* (James G. Speight, Marcel Dekker, Inc., 1980). The book introduces the reader to the science of coal, beginning with the formation of coal in the ground and progresses through the various chemical and analytical aspects of coal science to the established and proposed processes for the production of a variety of gaseous and liquid fuels. These latter aspects of coal technology are actually quite complex insofar as the technology is still evolving. Thus, processes that were of major interest at, say, manuscript preparation may, at the time of manuscript publication, no longer be in contention as a serious process option.

There is no satisfactory method by which such changes in technology and process planning can be satisfied, and the only way to inform the reader of the various process options is to present them in outline with the cautious corollary "here today—gone tomorrow." In more general terms, the book is written, like *The Chemistry and Technology of Petroleum*, as a teaching book from which the reader can gain a broad overview (with some degree of detail) of the concepts involved in coal science and technology. The book will, therefore, satisfy those who are just entering into this fascinating aspect of science and engineering as well as those (scientists and engineers) who are already working with coal but whose work is so specific that they also require a general overview. It will also be of assistance to petroleum refinery personnel who may, one day, be called upon to handle large supplies of liquid fuels from coal as a feedstock for the refinery system. For those readers who may require more detail in certain of the subject areas, bibliographies have been appended to each chapter. These will either directly provide the reader with the desired detail or will provide a compilation of literature (references for further consultation). The nature of the subject virtually dictates that any book on coal must include some chemistry (and the present book is not delinquent in this respect), but attempts have been made for the benefit of those readers without any formal post–high school training in chemistry (and who may, therefore, find chemistry lacking in any form of inspiration) to maintain the chemical sections in the simplest possible form. Wherever possible, simple chemical formulas have been employed to illustrate the book. However, for the reader with an in-depth knowledge of organic chemistry, a chapter that describes in detail the organic structures found in coal has been included.

The book contains both the metric and nonmetric measures of temperature (Celsius and Fahrenheit). However, it should be noted that exact conversion of the two scales is not often possible and, accordingly, the two temperature scales are interconverted to the nearest 5. At the high temperatures often quoted in the process sections, serious error will not arise from such a conversion. With regard to the remaining metric/nonmetric scales of measurement, there are also attempts to indicate the alternate scales. For the sake of simplicity and clarity, simple illustrations (often line drawings) are employed for the various process options, remembering, of course, that a line between two reactors may not only be a transfer pipe but also a myriad of valves and control equipment.

Dr. James G. Speight
Laramie, Wyoming

Author

Dr. James G. Speight is a senior fuel and environmental consultant with more than 45 years of experience in thermal/process chemistry, thermodynamics, refining of petroleum, heavy oil, and tar sand bitumen, and physics of crude with emphasis on distillation, visbreaking, coking units, and oil–rock or oil catalyst interactions. He received his BSc and PhD from the University of Manchester, England. He is the author of more than 45 books in petroleum science, petroleum engineering, and environmental sciences and has considerable expertise in evaluating new technologies for patentability and commercial application.

Although his career was focused predominantly in the commercial world, Dr. Speight has served as adjunct professor in the Department of Chemical and Fuels Engineering at the University of Utah and in the Departments of Chemistry and Chemical and Petroleum Engineering at the University of Wyoming. In addition, he was a visiting professor in chemical engineering at the following universities: the University of Missouri–Columbia, the Technical University of Denmark, and the University of Trinidad and Tobago.

Dr. Speight is recognized internationally as an expert in the characterization, properties, and processing of conventional and synthetic fuels. He is currently editor in chief of the *Journal of Petroleum Science and Technology* (Taylor & Francis Publishers), *Energy Sources—Part A: Recovery, Utilization, and Environmental Effects* (Taylor & Francis Publishers), *Energy Sources—Part B: Economics, Planning, and Policy* (Taylor & Francis Publishers), and *Journal of Sustainable Energy Engineering* (Scrivener publishing).

As a result of his work, he was awarded the Diploma of Honor, National Petroleum Engineering Society, for outstanding contributions to the petroleum industry in 1995 and the Gold Medal of the Russian Academy of Sciences (Natural) for outstanding work in the area of petroleum science in 1996. He also received the Specialist Invitation Program Speakers Award from NEDO (New Energy Development Organization, Government of Japan) in 1987 and again in 1996 for his contributions to coal research. In 2001, he was awarded the Einstein Medal of the Russian Academy of Sciences in recognition of outstanding contributions and service in the field of geologic sciences. In 2005, he was awarded the Gold Medal—Scientists without Frontiers, Russian Academy of Sciences, in recognition of continuous encouragement of scientists to work together across international borders. More recently (2012), he was awarded a doctorate in petroleum engineering from Dubna University (Moscow, Russia) for his outstanding contributions in the field of petroleum engineering.

Common Conversion Factors Used in Coal Technology

To Convert	To	Multiply by
Acres	Hectares	0.4047
Acres	Square feet	43,560
Acres	Square miles	0.001562
Acre feet	Barrels	7,758.0
Atmosphere	Centimeters of mercury	76
Atmosphere	Torr	760
Atmosphere	mm Hg	760
Atmosphere	Psia	14.686
Atmosphere	Inches Hg	29.91
Atmosphere	Bar	1.0133
Atmosphere	Feet H_2O	33.899
Barrel (oil)	U.S. gallons	42
Barrel	Cubic feet	5.6146
Barrel	Lbs water at 60°F	350
Barrel per day	Cubic centimeters/second	1.84
Btu	Foot pounds	778.26
Btu/lb	kcal/kg	1.8
Btu/lb	kJ/kg	2.33
Btu/hour	Horsepower	0.0003930
Btu	Kilowatt-hour	0.0002931
Btu/hour	Watts	0.2931
Centimeters	Inches	0.3937
Centimeters	Feet	0.03281
Cubic foot	Cubic meters	0.0283
Cubic foot	Cubic centimeters	28,317
Cubic foot	Gallons	7.4805
Cubic meters	Cubic feet	35.3145
Cubic meters	Cubic yards	1.3079
Cubic yards	Cubic meters	0.7646
Density of water at 60°F	Gram/cubic centimeter	0.999
Density of water at 60°F	Lb/cu ft	62.367
Density of water at 60°F	Lb/U.S. gallon	8.337
Feet	Meters	0.3048
Feet	Miles (nautical)	0.0001645
Feet	Miles (statute)	0.0001894
Gallons (U.S.)	Liters	3.7853
Gallon	Cubic inches	231
Gallon	Cubic centimeters	3,785.4
Gallon	Cubic feet	0.13368
Grams	Ounces (avoirdupois)	0.0353
Grams	Pounds	0.002205
Hectares	Acres	2.4710

(continued)

(continued)

To Convert	To	Multiply by
Inches	Millimeters	25.4000
Inches	Centimeters	2.5400
Kilograms	Pounds (avoirdupois)	2.2046
Kilograms	Pounds (troy)	2.679
Kilometers	Miles	0.6214
Kilowatt-hour	Btu	3412
Liters	Gallons (U.S.)	0.2642
Liters	Pints (dry)	1.8162
Liters	Pints (liquid)	2.1134
Liters	Quarts (dry)	0.9081
Liters	Quarts (liquid)	1.0567
Meters	Feet	3.2808
Meters	Miles	0.0006214
Meters	Yards	1.0936
Metric tons	Tons (long)	0.9842
Metric tons	Tons (short)	1.1023
Miles	Kilometers	1.6093
Miles	Feet	5,280
Miles (nautical)	Miles (statute)	1.1516
Miles (statute)	Miles (nautical)	0.8684
Millimeters	Inches	0.0394
Ounces (avoirdupois)	Grams	28.3495
Ounces (avoirdupois)	Pounds	0.0625
Ounces (liquid)	Pints (liquid)	0.0625
Ounces (liquid)	Quarts (liquid)	0.03125
Ounces (troy)	Ounces (avoirdupois)	1.09714
Pints (dry)	Liters	0.5506
Pints (liquid)	Liters	0.4732
Pints (liquid)	Ounces (liquid)	16
Pounds (troy)	Kilograms	0.3782
Pounds (avoirdupois)	Kilograms	0.4536
Pound	Grams	453.59
Pound	Ounces	16
1 psi	kPa	6.895
Quarts (dry)	Liters	1.1012
Quarts (liquid)	Liters	0.9463
Quarts (liquid)	Ounces (liquid)	32
Square feet	Square meters	0.0929
Square kilometers	Square miles	0.3861
Square meters	Square feet	10.7639
Square meters	Square yards	1.1960
Square miles	Square kilometers	2.5900
Square mile	Acres	640
Square yards	Square meters	0.8361
Tons (long)	Metric tons	1.016
Tons (short)	Metric tons	0.9072
Tons (long)	Pounds	2,240
Tons (short)	Pounds	2,000
Torr	Atmospheres	0.001316
Torr	mm Hg	1
Yards	Meters	0.9144
Yards	Miles	0.0005682

Part I

Character and Properties

Coal is an organic sedimentary rock that is formed from the accumulation and preservation of plant materials, usually in a swamp environment. Coal is a combustible rock, and along with petroleum and natural gas it is one of the three most important fossil fuels, such as for the generation of electricity.

Coal is one of the world's most important sources of energy, fuelling almost 40% of electricity worldwide. In many countries, this figure is much higher: Poland relies on coal for over 94% of its electricity, South Africa for 92%, China for 77%, and Australia for 76%.

Coal has been the world's fastest growing energy source in recent years—faster than gas, oil, nuclear, hydro, and renewables. It has played this important role for centuries—not only for providing electricity but also as an essential fuel for steel and cement production as well as other industrial activities.

The world currently consumes over 4 billion tons (4.0×10^9 ton) of coal. Coal is used by a variety of sectors, including power generation, iron and steel production, cement manufacturing, and as a liquid fuel. The majority of coal is either utilized in power generation (steam coal or lignite) or iron and steel production (coking coal).

As a result, coal will continue to play a key role in the world's energy mix, with demand in certain regions set to grow rapidly. Growth in both the steam and coking coal markets will be strongest in developing Asian countries, where demand for electricity and the need for steel in construction, car production, and demands for household appliances will increase as incomes rise.

COAL ORIGIN

Coal is a naturally occurring combustible material consisting primarily of the element carbon. It also contains low percentages of solid, liquid, and gaseous hydrocarbons and/or other materials, such as compounds of nitrogen and sulfur. It is usually classified into subgroups known as anthracite, bituminous coal, subbituminous coal, and lignite. The physical, chemical, and other properties of coal vary considerably from sample to sample.

The origins of coal lie in the realms of the distant past, but it is generally believed that coal takes its origins from dead plants and, possibly, animals. Coal has been formed at many times in the past, but most abundantly during the Carboniferous Age (about 300 million years ago) and again during the Upper Cretaceous Age (about 100 million years ago).

When plants and animals die, they normally decay and are converted to carbon dioxide, water, and other products that disappear into the environment. Other than a few bones, little remains of the dead organism. At some periods in Earth's history, however, conditions existed that made other forms of decay possible. The bodies of dead plants and animals underwent only partial decay. The products remaining from this partial decay are coal, oil, and natural gas (fossil fuels).

The initial stage of the decay of a dead plant is a soft, woody material known as peat, which, in some parts of the world, is still collected from boggy areas and used as a fuel. It is not a good fuel, however, as it burns poorly and produces a great deal of smoke.

As the processes of coalification (i.e., the transformation resulting from the increased temperatures and pressures) continue, there is a progressive transformation of the deposit: the proportion of carbon relative to oxygen rises and volatile substances and water are driven out. Continued compaction by overburden then converts lignite into bituminous (or soft) coal and finally into anthracite (or hard) coal.

Coal formed by these processes is often found layered between other layers of sedimentary rock. Sedimentary rock is formed when sand, silt, clay, and similar materials are packed together under heavy pressure. In some cases, the coal layers may lie at or very near Earth's surface. In other cases, they may be buried thousands of feet underground. Coal seams usually range from no more than 3–200 ft (1–60 m) in thickness. The location and configuration of a coal seam determines the method by which the coal will be mined.

The quality of each coal deposit is determined by temperature and pressure and by the length of time in formation, which is referred to as its organic maturity.

COAL RESOURCES

Coal is a nonrenewable resource, meaning it is not replaced easily or readily. Once a nonrenewable resource has been used up, it is gone for a very long time into the future, if not forever. Coal fits that description, since it was formed many millions of years ago but is not being formed in significant amounts any longer. Therefore, the amount of coal that now exists below Earth's surface is, for all practical purposes, all the coal available for the foreseeable future. When this supply of coal is used up, humans will find it necessary to find some other substitute to meet their energy needs.

It has been estimated that there are over 1 trillion tons (1.0×10^{12} ton) of proven coal reserves worldwide, which indicates (at current levels of use) there is enough coal to last for 200 years or more.

Most deposits of coal were formed during the Carboniferous and Permian periods. More recent periods of coal formation occurred during the early Jurassic and Tertiary periods. Coal deposits occur in all the major continents; the leading producers include the United States, China, Ukraine, Poland, the United Kingdom, South Africa, India, Australia, and Germany. China is also thought to have the world's largest estimated resources of coal, as much as 46% of all that exists. In the United States, the largest coal-producing states are Montana, North Dakota, Wyoming, Alaska, Illinois, and Colorado.

China produces the largest amount of coal each year, about 22% of the world's total followed by the United States (19%), the former members of the Soviet Union (16%), Germany (10%), and Poland (5%).

COAL CLASSIFICATION

Coal is a complex organic sedimentary rock and is classified according to its heating value and the percentage of carbon it contains.

Generally, coal is classified into two types: (1) *humic coal* (*woody coal*), which is derived from plant remains, and (2) *sapropelic coal*, which is derived from algae, spores, and finely divided plant material. However, coal is further classified according to its heating value and its relative content of elemental carbon. For example, anthracite contains the highest proportion of carbon (on the order of 86%–98% w/w) on a dry, ash-free basis and has the highest heat value of all forms of coal (13,500–15,600 Btu/lb) also on a dry, ash-free basis.

Bituminous coal generally has lower concentrations of pure carbon (46%–86%) and lower heat values (8,300–15,600 Btu/lb). Bituminous coals are often subdivided on the basis of their heat value, being classified as low, medium, and high volatile bituminous and subbituminous.

Subbituminous coal—also called black lignite—is generally dark brown to black and intermediate in rank between lignite and bituminous coal according to the coal classification used in the United States and Canada. In many countries, subbituminous coal is considered to be a brown coal. Subbituminous coal contains 42%–52% w/w and has calorific values ranging from 8,200 to 11,200 Btu/lb. Subbituminous coal is characterized by greater compaction than found in lignite as well as greater brightness and luster.

Lignite, the poorest of the true coals in terms of heat value (5500–8300 Btu/lb), generally contains about 46%–60% w/w carbon.

All forms of coal also contain other elements present in living organisms, such as sulfur, nitrogen, and oxygen, that may be low when compared to the carbon and hydrogen content but that have important environmental consequences when coal is used as a fuel or as a source of gaseous or liquid fuels.

COAL MINING, PREPARATION, AND TRANSPORTATION

Coal is extracted from Earth using one of two major methods: subsurface or surface (strip) mining. Subsurface mining is used when seams of coal are located at significant depths below Earth's surface. The first step in subsurface mining is to dig vertical tunnels into the Earth until the coal seam is reached. Horizontal tunnels are then constructed off the vertical tunnel. In many cases, the preferred way of mining coal by this method is called room-and-pillar mining. In room-and-pillar mining, vertical columns of coal (the pillars) are left in place as the coal around them is removed. The pillars hold up the ceiling of the seam, preventing it from collapsing on miners working around them. After the mine has been abandoned, however, those pillars may collapse, bringing down the ceiling of the seam and causing the collapse of land above the old mine.

Surface mining can be used when a coal seam is close enough to the Earth's surface to allow the overburden to be removed easily and inexpensively. In such cases, the first step is to strip off all of the overburden in order to reach the coal itself. The coal is then scraped out by huge power shovels, some capable of removing up to 100 m^3 at a time. Strip mining is a far safer form of coal mining for coal workers, but it presents a number of environmental problems. In most instances, an area that has been strip-mined is terribly scarred.

Coal mining generally takes place in rural areas where mining and the associated industries are usually one of, if not, the largest employers in the area. It is estimated that coal mining employs over 7 million people worldwide, 90% of whom are in developing countries.

Not only does coal mining directly employ millions worldwide, it generates income and employment in other regional industries that are dependent on coal mining. These industries provide goods and services into coal mining, such as fuel, electricity, and equipment, or are dependent on expenditure from employees of coal mines. Large-scale coal mines provide a significant source of local income in the form of wages, community programs, and inputs into production in the local economy.

However, mining and energy extraction can sometimes lead to land use conflicts and difficulties in relationships with neighbors and local communities. Many conflicts over land use can be resolved by highlighting that mining is only a temporary land use. Mine rehabilitation means that the land can be used once again for other purposes after mine closure.

Coal straight from the ground, known as run-of-mine (ROM) coal, often contains unwanted impurities such as rock and dirt and comes in a mixture of different-sized fragments. However, coal users need coal of a consistent quality. Coal preparation—also known as coal beneficiation or coal washing—refers to the treatment of ROM coal to ensure a consistent quality and to enhance its suitability for particular end uses.

The treatment depends on the properties of the coal and its intended use. It may require only simple crushing or it may need to go through a series of treatment processes to remove impurities.

To remove impurities, the raw ROM coal is crushed and then separated into various size fractions. Larger material is usually treated using *dense medium separation* (*heavy medium separation*) in which the coal is separated from other impurities by being floated in a tank containing a liquid of specific gravity, usually a suspension of finely ground magnetite. As the coal is lighter, it floats and can be separated off, while heavier rock and other impurities sink and are removed as waste.

The way that coal is transported to where it will be used depends on the distance to be covered. Coal is generally transported by conveyor or truck over short distances. Trains and barges are used for longer distances within domestic markets, or alternatively coal can be mixed with water to form a coal slurry and transported through a pipeline.

Ships are commonly used for international transportation, in sizes ranging from Handymax (40,000–60,000 DWT) and Panamax (approximately 60,000–80,000 DWT) to large Capesize vessels (>80,000 DWT). Approximately 700–800 million tons of coal is traded internationally annually and as much as 90% can be seaborne trade.

In all aspects of coal transportation, the costs can be very expensive—in some instances, it accounts for up to 70% of the delivered cost of coal.

COAL PROPERTIES

The properties of coal are affected by the composition of the coal. Furthermore, the natural constituents of coal can be divided into two groups: (1) the organic fraction, which can be further subdivided into microscopically identifiable macerals; and (2) the inorganic fraction, which is commonly identified as ash subsequent to combustion. The rank (thermal maturity) of the organic fraction of the coal is determined by the burial depth (pressure) and temperature. The composition of organic (non-mineral) fraction changes with rank with the main indicators of rank bringing the reflectance of the vitrinite, carbon, and volatile matter content on a dry ash-free basis.

Coal is composed of microscopically recognizable constituents, called macerals, which differ from one another in form and reflectance. Three principal maceral groups are identified and these are, in increasing order of carbon content, liptinite (exinite), vitrinite, and inertinite. Macerals and maceral groups differ in their chemical composition and thus their technical performance characteristics. The macerals of the liptinite group contain more hydrogen and are more generally reactive than the macerals of the inertinite group, while vitrinite group macerals range between the two.

In a single coal, vitrinite, which is usually the commonest maceral, has a higher reflectance than the associated liptinite, but a lower reflectance than inertinite. There is, therefore, a correlation between carbon content and reflectance, and this is used to precisely determine rank. The mean maximum reflectance of vitrinite in oil (Romax) as the level of organic maturity, or rank, of a coal sample.

The rank and proportions of liptinite, vitrinite, and inertinite do dictate the behavior of a coal during heating whether in a coke oven or a combustion flame. But other properties of the coal also have a significant influence, such as the following:

- The distribution of the inertinite and liptinite within vitrinite of the coal particle—a particle composed of only a pure maceral or distinctive gains of macerals will have different plastic properties (and therefore coke/char morphology) than a particle where the inertinite and/or liptinite is distributed throughout the vitrinite.
- The mineral matter can play a catalytic role, which will influence the decomposition of the coal and can also influence the reactivity of the resulting char or coke.

By far the most important properties of coal are the properties that relate to combustion. When coal is combusted, the two predominant products are carbon dioxide and water, with lesser amounts of sulfur oxides, nitrogen oxides, and unburned hydrocarbons. During this chemical reaction, a relatively large amount of heat energy is released. For this reason, coal has long been used by humans as a source of energy for heating homes and other buildings, running ships and trains, and in many industrial processes.

When coal is heated, it passes through a plastic stage, which is an indication of the initial softening, chemical reaction, gas liberation, and resolidification process within the coke oven and is an important requirement in the coke blend required for end product coke strength. In addition, the fluidity of the plastic stage is a major factor in determining what proportion of coal is used in a blend.

During the heating of coal, an unstable intermediate phase, called metaplast, is formed after the moisture is driven from the coal. The metaplast is responsible for the plastic behavior of coal. On further heating, a cracking process takes place in which tar is vaporized and nonaromatic groups are split off. This cracking process is accompanied by recondensation and formation of semicoke.

When a coal/blend is coked in slot-type ovens, two principal layers of plastic coal are formed parallel to the oven walls. They are linked near the sole and the top of the charge by two secondary plastic layers forming an envelope of plastic coal. As carbonization proceeds, the plastic layers move progressively inward, eventually meeting at the oven center. It is within these plastic layers that the processes that result in particulate coal being converted into porous, fused semi-coke take place. The semi-coke undergoes further devolatization and contracts, which results in fissures in the final coke.

The only small-scale methods that have stood the test of time and have been accepted as standard plasticity tests are the crucible swelling number, Gray-King coke type, dilatation characteristics, Gieseler plasticity, and, in some countries, the Rogas index. All of these are essentially empirical in nature and many are subjective, at least to some degree.

The Gieseler test is the only one that attempts to measure the actual extent of the plasticity of fluidity attained. The Gieseler test is used to characterize coals with regard to thermoplastic behavior and is sometimes an important tool used for coal blending for commercial coke manufacture. The maximum fluidity determined by the Gieseler is very sensitive to weathering (oxidation) of the coal.

The complete combustion of carbon and hydrocarbons rarely occurs in nature. If the temperature is not high enough or sufficient oxygen is not provided to the fuel, combustion of these materials is usually incomplete. During the incomplete combustion of carbon and hydrocarbons, other products besides carbon dioxide and water are formed. These products include carbon monoxide, hydrogen, and other forms of pure carbon, such as soot.

During the combustion of coal, minor constituents are also oxidized (i.e., they burn). Sulfur is converted to sulfur dioxide and sulfur trioxide, and nitrogen compounds are converted to nitrogen oxides. The incomplete combustion of coal and these minor constituents results in a number of environmental problems. For example, soot formed during incomplete combustion may settle out of the air and deposit an unattractive coating on homes, cars, buildings, and other structures. Carbon monoxide formed during incomplete combustion is a toxic gas and may cause illness or death in humans and other animals. Oxides of sulfur and nitrogen react with water vapor in the atmosphere and then settle out in the air as acid rain. Acid rain is responsible for the destruction of certain forms of plants and animals, especially fish in the acidified waterways.

In addition to these compounds, coal often contains a small percentage of mineral matter: quartz, calcite, or perhaps clay minerals. These components do not burn readily and so become part of the ash formed during combustion. This ash then either escapes into the atmosphere or is left in the combustion vessel and must be discarded. Sometimes coal ash also contains significant amounts of lead, barium, arsenic, or other elements. Whether airborne or in bulk, coal ash can therefore be a serious environmental hazard.

Restoring the area to its original state can be a long and expensive procedure. In addition, any water that comes in contact with the exposed coal or overburden may become polluted and require treatment.

Determining coal properties is an integral part of determining the most appropriate means of utilizing the coal (Part II) utility of the coal.

Depending on the rank and plastic properties, coals are divided into the following types:

- Hard coking coals are a necessary input in the production of strong coke. They are evaluated based on the strength, yield, and size distribution of coke produced, which is dependent on rank and plastic properties of the coal. Hard coking coals trade at a premium to other coals due to their importance in producing strong coke and as they are of limited resources.
- Semi-soft coking coal or weak coking coal is used in the coke blend but results in a low coke quality with a possible increase in impurities. There is scope for interchangeability between thermal coal and semi-soft coking coal, and thus, semi-soft coking coal prices have a high correlation with thermal prices.
- Coal used for pulverized coal injection reduces the consumption of coke per ton of pig iron as it replaces coke as a source of heat and, at high injection rates, as a reductant. PCI coal tends to trade at a premium to thermal coal depending on its ability to replace coke in the blast furnace.
- Thermal coals are mostly used for electricity generation. All coals can be combusted to release useful energy by the selection of suitable technology to match the coal rank and ash content. The majority of thermal coal traded internationally is fired as pulverized and varies in rank from subbituminous to bituminous.

For many centuries, coal was burned in small stoves to produce heat in homes and factories. As the use of natural gas became widespread in the latter part of the twentieth century, coal oil and coal gas quickly became unpopular since they were somewhat smoky and foul smelling. Today, the most important use of coal, both directly and indirectly, is still as a fuel, but the largest single consumer of coal for this purpose is the electrical power industry.

The combustion of coal in power-generating plants is used to make steam, which, in turn, operates turbines and generators. For a period of more than 40 years beginning in 1940, the amount of coal used in the United States for this purpose doubled in every decade. Although coal is no longer widely used to heat homes and buildings, it is still used in industries such as paper production, cement and ceramic manufacture, iron and steel production, and chemical manufacture for heating and steam generation.

Another use for coal is in the manufacture of coke. Coke is nearly pure carbon produced when soft coal is heated in the absence of air. In most cases, 1 ton of coal will produce 0.7 ton of coke in this process. Coke is valuable in industry because it has a heat value higher than any form of natural coal. It is widely used in steelmaking and in certain chemical processes.

In addition, for the more conventional uses of coal for power generation and production of coal tar and coke (Part II), a number of processes have been developed by which solid coal can be converted to a liquid or gaseous form for use as a fuel. Conversion has a number of advantages. In a liquid or gaseous form, the fuel may be easier to transport. Also, the conversion process removes a number of impurities from the original coal (such as sulfur) that have environmental disadvantages.

In the liquefaction process, the goal is to convert coal to a petroleum-like liquid that can be used as a fuel for motor vehicles and other applications. On the one hand, both liquefaction and gasification are attractive technologies in the United States because of its very large coal resources. On the other hand, the wide availability of raw coal means that expensive new technologies have been unable to compete economically with the natural product.

In the gasification process, crushed coal is forced to react with steam and either air or pure oxygen. The coal is converted into a complex mixture of gaseous hydrocarbons with heat values ranging from 100 to 1000 Btu. One day it may be possible to construct gasification systems within a coal mine, making it much easier to remove the coal (in a gaseous form) from its original seam.

ENVIRONMENTAL ISSUES

Coal is a globally abundant and a relatively cheap fuel and will continue to play a significant role in power generation and steel production. But coal use contributes significantly to carbon dioxide emissions. In the future, technologies that enable carbon dioxide management will become increasingly important.

Because of coal's high carbon content, increasing use will exacerbate the problem of climate change unless coal plants are deployed with very high efficiency and large-scale carbon capture and sequestration (CCS) is implemented. CCS is the critical enabling technology because it allows significant reduction in carbon dioxide emissions while allowing coal to meet future energy needs.

In fact, a significant charge on carbon emissions is needed in the relatively near term to increase the economic attractiveness of new technologies that avoid carbon emissions and specifically to lead to large-scale CCS in the coming decades. Appropriate measurement, monitoring, and verification are needed in the United States over the next decade with government support. This is important for establishing public confidence for the very large-scale sequestration program anticipated in the future.

Emissions will be stabilized only through global adherence to carbon dioxide emission constraints. China and India are unlikely to adopt carbon constraints unless the United States does so and leads the way in the development of CCS technology.

Key changes must be made to the current research programs and energy policies to successfully promote CCS technologies. A wider range of technologies should be explored, and modeling and simulation of the comparative performance of integrated technology systems should be greatly enhanced.

It is only through a thorough study of the properties and behavior of coal that such an understanding will be achieved.

FUTURE

It is estimated that over the next 30–50 years, global energy demand will increase by almost 60%. Two thirds of the increase will come from developing countries—by 2030, they will account for almost half of total energy demand.

With the uncertainties in the availability of other fossil fuels (such as petroleum, heavy oil, and natural gas), coal has a role to play in satisfying this demand.

1 Occurrence and Resources

1.1 INTRODUCTION

Coal (the term is used generically throughout the book to include all types of coal) is a black or brownish-black *organic sedimentary rock* of biochemical origin, which is combustible and occurs in rock strata (coal beds, coal seams) and is composed primarily of carbon with variable proportions of hydrogen, nitrogen, oxygen, and sulfur. Coal occurs in seams or strata.

The United States has a vast supply of coal, with almost 30% of world reserves and more than 1600 billion tons (1600×10^9 tons) as remaining coal resources. The United States is also the world's second largest coal producer after China and annually produces more than twice as much coal as India, the third largest producer (Höök and Aleklett, 2009, and references cited therein).

Deposits of coal, sandstone, shale, and limestone are often found together in sequences hundreds of feet thick. This period is recognized in the United States as the Mississippian and Pennsylvanian time periods due to the significant sequences of these rocks found in those states (i.e., Mississippi and Pennsylvania). Other notable coal-bearing ages are the Cretaceous, Triassic, and Jurassic Periods. The more recently aged rocks are not as productive for some reason, but lignite and peat are common in younger deposits, but generally, the older the deposit, the better the grade (higher rank) of coal.

In terms of *coal grade*, the grade of a coal establishes its economic value for a specific end use. Grade of coal refers to the amount of mineral matter that is present in the coal and is a measure of coal quality. Sulfur content, ash fusion temperature (i.e., the temperature at which measurement the ash melts and fuses), and quantity of trace elements in coal are also used to grade coal. Although formal classification systems have not been developed around grade of coal, grade is important to the coal user.

Coal has also been considered to be a *metamorphic rock*, which is the result of heat and pressure on organic sediments such as peat. However, the discussion in favor of coal as a sedimentary rock because most sedimentary rocks undergo some heat and pressure and the association of coal with typical sedimentary rocks and the mode of formation of coal usually keep low-grade coal in the sedimentary classification system. Anthracite, on the other hand, undergoes more heat and pressure and is associated with low-grade *metamorphic rocks* such as slate and quartzite. Subducted coal may become graphite in igneous rocks or even the carbonate-rich rocks (carbonatites).

The degree of metamorphosis results in differing coal types, each of which has different quality. However, peat is not actually a rock but no longer just organic matter and is major source of energy for many nonindustrialized countries. The unconsolidated plant matter is lacking the metamorphic changes found in coal. Thus, coal is classified into four main types, depending on the amount of carbon, oxygen, and hydrogen present. The higher the carbon content, the more energy the coal contains.

Lignite (brown coal) is the least mature of the coal types and provides the least yield of energy; it is often crumbly, relatively moist, and powdery. It is the lowest rank of coal, with a heating value of 4000–8300 Btu per pound. Most lignite mined in the United States comes from Texas. Lignite is mainly used to produce electricity.

Subbituminous coal is poorly indurated and brownish in color, but more like bituminous coal than lignite. It typically contains less heating value (8,300–13,000 Btu per pound) and more moisture than bituminous coal.

Bituminous coal was formed by added heat and pressure on lignite and is the black, soft, slick rock and the most common coal used around the world. Made of many tiny layers, bituminous coal looks smooth and sometimes shiny. It is the most abundant type of coal found in the United States and has 2–3 times the heating value of lignite. Bituminous coal contains 11,000–15,500 Btu per pound. Bituminous coal is used to generate electricity and is an important fuel for the steel and iron industries.

Anthracite is usually considered to be the highest grade of coal and is actually considered to be metamorphic. Compared to other coal types, anthracite is much harder, has a glassy luster, and is denser and blacker with few impurities. It is largely used for heating domestically as it burns with little smoke. It is deep black and looks almost metallic due to its glossy surface. Like bituminous coal, anthracite coal is a big energy producer, containing nearly 15,000 Btu per pound.

With an increasing coalification (i.e., progressing from lignite to subbituminous coal to bituminous coal to anthracite), the moisture content decreases while the carbon content and the energy content both increase.

Finally, *steam coal*, which is not a specific rank of coal, is a grade of coal that falls between bituminous coal and anthracite, once widely used as a fuel for steam locomotives. In this specialized use, it is sometimes known as *sea coal* in the United States. Small steam coal (*dry small steam nuts* [DSSN]) was used as a fuel for domestic water heating. In addition, the material known as *jet* is the gem variety of coal. Jet is generally derived from anthracite and lacks a crystalline structure, so it is considered to be a mineraloid. Mineraloid are often mistaken for minerals and are sometimes classified as minerals, but lack the necessary crystalline structure to be truly classified as a mineral. Jet, being one of the products of an organic process, remains removed from full mineral status.

Coal is the world's most abundant and widely distributed fossil fuel and possibly the least understood in terms of its importance to the world's economy. Currently, about 5 billion tons are mined in more than 40 countries. In spite of considerable publicity and push toward "clean fuels," coal continues to be second only to oil in meeting the world's energy needs. Estimate indicate that in excess of 5 billion tons will be required annually by the year 2050 just to generate electricity.

Coal provinces (clustering of deposits in one area) occur in regional sedimentary structures referred to as *coal basins*. More than 2000 sedimentary, coal-bearing basins have been identified worldwide but less than a dozen contain reserves of more than 200 billion tons.

However, any estimate of the amount or the lifetime of coal reserves (as for any estimates of the reserves of any natural resource) must incorporate a number of assumptions: (1) there will be no new discoveries of coal reserves, (2) hypothetical and speculative resources will not become reserves, (3) no new mining technologies will be developed (conditional resources will not become reserves), and (4) all available coal will be burned regardless of quality.

Although the majority of mined coal continues to be consumed within the country of production, the value of traded coal is increasing. The United States and Australia account for about 50% of world coal exports. This figure increases to 70% if exports from South Africa and Indonesia are included. Japan is the largest recipient of exported coal—approximately 25% of the world coal trade—and as such, agreements with coal suppliers and Japan can have a great influence on the world coal price. Japan, Taiwan, and South Korea together import about 45% of all coal exports and countries of the European Union accounts for another 30% of the total coal exports.

Coal is burned to produce energy and is used to manufacture steel. It is also an important source of chemicals used to make pharmaceuticals, fertilizers, pesticides, and other products. In fact, coal use is concentrated on power utility markets and the steel industry and has had a long history of being a primary energy source and still accounts for significantly larger reserve base than either petroleum or natural gas (Speight, 2007a,b). As an example, in the United States coal still accounts for over 50% of the domestic electricity generating industry requirements, all from domestic production. The European Union, on the other hand, must import approximately 50% of its energy requirements (in the form of oil, gas, uranium, and coal).

When burned, coal generates energy in the form of heat. In a power plant that uses coal as fuel, this heat converts water into steam, which is pressurized to spin the shaft of a turbine. This rotating shaft drives a generator that converts the mechanical energy of the rotation into electric power (Chapters 14 and 15).

Production of steel accounts for the second largest use of coal. Minor uses include cement manufacture, the pulp and paper industry, and production of a wide range of other products (such as coal tar and coal chemicals). The steel industry uses coal by first heating it and converting it into coke, a hard substance consisting of nearly pure carbon (Chapters 13, 16, and 17). The coke is combined with iron ore and limestone, and then the mixture is heated to produce iron. Other industries use different coal gases emitted during the coke-forming process to make fertilizers, solvents, pharmaceuticals, pesticides, and other products (Chapter 24).

Fuel companies convert coal into gaseous products or liquid products. Coal-based gaseous fuels are produced through the process of *gasification* (Chapters 20 and 21). In the gasification process, coal is heated in the presence of steam and oxygen to produce *synthesis gas*, a mixture of carbon monoxide, hydrogen, and methane used directly as fuel or refined into cleaner-burning gas.

Liquefaction processes convert coal into a liquid fuel that has a composition similar to that of crude petroleum (Chapters 18 and 19) (Speight, 2007a, 2008). In general, coal is liquefied by breaking hydrocarbon molecules into smaller molecules. Coal contains more carbon than hydrogen, so hydrogen must be added—directly as hydrogen gas or indirectly as water—to bond with the carbon chain fragments.

Through a process known as in situ *gasification* (Chapters 20 and 21), coal beds can be converted to gaseous products underground. To do this, the coal is ignited, air and steam are pumped into the burning seam, and the resulting gases are pumped to the surface.

As with industrial minerals, the physical and chemical properties of coal beds are as important in marketing a deposit as the *grade*.

The grade of a coal establishes its economic value for a specific end use. Grade of coal refers to the amount of mineral matter that is present in the coal and is a measure of coal quality. Sulfur content, ash fusion temperatures, that is, measurement of the behavior of ash at high temperatures, and quantity of trace elements in coal are also used to grade coal. Although formal classification systems have not been developed around grade of coal, grade is important to the coal user.

The simple percentage of carbon (gases and sulfur) is not sufficient to determine what an individual coal seam might be used for. The differences in the kinds of plant materials (called maceral) comprising the bed of coal (referred to as type), the degree of metamorphism (rank), and the range of impurities (grade) are used in classifying coal deposits. Coal rank is particularly important as the amount of fixed carbon, percentage of volatiles, and moisture in coal provide a measure of the Btus of energy that are available.

Production of coal is both by underground and open pit mining. Surface, large-scale coal operations are a relatively recent development, commencing as late as the 1970s. Underground mining of coal seams presents many of the same problems as mining of other bedded mineral deposits, together with some problems unique to coal. Current general mining practices include coal seams that are contained in beds thicker than 27 in. and at depths less than 1000 ft. Approximately 90% of all known coal seams fall outside of these dimensions and are, therefore, not presently economical to mine. Present coal mine technology in the United States, for instance, has only 220 billion tons (220×10^9 tons) of measured proven recoverable reserves out of an estimated total resource of 3–6 trillion tons (3–6 $\times 10^{12}$ tons).

Problems specific to coal mining include the fact that coal seams typically to occur within sedimentary structures of relatively moderate to low strengths. The control of these host rocks surrounding the coal seams makes excavation in underground mining a much more formidable task than that in hard, igneous rocks in many metal mines. Another problem is that coal beds can be relatively flat-lying, resulting in workings that extend a long distance from the shaft or adit portals. Haulage of large tonnages of coal over considerable distance, sometimes miles, is expensive.

Most coal seams emit off varying quantities of methane gas during the course of mining. Of the many molecular species trapped within coal, methane can be easily liberated by simply reducing the pressure in the bed. Other hydrocarbon components are tightly held and generally can be liberated only through different extraction methods.

Coal, being largely composed of carbonaceous material, can also catch fire, in some cases spontaneously (Chapter 14). Coal, for the miner, has not been an attractive occupation. Interestingly though, the problem of methane may in the future become a profitable by-product from closed coal mines. Many countries are reported to millions of cubic feet of coal bed methane (CBM) trapped in abandoned coal mines.

As coal contains both organic and inorganic components, run-of-mine (ROM) coal contains both these components in varying amounts. In many instances, coal beneficiation is required to reduce the inorganic matter (ash) so that a consistent product can be more easily marketed. Most coal beneficiation consists of crushing in order to separate out some of the higher ash content, or washing that exploits the difference in density between maceral and inorganic matter.

Coal is far from being a worn-out of faded commodity and offers much promise for future energy supply (Kavalov and Peteves, 2007; Speight, 2008, 2011; Malvić, 2011). Much research has gone into improving the efficiency of coal use, especially the implementation of coal-fired plants based on clean coal technology (pressurized fluidized bed combustion).

1.2 HISTORY

Coal is the most plentiful fuel in the fossil family and it has the longest and, perhaps, the most varied history (Francis, 1961; Freese, 2003; Höök and Aleklett, 2009). Coal has been used for heating since the cave man.

The lack of written records detracts from finding out the first use of coal. It can be assumed (tongue-in-cheek) that combustion of coal may have been used to provide warmth that enabled the human population to move into northern latitudes as earth emerged from the most recent ice age some 12,000 years ago. An early documented reference to the use of coal as fuel is the geological treatise *On Stones* by the Greek philosopher Theophrastus (ca. 371–287 BC). He described the certain black rocks once set on fire, burn like charcoal, and are used by those who work in metals (Mattusch, 2008).

Outcrop coal was used in Britain as forming part of the fuel for funeral pyres during the Bronze Age (3000–2000 BC) (Needham and Golas, 1999). In Roman Britain, exploitation of coalfields occurred throughout England and Wales at the end of the second century AD (Smith, 1997). Coal cinders have been found in the hearths of villas and military forts, particularly in Northumberland, that were built to protect the Roman Wall as well as at other sites (Forbes, 1966; Cunliffe, 1984; Smith, 1997; Salway, 2001).

After about 1000 AD, mineral coal came to be referred to as *sea coal*, initially because much coal was found on the shore, having fallen from the exposed coal seams on cliffs above or washed out of underwater coal outcrops but by the sixteenth century the name was corrupted because coal was carried to London from the north east of England by sea barges (Cantril, 1914). The alternative name was *pit coal* came from the recovery of coal from mines (pits), because it came from mines.

By the 1700s, the English craftsmen and foundry men found that coal could produce a fuel that burned cleaner and hotter than wood charcoal. However, it was the overwhelming need for energy to run the new technologies invented during the Industrial Revolution that provided the real opportunity for coal to fill the role of the dominant worldwide supplier of energy (Read, 1939).

In North America, the Hopi Indians, in what is now the U.S. Southwest, used coal for cooking and have been known (since the eleventh century) to use coal for heating and to bake the pottery from clay. Coal was later *rediscovered* in the United States by explorers in 1673—the first record of coal in the United States shows up in a map of the Illinois River prepared by Louis Joliet and Father Jacques Marquette in 1673–1674 (they labeled the coal deposits *charbon de terre*). In 1701, coal was found by Huguenot settlers on the James River in what is now Richmond, Virginia. In 1736, several "coal mines" were shown on a map of the upper Potomac River near what is now the border of Maryland and West Virginia.

The first *coal miners* in the American colonies were likely farmers who dug coal from beds exposed on the surface and sold it by the bushel. In 1748, the first commercial coal production began from mines around Richmond, Virginia. Coal was used to manufacture shot, shell, and other war material during the Revolutionary War.

The Industrial Revolution played a major role in expanding the use of coal when James Watt, a Scottish engineer, made improvements to the Newcomen steam engines which were fundamental to the changes brought by the Industrial Revolution in both the United Kingdom and throughout the world. The steam engine developed by Watt made it possible for machines to do work previously done by humans and animals.

During the first half of the 1800s, the Industrial Revolution spread to the United States. Steamships and steam-powered railroads were becoming the chief forms of transportation, and they used coal to fuel their boilers.

In 1816, the city of Baltimore, Maryland, began to light its streets with combustible gas made from coal. The first major boon for coal use occurred in 1830 when the Tom Thumb, the first commercially practical American-built locomotive, was manufactured. The Tom Thumb burned coal, and in rapid fashion, virtually every American locomotive that burned wood was converted to use coal. America's coal industry had begun taking shape. At this time, coal was being used to make glass in Fayette County, Pennsylvania. In the second half of the 1800s, more uses for coal were found. During the Civil War in the United States, weapons factories were beginning to use coal. In 1875, coke (which is made from coal) replaced charcoal as the primary fuel for iron blast furnaces to make steel.

The 1870s saw the next major surge in coal demand. In 1875, coke—a product of heating coal—replaced wood charcoal as the chief fuel for iron blast furnaces. Strip mining began in 1866 near Danville, Illinois, and in 1877, steam shovels were being used to reach a 3 ft thick coal bed in Kansas. With the rise of iron and steel, coal production increased by 300% during the 1870s and early 1880s.

In 1882 Edison built the first practical coal-fired electric generating station, supplying electricity to some residents of New York City. In 1901, General Electric Company built the first alternating current power plant at Ehrenfeld, Pennsylvania. The plant, designed to eliminate the difficulties in long-distance direct current transmission, was built for the Webster Coal and Coke Company.

When the United States entered the twentieth century, coal was the energy mainstay for the businesses and industries and remained the primary energy source for the United States—by the early 1900s, coal was supplying more than 100,000 coke ovens, mostly in western Pennsylvania and northwestern West Virginia—until the demand for petroleum products elevated petroleum to the primary energy source (Speight, 2007a). Automobiles needed gasoline and trains switched from coal to diesel fuel. Even homes that used to be heated by coal turned to oil or gas furnaces instead.

Coal production reached its low point in the early 1950s. Since 1973, coal production has increased by more than 95%, reaching record highs in 2008. Currently, coal supplies 23% of the nation's total energy needs, mostly for electricity production.

Burning coal to generate electricity is a relative newcomer in the long history of this fossil fuel. It was in the 1880s when coal was first used to generate electricity for homes and factories. Long after homes were being lighted by electricity produced by coal, many of them continued to have furnaces for heating and some had stoves for cooking that were fueled by coal.

In 1961, coal had become the major fuel used by electricity utilities to generate electricity, and a new era for coal began taking shape. Coal production in the United States had almost doubled, increasing from 520 million tons in 1970 to 1 billion tons for the first time in 1990 and to nearly 1.1 billion tons currently.

From its earliest use in colonial blacksmith shops, coal had helped wean larger cities from their reliance on imports from England and Nova Scotia, changed the nature of transportation, and powered an industrial revolution. Now, it is the dominant fuel for electric power generation in the United States and has become a major contributor to the modern energy mix of the United States. The real observation is the following: How far will coal progress and intrude into markets currently occupied by petroleum (Speight, 2007a) and natural gas (Speight, 2007b) as a major energy source of the United States?

1.3 ROCK ENVIRONMENT

Coal geology is a branch within the field of *geology* which is focused on coal, which is an abundant fossil fuel with a number of uses.

The study of coal geology includes a wide variety of topics, including coal formation, occurrence, and properties. Coal formation is of great interest because the process of formation can determine the geological composition of the coal, which in turn determines how coal can be used. Knowledge of coal formation can help geologists find new coal deposits and determine the extent and value of existing deposits.

Coal is believed to have been formed under anoxic conditions (i.e., conditions under which the oxygen level has been totally decreased). Such conditions exist in peat bogs or wetlands but for coal formation the peat bogs and wetlands existed millions of years ago. For example, coal has been reported in rocks as old as Proterozoic (possibly 2 billion years old) and as young as Pliocene (2 million years old) but the great majority of the world's coal was laid down during the Carboniferous Period, a 60 million year stretch (approximately 350–290 million years ago) (Table 1.1) when sea level was high and forests of tall ferns and cycads grew in gigantic tropical swamps.

Cycads are seed plants characterized by a large crown of compound leaves and a trunk. They are evergreen and have large pinnately (feather-like or multi-divided features arising from both sides of a common axis in plant) compound leaves. They are frequently confused with and mistaken for palms or ferns, but are only distantly related to both.

Coal formation began during the Carboniferous Period (known as the *first coal age*), which spanned 360 to 290 million years before present. The Carboniferous Period is divided into two parts. The Lower Carboniferous, also called the Mississippian, began approximately 360 million years ago and ended 310 million years ago. The Upper Carboniferous, or Pennsylvanian, extended from about 310 to 290 million years ago, the beginning of the Permian Period.

Coal formation continued throughout the Permian, Triassic, Jurassic, Cretaceous, and Tertiary Periods (known collectively as the *second coal age*), which spanned 290 to 1.6 million years before present. During that time, coal formation started from thick layers of dead plants that piled up in ancient swamps. The dead plant layers were buried deeper and deeper under even thicker layers of sand and mud. During burial, pressure and heat changed the plant material into coal. Coal from the western United States usually formed during or after the era of the dinosaurs (about 50–100 million years ago).

Coal formed during the first coal age is older, so it is generally located deeper in the earth's crust. The greater heat and pressures at these depths produces higher-grade coals such as anthracite and bituminous coals. Conversely, coals formed during the second coal age under less intense heat and pressure are generally located at shallower depths. Consequently, these coals tend to be lower-grade subbituminous and lignite coals.

In geological terms, coal is often defined as banded, streaky, black mineral which is readily combustible and contains more than 50% (by weight) or 70% (by volume) of carbonaceous material. However, to be considered as coal, the black rock must substantially be a deposit of compacted, decayed plant material.

The principal geological periods into which earth history has been divided (Table 1.1) have been established on the basis of the evolution of life forms recorded by fossils found in the successive

TABLE 1.1
Geological Timescale

Period	Epoch	Age	Dates (mya)
Permian	Lopingian	Chianghsingian	253.8–251.0
		Wuchiapingian	260.4–253.8
	Guadalupian	Capitanian	265.8–260.4
		Wordian	268.0–265.8
		Roadian	270.6–268.0
	Cisuralian	Kungurian	275.6–270.6
		Artinskian	284.4–275.6
		Sakmarian	294.6–284.4
		Asselian	299.0–294.6
Pennsylvanian (Carboniferous)	Late Pennsylvanian	Gzhelian	303.4–299.0
		Kasimovian	307.2–303.4
	Middle Pennsylvanian	Moscovian	311.7–307.2
	Early Pennsylvanian	Bashkirian	318.1–311.7
Mississippian (Carboniferous)	Late Mississippian	Serpukhovian	328.3–318.1
	Middle Mississippian	Visean	345.3–328.3
	Early Mississippian	Tourniaisian	359.2–345.3
Devonian	Late Devonian	Famennian	374.5–359.2
		Frasnian	385.3–374.5
	Middle Devonian	Givetian	391.8–385.3
		Eifelian	397.5–391.8
	Early Devonian	Emsian	407.0–397.5
		Pragian	411.2–407.0
		Lochkovian	416.0–411.2
Silurian	Pridoli		418.7–416.0
	Ludlow	Ludfordian	421.3–418.7
		Gorstian	422.9–421.3
	Wenlock	Homerian	426.2–422.9
		Sheinwoodian	428.2–426.2
	Llandovery	Telychian	436.0–428.2
		Aeronian	439.0–436.0
		Rhuddanian	443.7–439.0
Ordovician	Late Ordovician	Hirnantian	445.6–443.7
		Katian	455.8–445.6
		Sandbian	460.9–455.8
	Middle Ordovician	Darriwillian	468.1–460.9
		Dapingian	471.8–468.1
	Early Ordovician	Floian	478.6–471.8
		Tremadocian	488.3–478.6
Cambrian	Furongian	Stage 10	492–488.3
		Stage 9	496–492
		Paibian	499–496
	Series 3	Guzhangian	503–499
		Drumian	506.5–503
		Stage 5	510–506.5
	Series 2	Stage 4	515–510
		Stage 3	521–515
	Terreneuvian	Stage 2	528–521
		Fortunian	542–528

The dates shown on this geological timescale were specified by the International Commission on Stratigraphy in 2009.

layers of rocks deposited through geological time. There are also major breaks in the continuity of depositional environments that are recognized in the rock sequences.

Most sedimentary rocks were deposited as particles in water and the mud, silt, or sand deposited by streams ultimately formed shale (bedded mudstone), clay stone (non-bedded mudstone), siltstone, or sandstone. Early in geological time (although not yet generally recognized in the very earliest sedimentary rocks), animals and simple marine plants developed. Primitive algal remains, fungal remains, and wormlike burrows and trails have been recognized in pre-Cambrian sediments, but by earliest Cambrian time a relatively diverse assemblage of small marine animals had evolved. Many of the marine animals and plants extracted calcium carbonate from the seawater. When their remains accumulated on the sea floor, their shells and fine calcium carbonate muds formed deposits commonly from a few feet up to hundreds of feet thick that are now limestone and dolomite.

At various times in the geological past, large areas of the sea were apparently cut off from the rest, and evaporation first made the water too salty to support life and then resulted in deposits of salt and/or gypsum and anhydrite. Many other deposits are also recognized and include wind-blown deposits and deposits in river valleys and lakes. However, somewhat unique in character among the sedimentary rocks are the vast accumulations of plant material that formed at many different times to produce coal seams.

The major areas of coal distribution are principally in the Northern Hemisphere; with the exception of Australia, the southern continents are relatively deficient in coal deposits. This relatively uneven distribution is the result of the deposition and maturation of the plant at different times in the geological past in predominantly tropical latitudes, and the subsequent drift of the continents to their present-day positions. The oldest coals of any economic significance date from the Middle Carboniferous Period—the earliest geological strata in which coal has been identified are of Devonian age but they are generally of little economic significance. With the exception of parts of the Triassic Period, major coal deposits have been forming somewhere in the world throughout the last 320 million years. Sedimentary sequences of the last 2–3 million years do not contain coal— there has been insufficient time for them to develop from plant debris.

A broadband chain of large coalfields of Carboniferous age extends from the eastern United States, through Europe, Russia, and south into China. A second chain of Permian coalfields is found in the southern continents—South America, India, southern Africa, Australia, and Antarctica. The vast coalfields of the western United States and Canada are of Cretaceous and/or Tertiary age (Figure 1.1).

However, not all regions have coal reserves. Generally, regions where post-Devonian sediments were not deposited or have been completely eroded have no coal whatsoever, for example, Scandinavia and much of Africa. In addition, rocks with the potential for coal bearing, once deposited, may have been eroded away and some regions which do have rocks with the potential for coal bearing may be buried deeply under younger sedimentary sequences and cannot be worked economically.

In North America, extensive coal seams occur in the rocks of Mississippian and Pennsylvanian age and the equivalent carboniferous strata in Europe, although other occurrences of coal are not to be ignored and may even be prevalent in various countries (Figure 1.1) (Matveev, 1976; Hessley et al., 1986; Schobert et al., 1987; Hessley, 1990). Most coal mining in North America to date has focused on coal seams of Pennsylvanian/Carboniferous age in the eastern United States (Table 1.2). Of particular interest is the sulfur content of the various coals; sulfur oxides from coal use being recognized as major pollutants and, therefore, being of some concern in coal utilization (Chapters 14 and 15). The general character of coal sequences in Appalachian coals is found in all geological periods from Silurian through Quaternary time in various parts of the world (Ball, 1976; Ghosh, 1976; Matveev, 1976; Neto, 1976; Ensminger, 1977; Hessley et al., 1986; Hessley, 1990).

Commercially important coals are not usually found in strata older than Mississippian age, although Devonian coals have been mined in Europe (Table 1.3). Some Permian coals are mined in the United States, but coals of that age have much more important reserves in other parts of the

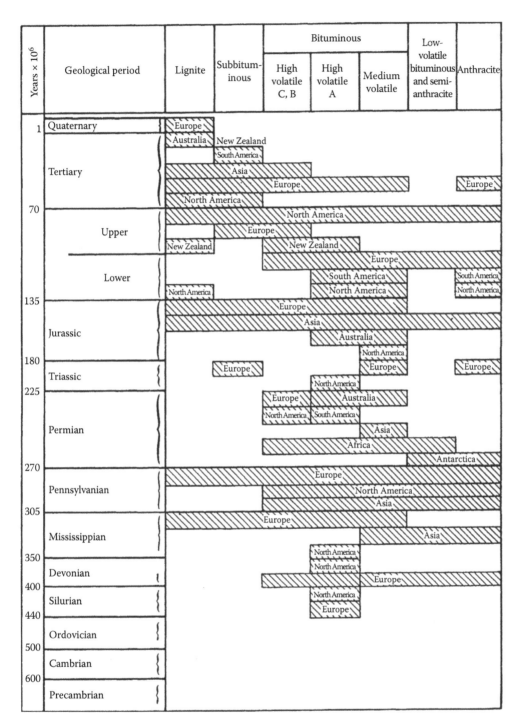

FIGURE 1.1 Geological age of the world coal resources.

world, particularly in various parts of Australia as well as in, what used to be, the USSR. Coals of Triassic age are known and have been mined in Europe but they have been relatively minor deposits. Jurassic coals also are in relatively minor deposits in most of the world, although they have been mined in some areas in Europe and Asia. However, Jurassic rocks contain important coal reserves in Siberia and in western Canada. Large reserves of coals of Cretaceous and Tertiary ages are found

TABLE 1.2

Coal Regions of the United States

Coal Province	Geographical Area	Range of Rank	Other Comments
Eastern (Appalachian)	PA, OH, WV, AL, TN, E. KY	HVB to anthracite	Variable S (medium to high)
Interior			
Eastern region	IL, IN, W. KY	HVB to HVA	High S
Western region	KS, OK, IA, AR, TX	HVC to HVA	High S
Northern great plains	ND and SD, parts of MT, WY, CO, and NM	Lignite to HVB	Low S, very large reserves of lignite
Rocky mountain	UT, CO, parts of WY and NM	Sbb to HVA	Low S, many separate basins
Pacific	WA (CA)	Sbb to HVA	Low S, much influenced igneous intrusions at depth
Gulf	TX, parts of MS, AL, LA	Lignite	Medium S
Alaskan	AK	Sbb to HVA	Low S

TABLE 1.3

Coal Regions of the World and Age of the Deposits

Period	North America	South America	Europe	Asia	Africa	Australia
Quaternary	l	l	l	l	l	l
Tertiary	a, lvb, SB, L	B, SB, L	B, L	SB, B, L	L	L
Cretaceous	a, LVB, SB, L	SB, B	B, L	b	SB	l
Jurassic	B	—	sb, b, l	a, LVB, b, sb, L	—	b
Triassic	b	—	a, B, sb, l	—	—	—
Permian	b	b	A, B	B	B, c	B
Pennsylvanian	A, LVB, B, c	b	A, LVB, B, sb, l, c	A, LVB, B	A, B, l	LVB, b
Mississippian	a, lvb, B	—	B, sb, l	A, LVB, b	—	—
Devonian	—	—	b	—	—	—

Key: L, l, lignite; SB, sb, subbituminous; B, b, high- and medium-volatile bituminous; LVB, lvb, low-volatile bituminous; A, a, anthracite; c, cannel. Capital letters indicate major deposits, and lowercase letters indicate smaller, less significant deposits.

in the western United States (Table 1.2), western Canada, and other parts of the world. Many of these Cretaceous/Tertiary deposits, especially in the western United States and western Canada, are of such a shallow nature that they are subject to surface-mining techniques.

Individual coal seams vary in thickness from a fraction of an inch up to many feet and may be as much as 300 ft (91.5 m) in some areas. Most of the bituminous coal seams are less than 20 ft (6.1 m) thick, and most mining has been in seams from 3 to 10 ft (0.9 to 3.5 m) thick.

The total sequence of sedimentary rocks may reach several thousand feet (but generally much less) and the coal-bearing seams that have been mapped throughout the world commonly occur at depths less than 5000 ft (1524 m). Most coal seams in the United States lie at depths up to 3000 ft (915 m) whereas most of the mining operations have been carried out at depths ranging from a few feet up to 1000 ft (305 m).

As a result of the formation processes, coal contains organic (carbon-containing) compounds transformed from ancient plant material (Chapters 3 and 10). The original plant material was composed of cellulose, the reinforcing material in plant cell walls; lignin, the substance that cements plant cells together; tannins, a class of compounds in leaves and stems; and other organic compounds, such as fats and waxes. In addition to carbon, these organic compounds contain hydrogen, oxygen, nitrogen, and sulfur. After a plant dies and begins to decay on a swamp bottom, hydrogen and oxygen (and smaller amounts of other elements) gradually dissociate from the plant matter, increasing its relative carbon content.

Coal also contains inorganic components—mineral matter (Chapter 7)—which are determined as *mineral ash* (Chapter 8). The mineral matter includes minerals such as pyrite (FeS_2) and the related mineral marcasite (FeS_2), as well as other minerals formed from metals that accumulated in the living tissues of the ancient plants. Quartz, clay, and other minerals are also added to coal deposits by wind and groundwater. The mineral matter lowers the fixed carbon content of coal, decreasing its heating value (Chapter 8).

1.4 SEAM STRUCTURE

Coal occurs in strata, which are more commonly referred to as *seams*, which occur at varying depths below the surface of the earth.

Many factors influence seam thickness, continuity, quality, and mining conditions. However, the focus here is on the geological factors which affect seam structure (Moore, 1940; Cargo and Mallory, 1974; Lindberg and Provorse, 1977; Cuff and Young, 1980; Hessley et al., 1986).

Some features were formed during the period of accumulation of the organic debris (or at some time shortly thereafter) whereas other features developed not only after burial but also after the onset of coalification. An entire coalfield or large parts of it may be affected by a geological feature or occurrence, though many geological features usually have only local significance, the recognition of the nature of many of these features (Eliot, 1978) is important to mining operations.

1.4.1 SHALE PARTINGS AND SPLIT COAL

Shale partings and *split coal* are terms applied to a coal seam that cannot be mined as a single unit because it is separated from the main stream by a parting of other sedimentary rock (Figure 1.2).

In the areas in which floral (vegetable) material accumulated (i.e., the *peat swamp*), streams periodically overflowed their banks and deposited mud (or silt) layers which eventually became bands of shale or siltstone after the vegetable material was coalified. In the close proximity of these rivers, relatively thick deposits were laid down, with some (at least partial) reestablishment of the swamp vegetation after the flood stage which resulted in partings or split coal.

During exploration, gamma-ray logs can be used to detect shale partings in a coal bed, but generally the thickness of thin partings may be exaggerated and varying thicknesses of parting may be recorded as similar partings (Wood et al., 1982, 2003).

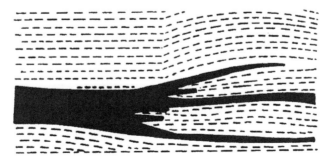

FIGURE 1.2 Splitting of a coal seam.

In underground mining methods, it is difficult if not impossible to separate the shale, clay, and other non-coal partings from the coal seam as the mining machinery removes the entire coal seam—partings included—to maintain a workable height for miners and sometimes to achieve maximum roof stability. The *raw coal* must be cleaned on the surface at the coal-processing plant.

1.4.2 Washouts

Washouts occur wherever a coal seam has been eroded away or where a river or stream channel has cut through the accumulating vegetable debris that will eventually for coal. After plant material has accumulated and is buried by various sediments, it may be removed by the downward erosive action of streams; such discontinuity in the rock sequence is termed a washout or cutout (Figure 1.3). The route of the washout channel is usually filled with the sand and/or gravel that was at the base of the stream/river bed and which often take the shape of elongated sandstone lenses.

Thus, washouts may happen shortly after deposition of the organic debris or much later, even after the coal has been covered by other sediments. Such channel washouts can usually be distinguished from channels that were present in the peat swamp and in which the peat may never have accumulated. Channels which existed in the swamps commonly have split coal immediately adjacent to the channel whereas channels or washouts that formed after burial of the peat normally cut sharply through the overlying strata and the coal and sometimes removed strata below the coal.

Some stream valleys may have been considerably deeper in the past than at the present because of partial filling by unconsolidated materials (Figure 1.3) causing discontinuities in coal seams because of the prior removal of the organic debris (coal precursors) from the stream bed.

Washouts are a major issue in mining operations—particularly in underground mining. Washouts can seriously reduce the workable area of coal and, therefore, the delineation of washouts is an important aspect of mine planning. Detailed interpretation of the sedimentary sequences may help predict the orientation of washouts.

1.4.3 Dipping and Folded Strata

Dipping strata usually involve strata that have a pronounced downdip. A *homocline* involves strata dipping in the same direction, though not necessarily exhibiting folding. The strata that form either side of the typical petroleum anticline are often referred to as dipping strata (Speight, 2007a,b).

The term *fold* is used in the geological sense when one or more originally flat and planar surfaces, such as sedimentary strata, are bent or curved as a result of permanent deformation. Throughout

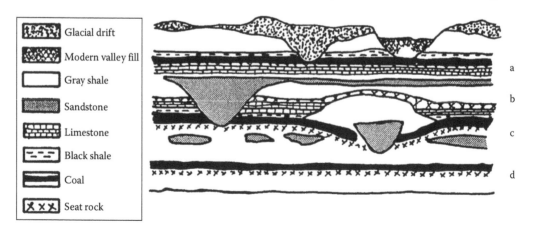

FIGURE 1.3 Coal seam discontinuities: (a) recent erosion, (b) preglacial erosion, and (c) and (d) stream beds at the time of debris accumulation.

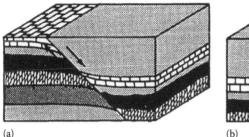 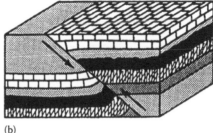

(a)   (b)

FIGURE 1.4 Types of fault: (a) a normal fault—the higher strata have moved downward—and (b) a reverse fault—the higher strata have moved upward.

geological time, horizontal and vertical stresses have resulted in folding of the rock sequences. Such folding may have resulted in localized variations in dip of the strata or may or may have affected very broad areas, such as an intermittently subsiding basin, or the strata may have been folded so intensely that the coal seam may be nearly vertical or actually overturned. Indeed, just as the strata are affected, so too are the coal seams.

1.4.4 FAULTS

Faults are fractures in the rock sequence along which strata on each side of the fracture appear to have moved, but in different directions. Such movement may be measured in inches, tens or even hundreds of feet, or (less commonly) miles; and the movement may be in any direction, from horizontal to vertical (Figure 1.4). It is necessary to note here that the term fault has often been (incorrectly) employed to indicate any type of discontinuity of a coal seam but, from a geological perspective, the term should be reserved to indicate differential movement of strata along a fracture surface.

   Where stresses are in opposite directions or tensional, rocks have "pulled apart" at the fracture surface and displacement may produce a normal fault or a thrust reverse fault (Figure 1.4). A borehole drilled very close to a normal fault may have a shortened interval between key strata, whereas in a reverse fault, part of the sequence may be repeated in the boring.

1.4.5 CLASTIC DIKES AND CLAY VEINS

Clastic rocks are composed of fragments (clasts) of preexisting rock. The term may be used in reference to sedimentary rocks as well as to transported particles whether in suspension or in deposits of sediment.

   A *clastic dike* is typically a vertical or near-vertical seam of sedimentary material that fills a crack in and cuts across sedimentary strata—the dikes are found in sedimentary basin deposits worldwide. Clastic dikes form rapidly by fluidized injection (mobilization of pressurized pore fluids) or passively by water, wind, and gravity (sediment swept into open cracks). Dike thickness varies from millimeters to meters and the length is usually many times the width.

   The irregular, vertical-to-inclined masses of certain materials (clay, silt, or sand) that interrupt a coal seam are called clastic dikes (Figure 1.5). They have also been called horsebacks, a name also applied (in some areas) to rolls. These dikes may vary from mere inches to several feet wide and usually extend laterally for distances up to hundreds of feet. They may also extend for some distance into the strata overlying the coal and may contribute to a weakened roof condition when the coal is mined.

   In addition to their contribution to roof instability, dikes increase the amount of waste material that must be removed from the coal and may also interfere with the drainage of mine gas (predominantly methane; see Chapter 5) during a mining operation.

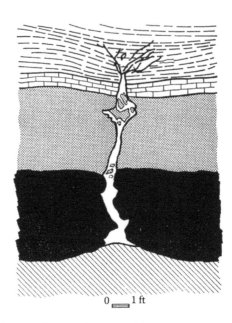

FIGURE 1.5   A clastic dike interrupting a coal seam and the overlying strata.

Clay veins are infilled fissures. Fissures may be infilled as a result of gravity, downward-percolating ground waters, or compactional pressures which cause unconsolidated clays or thixotropic sand to flow into the fissures (Chase and Ulery, 1987).

Clastic dikes and clay veins found in coal mines have caused numerous injuries and fatalities. These structures plague all phases of mining, including entry development, pillar recovery, and panel extraction. Clastic dikes and clay veins also increase production costs and may disrupt or halt mining.

1.4.6   FRACTURES AND CLEATS

Fracturing is the basic element of tectonic disturbance—the process of stratum folding gives rise to fracturing in beds. Cleats are natural fractures in coal. When the pressure in the coal seam is reduced through mining, gas begins to desorb and migrate through the coal matrix and through natural fractures such as cleat.

Fractures occur in nearly all coal beds, and can exert fundamental control on coal stability, min-ability, and fluid flow. It is therefore not surprising that coal fractures have been investigated since the early days of coal mining, and that published descriptions and speculation on fracture origins date from early in the nineteenth century (Laubach et al., 1998). Although various mining terms for systematic fractures in coal have been used over the years, they are still generally referred to by the ancient mining term "cleat" (Dron, 1925).

Cleats are natural opening-mode fractures in coal beds. They account for most of the permeability and much of the porosity of coal bed gas reservoirs and can have a significant effect on the success of engineering procedures such as cavity stimulations (Ayers and Kaiser, 1994; Ayers, 2002; Bachu and Karsten, 2003). Because permeability and stimulation success are commonly limiting factors in gas well performance, knowledge of cleat characteristics and origins is essential for successful exploration and production. Although the coal–cleat literature spans at least 160 years, mining issues have been the principal focus, and quantitative data are almost exclusively limited to orientation and spacing information. Few data are available on apertures, heights, lengths, connectivity, and the rela-tion of cleat formation to diagenesis, characteristics that are critical to permeability. Moreover, recent studies of cleat orientation patterns and fracture style suggest that new investigations of even these well-studied parameters can yield insight into coal permeability (Laubach et al., 1998).

A variety of factors have been cited as affecting cleat development. These factors include coal rank, coal composition, and layer thickness. To the extent that the impact of these factors on cleating has been explored quantitatively, average cleat spacing has been used to characterize the cleats. Other factors, such as mineral fill, degree of tectonic and compactional deformation, and coal age, have received little to no attention (Laubach et al., 1998).

There are two types of cleats related to the mechanisms for the origin: (1) endogenetic cleat and (2) exogenetic cleat.

The *endogenetic cleat* is formed during the process of physical changes in the properties of coal during the metamorphic process. Coal matter undergoes density changes and a decrease in its volume—processes are associated with the changes in the internal stress system, compaction and desiccation, and the formation of cleat planes.

The *exogenetic cleat* is formed as a result of the external stresses acting on the coal seam. These include tectonic stresses, fluid pressure changes, folding, and development of tensile stresses to which the coal seam is subjected during various time periods.

Endogenetic cleats are normal to the bedding plane of coal and generally occur in pairs. There are at least two sets of near perpendicular fractures that intersect the coal to form an interconnected network throughout a coal bed. These two fracture systems are known as face and butt cleats. The shorter butt cleat normally terminates at a face cleat, which is the prominent type of cleat.

Larger fractures (*joints*) are found to extend over the whole or part of the coal seam and are much less frequent than cleat. These are of the exogenetic origin and are related to the tectonic movement. The frequency of joints increases rapidly when approaching shear structures and faults, large joints running over several times the thickness of the seam and at low angle to the coal are of endogenetic origin. Joints can cut across the lithological boundaries in the seam, but are in general limited to the seam thickness. Increase in their frequency is an indication of an approaching geological structure.

Coals with bright lithotype layers, with a high percentage of vitrinite macerals, have greater amount of cleats than dull coals. It is commonly understood that cleats are formed due to the effects of the intrinsic tensile force, fluid pressure, and tectonic stress. The intrinsic tensile force arises from matrix shrinkage of coal, and the fluid pressure arises from hydrocarbons and other fluids within the coal. These two factors are considered to be the reasons for endogenetic cleat formation. On the other hand, the tectonic stress is regarded as extrinsic to cleat formation and is the major factor that controls the geometric pattern of cleats. Face cleats extend in the direction of maximum in situ stress, and butt cleats extend in the direction of minimum in situ stress which existed at the time of their formation. This is why regular cleats are formed in face and butt pairs.

Bituminous coal seams often exhibit numerous fractures (or cleats) that may be relatively closely spaced (from a fraction of an inch to a few inches apart). There are usually two preferred directions to the cleat which are at approximately 90° to each other. The more prominent cleat is termed the face cleat whereas the less prominent (or less well-developed) cleat is known as the butt cleat. The natural trend of the cleat may facilitate the breaking of the coal during the mining operations and may, therefore, also determine the direction of the mining operation.

Thin layers of pyrite, calcite, kaolinite, or other minerals are commonly found deposited on the relatively smooth surfaces of the cleat.

Cleats are fractures that usually occur in two sets that are, in most instances, mutually perpendicular and also perpendicular to bedding. Although pre-1990 geological literature and current mining usage distinguishes these sets on the basis of factors such as "prominence" that are difficult to quantify, abutting relations between cleats generally show that one set predates the other. This is a readily quantifiable distinction between sets in most outcrops and cores. Through-going cleats formed first and are referred to as face cleats; cleats that end at intersections with through-going cleats formed later and are called butt cleats (Laubach and Tremain, 1991). These fracture sets, and partings along bedding planes, impart a blocky character to coal. A hierarchy (or perhaps a continuum) of cleat sizes typify the population of cleats in a coal bed (Laubach and Tremain, 1991). This is most easily recognized in cleat heights, but size variations are also evident in cleat lengths, apertures, and spacings.

Owing to the increasing importance of coal beds as gas reservoirs, geologists are again becoming concerned with the characteristics and origins of cleat. For CBM extraction, knowledge of the properties of natural fractures is essential for planning exploration and development because of their influence on recovery of methane, and the local and regional flow of hydrocarbons and water. New mapping of cleat patterns, guided by recent conceptual advances in the description and theoretical understanding of fracture processes, is beginning to bring these patterns into focus. This, together with more rigorous study of the petrology of cleat development, may resolve uncertainties about causes of cleat that now hinder predictions of interwell-scale patterns of fractures in coal beds.

1.4.7 CONCRETIONS

A *concretion* is a volume of sedimentary rock in which mineral cement fills the spaces between the sediment grains. Concretions are often ovoid or spherical in shape, although irregular shapes also occur. Concretions form within layers of already-deposited sedimentary strata. They usually form early in the burial history of the sediment, before the rest of the sediment is hardened into rock.

Mine roof shale commonly contains concretions made up of calcite ($CaCO_3$), dolomite ($CaCO_3 \cdot MgCO_3$), siderite ($FeCO_3$), and pyrite (FeS_2) alone or in combination. In some cases, plant or animal fossils serve as nuclei around which the minerals accumulate.

The presence of large concretions in mine roof material may have considerable effect on the stability of the roof. In fact, differential compaction of the roof rocks frequently produces a slick surface that is relatively weak and may permit the larger concretions to drop from the roof coal is mined. Within the coal itself, pyritic concretions are common and may range from less than an inch (2.5 cm) to a few feet across. The larger masses probably are aggregations of smaller pyritic concretions and are commonly referred to by miners as sulfur balls.

On the other hand, *coal balls* are an aggregation of calcite and/or pyrite minerals and appear to represent the early onset of the petrification of the bed (plant) material in the peat swamp, prior to complete compaction and coalification (Francis, 1961). It appears that no marine remains have been found within a coal ball and although coal balls provide evidence of a marine intrusion, there is insufficient evidence that the original plant material grew in a totally marine environment (Stutzer, 1940).

1.4.8 IGNEOUS INTRUSIONS

Igneous intrusions form when magma cools and solidifies before it reaches the surface. In some districts, the coal and associated strata may have been intruded by once molten igneous rocks forcibly injected into the sedimentary sequence from below (Figure 1.6).

In coal seams, igneous rock is commonly seen as a dike, which is a nearly vertical tabular mass cutting across the bedding of the sediments. If a zone of weakness exists between the bedding planes, the molten material may have been forced to spread as a broad sheet parallel to the bedding plane to form a sill.

Dikes form as magma pushes up toward the surface through cracks in the rock. Dikes are vertical or steeply dipping sheets of igneous rock. Sills form as a horizontal or gently dipping sheet of igneous rock when magma intrudes between the rock layers. A *batholith* is a large deep-seated intrusion (sometimes called a *pluton*) that forms as thick, viscous magma slowly makes its way toward the surface, but typically does not reach the surface.

The size of such an igneous mass (and the temperature) may also play a prominent role in determining the rank of the coal. For example, coal adjacent to the intrusion may be either advanced in rank and has increased carbon content relative to the nonadjacent coal. The concept of higher-rank coal having

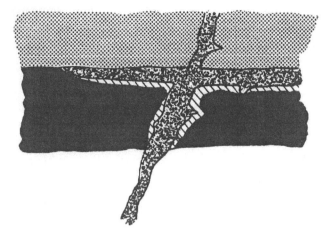

FIGURE 1.6 An igneous dike showing a thin zone adjacent to the igneous rock that has been thermally altered.

increased carbon content over lower-rank coal is very general, but not completely true (Chapters 2 and 8) and must be treated as a very general, somewhat less than precise term. On the other hand, the coal adjacent to the intrusion may have been *coked* having undergone a thermal transformation due to the heat liberated from the intrusion and, therefore, be similar to manmade process cokes.

The hardness of igneous rocks renders them a source of continued problems during the mining operations.

1.5 RESERVES

In any text dealing with coal, there must be recognition, and definition, of the terminology used to describe the amounts, or reserves, of coal available for recovery and processing (Figure 1.7). But the terminology used to describe coal (and for that matter any fossil fuel or mineral) resource is often difficult to define with any degree of precision.

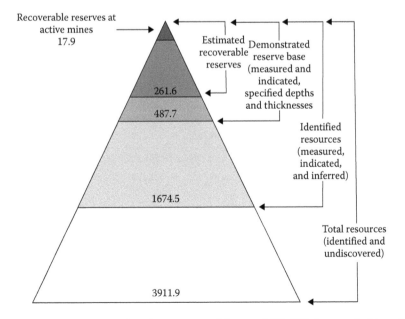

FIGURE 1.7 U.S. coal resources and coal reserves as of January 2009 (10^9 short tons). (From U.S. Energy Information Administration, Form EIA-7A, Coal Production Report, February 2009.)

Different classification schemes often use different words which should, in theory, mean the same but there will always be some difference in the way in which the terms can be interpreted. It might even be wondered that if the words themselves leave much latitude in the manner of their interpretation, how the resource base can be determined with any precision. The terminology used here is that more commonly found although other systems do exist and should be treated with caution in the interpretation.

Generally, when estimates of coal supply are developed, there must be a line of demarcation between coal *reserves* and *resources*. Reserves are coal deposits that can be mined economically with existing technology, or current equipment and methods. Resources are an estimate of the total coal deposits, regardless of whether the deposits are commercially accessible. For example, world coal reserves were estimated to be in excess of 1 trillion tons (1×10^{12} tons) and world coal resources were estimated to be approximately 10 trillion tons (10×10^{12} tons) and are geographically distributed in Europe, including all of Russia and the other countries that made up the Soviet Union, North America, Asia, Australia, Africa, and South America (Table 1.4).

However, there are definitions that go beyond *reserves* and *resources*. To begin at the beginning, the energy resources of the earth are subdivided into a variety of categories (Figure 1.8) and the resources of coal (as well as each of the other fossil fuel resources) can be further subdivided into different categories (Figure 1.9).

TABLE 1.4
Geographical Distribution of Recoverable Coal Reserves

By Country	
Country	Total Recoverable Coal ($\times 10^6$ tons)
United States	273,656
Russia	173,074
China	126,215
India	93,031
Australia	90,489
Germany	72,753
South Africa	54,586
Ukraine	37,647
Kazakhstan	37,479
Poland	24,427
By Region	
Region	Total Recoverable Coal ($\times 10^6$ tons)
Asia and Oceania	322,394
Eastern Europe	290,183
North America	282,444
Western Europe	101,343
Africa	61,032
Central and South America	23,977
Middle East	1,885

Source: EIA/DOE: http://www.eia.gov/pub/international/iea2002/ table82.xls. [Even though the data are from 2002, they are still valid estimates.]

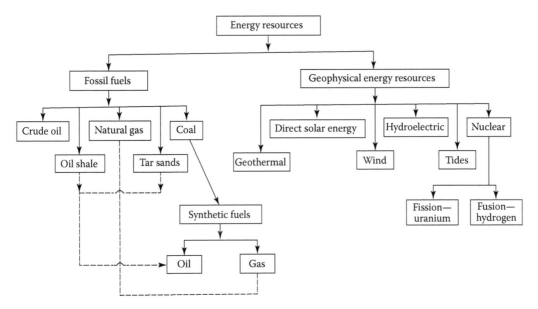

FIGURE 1.8 Energy resources of the earth.

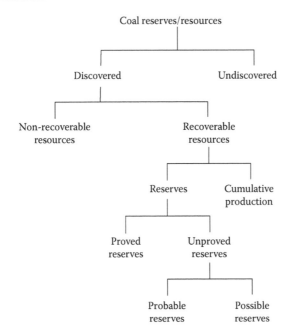

FIGURE 1.9 Resource and reserve terminology.

1.5.1 PROVEN RESERVES (PROVED RESERVES)

These reserves are those coal reserves that are actually found (proven), usually by drilling and coring. The estimates have a high degree of accuracy and are frequently updated as the mining operations proceed. However, even though the coal reserves may be proven, there is also the need to define the resources on the basis of what further amount of coal might be recoverable (using currently available mining technology without assuming, often with a very high degree of optimism) extravagantly, that new technology will miraculously appear or will be invented, and *non-recoverable* coal reserves will suddenly become *recoverable*.

1.5.2 INFERRED RESERVES (UNPROVED RESERVES)

The term *inferred reserves* is commonly used in addition to, or in place of, *potential reserves*. The inferred coal reserves are regarded as being of a higher degree of accuracy than the potential reserves and the term is applied to those reserves that are estimated using an improved understanding of seam structure (see Section 1.5.1). The term also usually includes those reserves that can be hoped to be recoverable by further development of mining technology.

1.5.3 POTENTIAL RESERVES (PROBABLE RESERVES, POSSIBLE RESERVES)

These reserves are the additional reserves of coal that are believed to exist in the earth. The data are estimated (usually from geological evidence) but have not been substantiated by any drilling or coring operations. Other terminologies such as *probable reserves* and *possible reserves* are also employed but fall into the subcategory of being unproven.

1.5.4 UNDISCOVERED RESERVES

One major issue in the estimation of coal (or for that matter any fossil fuel or mineral) resource is the all too frequent use of the term "undiscovered" resources. Caution is advised when using such data, as might be provided to "substantiate" such reserves, as they are very speculative and are regarded as many energy scientists as having little value other than being unbridled optimism. And perhaps, opportunism and/or charlatanism are being applied depending upon the resource under speculation and the potential market for the product.

The differences between the data obtained from these various estimates can be considerable but it must be remembered that any data about the reserves of natural gas (and, for that matter, about any other fuel or mineral resource) will always be open to questions about the degree of certainty (Figure 1.10).

There are three important items that counteract the guesswork applied to "undiscovered" resources: (1) the actual discoveries of new coal reserves, (2) the development of improved mining technologies for already known coal reserves, and (3) estimates of the coal resource base that are derived from known resource properties where the whole of the resource is not explored.

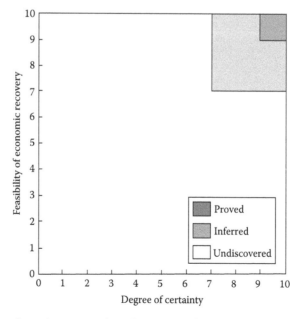

FIGURE 1.10 Degree of certainty–uncertainty of resource estimates.

It should also be remembered that the total resource base of any fossil fuel (or, for that matter, of any mineral) will be dictated by economics (Nederlof, 1988). Therefore, when coal resource data are quoted some attention must be given to the cost of recovering those reserves. And, most important, the economics must also include a cost factor that reflects the willingness to secure total, or a specific degree of, energy independence, including some serious consideration of the very, real environmental aspects of coal usage.

1.5.5 Coal Bed Methane

Methane adsorbed to the surface of coal is a very old issue, mainly because this explosive gas has made underground coal mines dangerous both from the risk of explosion and the possibility of an oxygen-poor atmosphere that would not support life. The main concern of miners with CBM has been how to get rid of it (Simpson et al., 2003).

In the past several decades, it had become clear that the CBM is a significant economic resource and capturing CBM for sale is often profitable even for exploiting coal seams that cannot be economically developed.

CBM is predominantly methane extracted from coal beds (coal seams) and has become an important source of energy in United States, Canada, and other countries. The gas is adsorbed into the solid matrix of the coal and is often referred to as *sweet gas* because of its lack of hydrogen sulfide. Unlike natural gas from conventional reservoirs (Mokhatab et al., 2006; Speight, 2007b), CBM contains very little higher-molecular-weight hydrocarbons such as propane or butane or condensate. It often contains up to a few percent carbon dioxide. Some coal seams contain little methane, with the predominant coal seam gas being carbon dioxide.

In fact, CBM has been suggested (Levine, 1991a,b) with sufficient justification that the materials comprising a coal bed fall broadly into the following two categories: (1) volatile low-molecular-weight materials (components) that can be liberated from the coal by pressure reduction, mild heating, or solvent extraction and (2) materials that will remain in the solid state after the separation of volatile components.

Techniques to detect and deal with CBM in mines have ranged from (1) the classic canary in a cage to detect an oxygen-poor atmosphere, (2) ventilation fans to force the replacement of a methane-rich environment with outside air, and (3) drilling CBM wells in front of the coal face to try to degas the coal prior to exposing the mine to the CBM. Each method has meant with a degree of success but none can prevent CBM from escaping from the coal seam and entering the air in an underground mine.

With the CBM's unique method of gas storage, the preponderance of the gas is available only to very low coal face pressures. The coal face pressure is set by a combination of flowing wellhead pressure and the hydrostatic head exerted by standing liquid within the wellbore. Effective removal of higher-molecular-weight hydrocarbons and other constituents (*liquids removal* or *deliquification*) techniques can reduce or remove the backpressure caused by accumulated liquid.

1.5.5.1 Reservoir Characteristics

Rock formations that are typically important to oil and gas production fall into three categories: (1) source rock, (2) reservoir rock, and (3) cap rock.

Source rock is a formation containing organic matter where the decomposition of the organic matter has resulted in the formation of complex hydrocarbonaceous products. *Reservoir rock* is a formation with pore volume capable of containing commercial quantities of hydrocarbons and/or hydrocarbonaceous material. *Cap rock* is a formation that is largely impermeable to the flow of liquids and gases and is located such that fluids that approach it from lower or adjacent formations cannot migrate any further (Bradley, 1992; Speight, 2007a,b).

In conventional petroleum and gas reservoirs, the formation of oil and/or gas is initiated in the source rock and migrates to the reservoir rock, and its migration is stopped by cap rock. CBM reservoirs do not follow this pattern. Coal meets the criteria for source rock since the very matrix of the coal is rich in organic matter. Coal fails the definition of reservoir rock since the pore volume of a coal bed is an order of magnitude less than conventional reservoirs. Because of its relative imperme-ability, coal is often likened to the cap rock for conventional reservoirs.

CBM is adsorbed to the surface of the coal and the adsorption sites can store commercial quantities of gas as *part* of the coal matrix. This must not be confused with conventional pore-volume storage. Gas within petroleum reservoir rock as a gas and the traditional pressure/temperature/volume relation-ships hold. Adsorbed gas molecules do not behave as a gas: (1) they do not conform to the shape of the container, (2) they do not conform to the modified ideal gas laws (i.e., $PV \neq ZnRT$), and (3) they take up substantially less volume than the same mass of gas would require within a pore volume.

1.5.5.2 Contamination

Adsorption sites within a coal seam will in a manner to catalysts for petroleum refining, accommo-date any molecule that fits within the site. The typical adsorption site diameter in a coal bed, while being suitable for methane adsorption, is also able to accommodate carbon dioxide and nitrogen. Higher-molecular-weight hydrocarbons do not fit well on the sites and it is rare to see a signifi-cant concentration of hydrocarbons of higher molecular weight than methane adsorbed on the coal. However, inter-bedded sand lenses do occur and they can hold any gas or liquid that their pore structure (i.e., pore size) can accommodate.

1.5.5.3 Recovery

A conventional reservoir is assumed to be approximately (if not fully) homogeneous and gas will tend to flow to a wellbore in a pore-to-pore Darcy flow. Coal had minimal porosity and, therefore, limited communication from one micro-cleat to the next. The best flow communication through the cleat system is along *face cleats*, which typically run vertically and tend to align themselves from cell to cell along the axis of maximum external stress on the coal. Butt cleats, which intersect the face cleats at 90° in the direction of least external stresses, do not tend to align, and they typically have a minimal contribution to gas production.

As a result, commercial flows within a CBM reservoir generally rely on a system of *channels* that have been created within the coal seam either through geological movements or designed stimula-tions. Wells where the stimulations have been aligned to cross the maximum number of face cleats have consistently outperformed wells that ignore the lay of the coal stresses.

These channels are analogous to pipe flow, and the pressure traverse from the wellbore out to considerable distances can be essentially constant. Since desorption is a pressure-swing phenom-enon, the extent of these channels has an overriding effect on the gas flow rate. If the system of channels and channel branches is extensive, the flow rate will be high but if the channels are absent, the ability of the well to allow a flow rate adequate for commercial production will be limited.

1.6 RESOURCES

In general terms, coal is a worldwide resource; the latest estimates (and these seem to be stable within minor limits of variation; Hessley, 1990) show that there is in excess of 1000 billion (10^9) tons of proven recoverable coal reserves throughout the world (Figure 1.11). In addition, consump-tion patterns give coal approximately 30% of the energy market share (Figure 1.12). Estimates of the total reserves of coal vary within wide limits, but there is no doubt that vast resources exist and are put to different uses (Horwitch, 1979; Hessley, 1990; Ogunsola, 1991). However, it is reasonable to assume that, should coal form a major part of any future energy scenario, there is sufficient coal for many decades (if not hundreds of years) of use at the current consumption levels. Indeed, coal is projected as a major primary energy source for power generation for at least the next several decades and could even surpass oil in use, especially when the real costs of energy are compared to the costs of using the indigenous coal resources of the United States (Hubbard, 1991; Speight, 2008).

In order to understand the politics of coal use and production, it is necessary to put coal into the per-spective of oil and gas. In the early days of the oil industry, the United States was the major producer and

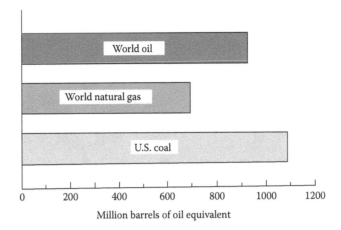

FIGURE 1.11 Estimates of world oil and gas resources compared to U.S. coal resources.

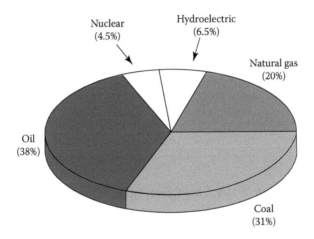

FIGURE 1.12 Estimate of world energy use form various sources.

was predominantly an exporter of crude oil thereby serving as the "swing" producer insofar as production was adjusted to maintain stability of world oil prices. However, oil production in the United States peaked in 1972 and has been in decline ever since and the mantle of oil power has shifted to the Middle East leading to new political and economic realities for the world. From the 1970s, oil prices have been determined more by international affairs (geopolitics) than by global economics (Yergin, 1991).

In contrast to current United States oil production and use patterns, the United States is not a significant importer of natural gas. Trade agreements with Canada and with Mexico are responsible for the import of natural gas but these are more of a convenience for the Border States rather than for the nation as a whole.

In the United States, the use of coal increased after World War II with the majority of the production occurring in the eastern states close to the population centers. The majority of the recovery methods used underground mining techniques in the seams of higher quality, that is, the minerals and water content of the coal, was relatively low and the coal had a high heat content (Chapters 8 and 9). However, by the late 1960s, oil and gas had captured most of the residential, industrial, and commercial leaving only power generation and metallurgical coke production as the major markets for coal.

On a global scale, the United States is a major source of coal (Figure 1.13) as well as a coal producer and coal exporter (Chadwick, 1992). There are many coal-producing states in the United States but the passage, and implementation, of the Clean Air Act in the early 1970s, opened up unforeseen markets for the easily (surface) mined low-sulfur coals (Tables 1.5 and 1.6 and Figure 1.14) from

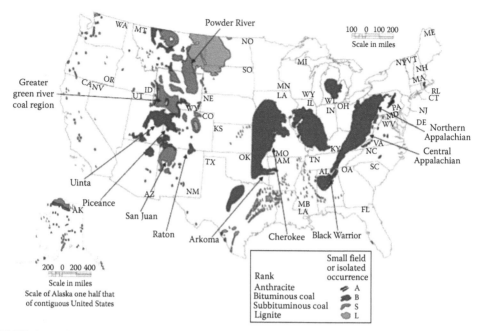

FIGURE 1.13 Coal reserves and distribution in the United States. (From Miller, C.L., in *2007 EIA Energy Outlook Modeling and Data Conference*, March 28, 2007. Office of Sequestration, Hydrogen, and Clean Coal Fuels, Office of Fossil Energy, U.S. Department of Energy, Washington, DC.)

TABLE 1.5
Distribution of U.S. Coal Reserves (% of Total)

Region[a]	Underground Minable (%)	Surface Minable (%)	Total (%)
Eastern United States	36.9	8.2	45.1
Western United States	30.8	24.1	54.9
Total United States	67.7	32.3	100.0

[a] Mississippi River is dividing line between East and West.

TABLE 1.6
Sulfur Content of U.S. Coals by Region

Region	No. of Samples	Organic S (%)	Pyritic S (%)	Total (%)
N. Appalachia	227	1.00	2.07	3.01
S. Appalachia	35	0.67	0.37	1.04
E. Midwest	95	1.63	2.29	3.92
W. Midwest	44	1.67	3.58	5.25
Western	44	0.45	0.23	0.68
Alabama	10	0.64	0.69	1.33

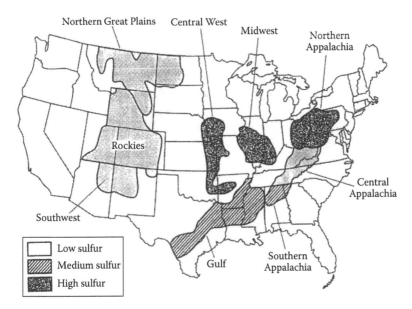

FIGURE 1.14   Distribution of sulfur content of U.S. coals.

the western United States have captured a substantial share of the energy (specifically, the electrical utility) markets. In addition, states such as Wyoming were the major beneficiaries of the trend to the use of low-sulfur coal (Sanami, 1991) and occupy a significant position in the coal reserves and coal production scenarios of the United States.

Furthermore, the two oil shocks of 1973 and 1979, as well as the political shock that occurred in Iran in 1980, brought about the rediscovery of coal through the realization that the United States and other Western countries had developed a very expensive habit insofar as they not only had a growing dependence upon foreign oil but also craved the energy-giving liquid. The discovery of the North Sea oil fields gives some respite to an oil-thirsty Europe but the resurgence of coal in the United States continued with the official *rebirth* of coal in 1977 as a major contributor to the United States' National Energy Plan. It is to be hoped that future scenarios foresee the use of coal as a major source of energy; the reserves are certainly there and the opportunities to use coal as a clean, environmentally acceptable fuel are increasing.

Coal usage in the United States is increasing even though in some countries (such as England where many coal mines have been closed) coal use appears to be on a decreasing curve as natural gas usage increases (Coal Voice, 1992; Mokhatab et al., 2006). Later governmental announcements suggested that the decision on 21 of the mines earmarked for closure was to be postponed and the future of the remaining 10 mines was to be reviewed. In most cases of such shut downs, it is not so much a question of economics but also a question of environmental costs of coal usage. Thus, major issues remain to be addressed on behalf of coal.

The question to be asked by any country, and Canada did ask this question in the early 1970s, is "What price are we willing to pay for energy independence?" There may never be any simple answer to such a question. But, put in the simplest form, the question states that if the United States (or, for that matter, any energy consumer) is to wean itself from imported oil (i.e., nonindigenous energy sources), there will be an economic and environmental cost if alternate sources are to be secured (NRC, 1979, 1990). In this light, there is a study which indicates that coal is by far the cheapest fossil fuel. However, the costs are calculated on a cost per Btu basis for electricity generation only and whilst they do show the benefits for using coal for this purpose, the data should not be purported to be generally applicable to all aspects of coal utilization.

Nevertheless, the promise for the use of coal is there insofar as the data do show the more stable price dependability of coal. If price stability can be maintained at a competitive level and the environmental issues can be addressed successfully, there is a future for coal. A long future and a bright future.

REFERENCES

Ayers, W.B. Jr. 2002. Coalbed gas systems, resources, and production and a review of contrasting cases from the San Juan and Powder River Basins. *American Association Petroleum Geologists Bulletin*, 86: 1853–1890.

Ayers, W.B. Jr. and Kaiser, W.A. (Eds.). 1994. *Coalbed Methane in the Upper Cretaceous Fruitland Formation, San Juan Basin, Colorado, and New Mexico*. Report of Investigations No. 218. Texas Bureau of Economic Geology, Austin, TX.

Bachu, S. and Karsten, M. 2003. Possible controls of hydrogeological and stress regimes on the producibility of coalbed methane in Upper Cretaceous–Tertiary Strata of the Alberta Basin, Canada. *American Association Petroleum Geologists Bulletin*, 87: 1729–1754.

Ball, C.W. 1976. In *Coal Exploration: Proceedings of the First International Coal Exploration Symposium*, W.L.G. Muir (Ed.). Miller Freeman Publications Inc., San Francisco, CA, Chapter 19.

Bradley, H.B. 1992. *Petroleum Engineering Handbook*, 3rd edn. Society of Petroleum Engineers, Richardson, TX.

Cantril, T.C. 1914. *Coal Mining*. Cambridge University Press, Cambridge, U.K., pp. 3–10.

Cargo, D.N. and Mallory, B.F. 1974. *Man and His Geologic Environment*. Addison-Wesley, Reading, MA.

Chadwick, J. 1992. *Mining Magazine*, 167(3): 168.

Chase, F.E. and Ulery, J.P. 1987. Clay veins: Their occurrence, characteristics, and support. Bureau of Mines Report of Investigations No. RI-9060. United States Department of the Interior, Washington, DC.

Coal Voice. 1992. 15(6): 9.

Cuff, D.J. and Young, W.J. 1980. *The United States Energy Atlas*. Free Press, New York.

Cunliffe, B.W. 1984. *Roman Bath Discovered*. Routledge, London, U.K., pp. 14–15.

Dron, R.W. 1925. Notes on cleat in the Scottish Coalfield. *Transactions of the Institution of Mining Engineers*, 70: 115–117.

Eliot, R.C. (Ed.). 1978. *Coal Desulfurization Prior to Combustion*. Noyes Data Corp., Park Ridge, NJ.

Ensminger, J.T. 1977. In *Environmental, Health and Control Aspects of Coal Conversion*, H.M. Braunstein, E.D. Copenhaver, and H.A. Pfuderer (Eds.). Oak Ridge National Laboratory, Oak Ridge, TN, Chapter 2.

Forbes, R.J. 1966. *Studies in Ancient Technology*. Brill Academic Publishers, Boston, MA.

Francis, W. 1961. *Coal: Its Formation and Composition*. Edward Arnold Ltd., London, U.K.

Freese, B. 2003. *Coal: A Human History*. Perseus Publishing, Perseus Books Group, Cambridge, MA.

Ghosh, P.K. 1976. In *Coal Exploration: Proceedings of the First International Coal Exploration Symposium*, W.L.G. Muir (Ed.). Miller Freeman Publications Inc., San Francisco, CA, Chapter 24.

Hessley, R.K. 1990. In *Fuel Science and Technology Handbook*, J.G. Speight (Ed.). Marcel Dekker Inc., New York.

Hessley, R.K., Reasoner, J.W., and Riley, J.T. 1986. *Coal Science*, John Wiley & Sons Inc., New York.

Höök, M. and Aleklett, K. 2009. Historical trends in American coal production and a possible future outlook. *International Journal of Coal Geology*, 78(3): 201–216.

Horwitch, M. 1979. In *Energy Future*, R. Stobaugh and D. Yergin (Eds.). Random House Inc., New York.

Hubbard, H.M. 1991. *Scientific American*, 264(4): 36.

Kavalov, B. and Peteves, S.D. 2007. The future of coal. Report No. EUR 22744 EN. Directorate-General Joint Research Center, Institute for Energy, Petten, the Netherlands.

Laubach, S.E., Marrett, R.A., Olson, J.E., and Scott, A.R. 1998. Characteristics and origins of coal cleat: A review. *International Journal of Coal Geology*, 35: 175–207.

Laubach, S.E. and Tremain, C.M. 1991. Regional coal fracture patterns and coalbed methane development. In *Proceedings, 32nd U.S. Symposium Rock Mechanics*, J.-C. Roegiers (Ed.), Balkema, Rotterdam, the Netherlands, pp. 851–859.

Levine, J.R. 1991a. New methods for assessing gas resources in thin-bedded, high-ash coals. In *Proceedings, 1991 Coalbed Methane Symposium*, Tuscaloosa, AL, Paper 9125, pp. 115–125.

Levine, J.R. 1991b. The impact of oil formed during coalification on generation and storage of natural gas in coalbed reservoir systems. In *Proceedings, 1991 Coalbed Methane Symposium*, Tuscaloosa, AL, Paper 9126, pp. 307–315.

Lindberg, K. and Provorse, B. 1977. In *Coal: A Contemporary Energy Story*, R. Conte (Ed.). Scribe Publishers Inc., New York, Chapter 1.

Malvić, T. 2011. Unconventional hydrocarbon gas sources as a challenge to renewable energies. *Energies*, 4: 1–14.

Mattusch, C. 2008. Metalworking and tools. In *The Oxford Handbook of Engineering and Technology in the Classical World*, J.P. Oleson (Ed.). Oxford University Press, Oxford, U.K., pp. 418–438.

Matveev, A.K. 1976. In *Coal Exploration: Proceedings of the First International Coal Exploration Symposium*, W.L.G. Muir (Ed.). Miller Freeman Publications Inc., San Francisco, CA, Chapter 2.

Miller, C.L. 2007. In *2007 EIA Energy Outlook Modeling and Data Conference*, March 28, 2007. Office of Sequestration, Hydrogen, and Clean Coal Fuels, Office of Fossil Energy, U.S. Department of Energy, Washington, DC.

Mokhatab, S., Poe, W.A., and Speight, J.G. 2006. *Handbook of Natural Gas Transmission and Processing*. Elsevier, Amsterdam, the Netherlands.

Moore, E.S. 1940. *Coal: Its Properties, Analysis, Classification, Geology, Extraction, Uses, and Distribution*. John Wiley & Sons Inc., New York.

Nederlof, M.H. 1988. *Annual Review of Energy*, 13: 95.

Needham, J. and Golas, P.J. 1999. *Science and Civilization in China*. Cambridge University Press, Cambridge, U.K., pp. 186–191.

Neto, J.N. 1976. In *Coal Exploration: Proceedings of the First International Coal Exploration Symposium*, W.L.G. Muir (Ed.). Miller Freeman Publications Inc., San Francisco, CA, Chapter 23.

NRC. 1979. *Energy in Transition 1985–2010*. National Research Council, National Academy of Sciences, W.H. Freeman & Company, San Francisco, CA.

NRC. 1990. *Fuels to Drive Our Future*. National Research Council, National Academy of Sciences, National Academy Press, Washington, DC.

Ogunsola, O.I. 1991. *Fuel Science and Technology International*, 9: 1211.

Read, T.T. 1939. The earliest industrial use of coal. *Transactions of the Newcomen Society*, 20: 119.

Salway, P. 2001. *A History of Roman Britain*. Oxford University Press, Oxford, U.K.

Sanami, Y. 1991. *Enerugi Keizei*, 17(11): 2.

Schobert, H.H., Karner, F.P., Olson, E.S., Kleesattel, D.P., and Zygarlicke, J. 1987. In *Coal Science and Chemistry*, A. Volborth (Ed.). Elsevier, Amsterdam, the Netherlands, p. 355.

Simpson, D.A., Lea, J.F., and Cox, J.C. 2003. Coal bed methane production. In *Proceedings. Production and Operations Symposium*. Society of Petroleum Engineers, Richardson, TX, March 2003, Paper SPE 80900.

Smith, A.H.V. 1997. Provenance of coals from Roman Sites in England and Wales. *Britannia*, 28: 297–324.

Speight, J.G. 2007a. *The Chemistry and Technology of Petroleum*, 4th edn. Taylor & Francis Group, Boca Raton, FL.

Speight, J.G. 2007b. *Natural Gas: A Basic Handbook*. GPC Books, Gulf Publishing Company, Houston, TX.

Speight, J.G. 2008. *Synthetic Fuels Handbook: Properties, Processes, and Performance*. McGraw-Hill, New York.

Speight, J.G. 2011. *Handbook of Industrial Hydrocarbon Processes*. Gulf Professional Publishing, Elsevier, Oxford, U.K.

Stutzer, O. 1940. *Geology of Coal*. University of Chicago Press, Chicago, IL.

Wood, G.H. Jr., Culbertson, W.C., Kehm, T.M., and Carter, M.D. 1982. Coal resources classification system of the United States Geological Survey. In *Proceedings, Fifth Symposium on the Geology of Rocky Mountain Coal 1982*, K.D. Gurgel (Ed.). Utah Geological and Mineral Survey, Bulletin No. 118, pp. 233–238.

Wood, G.H. Jr., Kehn, T.M., Carter, M.D., and Culbertson, W.C. 2003. Coal resource classification system of the U.S. Geological Survey. Geological Survey Circular No. 891. United States Geological Survey, Washington, DC. URL: http://pubs.usgs.gov/circ/c891/geophysical.htm

Yergin, D. 1991. *The Prize: The Epic Quest for Oil, Money, and Power*. Simon & Schuster, New York.

2 Classification

2.1 INTRODUCTION

Coal is a combustible dark-brown-to-black organic sedimentary rock that occurs in *coal beds* or *coal seams*. Coal is composed primarily of carbon with variable amounts of hydrogen, nitrogen, oxygen, and sulfur and may also contain mineral matter and gases as part of the coal matrix. Coal begins as layers of plant matter that has accumulated at the bottom of a body of water after which, through anaerobic metamorphic processes, changes the chemical and physical properties of the plant remains occurred to create a solid material.

Coal is the most abundant fossil fuel in the United States, having been used for several centuries (Table 2.1), and occurs in several regions (Figure 2.1). Knowledge of the size, distribution, and quality of the coal resources is important for governmental planning; industrial planning and growth; the solution of current and future problems related to air, water, and land degradation; and for meeting the short- to long-term energy needs of the country. Knowledge of resources is also important in planning for the exportation and importation of fuel.

The types of coal, in increasing order of alteration, are lignite (brown coal), subbituminous, bituminous, and anthracite. It is believed that coal starts off as a material that is closely akin to peat, which is metamorphosed (due to thermal and pressure effects) to lignite. With the further passing of time, lignite increases in maturity to subbituminous coal. As this process of burial and alteration continues, more chemical and physical changes occur and the coal is classified as bituminous. At this point, the coal is dark and hard. Anthracite is the last of the classifications, and this terminology is used when the coal has reached ultimate maturation.

The degree of alteration (or metamorphism) that occurs as a coal matures from peat to anthracite is referred to as the rank of the coal, which is the classification of a particular coal relative to other coals, according to the degree of metamorphism, or progressive alteration, in the natural series from lignite to anthracite (American Society for Testing and Materials [ASTM], 2011).

Low-rank coals (such as lignite) have lower energy content because they have low carbon content. They are lighter (earthier) and have higher moisture levels. As time, heat, and burial pressure all increase, the rank does as well. High-rank coals, including bituminous and anthracite coals, contain more carbon than lower-rank coals which results in a much higher energy content. They have a more vitreous (shiny) appearance and lower moisture content than lower-rank coals.

There are many compositional differences between the coals mined from the different coal deposits worldwide. The different types of coals are most usually classified by *rank* which depends upon the degree of transformation from the original source (i.e., decayed plants) and is therefore a measure of a coal's age. As the process of progressive transformation took place, the heating value and the fixed carbon content of the coal increased and the amount of volatile matter in the coal decreased.

The method of ranking coals used in the United States and Canada was developed by the ASTM, which is (1) heating value, (2) volatile matter, (3) moisture, (4) ash, and (5) fixed carbon (Chapters 8 and 9).

As a result of these differences, the terminology/nomenclature of coal is complex and variable and is the result of many schemes that have been suggested over the years since coal came into prominence as the object of scientific study.

TABLE 2.1

Timeline of Coal Discovery and Use in the United States

1000	Hopi Indians, living in what is now Arizona, use coal to bake pottery made from clay
1673	Louis Jolliet and Father Jacques Marquette discover *charbon de terra* (coal) at a point on the Illinois River during their expedition on the Mississippi River
1701	Coal is found by Huguenot settlers at Manakin on the James River, near what is now Richmond, Virginia
1748	The first recorded commercial coal production in the United States is recorded from mines in the Manakin area
1762	Coal is used to manufacture, shot, shell, and other war material during Revolutionary War
1816	Baltimore, Maryland, becomes the first city to light streets with gas made from coal
1830	The *Tom Thumb*—the first commercially practical American-built locomotive—is manufactured. Early locomotives that burned wood were quickly modified to use coal almost entirely
1839	The steam shovel is invented and eventually becomes instrumental in mechanizing surface coal mining.
1848	The first coal miners' union is formed in Schuylkill County, PA
1866	Surface mining, then called *strip mining*, begins near Danville, Illinois. Horse-drawn plows and scrapers are used to remove overburden so the coal can be dug and hauled away in wheelbarrows and carts
1875	Coke replaces charcoal as the chief fuel for iron blast furnaces
1881	An explosion and fire killed 38 men at a mine at Almy, Wyoming; this was not the first mine disaster in which there were fatalities but is served to emphasize the dangers of coal mining
1861	Thirteen more men died in another explosion and fire at Almy
1896	Steel timbering is used for the first time at the shaft mine of the Spring Valley Coal Co., where 400 ft of openings are timbered with 15 in. beams
1901	General Electric Co. builds the first alternating current power plant at Ehrenfeld, Pennsylvania, for Webster Coal and Coke Company, to eliminate inherent difficulties in long-distance direct-current transmission.
1903	June 30, an explosion in the Number One Mine in Hanna (Wyoming) killed 171 men
1907	United Mine Workers Union of American become active in seeking protection for miners
1908	March, an explosion in the Hanna (Wyoming) Mine #1 trapped 18 miners; the state mine inspector and 40 rescuers entered the mine and a second explosion occurred killing all 59 men
1912	The first self-contained breathing apparatus for mine rescue operations is used
1912	Between 1912 and 1938, 160 miners were killed in Wyoming in accidents involving five or more men; others continued to be killed in ones and twos
1923	August, 99 men were killed in an explosion in a mine near Kemmerer, Wyoming
1930	Molded, protective helmets for miners are introduced
1937	The shuttle car is invented
1961	Coal becomes the major fuel used by electric utilities to generate electricity
1973	Oil embargo by the Organization of Petroleum Exporting Companies (OPEC) focuses attention on the energy crisis and results in increased demand for coal in the United States
1977	Surface Mining Control and Reclamation Act (SMCRA) is passed
1986	Clean Coal Technology Act is passed
1990	U.S. coal production tops 1 billion tons in a single year for the first time
1996	Energy Policy Act goes into effect, opening electric utility markets for competition between fuel providers
2005	Congress passes and President signs into law the Energy Policy Act of 2005 that promotes increased use of coal through clean coal technologies

2.2 CLASSIFICATION SYSTEMS

Coal classification is almost 200 years old and was initiated by the need to establish some order to the confusion of different coals. Thus, several types of classification systems arose: (1) systems that were intended to aid scientific studies and (2) systems that were designed to assist coal producers and users. The scientific systems of classification are concerned with the origin, composition, and fundamental properties of coal, while the commercial systems are focused on market issues, utilization, technological properties, and the suitability of coal for certain end uses.

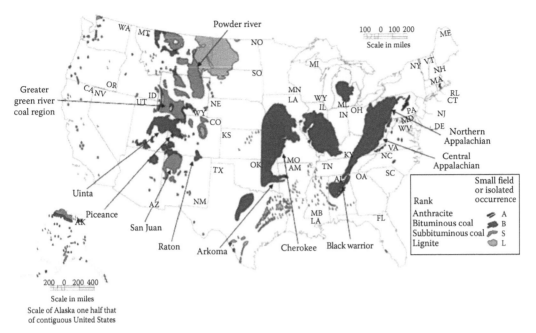

FIGURE 2.1 Coal reserves and distribution in the United States. (From Miller, C.L., in *2007 EIA Energy Outlook Modeling and Data Conference*, March 28, 2007, Office of Sequestration, Hydrogen, and Clean Coal Fuels, Office of Fossil Energy, U.S. Department of Energy, Washington, DC.)

Thus, the purpose of classifying coal in this way is to determine the best use of the various types of coal. The type and chemical composition must be matched to the most suitable end use.

In fact, it is worthy of note that even in the terminology of anthracite there are several variations which, although somewhat descriptive, do not give any detailed indications of the character of the coal. For example, some of the terms that refer to anthracite are black coal, hard coal, stone coal (which should not to be confused with the German *steinkohle* or the Dutch *steenbok*, which are terms that include all varieties of coal with a stone-like hardness and appearance), blind coal, Kilkenny coal, crow coal (from its shiny black appearance), and black diamond. Without seeing any of these particular coals, it would be difficult (perhaps impossible) to categorize the coal for any form of utilization (Thorpe et al., 1878; Freese, 2003).

European and American researchers in the nineteenth and early twentieth centuries proposed several coal classification systems. The earliest, published in Paris in 1837 by Henri-Victor Regnault (1810–1878), classified coal types according to their proximate analysis (determination of component substances, by percentage), that is, by their percentages of moisture, combustible matter, fixed carbon, and ash. This system is still favored, in modified form, by many American coal scientists (Speight, 2005).

Another widely adopted system, introduced in 1919 by the British scientist Marie Stopes (1880–1958), classifies types of coal according to their macroscopic constituents: clarain (ordinary bright coal), vitrain (glossy black coal), durain (dull rough coal), and fusain, also called mineral charcoal (soft powdery coal). Still another system is based on ultimate analysis (determination of component chemical elements, by percentage), classifying types of coal according to their percentages of fixed carbon, hydrogen, oxygen, and nitrogen, exclusive of dry ash and sulfur. Regnault had also introduced the concept of using ultimate analysis in 1837 and the British coal scientist Clarence A. Seyler developed this system in 1899 and greatly expanded it to include large numbers of British and European coals (Parr, 1922; Rose, 1934).

As the importance of the coal trade increased, however, it was realized that some more definite means of classifying coals according to their composition and heating value were desired because the lines of distinction between the varieties used in the past were not sufficiently definite for practical purposes.

Over the years, many classification systems have been proposed for coal. Finally, in 1929, with no universal classification system, a group of coal scientists based in the United States and Canada and working under guidelines established by the American Standards Association (ASA) and the ASTM developed a classification that became the standard in 1936. This system has remained in place since 1938.

Some of the classification systems are currently in use in several countries and included the ASTM system (used in North America), the National Coal Board (NCB) system (UK), the Australian system, and the German and International Systems (for both hard and soft coal) classifications (Carpenter, 1988). Each system involves the use of selected coal properties (chemical, physical, mechanical, and petrographic) as the determining factors leading to classification of coal but the variation of these properties can lead to a poor fit of a coal within the relevant classification system. It is not the purpose to enter into such details here but to give a general description of the ASTM system and the classification systems that relate to the United States through coal trade and other aspects of coal technology.

Generally and from the mineralogical aspects, coal can be defined (sometimes classified) as an organic sedimentary rocklike natural product. The resemblance of coal to a rock is due to its physical nature and composition (Chapters 3 and 4) and the inclusion of a "natural product" term in a general definition is not an attempt to describe coal as a collection of specific, and separate, natural product chemicals. Such chemical species are universally recognized (see, e.g., Fessenden and Fessenden, 1990) and are more distinct chemical entities than coal. But, the designation of coal as a natural product is no stretch of the truth and arises because of the oft-forgotten fact that coal is the result of the decay and maturation of floral remains (which are natural product chemicals) over geological time. Indeed, the organic origins of coal (Chapter 3) and its organic constituents (Chapter 10) are too often ignored. But, there are more appropriate definitions of coal and the manner by which this complex natural product can be classified.

Eventually, the need arose to describe each individual sample of coal in terms that would accurately (even adequately) depict the physical and/or chemical properties (Kreulen, 1948; Van Krevelen and Schuyer, 1957; Francis, 1961). Consequently, the terminology that came to be applied to coal essentially came into being as part of a classification system and it is difficult (if not impossible) to separate terminology from classification to treat each as a separate subject. This is, of course, in direct contrast to the systems used for the nomenclature and terminology of petroleum, natural gas, and related materials (Speight, 1990, 2007). Indeed, the coal classification stands apart in the field of fossil fuel science as an achievement that is second to none insofar as the system(s) allow classification of all the coals that are known on the basis of standardized parameters.

Of particular importance here is the carbon content of the coal, which is part of the basis for the modern classification system of coal. Thus, whereas the carbon content of the world's coals varies over a wide range (ca. 75.0%–95.0% w/w), petroleum, on the other hand, does not exhibit such a wide variation in carbon content; all of the petroleum, heavy oil, and bitumen (natural asphalt) that occur throughout the world fall into the range of 82.0%–88.0% w/w. Hence, little room is left for the design of a standardized system of petroleum classification and/or nomenclature based on carbon content. Before launching into a discussion of the means by which coal is classified, it is perhaps necessary to become familiar with the nomenclature (the equivalent terms nomenclature and terminology are used synonymously here) applied to coal even though this terminology may be based on a particular classification system.

Some mention will be made of the nomenclature and terminology of the constituent parts of coal, i.e., the lithotypes, the macerals, and the microlithotypes (Stopes, 1919, 1935; Spackman, 1958)

(Chapter 4). It is unnecessary to repeat this discussion except to note that this particular aspect of coal science (petrography) deals with the individual components of coal as an organic rock (Chapter 4) whereas the nomenclature, terminology, and classification systems are intended for application to the whole coal. Other general terms that are often applied to coal include rank and grade, which are two terms that describe the particular characteristics of coal.

The kinds of plant material from which the coal originated, the kinds of mineral inclusions, and the nature of the maturation conditions that prevailed during the metamorphosis of the plant material give ride to different coal types. The rank of a coal refers to the degree of metamorphosis; for example, coal that has undergone the most extensive change, or metamorphosis, has the highest rank (determined from the fixed carbon or heating value; Chapters 8 and 9). The grade of a coal refers to the amount and kind of inorganic material (mineral matter) within the coal matrix. Sulfur is, perhaps, the most significant of the non-hydrocarbon materials because of the potential to generate sulfur dioxide during combustion.

For example, within certain parts of the industry, coal can be classified into two main categories: (1) high-grade and (2) low-grade coals, which refer to the energy content of the coal. Among high-grade coals are anthracite and bituminous while low-grade coals include subbituminous coal and lignite. However, the grade is determined mainly by the sulfur content and the amount and type of ash, not recommended for use in coal resource estimations. Definitive statements as to the contents and types of sulfur and ash are preferable—statements indicating high, medium, or low grade are inappropriate without documentation.

Coal, in the simplest sense, consists of vestiges of various organic compounds that were originally derived from ancient plants and have subsequently undergone changes in the molecular and physical structure during the transition to coal (Chapters 3, 4, and 10).

There have been attempts to classify coal as one of the hydrocarbon resources of the earth (Figure 2.2), but the term *hydrocarbon* is used too loosely and extremely generally since coal is not a true hydrocarbon and contains atoms (nitrogen, oxygen, and sulfur) other than carbon and hydrogen–a hydrocarbon (by definition) contains carbon and hydrogen only. Furthermore, the classification of petroleum, heavy oil, and tar sand bitumen into various subgroups (Table 2.2) and the classification of petroleum, for example, by the use of density–gravity data (Table 2.3) offer some hope for the design of a classification system for petroleum (Speight, 2007).

One general system of nomenclature is a system by which coal can be divided into four major types: anthracite coal, bituminous coal, subbituminous coal, and lignite coal which show considerable variation in properties (Table 2.4). Peat is usually not classified as a coal and, therefore, is not included in this system.

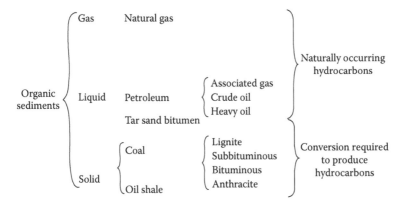

FIGURE 2.2 Classification of the various hydrocarbon and hydrocarbon-producing resources.

TABLE 2.2

Class Subgroups of Fossil Fuels and Derived Materials

Natural Materials	Manufactured Materials	Derived Materials
Asphaltite	Asphalt[b]	Asphaltenes
Asphaltoid	Coke	Carbenes
Bitumen (native asphalt)	Pitch	Carboids
Bituminous rock	Residuum[c]	Oils
Bituminous sand[a]	Synthetic crude oil	Resins
Heavy oil	Tar	
Kerogen	Wax	
Migrabitumen		
Mineral wax		
Natural gas		
Oil sand[c]		
Oil shale		
Petroleum		
Tar sand[c]		

[a] Tar sand—a misnomer; tar is a product of coal processing. Oil sand—also a misnomer but equivalent to the use of "oil shale." Bituminous sand—more correct; bitumen is a naturally occurring asphalt.

[b] Asphalt—a product of a refinery operation; usually made from a residuum.

[c] Residuum—the nonvolatile portion of petroleum and often further defined as "atmospheric" (bp >350°C [>660°F]) or vacuum (bp >565°C [>1050°F]).

TABLE 2.3

Attempted Classification of Petroleum, Heavy Oil, and Tar Sand Bitumen by Density–Gravity

Type of Crude Oil	Characteristics
1. Conventional or light crude oil	Density–gravity range less than 934 kg/m^3 (>20° API)
2. Heavy crude oil	Density–gravity range from 1000 kg/m^3 to more than 934 kg/m^3 (10° API to <20° API)
	Maximum viscosity of 10,000 mPa·s (cP)
3. Extra-heavy crude oil; may also include atmospheric residua (bp >340°C [>650°F])	Density–gravity greater than 1000 kg/m^3 (<10° API)
	Maximum viscosity of 10,000 mPa·s (cp)
4. Tar sand bitumen or natural asphalt; may also include vacuum residua (bp >510°C [>950°F])	Density–gravity greater than 1000 kg/m^3 (<10° API)
	Viscosity greater than 10,000 mPa·s (cP)

TABLE 2.4
Typical Properties of Coal

Sulfur content in coal
- Anthracite: 0.6%–0.77% w/w
- Bituminous coal: 0.7%–4.0% w/w
- Lignite: 0.4% w/w

Moisture content
- Anthracite: 2.8%–16.3% w/w
- Bituminous coal: 2.2%–15.9% w/w
- Lignite: 39% w/w

Fixed carbon
- Anthracite: 80.5%–85.7% w/w
- Bituminous coal: 44.9%–78.2% w/w
- Lignite: 31.4% w/w

Bulk density
- Anthracite: 50–58 (lb/ft^3), 800–929 (kg/m^3)
- Bituminous coal: 42–57 (lb/ft^3), 673–913 (kg/m^3)
- Lignite: 40–54 (lb/ft^3), 641–865 (kg/m^3)

Mineral matter content (as mineral ash)
- Anthracite: 9.7%–20.2% w/w
- Bituminous coal: 3.3%–11.7% w/w
- Lignite: 4.2% w/w

- Anthracite coal is coal of the highest metamorphic rank; it is also known as *hard coal* and has a brilliant luster being hard and shiny. It can be rubbed without leaving a familiar coal dust mark on the finger and can even be polished for use as jewelry. Anthracite coal burns slowly with a pale blue flame and may be used primarily as a domestic fuel.
- Bituminous ignites relatively easily coal burns with a smoky flame and may also contain 15%–20% w/w volatile matter. If improperly fired, bituminous coal is characterized with excess smoke and soot. It is the most abundant variety of coal, weathers only slightly, and may be kept in open piles with very little danger of spontaneous combustion, although there is evidence that spontaneous combustion (Chapter 14) is more a factor of extrinsic conditions such as the mining and storages practices (Chakravorty, 1984; Chakravorty and Kar, 1986) and the prevalent atmospheric conditions.
- Subbituminous coal is not as high on the metamorphic scale as bituminous coal and has often been called black lignite. Lignite is the coal that is lowest on the metamorphic scale. It may vary in color from brown to brown-black and is often considered to be intermediate between peat and the subbituminous coals.
- Lignite is often distinguished from the subbituminous coals by having lower carbon content and a higher moisture content. Lignite may dry out and crumble in air and is certainly liable to spontaneous combustion.

It is obvious that this general system of nomenclature offers little, if anything, in the way of a finite description of the various coals. In fact, to anyone but an expert (who must be presumed to be well versed in the field of coal technology) it would be extremely difficult to distinguish one piece of coal from another. Therefore, the terminology that is applied to different coals is much more logical when it is taken in perspective with the classification system from which it has arisen and becomes much more meaningful in terms of allowing specific definitions of the various types of coal that are known to exist.

Thus, there is a need to accurately describe the various coals in order to identify the end use of the coal and also to provide data which can be used as a means of comparison of the various worldwide coals. Hence, it is not surprising that a great many methods of coal classification have arisen over the last century or so (ASTM, 2011; ISO, 2011).

These methods were, of course, designed to serve many special interests ranging from the more academic classification systems geologists, paleobotanists, and chemists to the more commercial types of classification used by the utility company personnel. However, it became essential as coal rose to a prominent position as a fuel that a system be accepted which would be applied (almost) universally and the terminology of which would allow immediate recognition of a coal type irrespective of the place or country of origin of the coal.

The widespread occurrence of coal and its diversity of uses have, as stated earlier, resulted in the development of numerous classification systems. Indeed, these systems have invoked the use of practically every chemical and physical characteristic of coal. Consequently, the literature available for those wishing to make a thorough study of the various classification systems is voluminous, and it would be futile to enter upon a discussion of all the classification systems here.

Nevertheless, it will be useful to review the major classification systems in current use. In addition, several of the lesser known classification systems are also included because they often contain elements of coal terminology that may still be in current use by the various scientific disciplines involved in coal technology, although they may not be recognized as part of a more formalized classification system.

2.2.1 Geological Age

Coals have at various times been classified according to the geological age in which they were believed to have originated.

For example, coal paleobotanists have noted that three major classes of plants are recognizable in coal: coniferous plants, ferns, and lycopods. Furthermore, these plant types are not usually mixed in a random manner in a particular coal, but it has been observed that one particular class of these plant types usually predominates in a coal bed or seam. Thus, because of the changes in character and predominant types of vegetation during the 200 million years or so of the coal-forming period in the earth's history, it has often been found convenient and, perhaps, necessary to classify coal according to the age in which the deposit was laid down (Table 2.5).

TABLE 2.5
Coal Regions of the World and Age of the Deposits

Period	North America	South America	Europe	Asia	Africa	Australia
Quaternary	l	l	l	l	l	l
Tertiary	a, lvb, SB, L	B, SB, L	B, L	SB, B, L	L	L
Cretaceous	a, LVB, SB, L	SB, B	B, L	b	SB	l
Jurassic	B	—	sb, b, l	a, LVB, b, sb, L	—	b
Triassic	b	—	a, B, sb, l	—	—	—
Permian	b	b	A, B	B	B, c	B
Pennsylvanian	A, LVB, B, c	b	A, LVB, B, sb, l, c	A, LVB, B	A, B, l	LVB, b
Mississippian	a, lvb, B	—	B, sb, l	A, LVB, b	—	—
Devonian	—	—	b	—	—	—

Key: L, l, lignite; SB, sb, subbituminous; B, b, high- and medium-volatile bituminous; LVB, lvb, low-volatile bituminous; A, a, anthracite; c, cannel. Capital letters indicate major deposits, and lowercase letters indicate smaller, less significant deposits.

It should be noted, however, that deposits of vegetable matter are not limited to any particular era or period, but while these deposits occur even in pre-Cambrian rocks, the plants (i.e., terrestrial plants) that were eventually to become coal were not sufficiently abundant until the Devonian Period and it appears that such deposits really became significant during the Carboniferous Period.

2.2.2 BANDED STRUCTURE

Reference has been made elsewhere to the three general classes of coal: banded coal, non-banded coal, and impure coal (White, 1911; Thiessen, 1931) (Chapter 4) and further discussion is not warranted here. Nevertheless, the fact that many coals have a laminated structure consisting of layers which may vary considerably in thickness, luster, and texture has led to an attempt to classify coal by virtue of these differences (Table 2.6).

Since this banded structure persists in all types of coal from lignite to anthracite (although it is most obvious in the bituminous coals), there may, of course, be some merit in such a classification (Stach et al., 1982). However, the failure of such a classification system to take into account the elemental composition of the coal is a serious deficiency. Indeed, a similar statement may be made about all of the classification systems that involve the physical appearance of the coal. To all but the well-initiated, there is little, if any, difference between one piece of coal and another. Therefore, classification systems which rely on a physical property are not only difficult to rationalize but are even more difficult to accept.

Furthermore, the wide variation in the elemental (ultimate) composition of coals, irrespective of the banded structure, is the major objection to classification by physical methods alone.

2.2.3 RANK

An early method that attempted a definitive classification of coals on the basis of their composition and heating value was based on the ratio of the fixed carbon to the volatile combustible matter (C/V.Hc) (Frazer, 1877) in which the ratio of the volatile to fixed combustible matter was a logical basis for the classification of coals. After various attempts to make the fuel ratio of the different coals fit the descriptions of the varieties of coal, it was concluded that coal could be classified according to the fuel ratio within wide limits, and the following divisions were suggested:

Hard-dry anthracite: C/V.Hc = 100:12
Semianthracite: C/V.Hc = 12:8
Semibituminous coal: C/V.Hc = 8:5
Bituminous coal: C/V.Hc = 5:0

TABLE 2.6
Classification by Banded Structure

Designation	Thickness of Bands (mm)	Remarks
Coarsely banded	>2	
Finely banded or stripped	2–0.5	
Microbanded or striated	<0.5	Bands not visible to naked eye
Mixed banded		Both coarse and fine bands
Non-banded (little or no lamination)		Cannel and boghead coals that break with conchoidal fracture

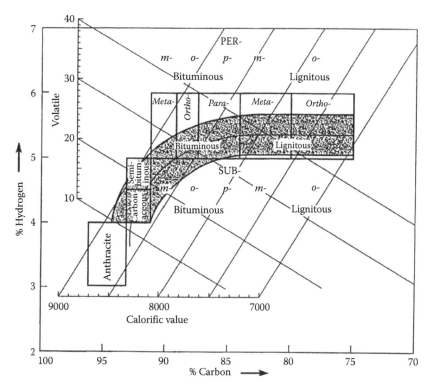

FIGURE 2.3 Classification by the Seyler system.

There are many compositional differences between the coals mined from the different coal deposits worldwide. The different types of coal are most usually classified by *rank* which depends upon the degree of transformation from the original source (i.e., decayed plants) and is therefore a measure of a coal's age. As the process of progressive transformation took place, the heating value and the fixed carbon value of the coal increased and the amount of volatile matter in the coal decreased.

Coal contains significant proportions of carbon, hydrogen, and oxygen with lesser amounts of nitrogen and sulfur. Thus, it is not surprising that several attempts have been made to classify coal on the basis of elemental composition. Indeed, one of the earlier classifications of coal, based on the elemental composition of coal, was subsequently extended. This system (Figure 2.3) offered a means of relating coal composition to technological properties and may be looked upon as a major effort to relate properties to utilization (Tables 2.7 and 2.8). Indeed, for coal below the anthracite rank, and with an oxygen content less than 15%, it was possible to derive relationships between carbon content (C% w/w), hydrogen content (H% w/w), calorific value (Q, cal g), and volatile matter (VM, % w/w):

$$C = 0.59(Q/100 - 0.367VM) + 43.4$$

$$H = 0.069(Q/100 + VM) - 2.86$$

$$Q = 388.12H + 123.92C - 4269T$$

$$VM = 10.61H - 1.24C + 84.15$$

Since these relationships only apply to specific coal types their application is limited, and it is unfortunate that composition and coal behavior do not exist in the form of simple relationships. In fact, classification by means of elemental composition alone is extremely difficult. Nevertheless,

TABLE 2.7
Classification by Degree of Carbonization

Classification	Heating Value (kcal/kg [Dry Basis])	Fuel Ratio	Caking Property
Anthracite	—	4.0 or greater	Noncaking
Bituminous coal	8400 or greater	1.5 or greater	Strong caking
		1.5 or less	
	8100 or greater	1.0 or greater	Caking
		1.0 or less	Weak caking
Subbituminous coal	7800 or greater	1.0 or greater	Weak caking
		1.0 or less	Noncaking
	7300 or greater	—	Noncaking
Brown coal	6800 or greater	—	Noncaking
	5800 or greater		

TABLE 2.8
Classification by Utilization

Anthracite		Anthracite		
Coking coal	Coking coal A	Bituminous coal	Strong-caking coal for coke	Ash content of 8% or less
	Coking coal B			Ash content exceeding 8%
	Coking coal C		Other coal for coke	Ash content of 8% or less
	Coking coal D			Ash content exceeding 8%
Steam coal	Steam coal A		Other	Ash content exceeding 8%
	Steam coal B		Other coal	Ash content of 8% or less
	Steam coal C			Ash content exceeding 8%

Seyler's attempt at coal classification should not be ignored or discredited as it offered an initial attempt at an introspective look at coal behavior.

Furthermore, it is worthy of note here that coal rank is often equated directly to carbon content. Even though this may actually appear to be the case, classification of coal by rank is a progression from high-carbon coal to low-carbon coal (or vice versa) but this is in conjunction with other properties of the coal.

The ASTM (2011) has evolved a method of coal classification over the years and is based on a number of parameters obtained by various prescribed tests for the fixed carbon value as well as other physical properties (Table 2.9) which can also be related to coal use (Figure 2.4). In the ASTM system, coal is classified based on certain gradational properties that are associated with the amount of change that the coal has undergone while still beneath the earth. The system uses selected chemical and physical properties that assist in understanding how the coal will react during mining, preparation, and eventual use.

The ASTM system is based on proximate analysis in which coals containing less than 31% volatile matter on the mineral matter–free basis (Parr formula) are classified only on the basis of fixed carbon, i.e., 100% volatile matter. Coal is divided into five groups: (1) >98% fixed carbon, (2) 98%–92% fixed carbon, (3) 92%–86% fixed carbon, (4) 86%–78% fixed carbon, and (5) 78%–69% fixed carbon. The first three of these groups are *anthracites*, and the last two are *bituminous coals*.

The *subbituminous coals* and lignite are then classified into groups as determined by the calorific value of the coals containing their natural bed moisture, i.e., the coals as mined but free from any moisture on the surface of the lumps. The classification includes three groups of bituminous coals with moist calorific value from above 14,000 Btu/lb (32.5 MJ/kg) to above 13,000 Btu/lb (30.2 MJ/kg);

TABLE 2.9

Coal Classification according to Rank (ASTM System)[a]

Class/Groups	Fixed Carbon Limits (Dry Mineral Matter–Free Basis), %		Volatile Matter Limits (Dry Mineral Matter–Free Basis), %		Gross Calorific Value Limits (Moist,[b] Mineral Matter–Free Basis), MJ/kg		Agglomerating Character
	Equal or Greater Than	Less Than	Greater Than	Equal or Less Than	Equal or Greater Than	Less Than	
Anthracitic							
Meta-anthracite	98			2			Agglomerating
Anthracite	92	98	2	8			
Semianthracite[c]	86	92	8	14			
Bituminous							
Low-volatile bituminous coal	78	86	14	22			Commonly agglomerating[e]
Medium-volatile bituminous	69	78	22	31			
High-volatile A bituminous coal			31		32.6[d]		
High-volatile B bituminous coal					30.2[d]	32.6	
High-volatile C bituminous coal					26.7	30.2	
					24.4	26.7	Agglomerating
Subbituminous							
Subbituminous A coal					24.4	26.7	Non-agglomerating
Subbituminous B coal					22.1	24.4	
Subbituminous C coal					19.3	22.1	
Lignitic							
Lignite A					14.7	19.3	
Lignite B						14.7	

[a] This classification does not apply to certain coals.

[b] Moist refers to coal containing its natural inherent moisture but not including visible water on the surface of the coal.

[c] If agglomerating classify in low-volatile group of bituminous class.

[d] Coals having 69% or more fixed carbon on the dry, mineral matter–free basis shall be classified according to fixed carbon, regardless of gross calorific value.

[e] It is recognized that there may be non-agglomerating varieties in these groups of the bituminous class, and that there are notable exceptions in the high-volatile C bituminous group.

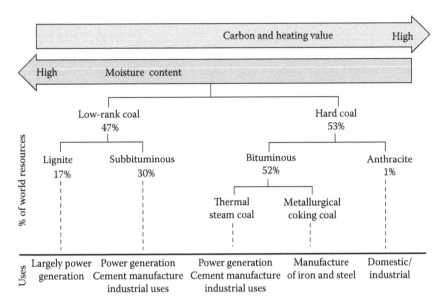

FIGURE 2.4   Coal rank and uses.

three groups of subbituminous coals with moist calorific value below 13,000 Btu/lb to below 8,300 Btu/lb (19.3 MJ/kg); and two groups of lignite coals with moist calorific value below 8,300 Btu/lb. The classification also differentiates between consolidated and unconsolidated lignite and between the weathering characteristics of subbituminous coals and lignite.

These test methods used for this classification system are (as already stated) based on proximate analysis and are (Luppens and Hoeft, 1992)

- *Heating value (calorific value)*, which is the energy released as heat when coal (or any other substance) undergoes complete combustion with oxygen. *Moist calorific value* is the calorific value of the coal when the coal contains its natural bed moisture. The natural bed moisture is often determined as the equilibrium moisture under prescribed standard conditions (Chapter 8). In addition, the agglomerating characteristics of coal are used to differentiate between certain adjacent groups.
- *Volatile matter*, which is the portion of a coal sample which, when heated in the absence of air at prescribed conditions, is released as gases and volatile liquids.
- *Moisture*, which is the water inherently contained within the coal and existing in the coal in its natural state of deposition. It is measured as the amount of water released when a coal sample is heated at prescribed conditions but does not include any free water on the surface of the coal; such free water is removed by air-drying the coal sample being tested.
- *Ash yield*, which is the inorganic residue remaining after a coal sample, is completely burned and is largely composed of compounds of silica, aluminum, iron, calcium, magnesium, and others. The ash may vary considerably from the mineral matter present in the coal (such as clay, quartz, pyrites, and gypsum) before being burned.
- *Fixed carbon value*, which is the remaining organic matter after the volatile matter and moisture have been released. It is typically calculated by subtracting from 100 the percentages of volatile matter, moisture, and ash. It is composed primarily of carbon with lesser amounts of hydrogen, nitrogen, and sulfur. It is often simply described as a coke-like residue. The value is calculated by subtracting moisture, volatile matter, and ash from 100% (Chapter 8).

The basis for such a classification is the proximate analysis (Chapter 8) of the coal resulting in the division of all coals into a series of classes and groups. Coal with a higher-rank than high-volatile B bituminous coal (class II, group 4), i.e., coal having less than 31% w/w volatile matter (daf), is classified according to the fixed carbon content; coal having more than 31% w/w volatile matter (daf) is classified according to the moist calorific value.

This system of classification, in fact, indicates the degree of coalification as determined by these methods of proximate; analysis with lignite being classed as low-rank coal; the converse applies to anthracite. Thus, coal rank increases with the amount of fixed carbon but decreases with the amount of moisture and volatile matter. It is, perhaps, easy to understand why coal rank is often (and incorrectly) equated to changes in the proportion of elemental carbon in coal (ultimate analysis; Chapter 8).

It is true, of course, that anthracite coal usually contains more carbon than bituminous coal which, in turn, usually contains more carbon than subbituminous coal, and so on. Nevertheless, the distinctions between the proportions of elemental carbon in the various coals are not so well-defined as for the fixed carbon and extreme caution is advised in attempting to equate coal rank with the proportion of elemental carbon in the coal.

There have been criticisms of this method of classification because of the variability of the natural bed moisture and the numbering system for the classes of coal. With regard to the natural bed moisture, the fact that it may vary over extremely wide limits has been cited as a distinct disadvantage to using this particular property as a means of classifying coal. In fact, the natural bed moisture is determined under a set of prescribed, and rigorously standardized, conditions, thereby making every attempt to offset any large variability in the natural bed moisture. With regard to the

---

TABLE 2.10

Coals of the United States Classified by Rank

	Coal Rank				Coal Analysis, Bed Moisture Basis						Rank	Rank
No.	Class	Group	State	County	M	VM	FC	A	S	Btu	FC	Btu
1	I	1	PA	Schuylkill	4.5	1.7	84.1	9.7	0.77	12,745	99.2	14,280
2	I	2	PA	Lackawanna	2.5	6.2	79.4	11.9	0.60	12,925	94.1	14,880
3	I	3	VA	Montgomery	2.0	10.6	67.2	20.2	0.62	11,925	88.7	15,340
4	II	1	WV	McDowell	1.0	16.6	77.3	5.1	0.74	14,715	82.8	15,600
5	II	1	PA	Cambria	1.3	17.5	70.9	10.3	1.68	13,800	81.3	15,595
6	II	2	PA	Somerset	1.5	20.8	67.5	10.2	1.68	13,720	77.5	15,485
7	II	2	PA	Indiana	1.5	23.4	64.9	10.2	2.20	13,800	74.5	15,580
8	II	3	PA	Westmoreland	1.5	30.7	56.6	11.2	1.82	13,325	65.8	15,230
9	II	3	KY	Pike	2.5	36.7	57.6	3.3	0.70	14,480	61.3	15,040
10	II	3	OH	Belmont	3.6	40.0	47.3	9.1	4.00	12,850	55.4	14,380
11	II	4	IL	Williamson	5.8	36.2	46.3	11.7	2.70	11,910	57.3	13,710
12	II	4	UT	Emery	5.2	38.2	50.2	6.4	0.90	12,600	57.3	13,560
13	II	5	IL	Vermilion	12.2	38.8	40.0	9.0	3.20	11,340	51.8	12,630
14	III	1	MT	Musselshell	14.1	32.2	46.7	7.0	0.43	11,140	59.0	12,075
15	III	2	WY	Sheridan	25.0	30.5	40.8	3.7	0.30	9,345	57.5	9,745
16	III	3	WY	Campbell	31.0	31.4	32.8	4.8	0.55	8,320	51.5	8,790
17	IV	1	ND	Mercer	37.0	26.6	32.2	4.2	0.40	7,255	55.2	7,610

Source: Considine, D.M., Ed., *Energy Technology Handbook*, McGraw-Hill, New York, 1977.

M, equilibrium moisture, %; VM, volatile matter, %; FC, fixed carbon, %; A, ash, %; S, sulfur, %; Btu, high heating value, Btu/lb; rank Btu, moist mineral matter–free Btu/lb; all calculations by Parr formulas.

numbering system, it has been indicated that the class numbering system should be reversed so that a high number would indicate a high rank.

Nevertheless, in spite of these criticisms, the method has survived and has been generally adopted for use throughout North America as the predominant of classification (Table 2.10).

2.2.4   Coal Survey (National Coal Board, UK)

The use of coal as a popular fuel in Britain, from the time of the industrial revolution, has led to the development of a system of classification in which code numbers are used to denote the different types of coal. The system relies heavily on the coke-forming characteristics of the various coals as well as on the types of coke produced by a standard of coking test (the Gray–King carbonization assay). The system also employs the amounts of volatile matter produced in percentage by the various coals (Table 2.11).

In this classification system, a three-digit code number is used to describe each particular coal (Table 2.11)—as with the ASTM system, the lower numbers are assigned to the higher-rank coal, i.e., anthracite. However, because of the various divisions of this particular system, such approximations have to be made with extreme caution and with careful cross-checking.

TABLE 2.11
National Coal Board (UK) System of Coal Classification

Class	Volatile Matter[a] (% w/w)	Comments	
101	<6.1	Anthracite	
102	3.1–9.0		
201	9.1–13.5	Dry steam coal	Low-volatile steam coal
202	13.6–15.0		
203	15.1–17.0	Coking steam coal	
204	17.1–19.5		
206	19.1–19.5	Low-volatile steam coal	
301	19.6–32.0	Prime cooking coal	Medium-volatile coal
305	19.6–32.0	Mainly heat altered coal	
306	19.6–32.0		
401	32.1–36.0	Very strongly coking coal	High-volatile coal
402	>36.0		
501	32.1–36.0	Strongly coking coal	
502	>36.0		
601	32.1–36.0	Medium coking coal	
602	>36.0		
701	32.1	Weakly coking coal	
702	>36.0		
801	32.1–36.0	Very weakly coking coal	
802	>36.0		
901	32.1–36.0	Noncoking coal	
902	>36.0		

[a] Volatile matter—dry mineral matter–free basis. In coal, those products, exclusive of moisture, given off as gas and vapor determined analytically.

Although the English system does appear to have some merit because of the dependence on two simple physical parameters (i.e., the volatile matter content of the coal and the Gray–King carbonization assay), there are, nevertheless, disadvantages to the method, not the least of which is the susceptibility of the Gray–King assay data to oxidation (weathering) of the coal and, apparently, the time required to conduct the assay.

2.2.5 International System

The International System of coal classification came into being after the World War II as a result of the greatly increased volume of trade between the various coal-producing and coal-consuming nations. This particular system, which still finds limited use in Europe, defines coal as two major types: hard coal and brown coal. For the purposes of the system, hard coal is defined as a coal with a calorific value greater than 10,260 Btu/lb (5700 kcal/kg; 23.9 kJ/kg) on a moist, but ash-free basis. Conversely, brown coal is defined as coal with a calorific value less than 10,260 Btu/lb (5700 kcal/kg; 23.9 kJ/kg). In this system, the hard coals (based on dry, ash-free volatile matter content and moist, ash-free calorific value) are divided into groups according to their caking properties (Chapters 8 and 9). These latter properties can be determined either by the free swelling test or by the Roga test and the caking property is actually a measure of how a coal behaves when it is heated rapidly.

The coal groups are then further subdivided into subgroups according to their coking properties (which may actually appear to be a paradox since the coking properties are actually a measure of how coal behaves when it is heated slowly).

Briefly, *coking coal* is a hard coal with a quality that allows the production of coke suitable to support a blast furnace charge. On the other hand, *steam coal* is all other hard coal not classified as coking coal. Also included are recovered slurries, middlings, and other low-grade coal products not further classified by type. Coal of this quality is also commonly known as thermal coal.

A three-digit code number is then employed to express the coal in terms of this classification system. The first figure indicates the class of the coal, the second figure indicates the group into which the coal falls, and the third figure is the subgroup (Table 2.12). Thus, a 523 coal would be a class 5 coal with a free swelling index (FSI) of 2.5–4 and an expansion (dilatation) falling in the range 0–50.

The class numbers derived by the International System of Coal Classification can be approximated to the coal rank derived by the ASTM system (Figure 2.5) except that the International System uses an ash-free calorific value while the ASTM system employs a mineral matter–free calorific value, which could cause a slight discrepancy in the assignments. The error lies in the fact that the weight of ash obtained by ignition is not the same as the weight of the mineral impurities present in the original coal. The expulsion of water of hydration from the clay/shale-type impurities and the expulsion of carbon dioxide from the carbonates introduce errors that are too large to be ignored. Consequently, a number of correction formulas have been derived which assume that the percentage of mineral matter in coal is numerically equal to 108% of the amount of ash plus 55% of the amount of the sulfur. This latter value takes into consideration the conversion of iron pyrites to iron oxide during the ashing procedure.

The system has been conveniently applied to various North American coals (Table 2.13) and can adequately express the character of these coals in terms of the three-digit code number.

Brown coals, like the hard coals, are also defined in terms of their calorific value (see the previous text) and were also recognized as potential fuels. The International System for the classification of brown coals is based on two inherent characteristics which indicate the value of brown coals as fuels: (1) the total moisture (ash-free basis) and (2) the yield of tar (daf basis). Thus, the six classes of brown coal based on the ash-free, equilibrium moisture content are divided into groups according to the yield of tar on a dry, ash-free basis (Table 2.14). This system indicates

TABLE 2.12

Coal Classification by the International System

Groups (Determined by Coking Properties)			Code Number	Subgroups (Determined by Coking Properties)		
	Alternative Group Parameters				Alternative Subgroup Parameters	
Group Number	Free Swelling Index (Crucible Swelling Number)	Roga Index		Subgroup Number	Audibert-Amu Dilatometer	Gray-King
3	>4	>45	435, 535, 635	5	>140	>G₈
			334, 434, 534, 634, 734	4	>50–140	G₅–G₈
			333, 433, 533, 633, 733	3	>0–50	G₅–G₈
			332a | 332b, 432, 532, 632, 732, 832	2	≈0	E–G
2	2 1/2–4	>20–45	323, 423, 523, 623, 723, 823	3	>0–50	G₅–G₈
			322, 422, 522, 622, 722, 822	2	≈0	E–G
			321, 421, 521, 621, 721, 821	1	Contraction only	B–D
1	1–2	>5–20	212, 312, 412, 512, 612, 712, 812	2	≈0	E–G
			211, 311, 411, 511, 611, 711, 811	1	Contraction only	B–D

The first figure of the code number indicates the class of the coal, determined by volatile matter content up to 33% volatile matter and by calorific parameter above 33% volatile matter.
The second figure indicates the group of coal, determined by caking properties.
The third figure indicates the subgroup, determined by coking properties

(continued)

TABLE 2.12 (continued)
Coal Classification by the International System

Groups (Determined by Coking Properties)			Code Number										Subgroups (Determined by Coking Properties)			
Group Number	Alternative Group Parameters: Free Swelling Index (Crucible Swelling Number)	Alternative Group Parameters: Roga Index	The first figure of the code number indicates the class of the coal, determined by volatile matter content up to 33% volatile matter and by calorific parameter above 33% volatile matter. The second figure indicates the group of coal, determined by caking properties. The third figure indicates the subgroup, determined by coking properties										Sub-group Number	Alternative Subgroup Parameters: Audibert-Amu Dilatometer	Alternative Subgroup Parameters: Gray–King	
0	0–1/2	>0–5	000	100 A	100 B	200	300	400	500	600	700	800	900	0	Non-softening	A
Class number →			0	1		2	3	4	5	6	7	8	9	As an indication, the following classes have an approximate volatile matter content of Class 6: 33%–41% volatile matter Class 7: 33%–44% volatile matter Class 8: 35%–50% volatile matter Class 6: 42%–50% volatile matter		
Class parameters	Volatile matter (dry, ash-free) →		0–3	>3–10: >3–6.5	>3–10: >6.5–10	>10–14	>14–20	>20–28	>28–33	>33	>33	>33	>33			
	Calorific parameter[a] →		—	—		—	—	—	—	>13,950	>12,960–13,950	>10,980–12,960	>10,260–10,980			
Classes: Determined by volatile matter up to 33% volatile matter and by calorific parameter above 33% volatile matter																

Source: Considine, D.M., Ed., *Energy Technology Handbook*, McGraw-Hill, New York, 1977.

Notes:

1. Where the ash content of coal is too high to allow classification according to the present system, it must be reduced by laboratory float-and-sink method (or any other appropriate means). The specific gravity selected for flotation should allow a maximum yield of coal with 5%–10% of ash.
2. 332a > 14%–16% volatile matter. 332b > 16%–20% volatile matter.

[a] Gross calorific value on moist ash-free basis (86°F/96% relative humidity) Btu per pound.

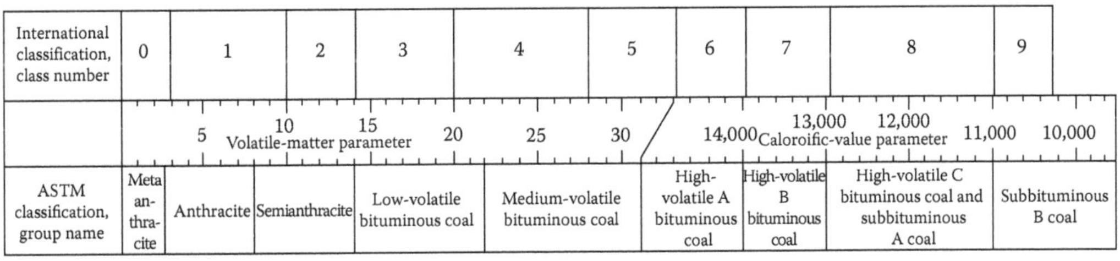

International classification, class number	0	1	2	3	4	5	6	7	8	9
		5 10 15 Volatile-matter parameter 20 25 30 / 14,000 13,000 12,000 Caloroific-value parameter 11,000 10,000									
ASTM classification, group name	Meta an- thra- cite	Anthracite	Semianthracite	Low-volatile bituminous coal	Medium-volatile bituminous coal	High-volatile A bituminous coal	High-volatile B bituminous coal	High-volatile C bituminous coal and subbituminous A coal		Subbituminous B coal

FIGURE 2.5 Comparison of class number (International System) with ASTM classification: (1) parameters in the International System are on an ash-free basis, ASTM parameters are mineral matter–free basis; and (2) no upper limit of calorific value for class 6 and high-volatile bituminous coals.

TABLE 2.13
Comparison of Selected US Coals by the ASTM System and the International System

			Classification	
State	County	Seam	ASTM[a]	International
West Virginia	McDowell	Pocahontas No. 3	LVB	333
Pennsylvania	Clearfield	Lower Kittaning	MVB	435
West Virginia	Monongah	Pittsburgh	HVAB	635
Illinois	Williamson	No. 6	HVBB	734
Illinois	Knox	No. 6	HVCB	821

Source: Considine, D.M., Ed., *Energy Technology Handbook*, McGraw-Hill, New York, 1977.

[a] LVB, low-volatile bituminous; MVB, medium-volatile bituminous; HVAB, high-volatile A bituminous; HVBB, high-volatile B bituminous; HVCB, high-volatile C bituminous.

TABLE 2.14
Classification of Brown Coal by the International System

Group No.	Group Parameter Tar Yield (Dry, Ash-Free) (%)[a]	Code Number					
40	>25	1040	1140	1240	1340	1440	1540
30	20–25	1030	1130	1230	1330	1430	1530
20	15–20	1020	1120	1220	1320	1420	1520
10	10–15	1010	1110	1210	1310	1410	1510
00	≤10	1000	1100	1200	1300	1400	1500
Class no.		10	11	12	13	14	15
Class parameter[b]		≤20	20–30	30–40	40–50	50–60	60–70

Source: Considine, D.M., Ed., *Energy Technology Handbook*, McGraw-Hill, New York, 1977.

[a] Gross calorific value below 10,260 Btu/lb. Moist ash-free basis (86°F/96% relative humidity).

[b] Total moisture, ash-free, %. The total moisture content refers to freshly mined coal. For internal purposes, coals with a gross calorific value over 10,260 Btu/lb (moist, ash-free basis), considered in the country of origin as brown coals but classified under this system to ascertain, in particular, their suitability for processing. When the total moisture content is over 30%, the gross calorific value is always below 10,260 Btu/lb.

the value of the brown coals either as a fuel or as a chemical raw material with the notation that brown coals with a propensity to produce high yields of thermal tar have found more general use in the chemical industry rather than as a fuel.

Critics of the International System note that the parameters used (as well as in some other national systems) to define degree of coalification (rank), i.e., volatile matter and calorific value, are dependent on variable maceral composition. On the other hand, the parameters used in the International Classification to determine the agglutinating and coking properties of coals are competing parameters, instead of following a hierarchy. In order to circumvent such conflicts, classification scheme has been proposed which is based on two primary parameters determined with microscopic techniques: (1) mean maximum reflectance of vitrinite, which is a good single measure of rank and (2) petrographic composition (vitrinite and exinite) as an indication of the type of coal.

A third parameter is chosen to qualify the different classes of coal: volatile matter for anthracitic coals; dilatation for semianthracite and bituminous coals; and calorific value for subbituminous coal and lignite. The scheme is expressed by mean of a code number of four digits, which refers to the rank (first digit), type (second and third digits), and qualification (fourth digit) of coal (Uribe and Pérez, 1985).

2.3   CORRELATION OF THE VARIOUS SYSTEMS

The earlier illustrations indicate that, with the exception of perhaps the geological classification and the banded structure classification system, coal classification can be a complex operation. While only the major classification systems (especially those relevant to the North American and British scenarios) have been illustrated in detail, there are, nevertheless, other systems which have in the past been used in various other countries. Indeed, these other systems for the classification of coal may still find use in the countries of their origin. No attempt has been made to illustrate these systems in the present context, but they are certainly very worthy of mention as an indication of the nomenclature and terminology applied to coals of the various types.

It was suggested earlier (Section 2.1) that it is difficult (if not impossible) to treat coal classification and coal terminology as separate entities. Indeed, the terms applied to the various coals as a result of a particular classification system invariably displace some of the older and less specific names of the coals. Thus, because of the various systems which have originated in the coal-producing (as well as the coal-consuming) nations, several names may have evolved for one particular type of coal (Table 2.15).

Such a profusion of names can make cross-referencing very difficult and it is beneficial for the coal scientist to become as familiar as possible with the various terminologies that exist. Of course, the only means by which this problem could be alleviated would be the establishment of a truly international system for the classification and nomenclature of coal.

2.4   EPILOGUE

Even though there have been serious, and successful, attempts to define coal by means of a variable series of classification systems, there is now, in the modern world, a potential lapse in the information. And that relates to the environmental issues. Other than comparing data (such as is provided by elemental analyses, Table 2.10), there is no other means by which the potential environmental liability of coal usage can be determined. Nor, for that matter, might there ever be, but such a possibility is always worthy of consideration as coal science and technology evolves and moves into the era of clean coal technology (Chapter 22).

TABLE 2.15

Comparison of Various Coal Classification Systems

Classes of the International System

	Parameters		Classes of National Systems							
	Volatile Matter Content	Calorific Value (Calculated to Standard Moisture Content)	Belgium	Germany	France	Italy	Netherlands	Poland	United Kingdom	United States
0	0–3		Maigre	Anthrazit	Anthracite	Antraciti speciali	Anthraciet	Meta-antracyt	Anthracite	Meta-anthracite
1A	3–6.5							Antracyt		
1B	6.5–10					Antraciti comuni		Polantracyt		Anthracite
2	10–14		1/4 gras	Magerkohle	Maigre	Carboni magri	Mager	Chudy	Dry steam	Semianthracite
3	14–20		1/2 gras, 3/4 gras	Esskohle	Demigras	Carboni semi-grassi	Esskool	Polkoksowy, Metakoksowy	Coking steam	Low-volatile bituminous
4	20–28			Feetkhole	Gras a courte flamme	Carboni grassi corta flamma	Vetkool	Ortokoksowy	Medium-volatile coking	Medium-volatile bituminous
5	28–33			Gaskohle	Grasproprement dit	Carboni grassi media flamma		Gazokoksowy	High-volatile	
6	>33 (33–40)	8450–7750				Carboni da gas	Gaskool			High-volatile bituminous A
7	>33 (32–44)	7750–7200		Gas-Flammkohle	Flambantgras	Carboni grassi da vapore	Gasvlamkool	Gazowy		High-volatile bituminous B
8	>33 (34–46)	7200–6100			Flambant see	Carboni secchi	Vlamkool	Gazowplomienny		High-volatile bituminous C
9	>33 (36–48)	<6100						Plomienny		Subbituminous

Source: Van Krevelen, D.W. and Schuyer, J., *Coal Science: Aspects of Coal Constitution*, Elsevier, Amsterdam, the Netherlands, 1957.

REFERENCES

ASTM. 2011. *Classification of Coals by Rank* (ASTM D388). Annual Book of ASTM Standards, Section 05.05. American Society for Testing and Materials, Philadelphia, PA.

Carpenter, A.M. 1988. Coal classification. Report No. IEACR/12. International Energy Agency, London, U.K.

Chakravorty, R.N. 1984. Report No. ERP/CRL 84-14. Coal Research Laboratories, Canada Centre for Mineral and Energy Technology, Ottawa, Ontario, Canada.

Chakravorty, R.N. and Kar, K. 1986. Report No. ERP/CRL 86-151. Coal Research Laboratories, Canada Centre for Mineral and Energy Technology, Ottawa, Ontario, Canada.

Considine, D.M. (Ed.). 1977. *Energy Technology Handbook*. Mc-Graw Hill, New York.

Fessenden, R.J. and Fessenden, J.S. 1990. *Organic Chemistry*, 4th edn. Brooks/Cole Publishing Company, Pacific Grove, CA.

Francis, W. 1961. *Coal: Its Formation and Composition*. Edward Arnold Ltd., London, U.K.

Frazer, P. Jr. 1877. Classification of coal. *Transactions of the American Institute of Mining Engineers*, 6: 430–451.

Freese, B. 2003. *Coal: A Human History*. Perseus Publishing, Cambridge, MA.

ISO. 2011. *Brown Coal and Lignites—Classification by Types on the Basis of Total Moisture Content and Tar Yield* (ISO 2950). International Standards Organization, Geneva, Switzerland.

Kreulen, D.J.W. 1948. *Elements of Coal Chemistry*. Nijgh and van Ditmar N.V., Rotterdam, the Netherlands.

Luppens, J.A. and Hoeft, A.P. 1992. *Journal of Coal Quality*, 10: 133.

Miller, C.L. 2007. In *2007 EIA Energy Outlook Modeling and Data Conference*, March 28, 2007. Office of Sequestration, Hydrogen, and Clean Coal Fuels, Office of Fossil Energy, U.S. Department of Energy, Washington, DC.

Parr, S.W. 1922. The classification of coal. *Industrial and Engineering Chemistry*, 14(10): 919–922.

Rose, H.J. 1934. Coal classification. *Industrial and Engineering Chemistry*, 26: 140–143.

Spackman, W. 1958. *Transactions of the New York Academy of Sciences*, Series 2. 20(5): 411.

Speight, J.G. 2005. *Handbook of Coal Analysis*. John Wiley & Sons Inc., Hoboken, NJ.

Speight, J.G. 1990. In *Fuel Science and Technology Handbook*. Marcel Dekker Inc., New York, Chapter 33.

Speight, J.G. 2007. *The Chemistry and Technology of Petroleum*, 4th edn. Taylor & Francis Group, Boca Raton, FL.

Stach, E., Taylor, G.H., Mackowsky, M-Th., Chandra, D., Teichmuller, M., and Teichmuller, R. 1982. *Textbook of Coal Petrology*, 3rd edn. Gebruder Bornfraeger, Stuttgart, Germany.

Stopes, M. 1935. *Fuel*, 14: 6.

Stopes, M. 1919. *Proceedings of the Royal Society*, 90B: 470.

Thiessen, R. 1931. Bulletin No. B344. United States Bureau of Mines, Washington, DC.

Thorpe, T.E., Green, A.H., Miall, L.C., Rücker, A.W., and Marshall, A. 1878. *Coal: Its History and Uses*. Macmillan & Co., London, U.K.

Uribe, C.A. and Pérez, F.H. 1985. Proposal for coal classification. *Fuel*, 64: 147–150.

Van Krevelen, D.W. and Schuyer, J. 1957. *Coal Science: Aspects of Coal Constitution*. Elsevier, Amsterdam, the Netherlands.

White, D. 1911. Bulletin No. B29. United States Bureau of Mines, Washington, DC.

3 An Organic Sediment

3.1 INTRODUCTION

In the chemical and geological senses, coal is an organic natural product or, more correctly, an organic rock formed from partially decomposed (and decomposing) plant debris (and, in some cases, animal debris) which had collected in regions where waterlogged swampy conditions prevailed (Terres, 1931; Lowry, 1945; Francis, 1961; Lowry, 1963; Murchison and Westoll, 1968; Elliott, 1981; Bend et al., 1991; Bend, 1992). These conditions prevented complete decay of the debris (to carbon dioxide and water) as it accumulated and was subsequently metamorphosed to eventually lead to the material now known as coal.

A sedimentary rock is formed by sedimentation at the surface of the earth and within bodies of water—*sedimentation* being the collective name for processes that cause minerals and organic particles (detritus) to settle and accumulate or minerals to precipitate from solution. In general terms, the debris consisted of trees, rushes, lycopods (a class of plants often loosely grouped as the *ferns* and *club mosses*), and several thousand plant species that have been identified in coal beds from their remnants (Francis, 1961; Van Krevelen, 1961). However, it does appear that none of the species identified in many different coals originated in brackish-water locales.

After sedimentation, diagenesis (the chemical, physical, and biological changes undergone by a sediment after its initial deposition) occurs. Some of these processes cause the sediment to consolidate and a compact, solid substance forms out of loose material. Burial of rocks due to ongoing sedimentation leads to increased pressure and temperature, which stimulates chemical reactions. An example is the reactions by which organic sediment (organic debris) becomes coal. When temperature and pressure increase still further, metamorphism follows diagenesis and the product is a metamorphic rock—in the coal series, anthracite is believed to be a metamorphic (organic) rock.

Similar types of botanical remains are found in all different types (ranks) of coal. But, as anticipated because of local and regional variations in the distribution of floral species (i.e., site specificity), the relative amounts can vary considerably from one site to another (Stutzer, 1940). In addition to variations in the types of flora, there is also the potential for regional variations in the physical maturation conditions (Table 3.1); these include differences such as variations in the oxygen content of the water as well as acidity/alkalinity and the presence (or absence) of microbial life forms. Variations of the plant forms (Table 3.2) due to climatic differences between the geological eras/periods would also play a role in determining the chemical nature of the constituents of the prenatal coal (Bend et al., 1991; Bend, 1992; Speight, 2007).

Thus, it is not surprising that coal differs markedly in composition from one locale to another. Indeed, pronounced differences in analytical properties (Chapter 8) of coal from one particular seam are not uncommon, due not only to the wide variety of plant debris that could have formed the precursor but also to the many different chemical reactions that can occur during the maturation process. Indeed, the development of maturation indices may enable scientists to determine the path (pathways) by which maturation occurred (Richardson and Holba, 1987; Ekwenchi et al., 1991).

Once plant debris has accumulated under the *correct* but difficult-to-define conditions, the formation of peat gradually occurs. Peat is not actually classified as coal and is not usually included in the *coal series*, but it is, nevertheless, believed to be that material that is formed at the initial

TABLE 3.1

Variation in the Conditions that Influence Coal Formation and Composition

Phase	Variables
Selective destruction of carbohydrates and lignin	Water level and degree of brackishness, oxygen level, temperature
Concentration of waxes, resins, cuticles, and spores	Acidity or alkalinity of water, oxidation–reduction potential, time
Accumulation of colloidal organic and inorganic debris	Presence and type of microbes, rate of land subsidence
Burial	

TABLE 3.2

Variations of Plant Forms with Geological Period

Period	Approximate Millions of Years before Present	Botanical Characteristics
Late Silurian	400	First appearance of lignified land plants
Carboniferous	270–350	Large, diverse flora of spore-dispersing plants, including ferns and slender trees with varying amounts of branching and leaf development
Permian	225–270	Seed-fern flora (glossopteris) flourishes all over Gondwanaland[a]
Triassic and Jurassic	180–225	Seed-bearing plants flourish, with conifers and cycadophytes prominent
Cretaceous	70–135	Flowering plants evolve (Angiosperms)
Late Cretaceous	~80	Essentially modern flora in most respects
Tertiary	50–70	Grasses and sedges appear

[a] Gondwanaland is the name given to a once-existing supercontinent comprising what are now South America, Africa, India, Australasia, and Antarctica.

step in the process (Francis, 1961; Bouska, 1981), assuming, of course, that there is a progression through the coal series from the lower-rank materials to the higher-rank coals (Table 3.3 and Figure 3.1).

In fact, peat might be considered to be almost analogous to the material often referred to as *protopetroleum*, that (at best) ill-defined (but more generally) unknown entity that is thought to be first-formed as organic material begins the first of another series of steps to form petroleum (Speight, 2007). There are indications or theories that coal or terrestrial (type III) organic matter can act as a source rock for petroleum (Snowdon, 1991). However, to become coal, peat being a complex chemical entity itself (Given, 1975) must progress through what is generally termed the coalification process—not it is by any means proven that peat is the precursor to coal.

Coalification can be described geochemically as consisting of three processes: (1) the microbiological degradation of the cellulose of the initial plant material, (2) the conversion of the lignin of the plants into humic substances, and (3) the condensation of these humic substances into larger coal molecules. The kind of decaying vegetation, conditions of decay, depositional environment, and the movements of the earth's crust are important factors in determining the nature, quality, and relative position of the coal seams.

However, the coalification process is subject to local chemical and geological influences, and the type of coal formed, being subject to these influences, is not consistent within any country or geological period. While it might be claimed that significant progress has been made in this area, the jury is still out leading to much speculation and the application of a plethora of characterization techniques.

TABLE 3.3
General Description of Different Coal Types and Peat

Rank (Excluding Peat)	Properties
Peat	A mass of recently accumulated to partially carbonized plant debris and is not classed as coal. Peat is organic sediment. It has a carbon content of less than 60% on a dry ash-free basis.
Lignite	Lignite is the lowest rank of coal and sometimes contains recognizable plant structures. By definition, it has a heating value of less than 8300 Btu/lb on a mineral matter–free basis. It has a carbon content of between 60% and 70% on a dry ash-free basis. In Europe, Australia, and the United Kingdom, some low-level lignite is called *brown coal*.
Subbituminous	Subbituminous coal that has been subjected to an increased level of organic metamorphism, which has driven off some of the oxygen and hydrogen in the coal. That loss produces coal with higher carbon content (71%–77% on a dry ash-free basis). Subbituminous coal has a heating value between 8,300 and 13,000 Btu/lb on a mineral matter–free basis. On the basis of heating value, it is subdivided into subbituminous A, subbituminous B, and subbituminous C ranks.
Bituminous	Bituminous is the most abundant rank of coal. It accounts for about 50% of the coal produced in the United States. Bituminous coal has a carbon content of between 77% and 87% on a dry ash-free basis and a heating value that is much higher than lignite or subbituminous coal. On the basis of volatile matter content, bituminous coals are subdivided into low-volatile bituminous, medium-volatile bituminous, and high-volatile bituminous. Bituminous coal is often referred to as *soft coal* but this designation is a Layman's term and has little to do with the hardness of the coal.
Anthracite	Anthracite is the highest rank of coal and has a carbon content of over 87% on a dry ash-free basis. Anthracite coal generally has the highest heating value per ton on a mineral matter–free basis. It is often subdivided into semianthracite, anthracite, and meta-anthracite on the basis of carbon content. Anthracite is often referred to as *hard coal* but this is a Layman's term and has little to do with the hardness of the coal.

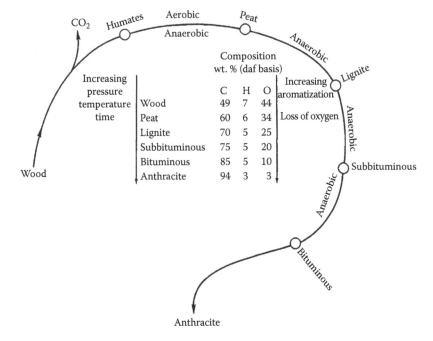

FIGURE 3.1   Simplified representation of the coalification process.

The ultimate goal to relate the structure of the native materials to the character and properties of the final coal is still unresolved and pending.

Indeed, since the resurgence of coal science in the 1980s, it is not obvious that there has been a corresponding increase in understanding the molecular makeup and molecular structure of coal. In fact, understanding the chemical and molecular makeup of coal is a challenge that may never be revolved. A more subtle challenge is to identify the chemical pathways followed during the formation of the combustible black rock that we know.

Indications can be given by tracing the *possible* chemical precursors in the original *mess of pottage* that can lead to a variety of hydrocarbon and heteroatom chemical functional groups in coal and which can be determined by various analytical techniques.

3.2 PRECURSORS OF COAL

Coal formation began during the Carboniferous Period, which is known as the *first coal age* and major coal deposits were formed in every geological period since the Upper Carboniferous Period, 270–350 million years ago.

Layers of plant debris were deposited in wet or swampy regions under conditions that limited exposure to air and complete decay as the debris accumulated, thereby resulting in the formation of peat. As the peat became buried by sediment, it was subjected to higher temperatures and pressures resulting in chemical and physical changes that, over time, formed coal. Cycles of plant debris accumulation and deposition were followed by diagenetic (i.e., biological) and tectonic (i.e., geological) actions, and, depending on the extent of temperature, time, and forces exerted, formed the different ranks of coal.

The concept that coal evolved from the organic debris of plant material is not new and came into being during the latter part of the eighteenth century (Moore, 1940; Hendricks, 1945). Precisely how this theory came about is still open to some speculation but the observation of plant *imprints* within the coal obviously would give rise to the idea of coal being associated with plant material. From that premise, it is one small step to the concept of plant material as the organic precursor to coal.

Indeed, in order to emphasize the complexity of coal and the complexity of the coal-forming processes, some consideration of the potential precursors are worthy of consideration. However, there is also cautious assumption that there are little, if any, differences (natural selection notwithstanding) between the natural product chemicals known now and those in existence at the time of deposition of the coal beds; perhaps a reasonable assumption, but it is fraught with uncertainties.

In fact, it may be better to garner a general understanding of the coal-forming processes through a general knowledge of natural product (naturally occurring) chemicals (Orheim, 1979). This is preferable to attempting to understand the detailed (and unintentionally misrepresented) chemistry that is often promoted as the result of a series of artificial (and potentially erroneous) conditions of coal formation, and the ensuing changes in the chemistry, all the while remembering the nature of the postulated precursors and the nature of coal.

In addition, although there are questions about the timing (sequential and concurrent) of the events which occurred during coal formation, there is the need to adhere to a convenient terminology. Hence, for the purposes of this discussion, some of the more convenient terminology will be employed. For example, the term "peat swamp" (in reality, a complex chemical and environmental entity) (Given, 1975) is used to mean the initial products of the anaerobic decay of the fallen trees (woody material). It is not, however, meant to show any compliance with the consecutive theory in which peat is gradually converted (*metamorphosed*) to coal by a series of sequential chemical and physical transformations. The converse is also true insofar as use of convenient, perhaps archaic, terminology does not mean rejection of a theory which advocates the concepts of concurrent chemical transformation without the insertion of coal types as intermediate products in the sequence.

On the other hand, the term organic debris is used globally in this text to mean any of the material after degradation or decay has commenced but before coalification proper has ensued.

TABLE 3.4
Types of Macromolecules (Other than Proteins) in Plants

Polymer	Structural Character	Plant Organ or Tissue
Lignin	Amorphous three-dimensional alkylphenolic	Wood
α-Cellulose	Partly crystalline β-1→4 glucan	Most tissues
Hemicelluloses	Various mixed sugar polymers; monomers include uronic acids	Wood
Suberin	Copolymer of long-chain fatty acids with phenolic acids	Bark, roots
Cutin	Polymers of ω-hydroxy fatty acids	Cuticle of leaf and stem
Sporopollenin	Ladder copolymer of carotenoids and fatty acids (?)	Exine of spores and pollen
Flavolans	Insoluble polymers of condensed tannins, phlobaphenes, polyphenolic	Inside cells of senescent leaf, inside some cells of stem

The makeup of the organic portion of plants consists of carbohydrates, lignin, and proteins, as well as other polymers (Table 3.4). Indeed, it is these higher polymers that are often considered as contributors to the organic substance of coal. But, of course, the relative amount of each of these constituents varies greatly with the particular species of plant as well as the relative stage of growth of the plant.

Furthermore, recognition of the fact that each of the earlier compound types is a very broad classification (and does not make any attempt to include individual molecular types) indicates the general complexity of the original plant debris. It is this fact alone (probably more than any other single contribution) that determines the complexity of the final *coal molecule*.

Nevertheless, it is possible (in an attempt to simplify an already complex situation) to acknowledge some generalities; the lignin content of plants may fall within the range 10%–35% w/w while in woody material the lignin may be within the narrower range of 25%–30% w/w (all of these approximations are given as percentages of the dry weight of the plant material). In fact, the protein content of grasses has been assigned to fall within the 15%–20% w/w while woody materials contain little, if any, protein constituents; the cellulose (carbohydrate) content of grasses falls in the 20%–55% w/w range (all of these approximate ranges are given as percentages of the dry weight of the plant material).

Thus, from these broad generalities alone, plant debris is complex and the original plant debris that formed the precursors to coal may have been even more complex (depending on location and the circumstances leading to the deposition). Even the most competent scientific detective must admit that the problem is a worthy one and that *the plot has thickened* with time.

Indeed, the concept of a *coal structure* (often referred to as an *average structure for coal*) has continued for several decades and it is very questionable, in the minds of many scientists and engineers, as to whether or any progress has been made down the highways and byways of uncertainty than was the case some 40 years ago. There are those who can, and will, argue convincingly for either side of this question. Or it might be wondered if (even denied that) there is a need to define coal in terms of a distinct molecular structure (Chapter 10).

Giving consideration to (1) the different chemical precursors of coal (as presented later), (2) the varied geological environment, (3) the maturation conditions, and (4) the variation of coal analysis within a seam, the rationale for an average structure of coal defeats itself.

3.2.1 LIGNIN

Lignin (Sarkanen and Ludwig, 1971) has been considered one of the most important substances involved in the transformation of plant constituents into coal (for discussion of this aspect of coal formation, please see Francis, 1961; Van Krevelen, 1961; Murchison and Westoll, 1968; Hyatsu et al., 1979, 1981; Lukoshko et al., 1979; Bredenberg et al., 1987; Hatcher and Lerch, 1991; Bend, 1992).

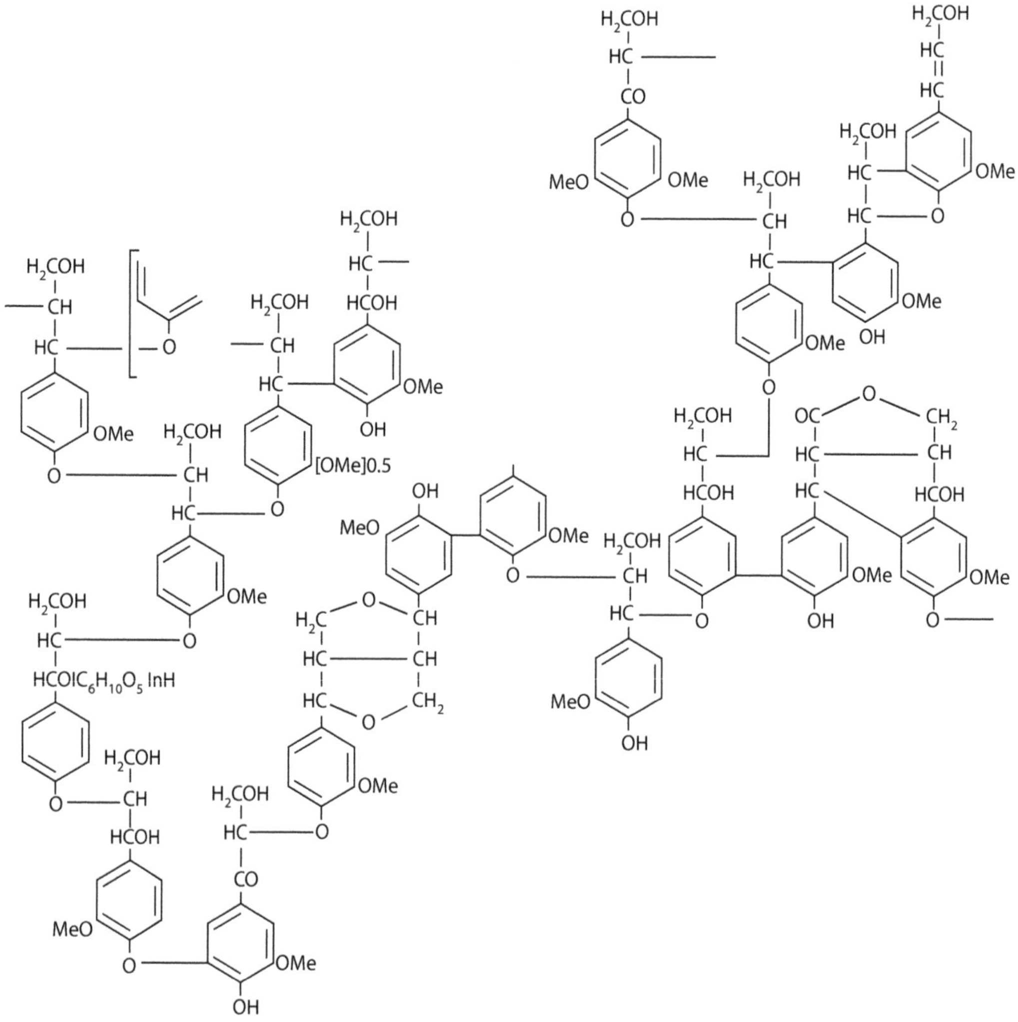

FIGURE 3.2   Segment of a hypothetical structure for lignin.

While the actual molecular structure of lignin (Figure 3.2) remains largely unknown and speculative, some information has been afforded by the identification of more specific molecules from the lignins of various plants, for example, coniferyl alcohol (Formula **1**; Figure 3.3) is the main constituent of the lignin of conifers (Masuku, 1992) while the lignin of deciduous trees contains sinapyl alcohol (Formula **2**; Figure 3.3) and the lignin of graminaceans (grasses) contains the somewhat less complex p-coumaryl alcohol (Formula **3**; Figure 3.3).

Lignin is regarded as a polymer (or mixed polymer) of any one (or more) of these alcohols. The extent of the amount of each particular alcohol has been presumed to be recognizable from determinations of the methoxyl (−OCH) content of various lignin species (Table 3.5), but the complication of nitrogen in the lignin is also evident. This latter problem has not been fully resolved and therefore it is not known whether the nitrogen is inherent in the lignin (i.e., chemically combined and part of the lignin substance) or whether the nitrogen arises by virtue of the presence of nitrogen-containing molecules that are difficult to separate from the lignin.

Nevertheless, it has been possible to make some general deductions about the nature of the linkages in this complex natural product. The ring carbon–carbon linkages are of the diphenyl type (Formula **1**; Figure 3.4) and there are also a variety of aliphatic carbon–carbon linkages between

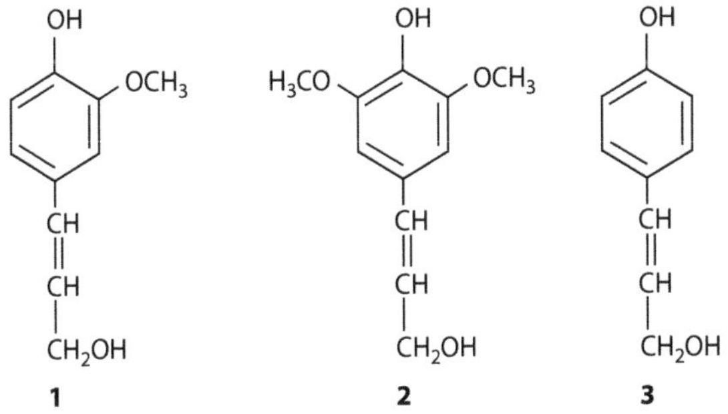

FIGURE 3.3 Monomeric units found in lignin.

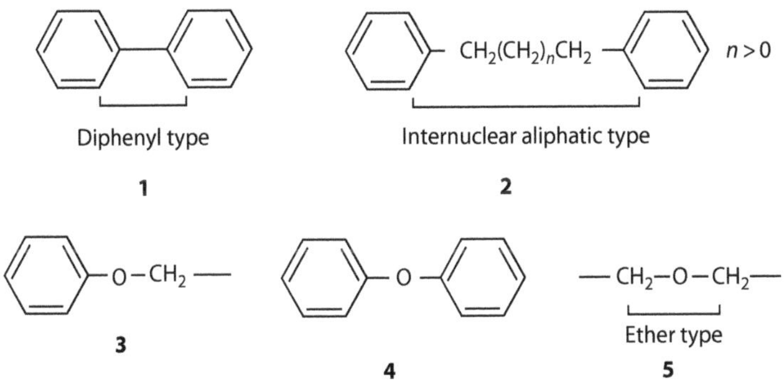

FIGURE 3.4 Various bond types that occur in lignin.

TABLE 3.5
Methoxyl and Nitrogen Content
(% w/w) in Lignin from Various Sources

Lignin Source	OCH$_3$	N
Hardwoods	20.0	0
Softwoods	15.0	0.2–0.3
Graminaceans	10.0	1.2–1.6
Leguminoseans	5.0	2.9–3.4

Source: Murchison, D. and Westoll, T.S., Eds., *Coal and Coal Bearing Strata*, Elsevier, New York, 1968.

---

the aromatic rings (Formula **2**; Figure 3.4). In addition, the molecule of lignin contains various ether linkages (Formulas **3** through **5**; Figure 3.4). In addition, investigations with lignin from conifers have shown that diphenyl-type linkages account for approximately 25% of the ring–ring linkages with approximately 50% of the monomers being connected through carbon–carbon linkages. The remainder is connected by means of ether bonds. Thus, while data of this type do not illustrate the molecular structure of a complex molecule such as lignin, they do illustrate the bonds which either may remain intact or may be ruptured during the maturation process.

There is little hard evidence to support the theory that lignin is the major organic precursor of coal. The nature of lignin has been well researched but there are still many unknowns. Thus, the concept of a largely unknown chemical precursor (lignin) forming, or being incorporated into, a largely unknown chemical product (coal) with the claims of an understanding of the precise chemistry may seem, and is, somewhat paradoxical.

3.2.2 CARBOHYDRATES

The simple sugars, or monosaccharides, are the building blocks of the more complex carbohydrates and they occur widely throughout nature, even in ancient sediments (Swain, 1969). Sugars are actually polyhydric alcohols containing five to eight carbon atoms, although the five- and six-carbon sugars are by far the most common. Although the simple formulas of the sugars indicate that they contain a carbonyl (aldehyde or ketone) function, this is not actually the case. The carbonyl functions usually occur in combination with one of the hydroxyl functions in the same molecule to form a cyclic hemiacetal or a hemiketal (Table 3.6).

Insofar as the monosaccharides do occur as such in nature, it is more common to find the sugars occurring naturally in pairs (disaccharides) or in threes (trisaccharides) and, more likely, as the high-molecular-weight polysaccharides (Table 3.7). It is the polysaccharides which most probably contribute to the source material, especially the two well-known polysaccharides cellulose and starch. The fibrous tissue in the cell wall of plants and trees contains cellulose and starch also occurs throughout the plant kingdom in various forms but usually as a food reserve. The chemical composition of starch vanes with the source but in any one starch there are two structurally different polysaccharides. Both usually consist entirely of glucose units but one is a linear structure (amylose) whereas the other is a branched structure (amylopectin).

Moreover, in contrast to the somewhat speculative chemistry of lignin, the structural chemistry of many polysaccharides (especially cellulose and starch) is generally more well-defined (Figure 3.5). For example, the molecular formula of cellulose is $(CHO_5)_n$, where n may be on the order of several thousand. Cellulose can be hydrolyzed by strong hydrochloric acid to give high (95%–96%) yields of glucose, indicating that glucose is the predominant monomeric unit of cellulose.

Hence, a possible molecular structure for cellulose is based on the glucose stereochemical arrangement (Formula 1; Figure 3.6). In addition, the cellulose molecule is not planar but has a "screw axis" where each glucose unit is virtually at right angles to the next, and although free rotation about the –C–O–C– bond might at first appear possible, steric effects appear to limit the amount of rotation that can occur. Furthermore, close packing of the atoms (in addition to the hindered rotation) gives rise to a rigid molecule and the hydrogen bonding that holds the long chains together gives cellulose a three-dimensional network.

While this description of cellulose presents a very simple description of the molecule for the present purposes, it is nevertheless not the complete structure. For example, cellulose very rarely gives 100% yields of glucose on hydrolysis, and the possibility of other groups which are not necessarily carbohydrate in nature for origin has also been raised. In short, the chemical structure of cellulose is much more complicated than that presented here but the structure outlined here should suffice to illustrate the general concepts of the structural chemistry of cellulose. It might do well to recall this anomaly of a well-defined structure being not so well-defined when the results of well-defined degradation experiments are considered as aids to defining the *structure of coal* (Chapter 10).

Being also a polysaccharide, the chemistry of starch (Figures 3.7 through 3.9) is often considered to parallel the chemistry of cellulose. Although this may not be entirely the case to the skilled biochemical investigator, it will suffice for the present text where the main point is to illustrate the complexity of the molecular entities that are postulated to have made some (perhaps significant) contributions to the plant debris, eventually forming coal (Francis, 1961; Van Krevelen, 1961).

Thus, speaking in very general terms, two high-molecular-weight polysaccharides (starch and cellulose) having the general formula $(C_6H_{10}O_5)_n$, where n may represent several hundred or even several thousand units, may have been incorporated into the plant debris and were eventually incorporated, albeit in an altered form, into the coal structure.

TABLE 3.6
Structure and Occurrence of Common
Five- and Six-Carbon Sugars

Sugar	Open Chain	Cyclic

A. Pentose

1. L-Arabinose
   Glycosides
   Polysaccharides

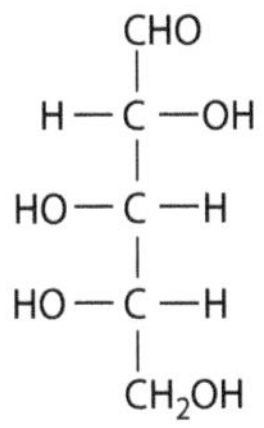

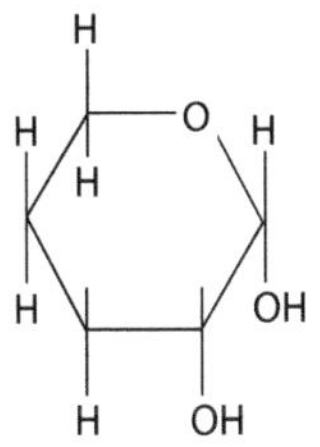

CHO
H—C—OH
HO—C—H
HO—C—H
CH$_2$OH

β-anomer

2. D-Ribose
   Nucleic acids
   Vitamins
   Coenzymes

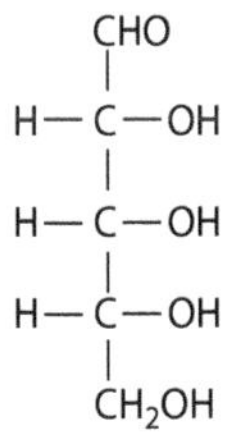

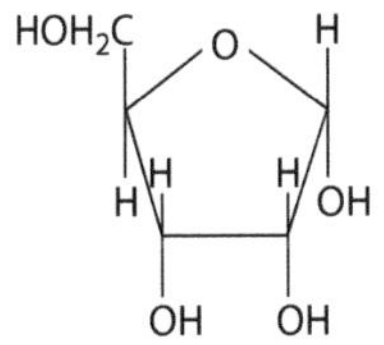

CHO
H—C—OH
H—C—OH
H—C—OH
CH$_2$OH

α-anomer

3. 2-Deoxy-D-Ribose
   Glycosides

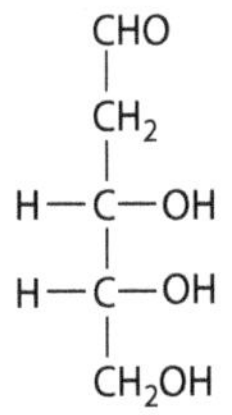

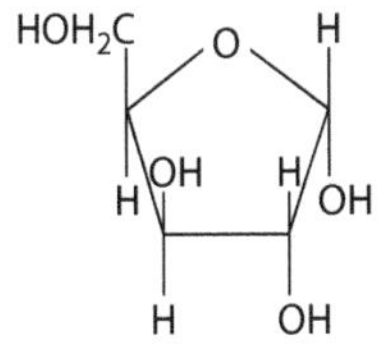

CHO
CH$_2$
H—C—OH
H—C—OH
CH$_2$OH

α-anomer

4. D-Xylose
   Polysaccharides

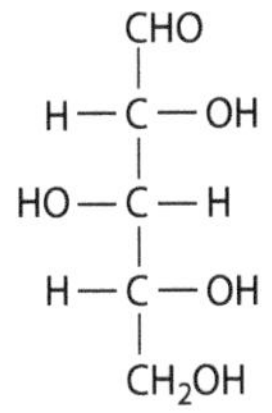

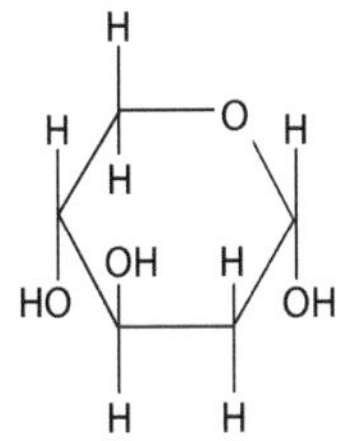

CHO
H—C—OH
HO—C—H
H—C—OH
CH$_2$OH

α-anomer

B. Hexose

1. D-Glucose

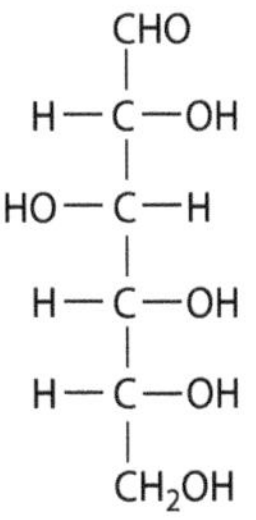

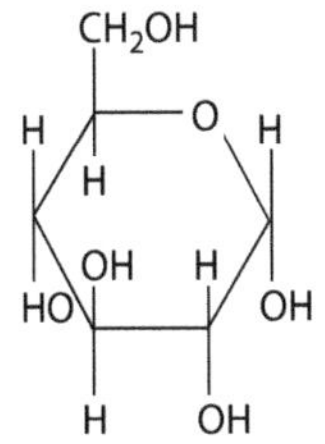

CHO
H—C—OH
HO—C—H
H—C—OH
H—C—OH
CH$_2$OH

(continued)

TABLE 3.6 (continued)
Structure and Occurrence of Common
Five- and Six-Carbon Sugars

Sugar	Open Chain	Cyclic
2. D-Fructose Honey Fruit juices		
3. D-Mannose Polysaccharides		
4. D-Galactose Oligosaccharides Polysaccharides		

3.2.3 PROTEINS

Proteins are nitrogen-containing organic substances which occur in the protoplasm of all plant and animal cells (Hare, 1969). The composition of proteins varies with the source, but a general range for protein composition can be deduced (Table 3.8) with other elements, for example, phosphorous (nucleoproteins) and iron (hemoglobin) also being present in trace amounts in certain proteins.

The characteristic structural feature of proteins is a chain (or ring) of amino acids joined together by amide linkages (Figure 3.10). Peptides are, in effect, a similar class of compounds but the distinction between a protein and a peptide is not always clear. One arbitrary choice is to use a molecular weight of 10,000 as a distinction between the two classes with proteins having a molecular weight in excess of 10,000 while peptides have a molecular weight below this value. The distinction might also be made in terms of differences in physical properties such as the degree of hydration and spatial arrangement of the molecules. For example, the naturally occurring peptides have relatively short, flexible chains and, although hydrated in aqueous solution, can be reversibly dehydrated. On the other hand, proteins have very long chains which have definite conformational character with water molecules filling the interstices.

TABLE 3.7
Natural Carbohydrates[a]

Monosaccharides
Pentoses ($C_5H_{10}O_5$)
    L-Arabinose
    D-Ribose
    D-Xylose
Hexoses ($C_6H_{12}O_6$)
    D-Glucose
    D-Fructose
    D-Galactose
    D-Mannose
Heptoses ($C_7H_{14}O_7$)
    D-Sedoheptulose

Oligosaccharides
Disaccharides
    Sucrose (D-glucose + D-fructose)
    Maltose (D-glucose + D-glucose)
    Lactose (D-galactose + D-glucose)
Trisaccharides
    Raffinose (D-glucose + D-fructose + D-galactose)

Polysaccharides
Plants
    Starch
    Cellulose
    Gum Arabic
    Agar
Animals
    Glycogen

[a] The prefixes D-(dextro) and L-(levo) indicate stereoisomers.

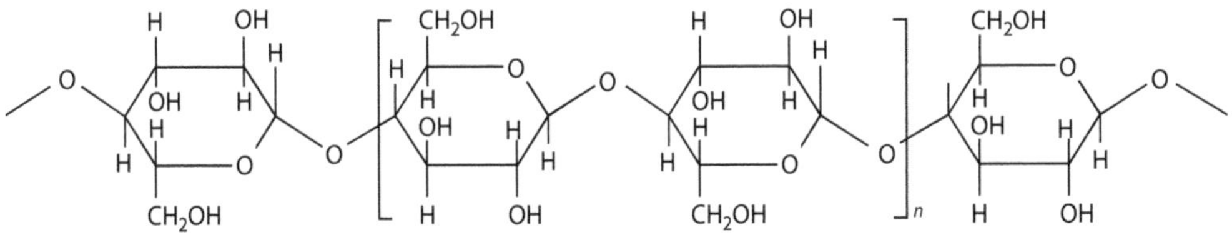

FIGURE 3.5 Simplified molecular structure of cellulose.

The special character of proteins derives not only from the length and complexity of the chains but also from the way in which they are synthesized by organisms in an often metastable form that can be easily and conveniently altered (denatured). As an aside, the boiling of an egg may be used as an example of such changes (denaturing) that can be brought about conveniently to proteins (in this case the egg albumin). In addition, the three-dimensional (spatial) character of proteins has an interesting arrangement that involves an α-helix.

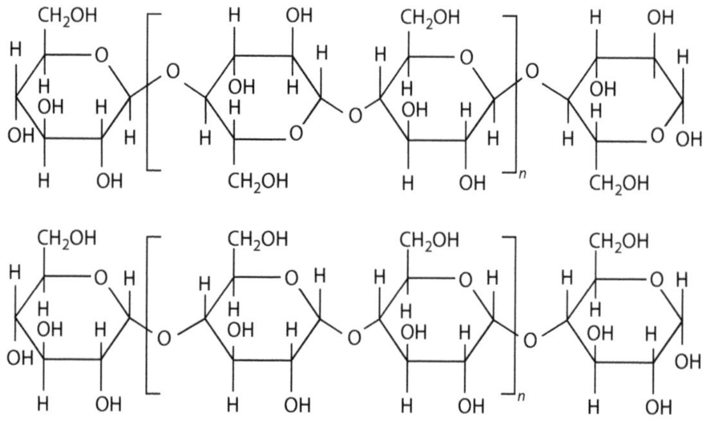

FIGURE 3.6 Simplified structures for cellulose and starch (see also Figures 3.7 through 3.9).

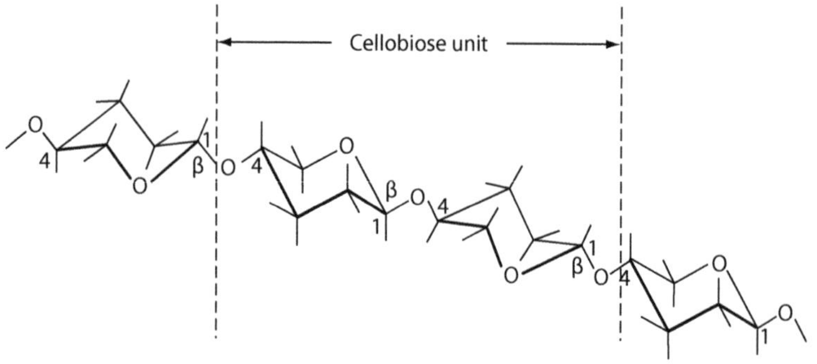

FIGURE 3.7 Three-dimensional representation of cellulose.

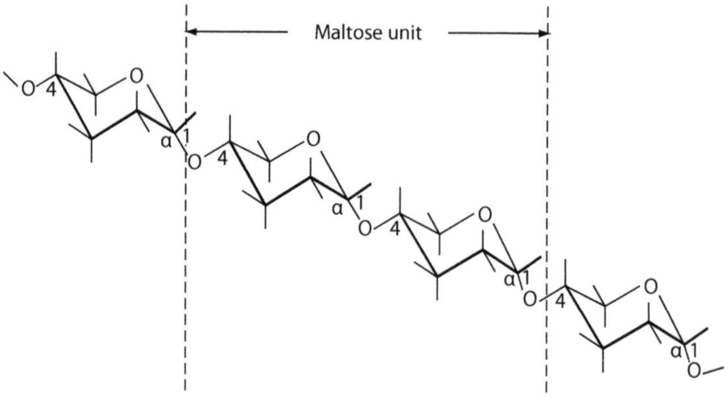

FIGURE 3.8 Three-dimensional representation of starch (amylose).

The principal feature of this helix is the coiling of peptide chains in such a way as to form hydrogen bonds between the amide hydrogen atom(s) and the carbonyl groups at spatially convenient points in the chain (Figure 3.11). Such bonds are due to the polarization of various organic functional groups where adjacent hydrogen is positively polarized and a heteroatom (an atom other than carbon and hydrogen, usually oxygen) is negatively polarized. The strengths of these bonds are usually of the order of 5–10 kcal (2.1–4.2 kJ) compared with 60–100 kcal (25.1–41.9 kJ) for the conventional organic σ-bonds.

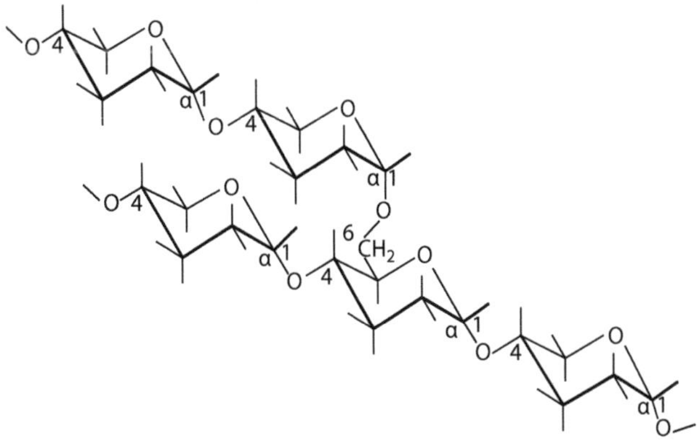

FIGURE 3.9   Three-dimensional representation of starch (amylopectin).

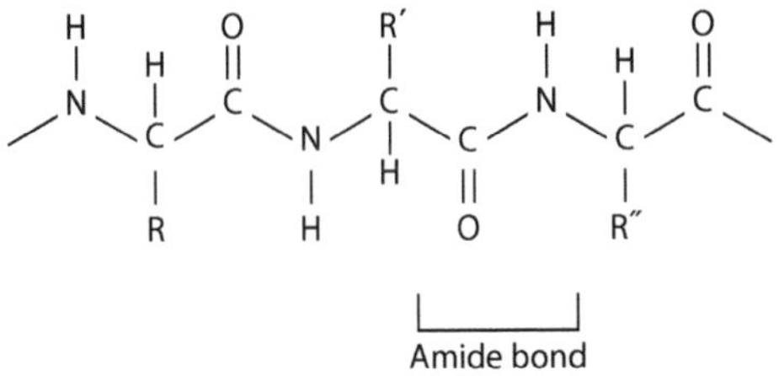

FIGURE 3.10   Representation of part of a protein (peptide) chain.

TABLE 3.8
Range of Protein Composition

Component	Percent by Weight
Carbon	47–50
Hydrogen	6–7
Nitrogen	15–18
Oxygen	23–26
Sulfur	<1

There is, however, the potential for the formation of much more stable hydrogen bonds within the protein helix resulting in the formation of *ring* systems which tend to confer an additional stability (10–20 kcal; 4.2–8.4 kJ) on the system.

The amino acid side chains lie outside the coil of the helix. It should be noted, however, that not all proteins are perfect α-helices since steric effects between certain amino acid side chains may be sufficient to influence the stability of the helix.

Proteins occur throughout nature in a wide variety of sizes and shapes (Table 3.9) and many proteins contain metals such as iron, zinc, and copper which, in turn, are intimately involved in the physiological (biological) functions of the molecules to which they are bound.

3.2.4   OILS, FATS, AND WAXES

Oils, fats, and waxes are common constituents of plants and usually consist of terpene hydrocarbons and their derivatives or alcohols, acids, and esters (Parker, 1969; Streibl and Herout, 1969; Wollrab and Streibl, 1969).

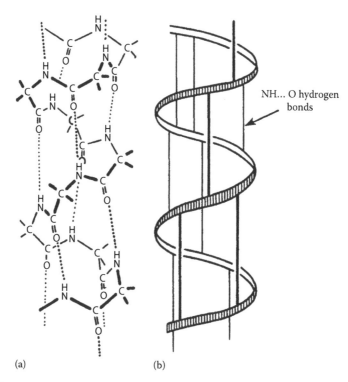

(a)   (b)

FIGURE 3.11   Representation of protein structure showing (a) atomic and (b) spatial arrangement of the helix. (From Roberts, J.D. and Caserio, M.C., *Basic Principles of Organic Chemistry*, Benjamin Book Publishers, Inc., New York, 1964.)

TABLE 3.9
Commonly Occurring Proteins

Name	Molecular Weight	Occurrence
Insulin	6,000	Pancreas
Ribonuclease	13,000	Pancreas
Lysozyme	14,600	Egg white
α-Chymotrypsin	25,000	Pancreas
Papain	21,000	Papaya
Hemoglobin	66,700	Red blood corpuscles
Catalase	250,000	Liver and kidney
Fibrogenin	330,000	Blood plasma
Tobacco mosaic virus	41,000,000	Infected tobacco plants

Source: Roberts, J.D. and Caserio, M.C., *Basic Principles of Organic Chemistry*, Benjamin Book Publishers, Inc., New York, 1964, p. 725.

However, oils occur in much greater concentrations in fruits and seeds. The so-called essential oils are usually sweet, or aromatic smelling organic compounds which may also contain sulfur and nitrogen in addition to the usual carbon and hydrogen. These oils are often of the terpene class with the general formula $(C_5H_8)_n$, where $n > 1$, or may be compound types similar to camphor or other oxidation products associated with the basic terpene structure.

TABLE 3.10
Fatty Acids and Their Natural Sources

Acid	Structure	Natural Source
Butyric	$CH_3CH_2CH_2CO_2H$	Cow butterfat
Hexanoic	$CH_3(CH_2)_4CO_2H$	Goat butterfat
Lauric	$CH_3(CH_2)_{10}CO_2H$	Laurel oil
Myristic	$CH_3(CH_2)_{12}CO_2H$	Nutmeg
Palmitic	$CH_3(CH_2)_{14}CO_2H$	Palm oil
trans-Crotonic	$CH_3CH=CHCO_2H$	Croton oil
Oleic	$CH_3(CH_2)_7CH=CH(CH_2)_7CO_2H$	Olive oil
Tariric	$CH_3(CH_2)_{10}C\equiv C(CH_2)_4CO_2H$	Lichens
Sorbic	$CH_3CH=CHCH=CHCO_2H$	Mountain ash berries
Linoleic	$CH_3(CH_2)_4CH=CHCH_2CH=CH(CH_2)_7CO_2H$	Cottonseed oil
Linolenic	$CH_3CH_2(CH=CHCH_2)_3(CH_2)_6CO_2H$	Linseed oil
Ricinoleic	$CH_3(CH_2)_5CH(OH)CH_2CH=CH(CH)\ COH$	Castor oil
Licanic	$CH_3(CH_2)_{13}CO(CH_2)_2CO_2H$	Oiticica oil
Eleostearic	$CH_3(CH_2)_3(CH=CH)_3(CH_2)_7CO_2H$	
Sterculic	$CH_3(CH_2)_7C=C(CH_2)_7CO_2HCH_2$	Lichens
Mycomycin	$HC=C-C\equiv C-CH-CH=C=CHCH=CHCH=CHCH_2CO_2H$	Lichens
Tuberculostearic	$CH_3(CH_2)_7CH(CH_3)(CH_2)_8CO_2H$	Tubercle bacillus

Source: Speight, J.G., *The Chemistry and Technology of Petroleum*, 1st edn., Marcel Dekker Inc., New York, 1980.

Alternately, oils found in plants may also be of the fatty acid (Table 3.10) or glyceride type or may even be esters of the higher monohydric alcohols and fatty acids. The acids most commonly found as glycerides, or esters in plants are

$$C_{15}H_{31}COOH \qquad C_{17}H_{35}COOH \qquad C_{17}H_{33}COOH$$
Palmitic acid   Stearic acid   Oleic acid

In the glycerides, any of the earlier fatty acids can combine with glycerol to yield a symmetrical glyceride in which the acid radicals are the same:

$CH_2OCOC_{15}H_{31}$ $CH_2OCOC_{17}H_{35}$
$CHOCOC_{15}H_{31}$ $CHOCOC_{17}H_{35}$
$CH_2OCOC_{15}H_{31}$ $CH_2OCOC_{17}H_{35}$
Tripalmitin   Tristearin

Different fatty acids may combine with one molecule of glycerin to produce a mixed glyceride.

$CH_2OCOC_{17}H_{35}$ $CH_2OCOC_{17}H_{35}$
$CHOCOC_{17}H_{35}$ $CHOCOC_{17}H_{35}$
$CH_2OCOC_{15}H_{31}$ $CH_2OCOC_{15}H_{33}$
Palmitodistearin   Oleodistearin

Vegetable oils and fats are usually liquids or semisolids at ambient temperature and may not actually possess a well-defined melting point. There are many other fatty acids (Table 3.10), other than palmitic, stearic, and oleic acids, which are found in plant tissue which may have

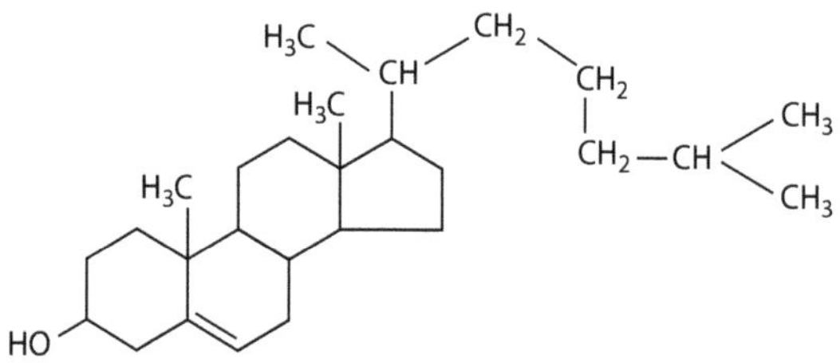

FIGURE 3.12 Cholesterol showing the four-ring steroid nucleus.

also been incorporated into the coal precursor thereby complicating the contribution from the fatty acids still further.

A group of chemical compounds which falls under the general classification *sterols* also occur in animal and plant oils and in fats. The structure of the sterols is based on the multi-ring perhydro-cyclopentaphenanthrene skeleton (Figure 3.12). The sterols are crystalline compounds and contain a hydroxyl group (e.g., cholesterol). They occur either in the uncombined form or as esters of the higher fatty acids and can be isolated from the nonsaponifiable portion of oils and fats. Sterols (or the related compounds, the steroids) also occur in animal tissue. In these cases, the compound may actually be cholesterol or the hydroxyl group or the alkyl side chain may have been altered to afford a structurally related compound (Table 3.11). In this latter case, it is presumed that the majority of the steroid-type material that contributed to the coal substance may not have arisen from animal remains and that the majority of the steroid-B-type material that eventually became part of the coal substance arose from plant debris.

Waxes often occur naturally as high-molecular-weight alcohols and esters, although free fatty acids may also occur within this group (Table 3.12). The constitution of the natural waxes in contrast to the so-called petroleum wax, which is actually a mixture of straight-chain paraffin hydrocarbons containing more than 17 carbon atoms and which are also solid at room temperature but which do not contain any oxygen functional groups (Speight, 2007).

The structural chemistry of the natural waxes is somewhat less well-defined than the structural chemistry of the oils and fats (glycerides). While the more common straight-chain materials are defined by their names, for example, octacosanol, 28-carbon alcohol, there is still some doubt about the structures of some of the less conventionally named natural waxes. Opinions have differed for many years as to the exact structures of some of these compounds (e.g., montanyl alcohol, which may actually be a mixture of alcohols having different carbon numbers) and thus the carbon number of some of these waxes with the less descriptive names may certainly be open to question.

3.2.5 MISCELLANEOUS ORGANIC MATERIALS

3.2.5.1 Resins

Resins are those materials which exude from wounds in the bark of many types of trees. Indeed, many types of resins exist and each type appears to be characteristic of the family of trees from which it originates.

Again, because of the potential number of compounds that could exist in a resin mixture, it is difficult to assign a definite structure to any particular resin. However, it is known that the resin colophony (or rosin, the nonvolatile residue from the steam distillation of turpentine) is actually a mixture of resin acids which are derived from the diterpenes. One of the best-known constituents of this mixture is the tricyclic diterpene *abietic acid* (Figure 3.13).

TABLE 3.11
Steroids and Their Natural Sources

Structure and Name		Occurrence
Ergosterol		Yeast
Stigmasterol		Plant sterol, soybean oil
Estrone		Humans
Testosterone		Male sex hormone
Cortisone		Adrenal cortex
Digitogenin		Plants
Ergosterol		Yeast

(*continued*)

TABLE 3.11 (continued)
Steroids and Their Natural Sources

Structure and Name	Occurrence

Stigmasterol — Plant sterol, soybean oil

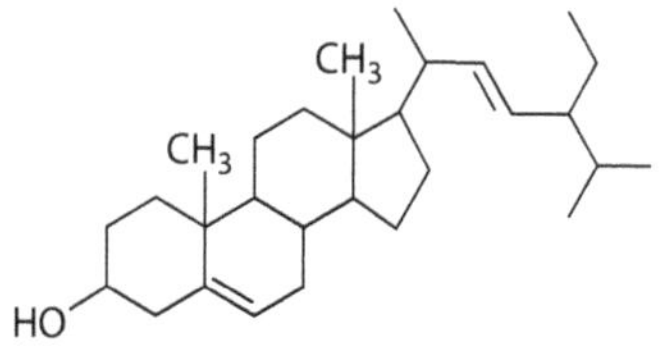

Estrone — Humans

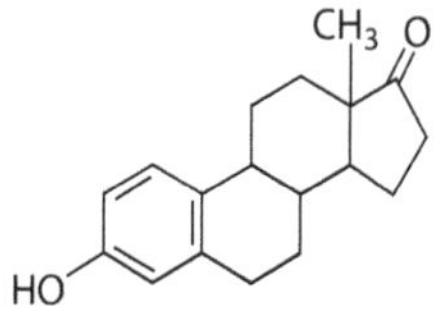

Testosterone — Male sex hormone

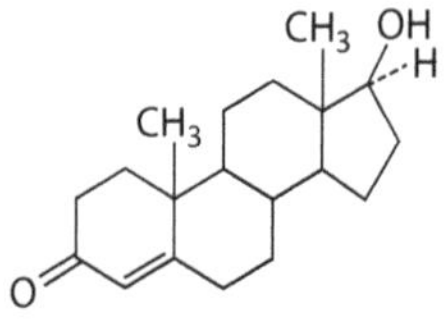

Cortisone — Adrenal cortex

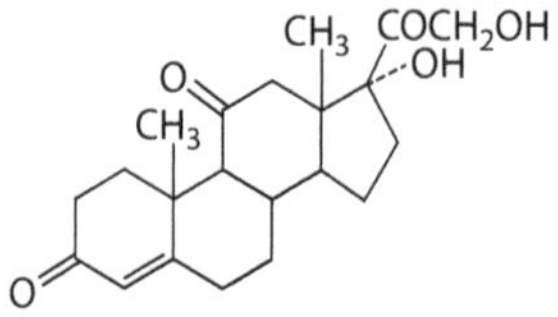

Digitogenin — Plants

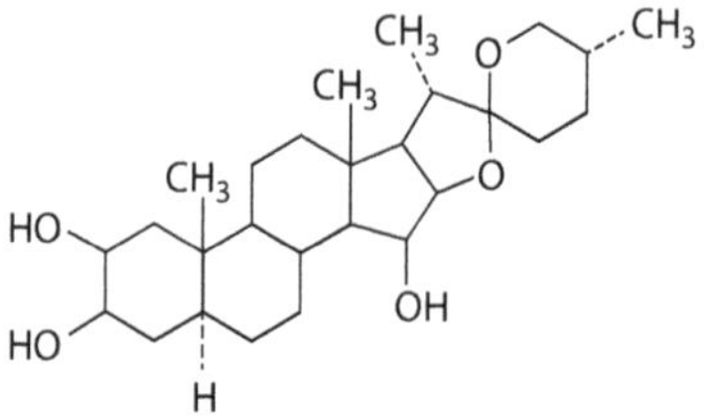

3.2.5.2 Tannins

Tannins are widely distributed in plants and, although they are present in most plant tissues, they are more often found in the bark. These compounds are usually glucosides that are based on the polyhydroxyphenol structure. For example, the so-called catechol tannins all evolve catechol on destructive distillation (thermal decomposition with the simultaneous removal of distillate) while the so-called pyrogallol tannins evolve pyrogallol on destructive distillation (Figure 3.14). However, this is not conclusive evidence for the existence of these two polyphenols alone in the tannins since the destructive distillation of other polyphenol systems could conceivably lead to catechol and/or pyrogallol. Nevertheless, the tannins are all considered to be derivatives of either gallic acid or ellagic acid (Figure 3.15) which may, in turn, be derived from simpler trihydroxyphenol and trihydroxyphenol derivatives. The catechol tannins usually contain 58%–60% w/w carbon and the pyrogallol tannins usually contain 51%–53% w/w carbon.

TABLE 3.12
Constituents of Natural Waxes

Compound Type	Name	Empirical Formula
Alcohol	Tetracosanol	$C_{24}H_{49}OH$
n-Aliphatic	Hexacosanol (ceryl)	$C_{26}H_{53}OH$
	Octacosanol	$C_{28}H_{57}OH$
	Nonacosanol (montanyl)	$C_{29}H_{59}OH$
	Triacontanol (myricyl)	$C_{30}H_{61}OH$
	Hentriacontanol (melissyl)	$C_{31}H_{63}OH$
	Dotriacontanol (lacceryl)	$C_{32}H_{65}OH$
Ketone	Carbocerotone	$C_{27}H_{55}CO$
	Montanone	$C_{29}H_{59}CO$
Acid, aliphatic, straight, or branched chain	Lignoceric	$C_{24}H_{48}O_2$
	Hexacosanic (cerotic)	$C_{26}H_{52}O_2$
	Heptacosanic (carbocerotic)	$C_{27}H_{54}O_2$
	Octacosanic	$C_{28}H_{56}O_2$
	Nonacosanic (montanic)	$C_{29}H_{58}O_2$
	Triacontanic (myricinic)	$C_{30}H_{60}O_2$
	Hentriacontanic (melissic)	$C_{31}H_{62}O_2$
	Dotriacontanol (lacceric)	$C_{32}H_{64}O_2$
	Hydroxymontanic	$C_{29}H_{58}O_3$
Ester	Miricyl cerotate	$C_{25}H_{51}COOC_{30}H_{61}$
	Octacosanyl cerotate	$C_{25}H_{51}COOC_{28}H_{57}$
	Octacosyl hydroxyoctacosanate	$C_{27}H_{55}(OH)COOC_{28}H_{57}$
	Ceryl octacosanate	$C_{27}H_{55}COOOC_{26}H_{53}$
	Montanyl montanate	$C_{28}H_{57}COOC_{29}H_{59}$
Hydroxy acid	2-Hydroxydocosanic (phellonic)	$C_{22}H_{42}O_3$
Resin	Abietic acid	$C_{20}H_{30}O_2$
	Montan resin	$C_{24}H_{34}O_2$

Source: Francis, W., *Coal: Its Formation and Composition*, Edward Arnold Ltd., London, U.K., 1961, p. 168.

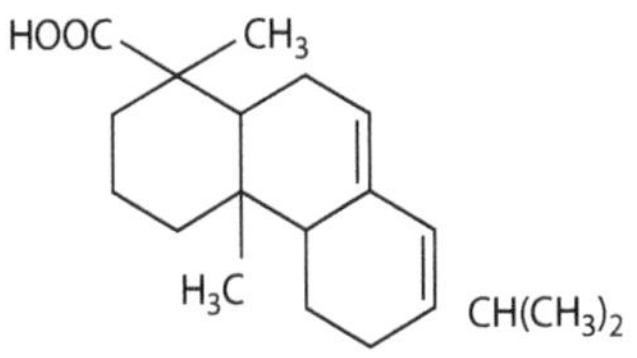

FIGURE 3.13 Abietic acid—a constituent of natural resins.

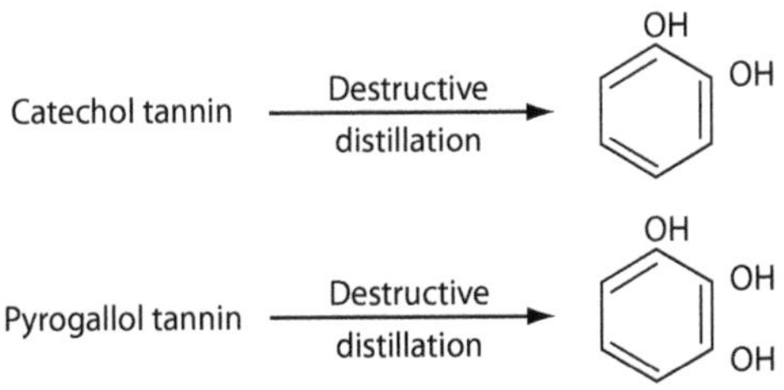

FIGURE 3.14 Representation of the thermal decomposition of tannins.

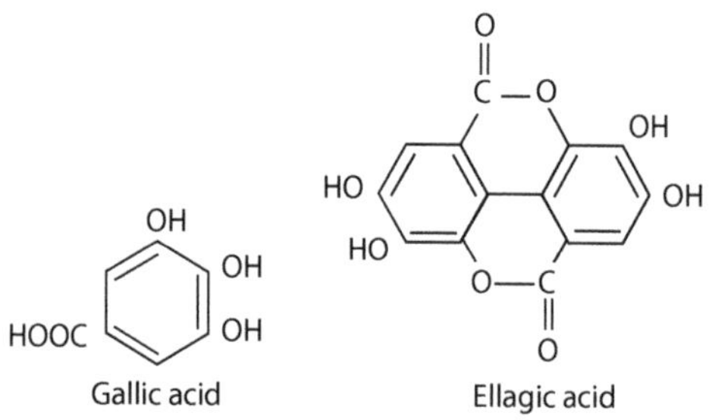

FIGURE 3.15 Gallic acid and ellagic acid—constituents of tannins.

3.2.5.3 Alkaloids

The alkaloids are perhaps the last class of organic compounds which originate in plants that are of any importance in the formation of the organic substance of coal. These compounds all contain basic nitrogen in the molecule with the nitrogen frequently occurring in a cyclic system; in addition, most of the alkaloids also contain oxygen. Alkaloids occur in the roots, bark, leaves, and within the cells of many plants (Dalton, 1979).

The structural chemistry of the alkaloids is variable because of the many locations in which nitrogen can occur in organic systems. However, it is generally recognized that the alkaloids may be based on any one of several individual (or even on a combination of two or more) systems (Table 3.13).

3.2.5.4 Porphyrins

One other class of nitrogen compounds exists which is based on the pyrrole system (Figure 3.16), the porphyrins. The porphyrins are actually based on a conjugated cyclic structure (porphine) consisting of four pyrrole rings linked together in the 2 and 5 positions by methine (=CH–) bridges (Figure 3.16).

Although porphine itself does not appear to exist in nature, structures based on this system are very important to the existence of plant and animal life (Baker, 1969). In the animal kingdom, the phosphine system occurs as hemoglobin (in the hemin part of the molecule) in the red corpuscles of the blood and acts as an oxygen carrier to the lungs and to the body tissue (Figure 3.17). In the plant kingdom (and perhaps more pertinent to the present case), porphyrin occurs in the chlorophyll found as the green coloring of leaves and stems which is essential for photosynthesis (Figure 3.17). The porphyrin constituents of coal have been proposed as being valuable markers in determining the extent of coalification (Bonnett et al., 1991) in a manner similar to the use of the porphyrin constituents of petroleum to trace the pathways of petroleum maturation (Branthaver, 1990).

3.2.5.5 Hydrocarbons

One final class of compounds that is worthy of mention is the polyenes because it is quite possible that it could make some contribution to the coal precursor. These compounds comprise bicyclic alkanes, the terpene hydrocarbons, and the carotenoid pigments (Figure 3.18) (Meinschein, 1969; Schwendinger, 1969; Luo et al., 1991).

In addition, the occurrence of picene derivatives as well as similar polynuclear aromatic systems (White, 1983) in the solvent extracts of coal might be cited as evidence for the inclusion of sterane-type materials in the precursor material to coal. This does assume that the skeletal structure remains intact throughout the maturation process and that there has been no degradation or even formation of new skeletal systems. Indeed, the biosynthesis of aromatic compounds is well documented, including the synthesis of aromatic species from nonaromatic precursors (Weiss and Edwards, 1980; Lee et al., 1981).

TABLE 3.13
Typical Alkaloid Systems

1. Indole

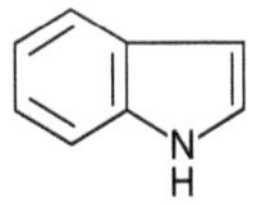

Example of indole alkaloids:

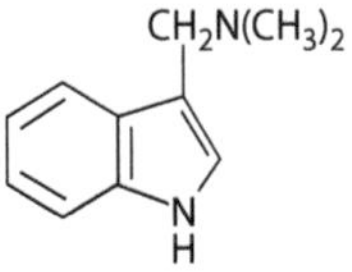

Gramine

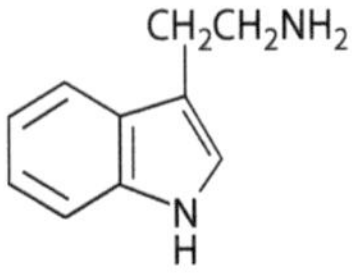

Tryptamine

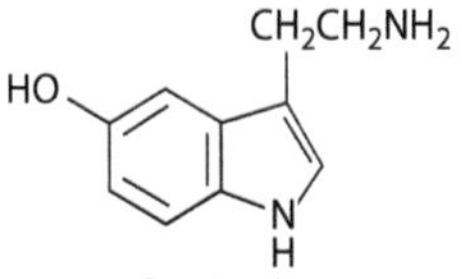

Serotonin

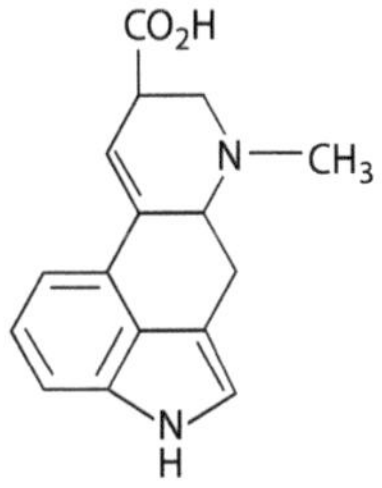

Lysergic acid

2. Pyridine

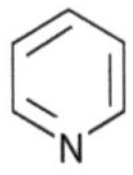

Example of pyridine alkaloids:

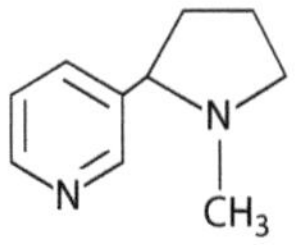

Nicotine

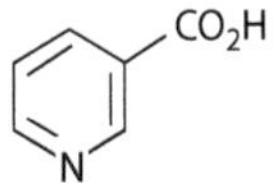

Nicotinic acid

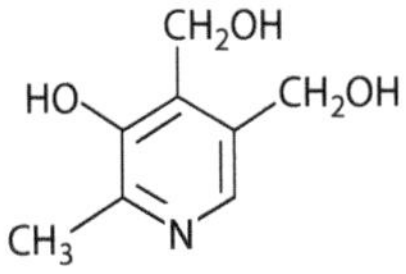

Pyridoxine

3. Quinoline

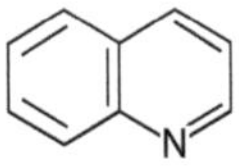

An example of a quinoline alkaloid:

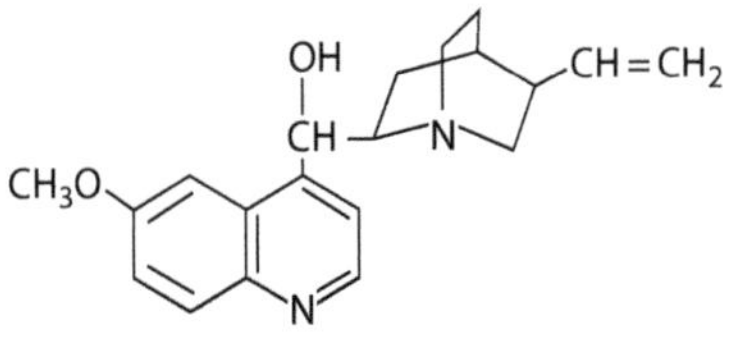

Quinine

(continued)

TABLE 3.13 (continued)
Typical Alkaloid Systems

4. Isoquinoline

Examples of isoquinoline alkaloids

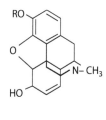

Papaverine   Narcotine   Morphine, R = H
Codeine, R = CH₃

Morphine, R = H
Codeine, R = CH₃

5. Pyrimidine

Examples of pyrimidine alkaloids:

Barbituric acid   Predominant
form

Veronal   Phenobarbital

6. Purine

Examples of purine alkaloids:

Caffeine   Theobromine

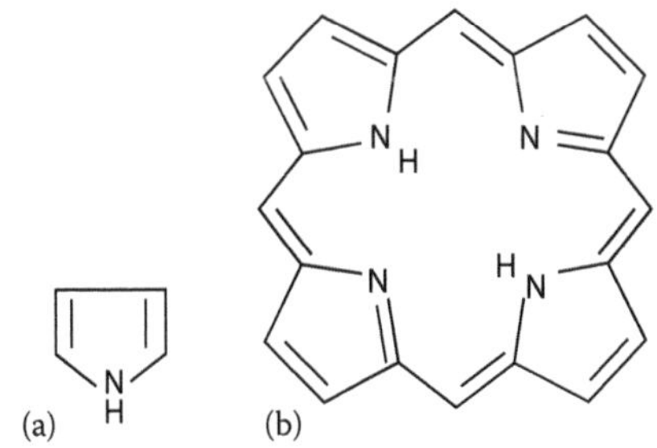

FIGURE 3.16 (a) Pyrrole, the major building block of porphyrins, and (b) porphine, the major structural unit of porphyrins.

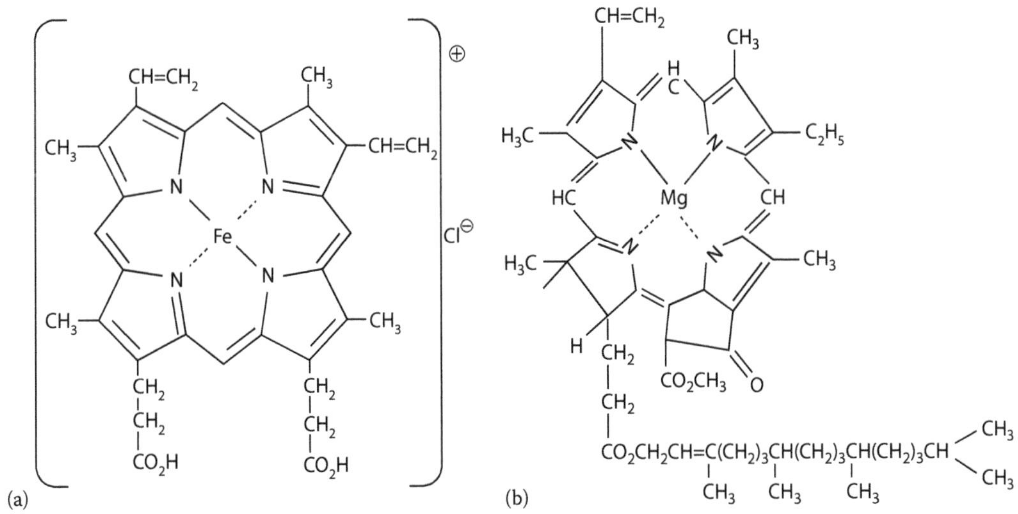

FIGURE 3.17 Two common porphyrins: (a) hemoglobin and (b) chlorophyll.

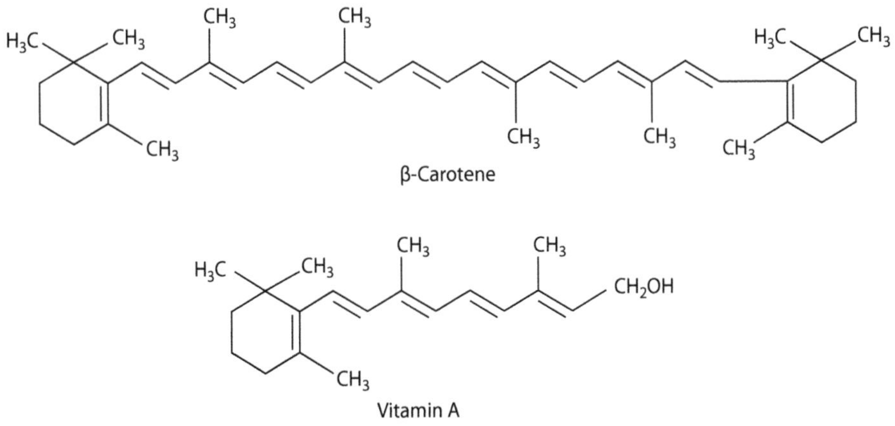

FIGURE 3.18 Carotenoid hydrocarbons.

3.2.5.6 Inorganic Constituents (Mineral Constituents) of Plants

The previous sections have dealt in some detail with the organic constituents of plants that could conceivably contribute to the source material. There is, however, one additional class of chemical material that originates in plants and eventually be included into the coal matrix. This, of course, is the mineral matter which arises from inorganic constituents of plants.

Mineral matter in coal (Chapter 7) is often classified as inherent mineral matter or as extraneous mineral matter. The inherent mineral matter is the inorganic material that originated as part of the plant material that formed the organic debris in the source bed. On the other hand, extraneous mineral matter is that inorganic material which was brought into the coal-forming deposit by various means from external sources.

Even though the major portion of mineral matter in the majority of coals is that which arises from extraneous sources, some recognition must still be given to the inorganic constituents of the plant source materials. The extraneous inorganic material which eventually becomes part of the coal can originate from a wide variety of sources. For example, phosphorus occurs in nucleic acids, magnesium occurs in chlorophyll, while calcium, sodium, iron, and potassium are also found in many plants. Indeed, mineral matter is incorporated into the cell walls of virtually all plants and may actually be chemically associated (or combined) with the organic constituents of the plants.

In fact, it is now generally believed that the majority of nickel and vanadium in petroleum results from the incorporation of chlorophyll (and chlorophyll-type) materials into the source bed; such materials are, of course, rich in these types of metals, which exist in a chemically combined form with the organic fragment. The same may be true for similar elements in coal.

The proportion of inorganic matter in the wood of trees is usually less than 1% w/w, whereas the amount of inorganic material in the leaves may be of the order of 2%–3% w/w with as much as 20% w/w of the outer tissues (cork) of trees being inorganic material. On an individual basis, iron is found principally in the leaves whereas the sap contains solutions of potassium, sodium, magnesium, and calcium as well as some iron.

Thus, although the average proportion of inherent inorganic matter in all coal deposits is relatively small, individual coal components may contain relatively high proportions of inherent inorganic constituents. For example, coal components containing high proportions of cuticle may also contain higher proportions of inherent iron compounds than coal components formed from other structural components of plants.

3.3 COAL-FORMING PROCESSES

Coal is the compacted and preserved remains of plant matter and when plant life containing cellulose-rich stems and leaves is highly abundant and special conditions exist (usually anaerobic conditions)—the plant matter does not totally decompose and is preserved in fossilized form. These types of plants had evolved by the Devonian Period and many coal deposits in Europe and North America date from the Carboniferous Periods of the Paleozoic when these areas were covered with forests dominated by large ferns and scale trees.

The *coalification process* (*coal-forming process*) is, simply defined, the progressive change in the plant debris as it becomes transformed from peat to lignite and then through the higher ranks of coal to anthracite (Francis, 1961; Sunavala, 1990).

In the peat swamp as dead plant matter accumulates, aerobic bacteria rapidly oxidize cellulose and other components producing methane (CH_4), carbon dioxide, and ammonia (from the nitrogen-containing components). The resulting decomposed material compacts about 50% and is largely composed of lignin, a complex, three-dimensional polymer rich in aromatic (benzene) rings. These bacteria quickly use up the available oxygen and die ending the first stage of the process. Anaerobic bacteria take over the decomposition process and they produce acids as metabolic waste products. When the pH of the medium is sufficient (pH ~ 4), these bacteria die. The product at this stage is a gel-like material called *Gytta* (sometime referred to as *proto-coal*). When the Gytta is buried to a depth of 2000–3000 ft, the temperature is approximately 100°C (212°F) at which water and other volatiles are driven off.

The coalification processes usually *assume* that there is a regular (or at least a near-regular) progression from lignite to anthracite and that the progression is consecutive (Figure 3.1), but concurrency (i.e., simultaneous formation of different coal types) must also be considered a possibility and may even be the norm under certain conditions and in a variety of locales. In summary, *consecutiveness* is not proven, although it is often claimed.

Whereas the degree of coalification generally determines the rank of the coal, the process is not a series of regular, or straightforward, chemical changes. Indeed, complexity is the norm insofar as the metamorphosis of the plant debris relies not only on geological time but also on other physical factors, such as temperature and pressure. Thus, the occurrence of a multitude of different and complex chemical reactions (no matter how simple they appear on paper or how simply they occur in laboratory simulations) is inherent in the coalification process.

When organic debris (which may be identified as peat) is buried underneath sedimentary cover (overburden), a variety of physicochemical processes occur as part of the metamorphosis. The major process parameters are believed to be the resulting heat and pressure developed because of the overlying sedimentary cover (Bouska, 1981; Stach et al., 1982). This leads to changes in the constituents of the debris such as an increase in the carbon content, alteration of the functional groups, alteration of the various molecular structures, ultimately resulting in the loss of water (i.e., hydrogen and oxygen). The water may be lost as liquid (or vapor) or in other forms and in the correct ratio of the elements with the general effect of a decrease in reactivity (i.e., an increased resistance of the residue to heat, oxidation) as well as a decrease in the response (from tests on the coal) to solvents.

Thus, the theories about the formation of coal require that the original plant debris eliminate hydrogen and oxygen either occasionally or continuously under the prevailing conditions, ultimately leading to a product containing approximately 90% carbon (i.e., anthracite). In order for this maturation to proceed, chemical principles require that oxidation reactions be completely inhibited; that is, the chemistry proceeds according to carbonaceous residue coal organic debris but not according to

$$C_nH_m + \rightarrow \text{carbonaceous residue} \rightarrow \text{coal}$$
Organic
debris

Rather than

$$C_nH_m + O_2 \rightarrow CO_2 + H_2O$$
Organic
debris

Furthermore, in the early stages of coalification, microorganisms may also play a role (possibly an important one); indeed, in a somewhat paradoxical manner, they may interact with the plant debris under aerobic (oxidation) conditions as well as anaerobic (reductive) conditions.

The formation of coal under the slow conditions generally referred to as geological time may, nevertheless, be regarded as occurring in the absence of oxygen, thereby promoting the formation of highly carbonaceous molecules through losses of oxygen and hydrogen from the original organic molecules (Figure 3.1).

Laboratory investigations into the nature of coal formation and the chemical transformations involved have been carried out for many years and perhaps a brief discussion is warranted here not only to indicate the complexity of the issue of coal-forming processes but also to add a caution of the speculation involved in this aspect of coal science. But first, a general comment, it is generally believed that there are, on the basis of laboratory "evidence" (which is also often found in the field of petroleum chemistry), indicators of the pathways (as generally elucidated earlier) of the maturation process. Second, since there is no way (other than time) to replicate geological time, the work often employs what are postulated to be precursors, pseudo-precursors, and model compounds that are coerced to react in a similar manner that they *interact during the coalification process* in temperature regions that can often be classed as excessive.

This, of course, is on the presumption that a high temperature is equivalent to, or a substitute for, geological time. Be that as it may, and as well as it may seem, increases in the temperature at which a reaction occurs not only increases the rate of the reaction (for the nonchemists, an increase in the temperature of 10°C, 18°F, usually doubles the rate of a chemical reaction) but also is more than capable, even likely, of changing the chemistry. Thus, caution is advised when applying these principles (no matter how sound the logic may appear) to coal formation.

The plant precursors that eventually formed coal were compacted, hardened, chemically altered, and metamorphosed by heat and pressure over geological time in anaerobic, aquatic environments where low oxygen levels prevented their conversion to carbon dioxide. Successive generations of this type of plant growth and death formed deep deposits of unoxidized organic matter that were subsequently covered by sediments and compacted into carboniferous deposits leading eventually to the various coal types. Evidence of the types of plants that contributed to carboniferous deposits can occasionally be found in the shale and sandstone sediments that overlie coal deposits or even within the coal itself.

The formation of coal is often explained by the assumption that plant material, collected over a period of time, underwent some, but generally incomplete, decay and the resulting product was then altered by various chemical and physical forces. It has also been generally advocated, often accepted, that one of the first-formed materials in the coalification process was peat, a carbonaceous material which is generally postulated to be the starting point for the beginning of the coalification process. Indeed, there are several postulates for the interrelationship of plant materials and the coalification process (Figures 3.19 and 3.20) (Kreulen, 1948; Christman and Oglesby, 1971) as well as the interrelationships among the formation pathways of the fossil fuels (Figure 3.21) (Breger, 1977).

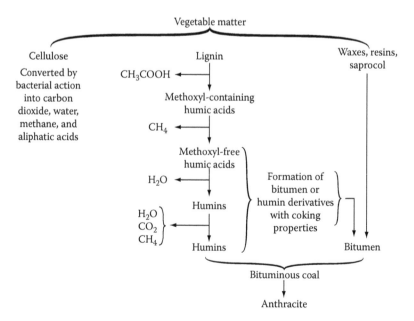

FIGURE 3.19 Hypothetical routes for coal formation from various plants.

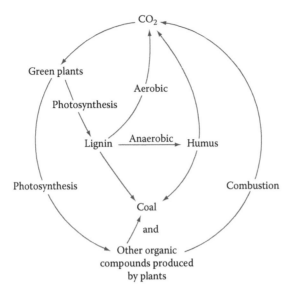

FIGURE 3.20   The carbon dioxide–plants–coal cycle.

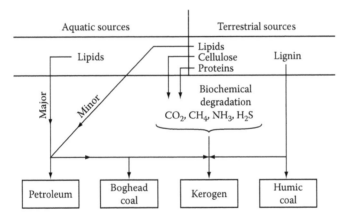

FIGURE 3.21   Routes for the formation of fossil fuels.

The interesting aspect of one of these examples (Figure 3.20) is the inclusion of carbon dioxide within the "cycle." Both postulates recognize the production of carbon dioxide from coal, thereby pointing out the relationship of carbon dioxide and coal but recognition of carbon dioxide as part of the *natural cycle* between plants, lignin, humus, and coal underscores the closeness of the relationship and the end product of the environmental aspects of coal use.

There is no doubt that as coal types *progress* through the *coal system*, there is a change in composition with the progression generally being to higher carbon content, lower hydrogen content, and lower oxygen content; sulfur and nitrogen contents can be variable (Table 3.14). Inspections of these analytical data have also raised the question of the role of another unknown natural product in the formation of coal, that is, kerogen.

Kerogen is a natural product of diverse character (Table 3.15) as well as origin (Scouten, 1990; Speight, 2007). There is the concept that kerogen plays a role in coal formation (Figure 3.22) but the role is very speculative, to say the least. Indeed, just as the role of kerogen in the genesis of petroleum is open to debate and question (Speight, 2007), so too must be the role of kerogen.

Thus, the path from kerogen to peat being unclear leaves the concept open to serious question and doubt. In fact, as long as there is some disagreement about the sequence of events that proceed

TABLE 3.14

Representative Ultimate Analysis for Peat and Various Coal Types

Rank	Percent C	Percent H	Percent O	Percent N	Percent S
Peat	55	6	30	1	1.3[a]
Lignite	72.7	4.2	21.3	1.2	0.6
Subbituminous	77.7	5.2	15.0	1.6	0.5
High-volatile B bituminous	80.3	5.5	11.1	1.9	1.2
High-volatile A bituminous	84.5	5.6	7.0	1.6	1.3
Medium-volatile bituminous	88.4	5.0	4.1	1.7	0.8
Low-volatile bituminous	91.4	4.6	2.1	1.2	0.7
Anthracite	93.7	2.4	2.4	0.9	0.6

[a] Ash and moisture content constitute remaining weight percent.

TABLE 3.15

Ultimate Analysis of Various Kerogen Concentrates

Organic Component	Green River (Rifle, Colorado)	Aleksinac (Yugoslavia)	Irati (Brazil)	Pumpherston (Scotland)
C	77.39	71.87	78.83	75.70
H	10.26	8.73	9.47	10.37
N	3.10	3.21	3.92	4.18
O + S (diff)	9.25	16.19	7.78	9.75
Atomic H/C ratio	1.59	1.46	1.44	1.64

from kerogen to a hypothetical precursor (i.e., peat) and then to coal (see, e.g., Hessley et al., 1986; Anderson and Mackay, 1990; Hessley, 1990; Bend, 1992), there are, or at least there should be, questions about the *progression* and *regularity* of the coal-forming process.

3.3.1 Establishment of Source Beds

The source bed concept of coal genesis is the outcome of an attempt over the past century or more to find answers to some of the outstanding problems of coal genesis—answers that would be of practical assistance in the difficult task of finding new coal seams.

The approach is to search for some phenomenon of coal habitat common to several major coal seams. One such phenomenon emerged from a study of many mining operations, namely, the correlation between stratigraphy and coal seams where the seams may be restricted to one particular sedimentary horizon.

The source bed concept postulates that all coal seams are derived from plant material that was deposited syngenetically at one particular horizon of the sedimentary basin constituting the field, and that changes to the plant material changed (evolved) to coal in varying degrees under the influence of rise in temperature and pressure of the rock environment.

The accumulated, compacted, and altered plants form the sedimentary rock that we know as coal. However, moving beyond the source bed concept, there are two theories that have been proposed to explain the formation of coal.

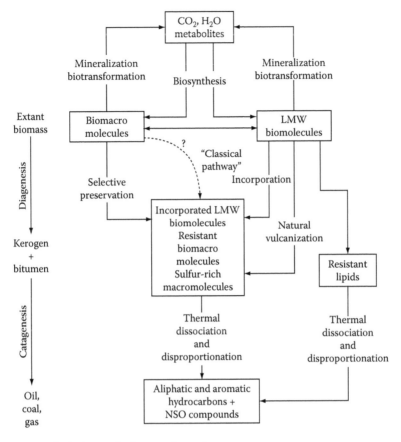

FIGURE 3.22   Hypothetical route for kerogen formation.

The first theory suggests that coal strata accumulated from plants which had been rapidly transported and deposited under flood conditions. This second theory which claims transportation of vegetable debris is called the *allochthonous theory*.

The second theory, which is favored by some geologists, is that the plants that compose the coal were accumulated in large freshwater swamps or peat bogs during many thousands of years. This theory, which supposes growth-in-place of vegetable material, is called the *autochthonous theory*.

3.3.1.1   Transportation Theory (Allochthonous Theory)

Throughout the last century or so, many scientists have subscribed to the theory that the coal source material has not formed by the degradation of plant material in situ in a peat swamp environment but rather accumulated from the deposition of transported material in aqueous environments such as lakes, seas, and estuaries (Moore, 1940; Stutzer, 1940; Francis, 1961). Indeed, the concept of transportation of the source material also led to the belief that not one but several processes were responsible for the different types of coal. Thus, there was much thought and discussion which actually led to serious questions about the autochthonous theory of coal formation and also brought discredit on the idea that coals form a prescribed and definite progression from peat to anthracite.

There is much to be said in favor of either the in situ theory or the transportation theory because the evidence in each ease must be circumstantial due to the nature of the subject. There will, undoubtedly, be coals for which the evidence will strongly favor the transportation theory, and even the stronger exponents of the in situ theory (the theory most generally favored by scientists) will recognize that some coals may not be the product of peat swamp deposits.

Thus, the nature of the problem of coal formation dictates that a compromise may need to be reached with, supposedly, the in situ theory predominating but recognizing that the transportation theory also has merit.

3.3.1.2 Peat Swamp Theory (Autochthonous Theory)

The concept of coal being formed from plant debris that had collected in a peat swamp environment, which is chemically very complex (Cooper and Murchison, 1969; Given, 1975), is not new and was, in fact, propounded in the latter part of the eighteenth century (see Section 3.2). Indeed, it was recognized at this time that the accumulation of the plant debris followed by the collection of later sediments in the same area and, thereby, effectively covering the plant debris was one of the prime requirements for the formation of coal from organic debris. It was also recognized that the organic debris must, at some stage of geological time, be altered to produce the sequence of coals beginning with peat and terminating with the high-rank anthracites.

In short, this early theory invoked the concept of the in situ origin of coal by noting that the low content of mineral matter as part of the coal matrix precluded the large-scale transportation of plant material since it was presumed that such transportation would necessarily involve the deposition of plant material in locales that could lead to the inclusion of considerable quantities of mineral matter within the coal.

The formation of petroleum should be compared here as it is generally believed (Speight, 2007) that the source material may be transported in a marine environment and deposited with mineral matter, but subsequent liquefaction of the source material allows further movement from the site. In the case of coal, the source material is (presumably) a solid and transportation from the source bed is prohibited; therefore, sizable amounts of mineral matter might be expected to occur within the resulting coal. Indeed, the occurrence of plant decay under marine conditions ultimately leading to the generation of coal (or, at least, a precursor to coal) should not be discarded. Marine plants and animals are rich sources of protein and generally contain more sulfur and nitrogen compounds than do freshwater species and offer a means by which depositional environment can be correlated with the mined coal (Stach et al., 1982; Hessley et al., 1986).

Other considerations to be noted in favor of an autochthonous theory are as follows: (a) There exist under clays (which are believed to be the remnants of the ancient soils) beneath a large number of banded coals; (b) the arrangement of plant debris (and its condition) is not believed to be that of transported material; and (c) the deposition of large quantities of plant material in aqueous sediments would be expected to be subject to bacterial decay as propounded for the theory of petroleum formation.

The debate over the origin of coal seams appears to have been decided in favor of in situ (or autochthonous) formation from peats formed slowly in swamps of various descriptions (Diessel, 1980; Stach et al., 1982). One of the key factors in this ascendancy of the peat swamp model over the various allochthonous (or transported) models was the recognition of so-called *fossil forests*—tree stumps with roots and logs in apparent growth positions on top of coal seams. The peat swamp model has not only become the basis of virtually all studies on coal seam formation but is now also the basis of studies on the coalification of the plant constituents to produce the various coal macerals. For this reason, considerable effort has been directed toward the study of modern peat-forming environments.

The ascendancy of the *peat swamp* model has led to some neglect of the evidence for the allochthonous, and catastrophic, deposition of coal seams. Even with abundant evidence for contemporaneous volcanism resulting in volcanically derived inter-seam sediments, such coals are still viewed as having formed in peat swamps that were periodically buried by volcanic debris.

3.3.2 GEOCHEMICAL AND METAMORPHIC PROCESSES

Coal is principally an organic substance and organic geochemistry is an important factor in the study of the origin, structure, quality, and utilization of coal. Determining the organic matter in coal is necessary for studying (1) the organic structure and the organic sulfur compounds in coal,

(2) modern peat-forming swamps as analogs of ancient peat-forming swamps, (3) the potential for coal gasification and liquefaction, and (4) the effects of coal mining and coal use on surface water, ground water, and air quality.

Once the source beds have been established, the onset of various processes occurs which ultimately results in the formation of coal. The first of these processes is often referred to as diagenesis which takes place under normal conditions of temperature and pressure and also includes the various biochemical processes (Teichmuller and Teichmuller, 1979; Bouska, 1981). There is also the suggestion that as a result of such processes, changes to the coal components may be continuous, insofar as the coal may be changing even in its current beds; changes to various coal macerals have been observed under various conditions that are equivalent to the coal-forming, or maturation, conditions (Smith and Cook, 1980).

The concept of *normal temperatures* and *normal pressures* is also a contradiction to the recent work whereby geological time has been simulated by substituting higher temperatures than might have ever been operative in the past. It is also a caution against the use of such data to project the precise chemistry of coal formation.

The end product of these early processes is (it is assumed) the familiar peat or, often, a soft lignite-type material.

Indeed, during the first stages of coalification, the various biological–bacterial processes may predominate over any other potential process(es), and, hence, this stage of coalification is often referred to as biochemical coalification and any alteration to the original plant material is usually only slight at this time. However, this stage is considered to be complete when bacterial and fungal activities have ceased and any residual microbial bodies have also ceased to function. This usually occurs when the deposit has been covered (and sealed) by an impervious sedimentary cover. The biochemical organisms may then exhaust the supply of plant substance that is necessary for survival or (in the case of aerobic organisms) exhaust the supply of oxygen.

Once the biochemical stage of coalification is effectively terminated, the changes in the plant debris (by this time it may be approaching peat or lignite in nature) are entirely physical or chemical and are usually determined by the prevailing conditions of temperature and/or pressure.

Obviously, other factors may also play an important part in determining the ultimate nature of the finished coal. These include the relative proportions of the chemical compounds that form the source material which become part of the coal substance and which are used as food by the microbial organisms. Indeed, the large variation of chemical compounds that can form the source material almost, even at this point, assure the complexity of coal.

Thus, the predominant coal-forming processes are believed to be brought about through the agency of increasing temperature and pressure. The effect of increase in the temperature is to increase the rate of chemical reaction while the effect of pressure is to bring reacting entities into sufficiently close proximity as to enable reaction to occur and, at the same time, to retard the reactions that result in the evolution of gaseous products.

The usual metamorphic processes are those which result from temperatures and pressures that arise solely as a result of the depth of burial of the organic material or perhaps even from the various movements in the earth that occur from time to time. The temperatures that are associated with depth in the earth's crust are generally assumed to be due to the natural thermal gradient which is, essentially, an increase in temperature with depth ($0.012°C$ or $0.022°F$ per 1 ft or 0.3 m of depth) which arises from the hot core of the earth.

There is also the possibility that the current temperature of a coal deposit may not necessarily be the maximum or minimum temperature to which that deposit has been subject; changes in depth because of earth movements or direct changes in temperature because of an igneous intrusion all serve to complicate the issue of temperature as well as pressure. Nevertheless, the use of high temperatures in laboratory studies is speculative and the use of the data therefrom to identify the precise chemistry of coal formation is always suspect. Indeed, the uncertainties that exist in the history of coals make the influence of temperature and pressure somewhat

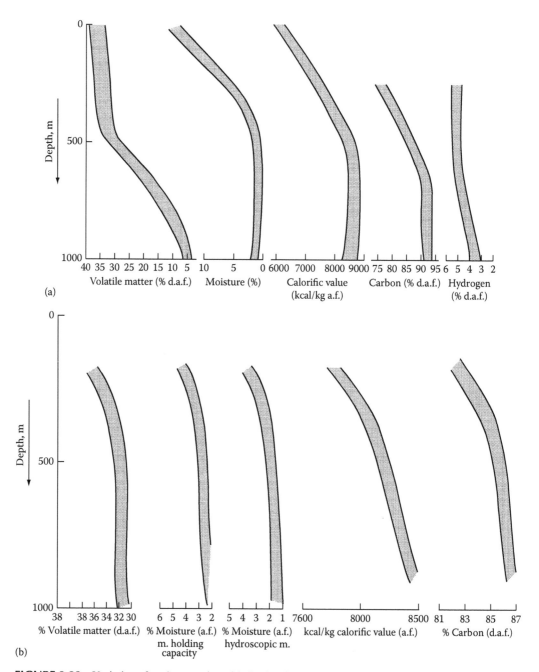

FIGURE 3.23 Variation of coal properties with depth using coals from (a) the Ruhr District (Germany) and (b) the Saar District (Germany). (From Murchison, D. and Westoll, T.S., Eds., *Coal and Coal Bearing Strata*, Elsevier, New York, 1968.)

circumstantial, but it has nevertheless been possible to note the influence of temperature and pressure in general terms.

For example, certain physical properties of coal (which are themselves a function of coal rank) change with the depth of burial (Figure 3.23) (Breger, 1958; Francis, 1961; Schmidt, 1979) but it should be noted here that the temperature gradient is, of course, influenced by the thermal conductivity of the rocks, which essentially makes comparisons of coals from different locales extremely

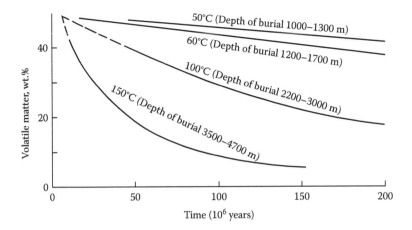

FIGURE 3.24 Variation of volatile matter yield with age of coal. (From Murchison, D. and Westoll, T.S., Eds., *Coal and Coal Bearing Strata*, Elsevier, New York, 1968.)

difficult if not prone to criticism and error. Hence, any such comparisons should be made with extreme caution since depth may not be the only factor which influences the metamorphic processes. In addition, some consideration should also be given to the duration of the exposure to heat as shown by laboratory studies (Figure 3.24) (Francis, 1961), although the cautions noted earlier about laboratory studies should be borne in mind. This is similar to the earlier notation of the relationship between coal rank (determined by volatile matter content) and to the duration of burial paleotemperature (Figure 3.25) (Huck and Karweil, 1955; Karweil, 1956).

With regard to the effects of pressure, it has generally been concluded that pressure does not usually promote the reactions associated with coalification, but even retards these reactions (Bouska, 1981; Stach et al., 1982).

For example, if methane formation is regarded as one of the coalification reactions, an increase in pressure will suppress excessive methane formation and the reaction will proceed more slowly at higher pressures. However, while pressure retards chemical reactions, it will cause changes in the physical structure of the material during the coalification process. For example, the decrease in

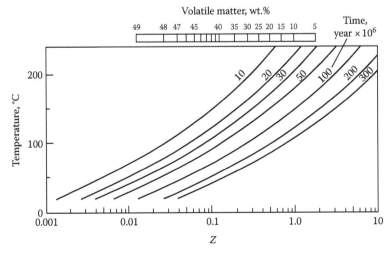

FIGURE 3.25 Variation of volatile matter yield to time of burial and temperature. (From Murchison, D. and Westoll, T.S., Eds., *Coal and Coal Bearing Strata*, Elsevier, New York, 1968.)

porosity (and the concurrent decrease in moisture content) and the increase in specific gravity of the lower-rank coals are caused mainly by the influence of pressure and can, in fact, be demonstrated quite readily under laboratory conditions.

Thus, it is generally accepted that the chemical changes which occur during coal metamorphosis are caused primarily by the maximum temperature to which the coal has been subjected over geological time. On the other hand, pressure retards many of these reactions but produces physical alterations in the texture of the material. It should be remembered, however, that the chemical and physical changes run parallel during the coalification process and are difficult to separate.

A third variable, *time*, also plays an important role in the coalification process and adds to the complexity of the overall scheme.

It is, at best, only possible to speculate on the routes of the coalification process on the basis of the current environment of the coal. Lack of substantiating data for the geological history of the sample as well as the lack of any data, other than speculative data, on the nature of the source material, make determination of the chemistry and physics of coalification extremely difficult if not impossible. Identifying the chemical reactions that occurred to form coal from the precursor material is somewhat more difficulty.

However, on the basis that there is some similarity of the polynuclear aromatic types produced (thermally) from coal under different conditions, there is the suggestion that the major compound types are either representative of similar aromatic moieties in the original coal or a result of the processing conditions involving complex reactions that lead to similar final products. It may also be that many of the same complex reactions occur during diagenesis (Nishioka and Lee, 1988).

3.4   HETEROATOMS IN COAL

Coal is used in processes such as combustion (Chapters 14 and 15), carbonization (Chapters 16 and 17), liquefaction (Chapters 18 and 19), and gasification (Chapters 20 and 21) as well as in the production of chemicals (Chapter 24). The origin of coal (Chapter 3) dictates the heterogeneous nature of coal and, thus, a coal must be characterized before it is used, whether as a single or blended coal.

The type of characterization that is performed in order to determine the properties of coal is used to predict its technological behavior. Typically, there are two characteristics that influence the use of coal: its composition and its rank. Coal composition is in turn represented by (1) type (nature of the organic components) and (2) grade, which is determined by the amount of mineral matter in the coal (or, in other words, by the dilution of the organic matrix by mineral matter).

At this point, it is pertinent to discuss the heteroatom species in coal (the non-hydrocarbon species in coal, i.e., nitrogen- oxygen-, and sulfur-containing species) insofar as these species can arise from the plant precursors as well as from external sources such as might occur during the maturation process. These species are also influential in determining the extent of atmospheric emissions that often have an adverse effect (and draw adverse comments) on the use of coal as an energy source (Chapters 22 and 25).

3.4.1   NITROGEN

Nitrogen has been the least studied of the principal elements in coal but, except in rare cases, the nitrogen present in coal is bound into the organic carbonaceous part. Fragmentary information is available concerning the nitrogen-containing compounds present in coal, but they do appear to be stable and are thought to be primarily heterocyclic. The original source of nitrogen in coal may have been both plant and animal protein. Plant alkaloids, chlorophyll, and other porphyrins contain nitrogen in cyclic structures stable enough to have withstood changes during the coalification process and thus to have contributed to the nitrogen content of coal. The amount of nitrogen in coal seems to be only weakly correlated with coal rank (Davidson, 1994).

Nitrogen is typically found in coal in the 0.5%–1.5% range by weight. As with oxygen, a number of different types of nitrogen-containing compounds have been isolated from coal-derived liquids (Schiller, 1977). Examples of these include anilines, pyridines, quinolines, isoquinolines, carbazoles, and indoles which carry alkyl and aryl substituents (Schweighardt et al., 1977; Palmer et al., 1992, 1994).

Five major nitrogen chemical structures, present in coals of varying ranks, have been quantitatively determined with the use of nitrogen x-ray absorption near-edge spectroscopy (XANES). Similar studies of the sulfur chemical structures of coals have been performed for the last 10 years; nitrogen studies on these fossil fuel samples have only recently been realized. XANES spectra of coals exhibit several distinguishable resonances which can be correlated with characteristic resonances of particular nitrogen chemical structures, thereby facilitating analysis of these complicated systems. All features in the XANES spectra of coals were accounted for; thus, all the major structural groups of nitrogen present in coals have been determined. A wide variety of aromatic nitrogen compounds are found in the coals; no evidence of saturated amines was found. Pyrrole derivatives, pyridine derivatives, pyridine derivatives, and aromatic amine derivatives are found in coal; of these, pyrrole structures are the most prevalent. Pyridine nitrogen is prevalent in all except low-rank coals. The low pyridine content in low-rank (high-oxygen) coals correlates with a high pyridone content. This observation suggests that, with increasing maturation of coal, the pyridone loses its oxygen and is transformed into pyridine. Aromatic amines are present at low levels in coals of all rank (Mullins et al., 1993).

The emissions of nitrogen oxides (NOx) from coal combustion may bear a relationship to the nitrogen types in the coal, although more nitrogen may remain in the char (Hindmarsh et al., 1994; Wang et al., 1994) compared to nitrogen remaining in petroleum coke (Speight, 2007).

The influence of coal properties on NOx emission levels is not well-defined. It is clear that several factors influence NOx formation, the roles and interaction of such factors as nitrogen partitioning (between the volatiles and chars), nitrogen functionality, coal rank, coal nitrogen content, and volatile content in the formation of NOx. However, it has been demonstrated repeatedly that the selection of coals on the basis of coal nitrogen content is an inappropriate and ineffective means of restricting NOx emissions. The selection of coals should be the sole preserve of the generating utility that has the best understanding of the influence of their combustion plant design and operation on the level of NOx emissions. Furthermore, the continued use of coal nitrogen as a controlling specification for coal supplies to an expanding utility market will not only fail to limit NOx emission levels but also result in the limitation on the development of new coal resources.

While the influence of nitrogen functionality on NOx emissions is not yet clearly defined, it has been found by a number of investigators that the pyrrole-type nitrogen structures are considerably less stable, and therefore more likely to react and oxidize to form NOx than the pyridine-type structures (Davidson, 1994), a factor which has been found to have influence on NOx generation from both volatile and char combustion. The coal nitrogen content, in isolation, gives no indication of the relative proportions of the different nitrogen functional groups in particular coal (Keleher, 1995).

Thus, it can be assumed that the amount of nitrogen in the coal, and the way in which it is bound into the coal structure, affects the amount and distribution of NOx emissions. The proportion of the coal sulfur volatilized on partial gasification could be related to the coal substance gasified, using one equation for all the coals. For coal nitrogen however, a single equation can describe the data from both pyrolysis and partial gasification (Middleton et al., 1997).

The suggestion that the nitrogen functionality exerts an indirect effect, possibly related to rank, is a reasonable explanation.

3.4.2 OXYGEN

Oxygen occurs in both the organic and inorganic portions of coal. In the organic portion, oxygen is present in hydroxyl (–OH), usually phenol groups, carboxyl groups (CO_2H), methoxyl groups (–OCH_3), and carbonyl groups (=C=O). In low-rank coal, the hydroxyl oxygen averages about 6%–9%, while high-rank coals contain less than 1%. The percentages of oxygen in carbonyl, methoxyl, and carboxyl groups average from a few percent in low-rank and brown coal to almost no measurable value in high-rank coal. The inorganic materials in coal that contain oxygen are the various forms of moisture, silicates, carbonates, oxides, and sulfates. The silicates are primarily aluminum silicates found in the shale-like portions. Most of the carbonate is calcium carbonate ($CaCO_3$), the oxides are mainly iron oxides (FeO and Fe_2O_3), and the sulfates are calcium and iron ($CaSO_4$ and $FeSO_4$).

The inclusion of oxygen in coal presumably arises from a variety of plant sources although there is also the strong possibility that oxygen inclusion occurs during the maturation by contact of the precursors with aerial oxygen and with the oxygenated waters that percolate through the detrital deposits. The oxygen content of coal ranges from a high of 20%–30% by weight for a lignite to a low of around 1.5%–2.5% by weight for an anthracite.

The oxygen content of a given coal is normally determined by difference:

$$\%Oxygen = 100 - (\%C + \%H + \%N + \%S + \%ash) \text{ (all calculated on a dry basis)}$$

Oxygen is known to occur in several different forms in coal, including phenolic hydroxyl, carboxylic acid, carbonyl, ether linkages, and heterocyclic oxygen (Chapter 10).

The presence of phenolic hydroxyl, and especially of carboxylic acid, is highly dependent on the rank or degree of coalification. The content of both these groups decreases rapidly with increasing carbon content (rank). Compounds containing the carboxylic acid group (–CO_2H) can be very readily extracted from peat and lignite, but seem to disappear in the subbituminous range. Likewise, the methoxyl group (–OCH_3) appears to be present in lignite, but has not been shown to be present in significant amounts in the bituminous and higher-rank coals (Deno et al., 1978).

The phenolic hydroxyl, ether linkages, and heterocyclic oxygen all appear to be present in bituminous coals and also to a smaller extent in the higher-rank coals. The phenolic hydroxyl content decreases almost linearly with increasing carbon content.

The most common form of heterocyclic oxygen is in furan ring systems. Substituted furan rings have been reported in coal extracts, pyrolysis tars, and oxidative degradation products (Duty et al., 1980; Bodzek and Marzek, 1981; see also White, 1983). However, to exactly what extent the dibenzofuran system exists in the parent coal is not clear. The isolation of dibenzofuran from oxidative degradation products and low-C temperature coal extracts supports the view that it is present in the parent coal.

The sulfur content of coal is quite variable, typically in the range 0.5%–5.0% w/w. This includes both inorganic and organic sulfur. Inorganic sulfur is present mainly in the form of iron pyrite. Sulfur contained in the precursor proteins and amino acids may be a source of sulfur in coal but hydrogen sulfide and pyrite are quite capable of reacting with the coal precursors to produce sulfur constituents as found in coal.

3.4.3 SULFUR

The inclusion of sulfur in coal is largely unknown and has been presumed to arise from the sulfur-containing plant constituents such as protein and oils although there are other means by which sulfur might be included into the coal (Casagrande and Seifert, 1977; Casagrande et al., 1979; Greer, 1979; Casagrande, 1987; Palmer et al., 1992, 1994). The predominant mode of sulfur incorporation into coal varies with the nature of the coal.

Sulfur is an important consideration in coal utilization, and, hence, there is a considerable amount of published work relating to the development of methods to improve the efficiency of the techniques as well as to improve the accuracy and precision of the sulfur determination (Speight, 2005 and references cited therein).

Sulfur is present in coal in three forms: (1) organically bound sulfur, (2) inorganic sulfur (pyrite or marcasite, FeS_2), and (3) inorganic sulfates (ASTM, 2011; ISO, 2011) (Wawrzynkiewicz and Sablik, 2002; Speight, 2005, and references cited therein). The amount of organic sulfur is usually <3% w/w of the coal, although exceptionally high amounts of sulfur (up to 11%) have been recorded. Sulfates (mainly calcium sulfate, $CaSO_4$, and iron sulfate [$FeSO_4$]) rarely exceed 0.1% except in highly weathered or oxidized samples of coal. Pyrite and marcasite (the two common crystal forms of FeS_2) are difficult to distinguish from one another and are often (incorrectly) designated simply as pyrite.

Some coals may show a general increase of total sulfur with both marine incursion and organic matter. Partitioning of this sulfur varies with organic content. In clay domains most of the sulfur is pyritic while in organic-rich samples organic sulfur is generally dominant, or present in concentrations approximately equal to pyritic sulfur (Bailey et al., 1990).

Free sulfur as such does not occur in coal to any significant extent. The amount of the sulfur-containing materials in coal varies considerably, especially for coals from different seams. In addition, pyrite is not uniformly distributed in coal and can occur as layers or slabs or may be disseminated throughout the organic material as very fine crystals. The content of sulfates, mainly gypsum ($CaSO_4 \cdot 7H_2O$) and ferrous sulfate ($FeSO_4 \cdot 7H_2O$), rarely exceeds trace amounts (i.e., <0.1%) except in highly weathered or oxidized coals.

When coal is mined, fresh sulfur-bearing minerals in the coal and rocks are exposed to air and water. The resulting chemical reactions produce sulfuric acid and precipitates. The acid water flowing from coal mines, if not treated, can damage life forms in the receiving streams. The iron and sulfate precipitates often discolor stream beds with yellow and orange stains. In a similar manner, when burned, sulfur escaping in the flue gases can combine with water in the atmosphere to produce acidic precipitation (*acid rain*). For the same reasons, burning high-sulfur coal can be corrosive to the metal equipment used in a power plant.

Organic sulfur can represent >50% of the total sulfur found in some coals and can arise from the interaction of peat-type precursors with hydrogen sulfide (Casagrande et al., 1979). On the other hand, there are several lines of evidence (Chou, 1990) which lead to the conclusion that sulfur in plant material is the principal source of sulfur in low-sulfur coal whereas in medium- and high-sulfur coals, seawater is also a predominant source of sulfur (Figure 3.26) (Chou, 1990).

The most abundant sulfur functional types found in coal are believed to be derivatives of the thiophene ring system. Other important sulfur types are aryl sulfides, alkyl sulfides, and acyclic sulfides. It should, however, be kept in mind that most of the information on sulfur compounds has been obtained by the analysis of small molecules obtained upon the depolymerization of the coal matrix. The more drastic depolymerization or degradation techniques are very likely to alter the chemical structure of the sulfur compounds originally present in the coal matrix.

3.4.4 OTHER ELEMENTS

A variety of other elements also occur in coal but their presence and amounts are strictly dependent upon the coal source (Speight, 2005 and references cited therein).

Chlorine occurs in coal (and is believed to be a factor not only in fouling problems but also in corrosion problems). The chlorine content of coal is normally low, usually only a few tenths of a percent or less. It occurs predominantly as sodium, potassium, and calcium chlorides, with magnesium and iron chlorides present in some coals. There is evidence that chlorine may also be combined with the organic matter in coal.

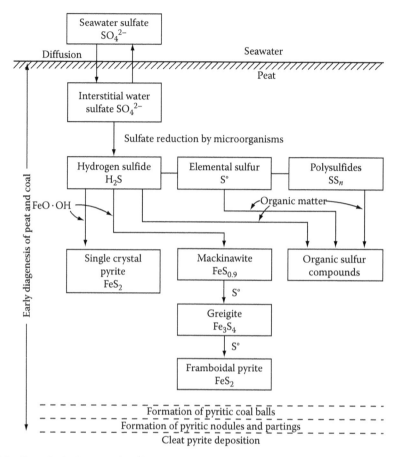

FIGURE 3.26 Hypothetical mechanism for sulfur incorporation into coal.

Mercury has been identified as a very dangerous environmental contaminant, largely by reason of the process of concentration in the food chain. Thus, the presence of mercury in coal is an extremely sensitive issue. The possible emission of mercury that may be found in coal is an environmental concern.

Mercury can go through a series of chemical transformations that convert elemental mercury (itself a toxin) to a highly toxic form (Tewalt et al., 2001). The most toxic form of mercury is methylmercury, an organic form created by a complex bacterial conversion of inorganic mercury and which enters and accumulates within the food chain, particularly in aquatic organisms such as fish and birds, causing various diseases in animals and humans. The creation of methylmercury in ecosystems is a function of mercury availability, bacterial population, nutrient load, acidity and oxidizing conditions, sediment load, and sedimentation rates. Methyl mercury in coal is known to be associated with iron (Fe), copper (Cu), and sulfur (S). In particular, it is often associated with sulfur in both pyrite and marcasite (FeS_2), where it is found in the form of a solid solution and in association with the mineral sphalerite (ZnS). In addition, some mercury may also be found organically bound in coal.

In the coal combustion process, mercury is released mainly as elemental mercury, since the thermodynamic equilibrium favors this state at coal combustion temperatures (Lu et al., 2007). In the combustion zone, mercury is vaporized from the coal as elemental mercury (Hg). As the flue gas temperature decreases, elemental mercury is oxidized to form mercuric oxide (HgO), as well as mercuric chloride ($HgCl_2$), and mercurous chloride (Hg_2Cl_2), subject to the presence of chlorine in the system.

Most coals contain small amounts of *mineral carbonates* made up primarily of calcium carbonate and to a lesser extent ferrous and other metal carbonates. Some coals contain a comparatively large amount of the inorganic carbonates, and the determination of carbon dioxide content is required in estimating the mineral matter content of these high-carbonate coals. Indeed, it is necessary to have a knowledge of the carbonate content of coal in order to correct the carbon figure and since, without resorting to very elaborate analyses, it would be impossible to express the carbonate content as definite quantities of calcium carbonate, magnesium carbonate, etc., it is customary, and sufficient for all analytical purposes, to express it simply in terms of carbon dioxide.

Arsenic and *selenium* occur in coal to the extent of several parts per million and, on combustion of the coal a varying quantity of these elements are released or retained in the ash, depending largely on the conditions under which the combustion takes place and on the nature of the coal ash.

3.5 EPILOGUE

It has been (and often still is) presumed that coal formation invokes the concept of a *regular* progression from peat (in a stepwise manner) to the higher-rank members of the coal series, ultimately leading to the anthracite coals. On the other hand, the possibility must also be considered whereby the formation of the various coals (from peat) may involve two or more different processes thereby leading to the various coal types.

It is difficult, however, to assemble evidence in favor of only one or the other of these theories insofar as each theory has its own particular merits and, moreover, the chemical makeup of the starting material and end product (coal) is, at best, extremely speculative making the task that much more difficult. Indeed, there may be some merit in accepting both theories as part of the coal-forming process.

In fact, it has been suggested that there exists a somewhat lesser dependence of the molecular chemistry of coal on rank and the chemistry of coal is heavily influenced by its source as well as by the early formation history (Berkowitz, 1988). It is also possible that coals of similar rank may, therefore, be chemically much more diverse than is usually supposed. Indeed, this concept is in agreement with similar conclusion about the formation of petroleum insofar as the chemical nature of the crude oil, with particular reference to the asphaltene constituents, is dependent not only on the types of precursors but also on the relative mix of these precursors that formed the *protopetroleum* (Speight, 2007) as well as on regional variations in the maturation conditions due to variations in climatic differences between various geological eras/periods (Bend, 1992; Speight, 2007).

The concept of source material and formation history playing a larger role in the determination of coal chemistry (Berkowitz, 1988) is quite logical; indeed the potential (and the more likely reality) that there are localized variations in precursor-type, mix of the precursors, as well as variation in maturation conditions are given little consideration as part of the coal-forming processes. The age-old concept of *high heat and pressure*, which one assumes would be a chemical equalizer insofar as all chemical and physical reactions proceed toward a graphite-type material, is given much more credence, for whatever reasons. Perhaps it is also time to cease considering coal as a graphitic material and look to coal as a natural product that is subject to local and regional variations in maturation conditions rather than as a conglomeration of large polynuclear aromatic sheets. Indeed, the perception of the chemical and physical behavior of coal has, for many years, been *guided* by the results of a series of tests that, at best, give only gross properties and afford no information whatsoever (or at the most optimistic, very little information) about the nature of coal.

Coal should be given its correct place in the natural order; it is a *secondary* (at least) natural product (in this case, a collection of chemical entities) that has undergone significant chemical and physical transformation in its history to produce a series of new chemical entities that still produce interesting chemical and physical phenomena.

REFERENCES

ASTM. 2011. *Test Method for Forms of Sulfur in Coal* (ASTM D2492). Annual Book of ASTM Standards, Section 05.05. American Society for Testing and Materials, West Conshohocken, PA.

Anderson, K.B. and Mackay, G. 1990. *International Journal of Coal Geology*, 16: 327.

Bailey, A.M., Sherrill, J.F., Blackson, J.H., and Kosters, E.C. 1990. *Sulfur and Pyrite in Precursors for Coal and Associated Rocks: A Reconnaissance Study of Three Modern Sites. Geochemistry of Sulfur in Fossil Fuels.* ACS Symposium Series. American Chemical Society, Washington, DC, Vol. 429, Chapter 10, pp. 186–203.

Baker, E.W. 1969. In *Organic Geochemistry: Methods and Results*, G. Eglinton and M.T.J. Murphy (Eds.). Springer-Verlag, New York, Chapter 19.

Bend, S.L. 1992. *Fuel*, 71: 851.

Bend, S.L., Edwards, I.A.S., and Marsh, H. 1991. *Fuel*, 70: 1147.

Berkowitz, N. 1988. In *Polynuclear Aromatic Compounds*, L.B. Ebert (Ed.). Advances in Chemistry Series No. 217. American Chemical Society, Washington, DC, Chapter 14.

Bodzek, D. and Marzek, A. 1981. *Fuel*, 60: 47.

Bonnett, R., Czechowski, F., and Hughes, P.S. 1991. *Chemical Geology*, 91: 193.

Bouska, V. 1981. *The Geochemistry of Coal*. Elsevier, Amsterdam, the Netherlands.

Branthaver, J.F. 1990. In *Fuel Science and Technology Handbook*, J.G. Speight (Ed.). Marcel Dekker, Inc., New York, Chapter 3 and references cited therein.

Bredenberg, J.B-son, Huuska, M., and Vuori, A. 1987. In *Coal Science and Chemistry*, A. Volborth (Ed.). Elsevier, Amsterdam, the Netherlands.

Breger, I.A. 1958. *Economic Geology*, 53: 823.

Breger, I.A. 1977. In *Future Supply of Nature-Made Petroleum and Gas*. Pergamon, New York, p. 913.

Casagrande, D.J. 1987. Sulfur in peat and coal. In *Coal and Coal-Bearing Strata: Recent Advances*, A.C. Scott (Ed.). Geological Society Special Publication No. 32, pp. 87–105. Presentation to Geological Society of London, London, U.K., April 1986.

Casagrande, D.J., Idowu, G., Friedman, A., Rickert, P., Siefert, K., and Schlenz, D. 1979. *Nature*, 282: 599.

Casagrande, D.J. and Seifert, K. 1977. *Science*, 195: 675.

Chou, C.L. 1990. In *Geochemistry of Sulfur in Fossil Fuels*. Symposium Series No. 429. American Chemical Society, Washington, DC, Chapter 2.

Christman, R.F. and Oglesby, R.T. 1971. In *Lignins: Occurrence, Formation, Structure, and Reactions*, K.V. Sarkanen and C.H. Ludwig (Eds.). Interscience/John Wiley & Sons, Inc., New York, Chapter 18.

Cooper, B.S. and Murchison, D.G. 1969. In *Organic Geochemistry: Methods and Results*, G. Eglinton and M.T.J. Murphy (Eds.). Springer-Verlag, New York, Chapter 29.

Dalton, D.R. 1979. *The Alkaloids: The Fundamental Chemistry, A: Biogenic Approach*. Marcel Dekker, Inc., New York.

Davidson, R. 1994. Nitrogen in coal. Report No. IEAPER/08. International Energy Agency, London, U.K.

Deno, N.C., Greigger, B.A., and Stroud, S.G. 1978. *Fuel*, 57: 455.

Diessel, C.F.K. 1980. *Coal Geology*. Australian Mineral Foundation, Adelaide, South Australia, Australia.

Duty, R.C., Hyatsu, R., Scott, R.G., Moore, L.P., Winans, R.E., and Studier, M.H. 1980. *Fuel*, 59: 97.

Ekwenchi, M.M., Imobighe, G.A., and Adejumo, M.A. 1991. *Fuel Processing Technology*, 28: 239.

Elliott, M.A. (Ed.). 1981. *Chemistry of Coal Utilization*, Second Supplementary Volume. John Wiley & Sons, Inc., New York.

Francis, W. 1961. *Coal: Its Formation and Composition*. Edward Arnold Ltd., London, U.K.

Given, P.H. 1975. In *Environmental Chemistry*, G. Eglinton (Senior Ed.). The Royal Society of Chemistry, London, England, Vol. 1, Chapter 3.

Greer, R.T. 1979. *Scanning Electron Microscopy*, (1): 377.

Hare, P.E. 1969. In *Organic Geochemistry: Methods and Results*, G. Eglinton and M.T.J. Murphy (Eds.). Springer-Verlag, New York, Chapter 18.

Hatcher, P.G. and Lerch, H.E. III. 1991. In *Coal Science II*, Symposium Series No. 461, H.H. Schobert, K.D. Bartle, and L.J.D. Lynch (Eds.). American Chemical Society, Washington, DC, Chapter 2.

Hendricks, T.A. 1945. In *Chemistry of Coal Utilization*, H.H. Lowry (Ed.). John Wiley & Sons, New York, Vol. 1, Chapter 1.

Hessley, R.K. 1990. In *Fuel Science and Technology Handbook*, J.G. Speight (Ed.). Marcel Dekker, Inc., New York.

Hessley, R.K., Reasoner, J.W., and Riley, J.T. 1986. *Coal Science*. John Wiley & Sons, Inc., New York.

Hindmarsh, C.J., Wang, W., Thomas, K.M., and Crelling, J.C. 1994. The release of nitrogen during the combustion of macerals, microlithotypes, and their chars. *Fuel*, 73: 1229–1234.

Huck, G. and Karweil, J. 1955. *Brennstoff Chemie*, 36: 1.

Hyatsu, R., Winans, R.E., and McBeth, R.L. 1979. Lignin-like polymers in coal. *Nature*, 278(5699): 31.

Hyatsu, R., Winans, R.E., McBeth, R.L., Scott, R.G., Moore, L.P., and Studier, M.N. 1981. In *Coal Structure*, M.L. Gorbaty and K. Ouchi (Eds.). Advances in Chemistry Series No. 192. American Chemical Society, Washington, DC, Chapter 9.

ISO. 2011. *Hard Coal—Determination of Forms of Sulfur* (ISO 157). International Standards Organization, Geneva, Switzerland, 331.

Karweil, J. 1956. *Z Deutsche. Geologische Gesellschaft*, 107: 132.

Keleher, 1995. *The Significance of Nitrogen in the Coal Combustion Process*. Department of Mines and Energy, Queensland Department of Minerals and Energy, Brisbane, Queensland, Australia.

Kreulen, D.J.W. 1948. *Elements of Coal Chemistry*. Nijgh and van Ditmar N.V., Rotterdam, The Netherlands.

Lee, M.L., Novotny, M.V., and Bartle, K.D. 1981. *Analytical Chemistry of Polycyclic Aromatic Compounds*. Academic Press, Inc., New York, Chapter 2.

Lowry, H.H. (Ed.). 1945. *Chemistry of Coal Utilization*. John Wiley & Sons, Inc., New York.

Lowry, H.H. (Ed.). 1963. *Chemistry of Coal Utilization*, Supplementary Volume. John Wiley & Sons, Inc., New York.

Lu, D., Anthony, E.J., and Tan, Y. 2007. Mercury removal from coal combustion by Fenton reactions. *Fuel*, 86: 2789–2797.

Lukoshko, E.S., Bambalov, N.H., and Smychnik, T.P. 1979. The change in composition of lignin in the process of peat formation. *Solid Fuel Chemistry (USSR)*, 13(3): 132.

Luo, B.J., Wang, Y.X., Meng, Q.X., Yang, X.H., Li, X.Y., and Cheng, N. 1991. *Science in China (Series B)*, 34: 363.

Masuku, C.P. 1992. *Journal of Analytical and Applied Pyrolysis*, 23: 195.

Meinschein, W.G. 1969. In *Organic Geochemistry: Methods and Results*, G. Eglinton and M.T.J. Murphy (Eds.). Springer-Verlag, New York, Chapter 13.

Middleton, S.P., Patrick, J.W., and Walker, A. 1997. The release of coal nitrogen and sulfur on pyrolysis and partial gasification in a fluidized bed. *Fuel*, 76: 1195–1200.

Moore, E.S. 1940. *Coal: Its Properties, Analysis, Classification, Geology, Extraction, Uses, and Distribution*. John Wiley & Sons, Inc., New York.

Mullins, O.C., Mitra-Kirtley, S., Van Elp, J., and Cramer, S.P. 1993. Molecular structure of nitrogen in coal from XANES spectroscopy. *Applied Spectroscopy*, 47: 1268–1275.

Murchison, D. and Westoll, T.S. (Eds.). 1968. *Coal and Coal Bearing Strata*. Elsevier, New York.

Nishioka, M. and Lee, M.L. 1988. In *Polynuclear Aromatic Compounds*, L.B. Ebert (Ed.). Advances in Chemistry Series D No. 217. American Chemical Society, Washington, DC, Chapter 14.

Orheim, A. 1979. In *Coal Exploration: Proceedings of the Second International Coal Exploration Symposium*, G.O. Argall, Jr. (Ed.). Miller Freeman Publications, Inc., San Francisco, CA, Chapter 14.

Palmer, S.R., Hippo, E.J., Crelling, J.C., and Kruge, M.A. 1992. Characterization and selective removal of organic sulfur from Illinois Basin coals. *Coal Preparation*, 10: 93–106.

Palmer, S.R., Kruge, M.A., Hippo, E.J., and Crelling, J.C. 1994. Speciation of heteroatoms in coal by sulfur- and nitrogen-selective techniques. *Fuel*, 73: 1167–1172.

Parker, P.L. 1969. In *Organic Geochemistry: Methods and Results*, G. Eglinton and M.T.J. Murphy (Eds.). Springer-Verlag, New York, Chapter 14.

Richardson, J.S. and Holba, A. 1987. In *Coal Science and Chemistry*, A. Volborth (Ed.). Elsevier, Amsterdam, the Netherlands, p. 343.

Roberts, J.D. and Caserio, M.C. 1964. *Basic Principles of Organic Chemistry*. Benjamin Book Publishers, Inc., New York.

Sarkanen, K.V. and Ludwig, C.H. 1971. *Lignins: Occurrence, Formation, Structure, and Reactions*. Interscience/John Wiley & Sons, Inc., New York.

Schiller, J.R. 1977. *Analytical Chemistry*, 49: 2292.

Schmidt, R.A. 1979. *Coal in America: An Encyclopedia of Reserves, Production, and Use*. McGraw-Hill, New York.

Schweighardt, F.K., White, C.M., Friedman, S., and Schultz, J.L. 1977. Preprints. *Division of Fuel Chemistry American Chemical Society*, 22(5): 124.

Schwendinger, R.B. 1969. In *Organic Geochemistry: Methods and Results*, G. Eglinton and M.T.J. Murphy (Eds.). Springer-Verlag, New York, Chapter 17.

Scouten, C.S. 1990. In *Fuel Science and Technology Handbook*, J.G. Speight (Ed.). Marcel Dekker, New York, Chapter 28.

Smith, G.C. and Cook, A.C. 1980. Coalification paths of exinite, vitrinite and inertinite. *Fuel*, 59: 641.

Snowdon, L.R. 1991. *Organic Geochemistry*, 17: 743.

Speight, J.G. 1980. *The Chemistry and Technology of Petroleum*, 1st edn., Marcel Dekker Inc., New York.

Speight, J.G. 2005. *Handbook of Coal Analysis*. John Wiley & Sons, Inc., Hoboken, NJ.

Speight, J.G. 2007. *The Chemistry and Technology of Petroleum*, 4th edn. Taylor & Francis Group, Boca Raton, FL.

Stach, E., Taylor, G.H., Mackowsky, M.-Th., Chandra, D., Teichmuller, M., and Teichmuller, R. 1982. *Textbook of Coal Petrology*, 3rd edn. Gebruder Bornfraeger, Stuttgart, Germany.

Streibl, M. and Herout, V. 1969. In *Organic Geochemistry: Methods and Results*, G. Eglinton and M.T.J. Murphy (Eds.). Springer-Verlag, New York, Chapter 16.

Stutzer, O. 1940. *Geology of Coal*. University of Chicago Press, Chicago, IL.

Sunavala, P.D. 1990. *URJA*, 28(2): 31.

Swain, F.M. 1969. In *Organic Geochemistry: Methods and Results*, G. Eglinton and M.T.J. Murphy (Eds.). Springer-Verlag, New York, Chapter 15.

Teichmuller, M. and Teichmuller, R. 1979. The diagenesis of coal (coalification). In *Diagenesis in Sediments and Sedimentary Rocks*, G. Larsen and G.V. Chilingarian (Eds.). Elsevier, New York, Chapter 5.

Terres, E. 1931. Contributions to the origin of coal and petroleum. *Proceedings. Third International Conference on Bituminous Coal*, 2: 797.

Tewalt, S.J., Bragg, L.J., and Finkelman, R. 2001. Mercury in US Coal: Abundance, Distribution, and Modes of Occurrence, Fact Sheet FS-095-01. United States Geological Survey, Washington, DC, September.

Van Krevelen, D.W. 1961. *Coal: Typology, Chemistry, Physics, Constitution*. Elsevier, Amsterdam, the Netherlands.

Wang, W., Brown, S.D., Thomas, K.M., and Crelling, J.C. 1994. Reactivity and nitrogen release from a rank series of coals during temperature programmed combustion. *Fuel*, 73: 341–347.

Wawrzynkiewicz, W. and Sablik, J. 2002. Organic sulfur in the hard coal of the stratigraphic layers of the Upper Silesian coal basin. *Fuel*, 81: 1889–1895.

Weiss, U. and Edwards, J.M. 1980. *The Biosynthesis of Aromatic Compounds*. John Wiley & Sons, Inc., New York.

White, C.M. 1983. In *Handbook of Polycyclic Aromatic Hydrocarbons*, A. Bjorseth (Ed.). Marcel Dekker, Inc., New York, Chapter 13.

Wollrab, V. and Streibl, M. 1969. In *Organic Geochemistry: Methods and Results*, G. Eglinton and M.T.J. Murphy (Eds.). Springer-Verlag, New York, Chapter 24.

4 An Organic Rock

4.1 INTRODUCTION

Coal and coal products will play an increasingly important role in fulfilling the energy needs of society. Future applications will extend far beyond the present major uses for power generation (Chapter 15) and chemicals production (Chapter 24). A key feature in these extensions will be the development of means to convert coal from its native form into useful gases and liquids in ways that are energy-efficient, nonpolluting, and economical.

The design of a new generation of conversion processes will require a deeper understanding of coal's intrinsic properties and the ways in which it is chemically transformed under process conditions. Coal properties such as the chemical form of the organic material, the types and distribution of organics, the nature of the pore structure, and the mechanical properties must be determined for coals of different ranks (or degrees of coalification) in order to use each coal type most effectively.

A second and more subtle challenge is to identify the chemical pathways followed during the thermal conversion of coal to liquids or gases (Chapters 18 through 21). This is accomplished by tracing the conversion of specific chemical functional groups in the coal and studying the effects of various inorganic compounds on the conversion process. Significant progress has been made in this area by combining test reactions with a battery of characterization techniques. The ultimate goal is to relate the structure of the native coal to the resulting conversion products.

Coal is an organic sedimentary rock that forms from the accumulation and preservation of plant materials (Chapter 3).

Briefly, sedimentary rocks are formed by the accumulation of sediments. There are three basic types of sedimentary rocks: (1) *clastic sedimentary rocks*, such as breccia, conglomerate, sandstone, and shale, which are formed from mechanical weathering debris, (2) *chemical sedimentary rocks*, such as rock salt and some limestone rocks, which form when dissolved materials precipitate from solution, and (3) *organic sedimentary rocks*, such as coal and some limestone rocks, which form from the accumulation of plant or animal debris.

Chemical sediments are created by a precipitation of low temperature/pressure minerals from water solution onto a depositional surface or within sediment pores. Depending on the acidity, oxidation, temperature, or salinity, a variety of chemical sedimentary rock may result. Examples of chemical sediments include carbonates, evaporites, opal, chert (may form in other ways too), iron oxides, and aluminum oxides. Chemical textures are usually crystalline with some special terms, such as oolitic or pisolitic. *Sandstone* is a clastic sedimentary rock made up mainly of sand-size (1/16 to 2 mm diameter) weathering debris.

Environments where large amounts of sand can accumulate include beaches, deserts, flood plains, and deltas. *Shale* is also a clastic sedimentary rock that is made up of clay-size (less than 1/256 mm in diameter) weathering debris. It typically breaks into thin flat pieces. *Siltstone* is a clastic sedimentary rock that forms from silt-size (between 1/256 and 1/16 mm diameter) weathering debris. *Conglomerate* is a clastic sedimentary rock that contains large (greater than 2 mm in diameter) rounded particles. The space between the pebbles is generally filled with smaller particles and/or chemical cement that binds the rock together. *Breccia* is a clastic sedimentary rock that is composed of large (over 2 mm diameter) angular fragments. The spaces between the large fragments can be filled with a matrix of smaller particles or mineral cement which binds the rock together.

Chert is a microcrystalline or cryptocrystalline sedimentary rock material composed of silicon dioxide (SiO_2). It occurs as nodules and concretionary masses and less frequently as a layered deposit. It breaks with a conchoidal fracture, often producing very sharp edges. Early people took advantage of how chert breaks and used it to fashion cutting tools and weapons.

Iron ore is a chemical sedimentary rock that forms when iron and oxygen (and sometimes other substances) combine in solution and deposit as a sediment. Hematite is the most common sedimentary iron ore mineral. *Rock salt* (*halite*) is also a chemical sedimentary rock that forms from the evaporation of ocean or saline lake. It is rarely found at Earth's surface, except in areas of very arid climate and is often mined for use in the chemical industry or for use as a winter highway treatment. *Limestone* is a rock that is composed primarily of calcium carbonate. It can form organically from the accumulation of shell, coral, algal, and fecal debris. It can also form chemically from the precipitation of calcium carbonate from lake or ocean water.

Coal is a sedimentary rock of biochemical origin. It forms from accumulations of organic matter, likely along the edges of shallow seas and lakes or rivers. Flat swampy areas that are episodically flooded are the best candidates for coal formation. During nonflooding periods of time, thick accumulations of dead plant material pile up. As the water levels rise, the organic debris is covered by water, sand, and soils. The water (often salty), sand, and soils can prevent the decay and transport of the organic debris. If left alone, the buried organic debris begins to go through the coal series as more and more sand and silt accumulates above it. The compressed and/or heated organic debris begins driving off volatiles, leaving primarily carbon behind.

The heterogeneity of coal arises because even during the formation of one coal seam, conditions vary and, hence, the types of coal vary depending upon the character of the original peat swamp (Chapter 3). Within a swamp, some areas might be shallow and other areas deep. Some areas might have woody plants and other areas grassy. The environment might be changing over time, making the bottom (older part) of the coal seam very different from the top. Varying water level and movement changes the degree of aeration and hence the activity of aerobic bacteria in bringing about decay. The different types of chemical substance present in plants (such as cellulose, lignin, resins, waxes, and tannins) are present in different relative proportions in living woody tissue, in dead cortical tissue as well as in seed and leaf coatings. In addition, these substances show differing degrees of resistance to decay.

Thus, as conditions fluctuate during the accumulation of plant debris, the botanical nature and chemical composition of the material surviving complete breakdown will fluctuate also, not only on a regional basis but also on a local basis (Chapter 3). This fluctuation is the origin of the familiar banded structure of coal seams, which is visible to the naked eye, and provides strong support case for the different chemical and physical behavior of coals.

Coal is also a combustible rock (Chapters 14 and 15) and along with oil and natural gas it is one of the three most important fossil fuels (Chapter 26). Coal was formed from accumulations of organic matter, likely along the edges of shallow seas and lakes or rivers, particularly swampy areas that are episodically flooded (Chapter 3). During nonflooding periods of time, thick accumulations of dead plant material pile up. As the water levels rise, the organic debris is covered by water, sand, and soils. The water (often salty), sand, and soils can prevent the decay and transport of the organic debris. The compressed and/or heated organic debris begins driving off volatile constituents.

The rate of plant debris accumulation required to produce a coal seam must be greater than the rate of decay. Once a thick layer of plant debris is formed, it must be buried by sediments such as mud or sand. These are typically washed into the swamp by a flooding river where the increasing weight of the mud and/or sand material compresses the plant debris and aids in its transformation into coal.

The ratio of plant debris to produced coal varies but has been assumed to be on the order of 10 to 1. That is, approximately 10 ft of plant debris will compact into just 1 ft of coal.

Plant debris accumulates very slowly; accumulating 10 ft of plant debris will take decades, if not centuries, and 50 ft of plant debris required to produce a 5 ft thick coal seam would require thousands of years to accumulate. During the time that the plant debris the water level of the swamp must remain stable, if the water level becomes too deep, the plants of the swamp will submerge and the coal-forming processes will be affected. On the other hand, if the water level is not maintained the plant debris will decay, eventually forming carbon dioxide and water.

Thus, in order to form a coal seam, the ideal conditions of perfect water depth must be maintained for considerable periods of time. It is apparent that the formation of a coal requires the coincidence of highly improbable events.

Deposits of coal, sandstone, shale, and limestone are often found together in sequences hundreds of feet thick. The key to large productive coal beds or seams seems to be long periods of time of organic accumulation over a large flat region, followed by a rapid inundation of sand or soil, and with this sequence repeating as often as possible. Such events happened during the Carboniferous Period—recognized in the United States as the Mississippian and Pennsylvanian time periods due to the significant sequences of these rocks found in several states (Chapters 1 and 3). Other coal-forming periods are the Cretaceous, Triassic, and Jurassic Periods.

Generally, lignite and peat are common in younger deposits—the older the deposit, the better the grade of coal.

Coal is also considered (perhaps without sufficient scientific foundation) to be a metamorphic rock—the result of heat and pressure on organic sediments such as peat—but most sedimentary rocks undergo some heat and pressure and the association of coal with typical sedimentary rocks and its mode of formation usually keep low-grade coal in the sedimentary classification system. On the other hand, anthracite, on the other hand, undergoes more heat and pressure and is associated with low-grade metamorphic rocks and is justifiably considered to be an organic metamorphic rock.

Thus, the degree of natural processing results in different quality of coal (Table 4.1) (see also Chapter 2). For example:

- Peat is considered (by some investigators) to be low-rank coal; it is not sufficiently natural to be classed as an organic rock but it is no longer merely a collection of organic debris.
- Lignite (brown coal) is the least mature of the true coals and the most impure; it is often relatively moist and can be crumbled to a powdery.

TABLE 4.1
Different Ranks of Coal

Rank	Properties
Lignite	The lowest rank of coal that contains recognizable plant structures; heating value <8300 Btu per pound on a mineral matter–free basis; carbon content: 60%–70% w/w (dry ash-free basis); often called *brown coal* in the UK, Europe, and Australia
Subbituminous	Relative to lignite, metamorphic processes have decreased the oxygen content and the hydrogen content to produce coal with a higher carbon content (71%–77% w/w on a dry ash-free basis) and a heating value from 8,300 to 13,000 Btu per pound (mineral matter–free basis); subbituminous coal is subdivided into the following: subbituminous A coal, subbituminous B coal, and subbituminous C coal on the basis of heating value
Bituminous	Bituminous coal has a carbon content of from 77% to 87% w/w (dry ash-free basis) and a heating value that is much higher than lignite or subbituminous coal; bituminous coal is subdivided into low-volatile bituminous coal, medium-volatile bituminous coal, and high-volatile bituminous coal on the basis of volatile matter production
Anthracite	Anthracite is the highest rank of coal and has a carbon content in excess of 87% w/w (dry ash-free basis); anthracite coal generally has the highest heating value per unit weight (mineral matter–free basis); it is often subdivided into semianthracite, anthracite, and meta-anthracite on the basis of carbon content

- Subbituminous coal is still poorly indurated and brownish in color, but is more closely related to bituminous coal than to lignite.
- Bituminous coal is the most commonly used coal; it occurs as a black, soft, shiny rock.
- Anthracite is the highest rank of coal and is considered to be a metamorphic organic rock; it is much harder and blacker than other ranks of coal, has a glassy luster, and is denser with few impurities.

4.2 PHYSICAL STRUCTURE

The behavior of coal during processing is determined by its physicochemical composition and structure. The examination of the physical and chemical structure of coal has been hampered by the inability to find techniques which measure any meaningful properties of such large complex structures. Most attacks on the problem have been by means of breaking down the structure into smaller, more tractable pieces, examining these and making inferences about the original structure. With many types of coal, the severity of the treatment needed to rupture the molecules raises doubts as to the validity of the method. It is claimed with some justification (but still with considerable doubt) that brown coal, which is geologically younger, may bear more resemblance to the molecules of classical organic chemistry.

Structurally, coal is a complex system (Chapters 3 and 10) (Thiessen, 1932) in which organic material dominates, typically representing 85%–95% w/w of a dry coal. These organic materials occur in various different petrographic types (macerals), which reflect the nature of the precursor plant material. Various inorganic materials, particularly aluminosilicates and pyrites (especially in high-sulfur coals), comprise 5%–15% of the coal. A third structural element, and perhaps its most distinctive feature when compared to other solid fossil fuel sources like petroleum and oil shale, is an extensive network of pores. These pores give coal a high surface area (>100 m²/g) for bituminous and subbituminous coals and lignites and an appreciable volume of pore space, allowing access to a significant fraction of the organic material.

The origin of coal has already been described in terms of the possible chemical constituents of the original plant material (Chapter 3). Coal from different geological time periods and different geographical areas may differ in composition and physical properties. It is often desirable to classify coals into groups with broadly common properties based on the degree of coalification, referred to as rank. Coal rank is partly determined from the percent carbon in the elemental analysis of the dry, mineral matter–free coal (Chapter 2). The percent carbon increases with increasing coalification. Thus, for example, a high-rank coal of 90% carbon is older and has undergone more extensive chemical change than a low-rank coal of 75% carbon.

Although such a description may serve to indicate the complexity of the coal by virtue of the almost limitless number of individual chemical compounds that can be included in the source material, it does not, however, present anything other than a speculative overview of the nature of coal.

There is a much more important aspect of coal composition that must be considered. For example, chemical constituents represent the molecular structure of coal but the form in which these chemical constituents eventually become incorporated into the coal matrix must also be given due consideration.

Thus, in terms of the chemical origin of coal, certain of the plant tissues contain chemical constituents of markedly similar molecular structures but there must be some attempt to recognize that incorporation into the coal precursor of a particular plant tissue (or a large part of the tissue of that particular plant) can also occur. While the maturation process may cause chemical changes to the individual chemical constituents of the tissue, this same maturation process(es) may also cause physical changes to the tissue (in to) that may cause it to appear as a recognizable entity in the coal (ASTM, 2011a–d).

Indeed, the acknowledgment that the tissues of the original plants can themselves contribute to the physical structure of coal has led to the development of the area of scientific investigation known as coal petrography (Murchison, 1991). Thus, coal petrography is an investigation (with subsequent identification) of coal with the macrostructures and microstructures that are the physical structure of coal.

There has, however, been a tendency to depart from these types of investigations over the last four decades and to concentrate on the so-called molecular structure of coal (Chapter 10) in the hope that elucidation of this elusive entity will assist in the understanding of the technological behavior of coal. However, as stated earlier, the large number of possible chemical constituents of the source material (Chapter 3) and, in addition, an incomplete knowledge of the prevailing maturation process(es) make the elucidation of the molecular structure (Chapter 9) of coal (if there is such a thing) very difficult, if not impossible.

Fortunately, there has been a surge of interest within the last decade in the physical macrostructure of coal with special emphasis on the relation of the petrographic properties of coal to its technological behavior. Indeed, equating technological behavior of coal (or, for that matter, any feedstock) with a definite entity should warrant far more attention than the attempts to equate technological behavior with a hypothetical (and, albeit, unknown) molecular structure.

4.3 PETROLOGY

Coal petrology, which dates back to the beginning of the twentieth century, is the fundamental discipline that deals with the origin, occurrence, physical and chemical properties, and utilization of coal (Bustin et al., 1983; Taylor et al., 1998).

Thus, coal petrology is the study of the organic and inorganic constituents of coal and their transformation via metamorphism. Coal petrology is applied to the studies of the depositional environments of coals, correlation of coals for geological studies, and the investigation of coals for their industrial utilization. Traditionally, the latter has been dominated by the use of coal petrology in the optimization of coal blends for the production of metallurgical coke, but can also include the use of petrology in evaluating coals for beneficiation (coal preparation for downstream utilization) and combustion.

Recently, significance of coal petrology has been demonstrated in CBM exploration and in potential sequestration of carbon dioxide in coal seams. Techniques developed in the study of coal are also used in the investigation of organic-rich rocks to evaluate source rocks in petroleum and natural gas exploration.

Coal is a combustible sedimentary rock and a versatile fossil fuel that has long been used for a variety of industrial and domestic purposes. Generally, it is used in processes such as combustion, gasification, liquefaction, and in carbonization for the manufacture of metallurgical coke and coal tar. It is also used as fuel for a range of manufacturing processes, such as the production of heat in cement kilns and other industrial plants, for gasification and petrochemical production, and for heating domestic and commercial buildings. However, its main use is for the generation of electricity.

Coal and its derivative products are used in a range of nonenergy applications such as the production of carbon anodes for the aluminum industry, graphite electrodes for the steel industry, and as a component for a number of other carbon-based materials and chemicals. Because coal is a heterogeneous material that is made of up organic and inorganic components, it must be characterized before it is used, to understand its properties, determine its quality, and predict its technological behavior.

Two main characteristics determine the properties of a coal: (1) composition and (2) rank (Table 4.1) (see also Chapter 2). Coal composition is in turn characterized by two essentially independent factors: type (nature of the organic components) and grade (presence of mineral matter) (Stach et al., 1982; Ward, 1984; Diessel, 1992; Van Krevelen, 1993; Taylor et al., 1998; Thomas, 2002; Suárez-Ruiz and Crelling, 2008):

- *Rank*. Coal rank reflects the degree of metamorphism (or coalification) to which the original mass of plant debris (peat) has been subjected during its burial history. This depends in turn on the maximum temperature to which it has been exposed and the time it has been held at that temperature. For most coals, rank also reflects the depth of burial and the geothermal gradient prevailing at the time of coalification in the basin concerned.

- *Type*. Coal type reflects the nature of the plant debris from which the original peat was derived, including the mixture of plant components (wood, leaves, algae, etc.) involved and the degree of degradation to which they were exposed before burial. The individual plant components occurring in coal, and in some cases fragments or other materials derived from them, are referred to as *macerals*. The kind and distribution of the various macerals are the starting point for most coal petrology studies.
- *Grade*. The grade of a coal reflects the extent to which the accumulation of plant debris has been kept free of contamination by inorganic material (mineral matter), before burial (i.e., during peat accumulation), after burial, and during coalification. A high-grade coal is therefore a coal, regardless of its rank or type, with a low overall content of mineral matter.

These parameters are the primary factors that influence a coal's specific physical and chemical properties and these properties in turn determine the overall quality of the coal and its suitability for specific purposes. Coal petrology, which dates back to the beginning of the twentieth century, is the fundamental discipline that deals with the origin, occurrence, physical and chemical properties, and utilization of coal.

Although various combinations of tests are used to evaluate the suitability of particular coals for industrial processes, the properties determined by these tests are ultimately related to coal composition and rank.

In addition to its chemical properties, the efficient use of a coal also requires a knowledge of its physical properties, such as its density (which is dependent on a combination of rank and mineral matter content), hardness, and grindability (both related to coal composition and rank). Other properties include its abrasion index (derived mainly from coarse-grained quartz) and the particle size distribution. Float-sink testing may also be included with the analysis process. This involves separating the (crushed) coal into different density fractions as a basis for assessing its response to coal preparation processes. Float-sink techniques may also be used to provide a coal sample that represents the expected end product of a preparation plant, in order to assess the quality of the coal that will actually be sold or used rather than the in situ or ROM material represented by an untreated (raw) coal sample.

Another factor that must be taken into account in determining coal quality is the degree of coal oxidation. Oxidation may affect both the organic and inorganic components and may lead to deterioration of the coal properties. Another possible consequence of coal oxidation is the spontaneous combustion (Misra and Singh, 1994; Beamish et al., 2001; Lyman and Volkmer, 2001; Beamish, 2005). The propensity to oxidation is mainly determined by the coal's rank in conjunction, perhaps, with the maceral and mineral (e.g., pyrite) content. Low-rank coals are particularly prone to spontaneous combustion. Other factors, such as the access of air to coal stockpiles, may need to be controlled in order to reduce the risk of spontaneous combustion. Petrographic examination may help to identify coals that have become oxidized.

Since the beginning of modern science, coal petrology has served as a powerful tool for the characterization of coals for both geological and industrial applications. As was mentioned earlier, the applications of coal petrology are wide-ranging. However, the applications of this science may sometimes be observed in apparently unrelated fields such as archaeological studies, materials science, and forensic geology.

4.4 PETROGRAPHY

Coal petrography (a branch of coal petrology) specifically deals with the analysis of the maceral composition and rank of coal and therefore plays an essential role in predicting coal behavior and should be an essential part of any coal analysis and testing program (Esterle and Ferm, 1986; Esterle et al., 1994). The fundamentals of organic petrography, maceral nomenclature, classification of coal components, and analytical procedures have been well established by the International Committee for Coal and Organic Petrology (ICCP, 1963, 1971a,b, 1975, 1993, 1998, 2001; Sýkorová et al., 2005).

Coal is a rock formed by geological processes and is composed of a number of distinct organic entities called macerals and lesser amounts of inorganic substances (mineral matter). The essence of the petrographic approach to the study of coal composition is the idea that coal is composed of macerals, which each have a distinct set of physical and chemical properties that control the behavior of coal. The organic units, composing the coal mass (macerals), are the descriptive equivalent of the inorganic units composing rock masses (minerals) and to which pathologists are accustomed to giving distinctive names (Stopes, 1919, 1935).

One of the most important aspects of coal petrography is an understanding of the effects that the petrographic constituents have on the behavior of coal during processing, in particular the processing sequences that are designed to produce gaseous and liquid fuels. While some data are available, a considerable amount of investigative work still remains to be carried out.

In keeping with the theme of this present chapter, a brief discussion of the influence of the petrographic composition of coal in relation to the technological properties of coal is warranted, although more specific reference will, as the data allow, be made to the various processes throughout the chapter.

The acceptance of coal as an organic sediment (Chapter 3) must lead to the realization and acceptance of the concept that, in a geological sense, coal is an organic rock. Thus, coal petrography is the science dealing with the description of coal as an organic mineral (Ting, 1978; Bend, 1992). On the other hand, coal petrology is a more general term which is intended as a description of the origin, history, occurrence, chemical composition, and classification of coal. Coal petrology is not as well advanced as the petrology of many of the minerals, perhaps because of the somewhat belated acceptance of coal as a rock in the geological sense.

Petrographically and chemically, coal is a complex material, and it is often convenient to describe coal in several ways. The most common way, of course, is in terms of the elemental (ultimate) composition (Chapter 8) where coal may actually be classified on the basis of the general formula: $C_nH_nN_xO_yS_z$ where n is the number of carbon atoms, y is the number of hydrogen atoms, x is the number of nitrogen atoms, y is the number of oxygen atoms, and z is the number of sulfur atoms.

Indeed, such was the case for many years and manipulations of these data leading to potential molecular structures for coal were the means by which coal was projected to behave in various processing and/or conversion scenarios; this, in fact, is still the case but the value of such mathematical manipulations must be questioned.

Although much valuable information can be obtained from such an examination, there may still remain considerable gaps in the data that are required for assessment of a particular coal sample as to the nature of the source material, contributions to the organic matrix of the various constituents, maturation conditions and pathways (Chapter 3), and (perhaps more pertinent to the text) the potential utility of the coal.

In the initial stages, coal petrography was a means of defining the coal as an organic rock but it is now elevated to a means by which the coal can be defined in terms of lithotype and maceral composition (Stopes, 1919, 1935; Spackman, 1958). It is worthy of note here that coal can be separated into the various maceral constituents by a variety of methods (Francis, 1961; Dyrkacz and Bloomquist, 1992a,b). This is of special importance in the study of coal behavior where it is known that the different petrographic constituents of coal can, and will, behave differently under various processing circumstances.

Indeed, one of the most important outcomes from the petrographic analysis of coal is the understanding of the effect which the petrographic composition has on the technological properties. In fact, the behavior of the petrographic constituents during coal utilization is now being documented with some regularity.

4.4.1 Lithotypes

Coal lithotypes represent the macrostructure of coal and are, in fact, descriptive of the coal. A piece of coal will usually exhibit a definite banded appearance due to the accumulation of different types of plant debris during the formation of the organic sediment (White, 1911; Thiessen, 1931; Muller et al., 1990;

Smith and Smoot, 1990). This may be due to not only the deposition of a variety of different organic compounds that are believed to be the precursors of coal (Chapter 3) but also the accumulation of different parts of the plant (as well as different plants, e.g., trees, ferns, mosses, etc.) during the formation of the sediment (Murchison, 1991; Puttmann et al., 1991). Each compound type, tissue type, or plant type then progresses through the various maturation stages that eventually lead to coal.

Thus, initial examination of coals that are higher in rank (Chapters 2 and 8) than lignite shows that two main features usually persist throughout the range of the coals. These features, which can be conveniently recognized by their general appearance as bright bands or as dull bands, can be attributed to the nature of the source material. The bright bands are believed to result from the main structural portions of plants, some of which may have retained their outward form and are recognizable (usually macroscopically or, if not, microscopically) as being derived from wood or cortex. On the other hand, the dull bands are believed to arise from a miscellaneous assortment of floral debris such as fragmentary cellular tissue, leaves, spores, pollen grains, cuticle, and other amorphous materials. The bright components of coals, because of their origin from wood, have been termed anthraxylon while the duller components, consisting of a miscellaneous assortment of debris, have been termed attritus (Table 4.2) (Thiessen, 1920).

Obviously, this system of nomenclature of the macrostructure of coal does not take into account the rank of the coal or any potential differences in chemical properties that may exist between the various coal types since the basis of such *macroscopic* names is the origin of the material.

Further close macroscopic examination of coals, however, has shown that they are much more complex and can actually be defined in terms of four macroscopically different bands. Thus, bright coal has been further subdivided into vitrain (Latin "vitrum": glass; brilliant to vitreous luster, conchoidal fracture) and clarain (Latin "clarus": bright pearly or near-vitreous luster, striated) while the dull coals are subdivided into fusain (Latin "fusus": spindle; silky luster, fibrous, and friable) and durain (Latin "durus": hard; generally dull or slight luster, granular fracture). Each of the lithotypes has individual characteristics that are easily distinguishable.

Vitrain occurs as well-defined narrow bands which are seldom greater than 0.5 in. (1.0 cm) thick. The single band does not exhibit any fine layering but is coherent, uniform, brilliant, and glossy (in fact, vitreous) in its texture.

Clarain, the second constituent of bright coal, occurs in bands of variable thickness that are horizontal to the bedding plane. Clarain bands are widely extended lenticular masses which have a well-defined smooth surface when broken at right angles to the bedding plane. These surfaces exhibit a very pronounced gloss and may be inherently banded as well as having bands of durain intercalated between its own bands.

TABLE 4.2
Microscopic Composition of Coal

Anthraxylon (bands >14 μm thick)
Attritus
   Translucent
      Humic matter (anthraxylon-like shreds <14 μm thick)
      Brown matter
      Spores, pollen, cuticles
      Resinous bodies
      Algal remains
   Opaque
      Granular and amorphous opaque matter
      Fusain splinters (<37 μm)
      Fusain (bands >37 μm)

TABLE 4.3

**Ultimate Composition and Gray–King Assay
of Different Lithotypes**

Component	Vitrain	Clarain	Durain	Fusain
Moisture, % w/w	1.7	1.4	1.2	0.9
Volatile matter, % w/w	34.6	37.6	32.2	19.1
Ash, % w/w	0.6	3.5	4.6	9.6
Carbon, % w/w	84.4	82.2	85.8	88.7
Hydrogen, % w/w	5.4	5.7	5.3	4.0
Sulfur, % w/w	1.0	2.3	0.9	1.0
Nitrogen, % w/w	1.5	1.9	1.4	0.7
Oxygen, % w/w	7.7	7.9	6.6	5.6
HHV, Btu/lb (dry ash-free)	14,790	14,790	15,100	14,840
Low-temperature assay (Gray–King) yields				
Coke, % w/w	69.0	71.5	76.2	—
Tar, % w/w	16.2	14.5	11.2	—
Liquor, % w/w	4.6	4.2	3.6	—
Gas, cm^3	13,440	12,800	11,340	—

Fusain, a constituent of dull coal, occurs predominantly as patches (or wedges) that are parallel to the bedding plane. Fusain consists of powdery, fibrous strands which can be easily fractured and separated from the coal mass. On the other hand, durain is hard with a close, firm texture and may even appear granular to the naked eye. It appears as bands of variable thickness which may be parallel to the bedding plane and which may contain intercalated bands of clarain.

Thus, there are consistent physical differences (as well as differences in elemental composition; Table 4.3) which allow distinctions to be made between the various lithotypes that constitute the macrostructure of coal. It is, however, necessary to recognize the difficulties that arise when a system of nomenclature devised for a series of coals from one geographical location (i.e., British subbituminous coals) is applied on an international basis. In fact, several such problems were encountered but they were perhaps vainly due to misinterpretation of the available local data and, most of all, the inability to recognize the individual lithotypes in indigenous coals. It is also worthy of mention here that there is a system which defines three major lithological classes of coal as (1) banded, (2) non-banded, and (3) impure (Table 4.4) (Spackman, 1975).

Banded coal contains visible bands of vitrain, which is the remains of single large fragments of ancient plants (see the earlier text), as well as bands of fusain, the remains of smaller fragments of ancient plants (see earlier text). Non-banded coal, however, displays a uniform texture which appears to be fine-grained, lacks a brilliant luster, and consists of comminuted, compacted sediments derived from plant detritus. This type of coal is much less common than banded coal in North America. There are two major types of non-banded coal: cannel coal and boghead coal (Stach et al., 1982).

Finally, impure coal is that coal which contains between 25% and 50% w/w ash (after ignition) on a dry basis. Two types of impure coal are bone coal and mineralized coal. The former (bone coal) contains clay or other fine-grained mineral matter and, moreover, if the bone coal contains more than 50% w/w of ash, it is often (and properly) termed carbonaceous shale or siltstone. Mineralized coal is impure coal which is heavily impregnated with mineral matter. The inorganic material can be dispersed and/or localized along fissures or cleat joints (Law, 1993; Laubach et al., 1998).

Other terms that are often employed in this chapter are splint coal and semisplint coal (Table 4.5) (Parks and O'Donnell, 1948). Splint coal contains more than 30% opaque matter together with some anthraxylon (more than 5% but usually only in minor amounts) while attritus is always the

TABLE 4.4
Nomenclature of Various Coal Types

(a) U.S. Bureau of Mines System
    I. Banded coals
        Characteristics: anthraxylon evident

Opaque matter:	<20%	Bright coal
	20%–30%	Semisplint coal
	>30%	Splint coal

    II. Non-banded coals
        Characteristics: anthraxylon not evident

Spore debris	Cannel coal
Algae debris	Boghead coal

(b) International System
    I. Humic (banded) coals

Charcoal-like	Fusain
Black, vitreous	Vitrain
Striated, glossy	Clarain
Nonstriated, matte	Durain

    II. Liptobiolithic (non-banded) coals

	Cannel coal
	Boghead coal

TABLE 4.5
Composition of Different Coal Types

Type	Components
Cannel coal	<5% anthraxylon and predominantly translucent attritus with little or no oil algae
Boghead coal	<5% anthraxylon and the translucent attritus predominantly oil algae
Bright coal	>5% anthraxylon and <20% opaque attritus
Semisplint coal	>5% anthraxylon and <20%–30% opaque attritus
Splint coal	>5% anthraxylon and >30% opaque attritus

predominant component. This type of coal acquired the name from the tendency to exhibit a hard, splintery fracture. On the other hand, semisplint coal contains 20%–30% opaque matter with some anthraxylon (usually more than 5% and occasionally equal to the amount of attritus) with minor amounts of fusain.

Semisplint coal is often referred to as block coal because of its tendency to exhibit a blocky fracture. In comparison with these two coals, bright coal has less than 20% opaque matter and the anthraxylon is always more than 5% and usually predominant; fusain is never abundant in bright coal (Table 4.4).

The coals described earlier are also referred to as humoliths (peat and lignite are also included under this definition), which are formed by way of one or more maturation paths from terrestrial vegetable matter. There are, however, a group of coals which are often classified as sapropelites. These coals originated from sapropelium, which is the organic matter originating from algae and other aquatic plants which had accumulated in marine, estuarine, or freshwater environments (sediments).

The sapropelites have often been considered to be a transition product between coal and petroleum. The validity of this comparison has often been questioned, and while there may be some merit in considering the sapropelites as a transition product, the complexities of the maturation

TABLE 4.6

Ultimate Analysis of Boghead Coal, Cannel Coal, and Banded Coal

	Composition (% w/w)						
	C	H	N	S	O	Moisture	Ash
Boghead coal	74.5	7.8	0.7	0.9	8.5	1.2	6.2
Boghead coal							
Dry basis	75.7	7.9	0.7	0.9	8.8	—	6.2
Daf basis	80.7	8.5	0.7	0.9	9.2	—	—
Cannel coal, daf basis	78.4	6.9	0.5	0.9	13.4	—	—
Banded coal, daf basis	77.0	5.1	1.2	0.9	15.2	—	—

Source: Drath, A., *Bull. Inst. Geol. Pologne*, 12, 1, 1932.

processes that are involved in the formation of coal and petroleum (and the lack of definite knowledge of these processes) (Speight, 2007) leave such statements open to very serious criticism.

The two major sapropelitic coals are cannel coal and boghead coal (the latter is also known as torbanite) (Table 4.5). The typical cannel coal consists essentially of the degradation products of an existing coal peat denuded to accumulate as an organic mud at the bottom of an aqueous, usually shallow, environment. Another important ingredient is a rich accumulation of spores. On the other hand, boghead coals are algal coals insofar as the whole mass of these coals originated from algal material without regard to the state of preservation (i.e., well preserved or completely decomposed) of the algal colonies. Boghead coals were believed to originate in the centers of the larger basins into which the transport of organic material was severely restricted or in the smaller basins where water entry was restricted or occurred through a filtering medium such as peat. In both cases, the water would be expected to be well aerated and have algal colonies flourishing on the surface.

Both boghead coal and cannel coal contain nitrogen, oxygen, and sulfur in addition to carbon and hydrogen (Table 4.6). They have a dull luster and fracture in a conchoidal pattern (i.e., the fracture surface has small rounded elevations and depressions). Cannel coal is black whereas boghead coal is in dull brown color, and splinters of these coals are easily ignited; in fact, cannel coal burns with a long and steady flame from which the name is derived (candle).

4.4.2 Macerals

Coal is the product of the deposition of the peat, its degradation (e.g., by insects or microorganisms), selective preservation of the surface litter; the growth of roots through the peat; the subsurface action of aerobic and anaerobic microorganisms; and the metamorphic changes of this organic mass through time.

As a result of the mode of formation, coal is an extremely complex heterogeneous material that is difficult to characterize in a chemical sense, although methods of classification have evolved based on bulk properties (Chapter 9).

To recap, coal is an organic rock formed by geological processes and is composed of a number of distinct organic entities called macerals and lesser amounts of inorganic substances—mineral matter, often incorrectly referred to as ash. The essence of the petrographic approach to the study of coal composition is the idea that coal is composed of macerals, which each have a distinct set of physical and chemical properties that control the behavior of coal. These organic units are the descriptive equivalent of the inorganic units (minerals) composing rock masses (Stopes, 1935).

The nomenclature employed to define coal as an organic rock distinguishes between the rock types (lithotypes) and their microscopic constituents. These constituents are called macerals by analogy with the minerals which occur in inorganic rocks. Identification of the macerals requires that the coal samples be polished and examined under vertically incident light using immersion objectives such as, for example, the oil immersion lens of a microscope. It should be noted, however, that this technique is not employed as such to enhance the resolution but to increase the contrast of the image.

The organic constituents in coal and non-coal organic-rich rocks are termed macerals (in the broader sense, and particularly for dispersed organic material, the term kerogen is also used). Macerals are the (optical) microscopically identifiable constituents in coal, somewhat analogous to minerals in an inorganic rock (Stopes, 1935). By convention, maceral names always have an *-inite* suffix. Thus, macerals are generally divided into the vitrinite (or *huminite* in lower-rank coals), inertinite, and liptinite groups.

Huminite/vitrinite macerals are derived from humic substances, the alteration products of lignin and cellulose. Huminite refers to macerals in lignite and subbituminous rank (see the following) coals (Sýkorová et al., 2005), and vitrinite to maceral of bituminous and anthracitic ranks.

The distinction with vitrinite is based on the division of the maceral group into three subgroups: (1) telovitrinite, (2) detrovitrinite, and (3) gelovitrinite, which are each further subdivided into two macerals. The dominant parameter for these newly ordered and in part newly defined subgroups is the degree of destruction (degradation), whereas the macerals can be further distinguished by their morphological characteristics and their degree of gelification (ICCP, 1994, 1998).

Huminite is divided into three subgroups based on the texture/morphology of the maceral: telo-huminite, with constituent macerals textinite and ulminite, the recognizably textured huminite macerals; detrohuminite, with macerals attrinite and densinite, the detrital huminite macerals; and gelohuminite, with macerals corpohuminite and gelinite, huminite macerals showing some degree of gelification.

Inertinite macerals are, to a certain degree, derived from the same starting materials as the huminite/vitrinite macerals (ICCP, 2001). In contrast to the latter maceral group, the inertinites have been oxidized, with fire thought to be the primary cause of their formation. The macerals fusinite and semifusinite are the products of such oxidation and, in most coals, are the most abundant inertinite macerals. Secretinite is a product of the oxidation of plant secretions. Macrinite is problematical, in part because it has been confused with what is now recognized as secretinite. Multiple pathways have been proposed for macrinite (ICCP, 2001). Funginite, with a fungus origin, is grouped with inertinites derived from plant cells by ICCP (2001). Fungi, however, are not plants but encompass two distinct eukaryote kingdoms, *Fungi* and *Protoctista*—kingdom slime molds. Micrinite is thought to have originated as a secondary maceral from the breakdown of hydrogen-rich liptinite.

The method of preparation of the samples for microscopic examination may vary according to the methods favored by a particular laboratory. However, there are standard methods described for the examination of coal samples by microscopic techniques.

For example, a crushed coal sample will be formed into a briquette with a cold-setting epoxy resin. When the resin is set, the surface is ground using water-resistant, adhesive-backed silicon carbide papers of grit size numbers 240, 320, 400, and 600. Then, polishing is carried out using aluminum oxide powders of specified sizes followed by treatment with a nap-free cloth of cotton and silk and chemo-textile material backed with water-resistant adhesive. There are, of course, suggested sequences for this procedure to produce a surface suitable for microscopic examination.

The macerals are actually identified microscopically by their form and reflectivity and are divisible into three basic groups: (1) the vitrinite group—sometimes called huminite in subbituminous coals—is derived from coalified woody tissue, (2) the exinite group—sometime called liptinite—is derived from the resinous and waxy parts of plants, and (3) the inertinite group is derived from charred and biochemically altered plant cell wall material and each group has a characteristic appearance and physical characteristics (Tables 4.7 and 4.8) (ICCP, 1963). The three groups are each further subdivided on the basis of individual maceral form, which is due in no small part to their mode of origin (Table 4.9) (Murchison et al., 1985).

TABLE 4.7

Subdivision of the Three Maceral Groups

Group	Subdivision	Characteristics
Vitrinite[a]	Collinite, telinite	Originates from humification and subsequent metamorphosis of cell wall materials from wood or cortex tissue. Translucent dark or light orange in transmitted light, dark, to light gray in reflected light
Inertinite	Macrinite	A totally structureless material probably evolved from humic mud and particles of diverse origin; opaque to transmitted light. Reflective and white in incident light. Particles 10–100 μm or more in diameter
	Micrinite	Derived from plant material that was macerated before coalification. Micrinite from lignite appears translucent yellowish brown to brown in transmitted light and dark gray in incident light, but micrinite from higher-rank coals is opaque to transmitted light and white in incident light. Particles 1–6 μm in diameter
	Semifusinite	Intermediate between fusinite and vitrinite with some of the characteristics of both
	Fusinite[b]	Fossil charcoal; exhibits cell structure; highly friable and hard. Formed by rapid alternative and charring of cell wall material before or soon after sedimentation. Opaque; polished faces highly reflective; white in vertically incident light
	Sclerotinite	Fossil bodies of fungal sclerotia; opaque, highly reflective
Exinite (or liptinite)	Resinite	High hydrogen content; formed from resinous material secreted by the plants; translucent dark or light orange in transmitted light; dark to light gray in reflected light
	Exinite	Hydrogen-rich; made up of fossil spores, pollens, cuticles (leaf surface materials and their excretions). Translucent yellow with low reflectivity
	Alginite	Fossil algal bodies making up boghead coal. Light yellow in transmitted light, dark in reflected light

Source: Spackman, W. *Proceedings,* Workshop on the Fundamental Organic Chemistry of Coal, National Science Foundation, University of Tennessee, Knoxville, TN, July 17–19, 1975, p. 12.

Note: See Figure 4.1 for a schematic representation.

[a] Pseudovitrinite has the characteristics of vitrinite but has a higher reflectance.

[b] Although there is some evidence for the formation of fusain during ancient forest fires, microscopic evidence (such as the preservation of cellular forms) does not logically support a forest fire origin. It is, therefore, quite conceivable that fusain is actually of dual origin, which could indeed be in keeping with the apparently conflicting evidence.

TABLE 4.8

General Appearance and Characteristics of Macerals

Maceral Group	Maceral Suite	Maceral Type	Range of Maximum Reflectance under Oil (%)	Macerals[a]
Vitrinite	Vitrinite suite	Anthrinoid	2.50–10.00	A_{25}–A_{100}
		Vitrinoid	0.40–2.40	V_4–V_{24}
		Xylinoid	0.10–0.39	X_1–X_3
Exinite	Liptinite suite	Exinoid	0.05–1.50	E_0–E_{15}
		Resinoid	0.05–1.50	R_0–R_{15}
Inertinite	Inertinite suite	Fusinoid	4.00–10.00	F_{40}–F_{100}
		Semifusinoid	0.20–3.99	SF_2–SF_{39}
		Micrinoid	0.20–8.00	M_2–M_{80}

[a] Provisionally differentiated into "entity types" and designated by type numbers ending in "-inite").

TABLE 4.9
Maceral Terminology and Origin

Maceral Group	Maceral	Origin
Vitrinite	Telinite	Humified plant remains typically derived from woody, leaf, or root tissue with well to poorly preserved cell structures
	Collinite	Humified material showing no trace of cellular structure, probably colloidal in origin
	Vitrodetrinite	Humified attrital or less commonly detrital plant tissue with particles typically being cell fragments
Liptinite (exinite)	Sporinite	Outer casing of spores and pollens
	Cutinite	Outer waxy coating from leaves, roots, and some related tissues
	Resinite	Resin filling in cells and ducts in wood; resinous exudations from damaged wood
	Fluorinite	Essential oils in part; some fluorinite may be produced during physicochemical coalification and represent nonmigrated petroleum
	Suberinite	Cork cell and related issues
	Bituminite	Uncertain but probable algal origin
	Alginite	Tests of some groups of green algae; material referred to alginite shows moderate to strong fluorescence
	Exudatinite	Veins of bitumen-related material expelled from organic matter during coalification
	Liptodetrinite	Detrital forms of liptinite that cannot be differentiated
Inertinite	Fusinite	Wood and leaf tissue oxidation
	Semifusinite	Wood or leaf tissue weakly altered by decay or by biochemical alteration
	Inertodetrinite	Similar to fusinite or semifusinite but occurring as small fragments
	Macrinite	Humic tissue probably first gelified and then oxidized by processes similar to those producing semifusinite
	Sclerotinite	Moderately reflecting tissue of fungal origin, largely restricted to Tertiary coals
	Micrinite	Largely of secondary origin formed by disproportionation of lipid or lipid-like compounds

The macerals have their own characteristic appearance (Figures 4.1 through 4.4) (Spackman, 1975). For example, vitrinite may vary from dark to light gray in appearance and may frequently show signs of botanical structure. On the other hand, stable exinite is much darker because it is composed of old megaspores, microspores, cuticles, and resin materials. Inertinite is frequently characterized by cellular structure and may appear, for example, as the cell structure in wood and may also exhibit intercellular spaces. It is noteworthy at this time that coal macerals exhibit a characteristic reflectance (ASTM, 2011a) which offers a means of differentiation between the individual macerals but the exinite macerals can exert a considerable influence on the vitrinite macerals (Raymond and Murchison, 1991). Thus, there is the need to state the amount and type of exinite present when estimating the maturity of organic-rich sediments (especially the sapropelic sediments) using vitrinite reflectance.

4.4.2.1 Vitrinite Macerals

Of the three maceral groups, vitrinite is more familiar as the brilliant black bands of coal. Vitrinite macerals are derived from the cell wall material (woody tissue) of plants, which are chemically composed of the polymers, cellulose and lignin. The vitrinite group is the most abundant group and typically makes up 50%–90% of most North American coals. However, most Gondwanaland coals and some western Canadian coals are vitrinite poor —the inertinite macerals dominate in these coals.

Two individual macerals form the major components of this group: telinite, which is composed of the cell wall material of the original plant(s), and collinite, which is derived from the substance that originally filled the cell cavities. Telinite appears as a structured maceral while collinite occurs relatively unstructured.

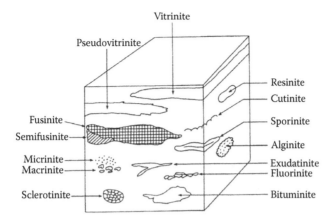

FIGURE 4.1   Representation of coal macerals (see Table 4.7).

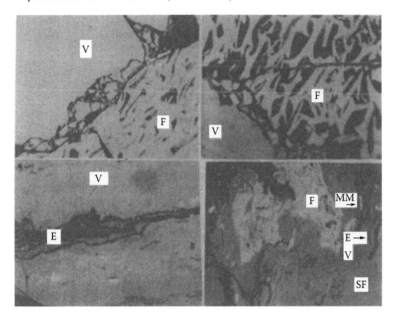

FIGURE 4.2   Photomicrographs of macerals from western Canadian low-volatile bituminous coal: V, vitrinite; F, fusinite; E, exinite; SF, semifusinite; and MM, massive micrinite or macrinite.

4.4.2.2   Liptinite Macerals

Liptinite (exinite) consists of macerals derived from spores, cuticles, resins, and algae and typically makes up about 5%–15% of most North American coals. The spores, which occur in tetrads, are virtually always compressed and show the tetrad scar (Figure 4.5); the spores are usually identified as the maceral sporite.

Liptinites generally make up about 5%–15% of most North American coals. They are usually more abundant in the Appalachian coals than any other US coals except cannel and boghead types where they dominate. At a reflectance of 1.35–1.40, most of the liptinite macerals disappear from coal. Cannel and boghead coals are petrographically distinguished from humic coals by both their maceral composition and texture. They have an abundance of liptinite macerals (sporinite in cannels and alginite in boghead coal) and a relative low level of vitrinite and inertinite macerals.

Sporinite is the most common of the liptinite macerals and is derived from the waxy coating of fossil spores and pollen. It generally has the form of a flattened spheroid with upper and lower hemisphere compressed until they come together. The outer surface of the sporinite macerals often shows various

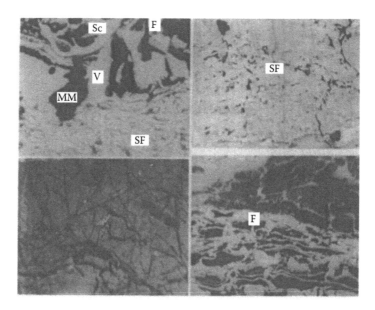

FIGURE 4.3 Photomicrographs of macerals from western Canadian low-volatile bituminous coal: V, vitrinite; F, fusinite; Sc, sclerotinite; SF, semifusinite; and MM—massive micrinite or macrinite.

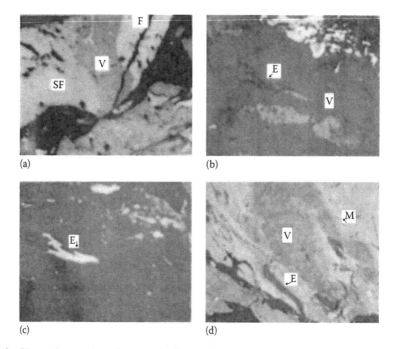

FIGURE 4.4 Photomicrographs of macerals from Canadian coals: (a) Alberta medium-volatile coal: SF, semifusinite; V, vitrinite; and F, fusinite. (b) Alberta subbituminous coal: V, vitrinite; F, fusinite; SF, semifusinite; and E, exinite. (c) Same as in (b) but under fluorescent light. (d) Sydney, Nova Scotia high-volatile bituminous coal: V, vitrinite; M, micrinite; and E, exinite.

kinds of ornamentation and in sections that are parallel or near parallel to the bedding plane of the coal, the sporinite macerals will appear to take on a disk or oval shape that can be confused with resinite. In Paleozoic coals, two sizes of spores are common—the smaller ones, usually <100 μm in size, are called microspores and the larger ones ranging up to several millimeters in diameter are called megaspores. Sporinite is also classified on the basis of the thickness of the spore walls—thin-walled

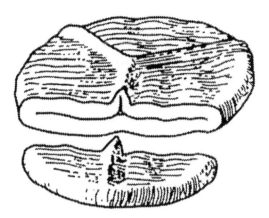

FIGURE 4.5 Representation of a cross-sectional view through a megaspore. (From Murchison, D. and Westoll, T.S., Eds., *Coal and Coal-Bearing Strata*, Elsevier, New York, 1968.)

spores (*tenuispores*) and thick-walled spores (*crassispores*). Spores formed in a sac (sporangium) on the original plants in which they were compressed into tetrahedral groups of four. Evidence of this formation can sometimes be seen under the microscope as a *trilete scar* on the surface of the sporinite.

The major petrographic feature of the liptinite group of macerals is that they all have a reflectance that is lower than the vitrinite macerals in the same coal. This group of macerals is very sensitive to advanced coalification and the exinite macerals begin to disappear in coals of medium-volatile rank and are absent in coals of low-volatile rank. When the exinite macerals are present in a coal, they tend to retain their original plant form and thus they are usually *plant fossils* (*phyterals*). The phyteral nature of the liptinite macerals is the main basis on which they are classified.

The leaves and needles of plants possess a protective coating skin of cutin often referred to as the cuticle and which is highly resistant to decomposition processes. These cuticles thus emerge as another maceral of the exinite group that is termed cutinite. While not very abundant, this maceral is commonly found in most coals and is derived from the waxy outer coating of leaves, roots, and stems.

Cutinite occurs as long stringers, which often have one surface that is fairly flat, and another surface which is crenulated. Cutinite usually has a reflectance that is equal to that of sporinite. Occasionally, the stringers of cutinite are distorted. Because cutinite occurs in sheetlike fragments and is very resistant to weathering, it is sometimes concentrated in weathered coals to which it gives a sheetlike appearance. Such coals are called *paper coals*.

As there are thin-walled and thick-walled cuticles, various profiles of this maceral may be observed in coals (Figure 4.6). In addition, the original plant contains leaf mesophyll between the cuticles which will, by the processes involved in maturation, be transformed into vitrinite. Thus, cutinite may occur as the thick-walled variety with the inner area(s) appearing as vitrinite. Exinite can appear as an elliptical or spindle-shaped body which, when well preserved, will be dark colored with the sporite and cutinite of the same coal appearing relatively lighter in color.

The third liptinite maceral is resinite, which is a term generally employed to include all of the resinous constituents of coals and also includes the resinified essential oils of lignites and bituminous coals. The fourth exinite maceral is alginite, which is formed from the remains of algal bodies and is, in fact, the principal component of boghead coals (torbanites).

Resinite macerals (Anderson et al., 1992) occur widely (often in minor amounts) throughout most American coals below medium-volatile bituminous rank. They are usually absent in coals of higher rank. Although resinite macerals usually make up less than 3% of most US coals, they are particularly abundant in coal of the Wasatch Plateau in Utah where they can account for as much as 15% of the macerals present. In most Appalachian and mid-western US coal seams, resinites occur as primary (present at the time of deposition) ovoid bodies with a long axis ranging from 25 to 200 µm. While primary ovoid bodies of resinite are also found in western US coals of Cretaceous/Tertiary

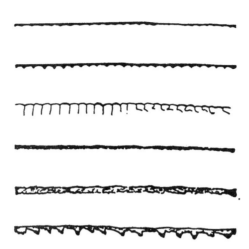

FIGURE 4.6 Representation of various thin-walled and thick-walled cuticles without auricular ledges. (From Murchison, D. and Westoll, T.S., Eds., *Coal and Coal-Bearing Strata*, Elsevier, New York, 1968.)

age, much resinite in these coals occurs as secondary cleat and void fillings. This secondary resinite shows an intrusive relationship to the host coal and often shows flow texture and carries xenoliths of coal in resinite veinlets. Fluorescence microscopy reveals that only the primary resinite ovoids commonly show oxidation or *reaction rims* that suggest a surface alteration.

The final members of the liptinite group have only recently been defined; they are fluorinite, bituminite, and exudatinite. These macerals are comparatively difficult to distinguish from the other constituents of the exinite group, hence their relatively recent identification. Fluorinite is believed to owe its origin to lipoids, fats, and proteins that were originally part of bacterial life forms. On the other hand, exudatinite is believed to owe its origin to exudates from the remaining exinite group and to the vitrinite group. All three of these *new* macerals are considered to have been mobilized and migrated during their history.

4.4.2.3 Inertinite Macerals

The third maceral group, inertinite, originates from plant remains similar to vitrinite but oxygen has usually played a stronger role during the biochemical stage of coalification and has been incorporated into the macerals either before or during their incorporation into peat.

The inertinite macerals are derived from plant material that has been strongly altered and degraded in the peat stage of coal formation. For example, fossil charcoal is the inertinite maceral, fusinite. In most North American coals, the inertinite macerals range from less than 5% to 40% with the highest amounts generally occurring in Appalachian coals. However, the inertinite macerals can make up over 50%–70% of some western Canadian coals. The inertinite macerals have the highest reflectance of all the macerals and are distinguished by their relative reflectance and structure.

Some of the cell structure of the wood may appear to be well preserved and, as is to be expected, transitions can exist between inertinite and vitrinite. Typically, the plant material in the inertinite macerals has been strongly altered and degraded in the peat stage of coal formation. For example, fossil charcoal is the inertinite maceral, fusinite.

In most North American coals, the inertinite macerals range from less than 5% to as much as 40% with the highest amounts generally occurring in Appalachian coals. However, the inertinite macerals can make up over 50%–70% of some western Canadian coals.

Two predominantly inertinite macerals are fusinite and semifusinite (Figure 4.3); fusinite is commonly referred to as fossil charcoal and usually shows well-defined cellular structures. Fusinite is seen in most coals and has a charcoal-like structure. Fusinite is always the highest reflecting maceral present and is distinguished by cell texture. It is commonly broken into small shards and fragments.

On the other hand, semifusinite is intermediate between fusinite and vitrinite showing the well-defined structure of wood, but the cell cavities (round, oval, or elongated) vary in size and are smaller and sometimes less well-defined than those of fusinite. Semifusinite has the cell texture and general features of fusinite except that it is of lower reflectance. In fact, semifusinite has the largest range of reflectance of any of the various coal macerals going from the upper end of the pseudovitrinite range to fusinite. Semifusinite is also the most abundant of the inertinite macerals.

Another maceral of the inertinite group is sclerotinite, which is the result of fungal remains. Sclerotinite occurs as ovoid bodies with cell structure, with reflectance covering the entire inertinite range. Sclerotinite is found mainly in coals of the Tertiary Period, particularly in the lignites, and also in bituminous coals of the Paleozoic era. Sclerotinite (Figure 4.3) is opaque, highly reflective, and, in the oval or round form, may vary in size from 20 to 300 μm (30 to 300×10^{-3} mm).

The final two macerals of the inertinite group are macrinite and micrinite. Originally, these two macerals were classified under the name micrinite, with macrinite being assigned the original terminology *massive micrinite* while micrinite was assigned the name *fine micrinite*. Macrinite (Figure 4.3) is composed of grains and is believed to arise from a sediment of inert detritus. Macrinite is a very minor component of most coals and usually occurs as structureless ovoid bodies with the same reflectance as fusinite.

On the other hand, micrinite is composed of smaller material and is believed to be a maturation product of protoplasm. Micrinite occurs as very fine granular particles of high reflectance and is commonly associated with the liptinite macerals but sometimes gives the appearance of actually replacing the liptinite. Micrinite is frequently, but not always, found in association with microspores. It is also found in the sapropelic (cannel and boghead) coals. Both macerals are virtually structureless.

It is also possible to make a general characterization of the three maceral groups on the basis of their chemical (ultimate) composition (Table 4.10). For example, the vitrinite macerals are relatively rich in oxygen whereas the exinite macerals are relatively rich in hydrogen. On the other hand, the inertinite macerals are relatively rich in carbon or, more in keeping with the other two maceral groups, are actually hydrogen-deficient. On the basis of these significant differences in composition, it is anticipated (and observed) that the three maceral groups show differing behavior in many of the technological processes for coal utilization.

Finally, there has been an effort by some petrographers to classify each maceral according to the major subdivisions of rank of coal. For example, it has been advocated that vitrinite found in lignite and in subbituminous coal should be called *xylinoid* while vitrinite found in bituminous coal would be called *vitrinoid* and vitrinite found in anthracite coal would be called *anthranoid*.

TABLE 4.10
Ultimate Analysis of Macerals of Similar Carbon Content

Component	Exinite	Vitrinite	Inertinite Others	Fusinite
C	82–83	83.0	83–85.0	94.0
H	8.7–9.0	5.5	2.7–4.0	2.8
O	6.0–7.3	9.0–10.1	9–12	2.3
N	0.5–1.4	1.3–2.0	1.3–1.9	0.9
S	0.5–0.6	0.5	10–15	
Volatile matter	80	33–40	10–15	5
Oil yield, wt%[a]	40–50	11–14	1–4	0

[a] From Fischer oil assay.

While there are, undoubtedly, advantages to such a further subdivision of the maceral terminology (of which the major advantage would be the specific coal from which the maceral was found), there are those workers who consider that such a subdivision of the maceral terminology would only add confusion to an area of coal technology that is only now becoming clear after years of misinterpretation and lack of a formalized system of nomenclature.

4.4.3 MICROLITHOTYPES

Microlithotypes are the microscopic analogs of the coal lithotypes and, hence, represent a part of the fine microstructure of coal(s). The microlithotypes are, in fact, associations of coal macerals with the proviso that the "associations" should occur within an arbitrary minimum bandwidth (50 μm, 50 × 10 mm). The composition of the microlithotypes in coal(s) appears to be limited by the types of organic matter present in the original coal precursor. Thus, even though there are theoretically many potential varieties of microlithotypes having differing compositions, the predominant microlithotypes are considerably less than would be anticipated on a random basis (Table 4.11 and Figure 4.7).

In more general terms, the microlithotype vitrite is composed predominantly of vitrinite and is a common constituent of humic coals (Table 4.4), especially vitrains. On the other hand, clarite (which also contains exinite in addition to vitrinite) is more commonly found as thick bands in clarains but does not occur in many other coals; clarite may also contain inorganic impurities such as clay, pyrite, and carbonates. Fusite consists predominantly of inertinite and also contains impurities such as kaolin (as well as other clays), pyrite, and carbonate minerals. Durite is composed predominantly of exinite and inertinite and has actually been classified as existing in two forms: durite E and durite I; the former (durite E) indicates that the durite is relatively rich in exinite whereas the latter (durite I) indicates the durite to be relatively rich in inertinite.

The mixed microlithotype clarodurite is intermediate in composition between durite and clarite with the proviso that while the amount of inertinite must be higher than the amount of vitrinite and the individual amounts of vitrinite, exinite, and inertinite should exceed 5%, the name structure also indicates a closer relationship to durite than to clarite. Clarodurite is also a common constituent of humic coals and the contaminants of this microlithotype are clays, pyrite, and siderite (ferrous carbonate). Conversely, duroclarite has maceral proportions that are closer to clarite than to durite. Thus, the proportion of vitrinite must exceed that of inertinite and the proportions of vitrinite, exinite, and inertinite present should exceed 5%. This microlithotype occurs in fairly thick bands and is a common constituent of most humic coals.

The microlithotype vitrinertite contains high (95%) proportions of vitrinite plus inertinite and, according to connotations employed earlier, vitrinertite V and vitrinertite I denote varieties if vitrinertite rich in vitrinite and in inertinite, respectively. It has also been suggested that this microlithotype may contain minute proportions (less than either the vitrinite or the inertinite) of the maceral exinite and that the inertinite may be present as fragments of fusinite or semifusinite, as sclerotinite, or as massive micrinite or as fine micrinite. Conflicting reports place this microlithotype in bituminous coals of high rank or in low-rank bituminous coals with increasing proportions in bituminous coals with less than 25% volatile matter in which exinite is not often distinguishable under the microscope.

Two other, but less well-known, microlithotypes that have also been identified are microite and sporite. Both are believed to be monomacerals but have received little attention.

4.4.4 INORGANIC CONSTITUENTS

Inorganic elements can be included in coal as minerals or as elements incorporated in the organic structure (Chapters 3, 7, and 10). The most common example of the latter is the incorporation of sulfur into macerals as organic sulfur. Minerals can be incorporated into the peat during deposition, result from epigenetic processes, or be the consequence of metamorphic changes within the coal.

TABLE 4.11
Composition of Microlithotypes

(a) Composition of the Various Microlithotypes

| Microlithotype | Macerals | | |
	Vitrinite	Exinite	Inertinite
Vitrite	>95	<5	
Clarite	<95	5–95	<5
Duroclarite	5–90	5–90	5–90
Clarodurite			
Vitrinertite	<95	<5	5–95
Durite	<5	5–95	5–95
Fusite		<5	>95
Sporite		>95 (Vt + I)	<5

(b) Relationship of the Various Groups in Coal Petrography

| | Maceral | | Microlithotype | |
Lithotype	Maceral	Maceral Group and Symbol	Microlithotype	Principal Groups of Constituent Macerals
Vitrain	Collinite, telinite	Vitrinite (Vt)	Vitrite	Vt
	Macrinite			Vt + I
	Micrinite			I
	Semifusinite			
Fusain	Fusinite	Inertinite (I)	Fusinite	
	Sclerotinite			
	Cutinite			
	Resinite	Exinite (E)	Sporite	E
	Sporinite			
	Alginite		Clarite	Vt + E
Clarain, durain			Durite	I + E
			Duroclarite	Vt + E + I
			Clarodurite	I + E + Vt

Sources: Francis, W., *Coal: Its Formation and Composition*, Edward Arnold Ltd., London, U.K., 1961; From Murchison, D. and Westoll, T.S., Eds., *Coal and Coal-Bearing Strata*, Elsevier, New York, 1968.

Note: Carbargilite (carbonaceous shale) contains 20%–60% by volume of inorganic mineral impurities. If the percentage of inorganic mineral is greater than 60%, the material is classified as shale.

Clay minerals, quartz, calcite, siderite, and pyrite/marcasite are the most common minerals in coals. All naturally occurring elements have been found in coal.

Coal petrography techniques, in addition to determining maceral/microlithotype composition, also include microscopic measurements of certain parameters that are indicators of the degree of coalification (metamorphism) termed coal rank. The coal rank series, from lowest to highest degree of metamorphism, is as follows: peat, lignite, subbituminous, high-volatile bituminous, medium-volatile bituminous, low-volatile bituminous, semianthracite, anthracite, and meta-anthracite (Table 4.1).

Coal metamorphism is the consequence of burial, particularly in the dewatering from peat to lignite to subbituminous coal, and enhanced temperatures. Temperature increase can be the result of burial at varying geothermal gradients, influx of thermal waters or brines through the coal, or, in rare cases, contact metamorphism in the vicinity of igneous intrusions.

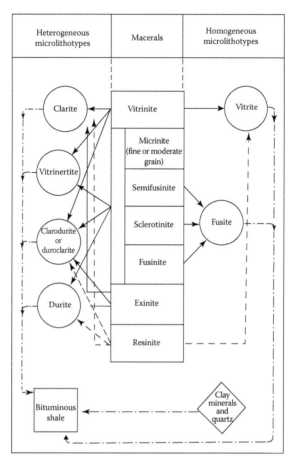

FIGURE 4.7 Relationships between macerals and microlithotypes. (From Francis, W., *Coal: Its Formation and Composition*, Edward Arnold Ltd., London, U.K., 1961.)

Various chemical parameters are used to delineate coal rank. At lower ranks, the equilibrium moisture is used as a measure of coalification. Within the bituminous coal series, heating value, carbon, and volatile matter, all as defined by standard organizations, can be used. At higher ranks, the hydrogen content is used. Petrographers often use the reflectance of vitrinite as a coal rank parameter, determined using a reflected light microscope, oil immersion optics, and a 546 nm band pass filter, as a broad standard. Vitrinite reflectance is also used in the evaluation of organic-rich petroleum source rocks.

4.5 PETROLOGY, PETROGRAPHY, AND BEHAVIOR

Although *coal rank* plays an important role in defining coal use, *coal type* and *coal grade* are also extremely important consideration (Table 4.12). These parameters are the primary factors that influence a coal's specific physical and chemical properties and these properties in turn determine the overall quality of the coal and its suitability for specific purposes.

In general terms, the behavior of coal during combustion, carbonization, pyrolysis, gasification, and direct conversion to liquids depends on the type and amount of the macerals present (Fryer et al., 1975; Falcon, 1978; Stach et al., 1982; Bend et al., 1991; Derbyshire, 1991; Kalkreuth et al., 1991; Murchison, 1991; Gagarin and Krichko, 1992; Sen, 1992; Speight, 2005,

---

TABLE 4.12

Differentiation of Coal Rank, Coal Type, and Coal Grade

Rank

- Indicative of the degree of metamorphism (or coalification) to which the original mass of plant debris (peat) has been subjected during its burial history
- Dependent on the maximum temperature to which the *proto-coal* has been exposed and the time it has been held at that temperature
- Also reflects the depth of burial and the geothermal gradient prevailing at the time of coalification in the basin concerned

Type

- Indicative of the nature of the plant debris (*proto-coal*) from which the coal was derived, including the mixture of plant components (wood, leaves, algae) involved and the degree of degradation before burial
- The individual plant components occurring in coal, and in some cases fragments or other materials derived from them, are referred to as *macerals*
- The kind and distribution of the various macerals are the starting point for most coal petrology studies

Grade

- Indicative of the extent to which the accumulation of plant debris has been kept free of contamination by inorganic material (mineral matter), before burial (i.e., during peat accumulation), after burial, and during coalification
- A high-grade coal is coal, regardless of its rank or type, with a low overall content of mineral matter

---

2008; Suárez-Ruiz and Crelling, 2008). Indeed, it would be most surprising if this were not the case in view of the differences in ultimate composition alone of the various macerals (Table 4.10). Furthermore, it appears that no two coals (no matter how close their relationship is in the various classification systems) have exactly the same petrographic composition and, consequently, the same reactive properties.

The basic chemical parameters of a coal are determined by proximate analysis (moisture, ash, volatile matter, and fixed carbon percentages) and ultimate analysis (carbon, hydrogen, nitrogen, sulfur, and oxygen contents). Other analyses that may be carried out include determining the forms of sulfur in the coal (pyritic, sulfate, organic) and the carbon (or CO_2) content which is derived from the carbonate fraction. Chlorine content (Spears, 2005), which is mainly associated with inorganic salts (relatively high proportions of chlorine may give rise to corrosion in coal utilization) and phosphorous content (an undesirable element in coals destined for use in the steel industry), may also be determined. The ash content of a coal may be analyzed to determine the presence of metal oxides (these influence coal and ash behavior during industrial usage) whereas the proportions of certain trace elements, some of which could be potentially hazardous to human health, are worth evaluating.

As well as proximate and ultimate analysis data, other coal quality parameters that need to be taken into account (e.g., in coal combustion) include information from the following tests:

- *The hardgrove grindability index (HGI)*. This indicates the ease with which the coal can be ground to fine powder. HGI is the test most directly related to the maceral and maceral group composition but is also dependent on rank and mineral content.
- *The heating value, calorific value, or specific energy*. This test indicates the amount of heat liberated per unit of mass of combusted coal. It is regarded as a rank-related parameter, but it is also dependent on the macerals in the coal and mineral composition.
- *The total sulfur content*. This may be derived from a combination of the organic constituents and the mineral matter.

- *The ash fusion temperatures.* They indicate the behavior of the ash residues from the coal at high temperatures and are mainly related to the chemical composition of the ash and the nature of the coal's mineral matter. They are used to indicate whether the ash will remain as a fine powder within the furnace system after the coal is burned or whether some of it might melt to form a slag on the boiler's heat exchange surfaces.

Other tests that provide information about the potential behavior of coals (e.g., in carbonization and coking processes) include the following:

- *The FSI or crucible swelling number* (*CSN*), which is a measure of the increase in volume of the coal when it is heated in the absence of air. This test is also used to characterize coals for combustion. The FSI is at least in part a rank-dependent parameter but also depends on the maceral composition of the coal, the vitrinite maceral group being the main contributor to the swelling properties.
- *The Roga index* provides information on the caking properties of the coal. The index itself is derived from the strength or cohesion of the coke produced in the crucible, as evaluated by a subsequent tumbler test.
- *The Gray–King and Fischer assays* determine the proportions of coke or char (carbonaceous solids), tar (organic liquids), liquor (ammonia-rich solutions), and gas produced when the coal is carbonized (heated in the absence of air) under laboratory conditions. Hence, they provide a basis for estimating the yields of coke and coke by-products obtained from the coal in an industrial coke oven or oil shale–processing plant.
- *Gieseler plastometer and Audibert-Arnu dilatomer tests* monitor how the coal behaves as the different macerals melt, devolatilize, and resolidify at different temperatures during the carbonization process. The Gieseler plastometer evaluates the coal's behavior by measuring the fluidity of a packed coal powder as it is heated, whereas the Audibert-Arnu dilatometer measures the contraction and expansion of a powdered sample pressed into a cylindrical coal "pencil." Such properties are significant when different coals are blended for coke production to ensure compatibility of the different blend components. Coal-blending strategies, for example, coke production are generally decided on the basis of a combination of rheological and petrographic parameters using individual coal samples. These parameters are used to select coals to make up a blend with specific coking properties.

The analytical procedures are used to determine the petrographic, physical, and chemical properties, and these tests are standardized in a number of international test methods (Peters et al., 1962; Karr, 1978a,b, 1979; Stach et al., 1982; Ward, 1984; Diessel, 1992; Van Krevelen, 1993; Taylor et al., 1998; Thomas, 2002; Speight, 2005, 2008; Suárez-Ruiz and Crelling, 2008).

Coal rank from the petrographic point of view is commonly expressed in terms of vitrinite reflectance which may act as an indicator that is independent of other factors (e.g., coal type or grade). Unlike other chemical parameters (e.g., carbon content, hydrogen content, volatile matter yield, and calorific value) it is not dependent on the overall composition of the coal. A number of coal properties progressively change with the advance in rank and the rank of a coal is therefore a major factor influencing its potential application.

The organic constituents of coal (liptinite, inertinite, and huminite/vitrinite) are individually and in combination (microlithotypes) fundamental to many coal properties. Vitrinite is the most common maceral group in many coals, especially in the Carboniferous coals of the Northern Hemisphere, and it is the properties of vitrinite, together with the variations in those properties with rank, that to a large extent determine the properties of the coal concerned.

The inorganic constituents of coal are often expressed using simple parameters such as ash yield, mineral matter, and sulfur content. Knowledge of the inorganic constituents including trace elements in the coal may take on a more complex role if emissions of the so-called hazardous air pollutants (HAPs) are regulated. HAPs generally include antimony (Sb), arsenic (As), beryllium (Be), cadmium (Cd), nickel (Ni), lead (Pb), selenium (Se), mercury (Hg), cobalt (Co), chromium (Cr), and manganese (Mn), with chlorine (Cl) and the radionuclides, thorium (Th) and uranium (U), also being included in some assessments.

Many of the elements that cause concern are trapped within the fly ash after coal combustion and sometimes in the bottom ash by-products. Coal beneficiation processes prior to utilization may serve as a means of reducing the levels of at least some trace elements. Elements of concern that occur at significant levels in the processing residues may give rise to waste disposal or control problems such as leaching into the natural environment via ground or surface water infiltration.

Early work on the relationship of petrographic composition of coal to its technological properties began in the 1920s and focused on the production of cokes and tars from various coal feedstocks. For example, in the coking process and in the low-temperature carbonization processes, exinite macerals were considered to be important constituents of the coal and any resulting tar. The microspores in this maceral group were recognized in the early days of coal petrography as precursors to tar and gas. In fact, yields of tar from spore- and cuticle-containing macerals may vary from 20% to as much as 40% w/w while the resinite maceral of the exinite group may produce as much as 80%–90% w/w of tar. In contrast, the relatively hydrogen-poor maceral groups (Table 4.11) produce much lower yields of tar. There are general claims or statements to the effect that an increase of 1% in the hydrogen content of a tar-producing maceral (i.e., an increase in the hydrogen content from, say, 7%–8%), will, on occasion, double the yield of tar.

In contrast to the exinite group of macerals, the inertinite macerals (which are relatively hydrogen-deficient) show little, or no, reaction during the coking process. Indeed, it is for this reason that this maceral group is given its name. On the other hand, vitrinite has been reported to be an excellent coke-producing maceral but its behavior is variable (see, e.g., Derbyshire, 1991). The rank (carbon content) of the vitrinite must fall within certain limits because the vitrinite found in the anthracite coals (92%–98% carbon, dry ash-free, or daf) and in the semianthracite coals (86%–92% carbon, daf) is virtually inert. However, if the volatile matter (in the lower-rank coals) amounts to 19%–33% w/w, the vitrinite constitutes the actual coking principle.

The carbon content of the two coal types are used here for simplification; describing coal in terms of carbon content alone is not recommended and can be misleading as there are several other factors which may be employed in addition to carbon content to describe coal rank (Chapter 2).

Obviously, in the field of metallurgical coke production, the coking potential of a coal can be determined within fairly accurate limits if the quantities and types of macerals in the coal are known. Consequently, as more data became available concerning the reactivity of the various macerals during gasification or pyrolysis (liquefaction), the more efficient will be the process control and degree of conversion.

In summary, certain of the waxy and resinous macerals are reactive to coal conversion processes while other macerals, notably those of the inertinite group, are notoriously unreactive. Thus, as petrography facilitates the qualitative and quantitative identification of the constituents of coal, it will, no doubt, find increasing application as a method for predicting and elucidating the behavior of any coal to conversion processes designed for the production of synthetic gaseous and liquid fuels (ASTM, 2011a).

REFERENCES

Anderson, K.B., Winans, R.E., and Botto, R.E. 1992. The nature and fate of natural resins in the geosphere: II. Identification, classification, and nomenclature of resinites. *Organic Geochemistry*, 18: 829–841.

ASTM. 2011a. *Standard Method for Microscopical Determination of the Reflectance of the Organic Components in a Polished Specimen of Coal* (ASTM D2793). Annual Book of ASTM Standards, Section 05.05. American Society for Testing and Materials, West Conshohocken, PA.

ASTM. 2011b. *Megascopic Description of Coal and Coal Beds and Microscopical Description and Analysis of Coal* (ASTM D2796). Annual Book of ASTM Standards, Section 05.05. American Society for Testing and Materials, West Conshohocken, PA.

ASTM. 2011c. *Preparing Coal Samples for Microscopical Analysis by Reflected Light* (ASTM D2797). Annual Book of ASTM Standards, Section 05.05. American Society for Testing and Materials, West Conshohocken, PA.

ASTM. 2011d. *Standard Method for Microscopical Determination of Volume Percent of Physical Components of Coal* (ASTM D2799). Annual Book of ASTM Standards, Section 05.05. American Society for Testing and Materials, West Conshohocken, PA.

Beamish, B.B. 2005. Comparison of the R_{70} self-heating rate of New Zealand and Australian coals to Suggate rank parameter. *International Journal of Coal Geology*, 64: 139–144.

Beamish, B.B., Barakat, M.A., and St. George, J.D. 2001. Spontaneous combustion propensity of New Zealand coals under adiabatic conditions. *International Journal of Coal Geology*, 45: 217–224.

Bend, S.L. 1992. *Fuel*, 71: 851.

Bend, S.L., Edwards, I.A.S., and Marsh, H. 1991. *Fuel*, 70: 1147.

Bustin, R.M., Cameron, A.R., Grieve, D.A., and Kalkreuth, W.D. 1983. *Coal Petrology: Its Principles, Methods, and Applications*. Geological Association of Canada, St. John's, Newfoundland, Canada.

Derbyshire, F. 1991. *Fuel*, 70: 276.

Diessel, C.F.K. 1992. *Coal-Bearing Depositional Systems*. Springer, Berlin, Germany.

Drath, A. 1932. *Bulletin of the Polish Geological Institute*, 12: 1.

Dyrkacz, G.R. and Bloomquist, C.A.A. 1992a. *Energy & Fuels*, 6: 357.

Dyrkacz, G.R. and Bloomquist, C.A.A. 1992b. *Energy & Fuels*, 6: 374.

Esterle, J.S. and Ferm, J.C. 1986. Relationship between petrographic and chemical properties and coal seam geometry, Hance seam, Breathitt formation, Southeastern Kentucky. *International Journal of Coal Geology*, 6: 199–214.

Esterle, J.S., O'Brien, G., and Kojovic, T. 1994. Influence of coal texture and rank on breakage energy and resulting size distributions in Australian coals. In *Proceedings. 6th Australian Coal Science Conference*. Australian Institute of Energy, Newcastle, New South Wales, Australia, pp. 175–181.

Falcon, R.M.S. 1978. Coal in South Africa, Part II: The application of petrography to the characterization of coal. *Minerals Science and Engineering*, 10(1): 28.

Francis, W. 1961. *Coal: Its Formation and Composition*. Edward Arnold Ltd., London, U.K.

Fryer, J.F., Campbell, J.D., and Speight, J.G. (Eds.). 1975. *Proceedings of a Symposium on Coal Evaluation*. Information Series No. 76. Alberta Research Council, Edmonton, Alberta, Canada.

Gagarin, S.G. and Krichko, A.A. 1992. *Fuel*, 71: 785.

ICCP. 1963. *International Handbook for Coal Petrology*, 2nd edn. International Committee for Coal Petrology, *International Handbook of Coal Petrography*, 1st Supplement to 2nd edn. Centre National de la Recherche Scientifique, Paris, France.

ICCP. 1971a. *International Handbook for Coal Petrology*, 2nd edn. International Committee for Coal Petrology, *International Handbook of Coal Petrography*, 2nd edn. Centre National de la Recherche Scientifique, Paris, France.

ICCP. 1971b. *International Handbook for Coal Petrology*, 2nd edn. International Committee for Coal Petrology, *International Handbook of Coal Petrography*, 1st supplement to 2nd edn. Centre National de la Recherche Scientifique, Paris, France.

ICCP. 1975. *International Handbook for Coal Petrology*, 2nd edn. International Committee for Coal Petrology, *International Handbook of Coal Petrography*, 2nd supplement to 2nd edn. Centre National de la Recherche Scientifique, Paris, France.

ICCP. 1993. *International Handbook for Coal Petrology*, 2nd edn. International Committee for Coal Petrology, *International Handbook of Coal Petrography*, 3rd supplement to 2nd edn. University of Newcastle on Tyne, U.K.

ICCP. 1994. The new vitrinite classification (ICCP System 1994). *Fuel*, 77: 349–358.

ICCP. 1998. New vitrinite classification (ICCP system 1994). *Fuel*, 77: 349–358.

ICCP. 2001. New inertinite classification (ICCP system 1994). *Fuel*, 80: 459–471.

Kalkreuth, W., Steller, M., Wieschenkamper, I., and Ganz, S. 1991. *Fuel*, 70: 683.

Karr, C. Jr. (Ed.). 1978a. *Analytical Methods for Coal and Coal Products*. Academic Press, Inc., New York, Vol. I.

Karr, C. Jr. (Ed.). 1978b. *Analytical Methods for Coal and Coal Products*. Academic Press, Inc., New York, Vol. II.

Karr, C. Jr. (Ed.). 1979. *Analytical Methods for Coal and Coal Products*. Academic Press, Inc., New York, Vol. III.

Laubach, S.E., Marrett, R.A., Olson, J.E., and Scott, A.R. 1998. Characteristics and origins of coal cleat: A review. *International Journal of Coal Geology*, 35: 175–207.

Law, B.E. 1993. The relation between coal rank and cleat spacing: Implications for the prediction of permeability in coal. In *Proceedings*. International Coalbed Methane Symposium, University of Alabama, Tuscaloosa, AL, pp. 435–442.

Lyman, R.M. and Volkmer, J.E. 2001. Pyrophoricity (spontaneous combustion) of powder river basin coals—Considerations for coalbed methane development. Coal Report No. CR 01-1. Wyoming State Geological Survey, Cheyenne, WY.

Misra, B.K. and Singh, B.D. 1994. Susceptibility to spontaneous combustion of Indian coals and lignites: An organic petrographic autopsy. *International Journal of Coal Geology*, 25: 265–286.

Muller, J.F., Abou Akar, A., and Kohut, J.P. 1990. In *Advanced Methodologies in Coal Characterization*, H. Charcosset and B. Nickel-Pepin-Donat (Eds.). Elsevier, Amsterdam, the Netherlands, Chapter 18.

Murchison, D.G. 1991. *Fuel*, 70: 296.

Murchison, D.G., Cook, A.C., and Raymond, A.C. 1985. *Philosophical Transactions of the Royal Society of London*, 315A: 157.

Parks, B.C. and O'Donnell, H.J. 1948. Bulletin No. 550. U.S. Bureau of Mines, Washington, DC.

Peters, J.T., Schapiro, N., and Gray, R.J. 1962. Know your coal. *Transactions American Institute of Mining and Metallurgical Engineers*, 223: 1–6.

Puttmann, W., Wolf, M., and Bujnowska, B. 1991. *Fuel*, 70: 227.

Raymond, A.C. and Murchison, D.G. 1991. *Fuel*, 70: 155.

Sen, S. 1992. *International Journal of Coal Geology*, 20(3/4): 263.

Smith, K.L. and Smoot, D.L. 1990. *Progress in Energy and Combustion Science*, 16: 1.

Spackman, W. 1958. *Transactions of the New York Academy of Sciences*, Series 2. 20(5): 411.

Spackman, W. 1975. *Proceedings. Workshop on the Fundamental Organic Chemistry of Coal*. National Science Foundation, University of Tennessee, Knoxville, TN, July 17–19, 1975.

Spears, D.A. 2005. A review of chlorine and bromine in some United Kingdom coals. *International Journal of Coal Geology*, 64: 257–265.

Speight, J.G. 2005. *Handbook of Coal Analysis*. John Wiley & Sons, Inc., Hoboken, NJ, 2005.

Speight, J.G. 2007. *The Chemistry and Technology of Petroleum*, 4th edn. Taylor & Francis Group, Boca Raton, FL.

Speight, J.G. 2008. *Synthetic Fuels Handbook: Properties, Processes, and Performance*. McGraw-Hill, New York.

Stach, E., Taylor, G.H., Mackowsky, M.-Th., Chandra, D., Teichmuller, M., and Teichmuller, R. 1982. *Textbook of Coal Petrology*, 3rd edn. Gebrüder Borntraeger, Stuttgart, Germany.

Stopes, M. 1919. *Proceedings of Royal Society*, 90B: 470.

Stopes, M. 1935. On the petrology of banded bituminous coals. *Fuel*, 14: 4–13.

Suárez-Ruiz, I. and Crelling, J.C. (Eds.). 2008. *Applied Coal Petrology. The Role of Petrology in Coal Utilization*. Elsevier, Academic Press, New York.

Sýkorová, I., Pickel, W., Christianis, K., Wolf, M., Taylor, G.H., and Flores, D. 2005. Classification of huminite—ICCP System 1994. *International Journal of Coal Geology*, 62: 85–106.

Taylor, G.H., Teichmüller, M., Davis, A., Diessel, C.F.K., Littke, R., and Robert, P. 1998. *Organic Petrology*. Berlin, Gebrüder Borntraeger, Stuttgart, Germany.

Thiessen, R. 1920. Bulletin No. 117. U.S. Bureau of Mines, Washington, DC.

Thiessen, R. 1931. Bulletin No. B344. U.S. Bureau of Mines, Washington, DC.

Thiessen, R. 1932. The physical structure of coal, cellulose fiber, and wood as shown by Spierer lens. *Industrial and Engineering Chemistry*, 24(9): 1032–1041.

Thomas, L. 2002. *Coal Geology*. John Wiley & Sons, Inc., New York.

Ting, F.T.C. 1978. In *Analytical Methods for Coal and Coal Products*, C. Karr, Jr. (Ed.). Academic Press, New York, Vol. I, Chapter 1.

Van Krevelen, D.W. 1993. *Coal: Typology, Chemistry, Physics, Constitution*, 3rd edn. Elsevier, Amsterdam, the Netherlands.

Ward, C.R. (Ed.). 1984. *Coal Geology and Coal Technology*. Blackwell Scientific Publications, Oxford, U.K.

White, D. 1911. Bulletin No. B29. United States Bureau of Mines, Washington, DC.

5 Recovery

5.1 INTRODUCTION

Recovery of coal (*coal mining, coal recovery*) is the act of removing coal from the ground that has been practiced throughout history in various parts of the world and continues to be an important economic activity today (Esterle, 2005). Historically, coal was used as a means of household heating but now it is mostly used in industry, especially in smelting and generation of electricity.

Coal is valued for its energy content, and since the 1880s, it has been widely used to generate electricity. Steel and cement industries use coal as a fuel for extraction of iron from iron ore and for cement production. In the United States, United Kingdom, and South Africa, a coal mine and its structures are a *colliery* whereas in Australia, the term *colliery* generally refers to an underground coal mine.

One of the earliest known references to coal was made by Greek philosopher and scientist Aristotle, who referred to charcoal-like rock found in Thrace (a region on the northeastern shore of the Aegean Sea) and in northeastern Italy. Although authentic records are unavailable, historians believe coal was first used commercially in China. Reports indicate the Fu-shun mine in northeastern China provided coal for smelting copper and for casting coins around 1000 BC.

Coal cinders found among Roman ruins in England suggest that the Romans harnessed energy from coal before 400 AD. In Roman Britain, coal was exploited in all major coalfields by the late second century AD (Smith, 1997). Much of the coal use remained localized—such as in hypocausts to heat public baths, the baths in military forts, and the baths in the villas of wealthy individuals—before trade developed along the North Sea coast supplying coal to London. After the Romans left Britain, in 410 AD, there are no records of coal being used in the country until the end of the twelfth century, which does not mean that coal was *not* used but merely that record keeping left the island with the Romans.

The written records of the monk Reinier of Liège from the early thirteenth century describe workers mining black earth in Europe. Blacksmiths used this black earth as fuel for metalworking. Other historical records contain numerous references to coal mining in England, Scotland, and continental Europe throughout the thirteenth century.

Shortly after the signing of the Magna Carta at Runnymede (near London), in 1215, coal began to be traded in areas of Scotland and the North East England, where the coal-bearing (carboniferous) strata were exposed on the sea shore, and thus, this type of coal became known as *sea coal* and as early as 1228 there are records that show that sea coal from the Durham coast (North East England) was being taken to London (Galloway, 1882; Freese, 2003). During the thirteenth century, the trading of coal increased across Britain and by the end of the century most of the coalfields in England, Scotland, and Wales were being worked on a small scale (Galloway, 1882; Flinn and Stoker, 1984). As the use of coal amongst the artisans became more widespread, it became clear that coal smoke was detrimental to health and the increasing pollution in London led to much unrest and agitation. As a result of this, a Royal proclamation was issued in 1306 prohibiting artificers of London from using sea coal in their furnaces and commanding them to return to the traditional fuels of wood and charcoal (Freese, 2003).

During the first half of the fourteenth century coal began to be used for domestic heating in coal-producing areas of Britain, as improvements were made in the design of domestic hearths (Freese, 2003). Edward III was the first king to take an interest in the coal trade of the northeast, issuing a number of writs to regulate the trade and allowing the export of coal to Calais (Freese, 2003).

The demand for coal steadily increased in Britain during the fifteenth century, but it was still mainly being used in the mining districts and coastal towns or being exported to continental Europe. However, by the middle of the sixteenth century supplies of wood were beginning to fail in Britain and the use of coal as a domestic fuel rapidly expanded (Freese, 2003). During the seventeenth century a number of advances in mining techniques were made, such as the use of test boring to find suitable deposits and pumps, driven by water wheels, to drain the collieries (Freese, 2003).

The Industrial Revolution, which began in Britain in the eighteenth century, was based on the availability of coal to power steam engines and, moreover, coal was cheaper and much more efficient than wood in most steam engines. International trade expanded exponentially when coal-fed steam engines were built for the railways and steamships in the 1810–1840 period.

In the early eighteenth century the demand for coal escalated when English iron founders John Wilkinson and Abraham Darby used coal, in the form of coke, to manufacture iron. An almost insatiable demand for coal was created by successive metallurgical and engineering developments, most notably the invention of the coal-burning steam engine by Scottish mechanical engineer James Watt in 1769.

Until the American Revolution (1775–1783), most of the coal consumed by the American colonies was imported from England or Nova Scotia. Wartime shortages and the need to manufacture munitions spurred the formation of small American coal-mining companies that mined Virginia's Appalachian bituminous field and other deposits. By the early 1830s, US mining companies had emerged throughout the Appalachian region and along the Ohio, Illinois, and Mississippi rivers. The construction of the first practical locomotive in 1804 in England by British engineer Richard Trevithick sparked a tremendous demand for coal. The growth of the railroad industry and the subsequent rise of the steel industry in the nineteenth century spurred enormous growth in the coal industry in the United States and Europe.

Coal mining essentially developed during the Industrial Revolution when coal provided the main energy source for industry and continuing in the Western countries as a means of providing energy for transportation, seeing it speak in the early to mid-years of the twentieth century (Galloway, 1882; Freese, 2003). However, coal remains an important energy source in the United States and many parts of the world, due to its low cost and abundance when compared to other fuels, particularly for electricity generation—providing much of the energy-based income for states such as Wyoming and West Virginia.

Increasing demand for coal worldwide combined with the depletion of many of the world's most easily accessible coal veins has led to an abundance of highly technical approaches to locating previously undiscovered coal deposits.

The first step in the recovery process is exploring and evaluating the coal deposit. Numerous techniques are employed to study the topography of the land, the extent of the coal bed, the thickness of the overburden, or overlying rock, and the conditions under which the coal must be mined.

While direct exploration methods such as core and rotary drilling remain the primary means of coal exploration, geophysical methods are increasingly being incorporated into broader coal exploration programs—particularly as precursors to expensive drilling operations. These methods consist of scanning the surface for chemical, electromagnetic, or topographic indicators of the material below. Typically, multiple instruments are used for geophysical scanning, as the readings from any one instrument are usually inconclusive.

5.2 EXPLORATORY DRILLING EQUIPMENT AND TECHNIQUES

Exploration for coal is emerging as a future potential source for thermal coal. The process involves discovering new regions and extracting coal economically from earth. Various coal-mining techniques include underground coal mining, surface coal mining, and mountain top removal method.

Coal reserves are available in almost every country worldwide, with recoverable reserves in around 70 countries. The biggest reserves are in the United States (Figure 5.1), Russia, China, and India.

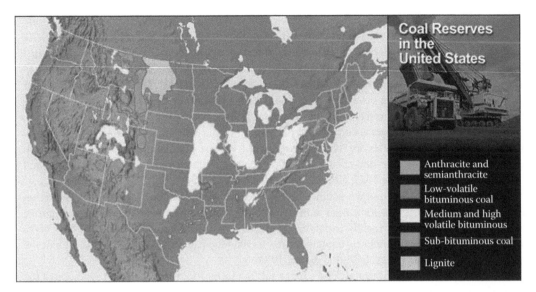

FIGURE 5.1 Coal reserves of the United States. (From American Coal Foundation, http://www.teachcoal. org/aboutcoal/articles/coalreserves.html)

After centuries of mineral exploration, the location, size, and characteristics of most countries' coal resources have been defined. What tends to vary much more than the assessed level of the *resource*—the potentially accessible coal in the ground—is the level classified as *proved recoverable reserves.*

Briefly, the *resource* is the amount of coal that may be present in a deposit or coalfield. This does not take into account the feasibility of mining the coal economically—not all resources are recoverable using current technology. The *reserves* are defined in terms of proved reserves (*measured reserves*) and probable reserves (*indicated reserves*) but probable reserves are estimated with a lower degree of confidence than proved reserves. Proved reserves (proven reserves) are reserves that are not only considered to be recoverable but can also be recovered economically, which takes into account what current mining technology can achieve and the economics of recovery. Proved reserves will therefore change according to the price of coal and advancements in mining technology.

Coal reserves are discovered through exploration activities. The process usually involves creating a geological map of the area followed by geochemical and geophysical surveys, and then exploration drilling. This allows an accurate picture of the area to be developed but the area will only be selected as a mining site if it is large enough and of sufficient quality that the coal can be economically recovered. Once this has been confirmed, mining operations begin.

5.2.1 EQUIPMENT AND TECHNIQUES

The equipment and techniques that have been applied to the search for new coal deposits to determine the extent of known deposits are many and varied (Muir, 1976; Argall, 1979). Thus, to enter upon a detailed discourse of such a collection of equipment and the accompanying techniques is well beyond the scope of this book.

However, in order to understand the nature of the material which the coal scientist is investigating, it is an advantage if he or she is aware of the effort that has already been put into defining the deposit. Similarly, some understanding of the effort that has gone into the mining of this organic rock may also be of value to the scientist. In addition, the reason that this section has been deemed necessary and its inclusion in this text will, hopefully, serve to remind the reader that the piece of coal which lies in the laboratory awaiting attention has been obtained at some great financial cost and (unfortunately, as happens too often) the human cost.

Exploration for coal has been carried out on a continuous basis for several decades but it was only the sudden realization that the so-called plentiful supply of oil and gas is limited (Speight, 1990, 2007 and references cited therein) that has caused a revitalization of coal exploration and expanded these operations to current levels of activity. The prior limited demand for coal exploration had caused the drilling industry to largely ignore the problems associated with coal drilling in order to concentrate efforts on the development of equipment and techniques for the detection of other mineral resources of which iron ore and the ores of lead, zinc, and copper may be cited as examples.

Four types of drilling systems employed for obtaining subsurface strata samples in coal exploration and development activities are the (1) rotary and (2) reverse circulation (noncore) systems and (3) conventional and (4) wire-line-coring systems. Each system is identifiable by its unique method of sample recovery (Landua, 1977; Chugh, 1985; Whittaker, 1985).

It is fortunate that the techniques acquired are of considerable value for all aspects of exploration drilling and many of the new tools used in mineral exploration are now being employed to improve drilling performance in the rapidly expanding search for, and delineation of, mineable coal beds. Indeed, the utilization of oilfield rig equipment and technology has added an extra dimension in the search for coal deposits (Ezra, 1976; Pidskalny, 1979).

As a consequence of the constantly changing trends in mining, there is a continued evolution of the types of tools and techniques that are necessary to meet the demands inherent in the search for new coal deposits (Galloway, 1882; Svendsen, 1976). For example, the expansion of strip-mining operations has dictated the need for more mobile drilling equipment which can be assembled and dismantled with minimal effort. Such equipment must not only be capable of drilling relatively shallow holes at relatively high speeds but also be capable of the maximum recovery of the core samples from the coal seam(s). In addition, core samples (Figure 5.2) of larger diameter are becoming of greater importance and drills (Figure 5.3) using tool diameters of the order of 3.5–4 in. (9–10 cm) or even larger must be capable of achieving high penetration rates in the drilling of either inclined or vertical holes.

Recognition of the many and varied types of drilling programs makes it extremely obvious that not one drill or drilling system can be expected to produce maximum results under all conditions.

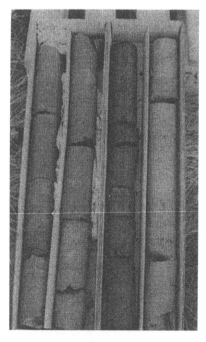

FIGURE 5.2   Recovered core samples (3 in. diameter) showing coal section (second from right) and shale section from above and below the coal seam.

FIGURE 5.3 Different types of drill bits capable of yielding a 3 in. core.

There are, however, certain features that are desirable in any drilling operation, including instrumentation, feed control, long feed and chucking, circulating pumps, and core bits and barrels.

It is also important that there be an understanding of best procedure and control features necessary to achieve the optimum results. If the bit pressure is too great, it is possible to block the bit and grind up the coal sample. Conversely, drilling too slowly in a coal seam may permit the circulating fluid to wash the coal away before it enters the core barrel. It is, therefore, necessary to drill cleanly into and through the coal seam with proper bit pressure.

Circulating pumps are an important part of the drilling program, most especially when drilling formations are soft and easily washed away. A drill pump must have adequate capacity for flushing of the hole during each phase of drilling and the pump should be fitted with a transmission or other means to allow adjusting the output flow without bypassing fluid. A closed-loop system from the pump to the bit will ensure that the fluid being pumped is reaching the bit. If the pump is not adjustable for output, the driller will be forced to bypass fluid and this creates a potentially dangerous situation that may result in a burned bit and possibly a lost drill hole.

Wire-line drilling is performed using thin-walled drill rods through which an inner tube can be lowered into position within the outer core barrel to receive the core as it is cut by the bit. After completion of the coring, the inner tube (containing the core samples) is disconnected so that it may be retrieved by hoisting to the collar of the hole; the drill rods, outer barrel, and bit remain in the hole until they need to be withdrawn to, say, replace a dull bit.

The use of the wire-line system results in an improved drilling process compared with the more conventional coring systems. In these latter operations, approximately 70% of the rig time involves pulling and lowering rods whereas only approximately 30% of the rig time is occupied in the pulling and lowering of the inner tube in the wire-line system. In addition, wire-line drilling is especially favorable for geological formations that are resistant to penetration by roller-rock bits.

In areas where the geology is less favorable, core drilling is performed with conventional or wire-line core barrel, using air, drill muds, or water, as may be required, for cooling the bit and removing the cuttings. As formations become harder and denser, the life of rock bits is shortened and, often due to lack of adequate bearing capacity, the smaller sizes of rock bits are totally unacceptable. If larger bit sizes are used, the size of drill rig and in-hole tools will also need to be enlarged at corresponding higher costs. In hard, dense formations, it is becoming common practice to employ diamond bits and wire-line drilling techniques, which has shown marked improvements in the economics of the operation as well as improvements in core recovery.

Offshore exploration for coal illustrates the successful employment of wire-line tools in a drilling operation, which undoubtedly is one of the more difficult undertakings; not only is it necessary to overcome the problems inherent in drilling through water in excess of 150 ft (48 m) depth into various types of unconsolidated material, but it is also necessary to operate from the deck of a ship on a body of water (Wallis, 1979). This, in itself, may seem a trivial matter to many, but it should be remembered that there are many offshore drilling operations that are extremely hazardous. For example, drilling off the coast east of Sunderland (on the northeast coast of England) is a major, usually hazardous, undertaking, especially when one considers that the North Sea is a body of water not noted for its quiet and dignified serenity.

Finally, a most important feature of the drilling operation is the counterweight system, which helps to maintain a relatively constant weight on the bit in spite of tidal, or wave, action. The close control of the weight allows the bit to cut cleanly and efficiently through the various formations with minimal blocking and a maximum core recovery.

5.2.2 Recent Advances

New techniques in coal exploration develop slowly. Most of the current work being done in this field relies heavily on the techniques and practices in common usage for the past five decades. In fact, since the mid-1970s, coal exploration has expanded and changed dramatically. The OPEC oil price variations focused attention on coal as a preferred alternative to petroleum, particularly in the area of use as electrical utility fuel. Sharply increased electrical power consumption has continued to spur exploration, although at a slower rate than in the mid-1980s through the early 1990s. In fact, the technological revolution of the 1990s ushered in new exploration technologies and techniques.

Moreover, developments in three areas are improving the geologist's ability to better quantify available information and to better predict the position and distribution of coal seams between and beyond drill holes: (1) improved geophysical techniques, (2) modeling of the deposition environments, and (3) manipulating available information with computer programs.

Several new geophysical techniques prove useful. These include improved resolution of downhole logging probes that more accurately indicate depth and thickness of seams and give coal quality information. High-resolution seismic equipment and techniques are now defining better the discontinuities in seams. Faults can be identified readily, but sedimentary cutouts are more difficult to define. New instrumentation in gravity and magnetic technology show some promise. These new geophysical methods lean heavily on manipulation of data by computers.

Modeling of depositional environments is gradually becoming more accepted as a better means of predicting what happens to the coal seam and adjacent rocks beyond the outcrops and drill holes. Not only does it allow the geologist to extrapolate the presence and thickness of seams, but also to predict the rock type that overlies and underlies the coal. All of this information is important for mine planning.

The area of modern coal exploration that has possibly seen the greatest change is in the collection and interpretation of geological data. An increased use of computers and accessories provides rapid handling of large amounts of data. Once the data are entered, the computer will construct a variety of maps, do statistical calculations, and tabulate requested information. Furthermore, the ability to handle vastly greater amounts of information coupled with the need for consistency has resulted in data collection standards that are much more rigorous than they were 25 years ago.

Once potential targets are identified, existing information is then assembled. Acquisition of publicly available information can be expedited by use of the Internet to identify resources and, in some cases, download reports and maps. Other critical information such as existing drill data, geological maps, and reserve estimates from preexisting exploration efforts are gathered from owners, operators, and companies that have prior experience in the area. These base data are then assembled and analyzed. Specific targets are then selected and preliminary budget forecasts made.

Not only have data collection standards changed, drilling techniques have improved as well. The use of high-density polymer foam, to facilitate cuttings removal and stabilize hole sidewalls, has largely replaced the use of bentonite mud, thereby allowing ready conversion of drill holes to water or monitoring wells. Specially constructed drill bits have tremendously increased drilling and coring penetration rates.

Increased pump capacity and the use of air compressors have also helped to increase drilling and coring penetration rates. This has helped to significantly reduce drilling and coring expenses on a cost-per-foot basis. Better bearings and the use of split tube core barrels have increased core recoveries to the point where anything less than 85% recovery is normally considered unacceptable. Standard coring procedures can recover samples usable for geotechnical testing.

Samples of drill cuttings are taken at regular intervals and described. Formal core descriptions are made and the core is frequently photographed in small increments. A recent development is the use of digital photographs that are then appended to computerized drilling reports. In fact over the past two-to-three decades, a virtual revolution has occurred in the way that coal exploration programs are conducted. These changes were caused by a number of factors including technology transfer from the petroleum and hard rock industries to the coal sector, and the advent of microelectronics-based technologies.

Readily available geological data, accurate maps, and reasonable projections of reserves quality and quantity, all serve to reduce operational costs, assist in mine planning, and improve financial forecasts (Ruppert et al., 1999; Watson et al., 2001; Rohrbacher et al., 2005; Heriawan and Koike, 2008). Coal exploration has become a much more vital and valuable part of coal-mining operations and will continue to serve these needs in the future.

5.3 MINING

The most economical method of coal extraction from coal seams depends on the depth and quality of the seams, the geology of the deposit, and environmental factors (Table 5.1). Coal-mining processes are differentiated by whether they operate on the surface or underground. Coal-mining operations

TABLE 5.1
Environmental Issues Related to Coal Mining

Issue	Comments	Source of Risk
Air quality	Emissions of particulates, nitrogen oxides, sulfur compounds, hydrocarbons, and carbon monoxide can be harmful to human health and terrestrial ecology	Surface mining Uncontrolled fired from abandoned mines and storage and waste piles
Water quality	Discharges from mines, runoff from storage and waste piles and reclaimed land, and groundwater changes during mining can affect pH and levels of pollutants, harming water quality and aquatic life	Surface-mine discharge Underground mine discharge Storage and waste pile runoff Aquifer modifications
Water quantity	Surface water and groundwater supplies can be reduced and redirected, affecting water availability for terrestrial and aquatic ecologies and other uses	Surface-mine drainage modification Aquifer modification Consumption
Land use	Mining causes temporary to permanent loss of wildlife habitat Restoration and protection of land uses before mining will require reclamation and subsidence control	Surface-mine reclamation Wildlife habitat destruction Subsidence
Waste disposal	Mine waste requires operations of disposal sites	Waste disposal from mines

can be described under two main headings: (1) underground (or deep) mining, in which the coal is extracted from a seam without removal of the overlying strata and (2) surface mining, in which the strata (overburden) overlying the coal seam are first removed after which the coal is extracted from the exposed seam or partially covered seam.

The most economical method of coal extraction from coal seams depends on the depth and quality of the seams, and the geology and environmental factors. Coal-mining processes are differentiated by whether they operate on the surface or underground. Many coals extracted from both surface and underground mines require washing in a coal preparation plant.

Each mining technique has its own individual merits and the method eventually employed to extract the coal and the technical and economic feasibility of coal recovery are based on (1) regional geological conditions, (2) overburden characteristics, (3) coal seam continuity, (4) seam thickness, (5) seam structure, (6) seam quality, (7) seam depth, (8) strength of materials above and below the seam for roof and floor conditions, (9) topography, especially altitude and slope, (10) climate, (11) land ownership as it affects the availability of land for mining and access, (12) surface drainage patterns, (13) groundwater conditions, (14) availability of labor and materials, (15) coal purchaser requirements in terms of tonnage, quality, and destination, and (16) capital investment requirements (Cassidy, 1973; Lindberg and Provorse, 1977; Martin, 1978).

Typically, seams relatively close to the surface, at depths less than approximately 200 ft, are usually surface-mined. Coal that occurs at depths of 200–300 ft is usually deep-mined but, in some cases, surface-mining techniques can be used. For example, some coals in the western United States occur at depths in excess of 200 ft are mined by open pit methods—the thickness of the seam (60–90 ft) renders the method economically feasible. Coal deposits occurring below 300 ft are usually deep-mined.

5.3.1 Underground Mining

To reach deeper coal beds, miners typically dig underground mines. Two or more shafts are tunneled down into a coal seam—typically, different shafts are used for the passage of miners and machinery and for the passage of mined coal.

There are three types of underground mines: (1) shaft mine, (2) slope mine, and (3) drift mine. The type excavated in a particular case depends on the depth of the coal deposit, the angle of the coal bed, and the thickness of the coal seam (Lindberg and Provorse, 1977; Euler, 1981).

A *drift mine* (Figure 5.4) is one in which a horizontal (or nearly horizontal) seam of coal outcrops to the surface in the side of a hill or mountain, and the opening into the mine, can be made directly into the coal seam. Thus, drift mines are used in cases where a coal seam outcrops on a hill or mountainside. A drift mine consists of a single passageway that follows the coal seam back into the mountain. Drift mines eliminate the need to tunnel through overlying rock to reach a coal deposit.

In the drift mine, mining is conducted using typically either longwall mining or room and pillar mining with continuous mining equipment. Coal is transported to the surface by conveyor belts. This method of mining is used when the coal seam outcrops at the surface, or when a bench has to be constructed on a mountain side to mine the coal. This type of mine is generally the easiest and most economical to open because excavation through rock is not necessary.

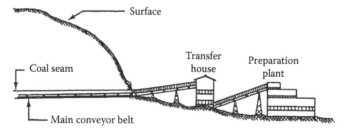

FIGURE 5.4   A drift mine.

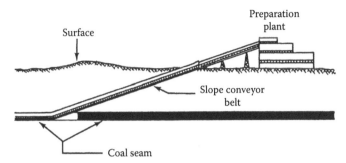

FIGURE 5.5 A slope mine.

A *slope mine* (Figure 5.5) is one in which an inclined opening is used to trap the coal seam or seams. The mine opening is made by tunneling from the surface down to the elevation of the coal seam. Mining is conducted using typically either longwall mining or room and pillar mining with continuous mining equipment. Coal is transported to the surface by conveyor belts. This method of mining coal is usually utilized when the coal seam is not far from the surface, and the outcrop of the coal seam is not exposed. A slope mine may follow the coal seam if the seam is inclined and outcrops, or the slope may be driven through rock strata overlying the coal to reach a seam which is below drainage.

Generally, the slope mines have not been under as much cover as shaft mines but, with the application of rock-tunneling machines, slopes can be extended to deeper coal seams. Coal transportation from a slope mine can be by conveyor or by track haulage (using a trolley locomotive if the grade is not severe) or by pulling mine cars up the slope using an electric hoist and steel rope if the grade is steep. The most common practice is to use a belt conveyor where grades do not exceed 18°.

A *shaft mine* (Figure 5.6) enters the coal seam by a vertical opening from the surface. Shaft mines are dug to reach deep coal beds, usually at least 660 ft (200 m) or more below the surface. A shaft mine uses two vertical shafts to reach the coal bed. Slope mines reach coal deposits that have been distorted or tilted by shifts in the earth's crust. A slope mine uses two angled shafts to reach the coal bed. The passageways of a slope mine typically begin where the inclined coal bed outcrops on the surface and follow the incline into the ground. Some slope mines angle down through the overburden to reach the sloping coal bed, then parallel the bed into the earth. If the grade of the slope mine passageway does not exceed 18°, the coal is usually transported from the mine by conveyor. For steeper grades, coal is typically removed by trolley or mine cars.

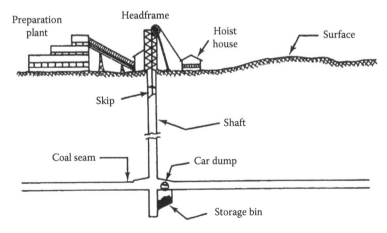

FIGURE 5.6 A shaft mine.

As a general rule, shafts are preferred to slopes for bringing coal out of the mine if the seam lies under very deep cover. The mine opening is made by sinking a shaft down to the elevation of the coal seam. Mining is conducted using typically either longwall mining or room and pillar mining with continuous mining equipment. Coal is transported to the surface by a skip hoist. This is the most expensive type of underground mine to build and operate and is only utilized when the coal seam is deep below the surface.

In both slope and shaft mines, ground support at the opening is dependent on several factors, such as the dimensions of the mine entry, the intended lifetime and use of the mine, the water and climatic conditions, as well as the nature of the exposed strata. In slope mines, it is common practice to use rock bolting only when the exposed rock tends to fragment but is otherwise sturdy; the general practice is to cover the sides and roof of the slope with a thin coating of cement sprayed on, to, say, a wire mesh. If the ground is *heavy* (badly fractured or overstressed), it may be necessary to support the slope with a poured concrete lining; the prevailing practice in shaft mines is to use such a concrete lining.

On occasion, any one mine may have all three types of openings (i.e., drift, slope, and shaft).

For example, miners might dig a drift mine to excavate coal from a hillside outcrop. As the drift mine follows the coal bed into the earth, the miners might dig angled passageways (a slope mine) from above down to the coal bed to shorten the transportation distance into the progressing mine. If the overburden becomes too deep above the progressing mine, the miners may dig more economical vertical passageways (a shaft mine) to provide transportation and ventilation.

In addition, coal haulage might come to the outside through a drift opening, especially if the cleaning plant is close to the outcrop. As the mine develops under heavier cover, additional openings become necessary at intervals for ventilation, and for portals to shorten the traveling time for people and supplies. A slope mine opening might be used for ingress/egress where the cover is not too great and a shaft for air supply. Eventually, a situation develops where the thickness of the overburden makes it more economical to use shaft openings for people, equipment, and supplies (Chadwick, 1992; Joy, 1992).

Once a coal deposit has been reached by a shaft, slope, or drift mine, workers mine the coal by one of two methods: the *room and pillar* method or the *longwall* method. Room and pillar mines extract coal at greater depths and are usually left standing when the mine is abandoned. Longwall mines are used at shallower depths and are allowed to collapse as the mine progresses.

Miners use two processes, known as *conventional mining* and *continuous mining*, to remove coal from room and pillar underground mines. Conventional coal mining replaced hand mining (mining with pick and shovel) in the 1930s.

5.3.1.1 Conventional Mining

In conventional mining, miners use power saws to slice a deep cut 10–12 ft (3–4 m) wide, into the bottom of a coal wall. Next, holes are drilled into the coal above this cut and fill the holes with explosives. The explosions dislodge chunks of coal from the wall. Conveyors or rubber-tired electric vehicles known as shuttle cars carry the coal chunks out of the mine.

Conventional mechanized mining (also called *cyclic mining*) involves a sequence of operations in the order (1) supporting the roof, (2) cutting, (3) drilling, (4) blasting, (5) coal removal, and (6) loading (Figure 5.7). After the roof above the seam has been made safe by timbering or by roof bolting, one or more slots (a few inches wide and extending for several feet into the coal) are cut along the length of the coal face by a large, mobile cutting machine. The cut, or slot, provides a free face and facilitates the breaking up of the coal, which is usually blasted from the seam by explosives.

These explosives (*permissible explosives*) produce an almost flame-free explosion and markedly reduce the amount of noxious fumes relative to the more conventional explosives. Other methods of breaking the coal consist of placing steel cartridges into the blast holes; the cartridges are then filled with compressed air (up to 2000 psi; 13.8 mp) or with liquid carbon dioxide. Release of the

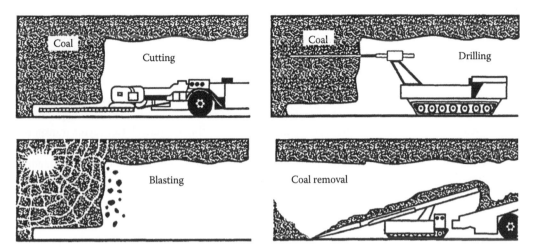

FIGURE 5.7   Underground mining operations.

gases brings about expansion and subsequent breaking of the coal in a manner analogous to the use of explosives. The coal may then be transported by rubber-tired electric vehicles (shuttle cars) or by chain (or belt) conveyor systems.

5.3.1.2   Continuous Mining

Continuous mining (Figure 5.8) was introduced during the late 1940s and began to replace the sequential and cyclic operations of cutting, drilling, blasting, and loading coal. The method makes use of a machine known as a *continuous miner* (Figure 5.9) that can be operated by remote control.

FIGURE 5.8   Continuous mining and loading.

FIGURE 5.9   A continuous miner.

This mobile machine has a series of metal-studded rotating drums that gouge coal from the face of the coal seam (known as the *wall face*). One continuous miner can mechanically break apart approximately 2 tons of coal per hour. After a wall face has been mined to a certain depth, miners stabilize the adjacent roof by bolting long rods into the mine ceiling, advance the ventilation, and begin a new continuous mining cycle.

In the method, the mine is divided into a series of 20–30 ft (6–9 m) rooms or work areas cut into the coal bed, and coal production can be as much as 5 tons of coal a minute. In fact, continuous mining is not necessarily *continuous* since room and pillar coal mining is cyclical. Conveyors transport the removed coal from the seam.

The weak link in this system was often the method of secondary transportation located immediately behind the continuous miner. The high rate of advance and loading of the mining equipment meant that normal secondary haulage systems were inadequate and mobile belt conveyors had to be introduced. These high-capacity conveyors have self-propelled drive and tail sections and sufficient belt storage to permit advances of up to 100 ft (30.5 m) without stopping the conveyor.

Cornering devices were also developed to carry coal around a series of right angle turns thereby removing the need for transference to another conveyor.

5.3.1.3 Longwall Mining

The longwall mining system, used for many years in Europe, involves the use of a mechanical self-advancing roof in which large blocks of coal are completely extracted in a continuous operation (Figure 5.10). Hydraulic or self-advancing jacks (*chocks*) support the roof at the immediate face as the coal is removed; as the face advances, the strata are allowed to collapse behind the support units.

Thus, instead of coal pillars, the longwall mining system uses a line of moving hydraulic jacks to support the roof temporarily in the mining area. No coal pillars are present to obstruct work, so a large coal-cutting machine cuts coal continuously along a wall face typically about 590 ft (180 m) wide. The coal-cutting machine works like a wood power saw, shredding coal from the wall in strips about 20–30 in. (50–75 cm) wide. As the coal-cutting machine strips layers of coal from the wall face like a meat cutter, the line of roof-supporting hydraulic jacks moves automatically behind the machine. As the hydraulic jacks move forward, the roof is allowed to collapse behind the equipment.

Longwall mining produces four to five times more coal from a given deposit than the room and pillar method because coal pillars are not built. But because longwall mining causes the land to sink, land use regulations prohibit this practice in many areas. Despite this prohibition, longwall mining still accounts for about 30% of the coal mined in the United States.

Thus, the method offers the advantages of coal recovery near that attainable with the conventional or continuous systems. In addition, longwall mining offers efficient method of coal recovery under extremely deep cover or overburden or when the roof is weak. Furthermore,

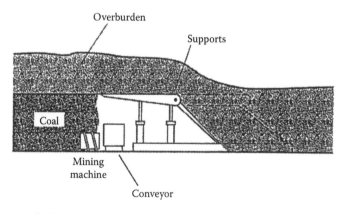

FIGURE 5.10 Longwall mining.

the surface effects of mining are minimized by allowing uniform or controlled subsidence as opposed to the more detrimental and dangerous uncontrolled subsidence. Rock dusting is not necessary along the longwall face and it appears that productivity (tons/person or tons/shift) is also increased.

Longwall systems allow a 60%–100% coal recovery rate when surrounding geology allows their use. Once the coal is removed, usually 75% of the section, the roof is allowed to collapse in a safe manner (WCA, 2009).

5.3.1.4 Shortwall Mining

The shortwall mining system originated in Australia and is, in essence, a combination of the continuous mining and longwall mining concepts.

Continuous, or conventional, mining equipment may be employed for field development after which a continuous miner, in conjunction with longwall-type roof supports, can be used to extract the bulk of the coal. The continuous miner shears coal panels 150–200 ft (40–60 m) wide and more than a half-mile (1 km) long, having regard to factors such as geological strata. Thus, the shortwall mining system appears to offer good recovery of the in-place coal with a marked decrease in the costs for roof support.

In fact, equipment costs may also be reduced by employing the development equipment during the retreat from the face. However, it should be noted that the passes along the coal face are restricted to one direction only (cf. the longwall mining system) since the mining machine is not usually adaptable for cutting in both directions.

5.3.1.5 Room and Pillar Mining

Some mention should also be made here of a technique known as room and pillar mining (Figure 5.11) which has been employed for many years in North America as a means of developing a coal face and, at the same time, retaining supports for the roof. As the name implies, the room and pillar method uses rows of large pillars of coal to support the roof of a mine.

Miners tunnel parallel passageways through the coal seam, and then cut 40–80 ft (12–24 m) wide pillars at regular intervals out of the separating walls of coal. The percentage of coal recovered from a coal seam mined by the room and pillar system depends on the economic incentive to remove as much coal as possible versus the number and size of coal pillars necessary to support the roof. Workers leave the pillars standing in areas where environmental regulations prohibit land subsidence (sinking or settling of land). In areas where land subsidence is acceptable, workers may remove some pillars just before closing the mine.

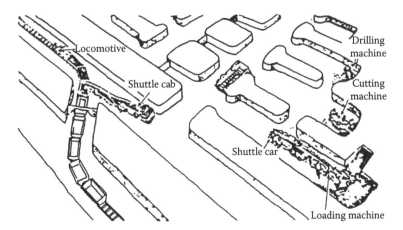

FIGURE 5.11   Room and pillar mining.

The percentage of coal recovered from a mineable seam depends on several factors, such as the number and size of protective pillars of coal thought necessary to support the roof safely, and of the percentage of pillar recovery. In general, the less coal that is extracted the less is the need for costly roof supports. The necessity to protect valuable surface land also has a bearing on the amount of coal mined. In some areas where the surface land is owned by the coal producer, most of the available coal can be mined, whereas in other areas a lesser percentage of the coal is mined to prevent surface damage from subsidence.

On the other hand, in certain heavily industrialized areas, such as Great Britain and Western Europe, the need to mine every possible ton of coal may override consideration for the surface. In such cases, and every care is taken to minimize damage to the surface, total extraction inevitably gives rise to some surface subsidence.

The percentage of coal that is removed may also be determined by the need to protect the mine workings from overhead water. For example, when mining operations are carried out below the sea (such as, as has already been mentioned, in the North Sea off the coast from Sunderland in North East England), only narrow working places may be allowed with large protective pillars of coal being left permanently between the rooms.

5.3.1.6 Miscellaneous Methods

Blast mining (sometime referred to as *conventional mining*) is an older practice in which explosives (such as dynamite) are used to fracture the coal seam, after which the coal is gathered and loaded on to shuttle cars or conveyors for removal to a central loading area. This process consists of a series of operations that begins with "cutting" the coal bed so it will break easily when blasted with explosives.

Retreat mining is a method in which the pillars or coal ribs used to hold up the mine roof are extracted, allowing the mine roof to collapse as the mining works back toward the entrance. This is one of the most dangerous forms of mining owing to imperfect predictability of when the ceiling will collapse and possibly crush or trap workers in the mine.

5.3.2 Mine Environment

Coal mines are hazardous operations. In the twentieth century, more than 100,000 miners died working in coal mines. Many accidents were caused by the mine structure failing through roof collapse or *rock bursts* (coal pillars exploding from the weight of excessive overburden). Other dangers that miners face include toxic or explosive gases released as the coal is mined, dangerous coal dust, and fires.

5.3.2.1 Mine Floor and Roof Strata

The nature of the overlying and underlying rocks in a coal mine may be a significant factor in mining coal. Very commonly, coal seams are underlain by clay stone (also referred to as seat earth, seat rock, underclay, or fire clay), which may vary considerably in physical properties. It may be relatively soft and, when wet, may hinder the use of heavy mining equipment. In many cases after rooms and entries have been mined out, the floor material is not strong enough to support the weight of the overburden, which is now concentrated in the pillars. In such an instance, the clay beneath the pillars flows laterally outward and upward into the void spaces of the rooms and entries; such movement is termed floor heaving squeezing. As the relatively plastic clay moves out from under the pillars, the roof moves down, adding to the general closing or squeezing of the roof and floor in the mined-out rooms and entries.

The type of strata in the first few feet above the coal also has a very important effect on mining. The material must be strong enough to remain stable for a required time when it is supported by either timbering or roof bolts. Shale and siltstone are the most common roof types, and probably the most varied in composition and character. Some of the shales and siltstones are relatively hard and make strong, stable roofs; others are relatively soft, fractured, or characterized by slips and are difficult to support. Limestone and sandstone generally make better roof strata, but variation in these rocks also results in variation of quality of the roof.

In room and pillar mines deep underground, the extreme weight of the overlying rock can break the pillars down, either gradually or in a violent collapse. If the mine roof or floor consists of softer material such as clay, the massive weight of the overburden can slowly push the pillars into the floor or ceiling, endangering the stability of the mine. When the mine roof and floor consist of particularly hard rock, massive overhead weight can overload the pillars, sometimes causing a spontaneous collapse, or rock burst.

5.3.2.2 Gas Evolution

During the coalification process, labile parts of the organic precursors are driven from the coal-generating matrix by heat and pressure (Chapter 3). The process usually results in the formation of volatile materials rather than the formation of a liquid material, as is presumed to occur during the generation of petroleum (Speight, 2007). Coals of increasingly higher rank may contain more free gas because deeper burial has not permitted escape of the gases, and they are trapped within fractures and pores of the coal seam. It is these gases which escape, quite often with violent and explosive consequences, when the coal is mined.

Gases trapped in deeper coal beds have a harder time escaping. Consequently, high-grade coals, which are typically buried deeper than low-grade coals, often contain more methane in the pores and fractures of the deposit. As coal miners saw or blast into a coal deposit, they can release these methane pockets, which may explode spontaneously, often with deadly results. Miners use a technique called *methane drainage* to reduce dangerous releases of methane. Before mining machinery cuts into the wall face, holes are drilled into the coal and methane is drawn out and piped to the surface.

Coal miners also risk being exposed to other deadly gases, including carbon monoxide, a poisonous by-product of partially burned coal. Carbon monoxide is deadly in quantities as little as 1%. It is especially prevalent in underground mines after a methane explosion. In the early 1800s, after a gas explosion, coal miners used canaries to test for carbon monoxide. If the canary died, the miners increased ventilation in the mine to remove the carbon monoxide. The miners then conducted the same test with another canary and repeated the process until a bird survived. Miners also tested for carbon monoxide and methane with a small flame. If the flame's size increased, methane was present in the air; if the flame went out, carbon monoxide was present.

Other dangerous gases locked inside coal deposits include hydrogen sulfide, a poisonous, colorless gas with an odor of rotten eggs, and carbon dioxide, a colorless, odorless gas. To prevent injury from inhaling these gases, underground coal mines must be sufficiently ventilated. Often, the total weight of air pumped through the mine exceeds the total weight of coal removed.

5.3.2.3 Rock Bursts

In areas of very deep mining, as additional weight is supported by the pillars on the coal face, stresses may build up such that the yield point is reached. If the coal pillars, the roof, or the floor is able to yield gradually, the effects of the stresses are relieved by floor heaving, or by pillar or roof failure. When the coal or the associated rocks are particularly strong and depths are great, the stresses may be relieved by rock bursts, which are relatively violent outbursts of the coal into the rooms and entries (Dumpleton, 1990; Beamish and Crosdale, 1998). Frequently, a large volume of gas (usually methane) is suddenly released during a rock burst in a coal mine; a rock burst has also been termed a bump.

5.3.2.4 Coal Dust

As coal is blasted, shredded, and hauled in a mine, large amounts of coal dust are produced. Coal dust is extremely flammable, and if ignited it can be more violently explosive than methane. Miners can reduce the buildup of coal dust by injecting pressurized water into coal beds before the coal is blasted or cut. Other methods to reduce coal dust include washing the dust from mining surfaces with water or covering it with dust from incombustible (nonflammable) rock (Colinet et al., 2010).

Miners who inhale coal dust over a prolonged period can damage their lung tissue. Often, these miners develop spots, lumps, or fibrous growths in their lungs, a condition known as pneumoconiosis, or black lung disease. Furthermore, black lung disease can develop into other, often fatal illnesses, including heart disease, emphysema, and cancer. To protect miners from black lung, many mines are equipped with coal dust filtering units.

5.3.2.5 Mine Fires

Coal mine fires can be triggered during routine mining operations. For example, sparks generated by mining equipment can ignite explosive gases, coal dust, and even the coal bed itself. Because coal beds provide an almost inexhaustible fuel source, once a coal seam is ignited, it can be extremely difficult to extinguish. The intense heat generated by burning coal can rupture the overlying rock strata, sometimes causing the roof to collapse. Uncontrollable fires in some coal deposits have continued burning for years, posing a danger to local communities.

Regulations to prevent and control mine fires have been enacted in the United States and other countries. The U.S. regulations require mining companies to install fire-fighting equipment and automatic fire-suppression systems in mines. Internal combustion engines, which power mining machinery, such as trucks, shuttles, tractors, and scoops, reach high temperatures that can pose a fire hazard. As a result, this equipment must be designed and operated according to the guidelines of the U.S. Mine Safety and Health Administration.

5.3.2.6 Mine Ventilation

Because of the occurrence of gases and dust, provision of adequate ventilation is, amongst other aspects (Ellendt, 1992; English, 1992), an essential safety feature of underground coal mining. In some mines, the average weight of air passing daily through the coal mines is about six times the total daily weight of coal produced, and many mines require the circulation of more than 500,000 ft^3 of air per minute. Obviously, not all of this vast quantity of air is required to enable miners to work in comfort. Most of it is required in order to adequately dilute the harmful gases, frequently termed damps (German damp, vapor), which are produced during mining operations.

5.3.3 Underground Transportation

The success of any mining operation, especially an underground operation, depends to a large extent on the quick and efficient removal of coal from the face. In order to accomplish this end, the coal is loaded at the face into shuttle cars, or on to conveyor belts, in the production area and is thus transferred to the main haulage system. Many miles of belt conveyors are used, most of them automatically controlled so as to stop the flow of coal in the event of any unforeseen condition (Mullen and Rinner, 1992).

Approximately 4000 miles (6400 km) of railroad are in use in North American underground coal mines in the second half of the twentieth century, with track equal in quality to that of many surface railroads. Trains carrying up to 26 tons (26×10^3 kg) of coal in each car were hauled by 35–50 tons ($35–50 \times 10^3$ kg) locomotives, either completely out of the mine or as far as the shaft or slope foot where the coal was transferred to the shaft hoist or slope conveyor.

5.3.4 Mine Gases

5.3.4.1 Firedamp (Methane)

The gas which occurs naturally in the coal seams is virtually always methane; it is a highly flammable gas and forms explosive mixtures with air (5%–14% v/v methane). The explosion can then cause the combustion of the ensuing coal dust (Torrent et al., 1991) thereby increasing the extent of the hazard.

Methane had frequently been termed (in Great Britain) marsh gas or firedamp, although the latter term should (more correctly) be confined to the flammable mixture(s) of methane and air.

In order to render the gas harmless, it is necessary to circulate large volumes of air to maintain the proportion of methane below the critical levels. In this context, it should be noted that the permissible maximum content of methane may vary from country to country but is usually of the order of 1%–2% v/v.

Methane is produced from the coal seam either gradually or as an "outburst"; the latter is, as the name implies, a sudden, violent discharge of short duration, usually accompanied by the displacement of large quantities of broken strata and dust. Fortunately, such outbursts are rare but firedamp (however formed) is a major hazard and, despite improvements in detection and ventilation techniques, dangerous accumulations still occur.

To reduce the danger from flammable gas underground, long boreholes may be drilled in the strata ahead of the working face and the methane drawn out of the workings and piped to the surface, a technique known as methane drainage that is common in Europe. The extracted gas is frequently utilized on the surface and pure methane has a calorific value of 1012 Btu/ft.

5.3.4.2 Whitedamp (Carbon Monoxide)

Carbon monoxide is a particularly harmful gas; as little as 1% in the air inhaled can cause death. It is the product of the incomplete combustion of carbon and is formed in coal mines chiefly by oxidation of coal, particularly in those mines where spontaneous combustion occurs. It is often found after explosions and occurs in the gases evolved by explosives.

5.3.4.3 Blackdamp (Carbon Dioxide)

Carbon dioxide (blackdamp, chokedamp, or stythe, as it is variously called) is found chiefly in old workings or badly ventilated headings. The actual percentage composition of blackdamp depends on its source and its physiological effect varies with composition.

5.3.4.4 Stinkdamp (Hydrogen Sulfide)

This is the name given by miners to hydrogen sulfide because of its characteristic smell. It is one of the first gases to be produced when coal is heated out of contact with air. It occasionally occurs in small quantities along with the methane given off by outbursts and is sometimes present in the fumes resulting from blasting.

5.3.4.5 Afterdamp (Mixed Gases)

This is the term applied to the mixture of gases found in a mine after an explosion or fire. The actual composition varies with the nature and amount of the materials consumed by the fire or with the extent to which firedamp or coal was involved in the explosion. Afterdamp is deficient in oxygen and has high carbon monoxide content.

5.3.5 MINE HAZARDS

The author, having been born and raised in a village that depended on a coal mine for its existence and without even referencing news media reports written as recently as the previous edition of this book, has no doubt that coal mining was and continues to be an extremely dangerous activity.

Mines are unique structures because they are not constructed of manmade materials, such as reinforced concrete, but are built of rock. Thus, integrity of a mine structure is greatly affected by the natural weaknesses or discontinuities that disrupt the continuity of the roof and rib. Geological discontinuities can originate while the material is being deposited by sedimentary or intrusive processes, or later when it is being subjected to tectonic forces. Depositional discontinuities include (1) slips, (2) clastic dikes, (3) fossil remains, (4) bedding planes, and (5) transition zones. Structural discontinuities include (1) faults, (2) joints, and (3) igneous dikes (Chapter 1).

Slips are breaks or cracks in the roof, and they are the features most often cited in underground coal mine fatality reports. When slips are more than several feet long and are steeply dipping, they form a ready-made failure surface. Their surfaces are usually slicken-sided, that is, smooth, highly polished, and striated. Two slips that intersect form an unsupported wedge that is commonly called a horseback. Undetected slips that do not fail during development have a tendency to pop out when subjected to abutment pressures generated during pillar recovery operations. Longer or angled bolts may be used to support slips, and straps or truss bolts can be even more effective.

Joints are fractures commonly found in hard rocks. They occur in sets with similar orientation. Often several sets of joints occur at angles to each other, creating unstable blocks that must be supported by roof bolts.

Fossil remains are the remnants of plants and animals that lived during the time when the sediments that later became rocks were being deposited. For example, kettle bottoms are fossil trees that grew in ancient peat swamps and they occur in every U.S. coal basin, but are especially abundant in southern West Virginia and eastern Kentucky. Dinosaur footprints are another fossil remain found in the roof rocks of coal mines in Utah, Wyoming, and Colorado. Fossil remains can fall without warning and should always be carefully supported. Roof bolt holes should never be drilled directly in fossil remains, because the vibrations could cause them to be dislodged.

Bedding planes are typically found in sedimentary rocks and can extend great lateral distances. Bedding planes represent sharp changes in deposition (such as limestone to clay, or sandstone to coal). These planes can separate readily and are frequently involved in roof falls.

Transition zones occur in many types of strata but are particularly common in sedimentary rocks. A transition zone occurs when some change in deposition causes a change in sedimentation. Different types of sediments compact at different rates. Discontinuities are abundant in the transition zones between distinct strata. For example, where ancient streambeds eroded the adjacent sediments, remnants of the stream channel disrupt the continuity of the normal roof beds, resulting in large slip planes.

Faults are structural displacements within the rock. Tectonic forces can cause rocks to break and slip. Faults often contain weak gouge material, and the country rock around them can be distorted, fractured, and hazardous. Faults are often cited as contributing to surface-mine high wall failures.

Discontinuities occur in many forms and are generally difficult to recognize in advance of mining. They often contribute to fatal accidents, frequently in combination with other factors.

As a result, hazards such as mine wall failure, roof collapse, and the release of gases such as methane (firedamp) as well as coal dust which can result in explosions are endemic to coal mines. Although most of these risks have been reduced in the past five decades, multiple fatality accidents still occur.

The menace of methane (firedamp) explosions in coal mines was widely recognized early in the nineteenth century and although the appropriate steps were taken to prevent them, these measures progressed slowly (Howe, 1992; Michelis and Margenburg, 1992). In the early years, miners took a canary underground to test for carbon monoxide immediately after an explosion. If the bird died, the miners would increase the ventilation until the gas was (apparently) removed and another bird survived. Alternatively, a miner would use his (live flame) lamp as a guide/test for the presence of methane and for carbon monoxide. Thus, if in a working area, the lamp flame increased, methane was present; if the flame extinguished, carbon monoxide was present. Not a reliable method in terms of health and safety—but more important, the danger of coal dust in mines was not, however, appreciated until the early years of this century and few people heeded Michael Faraday's thesis in 1845 that the ignition and explosion of firedamp would raise and then kindle coal dust.

Eventually it was realized that coal dust presents a far greater danger than methane insofar as dust explosions are far more violent in character and widespread in effect than firedamp explosions (Torrent et al., 1991). These latter are usually fairly localized in their effect but the danger lies in the fact that a relatively minor firedamp explosion may initiate a far more serious coal dust explosion. Precautions taken against the occurrence of coal dust explosions include the deposition of incombustible rock dust on the mine surfaces to suppress propagation of the explosion.

Fires are particularly dangerous in coal mines (Noack and Eicker, 1992) as, apart from the fact that a naked flame may initiate an explosion, there is the added complication that the coal itself provides a virtually inexhaustible supply of fuel. And, much of the equipment used underground is capable of initiating or sustaining a fire, for example, flammable conveyor belting.

It is recognized that continued inhalation of certain dusts is detrimental to health and may lead to reticulation of the lungs and eventually to fatal disease included under the general term pneumoconiosis (Freedman, 1978). Coal and silica dusts are particularly harmful when the particle size is below 2×10^{-8} in. (5×10^{-6} m or 5 μm) and, unfortunately, much of the dust produced during mining operations is below this size. Methods adopted to combat the dust hazard include the infusion of water under pressure into the coal before it is broken down; the spraying of water at all points where dust is likely to be formed; the installation of dust extraction units at strategic points; and the wearing of masks by miners operating drilling, cutting, and loading machinery.

5.3.6 SURFACE MINING

In surface mining, or strip mining, earth-moving equipment is used to remove the rocky overburden and then huge mechanical shovels scoop coal up from the underlying deposit. The modern coal industry has developed some of the largest industrial equipment ever made, including shovels (part of a piece of equipment known as a dragline) capable of holding in excess of 300 tons of coal.

To reach the coal, bulldozers clear the vegetation and soil. Depending on the hardness and depth of the exposed sedimentary rocks, these rocky layers may be shattered with explosives. To do this, workers drill *blast holes* into the overlying sedimentary rock, fill these holes with explosives, and then blast the overburden to fracture the rock. Once the broken rock is removed, coal is shoveled from the underlying deposit into giant earth-moving trucks for transport.

Surface mining developed as a natural extension of the early mining techniques by which man recovered coal from a seam. In the early stages, the recovered coal would come from exposed ledges or outcrops. In time, this supply would be exhausted and the earth would be scraped away to lay bare more of the seams that led to these outcroppings, thereby exposing even larger amounts of coal. However, as the overburden to be removed became too much for the primitive equipment then in use, the workings would be abandoned and fresh outcrops sought.

It had become obvious (presumably through accidents) to the early miners (probably individuals gathering fuel for their domestic fires) that coal lying close to the surface would not be safely mined from underground. Unless there was a solid roof above them which could be propped up, the tops of the shallow tunnels would collapse. Large reserves of coal were therefore left, too deeply buried to be exploited by the primitive equipment then available, and yet too shallow to be safely won from underground workings. Thus, surface mining as it is known today (Gambs and Walsh, 1981) was only developed further when improved earth-moving equipment became available, leading to the confident development of these shallower coal reserves (Ingle, 1992; Nies and Jurisch, 1992).

5.3.6.1 Strip Mining

Strip mining is a method of coal recovery in which the coal seam is exposed by removing the overburden (the earth above the coal seam) in long cuts or strips. The soil from the first strip is deposited in an area outside the planned mining area. Spoil from subsequent cuts is deposited as fill in the previous cut after coal has been removed.

Although certain coal deposits can be strip-mined without the use of explosives, some means of breaking the overburden is often necessary. This is usually accomplished by drilling horizontal or vertical blast holes (up to 14 in., 36 cm, diameter) into the overburden whereupon explosive charges are employed to break the rock fragments for subsequent easy removal. This latter part of the operation can be achieved by the use of a dragline (Figure 5.12), power shovel, excavators, bucket wheels, and trucks.

FIGURE 5.12 A dragline engaged in overburden removal. Weight: 7,000,000 lb, boom length: 425 ft, boom vertical height: 210 ft, and bucket capacity: 60 yd^3.

The overburden is then removed to a previously mined (and now empty) strip. When all the overburden is removed, the underlying coal seam will be exposed (a block of coal). This block of coal may be drilled and blasted (if hard) or otherwise loaded onto trucks or conveyors for transport to the coal preparation (or wash) plant. Once this strip is empty of coal, the process is repeated with a new strip being created next to it. This method is most suitable for areas with flat terrain.

The removed overburden (*spoil*) may, at some future time, be returned to the mined-out areas and, in fact, legislators in the various countries where strip mining is in operation are seriously considering the passage of laws relating to the return of mine spoil to mined-out areas so that the land can be restored for future productivity. On the other hand, caution is advised for such ventures since there are also laws which prohibit the "injection" or "burial" of any material that is not indigenous to the particular formations. While mine spoil might be cited as being indigenous to the geological formation from which it came, the effects of weathering (oxidation) and the interaction with acid materials (acid rain) must be given considerations and the extent of any chemical changes should be determined prior to return to the ground. It only takes one failure, or adverse environmental effect, to doom a concept to extinction.

5.3.6.2 Open Pit Mining

The characteristic that distinguishes this type of mine is essentially the thickness of the coal seam insofar as it is virtually impossible to backfill the immediate mined-out area with the original overburden when extremely thick seams of coal are involved. Thus, the coal is removed either by taking the entire seam down to the seam basement (i.e., floor of the mine) (Figure 5.13) or by benching (Figure 5.14). This latter technique involves the staged mining of the coal seam and has been employed with considerable success not only in many coal mines but also for the mining of the Athabasca oil sand deposits in northern Alberta (Canada), where the tar sand pay zone (ore body) may be some 200 ft (60 m) deep (Speight, 1990, 2008). If, however, significant parting exists on the floor of the mine from where the coal has recently been removed, this immediate worked out section will not be available for receipt of the overburden and in such a case the overburden may have to be transported from the area.

Similar types of equipment are used for coal removal as are used for overburden removal, although it is very rare that the same dragline or bucket system would be used for both tasks.

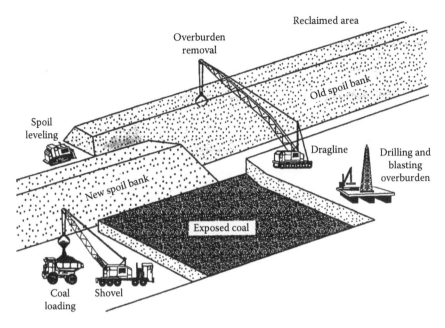

FIGURE 5.13 Open pit mining.

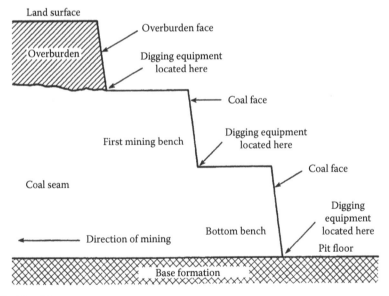

FIGURE 5.14 Benching.

The coal removal equipment (Figure 5.15), which uses *man-sized* buckets (Figure 5.16), will be supported by large trucks (Figure 5.16) for conveyance of the coal to a cleaning plant and/or a transportation point.

5.3.6.3 Contour Mining

Contour mining occurs on hilly or mountainous terrain, where workers use excavation equipment to cut into the hillside along its contour to remove the overlying rock and then mine the coal.

FIGURE 5.15 A surface-mining operation.

FIGURE 5.16 Relative size of dragline buckets.

The depth to which workers must cut into the hillside depends on factors such as hill slope and coal bed thickness. For example, steeper slopes require cutting away more overburden to expose the coal bed (Figure 5.17).

The contour mining method consists of removing overburden from the seam in a pattern following the contours along a ridge or around a hillside and prevails in mountainous and hilly terrain. If a coal seam is visualized as lying level at an elevation of 100 ft (30.5 m) above sea level, and the land surface elevation varies from 600 to 1400 ft (183–428 m) above sea level, a contour-stripping situation exists. Mining commences where the coal and surface elevations are the same (the cropline) and proceeds around the side of the hill on the cropline elevation.

The earth (overburden) overlying the coal may be removed by shovel, dragline, scraper, or bulldozer, depending on the depth and overburden character. The overburden is cast (spoiled) downhill from this first pit, after which the exposed coal is then loaded into trucks and removed from the pit. A second pit can be excavated by placing the overburden from it into the first pit.

Succeeding pits, if any, would follow in the same sequence, with the amount of overburden increasing for each succeeding pit until the economic limit of the operation, or the maximum depth limit of the overburden machine, is reached. This economic limit is determined by many variables,

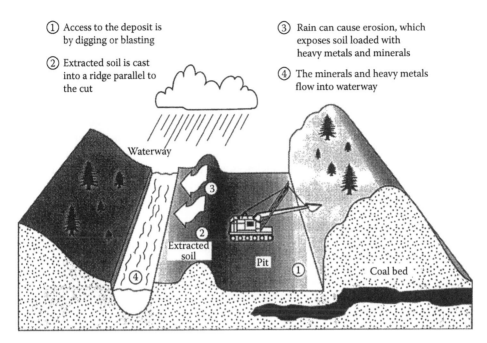

① Access to the deposit is
by digging or blasting

② Extracted soil is cast
into a ridge parallel to
the cut

③ Rain can cause erosion, which
exposes soil loaded with
heavy metals and minerals

④ The minerals and heavy metals
flow into waterway

Waterway

③

②
Extracted
soil

Pit

①

④

Coal bed

FIGURE 5.17   Potential for water contamination from mining operations.

some of which are thickness of the coal seam, quality, and marketability of the coal, nature of the overburden, capabilities of the equipment, and reclamation requirements.

The practice of depositing the spoil on the downslope side of the bench thus created has, for the most part, been terminated because the spoil consumed much additional land and created severe landslide and erosion problems. To alleviate these problems, a variety of methods were devised to use freshly cut overburden to refill mined-out areas. These haul-back or lateral movement methods generally consist of an initial cut with the spoil deposited downslope or at some other site and spoil from the second cut refilling the first. A ridge of undisturbed natural material is often left at the outer edge of the mined area to add stability to the reclaimed slope by preventing spoil from slumping or sliding downhill.

When the mining operation reaches a predetermined stripping ratio (tons of overburden/tons of coal), it is not profitable to continue and depending on the equipment available it may not be technically feasible to exceed a certain height of high wall.

Once the hill slope prevents further cutting into the hillside, miners often switch to a technique known as *auger mining* to extract more coal along the contour.

5.3.6.4 Auger Mining

Auger mining analogous to wood drilling is employed and the coal is extracted from a seam (which outcrops to the surface) by the use of large-diameter augers.

This particular method for the recovery of coal was developed mainly in the period following World War II. An attractive aspect of the method is the relatively low cost of the equipment since the technique essentially involves boring a series of parallel horizontal holes into a coal seam.

The technique is frequently employed in open pit mines where the thickness of the overburden at the high wall section of the mine is too great for further economic mining. This, however, should not detract from the overall concept and utility of auger mining as it is also applicable to underground operations. As the coal is discharged from the auger spiral, it is collected for transportation to the coal preparation plant or to the market.

In this technique, the miners drill a series of horizontal holes into the coal bed with a large auger (drill) powered by a diesel or gasoline engine. These augers are typically about 200 ft (60 m) long and drill holes between 2 and 7 ft (0.6–2.1 m) in diameter to depths of up to 300 ft (91 m) in the coal seam and the coal is moved out as the drill turns farther into the seam. As these enormous drills bore into the coal seam, they discharge coal like a wood drill producing wood shavings.

Additional auger lengths are added as the cutting head of the auger penetrates further under the high wall into the coal. Penetration continues until the cutting head drifts into the top or bottom, as determined by the cuttings returned, into a previous hole, or until the maximum torque or the auger is reached. Penetration in an auger operation may vary from a few feet up to 200 ft (61 m), depending on the pitch of the coal seam, the seam thickness, and the physical characteristics of the strata immediately above the coal seam. The better the roof strata and the more level the coal seam, the deeper the penetration.

5.3.6.5 Mountain Top Removal

Mountaintop removal/valley fill coal mining is, as the name implies, the removal of mountain top to recover coal from the seam(s) underneath. In the method, power shovels are used to dig into the soil for trucks to haul away after which a dragline is used to dig into the rock to expose the coal—explosives may also be used to make the overburden more amenable to removal. Other machines scoop out the layers of coal.

Unfortunately, the mountaintop method generates large amounts of solid waste that must be disposed in a sound environmental manner. Association of the waste with water (through heavy rains or river flooding) can cause the waste to move as a live mud stream causing serious environmental damage and danger to flora and fauna (including human life).

REFERENCES

Argall, G.O. Jr. (Ed.). 1979. *Coal Exploration: Proceedings of the Second International Coal Exploration Symposium*. Miller Freeman, San Francisco, CA.
Beamish, B.B. and Crosdale, P.J. 1998. Instantaneous outbursts in underground coal mines: An overview and association with coal type. *International Journal of Coal Geology*, 35: 27–55.
Cassidy, S.M. (Ed.). 1973. Elements of practical coal mining. *American Institute of Mining*. Metallurgical and Petroleum Engineers Inc., New York.
Chadwick, J. 1992. *Mining Magazine*, 166(6): 349.
Chugh, C.P. 1985. In *Manual of Drilling Technology*, A.A. Balkema (Ed.). Rotterdam, the Netherlands.
Colinet, J.F., Rider, J.P., Listak, J.M., Organiscak, J.A., and Wolfe, A.L. 2010. Best practices for dust control in coal mining. Information Circular 9517. CDC/NIOSH Office of Mine Safety and Health Research. US Department of Health and Human Services, Washington, DC.
Dumpleton, S. 1990. Outbursts in the South Wales Coalfield: Their occurrence in three dimensions and a method for identifying potential outburst zones. *Mining Engineer (London)*, 149: 322–329.
Ellendt, K.P. 1992. *Glueckauf-Forschung*, 53(1): 37.
English, W. 1992. *Glueckauf*, 128(4): 282.
Esterle, J.S. 2005. Mining and beneficiation. In *Applied Coal Petrology: The Role of Petrology in Coal Utilization*, I. Suárez-Ruiz and J. Crelling (Eds.). Elsevier, Amsterdam, the Netherlands, Chapter 3.
Euler, W.J. 1981. In *Coal Handbook*, R.A. Meyers (Ed.). Marcel Dekker, New York, Chapter 3.
Ezra, D. 1976. In *Coal Exploration: Proceedings of the First International Coal Exploration Symposium*, W.L.G. Muir (Ed.). Miller Freeman, San Francisco, CA.
Flinn, M.W. and Stoker, D. 1984. *History of the British Coal Industry: Volume 2. 1700–1830: The Industrial Revolution*. Oxford University Press, Oxford, U.K.
Freedman, R.W.D.W. 1978. In *Analytical Methods for Coal and Coal Products*, C. Karr, Jr. (Ed.). Academic Press, New York, Vol. II, Chapter 28.
Freese, B. 2003. *Coal: A Human History*. Perseus Publishing, Cambridge, MA.
Galloway, R.L. 1882. *A History of Coal Mining in Great Britain*. Macmillan Ltd., London, U.K.
Gambs, G.C. and Walsh, M.A. 1981. In *Coal Handbook*, R.A. Meyers (Ed.). Marcel Dekker Inc., New York, Chapter 4.

Heriawan, M.N. and Koike, K. 2008. Uncertainty assessment of coal tonnage by spatial modeling of seam distribution and coal quality. *International Journal of Coal Geology*, 76: 217–226.

Howe, M.B. 1992. *American Heritage of Invention & Technology*, 8(1): 54.

Ingle, J.H. 1992. *World Mining Equipment*, 16(3): 12.

Joy, A.D. 1992. *Journal of Mines, Metals and Fuels*, 40(1): 24.

Landua, H.L. 1977. Coring techniques and applications. In *Subsurface Geology*, L.W. Le Roy, D.O. LeRoy, and J.W. Raese (Eds.). Colorado School of Mines, Golden, CO, pp. 385–388.

Lindberg, K. and Provorse, B. 1977. In *Coal: A Contemporary Energy Story*, R. Conte (Ed.). Scribe Publishers Inc., New York, Chapter 2.

Martin, P.S. (Ed.). 1978. *A Report on the Current Status of the Canadian Coal Mining Industry*. Dames and Moore Consulting Engineers, Toronto, Ontario, Canada.

Michelis, J. and Margenburg, B. 1992. *Glueckauf-Forschung*, 53(1): 9.

Muir, W.L.G. (Ed.). 1976. *Coal Exploration: Proceedings of the First International Coal Exploration Symposium*. Miller Freeman, San Francisco, CA.

Mullen, R. and Rinner, A. 1992. *Pit and Quarry*, 84(11): 36.

Nies, G. and Jurisch, H. 1992. *Journal of Mines, Metals and Fuels*, 40(1): 46.

Noack, K. and Eicker, H. 1992. *Glueckauf-Forschung*, 53(1): 5.

Pidskalny, R.W. 1979. In *Coal Exploration: Proceedings of the Second International Coal Exploration Symposium*, G.O. Argall, Jr. (Ed.). Miller Freeman, San Francisco, CA, Chapter 4.

Rohrbacher, T.J., Luppens, J.A., Osmonson, L.M., Scott, D.C., and Freeman P.A. 2005. An external peer review of the U.S. Geological Survey energy resource program's economically recoverable coal resource assessment methodology—Report and comments. U.S. Geological Survey Open-File Report No. 2005-1076.

Ruppert, L.F., Tewalt, S.J., Bragg, L.J., and Wallack, R.N. 1999. A digital resource model of the Upper Pennsylvanian Pittsburgh coal bed, Monongahela Group, northern Appalachian basin coal region, USA. *International Journal of Coal Geology*, 41: 3–24.

Smith, A.H.V. 1997. Provenance of coals from Roman Sites in England and Wales. *Britannia*, 28: 297–324.

Speight, J.G. 1990. In *Fuel Science and Technology Handbook*. Marcel Dekker Inc., New York, Vol. 3, Chapter 12.

Speight, J.G. 2007. *The Chemistry and Technology of Petroleum*, 4th edn. Taylor & Francis Group, Boca Raton, FL.

Speight, J.G. 2008. *Synthetic Fuels Handbook: Properties, Processes, and Performance*. McGraw-Hill, New York.

Svendsen, W.W. 1976. In *Coal Exploration: Proceedings of the First International Coal Exploration Symposium*, W.L.G. Muir (Ed.). Miller Freeman, San Francisco, CA, Chapter 15.

Torrent, J.C., Fuchs, J.C., and Borrajo, J.L. 1991. *Combustion and Flame*, 87(3/4): 371.

Wallis, G.R. 1979. In *Coal Exploration: Proceedings of the Second International Coal Exploration Symposium*, G.O. Argall, Jr. (Ed.). Miller Freeman, San Francisco, CA, Chapter 6.

Watson, W.D., Ruppert, L.F., Tewalt, S.J., and Bragg, L.J. 2001. The Upper Pennsylvanian Pittsburgh coal bed: Resources and mine models. *Natural Resources Research*, 10: 21–34.

WCA. 2009. *Coal Mining*. World Coal Association, London, U.K.

Whittaker, A. (Ed.). 1985. *Field Geologist's Training Guide*. International Human Resources Development Corporation, Boston, MA.

6 Preparation, Transportation, and Storage

6.1 INTRODUCTION

Coal preparation (coal cleaning) is the means by which impurities such as sulfur, ash, and rock are removed from coal to upgrade its value (Anonymous, 1991, 1993, 1994). Coal preparation processes are categorized as either physical cleaning or chemical cleaning. Physical coal cleaning processes, the mechanical separation of coal from its contaminants using differences in density, are by far the major processes in use today. Chemical coal cleaning processes are currently being developed, but their performance and cost are undetermined at this time.

The direct objectives of coal cleaning practices are reduction (within predetermined limits) of size, moisture, ash, as well as sulfur (Williams, 1981; Couch, 1991). However, coal properties have a direct bearing not only on whether but also on how coal should be cleaned. Indeed, coal rank (rank being a complex property that is descriptive of the nature of the coal and its properties) (Chapters 2, 8, and 9) can, and usually does, play an important role in determining the feasibility and the extent of cleaning.

Coal quality is now generally recognized as having an impact, often significant, on coal combustion, especially on many areas of power plant operation (Parsons et al., 1987; Leonard, 1991). The parameters of rank, mineral matter content (ash content), sulfur content, and moisture content are regarded as determining factors in combustibility as it relates to both heating value and ease of reaction. In addition, although not always recognized as a form of cleaning or beneficiation, the size of the coal can also make a difference to its behavior in combustion (power plant) operations, hence the need for one or more forms of cleaning (pretreatment) prior to use.

The principal consideration of coal beneficiation is to upgrade the quality of coal for direct use in steam and power generation, or for special uses such as chemical feedstock, feed to liquefaction, and gasification. The properties and quantities of impurities in coal are known to be the major factors that place limitations on coal utilization. All coals are not the same (Chapters 1 and 2). Thus, the type of coal beneficiation technology and the extent of beneficiation depend mostly on the type of coal, the means of mining, and the clean coal utilization.

The output from any coal mine usually consists of raw coal (run-of-mine [ROM] coal) that is wetter and finer, and also contains more impurities than in the past due to (1) mining of lower quality coals, (2) advanced, continuous, and nonselective mining techniques that cause more coal being broken up, (3) more impurities being included, and (4) extensive water utilization for minimizing dust (Lockhart, 1984). While the quality of ROM coals is generally decreasing, the necessity for efficient coal beneficiation technology is significantly increasing, resulting in an increased demand for high-quality coals that meet both market and environmental standards.

Thus, the coal delivered to the coal preparation plant consists of coal, rocks, minerals, and any other form of material that is not coal. The coal also varies widely in size, ash content, moisture content, and sulfur content.

ROM coal has no size definition and consists of pieces ranging from fine dust to lumps as large as 2 ft (0.6 m) square, or larger depending on the rank of coal (Baafi and Ramani, 1979). It is often wet and may be contaminated with rock and/or clay; as such, the coal is unsuitable for commercial use. At best, the coal seams being worked may be relatively thick, unfaulted and uniform, free of associated rock partings, and dry. In such cases, the coal may require only some breaking or crushing and screening to produce a *pure* coal.

Mineral matter content ranges from 3% to 60% (mineral ash) at different mines. Most of the ash is introduced for the roof or bottom of the mine or from partings (small seams of slate) in the coal seam. This mineral matter (extraneous ash) is heavier than 1.80 specific gravity; the remaining mineral matter is inherent in the coal. The density of coal increases with the amount of mineral matter ash present.

The moisture content of the coal is also of two types. The surface moisture, which was introduced after the coal was broken loose from the seam, is easier to remove. This moisture is introduced by exposure to air, wet mining conditions, rainfall (in stockpiles), and water sprays. The remaining moisture (*bed moisture, cellular moisture, inherent moisture*) can be removed only by coking or combustion. This moisture was included during formation of the coal.

Sulfur in coal occurs as sulfates, organic sulfur, and pyrites (sulfides of iron). The sulfates usually are present in small quantities and are not considered a problem. Organic sulfur is bound molecularly into the coal and is not removable by typical coal preparation processes. Pyrites generally are present in the form of modules or may be more intimately mixed with the coal. Coal preparation plants remove only a portion of the pyritic sulfur; therefore, the degree of sulfur reduction depends on the percentage of pyrites in the coal, the degree to which this is intimately mixed with the coal, and the extent of coal preparation.

The size of the pieces of coal ranges upward to that of the size of foreign materials, such as a chunk of rock that has fallen from the mine roof or a metal tie; large pieces of coal from a very hard seam are sometimes included. Foreign materials are introduced into the coal during the mining process, the most common being roof bolts, ties, car wheels, timber, shot wires, and cutting bits.

All of the materials described earlier are combined with the coal to form the ROM feed. Coal, as referred to earlier, denotes the portion of the feed that is desired for utilization. These are the characteristics that can be controlled by preparation.

Thus, coal preparation serves several purposes. One important purpose is to increase the heating value of the coal by mechanical removal of impurities. This is often required in order to find a market for the product. ROM coal from a modern mine may incorporate as much as 60% reject materials.

Air pollution control often requires partial removal of pyrites with the ash to reduce the sulfur content of the coal (Godfrey, 1994). Ash content often must be controlled to conform to a prescribed quality stipulated in contractual agreements. Because of firing characteristics, it is often as important to retain the ash content at a given level as it is to reduce it. Freight savings are substantial when impurities are removed prior to loading. Finally, the rejected impurities are more easily disposed of at the mine site remote from cities than at the burning site, which is usually in a populated area.

Coal preparation is carried out at a facility that washes coal to remove soil and rock, preparing it for transport to market—a *coal preparation facility* may also be called a *coal-handling and preparation plant*. During the *preparation process*, as much waste as possible is removed from the coal to increase the market value of the coal and reduce the transportation costs.

Coal needs to be stored at various stages of the preparation process and conveyed around the preparation plant facilities. Stored coal (*stockpiled coal*) provides surge capacity to various parts of the preparation plant. A simple stockpile is formed by machinery dumping coal into a pile, either from dump trucks, pushed into heaps with bulldozers, or from conveyor belt booms. More controlled stockpiles are formed using a stacker (a large machine used to pile the coal into a stockpile) or multiple stackers to form piles along the length of a conveyor, and reclaimers (a large machine used to recover coal from a stockpile) to retrieve the coal when required for product loading. Taller and wider stockpiles reduce the land area required to store a set tonnage of coal. Larger coal stockpiles have a reduced rate of heat lost, leading to a higher risk of spontaneous combustion.

Briefly, the mined coal is loaded into a stockpile, with a reclaim tunnel beneath it. Then, the coal is transported to a raw coal silo, usually 10,000 tons capacity, for feed to the plant at a constant rate. Generally, the first stage is a crushing/screening plant (Figure 6.1), with heavy-media processing

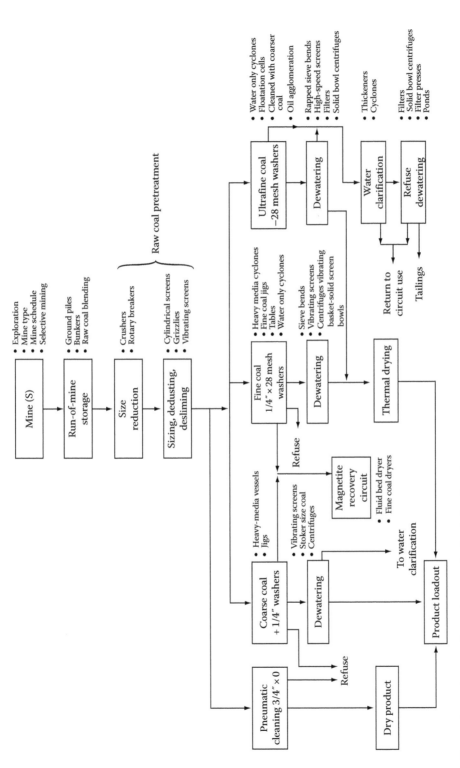

FIGURE 6.1 General layout of a coal preparation/coal cleaning plant.

(for coarse coal sizes—2 in. × 10 mesh), spirals for the middling sizes (10 mesh × 60 mesh), flotation for the −60-mesh fine coal feed.

Most conventional coal cleaning operations utilize gravity methods for the coarser size fractions and surface treatment methods for the finest particle sizes (Riley and Firth, 1993). The selection of equipment, especially for the finer sizes, depends on the mining method, coal hardness, and size distribution and amounts thereof. Typical of these is a dense-media cleaning process (Fourie, 1980) which use dense-media vessels or jigs for the coarsest size usually +3/8 in., dense-media cyclones, concentrating tables or jigs for the 3/8 in. × 28-mesh size, water-only cyclones or spirals, and sometimes flotation for the 28 × 100-mesh size and flotation for the −100 mesh.

Screening and centrifugal dryers dewater the coarser products while screen-bowl centrifuges and sometimes thermal dryers are utilized to reduce the moisture content of the finest sizes.

Metallurgical coal cleaning plants utilize thermal dryers—the coal is softer and more friable and thus has a finer size distribution after extraction by the mining machines. Coals for metallurgical use must be thoroughly processed and dried to meet the end user requirements. Additionally, flotation is typically utilized in these circuits due to the quantity of coal and quality of the needed end product (low ash, low sulfur) (Aplan, 1993; Burchfield, 1993).

On the other hand some steam coals, especially the harder ones (low hardgrove index), and some coals produced from surface mines have smaller quantities of the −100-mesh size. In many plants, there is such a small quantity of the −100-mesh size that this material is sent to disposal and is considered uneconomical to recover.

Coal flotation is a physiochemical process which exploits the differences in the wettability of hydrophobic clean coal and that of hydrophilic foreign particles (Arnold and Aplan, 1989; Fecko et al., 2005). It is, therefore, subject to the surface properties of coal pyrite and other types of commercially worthless material present in coal which plays a major role in determining separation of such material from coal (Luttrell and Yoon, 1994; Luttrell et al., 1994).

Oxidation also leads to the formation of various oxygen functional groups and soluble inorganic that can adsorb on the coal surface and modify its wettability and floatability. These groups have remarkable impacts on surface charge, which controls film-thinning process and thus flotation kinetics (Sokolović et al., 2006). Decreased coal recovery and increased concentrate ash content may be explained by oxidation of coal. In fact, a good correlation exists between the zeta potential and floatability and electrochemical tests confirm negatively effect of oxidation on the coal recovery and also the final effect of coal flotation process (Fonseca et al., 1993; Sokolović et al., 2006).

6.2 COAL PREPARATION

Coal preparation serves several purposes. One important purpose is to increase the heating value of the coal by mechanical removal of impurities. This is often required in order to find a market for the product. ROM coal from a modern mine may incorporate as much as 60% reject materials.

The energy content of coal is related to its rank (degree of coalification) (Table 6.1) which is influenced by the content of nonfuel components (e.g., minerals and moisture) (Chapters 7 through 9). Thus, a primary objective of coal cleaning is to maximize the recovery of the heat value of the coal, consistent with achieving standard specifications for ash, moisture, and sulfur contents.

Furthermore, since transportation costs are usually charged on a ton–mile basis (which does not distinguish between coal substance and moisture content), it is preferential to remove as much as possible of the extraneous mineral matter and water prior to shipping thereby reducing transportation costs for an "inferior" grade of coal and providing a higher energy material to the consumer.

Briefly, the grade of a coal establishes its economic value for a specific end use. Grade of coal refers to the amount of mineral matter that is present in the coal and is a measure of coal quality. Sulfur content, ash fusion temperature (measurement of the behavior of ash at high temperatures),

TABLE 6.1
Typical Properties of Different Rank Coals

Coal Type	Carbon (%)	Hydrogen (%)	Limits of Volatile Matter (%)	Fixed Carbon (%)	Calorific Value (Btu/lb)
Lignites	73–78	5.2–5.6	45–50	50–55	<8,300
Subbituminous	78–82.5	5.2–5.6	40–45	55–60	8,300–11,500
High-volatile bituminous	82.5–87	5.0–5.6	30–40	60–70	11,500–14,000
Medium-volatile bituminous	87–92	4.6–5.2	20–30	70–80	>14,000
Low-volatile bituminous	91–92	4.2–4.6	15–20	80–85	>14,000
Anthracite	95–98	2.9–3.8	5–10	91–95	>14,000

and quantity of trace elements in coal are also used to grade coal. Although formal classification systems have not been developed around grade of coal, grade is important to the coal user.

In fact, long-range transportation of lignites, more than one-third of which consists of water, can more than triple the initial mine-mouth costs calculated on an energy basis. There may however be some trade-off in transportation costs if the low-rank coal is sufficiently low in sulfur which, in turn, means a lower cost in terms of stack gas cleanup (Nowacki, 1980).

The need for coal cleaning can be reduced by choice of suitable mining methods; many mines include the methods by which oversize coal is reduced in size but the cleaning of ROM coal is, more often than not, a separate operation which is performed as a surface operation that is usually close to the mine-mouth. However, the term *coal preparation* includes, by definition, not only sizing (i.e., crushing and breaking) methods but also all of the handling and treatment methods that are required to prepare the coal for the market.

The scheme used in physical coal cleaning processes varies among coal cleaning plants but can generally be divided into four basic phases: initial preparation, fine coal processing, coarse coal processing, and final preparation (Figure 6.1).

In the initial preparation phase of coal cleaning, the raw coal is unloaded, stored, conveyed, crushed, and classified by screening into coarse and fine coal fractions. The size fractions are then conveyed to their respective cleaning processes.

Fine coal processing and coarse coal processing use similar operations and equipment to separate the contaminants. The primary difference is the severity of operating parameters. The majority of coal cleaning processes use upward currents or pulses of a fluid such as water to fluidize a bed of crushed coal and impurities. The lighter coal particles rise and are removed from the top of the bed. The heavier impurities are removed from the bottom. Coal cleaned in the wet processes then must be dried in the final preparation processes.

Final preparation processes are used to remove moisture from coal, thereby reducing freezing problems and weight and raising the heating value. The first processing step is dewatering, in which a major portion of the water is removed by the use of screens, thickeners, and cyclones (Hee and Laskowski, 1994; Nowak, 1994). The second step is normally thermal drying, achieved by any one of three dryer types: fluidized bed, flash, and multi-louvered. In the fluidized bed dryer, the coal is suspended and dried above a perforated plate by rising hot gases. In the flash dryer, coal is fed into a stream of hot gases for instantaneous drying. The dried coal and wet gases are both drawn up a drying column and into a cyclone for separation. In the multi-louvered dryer, hot gases are passed through a falling curtain of coal, which is then raised by flights of a specially designed conveyor.

Coal preparation processes can improve the ROM coal to meet market demands, as limited by the inherent characteristics of a given coal. The top size of the ROM can be reduced to any size specified, although control of the varying size increments can be poor, depending on the amount of crushing required.

Although inherent moisture cannot be changed, the surface moisture can be reduced to any level that is economically practicable. Considerations include the possibility of reexposure to moisture during transportation and subsequent storage and the fact that intense thermal drying increases ideal conditions for re-adsorption of moisture.

The free sulfur in the coal is subject to removal only by chemical treatment, which is not a coal preparation process, or by combustion. The reason that the pyrites can be partially removed in washing processes is that they are heavy enough to be removed with the ash. The processes can remove only 30%–60% of the pyrites, however, because some pyrites are not broken free of the coal and are present in a given piece in a quantity too small to increase its weight enough to be rejected.

Foreign metals can be removed relatively easily. Most wood fragments can be removed although small a few small pieces of wood cause no particular harm because they are combustible.

Thus, coal preparation is, of necessity, an integral part of the production and use of coals. The effect on costs can be as important as the planning of mine layouts; decisions concerning mining systems should be an essential element in all mining feasibility studies, especially in view of new (and/or renewed) environmental regulations such as the Clean Air Act Amendments in the United States (Elliott, 1992; Tumati and DeVito, 1992; Rosendale et al., 1993; Paul et al., 1994).

In more general terms, the primary aims of preparing coal for the market depend upon the nature of the raw coal (Table 6.2 and Figure 6.1) but, essentially, are (1) the reduction in size and control of size within the limits determined by the needs of transportation, handling, and utilization and

TABLE 6.2
General Methods of Coal Preparation and Levels of Cleaning

Level	Raw Coal Weight (%)	Raw Coal Btu Content (%)	Reduction Potential		Comments
			Ash	Sulfur	
1	98–100	99–100	None to minor	None	Crushing and breaking of raw coal to 3 in. size
2	75–85	90–95	Fair to good	None to minor	Coarse coal cleaning of 3 × 3/8 in. coal
3	60–80	80–90	Good	Fair	Moderate coal cleaning of 3 in. × 28-mesh coal
4	60–80	80–90	Good to excellent	Fair to good	Fine coal cleaning of 3 in. × 0-mesh coal
5	60–80	85–95	Deep-cleaned coal: excellent; middle-cleaned coal: none to fair	Multiple-stream coal preparation of two cleaned coal products: "deep-cleaned" coal and "middle grade" coal	

(2) the removal of extraneous mineral matter to a point that is satisfactory for the customer and specifications are met. This latter operation is more often referred to as control of ash content.

In the early days of the industry, coal was sold as it came from the ground but as the century advanced there was always the possibility of dispute about payment if the coal contained visible impurities. Thus, in the early-to-middle decades of the last century, some effort was made to physically remove the impurities from the coal as evidenced by the employment of *belt boys*. It was the sole purpose of these boys, hired straight from school, at age 12–15 (sometimes even younger) to stand at the side of the underground conveyer belts and remove (by hand) the pieces of rock and slate as the coal passed to the mine cars. By this means, much of the large-sized impurities were sorted and left in the waste area.

In many mines, the waste areas were called tipples because of the operation which transferred the coal from the mine cars to picking or sorting screens and where visible impurities were removed by hand. Tipples also segregated the ROM coal into size groups and, as the larger sizes could be more carefully hand-cleaned and were burned with greater ease and cleanliness in fireplaces and hand-stoked furnaces, size became associated with quality. The tipples grew to become what is now looked upon as environmentally unsuitable mountainous heaps of rock which still disfigure many coal-mining areas. Recently, there have been efforts to take back much of the tipple rock into the worked-out underground seams for storage.

Many coals also contain (in fact, it is a rare coal that does not contain any) sulfur which must also be removed before, or at some stage during coal use (see Section 6.5). In addition to sulfur, it is also of increasing importance to address the environmental hazards afforded by trace elements (Swaine, 1992).

Coal, and its associated mineral matter, have long been known to contain minute quantities of elements such as arsenic, cadmium, germanium, mercury, selenium, and even radionuclides, some of which are believed to be carcinogenic and all of which may otherwise be toxic (Chapter 7). Many trace elements are concentrated in particulate emissions; it is these extremely small particles that are the most difficult to eliminate and which are potentially the most harmful since they have the potential to enter, directly or indirectly, the human physiological system.

In more general terms, coal cleaning is a balance between recovery of the maximum energy content of the coal and rejection of extraneous mineral matter. Indeed, it is difficult (perhaps impossible) not to reject some coal with the rock and slate. Thus, every coal refuse pile in existence will contain some measure of recoverable energy. One major challenge is to reduce this potential loss of energy to the minimum practical amount.

The importance of adequate coal pretreatment technologies must be emphasized; many of the operating problems in cleaning plants are attributed to inadequate (inefficient) pretreatment, which results in large quantities of oversize (or undersize) material in the feeds to the various cleaning units which cause loss of cleaning efficiency, blockages, and even plant shutdown.

Conventional coal cleaning plants are quite efficient for Btu recovery, as well as for ash and pyritic sulfur reduction. Btu recovery is generally between 85% and 90% and the ash reductions on a lb of ash/MM Btu basis are usually in the 70%–80% range for Pittsburgh seam coals, and in the 85%–90% range for Illinois and central Appalachian coals (Rosendale et al., 1993).

6.3 SIZE REDUCTION

Pretreatment is, simply, breaking, crushing, and screening of the ROM coal in order to provide a uniform raw coal feed of predetermined top size thereby minimizing the production of material of ultrafine size by excessive crushing or handling. Thus, pretreatment and size reduction are terms that are often used synonymously.

Most conventional coal cleaning facilities utilize gravity methods for the coarser size fractions and surface treatment methods for the finest particle sizes (Riley and Firth, 1993). The selection of equipment, especially for the finer sizes, depends on the mining method, coal hardness, and size distribution and amounts thereof.

The first operations performed on ROM coal are removal of tramp iron and reduction of size to permit mechanical processing. The ROM coal is first exposed to a high-intensity magnet, usually suspended over the incoming belt conveyor which pulls the iron impurities out of the coal. This magnet sometimes follows the breaker, but always precedes a screen-crusher. The coal then goes to the breaker, which is a large cylindrical shell with interior lifting blades; the shell is perforated with holes (2–8 in. in diameter) to permit passage of small material.

The breaker rotates on a horizontal axis, receiving material in one end, tumbling it as it passes through the holes in the shell, and permitting the hard, large, unbroken material to pass out the rear of the machine. The small material (4 in.) goes to the cleaning plant, and the large rejected material falls into a bin to be hauled away.

Most commercial circuits utilize dense-media vessels of jigs for the coarsest size usually +3/8 in., dense-media cyclones, concentrating tables or jigs for the 3/8 in. × 28-mesh size, water-only cyclones or spirals, and sometimes flotation for the 28 × 100-mesh size and flotation for the −100 mesh.

Heavy-medium cyclones are extensively used for cleaning medium size coals, for example, 0.2–10 mm. Screening and centrifugal dryers dewater the coarser products while screen-bowl centrifuges and sometimes thermal dryers are utilized to reduce the moisture content of the finest sizes. Most metallurgical coal cleaning plants utilize thermal dryers. The coal is softer and more friable and thus has a finer size distribution after extraction by the mining machines. Coals for metallurgical use must be thoroughly processed and dried to meet the end user requirements. Additionally, flotation is always utilized in these circuits due to the quantity of coal and quality of the needed end product (low ash, low sulfur). On the other hand some steam coals, especially the harder ones (low hardgrove index), and some coals produced from surface mines have smaller quantities of the −100-mesh size. In many plants, there is such a small quantity of the −100-mesh size that this material is sent to disposal and is considered uneconomical to recover.

Size reduction of coal plays a major role in enabling ROM coal to be utilized to the fullest possible extent for power generation, production of coke, as well as other uses such as the production of synthetic fuels (Bevan, 1981). ROM coal is the *as-received* coal from the mine, whether the mining process is stripping, auger mining, continuous mining, short- or longwall mining (Chapter 5), or any other method currently practiced.

Since the mining processes differ in operation and since size reduction actually begins at the coal face (i.e., during mining), the mined coal will exhibit different characteristics. In fact, the mining process has a direct bearing on the size and on the size consistency of the coal. Thus, prior to final utilization of the coal, some degree of size reduction, or size control, is usually required. The number of stages in the size reduction process depends upon the specific utilization of the coal as well as the condition of the coal.

For example, coal which is destined for power generation may undergo size reduction to produce a product with a top size of 0.04 in. (1 mm). On the other hand, the size of the coal needed for a coking operation is coarse and the number of stages of *size reduction* involved in preparing a coal feed for a coking is somewhat less than required to prepare coal as the feedstock for power generation utilization.

Size reduction of coal plays a major role in enabling ROM coal to be utilized to the fullest possible extent for power generation, production of coke, and production of synthetic fuels. ROM coal is the *as-received* coal from the mining process prior to any type of treatment, whether the mining process be stripping, augering, continuous mining, short- or longwall mining, or any other method currently practiced. Once the ROM coal has been treated by screening, or crushing, the material is referred to as raw coal (Bevan, 1981).

Because the types of mining processes are varied, and size reduction actually begins at the face in the mining operation, it is quite understandable that the characteristics of the products from these various processes differ widely. The type of mining process directly affects the top size of the mine product. In addition, the amount of out-of-seam dilution in the mine product is directly related to the mining process employed.

The size of a coal particle is defined in terms of a surface opening through which the particle will barely pass or will not pass at all. In sizing a material, the individual particles are presented to the surface openings numerous times. However, it must be recognized that a three-dimensional particle is presented to a two-dimensional opening. In the case of a narrow, elongated particle, the particle will be sized according to the orientation of the particle with respect to the surface opening (ASTM, 2011c) that defines the top size or upper limit of a quantity of material as the smallest sieve (smallest opening) upon which is retained a total of less than 5% of the sample. In practice, the term *nominal top size* is used extensively when describing the output range of a size reduction device. The nominal top size of a quantity of material is the smallest sieve (smallest opening) upon which is retained a total of 5%–20% of the sample.

Reduction ratio is the ratio between the feed top size and the product top size, or the ratio between the feed and product sizes at which a specific percentage of the material passes. For size reduction units that produce a product top size larger than 1 mm, size reduction ratios are normally on the order of 8–1 while for units where the product top size is smaller than 1 mm, size reduction ratios can range from 200 to 1 and higher.

In general, four types of equipment are used for the size reduction of ROM coal: rotary breakers, roll crushers, hammer mills (ring mills), and impactors. At best, the crushing operation produces the desired (sized) product in a single stage. However, there are cases where the size reduction of coal entails multiple stage size reduction units. The final stage of coal pretreatment is to screen it into size fractions convenient for handling by the various process streams.

At this time, a note on the various terms that relate to coal pretreatment is pertinent.

The *top size* upper limit of the crushed (sized) material is defined by the smallest sieve opening upon which is retained a total of less than 5% of the sample (ASTM, 2011c). The term nominal top size is used to describe the product of a size reduction operation and is the smallest sieve opening upon which is retained a total of 5%–20% of the sample. The reduction ratio is the ratio between the feed top size and the product top size or the ratio between the feed and product sizes at which a determined percentage of the material passes. The relative ease with which a coal can be comminuted (reduced in top size) by mechanical action affects the design of a size reduction unit or operation. The term used to refer to this relative ease is the grind ability index of the coal.

The most common grindability index used in conjunction with coal size reduction is the HGI. This index is determined by grinding 50 g of 16 × 30-mesh dried coal in a standardized ball and race mill after which the sample is removed and sieved to determine the amount of material (W) which will pass through a 200-mesh sieve (ASTM, 2011b); as the resistance of the coal to grinding increases, the Hardgrove grindability index (HGI) decreases. The specific energy is the energy per unit of throughput that is required to reduce a feed coal to the desired product size.

The *specific energy* for size reduction is proportional to the *grindability index* of the coal and is also a function of the reduction ratio. Thus, the energy required for the size reduction of coal increases with increasing throughput as well as with the reduction ratio. For a "standard" throughput of coal, the energy required varies with the reduction ratio (Figures 6.2 and 6.3) and the relationship between the specific energy and the grindability is dependent upon the type of device (Figure 6.4).

Moisture in coal is not just an issue in terms of whether the coal is dry or not and whether the transportation costs warrant coal drying. Indeed, moisture content is a factor that must be taken into account when considering the energy requirements of a size reduction unit (Figure 6.5). Excessive moisture, but more particularly excessive surface moisture, can cause a lesser efficiency in fines removal (due to fines agglomeration) and the formation of emulsions can also be a problem in the selective agglomeration process (Bensley et al., 1977).

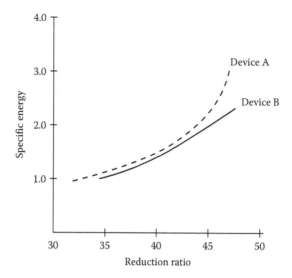

FIGURE 6.2 Relationship of specific energy to size reduction for two coal cleaning devices. (From Meyers, R.A., Ed., *Coal Handbook*, Marcel Dekker, New York, 1981, p. 176.)

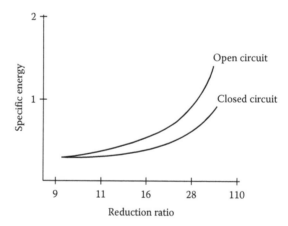

FIGURE 6.3 Relationship of specific energy to size reduction for open and closed circuits. (From Meyers, R.A., Ed., *Coal Handbook*, Marcel Dekker, New York, 1981, p. 176.)

6.3.1 ROTARY BREAKER

The rotary breaker (Figure 6.6) is a size reduction device that causes breakage of the coal material along natural cleavages by lifting the material to a given height and then dropping the material against a hard surface. The rotary breaker is used in reducing the top size of ROM or raw coal by causing breakage of coal along natural cleavages by lifting the material to a given height and then dropping the material against a hard surface. The raw coal from the rotary breaker can be utilized directly, undergo further size reduction, or the quality of the product can be upgraded in a coal-washing facility.

The rotary breaker (Figure 6.6) essentially consists of an outer cylinder with an inner rotating perforated cylinder fitted with lifters which is operated at the lowest speed that will cause a small particle on the liner to centrifuge. Coal is fed into one end of the rotating cylinder where it breaks because of the tumbling action due to gravity fall, and pieces less than the hole sizes (typically 1.5 in. [38 mm])

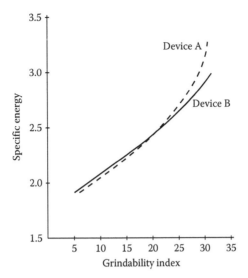

FIGURE 6.4 Relationship of specific energy to hardgrove grindability index for two size reduction devices. (From Meyers, R.A., Ed., *Coal Handbook*, Marcel Dekker, New York, 1981, p. 177.)

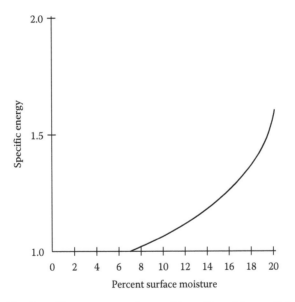

FIGURE 6.5 Relationship of specific energy to surface moisture. (From Meyers, R.A., Ed., *Coal Handbook*, Marcel Dekker, New York, 1981, p. 178.)

pass out to a bottom collection trough, giving a product with positive upper size control. The flow rate through the breaker is adjusted so that pure coal has broken through and rock passes through the exit; in practice, some coal will pass out with the rock. Large pieces of hard, dense rock will also act as grinding media. Because of the relatively mild breakage action, the production of fines is minimized. For harder coal, the length-to-diameter ratio is increased, and the diameter is increased to give a higher force due to the greater fall.

The liners of a rotary breaker are screen plates that allow peripheral discharge of properly sized material (Hicks, 1951). The unit size is based upon the diameter and length (Roman, 1967);

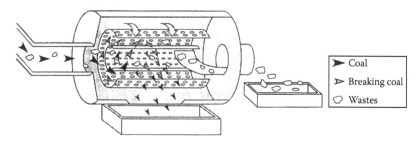

FIGURE 6.6 A rotary breaker.

the diameter ranges from 6 to 14 ft, with the length in the range of 8–28 ft; a length/diameter ratio between 1.33 and 2.00.

6.3.2 Roll Crusher

Roll crushers are size reduction devices that shear (or compress) the material that is to be reduced in size. The surfaces used can take the form of a rotating roll and a stationary anvil, or two equal-speed rolls rotating in opposite directions. The surfaces of the rolls can be smooth, corrugated, or toothed. The particular design of the teeth and corrugations is critical in keeping the production of fines and dust to a minimum (Roman, 1967).

The single-roll crusher (Figure 6.7) is a popular piece of equipment and primarily it consists of a heavy cast iron or steel-fabricated frame on which are mounted the crushing roll and the stationary breaker plate. The breaker plate is provided with renewable wear plates bolted to the breaker plate. The roll usually has a series of long teeth spaced at intervals and various short teeth covering the entire crushing surface. The coal is squeezed between the revolving roll and the breaker plate. The long teeth act as feeders and also penetrate the lumps of coal, splitting them into smaller pieces, while the smaller teeth make the proper size reduction.

For a given reduction ratio, single-roll crushers are capable of reducing ROM coal to a product with a top size in the range of 8 (200 mm) to 3/4 in. (19 mm) A single-roll crusher can usually accept up to 36 in. (0.9 m) cubes in the feed. Single-roll crushers can vary in size from 18 × 18 in. (0.5 × 0.5 m) to 36 × 100 in. (0.9 × 2.6 m) and have capacities up to 1500 tons/h (1.5 × 10^6 kg/h).

Double-roll crushers (Figure 6.8) consist of two rolls that rotate in opposite directions. A double-roll crusher can reduce ROM coal (maximum top size: 36 in., 0.9 m) to a product with a top size in the range of 14 in. (350 mm) to 3/4 in. (19 mm). The roll size can vary from 12 × 18 in. (0.3 × 0.5 m)

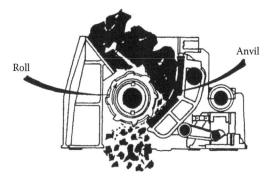

FIGURE 6.7 A single-roll crusher. (From Meyers, R.A., Ed., *Coal Handbook*, Marcel Dekker, New York, 1981, p. 187.)

FIGURE 6.8   A double-roll crusher. (From Meyers, R.A., Ed., *Coal Handbook*, Marcel Dekker, New York, 1981, p. 188.)

to 60 × 60 in. (1.6 × 1.6 m) and capacities may be up to 2000 tons/h (2 × 10 kg/h). Double-roll crushers generally produce minimal amounts of fines (Roman, 1967).

This double-roll crusher has the advantage of greater flexibility due to a variable gap setting, less fines production as compared to the single-roll crusher since the tooth–plate contact is absent, and a smaller limit in product size, depending on the type of teeth used. However, double-roll crushers have been cited to be more difficult to maintain in this heavy service, are more expensive, and may offer no particular advantage.

6.3.3   HAMMER MILL

The hammer mill is a variant of the conventional rotary breaker uses a closed far end with a series of hammers in a hammer mill rotating at this end, which breaks the harder coal and rock so that it is less than hole size and leaves through the holes.

The hammer mill (Figure 6.9) is a device in which the feed coal is contacted (impacted) by rotating hammers and then further impacted by contact with grid breaker plates. The grinding that occurs in the hammer mill results from the rubbing action of feed material between two hardened surfaces, i.e., the hammers and the grid or screen plates). The hammer mill usually produces a relatively high proportion of fine material (Roman, 1967). The fineness obtained can be varied by adjustments of revolutions per minute or the spacing between the hammer tips and the grate bars.

The ring-type hammer mill (Figure 6.10) is a modification of the hammer mill which was developed to help minimize the amount of fines in the product. In these units, the hammers have been

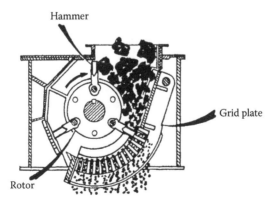

FIGURE 6.9   A hammer mill. (From Meyers, R.A., Ed., *Coal Handbook*, Marcel Dekker, New York, 1981, p. 189.)

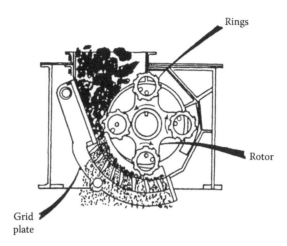

FIGURE 6.10 A ring-type hammer mill. (From Meyers, R.A., Ed., *Coal Handbook*, Marcel Dekker, New York, 1981, p. 190.)

replaced by rings (alternately toothed and plain) which revolve and cause size reduction by a rolling compression action rather than by a grinding action (Roman, 1967). The unit emphasizes *nipping* particles between the hammer and the grate bars. The hammer mill is a versatile unit and has a capacity up to 2500 tons/h for industrial units.

The rapid wear of the edges which occurs with straight-faced hammer mills when used at low gap settings is avoided by the circular shapes of the hammers (alternately toothed and plain) that are free to move around on their retaining pins. The mills are best suited for friable coals; otherwise, excessive abrasive wear occurs.

Slow-speed hammer mills or impactors are more difficult to maintain, and jaw crushers have not been required.

6.3.4 IMPACTORS

Impactors are size reduction devices that strike (impact) the coal which is then thrown against a hard surface or against other coal particles. The coal is typically contained within a cage, with openings on the bottom, end, or side of the desired size to allow pulverized material to escape.

The rotor-type impact mill (Figures 6.11 through 6.13) uses rotors to effect size reduction. In the mill, the coal drops into the path of the rotor, where it is shattered, driven against the impact surface,

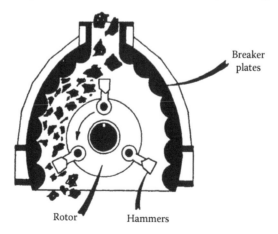

FIGURE 6.11 A reversible rotor-type impact mill. (From Meyers, R.A., Ed., *Coal Handbook*, Marcel Dekker, New York, 1981, p. 192.)

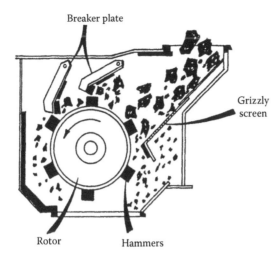

FIGURE 6.12 A single-rotor impact mill. (From Meyers, R.A., Ed., *Coal Handbook*, Marcel Dekker, New York, 1981, p. 193.)

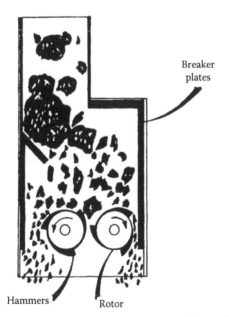

FIGURE 6.13 A double-rotor impact mill. (From Meyers, R.A., Ed., *Coal Handbook*, Marcel Dekker, New York, 1981, p. 194.)

and further reduced in size. The material rebounds into the path of the rotor and the cycle repeats itself until the product is discharged from the base.

There are two types of impact crushers: horizontal shaft impactor and vertical shaft impactor.

6.3.5 TUMBLING MILLS

The tumbling mill (Figures 6.14 and 6.15) is a grinding and pulverizing machine consisting of a shell or drum rotating on a horizontal axis. The material coal is fed into one end of the mill where it comes into contact with grinding material, such as iron balls. As the mill rotates, the material and grinding balls tumble against each other, the material being broken chiefly by attrition.

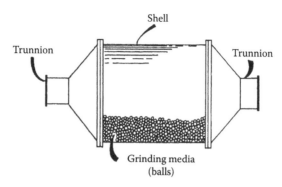

FIGURE 6.14 A ball mill. (From Meyers, R.A., Ed., *Coal Handbook*, Marcel Dekker, New York, 1981, p. 196.)

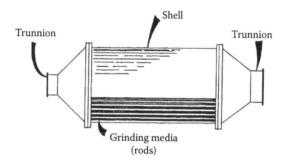

FIGURE 6.15 A rod mill. (From Meyers, R.A., Ed., *Coal Handbook*, Marcel Dekker, New York, 1981, p. 196.)

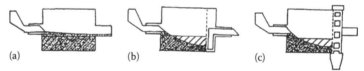

FIGURE 6.16 Ball and rod mill discharge methods: (a) trunnion overflow, (b) grate discharge, and (c) peripheral discharge. (From Meyers, R.A., Ed., *Coal Handbook*, Marcel Dekker, New York, 1981, p. 197.)

Tumbling mills are cylindrical size reduction devices and are essentially lined drums supported by hollow trunnions at each end. The units are manufactured with overflow, grate, or peripheral, discharge arrangements (Figure 6.16).

6.4 COAL CLEANING

ROM coal generally has 5%–40% of ash content and 0.34% of sulfur content depending on the geological conditions and mining technique used. Coal cleaning, therefore, is often required to remove excessive impurities for efficient and environmentally safe utilization of coal. In the United States, the coal cleaning is most common at Eastern and Midwestern mines.

Current commercial coal cleaning methods are invariably based on physical separation; chemical and biological methods tend to be too expensive. Typically, density separation is used to clean coarse coal while surface property-based methods are preferred for fine coal cleaning. In the density-based processes, coal particles are added to a liquid medium and then subjected to gravity or centrifugal forces to separate the organic-rich (float) phase from the mineral-rich (sink) phase.

Current commercial coal cleaning methods are invariably based on physical separation; chemical and biological methods tend to be too expensive. Typically, density separation is used to clean coarse coal while surface property-based methods are usually preferred for fine coal cleaning (Davis, 1993; Dodson et al., 1994). In the density-based processes, coal particles are added to a liquid medium and then subjected to gravity or centrifugal forces to separate the organic-rich (float) phase from the mineral-rich (sink) phase.

Density-based separation is commercially accomplished by the use of jigs, mineral spirals, concentrating tables, hydrocyclones, and heavy-media separators. The performance of density-based cleaning circuits is estimated by using laboratory float-sink tests. In the surface property-based processes, ground coal is mixed with water and a small amount of collector reagent is added to increase the hydrophobicity of coal surfaces. Subsequently, air bubbles are introduced in the presence of a frother to carry the coal particles to the top of the slurry, separating them from the hydrophilic mineral particles. Commercial surface property-based cleaning is accomplished through froth or column flotation.

Other physical cleaning methods include selective agglomeration, heavy-medium cycloning, and dry separation with electrical and magnetic methods (Couch, 1991, 1995). Selective agglomeration and advanced cycloning have the high probability of commercialization, particularly for reducing the sulfur content of coal (Couch, 1995). In selective agglomeration, the coal is mixed with oil. The oil wets the surface of coal particles and thus causes them to stick together to form agglomerates. The agglomerated coal particles are then separated from the mineral particles that stay in suspension because they do not attract oil to their surfaces.

6.4.1 EFFECT OF COMPOSITION AND RANK

Pure coal consists mainly of carbon with lesser quantities of hydrogen, nitrogen, oxygen, and sulfur within the organic matrix (Chapter 10); there are also mineral impurities (Chapter 7). When coal is burned, the organic matrix is converted to gaseous combustion products and the mineral impurities generally remain as an "ash" residue.

It is important to note here that the ash produced by combustion is not the mineral matter originally present in the coal, but the residues remaining after the minerals have been converted to the various elemental oxides.

Changes resulting from decomposition of mineral matter in this manner do not make any contribution to the calorific value of the coal but usually detract from it and where calorific values must be measured accurately, require corrections of the measured values. These corrections are not large and it is often considered sufficient (but not quite precise) to use the ash content, which is easily and rapidly determined, as a measure of the mineral impurities present in a coal.

When coals are pyrolyzed, this process is continued by the near complete elimination of hydrogen and oxygen. The carbon residue that remains, however, still contains small quantities of nitrogen and sulfur. Typical values for carbon, hydrogen, volatile matter (i.e., low-molecular-weight products of the pyrolysis), and residual (fixed, nonvolatile) carbon of the various classes of coal (Table 6.1) provide the basis of a system for describing coal rank (Chapters 2 and 8).

The chemical structure of coal is complex (Chapter 10), but there are strong indications (and little, or no, evidence to the contrary) that the nitrogen is entirely contained within the organic matrix. On the other hand, oxygen and sulfur occur in both organic and inorganic combinations.

In the context of coal cleaning, oxygen is often considered unimportant because it is nonpolluting. But some consideration should be given to the effect of oxygen on the fate of its nitrogen and sulfur compatriots as well as its effect when water (a product of the combustion of hydrocarbons which also contain oxygen) condenses with other by-products on the cooler parts of combustion systems; corrosive, aqueous acids can be the result.

Sulfur is a special case (see Section 6.5) because it is considered to be, and actually is, a more serious pollutant than oxygen. Sulfur occurs in various forms and is distributed throughout the organic matrix and in the minerals. As organic sulfur, it occurs in the organic structure of the coal and as pyritic sulfur, it occurs as discrete particles of pyrite (Fe). In addition, sulfates are occasionally found in the minerals. In summary, the bulk of the sulfur (organic and/or inorganic) present in coal has the potential to occur as gaseous combustion products.

Rank has been assumed to have an effect on the extent of hydrophobic character of coal (Chapters 2 and 9), there being a maximum for medium- and low-volatile bituminous grades. However, recent work on the prediction of coal hydrophobicity indicates that this property correlates better with the moisture content than with the carbon content and better with the moisture/carbon molar ratio than with the hydrogen/carbon or oxygen/carbon atomic ratio (Labuschagne et al., 1988). Thus, it appears that there is a relationship between the hydrophobicity of coal and the moisture content. But, there are differences in the behavior of coals and the differences are usually referred to as wettability, which can be quantified by measurement of the contact angle of the solid with water. Methods for calculating a floatability index of solid hydrocarbon materials have been suggested which utilize the elemental composition (Sun, 1968).

6.4.2 Methods

Methods employed for the removal of impurities from coal depend on the physical differences between the impurity and the coal; one such example is the difference in specific gravity.

Coal preparation plants generally use gravity process equipment to separate the refuse from the coal. Most of the extraneous impurities mined with coal are much heavier than the coal itself—coal has a specific gravity between 1.35 and 1.5, while the refuse rock has a specific gravity on the order of 2.1–2.3—and separation can be effected by immersing the ROM coal in a fluid having a specific gravity greater than that of the coal but less than that of the impurity (*heavy-media process*). This allows the coal to float and the heavy waste material to sink and the two products are collected separately (Figure 6.17) (Couch, 1991).

6.4.2.1 Dense-Media Washing

The *heavy-media process (dense-media process)* is the most popular method of cleaning coarse sizes; jig plants are probably the second most common method used for coarse coal.

In a heavy-media washing plant, all the cleaning is done by flotation in a medium of selected specific gravity, maintained by a dispersion of finely ground magnetite in water. The incoming raw coal is separated at 1/4 in. on an inclined screen. The *overs* proceed to a flat pre-wet screen, where the fine dust particles are sprayed off from the +1/4 in. coal. This increment is discharged into a heavy-medium vessel or bath, where the refuse is separated from the coal. The refuse is discharged to a "refuse rinse" screen, where it is dewatered. The use of magnetite has also been investigated in cyclone cleaning of coal (Klima et al., 1990).

The freed medium is divided into two parts: one returning directly to circulation via the heavy-medium sump and the other pumped to magnetite recovery. The refuse is discharged from the screen for disposal. The coal is discharged from the washer to a coal-rinse screen, where the coal is

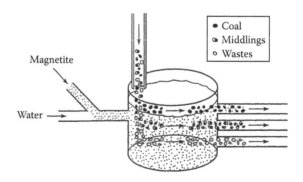

FIGURE 6.17 A dense-medium separation method.

dewatered and the medium is treated as from the refuse screen. The clean coal is then centrifuged, crushed, and loaded. The fine coal (less than 1/4 in.) from the raw coal screens is combined with magnetite and water and pumped to a heavy-media vessel in that the magnetite is finer and the effective specific gravity is different.

The refuse is dewatered and the medium is recovered, as in the coarse coal selection. The coal is discharged over a sieve bend and then proceeds to a centrifuge for final dewatering prior to transfer to a thermal dryer or to loading.

Because the magnetite recovered from the rinse screens is diluted by sprays, it is processed in magnetic separators for recovery of the solid mineral. Each washer (bath and cyclone) retains its own recovery system, which includes sumps, pumps, and magnetic separators. The separator is a shaft-mounted steel drum containing an interior fixed magnet. The cylinder rotates within a vessel containing coal slurry and magnetite, retrieving solid magnetite from the slurry by virtue of the magnetic qualities of the magnetite and the magnetic field within the drum.

The effluent from the centrifuges contains less than 28-mesh coal broken from larger pieces of clean coal. This material is thickened in a cyclone, de-slimed on a screen, and centrifuged prior to loading.

Heavy-media cyclones are being used more often for fine size fractions. Flotation is generally used to clean the −28-mesh size fraction, although spirals and heavy-media cyclones have shown success in cleaning down to 100-mesh coal feed. Spirals are generally used for middling sizes (−10 to 60 mesh).

Sand processes employ suspensions, often unstable, of sand in liquids whose effectiveness can be maintained only by high rates of agitation and recirculation. Specific gravity separations up to 1.90 are obtainable, but specific gravities of 1.45–1.60 are more commonly employed for bituminous coals.

The most successful commercial application of this process has been the Chance sand process (Figure 6.18) which is one of the most widely adopted heavy-medium cleaning processes and consists of a large, inverted conical vessel in which sand is maintained in suspension in an upward current of water. The density of the fluid can be varied by increasing or decreasing the amount of sand held in suspension. The process can be arranged so as to give an intermediate product between the pure coal and the pure rock. This is achieved by adjusting the speed of the upward water current through control of the volumes of water admitted at the several levels.

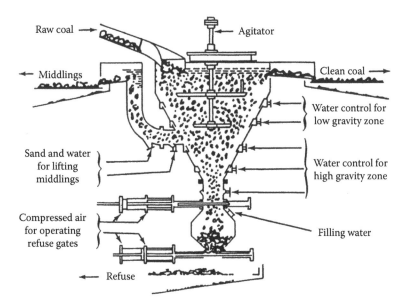

FIGURE 6.18 The chance sand flotation process.

6.4.2.2 Pneumatic Cleaning

Pneumatic cleaning devices, or air tables, are applied to the small fractions (less than 3/8 in.). In these devices, currents of air flow upward through a perforated bottom plate over which a layer of coal passes. The extreme fines are entrapped in the air and must be recaptured by cyclones and bag filters for return without quality improvement. As the coal reaches the end of the tables, the bottom layer is heavy (high-ash) material, a center layer is medium-weight coal and bone (high-ash), and the top layer is coal (low-ash). The middle layer must be incorporated with the refuse (and rewashed) or with the coal.

The efficiency of these devices is poor. Their ability to remove ash is limited to 2%–3%, regardless of how much is present. These devices represent the lowest capital investment of all cleaning devices, and they entail no problems of water supply and disposal.

The incoming coal must be screened, and, because feed to the tables must be dry, thermal drying of the raw feed is required at some plants. The thermal dryers, in turn, require cyclones and scrubbers for control of particulate emissions. Thermal dryers are fired with coal, oil, or gas.

6.4.2.3 Jig-Table Washing

Jig-table washing plants are so named because jigs are used to clean the >0.25 in. increment and Diester tables (oscillating table-sized sluices with a flat, riffled surface, approximately 12 ft², which oscillates perpendicular to the riffles, in the direction of the flow of coal) are used to clean the <0.25 in. increment. Froth cells and/or thermal dryers may be used in conjunction with this equipment.

The raw coal, restricted to sizes smaller than 8 in., is separated on a wet screen (usually 0.25 in. mesh). The large-sized increment goes into the jig; the remaining coal is sent to a separate cleaning circuit. The coal is dewatered on screens and in centrifuges, crushed to the desired size, and loaded. The jig makes the "equivalent" gravity separation on the principles of settling in rising and falling currents. The small-sized coal (less than 1/4 in.) is combined with the proper amount of water and distributed to the tables, where the refuse is separated from the coal. The refuse is dewatered on a screen and discarded. The clean coal is dewatered on a sieve bend (a stationary gravity screen), where the extreme fines are removed and discharged into a centrifuge for final dewatering and removal of the fines. The clean coal (+28 mesh) is then loaded or conveyed to a thermal dryer. The heavy rejects are discharged off one end of the discharge side of the table, the light coal is discharged from the opposite end, and the *middlings* are distributed between.

The slurry produced, along with the fines, requires clarification before recirculation is feasible.

6.4.2.4 Water Clarification

The water clarification plant receives all the slurry from the washing plant, separates the −48-mesh fraction for cleaning, and returns the water for reuse. The 48-mesh fraction flows to froth flotation cells, where it is mixed thoroughly with a reagent (light oil). The coal accepts a coating of oil and floats off the top of the liquid to a disk filter, where the excess water is drawn through a fabric by a vacuum. The water is recirculated to the washer, and the fine coal is transported to loading or to a dryer.

The refuse does not accept the oil coating and sinks, to be removed with most of the incoming water to a static thickener. The thickener is a large, circular, open tank, which retains the water long enough to permit the particles of refuse to sink to the bottom. Clarified water is removed from the surface by "skimming troughs" around the perimeter of the tank and is recirculated to the cleaning plant.

The tank is equipped with a rotating rake, which rakes the fine refuse from the bottom of the tank to the center of the tank, where it is collected by a pump and transferred to a disk filter. The filter removes part of the water for recirculation and discharges the solids as refuse.

6.4.2.5 Other Processes

Washability is a concept that exploits the differences that exist between the specific gravity of different coals and the associated minerals as a basis for predicting the yields and qualities of the products obtained for any given partition density (Mazumdar et al., 1992; Ryan, 1992). Washability

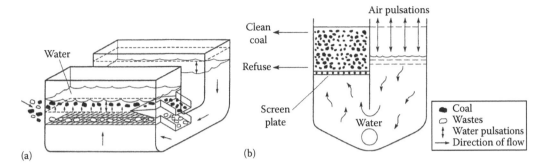

FIGURE 6.19 A jig: (a) sectional view and (b) end-on view.

data are always reported in terms of mean specific gravities of particles and those of the liquids used to effect separation. Of late, some degree of predictability during washing operations has become available (Vassallo et al., 1990) which affords a degree of luxury in the determination of recovery and overall behavior of the coal.

Laboratory washability data are determined using true solutions and it is customary to screen the coal sample into convenient size fractions, usually related to the particular application, and to test each size fraction in a series of solutions of incremental specific gravity. The material sinking at any particular specific gravity is recovered and tested at the next higher gravity level. The material floating is collected and analyzed as representing the particular size and gravity fraction. In practice, coal washing using water in a jig (Figure 6.19) is a common method of coal cleaning.

Separation of coal and mineral matter can also be achieved by exploiting differences in the surface properties; certain minerals, especially clays, have polar, hydrophilic surfaces although pyrite may exhibit markedly hydrophobic behavior and cause difficulties when present in discrete particles that are selectively floated with coal particles. Froth flotation and oil agglomeration methods (Mehrotra et al., 1983; Franzidis, 1987; Schlesinger and Muter, 1989; Vettor et al., 1989; Xiao et al., 1989; Couch, 1991; Carbini et al., 1992) are examples of how such separations can be achieved and although differences exist in the surface properties of the coal components (macerals; Chapters 4 and 9), these are generally small in relation to those of the minerals present.

Addition of small quantities of hydrocarbon oils to raw coal pulps causes preferential wetting of the coal by oil, tending to increase the surface tension forces operating at the coal/water interface. If air is admitted to the system, bubbles attach preferentially to the oil-wet coal surfaces thereby increasing the buoyancy of the coal particles, hence causing rapid separation from the heavier, unoiled mineral particles. The coal is collected as froth on the surface of the system and recovered by skimming.

Some heavy-medium flotation plants use finely crushed barium sulfate or magnetite in suspension in water. In some processes, termed dry, or pneumatic, cleaning air is used as the separating medium. Because densities are additive properties, the specific gravity of a suspension may be calculated from the concentration of solids in unit weight of suspension and the true density of the solids are known (Figure 6.20). However, the effective specific gravity of a suspension is strongly dependent on the stability of the suspension, which is, in turn, dependent on the fineness of the suspended particles.

In coal cleaning by cyclone (Figure 6.21), raw coal and fluid enter tangentially close to the top of the cylindrical section, forming a powerful vertical flow (Couch, 1991). The separation of the impurity from the coal occurs due to a buoyancy effect which arises because of the difference between the mass of any particle and mass of an equivalent volume of displaced fluid. When the fluid is water, this factor is equivalent to the specific gravity. The densest particles therefore move to the

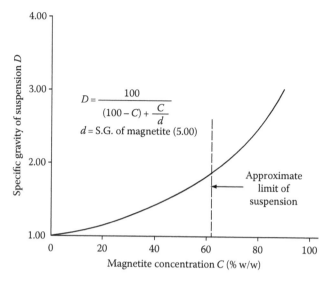

$$D = \frac{100}{(100 - C) + \dfrac{C}{d}}$$

$d = $ S.G. of magnetite (5.00)

FIGURE 6.20 Relationship of specific gravity of a magnetite suspension to concentration in water. (From Meyers, R.A., Ed., *Coal Handbook*, Marcel Dekker, New York, 1981, p. 268.)

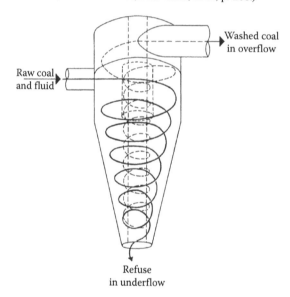

FIGURE 6.21 Cyclone separation. (From Meyers, R.A., Ed., *Coal Handbook*, Marcel Dekker, New York, 1981, p. 274.)

wall and downward, and exit with the cyclone underflow. Conditions can be adjusted so that less dense particles remain suspended in the vertical flow and exit from the top of the cyclone. This density separation can be controlled with great precision.

In practice, a material such as magnetite is reduced to very fine sizes, usually in ball mills. The viscosity of the suspension increases with increasing fineness of particle size and particle concentration. The settling rates may be reduced and the stability of the suspension improved by the presence of clays and it may be necessary to recondition the suspension by bleeding off part of the circulating volume and recovering the magnetite in magnetic drums and rejecting the clays. Indeed, there is a cause for optimism that magnetic methods of coal cleaning (pyrite removal) will be successful and be applicable as a complement to other methods (Kester et al., 1967; Ergun and Bean, 1968; Trindade et al., 1974; Oder, 1978, 1984, 1987).

In some countries, loess (being readily available and usually associated with the coal being mined) is used as the suspended medium, particularly for coals of lower rank. Medium conditioning in these cases is carried out by flocculation and concentration in settling cones. Dense-medium operations have been greatly assisted by the development of reliable instrumental methods for measuring the densities of suspensions, which have enabled heavy media to be automatically monitored and controlled.

Various polymeric flocculants exhibit some degree of selectivity for coal against mineral matter and include chemicals such as partially hydrolyzed polyacrylamide, nonionic polyacrylamide, polystyrene sulfonate, and polyacrylamide containing chelating and complexing groups. In some cases, selective flocculation processes suffer from relatively low selectivity. Thus, selective flocculation processes are usually run in multiple stages to remove the entrained ash-forming minerals.

Traditionally, precombustion cleaning has been concentrated on two major categories of cleaning technology: physical cleaning and chemical cleaning (Wheelock, 1977). A new category of coal cleaning, biological cleaning, has recently attracted much interest as advances have been made in microbial and enzymatic techniques for liberating sulfur and ash from coal (Couch, 1987, 1991; Dugan et al., 1989; Beier, 1990).

Microbes are effective in converting organic sulfur compounds such as thiophene derivatives and dibenzothiophene derivatives. Organics sulfur removal is in the neighborhood of 25% (there are claims of higher removal of sulfur) and the combined use of microbes, either simultaneously or sequentially, could potentially improve organic sulfur rejection. The limiting factors appear to be those of accessibility and residence time. Therefore, finer size coal should be used not only to improve accessibility of microbes to coal particle surfaces but also to reduce the overall retention time in the bioreactor.

6.5 COAL DRYING

The water content of a coal reduces its heating value, causes handling difficulties, increases handling and transportation costs, and reduces yields in carbonization and other conversion processes. Reduction of the water content is often desirable.

Water occurs in coals in three ways: (1) as inherent moisture contained in the internal pores of the coal substance, including water associated with the mineral impurities; (2) as surface moisture wetting the external surfaces of the coal particles in which adsorption may play a small part; and (3) as free water held by capillary forces in the interstices between the coal particles.

Inherent moisture is related to coal rank, being greatest for low-rank lignites and brown coals. However, the inherent moisture in coal can make a significant impact on the performance of low-rank coals (which are primarily used for electricity generation) and drying is generally restricted to (1) washed bituminous coals that are required to meet user specifications and (2) lignite or sub-bituminous coals that are employed for the manufacture of briquettes or of other specialty products such as absorbent carbon.

Surface moisture is related to the amount of available surface and the wettability of the coal but moisture contents are lower than might be expected from the available surface area because of the low wettability of coal compared with the surfaces of minerals. Surface moisture can be removed from washed higher-rank (bituminous and anthracite) coals by drainage on a screen while the dewatering of washed small coal or coal fines can be accomplished by the use of cyclones or centrifuges. If the moisture content must be reduced to lower levels, there are two alternate methods which involve the use of rotary kilns or fluidized bed dryers. But, such dried coal will reabsorb moisture on exposure to the atmosphere which may give rise to fire hazards as well as explosion hazards.

Thermal drying, particularly of metallurgical coals, has found extensive application in some parts of the "coal" world and there is growing interest in the development of efficient methods for drying low-rank coals. Thermal drying entails contacting wet coal particles with hot gases, usually combustion products, under conditions that promote evaporation of the surface moisture without causing degradation or incipient combustion of the coal.

In the process, the clean coal from various wet cleaning processes is wet and requires drying to make it suitable for transportation and final consumption. Thermal drying is employed to dry the wet coal.

Drying in the thermal dryer is achieved by a direct contact between the wet coal and currents of hot combustion gases. Various dryers marketed by different manufacturers work on the same basic principle.

The fluid-bed dryer operates under negative pressure in which drying gases are drawn from the heat source through a fluidizing chamber. Dryer and furnace temperature controllers are employed in the control system to readjust the heat input to match the evaporative load changes.

The multi-louver dryer is suitable for large volumes and for the coals requiring rapid drying. The coal is carried up in the flights and then flows downward in a shallow bed over the ascending flights. It gradually moves across the dryer, a little at each pass, from the feed point to the discharge point.

In the cascade dryer, wet coal is fed to the dryer by a rotary feeder; as the shelves in the dryer vibrate, the coal cascades down through the shelves and is collected in a conveyor at the bottom for evacuation. Hot gases are drawn upward through and between the wedge wire shelves.

In the flash dryer, the wet coal is continuously introduced into a column of high-temperature gases and moisture removal is practically instantaneous.

6.6 DESULFURIZATION

The total sulfur content of a raw coal is distributed throughout the macerals and minerals present and may occur as elementary sulfur, as sulfates, as sulfides, or in organic combination in the coal (Table 6.3) (Chapters 7 and 10).

It has been customary to classify the forms of sulfur as inorganic, pyritic, and organic; inorganic sulfur comprising the sulfates, pyritic the sulfides, and organic the remainder, including any

TABLE 6.3
Sulfur Distribution in Selected Coals

Region and Country	Location or Mine	Sulfur (% w/w)[a]			Ratio of Pyritic to Organic Sulfur
		Total	Pyritic	Organic	
Asia					
USSR	Shakhtersky	0.38	0.09	0.29	0.031
China	Taitung	1.19	0.87	0.32	2.7
India	Tipong	3.63	1.59	2.04	0.78
Japan	Miike	2.61	0.81	1.80	0.45
Malaysia	Sarawak	5.32	3.97	1.35	2.9
North America					
United States	Eagle No. 2	4.29	2.68	1.61	1.7
Canada	Fernie	0.60	0.03	0.57	0.053
Europe					
Germany	—	1.78	0.92	0.76	1.2
United Kingdom	Derbyshire	2.61	1.55	0.87	1.8
Poland	—	0.81	0.30	0.51	0.59
Africa					
South Africa	Transvaal	1.39	0.59	0.70	0.84
Australia	Lower Newcastle	0.94	0.15	0.79	0.19
South America					
Brazil	Santa Caterina	1.32	0.80	0.53	1.5

[a] Moisture-free basis, pyrite + sulfate reported as pyrite.

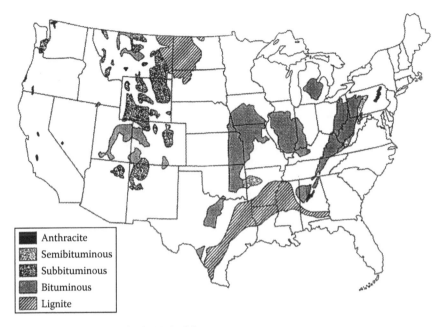

FIGURE 6.22   Coal occurrence in the United States.

elemental sulfur that may be present. Standard methods of analysis have been devised for direct determination of total sulfur, sulfate sulfur, and pyritic sulfur, the organic sulfur being reported as the difference between the total sulfur content and the sum of sulfate content plus pyritic sulfur content (Schmeling et al., 1978). On a more localized basis such as in the United States (Figure 6.22), the sulfur content will vary significantly by region (Table 6.4), thereby creating issues such as those arising from sales to regional power producers.

Even when coal has been prepared to meet the specifications for size, mineral (ash), and moisture contents, it may still be *dirty* by environmental standards. In this case, the important contaminant is sulfur which is converted, during combustion, to gaseous product(s):

$$S_{coal} + O_2 \rightarrow SO_2$$

$$2SO_2 + O_2 \rightarrow SO_3$$

Unless removed, the sulfur oxides end up as stack gas emissions.

TABLE 6.4
Sulfur Distribution in Selected U.S. Coals

Region	Total Sulfur (%)	Inorganic Sulfur (%)	Organic Sulfur (%)	Inorganic/Total Sulfur (%)
Northern Appalachia	3.01	2.01	1.00	67
Southern Appalachia	1.04	0.37	0.67	36
Alabama	1.33	0.69	0.64	52
Eastern Midwest	3.92	2.29	1.63	58
Western Midwest	5.25	3.58	1.67	68
Western	0.68	0.23	0.45	34
Total United States	3.02	1.91	1.11	63

In recent years, deliberate attempts have been made to bring about desulfurization on an industrial scale by modification of the coal preparation practices. In part, this has been due to development of more precise methods for the separation of coal and minerals. However, these practices are usually dependent upon the coal properties. Thus, where low-sulfur coal feedstock is a necessity, the deliberate selection of naturally occurring low-sulfur coal has been the most effective solution and has been the practice followed in producing metallurgical coals, political aspects notwithstanding as evidenced by the selection of higher sulfur (and inappropriate) coal for politically sensitive, rather than market, satisfaction.

There are strong incentives to develop processes for removing sulfur from coal before combustion (precombustion cleaning), during combustion, or after combustion (postcombustion cleaning) (Chapters 14, 15, and 23).

Indeed, since the passage of the original Clean Air Act of 1970, subsequently amended in November 1990, coal preparation efforts in the United States have emphasized development of technology for the reduction of sulfur. One approach to reducing sulfur emissions is accomplished by coal conversion, the direct thermal and/or chemical treatment of coal to produce virtually sulfur-free liquid fuels (Chapters 18 and 19) or gaseous fuels (Chapters 20 and 21); the sulfur can be recovered as a by-product of the conversion process (Chapter 23).

Coal desulfurization can be achieved on a commercial scale by means of physical or physicochemical methods which generally use the principal of density separation techniques or other techniques that exploit the surface properties of coals and minerals. For example, the methods exploit the difference in properties that exist between the various forms of pyritic sulfur (pyrite and/or marcasite, FeS_2, and occasionally including galena, PbS) and the organic matrix of the coal.

Desulfurization by chemical techniques is somewhat less well-developed than desulfurization by physical methods. However, a number of methods are under serious consideration and they can be divided into three general groups: (1) those which remove pyritic sulfur; (2) those which remove organic sulfur; and (3) those methods which remove either the pyritic sulfur or the organic sulfur or both (Couch, 1991; Ali and Srivastava, 1992).

Therefore, effective desulfurization requires that three criteria should be satisfied: (1) the reagent must be highly selective to either pyritic or organic sulfur (or both) and not significantly reactive with other coal components; (2) the reagent must be regenerable so that once-through reagent cost is not a major factor; and (3) the reagent should be either soluble or volatile in both its unreacted and reacted form so that it can be near totally recovered from the coal matrix.

The use of strong bases (alkali, caustic) appears to offer some solution to the problem of organic sulfur removal and this approach continues to be investigated (Schmidt, 1987; Chatterjee and Stock, 1991).

6.7   COAL SAMPLING

Coal sampling is an important part of the process control in a coal preparation plant. A *grab sample* is a one-off sample of the coal at a point in the process stream, and tends not to be very representative. A routine sample is taken at a set frequency, either over a period of time or per shipment. Coal sampling consists of several types of sampling devices such as a crosscut sampler to mimic the "stop belt" sample (Speight, 2005) which designate the manner in which coal must be sampled. A crosscut sampler mounts directly on top of the conveyor belt. The falling stream sampler is placed at the head section of the belt.

There are several points in the wash plant that many coal operations choose to sample the raw coal: (1) before it enters the plant, (2) any coal-related refuse, to determine the character of the coal reject, and (3) the cleaned coal, to determine the quality of the coal to be shipped.

Once a gross sample has been taken, it is crushed, and then quartered to obtain a net sample which is then sent to an independent laboratory for testing where the results will be shared with the buyer as well as the supplier. In many cases, the buyer may request a repeat analysis or a second analysis by another laboratory to assure the quality of the data. Continuous measurement of ash, moisture, heat content (Btu/lb), sulfur iron, calcium sodium, and other elemental constituents of the coal is reported.

6.8 TRANSPORTATION

There are many occasions when coal is transported by rail, road, and water in its journey from mine to market. In some mining areas near the coast, the coal was taken by conveyors directly from the mine to the holds of large coastal vessels. For example, in Britain, much of the coal from the northern coalfields is taken to the south in coastal cargo vessels called colliers. Large-scale haulage of coal by truck is normally economic only over relatively short distances.

There has, however, been the tendency during recent years to construct large industrial (chemical or power) plants close to the mine site in order to reduce coal-hauling costs and the coal is carried directly to the plant either by high-capacity truck (Figure 6.23) or by conveyor belts. In fact, the oil sand processing plants in northern Alberta employ this concept and transport the sand on several miles of conveyor belt to the processing plant (Speight, 1990).

It is easier (and more economical) to transport synthetic crude oil to market than to transport low-value oil sand. Similarly, it is much cheaper to transmit electricity over long distances by means of high-voltage wires than to move the equivalent tonnage of coal. However, the capital costs (and inconvenience) of constructing a plant in a remote area with a hostile environment near to a mine may dictate that this concept be impractical. Thus, the suggestion that coal be transported from the mine site completely or, in part, as a coal/water or coal/oil slurry in pipeline systems may have some merit and could afford a ready means of moving coal to markets using already existing pipeline system(s).

Generally speaking, the majority of mined coal is transported to market by railroad, the remainder being shipped or trucked to its destination or used at the mine. Shipping coal by rail has become a major industry in many parts of the world. In the western United States, where the wagon trails (Figure 6.24) once crossed the prairie, the trails are now in the form of railroads with the ever-present coal trains.

6.8.1 UNIT TRAIN

The *unit train* is the most common form of long-distance coal transportation. A unit train is group of railcars that operate in a dedicated shuttle service between a coal mine and a power plant.

A typical unit train consists of 10–120 railcars and 3–5 locomotives, with each railcar holding about 100 tons (105 kg) of coal. Carefully coordinated loading and unloading terminals are necessary to minimize costs. A unit train making a round trip from mine to plant has a typical turnaround time of 72 h, including a 4 h loading and 10 h unloading and servicing time per train.

FIGURE 6.23   Coal haulage in a multi-ton truck.

FIGURE 6.24 Stone marker showing the position of the overland trail (seen to the right of the marker).

FIGURE 6.25 Coal cars in Laramie, Wyoming.

The system is designed so that the trains can be loaded and unloaded can be achieved without stopping the train thereby providing a continuous means of shipping the coal as well as an increase in the rate at which the coal can be moved from the mine site to the consumer (Hanson, 1976; Lindberg and Provorse, 1977). As a typical example, it is difficult to drive on many roads in eastern Wyoming, on any given day, without passing several such trains carrying coal to market. In fact, rail service is the lifeline of the large majority (95%) of the western coal mines, whether it is on the high plains of Wyoming (Figure 6.25) or in the more northern areas such as Alaska (Figure 6.26).

The cost of shipping coal by train is often more than the mining costs. Using a barge or ship to move coal is a lot less expensive. In the United States, there are 25,000 miles of waterways, but not enough to reach all destinations in the country. To reduce transportation costs, power plants are sometimes constructed near coal mines.

6.8.2 COAL BARGE

Barges on rivers and lakes play an important role in coal transport in the United States and Europe. Coal-carrying barges move in tows of 15–40 barges, pulled by a single towboat of 2,000–10,000 hp.

FIGURE 6.26   Coal transportation by rail in Alaska.

A "jumbo" size barge carries 1,800 tons of coal, so a large tow can move 72,000 tons of coal, as much as five unit trains. These large volumes result in significant economies of scale and lower costs. Barge rates can run (on a cost-per-mile or cost-per-kilometer basis) a quarter or less of rail rates. However, waterways often follow circuitous routes, resulting in slow delivery times.

6.8.3   Coal Pipeline

Another method to transport coal is through a slurry pipeline. This connects a mine with a power plant where the coal is used to generate electricity. *Coal slurry pipelines* use a slurry of water and pulverized coal. The ratio of coal to water is about 1:1.

The coal removed from the mine is crushed to a diameter of around 1 mm, and is mixed with water in holding tanks with agitators, which keep the coal in suspension in the water. The pipeline consumes around a billion gallons of water annually. After 3 days, the slurry reaches the end of the pipeline, at the Mohave power plant, where it is held in agitated tanks, for immediate use, and in drying ponds, for later use. Heated centrifuges are used to get the water out. As of 2006, the plant is shut down because the coal and water supply terms are being renegotiated.

Coal slurry pipelines are potentially the least costly available means for transporting coal to any location, measured in economic terms. Whether this is true with reference to any particular pipeline can only be determined by detailed evaluation of the conditions of the route. The current coal transportation scenario does not offer any choices between slurry pipelines and railroad, which undoubtedly will necessarily minimize the cost of transporting coal. In this context, the present times warrant assessment of the potential economic, environmental, and social implications of coal slurry pipeline development and transportation of coal through it.

However, there is need for caution. The large water and energy requirements for coal slurry pose a significant barrier to further deployment, especially in arid regions of Australia and the western United States.

The *coal log pipeline* is another technology for transporting coal in which coal at the mine site is treated and compacted into cylindrical shapes (coal logs) (Liu et al., 1993). Then the coal logs are injected into an underground pipeline filled with water for transportation to destination which may be one or more than one power plants, or to a train station, a barge terminal, or a seaport, for intermodal transportation.

The coal must have been cleaned and crushed, with a binding agent comprised of coal pitch, bitumen, or wax. The coal mixture is then tightly compressed and compacted as coal logs that are 5%–10% thinner than the transportation pipeline. The logs are injected into a pipeline and pumped along using water. The pipeline can deliver the coal to coal-fired electric power stations or coal storage areas. The coal logs must then be crushed for use in fluidized bed, cyclone, or chain-grate stoker coal-burning boilers or pulverized for use in pulverized coal combustors.

Proponents of the coal log technology claim that in addition to being more cost-effective than coal slurry, the capsule pipeline is also more environmentally sound because the coal logs eliminate coal dust erosion of the pipe interior and erosion of coal fines by rain at the power plant storage site.

Since coal must be relatively dry before it can be burned efficiently, the coal must be dried after it arrives at the power plant. Coal transported as slurry requires drying and electricity generation will be substantially less if it is not dried effectively.

Coal logs do not require as much drying because they are packed so tightly that they do not absorb much water, and any water originally in the coal is squeezed out during compression.

6.8.4 TRUCK TRANSPORT

Coal-carrying vehicles are typically end-dump trucks with a carrying capacity of roughly 25–50 tons. Truck delivery is used extensively for small power plants in the eastern United States.

Coal can be moved by truck over regular highways in vehicles with 15–30 tons capacity. Coal can also be transported by large off-road trucks with capacities ranging from 100 to 200 net tons. These trucks are almost always diesel-powered with back or bottom dump.

Specially constructed roads for coal hauling are extensively used for mine-mouth power plants in the west, south, and east, while the hauling of coal by trucks on highways is more concentrated at surface mines. Truck hauls on public highways in the United States typically range from approximately 50–75 miles while off-road hauls are approximately 5–20 miles.

Trucks are the most versatile of all transportation modes for coal hauling because they can operate over the widest areas where roads are available.

However, adverse environmental impacts resulting from truck coal hauling are coal dust particle releases during coal loading or unloading, and coal dust entrainment during transport. Some coal will escape from the trucks during transport because the loads are normally uncovered. The coal dust tends to wash off roadways during rainstorms, causing aesthetic unsightliness and contamination of runoff waters. The air pollutant emissions from diesel fuel combustion add to the emissions.

6.8.5 OCEAN TRANSPORT

South Africa's Richards Bay Coal Terminal, the world's biggest coal-export facility. Global trade in coal increased from 498 Mt in 1990 to 917 Mt in 2007. Coal exports are expected to continue to rise, and, therefore, the importance of ocean transport will increase. Ocean transport of coal requires a system of (1) transportation from the mine to the port; (2) coal-handling facilities at the export port; (3) ocean carrier networks with adequate number and size of ships, contractual obligations, management of the fleet, and route decisions; (4) coal-handling facilities at the importing port; and (5) transportation from the port to the customer.

Ships are commonly used for international transportation, in sizes ranging from

- Handy size (40,000–45,000 dead weight tons [DWT]). A term normally taken to mean a vessel of about 10,000–40,000 DWT.
- Panamax (approximately 60,000–80,000 DWT). Technically, the maximum size vessel that can transit the Panama Canal—restriction of 32.2 M beam.
- Cape-size vessels (>80,000 DWT). A vessel that is too large to transit the Panama Canal and thus has to sail via Cape of Good Hope from Pacific to Atlantic, and vice versa.

However, the ability of coal to variously self-heat (spontaneous ignition), emit flammable gases, corrode and deplete oxygen levels has made the ocean transport of this commodity a particularly hazardous exercise. This is particularly the case in situations where loading is staggered or delayed and the potentially disastrous consequences of a shipboard coal fire can be realized.

6.8.6 CONVEYER BELT TRANSPORT

Conveyor belts are normally used in mine-mouth power plants to bring coal from the mining area to the storage or usage area. Conveyor belts can be used for coal transport in hilly terrain where roads are relatively inaccessible, typically being used to move coal over distances of 5–15 miles.

Conveyors have the advantage of being relatively maintenance free but have the disadvantage of location inflexibility, making a truck haul still necessary. Movable conveyor belts have been developed and used (U.S. DOE, 1979). The only adverse environmental impacts of conveyor belts for coal transport are coal dust losses during loading, unloading, or transport.

Conveyor belts do not use water, except for belt cleaning; they can use plant electricity and do not require petroleum as the energy source. However, conveyor belts tend to be very energy-intensive. As a result, conveyor belt transport of coal has been limited to shorter distances.

6.9 STORAGE

Storage of coal is an important part of coal-handling systems at mines, particularly since the advent of the unit-train concept in transportation.

With this in mind, coal storage is generally practiced in order to accomplish one, or a combination, of the following objectives: (1) to be ready for transportation promptly, (2) to facilitate blending in order to even out chemical and physical inconsistencies that exist in such a heterogeneous material and to produce a saleable product that has (at best) uniformity, and (3) to store coal of preferential sizes where the demand is seasonal (Berkowitz and Speight, 1973).

Thus, coal storage is necessary for (1) the production aspect and (2) the utilization aspect. However, coal storage may be accompanied by one or more undesirable events, the most important of which are (1) oxidation and spontaneous combustion, (2) changes in coal properties which may affect utilization, (3) degradation of coal due to multi-handling steps, and (4) the cost of the multi-handling steps as well as handling as well as the cost of the storage facilities.

However, coal reacts with ambient oxygen, even at ambient temperatures and the reaction is exothermic (Chapter 12). If the heat liberated during the process is allowed to accumulate, the rate of the earlier reaction increases exponentially and there is a further rise in temperature. When this temperature reaches the ignition temperature of coal, the coal ignites (*spontaneous ignition*—the onset of an exothermic chemical reaction and a subsequent temperature rise within the combustible material, without the action of an additional ignition source) and starts to burn (*spontaneous combustion*).

Generally, self-ignition occurs when the thermal equilibrium between the two counteracting effects of heat release due to the oxidation reaction and heat loss due to the heat transfer to the ambient is disturbed. When the rate of heat production exceeds the heat loss, a temperature rise within the material will consequently take place including a further acceleration of the reaction.

The temperature at which the coal oxidation reaction becomes self-sustaining and at which spontaneous combustion occurs varies generally depending on the type (nature and rank) of coal and the dissipation (or lack thereof) of the heat. For low-quality coal and where the heat retention is high, the coal starts burning at temperatures as low as 30°C–40°C (86°F–104°F).

6.9.1 OXIDATION AND SPONTANEOUS COMBUSTION

The tendency of a coal to heat spontaneously in storage is primarily dependent upon the tendency of the coal to oxidize. This in turn is closely correlated with (1) coal rank (the higher the rank, the lower the tendency to oxidize), (2) the size consistency or distribution of the coal in the pile (small pieces of coal have a higher surface area available for oxygen to react), (3) the method by which the coal is stockpiled, (4) the temperature at which the coal is stockpiled, (5) the amount and size of pyrite present, (6) moisture content and ventilation conditions in the pile, (7) time in storage, and (8) the presence of foreign materials. In addition, the variability of coal, added to these factors, does not allow accurate prediction of when spontaneous ignition (spontaneous combustion) will occur (Fieldner et al., 1945; Yoke, 1958; Feng, 1985).

6.9.1.1 Oxidation

Oxidation is an exothermic reaction and, since the rate of a chemical reaction increases for each 10°C (18°F), the reaction will generate heat at a faster rate than can be dissipated or expelled from the stockpile by natural ventilation. Hence, the temperature will rise to appoint where spontaneous ignition occurs and combustion ensues.

Complications may also arise in the case of coals with high moisture and sulfur content and those with tendencies to degrade when exposed to aerial oxygen. This is a critical issue in the case of low-rank, high-sulfur coals. Lignite and subbituminous coal are difficult to store without occurrence of spontaneous combustion, in contrast to anthracite where the potential for spontaneous ignition to occur is minimal.

Thus, oxidation of the coal substance proper is the primary cause of spontaneous heating. This heating, however slight, is caused by slow oxidation of coal in an air supply sufficient to support oxidation but not sufficient to carry away all heat formed and proceeds whenever a fresh coal surface is exposed to air (Berkowitz and Speight, 1973).

6.9.1.2 Oxidation and Rank

Coal is highly variable in the ability to absorb oxygen (thereby weathering or causing combustion) and oxygen absorption generally decreases with increasing rank, that is, low for anthracite and high for subbituminous coal and lignite (Fieldner et al., 1945). Oxygen absorption is also higher for those coals with high bed moisture (natural bed moisture), determined as *capacity moisture, natural bed moisture, equilibrium moisture* (ASTM, 2011a), oxygen content, and volatile content, that is, the low-rank coals (Speight, 2005).

Spontaneous combustion is a rank-related phenomenon. The tendency of coal for self-heating decreases as the rank increases, with lignite and subbituminous coals being more susceptible to self-heating than bituminous coals and anthracite (Pis, 1996). As rank decreases, inherent moisture, volatile matter, and oxygen and hydrogen contents increase. Medium- to high-volatile coals with a volatile matter content higher than 18% w/w daf perform a faster oxidation rate than low-volatile coals and are therefore more prone to spontaneous combustion than low-volatile ones. Furthermore, lower-rank coals often have a greater porosity than higher-rank coal and therefore more surface area is available for oxidation. Low-rank coals also contain long-chain hydrocarbons making them less stable than, for example, anthracite which has a lower hydrocarbon component. However, the oxidation rate for coals of the same rank may show variety within a wide range.

It is generally (but no totally) accepted that the mechanism of the oxidation of coal takes place in five steps, each one chemically dependent upon the temperature (Barkley, 1942; Parry, 1942; Roll, 1963):

1. The coal begins to oxidize slowly until a temperature of approximately 50°C (122°F) is reached.
2. At this point, the oxidation reaction increases at an increasing rate until the temperature of the coal is approximately 100°C–140°C (212°F–285°F).
3. At approximately 140°C (285°F), carbon dioxide and water vapor are produced and expelled from the coal.
4. Carbon dioxide liberation increases rapidly until a temperature of 230°C (445°F) is reached, at which stage spontaneous ignition may occur and spontaneous combustion may take place.
5. At 350°C (660°F), the coal ignites and combustion takes place vigorously.

In general, the critical temperature for bituminous coal in storage is approximately 50°C–66°C (122°F–150°F). From this temperature, heating will usually increase rapidly and may be unstable after which ignition occurs, unless preventive steps are taken.

The petrographic composition of a coal is determined by the nature of the original plant material from which it was formed and the environment in which it was deposited rather than the degree of coalification (i.e., rank) (Chapters 3 and 4). The homogenous microscopic constituents are called macerals on the analogy of minerals in inorganic rocks and can be distinguished in three groups. Vitrinite consists of woody material while exinite consists of spores, resins, and cuticles, and inertinite, the third maceral group, consists of oxidized plant material.

At constant rank, as the inertinite content of a coal increases, the self-heating propensity of the coal decreases. The general trend also indicates an increase in self-heating propensity with either increasing vitrinite or liptinite content.

Liptinite > Vitrinite ≫ Inertinite

Coal rank seems to play a more significant role in self-heating than the petrographic composition of coal.

6.9.1.3 Pyrite

Many chemicals in mineral form affect the oxidation rate to some extent, either accelerating or inhibiting it. Alkalis may have an influence of acceleration while borates and calcium chloride can act as retardants. The oxidation process is also promoted if ankerite is a constituent of the coal mineral matter. In contrast to this, silica and alumina retard the oxidation.

Pyrite (FeS_2) evolves heat from aerial oxidation and was believed to be the cause of the spontaneous heating of coal. The heat generation locally promotes the self-heating process of coal but the reaction products have a greater volume than the original pyrite, with the result of breaking open any coal in which they are embedded and thus exposing a greater surface of coal to the air.

The interaction of pyrite (FeS_2) with water and oxygen is also an exothermic reaction and results in the formation of iron sulfate ($FeSO_4$) and sulfuric acid (H_2SO_4). Thus, if coal is stored in the open, rain will most likely increase the rate of speed of this reaction and for the same reason water-flooding (to extinguish fires) may also increase the rate of the reaction.

Pyrite, through its transformation to bulkier materials, has also been cited as responsible in some cases for slacking and the resultant production of fines. For these reasons, coal users are generally reluctant to stockpile high-sulfur coal for extended periods of time (Berkowitz and Schein, 1951). However, low-sulfur content of coal is no guarantee of safe storage—coal with low-sulfur content can also ignite.

6.9.1.4 Coal Size and Stockpile Ventilation

Oxidation increases with increasing fineness of coal and the rate of oxidation of coal with oxygen of air is proportional to the specific internal surface. The proportional coefficient at low temperatures is the cube root but analysis also shows that both rate and extent of oxidation increase with the decrease in particle size, until a critical particle diameter is reached, below which the rate remains fairly constant.

The natural ventilation in coal storage piles is generally adequate to remove sensible heat as fast as it is liberated in the oxidation process. However, in situations where the ventilation is adequate to maintain oxidation but not adequate to dissipate the heat produced, the coal absorbs the heat causing a rise in the internal temperature. A chain reaction follows in that the oxidation rate increases with increasing temperature which, if allowed to proceed unchecked in a storage pile, the ignition temperature of the coal will eventually be reached and the pile will begin to smolder.

ROM coal is difficult to store because of large percentages of fines mixed with the lump. On the other hand, there is little danger in storing lump coals that have been double-screened or closely sized. The uniform pieces of coal are honeycombed with passages through which air can circulate freely and carry off the heat generated. However stockpiles of coal fines sufficiently compacted so as to exclude air, the potential for spontaneous ignition is diminished.

If coal is to be stored for prolonged periods of time, the pile should be constructed so that air (in the case of fine coal or mixed sizes such as ROM coal) is excluded. On the other hand, if the coal is to be stored as lump coal, air should be allowed to circulate freely through the pile.

The total exposed surface area of the coal is of importance in that the more area exposed, the better the chance of oxygen interacting or reacting uniting with the coal and any heat liberated in a given time for a given weight of coal will be higher (Elder et al., 1945; Berkowitz and Speight, 1973).

When coal stockpiles are constructed by allowing mixed varied size coal to fall, roll, or slide, the larger pieces tend to collect at the bottom outside of the pile and the fines will collect at the top and inside of the stockpile. As a result, air will move easily through the outer parts of the stockpile but with much led freedom in the interior of the stockpile. Such a pile will allow the development of *hot spots* which can lead to spontaneous ignition of the coal with subsequent combustion of stockpile.

6.9.1.5 Moisture Content

Moisture present in the coal is known to influence spontaneous heating in a stockpile insofar as the moisture affects ventilation (air flow) and pyrite reactivity.

The effect of moisture on the self-ignition is twofold: on the one hand, the vaporization of moisture consumes energy and hence the ignition process is impeded. On the other hand, a promotion of self-ignition by the wetting of materials prone to this has been observed (Gray, 1990).

$$\text{Dry coal} + \text{moisture} \rightarrow \text{wet coal} + \text{heat}$$

In addition to the heat of wetting moisture simply blocks the access of oxygen through the coal pores. The water vapor diffusing outward through the pores reduces the oxygen partial pressure and hence lowers the rate of the reaction or the polar water molecules attach to the reactive sites in coal (Jones, 1998).

The heat of condensation of coal in a stockpile can cause a rise in temperature in the pile, which is dependent upon the coal rank (Berkowitz and Schein, 1951). In addition, if dry screened coal is used as a storage-pile base for a shipment of wet coal, ignition can (or will) occur at the wet–dry interface of the two loads (Berkowitz and Speight, 1973). However, the more rapid oxidation occurring in high-moisture coals may be basically a function of coal rank rather than moisture content, since low-rank (high-oxygen) coal is usually also higher in moisture content.

6.9.1.6 Time Factor

The oxidation process takes place once a fresh coal surface is exposed to air; however, the oxygen absorption rate is inversely proportional to time if the temperature remains constant. Therefore, if the coal is stockpiled so that the temperature in the pile does not rise appreciably insofar as the heat is removed at least as fast as it is generated by the oxidation process, the oxidation rate and, thus, the deterioration or weathering rate of the coal will lessen with time (Vaughn and Nichols, 1985).

6.9.2 EFFECT OF STORAGE ON COAL PROPERTIES

Coal in storage has a tendency to lose heating value and coking quality. In general, high-rank coal (properly stored so as to limit oxidation to a minimum) will lose only about 1% of the heat value per year. On the other hand, improper storage can result in a 3%–5% loss in the heat value during the first year (Rees et al., 1961).

In addition, the coking characteristics of many coals and coal blends are so seriously affected by aging in storage that they may be totally worthless as a coke oven charge (Landers and Donoven, 1961). However, the data and claims can vary and range from (1) no effect on the coking properties after months and years of storage to (2) significant loss of coking properties in as little as 1 month of storage time. Storage of low-rank coal presents particular problems in that it is usually accompanied by loss of strength, degradation, and some loss of heating value (Jackman, 1957; Mitchell, 1963).

6.9.3 Elements of Safe Mine Storage and Reclaiming of Coal

6.9.3.1 Long-Term Storage

When coal is stockpiled in the open and is to remain in storage for long periods of time, such as through a winter or during extended periods of diminished sales, the area(s) selected for storage should be dry and either lie or be constructed to permit good drainage. The area must then be made free of all combustible material having low ignition temperatures, such as wood, rags, dry hay, and the like (Allen and Parry, 1954).

Clay or firmly packed earth, upon which fine coal is rolled, should form the base of the storage pile and the coal should be spread over the entire area in thicknesses of approximately 1–2 ft and compacted. The formation of conical piles should be avoided and the top and sides of the pile should be compacted or rolled to form a seal and exclude air. An effective seal for a coal pile is afforded by a continuous layer of fine coal followed by a covering of lump coal to prevent the loss of the seal through the action of wind and rain.

Larger sizes of screened coal can be stored with little difficulty. Loose storage that allows natural ventilation to dissipate the small amount of heat produced is usually adequate. In addition, seals of compacted fine coal may be employed. This can minimize the undesirable production of fines if the coal has a tendency to slack. ROM coal and stoker-sized coal should be stored by the layering method, with sides sloped for drainage (Chapter 6). Oil treatment of smaller sizes of coal in storage is at times desirable as it slows the absorption of moisture and oxygen.

When heating and fires develop in storage piles and it is impractical to spread out the coal to cool, smothering by compaction with heavy equipment is generally the best procedure since water-flooding may wash out voids in the pile, allowing the fire to spread. Heating in storage piles can be detected, before it becomes serious, by driving small-diameter pipes at intervals vertically into the pile and using thermometers or thermocouples to measure the temperature. The pipes should be driven completely through the pile to avoid a chimney effect.

6.9.3.2 Short-Term Storage

Similar actions to those described for long-term storage can (should) also be applied to the short-term storage of coal in stockpiles, particularly at unit train loading facilities with reclaiming tunnels. The major hazard associated with coal recovery tunnels is the possible formation of an explosive atmosphere originating from accumulation of methane and coal dust (Stahl and Dalzell, 1965). Methane often will accumulate despite what appears to be adequate ventilating practice; dust accumulations vary with the surface moisture of the coal.

The release of emanation of methane from coal forms a sluggish atmosphere and may inhibit low-temperature oxidation, exceptionally in coals with high content of gas but methane is also a potential source of energy (Thomas, 1992). Furthermore, as the methane desorption decreases sharply with time, more of the coal surface will be exposed to oxidation.

Closed-end coal recovery tunnels should be equipped with adequate escape passages that, if properly constructed, can also serve as ventilation ducts. Tunnel walls should be washed down frequently to prevent dust accumulation and welding, and electrical repair work should not be conducted in the tunnel during reclaiming operations or if gas or dust is present in the tunnel. Fire-fighting and respiratory protective equipment should be readily available.

REFERENCES

Ali, A. and Srivastava, S.K. 1992. *Fuel*, 71: 835.

Allen, R.R. and Parry, V.F. 1954. Storage of low-rank coals. Report of Investigations No. 5034. U.S. Bureau of Mines, Washington, DC.

Anonymous. 1991. Coal preparation technology—Influence and trends. *World Mining Equipment*, March, pp. 20–22.

Anonymous. 1993. *Facts about Coal*. National Coal Association, Washington, DC, p. 78.

Anonymous. 1994. An overview of U.S. Federal Coal Preparation Research. U.S. Department of Energy, Pittsburgh Energy Technology Center, Pittsburgh, PA, March, p. 24.

Aplan, F.F. 1993. Coal properties dictate coal flotation strategies. *Mining Engineering*, 45: 83–96.

Arnold, B.J. and Aplan, F.F. 1989. The hydrophobicity of coal macerals. *Fuel*, 68: 651–658.

ASTM. 2011a. *Standard Test Method for Equilibrium Moisture of Coal at 96 to 97 Percent Relative Humidity and 30°C* (ASTM D1412). Annual Book of Standards. American Society for Testing and Materials, West Conshohocken, PA.

ASTM. 2011b. *Standard Method of Test for Grindability of Coal by Hardgrove Machine Method* (ASTM D409). Annual Book of ASTM Standards, Section 05.05. American Society for Testing and Materials, West Conshohocken, PA.

ASTM. 2011c. *Standard Methods for Designating the Size of Coal from Its Sieve Analysis* (ASTM D431). Annual Book of ASTM Standards, Section 05.05. American Society for Testing and Materials, West Conshohocken, PA.

Baafi, E.Y. and Ramani, R.V. 1979. Rank and maceral effects on coal dust generation. *International Journal of Rock Mechanics and Mining Sciences and Geomechanics*, 16: 107–115 [Abstracts].

Barkley, J.F. 1942. Pointers on the Storage of Coal. Information Circular No. 7121. U.S. Bureau of Mines, Washington, DC.

Beier, E. 1990. In *Bioprocessing and Biotreatment of Coal*, D.L. Wise (Ed.). Marcel Dekker Inc., New York, p. 549.

Bensley, C.N., Swanson, A.R., and Nicol, S.K. 1977. *International Journal of Mineral Processing*, 4: 173.

Berkowitz, N. and Schein, H.G. 1951. Heats of wetting and the spontaneous ignition of coal. *Fuel*, 30: 94.

Berkowitz, N. and Speight, J.G. 1973. Prevention of spontaneous heating of coal. *Bulletin of the Canadian Institute of Mining and Metallurgy*, 66(736): 109.

Bevan, R.R. 1981. In *Coal Handbook*, R.A. Meyers (Ed.). Marcel Dekker Inc., New York, Chapter 5.

Burchfield, J.W. 1993. Automated flotation control at Jim Walter resource, Mining Division. In *Coal Prep'93, Proceedings of Tenth International Coal Preparation Exhibition and Conference*. Lexington, KT, May, pp. 159–168.

Carbini, P., Ciccu, R., Ghiani, M., and Satta, F. 1992. *Coal Preparation*, 11(1/2): 11.

Chatterjee, K. and Stock, L.M. 1991. *Energy & Fuels*, 5: 704.

Couch, G.R. 1987. Biotechnology and coal. Report No. ICTRS/TR38. A Coal Research, International Energy Agency, London, U.K.

Couch, G.R. 1991. Advanced coal cleaning technology. Report No. IEACR/44. IEA Coal Research, International Energy Agency, London, U.K.

Couch, G.R. 1995. Power from coal—Where to remove impurities? Report No. 1EACR/82. IEA Coal Research, International Energy Agency, London, U.K.

Davis, V.L. Jr. 1993. Implementation of microcell column flotation for processing fine coal. In *Coal Prep'93, Proceedings from the tenth International Coal Preparation Exhibition and Conference*. Lexington, KT, May, pp. 239–249.

Dodson, K., Ehlert, J., and Hausegger, S. 1994. Why are hyperbaric filters so popular in Europe for filtration of fine coal. In *Proceedings of the Twelfth Pittsburgh Coal Conference*. Pittsburgh, PA, September, pp. 1510–1515.

Dugan, P.R., McIlwain, M.E., Quigley, D., and Stoner, D.L. 1989. In *Proceedings. Sixth Annual International Pittsburgh Coal Conference*. University of Pittsburgh, Pittsburgh, PA, Vol. 1, p. 135.

Elder, J.L., Schmidt, L.D., and Steiner, W.A. 1945. Relative spontaneous heating tendencies of coals. Technical Paper No. 681. U.S. Bureau of Mines, Washington, DC.

Elliott, T.C. 1992. *Power*, January: 17.

Ergun, S. and Bean, E.H. 1968. Report of Investigations No. 7181. U.S. Bureau of Mines, Washington, DC.

Fecko, P., Pectova, I., Ovcari, P., Cablik, V., and Tora, B. 2005. Influence of petrographical composition on coal flotability. *Fuel*, 84: 1901–1904.

Feng, K.K. 1985. Spontaneous combustion of Canadian coals. *CIM Bulletin*, 78(877): 71–75.

Fieldner, A.C., Fisher, P.L., and Brewer, R.E. 1945. Annual Report of Research and Technologic Work on Coal-Fiscal Year 1944, Information Circular No. 7322. U.S. Bureau of Mines, Washington, DC.

Fonseca, A.G., Meenan, G.F., and Oblad, H.B. 1993. Automatic control coal flotation and dewatering processes. *Coal Preparation*, 17: 73–83.

Fourie, P.J.F. 1980. Dense media beneficiation of −0.5 mm coal in the Republic of South Africa. In *Proceedings. Fifth International Conference on Coal Research*. Dusseldorf, Germany, Vol. 2, pp. 737–747.

Franzidis, J.P. 1987. In *Proceedings. Fourth Annual Pittsburgh Coal Conference*. University of Pittsburgh, Pittsburgh, PA, p. 719.

Godfrey, R.L. 1994. Fuel upgrading as a method for pollution control. In *Proceedings. Eleventh Annual Pittsburgh Coal Conference*. Pittsburgh, PA, September.

Gray, B.F. 1990. The ignition of hygroscopic materials by water. *Combustion and Flame*, 79: 2–6.

Hanson, V.D. 1976. In *Coal: Scientific and Technical Aspects*, Part 2, M.E. Hawley (Ed.). Dowden, Hutchinson & Ross, Stroudsburg, PA, p. 284.

Hee, Y.B. and Laskowski, J.S. 1994. Effect of dense medium properties on the separation performance of a dense medium cyclone. *Minerals Engineering (United Kingdom)*, 7(2/3): 209–221.

Hicks, T. 1951. *Power*, July: 73.

Jackman, H.W. 1957. Weathering of Illinois coal during storage. Circular 2271, Illinois State Geological Survey, Champaign, IL.

Jones J.C. 1998. Temperature uncertainties in oven heating tests for propensity to spontaneous combustion. *Fuel*, 77(13): 1517–1519.

Kester, W.M., Leonard, J.W., and Wilson, E.B. 1967. *Mining Congress Journal*, July: 70.

Klima, M.S., Kilmeyer, R.P., and Hucko, R.E. 1990. Development of a micronized-magnetite cyclone process. In *Proceedings of the Eleventh International Coal Preparation Congress*. Tokyo, October, pp. 145–149.

Labuschagne, B.C.J., Wheelock, T.D., Guo, R.K., David, H.T., and Markuszewski, R. 1988. In *Proceedings. Fifth Annual International Pittsburgh Coal Conference*. University of Pittsburgh, Pittsburgh, PA, p. 417.

Landers, W.S. and Donoven, D.J. 1961. In *Storage, Handling and Transportation. Chemistry of Coal Utilization*, H.H. Lowry (Ed.). John Wiley & Sons Inc., New York, Chapter 7.

Leonard, J.W. III. (Ed.). 1991. *Coal Preparation*, 5th edn. Society of Mining Metallurgy and Exploration, Littleton, CO.

Lindberg, K. and Provorse, B. 1977. *Coal: A Contemporary Energy Story*, R. Conte (Ed.). Scribe Publishers Inc., New York.

Liu, H., Zuniga, R., and Richards, J.L. 1993. *Economic Analysis of Coal Log Pipeline Transportation of Coal*. CPRC Report No. 93-1. University of Missouri, Columbia, MO.

Lockhart, N.C. 1984. Dry beneficiation of coal. *Powder Technology*, 40: 17.

Luttrell, G.H., Venkatraman, P., and Yoon, R.H. 1994. Combining flotation and enhanced gravity for improved ash and sulfur rejection. In *Proceedings. Eleventh Pittsburgh Coal Conference*. Pittsburgh, PA, September, pp. 1273–1278.

Luttrell, G.H. and Yoon, R.H. 1994. Commercialization of the microcell column flotation technology. In *Proceedings. Eleventh Pittsburgh Coal Conference*. Pittsburgh, PA, September, pp. 1503–1508.

Mazumdar, M., Jacobsen, P.S., Killmeyer, R.P., and Hucko, R.E. 1992. *Journal of Coal Quality*, 11(1/2): 20.

Mehrotra, V.P., Sastry, K.V.S., and Morey, B.W. 1983. *International Journal of Mineral Processing*, 11(3): 175.

Meyers, R.A. (Ed.). 1981. *Coal Handbook*. Marcel Dekker Inc., New York.

Mitchell, J. 1963. Stockpiling of coking coals. *Blast Furnace and Steel Plant*, July: 540.

Nowacki, P. 1980. *Lignite Technology*. Noyes Data Corporation, Park Ridge, NJ.

Nowak, Z.A. 1994. Overview of fine coal dewatering in Europe. In *Coal Prep'94, Proceedings. Eleventh International Coal Preparation Exhibition and Conference*. Lexington, KY, May, pp. 71–79.

Oder, R.R. 1978. In *Novel Concepts, Methods and Advanced Technology in Particulate-Gas Separation*, T. Ariman (Ed.). University of Notre Dame, South Bend, IN, p. 108.

Oder R.R. 1984. *Separation Science and Technology*, 19(11/12): 761.

Oder, R.R. 1987. In *Proceedings. Fourth Annual Pittsburgh Coal Conference*. University of Pittsburgh, Pittsburgh, PA, p. 359.

Parry, V.F. 1942. Questions and answers on storage of coal in the rocky mountain area. Information Circular No. 7124. U.S. Bureau of Mines, Washington, DC.

Parsons, T.H., Higgins, S.T., and Smith, S.R. 1987. In *Proceedings. Fourth Annual Pittsburgh Coal Conference*. University of Pittsburgh, Pittsburgh, PA, p. 53.

Paul, B.C., Honaker, R.Q., and Ho, K. 1994. Production of Illinois basin compliance coal using enhanced gravity separation. Final Technical Report for Illinois Clean Coal Institute, DOE/ICCI Report DE-FC22-92PC92521. United States Department of Energy, Washington, DC, August.

Pis, J.J. 1996. A study of the self-heating of fresh coals by differential thermal analysis. *Thermochimica Acta*, 279: 93–101.

Rees, O.W., Coolican, F.C., Pierron, E.D., and Beele, C.W. 1961. Effects of outdoor storage on Illinois steam coal. Circular No. 313. Illinois State Geological Survey, Champaign, IL.

Riley, D.M. and Firth, B.A. 1993. Application of an enhanced gravity separator for cleaning fine coal. In *Coal Prep'93. Proceedings. Tenth International Coal Preparation Exhibition and Conference*. Lexington, KY, May, pp. 46–72.

Roll, W.H. 1963. Methods of coal storage. *Mechanization*, October: 28.

Roman, G.H. 1967. *Coal Age*, February: 107.

Rosendale, L.W., DeVito, M.S., Conrad, V.B., and Meenan, G.F. 1993. The effects of coal cleaning on trace element concentration. Air and Waste Management Association, Annual Meeting, Denver, CO, June.

Ryan, B. 1992. *Journal of Coal Quality*, 11(1/2): 13.

Schlesinger, L. and Muter, R.B. 1989. In *Proceedings. Sixth Annual International Pittsburgh Coal Conference*. University of Pittsburgh, Pittsburgh, PA, Vol. 2, p. 1163.

Schmeling, W.A., King, J., and Schmidt-Collerus, J.J. 1978. In *Analytical Chemistry of Liquid Fuel Sources: Tar Sands, Oil Shale, Coal, and Petroleum*, P.C. Uden, S. Siggia, and H.B. Jensen (Eds.). Advances in Chemistry Series No. 170. American Chemical Society, Washington, DC, Chapter 1.

Schmidt, C.E. 1987. In *Proceedings. Fourth Annual Pittsburgh Coal Conference*. University of Pittsburgh, Pittsburgh, PA, p. 143.

Sokolović, J., Stanojlović, R., and Marković, Z.S. 2006. Effect of oxidation on flotation and electro kinetic properties of coal. *Journal of Mining and Metallurgy A: Mining*, 42(1): 69–81.

Speight, J.G. 1990. *Fuel Science and Technology Handbook*, Marcel Dekker Inc., New York.

Speight, J.G. 2005. *Handbook of Coal Analysis*. John Wiley & Sons Inc., Hoboken, NJ.

Stahl, R.W. and Dalzell, C.J. 1965. Recommended safety precautions for active coal stockpiling and reclaiming operation. Information Circular No. 8256. U.S. Bureau of Mines, Washington, DC.

Sun, S.C. 1968. In *Coal Preparation*, J.W. Leonard and D.R. Mitchell (Eds.). AIME, New York, p. 10.

Swaine, D.J. 1992. *Energeia*, 3(3): 1.

Thomas, L. 1992. *Handbook of Practical Coal Geology*. John Wiley & Sons, Chichester, U.K.

Trindade, S.C., Howard, J.B., Kolm, H.H., and Powers, G.J. 1974. *Fuel*, 53: 178.

Tumati, P.R. and DeVito, M.S. 1992. Trace element emissions from coal combustion—A comparison of baghouse and ESP collection efficiency. In *Proceedings. EPRI Conference on the Effect of Coal Quality on Power Plant Performance*. San Diego, CA, August.

U.S. DOE. 1979. Final environmental impact statement fuel use act. Report No. DOE/EIS-0038. Office of Environment, United States Department of Energy, Washington, DC, April.

Vassallo, A.M., Fraser, B., and Palmisano, A.J. 1990. *Journal of Coal Quality*, 9(4): 105.

Vaughn, R.H. and Nichols, D.G. 1985. The estimation of risk of self heating and spontaneous combustion for coal in storage or transport. In *Proceedings. Coal Technology '85*. Pittsburgh, PA, November, pp. 222–231.

Vettor, A., Pellegrini, L., and Valentine, D. 1989. In *Proceedings. Sixth Annual International Pittsburgh Coal Conference*. University of Pittsburgh, Pittsburgh, PA, Vol. 2, p. 1214.

Wheelock, T.D. 1977. *Coal Desulfurization: Chemical and Physical Methods*. Symposium Series No. 64. American Chemical Society, Washington, DC.

Williams, D.G. 1981. In *Coal Handbook*, R.A. Meyers (Ed.). Marcel Dekker Inc., New York, Chapter 6.

Xiao, L., Somasunduran, P., and Vasudevan, T.V. 1989. In *Proceedings. Sixth Annual International Pittsburgh Coal Conference*. University of Pittsburgh, Pittsburgh, PA, Vol. 2, p. 1173.

Yoke, G.R. 1958. Oxidation of coal. Report of Investigations No. 207. Illinois State Geological Survey, Champaign, IL.

7 Mineral Matter

7.1 INTRODUCTION

As well as organic matter coal also contains a significant proportion of minerals and other inorganic contaminants, collectively referred to as mineral matter. Some of these materials may be removed during coal preparation, but a certain amount is intimately associated with the organic components and becomes involved in various ways when the coal is finally used.

The inorganic constituents of coal are often expressed in the form of ash yield and sulfur content. It is, however, often necessary to express an inorganic constituent relative to environmental emissions, such as the amount of sulfur dioxide produced during combustion (Chapters 14 and 15). Furthermore, many of the elements that may be of environmental concern occur in the fly ash after coal combustion (Chapter 14) and, in the case of power plants using flue gas desulphurization (FGD) systems, in the FGD (or scrubber) by-products (Chapter 15).

Coal beneficiation processes prior to utilization (Chapter 6) may also serve as a means of reducing the levels of at least some mineral elements. Such elements, especially those that occur at significant levels in the residue from utilization processes, can cause waste disposal or control problems, such as leaching into the environment (Chapter 25) following ground or surface water infiltration.

Thus, minerals in coal have been, and continue to be, a subject of much interest. Early studies approached the subject somewhat indirectly by means of chemical analysis of high-temperature ash and back calculation to obtain estimates of the mineral matter. Others supplemented chemical studies by handpicking the coarser minerals or performing density separations for chemical tests and optical microscopic studies. With the introduction of radio frequency ashing at low temperature (<150°C), it became possible to directly investigate all of the mineral constituents (Gluskoter, 1965; Miller, 1984).

The general term mineral matter as it applies to coal science and technology is a widely used, but often misrepresented (and even misinterpreted) expression insofar as the terms mineral matter and ash are often used interchangeably. This is, of course, incorrect since ash is, in reality, the residue remaining after complete combustion of the organic portion of the coal matrix to carbon dioxide and water, etc. Thus, the constituents of ash do not occur as such in coal but are formed as a result of chemical changes which take place in the mineral matter during the combustion (*ashing*) process. These changes usually involve the breakdown of complex chemical structures (such as can occur in clays and many minerals) with the formation of the metal oxides.

Coal is a sedimentary rock composed of three categories of substances: (1) organic carbonaceous matter—macerals, (2) inorganic (mainly crystalline) minerals, and (3) fluids. The latter occur in pores within and between the other two solid constituents. The fluids in coal prior to mining are mainly moisture and methane. Applied to coal, the term mineral matter is an inclusive term that refers to the mineralogical phases as well as to all other inorganic elements in coal, that is, the elements that are bonded in various ways to the organic (C, H, O, N, S) components.

The mineral matter content of coal varies considerably and may even be as high as 35% w/w of the coal. The composition of the mineral matter in the coal (or the composition of the mineral ash after combustion) is of importance for, as examples, the performance of design of postcombustion cleanup equipment, such as electrostatic precipitators and FGD units (Kelly and Spottiswood, 1982, 1989; Hjalmarsson, 1992). The alkali metals (sodium, potassium, and lithium) affect (decrease) the resistivity of the ash and can influence sulfur removal.

Aluminum, silicon, and iron can influence the size of the electrostatic precipitator. But, moreover, the fly ash can cause degradation of any catalyst used downstream by blocking the pores as well as causing erosion. This is in addition to any possible detrimental effects of the mineral matter in a coal conversion plant. Again, catalyst poisoning as well as adverse catalytic effects of the mineral constituents on the process may occur (Jenkins and Walker, 1978). On the other hand, any potential beneficial catalytic effects of the mineral constituents of coal (especially in relation to conversion processes) (Bredenberg et al., 1987) also need evaluation and precise definition.

Furthermore, ash fusibility can be strongly influenced by differences in calcium, magnesium, and iron content. Too high a magnesium content can cause clinkering troubles leading to magnesia swelling (Mackowsky, 1968).

While many environmental issues focus on the discharge of gaseous material such as sulfur oxides and nitrogen oxides (Chapters 22, 23, and 25), the discharge of mineral slag to the surrounding environment is also cause for concern. The constituents of these slags and their relative toxicity to the flora and fauna, not only in the immediate vicinity of the plant but also in areas quite remote from the plant, are also cause for concern. In addition, it is quite possible that the toxic materials can be transported by surface and ground waters as well as by the prevailing winds (e.g., sulfur dioxide, fly ash).

Coal may also be recognized as a source of valuable elements and inorganic materials. For example, coal may provide a supply of uranium or other desired elements while the production of sulfur from coal (by conventional gas-cleaning methods) is also feasible.

Hence, it is now considered essential that the nature of the mineral constituents of coal be more thoroughly understood in order to evaluate the nature of the inorganic material, for example., fly ash and discharge, produced when coal is employed as a fuel or feedstock (Babu, 1975; Braunstein et al., 1977; Given and Yarzab, 1978; Quast and Readett, 1991; Straszheim and Markuszewski, 1991).

7.2   ORIGIN OF MINERAL MATTER IN COAL

Mineral matter originates from the inorganic constituents of the vegetation which acted as the precursor to coal and from the mineral matter that was transported to the coal bed from a remote site (Chapters 3 and 4). Thus, mineral matter in coal has often been classified as inherent and extraneous mineral matter (Francis, 1961; Stach et al., 1982; Spears and Zheng, 1999).

The *inherent mineral matter* is that mineral matter which had its origin in the organic constituents of the plant giving rise to the coal bed but the extraneous mineral matter was brought into the coal-forming deposit by mechanical means from outside, for example, as dust by air or as suspended or dissolved material carried by water. The inherent mineral matter is also sometimes defined as the inorganic material combined with the organic coal substance, but such material need not be derived from the coal-forming plants.

The inherent mineral matter is usually much smaller in quantity than the extraneous mineral matter and can be expected to differ quite markedly in composition from the inorganic residue of the major coal-forming plant types. This is due to reuse (in no small part) of the inorganic elements by succeeding vegetation and to leaching of inorganic constituents by the waters percolating and flowing through the peat bog during coalification. The percolating waters may be presumed to have an increased dissolving action on the inorganic constituents because of their content of humic acids (Stevenson and Butler, 1969), carbon dioxide, and other products of decay. Furthermore, the differences in solubility and reactions of the inorganic elements present in the plants dictate that these elements not contribute to the coal mineral matter in proportion to their presence in the plant material.

There has been considerable discussion and conjecture about the origin of the so-called inherent and extraneous mineral matter in coal. Indeed, it is in regard to this particular aspect of coal technology that many researchers find this terminology not only misleading but also difficult to apply, especially to those minerals that are contemporaneous with the peat bog in which the coal was formed; such minerals may not have been incorporated into the plant matrix prior to the formation of the coal bed.

The intimate interrelationship between coal and mineral matter can only be adequately explained if it is accepted that the mineral matter became part of the coal matrix either during the early stages of coalification in the peat bog or at some early stage of the subsequent maturation process.

Mineral matter that originated within the immediate environs of the peat bog is often referred to as *authentic mineral matter* whereas mineral matter which was transported by water or by wind is often referred to as *allogenic mineral matter*. The additional term *syngenetic mineral matter* (Table 7.1) is also applied to mineral matter that may have been transported into future coal deposits (i.e., this mineral matter was not a part of the plant substance that contributed to the coal precursor nor was it introduced during the coalification process) by water or wind. On the other hand, *epigenetic mineral matter* refers to that material which was deposited into the peat bog by descending for ascending solutions in cracks or fissures or in the bedding planes of the coal; such mineral matter may often be found as cleat fillings (Spears and Caswell, 1986).

TABLE 7.1
Types of Coal Minerals and Their Origins

| Mineral Group | Syngenetic Formation (Intimately Intergrown) | | Epigenetic Formation | |
	Transported by Water or Wind	Newly Formed	Deposited in Fissures, Cleats, and Cavities (Coarsely Intergrown)	Transformation of Syngenetic Minerals (Intimately Intergrown)
Clay minerals	Kaolinite, illite, sericite, clay minerals with mixed-layer structure "tonstein"		—	Illite, chlorite
Carbonates	—	Siderite–ankerite concentrations, dolomite, calcite, ankerite	Ankerite, calcite, dolomite	—
	Siderite, calcite, ankerite in fusite			
Sulfide ores	—	Pyrite concretions, melnikowite-pyrite, coarse pyrite (marcasite), concretions of FeS_2–$CuFeS_2$–ZnS	Pyrite, marcasite, zinc sulfide (sphalerite), lead sulfide (galena), copper sulfide (chalcopyrite)	Pyrite from the transformation of syngenetic concretions of $FeCO_3$
	Pyrite in fusite			
Oxide ores	—	Hematite	Goethite, lepidocrocite (needle iron ore)	—
Quartz	Quartz grains	Chalcedony and quarts from the weathering of feldspar and mica	Quartz	—
Phosphates	Apatite	Phosphorite	—	—
Heavy minerals and accessory minerals	Zircon, rutile, tourmaline, orthoclase, biotite	—	Chlorides, sulfates, nitrates	—

Source: Murchison, D. and Westoll, T.S., Eds., *Coal and Coal-Bearing Strata*, Elsevier, New York, 1968.

7.3   OCCURRENCE

Mineral matter is common to all types of coal, and it has been recognized from the time when coal was first mined for general use that coal contained some material other than the main (organic) coal substance. The presence of the mineral matter is even more apparent when the coal is examined in place in the bed, after being mined or finally prepared for market, and, finally, when the residue from its combustion is examined.

In the bed, extraneous matter is apparent as definite horizontal layers of varying thickness and extent, as surface deposits or fillings in the vertical cleats, and more unusually as intrusions of clay-like (or other) material which occurs irregularly throughout the coal deposit. Furthermore, these horizontal layers that attain a considerable thickness may serve as a means of separating different coals or simply separating the different phases of formation of one particular coal. However, bands of this type are always regarded as being formed during the deposition of the coal precursors.

The most common minerals in coal (e.g., illite clay, pyrite, quartz, and calcite) are made up of these most common elements (in rough order of decreasing abundance): oxygen, aluminum, silicon, iron, sulfur, and calcium. These minerals and other less common minerals usually contain the bulk of the trace elements present in coal (Finkelman, 1982, 1993; Davidson and Clarke, 1996; Davidson, 2000).

Minerals in coal commonly occur as single crystals or clusters of crystals that are intermixed with organic matter or that fill void spaces in the coal; sizes of mineral grains range from submicroscopic to a few inches. Some clusters of mineral grains, however, such as fracture fillings or coal balls, may reach sizes that range from fractions of an inch up to several feet across. Coal balls result when mineral matter (such as calcite, pyrite, or siderite) infuses peat before it is compressed. Mineral grains in coal also often occur as discrete particles, such as the blebs of pyrite (framboids), which may be identified in the photomicrographs of coal. The more finely divided the mineral grains, the higher the magnification needed for identification.

Although much is known about the minerals in coal, much remains to be learned about their occurrence, abundance, origin, and composition. For example, the type of clay mineral in a coal, whether montmorillonite or illite, determines how a coal will react when burned.

Montmorillonite may or may not break down (dissociate) into its constituent parts when coal is burned; if it does dissociate, then, upon cooling, it may recombine with other elements or minerals to form mineral deposits on the inside surfaces of furnaces and boilers. This process (*slagging* or *fouling*) produces barriers to heat exchange in the affected equipment, which can substantially reduce its efficiency and require costly repairs. Illite, however, with its simpler composition, does not cause such problems under normal furnace operating conditions. Where these two clay minerals and others occur, their relative abundances, relationships to other minerals, and exact compositions are subjects for continued research in coal quality.

The deposits in the shrinkage cracks or cleats, which may, for example, be kaolinite, calcite, gypsum, or pyrite, are usually thick surface deposits and may not always be present and were presumably deposited later in the coal-forming process. On the other hand, the larger adventitious bodies of foreign matter in the coal (clay veins, sandstone intrusions, washouts, and the like) may be the result of rather unusual geological conditions during and after the time the coal was laid down (Chapters 3 and 4).

However, a large part of the mineral matter in marketed coal arises from the inclusion during mining of roof (or floor) rock of the impure streaks or of some of the large impurities. Such inclusions may be deliberate insofar as little effort is made to remove them during the mining operation, or they may be accidental in that their elimination during mining was not complete. Indeed, with the increasing use of mechanical methods of mining (particularly in large-scale strip operation), it is becoming the practice to do little exclusion during mining and to subject the coal to a cleaning process afterward (Chapter 6).

Obviously, the amount and type of minerals found in coal varies widely (Table 7.2) and depends on the coal history. The most abundant minerals are the clay minerals of which illite, kaolinite, and

TABLE 7.2

Indication of Minerals in Coal Expressed as a Percentage of the Total Mineral Matter

Classification	Mineral Constituents	Elkhorn No. 3 Seam, Kentucky	Hartshorne Seam, Kentucky
Silicates	Kaolinite	3–40	1–10
	Illite	Trace	1–10
	Chlorite	Trace	1–10
	Mixed-layer illite, montmorillonite	Trace	
Carbonates	Siderite		30–40
Oxides	Quartz	40–50	1–10
	Hematite		
	Rutile	1–10	
Sulfates	Gypsum	1–10	1–10
	Thernardite		
Sulfides	Pyrite	1–10	1–10

Source: Braunstein, H.M. et al. (Eds.), Environmental health and control aspects of coal conversion: An information overview, Report ORNL-EIS-94, Oak Ridge National Laboratory, Oak Ridge, TN, 1977, pp. 2–30.

TABLE 7.3

Iron Sulfate Minerals Identified in Weathered Illinois Coal

Mineral	Chemical Formula
Szomoluokite	$FeSO_4 H_2O$
Rozenite	$FeSO 4H_2O$
Melanterite	$FeSO_4 \cdot 7H_2O$
Coquimbite	$Fe(SO_4) 9H_2O$
Rosmerite	$Fe_2(SO_4) Fe_2(SO_4)_3 12H_2O$
Jarosite	$(Na,K)Fe_3(SO_4)_2(OH)_6$

Source: Gluskoter, H.J., *Trace Elements in Fuel*, S.P. Babu (Ed.), Advances in Chemistry Series No. 141, American Chemical Society, Washington, DC, Chapter 1, 1975.

montmorillonite occur most frequently. Pyrite is the common sulfide mineral while sulfates are relatively rare but increase with weathering (Table 7.3). Carbonates form readily in nonacid areas (dolomite and ankerite are encountered frequently) and quartz (which is found virtually in all coals) may occur in concentrations as high as 20% w/w of the total mineral matter. Sulfide minerals often constitute as much as 25% of the coal mineral matter.

There may also be regional variations in the mineral distribution in coal, and even a cursory examination of the major minerals that occur within the coal matrix presents only a general indication of the vast number of metals that can occur in coal. Even though less than 12 elements usually constitute the major portion of coal ash (Table 7.4), there are many individual mineral elements present which (on the basis that a *trace element* is a particular element which occurs in

TABLE 7.4
Major Inorganic Constituents
of Coal Ash

Constituents	Representative Percentage
SiO_2	40–90
Al_2O_3	20–60
Fe_2O_3	5–25
CaO	1–15
MgO	0.5–4
Na_2O	0.5–3
K_2O	0.5–3
SO_3	0.5–10
P_2O_5	0–1
TiO_2	0–2

concentrations of less than 0.1%; <1000 ppm) in the earth's crust are classified as trace elements (Table 7.5) (Finkelman, 1982, 1993).

Extensive chemical analyses on coals have shown that elements with relatively large ranges of concentrations (e.g., arsenic, As; barium, Ba; cadmium, Cd; iodine, I; lead, Pb; antimony Sb; and zinc, Zn) include those that are found in coals within sulfate and sulfide minerals. In addition, many elements appear to be positively correlated in coals such as calcium and manganese, zinc and cadmium, the chalcophile elements (i.e., those elements which commonly form sulfides, such as cobalt, Co; nickel, Ni; lead, Pb; and antimony, Sb) and the lithophile elements (i.e., those elements which commonly occur in silicate phases, such as silicon, Si; titanium, Ti; aluminum, Al; and potassium, K).

7.4   MINERAL TYPES

A variety of minerals have been reported to be present in coals although many occur infrequently and are not regularly found in all coals or may be detectable only in trace amounts in various suites of coals (O'Gorman and Walker, 1971; Gluskoter, 1975; Gluskoter et al., 1977; Finkelman, 1982, 1993; Roscoe and Hopke, 1982).

The complex environmental conditions which occur at the time the organic detritus is laid down of and the geological upheaval which may have occurred during the maturation process can play an important role in determining which minerals survive in coal (Cecil et al., 1982). Nevertheless, the large majority of minerals found in coal can be classified into several types, such as shale, kaolin compounds, sulfide derivatives, and carbonate derivatives (Table 7.6).

7.4.1   CLAY MINERALS (ALUMINOSILICATES)

The name clay is often used in various contexts: (1) to indicate particle size (<2 in. [<0.005 mm]), (2) to indicate a rock composed predominantly of clay minerals, and (3) to indicate a name for a group of minerals, the clay minerals.

These minerals are hydrated aluminosilicates which are characterized by a sheetlike structure and can be conveniently divided into three groups: (1) the kaolinite group, (2) the montmorillonite group, and (3) the potash clay (or hydrous mica) group (Table 7.7). In the kaolinite group, all have the same chemical composition and differ only in individual crystal structures. The montmorillonite group can be represented by means of ion substitutions in the general chemical formula. For example, in montmorillonite itself, approximately 16% of the aluminum

TABLE 7.5
Elements and Trace Elements in Coal

Constituent	Range
Arsenic	0.50–93.00 ppm
Boron	5.00–224.00 ppm
Beryllium	0.20–4.00 ppm
Bromine	4.00–52.00 ppm
Cadmium	0.10–65.00 ppm
Cobalt	1.00–43.00 ppm
Chromium	4.00–54.00 ppm
Copper	5.00–61.00 ppm
Fluorine	25.00–143.00 ppm
Gallium	1.10–7.50 ppm
Germanium	1.00–43.00 ppm
Mercury	0.02–1.60 ppm
Manganese	6.00–181.00 ppm
Molybdenum	1.00–30.00 ppm
Nickel	3.00–80.00 ppm
Phosphorus	5.00–400.00 ppm
Lead	4.00–218.00 ppm
Antimony	0.20–8.90 ppm
Selenium	0.45–7.70 ppm
Tin	1.00–51.00 ppm
Vanadium	11.00–78.00 ppm
Zinc	6.00–5350.00 ppm
Zirconium	8.00–133.00 ppm
Aluminum	0.43%–3.04%
Calcium	0.05%–2.67%
Chlorine	0.01%–0.54%
Iron	0.34%–4.32%
Potassium	0.02%–0.43%
Magnesium	0.01%–0.25%
Sodium	0.00%–0.20%
Silicon	0.58%–6.09%
Titanium	0.02%–0.15%
Organic sulfur	0.31%–3.09%
Pyritic sulfur	0.06%–3.78%
Sulfate sulfur	0.01%–1.06%
Total sulfur	0.42%–6.47%
Sulfur by x-ray fluorescence	0.54%–5.40%

Source: Ruch, R.R. et al., Environmental Geology Note No. 72, Illinois State Geological Survey, Urbana, IL, 1974, p. 18.

is substituted by magnesium and by other ions such as calcium, sodium, potassium, and hydrogen (as well as some additional magnesium).

Clay minerals are the most commonly occurring inorganic constituents of coals (Gluskoter, 1975) (as well as the strata associated with coals) and, therefore, can act as the source of a wide variety of metals in substantial or trace amounts. The most common clay minerals found in coals are kaolinite and illite while montmorillonite, chlorite, and sericite have also been regularly reported to occur in various coals.

TABLE 7.6
Minerals Commonly Associated with Coal

Group	Species	Formula
Shale	Muscovite	$(K, Na, H_2O, Ca)_2(Al, Mg, Fe, Ti)_4$
	Hydromuscovite	$(Al, Si)_8O_{20}(OH, F)_4$ (general formula)
	Illite	$(HO)_4K_2(Si_6 \cdot Al_2)Al_4O_{20}$
	Montmorillonite	$Na_2(Al\ Mg)Si_4O_{10}(OH)_2$
Kaolin	Kaolinite	$Al_2(Si_2O_5)(OH)_4$
	Livesite	$Al_2(Si_2O_5)(OH)_4$
	Metahalloysite	$Al_2(Si_2O_5)(OH)_4$
Sulfide	Pyrite	FeS_2
	Marcasite	FeS_2
Carbonate	Ankerite	$CaCO_3 (Mg, Fe, Mn)CO_3$
	Calcite	$CaCO_3$
	Dolomite	$CaCO_3\ MgCO_3$
	Siderite	$FeCO_3$
Chloride	Sylvite	KCl
	Halite	$NaCl$
Accessory minerals	Quartz	SiO_2
	Feldspar	$(K, Na)_2O\ Al_2O_3\ 6SiO_2$
	Garnet	$3CaO\ Al_2O_3\ 3SiO_2$
	Hornblende	$CaO\ 3FeO\ 4SiO_2$
	Gypsum	$CaSO_4\ 2H_2O$
	Apatite	$9CaO\ 3P_2O_5\ CaF_2$
	Zircon	$ZrSiO_4$
	Epidote	$4CaO\ 3Al_2O_3\ 6SiO_2\ H_2O$
	Biotite	$K_2O\ MgO\ Al_2O_3\ 3SiO_2\ H_2O$
	Augite	$CaO\ MgO\ 2SiO_2$
	Prochlorite	$2FeO\ 2MgO\ Al_2O_3\ 2SiO_2\ 2H_2O$
	Diaspore	$Al_2O_3\ H_2O$
	Lepidocrocite	$Fe_2O_3\ H_2O$
	Magnetite	Fe_3O_4
	Kyanite	$Al_2O_3\ SiO_2$
	Staurolite	$2FeO\ 5Al_2O_3\ 4SiO_2\ H_2O$
	Topaz	$2AlFO\ SiO_2$
	Tourmaline	$3Al_2O_3\ 4BO(OH)\ 8SiO_2\ 9H_2O$
	Hematite	Fe_2O_3
	Penninite	$5MgO\ Al_2O_3\ 3SiO_2\ 2H_2O$
	Sphalerite	ZnS
	Chlorite	$10(Mg, Fe)O\ 2Al_2O_3\ 6SiO_2\ 8H_2O$
	Barite	$BaSO_4$
	Pyrophillite	$Al_2O_3\ 4SiO_2\ H_2O$

7.4.2 QUARTZ (SILICA)

Quartz is a widely distributed mineral species consisting of silicon dioxide (silica, SiO_2) (Table 7.6). It is one of the most common minerals and is found in many varieties with very diverse modes of occurrence. Quartz is a primary constituent of rocks such as granite, quartz, porphyry, and rhyolite. It is also a common constituent in many gneisses (laminate rocks) and crystalline schists (foliated rocks). It is also, in a sense, mobile since by the weathering of silicates, silica passes into solution and is redeposited in cavities, crevices, and along joints of rocks of all types. Thus, it is not

TABLE 7.7
Subdivision of Clay Minerals
in Various Groups

Clay Mineral Group	Individual Members
Kaolinite	Kaolinite
	Dickite
	Nacrite
	Halloysite
Montmorillonite	Montmorillonite
	Beidellite
	Nontronite
	Saponite
	Hectorite
	Sauconite
Potash (hydrous mica)	Hydromica
	Hydrous mica
	Illite
	Glimmerton
	Bravaisite

surprising that quartz also occurs in all coals either by virtue of its proximity to the coal bed or by deposition at a later stage of coal formation.

7.4.3 CARBONATE MINERALS

Carbonate minerals are the salts of carbonic acid (H_2CO_3), and the extensive possibilities for interchanging the more common metals, such as calcium, magnesium iron, and manganese, are realized in the wide variety of carbonate minerals that occur in nature. Consequently, it is not surprising that a wide variety of carbonate minerals also occurs in coals (Table 7.6).

Calcite ($CaCO_3$) and siderite ($FeCO_3$) are commonly reported as constituents of the mineral matter in coals while the more complex carbonates dolomite ($CaCO_3 \cdot MgCO_3$) and ankerite ($2CaCO_3 \cdot MgCO_3 \cdot FeCO_3$) are also frequently reported.

7.4.4 SULFUR (SULFIDE AND SULFATE) MINERALS

The sulfide minerals of particular importance here are the pyrite group (Table 7.8), which are essentially metal disulfides occurring in widespread locations and different crystalline forms; pyrite itself is found in large deposits in metamorphic rocks.

The dominant sulfide mineral in coal is pyrite while marcasite has also been reported to be present in many coals. These minerals are dimorphic in that they are identical in chemical composition (FeS_2)

TABLE 7.8
The Pyrite Group of Minerals

Mineral	Formula (Approx.)
Pyrite	FeS_2
Marcasite	FeS_2
Braoite	$(Ni\,Fe)S_2$
Laurite	RuS_2

but differ in crystalline form—pyrite is cubic whereas marcasite is orthorhombic. Other sulfide minerals which have been found (but to a lesser extent) in coals are galena (lead sulfide, PbS) and sphalerite (zinc sulfide, ZnS).

The related sulfate minerals (Table 7.3) are not as common as the sulfides and are not usually present in unweathered fresh coals. Obviously, the anaerobic maturation of (the majority of) coals are not conducive to the formation of sulfates. For example, pyrite is markedly susceptible to oxidation and will decompose to iron sulfate minerals even under ambient conditions.

$$Fe_2S_3 + O_2 \rightarrow FeSO_4$$

7.5 CLASSIFICATION

Coal is a sedimentary rock composed of three categories of substances: (1) organic carbonaceous matter, termed macerals, (2) inorganic (mainly crystalline) minerals, and (3) fluids. The latter occur in pores within and between the other two solid constituents. The fluids in coal prior to mining are mainly moisture and methane.

Applied to coal, the term mineral matter is an inclusive term that refers to the mineralogical phases as well as to all other inorganic elements in the coal; that is, the elements that are bonded in various ways to the organic (C, H, O, N, S) components. The term mineral refers only to the discrete mineral phases.

Because coal is a type of sedimentary rock, 100 or so mineral can occur in coal; however, only about 15 are abundant enough to have high importance. It must be noted that minor impurities commonly substitute for the major cations as well as some anions which account for a considerable fraction of the minor and trace elements reported in coals.

More than 125 different minerals have been reported in coal and occur as discrete grains, flakes, or aggregates in one of five physical modes:

1. As microscopically disseminated inclusions within macerals (distinct organic designations such as vitrinite, liptinite, and inertinite)
2. As layers of partings wherein fine-grained clay minerals usually predominate
3. As nodules including lenticular and spherical concretions
4. As fissures including cleat and other fracture or void fillings
5. As rock fragments found within the coal bed as a result of faulting, slumping, or related disturbances

Useful also is a genetic classification of minerals in coal, wherein they are classified as (1) detrital, (2) syngenetic, or (3) epigenetic.

Detrital grains were introduced into a coal-forming basin (such as swamp land) by rivers, tidal waves, and wind. Syngenetic minerals were formed during the peat stage of coal formation and include minerals formed by crystallization of inorganic elements in plants (intrinsic mineral matter). Epigenetic minerals are those found as filling of fissures and voids after the peat was formed.

More generally, the minerals present in coals are grouped as either detrital or diagenetic. The detrital minerals undergo diagenetic modifications however, and provide one of the sources for elements incorporated into diagenetic minerals. The detrital minerals are present either dispersed or concentrated in the coal seams. The overall composition of the detrital minerals is a function of grain size and thus of current velocity.

Clay mineral composition varies with the maturity of the sediment, and in particular whether this is reworked from within the basin or is more representative of the upland source. Dirt partings (intraseam mudrocks) represent either an increase in detrital sediment input or a reduction in the rate of organic accumulation, or even loss of organic matter.

Wind-blown sediment in coals are best represented by altered volcanic ashes, tonsteins, providing a means of correlation and minerals for absolute age determinations. The alteration of ash to kaolinite both in the peat and in the associated sediments demonstrates which clay mineral is stable in the pore waters. Diagenetic kaolinite, probably formed from elements organically derived, is also recognized, particularly in low-ash coals. The elements present in diagenetic silicates, sulfides, and carbonates originated from within the peat and also from mud rocks in the sequence. The diagenetic sequence of minerals in the cleat is equated with that of other clastic sediments and is similarly interpreted as a burial depth sequence. The residual, diagenetic pore fluids resulting from burial diagenesis are the source of economically important, high chlorine concentrations in the coal (Spears, 2005).

7.6   EVALUATION OF MINERAL MATTER

The use of coal as a fuel and the varied, often detrimental, effects of the mineral matter on the fuel properties of coal has been a source of concern for consumers (Alpern et al., 1984). Even the interactions of the mineral constituents themselves (Table 7.9) have directed interest toward the mineral constituents of coals but with the tendency to an increased use of coal for power generation as well as for gasification and conversion plants that will enable coal to act as a source of liquid and gaseous fuels. Therefore, methods by which the mineral matter can be evaluated have been a constant target for quantitative and qualitative improvement.

Furthermore, the broad definition which defines mineral matter as all elements in coal except carbon, hydrogen, nitrogen, oxygen, and sulfur needs some modification since four of these five elements occur in inorganic locations and are, therefore, part of the mineral matter. For example, (1) carbon occurs in carbonates; (2) hydrogen occurs in water of hydration, as well as in any free water in the coal; (3) oxygen occurs in carbonates, sulfates, and silicates as well as in water as well as in water of hydration and free water in the coal; and (4) sulfur occurs in sulfates and sulfides, predominantly pyrite and marcasite.

Finally, mineral matter in coal is the parent material in coal from which ash is derived and which comes from minerals present in the original plant materials that formed the coal, or from extraneous sources such as sediments and precipitates from mineralized water. Mineral matter in coal cannot be analytically determined and is commonly calculated using data on ash and ash-forming constituents. Coal analyses are calculated to the mineral matter–free basis by adjusting formulas used in calculations in order to deduct the weight of mineral matter from the total coal.

TABLE 7.9
General Behavior of Minerals of High Temperatures

Inorganic Species	Behavior on Heating
Clays	Loose structural OH groups with rearrangements of structure and release of H_2O
Carbonates	Decompose with loss of CO_2; residual oxides fix some organic and pyritic S as sulfate
Quartz	Possible reaction with iron oxides from pyrite and organically held Ca in lignites; otherwise no reaction
Pyrite	In air, burns to Fe_2O_3 and SO_2; in VM test, decomposes to FeS
Metal oxides	May react with silicates
Metal carboxylates (lignites and subbituminous only)	Decompose; carbon in carboxylate may be retained in residue

Note: VM, volatile matter.

7.6.1 ASHING TECHNIQUES

There are many minerals in coal which will produce as ash residue (Table 7.6) and, although the ash composition varies around wide limits, the ash may appear to be composed of only a few of the more common elements (Table 7.4). But it is often the trace elements (Table 7.5) that can give ash some of its more obnoxious properties.

The determination of the mineral matter content of coal (determined as mineral ash after combustion) has been an essential part of coal evaluation for many years (Shipley, 1962; Rees, 1966; Given and Yarzab, 1978; Huggins et al., 1982; Nadkarni, 1982) (Chapter 8). For example, when coal is cleaned by various processes to reduce the sulfur and mineral content, it is an advantage to be aware of the mineral (ash) content to determine the best cleaning method insofar as the various cleaning methods have different tolerance levels for the mineral constituents of coal. Furthermore, the ash content is also a means of assessing the adequacy of the various sampling procedures (Chapter 8), and it is one of the criteria normally specified in contracts between a purchaser and a supplier of coal.

The amount of mineral matter in coking coal (Chapter 16) is an indication of the amount of ash that will eventually be part of the coke made from the coal. Thus, the higher the ash content of the coal, the lower the proportion of usable carbon in the coke and the more the fluxing limestone that must be added to the furnace to assist in ash (mineral) removal. However, the fouling tendency of a coal is dependent upon several factors, not the least of which is the nature of the mineral matter and the resulting ash (Beer et al., 1992).

Similarly, the higher the mineral matter content, the lower the heat of combustion obtainable from a unit sample of coal, hence the need for removal of mineral matter during cleaning and preparation operations (Chapter 6). A high mineral content also introduces additional problems such as a loss in the combustion efficiency as well as problems related to handling and disposing of larger amounts of mineral ash. Obviously, mineral matter in coal will (and often does) cause problems during utilization and measures to counteract any adverse effects that will arise from the presence of the mineral matter are necessary.

On the other hand, the potential benefits that could arise from the presence of this same mineral matter should not be ignored; catalytic effects in processes designed for the liquefaction (Chapters 18 and 19) and the gasification (Chapters 19 and 20) of coal may be cited as examples.

Mineral matter in coal is often determined indirectly with the ash analysis (determined by direct combustion of the sample) (Chapter 8) forming the basis of the calculation. However, determination and chemical analysis of the ash content of coal gives the average content of the inorganic elements in a particular coal but is not an indication of the nature or distribution of the mineral matter in coal. Nevertheless, ash analysis can provide valuable data which, when used with data from other sources, may give a representation of the mineral content of coal.

However, it must be emphasized that there has to be some attempt to recognize the limitations of the method before any projection relating to the mineral composition of coal is possible. For example, the high temperature required for the "ashing" may result in the loss of the volatile constituents of the minerals or the mineral constituents will undergo a chemical change. In the former case, certain of the mineral elements will escape detection while in the latter case the constituents of clays or shale (to cite an example) will lose water of hydration or the carbonate minerals will lose carbon dioxide and the oxides so produced may even undergo further reaction with sulfur oxides or with silica to produce completely different mineral species:

$$CaCO_3 \rightarrow CaO + CO_2$$

$$CaO + SO_3 \rightarrow CaSO_4$$

$$CaO + SiO_2 \rightarrow CaSiO_3$$

At the same time, pyrite will be converted to ferric oxide and oxides of sulfur:

$$4FeS_2 + 11O_2 \rightarrow 2Fe_2O_3 + 8SO_2$$

Any organically combined inorganic elements may be converted to the respective oxides and eliminated (e.g., nitrogen and sulfur) or may be retained in the ash (e.g., various metals that may be combined with an organic fragment).

The pathways of these various reactions can be followed by differential thermal analysis (DTA) which allows for the identification of the various minerals in more complex mixtures, as might be found in coal (Warne, 1979).

Several formulae have been proposed for calculating the amount of mineral matter originally in the coal using the data from ashing techniques as the basis of the calculations. Of these formulae, two have survived and have been used regularly to assess the proportion of mineral matter in coal and these are the Parr formula and the King–Mavies–Crossley formula.

In the Parr formula, the mineral matter content of coal is derived from the expression:

$$\% \text{ Mineral matter} = 1.08A + 0.55S$$

where
  A is the percentage of ash in the coal
  S is the total sulfur in coal

The King–Mavies–Crossley formula is a little more complex:

$$\% \text{ Mineral matter} = 1.09A + 0.5S_{pyr} + 0.8CO_2 - 1.1SO_{3ash} + SO_{3coal} - 0.5Cl$$

where
  A is the percentage of ash in the coal
  S_{pyr} is the percentage of pyritic sulfur in the coal
  CO_2 is the percentage of "mineral" carbon dioxide in the coal
  SO_{3ash} is the percentage of sulfur trioxide in the ash
  SO_{3coal} is the percentage of sulfur trioxide in the coal
  Cl is the percentage of chlorine in the coal

The Parr formula (which has been widely used in the United States) is obviously considerably simpler than the King–Mavies–Crossley formula (which has been in common use in Great Britain and Europe) and requires less analytical data. However, the King–Mavies–Crossley formula will, because of the detail, provide more precise mineral matter values. It is, however, a matter of assessing whether the slight improvement in precision is justifiable on the basis of the additional analytical effort.

There are, of course, limitations to the use of the data derived from the ashing techniques. For example, the indefinite amount of sulfur which may be retained in the ash will reduce the reliability of the *ash* values. In fact, the ash value is only an approximation of the noncombustible material in coal, and the relationship of ash composition to clinkering, boiler tube slagging, and other high-temperature behavior of the mineral constituents is complex and can be difficult to interpret. The disadvantages of using the ashing technique to evaluate the mineral matter in coals are evident from the earlier text but these disadvantages should not detract from a method that certainly has some merits.

7.6.2   ASH ANALYSIS AND/OR DIRECT MINERAL ANALYSIS

The evaluation of coal mineral matter by the ashing technique can be taken further insofar as attempts can then be made to determine the individual metal constituents of the ash. On the occasion when the mineral matter has been successfully separated from the coal, it is then possible to apply any one (or more) of several techniques (such as x-ray diffraction, x-ray fluorescence, scanning electron microscopy, and electron probe microanalysis) not only to investigate the major metallic elements in coal but also to investigate directly the nature (and amount) of the trace elements in the coal (Jenkins and Walker, 1978; Prather et al., 1979; Raymond and Gooley, 1979; Russell and Rimmer, 1979; Jones et al., 1992).

However, when coal ash is prepared for complete analysis, it has been considered necessary to ensure that sulfur is not retained in the ash. If sulfur is retained in the ash, the analysis will most likely be inaccurate unless corrections to a sulfur dioxide–free basis are made. In addition, attempts must be made to ensure that important elements are not lost during the ashing procedure by virtue of the higher temperatures (ca. 850°C, 1560°F) that are used. In fact, a low-temperature ash sample can be of the order of 150% w/w of the high-temperature ash sample.

To alleviate the problem of element loss by the use of high temperatures, various low-temperature ashing techniques have been developed. One method involves passage of "activated" oxygen (generated by passing commercial grade oxygen through the high-energy electromagnetic field produced by a radio-frequency oscillator) over a dry, finely ground coal sample. Throughout this whole procedure, the temperature has been observed to remain below 200°C (390°F) and is usually in the range of 150°C–160°C (300°F–320°F).

One other technique for the investigation of trace elements in coal applies the float-sink principle in which comminuted coal can be separated into several specific gravity fractions by flotation in mixtures of liquids such as perchloroethylene and naphtha; a liquid such as bromoform ($CHBr_3$) can also be employed for further subfractionation. This procedure can virtually be employed as a fractionation technique to remove significant proportions of the mineral matter from coal without resorting to the use of heat.

There are several options available for the choice of an analytical technique, although in practice it may be necessary to use more than one technique because of the limitations of the various methods with respect to concentrations of the individual elements, or the method may be determined by the amount of available sample. For example, application of any one of the conventional "wet" analytical methods (such as for the determination of silicates) may require up to 10 g of sample ash whereas a variety of spectrophotometric techniques may require only micrograms (1×10^{-6} g) of sample.

Spectrophotometric techniques have been the basis of many coal analysis methods. One of the most widely used techniques for analysis of trace elements is atomic absorption spectrometry, in which the standards and samples are aspirated into a flame. A hollow cathode lamp provides a source of radiation that is characteristic of the element of interest and the absorption of characteristic energy by the atoms of a particular element. X-ray fluorescence is also employed as a quantitative technique for trace element determination and depends on election of orbital electrons from atoms of the element when the sample is irradiated by an x-ray source.

Another method for trace element determination is neutron activation analysis (Weaver, 1978), which is a highly sensitive, nondestructive method for the analysis of many elements. Briefly, the samples are irradiated in a nuclear reactor which produces radioisotopes of the elements and appropriate measurements of the activities of the generated (daughter) species afford a means of calculating the concentrations of the parent isotopes (elements).

Although these three spectrophotometric methods are mentioned here as a means of examining the variety of trace elements which occur in coals, they are by no means the only applicable methods but are used as examples. Of course, each method has various detection limits for the different elements (Tables 7.10 through 7.12) and it is also essential to recognize the limitations of the

TABLE 7.10

Sensitivity and Detection Limits for Elements by Atomic Absorption Spectroscopy

Element	Sensitivity (µg/mL)	Range (µg/mL)	Detection Limits	
			(µg/mL)	(µg/m³)
Ag	0.036	0.5–5.0	0.003	0.1
Al	0.76	5–50	0.04	2
Ba	0.20	1–10	0.01	0.4
Be	0.017	0.1–1.0	0.002	0.08
Bi	0.22	1–10	0.06	3
Ca	0.021	0.1–1.0	0.005	0.02
Cd	0.011	0.1–1.0	0.006	0.03
Co	0.066	0.5–5.0	0.007	0.3
Cr	0.055	0.5–5.0	0.005	0.2
Cu	0.040	0.5–5.0	0.003	0.1
Fe	0.062	0.5–5.0	0.005	0.2
In	0.38	5–50	0.05	2
K	0.010	0.1–1.0	0.003	0.1
Li	0.017	0.1–1.0	0.002	0.08
Mg	0.003	0.05–0.50	0.0003	0.01
Mn	0.026	0.5–5.0	0.003	0.1
Na	0.003	0.05–0.50	0.0003	0.01
Ni	0.066	0.5–5.0	0.008	0.3
Pb	0.11	1–10	0.02	0.8
Rb	0.042	0.5–5.0	0.003	0.1
Sr	0.044	0.5–5.0	0.004	0.2
Tl	0.28	5–50	0.02	0.8
V	0.88	10–100	0.1	4
Zn	0.009	0.1–1.0	0.002	0.08

Source: Kneip, T.J. et al., *Health Lab. Sci.*, 12, 383, 1975.

TABLE 7.11

Detection Limits[a] for Various Elements by X-Ray Fluorescence

Ag, 1.2	Cs, 0.15	Nd, 0.30	Sr, 0.00007
Al, 5.0	Cu, 0.00002	Ni, 0.06	Tb, 159 mL^{-1}
As, 0.11	Eu, 0.66	P, 0.001	Te, 0.12
Au, 0.001 cm^{-2}	Fe, 0.0085	Pb, 0.0003	Th, 6.5 mL^{-1}
Ba, 0.12	Ga, 0.01	Rb, 0.0075	Ti, 0.001
Bi, 0.61	Hg, 0.24	Rh, 103 mL^{-1}	U (as UO$_2$), 0.72
Ca, 0.100	In, 1.1	Sc, 0.38	U, 0.00002
Cd, 0.40	K, 0.52	Se, 0.020 cm^{-2}	Y, 0.22
Ce, 0.17	La, 0.12	Si, 170 ppm	Yb, 6.8 mL^{-1}
Co, 0.05	Mn, 0.0015	Sm, 4.1 mL^{-1}	Zn, 0.00004
Cr, 0.00006	Mo, 0.072	Sn, 3.9 ppm	Zr, 0.00002

Source: Adapted from Braunstein, H.M. et al., Eds., Environmental health and control aspects of coal conversion: An information overview, Report ORNL-EIS-94, Oak Ridge National Laboratory, Oak Ridge, TN, 1977, pp. 5–13.

[a] In micrograms except as noted.

TABLE 7.12
Sensitivity of Various Elements to Neutron Activation Analysis

Sensitivity (g)	Elements
10^{-13}–10^{-12}	Dy, Eu
10^{-12}–10^{-11}	Au, In, Mn
10^{-11}–10^{-10}	Hf, Ho, Ir, La, Re, Rh, Sm, V
10^{-10}–10^{-9}	Ag, Al, As, Ba, Co, Cu, Er, Ga, Hg, Lu, Na, Pd, Pr, Sb, Sc, U, W, Yb
10^{-9}–10^{-8}	Cd, Ce, Cs, Gd, Ge, Mo, Nd, Os, Pt, Ru, Sr, Ta, Tb, Th, Tm
10^{-8}–10^{-7}	Bi, Ca, Cr, Mg, Ni, Rb, Se, Te, Ti, Tl, Zn, Zr
10^{-7}–10^{-6}	Pb
10^{-6}–10^{-5}	Fe

Source: Adapted from Braunstein, H.M. et al., Eds., Environmental health and control aspects of coal conversion: An information overview, Report ORNL-EIS-94, Oak Ridge National Laboratory, Oak Ridge, TN, 1977, pp. 5–13.

individual methods before any attempt is made to interpret the data. Nevertheless, the methods have found considerable use as means by which trace elements in coal can be identified as well as giving indications of the concentrations of the various elements.

7.7 CHEMISTRY OF ASH FORMATION

Ash-forming elements, that is, Al, Ca, Fe, K, Mg, Na, and Si, occur in fossil fuels or biofuels as internal or external mineral grains or simple salts such as NaCl or KCl, or are associated with the organic matrix of the fuel (Benson et al., 1993). In pulverized coal combustion, approximately 1% (w/w) of the inorganic metals is vaporized, while the rest remains in a condensed form as mineral inclusions (Flagan and Friedlander, 1978). Depending on the gas/particle temperature and local stoichiometry during coal particle heat-up, devolatilization, and char burnout, these mineral inclusions will undergo phase transformations and approach each other to form a fly ash fraction: the residual ash. The vaporized metal species may undergo several transformations: nucleation, subsequent coagulation, scavenging, heterogeneous condensation, and/or interactions with mineral inclusions in the burning char or residual fly ash particles. The extent of transformation depends on the total specific surface area of the residual fly ash particles, the cooling rate of the flue gas, the local stoichiometry, and the mixing in the gas phase.

When coal burns in air, as in the determination of ash in proximate analysis, all the organic material is oxidized or decomposed to give volatile products and the inorganic material associated with the coal is subjected to the combined effects of thermal decomposition and oxidation. As a result, the quantity and the composition of the resulting ash differ considerably from those of the inorganic materials originally associated with the pure coal substance.

It is therefore impossible to determine accurately the composition of the pure coal substance from the usual ultimate analysis simply by making allowance for the quantity of ash left behind as a residue when the coal is burned. Results obtained in this fashion are, as a consequence, quoted as being on the *dry, ash-free basis* and no claim is therefore made that these results do in fact represent the composition of the pure coal substance. If, however, it were possible to calculate accurately the quantity of mineral matter originally present in the coal sample, then by making due allowance for this material, the composition of the pure coal material could be deduced with reasonable precision and certainly with a greater accuracy than could be obtained by adopting the analytical figures calculated to a dry, ash-free basis.

The thermal and oxidation changes in which the mineral matter takes part during the combustion of coal are very complex but, fortunately, the majority of the reactions are fairly clearly understood (Table 7.9) and it has been possible to derive a formula which enables the mineral matter content of the original coal to be calculated from a knowledge of the quantity of ash produced on combustion, together with the quantities of pyrite sulfur, chlorine, carbonate present in the coal. The formula has been deduced from a consideration of the stoichiometric relationships existing between the reactants and products in reactions of the following types, all of which take place during combustion of coal and which together account for practically all the inorganic reactions involved in the process.

It is difficult to determine, either qualitatively or quantitatively, the mineral matter content of a coal from the high-temperature ash. During high-temperature ashing, as designated by various standards (usually 750°C [1382°F]), a series of reactions takes place involving the minerals in the coals. Of the major mineral groups, only quartz is not significantly altered during high-temperature ashing.

The clay minerals in coal contain water that is bound within their lattices. Kaolinite contains 13% bound water; illite contains 4.5% bound water; and montmorillonite contains 5% bound water. In addition, montmorillonite that occurs in mixed-layer clays also contains interlayer or adsorbed water. All the water is lost during the high-temperature ashing, for example,

$$Al_2O_3 \cdot 2SiO_2 \cdot xH_2O \rightarrow Al_2O_3 \cdot 2SiO_2 + xH_2O$$

During high-temperature ashing, pyrite and marcasite (FeS_2) are oxidized to ferric sulfate ($Fe_2(SO_4)_3$) and sulfur dioxide (SO_2). Some of the sulfur dioxide may remain in the ash in combination with calcium, but much is lost. If all of the possible sulfur dioxide is emitted during ashing, there would be a 33% loss in weight with respect to the weight of pyrite or of marcasite in the original sample:

$$4FeS_2 + 11O_2 \rightarrow 2Fe_2O_3 + 8SO_2$$

Calcium carbonate (calcite, $CaCO_3$) is calcined to lime (CaO) during high-temperature ashing and carbon dioxide is evolved, resulting in a 44% reduction in weight from the original calcite, for example,

$$CaCO_3 \rightarrow CaO + CO_2$$

Other metal carbonates behave similarly, that is, the oxides are formed during the ashing procedure.

The stable mineral quartz (silicon dioxide, silica, SiO_2) is the only major mineral found in coal that is inert during high-temperature ashing.

Sulfates in coal ash are derived from two sources: (1) sulfates (generally of calcium or magnesium) present as such in the coal sample and (2) sulfates formed by the absorption of sulfur oxides by the basic constituents of the ash during incineration of the coal and occurring mainly as calcium sulfate but also, to some extent, as alkali sulfates. The quantity of sulfates from the first source is usually very small and, in normal cases, can usually be neglected in comparison with those derived by absorption. Consequently, a determination of the sulfate in the ash enables a measure of the quantity formed during incineration to be established.

During combustion of the coal, organic chlorides are decomposed and liberate the chlorine atom as hydrochloric acid, while inorganic chlorides decompose with evolution of hydrochloric acid and ultimately leave a residue of the metallic oxide. Since approximately half the chlorine in coal occurs as inorganic chloride and the remainder as inorganic chloride, a correction for the chlorine present originally as inorganic material is easily applied when the total chlorine is known.

Except for oxygen and sulfur, elements that normally constitute the ash residues derived from coal combustion can arbitrarily be grouped as follows: (1) *major elements*, that is, elements in concentrations greater than 0.5% in the whole coal and these normally include aluminum, calcium, iron, and silicon; (2) *minor elements*, that is, those in the range of concentration of about 0.02 in the whole coal and these usually include potassium, magnesium, sodium, and titanium, and sometimes phosphorus, barium, strontium, boron, and others, depending on the geological area; and (3) *trace elements*, that is, all other inorganic elements usually detected in coal at less than 0.02% (200 ppm) down to parts per billion and below. Most nonmetallic elements, even though they are more volatile than metals, leave a detectable residue in coal ash.

7.8   EFFECT OF MINERAL MATTER IN COAL

Mineral matter in coal is one of the most important sources of problems in coal combustion, including fouling, slagging, corrosion, among others. Mineral matter transformation and slag formation are specific properties of coal that provide more information on the suitability for coal combustion or gasification.

The different states of iron (Fe^{2+} or Fe^{3+}) are the dominant reason for different sintering behaviors under different conditions. The iron-bearing minerals in ash, such as wustite (FeO), almandite ($3FeO \cdot Al_2O_3 \cdot 3SiO_2$) and fayalite (among others), are the most important factors influencing ash-sintering behavior because of the initial melting behavior during coal combustion.

Advanced industrial and utility power systems typically use direct fired gas turbine engines. Using coal to directly power a gas turbine has yet to be accomplished commercially, primarily because the ash causes erosion of the blades and deposition on the blades. If the combustion products contain a significant fraction of molten ash particles, deposition on the turbine blades occurs which blocks the flow path and degrades performance. If the ash particles are solid, erosion of the blades occurs which also degrades performance. In addition, mineral matter can cause corrosion of the blades. The size distribution, concentration, and composition of the ash, as well as the turbine design, determine the lifetime of the turbine blades.

Mineral matter in coal, during combustion, transforms into fly ash, and results in the buildup of ash deposits on heat transfer surfaces in PC-fired boilers. The ash formation process determines the ash character, that is, its particle size distribution and variation in chemistry. There are two models that can be used to represent ash formation from mineral matter: (1) coalescence of included mineral grains and (2) fragmentation of excluded mineral grains during combustion.

When coal is burned, most of the mineral matter and trace elements generally form ash; however, some minerals break down into gaseous compounds which go out the furnace's flue. Pyrite, for example, breaks down into the individual elements: iron and sulfur. Each element then combines with oxygen to become, respectively, iron oxide and sulfur dioxide. Iron oxide, a heavy solid, becomes part of the ash and sulfur dioxide is emitted as part of the flue gas. Some trace elements also dissociate from their organic or mineral hosts when coal is burned and follow separate paths. Most become part of the ash, but a few of the more volatile elements, such as mercury and selenium, may be emitted in the flue gas.

The mineral content of coal determines what kind of ash will be produced when it is burned. The fusion temperature (melting point) of the ash dictates the design of furnaces and boilers. In general, if the fusion temperature is relatively low, then the molten ash is collected at the bottom of the furnace as bottom ash, requiring one design; however, if the fusion temperature is relatively high, then the part of the ash that does not melt easily (*fly ash*) is blown through the furnace or boiler with the flue gas and is collected in giant filter bags, or electrostatic precipitators, at the bottom of the flue stack, requiring a different design.

Coal that is relatively rich in iron-bearing minerals (such as pyrite or siderite) has a low fusion temperature while coal relatively rich in aluminum-bearing minerals (such as kaolinite or illite) tends to have a high fusion temperature. If an electric generating or heating plant is designed to burn one type of coal, then it must continue to be supplied with a similar coal or undergo an extensive and costly redesign in order to adapt to a different type of coal. Similarly, furnaces designed to use coal that produces high amounts of heat will suffer severe losses in efficiency if they must accept coal that burns with substantially less heat.

Ash particulates are formed from mineral matter due to three different mechanisms. First, part of the mineral matter is found in layers or bands separate from the organic matter. This is called *adventitious mineral matter*, and it can be partially separated from the coal after crushing and fine grinding. The adventitious mineral matter is transformed directly to ash in the combustor, and the shape is semirounded. Second, mineral matter in coal is contained within the organic matrix in the form of chemically bound molecules and submicron crystals. Grinding does not liberate this intrinsic mineral matter. Third, some of the mineral matter is vaporized during combustion and condenses on cooler surfaces such as turbine blades.

Adventitious mineral matter is transformed directly to ash in the combustion zone. Depending on the temperature–time history, the ash particles will be spherical or semirounded. The size distribution of this ash depends on the size distribution of the adventitious mineral matter. Intrinsic mineral matter forms ash nodules in the pores of the char, and as char burnout proceeds, the ash nodules coalesce on the surface of the char.

Frequently, cenospheres (hollow glass-like spheres) are formed. The size distribution of this ash depends on the temperature–time history in the combustor. In PFBC tests, three hot cyclones yield an ash size distribution of 98% less than 10 μm and 80% less than 4 μm.

The ash loading and size distribution depend on (1) the extent of grinding and fuel cleaning and (2) the combustion time–temperature history. Regardless of the type of combustion system, much of the ash is in the 1–20 μm size range. Particles down to about 10 μm can be efficiently removed by hot cyclone collectors. Particles in the 1–10 μm range can be removed by other advanced methods; however, this tends to cause additional pressure drop and heat loss.

REFERENCES

Alpern, B., Nahuys, J., and Martinez, L. 1984. Mineral matter in ashy and non-washable coals: Its influence on chemical properties. *Community Service in Geology Portugal*, 70(2): 299–317.

Babu, S.P. (Ed.). 1975. *Trace Elements in Fuel*. Advances in Chemistry Series No. 141. American Chemical Society, Washington, DC.

Beer, J.M., Sarofim, A.F., and Barta, L.E. 1992. *Journal of the Institute of Energy*, 65(462): 55.

Benson, S.A., Jones, M.L., and Harb, J.N. 1993. Ash formation and deposition. In *Fundamentals of Coal Combustion—For Clean and Efficient Use*, L.D. Smoot (Ed.). Coal Science and Technology Series, Elsevier, Amsterdam, the Netherlands, Vol. 20.

Braunstein, H.M., Copenhaver, E.D., and Pfuderer, H.A. (Eds.). 1977. Environmental health and control aspects of coal conversion: An information overview. Report ORNL-EIS-94. Oak Ridge National Laboratory, Oak Ridge, TN.

Bredenberg, J.B-son, Huuska, M., and Vuori, A. 1987. In *Coal Science and Chemistry*, A. Volborth (Ed.). Elsevier, Amsterdam, the Netherlands, p. 1.

Cecil, C.B., Stanton, R.W., Dulong, F.T., and Renton, J.J. 1982. In *Atomic and Nuclear methods in Fossil Energy Research*, R.H. Filby (Ed.). Plenum Press, New York, p. 323.

Davidson, R.M. 2000. Modes of occurrence of trace elements in coal. Report CCC/36. International Energy Agency Coal Research, London, U.K.

Davidson, R.M. and Clarke, L.B. 1996. *Trace Elements in Coal*. International Energy Agency Coal Research, London, U.K.

Finkelman, R.B. 1982. In *Atomic and Nuclear methods in Fossil Energy Research*, R.H. Filby (Ed.). Plenum Press, New York, p. 141.

Finkelman, R.B. 1993. Trace and minor elements in coal. In *Organic Geochemistry*, M.H. Engel and S.A. Macko (Eds.). Plenum Press, New York, pp. 593–607.

Flagan, R.C. and Friedlander, S.K. 1978. Particle formation in pulverized coal combustion—A review. In *Recent Developments in Aerosol Science*, D.T. Shaw (Ed.). John Wiley & Sons, New York.

Francis, W. 1961. *Coal: Its Formation and Composition*. Edward Arnold Ltd., London, U.K.

Given, P.H. and Yarzab, R.F. 1978. In *Analytical Methods for Coal and Coal Products*, C. Kar, Jr. (Ed.). Academic Press, New York, Vol. II, Chapter 20.

Gluskoter, H.J. 1965. *Fuel*, 44: 285–291.

Gluskoter, H.J. 1975. Mineral matter and trace elements in coal. In *Trace Elements in Fuel*, S.P. Babu (Ed.). Advances in Chemistry Series No. 141. American Chemical Society, Washington, DC, Chapter 1.

Gluskoter, H.J., Ruch, R.R., Miller, W.G., Cahill, R.A., Dreher, E.C.B., and Kuhn, J.K. 1977. Trace Elements in coal: Occurrence and distribution. Circular No. 499. Illinois State Geological Survey, Urbana, IL.

Hjalmarsson, A.-K. 1992. Interactions in emissions control for coal-fired plants. Report No. IEACR/47. IEA Coal Research, International Energy Agency, London, U.K.

Huggins, F.E., Huffman, G.P., and Lee, R.J. 1982. In *Coal and Coal Products: Analytical Characterization Techniques*, E.L. Fuller, Jr. (Ed.). Symposium Series No. 205. American Chemical Society, Washington, DC, Chapter 12.

Jenkins, R.G. and Walker, P.L. Jr. 1978. In *Analytical Methods for Coal and Coal Products*, C. Kar, Jr. (Ed.). Academic Press Inc., New York, Vol. II, Chapter 26.

Jones, M.L., Kalmanovitch, D.P., Steadman, E.N., Zygarlicke, C.J., and Benson, S.A. 1992. In *Advances in Coal Spectroscopy*, H.L.C. Meuzelaar (Ed.). Plenum Press, New York, Chapter 1.

Kelly, E.G. and Spottiswood, D.J. 1982. *Introduction to Mineral Processing*. John Wiley & Sons, Inc., New York.

Kelly, D.G. and Spottiswood, D.J. 1989. The Theory of Electrostatic Separations: A Review, Part I. Fundamentals, Part II. Particle Charging, Part III. The Separation of Particles. *Minerals Engineering*, Vol. 2.

Kneip, T.J., Ajemian, R.S., Driscoll, J.N., Grundner, F.I., Kornreich, L., Loveland, J.W., Moyers, J.L., and Thompson, R.L. 1975. *Health Laboratory Science*, 12: 383.

Mackowsky, M.Th. 1968. In *Coal and Coal-Bearing Strata*, D. Murchison and T.S. Westoll (Eds.). Elsevier, New York, Chapter 13.

Miller, R.N. 1984. In *The Methodology of Low-Temperature Ashing. Interlaboratory Comparison of Mineral Constituents in a Sample from the Herrin (No. 6) Coal Bed from Illinois*, R.B. Finkelman, F.L. Fiene, R.N Miller, and F.O. Simon (Eds.). U.S. Geological Survey Circular No. 932, U.S. Geological Survey, Washington, DC, pp. 9–15.

Murchison, D. and Westoll, T.S. (Eds.). 1968. *Coal and Coal-Bearing Strata*. Elsevier, New York.

Nadkarni, R.A. 1982. In *Coal and Coal Products: Characterization Techniques*, E.L. Fuller Jr. (Ed.), Symposium Series No. 205, American Chemical Society, Washington, DC, Chapter 6.

O'Gorman, J.V. and Walker, P.L. 1971. Mineral characteristics of some American coals. *Fuel*, 50: 135.

Prather, J.W., Guin, J.A., and Tarrer, A.R. 1979. In *Analytical Methods for Coal and Coal Products*, C. Karr, Jr. (Ed.). Academic Press Inc., New York, Vol. III, Chapter 49.

Quast, K.B. and Readett, D.J. 1991. In *Coal Science II*, H.H. Schobert, K.D. Bartle, and L.J. Lynch (Eds.). Symposium Series No. 461. American Chemical Society, Washington, DC, Chapter 3.

Raymond, R. Jr. and Gooley, R. 1979. In *Analytical Methods for Coal and Coal Products*, C. Karr, Jr. (Ed.). Academic Press Inc., New York, Vol. III, Chapter 48.

Rees, O.W. 1966. Chemistry, uses and limitations of coal analyses. Report of Investigations No. 220. Illinois State Geological Survey, Urbana, IL.

Roscoe, B.A. and Hopke, P.K. 1982. In *Atomic and Nuclear Methods in Fossil Energy Research*, R.H. Filby (Ed.). Plenum Press, New York, p. 163.

Ruch, R.R., Gluskoter, H.J., and Shimp, N.F. 1974. *Occurrence and Distribution of Potentially Volatile Trace Elements in Coals*, Environmental Geology Note No. 72. Illinois State Geological Survey, Urbana, IL.

Russell, S.J. and Rimmer, S.M. 1979. In *Analytical Methods for Coal and Coal Products*, C. Karr, Jr. (Ed.). Academic Press Inc., New York, Vol. III, Chapter 42.

Shipley, D.E. 1962. Monthly Bulletin. British Coal Utilization Research Association. 26: 3.

Spears, D.A. 2005. A review of chlorine and bromine in some United Kingdom coals. *International Journal of Coal Geology*, 64: 257–265.

Spears, D.A. and Caswell, S.A. 1986. Mineral matter in coals: Cleat minerals and their origin in some coals from the English Midlands. *International Journal of Coal Geology*, 6: 107–125.

Spears, D.A. and Zheng, Y. 1999. Geochemistry and origin of some elements in UK coals. *International Journal of Coal Geology*, 38: 161–179.

Stach, E., Taylor, G.H., Mackowsky, M.-Th., Chandra, D., Teichmuller, M., and Teichmuller, R. 1982. *Textbook of Coal Petrology*, 3rd edn. Gebruder Bornfraeger, Stuttgart, Germany.

Stevenson, F.J. and Butler, J.H.A. 1969. In *Organic Geochemistry: Methods and Results*, G. Eglinton and M.T.J. Murphy (Eds.). Springer-Verlag, New York, Chapter 22.

Straszheim, W.E. and Markuszewski, R. 1991. In *Coal Science II*, H.H. Schobert, K.D. Bartle, and L.J. Lynch (Eds.). Symposium Series No. 461. American Chemical Society, Washington, DC, Chapter 3.

Warne, S.St.J. 1979. In *Analytical Methods for Coal and Coal Products*, C. Karr, Jr. (Ed.). Academic Press Inc., New York, Vol. III, Chapter 52.

Weaver, J.N. 1978. In *Analytical Methods for Coal and Coal Products*, C. Kar, Jr. (Ed.). Academic Press Inc., New York, Vol. I, Chapter 11.

8 Coal Analysis

8.1 INTRODUCTION

Although coal is an extremely complex material, the rapidly expanding use of coal throughout the nineteenth century and the early part of the twentieth century necessitated the design of acceptable methods for coal analysis with the goal of correlating fuel composition and properties with behavior (Campbell and Gibb, 1951; Baughman, 1978; Montgomery, 1978; Speight, 1990, 2005; Sen et al., 2009).

The primary reason for analyzing coal is to determine whether it will meet the needs of a specific application, or to characterize the general quality of the coal for future reference. For instance, coal may be analyzed to determine how much sulfur (or other element) is present, its form, and how it is distributed. If the sulfur is present in discrete pyrite grains, then much of it may be cleaned out of the coal; but, if it is organically bound, then the sulfur may be released only by burning, by using a solvent, or by employing a bacteriologic technique (although the latter two techniques are still largely experimental). If the coal has a high organic sulfur content, then it may have to be mixed or blended with a coal of lower sulfur content in order to meet sulfur emissions standards, or the sulfur may have to be cleaned out of the flue gas by FGD, which is an expensive procedure. Similarly, analysis may determine whether a trace element, such as arsenic, may be eliminated from a coal by washing or whether it must be trapped in the flue gas. Finally, in extreme cases, analysis may determine that the coal cannot be used.

In general, the coal-mining and coal-consuming industries determine coal quality for immediate or near-term use, whereas federal and state governments analyze coal (1) to characterize large areas of unmined coal (resources) for future reference, (2) to support policy decisions related to future coal use, and (3) to supply federal and state regulatory agencies with information about the quality or special characteristics of future coal supplies, such as the presence, amounts, and modes of occurrence of sulfur and potentially toxic trace elements (Table 8.1).

Coal sampling is performed whenever there is a need to determine or verify the analysis or content of some constituent in the coal and sampling the coal may be done either manually or automatically. In either case, the method of sampling must provide a representative sample of a relatively large amount of coal without bias. If moisture content determinations are to be made, the sample must be collected in a container which can be immediately sealed following the sample collection. Such endeavors were also necessary as coal performance became a more relevant factor to the users and economical aspects began to play a larger role in determining the position of coal as petroleum and natural gas offered strong competition.

Coal samples may be received at the laboratory in a variety of conditions. Accordingly, analytical results may be reported in a variety of ways, depending on the condition of the samples when received and the use of the analysis. Some samples arrive in a fresh condition, not long after sampling; these may be analyzed without further processing, except grinding and mixing. The analytical results for these samples are reported on an "as-received" basis. Some samples arrive dried out due to long storage, transportation over long distances, or mishandling. The results of analyses on dried-out samples are reported on a "dry" basis. The opposite conditions may also occur if the sample arrives wet because of excess moisture in the sample that may or may not be present in the coal bed; in this case, the analyses may be reported on a "moist" basis.

Analyses may also be reported on a "mineral matter–free" or "dry, mineral matter–free" basis for use in calculating coal rank because rank is a function only of the maturity of the organic matter. "Mineral matter–free" means that the amount of mineral matter in the sample has been subtracted from

TABLE 8.1

Sampling and Analytical Methods Used for Coal Analysis

Test/Property	Results/Comments
Sample information	
Sample history	Sampling date, sample type, sample origin (min, location)
Sampling protocols	Assurance that sample represents gross consignment
Chemical properties	
Proximate analysis	Determination of the "approximate" overall composition, i.e., moisture, volatile matter, ash, and fixed carbon content
Ultimate analysis	Absolute measurement of the elemental composition, i.e., carbon, hydrogen, sulfur, nitrogen, and oxygen content
Sulfur forms	Chemically bonded sulfur: organic, sulfide, or sulfate
Ash properties	
Elemental analysis	Major elements
Mineralogical analysis	Analysis of the mineral content
Trace element analysis	Analysis of trace elements; some enrichment in ash
Ash fusibility	Qualitative observation of temperature at which ash passes through defined stages of fusing and flow

the total analytical results to provide only the amount that is organic. *Dry, mineral matter–free* means that the sample was received in a dry or nearly dry state, or was dried out before an analysis was made.

An analysis on a mineral matter–free basis may also be used for comparison with the results of an as-received analysis of "ROM" coal (as it comes from the mine) if there is to be further processing, such as coal cleaning. ROM coal frequently contains (1) large amounts of mineral matter from partings in the coal bed, (2) rocks from above and below the coal bed that are mined along with the coal, (3) minerals that fill veins in the coal bed, and (4) mineral matter dispersed in the coal (such as pyrite grains). Mineral matter in ROM coal may cause the coal to exceed the ash and sulfur limits set by contract requirements and environmental regulations. A mineral matter–free analysis may indicate whether or not it will be worthwhile to clean the coal; however, the need for cleaning depends on the intended use of the coal and normal cleaning practices may not bring a coal to a completely mineral-free condition.

Most chemical analyses require the destruction of the coal sample in order to determine calorific value, volatile matter, and mineral matter content. Destruction is achieved by burning the coal sample in a high-temperature furnace. Although that procedure is fine for determining the calorific value of coal, it is not suitable for retention of some of its volatile inorganic elements, such as mercury. Subjecting the sample to a process called "low-temperature ashing," in which the sample is exposed to high-frequency radiation in a high-oxygen atmosphere, permits the organic matter to be driven off while preserving most of the mineral matter. Nevertheless, some of the very volatile inorganic elements (such as chlorine, mercury, or selenium) may be lost in this process and are best measured in whole-coal samples by x-ray fluorescence or other techniques. Most microscopic and microprobe techniques require whole-coal samples because the objective is to determine the interrelations between the constituent parts of the coal.

Assessment of any traits in coal properties is vital to ensuring a particular supply of coal is used in the most effective way. Thus, the data obtained from coal analyses (Smith and Smoot, 1990) are valuable not only for laboratory work but also, and perhaps more importantly, to establish the price of the coal by allocation of production costs as well as to control mine and cleaning operations and to determine plant efficiency.

In process-oriented terms, the goal of coal utilization is also to take a new concept from a small (laboratory or bench) scale to the pilot scale, thence to a demonstration scale unit, and finally to commercialization. It is only by assiduously following each step in this path by careful analyses of the coal feedstocks and the products that such goals can be achieved.

Some of the principal variables in terms of which coal quality is expressed are measured by empirical tests that do not directly measure the variables themselves but the behavior or properties of coal under standard conditions. Standardization of the conditions is essential for obtaining results that are comparable within any one laboratory and between different laboratories.

Much work, and the formation of various national standards associations, has led to the development of methods for coal evaluation. For example, the ASTM has carried out uninterrupted work in this field for many years while investigations on the development of the standardization of methods for coal evaluation has occurred in all of the major coal-producing countries (Table 8.2) (Montgomery, 1978; for coke analyses, see Patrick and Wilkinson, 1978).

As a noteworthy point, there are (in addition to the ASTM) other organizations for methods development and standardization which operate on a national level; examples are the British Standards Organization (BS) and the German Standards Organization (DIN). Furthermore, the increased trade between various coal-producing countries that followed World War II meant that cross-referencing of the already accepted standards was a necessity and the mandate for such work fell to the International Standards Organization (ISO), located in Geneva, Switzerland; membership in this organization is allocated to participating (and observer) countries.

It is also appropriate that in any discussion of the particular methods used to evaluate coal for coal products, reference should be made to the relevant test. Accordingly, the necessary ASTM test numbers have been included as well as those, where known, of the test numbers from the standards organizations of other countries. As a part of the multifaceted program of coal evaluation, new methods are continually being developed and the already-accepted methods may need regular modification to increase the accuracy of the technique as well as the precision of the results.

TABLE 8.2
Procedures and Purposes of American Society for Testing and Materials (ASTM) Methods for Testing Coal

Procedure	Purpose
Classification of coal by rank	Classify coal into categories indicating ranges of physical and chemical characteristics that are useful in making broad estimates of the behavior of coal in mining, preparation, and use
Coal ash	Determine amount of ash at a given temperature
Equilibrium moisture	Determine moisture-holding capacity of coal
Forms of sulfur	Determine whether sulfur is in pyrite, organically bound, or in sulfates
Free swelling index	Determine how well a coal will form coke
Fusibility of ash	Predict whether the coal ash will perform properly in the process for which it was chosen
Gross calorific value	Determine potential for energy production, Btu/lb
Hardgrove grindability	Determine resistance to grinding
Major and minor elements	Identify major and minor elements
Proximate analysis	Determine amount of ash, fixed carbon, moisture, and volatile matter
Reflectance of organic matter	Determine rank and how well a coal will form coke
Total moisture	Determine inherent water and any other water present
Trace elements	Identify trace elements
Ultimate analysis	Determine amount of ash, carbon, hydrogen, nitrogen, oxygen, and sulfur
Volatile matter	Identify products given off as gases or vapors
Maceral analysis	Determine kinds and amounts of macerals in coal

A complete discussion of the large number of tests that are used for the evaluation of coal (and coal products) would fill a considerable volume, indeed several volumes (see, e.g., Ode, 1963; Karr, 1978, 1979; Montgomery, 1978; Zimmerman, 1979; Gluskoter et al., 1981; Smith and Smoot, 1990; Speight, 2005). It does, however, seem appropriate that any text relating to the chemistry and technology of coal should present some indication of the methods available for the evaluation, and therefore, the scope of this chapter is limited to tests in common use.

At this point, it is advisable to note the differences which are inherent in the terms *accuracy* and *precision*.

The term *accuracy* is used to indicate the reliability of a measurement, or an observation; but it is, more specifically, a measure of the closeness of agreement between an experimental result and the true value.

On the other hand, the term *precision* indicates a measure of the degree to which replicate data and/or measurements conform to each other. Thus, it is possible that data can be very precise without necessarily being correct or accurate. These terms will be found throughout any text which is devoted to a description of standard methods of analysis and/or testing and have been (incorrectly) used interchangeably.

8.2   SAMPLING

Most analyses of coal for both standard and research purposes are conducted on carefully collected samples of whole coal. Coal samples are taken from a variety of places—individual coal beds in place, coal-mine conveyor systems, trucks, train cars, or stockpiles—depending on the needs of the sampling and analytical programs. The goal of this process is to collect a sample that will be as representative as possible of the coal bed or other source from which it is taken. In order to produce reliable results from testing programs, sampling must be done very carefully. Samples must be taken without contamination from extraneous material, and location and orientation must be documented.

There are also a few techniques for analyzing coal beds in place. Geophysical logging is conducted in drill holes during exploration and mine planning programs. This technique measures electrical resistivity, transmissivity of sound, inherent electrical properties (such as self-potential), and the reaction of coal to bombardment by atomic particles (such as neutrons) (Wood et al., 1983). These methods accurately identify coal beds and their boundaries, and help determine whether a coal bed merits additional exploration and mining; however, they are only good for an approximate analysis of coal quality.

Channel and drill-hole core samples are two of the most common types of samples used for characterizing a coal bed. A channel sample is taken from a freshly exposed coal bed (usually in an underground or surface coal mine) by cutting a 2–4 in. channel into the bed from top to bottom and collecting all the coal from the channel. Drill-hole core samples are taken from subsurface coal beds where the coring bit does the cutting. If the coal bed is divided horizontally into benches (individual beds of coal separated by mineral partings), or if the bed is layered by distinct types of coal, then the benches or beds may be sampled separately, which allows for a more detailed characterization of the individual parts. The ASTM recently has published standards for the collection of channel and core coal samples (ASTM International, 2011ag).

The number of channel or core samples taken depends on how well the coal bed needs to be characterized. If any two samples of the same bed show considerable variation in quality between them, then more samples may be taken from the area between the two initial samples; however, cost may be a factor in the decision to collect more samples.

The purpose of the analysis often determines the type of sample that is collected. If the purpose is to characterize a stock of already mined coal, then gross samples may be taken at regularly or randomly spaced intervals from the stockpile, coal car, or conveyor belt. If the purpose is to examine some specific property of a coal bed, then a sample of only a small part of the bed may be taken. If the purpose of the sample is to study some directionally controlled characteristic, such as the orientation of mineral grains, the sample must be labeled to show top and bottom and, in some cases, compass directions. For some types of analyses (e.g., by either petrographic or scanning electron microscope), small blocks of coal are cut out of larger samples and one side is polished and examined. This technique is used when it is necessary to see the microscopic details of the types, distributions, and arrangements of the macerals and minerals in a coal sample.

When a coal sample arrives at the lab, it is crushed (unless it is to be examined in block form) and its particles are thoroughly mixed to ensure that the sample is homogeneous. The crushed sample then is divided into appropriately sized subsamples for each analytical technique to be used.

Generally, there are two methods of coal sampling: (1) sampling in the mine, often referred to as *in situ sampling* or *seam sampling*, and (2) *ex situ sampling* which occurs after the coal has been mined. In situ sample is often carried out by geologists who are investigating the nature of the coal before development of the coal seam by mining methods.

8.2.1   IN SITU SAMPLING

Channel sampling is one of the best methods for sampling in-seam coal. When the coal sample is collected from an outcrop, the exposed area should be cleaned to avoid the weathered exposed coal surface. Normally, a small box cut is made at the coal outcrop exposing the entire thickness of the coal seam. For a relatively thin seam, only one coal section is recommended. However, if the seam is thick, two or more coal sections may be necessary to sample entire seam.

The channel sample should be cut perpendicular to the bedding plane and stored in plastic bags. Sampling of the full seam provides overall quality of the seam including all boney and mineral matters within coal. Sometimes, a more detailed analysis is required where the coal seams are collected separately from bone and rock partings. However, the bone and rock partings are collected separately. Then the position and thickness of the coal–bone–rock samples are recorded and sent to the lab for detailed analysis (*channel ply sampling*). Many times during channel ply sampling, a small portion of the roof and floor rock of the coal seam are also included in the coal analysis to allow *out-of-seam-dilution*.

In some cases, the strength of the coal becomes important mainly in the underground mining such as from the pillars when the seam has been mined using the *room-and-pillar mining* method. A large block of undisturbed coal is usually sampled (*pillar sampling*) from some specific areas of potential problems or areas with known problems; the sampling scheme is similar to the channel sampling method.

Core sampling is mainly a part of the exploration and reserve evaluation stage. This is however very important for the development of a future mine. A geologist is usually assigned to supervise a drilling program. Coal samples are collected in wooden box carefully in the field if not sampled at the field. Most of the time, an e-log is prepared for each completed hole in recent time. A geologist checks the e-log for the coal thickness and adjusts the "core recovery" for the collected coal seam.

Core recovery is usually >90% for the coal seams; however, a poor core recovery is possible if the driller is not very much experienced with coal drilling. The sampling method is similar to channel ply sampling. On the logging sheet, the depths of each ply sample intervals are recorded before sending to the lab. Total seam, total coal, and the % of coal recovery are also recorded on the sampling tags as a note to the coal analytical lab.

Cutting sampling (*chip sampling*) is a much less accurate sampling scheme than the core sampling. Cuttings are generated by rotary-type drilling where no core is recovered except chips. Air flush or mud-flush rotary drilling is a much faster drilling and mostly used for gas wells. This kind of sampling can only give a very general analysis of the coal. It is very difficult to collect samples and most of the time we have lots of impurities mixed in it. Also, the exact depth of coal cannot be accurately recorded unless generated from a geophysical log after drilling is completed.

Bulk samples are collected mainly for larger-scale tests, to check swelling the properties of various coal seams, to rank coal as *high pressure coal* and *low pressure coal*.

8.2.2 Ex Situ Sampling

Ex situ sampling (*non-in situ sampling*) is the method (or methods) of sampling coal from a stock-pile, a coal train (or other means of transportation), or at the time of entry into the preparation plant, or power plant.

Such methods are often not representative of the coal seam that is mined. The coal may be blended with *out-of-seam* products from the roof and floor strata or even with coal from two or more seams to meet certain quality standard specified by the client.

The basic purpose of collecting and preparing a sample of coal is to provide a test sample which when analyzed will provide the test results representative of the lot sampled. In order that the sample represents the coal from which it is taken, it is collected by taking a definite number of increments distributed throughout the whole volume of coal.

The procedure for sampling will, however, differ with the purpose and method of sampling. Samples may be required for technical evaluation, process control, quality control, or for commercial transactions. For quality assessment of coals from new sources, samples are to be drawn from in situ coal seams, either as rectangular blocks or pillars cut from full seam height, or from seam channels or from borehole cores.

The major consumers of indigenous coals belong to the core sectors such as steel, power, cement, chemicals, and domestic sectors. Quality monitoring of coal is an important activity for any commercial transactions between the consumers and the producers. As already mentioned, the method of sampling for quality monitoring differs and is governed by many factors. The sampling procedure will depend mainly on the nature of sample collection, that is, by mechanical or manual means, from moving belt or from stationary lots such as wagons and stockpiles. Normally, any sampling scheme is supposed to conform to relevant national or international standards. However, due to technical, cost, and time constraints, very often some modifications are made in the method of sampling jointly by the seller and the purchaser. It is a known fact that about 80% of the total variances involved at the different stages of sample collection, preparation, and analysis comes from errors during its collection only.

Thus, in order to test any particular coal, it is necessary (1) to obtain a sample of the coal and (2) to endure that the sample is representative of the bulk material. Thus, sampling is, by convention, the operation of removing a portion (the sample) from the greater bulk (the whole) of the material. However, the removal of the sample must be such that it has the same qualities (properties) as the bulk (Visman, 1969; Gould and Visman, 1981; Rose, 1992). In addition, there must be a clear understanding of the methods which constitute good (and correct) sampling practice. In more general terms, the effectiveness of a sampling method is the degree to which the composition and properties of the bulk coal is sampled. The fundamental requirements of sampling are as follows:

- All particles of coal in the lot to be sampled are accessible to the sampling equipment and each individual particle shall have an equal probability of being selected and included in the sample.
- The dimension of the sampling device used should be sufficient to allow the largest particle to pass freely into it.
- The first stage of sampling known as primary increments is the collection of an adequate number of coal portions from positions distributed over the entire lot to take care of the variability of the coal. The primary increments are then combined into a sample, as taken or after reducing the mass of the sample to a manageable size. From this gross sample, the required number and types of test samples are prepared by a series of processes jointly known as sample preparation.
- The minimum mass of the gross sample should be sufficient to enable particles to be present in the same proportions as in the lot of coal from which it is taken.
- To ensure that the result obtained has the *required precision*, the following issues are to be considered: (1) variability of coal, (2) number of samples from a lot, (3) number of increments comprising each sample, and (4) mass of sample relative to the nominal top size.

While drawing increments, great care should be taken to avoid the occurrence of bias in the results. The ideal method of sampling is the *stopped belt method*, which is considered free of bias. As implementation of such a method will affect the continuity of plant operations, it is not always practicable for routine sampling. However, any mechanical sampling device needs to be checked for bias by comparing with the results from the stopped belt reference method. In fact, the method of *stopped belt* sampling is often implemented to standardize any other mechanical automatic sampling systems.

Detailed documented procedures are laid down in various national and international standards for executing the job of representative sampling pertaining to different methods of execution.

Briefly, a *grab sample* is a one-off sample of the coal at a point in the process stream, and tends not to be very representative. A *routine sample* is taken at a set frequency, either over a period of time or per shipment. In fact, coal sampling consists of several types of sampling devices. A *crosscut* sampler to mimic the *stop belt* sample—a crosscut sampler mounts directly on top of the conveyor belt. The *falling stream* sampler is placed at the head section of the belt.

Joint sampling is carried out at the loading end by the representatives of the producer and the customer, following a methodology mutually agreed upon by both parties. Depending on the nature of the agreement, the loading point results can be taken exclusively for commercial transactions. In some cases, the mean value of the results of joint sampling at both the loading and unloading ends is considered. The tolerance values in the quality parameters are often defined, beyond which several bonus/penalty clauses are imposed but the tolerance value identified is compatible with the sampling scheme. Whether or not the tolerance value lies within the precision limit can be achieved by the implementation of a particular sampling scheme involving periodic testing.

There are several points in coal operations where samples are taken: (1) the raw coal, (2) the refuse, to determine what the plant missed, and (3) the clean coal, to determine exactly what is being used, sold, or shipped. The sampler is set according to *tons per hour, feet per minute*, and *top size of the product* on the actual belt. A sample is taken, crushed, and sent to a laboratory for testing where the results will be shared with the buyer as well as the supplier. The buyer in many cases will also sample the coal again once it is received to *double check* the results.

For homogeneous materials, sampling protocols are relatively simple and straightforward, although caution is always advised; but the heterogeneous nature of coal complicates sampling procedures. In fact, apart from variations in rank (Chapter 2) most coals are visibly heterogeneous (Chapters 4 and 7) and there is strong emphasis on need to obtain representative samples for testing and analysis (Gould and Visman, 1981).

When a property of coal (which exists as a large volume of material) is to be measured, there will almost always ("always" being a term like "never") be differences between the data from a gross lot or gross consignment and the data from the sample lot. This difference, called the sampling error, has a frequency distribution with a mean value and a variance. Variance is a statistical term defined as the mean square of errors; the square root of the variance is more generally known as the *standard deviation* or *standard error* of sampling.

Furthermore, recognition of the issues involved in obtaining representative samples of coal has resulted in the designation of methods which dictate the correct manner for the sampling of coal (ASTM, 2011b,f,h; ISO, 2011w,x). The number of riffling stages required to prepare the final sample depends on the size of the original gross lot. However, it is possible by use of these methods to reduce an extremely large consignment (that may be on the order of several thousand pounds) to a representative sample that can be employed as a laboratory test sample.

Every sampling operation consists of either extracting one sample from a given quantity of material or extracting from different parts of the lot a series of small portions or increments that are combined into one gross sample without prior analysis; the latter method is known as sampling by increments. The precision of sampling improves with the size of each of the increments collected and with the number of increments included in a gross sample, and manual sampling involves the

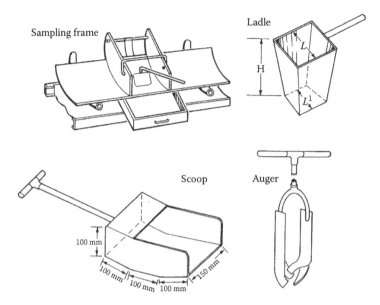

FIGURE 8.1   Sampling tools. (From Meyers, R.A., Ed., *Coal Handbook*, Marcel Dekker, New York, 1981.)

principle of ideal sampling (that every particle in the entire mass to be sampled has an equal opportunity to be included in the sample): (1) the dimensions of the sampling device and (2) proper use of the sampling device.

On a technical note, the opening of the sampling device must be 2.5–3 times the top size of the coal to meet requirements (ASTM, 2011h) and the ISO has established design criteria for several types of hand tools (Figure 8.1) that can be used for manual sampling. One consideration is that the device be able to hold the minimum increment weight specified without overflowing.

Stream sampling and flow sampling are terms usually reserved for the collection of sample increments from a free-falling stream of coal as opposed to the collection of increments from a stopped conveyor belt. Coal that passes from one belt to another at an angle tends to become segregated, with a predominance of coarse particles on one side and a predominance of fine particles on the other side. There are also situations where coal must be sampled when there is no motion (sampling at rest) and it may be difficult, if not impossible, to ensure that the sample is truly representative of the gross consignment.

An example of coal being sampled at rest is when samples are taken from railcars (car-top sampling), and caution is advised both in terms of the actual procedure and in the interpretation of the data. Again, some degree of segregation can occur as the coal is loaded into hopper cars. In addition, heavy rainfall can cause the moisture content of the coal to be much higher at the top and sides of the railcar than at the bottom. Similarly, the onset of freezing conditions can also cause segregation of the moisture content.

Thus, in car-top sampling, only the coal near the surface (as the name implies) has an opportunity to be included in the sample. Alleviation of this problem can be achieved by spacing the increments (uniformly and systematically) throughout the entire consignment. An alternate operation is to sample the coal as it is discharged from the bottom of hopper cars and, since the coal is sampled in motion, bottom sampling is considered to be an improvement over car-top sampling.

Taking samples of coal from sampling storage piles also can raise problems. For example, in conical-shaped piles segregation effects result in fines predominating in the central core as well as a gradation of sizes down the sides of the pile from generally fine material at the top of the pile to coarser coal at the base of the pile. If at all possible, coal piles should be moved to be sampled which, in turn, will determine how the coal is sampled. Where it is not possible to move a pile, there is no choice but to sample it "as is" and the sampling regime usually involves incremental spacing of the samples over the entire surface. Alternatively, the sampling of large coal piles can be achieved by core drilling or augering equipment or the coal can be exposed at various depths and locations (by means of heavy equipment such as a bulldozer) so that manual sampling can be performed.

There are a wide variety of devices that are available for machine sampling (mechanical sampling) and include flow-through cutters, bucket cutters, reciprocating hoppers, augers, slotted belts, fixed-position pipes, and rotating spoons (Figures 8.2 through 8.4). A major advantage of these systems is that they sample coal from a moving stream.

Sample preparation (ASTM, 2011f) includes drying (in air), as well as crushing, dividing, and mixing a gross sample to obtain an unbiased analysis sample. However, the procedure is usually accompanied by loss of moisture unless the sample increments are weighed as they are collected and then air-dried and reweighed before crushing and dividing.

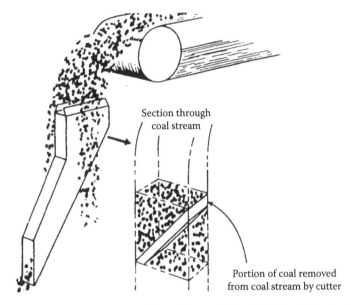

FIGURE 8.2 A cross-stream primary cutter. (From Meyers, R.A., Ed., *Coal Handbook*, Marcel Dekker, New York, 1981.)

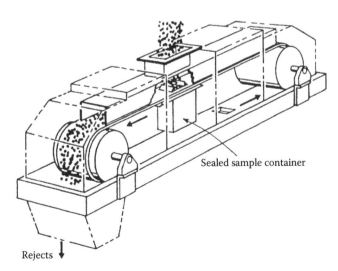

FIGURE 8.3 A slotted-belt secondary cutter. (From Meyers, R.A., Ed., *Coal Handbook*, Marcel Dekker, New York, 1981.)

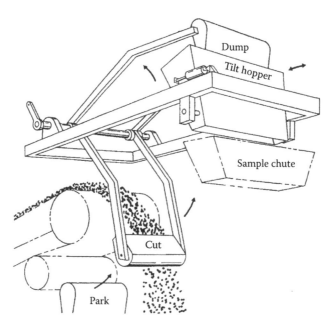

FIGURE 8.4 A swing-arm primary cutter. (From Meyers, R.A., Ed., *Coal Handbook*, Marcel Dekker, New York, 1981.)

Sampling plays a role in all aspects of coal technology. For example, one important objective for which there are no known specific standard methods relates to exploration, and sampling, of coal reserves as they exist in the ground. More specifically, the issues relate not only to determining the extent of the coal resource but also to the quality of the coal so that the amount may be determined (see, e.g., Averitt, 1981). This is achieved, for the most part, by core drilling but along outcrops, coal is sometimes sampled by channel sampling techniques.

The performance of coal preparation plants (see, e.g., Deurbrouck and Hucko, 1981) as well as the testing and routine quality control in mining operations and preparation plants requires sampling coal both in situ and at various stages of processing following removal from the seam. Monitoring of preparation plant performance, however, can be quite complex insofar as the feedstock to the plant may be taken from several streams with widely different sampling protocols. Such sampling may involve various manual sampling techniques.

The standard methods of sampling (ASTM, 2011h) usually apply to coal sales in which the purpose of the sampling is actually a method for valuation as a determinant of price or conformance with specifications (Janus and Shirley, 1973, 1975).

Coal sales are carried on in consignments or lots that are clearly definable as distinct units and can be sampled as such, but the size of such lots varies widely, from single truckloads to shiploads of $30–50 \times 10^3$ tons ($30–50 \times 10^6$ kg). The methods employed stipulate the minimum weight and minimum number of increments required, together with precautions, increment types, conditions of increment collection, and increment spacing.

Optimization of coal combustion in power plants and processes is a function of the many variable constituents of coal. Thus, it is not surprising, perhaps even anticipated, that sampling is conducted to determine efficiency, heat inputs, and operating needs.

The effect of fines content on the combustion of pulverized coal is quite dramatic (Field et al., 1967; Essenhigh, 1981), and the problems associated with collection of an unbiased sample of pulverized coal is a special situation (ASTM, 2011a). Operating samples are often collected from the feedstocks to power plant boilers on a shift or daily basis for calculation of heat balances and operating efficiencies. Another objective of operating samples is to document compliance with air pollution emission regulations based on fuel composition.

In summary, coal is an extremely complex material and there are methods which describe, in considerable detail, the process by which coal may be sampled for testing. Strict adherence to the specified methods are required if sampling of the coal is to be within the precision limits.

8.3 PROXIMATE ANALYSIS

The proximate analysis of coal was developed as a convenient and effective means for determining the distribution of products obtained by heating coal under a set of standard conditions. This particular group of tests has been used widely as the basis for coal characterization in connection with coal utilization. The proximate analysis (ASTM, 2011l) of coal may also be considered as the determination of the general properties of coal and is, in reality, the determination of moisture content, volatile matter content, ash yield, and (by difference) fixed carbon yield in contrast to the ultimate analysis of coal which provides the elemental composition (Figure 8.5).

The moisture, volatile matter, and ash results are typically among the primary parameters used for assessing the quality of coal. The moisture result is utilized for calculating the dry basis results of other analytical results. The ash result is utilized in the ultimate analysis calculation of oxygen by difference (ASTM, 2011p) and for calculating material balance and ash load purposes in industrial boiler systems. The volatile matter result indicates the coke yield on the carbonization process providing additional information on combustion characteristics of the materials, and establishes a basis for purchasing and selling coal. Fixed carbon is a calculated value of the difference between 100 and the sum of the moisture, ash, and volatile matter where all values are on the same moisture reference base.

Thus, the objective of the proximate analysis is to determine the amount moisture, volatile matter yield, ash yield, and fixed carbon from the coal sample. Mineral matter is not directly measured but may be obtained by one of a number of empirical formula either from the yield of mineral ash or from data derived from the ultimate analysis.

The variables are measured in percent by weight (% w/w) and are calculated in several different bases:

- AR (as-received) basis is the most widely used basis in industrial applications and puts all variables into consideration and uses the total weight as the basis of measurement.
- AD (air-dried) basis neglect the presence of moistures other than inherent moisture.
- DB (dry basis) leaves out all moistures, including surface moisture, inherent moisture, and other moistures.
- DAF (dry, ash-free) basis neglect all moisture and ash constituent in coal.
- DMMF (dry, mineral matter–free) basis leaves out the presence of moisture and mineral matter in coal, such as quartz, pyrite, calcite, and clay.

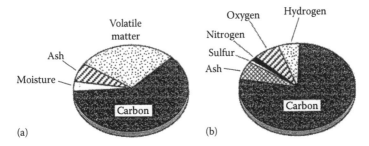

FIGURE 8.5   Data types obtained from (a) proximate analysis and (b) ultimate analysis.

Proximate Analysis	AR	AD	DB	DAF
Moisture (% w/w)	3.3	2.7		
Ash (% w/w)	22.1	22.2	22.8	
Volatile matter (% w/w)	27.3	27.5	28.3	36.6
Fixed carbon (% w/w)	47.3	47.6	48.9	63.4

8.3.1 MOISTURE

Moisture is an important property of coal, as all coals are mined wet. Groundwater and other extraneous moisture are known collectively as *adventitious moisture* and are readily evaporated. Moisture held within the coal itself is known as inherent moisture and is analyzed. Moisture may occur in four possible forms within coal:

- *Surface moisture*: water held on the surface of coal particles or macerals
- *Hydroscopic moisture*: water held by capillary action within the micro-fractures of the coal
- *Decomposition moisture*: water held within the coal's decomposed organic compounds
- *Mineral moisture*: water that comprises part of the crystal structure of hydrous silicates such as clay

As a result of these several forms, the quantitative measurement of water is complicated because the water is present within the coal matrix in more than one form (Allardice and Evans, 1978). The total moisture in coal may be determined by means of a single-stage method or by means of a two-stage method in which the as-received sample is air-dried at approximately room temperature and the residual moisture is determined in the sample (ASTM, 2011v; ISO, 2011l).

The air-drying step reduces the water in the sample to an equilibrium condition depending on the laboratory humidity and thereby minimizes any potential changes in the moisture content that might occur when the sample is prepared (say, by crushing) for further analysis. This two-stage method is particularly applicable for preparing samples that are to be submitted for extremely accurate analysis.

The methods for the determination of moisture in coal have been placed into the following categories: (1) the thermal methods, (2) a desiccators method, (3) the distillation method, (4) the extraction and solution methods, (5) the chemical methods, and (6) an electrical method. In the thermal methods, moisture may be determined either as the loss in weight when coal is heated to various temperatures (with the atmosphere and pressure variable) or as the weight gain of a vessel containing a desiccant through which passes the volatile materials evolved when the coal is heated. Similarly, the desiccator method involves a measure of the weight loss (of the coal) which occurs when the coal is maintained in a desiccator (in the presence of a desiccant) either at atmospheric pressure or at reduced pressure, but at ambient temperature.

The distillation method of moisture determination requires the collection and determination of the water evolved from the coal when the sample is heated in a boiling solvent which is itself immiscible with water. The solution and extraction methods require either solvent extraction of the water from the coal (followed by subsequent determination of the water content of the solvent) or use of a standard reagent which will exhibit differences in concentration by virtue of the water in the coal. The chemical methods of water determination invoke the concepts of direct chemical titration of the water or chemical reaction between the water and specific reagents which causes the evolution of gases and the water is determined by measurement of the volume produced. Finally, the electrical method requires the measurement of the dielectric constant of the coal from which the water content can be determined.

Obviously, each of these methods has merit and the advantages or disadvantages of any particular method must be considered prior to the application or acceptance of the method. Indeed, it must be remembered that the applicability of any one method is dependent not only on accuracy (as well as the reproducibility of that accuracy) but also on the applicability of that method to the whole range of coal types.

On a regular basis, the moisture content of coal is usually determined by measuring the percentage weight loss of a sample comminuted to pass through a 250 μm (60-mesh) sieve (ASTM, 2011l; ISO, 2011b,f,p). The sample (ca. 1 g) is maintained under controlled conditions (107°C [225°F]) in an inert atmosphere for 1 h. An alternate procedure, which involves distilling the coal with a water-immiscible liquid (such as toluene or xylene), is also available (BS, 2011; DIN, 2011b; ISO, 2011f,q) and is particularly applicable for use with high-moisture, low-rank coals that may also be easily oxidized.

For example, with lignite (and other low-rank coals), oxidation during the 1 h heating could conceivably cause a small increase in weight that will reduce the amount of moisture found in the coal. In addition, although the moisture as determined (particularly by the ASTM test) will, indeed, be mostly moisture, it may, however, include some adsorbed gases while some strongly adsorbed moisture will not be included. Caution is also necessary to ensure that the coal sample is not liable to thermal decomposition at the temperature of the moisture determination.

8.3.2 NATURAL BED MOISTURE (EQUILIBRIUM MOISTURE, CAPACITY MOISTURE)

In the ASTM system for the classification of coals by rank and in the International System for the classification of hard coals (Chapter 2), high-volatile coals are classified according to their calorific value on a moist basis. In this instance, the calorific value is quoted for the coal containing its natural bed moisture.

Indeed, at this point a distinction should be made between the natural bed moisture of coal and the as-received moisture of coal. The natural bed moisture of coal is the amount of water that a particular coal will hold when it is fully saturated at ca. 100% relative humidity (i.e., at the conditions approximating those of an undisturbed coal seam). It is also considered to be an indication of the total pore volume of the coal that is accessible to water. On the other hand, the as-received moisture content of coal relates to the amount of water in the coal at the time the coal is received for analysis. This may be somewhat smaller than the natural bed moisture, especially if the coal has been allowed to dry partially. In contrast, excessive amounts of surface moisture on the coal could lead to values for as-received moistures that are in excess of the natural bed moisture.

The natural bed moisture of coal is determined (ASTM, 2011d; ISO, 2011r; Luppens and Hoeft, 1992) by wetting the coal, removing the excess water by filtration, and then allowing moisture equilibration to occur by standing the coal over a saturated solution of potassium sulfate in a closed vessel, thereby maintaining the relative humidity at 96%–97%. The vessel must be evacuated to about 30 mm mercury and the whole maintained at 30°C (86°F) for 48 h for coals of higher rank than lignite while lignite will require 72 h to reach equilibrium. The method can also be employed to estimate the surface for extraneous moisture of wet coal; such moisture is the difference between the total moisture of the coal and the natural bed moisture.

8.3.3 VOLATILE MATTER

Volatile matter in coal refers to the components of coal, except for moisture, which are liberated at high temperature in the absence of air. The volatile matter obtained during the pyrolysis of coal consists mainly of combustible gases such as hydrogen, carbon monoxide, methane plus other hydrocarbons, tar as well as incombustible gases such as carbon dioxide and steam.

The volatile matter of coal is determined under rigidly controlled standards. In Australian and British laboratories, this involves heating the coal sample to 900°C ± 5°C (1650°F ± 10°F) for 7 min in a cylindrical silica crucible in a muffle furnace. The ASTM standard method of analysis involves heating coal to 950°C ± 25°C (1740°F ± 45°F) in a vertical platinum crucible.

The moisture that can be evolved by heating to temperatures only slightly higher than the boiling point of water is not included here but the water formed as part of the thermal decomposition process (and that originally existed as part of the coal substance but in a form other than water) is included.

The composition of the volatile matter evolved from coal is, of course, substantially different for the different ranks of coal and the proportion of incombustible gases increases as the coal rank decreases. Furthermore, in macerals isolated from any one particular coal, the volatile matter content decreases in a specific order; thus, exinite produces more volatile matter than vitrinite which, in turn, yields more volatile matter than inertinite.

The determination of the volatile matter content of coal (ASTM, 2011o; ISO, 2011j) is an important determination because volatile matter data are an integral part of coal classification systems (Chapter 2) and also form the basis of evaluating coals for their suitability for combustion and carbonization. The methods for determining volatile matter content are based on the same principle and consist of heating a weighed sample of coal (usually ca. 1 g) in a covered crucible to a predetermined temperature; the loss in weight (excluding losses due to water) is the volatile matter content (expressed as a weight percent).

The chief differences in the methods are (1) variations in the size, weight, and materials of the crucibles used, (2) the rate of temperature rise, (3) the final temperature, (4) the duration of heating, and (5) any modifications that are required for coals which are known to decrepitate or which may lose particles as a result of the sudden release of moisture or other volatile materials. In essence, all of these variables are capable of markedly affecting the result of the tests and it is, therefore, very necessary that the standard procedures be followed closely.

The arbitrary nature of the test for volatile matter content precludes a detailed discussion of the various national standards. However, in general terms, the temperature is in the range of 875°C–1050°C (1605°F–1920°F); the duration of heating is 3–20 min; and the crucibles may be platinum, silica, or ceramic material. The German standard specifies a temperature of 875°C (1605°F) to be in accord with the industrial coking practice while other standards specify temperatures of 1000°C–1050°C (1830°F–1920°F) to ensure maximum evolution of volatile matter under the test conditions; a temperature of 950°C (1740°F) is specified by the ASTM standard (ASTM, 2011o).

The various standards usually specify that a single crucible be employed although there has been a tendency in France and Belgium to advocate the use of two crucibles. In this method, the coal is heated in a crucible which is enclosed in a larger crucible with the space between the two crucibles filled with carbon (charcoal).

As an interesting comparison, the test for determining the carbon residue (Conradson), that is, the coke-forming propensity of petroleum fractions and petroleum products (ASTM, 2011ah) advocates the use of more than one crucible. A porcelain crucible is used to contain the sample and this is contained within two outer iron crucibles. This corresponds to the thermal decomposition of the sample in a limited supply of air (oxygen) and the measurement of the carbonaceous residue left at the termination of the test.

The advantage of using two crucibles in methods of this type is believed to arise because of the need to prevent partial oxidation and, hence, the reduction of accrued errors due to loss of material as carbon oxides. Indeed, the very nature of the test for volatile matter content (which is not a determination of low-molecular-weight volatile matter in the coal but is more a test for volatile matter formed as a result of a wide variety of decomposition reactions during the test) dictates that partial oxidation reactions be excluded.

Furthermore, the lower rate of heating (resulting from the insulating effect of the charcoal) assists in preventing the election of solid particles which can occur when lower-rank coals (in 10–20 mL platinum crucibles) are heated at the higher rate by direct insertion into the hot furnace. There is, therefore, no need for a modification of the method when it is applied to the lower-rank coals (an advantage) since, in borderline cases, the result may be up to 2% lower (i.e., 32% volatile matter instead of 34% volatile matter) when the modification (the lower heating rate) is employed.

Mineral matter may also contribute to the volatile matter by virtue of the loss of water from the clays, the loss of carbon dioxide from carbonate minerals, the loss of sulfur from pyrite (FeS_2), and the generation of hydrogen chloride from chloride minerals as well as various reactions that occur within the minerals thereby influencing the analytical data (Given and Yarzab, 1978).

The characterization of coal either as agglomerating or as non-agglomerating for the purposes of rank classification (Chapter 2) is carried out in conjunction with the determination of the volatile matter content. Thus, if the residue remaining from the determination is in the form of an agglomerate button that is capable of supporting a 500 g weight without pulverization of the button or if the button shows swelling or cell structure, then the coal is classified as agglomerating.

8.3.4 Ash

The *ash content* (more correctly, the *ash yield*) of coal is the noncombustible residue (usually metal oxides) left after coal is burned and represents the bulk mineral matter after carbon, oxygen, sulfur, and water (including from clays) has been driven off during combustion (Given and Yarzab, 1978). The mineral portion of coal has been ignored for too many years and it must be considered an integral part of the coal structure and so the ash (and its contents) must also be given consideration when a use is designed for coal.

The analysis is fairly straightforward, with the coal thoroughly burnt and the ash material expressed as a percentage of the original weight.

Ash is quantitatively and qualitatively different from the mineral matter originally present in coal (Chapter 7) because of the various changes that occur, such as loss of water from silicate minerals, loss of carbon dioxide from carbonate minerals, oxidation of iron pyrite to iron oxide, and fixation of oxides of sulfur by bases such as calcium and magnesium. In fact, incineration conditions determine the extent to which the weight changes take place and it is essential that standardized procedures be closely followed to ensure reproducibility.

The chemical composition of coal ash is an important factor in fouling and slagging problems and in the viscosity of coal ash in wet bottom and cyclone furnaces. The potential for the mineral constituents to react with each other (Given and Yarzab, 1978) as well as undergo significant mineralogical changes is high (Helble et al., 1989). In addition, coal with high-iron (20% w/w ferric oxide) ash typically exhibit ash-softening temperatures under 1205°C (2200°F). The use of coal with mineral matter that gives a high alkali oxide ash often results in the occurrence of slagging and fouling problems. As oxides, most ash elements have high melting points, but they tend to form complex compounds (often called eutectic mixtures) which have relatively low melting points. On the other hand, high-calcium low-iron ash coals tend to exhibit a tendency to produce low-melting range slags, especially if the sodium content of the slag exceeds about 4%.

The amount of ash is usually determined by heating (burning) a sample (1–2 g) of the coal in an adequately ventilated muffle furnace at temperatures in the range of 700°C–950°C (1290°F–1740°F). The temperature prescribed by the ASTM is 725°C (1335°F) for coal and up to, but not more than, 950°C (1740°F) for coke (ASTM, 2011n). Other standards (ISO, 2011t) may vary and require somewhat higher temperatures for the determination of the ash in coal. There are also other methods, predominantly thermal, which can be used for the analysis of coal ash (Voina and Todor, 1978).

Significant variations in the amount of ash can arise from the retention of sulfur. Thus, for high-rank coals, if the amount of pyrite and carbonate minerals is low, sulfur retention is not critical and ashing may be carried out rapidly. However, for coals with considerable amounts of pyrite and calcite, the procedure is to burn the coal at low temperatures to decompose the pyrite before the decomposition point of the carbonate minerals is reached. Thus, less sulfur remains in the coal to react with the oxides that are formed at higher temperatures. Another method consists of first ashing the coal, then treating the ash residue with sulfuric acid and igniting to constant weight. The amount of ash is calculated back to the calcium carbonate basis by subtracting three times the equivalent of carbon present as mineral carbonate from the ash as weighed.

Several formulae have been proposed for calculating the amount of mineral matter originally in the coal using the data from ashing techniques as the basis of the calculations. Of these formulae, two have survived and have been used regularly to assess the proportion of mineral matter in coal and these are the Parr formula and the King–Mavies–Crossley formula.

In the Parr formula, the mineral matter content of coal is derived from the expression:

$$\% \text{ Mineral matter} = 1.08A + 0.55S$$

where
    A is the percentage of ash in the coal
    S is the total sulfur in coal

On the other hand, the King–Mavies–Crossley formula is a little more complex:

$$\% \text{ Mineral matter} = 1.09A + 0.5S_{pyr} + 0.8CO_2 - 1.1SO_{3(in\ ash)} + SO_{3(in\ coal)} + 0.5Cl$$

where
    A is the percentage of ash in the coal
    S_{pyr} is the percentage of pyritic sulfur in the coal
    CO_2 is the percentage of "mineral" carbon dioxide in the coal
    $SO_{3(in\ ash)}$ is the percentage of sulfur trioxide in the ash
    $SO_{3(in\ coal)}$ is the percentage of sulfur trioxide in the coal
    Cl is the percentage of chlorine in the coal

The Parr formula (which has been widely used in the United States) is obviously considerably simpler than the King–Mavies–Crossley formula (which has been in common use in Great Britain and Europe) and requires less analytical data. However, the King–Mavies–Crossley formula will, because of the detail, provide more precise mineral matter values. It is, however, a matter of assessing whether the slight improvement in precision is justifiable on the basis of the additional analytical effort.

There are, of course, limitations to the use of the data derived from the ashing techniques. For example, the indefinite amount of sulfur which may be retained in the ash will reduce the reliability of "ash" values. In fact, the ash value is only an approximation of the noncombustible material in coal, and the relationship of ash composition to clinkering, boiler tube slagging, and other high-temperature behavior of the mineral constituents is complex and difficult to interpret.

At this point is worthy of note, in light of the fact that coal ash is measured by removal (combustion) of the organic part of the coal, that there are methods for measuring the mineral matter content of coal. Such a method involves demineralization of the coal which depends upon the loss of weight of a sample when it is treated with aqueous hydrofluoric acid at 55°C–60°C (130°F–140°F) (Radmacher and Mohrhauer, 1955; Bishop and Ward, 1958; Given and Yarzab, 1978; ISO, 2011m). However, pyrite is not dissolved by this treatment and must be determined separately. Other methods include the use of physical techniques such as scanning electron microscopy and x-ray diffraction (Russell and Rimmer, 1979).

There are also test methods (ASTM, 2011k,w) for the analysis of coal (and coke) ash which cover the determination of nine major constituents of ash: silicon dioxide, aluminum oxide, ferric oxide, titanium dioxide, phosphorous pentoxide, calcium oxide, magnesium oxide, sodium oxide, and potassium oxide (ASTM, 2011k) by a combination of spectrophotometric techniques, chelatometric techniques, and flame photometry. Determination of these same constituents, including manganese dioxide, can be achieved by atomic absorption (ASTM, 2011w).

The release of trace elements, associated with the combustion of coal, to the environment and disposal of coal ash, which often contains a wide range of trace elements, has become a matter of concern. The determination of these elements in coal (and coke) ash is a very important aspect of coal analysis and involves the use of atomic absorption (ASTM, 2011x). The methods cover the determination of beryllium, chromium, copper, manganese, nickel, lead, vanadium, and zinc. The use of x-ray fluorescence (Prather et al., 1979), the electron probe microanalyzer (Raymond and Gooley, 1979), for determination of trace elements in coal has also been reported.

Slagging, fouling, and clinkering difficulties have been found to correlate with the fusibility of the coal ash (ASTM, 2011e; BS, 2011; DIN, 2011e; ISO, 2011i) and there have been attempts to predict the fusibility of coal ash from compositional data (Vorres, 1979); temperatures are observed at which triangular cones prepared from the ash begin to deform and then pass through specified stages of fusion when heated at a specified rate. The test procedure provides for a controlled atmosphere; the test is first performed in a reducing atmosphere and is then repeated with a second set of cones in an oxidizing atmosphere. During combustion, there are zones in which the supply of oxygen is depleted or carbon dioxide is reversibly reduced to carbon monoxide in the presence of excess carbon. This can produce a reducing atmosphere in a hot zone where ash particles in an incipient state of fusion begin to melt. The significance of performing the test in reducing and oxidizing atmospheres is that most oxides of metals exhibit higher fusion temperatures in their highest state of oxidation.

The critical temperature most commonly referenced in the evaluation of the properties of coal ash is the softening temperature. This is the temperature at which the cone has fused down to a spherical lump in which the height is equal to the width at the base.

An analysis of coal ash may also be carried out to determine not only the composition of coal ash but also the levels at which trace elements occur in ash. These data are useful for environmental impact modeling, and may be obtained by spectroscopic methods such as inductively coupled plasma (ICP) or atomic absorption spectroscopy (AAS).

Beside composition of coal ash, ash fusion point is also one significant parameter in ash analysis. The optimum operating temperature of coal processing will depend on the gas temperature and also the ash fusion point.

In fact, the behavior of the constituents of coal ash at high temperature is a critical factor in selecting coals for steam power generation. Melting of the ash constituents may cause them to stick to the walls of the reactor and result in a buildup. However, most furnaces are designed to remove ash as a powdery residue. Coal which has ash that fuses into a hard glassy slag known (*clinker*) is usually unsatisfactory in furnaces as it requires cleaning. However, furnaces can be designed to handle the clinker, generally by removing it as a molten liquid.

Ash fusion temperatures are determined by viewing a molded specimen of the coal ash through an observation window in a high-temperature furnace (ASTM, 2011e). The ash, in the form of a cone, pyramid, or cube, is heated steadily past 1000°C (1832°F) to as high a temperature as possible, preferably 1600°C (2910°F). The following temperatures are recorded:

- *Deformation temperature*: the temperature when the corners of the mold first become rounded
- *Softening (sphere) temperature*: the temperature when the top of the mold takes on a spherical shape
- *Hemisphere temperature*: the temperature when the entire mold takes on a hemisphere shape
- *Flow (fluid) temperature*: the temperature when the molten ash collapses to a flattened button on the furnace floor

The ash fusibility temperatures are used to predict whether the ash will perform properly in the process for which the coal was chosen.

8.3.5 FIXED CARBON

The *fixed carbon content* (more correctly, the *fixed carbon yield* or *carbonaceous residue yield*) of the coal is the carbon found in the material which is left after volatile materials are driven off. This differs from the ultimate carbon content of the coal because some carbon is lost in hydrocarbons with the volatiles.

The fixed carbon yield differs from the ultimate carbon content of the coal because some carbon is lost in hydrocarbons with the volatiles. The fixed carbon value is determined by subtracting from 100 the resultant summation of moisture, volatile matter, and ash with all percentages on the same moisture reference base (ASTM, 2011l).

The fixed carbon yield—if the residue also contains mineral ash—represents the approximate yield of coke from coal under standard test conditions (Zimmerman, 1979; see also *carbon residue* of petroleum and petroleum products, Speight, 2007).

8.4 ULTIMATE ANALYSIS (ELEMENTAL ANALYSIS)

The classification systems for coal variously involve either proximate analysis or ultimate analysis or a combination of both (Chapter 2). Whereas proximate analysis is essentially an examination of the suitability of coal for combustion or for coking purposes, ultimate analysis is, in fact, an absolute measure of the elemental composition of coal.

For example, all of the carbon in coal is determined by ultimate analysis and it is not an indication (or determination) of the carbon-forming propensity (i.e., the coke-producing ability) of the coal as is the case with the test for the volatile matter content of coal. Thus, just as there has been the need to develop standard methods for the proximate analysis of coal, there has also been the necessity to develop standard methods for the ultimate analysis of coal.

The objective of ultimate analysis is to determine the constituents of coal in the form of the proportions of the chemical elements. Thus, the ultimate analysis (Figure 8.5) (ASTM, 2011p) determines the amount of carbon (C), hydrogen (H), oxygen (O), sulfur (S), and other elements within the coal sample.

The carbon so determined includes that present in the organic coal substance as well as that originally present as mineral carbonates and, similarly, the hydrogen includes that of the organic coal substance and the hydrogen present in the form of moisture and the water of constitution of the silicate minerals.

In some standards, the reported data for hydrogen in the as-received and air-dried samples includes hydrogen both in the coal substance and in the moisture while in other standards, the value for the determined hydrogen is corrected hydrogen; moisture is reported separately. All of the nitrogen is present as part of the organic coal substance but the sulfur is recognized as being present in three different forms: (1) organic sulfur compounds within the coal structure, (2) pyrite or marcasite (FeS_2), and (3) inorganic sulfates. If suitable corrections are made to the data which will account for the carbon, hydrogen, and sulfur derived from the inorganic material and for the conversion of ash to mineral matter, the ultimate analysis will represent the elemental composition of the organic portion of the coal matrix in terms of the proportion of carbon, hydrogen, nitrogen, oxygen, and sulfur.

The ultimate analysis of coal (ASTM, 2011aa,ad) gives the amounts of carbon, hydrogen, nitrogen, sulfur, and oxygen comprising the coal. Oxygen is determined by difference, that is, subtracting the total percentages of carbon, hydrogen, nitrogen, and sulfur from 100 because of the complexity in determining oxygen directly. However, this technique does accumulate all the errors in determining the other elements into the calculated value for oxygen.

The ultimate analysis is used with the heating value of the coal to perform combustion calculations including the determination of coal feed rates, combustion air requirements, weight of products of combustion to determine fan sizes, boiler performance, and sulfur emissions.

8.4.1 CARBON AND HYDROGEN

Carbon and hydrogen, respectively, account for 70%–95% and 26% (daf) of the organic substance of coal and are often thought to be the most important constituents of coal. The methods for determining carbon and hydrogen involve combustion of an exact amount of the coal in a closed system and the products of the combustion (CO_2/H_2O) determined by absorption (ASTM, 2011r; ISO, 2011n,o). The combustion is usually accomplished by placing the ground coal (to pass through, e.g., a 60-mesh/250 μm sieve) in a stream of dry oxygen at temperatures on the order of 850°C–900°C (1560°F–1650°F) (ASTM, 2011r). Complete conversion of the combustion gases to carbon dioxide and water can be achieved by passing the gases through heated cupric oxide.

This particular method of combustion analysis for carbon and hydrogen (Liebig method) is used internationally although some modifications may have been made by the various national standards organizations. The sample size can vary from as little as 1–3 mg (microanalysis) to 100–500 mg (macroanalysis) with combustion temperatures as high as 1300°C (2370°F). However, the method must ensure complete conversion of all the carbon to carbon dioxide and all of the hydrogen to water. Oxides of sulfur and chlorine are removed by passing the products of combustion over heated lead chromate and silver gauze. The carbon and hydrogen are calculated from the increase in weight of the absorbants used to collect the water and carbon dioxide. Oxides of sulfur (and chlorine, which may be released in significant amounts) are usually removed from the combustion gases by passage over silver at ca. 600°C (1110°F) while nitrogen dioxide is removed by lead chromate or manganese dioxide.

8.4.2 NITROGEN

Until recently, there was little, if any, information about the nature of nitrogen in coals. Although a variety of nitrogen types are believed to exist in the coal matrix, it was thought to exist mainly in condensed heterocyclic structures (Kirner, 1945) and only fragmentary interest has been shown in this element as part of the coal substance. More recent work (Wallace et al., 1989; Kirtley et al., 1992) have shown that nitrogen does, in fact, exist in ring systems in coal, particularly pyridine-type and pyrrole-type nitrogen.

The determination of nitrogen in coal is based on the principles of decomposition, oxidation, and reduction. The decomposition principle was introduced by Kjeldahl in 1883 when it was noted that nitrogen-containing compounds are attacked by boiling sulfuric acid with the liberation of nitrogen as ammonium sulfate. The Kjeldahl method consists of boiling the pulverized coal with concentrated sulfuric acid containing potassium sulfate and a suitable catalyst to reduce the time for digestion. The catalyst is usually a mercury salt, selenium itself, or a selenium compound, or a mixture of the two. Selenium is regarded as being particularly advantageous.

Dumas introduced the oxidation method in which a mixture of nitrogen-containing organic material and copper oxide is heated in an inert atmosphere to form carbon dioxide, water, and nitrogen. The carbon dioxide is absorbed, the water condensed, and the nitrogen determined volumetrically. Although the Dumas method has been employed for many years, various modifications have been made to increase the accuracy and the precision.

The gasification method consists of mixing the coal for coke with a mixture of two parts of the Eschka mixture (i.e., 67% w/w light calcined magnesium oxide and 33% w/w anhydrous sodium carbonate), six parts of soda lime, and one part of molybdenum oxide. The sample is then placed in a porcelain boat, covered with platinum gauze, and heated (in a quartz tube) to 200°C–250°C (390°F–480°F). The sample is then heated in steam at 850°C–950°C (1560°F–1740°F) depending on whether the sample is coal or coke, respectively. The gases pass into a solution of 0.1 N sulfuric acid where the ammonia is absorbed and determined by one of the usual techniques.

The standard procedure in many laboratories is the Kjeldahl method (ASTM, 2011s; ISO, 2011c,d), although there are the standard methods which involve the Dumas technique (DIN, 2011a) and the gasification procedure (DIN, 2011a). Neutron activation analysis has also been proposed for the determination of nitrogen in coal, coal ash, and related products (Volborth, 1979a,b).

8.4.3 SULFUR

Sulfur is present in coal either as organically bound sulfur or as inorganic sulfur (pyrite or marcasite and sulfates) (Kuhn, 1977). The amount of organic sulfur is usually 3% w/w of the coal, but exceptional amounts of sulfur (up to 11%) have been recorded. The sulfates (mainly calcium and iron) rarely exceed 0.1% except in highly weathered or oxidized samples of coal; pyrite and marcasite (the two common crystal forms of FeS_2) are difficult to distinguish from one another and are often (incorrectly) designated simply as pyrite.

Sulfur is an important consideration in coal utilization and, hence, there is a considerable amount of published work relating to the development of methods to improve the efficiency of the techniques as well as to improve the accuracy and precision of the sulfur determination (Ahmed and Whalley, 1978; Chakrabarti, 1978a; Attar, 1979; Raymond, 1982; Gorbaty et al., 1992).

The three most widely used methods for sulfur determination are (1) the Eschka method, (2) the bomb combustion method, and (3) the high-temperature combustion method, and all are based on the combustion of the sulfur-containing material to produce sulfate, which can be measured either gravimetrically or volumetrically.

The Eschka method for the determination of sulfur originated more than 100 years ago and has been accepted as a standard method in several countries. The method has distinct advantages in that the equipment is relatively simple and only the more convenient analytical techniques are employed.

Methods involving combustion of the organic material in a "bomb" are well-known as a means of sulfur determination in solid fuels and are often cited as alternate methods for sulfur determination in several national standards. The method has the advantage because the sulfur is not "lost" during the process and is particularly favored when the calorific value of the coal is also required. The method of decomposition involves the Parr fusion procedure in the presence of sodium peroxide and oxygen at high pressures (300–450 psi).

A high-temperature combustion method for the determination of sulfur in solid fuels has also been adopted in many laboratories and is advantageous insofar as chlorine can be determined simultaneously. The method requires that the coal sample be heated to 1250°C–1350°C (2280°F–2460°F) in the presence of excess kaolin, ferric phosphate, or aluminum oxide to enhance the removal of the sulfur (as sulfate) from the ash. Oxygen is also included to produce oxides of sulfur which are absorbed in hydrogen peroxide (whereby sulfur dioxide is converted to sulfur trioxide). The solution is then titrated with standard alkali solution which gives the total acidity (due to the hydrochloric and sulfuric acids that are formed).

The three methods of sulfur determination have all found application in various standards (ASTM, 2011q; BS, 2011; DIN, 2011d; ISO, 2011e,g) with the modern variation of the Eschka procedure often being favored because of its relative simplicity. In all cases, the proportion of organically bound sulfur can be obtained by subtracting the inorganic sulfur (sulfate plus pyrite sulfur; which is determined by the conventional Powell and Parr method; BS, 2011) from the total sulfur content, although it is possible (ASTM, 2011j; ISO, 2011a) to distinguish between the types of sulfur, thereby allowing the separate determination of each sulfur type in coal.

Thus, sulfate sulfur can be determined as the amount of material soluble in dilute hydrochloric acid while pyrite and marcasite are determined as the amount soluble in dilute nitric acid. However, the nitric acid extract is analyzed for iron, which is a more reliable measure of pyritic sulfur than sulfur itself. The nitric acid treatment may be carried out either on a sample which has been treated with dilute hydrochloric acid or on a separate sample; in the former case, the iron content must be corrected by the amount of iron dissolved in the hydrochloric acid. Organic sulfur (which is insoluble in either hydrochloric acid or in nitric acid) is determined by difference from the data for total sulfur and the sulfate plus pyrite sulfur.

The determination of the *forms of sulfur* forms (ASTM, 2011j) measures the quantity of sulfate sulfur, pyritic sulfur, and organically bound sulfur in the coal. The sulfur forms test measures sulfate and pyritic sulfur and determines organic sulfur by difference. Pyritic sulfur is an indicator of potential coal abrasiveness and is used for assessing coal cleaning because sulfur in the pyritic form is readily removed using physical coal-cleaning methods.

8.4.4 Oxygen

The lack of a satisfactory direct method for determining oxygen in coal and similar carbonaceous materials has dictated, historically (some would say hysterically), that oxygen be determined by subtracting the sum of the other components of ultimate analysis from 100 (ASTM, 2011p):

$$\% \text{ Oxygen} = 100 - (\%C + \%H + \%N + \%S_{organic})$$

The disadvantage of such an indirect method is that all of the errors of the other determinations are accumulated in the oxygen data. However, there are several methods for the direct determination of oxygen that have met with some success when applied to coal and, therefore, deserve some mention here because it is conceivable that at some future date one of these methods (or a modification thereof) could find approval as a recognized standard method for the direct determination of oxygen in coal.

An oxidation method that has been applied to coal involves the combustion of a weighed amount of coal with a specific quantity of oxygen. The oxygen in the coal is determined by deducting the added quantity of oxygen from the sum of the residual oxygen and the oxygen of the oxidation products. The disadvantages of the method arise because of the difficulties associated with the accurate measurement of the substantial volume of oxygen used and the quantity and quality (i.e., composition) of the combustion products.

Another method for the determination of oxygen in coal involves reduction of the coal by pyrolysis in the presence of hydrogen whereupon the oxygen is converted catalytically to water. However, the procedure is relatively complex and, furthermore, the catalyst may be poisoned by sulfur and by chlorine.

The most widely used procedure for oxygen determination consists of pyrolyzing coal in the presence of nitrogen and subsequent passage of the products over hot (1100°C [2010°F]) carbon or platinized carbon (ISO, 2011v). The oxygen in the volatile products is thereby converted to carbon monoxide, which can be determined by a variety of techniques (Gluskoter et al., 1981; Frigge, 1984). However, this particular method suffers from the errors induced by the mineral matter in the coal (especially the carbonate minerals and the water of hydration that is an integral part of many of the minerals) which can "increase" the determined oxygen. This particular problem may be resolved by first demineralizing the coal with a mixture of hydrofluoric and hydrochloric acids, but it must be presumed that the acid moieties are not incorporated into the coal.

However, in spite of the reliance on the difference method, over the last two decades the direct determination of oxygen in coal has been achieved by use of neutron activation (Volborth, 1979a,b; Mahajan, 1985; Volborth et al., 1987). The concentration of oxygen is determined by measuring the radiation from the sample. The method is nondestructive and rapid, but if only the organic oxygen is to be determined, then the sample must first be demineralized.

8.4.5 Chlorine

Chlorine occurs in coal (Chakrabarti, 1978b, 1982; Hower et al., 1992) and is believed to be a factor not only in fouling problems but also in corrosion problems (Canfield et al., 1979; Slack, 1981).

Although the validity of the values is uncertain, generally accepted fouling classification of coal, according to total chlorine content (ASTM, 2011i; ASTM, 2011z; ISO, 2011h,k), is as follows:

Chlorine (% w/w)	Fouling Type
<0.2	Low
0.2–0.3	Medium
0.3–0.5	High
>0.5	Severe

In terms of corrosion, the occurrence of chlorine in coal leads to the formation of hydrogen chloride and the condensation of water containing hydrogen chloride (hydrochloric acid) on the cooler parts of combustions equipment can lead to severe corrosion of the metal surfaces.

The test for the determination of chlorine is performed by burning the coal mixed with Eschka mixture in an oxygen bomb followed by potentiometric titration with silver nitrate solution or by a modified Volhard colorimetric titration. The repeatability interval for either procedure is the same.

8.4.6 MERCURY

Mercury occurs in coal (Chapter 7) has been identified as a very dangerous environmental contaminant, largely by reason of the process of concentration in the food chain.

In the United States, Appalachian bituminous coal and Western subbituminous coal accounted for considerable amounts of mercury entering coal-fired power plants (Pavlish et al., 2003). The composition of these coals is quite different, which can affect their mercury emissions. Appalachian coals typically have high mercury, chlorine, and sulfur contents and low calcium content, resulting in a high percentage of oxidized mercury (i.e., mercuric oxide, HgO); in contrast, Western subbituminous coals typically have low concentrations of mercury, chlorine, and sulfur contents and high calcium content, resulting in a high percentage of elemental mercury (Hg).

The test for mercury (ASTM, 2011y) consists of burning the sample in an oxygen bomb with dilute nitric acid and determination of the mercury by flameless cold vapor atomic absorption.

8.5 CALORIFIC VALUE

The calorific value of coal is a direct indication of the heat content (energy value) of the coal and this particular property is considered to be one of the most important means by which coal can be evaluated. Strictly speaking, the calorific value is neither part of the proximate analysis nor part of the ultimate analysis; it is, in fact, one of the many physical properties of coal and, as such, is often found in the various sections that deal with the physical properties of coal. In the present context, the importance of the calorific value as one of the means by which coal can be evaluated dictates that it be included in this particular section as well as in the section describing the general thermal properties of coal (Chapter 11 and Chapter 12).

The calorific value of coal is actually a complex function of the elemental composition and is usually reported as the gross calorific value (GCV) with a correction applied if the net calorific value (NCV) is of interest. For the analysis of coals, the calorific value is determined in a bomb calorimeter either by a static (isothermal) or an adiabatic method.

In the isothermal method (ASTM, 2011u; ISO, 2011u), the calorific value is determined by burning a weighed sample of coal in oxygen under controlled conditions and the calorific value is computed from temperature observations made before, during, and after combustion with appropriate allowances made for the heat contributed by other processes. The adiabatic method (ASTM, 2011g; ISO, 2011u) consists of burning the coal sample in an adiabatic bomb calorimeter under controlled conditions. The calorific value is calculated from observations made before and after the combustion.

The computed value for the calorific value of coal is usually expressed in British thermal units per pound, kilocalories per kilogram, or in kilojoules per kilogram (1.8 Btu/lb = 1.0 kcal/kg = 4.187 kJ/kg).

In either form of measurement, the recorded calorific value is the GCV whereas the NCV is calculated from the GCV (at 20°C; 68°F) by making a suitable subtraction (=1030 Btu/lb = 572 cal/g = 2.395 MJ/g) to allow for the water originally present as moisture as well as that moisture formed from the coal during the combustion. The deduction, however, is not equal to the latent heat of vaporization of water (1055 Btu/lb, 2.4 MJ/g, at 20°C, 68°F) because the calculation is made to reduce from the gross value at constant volume to a net value at constant pressure for which the appropriate factor under these conditions is 1030 Btu/lb (2.395 MJ/g).

If a coal does not have a measured heat content (calorific value), it is possible to make a close estimation of the calorific value (cv) by means of various formulae, the most popular of which are (Selvig, 1945)

The Dulong formula:

$$cv = 144.4(\%C) + 610.2(\%H) - 65.9(\%O) - 0.39(\%O)$$

The Dulong–Berthelot formula:

$$cv = 81370 + 345\ [\%H - (\%O + \%N - 1)/8] + 22.2(\%S)$$

%C, %H, %N, %O, and %S are the respective carbon, hydrogen, nitrogen, oxygen, and organic sulfur contents of the coal (all of which are calculated to a dry, ash-free basis). In both cases, the calculated values are in close agreement with the experimental calorific values.

In some instances, an adiabatic bomb calorimeter may not be available or the sample may even be too small accurate use. To combat such problems, there is evidence that DTA is applicable to the determination of calorific value of coal. Data obtained by use of the DTA method are in good agreement with those data obtained by use of the bomb calorimeter (Munoz-Guillena et al., 1992).

8.6 REPORTING COAL ANALYSES

Analyses may be reported on different bases (ASTM, 2011t; ISO, 2011s) with regard to moisture and ash content. Indeed, results that are "as determined" refer to the moisture condition of the sample during analyses in the laboratory; a frequent practice is to air-dry the sample, thereby bringing the moisture content to approximate equilibrium with the laboratory atmosphere in order to minimize gain or loss during sampling operations (ASTM, 2011f; ISO, 2011w). Loss of weight during air-drying is determined to enable calculation on an "as-received" basis (the moisture condition when the sample arrived in the laboratory). This is, of course, equivalent to the "assampled" basis if no gain or loss of moisture occurs in the interim.

Analyses reported on a *dry* basis are calculated on the basis that there is no moisture associated with the sample. The moisture value (ASTM, 2011m) is used for converting the "as-determined" data to the "dry" basis. Analytical data that are reported on a "dry, ash-free" basis are calculated on the theory that there is no moisture or ash associated with the sample. The values obtained for moisture determination (ASTM, 2011m) and ash determination (ASTM, 2011n) are used for the conversion. Finally, data calculated on an "equilibrium moisture" basis are calculated to the moisture level determined (ASTM, 2011d) as the equilibrium (capacity) moisture.

Hydrogen and oxygen reported on the moist basis may or may not contain the hydrogen and oxygen of the associated moisture, and the analytical report should stipulate which is the case because of the variation in the conversion factors (Table 8.3). These factors apply to calorific values as well as to proximate analysis (Table 8.4) and to ultimate analysis (Table 8.5).

TABLE 8.3

Conversion Factors for Components Other than Hydrogen and Oxygen (ASTM, 2011t)[a]

Given	As Determined (ad)	As Received (ar)	Dry (d)	Dry Ash-Free (daf)
As determined (ad)	—	$\dfrac{100 - M_{ar}}{100 - M_{ad}}$	$\dfrac{100}{100 - M_{ar}}$	$\dfrac{100}{100 - M_{ad} - A_{ad}}$
As received (ar)	$\dfrac{100 - M_{ad}}{100 - M_{ar}}$	—	$\dfrac{100}{100 - M_{ar}}$	$\dfrac{100}{100 - M_{ar} - A_{ar}}$
Dry (d)	$\dfrac{100 - M_{ad}}{100}$	$\dfrac{100 - M_{ar}}{100}$	—	$\dfrac{100}{100 - A_{d}}$
Dry, ash-free (daf)	$\dfrac{100 - M_{ad} - A_{ad}}{100}$	$\dfrac{100 - M_{ar} - A_{ar}}{100}$	$\dfrac{100 - A_{d}}{100}$	—

Note: M, percent moisture by weight; A, percent ash by weight.

[a] For example, given ad, to find ar use the formula:

$$ar = ad \times \frac{100 - M_{ar}}{100 - M_{ad}}$$

TABLE 8.4

Examples of Data (% w/w) Obtained by Proximate Analysis (ASTM, 2011l)

Gain	Moisture	Ash	Volatile	Fixed Carbon
Air-dried	8.23	4.46	40.05	47.26
Dry	—	4.86	43.64	51.50
As received[a]	23.24	3.73	33.50	39.53

[a] Air-dry loss in accordance with ASTM (2011f) Method = 16.36%.

TABLE 8.5

Examples of Data (% w/w) Obtained by Ultimate Analysis (ASTM, 2011p)

Basis	Component (% w/w)							Total (%)
	Carbon	Hydrogen	Nitrogen	Sulfur	Ash	Oxygen[a]	Moisture	
As determined[b,c]	60.08	5.44	0.88	0.73	7.86	25.01	9.00	100.0
Dry	66.02	4.87	0.97	0.80	8.64	18.70	0.00	100.0
As received[d]	46.86	6.70	0.69	0.57	6.13	39.05	(29.02)	100.0
As received[e]	46.86	3.46	0.69	0.57	6.13	13.27	29.02	100.0

[a] By difference.

[b] After air-dry loss (22.00%) in accordance with ASTM (2011f).

[c] Hydrogen and oxygen include hydrogen and oxygen in sample moisture, M_{ad}.

[d] Hydrogen and oxygen include hydrogen and oxygen in sample moisture, M_{ar}.

[e] Hydrogen and oxygen do not include hydrogen and oxygen in sample moisture, M_{ar}.

When hydrogen and oxygen percentages do contain hydrogen and oxygen of the moisture, values on the dry basis may be calculated according to the formulae:

$$H_d = (H1 - 0.1111M1) \times 100/(100 - M1)$$

$$O_d = (O1 - 0.8881M1) \times 100/(100 - M1)$$

H_d and O_d are weight percent of hydrogen and oxygen on the dry basis, and H1 and O1 are the given or determined weight per cent of hydrogen and oxygen, respectively, for the given or determined weight percent of moisture M1. Rearrangement of these equations to solve for H1 and O1 yields equations for calculating moisture containing hydrogen and oxygen contents H1 and O1 at any desired moisture level M1.

The mineral matter in coal loses weight during ashing because of the loss of water of constitution of clays, the loss of carbon dioxide from carbonate minerals such as calcite, and the oxidation of pyrite (FeS_2) to ferric oxide (Fe_2O_3). In addition, any chlorine in the coal is converted to hydrogen chloride but the change in weight may not be significant.

Analyses and calorific values are determined on a mineral matter–free basis by the Parr formulae (ASTM, 2011c) with corrections for pyrite and other mineral matter. The amount of pyrite is taken to be that equivalent to the total sulfur of the coal, which in spite of the potential error has been found to correlate well in studies of mineral matter. The remaining mineral matter is taken to be 1.08 times the weight of the corresponding (iron oxide–free) ash:

$$MM = 1.08A + 0.55S$$

MM, A, and S are weight percent of mineral matter, ash, and total sulfur, respectively.

Such data are necessary for calculation of parameters in the classification of coal by rank (Chapter 2) and which are dry, mineral matter–free volatile matter (or fixed carbon) as well as moist mineral matter–free GCV. For volatile matter and fixed carbon data, it is also necessary to assume that 50% of the sulfur is volatilized in the volatile matter test and therefore should not be included as part of the organic volatile matter (nor should the loss from clays and carbonate minerals):

$$FC_{dmmf} = [(FC - 0.15S)]/[100 - (M + 1.08A + 0.55S)]$$

$$VM_{dmmf} = 100 - FC_{dmmf}$$

where
  FC_{dmmf} and VM_{dmmf} are the fixed carbon and volatile matter, respectively, on a dry, mineral matter–free basis
  FC, M, A, and S are the determined fixed carbon, moisture, ash, and total sulfur, respectively

In the Parr formula for moist, mineral matter–free calorific value, the moisture basis used is that of the inherent moisture of the coal in the seam (natural bed moisture, capacity moisture):

$$\text{Moist, MM} - \text{free Btu} = [100(Btu - 50S)]/[100 - (1.08A + 0.55S)]$$

where
  Btu is the calorific value (Btu/lb)
  A is the ash (% w/w)
  S is sulfur (% w/w)
  All are on the moist (natural bed) basis

8.7 PRECISION AND ACCURACY

Reference has already been made to the terms precision and accuracy (Table 8.6) and further definition is not warranted here. Nevertheless, a comment is required to relate these terms to the various aforementioned tests for coal. In commercial enterprises, unit prices of coal reflect the relationship between a desirable property or a combination of properties and quantity of the coal.

TABLE 8.6
Precision of Analyses and Tests

Component	Repeatability	Reproducibility
Proximate Analysis		
Moisture		
Less than 5%	0.2	0.3
Greater than 5%	0.3	0.5
Ash		
No carbonates present	0.2	0.3
Carbonates present	0.3	0.5
Coals with more than 12% ash containing carbonate and pyrite		
Volatile matter		
High-temperature coke	0.2	0.4
Anthracite	0.3	0.6
Semianthracite, bituminous, low-temperature coke and char	0.5	1.0
Subbituminous	0.7	1.4
Lignite	1.0	2.0
Ultimate Analysis		
Carbon	0.3	—
Hydrogen	0.07	—
Sulfur by Eshka or bomb-washing method		
Less than 2% sulfur	0.05	0.10
Equal to or greater than 2% sulfur	0.10	0.20
Coke	0.03	0.05
Sulfur by high-temperature combustion		
Less than 2% sulfur	0.05	0.15
Equal to or greater than 2% sulfur	0.05	0.25
Coke	0.05	0.15
Nitrogen	0.05	—
Forms of sulfur		
Sulfate	0.02	0.04
Pyritic, less than 2%	0.05	0.30
Pyritic, equal to or greater than 2%	0.10	0.40
Calorific value, Btu/lb	50	100
Hardgrove grindability index	2	3

Source: Preparation of a coal conversion technical systems data book, ERDA contract No. E(49-18)-1730, Report No. FE-1730-21, 1976.

Thus, the *accuracy* of the measurement of coal properties is extremely important. The data need to be accurate values insofar as any levels of error are not a matter of any economic consequence. Furthermore, the word *accuracy* is used to indicate the reliability of a measurement or an observation, but it is, more specifically, a measure of the closeness of agreement between an experimental result and the true value. Thus, the accuracy of a test method is the degree of agreement of individual test results with an accepted reference value. Accuracy, similar to precision, is often expressed inversely in terms of the standard deviation or variance.

Precision, by definition, does not include any systematic error or bias, but accuracy, by definition, does. Precision is a measure of the degree to which replicate data and/or measurements conform to each other; the degree of agreement among individual test results obtained under prescribed similar conditions. Hence, it is possible that data can be very precise without necessarily being correct or accurate. Precision is commonly expressed inversely by the imprecision of results in terms of their standard deviation or their variance. Precision, by definition, does not include systematic error or bias.

In any form of analysis, accuracy and precision are required; otherwise, the analytical data are suspect and cannot be used with any degree of certainty (Speight, 2005). This is especially true of analytical data used for commercial operations where the material is sold on the basis of purity and, being a complex heterogeneous material, *coal purity* refers to the occurrence (or lack thereof) of foreign constituents (water, pyrite, and mineral matter) within the organic coal matrix.

For coal that is sampled in accordance with standard methods (ASTM, 2011h,ab,ac,ae,af; ISO, 2011y) and with the standard preparation of the samples for analysis (ASTM, 2011b,f), the overall variance of the final analytical data is minimized and falls within the limits of anticipated experimental difference.

Estimation of the limits of accuracy (deviation from a true or theoretical value) is not ordinarily attempted in coal analysis. Precision, on the other hand, is determined by means of cooperative test programs.

Both (1) repeatability, the precision with which a test can be repeated in the same laboratory, usually but not always by the same analyst using the same equipment and following the prescribed method(s), and (2) reproducibility, the precision expected of results from different laboratories, are determined. Values quoted in test methods are the differences between two results that should be exceeded in only 5 out of 100 pairs of results, equal to 2/2 times the standard deviation of a large population of results.

8.8 INTERRELATIONSHIPS OF ANALYTICAL AND PHYSICAL DATA

Various interrelationships exist between the different properties of coal. For example, variations in hydrogen content with carbon content (Figure 8.6) or oxygen content with carbon content (Figure 8.7) and with each other (Figure 8.8) have been noted. However, it should be noted that many of the published reports cite the variation of analytical data or test results not *with* rank in the true sense of the word but with elemental carbon content that can only be approximately equated to rank (Chapters 2 and 9).

Other relationships also exist, such as variations of natural bed moisture with depth of burial (Figures 8.9 and 8.10) as well as the variations in volatile matter content of vitrinites obtained from different depths (Figure 8.11). This latter observation (i.e., the decrease in volatile matter with the depth of burial of the seam) is a striking contrast to parallel observations for petroleum where an increase in the depth of the reservoir is accompanied by an increase in the proportion of lower-molecular-weight (i.e., more volatile) materials (Figure 8.12).

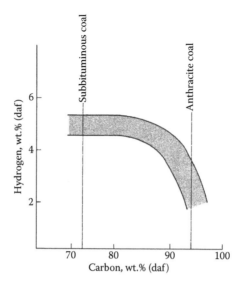

FIGURE 8.6 Variation of hydrogen content with carbon content. (Adapted from Berkowitz, N., *An Introduction to Coal Technology*, Academic Press, New York, 1979.)

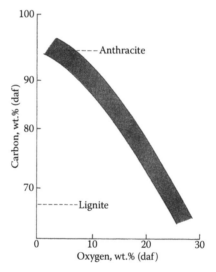

FIGURE 8.7 Variation of oxygen content with carbon content. (Adapted from Berkowitz, N., *An Introduction to Coal Technology*, Academic Press, New York, 1979.)

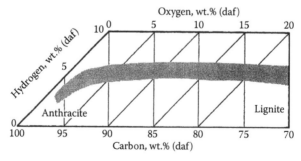

FIGURE 8.8 Relationship of carbon, hydrogen, and oxygen contents. (From Lowry, H.H., Ed., *Chemistry of Coal Utilization*, John Wiley & Sons, New York, 1945.)

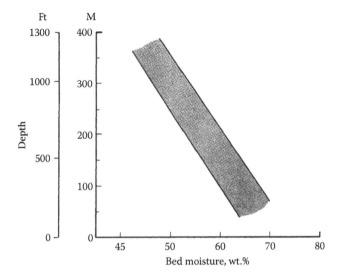

FIGURE 8.9   Variation of natural bed moisture with depth for German low-rank coals. (From Murchison, D. and Westoll, T.S., Eds., *Coal and Coal bearing Strata*, Elsevier, New York, 1968.)

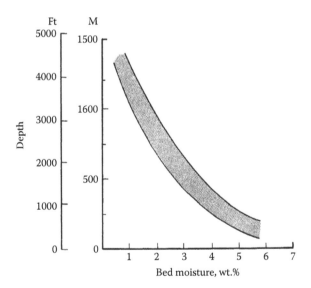

FIGURE 8.10   Variation of natural bed moisture with depth for German high-rank coals. (From Murchison, D. and Westoll, T.S., Eds., *Coal and Coal bearing Strata*, Elsevier, New York, 1968.)

Similarly, the tendency to a carbon-rich material in the deeper coal seams (Figure 8.13) appears to be in direct contrast to the formation of hydrogen-rich species (such as the constituents of the gasoline fraction) in the deeper petroleum reservoirs. Obviously, the varying maturation processes play an important role in determining the nature of the final product as does the character of the source material (Chapter 3).

Finally, it is also possible to illustrate the relationship of the data from proximate analysis and the calorific value to coal rank (Figures 8.14 and 8.15).

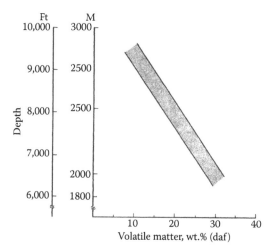

FIGURE 8.11 Variation of volatile matter yield with depth for vitrinites. (From Murchison, D. and Westoll, T.S., Eds., *Coal and Coal bearing Strata*, Elsevier, New York, 1968.)

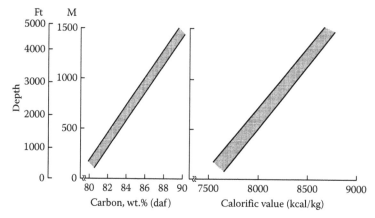

FIGURE 8.12 Variation of carbon content and calorific value with depth for vitrinites. (From Murchison, D. and Westoll, T.S., Eds., *Coal and Coal bearing Strata*, Elsevier, New York, 1968.)

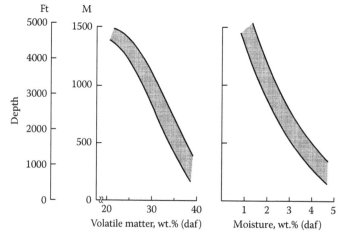

FIGURE 8.13 Variation of volatile matter yield and moisture content with depth for German coals. (From Murchison, D. and Westoll, T.S., Eds., *Coal and Coal bearing Strata*, Elsevier, New York, 1968.)

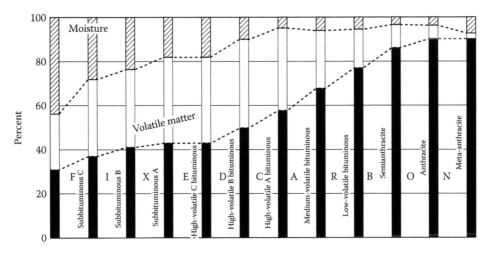

FIGURE 8.14 Proximate analysis and coal rank.

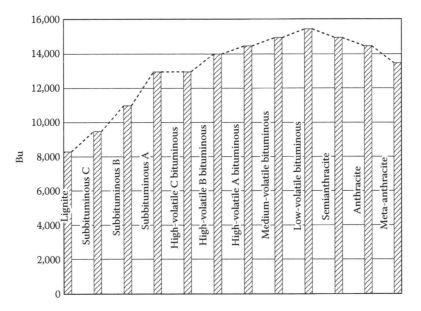

FIGURE 8.15 Calorific value and coal rank.

REFERENCES

Ahmed, S.M. and Whalley, B.J.P. 1978. In *Analytical Methods for Coal and Coal Products*, C. Karr, Jr. (Ed.). Academic Press, Inc., New York, Vol. I, Chapter 8.

Allardice, D.J. and Evans, D.G. 1978. In *Analytical Methods for Coal and Coal Products*, C. Karr, Jr. (Ed.). Academic Press, Inc., New York, Vol. I, Chapter 7.

ASTM. 2011a. *Method of Sampling and Fineness Test of Pulverized Coal* (ASTM D197). Annual Book of ASTM Standards, Section 05.05. American Society for Testing and Materials, West Conshohocken, PA.

ASTM. 2011b. *Practice for Collection and Preparation of Coke Samples for Laboratory Analysis* (ASTM D346). Annual Book of ASTM Standards, Section 05.05. American Society for Testing and Materials, West Conshohocken, PA.

ASTM. 2011c. *Classification of Coals by Rank* (ASTM D388). Annual Book of ASTM Standards, Section 05.05. American Society for Testing and Materials, West Conshohocken, PA.

ASTM. 2011d. *Test Method for Equilibrium Moisture of Coal at 96 to 97 Percent Relative Humidity and 30°C* (ASTM D1412). Annual Book of ASTM Standards, Section 05.05. American Society for Testing and Materials, West Conshohocken, PA.

ASTM. 2011e. *Test Method for Fusibility of Coal and Coke Ash* (ASTM D1857). Annual Book of ASTM Standards, Section 05.05. American Society for Testing and Materials, West Conshohocken, PA.

ASTM. 2011f. *Method for Preparing Coal Samples for Analysis* (ASTM D2013). Annual Book of ASTM Standards, Section 05.05. American Society for Testing and Materials, West Conshohocken, PA.

ASTM. 2011g. *Test Method for Gross Calorific Value of Coal and Coke by the Adiabatic Bomb Calorimeter* (ASTM D2015). Annual Book of ASTM Standards, Section 05.05. American Society for Testing and Materials, West Conshohocken, PA.

ASTM. 2011h. *Method for Collection of a Gross Sample of Coal* (ASTM D2234). Annual Book of ASTM Standards, Section 05.05. American Society for Testing and Materials, West Conshohocken, PA.

ASTM. 2011i. *Test Method for Chlorine in Coal* (ASTM D2361). Annual Book of ASTM Standards, Section 05.05. American Society for Testing and Materials, West Conshohocken, PA.

ASTM. 2011j. *Test Method for Forms of Sulfur in Coal* (ASTM D2492). Annual Book of ASTM Standards, Section 05.05. American Society for Testing and Materials, West Conshohocken, PA.

ASTM. 2011k. *Method for Analysis of Coal and Coke Ash* (ASTM D2795). Annual Book of ASTM Standards, Section 05.05. American Society for Testing and Materials, West Conshohocken, PA.

ASTM. 2011l. *Practice for Proximate Analysis of Coal and Coke* (ASTM D3172). Annual Book of ASTM Standards, Section 05.05. American Society for Testing and Materials, West Conshohocken, PA.

ASTM. 2011m. *Test Method for Moisture in the Analysis Sample of Coal and Coke* (ASTM D3173). Annual Book of ASTM Standards, Section 05.05. American Society for Testing and Materials, West Conshohocken, PA.

ASTM. 2011n. Test *Method for Ash in the Analysis Sample of Coal and Coke from Coal* (ASTM D3174). Annual Book of ASTM Standards, Section 05.05. American Society for Testing and Materials, West Conshohocken, PA.

ASTM. 2011o. *Test Method for Volatile Matter in the Analysis Sample of Coal and Coke* (ASTM D3175). Annual Book of ASTM Standards, Section 05.05. American Society for Testing and Materials, West Conshohocken, PA.

ASTM. 2011p. *Practice for Ultimate Analysis of Coal and Coke* (ASTM D3176). Annual Book of ASTM Standards, Section 05.05. American Society for Testing and Materials, West Conshohocken, PA.

ASTM. 2011q. *Test Method for Total Sulfur in the Analysis Sample of Coal and Coke* (ASTM D3177). Annual Book of ASTM Standards, Section 05.05. American Society for Testing and Materials, West Conshohocken, PA.

ASTM. 2011r. *Test Method for Carbon and Hydrogen in the Analysis Sample of Coal and Coke* (ASTM D3178). Annual Book of ASTM Standards, Section 05.05. American Society for Testing and Materials, West Conshohocken, PA.

ASTM. 2011s. *Test Method for Nitrogen in the Analysis Sample of Coal and Coke* (ASTM D3179). Annual Book of ASTM Standards, Section 05.05. American Society for Testing and Materials, West Conshohocken, PA.

ASTM. 2011t. *Practice for Calculating Coal and Coke Analyses from As-Determined to Different Bases* (ASTM D3180). Annual Book of ASTM Standards, Section 05.05. American Society for Testing and Materials, West Conshohocken, PA.

ASTM. 2011u. *Test Method for Gross Calorific Value by the Isoperibol Bomb Calorimeter* (ASTM D3286). Annual Book of ASTM Standards, Section 05.05. American Society for Testing and Materials, West Conshohocken, PA.

ASTM. 2011v. *Test Method for Total Moisture in Coal* (ASTM D3302). Annual Book of ASTM Standards, Section 05.05. American Society for Testing and Materials, West Conshohocken, PA.

ASTM. 2011w. *Test Method for Major and Minor Elements in Coal and Coke Ash by the Atomic Absorption Method* (ASTM D3682). Annual Book of ASTM Standards, Section 05.05. American Society for Testing and Materials, West Conshohocken, PA.

ASTM. 2011x. *Test Method for Trace Elements in Coal and Coke Ash by the Atomic Absorption Method* (ASTM D3683). Annual Book of ASTM Standards, Section 05.05. American Society for Testing and Materials, West Conshohocken, PA.

ASTM. 2011y. *Test Method for Mercury in Coal by the Oxygen Bomb Combustion/Atomic Absorption Method* (ASTM D3684). Annual Book of ASTM Standards, Section 05.05. American Society for Testing and Materials, West Conshohocken, PA.

ASTM. 2011z. *Standard Test Method for Total Chlorine in Coal by the Oxygen Bomb Combustion/Ion Selective Electrode Method* (ASTM D4208). Annual Book of Standards, American Society for Testing and Materials, West Conshohocken, PA.

ASTM. 2011aa. *Standard Test Method for Sulfur in the Analysis Sample of Coal and Coke Using High-Temperature Tube Furnace Combustion* (ASTM D4239). Annual Book of Standards, American Society for Testing and Materials, West Conshohocken, PA.

ASTM. 2011ab. *Standard Practice for Collection of Channel Samples of Coal in a Mine* (ASTM D4596). Annual Book of Standards, American Society for Testing and Materials, West Conshohocken, PA.

ASTM. 2011ac. *Standard Practice for Mechanical Auger Sampling* (ASTM D4916). Annual Book of Standards, American Society for Testing and Materials, West Conshohocken, PA.

ASTM. 2011ad. *Standard Test Methods for Instrumental Determination of Carbon, Hydrogen, and Nitrogen in Laboratory Samples of Coal* (ASTM D5373). Annual Book of Standards, American Society for Testing and Materials, West Conshohocken, PA.

ASTM. 2011ae. *Standard Practice for Manual Sampling of Coal from Tops of Barge* (ASTM D6315). Annual Book of Standards, American Society for Testing and Materials, West Conshohocken, PA.

ASTM. 2011af. *Standard Practice for Bias Testing a Mechanical Coal Sampling System.* Annual Book of Standards, American Society for Testing and Materials, West Conshohocken, PA.

ASTM International. 2011ag. Annual Book of Standards, American Society for Testing and Materials, West Conshohocken, PA.

ASTM. 2011ah. *Test Method for Conradson Carbon Residue of Petroleum Products* (ASTM D189). Annual Book of ASTM Standards, Section: 05.05. American Society for Testing and Materials, West Conshohocken, PA.

Attar, A. 1979. In *Analytical Methods for Coal and Coal Products*, C. Karr, Jr. (Ed.). Academic Press, Inc., New York, Vol. III, Chapter 56.

Averitt, P. 1981. In *Chemistry of Coal Utilization*, Second Supplementary Volume, M.A. Elliott (Ed.). John Wiley & Sons, Inc., New York, Chapter 2.

Baughman, G.L. 1978. *Synthetic Fuels Data Handbook.* Cameron Engineers, Denver, CO.

Berkowitz, N. 1979. *An Introduction to Coal Technology.* Academic Press, Inc., New York.

Bishop, M. and Ward, D.L. 1958. The direct determination of mineral matter in coal. *Fuel*, 27: 191.

BS. 2011. British Standards Organization (BS 1016). London, U.K.

Campbell, J.R. and Gibb, W. 1951. *Methods of Analysis of Fuels and Oils.* Constable and Co. Ltd., London, U.K.

Canfield, D.R., Ibarra, S., and McCoy, J.D. 1979. *Hydrocarbon Processing*, 58(7): 203.

Chakrabarti, J.N. 1978a. In *Analytical Methods for Coal and Coal Products*, C.K. Karr, Jr. (Ed.). Academic Press, Inc., New York, Vol. 1, Chapter 9.

Chakrabarti, J.N. 1978b. In *Analytical Methods for Coal and Coal Products*, C.K. Karr, Jr. (Ed.). Academic Press, Inc., New York, Vol. 1, Chapter 10.

Chakrabarti, J.N. 1982. In *Coal and Coal Products: Analytical Characterization Techniques*, E.L. Fuller, Jr. (Ed.). Symposium Series No. 205. American Chemical Society, Washington, DC, Chapter 8.

Deurbrouck, A.W. and Hucko, R.E. 1981. In *Chemistry of Coal Utilization*, Second Supplementary Volume, M.A. Elliott (Ed.). John Wiley & Sons, Inc., New York, Chapter 10.

DIN. 2011a. Deutsches Institut für Normung e. V (DIN 15722), Berlin, Germany.

DIN. 2011b. Deutsches Institut für Normung e. V (DIN 51718), Berlin, Germany.

DIN. 2011c. Deutsches Institut für Normung e. V (DIN 51722), Berlin, Germany.

DIN. 2011d. Deutsches Institut für Normung e. V (DIN 51724), Berlin, Germany.

DIN. 2011e. Deutsches Institut für Normung e. V (DIN 51730), Berlin, Germany.

Essenhigh, R.H. 1981. In *Chemistry of Coal Utilization*, Second Supplementary Volume, M.A. Elliott (Ed.). John Wiley & Sons, Inc., New York, Chapter 19.

Field, M.A., Gill, D.W., Morgan, B.B., and Hawksley, P.G.W. 1967. *Combustion of Pulverized Coal.* British Coal Utilization Research Association, Leatherhead, Surrey.

Frigge, J. 1984. Erdol Kohle, *Erdgas Petrochemie*, 37: 267.

Given, P.H. and Yarzab, R.F. 1978. In *Analytical Methods for Coal and Coal Products*, C. Karr, Jr. (Ed.). Academic Press, Inc., New York, Vol. II, Chapter 20.

Gluskoter, H.J., Shimp, N.F., and Ruch, R.R. 1981. In *Chemistry of Coal Utilization*, Second Supplementary Volume, M.A. Elliott (Ed.). John Wiley & Sons, Inc., New York, Chapter 7.

Gorbaty, M.L., Kelemen, S.R., George, G.N., and Kwiatek, P.J. 1992. *Fuel*, 71: 1255.

Gould, G. and Visman, J. 1981. In *Coal Handbook*, R.A. Meyers (Ed.). Marcel Dekker, Inc., New York, Chapter 2.

Helble, J.J., Srinivasachar, S., Boni, A.A., Kang, S.-G., Sarofim, A.F., Beer, J.M., Gallagher, N., Bool, L., Peterson, T.W., Wendt, J.O.L., Shah, N., Huggins, F.E., and Huffman, G.P. 1989. In *Proceedings Sixth Annual International Pittsburgh Coal Conference*. University of Pittsburgh, Pittsburgh, PA, Vol. 1, p. 81.

Hower, J.C., Riley, J.T., Thomas, G.A., and Griswold, T.B. 1992. *Journal of Coal Quality*, 10: 152.

ISO. 2011a. *Hard Coal—Determination of Forms of Sulfur* (ISO 157). International Standards Organization, Geneva, Switzerland, 331.

ISO. 2011b. *Determination of Moisture in the Analysis Sample* (ISO 331). International Standards Organization, Geneva, Switzerland.

ISO. 2011c. *Coal—Determination of Nitrogen by the Macro Kjeldahl Method* (ISO 332). International Standards Organization, Geneva, Switzerland.

ISO. 2011d. *Coal and Coke—Determination of Nitrogen, Semi-Micro Method* (ISO 333). International Standards Organization, Geneva, Switzerland.

ISO. 2011e. *Coal and Coke—Determination of Total Sulfur, Eschka Method* (ISO 334). International Standards Organization, Geneva, Switzerland.

ISO. 2011f. *Hard Coal—Determination of Moisture in the Analysis Sample* (ISO 348). International Standards Organization, Geneva, Switzerland.

ISO. 2011g. *Coal and Coke—Determination of Sulfur, High-Temperature Method* (ISO 351). International Standards Organization, Geneva, Switzerland.

ISO. 2011h. *Coal and Coke—Determination of Chlorine (High-Temperature Combustion Method)* (ISO 352). International Standards Organization, Geneva, Switzerland.

ISO. 2011i. *Coal and Coke—Determination of Fusibility of Ash* (ISO 540). International Standards Organization, Geneva, Switzerland.

ISO. 2011j. *Coal and Coke—Determination of Volatile Matter* (ISO 562). International Standards Organization, Geneva, Switzerland.

ISO. 2011k. *Coal and Coke—Determination of Chlorine Using Eschka Mixture* (ISO 587). International Standards Organization, Geneva, Switzerland.

ISO. 2011l. *Hard Coal—Determination of Total Moisture* (ISO 589). International Standards Organization, Geneva, Switzerland.

ISO. 2011m. *Coal—Determination of Mineral Matter* (ISO 602). International Standards Organization, Geneva, Switzerland.

ISO. 2011n. *Coal and Coke—Determination of Carbon and Hydrogen, Liebig Method* (ISO 609). International Standards Organization, Geneva, Switzerland.

ISO. 2011o. *Coal and Coke—Determination of Carbon and Hydrogen, High-Temperature Combustion Method* (ISO 625). International Standards Organization, Geneva, Switzerland.

ISO. 2011p. International Standards Organization (ISO 687), Geneva, Switzerland.

ISO. 2011q. International Standards Organization (ISO 1015), Geneva, Switzerland.

ISO. 2011r. *Hard Coal—Determination of Moisture-Holding Capacity* (ISO 1018). International Standards Organization, Geneva, Switzerland.

ISO. 2011s. International Standards Organization (ISO 1170), Geneva, Switzerland.

ISO. 2011t. *Solid Mineral Fuels—Determination of Ash* (ISO 1171). International Standards Organization, Geneva, Switzerland.

ISO. 2011u. *Solid Mineral Fuels—Determination of Gross Calorific Value and Calculation of Net Calorific Value* (ISO 1928). International Standards Organization, Geneva, Switzerland.

ISO. 2011v. *Hard Coal—Determination of Oxygen* (ISO 1944). International Standards Organization, Geneva, Switzerland.

ISO. 2011w. International Standards Organization (ISO 1988), Geneva, Switzerland.

ISO. 2011x. International Standards Organization (ISO 2309), Geneva, Switzerland.

ISO. 2011y. International Standards Organization (ISO 13909), Geneva, Switzerland.

Janus, J.B. and Shirley, B.S. 1973. Report of Investigations No. 7848. United States Bureau of Mines, Washington, DC.

Janus, J.B. and Shirley, B.S. 1975. Report of Investigations No. 7848. United States Bureau of Mines, Washington, DC.

Karr, C. Jr. 1978. (Ed.). *Analytical Methods for Coal and Coal Products*. Academic Press, Inc., New York, Vols. I and II.

Karr, C. Jr. 1979. (Ed.). *Analytical Methods for Coal and Coal Products*. Academic Press, Inc., New York, Vol. III.

Kirner, W.R. 1945. In *Chemistry of Coal Utilization*, H.H. Lowry (Ed.). John Wiley & Sons, Inc., New York, Chapter 13.

Kirtley, S.M., Mullins, O.C., van Elp, J., and Cramer, S.P. 1992. Preprints. Division of Fuel Chemistry, American Chemical Society, Washington, DC, 37(3): 1103.

Kuhn, J.K. 1977. In *Coal Desulfurization*, T.D. Wheelock (Ed.). Symposium Series No. 64. American Chemical Society, Washington, DC, p. 16.

Lowry, H.H. (Ed.). 1945. *Chemistry of Coal Utilization*. John Wiley & Sons, Inc., New York.

Luppens, J.A. and Hoeft, A.P. 1992. *Journal of Coal Quality*, 10: 133.

Mahajan, O.P. 1985. *Fuel*, 64: 973.

Meyers, R.A. (Ed.). 1981. *Coal Handbook*. Marcel Dekker, Inc., New York.

Montgomery, W.J. 1978. In *Analytical Methods for Coal and Coal Products*, C. Karr, Jr. (Ed.). Academic Press, Inc., New York, Vol. I, Chapter 6.

Munoz-Guillena, M.J., Linares-Solano, A., and Salinas-Martinez de Lecea, C. 1992. *Fuel*, 71: 579.

Murchison, D. and Westoll, T.S. (Eds.). 1968. *Coal and Coal bearing Strata*. Elsevier, New York.

Ode, W.H. 1963. In *Chemistry of Coal Utilization*, Supplementary Volume, H.H. Lowry (Ed.). John Wiley & Sons, Inc., New York, Chapter 5.

Patrick, J.W. and Wilkinson, H.C. 1978. In *Analytical Methods for Coal and Coal Products*, C. Karr, Jr. (Ed.). Academic Press, Inc., New York, Vol. II, Chapter 29.

Pavlish, J.J., Sondreal, E.A., Mann, M.D., Olson, E.S. Galbreath, K.C., Laudal, D.L., and Benson, S.A. 2003. Status review of mercury control options for coal-fired power plants. *Fuel Processing Technology*, 82: 89–165.

Prather, J.W., Guin, J.A., and Tarrer, A.R. 1979. In *Analytical Methods for Coal and Coal Products*, C. Karr, Jr. (Ed.). Academic Press, Inc., New York, Vol. III, Chapter 49.

Radmacher, W. and Mohrhauer, P. 1955. The direct determination of the mineral matter content of coal. *Brennstoff Chemie*, 36: 236.

Raymond, R. Jr. 1982. In *Coal and Coal Products: Analytical Characterization Techniques*, E.L. Fuller, Jr. (Ed.). Symposium Series No. 205. American Chemical Society, Washington, DC, Chapter 9.

Raymond, R. Jr. and Gooley, R. 1979. In *Analytical Methods for Coal and Coal Products*, C. Karr, Jr. (Ed.). Academic Press, Inc., New York, Vol. III, Chapter 48.

Rose, C.D. 1992. *Journal of Coal Quality*, 11(1/2): 6.

Russell, S.J. and Rimmer, S.M. 1979. In *Analytical Methods for Coal and Coal Products*, C. Karr, Jr. (Ed.). Academic Press, Inc., New York, Vol. III, Chapter 42.

Selvig, W.A. 1945. In *Chemistry of Coal Utilization*, H.H. Lowry (Ed.). John Wiley & Sons, Inc., New York, Chapter 4.

Sen, R., Srivastava, S.K., and Singh, M.M. 2009. Aerial oxidation of coal-analytical methods, instrumental techniques and test methods: A survey. *Indian Journal of Chemical Technology*, 16: 103–135.

Slack, A.V. 1981. In *Chemistry of Coal Utilization*, Second Supplementary Volume, M.A. Elliott (Ed.). John Wiley & Sons, Inc., New York, Chapter 22.

Smith, K.L. and Smoot, L.D. 1990. *Progress in Energy and Combustion Science*, 16: 1.

Speight, J.G. 1990. In *Fuel Science and Technology Handbook*, J.G. Speight (Ed.). Marcel Dekker, Inc., New York, Chapters 2 and 12.

Speight, J.G. 2005. *Handbook of Coal Analysis*. John Wiley & Sons, Inc., Hoboken, NJ.

Speight, J.G. 2007. *The Chemistry and Technology of Petroleum*, 4th edn. Taylor & Francis Group, Boca Raton, FL.

Visman, J. 1969. *Material Research and Standards*, 9(11): 9.

Voina, N.I. and Todor, D.N. 1978. In *Analytical Methods for Coal and Coal Products*, C. Karr, Jr. (Ed.). Academic Press, Inc., New York, Vol. II, Chapter 37.

Volborth, A. 1979a. In *Analytical Methods for Coal and Coal Products*, C. Karr, Jr. (Ed.). Academic Press, Inc., New York, Vol. III, Chapter 47.

Volborth, A. 1979b. In *Analytical Methods for Coal and Coal Products*. Academic Press, Inc., New York, Vol. III, Chapter 55.

Volborth, A., Dahy, J.P., and Miller, G.E. 1987. In *Coal Science and Chemistry*, A. Volborth (Ed.). Elsevier, Amsterdam, the Netherlands.

Vorres, K. 1979. In *Analytical Methods for Coal and Coal Products*, C. Karr, Jr. (Ed.). Academic Press, Inc., New York, Vol. III, Chapter 53.

Wallace, S. Bartle, K.D., and Perry, D.L. 1989. *Fuel*, 68: 1450.

Wood, G.H., Jr., Kehn, T.M., Carter, M.D., and Culbertson, W.C. 1983. Coal resource classification system of the U.S. Geological Survey. U.S. Geological Survey Circular 891. U.S. Department of the Interior, Washington, DC.

Zimmerman, R.E. 1979. *Evaluating and Testing the Coking Properties of Coal*. Miller Freeman & Co., San Francisco, CA.

9 Coal Properties

9.1 INTRODUCTION

Coal is a naturally occurring combustible material with varying composition and it is not surprising that the properties of coal vary considerably from coal type to coal type and even from sample to sample within a specific coal types. This can only be ascertained by application of a series of standard test methods (Zimmerman, 1979; Speight, 2005).

The constituents of coal can be divided into two groups: (1) the organic fraction, which can be further subdivided into soluble and insoluble fractions (Chapters 10 and 11) as well as microscopically identifiable macerals (Chapter 4), and (2) the inorganic fraction, which is commonly identified as ash subsequent to combustion (Chapter 7). Because of this complex heterogeneity it might be expected that the properties of coal can vary considerably, even within a specific rank of coal.

Furthermore, depending on the rank and plastic properties coals are divided into the following types:

- Hard coking coals are necessary for the production of strong coke; they are evaluated based on the strength, yield, and size distribution of coke produced which is dependent on rank and plastic properties of the coal.
- Semisoft coking coal (weak coking coal) is used in the coke blend, but results in a low coke quality with a possible increase in impurities; there is scope for interchangeability between thermal coal and semisoft coking coal.
- Coal used for pulverized coal injection reduces the consumption of coke per ton of pig iron as it replaces coke as a source of heat and, at high injection rates, as a reductant.
- Thermal coals are mostly used for electricity generation; the majority of thermal coal traded internationally is fired as pulverized fuel and vary in rank from subbituminous coal to bituminous coal.

Evaluation of coal for any of the earlier uses can be achieved by the determination of several noteworthy properties (Chapter 8) and there are also various other properties (Table 9.1) which provide even more valuable information about the potential use for coal (Van Krevelen, 1957). Indeed, there are also those properties of organic materials which offer valuable information about environmental behavior (Lyman et al., 1990). Hence, environmental issues lead to an additional need to investigate the properties of coal.

In the broadest sense, it has been suggested that the granular nature of high-rank coals is of importance in understanding the physical nature of coal if coal is modeled in terms of a granular medium which consists of graphite-like material embedded in an insulating organic matrix (Cody et al., 1990). Indeed, there have been several earlier suggestions of the graphite-like nature of coal, particularly from x-ray diffraction studies (Speight, 1978, and references cited therein) and perhaps this is a means by which the behavior of coal can be modeled. But, if this be the case, the precise role of the smaller aromatic systems needs also to be defined more fully. Nevertheless, it certainly offers new lines of thinking about coal behavior.

TABLE 9.1

Selected Properties Used for Coal Evaluation and Use

Test/Property	Results/Comments
Physical properties	
Density	True density as measured by helium displacement
Specific gravity	Apparent density
Pore structure	Specification of the porosity or ultrafine structure of coals and nature of pore structure between macro, and transitional pores
Surface area	Determination of total surface area by heat of absorption
Reflectivity	Useful in petrographic analyses
Mechanical properties	
Strength	Specification of compressibility strength
Hardness/abrasiveness	Specification of scratch and indentation hardness; also abrasive action of coal
Friability	Ability to withstand degradation in size on handling, tendency toward breakage
Grindability	Relative amount of work needed to pulverize coal
Dustiness index	Amount of dust produced when coal is handled
Thermal properties	
Calorific value	Indication of energy content
Heat capacity	Measurement of the heat required to raise the temperature of a unit amount of coal 1°
Thermal conductivity	Time rate of heat transfer through unit area, unit thickness, unit temperature difference
Plastic/agglutinating	Changes in a coal upon heating; caking properties of coal
Agglomerating index	Grading on nature of residue from 1 g sample when heated at 950°C (1550°F)
Free swelling index	Measure of the increase in volume when a coal is heated without restriction
Electrical properties	
Electrical resistivity	Electrical resistivity of coal measured in ohm-centimeters
Dielectric constant	Measure of electrostatic polarizability

9.2 PHYSICAL PROPERTIES

The physical properties and the behavior of coal play an important part in dictating the methods by which coal should be handled and utilized. This section considers those properties such as, for example, density and hardness, which are definitely physical in nature, in contrast to the properties (e.g., spectroscopic properties) which arise by virtue of the molecular structural types that occur within the coal (Chapter 10).

At first consideration, there may appear to be little, if any, relationship between the physical and chemical behavior of coal but in fact the converse is very true. For example, the pore size of coal (which is truly a physical property) is a major factor in determining the chemical reactivity of coal (Walker, 1981). And chemical effects which result in the swelling and caking of coal(s) have a substantial effect on the means by which coal should be "handled" either prior to or during a coal conversion operation.

9.2.1 DENSITY (SPECIFIC GRAVITY)

For porous solids, such as coal, there are three different density measurements: true density, particle density, and apparent density.

The *true density* is usually determined by displacement of a fluid, but because of the porous nature of coal and also because of physicochemical interactions the observed density data vary with the particular fluids employed (Agrawal, 1959; Mahajan and Walker, 1978).

The apparent density of coal is determined by immersing a weighed sample of coal in a liquid followed by accurate measurement of the liquid that is displaced. For this procedure, the liquid should (1) wet the surface of the coal, (2) not absorb strongly to the coal surface, (3) not cause swelling, and (4) penetrate the pores of the coal.

It is difficult (if not impossible) to satisfy all of these conditions as evidenced by the differing experimental data obtained with solvents such as water, methanol, carbon tetrachloride, benzene, and other fluids. Thus, there is the need to always specify the liquid employed for the determination of density by means of this (pycnometer) method. Furthermore, a period of 24 h may be necessary for the determination because of the need for the liquid to penetrate the pore system of the coal to the maximum extent.

The true density of coal is usually determined by helium displacement and, therefore, is often referred to as the "helium density." Helium is used because it has the ability to penetrate all of the pores of a given sample of coal without (presumably) any chemical interaction. Thus, in the direct-pressure method, a known quantity of helium and a weighed sample of coal are introduced into an apparatus of known volume whereupon the pressure of the helium at a given temperature allows calculation of the volume of the coal. In the indirect method, mercury is used to compensate for the helium displaced by the introduction of the coal.

It is generally believed that the use of helium gives a more accurate determination of coal density but there is evidence (Kotlensky and Walker, 1960) that part of the pore system may be inaccessible to the helium. Thus, when helium is used as the agent for determining coal density, the density (helium density) may differ from the true density and may actually be lower than the true density.

The particle density is the weight of a unite volume of solid including the pores and cracks (Mahajan and Walker, 1978). The particle density can be determined by any one of three methods, viz, (1) mercury displacement (Gan et al., 1982), (2) gas flow (Ergun, 1951), and (3) silanization (Ettinger and Zhupakhina, 1960).

The density of coal shows a notable variation with rank for carbon content (Figure 9.1) and, in addition, the methanol density is generally higher than the helium density because of the contraction of adsorbed helium in the coal pores as well as by virtue of interactions between the coal and the methanol which result in a combined volume that is notably less than the sum of the separate volumes. Similar behavior has been observed for the water density of coals having 80%–84% carbon.

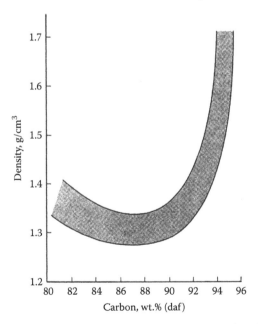

FIGURE 9.1 Variation of coal density with carbon content. (From Berkowitz, N., *An Introduction to Coal Technology*, Academic Press, New York, 1979.)

Coal with more than 85% carbon usually exhibits a greater degree of hydrophobic character than the lower-rank coals with the additional note that the water density may be substantially lower than the helium density; for the 80%–84% carbon coals, there is generally little, if any, difference between the helium density and the water density.

However, on the issue of the hydrophobicity of coal, recent work on the prediction of this characteristic indicates that the hydrophobicity of coal correlates better with the moisture content than with the carbon content and better with the moisture/carbon molar ratio than with the hydrogen/carbon or oxygen/carbon atomic ratios. Thus, it appears that there is a relationship between the hydrophobicity of coal and the moisture content (Labuschagne, 1987; Labuschagne et al., 1988).

An additional noteworthy trend is the tendency for the density of coal to exhibit a minimum value at approximately 85% carbon. For example, a 50%–55% carbon coal will have a density of approximately 1.5 g/cm and this will decrease to, say, 1.3 g/cm for an 85% carbon coal followed by an increase in density to ca. 1.8 g/cm for a 97% carbon coal. On a comparative note, the density of graphite (2.25 g/cm) also falls into this trend.

Determination of the density of various coal macerals have also been reported (Table 9.2) and, although the variations are not great, the general order of density for macerals (having the same approximate carbon content) is

Exinite < vitrinite < micrinite

However, it should also be noted that the density of a particular maceral does vary somewhat with the carbon content (Figure 9.2) (Van Krevelen, 1957).

The *in-place density* of coal is the means by which coal in the seam can be expressed as tons per acre per foot of seam thickness and/or tons per square mile per foot of seam thickness (Table 9.3).

The bulk density of the solid is the mass of the solid particles per unit of volume they occupy. Major factors influencing the bulk density of coal are moisture content, particle surface properties, particle shape, particle size distribution, and particle density (Leonard et al., 1992).

TABLE 9.2
Maceral Density

Macerals	% C	D_{CH_3OH}	$d_{He(calc)}$
Exinites	85.49	1.201	1.187
	87.41	1.213	1.193
	89.10	1.288	1.267
	89.29	1.347	1.325
	83.5	1.345	1.304
Vitrinites	88.36	1.317	1.295
	88.84	1.368	1.338
	86.77	1.463	1.435
	87.98	1.415	1.368
Micrinites	89.59	1.414	1.389
	89.78	1.413	1.385

Sources: Braunstein, H.M. et al., Eds., Environmental health and control aspects of coal conversion: An information overview, Report ORNL-EIS-94. Oak Ridge National Laboratory, Oak Ridge, TN, 1997; Kroger, C. and Badenecker, J., *Brennstoff Chem.*, 38, 82, 1957.

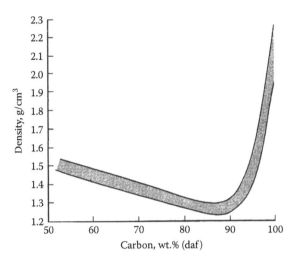

FIGURE 9.2   Variation of vitrinite density with carbon content. (From Berkowitz, N., *An Introduction to Coal Technology*, Academic Press, New York, 1979.)

TABLE 9.3
In-Place Density of Coal

	Weight of In-Place Coal	
Rank	Tons per Acre per Foot of Thickness	Tons per Square Mile per Foot of Thickness
Anthracite	2310	1479×10^6
Semianthracite	2039	1305×10^6
Bituminous	1903	1218×10^6
Subbituminous	1767	1131×10^6
Lignite	1631	1044×10^6

Source: Baughman, G.L., *Synthetic Fuels Data Handbook*, 2nd edn., Cameron Engineers, Inc., Denver, CO, 1978, p. 166.

Generally, the manner the bulk sample packs into a confined space is reflected by its bulk density and is related to the size distribution as well as the effects of moisture on the packing ability of the particles. There will always be such a mixture of fine and coarse particles at which the bulk density will assume the highest value, higher than assumed by any of these fractions when packed separately. As the range of particle size is increased, the bulk density is also increased (Wakeman, 1975).

The *bulk density* is not an intrinsic property of coal and varies depending on how the coal is handled. Bulk density is the mass of many particles of coal divided by the total volume occupied by the particles—the total volume includes particle volume, interparticle void volume, and internal pore volume. This allows the density of coal to be expressed in terms of the cubic foot weight of crushed coal (ASTM, 2011c), which varies with particle size of the coal and packing in a container (Table 9.4). In addition, the bulk density of coal varies with rank:

- Anthracite: 50–58 lb/ft^3 (800–929 kg/m^3)
- Bituminous coal: 42–57 lb/ft^3 (673–913 kg/m^3)
- Lignite: 40–54 lb/ft^3 (641–865 kg/m^3)

TABLE 9.4
Bulk Density of Coal[a]

Particle Size of Coal Sample	Bulk Density (lb/ft³)	Percent Voids
Mine run	55	37
Lump (plus 6 in.)	50	43
6 × 3 in.	48	45
3 × 2 in.	46	47
2 in. screenings	49	44
2 × 1 1/2 in.	45	48
3/4 × 7/16 in.	42	52
7/16 in. × 0	47	46
5/16 in. × No. 10	45	48
1/16 in. × 48 M	42	52
No. 48 × 0	35	60

Source: Baughman, G.L., *Synthetic Fuels Data Handbook*, 2nd edn., Cameron Engineers, Inc., Denver, CO, 1978.

[a] Specific gravity of coal: 1.4.

General formula used for soil samples (Tisdall, 1951; Birkeland, 1984) may also be applied to coal. Thus,

$$\text{Bulk density} = \left(\frac{\text{Weight}}{\text{Volume}} \right) = \left(\frac{63}{35} \right)$$

$$\text{Porosity (\%)} = \left(1 - \frac{\text{Bulk density}}{\text{Particle density}} \right) \times 100$$

$$\text{Porosity (\%)} = \left(1 - \frac{1.8}{2.65} \right) \times 100$$

The bulk density of coal decreases with increasing moisture content until a minimum is reached (at approximately 6%–8% moisture) but then increases with moisture content. With the use of aqueous solutions of wetting additives, the coal bulk density can be increased by up to 15% (Leonard et al., 1993).

Due to the hydrophobic nature of higher-rank coals, their bulk density rapidly decreases with increasing moisture, as the water stays on the surface of the coal in between the particles, increasing the volume of the bulk solid (decreasing bulk density). Further increase in moisture leads to the minimum bulk density attained at about 6%–8% moisture content (Leonard et al., 1992). Beyond this level of moisture, more water penetrates the spaces in between the particles and forms a water layer around them allowing for aggregation which leads to tighter packing configuration. As a result particles are packed into smaller volume which further leads to an increase in bulk density of the sample. Furthermore, small additions of chemical reagents reducing surface tension of water resulted in an increased bulk density. These experiments showed that the bulk density could be increased by 13%–15% with the use of such additives (Leonard et al., 1992).

The aggregation of fines, leading to tight packing of particles, affects the bulk density of coal samples. Wettability appears to be a controlling factor in the aggregation of fine coal particles.

As a result, different patterns of bulk density can be anticipated with moisture increase for hydrophilic coal samples and also for hydrophobic coal samples (Holuszko and Laskowski, 2010).

9.2.2 POROSITY AND SURFACE AREA

Coal is a porous material; thus, the porosity and surface area of coal (Mahajan and Walker, 1978) have a large influence on coal behavior during mining, preparation, and utilization.

Although porosity dictates the rate at which methane can diffuse out of the coal (in the seam) and there may also be some influence during preparation operations in terms of mineral matter removal, the major influence of the porous nature of coal is seen during the utilization of coal. For example, during conversion chemical reactions occur between gas (and/or liquid) products and surface features, much of which exist within the pore systems.

The pore systems of coal have generally been considered to consist of micropores having sizes up to approximately 100 and macropores having sizes greater than 300 Å (Gan et al., 1972; Mahajan and Walker, 1978). Other work (Kalliat et al., 1981), involving a small-angle x-ray investigation of porosity in coals, has thrown some doubt upon this hypothesis by bringing forward the suggestion that the data are not consistent with the suggestion that many pores have dimensions some hundreds of angstrom units in diameter but have restricted access due to small openings which exclude nitrogen (and other species) at low temperatures. Rather, the interpretation that is favored is that the high values of surface area obtained by adsorption studies are the result of a large number of pores with minimum pore dimensions which are not greater than ca. 30.

There are also indications that the adsorption of small molecules on coal, such as methanol, occurs by a site-specific mechanism (Ramesh et al., 1992). In such cases, it appears that the adsorption occurs first at high-energy sites but with increasing adsorption the (methanol) adsorbate continues to bind to the surface rather than to other (polar) methanol molecules and there is evidence for both physical and chemical adsorption. In addition, at coverages below a monolayer, there appears to be an activation barrier to the adsorption process. Whether or not such findings have consequences for surface area and pore distribution studies remains to be seen. But there is the very interesting phenomenon of the activation barrier which may also have consequences for the interpretation of surface effects during coal combustion (Chapters 14 and 15). As an aside, adsorption studies of small molecules on coal has been claimed to confirm the copolymeric structure of coal (Milewska-Duda, 1991).

As already noted with respect to coal density, the porosity of coal decreases with carbon content (Figure 9.3) (King and Wilkins, 1944) and has a minimum at approximately 89% carbon coals followed by a marked increase in porosity. The "nature" of the porosity also appears to vary with carbon content (rank); for example, the macropores are usually predominant in the lower carbon (rank) coals while higher carbon (rank) coals contain predominantly micropores. Thus, pore volume can be calculated from the relationship:

$$V_p = 1/\rho_{Hg} - 1\rho_{He}$$

In this equation, ρ_{Hg} is the mercury density and ρ_{He} is the helium density, decreases with carbon content (Figure 9.4). In addition, the surface area of coal varies over the range of 10–200 m^2/g and also tends to decrease with the carbon content of the coal.

Porosity and surface area are two very important properties with respect to the gasification of coal since the reactivity of coal increases as the porosity and surface area of the coal increases. Thus, the gasification rate of coal is greater for the lower-rank coals than for the higher-rank coals. The porosity of coal is calculated from the relationship:

$$\rho = 100\rho_{Hg}(1/\rho_{Hg} - 1\rho_{He})$$

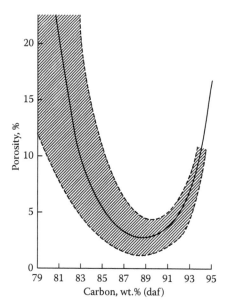

FIGURE 9.3 Variation of porosity with carbon content. (From Berkowitz, N., *An Introduction to Coal Technology*, Academic Press, New York, 1979.)

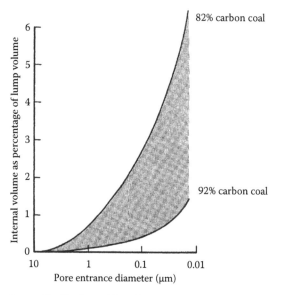

FIGURE 9.4 Variation of pore distribution with carbon content. (From Berkowitz, N., *An Introduction to Coal Technology*, Academic Press, New York, 1979.)

By determining the apparent density of coal in fluids of different, but known, dimensions, it is possible to calculate the pore size (pore volume) distribution. The open pore volume (V), that is, the pore volume accessible to a particular fluid, can be calculated from the relationship:

$$V = 1\rho_{Hg} - 1\rho_a$$

where ρ_a is the apparent density in the fluid.

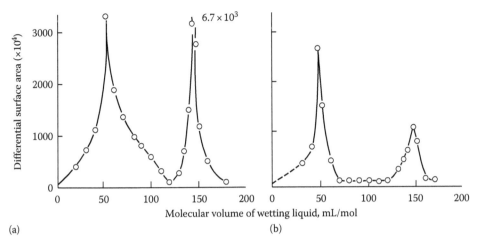

(a) (b)

FIGURE 9.5 Variation of surface area with pore size for (a) a low-rank coal and (b) a high-rank coal. (From Berkowitz, N., *An Introduction to Coal Technology*, Academic Press, New York, 1979.)

The size distribution of the pores within a coal can be determined by immersing the coal in mercury and progressively increasing the pressure. Surface tension effects prevent the mercury from entering the pores having a diameter smaller than a given value d for any particular pressure p such that

$$p = 4\sigma \cos \theta / d$$

where
    σ is the surface tension
    θ is the angle of contact

By recording the amount of mercury entering the coal for small increments of pressure, it is possible to build up a picture of the distribution of sizes of pores (Figure 9.5) (Van Krevelen, 1957). However, the total pore volume accounted for by this method is substantially less than that derived from the helium density, thereby giving rise to the concept that coal contains two pore systems: (1) a macropore system accessible to mercury under pressure and (2) a micropore system that is inaccessible to mercury but accessible to helium. By using liquids of various molecular sizes, it is possible to investigate the distribution of micropore sizes. However, the precise role or function of these micropores as part of the structural model of coal is not fully understood, although it has been suggested that coal may behave in some respects like a molecular sieve.

9.2.3 REFLECTANCE

It is often a surprise to many workers that coal, being often regarded as a black solid which appears to be impervious to light, should be recognized as having optical properties. That some preparation or conditioning of the coal is necessary is also of importance.

Thus, coal may be examined in visible light by either transmission or reflectance (Tschamler and de Ruiter, 1963). The former is a measure of light absorbance at various wavelengths and may be determined for thin sections of coal or finely divided coal pressed into a potassium bromide disk or solutions of coal extracts in solvents such as pyridine or films of coal that have been deposited by evaporation of a dispersing liquid.

Coal reflectance (ASTM, 2011k) is very useful because it indicates several important properties of coal (Davis, 1978). Among these are the content of several macerals (ASTM, 2011l) and carbonization temperature. Coal reflectance is determined by the relative degree to which a beam of polarized light is reflected from a polished coal surface. The coal is crushed to pass a number 20 screen (850 μm screen) with minimal fines production and the particles are formed into a briquette held together with a binder. One side of the briquette is polished using successively finer abrasives until a smooth surface that is scratch-free, smear-free, and char-free is obtained. A metallurgical, or opaque-ore, microscope is employed to determine the reflectance with vertical illumination using polarized light.

Prior to measurement of reflectance, the sample face is covered with cedar oil or a commercial immersion oil and then multiple readings are taken of the maximum reflectance of the coal component (e.g., vitrinite) of interest. These values are compared with readings of high-index glass standards (of known reflectance) which are available with reflectance values typically ranging from 0.302% to 1.815%.

Thus, although coal always appears as a black mass, thin layers and polished faces exhibit a variety of colors. For example, in incident light, fusinite and micrinite are white whereas exinite is a translucent yellow color; on the other hand, exinite is orange in transmitted light. Obviously, these color differences are employed for differentiating maceral types. In addition, the reflectance of coal varies with rank (Figure 9.6) and the reflectance data for air are considerably higher than those obtained using an oil medium (Table 9.5).

Coal reflectance is important in aiding the determination of the maceral composition of coal which, in turn, is helpful for the prediction of behavior in processing (Davis et al., 1991) (Chapter 4). For example,

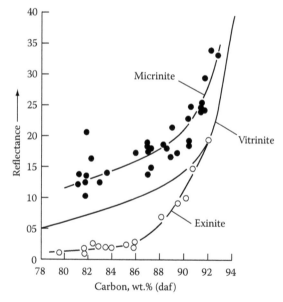

FIGURE 9.6 Variation of reflectance with carbon content for different macerals. (From Dormans, H.N.M. et al., *Fuel,* 26, 321, 1957.)

TABLE 9.5

Reflectance of Coal in Air and Oil

Carbon Content of Coal (%)	Maximum Reflectance, Approximate (%)	
	Air	Oil
60	0.4	6
90	1.0	8
96	6.5	17

the basic use for which determination of vitrinite reflectance is applied is the measurement of the degree of metamorphism or rank of coals and the organic components of sediments. It is possible to relate the reflectance of a coal to parameters such as carbon content, volatile matter, and calorific value.

Petrographic analyses are also used to predict coke properties (such as strength) which would be produced from the coal. Other uses include the estimation of the chemical properties of fresh coals from the reflectance of weathered specimens.

9.2.4 REFRACTIVE INDEX

The refractive index of coal can be determined by comparing the reflectance in air with that in cedar oil (Cannon and George, 1943; Speight, 2005). For vitrinite, the refractive index usually falls within the range of 1.68 (58% carbon coal) to 2.02 (96% carbon coal).

9.3 MECHANICAL PROPERTIES

In contrast to the proximate analysis, ultimate (elemental) analysis and certain of the physical properties, the mechanical properties of coal have been little used. But these properties are of importance and should be of consideration in predicting coal behavior during mining, handling, and preparation.

For example, the mechanical properties of coal are of value as a means of predicting the strength of coal and its behavior in mines when the strength of coal pillars and stability of coal faces are extremely important factors. The mechanical properties of coal are also of value in areas such as coal winning (for the design and operation of cutting machinery), comminution (design and/or selection of mills), storage (flow properties, failure under shear), handling (shatter and abrasion during transport), as well as in many other facets of coal technology (Yancey and Geer, 1945; Van Krevelen, 1957; Brown and Hiorns, 1963; Trollope et al., 1965; Evans and Allardice, 1978).

9.3.1 STRENGTH

Interest in the compressive strength of coal arises mainly from the relationship to the strength of coal pillars used for roof support in mining operations (Chapter 5).

There are different methods for estimating coal strength and hardness: compressive strength, fracture toughness, or grindability, all of which show a trend relative to rank, type, and grade of the coal. The measurement of coal strength is affected by the size of the test specimen, the orientation of stress relative to banding, and the confining pressure of the test (Hobbs, 1964; Zipf and Bieniawski, 1988; Medhurst and Brown, 1998).

By its nature, coal is a banded material which makes it weak in comparison to most other rocks. Intact rock strength is commonly defined as the strength of the rock material that occurs between discontinuities, which in coal are closely spaced and related to lithotype banding and cleat. For a given rank, individual lithotypes can have large compressive strength differences owing to wide ranges in maceral composition, banding texture, and cleat density (Medhurst and Brown, 1998).

Thus, the strength of a bituminous coal specimen is influenced also by its lateral dimension, the smaller specimens showing greater strength than the larger (Table 9.6), which can be attributed to the presence in the larger specimen of fracture planes or cleats. In fact, it is the smaller samples which present a more accurate indication of the strength of the coal. A factor that probably contributes to the lower strength of large blocks of coal tested in the laboratory is the difficulty of mining large blocks and transporting them from the mine to the laboratory without imposing strains that start disintegration along the cleats.

The variation of compressive strength with rank of coals has been noted and a plot of strength against volatile matter shows the customary minimum to be 20%–25% dry, ash-free volatile matter (Table 9.7 and Figure 9.7) for compression both perpendicular and parallel to the bedding plane.

TABLE 9.6
**Variation of Compressive Strength
of Coal with Size**

Size	Average Maximum Load
2.5–4 in. cubes	2486
7–8 in. cubes	2170
10–12 in. cubes	2008
12 × 12 × 18 in. high	1152
Approximately 30 in. cube	817
Approximately 54 in. cube	306

TABLE 9.7
**Compressive Strength of Various
Coals**

Coal Rank	Compressive Strength (psi)
Anthracite	2370
	3390
	2000
	1740
Bituminous	310–2490
	500
Lignite	2190–6560

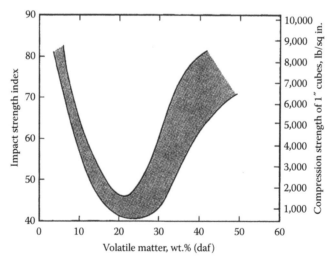

FIGURE 9.7 Variation of coal strength with volatile matter yield. (From Brown, R.L. and Hiorns, F.J., *Chemistry of Coal Utilization*, Supplementary Volume, H.H. Lowry, Ed., John Wiley & Sons, New York, 1963, Chapter 3.)

TABLE 9.8
Scratch Hardness of Various Coals

Material	Scratch Hardness Relative to Barnsley Soft Coal
Anthracite, Great Mountain	1.70
Anthracite, Red Vein	1.75
Welsh steam	0.29
Barnsley hards	0.85
Barnsley softs	1.00
Illinois coal	1.10
Cannel	0.92
Carbonaceous shale	0.69
Shale	0.32
Pyrite	5.71
Calcite	1.92

9.3.2 HARDNESS

Considerable information is available on the properties of coal that relate to its hardness, such as friability and grindability, but little is known of the hardness of coal as an intrinsic property.

The scratch hardness of coal can be determined by measuring the load on a pyramidal steel point required to make a scratch 100 μm in width on the polished surface of a specimen. The scratch hardness of anthracite is approximately six times that of a soft coal whereas pyrite is almost 20 times as hard as a soft coal (Table 9.8). Similar data were noted for anthracite and cannel coal but durain, the reputedly hard component of coal, was found to be only slightly harder than vitrain and cannel coal.

Although the resistance of coal to abrasion may have little apparent commercial significance, the abrasiveness of coal is, on the other hand, a factor of considerable importance. Thus, the wear of grinding elements due to the abrasive action of coal results in maintenance charges that constitute one of the major items in the cost of grinding coal for use as pulverized fuel. Moreover, as coals vary widely in abrasiveness, this factor must be considered when coals are selected for pulverized fuel plants. A standardized, simple laboratory method of evaluating the abrasiveness of coal would assist, like the standard grindability test now available, in the selection of coals suitable for use in pulverized form.

Actually, the abrasiveness of coal may be determined more by the nature of its associated impurities than by the nature of the coal substance. For example, pyrite is 20 times harder than coal, and the individual grains of sandstone, another common impurity in coal, also are hard and abrasive.

9.3.3 FRIABILITY

One measure of the strength of coal is its ability to withstand degradation in size on handling. The tendency toward breakage on handling (friability) depends on toughness, elasticity, and fracture characteristics as well as on strength, but despite this fact the friability test is the measure of coal strength used most frequently.

Friability is of interest primarily because friable coals yield smaller proportions of the coarse sizes which may (depending on use) be more desirable and there may also be an increased amount of surface in the friable coals. This surface allows more rapid oxidation; hence, conditions are more favorable for spontaneous ignition (Chapter 14), loss in coking quality in coking coals, and other changes that accompany oxidation. These economic aspects of the friability of coal have provided the incentive toward development of laboratory friability tests.

The tumbler test for measuring coal friability (ASTM, 2011f) employs a cylindrical porcelain jar mill (7.25 in. [18.4 cm] in size) fitted with three lifters that assist in tumbling the coal. A 1000 g

TABLE 9.9
Variation of Friability with Rank

Rank of Coal	Number of Tests	Friability (%)
Anthracite	36	33
Bituminous (lv)[a]	27	70
Bituminous	87	43
Subbituminous A	40	30
Subbituminous B	29	20
Lignite	16	12

Source: Yancey, H.R. and Geer, M.R., *Chemistry of Coal Utilization*, H.H. Lowry (Ed.), John Wiley & Sons, New York, 1945, p. 3.

[a] lv, low volatile.

sample of coal sized between 1.5 and 1.05 in. (38.1 and 26.7 mm) square-hole screens is tumbled in the mill (without grinding medium) for 1 h at 40 rpm. The coal is then removed and screened on square-hole sieves with openings of 1.05 in. (26.7 mm), 0.742 in. (18.8 mm), 0.525 in. (13.3 mm), 0.371 in. (9.42 mm), 0.0369 in. (1.19 mm), and 0.0117 in. (0.297 mm).

Friability is reported as the percentage reduction in the average particle size during the test. For example, if the average particle size of the tumbled coal were 75% that of the original sample, the friability would be 25%.

A drop shatter test has also been described for determining the friability of coal (ASTM, 2011e) which is similar to the standard method used as a shatter test for coke (ASTM, 2011m). In this method, a 50 lb sample of coal (2–3 in. [4.5–7.6 cm]) is dropped twice from a drop-bottom box onto a steel plate 6 ft below the box. The materials shattered by the two drops are then screened over round-hole screens with 3.0 in. (76.2 mm), 2.0 in. (50.8 mm), 1.5 in. (38.1 mm), 1.0 in. (25.4 mm), 0.75 in. (19.05 mm), and 0.5 in. (12.7 mm) openings and the average particle size is determined.

The average size of the material, expressed as a percentage of the size of the original sample, is termed the sizeability, and its complement, the percentage of reduction in average particle size, is termed the percentage friability. Provision is made for testing sizes other than that stipulated for the standard test to permit comparison of different sizes of the same coal.

Attempts have been made to correlate the friability of coal with rank (Table 9.9). Lignite saturated with moisture was found to be the least friable and friability increased with coal rank to a maximum for coals of the low-volatile bituminous coal. The friability of anthracites is comparable with that of subbituminous coals; both are stronger than bituminous coals and decidedly more resistant to breakage than some of the extremely friable semibituminous coals.

The relationship between the friability of coal and its rank has a bearing on its tendency to undergo spontaneous heating and ignition (Chakravorty, 1984; Chakravorty and Kar, 1986). The friable, low-volatile coals, because of their high rank, do not oxidize readily despite the excessive fines and the attendant increased surface they produce on handling. Coals of somewhat lower rank, which oxidize more readily, usually are relatively nonfriable; hence, they resist degradation in size with its accompanying increase in the amount of surface exposed to oxidation. But above all, the primary factor in coal stockpile instability is unquestionably oxidation by atmospheric oxygen whilst the role of any secondary factors such as friability is to exacerbate the primary oxidation effect (Jones and Vais, 1991).

9.3.4 GRINDABILITY

The grindability of coal (i.e., the ease with which coal may be ground fine enough for use as pulverized fuel) is a composite physical property embracing other specific properties such as hardness, strength, tenacity, and fracture. Several methods of estimating relative grindability utilize a porcelain jar mill in which each coal may be ground for, say, 400 revolutions and the amount of new surface is estimated from screen analyses of the feed and of the ground product. Coals are then rated in grindability by comparing the amount of new surface found in the test with that obtained for a standard coal.

The test for grindability (Hardgrove, 1932; Edwards et al., 1980; ASTM, 2011d; ISO, 2011d) utilizes a ball-and-ring-type mill in which a 50 g sample of closely sized coal is ground for 60 revolutions after which the ground product is screened through a 200-mesh sieve and the grindability index (GI) is calculated from the amount of undersize produced using a calibration chart (Table 9.10).

TABLE 9.10
Hardgrove Grindability Indexes of Selected U.S. Coals

State	County	Bed	Mine	Hardgrove Grindability Index
Alabama	Walker	Black Creek	Drummond	46
Colorado	Fremont		Monarch No. 4	46
	Mesa		Cameo	47
Illinois	Fulton	No. 2	Sun Spot	53
	Stark	No. 6	Allendale	61
	Williamson	No. 6	Utility	57
Indiana	Pike	No. V	Blackfoot	54
Iowa	Lucas	Cherokee	Big Ben	61
Kansas	Crawford	Bevier	Clemens	62
Kentucky	Bell	High Splint	Davisburg	44
	Muhlenberg	No. 11	Crescent	55
	Pike	Elkhorn Nos. 1 and 2	Dixie	42
Missouri	Boone	Bevier	Mark Twain	62
Montana	Richland		Savage	62
New Mexico	McKinley	Black Diamond	Sundance	51
North Dakota	Burke		Noonan	38
Ohio	Belmont	No. 9	Linda	50
	Harrison	No. 8	Bradford	51
Pennsylvania	Cambria	Lower Kittanning (bituminous)	Bird No. 2	109
	Indiana	Lower Freeport (bituminous)	Acadia	83
	Schuylkill	Various (anthracite)		38
	Washington	Pittsburgh	Florence	55
	Westmoreland	Upper Freeport	Jamison	65
Tennessee	Grundy	Sewanee	Ramsey	59
Utah	Carbon	Castle Gate	Carbon	47
Virginia	Buchanan	Splash Dam	Harman	68
	Dickenson	Upper Banner		84
	Wise	Morris	Roda	43
West Virginia	Fayette	Sewell	Summerlee	86
	McDowell	Pocahontas No. 3	Jacobs Fork	96
	Wyoming	Powellton	Coal Mountain	58
	Wyoming	No. 2 Gas	Kopperston	70
Wyoming	Campbell	Smith/Rowland	Wyodak	59

Source: Baughman, G.L., *Synthetic Fuels Data Handbook*, 2nd edn., Cameron Engineers, Inc., Denver, CO, 1978, p. 169.

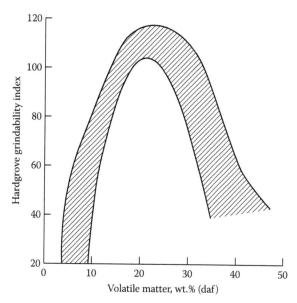

FIGURE 9.8   Variation of the hardgrove grindability index with volatile matter yield. (From Berkowitz, N., *An Introduction to Coal Technology*, Academic Press, New York, 1979.)

The results are converted into the equivalent HGI. High GI numbers indicate easy-to-grind coals. There is a rough relationship between volatile content and grindability in the low-, medium-, and high-volatile bituminous rank coals. Among these the low-volatile coals exhibit the highest GI values, often over 100. The high-volatile coals range in GI from the mid-50s down to the high 30s. There is also a correlation of sorts between friability and grindability. Soft, easily fractured coals generally exhibit relatively high GI values. ASTM has two methods for measuring friability, the drop shatter test, D440, and the tumbler test, D441, which should be used where more than a rough indication of friability is needed.

A general relationship exists between the grindability of coal and rank (Figure 9.8) insofar as the easier-to-grind coals are in the medium- and low-volatile groups but, nevertheless, the relationship between grindability and rank is far too approximate to permit grindability to be estimated from coal analysis.

9.3.5   DUSTINESS INDEX

The concept of a *dustiness index* was proposed to enable comparison among dust-producing capacities of different bulk materials, such as coal. Dustiness estimation methods were developed with a view to establishing relative *dustiness indices* (BOHS, 1985; Upton et al., 1990; Lyons and Mark, 1994; Vincent, 1995; Breum et al., 1996). The objective is to provide criteria for the selection of products that will lead to less dust emissions. It is important to note, however, that different dustiness methods will produce different rank orderings.

Dust removal from mine workings and coal preparation plants is an important aspect of safety (Chapters 5 and 6) and there have been constant attempts to improve dust removal technology (Henke and Stockmann, 1992). As the techniques have been developed, the predictability of how coal will behave, in terms of the dust produced, under certain conditions has also been sought. The "dustiness index" of coal is a means of determining the relative values which represent the amount of dust produced when coal is handled in a standard manner.

For the test method, a 50 lb sample of coal is placed on a slide plate in a metal cabinet of prescribed size. When the plate is withdrawn, the sample falls into a drawer and, after 5 s, two slides are inserted into the box. The slides collect suspended dust particles for 2 min (coarse dust) or for 10 min (fine dust). The dustiness index is reported as 40 times the gram weight of dust that has settled after either 2 min or an additional 8 min.

9.3.6 Cleat Structure

A system of joint planes is often observed in coal formations and these joint planes (cleats) are usually perpendicular to the bedding planes (Chapters 1 and 7); thus, cleat joints are usually vertical. The main system of joints is more commonly called the face cleat whereas a cross system of jointing is called the butt cleat.

Cleats are natural opening-mode fractures in coal beds. They account for most of the permeability and much of the porosity of coal bed gas reservoirs and can have a significant effect on the success of engineering procedures such as cavity stimulations. Because permeability and stimulation success are commonly limiting factors in gas well performance, knowledge of cleat characteristics and origins is essential for successful exploration and production.

Few data are available on apertures, heights, lengths, connectivity, and the relation of cleat formation to diagenesis, characteristics that are critical to permeability (Laubach et al., 1998). Moreover, recent studies of cleat orientation patterns and fracture style suggest that new investigations of even these well-studied parameters can yield insight into coal permeability.

More effective predictions of cleat patterns will come from advances in understanding cleat origins. Although cleat formation has been speculatively attributed to diagenetic and/or tectonic processes, a viable mechanical process for creating cleats has yet to be demonstrated. Progress in this area may come from recent developments in fracture mechanics and in coal geochemistry.

It has been reported that the cleat system in coal has a pronounced effect on the properties of a coal deposit. For example, holes drilled into coal perpendicular to the face cleat are said to yield from 2.5 to 10 times the amount of methane gas from the formation as holes drilled perpendicular to the butt cleat. Also, the cleat system of fracture and the frequency of cleats may determine the size of ROM coal. In general, pair of cleats will be oriented at approximately 90° to each other and the orientation of the cutting elements influence the output of coal-mining machines (Figure 9.9).

Directional properties also occur in coal and these properties can affect the direction of flow of gases. For a particular coal seam, analyses of natural fracture orientation, directional permeability, directional ultrasonic velocity, and directional tensile strength (Berenbaum and Brodie, 1959; Ingles, 1961) may disclose distinct coal structures that are important for gas flow through the coal (Skawinski et al., 1991; Puri and Seidle, 1992).

Cleat structure is of special importance in the passage of methane through, and from, coal beds (Chapters 1 and 7) as well as in the planning of underground coal gasification tests (Chapters 20 and 21).

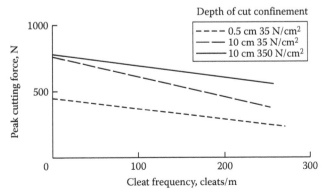

FIGURE 9.9 Effects of cleats on the cutting force. (From Baughman, G.L., *Synthetic Fuels Data Handbook*, 2nd edn., Cameron Engineers, Inc., Denver, CO, 1978.)

9.3.7 Deformation and Flow under Stress

The rheology (deformation and flow) of coal has been studied in an effort to apply it to characteristics of coal in coal mines; coal elasticity (quality of regaining original shape after deformation under strain) has also been studied. However, it may be quite difficult to obtain meaningful measurements of coal elasticity.

The heat that is generated in cutting a uniform sample (usually by means of a grinding wheel consisting of grit embedded in rubber) can cause plastic deformation of the coal surface and thereby affect the plastic properties. Water used to cool the grinding interface and to carry away particles may be absorbed by the coal and affect the elastic properties. Furthermore, discontinuities in the coal structure give a wide sample-to-sample variation.

9.4 THERMAL PROPERTIES

The thermal properties of coal are important in determining the applicability of coal to a variety of conversion processes. For example, the heat content (also called the heating value or calorific value) (Chapter 8) is often considered to be the most important thermal property. However, there are other thermal properties that are of importance insofar as they are required for the design of equipment that is to be employed for the utilization (conversion, thermal treatment) of coal in processes such as combustion, carbonization, gasification, and liquefaction. Plastic and agglutinating properties as well as phenomena such as the agglomerating index give indications of how coal will behave in a reactor during a variety of thermal processes (Chan et al., 1991).

9.4.1 Calorific Value

The calorific or heating value of a coal is a direct indication of the energy content and therefore is probably the most important property for determining the usefulness of coal (Figure 9.10). It is the amount of energy that a given quantity of coal will produce when burned. Heating value is used in determining the rank of coals. It is also used to determine the maximum theoretical fuel energy available for the production of steam. It is used to determine the quantity of fuel that must be handled, pulverized, and fired in the boiler.

The calorific value is reported as GCV, with a correction made if NCV is of interest (ASTM, 2011b,i,o,p; ISO, 2011c). For solid fuels such as coal, the gross heat of combustion is the heat produced by the combustion of a unit quantity of the coal in a bomb calorimeter with oxygen and under a specified set of conditions. The unit is calories per gram, which may be converted to the alternate units (1.0 kcal/kg = 1.8 Btu/lb = 4.187 kJ/kg).

The experimental conditions require an initial oxygen pressure of 300–600 psi and a final temperature in the range of 20°C–35°C (68°F–95°F) with the products in the form of ash, water, carbon dioxide, sulfur dioxide, and nitrogen. Thus, once the GCV has been determined, the NCV (i.e., the net heat of combustion) is calculated from the GCV (at 20°C; 68°F) by deducting 1030 Btu/lb (2.4×10^3 kJ/kg) to allow for the heat of vaporization of the water. The deduction is not actually equal to the heat of vaporization of water (1055 Btu/lb; 2.45×10^3 kJ/kg) because the calculation is to reduce the data from a gross value at constant volume to a net value at constant pressure. Thus, the differences between the GCV and the NCV are given by

$$NCV \text{ (Btu/lb)} = GCV - (1030 \times \text{total hydrogen} \times 9)/100.$$

The enthalpy, or heat content, of various coals has also been reported (Table 9.11) but has actually received somewhat less attention than the calorific value.

There are also reports of the use of DTA for the determination of the calorific value of coal (Munoz-Guillena et al., 1992).

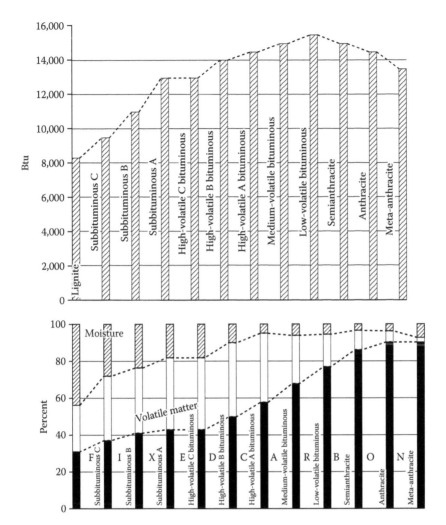

FIGURE 9.10 Variation of calorific value with rank. (From Baughman, G.L., *Synthetic Fuels Data Handbook*, 2nd edn., Cameron Engineers, Inc., Denver, CO, 1978.)

9.4.2 HEAT CAPACITY

The heat capacity of a material is the heat required to raise the temperature of one unit weight of a substance 1° and the ratio of the heat capacity of one substance to the heat capacity of water at 15°C (60°F) in the specific heat. The heat capacity of coal can be measured by standard calorimetric methods for mixtures (e.g., see ASTM, 2011a).

The units for heat capacity are Btu per pound per degree Fahrenheit (Btu/lb/°F) or calories per gram per degree centigrade (cal/g/°C), but the specific heat is the ratio of two heat capacities and is therefore dimensionless. The heat capacity of water is 1.0 Btu/lb/°F (=4.2 × 10³ J/kg/K), and, thus, the heat capacity of any material will always be numerically equal to the specific heat. Consequently, there has been the tendency to use the terms heat capacity and specific heat synonymously.

The specific heat of coal (Table 9.12) usually increases with its moisture content (Figure 9.11), decreases with carbon content (Figure 9.12), and increases with volatile matter content (Figure 9.13), with ash content exerting a somewhat lesser influence.

TABLE 9.11
Heat Content of Coal at Various Temperatures

Temperature		Heat Content		
		As Tested		Ash-Free Basis
°C	°F	cal/g	Btu/lb	cal/g
Lignite (Texas)				
32.7	90.9	11.8	21.2	13.5
69.3	156.7	20.2	36.4	22.5
95.3	203.5	25.4	45.7	27.7
34.4	273.9	39.2	70.6	42.5
Subbituminous B (Wyoming)				
42.3	108.1	14.1	25.4	14.5
65.0	149.0	19.4	34.9	19.8
89.7	193.5	26.4	47.5	26.9
112.6	234.7	34.0	61.2	34.6

Source: Adapted from Baughman, G.L., *Synthetic Fuels Data Handbook*, 2nd edn., Cameron Engineers, Inc., Denver, CO, 1978, p. 173.

TABLE 9.12
Specific Heat of Air-Dried Coals

Source	Proximate Analysis (% w/w)			
	Moisture	Volatile Matter	Fixed Carbon	Ash
West Virginia	1.8	20.4	72.4	5.4
Pennsylvania (bituminous)	1.2	34.5	58.4	5.9
Illinois	8.4	35.0	48.2	8.4
Wyoming	11.0	38.6	40.2	10.2
Pennsylvania (anthracite)	0.0	16.0	79.3	4.7

Source	Mean Specific Heat			
	28°C–65°C	25°C–130°C	25°C–177°C	25°C–227°C
West Virginia	0.261	0.288	0.301	0.314
Pennsylvania (bituminous)	0.286	0.308	0.320	0.323
Illinois	0.334			
Wyoming	0.350			
Pennsylvania (anthracite)	0.269			

Source: Baughman, G.L., *Synthetic Fuels Data Handbook*, 2nd edn., Cameron Engineers, Inc., Denver, CO, 1978, p. 172.

The values for the specific heats of various coals fall into the general range 0.25–0.37 but, as with other physical data, comparisons should only be made on an equal (moisture content, ash content, etc.) basis.

Estimates of the specific heat of coal have also been made on the assumption that the molecular heat of a solid material is equal to the sum of the atomic heats of the constituents (Kopp's law); the atomic heat so derived is divided by the atomic weight to give the (approximate) specific heat.

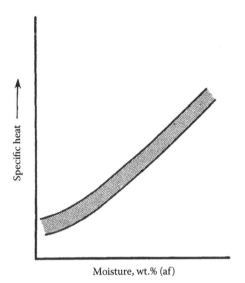

FIGURE 9.11   Variation of specific heat with moisture content. (From Baughman, G.L., *Synthetic Fuels Data Handbook*, 2nd edn., Cameron Engineers, Inc., Denver, CO, 1978.)

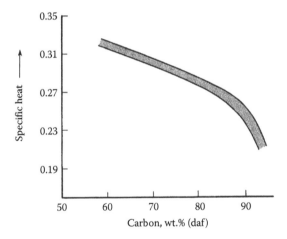

FIGURE 9.12   Variation of specific heat with carbon content. (From Baughman, G.L., *Synthetic Fuels Data Handbook*, 2nd edn., Cameron Engineers, Inc., Denver, CO, 1978.)

Thus, from the data for various coals, it has been possible to derive a formula which indicates the relationship between the specific heat and the elemental analysis of coal (mmf basis):

$$C_p = 0.189C + 0.874H + 0.491N + 0.3600 + 0.215S$$

C, H, N, O, and S are the respective amounts (% w/w) of the elements in the coal.

9.4.3   THERMAL CONDUCTIVITY

Thermal conductivity is the rate of transfer of heat by conduction through a unit area across a unit thickness for a unit difference in temperature:

$$Q = \frac{kA(t_2 - t_1)}{d}$$

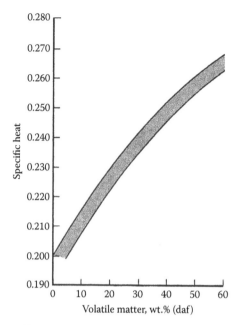

FIGURE 9.13 Variation of specific heat with volatile matter yield. (From Baughman, G.L., *Synthetic Fuels Data Handbook*, 2nd edn., Cameron Engineers, Inc., Denver, CO, 1978.)

where

Q is the heat, expressed as kcal/s·cm·°C or as Btu/ft·h·°F (1 Btu/ft·h·°F = 1.7 J/s·m·K)

A is the area

$t_2 - t_1$ is the temperature differential for the distance (d)

k is the thermal conductivity (Carslaw and Jaeger, 1959)

However, the banding and bedding planes in coal (Chapters 1 and 4) can complicate the matter to such an extent that it is difficult, if not almost impossible, to determine a single value for the thermal conductivity of a particular coal. Nevertheless, it has been possible to draw certain conclusions from the data available. Thus, monolithic coal is considered to be a medium conductor of heat with the thermal conductivity of anthracite being on the order of $5-9 \times 10^{-4}$ kcal/s·cm·°C while the thermal conductivity of monolithic bituminous coal falls in the range of $4-7 \times 10^{-4}$ kcal/s·cm·°C. For example, the thermal conductivity of pulverized coal is lower than that of the corresponding monolithic coal. For example, the thermal conductivity of pulverized bituminous coal falls into the range of $2.5-3.5 \times 10^{-4}$ kcal/s·cm·°C.

The thermal conductivity of coal generally increases with an increase in the apparent density of the coal as well as with volatile matter yield, ash yield, and temperature. In addition, the thermal conductivity of the coal parallel to the bedding plane appears to be higher than the thermal conductivity perpendicular to the bedding plane.

There is little information about the influence of water on the thermal conductivity of coal, but since the thermal conductivity of water is markedly higher than that of coal (about three times) the thermal conductivity of coal could be expected to increase if water is present in the coal.

9.4.4 PLASTIC AND AGGLUTINATING PROPERTIES

Plasticity refers to the melting and bonding behavior of the coal and (1) is an indication of the initial softening, chemical reaction, gas liberation, and resolidification process within the coke oven,

(2) is an important requirement in the coke blend and is required for end product coke strength, and (3) the fluidity of the plastic stage is a major factor in determining what proportion of a coal is used in a blend.

When coal is heated, an unstable intermediate phase (*metaplast*) is formed after the moisture is driven from the coal. This intermediate phase is responsible for the plastic behavior of coal. On further heating, a cracking process takes place in which tar is vaporized and nonaromatic groups are split off followed by recondensation and formation of semicoke.

When coal or a blend of coals is coked in slot-type ovens, two principal layers of plastic coal are formed parallel to the oven walls. As carbonization proceeds, the plastic layers move progressively inward eventually meeting at the oven center. It is within these plastic layers that the processes which result in particulate coal being converted into porous, fused semicoke take place. The semicoke undergoes further devolatization and contracts which result in fissures in the final coke product.

All coals undergo chemical changes when heated but there are certain types of coal which also exhibit physical changes when subjected to the influence of heat. These particular types of coals are generally known as "caking" coals whereas the remaining coals are referred to as *noncaking coals*.

The caking coals pass through a series of physical changes during the heating process insofar as they soften, melt, fuse, swell, and resolidify within a specific temperature range. This temperature has been called the "plastic range" of coal and thus the physical changes which occur within this range have been termed the *plastic properties* (*plasticity*) of coal.

The caking tendency of coals increases with the volatile matter content of the coal and reaches a maximum in the range of 25%–35% w/w volatile matter but then tends to decrease. In addition, the caking tendency of coal is generally high in 81%–92% w/w carbon coals (with a maximum at 89% carbon); the caking tendency of coal also increases with hydrogen content but decreases with oxygen content and with mineral matter content.

When noncaking (nonplastic) coals are heated, the residue is pulverent and noncoherent. On the other hand, the caking coals produce residues that are coherent and have varying degrees of friability and swelling. In the plastic range, caking coal particles have a tendency to form agglomerates (cakes) and may even adhere to surfaces of process equipment, thereby giving rise to reactor plugging problems.

Thus, the plastic properties of coal are an important means of projecting and predicting how coal will behave under various process conditions as well as assisting in the selection of process equipment. For example, the plasticity of coal is a beneficial property in terms of the production of metallurgical coke but may have an adverse effect on the suitability of coal for conversion to liquids (Chapters 18 and 19) and gases (Chapters 20 and 21). As an example, the Lurgi gasifier is unable to adequately gasify caking coals unless the caking properties are first nullified by a prior oxidative treatment.

The plastic behavior of coal is of practical importance for semiquantitative evaluation of metallurgical coal and coal blends used in the production of coke for the steel industry. When bituminous coals are heated in the absence of air through over the range of 300°C–550°C (570°F–1020°F), volatile materials are released and the solid coal particles soften (to become a plastic-like mass) which swells and eventually resolidifies.

The Gieseler test is a standard test method in which attempts the actual extent of the plasticity of fluidity attained is measured. The Gieseler test is used to characterize coals with regard to thermoplasticity and is an important method used for coal blending for commercial coke manufacture. The maximum fluidity determined by the Gieseler is very sensitive to weathering (oxidation) of the coal.

Briefly, the Gieseler plastometer (ASTM, 2011h,j) is a vertical instrument consisting of a sample holder, a stirrer with four small rabble arms attached at its lower end with the means of (1) applying a torque to the stirrer, (2) heating the sample that includes provision for controlling temperature and rate of temperature rise, and (3) measuring the rate of turning of the stirrer.

A sample of −60-mesh coal is packed in the sample holder; the holder is completely filled and the rabble arms of the stirrer are all in contact with the coal. The apparatus is then immersed in the heating bath and a known torque applied to the stirrer. During the initial heating no movement of the stirrer occurs but as the temperature is raised the stirrer begins to rotate. With increasing temperature, the stirrer speed increases until at some point the coal resolidifies and the stirrer is halted (Berkowitz, 1979).

Data usually obtained with the Gieseler plastometer are (1) softening temperature (the temperature at which stirrer movement is equal to 0.5 dial divisions per minute) which may be characterized by other rates but if so the rate must be reported, (2) maximum fluid temperature (the temperature at which stirrer movement reaches maximum rate in terms of dial divisions per minute), (3) solidification temperature (the temperature at which stirrer movement stops), and (4) maximum fluidity (the maximum rate of stirrer movement in dial divisions per minute).

Plastic properties of coal, as determined by the Gieseler plastometer, appear to be sensitive to oxidation, which can have a marked effect in decreasing the maximum fluidity. In fact, prolonged oxidation may completely destroy the fluidity of a coal. To reduce oxidation, samples should be tested soon after collection or, if delay is unavoidable, storage under water or in a nonoxidizing atmosphere such as nitrogen is advisable.

Gieseler data are used primarily for assessing coking properties and, furthermore, there is the additional capability that maximum fluidity data can also be used in selecting coals for blending to produce desirable metallurgical coke. For example, if a coal of low maximum fluidity is to be used for making coke, it may be necessary to blend with it a coal of higher fluidity.

The plasticity of individual coal macerals has also received some attention but the investigations can be complicated by the difficulties encountered in the isolation of the macerals in the "pure" state. In spite of this, there are reports of the behavior of the macerals. Thus, exinite tends to be quite plastic, with a low softening point, wide plastic range, high fluidity in the plastic state, and a high degree of swelling. This is understandable because exinite is a hydrogen-rich maceral (Chapter 4) and contains up to 70% volatile matter when isolated from high-volatile coal. It is almost impossible to plasticize fusinite insofar as this maceral is a fossil charcoal which is hard and friable (Chapter 4); vitrinite generally shows an intermediate plasticity behavior.

Although it is possible to formulate the general stages that occur up to and during the plastic stage of coal, the exact mechanism of coal fusion is not completely understood. There seems to be little doubt that the process is concerned with the production and/or liberation of liquid tars within the coal.

In terms of the elemental composition of coal, there is a relative hydrogen deficiency, but there are theories that admit to the presence of hydrogen-rich liquid (and mobile) hydrocarbons (Chapter 10) that are enclosed within the coal matrix and which are often (erroneously) called bitumen (Chapters 10 and 11) and which should not be confused with the bitumen (natural asphalts) that occur in various deposits throughout the world (Speight, 1990, 2007). The application of heat results in the liberation of these hydrocarbon liquids and forms other hydrocarbons (*thermobitumen*) by scission of hydrocarbon fragments from the coal structure, and the overall effect is the formation of a high-carbon coke and a hydrocarbon tar, the latter being responsible for the fluidity of the mass. With increased heating, the tar partly volatilizes and partly reacts to form nonfluid material ultimately leading to the coke residue (Figure 9.14).

When coal is heated in a vacuum, the plastic range is generally reduced substantially, perhaps because of the rapid evaporation of the bituminous hydrocarbons that are reputedly responsible for the fluidity of the plastic coal. Heating coal to the plastic range followed by rapid cooling yields coal with a lower softening point (if plasticized a second time) and this has been ascribed to the presence of liquid in the coal that arose from the first heating.

An additional property of coal that is worthy of mention at this time is the softening point, which is generally defined as the temperature at which the particles of coal begin to melt and become rounded. The softening point indicates the onset of the plasticity stage and is (as should be

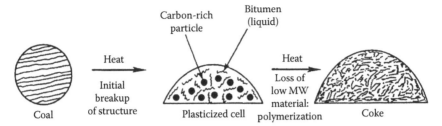

FIGURE 9.14 Coal behavior during heating through the plastic range.

anticipated) a function of the volatile matter content of coal. For example, coal with 15% volatile matter will have a softening point of the order of 440°C (825°F) which will decrease to a limiting value of ca. 340°C (645°F) for coal with 30% volatile matter.

The material thought to be responsible for conferring the plastic properties on coal can be removed by solvent extraction to leave a nonplastic residue (Pierron and Rees, 1960). Plastic properties can be restored to the coal by recombining the solvent extracts with the insoluble residue.

9.4.5 AGGLOMERATING INDEX

The agglomerating index is a grading index based on the nature of the residue from a 1 g sample of coal when heated at 950°C (1740°F) in the volatile matter determination (ASTM, 2011n).

The agglomerating index has been adopted as a requisite physical property to differentiate semi-anthracite from low-volatile C bituminous coal and also high-volatile C bituminous coal from sub-bituminous coal (Table 9.13). From the standpoint of the caking action of coal in coal-burning equipment, the agglomerating index has some interest. For example, coals having indexes NAa or NAb, such as anthracite or semianthracites, certainly do not give any problems from caking while those coals having a C_g index are, in fact, the high-caking coals.

The agglomerating (or agglutinating) tendency of coal may also be determined by the Roga test (ISO, 2011a), and the Roga index (calculated from the abrasion properties when a mixture of a specific coal and anthracite is heated) is used as an indicator of the agglomerating tendencies of coals (Table 9.14).

9.4.6 FREE SWELLING INDEX

The FSI of coal is a measure of the increase in volume of a coal when it is heated (without restriction) under prescribed conditions (ASTM, 2011g; ISO, 2011a). The ISO test (ISO, 2011a), Roga test, measures mechanical strength rather than size profiles of coke buttons; another ISO test (ISO, 2011b) gives a CSN of coal.

The nature of the volume increase is associated with the plastic properties of coal (Loison et al., 1963) and, as might be anticipated, coals which do not exhibit plastic properties when heated do not, therefore, exhibit free swelling. Although this relationship between free swelling and plastic properties may be quite complex, it is presumed that when the coal is in a plastic (or semifluid) condition the gas bubbles formed as a part of the thermal decomposition process within the fluid material cause the swelling phenomenon which, in turn, is influenced by the thickness of the bubble walls, the fluidity of the coal, and the interfacial tension between the fluid material and the solid particles that are presumed to be present under the test conditions.

The test for the FSI of coal (ASTM, 2011g) requires that several 1 g samples of coal be heated to 820°C (1508°F) within a specified time to produce buttons of coke. The shape, or profile, of the buttons (Table 9.15) determines the FSI of the coal (BSI, 2011). Anthracites do not usually fuse or exhibit a FSI whereas the FSI of bituminous coals will increase as the rank increases from the high-volatile C bituminous coal to the low-volatile bituminous coal.

TABLE 9.13

Agglomerating and Caking Properties of Coals as Indicated by the Residue from the Volatile Matter Determination

Designation		
Class[a]	Group	Appearance of Residue
Non-agglomerating (button shows no swelling or cell structure and will not support a 500 g weight without pulverizing)	NA, non-agglomerate	NA[a], noncoherent residue Nab, coke cotton shows no swelling or cell structure and after careful removal from the crucible will pulverize under a 500 g weight carefully lowered onto the button
	A, agglomerate (button dull black, sintered, shows no swelling, or cell structure will support a 500 g weight without pulverizing)	Aw, weak agglomerate (buttons come out of crucible in more than one piece) Af, firm agglomerate (buttons come out of crucible in one piece)
Agglomerating (button shows swelling or cell structure or will support a 500 g weight without pulverizing)	C, caking (button shows swelling or cell)	Cp, poor caking (button shows slight swelling with small cell, has slight gray luster) Cf, fair caking (button shows medium swelling and good cell, structure; has characteristic metallic luster) Cg, good caking (button shows strong swelling and pronounced cell structure with numerous large cell and cavities; has characteristic metallic luster)

[a] Agglomerating index: Coals that in the volatile matter determination produce either an agglomerate button that will support a 500 g weight without pulverizing or a button showing swelling or cell structure shall be considered agglomerating from the standpoint of classification.

TABLE 9.14

Data for the Plastic Properties of Coals

Coal Type	Swelling Index	Dilation (%)	Roga Index
Noncaking	0	0	0–5
Weakly caking	1–2	0	5–20
Medium caking	2–4	0–40	20–50
Strongly caking	>4	>50	>50

Other effects which can influence the FSI of coal include the weathering (oxidation) of the coal. Hence, it is advisable to test coal as soon as possible after collection and preparation. There is also evidence that the size of the sample can influence the outcome of the free swelling test; an excess of fine (−100-mesh) coal in a sample has reputedly been responsible for excessive swelling to the extent that the FSI numbers can be up to two numbers higher than is the true case.

The FSI of coal is believed to be of some importance in assessing the coking properties of coal, but absolute interpretation of the data is extremely difficult. In general terms, the FSI of bituminous coals generally increases with an increase in rank (Rees, 1966) but the values for individual coals within a rank may vary considerably. The values for the lower-rank coals are normally less than those for bituminous coals; anthracite does not fuse and shows no swelling value. Furthermore,

---

TABLE 9.15

Free Swelling Indexes for Selected U.S. Coals

Rank	Coal	Free Swelling Index
High-volatile C	Illinois No. 6	3.5
High-volatile B	Illinois No. 6	4.5
High-volatile B	Illinois No. 5	3.0
High-volatile A	Illinois No. 5	5.5
High-volatile A	Eastern	6.0–7.5
Medium-volatile	Eastern	8.5
Low-volatile	Eastern	8.5–9.0

Source: Baughman, G.L., *Synthetic Fuels Data Handbook*, 2nd edn., Cameron Engineers, Inc., Denver, CO, 1978, p. 176.

---

a coal exhibiting a FSI of 2, or less, will most likely not be a good coking coal whereas a coal having a FSI of 4, or more, may have good coking properties.

9.5 ELECTRICAL PROPERTIES

Knowledge of the electrical properties of coal is also an important aspect of coal characterization and behavior. Electrical properties are useful for cleaning, mining, pyrolysis, and carbonizing processes. They are also of special interest in the electro-linking process for permeability enhancement and as a means to locate regions with different physical properties during in situ coal gasification.

9.5.1 SPECIFIC RESISTANCE (RESISTIVITY)

Specific resistance (resistivity) is the electrical resistance of a body of a unit cross section and of unit length and is expressed in ohm-centimeters:

$$p = RA/L$$

where
p is the specific resistance
R is the resistance of the substance
A is the cross-sectional area
L is the length

The specific resistance of coals (Table 9.16) may vary from thousands of ohm-centimeters to millions of ohm-centimeters depending on the direction of measurement.

9.5.2 ELECTRICAL CONDUCTIVITY

The electrical conductivity of coal is generally discussed in terms of specific resistance, p (units of p are ohm-centimeters), and is defined as the resistance of a block of coal 1 cm long and having a 1 cm^2 cross section. Substances having a specific resistance greater than approximately 1×10^{15} ohm-cm are classified as insulators while those with a specific resistance of less than 1 ohm-cm are conductors; materials between these limits are semiconductors.

Electrical conductivity depends on several factors, such as temperature, pressure, and moisture content of the coal. The electrical conductivity of coal is quite pronounced at high temperatures

TABLE 9.16
Specific Resistance of Coal

Material	Specific Resistance (ohm-cm)
Graphite	$(0.8-10) \times 10^{-3}$
Graphite	4×10^{-3}
Anthracite[a]	
Parallel to bedding	$(7-90) \times 10^3$
Perpendicular to bedding	$(17-34) \times 10^3$
Bituminous[a]	
Parallel to bedding	$(0.004-360) \times 10^8$
Perpendicular to bedding	$(3.1-530) \times 10^9$
Brown coal	
20%–25% H_2O	10^4
Dry	$10^{10}-10^{13}$
Copper	1.7×10^{-6}
Water (distilled)	$8.5-25 \times 10^6$

Source: Baughman, G.L., *Synthetic Fuels Data Handbook*, 2nd edn., Cameron Engineers, Inc., Denver, CO, 1978, p. 170.

[a] Dry

(especially above 600°C [1110°F]), where the coal structure begins to break down. Moisture affects electrical conductivity to a marked extent, resulting in a greatly increased conductivity. To prevent any anomalies from the conductance due to water, the coal is usually maintained in a dry, oxygen-free atmosphere and, to minimize the problems that can arise particularly because of the presence of water, initial measurements are usually taken at approximately 200°C (390°F) and then continued to lower temperatures.

Coal is considered to be a semiconductor since the specific resistance of bituminous coal ranges from 1×10^{10} to 1×10^{14} ohm-cm. Anthracite has values ranging from 1 to 1×10^4 ohm-cm but exhibits some directional dependence (anisotropy).

The conductivity of coal is explained in part by the partial mobility of electrons in the coal structure lattice which occurs because of unpaired electrons (free radicals) (Chapter 10). Mineral matter in coal may have some influence on electrical conductivity. The conductivities of coal macerals show distinct differences; fusains conduct electricity much better than clarains, durains, and vitrains.

9.5.3 DIELECTRIC CONSTANT

Dielectric constant is more useful than electrical conductivity in characterizing coal and is a measure of the electrostatic polarizability of the dielectric coal (Chatterjee and Misra, 1989). The dielectric constant of coal is believed to be related to the polarizability of the π-electrons in the clusters of aromatic rings within the chemical structure of coal (Chapter 10).

Like conductivity, dielectric constant is strongly dependent on water content. Indeed, the dielectric constant can even be used as a measure of moisture in coal. Meaningful dielectric constant measurements of coal require drying to a constant dielectric constant and several forms of coal are used for dielectric constant measurements. These include precisely shaped blocks of coal, mulls of coal in solvents of low dielectric constant, or blocks of powdered coal in a paraffin matrix.

The dielectric constant varies with coal rank (Chatterjee and Misra, 1989). The theorem that the dielectric constant is equal to the square of the refractive index (which is valid for nonconducting, nonpolar substances) holds only for coal at the minimum dielectric constant. The decreasing value of dielectric constant with rank may be due to the loss of polar functional groups (such as hydroxyl or carboxylic acid functions) but the role of the presence of polarizable electrons (associated with condensed aromatic systems) is not fully known. It also appears that the presence of intrinsic water in coal has a strong influence on the dielectric properties (Chatterjee and Misra, 1989).

9.6 EPILOGUE

Knowledge of coal properties is an important aspect of coal characterization and has been used as a means of determining the suitability of coal for commercial use for decades, perhaps even centuries. In fact, the molecular characterization of coal (Chapter 10) is seen to be of little, or no, by some consumers.

Therefore, the properties outlined in this chapter (and also in Chapter 8) must always be borne in mind when consideration is being given to the suitability of coal for a particular use. It must also be borne in mind that a coal which at first appears unsuitable for use by a consumer might become eminently suitable by a "simple" or convenient pretreating step almost analogous to the "conditioning" of asphalt by air blowing (Speight, 1991), in light of various environmental regulations and how the data might be used to predict the suitability of coal for use in an environmentally cleaner and safer manner.

The analysis of coal ash (Chapter 8) for environmentally hazardous trace elements is but one example of how the data might be used. The mineral portion of coal has been ignored for too many years. It must be considered an integral part of the coal structure and so the ash (and its contents) must also be given consideration when a use is designed for coal.

REFERENCES

Agrawal, P.L. 1959. *Proceedings*. Symposium on the Nature of Coal. Central Fuel Research Institute, Jealgora, India, p. 121.

ASTM. 2011a. *Test Method for Mean Specific Heat of Thermal Insulation* (ASTM C351). Annual Book of ASTM Standards, Section 04.06. American Society for Testing and Materials, West Conshohocken, PA.

ASTM. 2011b. *Terminology of Coal and Coke* (ASTM D121). Annual Book of ASTM Standards, Section 05.05. American Society for Testing and Materials, West Conshohocken, PA.

ASTM. 2011c. *Test Method for Cubic Foot Weight of Bituminous Coal* (ASTM D291). Annual Book of ASTM Standards, Section 05.05. American Society for Testing and Materials, West Conshohocken, PA.

ASTM. 2011d. *Standard Test Method for Grindability of Coal by the Hardgrove-Machine Method* (ASTM D409). Annual Book of Standards, American Society for Testing and Materials, West Conshohocken, PA.

ASTM. 2011e. *Method for Drop Shatter Test for Coal* (ASTM D440). Annual Book of ASTM Standards, Section 05.05. American Society for Testing and Materials, West Conshohocken, PA.

ASTM. 2011f. *Method for Tumbler Test for Coal* (ASTM D441). Annual Book of ASTM Standards, Section 05.05. American Society for Testing and Materials, West Conshohocken, PA.

ASTM. 2011g. *Test Method for Free-Swelling Index of Coal* (ASTM D720). Annual Book of ASTM Standards, Section 05.05. American Society for Testing and Materials, West Conshohocken, PA.

ASTM. 2011h. *Method of Test for Plastic Properties of Coal by Gieseler Plastometer* (ASTM D1812). Annual Book of Standards, American Society for Testing and Materials, West Conshohocken, PA.

ASTM. 2011i. *Test Method for Gross Calorific Value of Coal and Coke by the Adiabatic Bomb Calorimeter* (ASTM D2015). Annual Book of ASTM Standards, Section 05.05. American Society for Testing and Materials, West Conshohocken, PA.

ASTM. 2011j. *Test Method for Plastic Properties of Coal by the Constant-Torque Gieseler Plastometer* (ASTM D2639). Annual Book of ASTM Standards, Section 05.05. American Society for Testing and Materials, West Conshohocken, PA.

ASTM. 2011k. *Method for Microscopical Determination of the Reflectance of the Organic Components in a Polished Specimen of Coal* (ASTM D2798). Annual Book of ASTM Standards, Section 05.05. American Society for Testing and Materials, West Conshohocken, PA.

ASTM. 2011l. *Method for Microscopical Determination of Volume Percent of Physical Components of Coal* (ASTM D2799). Annual Book of ASTM Standards, Section 05.05. American Society for Testing and Materials, West Conshohocken, PA.

ASTM. 2011m. *Method for Drop Shatter for Coke* (ASTM D3038). Annual Book of ASTM Standards, Section 05.05. American Society for Testing and Materials, West Conshohocken, PA.

ASTM. 2011n. *Test Method for Volatile Matter in the Analysis Sample of Coal and Coke* (ASTM D3175). Annual Book of ASTM Standards, Section 05.05. American Society for Testing and Materials, West Conshohocken, PA.

ASTM. 2011o. *Test Method for Gross Calorific Value of Coal and Coke by the Isoperibol Bomb Calorimeter* (ASTM D3286). Annual Book of ASTM Standards, Section 05.05. American Society for Testing and Materials, West Conshohocken, PA.

ASTM. 2011p. *Standard Test Method for Gross Calorific Value of Coal and Coke* (ASTM D5865). Annual Book of Standards, American Society for Testing and Materials, West Conshohocken, PA.

Baughman, G.L. 1978. *Synthetic Fuels Data Handbook*, 2nd edn. Cameron Engineers, Inc., Denver, CO.

Berenbaum, R. and Brodie, J. 1959. *British Journal of Applied Physics*, 10: 281.

Berkowitz, N. 1979. *An Introduction to Coal Technology*. Academic Press, Inc., New York.

Birkeland, P.W. 1984. *Soils and Geomorphology*. Oxford University Press, New York, pp. 14–15.

BOHS. 1985. Dustiness estimation methods for dry materials: Part 1, their uses and standardization; Part 2, towards a standard method. Technical Report No. 4, British Occupational Hygiene Society Technology Committee, Working Group on Dustiness Estimation. H&H Scientific Reviews, Leeds, U.K.

Braunstein, H.M., Copenhaver, E.D., and Pfuderer, H.A. (Eds.). 1977. Environmental health and control aspects of coal conversion: An information overview. Report ORNL-EIS-94. Oak Ridge National Laboratory, Oak Ridge, TN.

Breum, N.O., Nielsen, B.H., Nielsen, E.M., Midtgaard, U., and Poulsen, O.M. 1996. Dustiness of compostable waste: A methodological approach to quantify the potential of waste to generate airborne micro-organisms and endotoxin. *Waste Management & Research*, 15: 169–187.

Brown, R.L. and Hiorns, F.J. 1963. In *Chemistry of Coal Utilization*, Supplementary Volume, H.H. Lowry (Ed.). John Wiley & Sons, Inc., New York, Chapter 3.

BSI. 2011. EN-analysis and testing of coal and coke. Part 107: Caking and swelling properties of coal, Section 107.1: Determination of crucible swelling number (BSI BS 1016–107.1). British Standards Institution, London, U.K.

Cannon, C.G. and George, W.H. 1943. Refractive index of coals. *Nature*, 151(January 9): 53–54.

Carslaw, H.S. and Jaeger, J.C. 1959. *Conduction of Heat in Solids*, 2nd edn. Oxford University Press, Oxford, U.K., p. 189.

Chakravorty, R.N. 1984. Report No. ERP/CRL 84-14. Coal Research Laboratories. Canada Centre for Mineral and Energy Technology, Ottawa, Ontario, Canada.

Chakravorty, R.N. and Kar, K. 1986. Report No. ERP/CRL 86-151. Coal Research Laboratories. Canada Centre for Mineral and Energy Technology, Ottawa, Ontario, Canada.

Chan, M.-L., Parkyns, N.D., and Thomas, K.M. 1991. *Fuel*, 70: 447.

Chatterjee, I. and Misra, M. 1989. *Proceedings Materials Research Symposium*. Materials Research Society, Vol. 189, Warrendale, PA, p. 195.

Cody, G.D., Jr., John W. Larsen, J.W., Siskin, M., and Cody, G.D., Sr. 1990. The granular nature of high rank coals. In *Proceedings, 1990 MRS Spring Meeting*. Materials Research Society, Warrendale, PA.

Davis, A. 1978. In *Analytical Methods for Coal and Coal Products*, C. Karr, Jr. (Ed.). Academic Press, Inc., New York, Vol. I, Chapter 2.

Davis, A., Mitchell, G.D., Derbyshire, F.J., Rathbone, R.F., and Lin, R. 1991. *Fuel*, 70: 352.

Dormans, H.N.M., Huntjens, F.J., and van Krevelen, D.W. 1957. *Fuel*, 26: 321.

Edwards, G.R., Evans, T.M., Robertson, S.D., and Summers, C.W. 1980. *Fuel*, 59: 826.

Ergun, S. 1951. *Analytical Chemistry*, 23: 151.

Ettinger, I.L. and Zhupakhina, E.S. 1960. *Fuel*, 39: 387.

Evans, D.G. and Allardice, D.J. 1978. In *Analytical Methods for Coal and Coal Products*, C. Karr, Jr. (Ed.). Academic Press, Inc., New York, Vol. I, Chapter 3.

Gan, H., Nandi, S.P., and Walker, P.L., Jr. 1972. *Fuel*, 51: 272.

Hardgrove, R.M. 1932. *Transactions of the American Society of Mechanical Engineers*, 54: 37.

Henke, B. and Stockmann, H.W. 1992. *Glueckauf-Forschungshefte*, 53(1): 249.

Hobbs, D.W. 1964. Strength and stress–strain characteristics of coal in triaxial compression. *Journal of Geology*, 72: 214–231.

Holuszko, M.E. and Laskowski, J.S. 2010. *Proceedings. International Coal Preparation Conference*. International Coal Preparation Congress (ICPC). Lexington, KT, April 25–29.

Ingles, O.G. 1961. In *Agglomeration*, W.A. Knepper (Ed.). John Wiley & Sons, Inc., New York, p. 29.

ISO. 2011a. *Hard Coal—Determination of Coking Power, Roga Test* (ISO 335). International Standards Organization, Geneva, Switzerland.

ISO. 2011b. *Coal—Determination of Crucible Swelling Number* (ISO 501). International Standards Organization, Geneva, Switzerland.

ISO. 2011c. *Solid Mineral Fuels—Determination of Gross Calorific Value and Calculation of Net Calorific Value* (ISO 1928). International Standards Organization, Geneva, Switzerland.

ISO. 2011d. *Hard Coal—Determination of Grindability* (ISO 5074). International Standards Organization, Geneva, Switzerland.

Jones, J.C. and Vais, M. 1991. *Journal of Hazardous Materials*, 26: 203.

Kalliat, M., Kwak, C.Y., and Schmidt, P.W. 1981. In *New Approaches in Coal Chemistry*, B.D. Blaustein, B.C. Bockrath, and S. Friedman (Eds.). American Chemical Society, Washington, DC, Chapter 1.

King, J.G. and Wilkins, E.T. 1944. *Proceedings. Conference on the Ultrafine Structure of Coal and Cokes*. British Coal Utilization Research Association, London, U.K., p. 45.

Kotlensky, W.V. and Walker, P.L. Jr. 1960. *Proceedings. Fourth Carbon Conference*. Pergamon, Oxford, England, p. 423.

Kroger, C. and Badenecker, J. 1957. Physical and chemical properties of the components of hard coal (Macerals). *Brennstoff Chemie*, 38: 82.

Labuschagne, B.C.J. 1987. *Proceedings. Fourth Annual Pittsburgh Coal Conference*. University of Pittsburgh, Pittsburgh, PA, p. 417.

Labuschagne, B.C.J., Wheelock, T.D., Guo, R.K., David, H.T., and Markuszewski, R. 1988. *Proceedings. Fifth Annual International Pittsburgh Coal Conference*. University of Pittsburgh, Pittsburgh, PA, p. 417.

Laubach, S.E., Marrett, R.A., Olson, J.E., and Scott, A.R. 1998. Characteristics and origins of coal cleat: A review. *International Journal of Coal Geology*, 35: 175–207.

Leonard, J.W., Paradkar, A., and Groppo, J.G. 1992. Handle more coal by increasing its bulk density. *Mining Engineering*, April: 317–318.

Leonard, J.W., Paradkar, A., and Groppo, J.G. 1993. Are big gains in coal loading productivity possible? A look at some of the fundamentals. *Proceedings*. SME Annual Meeting, Reno, NV, February 15–18, Preprint number 93-58.

Loison, R., Peytavy, A., Boyer, A.F., and Grillot, R. 1963. In *A Chemistry of Coal Utilization*, Supplementary Volume, H.H. Lowry (Ed.). John Wiley & Sons, Inc., New York, Chapter 4.

Lyman, W.J., Reehl, W.F., and Rosenblatt, D.H. 1990. *Handbook of Chemical Property Estimation Methods: Environmental Behavior of Organic Compounds*. McGraw-Hill, New York.

Lyons C.P. and Mark, D. 1994. Development and testing of a procedure to evaluate the dustiness of powders and dusts in industrial use, Report LR 1002. Environmental Technology Executive Agency of the U.K. Department of Trade and Industry, London, U.K.

Mahajan, O.P. and Walker, P.L. Jr. 1978. In *Analytical Methods for Coal and Coal Products*, C. Karr, Jr. (Ed.). Academic Press, Inc., New York, Vol. I, Chapter 4.

Medhurst, T. and Brown, E.T. 1998. A study of the mechanical behavior of coal for pillar design. *International Journal of Rock Mechanics and Mining Sciences*, 35: 1087–1105.

Milewska-Duda, J. 1991. *Archives of Mining Sciences*, 36: 369.

Munoz-Guillena, M.J., Linares-Solano, A., and Salinas-Martinez de Lecea, C. 1992. *Fuel*, 71: 579.

Pierron, E.D. and Rees, O.W. 1960. Solvent extract and the plastic properties of coal. Circular No. 288. Illinois State Geological Survey, Urbana, IL.

Puri, R. and Seidle, J.P. 1992. *In Situ*, 16: 183.

Ramesh, R., Francois, M., Somasundaran, P., and Cases, J.M. 1992. *Energy & Fuels*, 6: 239.

Rees, O.W. 1966. Chemistry, uses, and limitations of coal analysis. Report of Investigations No. 220. Illinois State Geological Survey, Urbana, IL.

Skawinski, R., Zolcinska, J., and Dyrga, L. 1991. *Archives of Mining Sciences*, 36: 227.

Speight, J.G. 1978. In *Analytical Methods for Coal and Coal Products*, C. Karr, Jr. (Ed.). Academic Press, Inc., New York, Chapter 22.

Speight, J.G. 1990. In *Fuel Science and Technology Handbook*, J.G. Speight (Ed.). Marcel Dekker, New York, Chapters 2 and 12.

Speight, J.G. 2005. *Handbook of Coal Analysis*. John Wiley & Sons, Inc., Hoboken, NJ.

Speight, J.G. 2007. *The Chemistry and Technology of Petroleum*, 4th edn. Taylor & Francis Group, Boca Raton, FL.

Tisdall, A.L. 1951. Comparison of methods of determining apparent density of soils. *Australian Journal of Agricultural Research*, 2: 349–354.

Trollope, D.H., Rosengren, K.J., and Brown, E.T. 1965. *Geotechnique*, 57: 363.

Tschamler, H. and de Ruiter, E. 1963. In *Chemistry of Coal Utilization*, Supplementary Volume, H.H. Lowry (Ed.). John Wiley & Sons, Inc., New York, Chapter 2.

Upton, S.L., Hall, D.J., and Marsland, G.W. 1990. Some experiments on material dustiness. *Proceedings. Aerosol Society Annual Conference 1990*. Surrey, U.K.

Van Krevelen, D.W. 1957. *Coal: Aspects of Coal Constitution*. Elsevier, Amsterdam, the Netherlands.

Vincent, J.H. 1995. *Aerosol Science for Industrial Hygienists*. Pergamon/Elsevier, Oxford, U.K.

Wakeman, R.J. 1975. Packing densities of particles with log-normal distribution. *Powder Technology*, 11: 297–299.

Walker, P.L. Jr. 1981. *Philosophical Transactions of the Royal Society (London)*, 300A: 65.

Yancey, H.R. and Geer, M.R. 1945. In *Chemistry of Coal Utilization*, H.H. Lowry (Ed.). John Wiley & Sons, Inc., New York, p. 145.

Zimmerman, R.E. 1979. *Evaluating and Testing the Coking Properties of Coal*. Miller Freeman & Co., San Francisco, CA.

Zipf, R.K. and Bieniawski, Z.T. 1988. Estimating the crush zone size under a cutting tool in coal. *International Journal Mining Geology and Engineering*, 6: 279–295.

10 Organic Constituents

10.1 INTRODUCTION

Coal is a sedimentary rock that was formed from the accumulation of vegetative debris that has undergone physical and chemical changes over millennia. These changes include decaying of the vegetation, deposition, and burying by sedimentation, compaction, and transformation of the plant remains. As a result, coal is composed of both organic and inorganic material and manifests itself in the form of macerals (Chapter 4), discrete minerals (Chapter 7), inorganic elements (some as mineral matter others in molecular bonding arrangements with the organic matrix), and water and gases contained in submicroscopic pores.

Organically, coal is a complex heterogeneous mixture (Chapters 3, 4, and 8) that consists primarily of carbon, hydrogen, and oxygen, and lesser amounts of sulfur and nitrogen. Inorganically, coal consists of a diverse range of ash-forming compounds distributed throughout the coal. The inorganic constituents can vary in concentrations from several percentage points to parts per billion of the coal.

The organic constituents of coal, including both the maceral groups (liptinite, inertinite, and huminite/vitrinite) and the individual macerals in those groups (Chapter 4), are the basic elements of many properties of coal. Vitrinite is the most common maceral group in many coals and it is the properties of the vitrinite in such coals, together with the variations in those properties with rank, that to a large extent determine the properties of the coal concerned.

The presence of inorganic matter is an additional facet affecting coal in all of its uses. Whether as discrete minerals or chemically associated with the organic portion of coal, mineral matter can be troublesome, by forming ash during combustion, or beneficial, by catalyzing reactions during direct coal liquefaction or gasification. Furthermore, since coal is solid, inherent porosity can trap or release gas, water, or solvent-soluble material, and influence accessibility of reagents. The interplay among the organic, inorganic, and physical components ultimately determines many of the applications for which coal can be used.

The determination of the properties of coal as a means of coal evaluation and with the goal of predicting behavior during utilization is well documented (Chapters 8 and 9). However, there is another area of coal characterization that has been at least equally well documented, but perhaps less well recognized as a means of evaluation, which involves studies of the molecular constituents of coal.

In recent decades, there has been a strong tendency to characterize coals by structural analysis. Thus, coal can be characterized in terms of aromaticity (percentage of total carbon atoms existing in aromatic structures), number of aromatic atoms (or rings per cluster), number and types of functional groups, locations of oxygen atoms, location and types of heteroatoms. This has led to the postulation of speculative structure for coal which may (questionably) lead to a better understanding of utilization processes such as combustion.

On the one hand, identification of the molecular constituents of coal is an exceptionally formidable, if not impossible, task. And yet, significant advances have been made recently in bringing about an understanding of the molecular nature of coal. As always, there are, and will be, serious questions about the need for such an understanding, relative to the use of coal, but the knowledge gained can often help offset a difficult-to-understand aspect of a processing sequence (Snape, 1987). The information may also be an in-road to predicting coal behavior during utilization.

Thus, it can be argued (often successfully to a point but not *ad nauseam*) that an understanding of the chemical nature of coal constituents is, just like an understanding of the chemical and thermal behavior of coal (Chapters 13 and 14), a valuable part of projecting the successful use of coal for conversion and/or utilization processes such as combustion (Chapters 14 and 15), carbonization an briquetting (Chapters 16 and 17), liquefaction (Chapters 18 and 19), gasification (Chapters 20 and 21), or as a source of chemicals (Chapter 24).

At this point, it should be noted that there are differences (which are more than semantic) between understanding the nature of the chemical constituents of coal and delineating the molecular structure of coal. The former is not insurmountable; the latter is certainly of questionable validity in view of the complex nature of the material under study. Dogmatic statements relating to the structure of coal are to be deplored. Indeed, perhaps the safest (and the most sane) statement relates to coal being an organic rock. But nothing ventured, nothing gained.

Indeed, statements relating to the derivation of hypothetical and representative models are of some value. In fact, the value of any structural model of a complex molecular entity lies in its use as a means of not only understanding process chemistry and physics but also in its use as a means by which processes can be understood and predictions (perhaps tongue-in-cheek and hopefully near to reality) can be made. Of course, such visions are always subject to the willingness of the chemical modelist to learn from experience and also to subject the model to the necessary changes to render it workable (Speight, 2007).

As an aside, but certainly worthy of note, the identification of many of the constituents of petroleum have been achieved as a result of the volatility of these constituents and subsequent application of methods such as gas–liquid chromatography and mass spectroscopy. But in the case of petroleum residua, which are the nonvolatile constituents of petroleum (boiling point: 345°C [650°F]), identification of the individual constituents is much more difficult and heavy reliance has to be put on identification by molecular type (Speight, 2007). The same is essentially true for coal.

Identification of the constituents of complex mixtures (such as petroleum residua and coal) by molecular type may proceed in a variety of ways but generally can be classified into three methods: (1) chemical techniques, (2) spectroscopic techniques, and (3) physical property methods whereby various structural parameters are derived from a particular property by a sequence of mathematical manipulations. The end results of these methods are indications of the structural types present in the material (Stadelhofer et al., 1981).

However, the unfortunate tendency of researchers is to attempt to interrelate these structural types into a so-called *average structure* but the pronounced heterogeneity of coal makes the construction of "average" structures extremely futile and, perhaps, misleading. In fact, the heterogeneous chemical structures of the wide range of plant chemicals which formed the starting material for coal (Chapter 3) promise, but do not guarantee, an almost unlimited range of chemical structures within the various types of coal. Thus, it is perhaps best to consider coal as a variety of chemical entities which virtually dictate coal properties and reactivity under specific conditions.

10.2 SOLVENT EXTRACTION

From a more practical viewpoint, the solvent extraction of bituminous coals has been used as a means of coproducing clean liquid transportation fuels as well as solid fuels for gasification. Coal solvents are created by hydrogenating coal tar distillate fractions to the level of a fraction of a percent, thus enabling bituminous coal to enter the liquid phase under conditions of high temperature (above 400°C [750°F]). The pressure is controlled by the vapor pressure of the solvent and the cracked coal. Once liquefied, mineral matter can be removed via centrifugation, and the resultant heavy oil product can be processed to make pitches, cokes, as well as lighter products.

Solvents for coal extraction (Chapters 10 and 11) have been classified into four types:

1. *Nonspecific solvents*: They extract a small amount (up to approximately 10%) of coal at temperatures up to 100°C (212°F). These are low boiling liquids like methanol, ethanol, benzene, acetone, and ether. The extract is believed to arise from the resins and waxes occluded in the coal matrix.

2. *Specific solvents*: They extract 20%–40% of coal at temperatures below 200°C. The nature of the coal extract and the parent coal is believed to be similar. Hence, such solvents can be in fact considered nonselective in their action on coal. Pyridine, *N*-methylpyrrolidone, dimethylformamide, and dimethylacetamide are examples of this type of solvent. They are mostly nucleophilic in nature due to the presence of a lone pair of electrons on the nitrogen atom.

3. *Degrading solvents*: They extract up to 90% of the coal at temperatures up to 400°C (750°F). The mechanism of solvent action is by thermal degradation of coal into smaller fragments. At the end of the extraction, the solvent can be almost completely recovered without change in its chemical form. Examples of this type of solvents include phenanthrene, diphenyl, phenanthridine, and tar oil fractions.

4. *Reactive solvents*: These solvents interact with the coal chemically. Extraction of coal with such solvents is referred to as "extractive chemical disintegration." Reactive solvents are generally hydrogen donors. The smaller coal fragments formed by thermal disintegration are stabilized by hydrogen, which is donated by the solvent. The coal—as well as the solvent—changes appreciably during the extraction. Hydroaromatic compounds are good hydrogen-donor solvents, which are converted to their corresponding aromatic counterparts during extraction. For example, tetralin (1,2,3,4-tetrahydronaphthalene) is converted to naphthalene upon donation of four hydrogen atoms.

Compressed gases (such as carbon dioxide) can also be used effectively as solvents for extraction of coal. Toluene, dodecane, *p*-cresol, etc. can be applied under supercritical conditions. Some advantages of using supercritical fluid as solvents are (1) an extract with low molecular weight (approximately 500) and higher hydrogen content can be obtained, (2) solvent recovery is easy, and (3) the reduction in the operating pressure or temperature precipitates the extract.

The purpose of many solvent extraction investigations is to develop continuous processes for solvent extraction of coal for the production of carbon products, which include materials used in metals smelting, especially in the aluminum and steel industries, as well as porous carbon structural material (*carbon foam* and *carbon fibers*).

Current sources of materials for these processes generally rely on petroleum distillation products or coal tar distillates obtained as a by-product of metallurgical coke production facilities. In the former case, the industry, just as the energy industry in general, is dependent upon foreign sources of petroleum. In the latter case, metallurgical coke production is decreasing due to the combined difficulties associated with poor economics and a significant environmental burden. Thus, a significant need exists for an environmentally clean process which can use domestically obtained raw materials and which can still be very competitive economically. Continuous processes using solvent extraction of coal offer the potential to accomplish all of these objectives.

Low emission liquefaction processes are particularly important in a scenario in which greenhouse gas mitigation is essential. Likely such scenarios will emphasize the use of technologies such as wind and nuclear power for central station power, while hydrocarbons will be increasingly reserved for liquid transportation fuel applications.

Lower-rank coals such as subbituminous coal and lignite are desirable feedstocks for this process due to their low cost, high hydrogen-to-carbon ratio, and highly aliphatic compared to bituminous coals, which can result in superior transportation fuels. However, these advantages are partially offset by the high moisture content and high ash content which typically accompany lignite and subbituminous coals. In particular, then high mineral matter content (determined as mineral ash) on the order of 20% w/w is problematic because centrifugation might not succeed in increasing the ash content of the tails (Kennel et al., 2009).

It is, therefore, the purpose of this chapter to present some indication of the methods which allow coal to be defined in terms of specific structural entities and also to include an assessment of the various *molecular* structures proposed for coal.

10.2.1 SOLUBLE CONSTITUENTS

Coal is usually considered to be a high-molecular-weight organic material composed of a variety of carbon systems. Indeed, many structures have been postulated for coal and all of these postulates invoke the concept of an extremely complex system having a high molecular weight.

Solvent extraction under mild conditions has been and continues to be actively pursued, primarily as a tool to probe the molecular properties and structure of coal and, in some instances, for commercial applications. Research shows that the effectiveness of solvents for coal extraction is influenced by coal type, rank, and petrographic composition, and by the chemical nature of the solvent.

Whether the vitrinite in coal is a micellar complex (Van Krevelen, 1993) or three-dimensional network (Given, 1984a) remains to be proven, and whether solvent effectiveness can be explained entirely by solubility parameter (Van Krevelen, 1965), donor–acceptor number (Szeliga and Marzec, 1983), hydrogen bonding (Larsen et al., 1985), solvent synergism (Iino et al., 1985), or other associative forces (Nishioka and Larsen, 1990) is still open to debate.

Though still relevant today, solvents used to extract coal have been classified into four groups (Oele et al., 1951). Solvents in the first category, nonspecific extraction, extract only a few percent of the coal at temperatures below about 100°C (212°F). These solvents (e.g., benzene, hexane, carbon tetrachloride, and ethanol) extract only a few percent of the coal thought to be primarily resins and waxes. In specific extraction, the coal material solubilized can typically amount to about 20% to 40% w/w at temperatures up to about 200°C (390°F). Many of these solvents are (1) nucleophilic, (2) have electron donor–acceptor properties, and (3) high internal pressure. However, it must be recognized in any such system that the efficiency of solvents in this system is closely related (even dictated) by temperature.

Pyridine, quinoline, and *N*-methyl pyrrolidone are examples of solvents used for specific extraction. It is noteworthy that remarkable ambient temperature extraction efficiencies have been reported using mixed carbon disulfide–*N*-methyl pyrrolidone solutions (Iino et al., 1988). During extractive disintegration temperatures approaching the thermal decomposition of coal, up to 350°C (660°F), it can be (should be) anticipated that solvent components which can redistribute hydrogen from and to reactive coal components by a hydrogen exchange mechanism appear to be most effective (Golumbic et al., 1950; Heredy and Fugass, 1966; Derbyshire and Whitehurst, 1981; Derbyshire etal., 1982).

The extractive chemical disintegration process can be called direct coal liquefaction. Here, solvents rich in hydroaromatic components are especially suited in extracting nearly all of the reactive coal macerals. These types of solvents actively participate chemically in bond breakage and stabilization, are consumed or structurally changed, and are normally used at temperatures considerably in excess of 300°C (570°F). On the other hand, because of the heterogeneous nature of coal it is a distinct possibility there may be/could be no clear operational or mechanistic distinction between *extractive disintegration* and *extractive chemical disintegration* processes.

When certain bituminous coals are heated, they soften and become fluid commensurate with the evolution of gas and tar (Van Krevelen et al., 1956; Waters, 1962). The plastic behavior is transient and the mass eventually thickens, swells, and fuses to form a porous solid or coke. This phenomenon is of the utmost importance to the production of metallurgical coke, and in other processes sensitive to caking and agglomeration of coal. Because of the impact of plastic behavior on industrial processes, the solvent extraction of coal has been studied in the past in an attempt to isolate the *coking principle* from the coal (Burgess and Wheeler, 1911; Illingworth, 1922; Dryden and Pankhurst, 1955).

However, coal–solvent interactions are very complex and there have been various attempts to correlate the extraction efficiency with the nature of the solvent, nature of the coal, and extraction conditions.

The plastic behavior in coals has been explained by the hypothesis that the development of plasticity is a transient, in situ, hydrogen-donor process (Neavel, 1982). The solvent-soluble material contains more hydrogen than the parent coal, and is thought to be responsible for solvation and hydrogen stabilization of the molecular units in coal as they become mobile with increasing temperature. It is believed that the stability of the metaplast produced by coking coals during their softening and resolidification, which is emulated in their plastic range, is responsible for the quality and quantity of extract produced. These extractable components are also influential in other processes such as direct coal liquefaction (Derbyshire et al., 1986).

It is well established that there is a correspondence between the amount of material extractable from coal using specific solvents and rank. The quantity of extracts in pyridine increases as the carbon content increases, peaks around 86% w/w carbon, and then drops precipitously. There also is a corresponding relationship between the amount of material extractable from bituminous coal and thermoplasticity. The significance of the extracts on plasticity is made clear when extracts from good coking coals are blended with noncoking coals to induce fluidity (Brown and Waters, 1966a; Stansberry et al., 1987) and extracting prime coking coals leaves a residue devoid of plastic behavior (Brown and Waters, 1966b).

The nondestructive extraction of coal using a variety of solvents has shown that coal does, in fact, contain low-molecular-weight species (Chapters 10 and 11) that appear to have been trapped within the pore system of the coal (Vahrman, 1970; Bartle et al., 1978, 1982; Davis et al., 1989; Litke et al., 1990; Smith and Smoot, 1990; Pickel and Gotz, 1991).

There is evidence for and against the chemical relationship of these materials to the non-extractable portion of coal. Arguments pass back and forth on the issue of the extractable species being lower-molecular-weight chemical brothers, sisters, and cousins of the non-extractable material. Then there are those molecular genealogists who cannot see any chemical relationship whatsoever between the extractable and non-extractable species. And yet, the dilemma is that the scientists and engineers who fall on both sides of the issue may be correct. Obviously, caution is advised here and whether or not the extracts resemble the coal in terms of structural types is certainly open to question because of the diversity of the coal examined.

The volatile matter test for coal (Chapter 8) has been cited as an indication that coal contains lower-molecular-weight species, but it must be noted that the material produced in this test is more likely to be an artifact of the high temperatures (950°C [1740°F]) employed than the release of these materials by the selective thermal destruction of the surrounding coal matrix. The most obvious lower-molecular-weight constituent of coal is methane (Hamming et al., 1979), which has long been recognized as a lower-molecular-weight constituent that is often released in substantial quantities during mining operations.

However, there are other more complex organic constituents of coal which may not be volatile but can be extracted by various solvents (Chapters 10 and 11). It is the nongaseous low-molecular-weight species with which the present chapter is concerned. Thus, for the purposes of this chapter, the lower-molecular-weight species of coal are generally defined as those materials which are produced from coal by nondestructive solvent extraction.

Indeed, evidence accumulated over the last five decades indicates that those constituents of coal which have molecular weights greater than 1000 form a greater part of the coal matrix than had previously been acknowledged (Bartle et al., 1978; Hamming et al., 1979; Raj, 1979; White, 1983). Furthermore, the importance of these low-molecular-weight constituents lies in the fact that they can, and presumably do, influence the properties of the whole coal. Thus, properties such as the atomic hydrogen/carbon ratio, volatile matter content, and coking (or caking) properties will certainly fall under the sphere of influence of the lower-molecular-weight constituents of coal.

In more specific terms, alkanes can be extracted from coal(s) by a mixture of hot benzene and ethanol over a period of 250 h. The alkanes are actually isolated from the extract by adsorption chromatography and those alkanes that have received considerable attention are that portion which

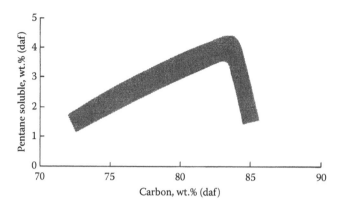

FIGURE 10.1 Variation of pentane solubility with carbon content. (From Raj, S., Ph.D. Thesis, The Pennsylvania State University, University Park, PA, 1976.)

elutes first from the adsorbent in pentane. The yields of this fraction varied markedly with the rank of the coal (Figure 10.1) (Raj, 1976), but all of the coals that have been studied were noted to yield n-alkanes in the series C_{13}–C_{33}. The alkanes from the lower-rank coals were highly branched and tended to predominate over the n-alkanes; in some coals, noticeable amounts of cyclic alkanes have also been identified. In addition, isoprenoid hydrocarbons as well as sterane-type and terpane-type materials have been extracted from coal (Bartle et al., 1978).

There have been many studies of the low-molecular-weight extracts of coal(s) and it is not surprising that there have also been several attempts to relate the structures of these constituents to the organic material that constitutes the bulk of the coal. For example, branched and cyclic alkanes are believed to be derived from substances in plants or in peat that either were the alkanes themselves or carried oxygen functions (such as some of the terpenoid materials), and the variation of these lower-molecular-weight constituents is presumed to be due to the diversity of the original precursors (Chapter 3) rather than the different conditions acting on similar starting materials during the coalification process. In relation to the polycyclic alkanes that have been isolated from coal, polycyclic aromatic hydrocarbons have also been isolated from coal and these compounds bear a diagenetic relationship to possible triterpenoid precursors (Chaffee and Fookes, 1988).

In general terms, lignite and anthracite coals appear to contain much lower proportions of the volatile organic compounds (10) than the bituminous coals, although alkaline extraction of lignite (and bituminous coal) will yield organic acids of various types that can be characterized. On the other hand, the hexane-soluble portion of the pyridine extracts that were obtained at 50°C (120°F) from coal (carbon content: 83.6% w/w) have been identified as alkylated chrysenes as well as alkylated picenes in addition to a mixture of C_{28}, C_{29}, and C_{30} paraffins. There were also indicates of the presence of an alicyclic or methyl-substituted five-ring (cat-condensed) system as well as 1,2,5,6-dibenzanthracene(s) (Table 10.1).

Attempts to isolate low-molecular-weight materials from coal have centered on the use of flash heating and supercritical gas extraction. In the former case, the products indicated that the low-energy flash heating brought about the production of high yields of aliphatic hydrocarbons. Of course, it is necessary to assume that the low-molecular-weight species are the primary products of the process insofar as they are rapidly removed from the reaction zone with little, if any, further molecular alteration.

Application of supercritical gas extraction (Schneider et al., 1980; Bright and McNally, 1992; Kiran and Brennecke, 1993) has lately received considerable attention when applied to coal. For example, studies have been reported relating to the chemical nature of extracts of coal (volatile matter content in excess of 36%) obtained using toluene (under pressure) at 350°C (660°F). The extracts contained aromatics (benzene derivatives, naphthalene derivatives, and phenanthrene derivatives) as well as n-paraffins, sterane, and materials such as phytane, pristane, and farnesane (Table 10.2) (Bartle et al., 1975; Smith and Smoot, 1990). There was a predominance of n-alkanes.

TABLE 10.1
Higher-Ring Polynuclear
Aromatic Systems in Coal Extracts

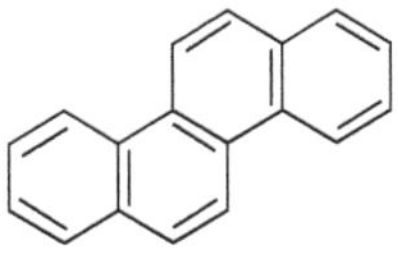

Chrysene

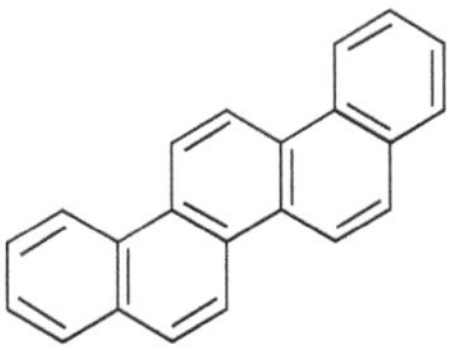

Picene

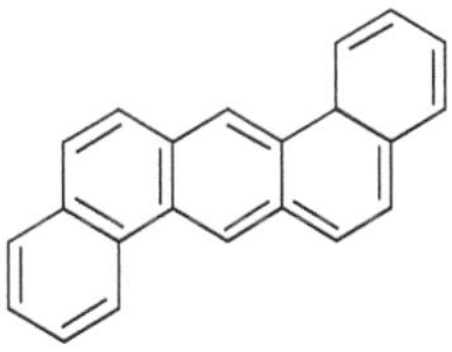

1,2,5,6-Dibenzanthracene

TABLE 10.2
Low-Molecular-Weight Constituents
in a Supercritical Gas Extract of Coal

Component	Amount in Eluate (% w/w)
n-Octane	0.2
n-Nonane	0.6
n-Decane	0.2
n-Undecane	0.5
Branched paraffins between C_{11} and C_{12}	5.5
n-Dodecane	0.6
2,6,10-Trimethylundecane	0.1
n-C_{13}–C_{31}	58.1
Farnesane	0.4
2,6,10-Trimethyldodecane	1.5
Norpristane	2.5
Pristane	8.4
Phytane	1.5

Source: Bartle, K.D. et al., *Fuel*, 54, 226, 1975.

Structural parameters derived by nuclear magnetic resonance studies were employed to calculate the conformation and size of the hypothetical aromatic systems (Figure 10.2) (Bartle et al., 1975). And it is interesting to note the small size of the systems coincide with the natural product origins of coal but offer an attractive alternative to the graphite-type layers that were (even still are) perceived to exist in coal. Such products could arise from the former, but not from the latter.

Further work on the supercritical extraction of coal at 400°C (750°F) in the presence of hydrogen and zinc chloride produced an extract of which approximately 50% was soluble in pentane and had an *average* (that word again) structure consisting of three aromatic rings which carried two methyl groups and an alicyclic ring. The absence of hydrogen produced less extract having a more complex structure. Other studies using Turkish lignite have indicated the presence of aromatic ether oxygen and single aromatic ring systems that carry alkyl substituents which themselves have at least eight carbon atoms.

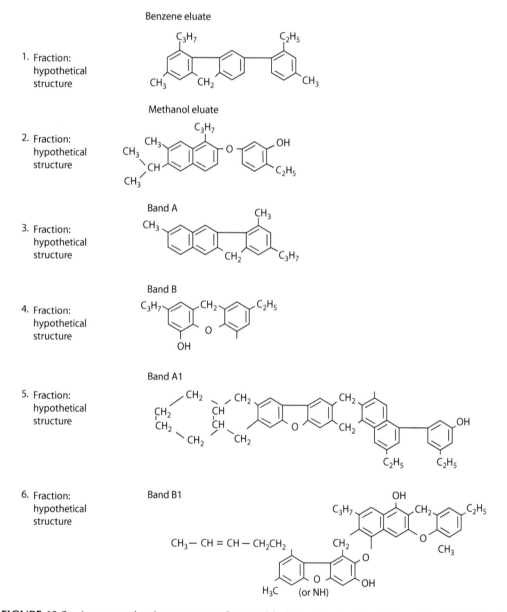

FIGURE 10.2 Average molecular structures of supercritical gas extracts of coal. (From Bartle, K.D. et al., *Fuel*, 54, 226, 1975.)

Studies of the smaller molecules of coal have also been made by solvating coals in solvents that are capable of donating hydrogen to the reaction. For these purposes, solvation was defined as thermal bond breaking at moderate temperatures under conditions of bond stabilization (such as the donation of hydrogen) and the method is actually analogous to various solvent extraction techniques (Chapters 10 and 11), but using much higher temperatures.

For example, subbituminous coal will "dissolve" in anthracite oil at temperatures equal to, or in excess of, 325°C (800°F) with the production of three-ring aromatic systems that have a limited molecular weight range (175–200). In the absence of hydrogen or "added" catalysts, differences in the yield and ring distribution of the aromatic systems brought about the suggestion that the process involved the scission of aryl-oxygen bonds as well as carbon–nitrogen bonds and some carbon–carbon (aliphatic) bonds, thereby giving rise to the "monomer" systems of the coal as the reaction products. Furthermore, these products contained condensed ring systems that appeared to be of the order of four-ring (pyrene) structures at the most.

Other studies of material extracted from high-volatile bituminous coal (in the presence of hydrogen at 450°C [840°F]) indicate the occurrence of dihydrophenanthrene units in the extract and, therefore (by extrapolation and inference), in the coal. This, of course, is analogous to the earlier postulate of such units as an integral part of the structure of coal.

However, it must be emphasized that the high temperatures used in the pursuance of such studies can at best only provide indirect evidence for the presence of any of the postulated units in the parent coal and such postulates must, therefore, be assessed accordingly.

10.2.2 INSOLUBLE CONSTITUENTS

The high-molecular-weight constituents of coal are much more difficult to define than the low-molecular-weight constituents insofar as they (1) are difficult, if not impossible, to extract with solvents and (2) cannot be "distilled," either individually or as various compound classes, from coal without the onset of thermal decomposition to low-molecular-weight fragments or, as is often the case, to a carbonaceous residue (Davidson, 1980).

Indeed, the identity of the higher-molecular-weight species in coal is often inferred from a variety of chemical, physicochemical, and physical studies. Thus, coal itself (with the exception of the extractable material) may simply be looked upon as a conglomeration of high-molecular-weight species that may, or may not, be covalently (or, e.g., hydrogen-bonded) together to form a complex "macromolecule." But, in terms of coal science, nothing is simple and the term "macromolecule" is used here very loosely and it is not used with the object of defining any coal as a single molecular entity.

There are many reports scattered throughout the scientific literature over the last century, at least, which relate to attempts to formulate a structure for coal. These efforts have been founded on the presumption that if the structure of a widely used fuel, such as coal, is known and understood, then it may then be possible to use the material much more efficiently. A similar line of thinking has been applied to the search for the structures that exist in the higher-molecular-weight fractions of petroleum which are particularly prone to produce coke and/or reduce catalyst activity during the processing of these materials (Speight, 2007). In the present context, such studies have been more often over the past two decades since coal has, once again, come into increased popularity as a potential source of liquid fuels.

As a result of these studies, it has been possible to derive data that contribute to the knowledge of the higher-molecular-weight constituents of coal but the methods are often criticized because of the repeated attempts by investigators to treat the data too literally and, thus, define coal in terms of a molecular structure. This, of course, leaves the whole area represented by these investigations open to extremely severe criticism and debate as to the usefulness of the exercise.

Nevertheless, such investigations must not be wholly rejected and discredited and are actually praiseworthy insofar as they provide valuable information about the structural types present in coal. But the danger lies in attempting to present the data in terms of an *average* structure.

As a side note, throughout this chapter, the term *structure of coal* will be employed to indicate the structure of the higher-molecular-weight species, in fact the macromolecular part, of coal to the exclusion of those lower-molecular-weight materials that can be extracted according to the methods outlined in the preceding section.

10.3   CHEMICAL METHODS

The application of chemical methods as attempts to determine the structure of coal has a long history insofar as coal has been known to be reactive to a variety of chemicals agents (Chapter 12). Such attempts were, of course, originally applied with the object of degrading coal to smaller molecules that would not only have a greater degree of utility but also be much easier to identify.

In fact, application of chemical methods to the elucidation of coal structure is analogous to structural determinations of a variety of natural products (i.e., alkaloids, cellulose, proteins, etc.) but is complicated by virtue of the complex nature of coal and the inability to recognize any of the individual molecular types in their true molecular context in this particular "natural product." Nevertheless, several very valiant, and noteworthy, attempts have been made to define the structure of coal by the application of such methods, with each attempt usually adding a little more to the data bank of molecular imagination. And to make matters even more confusing, there is often the assumption in many of the investigations, that very little, if any, molecular alteration occurs during the chemical fragmentation and that the low-molecular-weight fragments formed are truly representative of the structural types that are present in the original coal.

The oxidation of coal using reagents other than oxygen has been extensively studied and, the "common" oxidants such as nitric acid, permanganate, dichromate, various peroxides, as well as hypochlorites have all been applied to the oxidation of coal (Hayatsu et al., 1982; Speight, 1987).

In the early years of coal science, one of the prime motives for investigating the oxidation of coal was the production of chemicals from coal. More recently, the oxidation of coal using specific oxidants has become a prime means by which structural entities in coal have been identified. This has, of course, led to postulates of the molecular structure of coal.

However, the diversity of the oxidants renders the oxidation of coal very complex because the experimental parameters can vary widely (Speight, 1987). The diversity in the structural types in coal (which vary not only with rank but also within the same rank) causes many problems associated with studies of the oxidative degradation of coal. For example, optimal conditions of time, temperature, and ratio of oxidant to coal can only be determined when several experiments are performed for each oxidant. Furthermore, the presence of the mineral matter must also be considered to be an integral part of coal oxidation since mineral constituents may change the chemistry of the oxidation. If coal is pretreated with hydrochloric acid to remove mineral matter prior to oxidation, the actual oxidation reaction may be more facile.

At this point, it needs to be stated that the studies of the oxidation of coal have resulted in valuable contributions toward understanding the nature of (a) the polynuclear aromatic systems, (b) the aliphatic systems, and (c) the nature of the heteroatomic systems (Hayatsu et al., 1982; Speight, 1987; Winans et al., 1988). However, considerable gaps still remain in the knowledge of coal structure.

The concept of aliphatic structures being a predominant part of coal structure has been advocated on the basis of the oxidation of coal by sodium hypochlorite (Chakrabartty and Berkowitz, 1974, 1976). This particular concept requires that part or all of the coal carbon exist as sp carbon and invokes the three-dimensional adamantane system as the major building block of the coal structure.

The concept was derived on the presumption that when organic structures are degraded by means of sodium hypochlorite the occurrence of carbon dioxide in the products indicates the presence of sp^3 carbon. In addition, the requirement is that the majority of the carbon in coal (the exact proportion varying according to the rank of the coal) exists in the diamondoid aliphatic structure (Chakrabartty and Kretschmer, 1972). Thus, it was estimated that 59%–71% of the total carbon in a variety of coals (carbon content 76.1%–90.2%) occurred in the sp^3 valence state (Table 10.3)

TABLE 10.3
Occurrence of sp^3 Carbon
in Various Coals

Carbon in Coal (% w/w)	$\dfrac{sp^3 \text{ Carbon}}{\text{Total Carbon}} \times 100$
76.1	68
80.5	70
83.1	68
85.0	63
86.4	71
90.2	59

Source: Chakrabartty, S.K. and Kretschmer, H.O., *Fuel*, 51, 160, 1972.

(Chakrabartty and Kretschmer, 1972) on the assumption that, in each case, carbon dioxide is produced only by the oxidation of sp^3 (fully saturated) carbon.

The postulate of a diamondoid (adamantane-type) system in coal (but not necessarily as the main building block of coal) was first advocated in the late 1950s on the basis of the spectroscopic properties (especially x-ray diffraction profiles) of coal but only required that a small part of the carbon exist in such structures. This postulate was truly novel insofar as it offered the attractive option of a three-dimensional structure for coal, an option that has only been acknowledged once in the chemical formulae of that era.

However, these "hypochlorite" studies have been criticized on the basis of the claim that sodium hypochlorite only oxidize sp^3 carbon under the reaction conditions. For example, there is the claim that coal oxidation, by sodium hypochlorite under rigidly controlled experimental conditions, produces data which show that aromatic systems also undergo oxidation by hypochlorite. Thus, there are legitimate question about the occurrence of adamantane-type systems in coal. It is suggested that the main structural unit of bituminous coal is a three-dimensional cross-linked system (Mayo, 1975; Lucht and Peppas, 1981) which contains condensed aromatic, as well as alicyclic, systems that are connected by ether (or thioether) linkages as well as by methylene (or polymethylene) linkages.

Oxidizing agents, such as nitric acid, sodium dichromate, and oxygen, attack aromatic/aliphatic substrates predominantly at the benzylic carbon thereby bringing about oxidative cleavage, but the hydrogen peroxide/trifluoroacetic acid/sulfuric acid reagent attacks the aromatic rings and leaves the benzylic positions comparatively unconverted (Hayatsu et al., 1981). The products of this oxidation contain a substantial part of the original hydrogen and ca. 75% of this hydrogen appears as acetic acid, succinic acid, glutaric acid, and methanol (Table 10.4) (Deno et al., 1978). The results of such studies have been interpreted in terms of methyl groups attached to aryl rings producing the acetic acid, and the glutaric acid is presumed to arise from the presence of 1,3-diarylpropane (or indane) structures in the coal(s).

On the other hand, the succinic acid is believed to originate from $ArCH_2CH_2Ar$ systems, indane systems, $-CH_2CH_2CHOH$ systems, and $-CH_2CH_2CH(R)-$ systems. The absence of isobutyric acid was regarded as strong evidence for the absence of isopropyl groups in the original coal while the absence of ethanol and the higher alcohols was presumed to indicate that the methoxy function was the only alkoxy function present in the coal.

Finally it should be noted that in these oxidation reactions (as for many other reactions of coal), the yield of the oxidation products were not quantitative and therefore cannot be construed to represent the numerical distribution of the various functions within coal. However, it does seem that this technique provides a means of characterizing parts of coal structure that are often misrepresented or even not represented when other chemical reagents or physical methods are employed.

TABLE 10.4
**Proportions of Hydrogen in the Products
from Coal Oxidation**

Component	Type of Coal		
	Pittsburgh Seam	Illinois No. 6	Lignite
Carbon (% w/w)	79.6	70.8	65.3
Hydrogen (% w/w)	5.3	5.2	4.4
Moisture (% w/w)	1.5	17.7	35.7
Hydrogen (%) as			
Acetic acid	0.9	6.1	4.2
Succinic acid	4.3	13.4	6.0
Glutaric acid	0.5	2.2	—
Methanol	0	0	16.2

Source: Deno, N.C. et al., *Fuel*, 57, 455, 1978.

Just as oxidation methods have been applied to delineation of the aliphatic (and the alicyclic) structures in coal, they have also been applied to the determination of the aromatic systems in coal. In fact, a particular oxidizing agent may produce data relating to either the aliphatic or the aromatic moieties in coal (or even relating to the lack of such systems).

An oxidizing agent which has been claimed to show reasonable selectivity is sodium dichromate ($Na_2Cr_2O_4$) insofar as it appears to oxidize substituted polycyclic aromatic systems with a minimum of ring degradation (Hayatsu et al., 1981). Thus, bituminous coal can be oxidized, by sodium dichromate, to acid products which constitute approximately 53% w/w of the original coal. In fact, as a result of careful analysis, 35 aromatic acids (as well as heterocyclic systems) have been among the oxidation products and, furthermore, the carboxyl functions numbered as many as five on benzene nuclei and as many as four on polynuclear aromatics; these functions are thought to arise from the aliphatic or alicyclic linkages between the aromatic units (Figure 10.3) (Hayatsu et al., 1975).

In spite of disputed claims that the sodium dichromate oxidation of coal is not indicative of the aromatic species indigenous to coal, the method has been used to compare the structures of lignite, bituminous coal, and anthracite. As a result, the data indicated that the ring systems increased in size from the lower-rank lignite coal through bituminous coal to anthracite (Figure 10.4) (Hayatsu et al., 1978).

There is also evidence for the presence of fluorene-type derivatives in coal. The technique of selective methylation whereby an oxygen-methylated bituminous coal was treated with a series of carbanion bases followed by quenching with labeled methyl iodide appears to be a promising means of estimating the fluorene-type structural units in coal (Chambers et al., 1988) (Chapter 12). The data suggest that any molecular representations of coal structure should include the five-membered cyclopentadiene ring systems which are common to all fluorene derivatives.

10.4 SPECTROSCOPIC METHODS

Spectroscopic methods have been applied to the elucidation of the structures in coal from very early in the development of the methods (Speight, 1978). In the initial stages of the evolution of the spectroscopic methods, the data derived by their application to coal were more of a diagnostic nature as, for example, determination of functional entities or carbon–hydrogen bonds by means of infrared spectroscopy or determination of aromatic and aliphatic hydrogen by proton magnetic resonance spectroscopy. However, virtually all of the methods have at one time or another been applied to coal as a means of deriving more detailed information about coal structure with special emphasis on the

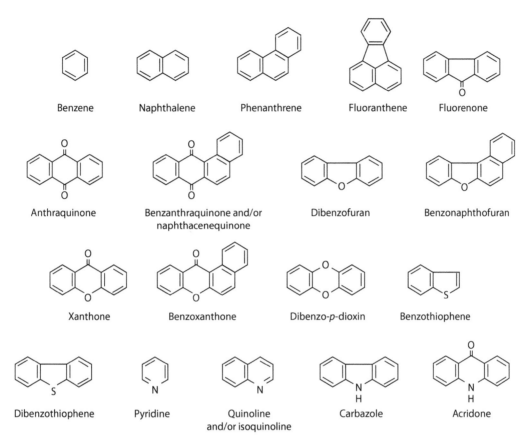

FIGURE 10.3 Structures of aromatic units in coal based on oxidation studies. (From Hayatsu, R. et al., *Nature*, 261, 77, 1975.)

higher-molecular-weight portion of coal that is unavailable for closer examination by initial extraction from the coal matrix and subsequent examination. In the majority of the cases, these investigations culminated in the derivation of so-called average structures which essentially represented efforts by the investigators to all of the findings about the various functions into one structure.

However praiseworthy this may be, it is still difficult to portray the various maturation paths of coal (or, in other instances, petroleum) as occurring regularly so that, for example, the heteroatoms are scattered evenly throughout the model structure. Nature does not always behave in such regular fashion and there may be a tendency to concentrate, say, the heteroatoms only in certain parts of a molecule. Recognition of this fact alone detracts markedly from, and even nullifies, many of the structures that have been proposed for coal. For the present purposes, it is sufficient to note that spectroscopic methods of estimating the structure of coal also (as do chemical methods of structural determination) suffer from several, sometimes severe, limitations. Thus, while some useful information can be derived about the structure of coal, the concept that the data will eventually lead to an "average" or even to a "representative" structure can only be cautiously (if at all) accepted.

10.4.1 INFRARED ABSORPTION SPECTROSCOPY

Of all the physical techniques, infrared spectroscopy gives perhaps the most valuable information about the constitution of organic materials. Indeed, qualitative information about specific structural elements can often be surmised even though the spectra are too complex for individual compound analysis.

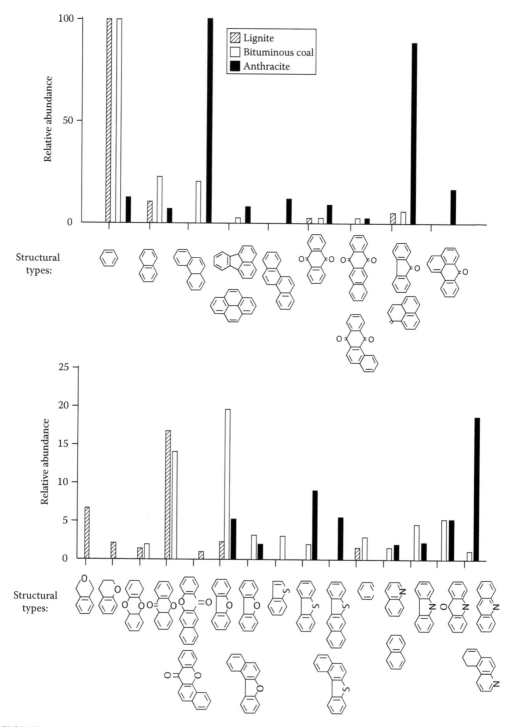

FIGURE 10.4 Relative abundances of aromatic and heteroaromatic units from oxidation of lignite, bituminous coal, and anthracite. (From Hayatsu, R. et al., *Fuel*, 57, 541, 1978.)

TABLE 10.5
Infrared Analysis of Coal

cm⁻¹	Assignment
3030	Aromatic CH
2978	CH$_3$
2940	Aliphatic CH
2925; 2860	CH$_3$, CH$_2$, CH
1600	Aromatic ring C==C
1575	Condensed aromatic ring C==C
1460	Aliphatic CH$_2$ and CH$_3$ groups
1370	CH$_3$ group, cyclic CH$_2$ group
870; 814; 760	Hydrogen atoms on substituted benzene rings

Source: Speight, J.G., *Analytical Methods for Coal and Coal Products*, C. Karr, Jr., Ed., Academic Press, Inc., New York, 1978, Vol. II, Chapter 22.

Infrared spectroscopic studies of coal have been the subject of many publications and attempts have been made to diagnose the functional groups and carbon skeletons of the coals. From the results of these investigations, it was generally established (Table 10.5) (Speight, 1978) that coal contained various aliphatic and aromatic carbon–carbon and carbon–hydrogen functions but few, if any, isolated unsaturated carbon–carbon bonds. It has been reported for a variety of coal fractions (in particular, the optical densities of the two peaks at 3030 and 2929 cm⁻¹) that the ratio of aromatic hydrogen to total hydrogen increased with rank; this phenomenon has also been observed in a series of Japanese coals. In addition it has also been reported that bituminous coals display strong absorption in the infrared which arises from nonaromatic carbon–hydrogen functions, as well as intense absorption from aliphatic or alicyclic groups, and the intensity of the 1600 cm⁻¹ absorption typifies lower-rank coals.

The three absorption bands at 760, 814, and 870 cm⁻¹ have been assigned to out-of-plane vibrations of one isolated, two adjacent, and four adjacent aromatic CH groups, respectively, while the relative intensity of these bands was suggested to give an indication of the degree of condensation of the aromatic clusters (Figure 10.5).

The absorption band at 3030 cm⁻¹ (aromatic CH), which appears in a coal with 81% carbon, becomes more pronounced as coal rank increases while the band at 2920 cm⁻¹ generally increases with rank to 86% carbon but decreases sharply thereafter. The assignment of absorptions in the infrared spectra to various oxygen functions has also received some attention (Table 10.6) (Speight, 1978).

Pyridine extraction of coal appears to afford, as indicated by infrared absorption studies, a two-component system, one rich in carbonyl and hydroxyl functions and the other in nonaromatic

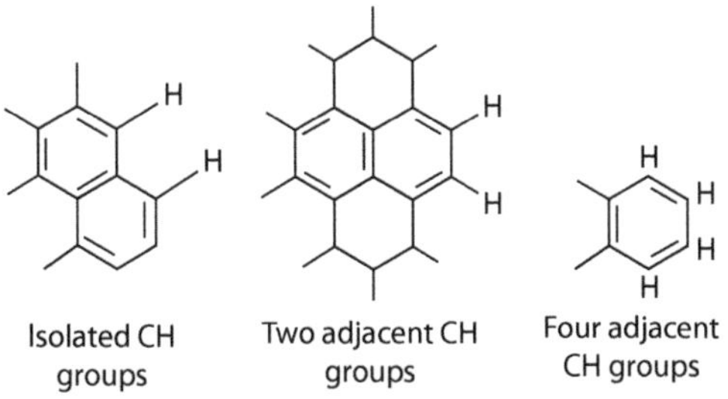

Isolated CH groups Two adjacent CH groups Four adjacent CH groups

FIGURE 10.5 Proton location on polynuclear aromatic rings.

TABLE 10.6
Infrared Analysis of Functional
Groups in Coal

Spectral Line (cm^{-1})	Assignment
3300	Associated OH and NH
1700	C==O
1600	Hydrogen-bonded C==O
1300–1000	C–O (phenols)
	C_{ar}–O–C_{ar}
	C–O (alcohols)
	C_{ar}–O–C_{al}
	C_{al}–O–C_{al}

Source: Speight, J.G., *Analytical Methods for Coal and Coal Products*, C. Karr, Jr., Ed., Academic Press, Inc., New York, 1978, Vol. II, Chapter 22.
Notes: C_{ar}, Aromatic carbon; C_{al}, aliphatic carbon.

carbon–hydrogen bonds. Another report on the infrared spectral analysis of coal indicates that the spectra are generally similar in higher-rank coals but different in the lower-rank coals. Hydrogen-bonded hydroxyl and carbonyl, or polycyclic quinones were prominent in some fractions while other fractions contained higher proportions of aliphatic and alicyclic ethers, epoxides, sulfoxides, and sulfones.

The presence of quinones in coal and oxidized coals and the products of the reaction of the latter with bisdiazonium compounds have also received some attention. An absorption band at 1640 cm^{-1}, observed in the spectra of model compounds obtained from the polycondensation of phenanthrene, pyrene, and chrysene with formaldehyde and subsequent pyrolysis and nitric acid oxidation, was also assigned to nonchelated quinoid carbonyl groups. This band was absent from the spectra of coal, oxidized coals and their products after reaction with bisdiazonium compounds after pyrolysis and oxidation with nitric acid.

A quantitative study of the changes occurring in the infrared spectra of coals during acetylation indicated that part of the oxygen is in the form of hydroxyl groups which are sterically hindered from acetylation and are thermally stable up to 450°C (840°F).

Many research groups have focused their attention on an assignment of the 1600 cm^{-1} band in the infrared spectra of coals. This band also occurs in the asphaltene fraction of petroleum, even in those with less than 1% oxygen, and has been reported to be unaffected either by acetylation or by oxidation reactions.

From the lithium aluminum hydride (Leah) reduction of coals, it was concluded that the 1600 cm^{-1} band is due predominantly to the presence of hydrogen-bonded carbonyl groups in the coals. It has also been suggested that the band at 1600 cm^{-1} could not be assigned principally to carbonyl absorption and an interpretation can be made on the basis of a polynuclear aromatic system. There is also the contention that the 1600 cm^{-1} band arises through the occurrence of donor–acceptor interactions between the aromatic sheets in the molecular species or that noncrystalline graphite-type structures may be responsible for the 1600 cm^{-1} band.

It has also been reported that ratio of aromatic hydrogen to total hydrogen increases with increasing rank and that at a 94% carbon content coal is completely aromatic. It has also been suggested that the percentage of the hydrogen contained in methyl groups probably lies in the range of 15%–25% and the methyl content decreases with increasing rank of the coal.

10.4.2   Nuclear Magnetic Resonance Spectroscopy

Nuclear magnetic resonance spectroscopy has proved to be of great value in fossil fuel research because it allows rapid and nondestructive determination of the total hydrogen content and distribution of hydrogen among the chemical functional groups present in the sample (Bartle and Jones, 1978; Retcofsky and Link, 1978; Petrakis and Edelheit, 1979; Snape et al., 1979; Cookson and Smith, 1982, 1987; Miknis, 1982; Davidson, 1986; Botto and Sanada, 1992; Meiler and Meusinger, 1992).

In an early publication on the subject it was reported that in a high-rank coal, 33% of the hydrogen atoms occurred in methylene groups of bridge and alicyclic structures while in a low-rank coal 67% of the hydrogen atoms occurred in this form and/or in long chains; it was also concluded that the proportion of perimethyl groups in coal is very small. The nuclear magnetic resonance spectra of the products obtained from the vacuum carbonization of coal have also been studied and it was concluded that the nonaromatic hydrogen occurred almost exclusively on saturated carbon atoms as aliphatic, alicyclic, or hydroaromatic structures. Solvent extracts of coal have been shown to contain naphthene-type compounds having four or five rings.

Nuclear magnetic resonance spectroscopy has also been used to derive structural parameters for the distribution of aliphatic hydrogen in coal and its extraction products. A ratio of four methylene groups per methyl group was noted; also, the methyl group content was higher in the soluble material and varied with the solvent power of the solvent.

Studies of coals of different ranks have shown that the proportion of aromatic hydrogen varies from 20% to 30% of the total hydrogen and that no simple relationship appears to exist with rank. Beta-paraffinic naphthenic hydrogen comprised 38%–66% of the nonaromatic hydrogen while the most striking variation in the structures of the coals was in the number of methylene bridges (Calkins and Spackman, 1986).

In addition, studies of the constitution of the pyridine extracts of coal have led to the conclusion, assuming that there are structural relationships between the extracts and the insoluble material, that coal consists of molecules having four to six aromatic rings carrying aliphatic side chains of four or five carbon atoms; the average bituminous coal had molecules with four or five aromatic rings bearing side chains of three or four carbon atoms.

A study of the hydrogen distribution in the pyridine and chloroform extracts from coal has indicated that ca. 50% of the aliphatic hydrogen was present on carbon atoms directly attached to aromatic rings. Indeed, the aromatic carbon atoms directly attached to aromatic rings has, in part, been reaffirmed by use of proton magnetic resonance to examine hydrogen distribution in coal.

Carbon-13 magnetic resonance spectroscopy has found some use in determining the fraction of carbon atoms that are in aromatic locations (f_a) as well as in attempting to define the structure of the aromatic ring system, although the question of the aromatic nature of coal as determined by this method is certainly open to debate. And there is the possibility of serious under estimation of the aromatic nature of coal by this method (Snape et al., 1989).

Carbon-13 magnetic resonance has been applied to coal and to the elucidation of structures of soluble coal fractions (Bartle and Jones, 1978). The data obtained, used in conjunction with those obtained from proton magnetic resonance and from elemental analysis, indicated an typical molecule of the neutral oil to be 70% aromatic and to consist of a naphthalene ring system bearing two or three saturated side chains, each having less than three atoms. Data from the investigation of the supercritical gas extraction of coal point to a similar conclusion insofar as the structural entities can be envisaged as consisting of small aromatic ring systems, with the inclusion of hydroaromatic ring systems, and the attendant alkyl groups (predominantly methyl) as well as a variety of oxygen functions.

Other evidence from carbon-13 magnetic resonance studies of the carbon disulfide extract of coal also favors the occurrence of small-ring systems. The mean structural unit appears to consist of two-to-three ring condensed aromatic systems with 40% of the available aromatic carbons bearing alkyl, phenolic, and/or naphthenic groups. Indeed, the mass spectrum of the extract indicated the presence of alkyl aromatic compounds having from 1 to 10 or more alkyl carbons per molecule.

The method, usually called cross-polarization (CP) ^{13}C NMR, has several advantages over the conventional technique (Miknis et al., 1981; Smith and Smoot, 1990). The weak signals from the ^{13}C atoms (abundance 1.1%) are enhanced by transferring polarization from protons to carbons and then decoupling the C–H couplings by irradiation with high-power radiofrequency. This enables the use of the short proton relaxation time to be used instead of the longer ^{13}C relaxation time as the waiting time between experiments. The technique also offers valuable information about the oxygen distribution in coal (Franco et al., 1992).

The CP method also makes assumptions about proton–proton and carbon–proton distances; if there are extensive regions of nonprotonated materials there would be no ^{13}C polarization and hence no signal contribution. This was tested for by comparing the CP spectra with conventional spectra generated by ^{13}C relaxation. This has indicated that the higher-rank coals are indeed highly aromatic substances and the CP ^{13}C NMR could be used with confidence to determine the carbon aromaticity.

Individual macerals have also been studied by ^{13}C magnetic resonance spectroscopy; the macerals were from a West Virginia bituminous coal (85.8% C). The calculated values were vitrinite 85%, exinite 66%, micrinite 85%, and fusinite 93%–96%. These results were in agreement with earlier work which had indicated that the aromaticity increases in the order:

Exinite < vitrinite/micrinite < fusinite

Exinite is thought to be derived from the nonwoody tissue of plants whereas the others are from the woody and cortical tissues of the plant precursors (Chapters 3 and 4) which may offer a partial explanation for the lower value of exinite aromaticity.

The f_a data obtained from the CP spectra can also be used to estimate the size of the aromatic ring systems. There are assumptions, however, such as the nonaromatic carbons are predominantly methylene ($-CH_2-$) and that the oxygen and 50% the nonaromatic carbons are directly bonded to aromatic rings. Thus, the hypothetical unsubstituted aromatic nuclei can be estimated from the equation:

$$H_{aru}/C_{ar} = (H - H_{ali})(2 + O)/C_{ar}$$

The terms "aru" and "ali" denote atoms in aromatic and aliphatic groupings, respectively.

Thus, high-resolution ^{13}C and solid-state magnetic resonance data have provided possible values for the average aromatic ring size in vitrains (77.0% C and 90.3% C) (Table 10.7) (Gerstein et al., 1979).

TABLE 10.7

Structural Parameter for Two Vitrains

	Virginia Vitrain (Pocahontas No. 4)	Iowa Vitrain ("Star")
Carbon (% w/w)	90.3	77.0
Hydrogen (% w/w)	4.43	6.04
%C_a	86	71
%H_a	77	37
H_{ar}/C_{ar}	0.53	0.40

Source: Gerstein, B.C. et al., Preprints. Division Fuel Chemistry, American Chemical Society, 24(1), 90, 1979.

Notes: %C_a, % Carbon that is aromatic; %H_a, % hydrogen that is aromatic.

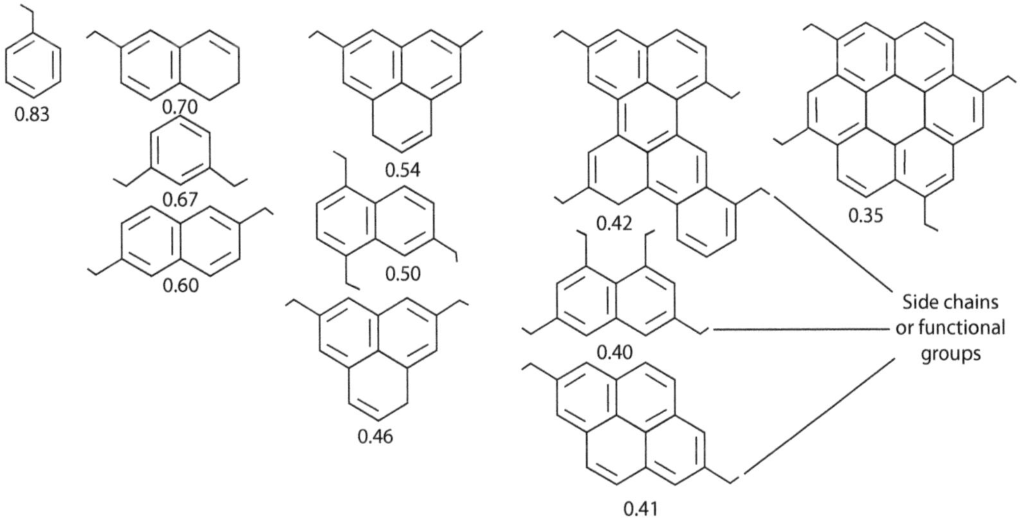

FIGURE 10.6 Aromatic cluster as a function of the H_{ar}/C_{ar} parameter. (From Gerstein, B.C. et al., Preprints, Division Fuel Chemistry, American Chemical Society, 24(1), 90, 1979.)

It has been pointed out that the lower H_{ar}/C_{ar} for the lower-rank coal may be surprising as one would expect larger ring systems in the older coal; however, the average ring size depends greatly on the number of side chains connected to the ring (i.e., on the degree of ring substitution) (Figure 10.6) (Gerstein et al., 1979).

In summary, it appears that magnetic resonance data support the assumption that the structural units increase in size with rank but the ring systems are smaller than the multi-ring plates, which were assumed from earlier studies.

10.4.3 X-Ray Diffraction

This method (Grigoriew, 1990; Lin and Guet, 1990) requires that within a given fraction the larger portion of the molecules contain within and among themselves certain repeated structural features as, for examples, sheets of condensed aromatic rings. With the realization that this assumption may have no real justification, the data must be treated with some reservation.

The x-ray scattering from coal was the subject of several early studies which led to the postulation that coal contains aromatic layers about 20–30 in. diameter, aligned parallel to near-neighbors at distances of about 3.5 (Hirsch, 1954). Later development of the method led to the conclusion that the aromatic carbon occurred in layers composed of four to five condensed rings for coal up to 87% carbon, to about 30 condensed rings in coal having 94% carbon. It has also been proposed that if coal (up to about 90% w/w carbon) contained condensed aromatic ring systems, the average number of rings could not exceed four, in keeping with other work which indicated that coal (up to about 90% carbon) contained an appreciable proportion of small layers consisting of one to three rings.

It has also been noted that these small condensed aromatic regions form part of larger units which may themselves be linked to other such units by aliphatic or alicyclic material or by five-membered rings to form large buckled sheets. Above 90% C, the condensed aromatic layers are assumed to increase rapidly in size with increasing rank. It was also concluded that alicyclic structures were present to a considerable degree in coals of medium rank.

Further studies have indicated considerable differences in the x-ray scattering characteristics of coal components and that certain high-rank coals gave rise to the three-dimensional crystalline reflections of graphite. After a further systematic study of several high-rank coals, single graphite crystals (up to 3 μm in size) were noted in one of the coals and it was concluded that coal ultimately becomes graphite, presumably if the maturation process were allowed to go unchecked.

In order to gain additional information about coal structure, a method was developed to derive the layer distribution in coal in terms of molecular size histograms. It was difficult to make an accurate estimate of the number of condensed aromatic rings per layer since edge groups may contribute to the diffraction of the layer, but the results indicate that all coals with up to 90% carbon may contain only very few condensed rings.

10.4.4 ELECTRON SPIN RESONANCE

Early work on a series of carbonized coals gave 3×10^{19} free radicals per gram (1 free radical per 1600 carbon atoms). It was also established that the free radical content of coal at first increases slowly (in the range of 70%–90% carbon), rises markedly (90%–94% w/w C), and then decreases to limits below detectability. Thus, in a coal having 70% carbon there is one radical per 50,000 carbon atoms but this is increased to one radical per 1,000 carbon (atoms in coal with 94% w/w carbon).

More recent electron spin resonance studies on vitrains and fusains from a large number of coals show that concentrations of unpaired electrons in vitrains increase with increasing carbon content up to ca. 94% after which they decrease rapidly (Figure 10.7) (Retcofsky et al., 1978). This is not observed for fusains where there does not seem to be any correlation with rank. For the vitrains an increase in ESR line width with increasing rank is observed; for lignite to low-rank bituminous, this decreases through the higher-rank bituminous coals to the early anthracite stages and then increases markedly (Figure 10.8) (Retcofsky et al., 1978).

The ESR g values, which arise from spin–orbit coupling, are higher than 2.0023, which is the free electron value in the absence of spin–orbit coupling. The g values for vitrains and fusains

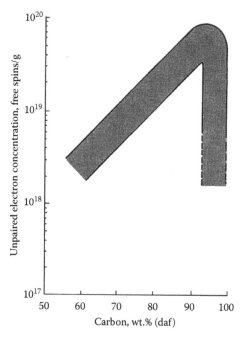

FIGURE 10.7 Variation of unpaired electron concentration with carbon content. (From Retcofsky, H.L. et al., *Organic Chemistry of Coal*, J.W. Larsen, Ed., Symposium Series No. 71, American Chemical Society, Washington, DC, 1978, p. 142.)

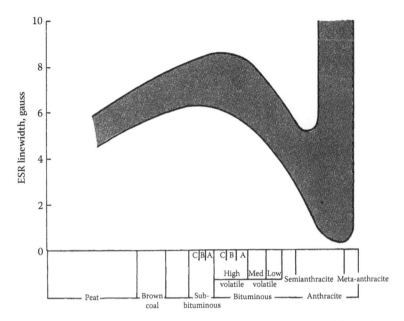

FIGURE 10.8 Variation of ESR line width with vitrain rank. (From Retcofsky, H.L. et al., *Organic Chemistry of Coal*, J.W. Larsen, Ed., Symposium Series No. 71, American Chemical Society, Washington, DC, 1978, p. 142.)

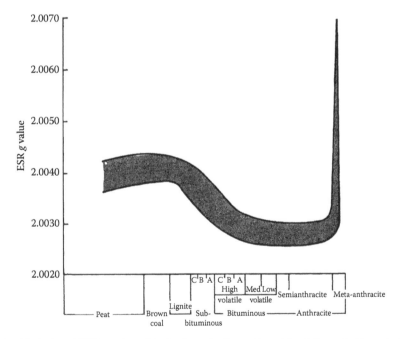

FIGURE 10.9 Variation of ESR *g* value with vitrain rank. (From Retcofsky, H.L. et al., *Organic Chemistry of Coal*, J.W. Larsen, Ed., Symposium Series No. 71, American Chemical Society, Washington, DC, 1978, p. 142.)

decrease from low to high rank with the exception of the values for vitrains from very high-rank coals (Figure 10.9) (Retcofsky et al., 1978); these are markedly higher.

Organic free radicals have higher *g* values if the unpaired electron is stabilized by atoms with high spin–orbit coupling constants. The high *g* values for low-rank coals can be explained by localization of unpaired electrons on heteroatoms. As the carbon content increases and the oxygen

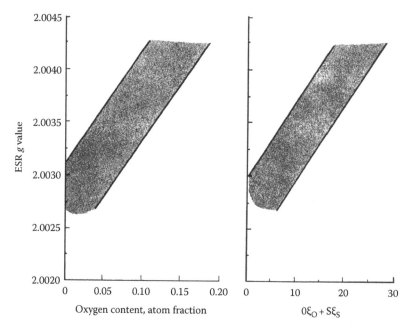

FIGURE 10.10   Variation of ESR g value with vitrain functionality. (From Retcofsky, H.L. et al., *Organic Chemistry of Coal*, J.W. Larsen, Ed., Symposium Series No. 71, American Chemical Society, Washington, DC, 1978, p. 142.)

content decreases, the g values decrease and the radicals become localized on aromatic hydrocarbons until the g values increase again with the formation of graphitic structures by condensation of aromatic rings (Figure 10.10) (Retcofsky et al., 1979).

There is the opinion that the g values are very important structural parameters and that, if used cautiously, the g values found in coal could be compared with the g values of pure compounds to obtain some indication of the paramagnetic species present in coal and perhaps other structural information.

Their comparisons show that for the higher-rank coals the g values are typical of those found for typical pi-type aromatic hydrocarbon radicals but the higher g values for subbituminous coal indicates the probability that oxygen atoms are important. Comparisons with pure compounds suggest that these radicals might be of the quinone-type in molecules containing four to nine rings or they may be due to methoxybenzene radicals, and/or related compounds. The greater line widths found for subbituminous coals imply that atoms with attached hydrogen atoms participate more in the molecular orbital of the unpaired electron in subbituminous radicals than in the higher-rank coals. It appears very likely, from the four coals studied, that the radicals in subbituminous coal are different from higher-rank coal radicals.

Electron spin resonance spectra of coals usually consist of a single line with no resolvable fine structure; however, the electron nuclear double resonance (ENDOR) technique can show hyperfine interactions not easily observable in conventional ESR spectra. Recently, this technique has been applied to coal and it is claimed that the very observation of an ENDOR signal shows interaction between the electron and nearby protons and that the results indicate that the interacting protons are twice removed from the aromatic rings on which, it is assumed, the unpaired electron is stabilized.

10.4.5   MASS SPECTROSCOPY

Mass spectroscopy adds further knowledge to the structural analysis of coal by allowing calculation of ring distribution as well as identification of individual molecular ions (Herod et al., 1991). For example, gas chromatography–mass spectroscopy as well as pyrolysis gas chromatography–mass spectroscopy of coal has enabled low-molecular-weight benzenes, phenols, and naphthalenes to be identified as well as the C_{27} and C_{29}–C_{30} hopanes in addition to several C_{15} sesquiterpenes (Gallegos, 1978; Smith and Smoot, 1990; Blanc et al., 1991).

The use of a pyrolysis technique masks the precise manner in which the thermal products are located within the coal matrix, that is, whether or not they were trapped ("caged") in the coal or whether they were an integral part of the organic structure of the coal. However, the identification of these materials does offer valuable information about the geochemical origin of the coal as well as some skeletal information about the coal (Chapter 3).

Application of this technique to the identification of methyl esters of the organic acids obtained by the controlled oxidation of bituminous coal allowed the more volatile benzene carboxylic acid esters to be identified (Studier et al., 1978). These were esters of benzene tetracarboxylic acid, terephthalic acid, toluic acid, and benzoic acid. Decarboxylation of the total acid mixture was shown to afford benzene, toluene, C_2-benzenes (i.e., ethylbenzene or xylenes), C_3-benzenes, butylbenzenes, C_5-benzenes, C_7-benzenes, naphthalene, methylnaphthalene, C_2-naphthalene, biphenyl, methylbiphenyl, C_3-biphenyl, indane, methylindane, C_2-indane, phenanthrene, and fluorene.

It was concluded that these nuclei occur in bituminous coal but are linked by more readily oxidizable structures. An examination of the solvent extracts of coal has produced evidence for the presence of short methylene chains which were part of neither an aliphatic chain nor an alicyclic ring. It was noted that a series of tetralin derivatives (or indane derivatives) and higher katacondensed aromatics were prevalent in the extracts of this same coal.

High-resolution mass spectroscopic analyses of pyridine extracts from reduced and untreated coals support the concept that ether linkages exist in the coal and are split during hydrogenation and that hydroaromatic compounds can be formed by addition of hydrogen to the aromatic nuclei.

10.4.6 ULTRAVIOLET SPECTROSCOPY

The ultraviolet spectra of coals, examined as suspensions in potassium bromide, show an absorption band at 2650 Å which becomes more pronounced with increasing rank of the coal. This band has been assigned to aromatic nuclei and, on the basis of data obtained from comparison between the specific extinction coefficients of coal and those of standard condensed aromatic compounds, it has been concluded that the concentration of aromatic systems in coal is lower than had previously been believed.

Other investigations have led to the conclusions that coal contains benzene and naphthalene rings, heterocyclic nitrogen, cyclic ethers, hydroxyl oxygen, as well as methylene groups and may even be fairly uniform in structure. In addition, the position of the maxima in the ultraviolet spectra of coal fractions appeared to indicate a mean cluster size comparable with that obtained by other methods. Thus, there was the suggestion that low-rank coals contained small aromatic nuclei of various kinds with a gradual coalescence of the units to form much larger layers of aromatic nuclei occurring as rank increases.

10.4.7 ELECTRON MICROSCOPY

Electron microscopy has also been employed as a means of elucidating coal structure; granules as small as 200 to about 1000 μm in diameter have been observed. In addition, it has been reported that particles range from 250 Å in a low-rank coal to 100 Å in a high-rank coal while two general ranges for the ultrafine structures have also been observed; one <100 μm and the other >100 μm with some of the particles in the form of polygonal platelets.

10.5 PHYSICAL PROPERTY METHODS

In addition to the chemical and spectroscopic methods that are applied to the identification of structural types in coal, it is also possible to make an evaluation of various structural parameters by application of various formulae derived from the physical property data. These particular techniques have received wide application to nonvolatile petroleum fractions because of the wider range

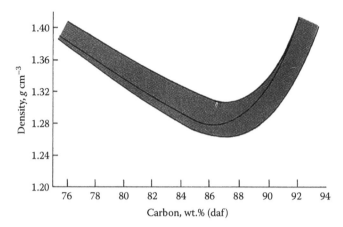

FIGURE 10.11 Variation of vitrinite density with carbon content. (From Davis, A., *Scientific Problems of Coal Utilization*, B.R. Cooper, Ed., United States Department of Energy, Washington, DC, 1978a.)

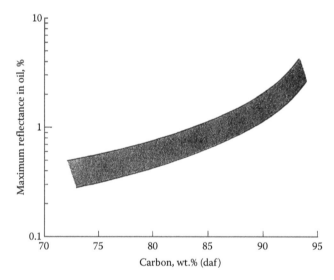

FIGURE 10.12 Variation of vitrinite reflectance with carbon content. (From Davis, A., *Scientific Problems of Coal Utilization*, B.R. Cooper, Ed., United States Department of Energy, Washington, DC, 1978.)

of physical properties than can be conveniently examined due to the easier examination of the liquid petroleum fractions (Speight, 2007). This, it is not surprising that they have also been of some value to determine the various structural parameters of coal and therefore deserved some mention at this point. In addition, although these methods are considered by some researchers to be of historical value only, there are other workers who consider these methods of continued value as a means of examining coal.

However, it should be noted here that the physical properties of coal are complicated by the fact that layering of the coal and mineral constituents at the time of bed deposition (as well as changes resulting from the subsequent pressure of the overburden) impact vertical and horizontal inequalities in the physical properties. Thus, any data derived by these methods must be expected to vary within certain limits (e.g., see Figures 10.11 and 10.12) (Davis, 1978a).

The molar volume is determined by summation of the single-bonded atomic volumes per gram atom minus a term to allow for unsaturation:

$$\text{Molar volume, } M/d = 9.9 + 3.1\text{H/C} + 3.8\text{O/C} + 1.5\text{N/C} + 14\text{S/C} - 9.1 - 3.6\text{H/C} \cdot R/\text{C}$$

where

M is the molecular weight

d is the specific gravity

The terms C, H, N, O, S represent the numbers of carbon, hydrogen, nitrogen, oxygen, and sulfur atoms per molecule, respectively.

R is the number of rings per molecule where

$$2(R-1)/C = 2 - f_a - H/C \ G$$

As before, f_a is the fraction of carbon atoms that are aromatic.

This relationship has been shown to be true for a variety of polymers but it is more likely now to be used in the abbreviated form:

$$\text{Molar volume} = 9.9C + 3.1H + 3.8O + - KM$$

KM is a correction term for multiple bonds. Thus, since

$$\%C/100 = 12(C/M)$$

The equation becomes (after division by C)

$$1200/d(\%C) = 9.9 + 3.1H/C + 3.8O/C + (\text{term for aromatic C=C bonds}).$$

An equation of this form has been found to be satisfied by data derived from 18 different polymers. By the use of a number of alkyl aromatic compounds and pitch fractions as reference substances, a calibration has been effected in terms of f_a, the fraction of the carbon in aromatic groups (Figure 10.13; terms for N/C, O/C, and S/C have been deleted by application of corrections) (Van Krevelen and Schuyer, 1957).

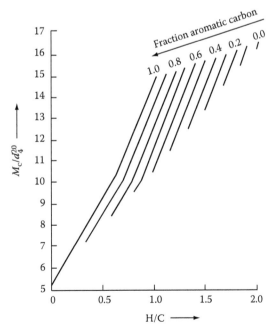

FIGURE 10.13 Variation of the fraction of aromatic carbon with the molar volume and atomic H/C ratio. (From Van Krevelen, D.W. and Schuyer, J., *Coal Science*, Elsevier, Amsterdam, the Netherlands, 1957.)

TABLE 10.8

Vitrinite Aromaticity by Various Methods

Carbon Content (%)	Basis of Fractional Aromaticity			
	Molar Volume	Sound Velocity	Heat of Combustion	UV Spectrum
70.5	0.70			
75.5	0.78			
80.0		0.79	~0.82	
81.5	0.83			
84.0				>0.78
85.0	0.85	0.82	~0.82	
87.0	0.86			
89.0	0.88	0.85		
90.0	0.90		~0.85	
91.2	0.92	0.90		
92.5	0.96	0.93		
93.4	0.99	0.97		
94.2	1.00			
95.0	1.00		1.00	
96.0	1.00			

Source: Van Krevelen, D.W., *Coal*, Elsevier, Amsterdam, the Netherlands, 1961.

Other properties can be used in a similar way including heat of combustion and sound velocity, giving f_a values in reasonable agreement (Table 10.8) (van Krevelen, 1961). There is still, however, some disagreement between different investigators over the correct values of f_a, and the available evidence can best be summarized by saying that a coal of 80% C has f_a between 0.7 and 90.8 and a coal of 90% C has f_a between 0.8 and 0.9.

Another additive property, namely, the molar refraction

$$(n^2 - 1)/(n^2 + 2)M/d$$

where n is the refractive index, has also been used for structural determinations of coal. Thus, if the reflectance, R, of coal (Davis, 1978b) is measured using two media with different refractive indices (n_o) in contact with the coal, the refractive index n can be derived from solving two simultaneous equations:

$$R = [(n - n_o)^2 + k^2]/[(n + n_o)^2 + k^2]$$

where k is the unknown absorption coefficient of coal. Elimination of the molecular weight term (division by C) gives

$$[(n^2 - 1)/(n^2 + 2)] 1200/\%C \cdot d = 2.6 + 1.0H/C + l_m$$

where l_m, the difference between the experimentally determined molar refraction and the value calculated from the atomic contributions, is assumed to be due to the aromatic C=C bonds. It can be shown that m is theoretically proportional to the number of aromatic carbon atoms per molecule (C_{ar})

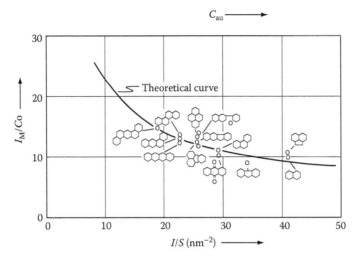

FIGURE 10.14 Variation of molar increment per gram atom of aromatic carbon with aromatic surface area. (From Schuyer, J. and van Krevelen, D.W., *Fuel*, 33, 176, 1954.)

and is a function of the surface area of the molecule (this arises from consideration of the polarizability associated with mobile electrons in an aromatic skeleton).

Experimental points for a number of pure compounds show satisfactory agreement with the theoretical relationship (Figure 10.14) (Schuyer and van Krevelen, 1954) and the surface area of the aromatic layers in coal, from which the number of carbon atoms in the aromatic layers can be calculated (Table 10.9) (van Krevelen, 1961). For reasonable agreement with the values derived from x-ray analysis for coals of higher rank, the values decrease possibly because of the development of charge-transfer characteristics and semiconducting properties.

There is also supporting evidence (Table 10.9) for the size of the aromatic layers from, for example, the elemental analysis. Very high-rank coals contain few hydrogen atoms to saturate the edge valencies

TABLE 10.9
Layer Sizes in Vitrinites (Expressed as Number of Carbon Atoms per Layer)

Carbon Content (%)	X-Ray Diffraction	Elementary Composition	Molar Refraction Increment	Semiconductivity	UV Spectrum
70.5			12		
75.5			13		
81.5	16		17		
84.0	17				>17
85.0	17		21		
87.0	17		(23)		
89.0	18				
90.0	18				
91.2	18				
92.5	18				
93.4	20	(20)		(45)	
94.2	(30)	22–40		(50)	
95.0		43–60		55	
96.0		85–100		>60	

Source: Van Krevelen, D.W., *Coal*, Elsevier, Amsterdam, the Netherlands, 1961.

of aromatic molecules. The semiconduction properties (the energy barrier, du, is a function of the number of aromatic carbon atoms per molecule in polycyclic aromatic compounds) and the ultraviolet spectrum (which can be synthesized from those of model compounds) also offer supporting evidence.

10.6 HETEROATOMS

During the investigations of coal structure, there has, with only minor exceptions, been the tendency to omit investigations of the heteroatoms (i.e., nitrogen, oxygen, and sulfur). Heteroatoms in coal are a very important aspect of coal structure (as those in the heavier petroleum fractions) and may be extremely important when considered in the light of coal conversion processes and design of catalysts for these processes. It is, therefore, essential that some consideration be given to these nonhydrocarbon species at this point (Table 10.10).

Most of the early evidence for the presence, or the absence, of functional groups in coal has been obtained from infrared spectra of solid coals (Table 10.6). However, more recent and detailed examination of the functional entities in coal has brought to light several more interesting features of the occurrence and the distribution of the heteroatoms.

10.6.1 NITROGEN

The nitrogen functionality in coal is believed to be representative of the nitrogen species of the original plant matter. This is consistent with the low degree of variation in the nitrogen content of coal (1%–2%). Metals also occur in coal and can be associated with clays or with the porphyrin system and can provide an index of coalification (Bonnett et al., 1991); metals also occur in coal as salts.

From a practical viewpoint (especially since coal combustion can produce nitrogen oxides), it is essential that more attention be given to the organic nitrogen compounds in coal. Data from the ultimate analysis of coal usually indicate nitrogen in coal to be within the range of 0.5%–2.0% and fragmentary evidence appears to favor nitrogen present in such heterocyclic locations as substituted acridine(s) and substituted benzoquinoline(s).

TABLE 10.10
Nitrogen and Oxygen Functional Types in Coal

There is indirect evidence for the presence of nitrogen ring systems in coal—the following compounds have been identified by mass spectroscopy as being present in coal tar pitch: pyridines, pyrroles, quinolines, azapyrenes, benzacridines, dibenzacridines, and substituted carbazoles. Allowances must be made for any structural alterations during the thermal production of the pitch as well as for aromatization of any hydroaromatic ring systems.

Perhaps one of the most interesting findings is that the total nitrogen content of a series of bituminous coals has been shown to peak at ca. 85% carbon and there is also a variation in nitrogen functionality. Nitrogen generally exists in a variety of heterocyclic forms with pyrrole-type nitrogen predominating throughout the bituminous coal range but the pyridine-type nitrogen increases with rank. It is also reported from this study that nitrogen is generally concentrated in the vitrinite group of macerals which also contain the highest proportion of pyridine-type nitrogen (Burchill, 1987).

10.6.2 OXYGEN

Of the three heteroelements in coal, oxygen has received the most attention and, consequently, some generalizations are possible. But first, the oxygen functionalities in coal can be divided into four categories: (1) carboxyl, (2) carbonyl, (3) hydroxyl, and (4) ether but only categories (1), (2), and (3) are amenable to quantitative analysis (Attar and Hendrickson, 1982).

Analyses of coals varying in rank have shown that rather surprising correlations exist between the carbon content of coal and the content of various functional groups (Schafer, 1970; Cronauer and Ruberto, 1977) and there appears to be a relationship between functionality and carbon content of coal (Figures 10.15 through 10.19) (Whitehurst et al., 1980).

Spectrometric and chemical methods have been applied to the qualitative analysis of ether bonds in coal (Barron and Wilson, 1981; Painter et al., 1981; Wender et al., 1981; Siskin and Aczel, 1983; Youtcheff and Given, 1982, 1984; Mallya and Zingaro, 1984; Yoshida et al., 1984; Deno et al., 1985; Franco et al., 1992) with some degree of success. Indeed, there is evidence to presume that the ether bond in coal is mainly of the diaryl (Ar–O–Ar) and benzyl–aryl ($C_6H_5CH_2$–O–Ar) types although methoxyl groups (CH_3O–) and dialkyl ethers (R–O–R$'$) are found in low-rank coals (Bredenberg et al., 1987).

In general, fresh (unweathered) higher-rank (bituminous and anthracite) coals appear to contain little, or no, aliphatic hydroxyl, carboxyl, or methoxyl functions while brown coals are reputed to contain aliphatic hydroxyl groups. Furthermore, there have been suggestions that, except for lignites, virtually all of the oxygen in coal can be accounted for as carboxyl,

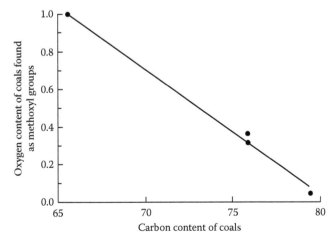

FIGURE 10.15 Methoxyl content and carbon content of selected coals. (From Whitehurst, D.D. et al., *Coal Liquefaction: The Chemistry and Technology of Thermal Processes*, Academic Press, Inc., New York, 1980.)

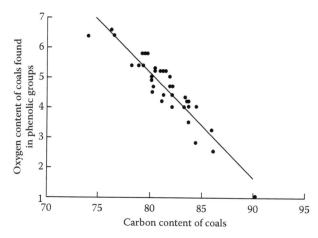

FIGURE 10.16 Phenolic hydroxyl content carbon content of selected coals. (From Whitehurst, D.D. et al., *Coal Liquefaction: The Chemistry and Technology of Thermal Processes*, Academic Press, Inc., New York, 1980.)

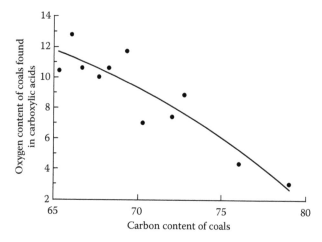

FIGURE 10.17 Carboxylic acid content carbon content of selected coals. (From Whitehurst, D.D. et al., *Coal Liquefaction: The Chemistry and Technology of Thermal Processes*, Academic Press, Inc., New York, 1980.)

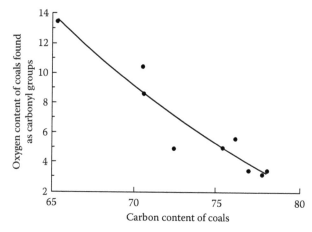

FIGURE 10.18 Carbonyl group content and carbon content of selected coals. (From Whitehurst, D.D. et al., *Coal Liquefaction: The Chemistry and Technology of Thermal Processes*, Academic Press, Inc., New York, 1980.)

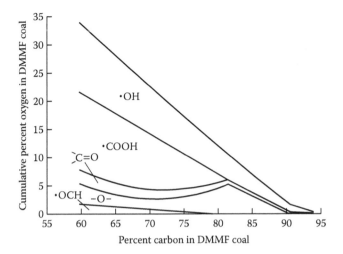

FIGURE 10.19 Distribution of oxygen functionality and carbon content of selected coals. (From Whitehurst, D.D. et al., *Coal Liquefaction: The Chemistry and Technology of Thermal Processes*, Academic Press, Inc., New York, 1980.)

hydroxyl, or carboxyl with a minor amount of heteronuclear oxygen, although there is now growing evidence for various ethers in coals.

However, it is essential to recognize that all, say, carbonyl groups are not chemically equivalent since a quinoid carbonyl function may be expected to behave differently from a carbonyl function conjugated with an aromatic ring.

A great deal of information about the location of oxygen in coal has been derived from infrared spectroscopic investigations. However, infrared spectroscopic data tend to suffer from their inability to be quantitative and can, at best, be used to illustrate trends throughout the progression from low-rank coals to high-rank coals (Table 10.11).

TABLE 10.11
IR Absorption Bands in Coal and Relative Strengths

Frequency (± 5 cm^{-1})	Assignment	Low-Rank Coals	High-Rank Coals
750–810	Substituted aromatic	Weak	Strong
850–870	Aromatic	Weak	Strong
1250	$\phi - O \diagup^H$	Strong	Weak
1370	$-CH_3$	Strong	Weak
1450	Aliphatic, alicyclic	Strong	Weak
1600[a]	$-C==C-/C==O$	Strong	Weak
1700	$C=O, -\overset{O}{\overset{\|}{C}}-OH$	Strong	Weak
2860	CH_2	Variable	Weak
2910	CH_2	Variable	Weak
3030–3050	$C_{ar}-H$	Weak	Strong
3300	$-O\diagup^H$	Strong	Weak

[a] Bands at 1360 and 1600 cm^{-1} have been reported for graphite.

TABLE 10.12

Analysis of Oxygen Functions in Coal

Coal	Burning Star (Bituminous)	Belle Ayr (Subbituminous)
Oxygen content as		
Hydroxylic (–OH)	2.4	5.6
Carboxylic (–COOH)	0.7	4.4
Carbonylic (==C==O)	0.4	1.0
Etheric (–O–)	2.8	0.9
Total	6.3	11.9
Oxygen by difference		
Ash basis	5.9	16.2
Mineral matter basis	—	16.0

Source: Ruberto, R.G. and Cronauer, D.G., *Organic Chemistry of Coal*, J.W. Larsen, Ed., Symposium Series No. 71, American Chemical Society, Washington, DC, 1978, p. 50.

Notable attempts have also been made to determine the oxygen functions (Table 10.12) (Ruberto and Cronauer, 1978) in a subbituminous coal and in a bituminous coal. The results are consistent with the generally agreed finding that the oxygen content of coal decreases with rank. From solvation studies of a subbituminous coal, it was concluded that a significant portion of the oxygen occurs in saturated ether functional groups alpha or beta to the aromatic moieties or a furan systems. In addition, significant quantities were found of xanthone in lignite, xanthone and dibenzofuran in bituminous coal, and dibenzofuran in anthracite.

The importance of phenolic hydroxyl and its characteristics has been discussed many times, and there is evidence to suggest that OH/coal is a rank parameter in its own right, it decreases with increasing degree of coalification, but that OH/O is a rather poor rank parameter. These results implied that the sum of the nonhydroxyl oxygen functional groups present (carbonyl, ether, heterocyclic oxygen) in coal changes in a random manner with rank.

10.6.3 SULFUR

Sulfur and oxygen in organic systems are often compared and similar behavioral patterns noted, but sulfur (perhaps because of the relative amounts of sulfur and oxygen in coal) has not received the same degree of attention in coal and, consequently, much less is known about the organic sulfur in coal than the organic oxygen. Nevertheless, recent studies have focused on the sulfur types in coal with some degree of success (Calkins et al., 1992; Sinninghe Damste and de Leeuw, 1992). In more general terms, there are two major classes of sulfur which predominate in coal: (1) inorganic sulfur (mostly pyrite) and (2) organic sulfur.

It is believed that the original decaying plant matter contained up to 0.5% sulfur and that any higher sulfur contents were caused by sulfate-reducing bacteria (Figure 10.20). The organic matter was submerged in saline or brackish water containing sulfate salts. The bacteria consumed a portion of the organic matter and used sulfate (SO_2) as an oxygen source, thus producing a variety of reduced- sulfur-containing ions (HS–, CHSX, etc.). If ionic iron was present in the water, iron sulfide precipitated and produced pyrite. It is also conceivable that the organic moieties in the coal were also changed as the oxygen functionality was exchanged for sulfur.

Since the kinetics of pyrite and organic sulfur production appear to be parallel, there is the distinct possibility that the ratio of organic to inorganic sulfur is unique to specific locations and was

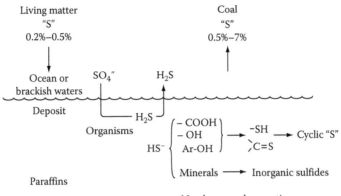

FIGURE 10.20   Potential routes to the origin of organic sulfur in coal. (From Whitehurst, D.D. et al., *Coal Liquefaction: The Chemistry and Technology of Thermal Processes*, Academic Press, Inc., New York, 1980.)

TABLE 10.13
Distribution of Organic Sulfur in Coal

Coal	Organic S (% w/w)	% Organic S Accounted	Thiolic	Thiophenolic	Aliphatic Sulfide	Aryl Sulfide	Thiophenes
Illinois	3.2	44	7	15	18	2	58
Kentucky	1.43	46.5	18	6	17	4	55
Martinka	0.60	81	10	25	25	8.5	21.5
Westland	1.48	97.5	30	30	25.5	—	14.5
Texas	0.80	99.7	6.5	21	17	24	31.5

Source: Attar, A. and Dupuis, F., Preprints, Division of Fuel Chemistry American Chemical Society, 24(1), 166, 1979.

dictated by the composition of the dissolved salts in the deposit and physical parameters such as time, temperature, and pressure.

Heterocyclic sulfur compounds have been proposed/identified in coals but, until recently, there have been very few studies concentrated on the organic sulfur functional groups in coal (Attar, 1979; Gorbaty et al., 1991). However, recently reported data on these functionalities in different coals (Table 10.13) (Attar and Dupuis, 1979) show the distribution of organic sulfur groups in coal and show that the content of thiols (–SH) is substantially larger in lignites and high-volatile bituminous coals than in low-volatile bituminous coals. The fraction of aliphatic sulfides (R–S–R) remains approximately constant at 18%–25%.

The data indicate that larger fractions of the organic sulfur are present as thiophene sulfur in higher-rank coals than in lower ones. The data on –SH, R–S–R, and thiophene sulfur are explained by suggesting that the coalification process causes the organic sulfur to change from –SH through R–S–R to thiophene in condensation reactions.

10.7   MOLECULAR WEIGHT

Determination of the molecular weight (size) of the higher-molecular-weight constituents of petroleum (the asphaltene constituents) is somewhat easier to investigate than the molecular weight of coal constituents by virtue of their solubility in aromatic (and other highly potent) solvents. In contrast, the molecular weight of coal is, because of the nature of the material, unknown.

There have been many attempts to define the molecular size of coal by means of molecular weight techniques. The majority of such attempts depend on the conversion of the largely insoluble coal material into soluble fractions under relatively mild conditions on the optimistic note that the coal "molecules" remain largely unchanged and a major attempt is made to avoid the thermal decomposition reactions which take place above 350°C (660°F).

10.7.1 DEPOLYMERIZATION

Depolymerization of coal is a technique which, it is claimed, solubilizes coal by cleaving methylene bridges in the coal. The methylene chains joining aromatic groups can be cleaved at the ring and the "free" alkyl group and then alkylate another aromatic substrate (Larsen et al., 1981; Mastral-Lamarca, 1987; Sharma, 1988):

$$Ar - (CH_2)_n - Ar' + C_6H_5OH = ArH + HOC_6H_4 - (CH_2)_n - Ar$$

$$HOC_6H_4 - (CH_2)_n - Ar' + C_6H_5OH = HOC_6H_4 - (CH_2)_n - C_6H_4OH$$

This technique has been used to estimate the molecular weight distribution in coal but the (number average) molecular weights (Mn) of the material soluble in pyridine after depolymerization can be affected by the presence of colloidal material. Whilst the reported value may be in the region of 400, removal of the colloidal material (by, say, centrifugation) may increase this value to ca. 1000.

However, there is evidence to indicate that this particular reaction does not produce the anticipated low-molecular-weight materials as the major products and, thus, the presence of colloidal material and the higher-molecular-weight products certainly casts some doubt on the validity of any conclusions drawn from the data acquired by this method.

10.7.2 REDUCTIVE ALKYLATION

In principle, the reductive alkylation of coal involves treatment of coal with an alkali metal in tetrahydrofuran in the presence of naphthalene whereby a coal *polyanion* is produced that is capable of undergoing further reaction with, say, an alkali halide. The resulting product has been presumed to be the alkylated coal but the "relatively straightforward" chemistry is, in fact, a complex sequence of reactions. For example, ether bridges are also cleaved under the conditions of the reaction as are carbon bonds. That the former can happen makes the resulting product mix somewhat more complex than if ethers were not cleaved but the cleavage of carbon–carbon bonds ensures the complexity of the product mix as well as the chemistry involved in the process.

It was originally anticipated that the reaction would not cleave coal "molecules" but an examination of the molecular weight distribution of the products from the ethylation of coal (carbon content: 90% w/w) produced data which indicated a number average molecular weight of the order of 3300 (Figure 10.21) (Sternberg et al., 1971).

In another series of studies, vitrinite was treated under the conditions prescribed for reductive alkylation. The number average molecular weight was determined for the benzene and pyridine extracts and it was concluded that the products were of two types: (1) those having a number average molecular weight below 500 and (2) those having a molecular weight (uncorrected for the added alkyl groups) of approximately 2000. Other data from the studies of coals of different rank (ranging from a 78% C to 92.6% C) have also produced alkylated products ranging in molecular weight from 500 to 800 for the lowest rank coal to products ranging in molecular weight from 1300 to 2000 for the highest rank coal.

Thus, even though the evidence appears to indicate that coal can be selectively cleaved by the technique of "reductive alkylation" to lower-molecular-weight fragments thereby allowing

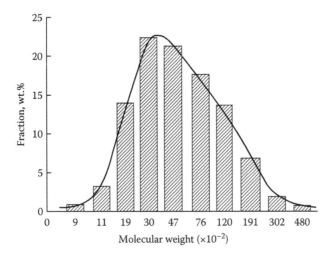

FIGURE 10.21 Molecular weight distribution of the product from coal ethylation. (From Sternberg, H.W. et al., *Fuel*, 50, 432, 1971.)

significant deductions to be made about the chemical nature of coal, there were other factors that needed consideration. For example, as noted earlier, it was originally believed that the alkylation reaction was quite specific and the chemistry was fairly simple and lacked any degree of complexity, but it has now been established that the chemistry is very complex and even side reactions between the tetrahydrofuran solvent and the alkali naphthalide will occur (in the absence of coal), thereby complicating the product mix still further.

Obviously, there are many facets to this particular reaction that need to be resolved before the data can be used with any degree of certainty to project to the chemical nature of coal.

10.7.3 ACYLATION

Acylation attempts to depolymerize coal by the introduction of acyl [RC(=O)–O–] groups have also been reported.

For example, Friedel-Crafts acylation of coals resulted in an increase in extractability which was dependent on the chain length of the inserted acyl groups. The molecular weights of the soluble products were only slightly higher than those reported for many of the alkylation experiments (ranging from 930°C to 4400°C for a coking coal and from 1900°C to 5200°C for a dry seam coal); the molecular weights increased with increasing chain length of the acyl substituent.

It has also been noted from work on the determination of the molecular weight of acylated coals that the molecular weight values were much higher than those reported from alkylation studies and were actually of the order of 100,000–1,000,000.

Again, there are questions of the interpretation of the data that are obtained by application of a "simple" chemical reaction sequence to coal.

10.7.4 BASE ALCOHOL HYDROLYSIS

Substantial increases in the extractability of coal with pyridine have been obtained by reacting coal with sodium hydroxide in ethyl alcohol. The yield of the extract rises with temperature and reaches a maximum at 450°C (840°F) after which the yield of extract decreases and there is also an observed increase in the yield of extract with coal rank.

One controversial point that arises from these investigations is the use of high temperatures, in excess of 350°C (660°F) that can cause substantial chemical and physical changes in the coal (Chapter 13).

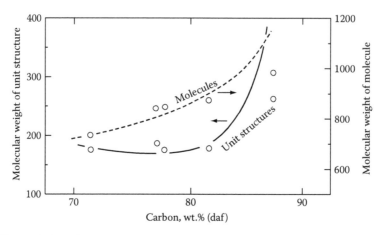

FIGURE 10.22 Variation of the molecular weight per molecule with the unit structure weight with coal carbon content. (From Makabe, M. and Ouchi, K., *Fuel*, 58, 43, 1979.)

As an example, the aromatization of naphthene rings, thermal dealkylation, thermal dehydrocyclization of alkyl chains, and subsequent aromatization, as well as some destruction of (oxygen) heterocyclic systems to name only four possibilities. Nevertheless, the data have been interpreted as indicating that the molecular weight of a coal molecule increases with rank of the coal but that of the "unit structures" remains constant up to approximately 82% carbon coal after which an increase is noted (Figure 10.22) (Makabe and Ouchi, 1979). There is no evidence to indicate that the unit structures are linked linearly (i.e., in the manner of a polymer chain) or that the unit structures are identical.

10.7.5 Miscellaneous Methods

A further approach to the determination of the molecular weight of coal invokes the concept of the time-dependent response of bituminous coals to constant stress and presents indications that these coals are cross-linked, three-dimensional molecules. There have also been attempts to apply the Flory-Huggins theory to coal but there is some question about the validity of such an approach. But coal is most unlikely to be a simple polymer network: its heterogeneity, its mineral content (e.g., clays might act as fillers in the polymer immobilizing any layers), and other factors indicate that a simple application of polymer theory is not justified. If some of the problems of the application of polymer theory to coals could be resolved, the method could provide a valuable aid to chemical methods of molecular weight determination.

In summary, there are many postulates of the structural entities present in coal (derived from a variety of experimental and theoretical methods) which involve estimations of the hydrocarbon skeleton, but although spectroscopic techniques may appear quite formidable as an aid to structural analysis of coal, the validity of any conclusions may always be suspect.

10.8 ASSESSMENT OF COAL STRUCTURE

Coal structure can be subdivided into two categories: (1) the physical structure and (2) the chemical structure. Coal contains extractable smaller molecules (Section 10.2) which are an integral part of coal and bear a physical and chemical relationship to the non-extractable material. In contrast to many earlier studies of coal "structure," both are an essential part of any consideration of the structural types in coal and one cannot exist without the other.

Therefore, in terms of physical structure, it is appropriate to note that coal is considered to be a two-component system, a mobile phase and a macromolecular network (Given, 1984a,b;

Given et al., 1986) which consists of aromatic ring clusters linked by bridges (Solomon et al., 1992). Thus, lower-molecular-weight species, the identifiable components of the mobile system, must also be given attention. How they exist in conjunction with the main body of the coal is another issue. They have been variously referred to as "guest molecules" (Redlich et al., 1985), clathrates (Given et al., 1986), and (perhaps even less correctly) as *bitumen* (Chapter 11) (Kreulen, 1948; Grint et al., 1985; Pickel and Gotz, 1991; Speight, 2007, 2008, 2009).

Recent work has used the yield of the chloroform extract of coal as an indication of the extent of the mobile phase (Derbyshire et al., 1991). In untreated coals, only a portion of the mobile phase appears to be removable but after "mild" preheating there are sharp increases in the yield of the extract. It has also been noted (Brown and Waters, 1966a,b), perhaps to no one's surprise, that there is an increase in the yield of chloroform-extractable material with treatment temperature. This has been interpreted to mean that there is a gradation in the manner in which the smaller molecules are associated with the network. Thermal studies also tell us that the thermal chemistry of chemical bonds can also vary with temperature.

Thus, caution is advised in the interpretation of the results of thermal investigations. There is a real need to recognize, or even differentiate between, the different chemical and physical events that can occur within the high-temperature domains, even though these events may differ from the preferred event.

Indeed, the use of high temperatures in a mass spectrometric examination of the species that constitute the mobile phase might also be suspect (Yun et al., 1991; Winans et al., 1992), even though the data appear to reveal the existence of a thermally extractable bitumen-like fraction which is chemically distinct from the remaining coal components.

Although such data might be cited as evidence that the weak bonding between the mobile phase and the network, consideration must also be given to the chemical changes that temperatures in excess of 300°C (570°F) and the presence of hydrogen can cause. Such conditions border on the conditions used for hydroprocessing petroleum fractions which, while the physical appearance of the fraction might not change, can bring about chemical changes such as the removal of a sigma-bonded heteroatom (nitrogen, sulfur, or oxygen) (Speight, 2007). Indeed, the effects of such reaction conditions on coal must be given serious consideration. The mere appearance that coal is not affected by high temperature (300°C [570°F]) is not indication of the chemistry that could occur on a molecular scale.

Thus, the results of such work, as mentioned here (and there are many other examples), may provide strong evidence that the thermal extraction (in many cases, *thermal decomposition* is a more appropriate term) of coal produces evidence for the molecular species that constitute the mobile phase, but, the thermal chemistry of coal is much more complex than these data would suggest.

In other words, the pertinent issues is the amount of the mobile phase, under prevailing ambient conditions and the amount of the mobile phase is generated by the application of thermal conditions that exceed the thermal decomposition threshold of the molecular species in coal. Part of the answer may lie in a thorough understanding of the plastic properties of coal (Grimes, 1982) (Chapter 9).

That coal contains low-molecular-weight extractable species is a fact. That these species may form a mobile phase within the macromolecular network of coal is effective in explaining some of the many facets of coal behavior, including physical phenomena such as porosity and solvent diffusion (Rodriguez and Marsh, 1987; Hall et al., 1992). However, that the constituents of this network can be extracted (unchanged) by thermal means or by solvent treatment after exposure of the coal to high temperatures where the stability of many organic species is suspect and that only the "weak" bonds are broken is open to question.

In the chemical sense, it should never be forgotten that coal (like petroleum; Speight, 2007) is a natural product and, as such, must be considered to contain vestiges (perhaps somewhat changed from the original) of the plant material. But to an extent, the nature of the coal must be influenced by the nature of the original material (Given, 1984b; Derbyshire et al., 1989). Studies of the maceral type and content of coal (Chapter 4) show this to be at least true in

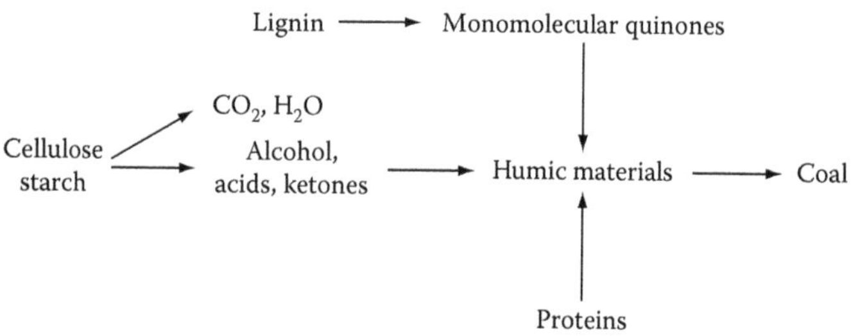

FIGURE 10.23 Routes to coal from natural product systems. (From Volborth, A., *Analytical Methods for Coal and Coal Products*, Academic Press, Inc., New York, Vol. III, Chapter 55, 1979.)

principle but these are physical signs of the coal constitution. Whether these vestiges remain as largely unchanged entities or whether they are completely different from their original form (Figure 10.23) is another question (Volborth, 1979).

As an illustration, it has been shown that the infrared and x-ray diffraction spectra of the aromatic acids and their methyl esters formed during oxidation of carbon black were almost identical to those of mellitic acid and its methyl ester, but the chemical and physical properties were markedly different and the ultraviolet spectra bore very little resemblance to those of the pure materials.

An examination of the solubility of coal in a variety of solvents and examination of the infrared spectra of the extracts indicated that coal consists of structures of basically similar chemical type and suggested that the coals closely related in rank may be homogeneous in chemical structure. Indeed, there are numerous examples cited in the literature which support this view and it is not surprising that material extracted from coal has been employed as being representative of that particular rank for structural determinations and differences exist predominantly in the molecular weights (i.e., degree of polymerization) of the structural entities. Caution is advised in interpretations of this nature because the pyridine extracts of coal may, other claims to the contrary because of the complex nature of coal, differ substantially with increasing extraction time and different constituents predominate at different stages of the extractions.

Nevertheless, the major drawback to the investigation of coal structure has been the incomplete solubility of the material, which has in many cases dictated that structural determinations be carried out on extracted material. Even then, the answer may not be complete. For example, coal structure is often considered to analogous to humic acid structure; humic acids are considered to be soluble molecular entities that are produced during the formation of coal. However, humic acid structure is not well known and has been, in the past, represented as involving large condensed nuclear systems (Figure 10.24).

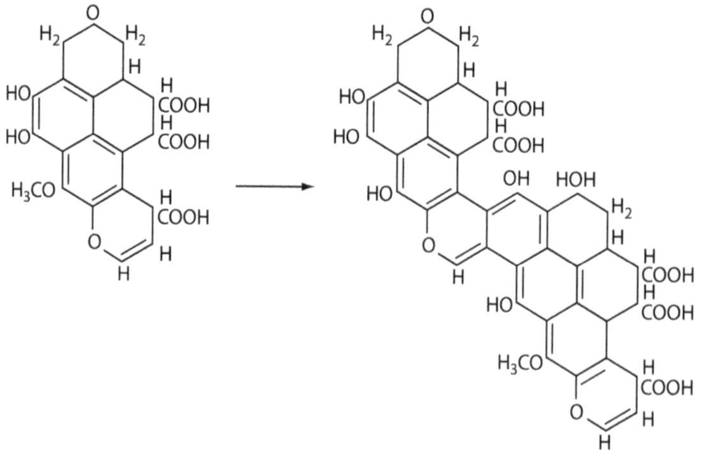

FIGURE 10.24 Hypothetical structures for humic acids and interconversional relationships.

In addition, an aspect of coal science that is often carried in the minds of those whose goal is structural elucidation is the thermal decomposition of coal to coke (Chapters 13 and 16).

Briefly, it has been assumed, on the basis of the behavior of the thermal decomposition of poly-nuclear aromatic systems, that coal must also consist of large polynuclear aromatic systems. Be that as it may, such assumptions are highly speculative and, to say the least, somewhat lacking in caution. As an example, similar lines of thinking have been applied to structural assumptions about petroleum asphaltene constituents when it is known from other pyrolysis studies that smaller, but polar systems, can produce as much thermal coke as the larger nonpolar highly condensed systems (Speight, 2007). Indeed, it is now recognized, on the basis of other studies (Winans et al., 1988), that coal structure (especially the structure of non-anthracitic coals) is not necessarily dominated by polycyclic/polynuclear aromatic species.

Thus, concepts of coal structure must also satisfy any of the relevant data from the variety of chemical manipulations of coal that have been described in a previous section and add positive knowledge about the nature of coal. In addition, a consensus of the aromaticity values reported would be that there appears to be a definite increase with rank. The values reported range from 40% to 50% for low-rank lignite and subbituminous coals to nearly 100% for anthracites.

There is some correlation between aromaticity and the weight percentage of carbon. As ^{13}C magnetic resonance is widely used as a method of estimating the aromatic carbon content of coal, it is worth noting that when line width is correlated with carbon content of coal(s) there is a pronounced inflection at ca. 93% carbon (Figure 10.25) (Retcofsky and Firedel, 1973). Presumably, this indicates a decrease in the diversity of carbon types in the high-rank (anthra-cite) coals. And there are many other similar correlations between the physical/chemical properties of coal but whether these correlations are genetic or generic is another issue (Berkowitz, 1988). Petroleum shows similar genetic/generic correlations and the issue is the same reality or perception (Speight, 2007).

Indeed, the issue of reality or perception has run amok in recent years in petroleum science where workers have laid claim to be the first with a particular structural postulate when their results were based on (at the time believable but proven and acknowledged since then to be) faulty labora-tory techniques which produced erroneous, and essentially fictitious, data. And yet we are asked to believe but the ethics of such claims leads to other issues. Back to the present issue.

An early model proposed for coal structure was based on ovalene (Figure 10.26) with the non-aromatic part represented as alkyl substituents. Yet another early attempt to illustrate coal structure involved the suggestion that ether linkages were a means of combining alkyl-substituted pyrene and/or coronene nuclei (Figure 10.26). In addition, coal was believed to be composed of an aromatic cluster unit bearing alkyl, as well as oxygen, substituents; the unit was approximately fused aromatic rings in low-rank coals, but 10–20 fused rings in high-rank coal. The evidence available from the infrared and x-ray methods appeared to suggest that coal exists as polymeric structures in which the monomer

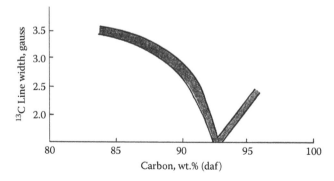

FIGURE 10.25 Variation of ^{13}C NMR line width with carbon content. (From Retcofsky, H.L. and Firedel, R.A., *J. Phys. Chem.*, 77, 68, 1973.)

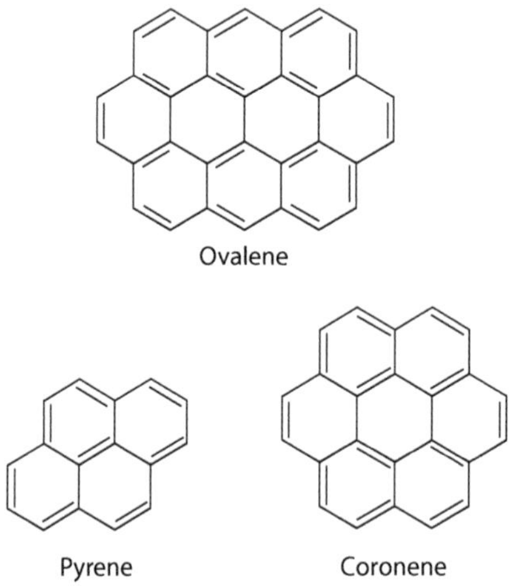

Ovalene

Pyrene   Coronene

FIGURE 10.26   Early structures of coal were based on ovalene, pyrene, and coronene.

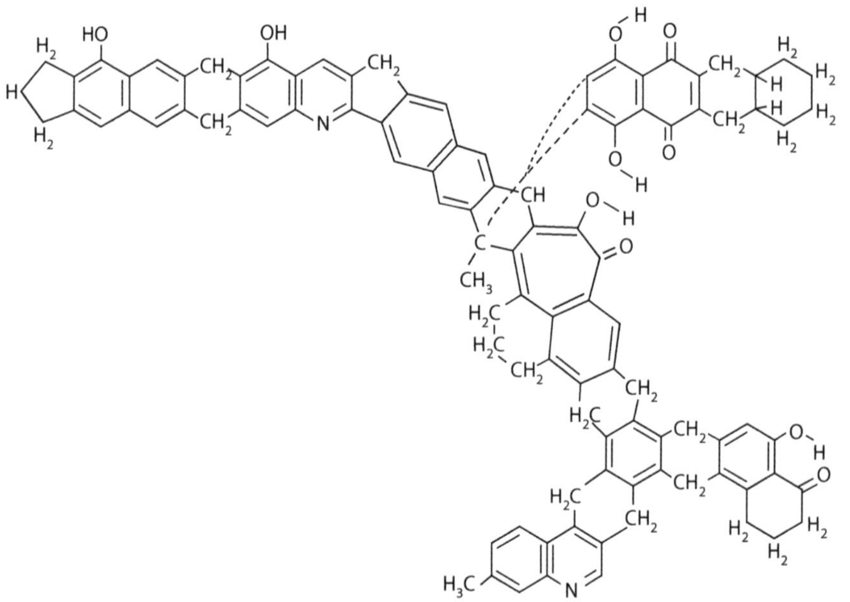

FIGURE 10.27   Introduction of a third dimension into coal structure. (From Given, P.H., *Fuel*, 39, 147, 1960.)

is based on 9,10-dihydroanthracene (Figure 10.27) (Given, 1960). The model was later modified, for various reasons, to convert the basic aromatic unit of coal from a dihydroanthracene unit to a related dihydrophenanthrene unit without seriously affecting the overall model for coal structure.

These postulates of coal structure also contained a novel element insofar as they invoke the concept of an extra dimension. This was derived by employing a triptycene moiety in the coal molecule which necessitated that not only the structure be buckled but also part of the coal molecule projects in an additional dimension to the rest. A structure, consistent with the x-ray data, represented the coal "molecule" as a buckled sheet consisting of condensed aromatic and hydroaromatic rings bearing only short alkyl substituents (methyl and ethyl) with bridging and nonbridging methylenes adjoining the aromatic rings (Figure 10.28) (Cartz and Hirsch, 1960); other similar formulae have also appeared for a vitrinite (Figure 10.29) (Gibson, 1978) and for vitrain with different carbon contents (Figure 10.30) (Pitt, 1979).

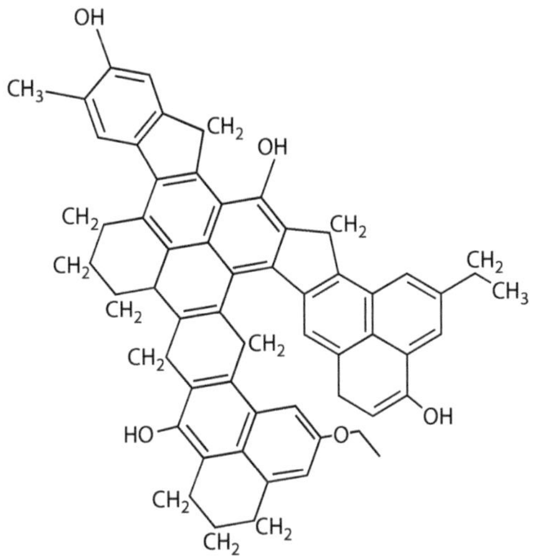

FIGURE 10.28 Introduction of the buckled sheet concept into coal structure. (From Cartz, L. and Hirsch, P.B., *Phil. Trans. R. Soc.*, A252, 557, 1960.)

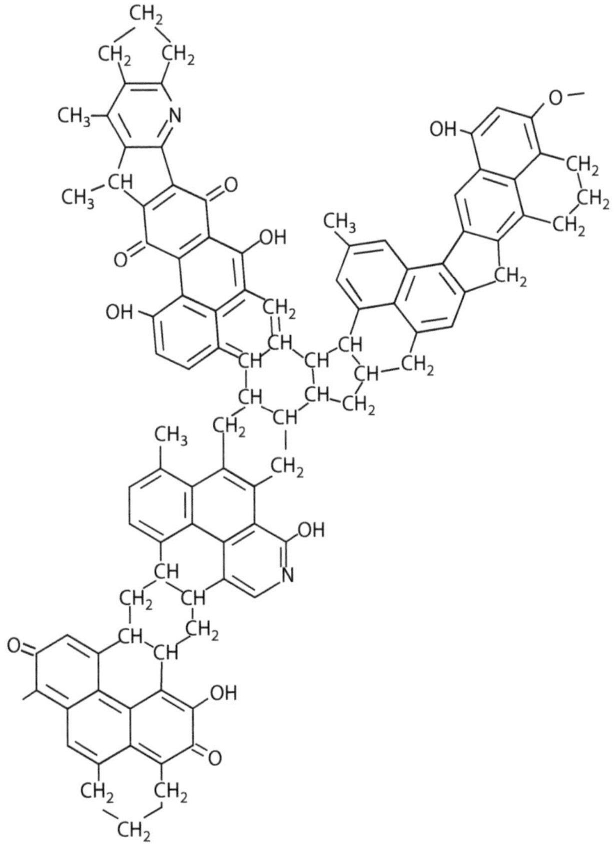

FIGURE 10.29 Hypothetical structure for vitrinite (82% w/w carbon). (From Gibson, J., *J. Inst. Fuel*, 51, 67, 1978.)

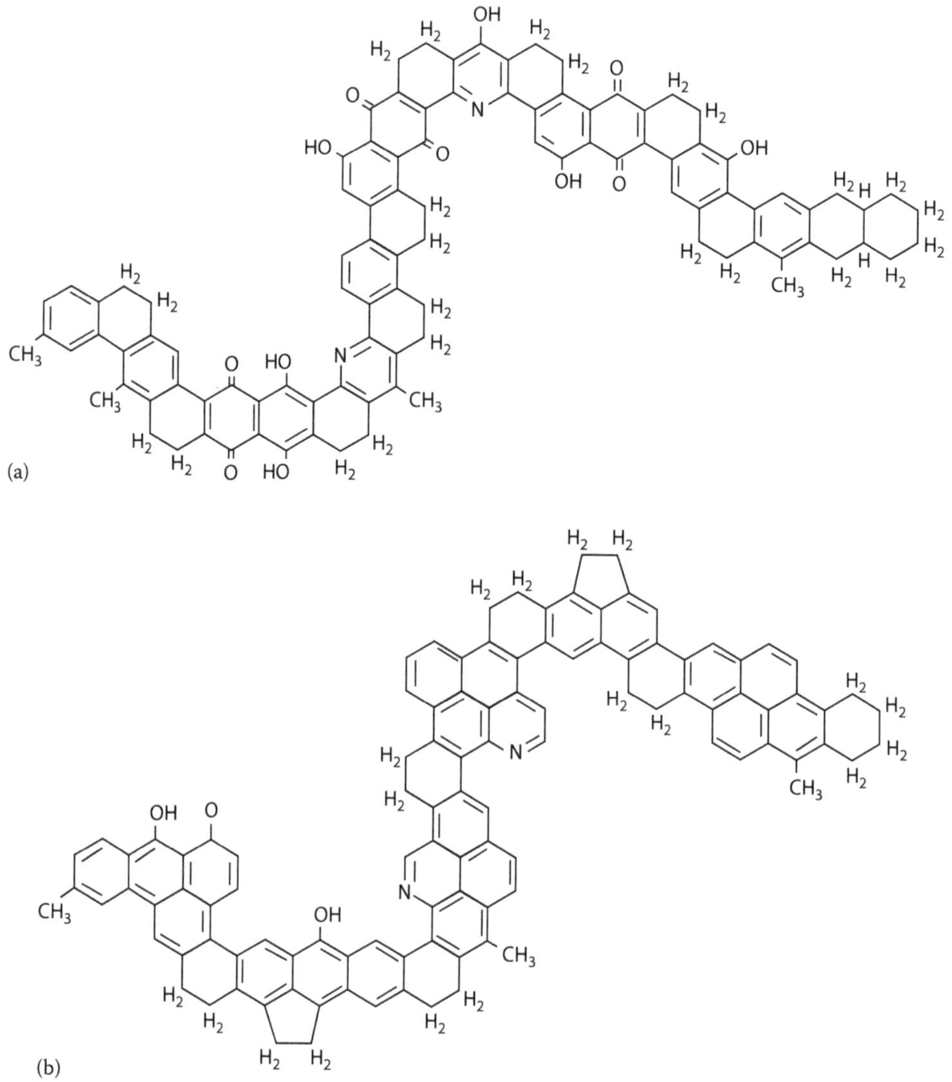

FIGURE 10.30 Hypothetic structures of (a) vitrain, 80% w/w carbon, and (b) vitrain, 90% w/w carbon. (From Pitt, G.J., *Coal and Modern Coal Processing: An Introduction*, G.J. Pitt and G.R. Millward, Eds., Academic Press, Inc., New York, 1979, Chapter 2.)

While chemical and spectroscopic evidence indicated a vast complex network of condensed aromatic and heteroaromatic nuclei bearing alkyl, as well as heteroatom, substituents, the aromatic centers varied from two to nine rings with the heteroatoms occurring as heterocyclic ring systems or as substituents on the aromatic and naphthenic nuclei (Figure 10.31) (Hill and Lyon, 1962). This postulate also invoked the concept that tetrahedral three-dimensional carbon–carbon bonds were present in the coal molecule, thus indicating an even higher degree of complexity than can be illustrated. However, the postulate is actually of forerunner of the current consistent thinking on coal structure in which coal is considered to be a two-component system, a mobile phase and a macromolecular network (Green et al., 1982; Given, 1984a,b; Given et al., 1986; Nishioka, 1992) which consists of aromatic ring clusters linked by covalent bridges or hydrogen-bonded structures (Peppas and Lucht, 1984; Solomon et al., 1992). But how the lower-molecular-weight species exist in conjunction with the main body of the coal is another issue. They have been variously referred to

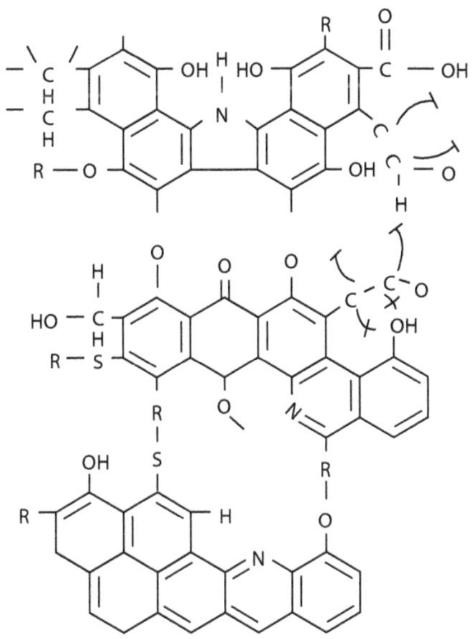

FIGURE 10.31 Section of a hypothetical coal structure using the concept of a network of condensed aromatic nuclei and heteroatomic nuclei. (From Hill, G.R. and Lyon, L.B., *Ind. Eng. Chem.*, 54, 36, 1962.)

as *guest molecules* (Redlich et al., 1985), "clathrates" (Given et al., 1986), and perhaps most incorrectly as *bitumen* (Chapter 11) (Grint et al., 1985; Pickel and Gotz, 1991; Speight, 2007, 2008, 2009).

Other workers considered coal to be composed of essentially small, but heterogeneous, condensed aromatic ring systems. It was also noted that the sum of the aromaticity and alicyclicity from lignite to high-rank bituminous coal was substantially constant and structures were proposed to illustrate the transition from lignite to the higher-rank coals (Figure 10.32) (Mazumdar et al., 1962). Another model which has been used to permit visualization of the chemistry of thermal decomposition (Figure 10.33) and invoke the concept of coal being composed of groups of fused aromatic and hydroaromatic ring systems linked by relatively weak aliphatic and ether bridges (Solomon, 1981).

It is apparent that the theory of coal structure has evolved from those theories which invoke the concepts of highly condensed aromatic clusters. In fact, through the agency of later studies involving an x-ray diffraction technique, the majority of the investigators deduced that low- to medium-rank coals consisted of clusters of condensed (three- to six-ring) aromatic nuclei. It was not fully understood what proportion of aliphatic structures occurred in coal and, in the majority of postulates, the aliphatic portion of the molecule was largely ignored. Although all of the postulated coal structures have been found satisfactory to explain the constitution and chemical behavior of coal to some extent, but not in its entirety, there still remain certain inexplicable facets of the chemical and physical behavior of coal.

In an early attempt to elucidate the role of aliphatic structures in coal, it was suggested that the principal entities in coal may not be predominantly polynuclear aromatics but could well be composed of significant proportions of tetrahedral carbon structures, including quaternary carbons. Indeed, progressive extraction of coal with pyridine affords significant quantities of benzenes, phenols, dihydric and/or alkoxyphenols, and naphthalenes with variously condensed aromatics being obtained in the later stages of the extraction. In addition, *n*-alkanes, branched alkanes, cycloalkanes, alkylated benzene, naphthalenes, and other condensed aromatics up to picenes as well as benzfluoranthrenes can be isolated in substantial quantities by exhaustive extraction of coal.

The role of the aromatic centers in coal structure has also been questioned insofar as clusters of tetrahedrally bonded carbon atoms give rise to x-ray diffraction bands in approximately the same angular region where the two-dimensional reflections of graphite-like layers occur. In fact,

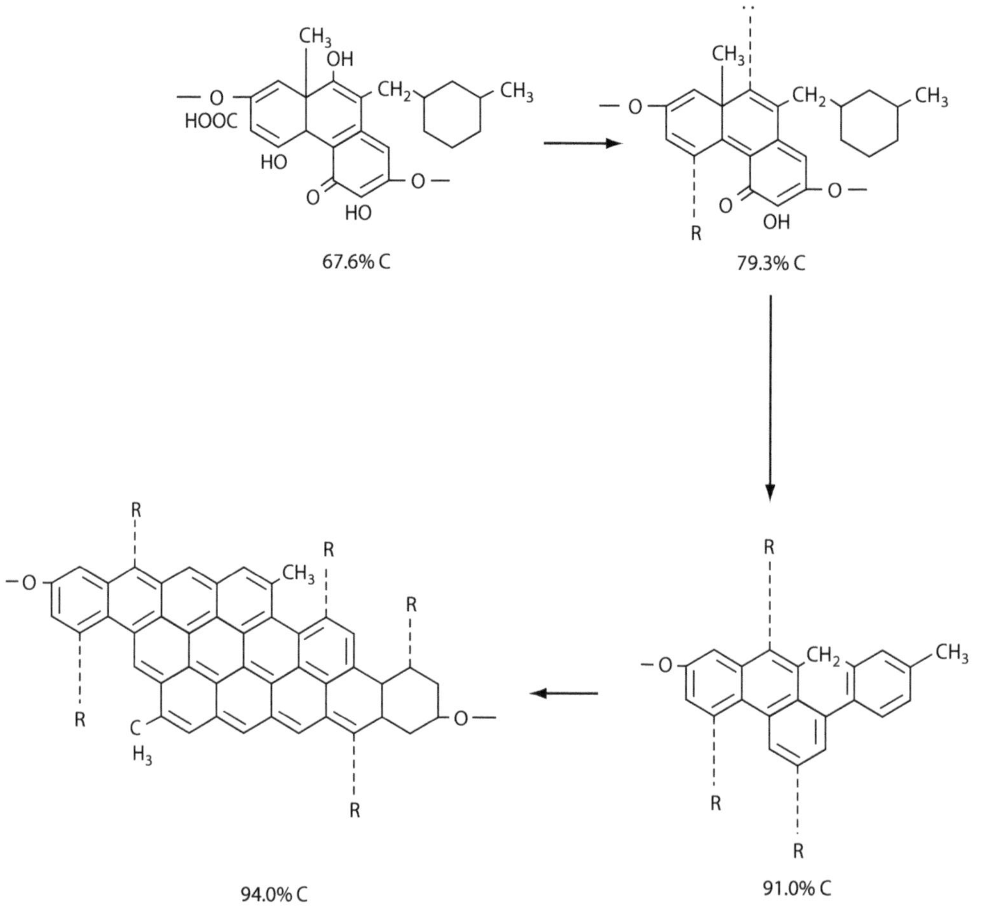

FIGURE 10.32 Hypothetical structures for various coals from lignite (68% w/w C) to anthracite (91% w/w C) showing the transformations during maturation. (From Mazumdar, B.K. et al., *Fuel*, 41, 129, 1962.)

it may be concluded from the diffuse diffraction peaks that many amorphous carbonaceous materials produced diffraction patterns in these particular regions and that it is very difficult to ascertain if graphite-like or diamond-like structures, or both, are present.

Thus, there has been the suggestion that coal may contain significant proportions of adamantane-type structures and is basically a modified bridged tricycloalkane (adamantane), or polyamantane, system (Figure 10.34) (Chakrabartty and Berkowitz, 1974, 1976). The model requires that the polyamantane units increase in size with increase in rank and that benzenoid carbons occupy the periphery of the units. The units would have to aromatize as bituminous coals pass to the anthracite stage. The coalification process was interpreted as first producing a stable aliphatic carbon skeleton from the original plant matter and then producing individual benzene rings.

Obviously, on a molecular basis and because of its complexity coal has been viewed as having many forms, only a few of which are reviewed here. The argument for any particular structure can sway back and forth but the concept of a "chemical structure" of coal or even a simple, repetitive unit cannot really be justified given the extensive heterogeneity of coals.

The origin of coal dictates that it be a very complex organic material product, having little volatility and containing very molecules which are of different chemical origins and, hence, different size and chemical structure (Haenel, 1992). Indeed, it must always be emphasized that coal, because of

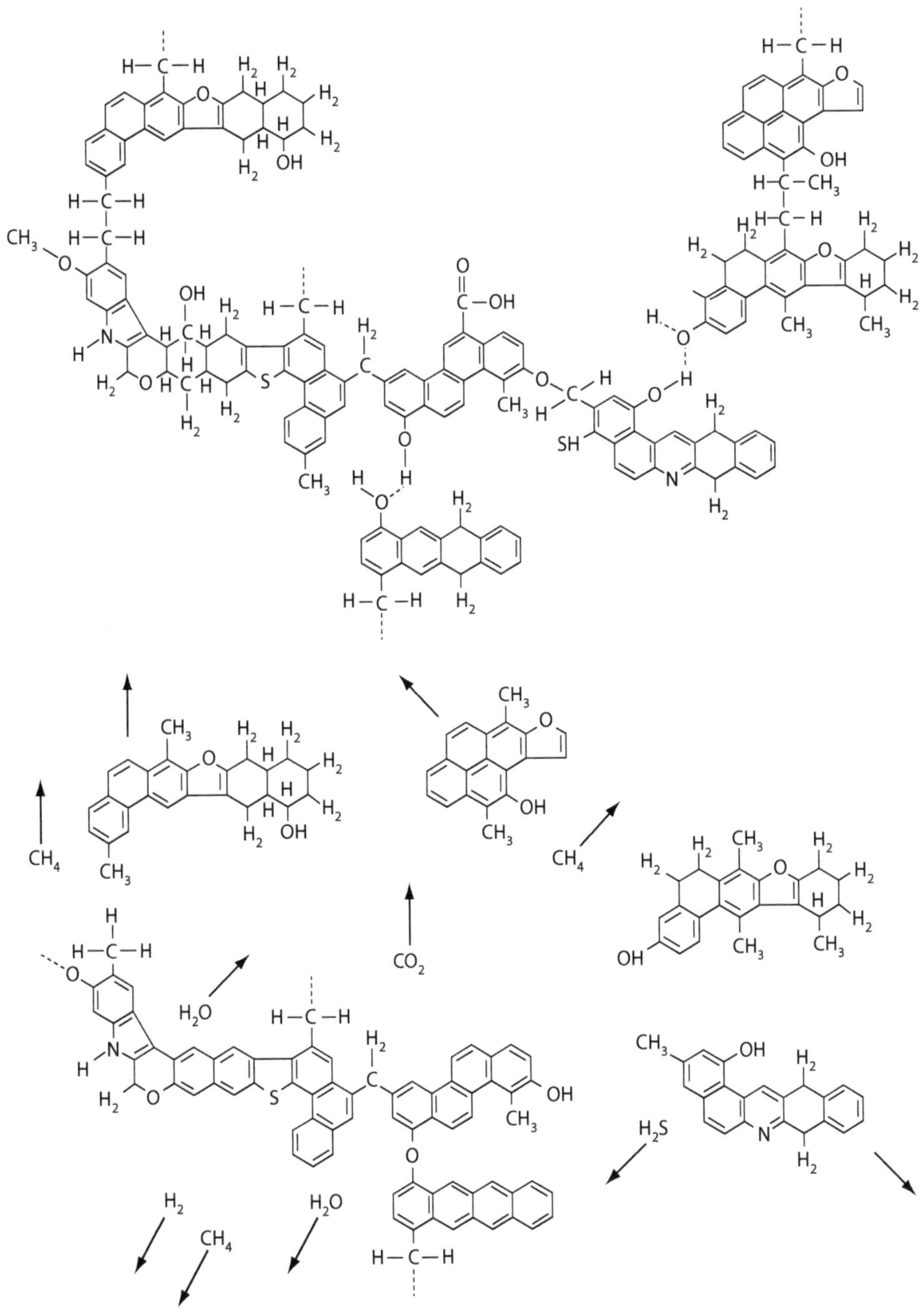

FIGURE 10.33 Hypothetical structures for coal and proposed thermal conversion pathways. (From Solomon, P.R., *New Approaches in Coal Chemistry*, B.D. Blaustein et al., Eds., Symposium Series No. 169, American Chemical Society, Washington, DC, 1981, p. 61.)

the complexity of its origins, cannot exist as a single molecular entity. Indeed, coal and petroleum, irrespective of their different phases (i.e., solid versus liquid), might be considered as similar entities, albeit formed from different mixes of the precursors and under different conditions. And no one would ever accuse petroleum of being a single molecular entity and the era of claiming asphaltene constituents (from petroleum) as being single molecular entities is, unfortunately, not yet over.

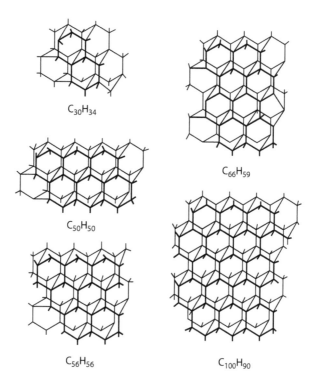

$C_{30}H_{34}$

$C_{66}H_{59}$

$C_{50}H_{50}$

$C_{56}H_{56}$

$C_{100}H_{90}$

FIGURE 10.34 Coals (70%–98% w/w C) as hypothetical polyamantane structures. (From Chakrabartty, S.K. and Berkowitz, N., *Fuel*, 53, 240, 1974.)

In fact, coal should be considered (on a molecular scale) as a complex heterogeneous organic rock and it is impossible to represent such a material by any single organic structure. What we obtain from oxidation studies (and for that matter from any other of the so-called structural studies) is a series of structural or functional types that occur in coal, subject of course to the above constraint. This must not be construed as a criticism that should bring an end to studies focused on the structural nature of coal but the limitations must be recognized in order to continue meaningful scientific endeavors that will aid in understanding the chemical and physical nature of coal.

Furthermore, some attention must also be paid to the physical aspects of coal structure which could involve charge-transfer complexes well as the potential for solvent-induced associations (Nishioka et al., 1991).

More recent application of computer modeling techniques to the issue of coal structure (Carlson and Granoff, 1991; Carlson, 1992) coupled with the potential of synthesizing a model of acceptable behavioral characteristics (Gunderman et al., 1989) might prove to be quite revealing but it should always be remembered that the computer is not a magician but can only feed upon the data in a somewhat limited (usually nonthinking) form. A sound diet of nutritional data could, however, prove to be quite revealing.

There are also indications that coal may be a system of peri-condensed polymeric structures in contrast to the suggestion of coal being predominantly kata-condensed. The occurrence of anthracene in the thermal products of coal processing has, on many occasions, been cited as evidence for the predominantly condensed nature of the aromatic systems in coal. Be that as it may, and there is some degree of truth to this supposition, there is also the distinct possibility that such anthracene systems in the thermal products are, to a degree, thermal artifacts that are formed by various dehyrocyclization reactions.

On a natural product basis, the occurrence of anthracene-type systems in nature and the occurrence of systems that could conceivably form anthracene are not unknown (Weiss and Edwards, 1980).

However, the occurrence of phenanthrene systems (the analogous peri-system to anthracene) is also well documented (Fieser and Fieser, 1949) but has often been ignored in terms of structural entities in coal. On the other hand, the phenanthrene system may be (or, at least, appears to be) prevalent in petroleum (Speight, 2007). Thus, the differences in precursor types and maturation paths notwithstanding, there is the distinct possibility of phenanthrene systems occurring in coal to an extent not previously recognized.

Furthermore, it is somewhat interesting to speculate on the similarity, or dissimilarity, between the structural types that are believed to exist in coal and those found in (or speculated to be in the higher-molecular-weight fractions of) petroleum. To date, there have not been any serious efforts to match the two. Nor were there any reasons to do this. Nevertheless, the concept of similarity is intriguing. Perhaps the reasons for the lack of comparison has been the complete differences in character of the two materials as well as the, apparently, overwhelming desire of coal to produce coke on heating. Indeed, the propensity for coal to form high yields of coke in thermal reactions has been a cause of question and puzzlement since the very early days of coal technology.

In fact, it is this aspect of coal technology that has, more than anything else, been cited as the reasons why coal consists of larger polynuclear aromatic systems. There being the erroneous consideration that such polynuclear aromatic systems are necessary to produce coke. And yet, the production of carbons from sources other than from nonaromatic precursors can give rise to materials that are eminently graphitizable.

It is now becoming more obvious (Speight, 2007), and perhaps has been obvious to polymer scientists for some time, that large polynuclear aromatic systems are not necessary to produce high yields of thermal coke. It is not only the elemental composition but also the chemical configuration of the system which plays a role in coke/carbon formation not heretofore considered in coal and petroleum science. Thus, the concept of relatively small polynuclear aromatic systems in coal is not at all outlandish and may (considering the natural product origins of coal) be expected, or certainly preferable.

However, it is the obvious physical differences between coal and petroleum that can raise questions when similarities are considered. Perhaps the most convenient approach is to consider the differences in dimension and space between the two. The properties of coal are very suggestive of a three-dimensional network. This is much less obvious in petroleum (asphaltene constituents) and may only occur to a very minor extent. Such a difference in spatial arrangement would certainly account for some, if not all, of the differences between the two. Serious consideration of such a proposition would aid physical/chemical/structural studies in both fields and would, hopefully, induce a more constructive thinking in terms of coal/petroleum behavior.

More than anything, the structural studies of coal are limited by the continued insistence that one molecule of unlimited size can be constructed which will explain all of the properties of coal. And there appears to be a more distinctive, and recent, trend to the belief that the representation of coal by an "average molecular structure" is inappropriate insofar as it does not reflect the molecular diversity of the components of coal and there is an overriding tendency to ignore the known diversity of coal which is evident from petrographic studies (Chapter 4) (Given, 1984b; Berkowitz, 1988; Haenel et al., 1989; Haenel, 1992).

An alternative choice is the representation of coal as a two-component system, thereby abandoning the concept of individual structures (Figure 10.35) (Haenel, 1992). Obviously during the use of such a model, the details of any chemical transactions may be missing (but they should always be borne in mind and diminished or ignored) and the model might be convenient to explain many, if not all, of the physicochemical aspects of coal behavior. A very worthy accomplishment, indeed.

The fact that coals are heterogeneous as a group and, indeed, heterogeneous individually does not mean that there cannot be a concept of a macromolecular structure. But such a concept should include a variety of molecular types, perhaps in a manner analogous to the formulation of the structural types in petroleum asphaltene constituents (Figure 10.36) (Long, 1979; Speight, 2007).

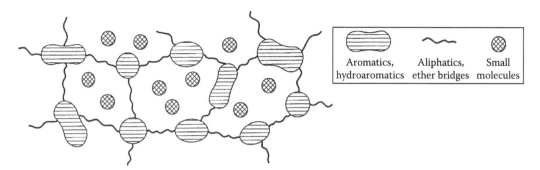

FIGURE 10.35 Coal as a two-phase system involving a three-dimensional network. (From Haenel, M.W., *Fuel*, 71, 1211, 1992.)

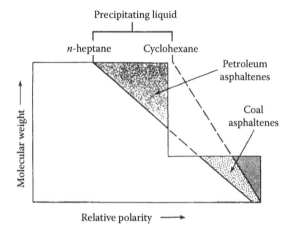

FIGURE 10.36 Representation of the petroleum asphaltene fraction and the coal asphaltene fraction on the basis of molecular weight and polarity.

There would need to be provisions made for the three-dimensional aspect of coal structure as well as for the presence of the lower-molecular-weight constituents.

Acceptance of these premises would presumably (or, at least, hopefully) facilitate a better understanding of the concepts of coal behavior during utilization, such as in beneficiation, combustion, and gasification processes as well as in liquefaction processes. However, the behavior of coal does not and cannot be represented by an average structure. At best, determining an average structure for a heterogeneous material such as coal is a paper exercise that may bear little relationship to reality.

The most appropriate manner in which to represent the structure of coal is through the use of the chemical and physical properties (Levine et al., 1982). Such a representation can be made in terms of both the chemical and physical bonding processes responsible for its structural integrity and the extensive network of pores that permeate the organic material. Information on the microscopic chemistry of coal and its relationship to coal's physical structure and reactivity is an essential component in the successful development of the next generation of coal conversion technologies.

Future applications of coal will extend beyond the present major uses for power generation, metals processing, and chemicals production (Speight, 2008). A key feature in these extensions will be the development of means to convert coal from its native form into useful gases and liquids. The design of a new generation of conversion processes will require a deeper understanding of the intrinsic properties of coal and the ways in which it is chemically transformed under process conditions. Coal properties such as the chemical form of the organic material, the types and distribution of organics, the nature of the pore structure, and the mechanical properties must be determined for coals of different ranks (or degrees of coalification) in order to use each coal type most effectively (Neavel, 1982).

However, it is more imperative to understand the chemical pathways followed during the thermal conversion of coal to liquids or gases. This is accomplished by tracing the conversion of specific chemical functional groups in the coal and studying the effects of various inorganic compounds on the conversion process. The ultimate goal is to relate the structure of the native coal to the resulting conversion products.

10.8.1 ELEMENTAL ANALYSIS

Elemental analysis provides a perspective of the macrochemical form of coal. Organic material typically represents more than 75% of coal, the balance being mineral matter and water. Low-sulfur coal is particularly desirable for power generation and the synthesis of environmentally acceptable conversion products. Coal has a significant chemical structure to petroleum, with higher levels of aromatic and other unsaturated species than found in petroleum. A second undeniable feature is the high level of organic oxygen in coal—many times the oxygen levels in petroleum—which strongly influences the structure and reactivity of coal.

10.8.2 OXYGEN FUNCTIONAL GROUPS IN COAL

Oxygen occurs in various chemical forms in coal—as carboxyl or phenol groups, as heterocyclic oxygen in aromatic or naphthenic rings, or as bridges between the nuclei as ethers or, to a lesser extent, as esters (Levine et al., 1982). These types of oxygen functional groups strongly influence coal reactivity and their relative numbers vary significantly as a function of coal rank.

To clarify oxygen chemistry in coal, there is a need to determine how many oxygen functional groups of a given type exist and how accessible they are for different reagents (Liotta, 1981). Phenols can be converted to ethers and carboxylic acids can be converted to esters selectively and under mild conditions by using methyl iodide (or any primary alkyl iodide or bromide) as the alkylating agent in the presence of a quaternary ammonium hydroxide.

A second major chemical issue concerns the important role oxygen plays in bridges connecting nuclei of the coal structure (Heredy and Neuworth, 1962; Taylor and Bell, 1980; Vernon, 1980) and the influence of the oxygen functions on reaction conditions similar to those used in coal conversion processes (Schlosberg et al., 1981a,b).

Furthermore, many of the bridging structures in coal are highly reactive—more so than the similar bridge structures in petroleum (Speight, 2007). The difficulty is that the products of such reactions depend sensitively on the availability of hydrogen to terminate the process, capping off the radicals that are thermally generated. In the presence of hydrogen, from a good donor solvent for example, radical processes are rapidly quenched and small molecules are formed. In a hydrogen-deficient system, these radicals propagate, forming large amounts of high-molecular-weight products.

10.8.3 PORE STRUCTURE

The thermal decomposition of coal (Chapter 13) emphasizes the need for access of reagents such as hydrogen to regions of the coal undergoing conversion. Thus, a key consideration in coal conversion is the mass transfer of reagents and products which occurs by means of coal's extensive pore network.

A wide variety of pore types exist in coals of different ranks—ranging from large macropores (2300 Å) to micropores (12 Å) (Gan et al., 1972). The total volume of such pores, the pore size distribution, and the effective surface area of the organic material are needed to anticipate coal's potential reactivity.

Understanding the pore chemistry and physics of complicated materials such as coal can make a significant long-term contribution to advanced coal conversion technology.

In conclusions, there is much to be gained by the analysis of coal and understanding the influence of coal composition on use (Chapters 8 and 9) but, as noted elsewise (Chapter 3), giving consideration to (1) the different chemical precursors of coal, (2) the varied geological environment, (3) the maturation conditions, and (4) the variation of coal analysis within a seam results in the rationale for an average structure of coal being of little value.

REFERENCES

Attar, A. 1979. In *Analytical Methods for Coal and Coal Products*, C. Karr, Jr. (Ed.). Academic Press, Inc., New York, Vol. III, Chapter 56.

Attar, A. and Dupuis, F. 1979. Preprints. Division of Fuel Chemistry. American Chemical Society, 24(1): 166.

Attar, A. and Hendrickson, G.G. 1982. In *Coal Structure*, R.A. Meyers (Ed.). Academic Press, Inc., New York.

Barron, P.F. and Wilson, M.A. 1981. *Nature*, 289: 275.

Bartle, K.D. and Jones, D.W. 1978. In *Analytical Methods for Coal and Coal Products*, C. Karr, Jr. (Ed.). Academic Press, Inc., New York, Vol. II, Chapter 23.

Bartle, K.D., Jones, D.W., and Pakdel, H. 1978. In *Analytical Methods for Coal and Coal Products*, C. Karr, Jr. (Ed.). Academic Press, Inc., New York, Vol. II, Chapter 25.

Bartle, K.D., Jones, D.W., and Pakdel, H. 1982. In *Coal and Coal Products: Analytical Characterization Techniques*, E.L. Fuller, Jr. (Ed.). Symposium Series No. 205. American Chemical Society, Washington, DC, Chapter 2.

Bartle, K.D., Martin, T.G., and Williams, D.F. 1975. *Fuel*, 54: 226.

Berkowitz, N. 1988. In *Polynuclear Aromatic Compounds*, L.B. Ebert (Ed.). Advances in Chemistry Series No. 217. American Chemical Society, Washington, DC, Chapter 13.

Blanc, P., Valisolalao, J., Albrecht, P., Kohut, J.P., Muller, J.F., and Duchene, J.M. 1991. *Energy & Fuels*, 5: 875.

Bonnett, R., Czechowski, F., and Hughes, P.S. 1991. *Chemical Geology*, 91: 193.

Botto, R.E. and Sanada, Y. (Eds.). 1992. *Magnetic Resonance of Carbonaceous Solids*. Advances in Chemistry Series No. 229. American Chemical Society, Washington, DC.

Bredenberg, J.B-son, Huuska, M., and Vuori, A. 1987. In *Coal Science and Chemistry*, A. Volborth (Ed.). Elsevier, Amsterdam, the Netherlands, p. 1.

Bright, F.V. and McNally, M.E.P. (Eds.). 1992. *Supercritical Fluid Technology: Theoretical and Applied Approaches in Analytical Chemistry*. Symposium Series No. 488. American Chemical Society, Washington, DC.

Brown, H.R. and Waters, P.L. 1966a. *Fuel*, 45: 17.

Brown, H.R. and Waters, P.L. 1966b. *Fuel*, 45: 41.

Burchill, P. 1987. In *Proceedings. International Conference on Coal Science*, J.A. Moulijn (Ed.). Elsevier, Amsterdam, the Netherlands, p. 5.

Burgess, M.J. and Wheeler, R.V.J. 1911. *Chemical Society*, 99: 649.

Calkins, W.H. and Spackman, W. 1986. *International Journal of Coal Geology*, 6: 1.

Calkins, W.H., Torres-Ordonez, R.J., Jung, B., Gorbaty, M.L., George, G.N., and Kelemen, S.R. 1992. *Energy & Fuels*, 6: 411.

Carlson, G.A. 1992. *Energy & Fuels*, 6: 771.

Carlson, G.A. and Granoff, B. 1991. In *Coal Science II*, H.H. Schobert, K.D. Bartle, and L.J.D. Lynch (Eds.). Symposium Series No. 461. American Chemical Society, Washington, DC, Chapter 12.

Cartz, L. and Hirsch, P.B. 1960. *Philosophical Transactions of the Royal Society*, A252: 557.

Chaffee, A.L. and Fookes, C.J.R. 1988. *Organic Geochemistry*, 12: 261.

Chakrabartty, S.K. and Berkowitz, N. 1974. *Fuel*, 53: 240.

Chakrabartty, S.K. and Berkowitz, N. 1976. *Fuel*, 55: 362.

Chakrabartty, S.K. and Kretschmer, H.O. 1972. *Fuel*, 51: 160.

Chambers, R.R. Jr., Hagaman, E.W., and Woody, M.C. 1988. In *Polynuclear Aromatic Compounds*, L.B. Ebert (Ed.). Advances in Chemistry Series No. 217. American Chemical Society, Washington, DC, Chapter 15.

Cookson, D.J. and Smith, B.E. 1982. *Fuel*, 61: 1007.

Cookson, D.J. and Smith, B.E. 1987. In *Coal Science and Chemistry*, A. Volborth (Ed.). Elsevier, Amsterdam, the Netherlands, p. 61.

Cronauer, D.C. and Ruberto, R.G. 1977. Report No. EPRI-AF-442. Electric Power Research Institute, Palo Alto, CA.

Davidson, R.M. 1980. Molecular structure of coal. Report No. ICTIS/TRO8. International energy Agency, London, U.K.

Davidson, R.M. 1986. Nuclear magnetic resonance studies of coal. Report No. ICTIS/TR32. International Energy Agency, London, U.K.

Davis, A. 1978a. In *Scientific Problems of Coal Utilization*, B.R. Cooper (Ed.). United States Department of Energy, Washington, DC.

Davis, A. 1978b. In *Analytical Methods for Coal and Coal Products*, C. Karr, Jr. (Ed.). Academic Press, Inc., New York, Vol. I, Chapter 2.

Davis, M.F., Quinting, G.R., Bronnimann, C.E., and Maciel, G.E. 1989. *Fuel*, 68: 763.

Deno, N.C., Greigger, B.A., and Stroud, S.G. 1978. *Fuel*, 57: 455.

Deno, N.C., Jones, A.D., Owen, B.O., and Weinschenk, J.I. 1985. *Fuel*, 64: 1286.

Derbyshire, F.J., Davis, A., and Lin, R. 1991. In *Coal Science II*, H.H. Schobert, K.D. Bartle, and D.L.J. Lynch (Eds.). Symposium Series No. 461. American Chemical Society, Washington, DC, Chapter 7.

Derbyshire, F.J., Davis, A., Stansberry, P.G., and Terrer, M.-T. 1986. *Fuel Processing Technology*, 1986, 12, 127.

Derbyshire, F.J., Marzec, A., Schulten, H.-R., Wilson, M.A., Davis, A., Tekely, P., Delpeuch, J.J., Jurkiewicz, A., Bronnimann, C.E., Wind, R.A., Maciel, G.E., Narayan, R., Bartle, K.D., Snape, and C.E. 1989. *Fuel*, 68: 1091.

Derbyshire, F.J., Odoerfer, G.A., Varghese, P., and Whitehurst, D.D. 1982. *Fuel*, 61: 899.

Derbyshire, F.J. and Whitehurst, D.D. 1981. *Fuel*, 60: 655.

Dryden, I.G.C. and Pankhurst, K.S. 1955. *Fuel*, 34: 363.

Fieser, L.F. and Fieser, M. 1949. *Natural Products Related to Phenanthrene*. Reinhold Publishing Corporation, New York.

Franco, D.V., Gelan, J.M., Martens, H.J., and Vanderzande, D.J.-M. 1992. *Fuel*, 71: 553.

Gallegos, E.J. 1978. In *Analytical Chemistry of Liquid Fuel Sources: Tar Sands, Oil Shale, Coal, and Petroleum*, P.C. Uden, S. Siggia, and H.B. Jensen (Eds.). Advances in Chemistry Series No. 170. American Chemical Society, Washington, DC, Chapter 2.

Gan, H., Nandi, S.P., and Walker, P.L. 1972. *Fuel*, 51: 272–277.

Gerstein, B.C., Ryan, L.M., and Murphy, P.D. 1979. Preprints. Division Fuel Chemistry. American Chemical Society, 24(1): 90.

Gibson, J. 1978. *Journal of the Institute of Fuel*, 51: 67.

Given, P.H. 1960. *Fuel*, 39: 147.

Given, P.H. 1984a. *Progress in Energy Combustion Science*, 10: 149.

Given, P.H. 1984b. In *Coal Science*, M.L. Gorbaty, J.W. Larsen, and I. Wender (Eds.). Academic Press Inc., New York, Vol. 3, p. 63.

Given, P.H., Marzec, A., Barton, W.A., Lynch, L.J., and Gerstein, B.C. 1986. *Fuel*, 65: 155.

Golumbic, C., Anderson, J.B., Orchin, M., and Storch, H.H. 1950. Report of Investigations No. 4664. Bureau of Mines, US Department of the Interior, Washington, DC.

Gorbaty, M.L., George, G.N., and Kelemen, S.R. 1991. In *Coal Science II*, H.H. Schobert, K.D. Bartle, and L.J. Lynch (Eds.). Symposium Series No. 461. American Chemical Society, Washington, DC, Chapter 10.

Green, T., Kovac, J., Brenner, D., and Larsen, J.W. 1982. In *Coal Structure*, R.A. Meyers (Ed.). Academic Press, Inc., New York, p. 199.

Grigoriew, H. 1990. *Fuel*, 69: 840.

Grimes, W.R. 1982. In *Coal Science*, M.L. Gorbaty, J.W. Larsen, and I. Wender (Eds.). Vol. 1, p. 21.

Grint, A., Mehani, S., Trewhella, M., and Crook, M.J. 1985. *Fuel*, 64: 1355.

Gunderman, K.-D., Humke, K., Emrich, E., and Rollwage, U. 1989. *Erdoel and Kohle*, 42(2): 59.

Haenel, M.W. 1992. *Fuel*, 71: 1211.

Haenel, M.W., Collin, G., and Zander, M. 1989. *Erdoel Erdgas Kohle*, 105: 71, 131.

Hall, P.J., Mark, T.F., and Marsh, H. 1992. *Fuel*, 71: 1271.

Hamming, M.C., Radd, F.J., and Carel, A.B. 1979. In *Analytical Methods for Coal and Coal Products*, C. Karr, Jr. (Ed.). Academic Press, Inc., New York, Vol. III, Chapter 39.

Hayatsu, R., Scott, R.G., Moore, L.P., and Studier, M.H. 1975. *Nature*, 261: 77.

Hayatsu, R., Scott, R.G., and Winans, R.E. 1982. Oxidation of coal. In *Oxidation in Organic Chemistry*, Part D. Academic Press, Inc., New York, p. 279.

Hayatsu, R., Winans, R.E., Scott, R.G., McBeth, R.L., and Moore, L.P. 1981. *Fuel*, 60: 77.

Hayatsu, R., Winans, R.E., Scott, R.G., Moore, L.P., and Studier, M.H. 1978. *Fuel*, 57: 541.

Heredy, L.A. and Fugass, P. 1966. In *Coal Science*. Advances in Chemistry Series No. 55. American Chemical Society, Washington, DC.

Heredy, L.A. and Neuworth, M.B. 1962. *Fuel*, 41: 221–231.

Herod, A.A., Stokes, B.J., and Radeck, D. 1991. *Fuel*, 70: 329.

Hill, G.R. and Lyon, L.B. 1962. *Industrial & Engineering Chemistry*, 54: 36.

Hirsch, P.B. 1954. *Proceedings of the Royal Society A*, 226: 143.

Iino, M., Ogawa, T., and Zeiger, E. 1985. Kinetic properties of the blue light response of stomata. *Proceedings of the National Academy of Sciences of the United States of America*, 82: 8019–8023.

Iino, M., Takanohash, H., Ohsuga, H., and Toda, K. 1988. *Fuel*, 67: 1639.

Illingworth, S.R. 1922. *Fuel*, 1: 17.

Kennel, E.B., Mukka, M., Stiller, A.H., and Zondlo, J.W. 2009. Solvent extraction of low grade coals for clean liquid fuels. *Proceedings. 2008 National Meeting*. American Institute of Chemical Engineers, Paper 123a.

Kiran, E. and Brennecke, J.F. (Eds.). 1993. Supercritical fluid engineering science: Fundamentals and applications. Symposium Series No. 514. American Chemical Society, Washington, DC.

Kreulen, D.J.W. 1948. *Elements of Coal Chemistry*. Nijgh & van Ditmar N.V., Rotterdam, the Netherlands.

Larsen, J W., Green, T.K., and Kovac, J. 1985. *Journal of Organic Chemistry*, 50: 4729.

Larsen, J.W., Urban, L., Lawson, G., and Lee, D. 1981. *Fuel*, 60: 267.

Levine, D.G., Schlosberg, R.H., and Silbernagel, B.G. 1982. Understanding the chemistry and physics of coal structure: A review. *Proceedings of National Academy of Sciences USA*, 79: 3365–3370.

Lin, Q. and Guet, J.M. 1990. *Fuel*, 69: 821.

Liotta, R. 1981. *Journal of American Chemical Society*, 103: 1735–1742.

Litke, R., Leythaeuser, D., Radke, M., and Schaefer, R.G. 1990. *Organic Geochemistry*, 16: 247.

Long, R.B. 1979. Preprints. Division of Petroleum Chemistry. American Chemical Society, 24(4): 891.

Lucht, L. and Peppas, N.A. 1981. In *New Approaches in Coal Chemistry*, B.D. Blaustein, B.C. Bockrath, and S. Friedman (Eds.). American Chemical Society, Washington, DC, Chapter 3.

Makabe, M. and Ouchi, K. 1979. *Fuel*, 58: 43.

Mallya, N. and Zingaro, R.A. 1984. *Fuel*, 63: 423.

Mastral-Lamarca, A.M. 1987. In *Coal Science and Chemistry*, A. Volborth (Ed.). Elsevier, Amsterdam, the Netherlands, p. 289.

Mayo, F.R. 1975. *Fuel*, 54: 273.

Mazumdar, B.K., Chakrabartty, S.K., and Lahiri, A. 1962. *Fuel*, 41: 129.

Meiler, W. and Meusinger, R. 1992. Annual report. *NMR Spectroscopy*, 24: 331.

Miknis, F.P. 1982. *Magnetic Resonance Review*, 7(2): 87.

Miknis, F.P., Sullivan, M.J., Bartuska, V.J., and Maciel, G.E. 1981. *Organic Geochemistry*, 3(1): 19.

Neavel, R.C. 1982. In *Coal Science*, M.L. Gorbaty, J.W. Larsen, and I. Wender (Eds.). Academic Press, Inc., New York, Vol. 1.

Nishioka, M. 1992. *Fuel*, 71: 941.

Nishioka, M., Gebhard, L.A., and Silbernagel, B.G. 1991. *Fuel*, 70: 341.

Nishioka, M. and Larsen, J.W. 1990. *Energy & Fuels*, 4: 100.

Oele, A.P., Waterman, H.I., Goedkoop, M.L., and van Krevelen, D.W. 1951. *Fuel*, 30: 169.

Painter, P.C., Coleman, M.M., Snyder, R.W., Mahajan, O.P., Komatsu, M., and Walker, P.L. Jr. 1981. *Applied Spectroscopy*, 35(1): 106.

Peppas, N.A. and Lucht, L.M. 1984. *Chemical Engineering Communications*, 30: 291.

Petrakis, L. and Edelheit, E. 1979. *Applied Spectroscopy Reviews*, 15(2): 195.

Pickel, W. and Gotz, G.K.E. 1991. *Organic Geochemistry*, 17: 695.

Pitt, G.J. 1979. In *Coal and Modern Coal Processing: An Introduction*, G.J. Pitt and G.R. Millward (Eds.). Academic Press, Inc., New York, Chapter 2.

Raj, S. 1976. Ph.D. Thesis. On the Study of the Chemical Structure Parameters of Coal. The Pennsylvania State University, University Park, PA.

Raj, S. 1979. *Chemical Technology (Houston)*, 2(4): 54.

Redlich, P., Jackson, W.R., and Larkins, F.P. 1985. *Fuel*, 64: 1383.

Retcofsky, H.L. and Firedel, R.A. 1973. *Journal of Physical Chemistry*, 77: 68.

Retcofsky, H.L., Hough, M., and Clarkson, R.B. 1979. Preprints. Division of Fuel Chemistry. American Chemical Society, 24(1): 83.

Retcofsky, H.L. and Link, T.A. 1978. In *Analytical Methods for Coal and Coal Products*, C. Karr, Jr. (Ed.). Academic Press, Inc., New York, Vol. II, Chapter 24.

Retcofsky, H.L., Thompson, G.P., Hough, M., and Friedel, R.A. 1978. In *Organic Chemistry of Coal*, J.W. Larsen (Ed.). Symposium Series No. 71. American Chemical Society, Washington, DC, p. 142.

Rodriguez, N.M. and Marsh, H. 1987. *Fuel*, 66: 1727.

Ruberto, R.G. and Cronauer, D.G. 1978. In *Organic Chemistry of Coal*, J.W. Larsen (Ed.). Symposium Series No. 71. American Chemical Society, Washington, DC, p. 50.

Schafer, H.N.S. 1970. *Fuel*, 49: 197.

Schlosberg, R.H., Ashe, T.R., Pancirov, R.J., and Donaldson, M. 1981a. *Fuel*, 60: 155–157.

Schlosberg, R.H., Davis, W.H., and Ashe, T.R. 1981b. *Fuel*, 60: 201–204.

Schneider, G.M., Stahl, E., and Wilke, G. 1980. *Extraction with Supercritical Gases*. Verlag Chemie, Weinheim, Germany.

Schuyer, J. and van Krevelen, D.W. 1954. *Fuel*, 33: 176.

Sharma, D.K. 1988. *Fuel*, 67: 186.

Sinninghe Damste, J.S. and de Leeuw, J.W. 1992. *Fuel Processing Technology*, 30: 109.

Siskin, M. and Aczel, T. 1983. *Fuel*, 62: 1321.

Smith, K.L. and Smoot, L.D. 1990. *Progress in Energy and Combustion Science*, 16: 1.

Snape, C.E. 1987. *Fuel Processing Technology*, 15: 257.

Snape, C.E., Axelson, D.E., Botto, R.E., Delpeuch, J.J., Tekely, P., Gerstein, B.C., Pruski, M., Maciel, G.E., and Wilson, M.A. 1989. *Fuel*, 68: 547.

Snape, C.E., Ladner, W.R., and Bartle, K.D. 1979. Survey of carbon-13 chemical shifts in aromatic hydrocarbons and its application to coal-derived materials. *Analytical Chemistry*, 51: 2189.

Solomon, P.R. 1981. In *New Approaches in Coal Chemistry*, B.D. Blaustein, B.C. Bockrath, and S. Friedman (Eds.). Symposium Series No. 169. American Chemical Society, Washington, DC, p. 61.

Solomon, P.R., Best, P.E., Yu, Z.Z., and Charpenay, S. 1992. *Energy & Fuels*, 6: 143.

Speight, J.G. 1978. In *Analytical Methods for Coal and Coal Products*, C. Karr, Jr. (Ed.). Academic Press, Inc., New York, Vol. II, Chapter 22.

Speight, J.G. 1987. In *Coal Science and Chemistry*, A. Volborth (Ed.). Elsevier, Amsterdam, the Netherlands, p. 183.

Speight, J.G. 2007. *The Chemistry and Technology of Petroleum*, 4th edn. Taylor & Francis Group, Boca Raton, FL.

Speight, J.G. 2008. *Synthetic Fuels Handbook: Properties, Processes, and Performance*. McGraw-Hill, New York.

Speight, J.G. 2009. *Enhanced Recovery Methods for Heavy Oil and Tar Sands*. Gulf Publishing Company, Houston, TX.

Stadelhofer, J.W., Bartle, K.D., and Matthews, R.S. 1981. *Erdoel und Kohle*, 34: 71.

Stansberry, P.G., Lin, L., Terrer, M.-T., Lee, C.W., Davis, A., and Derbyshire, F.J.J. 1987. *Energy & Fuels*, 1: 89.

Sternberg, H.W., Della Donne, C.L., Pantages, P., Moroni, E.C., and Markby, R.E. 1971. *Fuel*, 50: 432.

Studier, M.H., Hyatsu, R., and Winans, R.E. 1978. In *Analytical Methods for Coal and Coal Products*, C. Karr, Jr. (Ed.). Academic Press, Inc., New York, Vol. II, Chapter 21.

Szeliga, J. and Marzec, A. 1983. *Fuel*, 62: 1229.

Taylor, N.D. and Bell, A.T. 1980. *Fuel*, 59: 499–506.

Vahrman, M. 1970. *Fuel*, 49: 5.

Van Krevelen, D.W. 1961. *Coal*. Elsevier, Amsterdam, the Netherlands.

Van Krevelen, D.W. 1965. *Fuel*, 44: 229.

Van Krevelen, D.W. 1993. *Coal: Typology, Physics, Chemistry, Constitution*. Elsevier, New York.

Van Krevelen, D.W., Huntjens, F.J., and Dormans, H.N.M. 1956. *Fuel*, 35: 462.

Van Krevelen, D.W. and Schuyer, J. 1957. *Coal Science*. Elsevier, Amsterdam, the Netherlands.

Vernon, L.W. 1980. *Fuel*, 59: 102–106.

Volborth, A. 1979. In *Analytical Methods for Coal and Coal Products*. Academic Press, Inc., New York, Vol. III, Chapter 55.

Waters, P.L. 1962. *Fuel*, 41: 3.

Weiss, U. and Edwards, J.M. 1980. *The Biosynthesis of Aromatic Compounds*. John Wiley & Sons, Inc., New York.

Wender, I., Heredy, L.A., Neuworth, M.B., and Dryden, I.G.C. 1981. In *Chemistry of Coal Utilization*, Second Supplementary Volume, M.A. Elliott (Ed.). John Wiley & Sons, Inc., New York, Chapter 8.

White, C.M. 1983. In *Handbook of Polycyclic Aromatic Hydrocarbons*, A. Bjorseth (Ed.). Marcel Dekker, Inc., New York, Chapter 13.

Whitehurst, D.D., Mitchell, T.O., and Farcasiu, M. 1980. *Coal Liquefaction: The Chemistry and Technology of Thermal Processes*. Academic Press, Inc., New York.

Winans, R.E., Hayatsu, R., and McBeth, R.L. 1988. Preprints. Division of Fuel Chemistry. American Chemical Society, 33(1): 407.

Winans, R.E., Melnikov, P.E., and McBeth, R.L. 1992. Preprints. Division of Fuel Chemistry. American Chemical Society, 37(2): 693.

Yoshida, T., Tokuhasho, K., Narita, H., Hasegawa, Y., and Maekawa, Y. 1984. *Fuel*, 63: 282.

Youtcheff, J.S. and Given, P.H. 1982. *Fuel*, 61: 980.

Youtcheff, J.S. and Given, P.H. 1984. Preprints. Division of Fuel Chemistry. American Chemical Society, 29(5): 1.

Yun, Y., Meuzelaar, H.L.C., Simmleit, N., and Schulten, H.-R. 1991. In *Coal Science II*, H.H. Schobert, K.D. Bartle, and L.J. Lynch (Eds.). Symposium Series No. 461. American Chemical Society, Washington, DC, Chapter 8.

11 Solvent Extraction

11.1 INTRODUCTION

The problem of declining supplies of petroleum and natural gas has encouraged investigations into solvent refining of coal. The essential aspects of this type of conversion process are broadly described (1) as the thermal fragmentation of macromolecular units in the presence of a solvent or a hydrogen-donor solvent in the temperature range 375°C–425°C (710°F–800°F), associated with the intermolecular and intramolecular movement of hydrogen atoms in the reactants to produce pyridine-soluble solids and (2) as the progressive hydrogenation of the fragmented products to yield finally a transportable fuel or synthetic crude oils.

Solvent extraction is an operational definition for a solvent-based process to isolate soluble material from the complex and heterogeneous mixture, which is coal. Unless the separation process is specified, including filter pore size or centrifugation clearing factor, the material isolated is not well defined and may not be the same as that isolated in other laboratories. This is especially true for coals subject to weathering.

Solvent extraction is accomplished by contacting coal with a solvent and separating the residual coal material from the solvent and the extracts (Vorres, 1993 and references cited therein). However, extraction is, typically, mass transfer limited, so thorough mixing of the solvent and coal is required. Briefly, the extraction solvent is well mixed with the coal to allow potentially soluble constituents to transfer to the solvent. The residual coal and solvent are then separated by physical methods, such as gravity decanting, filtering, or centrifuging. Distillation regenerates the solvent from the extracted material.

Nondestructive solvent extraction of coal is the extraction of soluble constituents from coal under conditions where thermal decomposition does not occur. On the other hand, solvolysis (destructive solvent extraction) refers to the action of solvents on coal at temperatures at which the coal substance decomposes and in practice relates in particular to extraction at temperatures between 300°C and 400°C (570°F and 750°F). In the present context (i.e., the solvents extraction of coal), the solvent power of the extracting liquid appears to be solely determined by the ability of the solvent to alter the coal physically (by swelling). In this respect, the most effective solvents are aromatics, phenol derivatives, naphthol derivatives, anthracene, and phenanthrene.

Generally, coal should be dried before extraction in order to determine extract yields and material balances; however, this may collapse pore structure in low-rank and produce other changes in high-rank coals. Even drying at 100°C (212°F) can induce cross-linking reactions in low-rank coals, which reduce the tendency of the coal to swell in solvents such as pyridine.

Solvent extraction in coal research (Chapter 10) has been used for isolation and characterization of both soluble and insoluble coal fractions (Van Krevelen, 1957) and fall into four general areas: (1) improvement in extraction yields or selectivity, (2) correlation of solvent swelling and extraction behavior to structural models for the insoluble organic portion of coal, (3) analyses of extracts to identify and quantify organic compounds in the raw coal, and (4) use of solvent extraction to predict or influence coal behavior in some other process such as liquefaction.

Furthermore, from a more practical viewpoint, the solvent extraction of coal has been used as a means of coproducing clean liquid transportation fuels as well as solid fuels for gasification. Coal solvents are created by hydrogenating coal tar distillate fractions to the level of a fraction of a percent, thus enabling bituminous coal to enter the liquid phase under conditions of high temperature (above 400°C [750°F]). The pressure is controlled by the vapor pressure of the solvent and the cracked coal. Once liquefied, mineral matter can be removed via centrifugation, and the resultant heavy oil product can be processed to make pitches, cokes, as well as lighter products.

Methods of coal liquefaction have been available since the beginning of the twentieth century but the cost has initiated searches for more effective new processes. For example, in the Bergius process for direct coal liquefaction, the coal is treated with hydrogen under pressure at 450°C (840°F) in the presence of a solvent and an iron oxide catalyst. The activity of this catalyst is low, however, because the solid iron oxide cannot enter the macromolecular network structure of the insoluble coal. Semianthracite coal, which only contains a small amount of volatile components, cannot be converted by this process.

Thus, the solvent extraction of coal, far from being a theoretical study, is, in fact, very pertinent to the behavior of coal in a variety of utilization operations. For example, the liquefaction of coal (Chapters 18 and 19) often relies upon the use of solvent and, in addition, the thermal decomposition of coal can also be considered to be an aspect of the solvent extraction of coal. The generation of liquid products during thermal decomposition can be considered to result in the exposure of coal to (albeit coal-derived) solvents with the result that the solvent materials being able to influence the outcome of the process. And there are many more such examples.

The solvent extraction of coal has been employed for many years (Fremy, 1861; de Marsilly, 1862; Oele et al., 1951) as a means of studying the constitution of coal with the yield and the nature of the extract dependent on the solvent type, extraction conditions, and, last but not least, the coal type (Bedson, 1902; Kreulen, 1948; Van Krevelen, 1957; Francis, 1961; Dryden, 1963; Wise, 1971; Hombach, 1980; Pullen, 1983; Stenberg et al., 1983; Given, 1984; Litke et al., 1990; Blanc et al., 1991; Pickel and Gotz, 1991). The solvent extraction of coal should not be confused with the principles that lie behind coal cleaning (Chapter 6) even though solvents (more frequently referred to as *heavy liquids*) may be used. The approach uses a different perspective and the outcome is usually quite different.

Solvent extraction is very pertinent to the liquefaction of coal in the presence of a solvent, and especially to the more recent concept of coprocessing coal with a variety of other feedstocks such as heavy oils, tar sand bitumen, and residua (Chapters 18 and 19) (Speight, 2007, 2008). There have been many attempts to define solvent behavior in terms of one or more physical properties of the solvent, and not without some degree of success. However, it is essential to note that the properties of the coal also play an important role in defining behavior of a solvent, and it has been reported that the relative solvent powers of two solvents may be reversed from one coal type to another.

Two properties that have found some relevance in defining solvent behavior with coal (as well as with other complex carbonaceous materials such as petroleum asphaltene constituents) are the surface tension and the *internal pressure*. However, the solvent power of primary aliphatic amines (and similar compounds) for the lower-rank coals has been attributed to the presence of an unshared pair of electrons (on the nitrogen atom).

Early work on the solvent extraction of coal was focused on an attempt to separate from coal a "coking principle" (i.e., the constituents believed to be responsible for coking and/or caking properties). But solvent extraction has actually been used to demonstrate the presence in coal of material that either differed from the bulk of the coal substance or was presumed to be similar to the bulk material as, for example, in some of the earlier studies on the nature of the structural entities in coal (Chapter 10).

An example of the difference of the solvent extracts from the bulk material comes from a series of studies on the exhaustive extraction of coal by boiling pyridine and fractionation of the regenerated soluble solids by sequential selective extraction schemes (Figure 11.1). Subsequent analyses showed that the petroleum ether-soluble material was mostly composed of hydrocarbons (e.g., paraffins, naphthene derivatives, and terpene derivatives), while the ether-soluble, acetone-soluble, and acetone-insoluble fractions were resin-like substances with 80%–89% carbon and 8%–10% hydrogen. Indeed, this, and later work (Vahrman, 1970), led to the concept that coal is a two-component or two-phase system (Derbyshire et al., 1991; Yun et al., 1991). In general terms, solvents for coal extraction can be grouped into four general categories: (1) nonspecific, (2) specific, (3) degrading, and (4) reactive (Oele et al., 1951; Williams et al., 1987). Such solvents are used in a variety of solvent-enhanced liquefaction operations (Chapters 18 and 19).

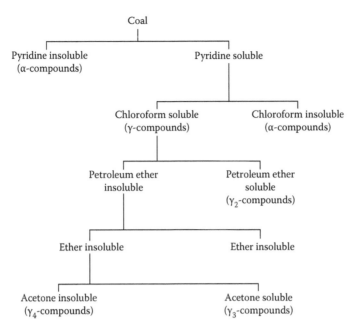

FIGURE 11.1   Fractionation of the pyridine extracts from coal. (From Berkowitz, N., *An Introduction to Coal Technology*, Academic Press Inc., New York, 1979.)

In general terms, solvents for the extraction of coal can be described and generally defined as follows:

1. Nonspecific solvents extract a small amount (on the order of 10% w/w) of coal at temperatures up to about 100°C (212°F). The extract is thought to arise from the so-called resins and waxes that do not form a major part of the coal substance and are typically aliphatic in nature. Low-boiling liquids such as methanol, ethanol, benzene, acetone, and diethyl ether are examples of the nonspecific solvents.

2. Specific solvents extract 20%–40% of the coal at temperatures below 200°C (390°F) and the nature of the extract is believed to be similar to, or even represent, the original coal. Specific solvents extract coal by a process of physical dissolution and the nature of the coal extract and the parent coal is believed to be similar. Hence, such solvents can be considered nonselective in their action on coal and usually contain a nitrogen atom and an oxygen atom with unshared electrons as a lone pair (Dryden, 1951a). Pyridine, *N*-methylpyrrolidone, dimethylformamide, and dimethylacetamide are examples of this type of solvent. They are mostly nucleophilic in nature due to the presence of a lone pair of electrons on the nitrogen atom.

3. Degrading solvents extract major amounts of the coal (up to more than 90%) at temperatures up to 400°C (750°F). This type of solvent can be recovered substantially unchanged from the solution and the solvent action is presumed to depend on mild thermal degradation of the coal to produce smaller soluble fragments. Examples of this type of solvents include phenanthrene, diphenyl, phenanthridine, and tar oil fractions.

4. Reactive solvents dissolve coal by active interaction. Reactive solvents are generally hydrogen donors. The smaller coal fragments formed by thermal disintegration are stabilized by hydrogen, which is donated by the solvent. The coal, as well as the solvent, changes appreciably during extraction. Hydroaromatic compounds are good hydrogen-donor solvents, which are converted to their corresponding aromatic counterparts during extraction. For example, tetralin (1,2,3,4-tetrahydronaphthalene) is converted to naphthalene upon donation of four hydrogen atoms. The products from the coal can vary in composition depending upon the reaction severity and the ratio of the solvent to the coal. In addition, the extracts differ markedly in properties from those obtained with degrading solvents.

Another class of solvents is the so-called super-solvents—typically dipolar aprotic solvents that are capable of dissolving a large amount of the organic material in coal (Stiller et al., 1981). Super-solvents can dissolve many substances, both polar and nonpolar.

More recently, bitumen from oil sand deposits and other heavy petroleum-based feedstocks have also been employed as reactive solvents with a degree of success that warrants further work and investigation (see, for example, Aitchison et al., 1990; Wang and Curtis, 1992).

Considerable attention has been paid to the use of compressed gases and liquids as solvents for extraction processes (Schneider et al., 1980; Dainton and Paul, 1981; Bright and McNally, 1992; Kiran and Brennecke, 1992) although the law of partial pressures indicates that when a gas is in contact with a material of low volatility, the concentration of "solute" in the gas phase should be minimal and decreases with increased pressure. Nevertheless, deviations from this law occur at temperatures near the critical temperature of the gas and the concentration of solute in the gas may actually be enhanced as well as increased with pressure.

The technique of extracting virtually nonvolatile substances is particularly useful for those materials that decompose before reaching boiling point and is, therefore, well suited to the extraction of the liquids formed when coal is heated to about 400°C (750°F). Thus, supercritical gas/fluid extraction affords a means of recovering the liquid products when they are first formed, avoiding undesirable secondary reactions (such as coke formation), and yields of extract up to 25% or 30% have been recorded.

While the yields of extract may be lower than that can be obtained with some liquid solvents and the use of high pressures might appear to be disadvantageous, there are nevertheless two positive features related to supercritical gas/fluid extraction of coal:

1. The extracts have lower molecular weights (about 500 compared to greater than 2000) and higher hydrogen content and may, presumably, be more readily converted to useful products.
2. Solvent removal and recovery is more efficient, that is, a pressure reduction has the ability to precipitate the extract almost completely.

Solvent extraction is often used to prepare the insoluble residue for solvent swelling studies that are used to infer structural information by reference to the literature on solvent swelling of cross-linked polymers. Access to this active field can be gained from recent papers (Larsen and Mohammadi, 1990; Fujiwara et al., 1992; Nishioka, 1992; Painter, 1992: McArthur et al., 1993b).

Pretreatment of coals with solvents prior to liquefaction often leads to increased yields of desirable products, lower conversion temperatures, or both. Studies utilizing both very polar and nonpolar solvents illustrate the complexity of the effects in that both swelling and non-swelling solvents can improve liquefaction depending on the process and coal used and the extent of molecular aggregation (Joseph, 1991; Larsen and Hall, 1991; McArthur et al., 1993a).

Brief thermal treatment of coal with solvents in which the coal is insoluble (such as water or chlorobenzene) causes irreversible changes in swelling and extraction yields with good solvents (defined as those such as pyridine that swell and extract coal to the greatest extent). Physical association with strong concentration and temperature dependence is suggested as a better representation than a cross-linked network (for the insoluble portion of most coals) and provides implications for coal liquefaction.

Finally, whatever the principles applied to coal science, it must not be forgotten that there is the likelihood that coal macromolecules are associated (interlocked and intermeshed) with each other so strongly that it generally requires stronger conditions for extraction or solubilization of a majority. Extraction of coal in an organic solvent is an essential requirement for the conversion of coal to value-added fuels and chemicals without a significant loss in original coal constitution and its energy.

11.2 PHYSICOCHEMICAL CONCEPTS

The solvent extraction of coal is, in essence, a mild form of chemical conversion because in addition to the purely solvent action, there may also be molecular alterations that are definite and irreversible. Coal–solvent interactions are complex (Szeliga, 1987) but, in more general terms, extraction is usually enhanced by temperature; in addition, the presence of hydrogen will significantly alter the molecular changes.

When coal is heated in a slurrying vehicle, it is liquefied at 400°C–500°C (750°F–930°F). Though the reaction mechanism involving conversion of coal to oil is very complex, it appears that the interaction of coal with solvent at the initial stage of the reactions plays the vital role to determine the sequential conversion of coal substances—first to a pyridine-soluble solid and thereafter to benzene-soluble liquid hydrocarbons and low-boiling products. Thus the isolation and identification of the products of coal–solvent interactions to yield pyridine-soluble matter may provide information regarding the suitability of the coal for liquefaction.

The rates of the fragmentation reactions in the coal feedstock are rapid and, for the most part, nonreversible (unless combination with other moieties to form coke occurs) and depend on the nature of coals. The microcomponents, as identified by the microscopic examination of coal particles as vitrinite and exinite, are the most reactive materials that undergo rapid fragmentation. The other components—semifusinite and fusinite—produce insignificant amounts of liquid products but may act as nuclei for coke formation, which might be deleterious to ultimate conversion.

Since the different microcomponents react differently, it would be necessary to separate them as several gravity-cut by a flotation technique and establish their reactivity separately. The reactivity of the as-mined coal (composite sample), along with the reactivity of the microcomponents, would be a valuable parameter. A suitable test method would also provide data for yield and composition of gaseous products and minimum hydrogen requirement that has to be fed in the reactor using solvent as a carrier. The spectroscopic characterization of the extractable matter would provide information helpful in downstream processing of the primary liquefaction products.

The extractive capabilities of various solvents have been correlated, amongst other phenomena (Given, 1984), with the *internal pressure* (P_i) of the solvent:

$$P_i = (dH_v - RT)/(M/r)$$

where
 dH_v is the latent heat of vaporization
 R is the gas constant, in consistent units
 T is the temperature (absolute)
 M is the molecular weight
 r is the density

The higher the internal pressure, the greater the solvent action on coal. Notable among the solvents that have been applied to the solvent extraction of coal are, in increasing effectiveness, naphthalene, tetralin, pyridine, aniline, cresols, and phenol. Tetralin and other cycloalkanes are somewhat unusual in that they are capable of transferring hydrogen to the coal or coal fragments during the extraction process and providing hydrogenation (Clarke et al., 1984). At temperatures above 400°C (750°F)—a temperature that is above the decomposition point of most coals—tetralin will also decompose to naphthalene.

In addition, coal tar fractions are effective solvents while paraffinic hydrocarbons are ineffective for the solvent extraction of coal. The former concept (i.e., the utilization of solvents similar to coal tar materials) is a feature of commercial hydrogenation recycle. Partially reacted liquid is recycled to form a paste with the ground coal fuel. The coal is partially dissolved and reacted while the recycle oil is also reacted to a further stage of conversion.

Temperature and pressure alone are capable of producing liquids from coals. The higher the rank, the more difficult the operation and, in general, the lower the rank (or the higher the hydrogen content) of a coal, the more easily it is dissolved in a solvent. For example, pyridine is a very efficient solvent for bituminous coals but has no effect on anthracite.

For a particular coal, there is an optimum temperature for solvent extraction that is usually recognized as the decomposition temperature of the coal. The effect of particle size apparently has little effect on solvent action, with or without accompanying hydrogenation, and has been interpreted in terms of an equilibrium condition rather than rate.

The coal solution process or solvent extraction process has been observed to involve several stages, including colloidal dispersion and, finally, true solution. This sequence involves the formation ("depolymerization") of relatively small organic fragments and the formation of higher-molecular-weight materials, such as char and coke, and is, as has been noted already, markedly affected by the presence of hydrogen in the system.

More recently, the mechanisms of solvent diffusion into coal have been shown to vary with the extent to which the diffusing solvent reacts with the coal (Hall et al., 1992). For basic (usually nitrogen-containing) solvents, this may be related to the number of hydrogen bonds that are disrupted by the diffusing solvent, which, in turn, is related to coal structure (Carlson, 1992; Nishioka, 1992).

Removal of the solvent from the extract phase leaves most products of the same type as the original but smaller in molecular size due to a mild pyrolyzing effect of the solvent, and thus may provide a method for estimating the structure of coal by projecting structural data derived for the extract(s) to the original coal. The extracts have often been (incorrectly) called bitumen (Chapter 10) but part of the soluble products may appear as oils, fats, hydrocarbons, acids, alcohols, esters, resins, and waxes, which may have been occluded within the solid structure of the coal system. A part of the soluble product may thus result from thermal depolymerization of the coal as well as from the solvent.

The bituminous extracts may also be separated into solid and oily materials in a manner similar to the fractionation of petroleum, tar sand bitumen (natural asphalt), refinery (manufactured) asphalts, and petroleum residua (Figure 11.2). As already noted (see also p. 183), the term bitumen is more correctly used for the natural asphalts that occur in various parts of the world and an artifact of the thermal process is often called tar or pitch (Speight, 2007).

In certain instances, the extracted bitumen will, depending on the solvent, have poorer coking properties than the original coal. However, if the bitumen extract is recombined with the residue and coked, the solid portion of the extract promotes swelling and the liquid portion of the extract promotes cementing. Furthermore, the bituminous extract from a poor coking coal may, when added to

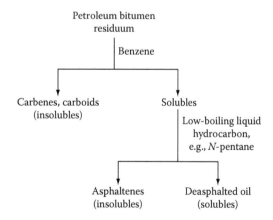

FIGURE 11.2 Fraction scheme for petroleum, heavy oil, and bitumen—pyridine-soluble tars from coal can be fractionated in a similar manner. (From Speight, J.G., *The Chemistry and Technology of Petroleum*, 4th edn., Taylor & Francis Group, Boca Raton, FL, 2007.)

the insoluble residue from a good coking coal, produce a good coke but the extract from a good coking coal may, when added to the insoluble residue from a poor coking coal, produce only a poor coke.

At temperatures below those normally required for the thermal decomposition of coal, the yields of extract vary directly with extraction temperature. This effect is usually most pronounced with the nonspecific solvents but it has been noted that a solvent such as ethylenediamine will produce from bituminous coal almost three times as much extract at its boiling point (115°C [240°F]) as at room temperature and also enhances the effect of extraction with other solvents such as N-methyl-2-pyrrolidone (Pande and Sharma, 2002). In fact, the yields of extracts obtained with a series of primary aliphatic amines have been found to vary with the extraction temperature rather than with any other solvent property.

Efficient contact between the solvent and the coal is also an important physical phenomenon because the process of coal solubilization is, even at the boiling point of the solvent, relatively slow; Soxhlet extraction of coal is a much more prolonged operation than shaking the coal with boiling (or near boiling for correct comparison) solvent. Furthermore, contact between the solvent and the coal is enhanced by using coal of a markedly reduced (–72 mesh) size; use of lump coal is always inefficient.

Thus, yields of extract tend to increase when the particle size of the treated coal is reduced; the greatest effect is observed when coals are milled to the order of size of 1 μm (0.001 mm). In extraction of bituminous coals with primary aliphatic amines, the effect of particle size over the range 72–240 mesh was always small but increased somewhat with increasing rank, as would be expected from the decrease in porosity. The effect of time on the yield of the extract has also been investigated and, like the effect of efficient contact between the coal and the solvent, may vary with the conditions prevalent during the extraction.

The effect of time of extraction on yields of extract (using bituminous coals of various ranks and primary aliphatic amines as solvents) differs according to whether (1) the coal is shaken in a tube with the solvent at room temperature or (2) it is extracted in a Soxhlet thimble near the boiling point of the solvent.

In fact, although the yield may approach an asymptotic limit over a period of days, there is no indication of an absolute limit over a greater period of time. However, when the coal and solvent are shaken together, the extraction may be more rapid and with small particles of coal (–72 mesh) much shorter periods may be adequate. Excessive shaking of the coal and solvent may have a deleterious effect on yield and, in addition, when a solvent and coal are heated together there is always a danger that the extraction yield will pass through a maximum because of aggregation or resorption of the extract, and a Soxhlet type extractor or a countercurrent extractor should be used when possible.

The effect of temperature is most marked for benzene-type solvents. With neutral tar oils, a sharp rise in yield from less than 3% at about 200°C (390°F) to more than 50% at between 350°C and 400°C (660°F and 750°F) has been reported. Ethylenediamine extracts at least half as much from low-rank bituminous coals at room temperature as at its boiling point (116°C [240°F]). Furthermore, for a series of the primary aliphatic amines, the limiting yield reached in a period of days has been found to be independent of the method of extraction and depends to a first order of approximation on the temperature of extraction rather than on the particular amine used.

An explanation of this in terms of the known solubility behavior of high polymers has been proposed. In a study of extraction of coal with tar oil at 300°C–400°C (570°F–750°F), it has been reported that the logarithm of the yield of extract varied linearly with the reciprocal of the absolute temperature, a relationship also noted for amine solvents at lower temperatures.

The presence of moisture in the solvent and of oxygen in the extraction apparatus both tend to decrease the yield obtained. The efficiency of extraction of coals by alcoholic potash is reduced by the presence of water. With amine solvents, the virtual absence of moisture is essential if good yields of extract are to be obtained because the specificity of the solvent is directly affected by water. However, the presence of moisture does not appear to be critically important with the majority of solvents that are immiscible with water. On the other hand, water has been reported to increase the yield of material that is extractable from coal by 2-naphthol.

The use of solvents at temperatures above their critical temperature can lead to enhanced yields of extracts due to changes in the solvent properties. Even nonspecific solvents, such as hydrocarbons, can give yields of extracts that are ca. 20% w/w of the dry, ash-free coal (Bartle et al., 1975; Whitehead and Williams, 1975; Williams, 1975).

11.3 ACTION OF SPECIFIC SOLVENTS

In general circumstances, unless solvolysis is involved, the more common organic solvents, such as benzene, alkylbenzenes, methanol, acetone, chloroform, and diethyl ether, dissolve little of the true coal substance and usually extract only that material that is occluded within the coal matrix.

The effect of pyridine on coal has been known since the late days of the nineteenth century (Bedson, 1899) and extensive follow-up studies were carried out to determine the comparative extractability of pyridine and chloroform (Cockram and Wheeler, 1927, 1931; Blayden et al., 1948; Wender et al., 1981).

Although pyridine is a good solvent for many coals and has been extensively studied, pyridine cannot be completely removed from coal or extracts by heating, either under reduced pressure or in a flow of nitrogen. Even though certain recipes for washing the remaining vestiges of pyridine out of the coal have been proposed, trace amounts of pyridine can still be detected by spectroscopic method. Indeed, placing the pyridine-treated-and-washed coal into a sealed container for 1 year will still give the odor of pyridine when the container is opened—the human nose is extremely sensitive to traces of pyridine on coal fractions.

These studies, as well as later work (Dryden, 1950, 1951a,b; Given, 1984), showed that significant yields of extracts, often as high as 35%–40%, can be obtained by using pyridine, certain heterocyclic bases, or primary aliphatic amines (which may, but need not, contain aromatic or hydroxyl substituents). Secondary and tertiary aliphatic amines are often much less effective insofar as more than one alkyl group on the amine appears to present steric problems that interfere with the interaction between the solvent and the coal.

There are reports, however, where the nature of the coal has been altered by chemical reaction prior to extraction in attempts to achieve solubilization (Given, 1984). For example, the addition of alkyl groups to the *coal molecule* provides an enhancement in the solubility of the coal (Given, 1984; Yoneyama et al., 1992). This may be ascribed either to *unknown/unpredictable* reactions of the coal by which smaller (soluble) molecules are produced or by the alkyl groups causing a reduction in the solubility parameter of the coal and bringing it closer to the solubility parameter of the solvent thereby effecting solubility, in a manner similar to the effects noted in petroleum science (Speight, 2007).

Other examples of the chemical alteration of coal to improve the extractability yield include lithium/amine reduction (Given et al., 1959; Given, 1984), sodium/alcohol reduction (Ouchi et al., 1981), and sodium/potassium/glycol ether reduction (Niemann and Hombach, 1979). But it must be remembered that even though the extractability of the coal is enhanced, the chemistry of these reactions is not well understood and is often subject to speculation leaving the precise reasons for solubility enhancement open to speculation also. Reactions that enhance coal solubility by "depolymerization" (Heredy et al., 1965) also suffer from the same unknowns.

Insofar as the ability of solvents to extract material from coal can be correlated with the presence of an unshared pair of electrons on a nitrogen atom or an oxygen atom in a solvent molecule, it is difficult to fully rationalize such a concept. For example, some nonsolvents (of which methanol may be cited as an example) have the ability to swell coal almost as much as the more specific solvents (such as pyridine) (Franz, 1979; Szeliga, 1987). Furthermore, the approximate linear relationship between extract yields and the internal pressures of the solvents no longer holds when the solvents are used at temperatures below their normal boiling points.

At higher temperatures (i.e., during extraction of coal at ca. 400°C [750°F]), a two-ring solvent containing a hydroxyl group attached to an aromatic ring and a hydroaromatic ring (a potential hydrogenating agent) has been found to be particularly effective. Furthermore, the extent of the extraction at atmospheric pressure from a bituminous coal with a number of aromatic solvents containing one or more rings has been found to be related to the boiling point of the solvent.

In the higher-rank bituminous coals, α- and β-naphthols are both effective solvents, but, for subbituminous coal, β-naphthol may produce even less extract than phenanthrene but α-naphthol may extract as much as 83% w/w of the coal. Ring compounds, such as phenanthrene, appear to be superior for extracting bituminous coals of medium rank. A study of extraction of Dutch bituminous coals at 200°C–400°C (390°F–750°F) illustrated the importance of simultaneous hydrogenation via the hydroaromatic portion of the solvents; diphenylamine has been found to be an effective catalyst for transfer of hydrogen in this manner because it increases the yield of extract.

Between 360°C (680°F) and 400°C (750°F), the solvent power of various solvent types has been found to decrease in the following order: amines, phenols, cyclic hydrocarbons, aliphatic hydrocarbons, but the substitution of hydrogen for functions such as imine (=NH) or amino (–NH$_2$) may increase the solvent power. With aliphatic amines at lower temperatures, the reverse is true. Insertion of a carbonyl group in the *para* position causes a sharp increase in effectiveness.

Many organic liquids have been suspected of exercising more than a solvent action on coal. An indication of chemical interaction is the observation that the total weight of products sometimes exceeds the original weight of coal, although up to about 5% may be the result of strong adsorption of solvents on the residue and extract. With mixed solvents, the potential for interaction may be increased.

Mixtures of the higher ketones and formamides have been reported to be considerably better solvents for coal (82% C; Northumberland, England) than either of the pure components. Mixtures worthy of note are acetophenone-monomethylformamide and ethylcyclohexanone-dimethylformamide (equimolecular proportions). Thus, some solvent pairs may show enhanced solvent power, whereas others may behave independently in admixture.

11.3.1 INFLUENCE OF COAL RANK

Coal rank has a considerable influence on the nature and the amount of extracts obtained by the solvent extraction of coal (Kiebler, 1945; Vorres, 1993). In addition, the soluble products of the extraction, whether they be called extracts or (incorrectly) bitumen, vary according to the means by which they are obtained.

The solubility of coal in solvents decreases rapidly as the carbon content of the coal increases from 85% to 89% (dry ash-free basis) and is negligible for anthracite (92%–93% w/w carbon). This phenomenon is generally applicable to all temperatures and types of solvents. However, a decrease in coal particle size (thereby increasing the surface if the coal available for solvent contact) or an increase in temperature increases the extent of extraction.

It is generally recognized that vitrain is the most soluble constituent of any particular coal whereas fusain is the least soluble. Indeed, early work on the liquefaction of coal by dissolution in a solvent (Chapters 18 and 19) showed that even at temperatures of the order of 400°C (750°F) fusain is, to all intents and purposes, insoluble. Under these aforementioned conditions, durain did show some response to the solvent but still did not match the solubility of vitrain.

11.3.2 INFLUENCE OF SOLVENT TYPE

The results of extracting coals with *benzene* and benzene/ethanol mixtures have been reported (Table 11.1), but only a broad general trend seems to emerge from these observations. Thus, it appears that for coals with more than 88% carbon content and less than 25% volatile matter, the amount of extract obtainable decreases rapidly; for coals of carbon content lower than this limit, no definite trend with rank appears to be evident.

The first systematic work on the extraction of coal using *nitrogen-containing solvents* (e.g., pyridine) resulted in the production of substantial amounts of extract, which was then further fractionated to produce a series of fractions based on their solubility/insolubility in different solvents (Figure 11.3) (Burgess and Wheeler, 1911).

TABLE 11.1

Influence of Coal Rank on the Degree of Extraction by Aromatic and Hydroaromatic Solvents

Yield	Solvent	Details
Increasing with decreasing rank	Naphthalene Hydrogenated pitch	
Exhibiting a maximum	Anthracene oil, retene, pyridine	Ten coals, 14.6%–43.7% volatile matter, pronounced peak in 30%–40% range with rapid decrease below about 20% volatile matter
	Neutral tar oil	Indication of flat maximum between 25% and 40% volatile matter, but points very scattered
	Tetralin	Yield increasing with increasing rank, then decreasing
	Phenanthrene	Coals from lignite to anthracite maximum at coal of 84.4% C

Source: Lowry, H.H., Ed., *Chemistry of Coal Utilization*, Supplementary Volume, John Wiley & Sons, Inc., New York, 1963.

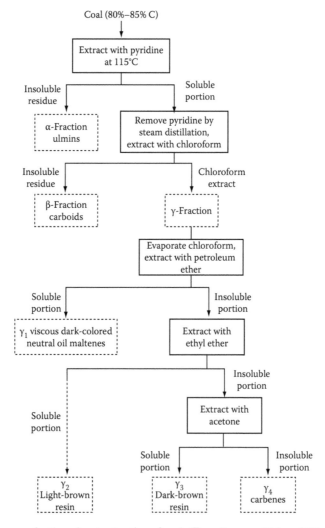

FIGURE 11.3 Sequence for the solvent extraction of coal. (From Burgess, M.J. and Wheeler, R.V., *J. Chem. Soc.*, 99, 649, 1911.)

TABLE 11.2

Influence of Coal Rank on the Degree of Extraction by Nitrogen-Containing Solvents

Yield	Solvent	Details
Increasing with decreasing rank	Pyridine	Two coals of 90.4% and 88.5% C
	Ethylenediamine	British bright bituminous and anthracite coals, 78%–94% C; particularly rapid decrease in yield between 85% and 88% C
Exhibiting no correlation	Pyridine	South Wales coals, maximum extraction at 30% volatile matter

Source: Lowry, H.H., Ed., *Chemistry of Coal Utilization*, Supplementary Volume, John Wiley & Sons, Inc., New York, 1963.

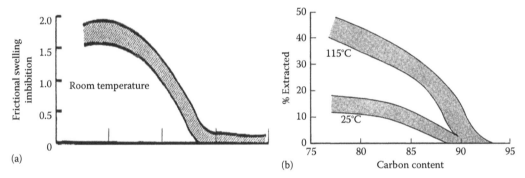

FIGURE 11.4 Variation of (a) imbibition of the solvent and (b) degree of extraction with carbon content. (From Dryden, I.G.C., *Fuel*, 30, 39, 1951.)

Since that time, a general correlation between yield of extract and carbon content has been reported for ethylenediamine (Table 11.2 and Figure 11.4). Various other amine solvents (e.g., monoethanolamine) show similar behavior insofar as the extract yield may decrease markedly with increase in rank for coals having less than 85% w/w carbon. On the other hand, the yield of extract using solvents such as benzylamine, piperidine, and pyridine may show much less variation with rank and the effectiveness of many solvents may decrease markedly for coal having more than 88% carbon (Figure 11.5).

Thus, it is possible to deduce several preferred options in the solvent extraction of coals using the so-called specific solvents. For example, the yield of extract usually decreases (but may, on occasion, increase) with an increase in the carbon content of the coal over the range 80% to ca. 87% carbon. However, use of a solvent such as pyridine may produce anomalous results and the petrographic composition of the coal may also have an effect. For coals having more than approximately 81% carbon the yield of extract is diminished to such an extent that only negligible yields of extracts are noted for coals having more than approximately 93% carbon.

There is a relationship between the yield of the extract and the saturation sorption or imbibition of solvent, which is independent of the rank coal or the particular amine solvent (Figure 11.4). An adsorption isotherm for ethylenediamine vapor on an 82% carbon coal exhibited three main features: (1) chemisorption up to 3%–6% adsorbed, (2) a fairly normal sorption isotherm from the completion of chemisorption up to a relative pressure of at least 0.8, and (3) a steeply rising indefinite region near saturation that corresponded to observable dissolution of the coal.

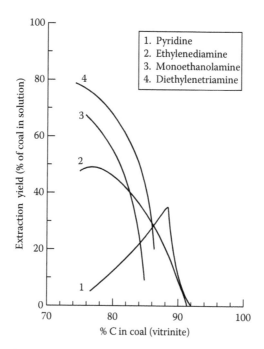

FIGURE 11.5 Extraction using different solvents with carbon content for vitrinites. (From Van Krevelen, D.W., *Fuel*, 44, 221, 1965.)

11.4 COMPOSITION OF THE EXTRACTS

Identification of the molecular species within coal extracts is an important step in determining not only the composition of the extracts but also the effectiveness of the extraction process.

There are many examples reported in the scientific literature in which the properties of coal extracts resemble the properties of the original coal. Whether or not this is fortuitous or to be anticipated because of the nature of the technique or because of a true resemblance has been the subject of many lengthy debates. The matter of "true" resemblance between coal extracts and the original coal is still not resolved, but for the present context it should be noted that when consideration is given to the comparative analyses of extracts and residues, it is essential to distinguish between low-temperature extractions and high-temperature extractions whereby (in the latter case) the extract may be substantially richer in hydrogen.

Size exclusion chromatography (SEC) is a powerful method for determining molecular weight distributions of mixtures. Because coal fractions usually contain both polar and nonpolar molecules, interpretation of SEC data is difficult and subject to error. Advantages and problems with tetrahydrofuran (THF), *N*-methyl-2-pyrrolidone, and pyridine have been observed (Lafleur and Nakagawa, 1989) but some investigators continue to employ these solvents, thereby ignoring any pitfalls.

Solvents (such as pyridine, dimethylformamide, and *N*-methyl-2-pyrrolidone) that prevent self-association of molecular species in coal extracts give more realistic molecular weights but limit the choice of SEC detectors. The common refractive index detector overestimates the contribution of oxygen-rich species such as phenols and gives average molecular weights that are too high unless calibrated with appropriate compounds. Changes in SEC molecular size distributions find use in studies of the progress of coal liquefaction and other processes.

Among the methods that have the possibility for identification of specific compounds in solvent extracts, GC/MS has been the most studied. However, a disadvantage to the use of nonpolar solvents is that potentially soluble molecules may not be accessible to the solvent and their absence may bias the conclusions drawn. For this purpose, mixed solvents such as benzene/methanol should be used with caution since the benzene and methanol can partition on the solid phase, thereby causing changes in the extract composition and errors in the data.

To avoid the problems associated with the use of mixed solvents, chromatographic fractionation of pyridine extracts is often used to isolate fractions that can be analyzed by mass spectroscopy. In the usual methods, elution from a silica gel column or alumina column will produce increasingly polar extract fractions that may be further fractionated on other supports for specific analyses.

The reader is well advised to examine not only the composition and character of coal extracts but also the means by which they were produced and defined.

Extract compositions depend on the particular coal/solvent system and on extraction conditions (Raj, 1979). For example, nonspecific solvents generally extract coal selectively and dissolve primarily waxy and resinous substances that, although derived from the original plant debris, may not be integral parts of the coal substance.

The extraction of peat(s) and lignite(s) with benzene or benzene–methanol mixtures followed by fractionation of the crude extract with diethyl ether or hot ethyl alcohol yields Montan waxes and resins (Table 11.3) and has actually been used commercially. The waxes consist mostly of aliphatic (C_{24}–C_{32}) acids and the corresponding esters as well as C_{24}–C_{32} alcohols, while the resins (which resemble colophy or hauri gums) are derivatives of mono- and dibasic C_{12}–C_{20} acids with aromatic/hydroaromatic nuclei (e.g., abietic acid).

The amount of extractable waxes and/or resins decreases markedly with increase in coal rank, and since nonspecific solvents can also take small quantities of undifferentiated coal-like material into solution, bituminous coal extracts are generally less "specialized." However, even these extracts are the result of solvent selectivity and have higher hydrogen/carbon atomic ratios than the parent coal.

In contrast, specific solvents, which dissolve a large fraction of the coal, do so nonselectively. Thus, while the differing solubility of the petrographic constituents will influence the composition of coal extracts, the extracted material may resemble the insoluble residue closely and, in fact, may be virtually indistinguishable from it. Indeed, material such as waxes and resins, unless present in phenomenally large amounts, is entirely masked by the remainder of the coal "substance." The elemental compositions of the extracts as well as the properties (especially those associated with the chemical aspects of the dissolved material, e.g., magnetic and spectroscopic properties) are all almost identical with those of the original coal.

As with any complex organic mixture, there is always the tendency to freely apply fractionation procedures in order to characterize the product. This is also true for coal extracts and the procedures applied vary from simple solvent procedures (Figure 11.2) to more complex physicochemical procedures. Thus, in the present context, several combinations of solvents have been developed for fractionation of the extracts from coal, many of which have been used by only one particular laboratory.

TABLE 11.3
Montan Waxes and Resins
Extracted from Coal

	Waxes	Resins
Carbon (% w/w)	80.5	80.0
Hydrogen (% w/w)	13.0	11.0
Oxygen (% w/w)	6.5	9.0
Melting point (°C)	80	65–70
Acid number	23	33
Saponification number	95	76

Among these are mentioned morpholine and hexachlorobiphenyl, anthracene oil, retene, carbon disulfide, carbon tetrachloride, and petroleum ether (Davies et al., 1977). It has been shown that the fractionation by pyridine and chloroform into various fractions does not yield consistent data when the method is transferred from a microscale investigation to a larger-scale operation.

Similar observations have been reported in the fractionation of the higher-molecular-weight constituents of petroleum (Speight, 2007) and there is obviously some degree of uncertainty in these methods of fractionation. On the other hand, there are many reports of attempts at the fractionation of coal extracts where some degree of success has been achieved. For example, the fractionation of the extracts obtained from bituminous coal with tetralin and *m*-cresol mixtures by adding ether, dioxane, and pyridine yielded fractions with different H/C ratios and molecular weights. Phenanthrene extracts may be fractionated into products having different degrees of aromaticity, molecular weights, and coking properties by the use of benzene and cyclohexane.

Extraction of lower-rank coals in ethylenediamine at room temperature shows several interesting features. The limiting yield of soluble material appears to depend only on the type of coal, the solvent, and the temperature of extraction; yields less than the limiting value may be obtained and it is essential to extract the coal with more than one successive portion of solvent. If, during extraction of coal by direct shaking of the solvent, the concentration of extract in solution exceeds a "critical" value (at which a solution tends to form a surface skin and to throw down sludge), some of the extract that was dissolved in solvent, which is itself imbibed in the pores of the residue, appears to be forced out thereby increasing the concentration of the free liquid phase; this is presumably an effect on the partition resulting from increased aggregation in the free solution.

When a sample of coal is successively extracted with fresh portions of ethylenediamine by shaking at room temperature, the amount of solvent imbibed by the residue undergoes periodic fluctuations, probably owing to rearrangements of physical structure as the amount extracted increases. In a given stage of extraction the ratio of solvent to coal exercises an important influence on the apparent yield of extract, but after sufficient stages the total result is the same for all ratios and sufficiently high to produce a mobile suspension of coal in the liquid.

A three-stage mechanism of extraction has been proposed and mathematical equations derived to fit this concept. In a study of the extraction of bituminous coals at 200°C–400°C (390°F–750°F) with polycyclic compounds, the kinetics of extraction up to 350°C (660°F) were computed; activation energies between 14 and 40 kcal/mol were derived and were dependent on the rank of the coal and on the particular solvent employed.

When lower-rank bituminous coals are extracted with ethylenediamine below its normal boiling point (i.e., 116°C [240°F]), the products consist entirely of solid extract and solid residue with no appreciable amount of oily material (although the extract can yield more volatile matter when heated). It has also been noted that the dissolution of coal in oils containing oleic acid affords residues (after 75% of the coal is dissolved) having proximate and ultimate analyses similar to those of the original coals.

Therefore, data of this type have been used to support the concept of a more uniform composition for coals and are further supported by a study of extraction of coals with anthracene oil at 350°C (660°F) in which the H/C atomic ratio of undissolved residue was observed to remain remarkably constant up to 80% extraction. Furthermore, the infrared spectra of extracts from coals in the range 78%–89% carbon resemble those of the parent coals.

Aromatic species are also extractable from coal by various solvents, especially benzene, and often can be related to the natural product origins of the coal (Table 11.4) (White, 1983).

11.5 SOLVOLYSIS

Since the 1970s, coal liquefaction has been extensively studied worldwide to produce petroleum substitutes (Speight, 2008). Major processes are operated under high or moderate hydrogen pressures with significant consumption of this element to *depolymerize* coal into a product similar to petroleum—it should be noted that the word depolymerize is used incorrectly since coal is not a

TABLE 11.4

Examples of Polynuclear Aromatic Compounds Isolated from Coal

Compound	Formula	Structure
1. Octahydro-2,2,4a,9-tetramethylpicene	$C_{22}H_{30}$	
2. Tetrahydro-1,2,9-trimethylpicene	$C_{25}H_{24}$	
3. Tetrahydro-2,2,9-trimethylpicene	$C_{25}H_{24}$	
4. 1,2,9-Trimethylpicene	$C_{25}H_{20}$	
5. 23,25-Bisnormethyl-2-desoxyallobetul-1,3,5-triene	$C_{26}H_{40}O$	
6. 23,24,25,26,27-Pentanormethyl-2-desoxyallobetul-1,3,5,7,9,11,13-hepta-ene	$C_{26}H_{28}O$	

Source: White, C.M., *Handbook of Polycyclic Aromatic Hydrocarbons*, A. Bjorseth, Ed., Marcel Dekker, Inc., New York, Chapter 13, 1983.

polymer in the true chemical sense and use of the word in this context can lead to serious misunderstanding about the chemical nature of coal (Chapters 2 through 4 and 7).

Nevertheless, in contrast to these processes, the degrading extraction (solvolysis) of various ranks of coal (including lignite) (Giralt et al., 1988) with stable solvents uses the hydrogen provided from dehydrogenation of the solvent or by the hydrogen-rich portions of coal (Chapters 10, 18, and 19).

Although yields obtained with this method are lower than when hydrogen gas is used, the lower pressure required makes this process suitable for coal liquefaction. Alternatively, it may be used as the first stage in coal hydrogenation processes.

Tetralin is normally used as a model solvent to study extraction processes due to its great hydrogen-donor capacity and well-known dehydrogenation reactions while anthracene oil is a solvent of industrial interest.

In spite of the large volume of work, it must be said that the mechanism of coal liquefaction is not clearly understood due to both the complexity of products and the vast differences found among coals of similar rank—in other words, the pronounced *heterogeneity* of coal. Furthermore, the heterogeneity of coal has not deterred investigators devising kinetic models for the liquefaction of bituminous and subbituminous coals in tetralin and anthracene oil. The issue is whether or not such work and models bear any relationship to the realty of the coal liquefaction process.

Coal macromolecules are associated (interlocked and intermeshed) with each other so strongly that it generally requires stronger conditions for extraction or solubilization of more than 10%–20% of coal. Extraction of coal in an organic solvent is an essential requirement for the conversion of coal to value-added fuels and chemicals without a significant loss in original coal constitution and its energy. Such a concept require that the coal be treated with a solvent under conditions such that the coal is decomposed into smaller molecular entities—the process is known as *solvolysis*, which refers to the action of solvents on coal at temperatures at which the coal substance decomposes and in practice relates in particular to extraction at temperatures between 200°C and 400°C (390°F and 750°F). This concept is often used as a means of coal liquefaction (Chapters 18 and 19).

The increased solubility of coal under these conditions (more specifically, at these temperatures) was first reported in 1916 when subbituminous and bituminous coals were extracted with benzene at 250°F (480°F) and yields of extracts were approximately 10–20 times greater than those obtainable by Soxhlet extraction at the normal boiling point of benzene. Since the nature and extent of the thermal degradation processes in coal depend on temperature, it must be anticipated that the yield and composition of soluble matter extracted under solvolytic conditions will also vary with extraction temperatures. However, extract yields are, as in milder forms of solvent extraction, also dependent on the choice of solvent.

For example, at temperatures below 350°C (660°F), the solvent power of the extracting liquid appears to be solely determined by the ability of the solvent to alter the coal physically (by swelling) prior to the onset of the degradation (depolymerization) process. In this respect, the most effective solvents are aromatics, phenols, naphthols, anthracene, and phenanthrene.

On the other hand, at temperatures in excess of 350°C (660°F) the coal undergoes decomposition without the aid of the solvent but the solvent is necessary insofar as the radicals generated during the decomposition process must be prevented from forming higher-molecular-weight materials such as hydrogen-deficient chars or cokes. Thus, if the solvent has the capacity of donating hydrogen to the coal, extract yields approaching 100% can be achieved. The formation of large amounts of soluble matter under severe solvolytic conditions can therefore be identified with the initial stages of coal hydrogenation in H-donor systems (Chapters 10, 18, and 19).

On the other hand, low-rank coals are currently used just as fuels near coal mines because such coals (e.g., brown coal and lignite) contain large amounts of water (up to 60% w/w) and dewatering or drying is essential when brown coals are transported and stored to be utilized. Dewatered brown coals unfortunately tend to have high spontaneous combustion tendency as compared to non-dewatered coals, which causes serious problems for transportation and storage (Xian et al., 2010).

Because the spontaneous combustibility of brown coal is associated with its high oxygen content coming from oxygen functional groups, it is required to reduce the amount of oxygen functional groups to suppress the spontaneous combustibility. Therefore, both dewatering and upgrading through the use of solvent degradation might be an option for the more efficient utilization of these coals as fuels or as chemical precursors.

Suitable solvents for this process include tetralin, anthracene oil, 5-hydroxytetralin, and *o*-cyclohexylphenol. However, solvents such as naphthalene, cresol, diphenyl, or *o*-phenylphenol, which will not donate hydrogen to the process, will generally only produce optimum yields of extracts of the order of 20%–30% w/w of the coal at 400°C (790°F).

The flexibility of coal structure and mobility of loosened molecules are responsible for the successive extraction of coal in solvents having different chemical characteristics. In fact, coal has a heterogeneous structure having different structural units, that is, polyaromatic, hydroaromatic, and paraffinic units linked through C–C, C–N–C, C–O–C, and C–S–C linkages.

The solvents having structural similarities with the structural units present in coal can easily dissociate and mobilize the coal molecules for their solution in that solvent. The extent of similarity between the structures of coal and that of the solvent and the boiling point of the solvent were generally found to determine the extent of extractability of coal.

The point or position of attack, that is, chemical depolymerization or physical dissociation or dispersion of coal, by different solvents can be different depending on the bond strength or the particular structural similarities between the coal and the solvent and, of course, depending on the boiling point of the solvent. Anthracene oil was found to be a wonder solvent for extraction of coal and for depolymerizing the coal (both chemically and physically) to render the enhanced amount of the treated residual coal in the next solvent. The actual three-dimensional shape and size of the solvent molecule (especially angular arrangement of condensed aromatic rings instead of linear) were reported to be important factors for the ability of solvents to dissolve and disperse coal. Similarity of coal structure with angular ring phenanthrene-type molecules allows the free passage (inside and outside) to these molecules inside the solid coal.

This similarity of structures also helps in breaking the coal–coal (molecular) physical interactions such as London, van der Waals forces, and hydrogen bonding by the solvents. Since the coal structure is heterogeneous, a successive multiple solvent attack has been found to be the best method to dissipate inter- and intramolecular forces and disperse the three-dimensional cross-linked gel structure of coal macromolecules. This allows enhanced extraction of coal without using any high-pressure method involving hydrogenation at elevated temperatures (thereby causing chemical changes to the system). Successive solvolytic extraction of coal affords a convenient and cost-effective technique for solvent refining of coal under ambient pressure conditions.

11.6 SOLVENT SWELLING OF COAL

Solvent swelling of coal occurs when the physical dimensions of coal increase due to the presence of a solvent. Coal swelling occurs with changes to other coal properties, such as coal extraction yield or coal surface area.

Coal begins to swell as it imbibed a solvent for which it has an affinity. As the coal absorbs solvent, the size increases but the coal maintains the original shape. When the solvent is removed, the coal shrinks to near the original size and shape—the process is thought to be reversible (Brenner, 1983, 1984), although this has been disputed for bituminous coal (Nishioka, 1993a,b). Some destruction of coal samples occurs after swelling and shrinking, but this destruction is thought to be due to mechanical stresses rather than chemical changes (Brenner, 1983, 1984). Such a phenomenon could be used to contradict the concept of the reversibility of the swelling process.

The overall implications of coal selling experiments suggest that the dissociation of the non-covalent bonds in coal is crucially important during coal swelling in solvents (Chen et al., 2011).

The amount of coal swelling was measured by the swelling ratio, represented by the symbol, Q:

$$Q = (\text{volume of swollen coal})/(\text{volume of original coal})$$

Solvents for which coal has a high affinity are referred to as good swelling solvents. Swelling in good swelling solvents was found to be independent of the solvent-to-coal weight ratio and grinding direction (Cody et al., 1988; Mastral et al., 1990). In good swelling solvents such as N-methyl-2-pyrrolidone and pyridine, coal was capable of swelling to over twice of its original volume, while still retaining its original shape (Takanohashi and Iino, 1995).

Good extraction solvents were usually good swelling solvents. The ability of a solvent to swell coal was a strong function of the electron-donating ability of the solvents (Takanohashi and Iino, 1995). Because of the anisotropic nature of coal, coal swelled preferentially in a direction perpendicular to the bedding plane of the coal seam. This directional swelling was observed because coal appears to have been more highly cross-linked in the bedding plane than perpendicular to it. This directional swelling of coal was not noted in most studies because traditionally only bulk swelling has been measured, not the swelling of individually oriented coal pieces. Measuring the swelling ratio of individual coal pieces yielded clues to the structure of coal not provided by the study of the bulk swelling behavior of coal. The perpendicular/parallel swelling ratios are highest in pyridine and lowest in chlorobenzene, indicating a highly anisotropic arrangement of covalent bonds. Also, the time to reach maximum swell parallel to the bedding plane is shorter than the time to reach maximum swell perpendicular to the bedding plane. Cody et al. (1988) also discovered that swelling measured as a function of time passed through a maximum due to the formation of a metastable state.

In addition, observing the swelling ratios of different ranks in different solvents has provided information about the structural changes across varying ranks. For example, as might be anticipated, swelling ratios were higher for lower-rank coals and swelling could be used to improve THF-soluble materials after liquefaction with H-donor solvents. The trend of increased THF-soluble materials correlated with coals of increased swelling ratios. As a result, it was postulated that liquefaction of coal by H-donor solvents is a surface area–dependent reaction and that pre-swelling the coal was a good method for producing greater penetration and diffusion of reactants, increasing the liquefaction yield (Rincón and Cruz, 1988).

The rate at which coals swell is dictated by the rate at which the solvent diffuses into the coal. This is controlled by solvent properties, the size of the coal particles, and the average molecular weight between the cross-links of the coal matrix (Olivares and Peppas, 1992). Coal is a glassy solid at room temperature, but transitions to a flexible state as it absorbs solvent and the flexible nature of the swollen coal suggested lower effective cross-link density, and suggested that the elasticity of the solvent swollen coal may be predominantly rubber-like. The transition from the glassy to rubbery state is generally very sharp (Olivares and Peppas, 1992).

11.7 EPILOGUE

Generally, with careful attention to coal history, solvent history, and solvent properties, the solvent extraction of coal may be a useful technique for application to coal science. Studies of the extraction process itself and the related solvent swelling can provide insights into coal matrix structure only with careful, well thought out, and reproducible analyses. The data may also provide detailed and valuable information about the chemical species present in both extract and raw coal.

Coal partially dissolves in a number of solvents and although a wide range of organic solvents can be used, dissolution is never complete and usually requires heating to temperatures sufficient for some thermal degradation or solvent reaction to take place. For some coals, dissolution on the order of 10%–40% (w/w of coal) can be achieved near room temperature (to a limiting value) (Landau and Asbury, 1938) and up to 90% (w/w of coal) at temperatures approaching 400°C (750°F). However, it must be recognized that this latter effect is not so much due to dissolution of the original coal but due to dissolution of the products of the thermal degradation of the reaction of the coal to lower-molecular-weight species and/or due to thermal reaction of the coal with the solvent.

In addition, the solvent extraction of coal can be improved with various types of pretreatment, one of which is thermal pretreatment. For example, when certain coals were heated at 200°C–400°C (390°F–750°F) in an inert atmosphere, cooled, and extracted with solvents, the yield of extract is often higher than the yield of extract obtained with untreated coals for a specific type of solvent.

The solubility of coal in solvents decreases rapidly as the carbon content of the coal increases from 85% to 89% (dry ash-free basis) and is negligible for anthracite (92%–93% w/w carbon). This phenomenon is generally applicable to all temperatures and types of solvents. However, a decrease in coal particle size (thereby increasing the surface if the coal available for solvent contact) or an increase in temperature increases the extent of extraction.

As expected, aromatic solvents are more effective for coal extraction than neutral aliphatic solvents. Whether the solvent action of coal is physical or chemical in nature has always been disputed. It is suspected that certain organic liquids perform more than a solvent action on coal. An indication of chemical interaction is obtained from the observation that the total weight of the products sometimes exceeds that of the original coal.

The presence of oxygen during the solvent extraction of coal decreases the yield for most solvents. The presence of moisture in coal has an unfavorable effect on extraction yield for solvents that are miscible with water. The moisture content of coal is extremely important, especially for any extraction process in which the solvent would be recycled.

The hydrogen transfer process is recognized as a stage in coal liquefaction. Furthermore, during the extraction process, the temperature at which the extraction is being conducted and the solvent play major roles in whether or not a hydrogenation reaction will take place. However, extraction conducted at relatively high temperatures (>300°C [>570°F]), which causes the hydrogen-donor ability of tetralin to increase the final hydrogen-to-carbon ratio in the extracts compared to that in the raw materials.

Furthermore, when solvent treatment is applied to coal as liquefaction process the challenge is to (1) reject the mineral matter, (2) reduce the sulfur content of the products, (3) control the microstructure or macrostructure, and (4) control the chemistry.

The flexibility of the coal matrix and mobility of the lower-molecular-weight species (Chapter 10) (as well as the solubility of any products) will be responsible for the successive extraction of coal in solvents having different chemical characteristics. Solvents having structural similarities with the structural units present in coal may have enhanced ability to dissociate and mobilize the coal molecules for their solution in that solvent. The extent of similarity between the structures of coal and that of the solvent and the boiling point of the solvent were generally found to determine the extent of extractability of coal.

REFERENCES

Aitchison, D.W., Clark, P.D., Hawkins, R.W., Logan, L.H., and Ohuchi, T. 1990. *Fuel*, 69: 97.

Bartle, K.D., Martin, T.G., and Williams, D.F. 1975. *Fuel*, 54: 226.

Bedson, P.P. 1899. Extraction of coal. *Journal of the Indian Chemical Society*, 21: 241.

Bedson, P.P. 1902. *Journal of Indian Chemical Society*, 21: 241.

Berkowitz, N. 1979. *An Introduction to Coal Technology*, Academic Press Inc., New York.

Blanc, P., Valisolalao, J., Albrecht, P., Kohut, J.P., Muller, J.F., and Duchene, J.M. 1991. *Energy & Fuels*, 5: 875.

Blayden, H.E., Gibson, J., and Riley, H.L. 1948. *Journal of Chemical Society*, 1693.

Brenner, D. 1983. In situ microscopic studies of the solvent-swelling of polished surfaces of coal. *Fuel*, 62: 1347.

Brenner, D. 1984. Microscopic in-situ studies of the solvent induced swelling of thin sections of coal. *Fuel*, 63: 1325.

Bright, F.V. and McNally, M.E.P. (Eds.). 1992. *Supercritical Fluid Technology: Theoretical and Applied Approaches in Analytical Chemistry*. Symposium Series No. 488. American Chemical Society, Washington, DC.

Burgess, M.J. and Wheeler, R.V. 1911. *Journal of Chemical Society*, 99: 649.

Carlson, G.A. 1992. *Energy & Fuels*, 6: 771.

Chen, L., Yang, J., and Liu, M. 2011. Kinetic modeling of coal swelling in solvents. *Industrial & Engineering Chemistry Research*, 50: 2562–2568.

Clarke, J.W., Rantell, T.D., and Snape, C.E. 1984. Reactivity of cycloalkanes during the solvent extraction of coal. *Fuel*, 63(10): 1476–1478.

Cockram, C. and Wheeler, R.V. 1927. *Journal of Chemical Society*, 700.

Cockram, C. and Wheeler, R.V. 1931. *Journal of Chemical Society*, 854.

Cody, G.D. Jr., Larsen, J.W., and Siskin, M. 1988. Anisotropic solvent swelling of coals. *Energy & Fuels*, 2(3): 340–344.

Dainton, A.D. and Paul, P.F.M. 1981. In *Energy and Chemistry*, R. Thompson (Ed.). The Royal Society of Chemistry, London, England, p. 32.

Davies, G.O., Derbyshire, F.J., and Price, R. 1977. *Journal of the Institute of Fuel*, 121.

de Marsilly, C. 1862. *Annales de chimie et de physique*, 66(3): 167.

Derbyshire, F.J., Davis, A., and Lin, R. 1991. In *Coal Science II*, H.H. Schobert, K.D. Bartle, and L.J. Lynch (Eds.). Symposium Series No. 461. American Chemical Society, Washington, DC, Chapter 7.

Dryden, I.G.C. 1950. Behavior of bituminous coals towards solvents, I. *Fuel*, 29: 197–207.

Dryden, I.G.C. 1951a. *Fuel*, 30: 39.

Dryden, I.G.C. 1951b. *Nature (London)*, 16: 561.

Dryden, I.G.C. 1963. In *Chemistry of Coal Utilization*, Supplementary Volume, H.H. Lowry (Ed.). John Wiley & Sons, Inc., New York, Chapter 6.

Francis, W. 1961. *Coal: Its Formation and Composition*. Edward Arnold Ltd., London, U.K., p. 452 et seq.

Franz, J.A. 1979. *Fuel*, 58: 405.

Fremy, E. 1861. *Comptes Rendus*, 52: 114.

Fujiwara, M., Ohsuga, H., Takanohashi, T., and Iino, M. 1992. *Energy & Fuels*, 6: 859–862.

Giralt, J., Fabregat, A., and Giralt, F. 1988. Kinetics of the solvolysis of a catalan lignite with tetralin and anthracene oil. *Industrial & Engineering Chemistry Research*, 27: 1110–1114.

Given, P.H. 1984. In *Coal Science*, M.L. Gorbaty, J.W. Larsen, and I. Wender (Eds.). Academic Press, Inc., New York, Vol. 3, p. 63.

Given, P.H., Lupton, V., and Peover, M. 1959. *Proceedings. Residential Conference on the Science in the Use of Coal*, Institute of Fuel, London, U.K., p. A38.

Hall, P.J., Thomas, K.M., and Marsh, H. 1992. *Fuel*, 71: 1271.

Heredy, L.A., Kostyo, A.E., and Neuworth, M.B. 1965. *Fuel*, 44: 125.

Hombach, H.P. 1980. General aspects of coal solubility, *Fuel*, 59: 465.

Joseph, J.T. 1991. *Fuel*, 70: 179, 459.

Kiebler, M.W. 1945. In *Chemistry of Coal Utilization*, H.H. Lowry (Ed.). John Wiley & Sons, Inc., New York, Chapter 19.

Kiran, E. and Brennecke, J.F. (Eds.). 1992. *Supercritical Fluid Engineering Science: Fundamentals and Applications*. Symposium Series No. 514. American Chemical Society, Washington, DC.

Kreulen, D.J.W. 1948. *Elements of Coal Chemistry*. Nijgh & van Ditmar N.V., Rotterdam, the Netherlands, Chapter 14.

Lafleur, A.L. and Nakagawa, Y. 1989. Multimode size exclusion chromatography with poly(divinylbenzene)/N-methylpyrrolidinone for the characterization of coal-derived mixtures. *Fuel*, 68: 741–752.

Landau, H.G. and Asbury, R.S. 1938. Ultimate yield of solvent extraction of coal. *Industrial & Engineering Chemistry Research*, 30(1): 117.

Larsen, J.E. and Hall, P.J. 1991. *Energy & Fuels*, 5: 228.

Larsen, J.W. and Mohammadi, M. 1990. *Energy & Fuels*, 4: 107.

Litke, R., Leythaeuser, D., Radke, M., and Schaefer, R.G. 1990. *Organic Geochemistry*, 16: 247.

Lowry, H.H. (Ed.). 1963. *Chemistry of Coal Utilization*, Supplementary Volume. John Wiley & Sons, Inc., New York.

Mastral, A.M., Izquierdo, M.T., and Rubio, B. 1990. Network swelling of coals. *Fuel*, 69: 892–895.

McArthur, C.A., Hall, P.J., MacKinnon, A.J., and Snape, C.E. 1993b. Preprints. Division of Fuel Chemistry. *American Chemistry Society*, 33(4): 1290–1296.

McArthur, C.A., Hall, P.J., and Snape, C.E. 1993a. Preprints. Division of Fuel Chemistry. *American Chemistry Society*, 33(2): 565–570.

Niemann, K. and Hombach, H.-P. 1979. *Fuel*, 58: 853.

Nishioka, M. 1992. *Fuel*, 71: 941–948.

Nishioka, M. 1993a. Irreversibility of the solvent swelling of bituminous coals. *Fuel*, 72: 997–1000.

Nishioka, M. 1993b. *Fuel*, 72: 1001–1011.

Oele, A.P., Waterman, H.I., Goldkoop, M.L., and van Krevelen, D.W. 1951. *Fuel*, 30: 169.

Olivares, J.M. and Peppas, N.A. 1992. The effect of temperature treatment on penetrant transport in coal. *Chemical Engineering Communications*, 115: 183–204.

Ouchi, K., Shivaishi, H, Itoh, H., and Makabe, M. 1981. *Fuel*, 60: 474.

Painter, P. 1992. *Energy & Fuels*, 6: 863–864.

Pande, S. and Sharma, D.K. 2002. Ethylenediamine-assisted solvent extraction of coal in *N*-methyl-2-pyrrolidone: Synergistic effect of ethylenediamine on extraction of coal in *N*-methyl-2-pyrrolidone. *Energy & Fuels*, 16(1): 194–204.

Pickel, W. and Gotz, G.K.E. 1991. *Organic Geochemistry*, 17: 695.

Pullen, J.R. 1983. In *Coal Science*, M.L. Gorbaty, J.W. Larsen, and I. Wender (Eds.). Academic Press, Inc., New York, Vol. 2, p. 173.

Raj, S. 1979. *Chemical Technology (Houston)*, 2(4): 54.

Rincón, J.M. and Cruz, S. 1988. Influence of pre-swelling on the liquefaction of coal. *Fuel*, 67: 1162–1163.

Schneider, G.M., Stahl, E., and Wilke, G. 1980. *Extraction with Supercritical Gases*. Verlag Chemie, Weinheim, Germany.

Speight, J.G. 2007. *The Chemistry and Technology of Petroleum*, 4th edn. Taylor & Francis Group, Boca Raton, FL.

Speight, J.G. 2008. *Synthetic Fuels Handbook: Properties, Processes, and Performance*. McGraw-Hill, New York.

Stenberg, V.I, Baltisberger, R.J., Patel, K.M., Raman, K., and Wollsey, N.F. 1983. In *Coal Science*, M.L. Gorbaty, J.W. Larsen, and I. Wender (Eds.). Academic Press, Inc., New York, Vol. 2, p. 125.

Stiller, A.H., Sears, J.T., and Hammack, R.W. 1981. Coal extraction process. US Patent 4,272,356, June 9.

Szeliga, J.S. 1987. In *Coal Science and Chemistry*, A. Volborth (Ed.). Elsevier, Amsterdam, the Netherlands, p. 405.

Takanohashi, T. and Iino, M. 1995. Investigation of associated structure of upper freeport coal by solvent swelling. *Energy & Fuels*, 9: 788–793.

Vahrman, M. 1970. *Fuel*, 49: 5.

Van Krevelen, D.W. 1957. *Coal Science: Aspects of Coal Constitution*. Elsevier, Amsterdam, the Netherlands, Chapter 5.

Van Krevelen, D.W. 1965. *Fuel*, 44: 221.

Vorres, K.S. 1993. *Users Handbook for the Argonne Premium Coal Sample Program*. Argonne National Laboratory, Argonne, IL; National Technical Information Service, United States Department of Commerce, Springfield, VA.

Wang, S.L. and Curtis, C.W. 1992. Preprints. Division of Fuel Chemistry. American Chemical Society, Washington, DC, 37(4): 1769.

Wender, I., Heredy, L.A., Neuworth, M.B., and Dryden, I.G.C. 1981. In *Chemistry of Coal Utilization*, Second Supplementary Volume, M.A. Elliott (Ed.). John Wiley & Sons, Inc., New York, Chapter 8.

White, C.M. 1983. In *Handbook of Polycyclic Aromatic Hydrocarbons*, A. Bjorseth (Ed.). Marcel Dekker, Inc., New York, Chapter 13.

Whitehead, J. and Williams, D. 1975. *Journal of the Institute of Fuel*, 48: 182.

Williams, D.F. 1975. *Indian Journal of Chemistry*, 10: 20.

Williams, J.M., Vanderborgh, N.E., and Walker, R.D. 1987. In *Coal Science and Chemistry*, A. Volborth (Ed.). Elsevier, Amsterdam, the Netherlands, p. 435.

Wise, W.S. 1971. *Solvent Treatment of Coal*. Mills and Boon, London, U.K.

Xian, L., Hasegawa, Y., Morimoto, M., Ashida, R., and Miura, K. 2010. Degradative solvent extraction of low-rank coal for converting them into extracts having similar chemical and physical properties. *Proceedings. 13th Asia Pacific Confederation of Chemical Engineering Congress*. Taipei, Taiwan, October 5–8.

Yoneyama, Y., Nakayama, H., Nishi, H., and Kato, T. 1992. *Bulletin of the Chemical Society of Japan*, 65: 987.

Yun, Y., Meuzelaar, H.L.C., Simmleit, N., and Schulten, H.-R., 1991. In *Coal Science II*, H.H. Schobert, K.D. Bartle, and L.J. Lynch (Eds.). Symposium Series No. 461. American Chemical Society, Washington, DC, Chapter 8.

12 Chemical Reactivity

12.1 INTRODUCTION

Coal is used in processes such as combustion (Chapters 14 and 15), carbonization (Chapters 16 and 17), liquefaction (Chapters 18 and 19), and gasification (Chapters 20 and 21). The products derived from coal are used as precursors of other materials and in the production of chemicals (Chapter 24). Thus, a coal must be characterized before it is used, whether as a single or blended coal, in order to find out the properties of a coal, to determine its quality, and to predict its technological behavior (Chapters 4, 8, and 9).

Furthermore, an understanding of the chemical behavior of coal is an essential part of projecting the successful use of coal, say, as a source of chemicals or for conversion and/or utilization processes.

Since the early days of coal science, a substantial amount of work has been carried out on the various aspects of the chemistry of coal. Furthermore, the trends toward the use of coal as a source of liquid and gaseous fuels have caused a marked increase in studies related to the chemical reactions of coal. It is, perhaps, unfortunate that the investigations related to the chemical behavior of coal have in the past been so diversified as to appear to be completely without any focus or even any mutually common goal.

It is difficult, if not impossible, to predict the reactivity of coal based on physical or chemical analyses, because coal is a complex mixture of macromolecules as well as a physical mixture of organic and inorganic constituents. Evidence can be accumulated to reason that reactions in the organic materials can be influenced by both clay minerals and pyrites found distributed within the organic coal matrices. One convenient basis for understanding the reactions in both systems is to examine reactive functional groups.

Nevertheless, the emphasis on the various aspects of coal conversion requires that investigators have a sound knowledge of the chemical behavior of coal. And there is an additional caveat. The reactivity of coal with various reagents is not just a matter of bring together the coal and the reagent; it also appears that many of the reactions of coal are dependent upon the diffusion of the reagents through the coal (Larsen et al., 1981).

Thus, even though the chemistry is often represented by one or more simple chemical equations, there is much more to coal chemistry (in addition to diffusional limitations) than these equations indicate. For example, coal combustion (using the combustion process as an example) (Chapters 14 and 15) is generally, and simply, represented as the reaction of the carbon of the coal with oxygen with the overall stoichiometry of the reaction always being established:

$$C_{coal} + O_2 \rightarrow 2CO$$

$$C_{coal} + O_2 \rightarrow CO_2$$

$$C_{coal} + O_2 \rightarrow CO_2$$

$$C_{coal} + H_2O \rightarrow CO + H_2$$

Thus

$$C_{coal} + O_2 \rightarrow 2CO$$

$$2CO + O_2 \rightarrow 2CO$$

Any reactions of hydrogen, for example, $2H_{coal} + O_2 \rightarrow H_2$, are considered to be of secondary importance. However, it must be recognized that coal is very complex and that the heteroatoms (nitrogen, oxygen, and sulfur) can exert an influence on the chemistry of the combustion process and it is this influence that can bring about serious environmental concerns.

For example, the conversion of nitrogen and sulfur to their respective oxides (represented by conveniently simple chemical equations) during combustion is a major issue:

$$S_{coal} + O_2 \rightarrow SO_2$$

$$2SO_2 + O_2 \rightarrow 2SO_3$$

$$N_{coal} + O_2 \rightarrow 2NO$$

$$2NO + O_2 \rightarrow 2NO$$

$$N_{coal} + O_2 \rightarrow NO$$

In addition, the sulfur dioxide that escapes into the atmosphere is either deposited locally or undergoes further conversion to sulfurous and sulfuric acids by reaction with moisture in the atmosphere:

$$SO_2 + H_2O \rightarrow H_2SO_3$$

$$2SO_2 + O_2 \rightarrow 2SO_3$$

$$SO_3 + H_2O \rightarrow H_2SO_4$$

Thus

$$2SO_2 + O_2 + 2H_2O \rightarrow 2H_2SO_4$$

Similarly, the chemistry of the nitrogen species in coal is also of some importance. Nitrogen oxides (Morrison, 1980) also contribute to the formation and occurrence of acid rain, in a similar manner to the production of acids from the sulfur oxides, yielding nitrous and nitric acids:

$$NO + H_2O \rightarrow H_2NO_2$$

$$2NO + O_2 \rightarrow 2NO$$

$$NO_2 + H_2O \rightarrow HNO_3$$

Thus

$$2NO + O_2 + H_2O \rightarrow 2HNO_3$$

The reactions of coal in combustors are far more complex than these simple equations would illustrate. Nevertheless they are a beginning. Whereas the understanding complex nature of coal as a molecular entity has improved (Berkowitz, 1979; Meyers, 1981; Hessley et al., 1986; Hessley, 1990; Haenel, 1992) (Chapter 10)—but not to the extent that a structure for coal has been identified with any degree of certainty—explanations of the chemical reactions of coal are generally simplified (to the point of over simplification) and are often misunderstood so that, at best, they are only given passing recognition. But coal chemistry is extremely important to coal utilization. If the chemistry can be controlled, more efficient conversion processes can result. And understanding the chemistry can only be achieved by a thorough understanding of coal behavior in various reactions systems.

It should also be noted here that many of the chemical reactions applied to coal have been applied as a means of determining (or estimating) the structural/functional types present in coal (Chapter 10), which is well worth bearing in mind when the chemistry of coal is under study insofar as any derived structures must fit the chemical (and physical) behavior of the coal.

Also, as precombustion sulfur removal becomes an attractive alternative to flue gas scrubbers, scientists have explored several methods of removing inorganic sulfur from coal and coal chars. Oxidation, microbial oxidation/desulfurization, and halogenation are avenues that are now, and have been, under investigation for some time (Thomas, 1995).

Therefore, it is with this in mind that the present chapter represents an attempt to give a pertinent summary of the chemistry of coal and to show how this knowledge might be applied to *mapping* the behavior of coal in the various conversion systems. It would have been inappropriate to attempt to include all of the reactions of coal in this chapter. It is necessary that many of the reactions be included elsewhere in this text (for example, see Chapter 13) and only the reactions pertinent to the meaning of this chapter are included here.

12.2   REACTIONS WITH OXYGEN (OR AIR)

Oxidation is a complex process that consists of parallel but competing/interacting reaction processes. Although at least four such processes are believed to exist, the exact number, nature, and kinetics of such processes are not very clearly understood. Although aerial oxidation of coal is essentially a chemical reaction process, it is influenced by, apart from its original chemical composition, other factors like temperature, moisture, catalytic effects of water, and components in the mineral matter. Furthermore, the effects of the physical and surface properties play a role that is not properly understood.

Oxidation of coal in air is a factor that must be taken into account in determining coal quality, which has an effect on the technological behavior of coal (Chapter 4). Coal oxidation may result from exposure to weathering processes during handling and transport or when the coal is stockpiled under different environmental conditions and can affect both the organic and inorganic components giving rise to the deterioration of coal properties (Chapter 9).

A different consequence of coal oxidation is the development of spontaneous combustion (Chapters 9 and 14), when the heat generated by in situ oxidation causes the coal to smolder and ultimately burn without any external heat source. The liability to oxidation is mainly determined by the coal's rank, in conjunction perhaps with the maceral content (Chapter 4) and the content of mineral matter (Chapter 7). Low-rank coal is particularly prone to spontaneous combustion and factors, such as access of air to coal stockpiles, may need to be controlled to reduce the ever-present risk of spontaneous ignition and combustion.

On the practical side, progress in the science of self-heating leading to useful preventive methodologies with wide and extensive applicability on industrial level is yet to be developed. Although the negative effects of aerial oxidation on the technologically useful properties, for example, coking properties, lowering of tar yield, combustion behavior, and spontaneous combustion of coal, are well documented in literature, again the exact mechanisms of these phenomena are yet to be firmly established and considerable difference of opinion still exists.

The oxidation of coal is of considerable interest as a means of modifying the physical and chemical properties of the coal. In particular, the caking of the coal on carbonization may be reduced or prevented by oxidation treatment. The degree of oxidation is important, about 1% of oxygen being adsorbed on, or combined with, the coal. It may be estimated by chemical analysis or trial carbonization, but for the control of oxidation plant, a more rapid indication is required of any departure from the correct operating conditions.

Because coal is a heterogeneous material, factors other than the organic matter and oxygen can affect the oxidation reaction. These factors, including minerals, moisture, thermal chemistry, and particle size and surface area, are considered. However, the parameters affecting the natural oxidation are far from fully understood and unresolved differences remain on the true nature of the structural changes oxidation causes in the organic matter. Oxidation generally has a deleterious effect on the way coal behaves in combustion, beneficiation, and processing (especially carbonization).

The reaction of coal with aerial oxygen has been the subject of many investigations (Van Krevelen and Schuyer, 1957; Dryden, 1963). Weathering studies, the incorporation of functional groups, and the mechanisms of spontaneous ignition of coal are only three of the many reasons of interest (Wender et al., 1981). Furthermore, a standard test method is available for determining the relative degree of oxidation in bituminous coal by alkali extraction (ASTM, 2011).

Generally, aerial oxidation renders coal increasingly soluble in alkali as "humic acid" with ultimate loss of carbon as carbon oxides and hydrogen as water. The solubilized coal is spectroscopically similar to the parent coal except for oxygen functional groups identified as phenolic –OH, –COOH, and =CO (ketone and/or quinone), respectively. The peripheral molecular changes associated with the formation of oxygen functions are explained by a scheme resembling the gas-phase oxidation of naphthalene to phthalic acid and maleic anhydride in sequential steps. The limiting elemental composition of the soluble coal (humic acids) depends on the temperature of oxidation, and tends to vary slightly with coal rank. Higher temperature and/or higher oxygen partial pressure accelerates the rate of the reaction. At temperatures beyond 250°C (480°F), a virtual low-temperature combustion process ensues (Chakrabartty, 1981).

Coal (with the exception of anthracite) and the coal macerals (with the exception of inertinite) react with oxygen at ambient temperature. However, even a low degree of oxidation can have a significant effect on the properties and behavior of the coal (Beafore et al., 1984). One such example is the loss of fluidity (of bituminous coals) by mild oxidation (Clemens et al., 1989; Seki et al., 1990a,b) that appears to affect both the coal extract and the solvent-insoluble residue (Seki et al., 1990a,b). Property changes are particularly important when coal is stockpiled before use, which can lead to oxidation by aerial oxygen. There is also evidence that the sulfur in coal be removed by oxidation as a means of coal cleaning (Beckberger et al., 1979). Thus, there is the opportunity for oxidation reactions to play a beneficial role in coal use.

The oxidation reaction is, in fact, a general occurrence over a wide range of conditions (Given, 1984) but, more specifically, increases readily with temperature and decreases with particle size. This latter phenomenon is an indication of the function of surface area as one of the reaction parameters. If the coal contains a substantial proportion of indigenous moisture, the oxidation reaction rate will be relatively high.

On the other hand, coal that has already been oxidized (weathered) is much more difficult to oxidize and any further oxidation occurs only with difficulty, if at all. In addition to having an adverse effect on the coking (carbonizing) properties of coal (Chapter 16), atmospheric oxidation is also perceived to reduce the amount of soluble material in coal but some coals have been noted to show an increase in extractable matter (humic acids) after weathering (which also improves the flotation response of coal and therefore provides a benefit for coal cleaning operations) (Chapter 6) (Garcia et al., 1991; Lalvani and Kang, 1992).

It is, in fact, apparently contradictory observations of this type (where, more often than not, direct comparisons of the experimental conditions are difficult, if not impossible) that also add many complications to coal chemistry. Of course, the apparent contrasting nature of the data is dependent upon the definition of extractable matter. Organic nonpolar material that is extractable by organic solvents might be expected to decrease with oxidation. On the other hand, polar material such as humic acids that are produced by coal oxidation should be expected to decrease. It may be, after all, a matter of definition.

The exhaustive, low-temperature oxidation of coal has also been reported to yield products that are completely soluble in aqueous alkali and this phenomenon may actually be used as an indication of the extent of coal oxidation.

The low-temperature oxidation of coal has been compared with the combustion of coal (Chapters 14 and 15) insofar as the gases (carbon monoxide, CO; carbon dioxide, CO_2; and water, H_2O) evolved during both processes are the same, but a notable difference occurs in the rate of reaction and, furthermore, up to 50% of the oxygen consumed during the oxidation may remain in the coal and may even appear as a gain in weight. The oxygen adsorbed by the coal has been considered to be retained by the involvement of a coal–oxygen complex, which then decomposes to produce carbon monoxide, carbon dioxide, and water.

The low-temperature oxidation of coal is an exothermic process and (as noted earlier) the rate of reaction increases markedly with temperature. Thus, under the appropriate conditions, the oxidation sequence can (and often does) lead to the spontaneous ignition of the coal.

The low-temperature oxidation of coal dramatically influences and alters inherent properties. The deleterious effect that oxidation has upon the coking and caking properties of coal, through the loss of plasticity and fluidity, is well known.

12.2.1 Chemical and Physicochemical Aspects

At ambient temperatures, coal is presumed to react with oxygen by means of the transient formation of a coal–oxygen ("oxy-coal") complex. The so-called oxy-coal is evident from the changes that can be observed to have taken place by microscopic examination of the coal (oxidation edges) (Figure 12.1) and the development of more reflective material (Van Krevelen and Schuyer, 1957).

The low-temperature oxidation of coal is presumed to proceed by way of three distinct steps: (1) the chemisorption of oxygen on to the surface of the coal to form an oxygen complex with the coal, (2) the decomposition of the coal–oxygen complex, and (3) the formation of oxy-coal (Petit, 1990; Petit and Boettner, 1990; Petit and Cheng, 1990). The chemisorption of oxygen on to the coal surface is believed to produce a hydroperoxide and/or a peroxide (Given, 1984) (the exact form of

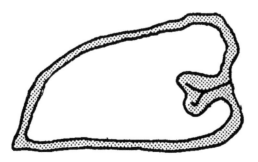

FIGURE 12.1 Representation of the oxidation edge (shaded area) of a coal particle.

$$(\text{Coal}){:}H + O_2 \rightarrow (\text{coal}){\cdot} + HO_2{\cdot}$$
$$(\text{Coal}){\cdot} + O_2 \rightarrow (\text{coal})O_2{\cdot}$$
$$(\text{Coal})O_2{\cdot} + (\text{coal}){:}H \rightarrow (\text{coal})OOH + (\text{coal}){\cdot}$$

FIGURE 12.2 Oxidation of coal by formation of a transient hydroperoxide.

which is still only presumptuous and which may vary with the particular site of attack) and the heat evolved by such a process is of the order of 100 kcal/mol oxygen.

The formation of an intermediate hydroperoxide (Figure 12.2) (Petit and Boettner, 1990) is postulated by virtue of the presence of free radical (i.e., unpaired electron) sites within the coal (Chapter 10). The ensuing decomposition of the coal–oxygen complex leads to the formation of oxygen-containing functional groups (such as carboxylic acid, carbonyl, phenolic hydroxyl) on the coal. With time the process will eventually lead to degradation of the coal substance and the formation of alkali-soluble products (often referred to as humic acids) (see Howard, 1945a; Van Krevelen and Schuyer, 1957; Stevenson and Butler, 1969), which will, in turn, be degraded to progressively lower-molecular-weight products.

However, oversimplifications notwithstanding, the aerial oxidation of coal is quite a complex process. In the initial stages, a slight weight gain (due to the incorporation of oxygen without the loss of any volatile oxidation products) may be noted followed by a more obvious decrease in the carbon and hydrogen contents of the product(s) eventually approaching the apparently "limiting" values of ca. 60%–65% carbon and 2%–3% hydrogen depending on the rank of the starting coal and the oxidation conditions (especially temperature) (Figure 12.3) (see also Fuerstenau and Diao, 1992).

While high temperatures (70°C [160°F]) usually accelerate the process of dehydrogenation or reactivity of the aromatic systems (Rausa et al., 1989), the thermal stability of the transient intermediates is often decisive in determining the product distribution. For example, at temperatures 72°C (160°F) the reaction does not usually proceed to the point where "massive" degradation of the coal substance is observed but usually terminates with the formation of the various oxygen functional groups within the coal.

However, in the high-temperature range (70°C [160°F]) the oxidation rate is increased markedly and also becomes dependent on the porosity of the coal. In this range, there may be a decrease in the oxidation rate with increasing particle size or with increasing rank. There may also be an increasing amount of carbon dioxide in the gaseous products.

The formation of humic acids only occurs readily at temperatures above 70°C (Estevez et al., 1990); at temperatures in the range 150°C–250°C (300°F–480°F), these acid products are reputed to be stable, which is in itself surprising since petroleum asphaltene constituents (perhaps a very distant cousin of humic materials but a complex natural product that also contains oxygen functions) evolve oxygen-containing gases (presumably from the oxygen-containing functions) quite readily over the temperature range 150°C–300°C (300°F–570°F) (Moschopedis et al., 1978).

In the "pure" state, humic acids are black, lustrous solids having ca. 55%–60% carbon and 2%–3% hydrogen with the balance of the elemental composition appearing predominantly as oxygen. The molecular weights of humic acids vary considerably and if a range was to be assigned for these molecular weights it would have to be 600–10,000. The reasons for this widespread observed molecular weights appear to lie in the tendency of these materials to associate even in dilute solution, and, therefore, the observed molecular weights depend to a large extent on the solvent used for the determination as well as on the methods employed.

Various attempts to elucidate the molecular structure of humic acids have been reported, but the complex nature of these oxidation products appears to defy resolution. Indeed, the number of potential species that may constitute "humic acid" lends little assistance to the problem and even complicates the matter still further. Thus, any deductions that have been reported with respect to a molecular structure for humic acids must be open to severe criticism because of the speculation and guesswork that are involved. In fact, as with the studies relating to the "molecular structure" of coal, such claims are only of very limited value and add to the general confusion that already exists.

12.2.2 Effect of Moisture

The presence of moisture does have a noticeable effect on the rate of the low-temperature oxidation of coal. Thus, if the coal is thoroughly dried, the rate of oxidation decreases (see, for example, Ogunsola and Mikula, 1992), but other factors may play a role in this otherwise simple observation. For example, water is a product of the oxidation reaction and if the rate of the oxidation appears to increase with time it may be that the water that is formed as a reaction product actually causes the

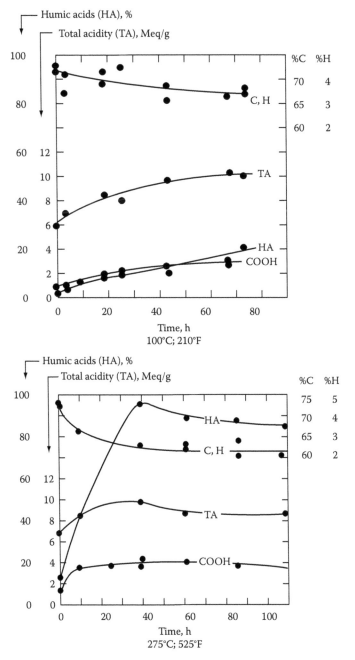

FIGURE 12.3 Aerial oxidation of subbituminous coal at various temperatures. (From Jensen, E.J. et al., Advances in Chemistry Series No. 55, American Chemical Society, Washington, DC, 1966, p. 621.)

(*continued*)

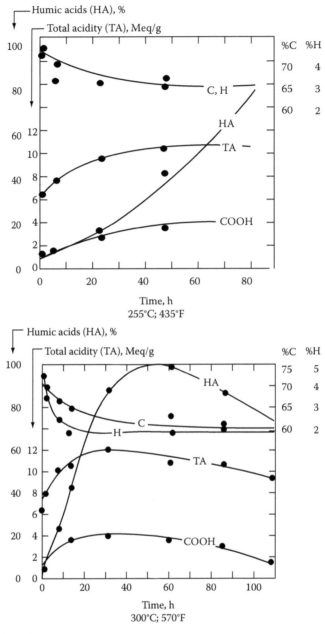

FIGURE 12.3 (continued)   Aerial oxidation of subbituminous coal at various temperatures. (From Jensen, E.J. et al., Advances in Chemistry Series No. 55, American Chemical Society, Washington, DC, 1966, p. 621.)

formation of the transient peroxide or hydroperoxide. On the other hand, severe drying of the coal may adversely affect the pore system of the coal thereby promoting a marked reduction in the reaction rate by virtue of a reduced surface area available for the oxidation.

12.2.3   EFFECT OF PYRITE

Pyrite (FeS_2) and the related mineral marcasite (FeS_2) occur frequently in coal and are recognized as the major mineral constituents of many coals (Chapter 7). Thus, it is not surprising that there has been some interest in the effects of these materials (but more particularly pyrite because of its more common occurrence) on the oxidation of coal.

The low-temperature oxidation of pyrite is exothermic:

$$2FeS_2 + 7O_2 + 2H_2O \rightarrow 2H_2SO_4 + 2FeSO_4 + 62 \text{ kcal/mol}$$

Thus, storing coal where it would be repeatedly wet (by rain) may favor pyrite oxidation.

If the coal is stored by complete immersion in water, then this would not be the case. It appears, however, that in spite of the exothermic value of the oxidation of pyrite, the pyrite may only play a minor part in the oxidation of coal but, on occasion, it has been suggested that pyrite oxidation will contribute to the spontaneous ignition of coal.

12.2.4 SPONTANEOUS IGNITION

The spontaneous ignition of coal (also variously referred to as the spontaneous combustion or autogenous heating of coal) has recognized as a hazard for some time to the extent that, in the early years of this century, guidelines were laid down for the strict purpose of minimizing the self-heating process (Haslam and Russell, 1926) and have been revised since that time (Allen and Parry, 1954). Indeed, the phenomenon of spontaneous ignition is not limited to coal but has also been observed in other piles of organic debris (Vovelle et al., 1983; Gray et al., 1984; Jones, 1990; Jones et al., 1990).

Spontaneous combustion, or self-heating, of coal is a naturally occurring process caused by the oxidation of coal. The self-heating of coal is dependent on a number of controllable and uncontrollable factors. Controllable factors include close management in the power plant, of coal storage in stockpiles, silos/bunkers, and mills and management during coal transport. Uncontrollable factors include the coal itself and ambient conditions.

The self-heating of coal is due to a number of complex exothermic reactions. Coal will continue to self-heat provided that there is a continuous air supply and the heat produced is not dissipated. The property of coal to self-heat is determined by many factors, which can be divided into two main types: properties of the coal (intrinsic factors) and environment/storage conditions (extrinsic factors). Self-heating results in degradation of the coal by changing its physical and chemical characteristics, factors that can seriously affect boiler performance.

The risk of spontaneous combustion during final preparation such as in silos/bunkers and mills also presents concerns in some cases. Properties that influence the propensity of coal to self-heat include volatile content, coal particle size, rank, heat capacity, heat of reaction, the oxygen content of coal, and pyrite content. The propensity of coal to self-heat and spontaneously combust tends to increase with decreasing rank. Thus, lignites and subbituminous coals are more prone to spontaneous combustion than bituminous coals and anthracites.

The temperature of coal increases due to self-heating until a plateau is reached, at which the temperature is temporarily stabilized. At this point, heat generated by oxidation is used to vaporize the moisture in the coal. Once all the moisture has been vaporized, the temperature increases rapidly. On the other hand, dry material can readily ignite following the sorption of water—dry coal in storage should not be kept in a damp place because this can promote self-heating. Therefore, it is recommended that dry and wet coal be stored separately.

Coal stockpiles are prone to spontaneous combustion especially where large quantities are stored for extended periods. Coals that exhibit the greatest tendency to self-heat (that is lignite, subbituminous coal, and brown coal) are rarely stored for any length of time at the power station. Self-heating occurs more commonly. Self-heating of coal is a naturally occurring process caused by the oxidation of coal. Natural oxidation is uncontrolled and can lead to emissions and spontaneous combustion. Unless handled correctly, the results can be catastrophic in damage to power plant equipment.

This is a reflection on the relative length of storage time involved at each stage. Spontaneous combustion in stockpiles poses significant safety, environmental, economic, and handling problems.

Thus, coal presents hazards between the time it is mined and its eventual consumption in boilers and furnaces. Below are listed some of the characteristics of spontaneous fires in coal that can be used to evaluate the potential for coal fires and as guidelines for minimizing the probability of a fire.

The higher the inherent (equilibrium) moisture, the higher the heating tendency. The lower the ash-free Btu, the higher the heating tendency. The higher the oxygen content in the coal, the higher the heating tendency.

Sulfur, once considered a major factor, is now thought to be a minor factor in the spontaneous heating of coal. There are many very low-sulfur Western subbituminous coals and lignite that have very high oxidizing characteristics and there are high-sulfur coals that exhibit relatively low oxidizing characteristics.

The oxidation of coal is a solid–gas reaction, which happens initially when air passes over the coal surface. Oxygen from the air combines with the coal, raising the temperature of the coal. As the reaction proceeds, the moisture in the coal is liberated as a vapor and then some of the volatile matter that normally has a distinct odor is released. The amount of surface area of the coal that is exposed is a direct factor in its heating tendency. The finer the size of the coal, the greater the surface area exposed to the air and the greater the tendency for spontaneous ignition.

Thus, the spontaneous ignition of coal is believed to center around the basic concept of the oxidation of carbon to carbon dioxide:

$$C + O_2 = CO_2$$

This particular reaction is exothermic (94 kcal/mol) and will be self-perpetuating especially since the rates of organic chemical reactions usually double for every 10°C (18°F) rise in temperature. Furthermore, there has also been the suggestion that the heat release that accompanies the wetting of dried (or partially dried) coal may be a significant contributory factor in the onset of burning.

Support for such a concept is derived from the observations that stored coal tends to heat up when exposed to rain after a sunny period (during which the coal has been allowed to dry) or when wet coal is placed on a dry pile (Berkowitz and Schein, 1951). Similar effects have been noted during the storage of hay in the conventional haystacks and ignition has been noted to occur. Thus, any heat generated by climatic changes will also contribute to an increase in the rate of the overall oxidation process. Obviously, if there are no means by which this heat can be dissipated, the continued oxidation will eventually become self-supporting and will ultimately result in the onset of burning.

Spontaneous ignition and the ensuing combustion of coal are usually the culmination of several separate chemical events (see Chapters 13 and 14) and although precise knowledge of the phenomenon is still somewhat incomplete but is gradually becoming known (Kreulen, 1948; Dryden, 1963; Gray et al., 1971; de Faveri et al., 1989; Vilyunov and Zarko, 1989; Jones and Wake, 1990; Shrivastava et al., 1992), there are means by which the liability of a coal to spontaneously ignite can be tested (Schmeling et al., 1978; Chakravorty, 1984; Chakravorty and Kar, 1986; Tarafdar and Guha, 1989; Jones and Vais, 1991; Ogunsola and Mikula, 1991; Chen, 1992).

For example, exposure of coal (freshly mined) to air will bring about not only loss of moisture but also oxidation. The latter process, often referred to as autooxidation or autoxidation (Joseph and Mahajan, 1991), commences when the coal reacts with oxygen (of the atmosphere). Both processes result in an alteration of the properties of the coal, that is, there is a decrease in the calorific value of the coal through the introduction of oxygen functions while there is also a very marked, adverse, effect on the caking properties of the coal.

The relationship between the friability of coal and its rank has a bearing on its tendency to heat or spontaneously ignite (Chakravorty, 1984; Chakravorty and Kar, 1986). The friable, low-volatile coals, because of their high rank, do not oxidize readily despite the excessive fines and the attendant increased surface they produce on handling. Coals of somewhat lower rank, which oxidize more readily, usually are relatively nonfriable; hence, they resist degradation in size with accompanying increase in the amount of surface exposed to oxidation. But above all, the primary factor in coal stockpile instability is unquestionably oxidation by atmospheric oxygen whilst the role of any secondary factors such as friability seems to exacerbate the primary oxidation effect (Jones and Vais, 1991).

More recent work has presented indications that the tendency for spontaneous ignition is reduced by thermal upgrading and further decreased with increase in treatment temperature (Ogunsola and Mikula, 1992). The decrease in the tendency to spontaneously ignite appears to be due to the loss of the equilibrium moisture as well as the loss of oxygen functional groups. The loss of the equilibrium moisture is an interesting comment because of the previous comment that the presence of indigenous moisture appears to enhance (i.e., increase the rate of) the oxidation reaction.

Coal tends to spontaneously ignite when the moisture within the pore system is removed leaving the pores susceptible to various chemical and physical interactions (Berkowitz and Speight, 1973) that can lead to spontaneous ignition. It is a question of degree and the correct order of reactions being in place. It is obvious that the system is complex and, as noted earlier, spontaneous ignition is the culmination of several interrelated chemical and physical events.

12.3   REACTIONS WITH OXIDANTS

In the very early days of coal science, the prime reason for the oxidation of coal was the production of chemicals but, economic considerations notwithstanding, there was always the underlying theme of the identification of the low-molecular-weight products and the ensuing attempts to relate these products to coal structure.

Thus, it is not surprising that, in view of the recent emphasis (perhaps, as some scientists and engineers would suggest) on the elucidation of the structure of the coal *molecule*, oxidation techniques have once again assumed a major role in coal science and, lately, as a means of projecting the product data to the various structural entities (Chapter 10) (Speight, 1987; Haenel, 1992). Thus, the oxidation of coal by materials other than oxygen is an extensive part of coal science. For example, the well-known oxidants such as nitric acid, permanganate, dichromate, hydrogen peroxide, performic acid, as well as hypochlorites have all been applied to coal (Mayo, 1975; Chakrabartty and Berkowitz, 1976; Studier et al., 1978; Duty et al., 1980; Liotta and Hoff, 1980; Deno et al., 1981; Wender et al., 1981).

Although studies of coal oxidation by various oxidants never really ceased, the apparent revival of interest may be ascribed to the concept that coal is a polyadamantane (polyamantane) system, as deduced from the products arising from the interaction of coal with sodium hypochlorite (Chakrabartty and Berkowitz, 1976). Nevertheless, it is not the intention here to enter into a discourse on the structure of coal as arrived at by the use of oxidants (see Chapter 10 for a discussion of such reactions) but it is the intention to present a brief indication of the means by which coal reacts with the various oxidizing agents.

Most oxidizing agents cause changes in the surface character of the coal and in this respect permanganate is particularly effective. However, in more general terms, oxidants appear to react with coal quite indiscriminately, in spite of repeated claims for the specificity of various oxidants. The conditions reported for such investigations do, however, vary considerably but the oxidants are nevertheless considered to attack coal substantially more rapidly than aerial oxygen or even pure oxygen.

Moreover, the diversity of the oxidants renders the oxidation of coal very complex because the experimental parameters can vary widely (Speight, 1987). The diversity in the structural types in coal (which vary not only with rank but also within the same rank) causes many problems associated with studies of the oxidative degradation of coal. For example, optimal conditions of time, temperature, and ratio of oxidant to coal can only be determined when several experiments are performed for each oxidant. Furthermore, the presence of the mineral matter must also be considered to be an integral part of coal oxidation since mineral constituents may change the chemistry of the oxidation. If coal is pretreated with hydrochloric acid to remove mineral matter prior to oxidation, the actual oxidation reaction may be more facile.

The oxidation of coal generally produces a complex mixture of products and any one particular solvent extraction procedure is not usually adequate for extracting all of the organic acids. For example, it may be necessary to use a sequence of solvents or solvent blends of different polarity or hot solvents

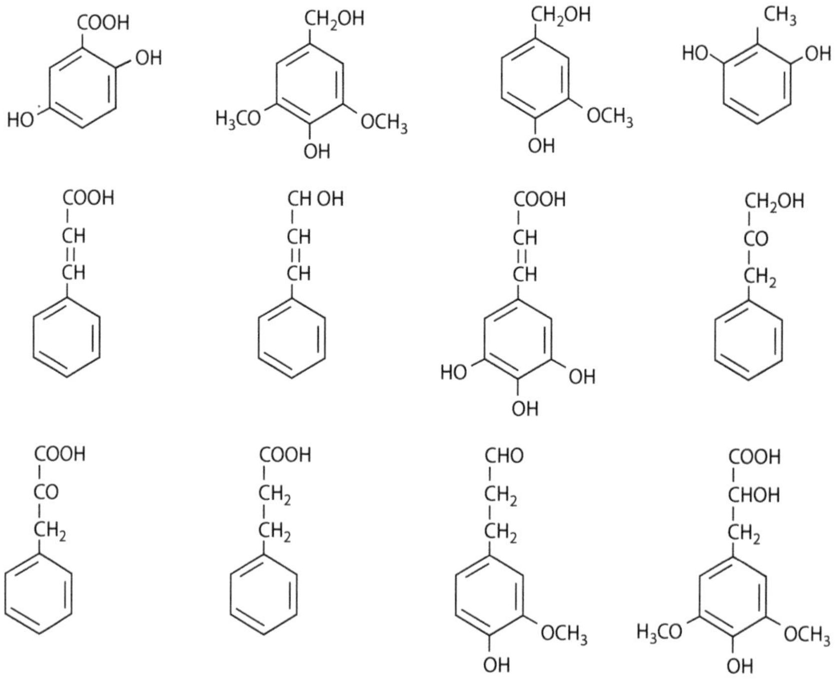

FIGURE 12.4 Products from the oxidation of coal with performic acid. (From Raj, S., Ph.D. Thesis, Pennsylvania State University, PA, 1976.)

to accomplish product isolation (Duty et al., 1980). Therefore, careful choice of the correct solvent or solvent blends is a major factor in isolation of the organic materials produced by the oxidation of coal.

In addition to the correct choice of solvent for separation of the oxidation products, it must also be recognized that humic acids are also produced by the oxidation of coal. Humic acids are complex molecular entities that are difficult to characterize but are considered to be an extremely heterogeneous mixture of moderately high-molecular-weight polymers that contain acidic (carboxylic and phenolic) functions (Stevenson and Butler, 1969). Humic acids have a low solubility in organic solvents and, therefore, may be difficult to separate from any coal that has not been oxidatively degraded to the lower-molecular-weight organic products. However, humic acids are soluble in aqueous alkaline solutions and should be isolated by this means (Stevenson and Butler, 1969).

Once formed, these acids then degrade to lower-molecular-weight species such as the more easily defined aliphatic and aromatic carboxylic acids. The final step is the formation of the carbon dioxide from these acids. Acids containing polycondensed ring systems are not often detected among the products of the oxidation of coal but there are claims that the mixed aromatic acids from the oxidation of coal with performic acid also contain a variety of phenols and alkyl-substituted phenols (Figure 12.4).

Although not usually recognized as a means of oxidation, the treatment of coal with concentrated sulfuric acid also gives rise to the incorporation of oxygen functions (phenolic hydroxyl, $-OH$, and carboxylic groups, $-CO_2H$) into the coal (Iyengar et al., 1959). Some incorporation of sulfonic acid functions ($-SO_3H$) occurs but usually to a lesser extent than the incorporation of the phenolic functions.

12.4 BACTERIAL OXIDATION OF COAL

The problem of stream pollution caused by acid mine drainage has prompted several studies (Leathen, 1952; Leathen et al., 1953a,b; Braley, 1954; Temple and Koehler, 1954; Ashmead, 1955; Moulton, 1957; Beck, 1960; Brant and Moulton, 1960; Hoffman et al., 1981; Atkins and

Pooley, 1982; Atkins and Singh, 1982; Wang et al., 2007). These investigators agree that the ultimate source of the sulfuric acid and iron pollutants is the pyritic material associated with coal and coal-bearing strata.

Oxidation of these iron disulfides results in the production of ferrous sulfate and sulfuric acid. Subsequent oxidation of ferrous sulfate yields additional sulfuric acid and hydrated oxides of iron. The latter forms the unsightly yellow-to-red muds characteristic of streams receiving acid mine drainage.

Iron-oxidizing and sulfur-oxidizing chemoautotrophs have been isolated repeatedly from acid waters, both in United States (Colmer and Hinkle, 1947; Leathen and Madison, 1949; Colmer et al., 1950; Bryner et al., 1954; Brant and Moulton, 1960) and in Europe (Ashmead, 1955; Kullerud and Yoder, 1959; Zarubina et al., 1959), and have been shown to accelerate the rate of oxidation of pyrite and other sulfide minerals. These workers measured pyrite oxidation by titrating for increased acidity, or by determining the release of soluble iron or sulfate S from the insoluble minerals.

Current interest in the problem of air pollution by sulfur dioxide from the combustion of sulfur-containing fuels and continued concern with the acid mine water problem have stimulated the study of the role of bacteria in the oxidation of the pyritic constituents of coal. Consequently, studies of the iron-oxidizing chemolithoautotroph *Ferrobacillus ferrooxidans* as an agent for the desulfurization of coal were started; supporting physiological studies of the organism were required to determine the bacteriological, chemical, and physical factors relating to its ability to oxidize the pyrites in coal.

The use of iron-oxidizing autotrophs in the removal of pyrite from coal (Ashmead, 1955; Zarubina et al., 1959) and in the secondary recovery of copper and molybdenum from their sulfide minerals has already been reported (Bryner et al., 1954; Bryner and Anderson, 1957; Bryner and Jameson, 1958).

Generally, (1) the rate of oxidation of pyrite increases with the exposed, that is, available, pyrite surface; the coal must be in a finely divided state to expose a maximum of the embedded pyrite. (2) *F. ferrooxidans* can accelerate the oxidation of diverse pyritic materials and coarsely crystalline marcasite, but not the oxidation of coarsely crystalline pyrite. (3) *Thiobacillus thiooxidans* does not enhance the oxidation of the experimental materials, with the possible exception of marcasite; this organism appears to play no role in pyrite oxidation. (4) Lattice imperfections or the presence of some impurity in the mineral affects the ability of the test organisms to accelerate oxidation—coarsely crystalline pyrite proved resistant to oxidation, whereas coarsely crystalline marcasite and pyrite concretions and concentrates from coal were oxidizable (Silverman et al., 1961).

Bioprocessing of coal is emerging as a new technology for coal cleaning and coal conversion processes (Attia, 1985; Dugan, 1986; Ehrlich and Holmes, 1986; Olson et al., 1986; Couch, 1987; Markuszewski and Wheelock, 1990). Coal conversion under milder conditions continues to offer the potential for improved liquefaction of coal. The overall success of liquefaction may entail improved coal pretreatment, disposable catalysts, upgrading of coal liquids in terms of removal of nitrogen and organic sulfur, and novel hydrogen generation and utilization techniques. The development of these techniques as applied to the aforementioned objectives offers an alternative approach if appropriate microbial cultures and environmental conditions can be established.

Under appropriate conditions, over 90% of the pyritic sulfur from coals can be removed by the mesophilic sulfur-oxidizing autotrophic bacteria *Thiobacillus ferroxidans* and *T. thiooxidans*, but these bacteria were incapable of removing organic sulfur (Khalid et al., 1989, 1990a,b; Bhattacharyya et al., 1990; Huffman et al., 1990). However, the thermophilic archaebacterium *Sulfolobus brierlayi* was able to remove over 95% of the pyritic sulfur and over 30% of the organic sulfur from the untreated coal when the cells of *S. brierleyi* were acclimatized. The aerobic biosolubilization of low-rank coal to polar water-soluble products has been demonstrated (Scott et al., 1986; Cohen et al., 1987; Pyne et al., 1987; Wilson et al., 1987).

12.5   HYDROGENATION

In the historical sense, the hydrogenation of coal is not new and has been known for at least 125 years since coal was first converted to liquid products by the reaction with hydrogen iodide in sealed glass tubes at 270°C (520°F) (Berthelot, 1868, 1869). In addition, the hydrogenation of coal under high temperature and high hydrogen pressure led to a process for commercial production of liquid fuels from coal by direct hydrogenation (for example, see Bergius, 1912, 1913, 1925), which culminated in the production of approximately 100,000 bbl/day (15.9 × 10⁶ L/day) of liquid fuels for the German war effort in the early to mid-1940s. Thus, the hydrogenation of coal was seen, from that time, as a means of producing liquid fuels (Chapters 18 and 19), depending upon the need and the various political influences that, on occasion, have a habit of affecting the modern world (Chapter 26).

The relationship between coal composition and hydrogenation reactivity has been studied extensively and coal with less than 85% w/w carbon (daf) made poor liquefaction feedstocks. It has been suggested that coal reactivity is related to rank (Francis, 1961). Coal is, in actual fact, a hydrogen-deficient organic natural product having an atomic hydrogen/carbon ratio of approximately 0.8 compared with an atomic hydrogen/carbon atomic ratio of 1.4:1.8 for various crude oils, heavy oil, and bitumen (Table 12.1).

So it is not surprising that the interest in the hydrogenation of coal has focused, and continues to focus, on the production of liquids as suitable alternates to petroleum feedstocks (Krichko and Gagarin, 1990) (Chapters 18, 19, and 26). Chemical hydrogenation has seen some use in structural interpretations but (unless there is some means by which it can be related to commercial interests) been largely ignored in terms of coal utilization.

From the low hydrogen/carbon atomic ratio, it is obvious that coal requires the addition of some hydrogen as a means of adequate conversion to liquid products. This concept does not ignore the formation of liquids from coal without the addition of external hydrogen since part of the coal organic matrix is capable of acting as a hydrogen donor but with the simultaneous formation (usually in substantial quantities) of hydrogen-deficient coke (H/C < 0.5).

TABLE 12.1
Analysis and Atomic H/C Ratios of Selected Fuels

Fuel	C	H	O	N	S	H/C[a]
Coals						
Low-volatile bituminous	90.5	4.7	2.8	1.3	0.7	—
Medium-volatile bituminous	88.4	5.0	4.1	1.7	0.8	0.68
High-volatile bituminous	76.4	5.3	15.8	1.6	0.9	0.83
Subbituminous A	75.8	5.3	15.5	1.9	1.5	0.84
Subbituminous B	75.3	5.1	17.4	1.5	0.7	0.81
Subbituminous C	73.7	5.3	19.1	1.3	0.6	0.86
Lignite	70.2	5.6	20.8	1.4	2.0	0.96
Wood	49.9	6.0	43.9	0.2	—	1.44
Bitumen						
Cold lake	83.7	10.4	1.0	0.2	4.7	1.49
Athabasca	83.1	10.3	1.6	0.4	4.6	1.49
Crude oil	85.8	13.0	—	0.2	1.0	1.82
Gasoline	86.0	14.0	—	—	—	1.95
Natural gas	75.8	24.2	—	—	—	3.83

[a]   Atomic ratio.

It must also be recognized that the hydrogenation of coal is not only a means of saturating hydrogen-deficient (e.g., aromatic) products but also a means of terminating the copious side reactions that can occur, ultimately leading to the formation of coke. The hydrogen is thus effective in controlling the molecular weight of the species that make up the liquid products. The hydrogen also acts as a scavenger insofar as it is effective in achieving removal of nitrogen, oxygen, and sulfur from the coal as the hydrogenated analogs (i.e., ammonia, NH_3, water, H_2O, and hydrogen sulfide, H_2S, respectively).

At this point, it is worthy of note that the formation of water by the interaction of hydrogen with the oxygen species in the coal is considered by many researchers to be wasteful in terms of hydrogen consumption and, therefore, alternate means for the pretreatment of coal to remove all (or part) of the oxygen prior to hydrogenation to liquids (Chapters 18 and 19) are always worthy of consideration. And yet, there may be benefits to such treatment.

For example, there are observations that noncaking coals acquire coking properties when heated with hydrogen at various temperatures just below that of active decomposition (usually 350°C [660°F]) (Crawford et al., 1931). It was concluded that the loss of oxygen and other functional groups, as well as the incorporation of hydrogen into the coal, was responsible for the acquired coking properties (Kukharenko, 1950; Kukharenko and Matveeva, 1950; Ahuja et al., 1959). Dehydrogenation by chemical means appears to reverse this effect as well as completely inhibits tar formation (Mazumdar et al., 1959).

Just as the complete oxidation of coal can be represented by conversion to carbon dioxide

$$C + O_2 = CO_2$$

the complete hydrogenation of coal can (in theory) be represented by conversion to methane:

$$C + 2H_2 \rightarrow CH_4$$

However, in the present context, the hydrogenation of coal is not considered to be the complete conversion of coal to methane by gasification (Chapters 20 and 21) but any means by which hydrogenation affects the raising of the natural atomic hydrogen/carbon ratio producing either partially hydrogenated products (such as tar) or even the more volatile liquid products.

Thus, unlike the oxidation of coal, which can occur with some degree of spontaneity and even lead to ignition at ambient temperatures, the hydrogenation of coal is relatively difficult (Weiler, 1945a,b) and the introduction of sufficient hydrogen to transform the coal into liquid hydrocarbons can only be achieved with more than a moderate degree of success at temperatures in excess of 350°C (660°F) when the coal is undergoing decomposition (Chapters 13 and 14). At temperatures below 350°C (<660°F), the hydrogenation reaction does not usually proceed beyond the point where the coal is reduced to a product that exhibits a marked increase in pyridine solubility.

Thus, the type of conversion that can be attained without substantial decomposition of the coal can be illustrated by the reaction of coal with lithium in a solvent such as ethylenediamine at ca. 100°C (212°F) (Given, 1984). Hence (aside from complications that include the incorporation of solvent into the product), up to 55 additional hydrogen atoms per 100 carbon atoms can be introduced into coal in this manner. For example, vitrains are capable of incorporating an additional 20 hydrogen atoms per 100 carbon atoms (at 91% C) to yield predominantly solid reaction products.

As with the oxidation of coal, the product characteristics and product disposition depend markedly on coal rank and reaction conditions. For example, hydrogenation products range from the highly carbonaceous cokes, chars, tars, and heavy oils to lighter naphtha fractions and even methane. Thus, at relatively mild conditions, asphalts are usually the predominant products while as the severity of the reaction conditions is increased, liquids boiling in the same range as petroleum naphtha

(30°C–200°C [85°F–390°F]) are produced. Under extremely severe conditions, methane becomes the predominant product and the operation is often referred to as hydrogasification (Chapters 20 and 21).

Thus, with coal utilization in mind, the hydrogenation of coal can be generally divided into two categories: (1) the liquid-phase hydrogenation of coal and (2) the vapor-phase hydrogenation of coal.

12.5.1 LIQUID-PHASE HYDROGENATION

The liquid-phase hydrogenation of coal has been a topic of interest since the early 1900s (for example, see Bergius, 1912, 1913, 1925), leading to one of the first commercial developments for the production of liquid fuels from coal. Of late, research efforts have been directed toward cheaper sources of hydrogen, better catalysts, less severe operating conditions, and simpler operating procedures. There has even been some emphasis on the elimination of the carrier liquid and more facile separation of the products (Chapters 18 and 19).

The liquid-phase hydrogenation of coal can occur at 370°C–540°C (700°F–1000°F) and at pressures of 1500–7500 psi. The catalyst composition can also vary (Table 12.2) and the liquid vehicle may be a coal tar fraction or other coal product—the product type and the product distribution depend largely on the operating conditions but yields (conversion) of the order of 70%–95% have been reported.

An increase in the process temperature usually leads to an increase in the yield of naphtha and gases and may even reduce the yield of middle distillates but only rarely appears to reduce the yield of the heavier products. The residence (or reaction) time may be of the order of 2 h and an increase in the residence time will not only increase the yield of gas, naphtha, and middle distillate but also serve to decrease the yield of the heavier oils. An increase in the pressure of the hydrogen (say, from 1200 to 1800 psi) can markedly affect the overall conversion and may even (under certain prime conditions) double the conversion yield; there may, however, be other factors, such as the product distribution, which make the increased conversion unfavorable and, therefore, not a true benefit to the process.

The chemistry of the liquid-phase hydrogenation of coal has been the focus of considerable attention and it has been suggested that hydroaromatic compounds are formed in the early stages of the process by interaction of the hydrogen and constituents of the liquid vehicle. These hydroaromatic compounds then furnish hydrogen to the hydrogen-deficient entities that are produced during the decomposition of the coal (Figure 12.5) and thus, like many other reactions of coal, may only proceed with any degree of efficiency at temperatures in excess of 350°C (660°F). Complex free-radical sequences have been postulated to be involved in which the coal fragments become stabilized by abstraction of hydrogen from the donor vehicle without the necessary requirement of catalysts or high pressures (for example, see Krichko and Gagarin, 1990).

Kinetic studies using tetrahydronaphthalene (tetralin) as the hydrogen donor have shown that sufficient hydrogen is available for liquefaction of coal when the proportions of vehicle coal are of the order of 2:1 and that the conversion is quite rapid at temperatures on the order of 400°C (750°F). Thus, under these particular conditions (Figures 12.6 and 12.7), coal becomes almost 100% soluble in pyridine within 10 min without any massive increase in hydrogen content of the product relative to the original coal.

TABLE 12.2
Catalysts for Liquid-Phase Hydrogenation of Coal

Catalyst	Comments
Tin oxalate + chlorides	Superior to iron; expensive and corrosive
Ammonium molybdate	Excellent catalyst yielding low-ash oil; expensive without recovery
Zinc chloride	Produces more aliphatic product; high proportions required
Co–Mo ("moly") on silica alumina	Promotes hydrodesulfurization with liquefaction

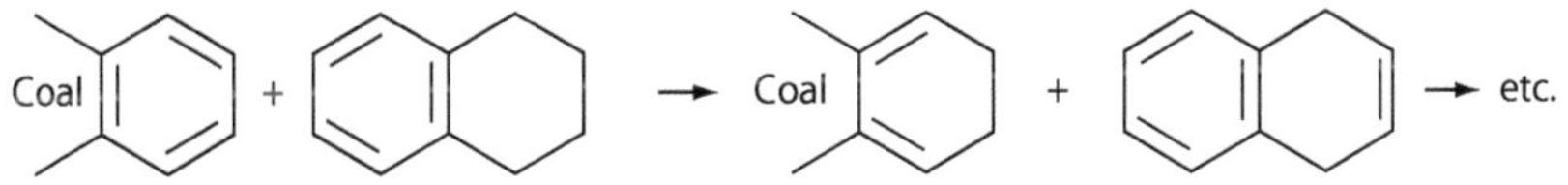

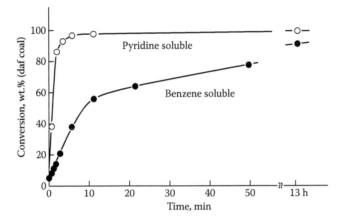

FIGURE 12.5 Hydrogen transfer reactions.

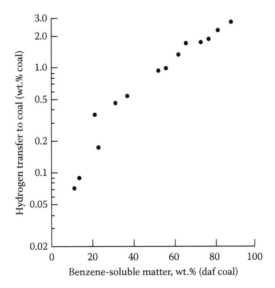

FIGURE 12.6 Variation of conversion of high volatile C bituminous coal with time. (From Neavel, R.C., *Fuel*, 55, 237, 1976.)

FIGURE 12.7 Variation of conversion of hvC bituminous coal with hydrogen transfer. (From Neavel, R.C., *Fuel*, 55, 237, 1976.)

Coals up to and including high-volatile C bituminous coals generally react at similar rates and usually disintegrate in the donor vehicle at the reaction temperature thereby minimizing any requirements for prior crushing of the coal. However, it should be remembered that coal is composed of different macerals (Chapter 4) and that maceral content may exert an unexpected influence on process efficiency. For example, while vitrinites and exinites may react completely, fusinites are generally inert and may, therefore, appear in the residue with the mineral matter.

Thus, coal composition in terms of the maceral content is a very important addendum to the generalized corollary that rank plays a significant role in coal conversion (Chapters 18 and 19).

12.5.2 Vapor-Phase Hydrogenation

The vapor-phase hydrogenation of coal is actually (as intended here) the direct reaction between hydrogen and coal without a vehicle. The reaction may require higher temperatures or may proceed at lower pressures than those employed for the liquid-phase hydrogenation of coal.

Obviously, the nature of the reaction under these conditions will differ from the reaction in the presence of a carrier insofar as there is a direct reaction between the hydrogen and the coal. In addition, vapor-phase (or secondary) hydrogenation may also follow the primary hydrogenation in which volatile products from the decomposition of the coal or from the reaction of the coal with hydrogen then react with more hydrogen to modify the slate of primary reaction products.

Commercially, the vapor-phase hydrogenation of coal finds application as the hydrogasification of coal (Chapters 20 and 21). The reactor temperature may be as high as 980°C (1800°F) with reactor pressures falling into the range 500–10,000 psi. As with the liquid-phase hydrogenation of coal, the process parameters are selected on the basis of product type as well as the product distribution that is desired. In addition, coal type and catalyst also play major roles in defining the process parameters. A preliminary hydrogenation for supplementary stage may also be employed and actually serves to hydrogenate unsaturated material as well as those oxygen species that could markedly decrease the life of the catalyst.

Lower-rank coals tend to react more readily than higher-rank coals or coke. While methane may result more readily from the direct reaction of coal carbon and hydrogen, it may also result from the thermal decomposition and hydrogenation (hydrocracking) of higher-molecular-weight compounds formed in the earlier stages of hydrogenation. The overall rate of methane formation can be regarded as first order with respect to hydrogen partial pressure.

In hydrogasification reactions, the conversion to methane is favored by high pressures and by temperatures at the lower end of the range. The reaction is rapid at temperatures above 480°C (900°F) and while the original coal is usually more reactive than the char produced, it may appear paradoxical that the char can often be converted to methane more easily than coal. There are also reports that sodium carbonate can increase the rate of the reaction of coal with hydrogen in the range 815°C–900°C (1500°F–1650°F). However, for those catalysts that are added to the reaction mix as a powder, tin(II) chloride has been noted as having exceptionally beneficial effects.

On the other hand, for catalysts that are added to the reaction mix by means of solution impregnation of the coal, ammonium molybdate, tin(II) chloride, nickel(II) chloride, and iron(II) sulfate have been noted to be beneficial.

12.5.3 Influence of Petrographic Composition

There are reports in which coal rank has been correlated with the extent of the catalytic liquefaction under batch conditions. Thus, while several of the correlations are questionable, there does appear to be a maximum with high-volatile C bituminous coal. However, the ability of coal to undergo liquefaction appears to correlate much more closely with maceral content than with coal rank. For example, vitrinite (Chapter 4) has been found to play a significant role in the liquefaction of coal and an almost linear relationship has been reported to exist between the percent conversion to liquid products and the vitrinite content.

On the other hand, fusinite offers some resistance to the low-temperature hydrogenation process, which may not be surprising considering the nature of this particular maceral (Chapter 4). And, microscopic examination of residual material (often referred to incorrectly as *unconverted material*) from coal liquefaction has indicated structures characteristic of fusinite thereby offering further evidence for the comparatively low reactivity of this maceral.

Other macerals, sporinite, resinite, and cutinite (Chapter 4), are all readily liquefied by hydrogenation while there is also evidence to indicate that the maceral micrinite may be more easily liquefied than has previously been thought possible, but some further substantiation is necessary.

12.5.4 DESULFURIZATION DURING HYDROGENATION

While the primary goal of coal hydrogenation can be seen as conversion of the coal to liquid products, the conversion of the coal to low-sulfur liquid products can also be achieved. Indeed, these two processes usually occur simultaneously but the extent to which each process occurs will vary with the catalyst and with the process parameters.

For example, molybdenum-containing catalysts are quite effective for coal desulfurization as well as for coal hydrogenation. The use of such catalysts has been based largely on the experience gained in the petroleum industry (Speight, 2000, 2007) and the most commonly used catalyst is cobalt molybdate, which is a mixture of cobalt oxide (CoO) and molybdenum oxide (MoO$_3$) on an alumina (Al$_2$O$_3$) support. The catalyst may be in the form of spheres or cylinders with dimensions in the range 2–5 mm; surface areas of these catalysts are often on the order of several hundred m^2/g with pore volumes on the order of 0.5 cm^3/g.

The activity of the catalyst depends on maintenance of the open-pore structure, and it is therefore essential to minimize the deposition of coke on the catalyst. This may be achieved by pretreating the catalyst with a sulfur-containing species that minimizes the side reactions that ultimately lead to coke formation; the treatment actually yields a surface that (in a simplified form) (Figure 12.8) appears as molybdenum atoms, sulfur atoms, and spaces for sulfur atoms.

A number of problems occur with catalysts used for petroleum desulfurization, and these are likely to be worse with catalysts used with coal. For example, the catalyst particles may become coated with iron sulfide (presumable from the indigenous pyrite). This deposit, along with any other deposits, decreases the efficiency of the catalyst and could eventually lead to plugging of the catalyst bed. In fact, microscopic examinations of various catalysts have shown that deposits containing iron, titanium, and silicon are present within the catalyst pore system.

The temperature stability of the desulfurization catalyst is also a major process consideration. Thus, hydrodesulfurization reactions are usually exothermic and can produce more than enough heat to raise the temperature of the catalyst bed by as much as 85°C (185°F) and, although cold hydrogen may be added at various stages to compensate for such a temperature increase, the so-called hot spots can (and often do) occur within the reactor system. Such an almost uncontrollable phenomenon can result in "runaway" reactions thereby leading to even higher temperatures as well as spreading of the hot spot; such a spot near a reactor wall can often lead to wall failure and even to explosion.

Steric (structural) factors are always important in hydrodesulfurization reactions, particularly where a catalyst is employed and where the intimate molecule–catalyst contact is necessary to achieve the desired degree of success. For example, dibenzothiophene undergoes hydrodesulfurization some 100 times faster than a derivative that contains two methyl groups attached in such a position as to make sulfur–catalyst contact difficult.

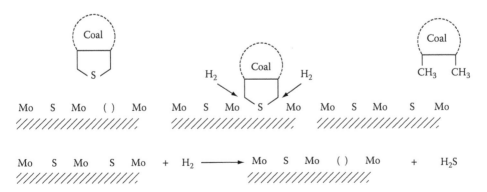

FIGURE 12.8 Interaction of coal-bound organic sulfur with a catalyst.

Finally, it should be noted that hydrodesulfurization does not occur alone; the sulfur removal reaction is inevitably accompanied by hydrocracking and hydrogenation reactions that effectively cause degradation of the coal into lower-molecular-weight fragments. At the same time, oxygen (removed as water) is eliminated but this particular reaction is considered to be wasteful because of the overall increased hydrogen consumption of the process. Hydrodenitrogenation removes pollutant nitrogen components from the coal and, in spite of the enriched hydrogen atmosphere, coke deposition on the catalyst is virtually impossible to eliminate and leads to additional catalyst fouling.

12.6 HALOGENATION

Coal reacts quite readily with the halogens and the ease with which coal can be halogenated was first demonstrated (Bevan and Cross, 1881) when it was shown that cannel coal (Chapter 4) afforded, in toto, alkali-soluble products when shaken with "chlorine water"; it was also noted that bituminous coal could also be rendered soluble by treatment with hydrochloric acid and potassium chlorate.

Subsequent investigations (Weiler, 1945a,b; Wender et al., 1981) have shown that chlorination can also be achieved by either exposing pulverized coal to chlorine or by passing chlorine gas through a suspension of coal in water or in carbon tetrachloride or even by reacting coal with chlorine dioxide in a pressure vessel (Pinchin, 1958; Bhowmik et al., 1959; Honda et al., 1959). In some cases, depending upon the procedure, the products from the chlorination of lower-rank coals are alkali soluble.

The amount of chlorine that appears in the products arising from any one of these methods may vary from a high of ca. 40% w/w for a lignite to ca. 25% w/w for a bituminous coal. There may also be an increase in the reaction temperature. However, a substantial portion (ca. 50%) of the chlorine "introduced" into the coal can be removed by heating the chlorinated material under reflux with aqueous sodium hydroxide or by extraction of the material with concentrated ammonia or with aqueous sodium carbonate; almost all of the chlorine is removed from the chlorinated material by heating to 550°C (1020°F).

Halogenation of dry coal with gaseous chlorine is an exothermic reaction and is accompanied by the evolution of hydrogen chloride; pyrolysis of the chlorinated material yields little, if any, tar. To explain these observations, it has been suggested that the chlorine is introduced into the coal by means of simultaneous addition and substitution reactions (Figure 12.9).

However, chlorination chemistry (as it applies to coal) is not that simple (Gonzalez de Andres et al., 1990a) and there are indications that the properties of the chlorinated products depend not only on the extent of the chlorination but also on the coal type as well as on the reagent. There is also the suggestion (as mentioned earlier) that the differences between coals are reflected in a higher uptake of the chlorine with a lignite than with the higher-rank coals.

In attempts to chlorinate different macerals, the products indicate that exinite has a greater proportion of disordered structure than fusinite and the number of condensed aromatic rings

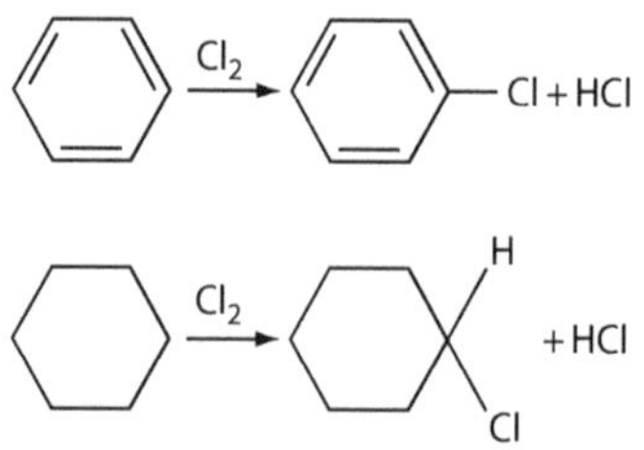

FIGURE 12.9 Chlorination (substitution) reactions.

is not more than two. In addition, sulfur is more firmly bound and the number of peripheral groups is greater. On the other hand, fusinite has more condensed structures, the number of hydroaromatic structures is negligible, and the number of condensed rings present is more than two (Roy, 1965).

The reactions of coal with bromine, iodine, and fluorine have also received some attention but not to the same extent as the reaction with chlorine (Weiler, 1945a,b; Dryden, 1963; Wender et al., 1981). Reactions of lignite with sulfur tetrafluoride are reported to be a reliable means of introducing fluorine into specific sites of the coal (Hagaman and Lee, 1992).

Thus, bromination (which can be achieved by shaking powdered coal with bromine water in carbon tetrachloride or in chloroform) is also believed to involve addition and substitution reactions; the products can contain up to 20% w/w bromine. A similar concept has been invoked for the interaction of iodine with coal but the reaction of fluorine with coal will ultimately yield a mixture of fluorocarbon oils.

Mixed halogens also react with coal; for example, chlorine trifluoride will react with coal at 205°C–400°C (400°F–750°F) to yield chlorofluorocarbon oils and gases. The former products (i.e., the oils) may contain the majority (ca. 60%) of the halogen reacted.

12.7 ALKYLATION

Coal is a complex macromolecule of predominantly aromatic character that lends itself to alkylation reactions in which an alkyl group is substituted into an aromatic ring for a hydrogen atom. Indeed, the alkylation of coal has received considerable attention as a means of increasing the solubility of coal in various organic solvents thereby enhancing the reactivity to liquefaction (Larsen and Kuemmerle, 1976; Davidson, 1982; Given, 1984; Stock, 1987; Baldwin et al., 1991; Sharma and Mishra, 1992).

For example, the reaction of coal with an alkyl chloride in the presence of aluminum chloride (in a solvent such as carbon disulfide) can increase the solubility of the coal from ca. 11% pyridine-soluble material in the original coal to as much as 39% pyridine-soluble material (Schlosberg et al., 1978). Alternatively, the reaction of coal with an alkene (cetene) or with alcohols (dodecanol/cetanol) in the presence of aluminum chloride renders more than 50% of the coal soluble in quinoline (Sharma and Mishra, 1992).

In fact, the reaction of coal under Friedel–Crafts conditions is not a simple alkylation but involves a series of complex reactions. The molecular dynamics of coal molecules under such conditions and the mineral matter in coal may interfere with or catalyze these reactions. Extraction can be used as a measure of the effectiveness of reactions such as degradation, disintegration, and transalkylation of coal. In fact, the reaction of coal with alkylating agents under Friedel–Crafts conditions may be a complicated series reaction involving alkylation–transalkylation–dehydrogenation–degradation–dissociation (Sharma and Mishra, 1992).

Other methods of alkylation include the use of alcohols (as the alkylating agents) with hydrogen fluoride (Flores et al., 1978) and the zinc chloride–alcohol system (Berlozecki and Bimer, 1991), and the use of an alkyl iodide in the presence of cadmium (Yoneyyama et al., 1992).

However, the reactions are much more complex than the "simple" alkylation would indicate; the reaction product may be much more aromatic than the original coal. Such a phenomenon could result as a simultaneous condensation that is catalyzed by the aluminum chloride (Figure 12.10) or even by condensation and subsequent aromatization of indigenous alkyl side chains. In fact, it is in this connection relevant to note that aluminum chloride has also been found to accelerate the formation of graphite from aromatic compounds.

The concept of coal alkylation has been expanded to a variety of studies relating to coal structure (Sternberg et al., 1971; Sternberg and Delle Donne, 1974; Wachowska, 1979; Alemany et al., 1981; Wender et al., 1981; Handy and Stock, 1982). This is, essentially, a method of "reductive alkylation," which, as a first step, necessitates reacting the coal with metallic potassium and naphthalene in

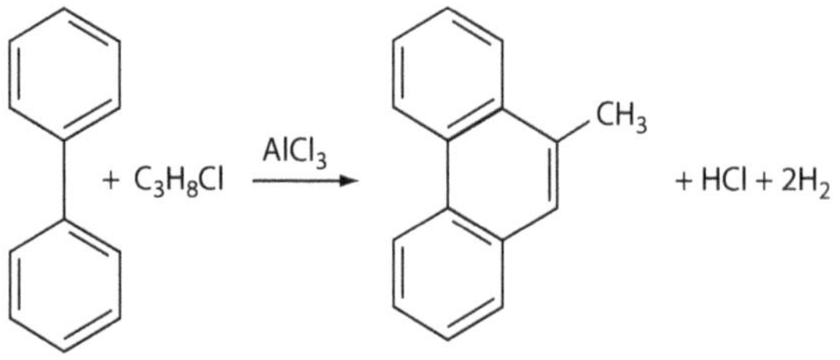

FIGURE 12.10 Friedel–Crafts cycloalkylation with aromatization.

tetrahydrofuran (THF). Through an electron transfer process, a coal "anion" is produced, which can then be alkylated by means of an alkyl halide:

$$\text{Coal} + n\text{K (in THF/naphthalene)} \rightarrow \text{coal}^{n-} \; n\text{K}^+$$

$$\text{Coal} + n\text{C}_2\text{H}_5\text{I} \rightarrow \text{coal}(\text{C}_2\text{H}_5)_n + n\text{I}^-$$

The concentration of coal anions can be determined by reacting the sample with water to form hydroxide ion

$$\text{Coal}^{n-} + n\text{H}_2\text{O} \rightarrow \text{coal}(n\text{H}) + n\text{OH}^-$$

and subsequent titration of the reaction product against acid. As originally conceived, the reductive alkylation of coal was presumed to involve simple and straightforward chemical reactions with only minimal side reactions occurring. However, many subsequent workers have raised serious questions as to the authenticity of the initially proposed mechanism of the reaction. The complex side reactions that occur throw some doubts upon the precise nature of the products as well as on the mechanism of the reaction (Ebert et al., 1981).

For example, it has been reported that aromatic hydrocarbons can cause the cleavage of THF in the presence of a metal complex and then generate polymeric substances that embody some of the cleaved residues. In addition, the gaseous by-products of the alkylation contain traces of butane (from the coal or from the solvent) as well as significant amounts of elemental hydrogen. And, THF is capable of participating in the reaction (as has been shown by parallel studies on petroleum asphaltene fractions) (Speight and Moschopedis, 1980), thereby adding more complications to the reaction chemistry and product distribution.

Obviously, there are many unexplored facets of the reductive alkylation of coal. Although there have been many attempts to elucidate the mechanism of what was first thought to be a relatively simple reaction, this has not yet been achieved to any degree of completeness and thus remains a very complex and little understood aspect of coal chemistry.

The technique of selective methylation whereby an oxygen-methylated bituminous coal was treated with a series of carbanion bases followed by quenching with labeled methyl iodide has also been employed as a means of determining/estimating the fluorene-type structural units in coal (Chambers et al., 1988). The data suggest that representations of coal structure should include the five-membered cyclopentadiene ring systems that are common to all fluorene derivatives.

Transalkylation has been used as a means of identifying the alkyl substituents on the aromatic rings of coal; although normal and branched alkyl chains having up to six carbon atoms per chain were identified, no alkyl chain having more than six carbon atoms was observed in the products. And, the solubility of the coal was not increased by the transalkylation. This is to be expected since, in a similar manner, a decrease in the number of alkyl groups (per molecule) has also been found to decrease the solubility of petroleum asphaltene constituents and is believed to be followed by phase separation as a prelude to coking in thermal processing (Speight, 2007).

Whilst not precisely the same classification as coal alkylation, the reactions of coal with dienophiles have been examined in some detail (Larsen and Lee, 1983; Zher'akova and Kochkan'an, 1990). The reactions are diffusion controlled and appear to be rank dependent, proceeding with the formation of diene synthesis products.

12.8 DEPOLYMERIZATION

In the true sense, coal is *not* a polymer and therefore the word *depolymerization* must be used with caution. As used in the current context, depolymerization refers to the conversion of the coal macromolecule to lower-molecular-weight products.

On the molecular level, coal is an extremely complex substance and in spite of extensive research in this area the exact structure of coal has yet to be determined (Chapter 10). Different authors have proposed various coal models from time to time. Broadly it is agreed that coal consists of heterogeneous polyaromatic clusters in a complex array resulting in a highly cross-linked macromolecular gel structure. Another opinion that coal has a highly associated structure also exists.

On the macrolevel too, coal is extremely heterogeneous consisting of various minerals and maceral components whose composition varies from coal to coal, which increases the complexities in the study on aerial oxidation of coal.

The concept of coal being an organic polymer received little attention in the early days of coal science but has, of late (for reasons that often defy justification), been the subject of several investigations. However, there is also the need to adequately, or appropriately, define *polymerization* as it is applied to coal chemistry.

For example, a polymer that is produced from *a lot of monomers* gives the monomer once again by depolymerization. The production of low-molecular-weight species by the reaction of coal with a *depolymerizing* agent does not mean that coal has been depolymerized, or that it even was a polymer. The misused term means that coal has been converted to lower-molecular-weight products and these products may or may not be representative of the structural entities within the coal.

Recalling the potential precursors that could be included in the original sour mash (*proto-coal*) that eventually became coal (Chapter 3), it means that coal has been converted to various low-molecular-weight, soluble (often, but not guaranteed) chemical species. And to consider coal a polymer of these materials may not be chemically correct. Nevertheless, mental meanderings and thoughtful speculation about the definition of *depolymerization* aside, valuable data have been produced from such work, thereby providing an enhanced understanding of the nature of coal, but not the *average structure* of coal.

Thus, the *depolymerization of coal* (Heredy et al., 1965; Heredy, 1981; Wender et al., 1981; Given, 1984) by means of a phenol/boron trifluoride mix is reputed to proceed by the cleavage of aryl–alkyl–aryl systems (Figure 12.11). On this basis, the extent of the polymeric nature of coal is proportional to the number of methylene bridges in the coal, which can be determined from the concentration of di(hydroxyphenyl)methane in the reaction products.

By means of this technique, it is possible to produce phenol-soluble material amounting to as much as 75% w/w of the starting coal (lignite; 70% carbon); on the other hand, a low-volatile bituminous coal (89% carbon) may only afford as little as 9% w/w phenol-soluble material. The molecular weights of the phenol-soluble products usually fall into the range 350–1050. Coal depolymerization is also reputed to occur when the coal is treated with an iron-based catalyst in the presence of hydrogen (Wang et al., 1992).

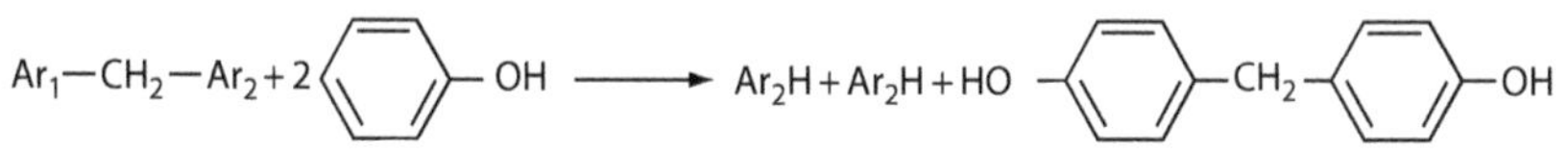

FIGURE 12.11   So-called depolymerization of coal.

12.9 HYDROLYSIS

The hydrolysis of coal with alkali produces low yields of alkali-soluble products that are usually phenolic or acidic in nature (Howard, 1945b; Wender et al., 1981). The lower-rank coals are more reactive to hydrolysis but as the rank (or age) of the coal increases the solubility is often noted to decrease.

Fused alkali at 650°C (1200°F) converts lignite or bituminous coals to brown, alkali-soluble products. On the other hand, aqueous potassium hydroxide at 280°C (535°F) will convert lignite to alkali-soluble products but the higher-rank coals are somewhat more stable to this treatment. It has also been reported that an alkali with an organic solvent may even be better for the degradation. For example, a mixture of 10% aqueous potassium hydroxide plus the monoethyl ether of ethylene glycol has been employed to "hydrolyze" coal and the extraction of peat or lignite with water–benzene mixtures enhances the recovery of bitumen-like products.

The use of strong caustic also provides the potential for coal cleaning (chemical demineralization of coal) (Chapter 6) leaving the organic portion of the coal relatively untouched, other than the dissolution of humic material (Bowling et al., 1990). The addition of sulfur appears to enhance the potential of the coal for dissolution thereby offering another means of cleaning through chemical reaction (Dorland et al., 1992) with the process appearing to be generally rank dependent. On the other hand, molten caustic is effective in the desulfurization of coal (Kusakabe et al., 1989; Balcioglu and Yilmaz, 1990; Majchrowicz et al., 1991) thereby offering an additional route to clean coal by the use of caustic (Schmidt, 1987; Chatterjee and Stock, 1991) (Chapter 6).

Other reagents such as hydrogen peroxide, oxidation in alkaline solution, a chlorinolysis/hydrothermal treatment, copper sulfate are also claimed to be effective for the desulfurization, and in some cases demineralization, of coal (Chuang et al., 1983; Ahnonkitpanit and Prasassarakich, 1989; Kin et al., 1990; Ali et al., 1992; Prasassarakich and Pecharanond, 1992). In addition, the hydrothermal treatment of coal at 350°C (660°F) is reputed to affect the cleavage of the ether moieties as well as produce a coal product that is free-flowing and non-agglomerating (Ross et al., 1991).

Whereas the use of bio-organic agents with coal may not be a *reaction* of coal in the strictest sense, insofar as it is not caused by a chemical agent but by a live biological agent, it is nevertheless a reaction of coal (involving chemical changes within the coal matrix) that needs to be mentioned here because of its application to coal cleaning (Chapter 6).

More recently, the biodesulfurization of coal by a variety of biological agents has also received considerable attention (Couch, 1987, 1991; Chandra and Mishra, 1988; Dugan et al., 1989; Beier, 1990; Dugan and Quigley, 1991; El-Sawy and Gray, 1991; Lu, 1992). In addition, the use of bio-organic reagents enhances the solubility of coal, which, in turn, effects demineralization (Sharma et al., 1992).

Generally, coal treated with bacteria shows an improvement in quality. Apart from the reduction of the sulfur content, the mineral matter content (measured as ash production) in bacteria-treated coals are substantially reduced. However, the removal of organic sulfur depends on the structure and type of the sulfur-containing organic compounds present in coal.

Other work (Malik et al., 2004) has focused on the various factors (which are suspected to limit the coal biodesulfurization process rate) on bacterial oxidation of ferrous iron. An attempt has been made to find whether this reaction is inhibited due to the presence of solid particles and/or the release of inorganic components from solids during coal biodesulfurization. The data showed that silicon was responsible for retarding the central step of iron oxidation during coal biodesulfurization. Consequently, operational strategies, which tend to minimize the concentration of toxic inorganic components in the leachate, would be a better option than conventional batch process for enhanced biodesulfurization.

12.10 MISCELLANEOUS REACTIONS

There are a host of chemical reactions of coal that are dispersed throughout the scientific and engineering literature but are too voluminous to report here. Many of these reactions are "miscellaneous" in nature and do not appear to have a bearing on chemistry and "coal utilization." However, there are other reactions of coal that do have importance in terms of coal utilization insofar as they provide valuable information about the behavior of coal under various conditions and are reported here.

12.10.1   Reactions of Oxygen-Containing Functional Groups

The predominant oxygen-containing functional groups in coals are carbonyl ($>C=O$), hydroxyl ($-OH$), and carboxylic ($-CO_2H$), although the latter is not usually present in significant amounts in coals that are higher than subbituminous rank (Dryden, 1963; Wender et al., 1981). Reactions of the functional groups are generally (correctly or incorrectly) considered to be of minimal importance in terms of coal conversion scenarios but are nevertheless significant insofar as they afford an insight into the total behavior of coal as well as into the molecular environments of the oxygen species. For example, the role of oxygen functions during the events that lead to the spontaneous ignition process (Chapters 13 and 14) has been noted and the role of these same functions in coal macromolecular structure (which can decide the behavior of coal) has been, and continues to be, investigated.

Oxygen functions in coal have been studied in terms of their reduction behavior. Thus, the most likely fate of the carbonyl groups during coal hydrogenation is reduction to the hydroxyl groups:

$$>C=O+H_2 \rightarrow > CH=OH$$

On the other hand, carbonyl oxygen occurring as aldehyde functions is very susceptible to oxidation whereas ketone functions are much less susceptible. The oxidation of aldehydes in air (autooxidation) occurs readily by way of a free-radical mechanism:

$$RCH=O+O_2 \rightarrow RCO_2H$$

Hydroxyl oxygen, especially in the higher-rank coals, occurs predominantly as phenolic oxygen and carbonyl oxygen. Most of the chemical reactions used for hydroxyl oxygen have been used with the object of estimating the proportion present. However, phenolic hydroxyl has a limited reactivity but the function is acidic and the reaction has been used as a means of determining the phenolic hydroxyl content of coal by titration of the excess barium hydroxide:

$$2ROH + Ba(OH)_2 \rightarrow (2RO^-)Ba^{2+} + H_2O$$

In addition, phenolic hydroxyl adjacent to a carboxyl group on an aromatic ring constitutes the major entity responsible for the chelation of metals in coal, particularly in the lower-rank coals. Phenolic hydroxyl is also determined by reaction with diazomethane (CH_2N_2) to form the methoxyl group ($-OCH_3$) with the release of elemental nitrogen.

Total hydroxyl can be determined by reaction with acetic anhydride in pyridine; this process may require prolonged periods (several days) at 90°C (195°F). Alcoholic hydroxyl can be determined by esterification with phthalic anhydride and this reaction also requires several days at 90°C (195°F) to ensure completion.

12.10.2   Reactions of Sulfur-Containing Functional Groups

Sulfur occurs in coal as both organic (Chapter 10) and inorganic sulfur (Chapter 7) and the reactions of both forms of sulfur are quite important for coal combustion and conversion. During coal combustion, both of the sulfur forms appear as sulfur dioxide, which is the most significant pollutant from sulfur-containing coals. Thus, one of the major objectives of many coal conversion programs is to bring about the facile separation of sulfur from the combustible fraction of the coal.

Organic sulfur in coal can undergo oxidation reactions and can be involved in reduction reactions. Thus sulfur is contained in a variety of groups (Chapter 10) that are susceptible to further reaction. Thus, under partially oxidizing conditions, organic sulfur adds oxygen to form sulfones (O=S=O) but if the sulfur is originally present as mercaptan (thiol) sulfur (–SH), the product is a sulfonic acid (–SO$_3$H).

The formation of sulfones, followed by elimination of sulfur as sulfuric acid, has been proposed as a means of removing the organic sulfur from coal prior to combustion. Unfortunately, the formation of a sulfone from sulfur in a thiophene group is rather difficult. However, up to 40% of the organic sulfur in bituminous coal can be removed when the coal is subjected to 1000 psi air at 200°C (390°F) for 1 h, implying that there exists within the coal an appreciable fraction of non-thiophene sulfur from which sulfuric acid is produced:

$$R^1CH_2SR^2 + 3O_2 + H_2O \rightarrow R^1CO_2H + R^2OH + H_2SO_4$$

Aromatic groups bound to the sulfur form phenols and aliphatic groups normally form carboxylic acids.

The reduction of organic sulfur by hydrogen to hydrogen sulfide is the prime route to sulfur removal through coal conversion and is one of the main objectives of many coal conversion processes. In general, organic sulfides (thioethers), mercaptans, and organic disulfides are relatively easy to reduce, whereas heterocyclic sulfur (thiophenes) normally requires a catalyst for significant conversion to hydrogen sulfide:

$$R^1CH_2SR^2 + 2H_2 \rightarrow R^1CH_3 + R^2H + H_2S$$

Because of the nature of coal, some consideration must also be given to the reactions of the inorganic sulfur (Chapter 7). This particular type of sulfur is often difficult to remove completely and therefore passes as part of the coal substance to the various utilization processes.

The sulfur in pyrite (or marcasite, FeS$_2$) oxidizes through microbial action when exposed to air to produce sulfates and sulfuric acid:

$$2FeS_2 + 7O_2 + 2H_2O \rightarrow 2FeSO_4 + 2H_2SO_4$$

Similar reactions are also possible with a variety of other sulfide minerals such as galena (PbS) and sphalerite (ZnS). These reactions produce acid mine water, a major pollution problem from coal mining.

During combustion, the sulfide minerals are oxidized to sulfur dioxide:

$$4FeS_2 + 11O_2 \rightarrow 8SO_2 + 2Fe_2O_3$$

As a result of this reaction, the various sulfide minerals are the most troublesome of the coal mineral components.

Pyrite reacts with ferric ion in solution:

$$FeS_2 + 2Fe^{3+} \rightarrow 3Fe^{2+} + 2S$$

Other reactions occur, such as that producing sulfate ion, SO$_4{}^{2-}$. The elemental sulfur by-product from the previous reaction may be removed for commercial use by extraction with toluene or kerosene, or by steam vaporization; a portion of the iron(II) produced may be oxidized to iron(III):

$$Fe^{2+} 4H^+ + O_2 \rightarrow 4Fe^{3+} + 2H_2O$$

The iron(III) produced may be used for further pyrite beaching; these reactions are the basis of a process for the removal of pyrite from coal.

Heating pyritic sulfur at 600°C–800°C (1110°F–1470°F) in a reducing atmosphere results in the formation of *pyrrhotite* (a mineral having the approximate formula FeS (but actually ranging in formula from Fe_5S_6 to $Fe_{16}S_{17}$) and hydrogen sulfide:

$$FeS_2 + H_2 \rightarrow FeS + H_2S$$

An external source of hydrogen is not an essential prerequisite for this reaction to occur. In an inert atmosphere, such as that present during coal pyrolysis in the absence of oxygen, hydrogen from the coal combines with the sulfur to give hydrogen sulfide. In the presence of hydrogen, pyrrhotite may be further reduced to form elemental iron as well as hydrogen sulfide:

$$FeS + H_2 \rightarrow Fe + H_2S$$

The rate of this reaction is much slower than the rate of the reaction for the reduction of pyrite to pyrrhotite.

In general, when mineral sulfur in coal is converted to hydrogen sulfide by a coal conversion process, reduction of the metal ion in the sulfide is required. This can occur through reduction of the metal to a lower oxidation state, reduction of the metal ion to the metallic state, or reduction of the metal to the hydride:

$$As_2S_3 + 6H_2 \rightarrow 2AsH_3 + 3H_2S$$

This reaction is partially responsible for the formation of volatile arsine, AsH_3, as a by-product of coal hydrogenation.

12.10.3 REACTIONS OF NITROGEN-CONTAINING FUNCTIONAL GROUPS

Little is known about the nature of nitrogen-containing functional groups in coal, and even less about their specific reactions under the various conditions to which coal is subjected. Coal-derived nitrogen compounds result in the formation of pollutants and toxic substances, which interfere with synthetic fuel processing and, more importantly, the environment. The former problem results from the fact that most of these compounds are basic. Therefore, they are poisons for the acidic catalysts used in synthetic crude oil refining. The latter problem arises because of the formation of nitrogen oxides during combustion and the subsequent "ex-utility plant" conversion of these oxides to nitrous and nitric acids that are contributors to acid rain.

Stated simply and more in terms with the current context, but without consideration of the chemistry, the combustion of coal (which invariably contains some nitrogen, or by the inclusion of aerial nitrogen in the combustor) produces nitrogen oxides (NO and NO_2) (Chapters 14, 15, 23, and 26), which cause a major pollution problem (Chapter 25). The irony is that the smoke yield, the swelling and caking power, the ignition and decomposition temperatures of bituminous coals are reduced by pretreatment with the oxides of nitrogen, presumably through the formation of oxygen-containing groups, nitration, as well as some dehydrogenation (Billington, 1954). This is essentially the reverse effect noted for the reaction of coal with hydrogen and coincides with the effects of dehydrogenation of coal (Kukharenko, 1950; Kukharenko and Matveeva, 1950; Mazumdar et al., 1959).

On the other hand, coal pyrolysis and coal hydrogenation produce ammonia (a useful by-product), which arises from the more labile nitrogen functional groups, such as amine ($–NH_2$); less reactive nitrogen functions are converted to ammonia by catalytic hydrogenation.

12.10.4 REACTIONS OF MINERAL MATTER AND TRACE ELEMENTS

There are only fragmentary reports about the reactions of coal mineral matter (for example, see Gonzalez de Andres et al., 1990a,b). Considering the large variety and substantial amounts of mineral matter and trace elements in coal (Chapter 7), little is known about their reactions and, more importantly, their participation in coal conversion processes. For example, the treatment of coal with methanol and hydrochloric acid removes virtually all of the calcium species (Shams et al., 1992) that are responsible for retrogressive reactions during the thermal treatment of coal leading to THF-insoluble materials (coke precursors) during liquefaction.

This general lack of interest in the chemistry of the mineral constituents is unfortunate in view of the increasing recognition of the importance of the minerals in conversion as well as the toxicity of some trace elements, especially the heavier metals. On this latter aspect, it is worthy of note that there is a high potential for the recovery of valuable mineral elements from coal conversion processes as added-value products in addition to their removal from the system prior to the onset of adverse effects on the environment.

REFERENCES

Ahnonkitpanit, E. and Prasassarakich, P. 1989. *Fuel*, 68: 819.
Ahuja, L.D., Kini, K.A., Basak, N.G., and Lahiri, A. 1959. In *Proceedings. Symposium on the Nature of Coal.* Central Fuel Research Institute, Jealgora, Jharkhand, India, p. 215.
Alemany, L.B., Handy, C.I., and Stock, L.M. 1981. In *Coal Structure*, M.L. Gorbaty and K. Ouchi (Eds.). Advances in Chemistry Series No. 192. American Chemical Society, Washington, DC, Chapter 14.
Ali, A., Srivastava, S.K., and Haque, R. 1992. *Fuel*, 71: 835.
Allen, R.R. and Parry, V.F. 1954. Report of Investigations No. 5034. United States Bureau of Mines, Washington, DC.
Ashmead, D. 1955. The influence of bacteria in the formation of acid mine waters. *Colliery Guardian*, 190: 694–698.
ASTM. 2011. *Standard Test Method for Determining the Relative Degree of Oxidation in Bituminous Coal by Alkali Extraction* (ASTM D5263). Annual Book of Standards, American Society for Testing and Materials, West Conshohocken, PA.
Atkins, A.S. and Pooley, F.D. 1982. The effects of bio-mechanisms on acidic mine draining in coal mining operations. *International Journal of Mine Water*, 1: 31–44.
Atkins, A.S. and Singh, R.N. 1982. A study of acid and ferruginous mine water in coal mining operations. *International Journal of Mine Water*, 2: 37–57.
Attia, Y.A. (Ed.). 1985. *Processing and Utilization of High-Sulfur Coals*. Elsevier, Amsterdam, the Netherlands.
Balcioglu, N. and Yilmaz, M. 1990. *Fuel Science and Technology International*, 8: 689.
Baldwin, R.M., Kenner, D.R., Nguanprasert, O., and Miller, R.L. 1991. *Fuel*, 70: 429.
Beafore, F.J., Cawiezel, K.E., and Montgomery, C.T. 1984. *Journal of Coal Quality*, 3(4): 17.
Beck, J.V. 1960. A ferrous-ion-oxidizing bacterium. I. Isolation and some general physiological characteristics. *Journal of Bacteriology*, 79: 502–509.
Beckberger, L.H., Burk, E.H. Jr., Grosboll, M.P., and Yoo, J.S. 1979. Report No. EPRI EM-1044. Research Project No. 833-1, p. 549.
Beier, E. 1990. In *Bioprocessing and Biotreatment of Coal*, D.L. Wise (Ed.). Marcel Dekker, Inc., New York.
Bergius, F. 1912. *Z Angewandte Chemie*, 23: 2540.
Bergius, F. 1913. *Journal of the Indian Chemical Society*, 32: 462.
Bergius, F. 1925. *Fuel*, 4: 458.
Berkowitz, N. 1979. *An Introduction to Coal Technology*. Academic Press, Inc., New York.
Berkowitz, N. and Schein, H.G. 1951. *Fuel*, 30: 94.
Berkowitz, N. and Speight, J.G. 1973. Prevention of spontaneous heating of coal. *Canadian Institute of Mining and Metallurgical Bulletin*, 66(736): 109.
Berlozecki, S. and Bimer, J. 1991. *Koks, Smola, Gaz*, 36(7): 169.
Berthelot, M. 1868. *Bulletin de la Societe Chimique de France*, 9: 8.
Berthelot, M. 1869. *Bulletin de la Societe Chimique de France*, 11: 278.
Bevan, E.J. and Cross, C.F. 1881. *Chemistry News*, 44: 185.
Bhattacharyya, D., Hsieh, M., Francis, H., Kermode, R.I., and Aleem, M.I.H. 1990. Biological desulfurization of coal by mesophilic and thermophilic microorganisms. *Resources Conservation and Recycling*, 3: 81–96.

Bhowmik, J.N., Mukherjee, P.N., Mukherjee, A.K., and Lahiri, A. 1959. In *Proceedings, Symposium on the Nature of Coal*. Central Fuel Research Institute, Jealgora, India, p. 242.

Billington, A.H. 1954. *Fuel*, 33: 295.

Bowling, K.M., Chandler, P.J., and Waugh, A.B. 1990. *Proceedings of the Eleventh International Coal Preparation Congress*. Mining and Minerals Processing Institute, Tokyo, Japan, p. 349.

Braley, S.A. 1954. A summary report of Commonwealth of Pennsylvania Department of Health Industrial Fellowship, No. 1–7. Mellon Institute of Industrial Research Fellowship No. 326B.

Brant, R.A. and Moulton, E.Q. 1960. Acid mine drainage manual. Ohio State University Studies. Eng. Expt. Sta. Bull. No. 179.

Bryner, L.C. and Anderson, R. 1957. Microorganisms in leaching sulfide minerals. *Industrial & Engineering Chemistry*, 49: 1721–1724.

Bryner, L.C., Beck, J.V., Davis, D.B., and Wilson, D.G. 1954. Micro-organisms in leaching sulfide minerals. *Industrial & Engineering Chemistry*, 46: 2587–2592.

Bryner, L.C. and Jameson, A.K. 1958. Microorganisms in leaching sulfide minerals. *Applied Microbiology*, 6: 281–287.

Chakrabartty, S.K. 1981. Oxidation of coal: A mechanistic puzzle. Preprints. Division of Fuel Chemistry. *American Chemical Society*, 26(1): 10–15.

Chakrabartty, S.K. and Berkowitz, N. 1976. *Fuel*, 55: 362.

Chakravorty, R.N. 1984. Report No. ERP/CRL 84-14. Coal Research Laboratories, Canada Centre for Mineral and Energy Technology, Ottawa, Ontario, Canada.

Chakravorty, R.N. and Kar, K. 1986. Report No. ERP/CRL 86-151. Coal Research Laboratories, Canada Centre for Mineral and Energy Technology, Ottawa, Ontario, Canada.

Chambers, R.R. Jr., Hagaman, E.W., and Woody, M.C. 1988. In *Polynuclear Aromatic Compounds*, L.B. Ebert (Ed.). Advances in Chemistry Series No. 217. American Chemical Society, Washington, DC, Chapter 15.

Chandra, D. and Mishra, A.K. 1988. Desulfurization of coal by bacterial means. *Resources Conservation and Recycling*, 1(3–4): 293–308.

Chatterjee, K. and Stock, L.M. 1991. *Energy & Fuels*, 5: 704.

Chen, X.D. 1992. *Combustion and Flame*, 90(2): 114.

Chuang, K.C., Markuszewski, R., and Wheelock, T.D. 1983. *Fuel Processing Technology*, 7: 43.

Clemens, A.H., Matheson, T.W., Lynch, L.J., and Sakurovs, R. 1989. *Fuel*, 68: 1162.

Cohen, M.S., Bowers, W.C., Aronson, H., and Gray, E.T. 1987. Cell-free solubilization of coal by *Polyporus versicolor*. *Applied Environmental Microbiology*, 53: 2840–2843.

Colmer, A.R. and Hinkle, M.E. 1947. The role of microorganisms in acid mine-drainage. *Science*, 106: 253–256.

Colmer, A.R., Temple, K.L., and Hinkle, M.E. 1950. An iron-oxidizing bacterium from the acid drainage of some bituminous coal mines. *Journal of Bacteriology*, 59: 317–328.

Couch, G.R. 1987. Biotechnology and coal. Report No. ICTRS/TR38. IEA Coal Research, International Energy Agency, London, U.K.

Couch, G.R. 1991. Advanced coal cleaning technology. Report No. IEACR/44. IEA Coal Research, International Energy Agency, London, U.K.

Crawford, A., Williams, F.A., King, J.G., and Sinnatt, F.S. 1931. Research Technical Paper No. 29. Department of Scientific and Industrial Research, London, U.K.

Davidson, R.M. 1982. In *Coal Science*, M.L. Gorbaty, J.W. Larsen, and I. Wender (Eds.). Academic Press, Inc., New York, Vol. 1, Chapter 2.

De Faveri, D.M., Zonato, C., Vidili, A., and Ferrailo, G. 1989. *Journal of Hazardous Materials*, 21: 135.

Deno, N.C., Curry, K.W., Jones, A.D., Keegan, K.R., Rakitsky, W.G., Richter, C.A., and Minard, R.D. 1981. *Fuel*, 60: 210.

Dorland, D., Stiller, A.S.H., and Mintz, E.A. 1992. *Fuel Processing Technology*, 30: 195.

Dryden, I.G.C. 1963. In *Chemistry of Coal Utilization, Supplementary Volume*, H.H. Lowry (Ed.). John Wiley & Sons, Inc., New York, Chapter 6.

Dugan, P.R. 1986. Microbiological desulfurization of coal and its increased monetary value. *Biotechnology, Bioengineering Symposium*, 16: 185–204.

Dugan, P.R., McIlwain, M.E., Quigley, D., and Stoner, D.L. 1989. *Proceedings. Sixth Annual International Pittsburgh Coal Conference*. University of Pittsburgh, Pittsburgh, PA, Vol. 1, p. 135.

Dugan, P.R. and Quigley, D.R. 1991. *Fuel*, 70: 571.

Duty, R.C., Hayatsu, R., Scott, R.G., Moore, L.P., Winans, R.E., and Studier, M.N. 1980. *Fuel*, 59: 97.

Ebert, L.B., Mills, D.R., Matty, L., Pancirov, R.J., and Ashe, T.R. 1981. In *Coal Structure*, M.L. Gorbaty and K. Ouchi (Eds.). Advances in Chemistry Series No. 192. American Chemical Society, Washington, DC, Chapter 15.

Ehrlich, H.L. and Holmes, D.S. (Eds.). 1986. *Workshop on Biotechnology for the Mining*. Metal Refining and Fossil Fuel Processing Industries. John Wiley & Sons, Inc., New York.

El-Sawy, A. and Gray, D. 1991. *Fuel*, 70: 591.

Estevez, M., Juan, R., Ruiz, C., and Andres, J.M. 1990. *Fuel*, 69: 157.

Flores, R.A., Geigel, M.A., and Mayo, F.R. 1978. *Fuel*, 57: 697.

Francis, W. 1961. *Coal: Its Formation and Composition*. Edward Arnold Ltd., London, U.K.

Fuerstenau, D.W. and Diao, J. 1992. *Coal Preparation*, 10(1/4): 1.

Garcia, A.B., Moinelo, S.R., Matinez-Tarazona, M.-R., and Tascon, J.M.D. 1991. *Fuel*, 70: 1391.

Given, P.H. 1984. In *Coal Science*, M.L. Gorbaty, J.W. Larsen, and I. Wender (Eds.). Academic Press, Inc., New York, Vol. 3, p. 63.

Gonzalez de Andres, A.I., Bermejo, J., Martinez-Alonso, A., Moinelo, S.R., and Tascon, J.M.D. 1990b. *Fuel*, 69: 873.

Gonzalez de Andres, A.I., Moinelo, S.R., Bermejo, J., and Tascon, J.M.D. 1990a. *Fuel*, 69: 867.

Gray, P., Boddington, T., and Harvey, D.I. 1971. *Philosophical Transactions of Royal Society*, A270: 467.

Gray, P., Griffiths, J.F., and Hasko, S.M. 1984. *Journal of Chemical Technology & Biotechnology*, 34A: 453.

Haenel, M.W. 1992. *Fuel*, 71: 1211.

Hagaman, E.W. and Lee, S.-K. 1992. Preprints. Division of Fuel Chemistry. American Chemical Society, Washington, DC, 37(3): 1166.

Handy, C.I. and Stock, L.M. 1982. *Fuel*, 61: 700.

Haslam, R.T. and Russell, R.P. 1926. *Fuels and their Combustion*. McGraw-Hill, New York, Chapter 4.

Heredy, L.A. 1981. In *Coal Structure*, M.L. Gorbaty and K. Ouchi (Eds.). Advances in Chemistry Series No. 192. American Chemical Society, Washington, DC, Chapter 12.

Heredy, L.A., Kostyo, A.E., and Neuworth, M.B. 1965. *Fuel*, 44: 125.

Hessley, R.K. 1990. In *Fuel Science and Technology Handbook*, J.G. Speight (Ed.). Marcel Dekker, Inc., New York.

Hessley, R.K., Reasoner, J.W., and Riley, J.T. 1986. *Coal Science*. John Wiley & Sons, Inc., New York.

Hoffmann, M.R., Faust, B.C., Panda, F.A., Koo, H.H., and Tsuchiya, H.M. 1981. Kinetics of the removal of iron pyrite from coal by microbial catalysis. *Applied and Environmental Microbiology*, 42(2): 259–271.

Honda, H., Hirose, Y., and Yamakawa, T. 1959. *Proceedings of the Symposium on the Nature of Coal*. Central Fuel Research Institute, Jealgora, India, p. 238.

Howard, H.C. 1945a. In *Chemistry of Coal Utilization*, H.H. Lowry (Ed.). John Wiley & Sons, Inc., New York, Chapter 9.

Howard, H.C. 1945b. In *Chemistry of Coal Utilization*, H.H. Lowry (Ed.). John Wiley & Sons, Inc., New York, Chapter 11.

Huffman, G.P., Huggins, F.P., Francis, H.E., Milra, S., and Shah, N. 1990. Structural characterization of sulfur in bioprocessed coal, R. Markuszewski and T.D. Wheelock (Eds.). *Third International Conference on Processing and Utilization of High-Sulfur Coals* 111. Elsevier, Amsterdam, the Netherlands, pp. 23–32.

Iyengar, M.S., Dutta, S.N., and Lahiri, A. 1959. *Proceedings of Symposium on the Nature of Coal*. Central Fuel Research Institute, Jealgora, India, p. 230.

Jensen, E.J., Melnyk, N., and Berkowitz, N. 1966. Advances in Chemistry Series No. 55, American Chemical Society, Washington, DC, p. 621.

Jones, J.C. 1990. *Fuel*, 69: 399.

Jones, J.C., Rahmah, H., Fowler, O., Vorasurajakart, J., and Bridges, R.G. 1990. *Journal of Life Sciences*, 8: 207.

Jones, J.C. and Vais, M. 1991. *Journal of Hazardous Materials*, 26: 203.

Jones, J.C. and Wake, G.C. 1990. *Journal of Chemical Technology & Biotechnology*, 48: 209.

Joseph, J.T. and Mahajan, O.P. 1991. In *Coal Science II*, H.H. Schobert, K.D. Bartle, and L.J. Lynch (Eds.). Symposium Series No. 461. American Chemical Society, Washington, DC, Chapter 23.

Khalid, A.M., Bhattacharyya, D., and Aleem, M.I.H. 1989. *Sulfur Metabolism and Coal Desulfurization Potential of Sulfolobus brierleyi and Thiobacilli*, S. Yunker (Ed.). *Proceedings. Symposium on Biological Processing of Coal and Coal-Derived Substances*. EPFU ER-6572. Palo Alto, CA, pp. 2–19; 2–36.

Khalid, A.M., Bhattacharyya, D., and Aleem, M.I.H. 1990a. *Coal Desulfurization and Electron Transport-Linked Oxidation of Sulfur Compounds by Thiobacillus and Sulfolobus Species*, G.E. Pierce (Ed.). Development in Industrial Microbiology, Society for Industrial Microbiology, 31: 115–126.

Khalid, A.M., Bhattacharyya, D., Hseih, M., Kermode, R.I., and Aleem, M.I.H. 1990b. Biological desulfurization of coal, R. Markuszewski and T.D. Wheelock (Eds.). *Third International Conference on Processing and Utilization of High-Sulfur Coals*. Elsevier, Amsterdam, the Netherlands, pp. 469–480.

Kin, B., Dincer, A., Kuyulu, A., and Piskin, S. 1990. *Fuel Science and Technology International*, 8: 271.

Kreulen, D.J.W. 1948. *Elements of Coal Chemistry*. Nijgh & van Ditmar N.V., Rotterdam, Chapter XII.

Krichko, A.A. and Gagarin, S.G. 1990. *Fuel*, 69: 891.

Kukharenko, T.A. 1950. *Journal of Applied Chemistry (USSR)*, 23: 655.

Kukharenko, T.A. and Matveeva, I.I. 1950. *Journal of Applied Chemistry (USSR)*, 23: 773.

Kullerud, G. and Yoder, H.S. 1959. Pyrite stability relations in the Fe–S system. *Economic Geology*, 54: 533–572.

Kusakabe, K., Orita, M., Kato, L., Morooka, S., Kata, Y., and Kusonoki, K. 1989. *Fuel*, 68: 396.

Lalvani, S.B. and Kang, J.W. 1992. *Fuel Science and Technology International*, 10: 1241.

Larsen, J.W., Green, T.K., Choudhury, P., and Kuemmerle, E.W. 1981. In *Coal Structure*, M.L. Gorbaty and K. Ouchi (Eds.). Advances in Chemistry Series No. 192. American Chemical Society, Washington, DC, Chapter 18.

Larsen, J.W. and Kuemmerle, E.W. 1976. *Fuel*, 56: 162.

Larsen, J.W. and Lee, D. 1983. *Fuel*, 62: 1351.

Leathen, W.W. 1952. Microbiological studies of bituminous coal mine drainage. Commonwealth of Pennsylvania Department of Health Industrial Fellowship No. 326B-6.

Leathen, W.W., Braley, S.A., and McIntyre, L.D. 1953a. The role of bacteria in the formation of acid from certain sulfuritic constituents associated with bituminous coal. I. *Thiobacillus thiooxidans*. *Applied Microbiology*, 1: 61–64.

Leathen, W.W., Braley, S.A., and McIntyre, L.D. 1953b. The role of bacteria in the formation of acid from certain sulfuritic constituents associated with bituminous coal. II. Ferrous iron oxidizing bacteria. *Applied Microbiology*, 1: 65–68.

Leathen, W.W. and Madison, K.M. 1949. The oxidation of ferrous iron by bacteria found in acid mine waters. *Bacteriological Proceedings*, p. 64.

Liotta, R. and Hoff, W.S. 1980. *Journal of Organic Chemistry*, 45: 287.

Lu, L.-K. 1992. *Fuel Science and Technology International*, 10: 1251.

Majchrowicz, B.B., Franco, D.V., Yperman, J., Reggers, G., Gelan, J., Martens, H., Mullens, J., and van Poucke, L.C. 1991. *Fuel*, 70: 434.

Malik, A., Dastidar, M.G., and Roychoudhury, P.K. 2004. Factors limiting bacterial iron oxidation in biodesulfurization system. *International Journal of Mineral Processing*, 73(1): 13–21.

Markuszewski, R. and Wheelock, T.D. (Eds.). 1990. *Processing and Utilization of High-Sulfur Coals*. Elsevier, Amsterdam, the Netherlands.

Mayo, F.R. 1975. *Fuel*, 54: 273.

Mazumdar, B.K., Chakrabartty, S.K., Chowdhury, S.S., and Lahiri, F.A. 1959. *Proceedings. Symposium on the Nature of Coal*. Central Fuel Research Institute, Jealgora, Jharkhand, India, p. 219.

Meyers, R.A. (Ed.). 1981. *Coal Handbook*. Marcel Dekker, Inc., New York.

Morrison, G.F. 1980. *Nitrogen Oxides from Coal Combustion—Abatement and Control*. IEA Coal Research, London, U.K.

Moschopedis, S.E., Parkash, S., and Speight, J.G. 1978. *Fuel*, 57: 431.

Moulton, E.Q. (Ed.). 1957. The acid mine-drainage problem in Ohio. Ohio State University Studies. Eng. Expt. Sta. Bull. No. 166.

Neavel, R.C. 1976. *Fuel*, 55: 237.

Ogunsola, O.I. and Mikula, R.J. 1991. *Fuel*, 70: 258.

Ogunsola, O.I. and Mikula, R.J. 1992. *Fuel*, 71: 3.

Olson, G.I., Brinckman, F.E., and Iverson, W.P. 1986. Processing of coal with microorganisms. EPRI Final Report. EPRI AP-4472. Palo Alto, CA.

Petit, J.C. 1990. *Fuel*, 69: 861.

Petit, J.C. and Boettner, J.C. 1990. In *Advanced Methodologies in Coal Characterization*, H. Charcosset and B. Nickel-Pepin-Donat (Eds.). Elsevier, Amsterdam, the Netherlands, Chapter 14.

Petit, J.C. and Cheng, Z.X. 1990. In *Advanced Methodologies in Coal Characterization*, H. Charcosset and B. Nickel-Pepin-Donat (Eds.). Elsevier, Amsterdam, the Netherlands, Chapter 15.

Pinchin, F.J. 1958. *Fuel*, 37: 293.

Prasassarakich, P. and Pecharanond, P. 1992. *Fuel*, 71: 929.

Pyne, J.W., Stewart, D.L., Fredricson, J., and Wilson, B.W. 1987. Solubilization of leonardite by an extracellular fraction from *Coriolcus versicolor*. *Applied Environmental Microbiology*, 53: 2844–2848.

Raj, S. 1976. PhD thesis, On the Study of the Chemical Structural Parameters of Coal. Pennsylvania State University.

Rausa, R., Calemma, V., Ghelli, S., and Girardi, E. 1989. *Fuel*, 68: 1168.

Ross, D.S., Loo, B.H., Tse, D.S., and Hirschon, A.S. 1991. *Fuel*, 70: 289.

Roy, M.M. 1965. Studies on coal macerals 4: Chlorination of exinite and fusinite. *Economic Geology*, 60(7): 1404–1410.

Schlosberg, R.H., Gorbaty, M.L., and Aczel, T. 1978. *Journal of American Chemical Society*, 100: 4188.

Schmeling, W.A., King, J., and Schmidt-Collerus, J.J. 1978. In *Analytical Chemistry of Liquid Fuel Sources: Tar Sands, Oil Shale, Coal, and Petroleum*, P.C. Uden, S. Siggia, and H.B. Jensen (Eds.). Advances in Chemistry Series No. 170. American Chemical Society, Washington, DC, Chapter 1.

Schmidt, C.E. 1987. *Proceedings. Fourth Annual Pittsburgh Coal Conference.* University of Pittsburgh, Pittsburgh, PA, p. 143.

Scott, C.D., Strandberg, G.W., and Lewis, N.S. 1986. Microbial solubilization of coal. *Biotechnological Progress*, 2: 131–139.

Seki, H., Ito, O., and Iino, M. 1990a. *Fuel*, 69: 317.

Seki, H., Ito, O., and Iino, M. 1990b. *Fuel*, 69: 1047.

Shams, K., Miller, R.L., and Baldwin, R.M. 1992. *Fuel*, 71: 1015.

Sharma, D.K. and Mishra, S. 1992. *Fuel Science and Technology International*, 10: 1601.

Sharma, D.K., Singh, S.K., and Behera, B.K. 1992. *Fuel Science and Technology International*, 10: 223.

Shrivastava, K.L., Tripathi, R.P., and Jangid, M.L. 1992. *Fuel*, 71: 377.

Silverman, M.P., Rogoff, M.H., and Wender, I. 1961. Bacterial oxidation of pyritic materials in coal. *Applied Microbiology*, 9: 491–496.

Speight, J.G. 1987. In *Coal Science and Chemistry*, A. Volborth (Ed.). Elsevier, Amsterdam, the Netherlands, Chapter 6.

Speight, J.G. 2000. *The Desulfurization of Heavy Oils and Residua*, 2nd edn. Marcel Dekker, Inc., New York.

Speight, J.G. 2007. *The Chemistry and Technology of Petroleum*, 4th edn. Marcel Dekker, Inc., New York.

Speight, J.G. and Moschopedis, S.E. 1980. *Fuel*, 59: 440.

Sternberg, H.W. and Delle Donne, C.L. 1974. *Fuel*, 53: 172.

Sternberg, H.W., Delle Donne, C.L., Pantages, P., Moroni, E.C., and Markby, R.E. 1971. *Fuel*, 50:432.

Stevenson, F.J. and Butler, J.H.A. 1969. In *Organic Geochemistry: Methods and Results*, G. Eglinton and M.T.J. Murphy (Eds.). Springer-Verlag, New York, Chapter 22.

Stock, L.M. 1987. Coal alkylation and pyrolysis. Preprints. Division of Fuel Chemistry. American Chemical Society, 32(1): 463–470.

Studier, M.H., Hyatsu, R., and Winans, R.E. 1978. In *Analytical Methods for Coal and Coal Products*, C. Karr, Jr. (Ed.). Academic Press, Inc., New York, Vol. II, Chapter 21.

Tarafdar, M.N. and Guha, D. 1989. *Fuel*, 68: 315.

Temple, K.L. and Koehler, W.A. 1954. Drainage from bituminous coal mines. West VA. Univ. Bull. Eng. Expt. Sta. Research Bull. No. 25.

Thomas, T. 1995. Developments for the precombustion removal of inorganic sulfur from coal. *Fuel Processing Technology*, 43(2): 123–128.

Van Krevelen, D.W. and Schuyer, J. 1957. *Coal Science: Aspects of Coal Constitution*. Elsevier, Amsterdam, the Netherlands, Chapter 15.

Vilyunov, V.N. and Zarko, V.E. 1989. *Ignition of Solids*. Elsevier, Amsterdam, the Netherlands.

Vovelle, C., Mellottee, H., and Delbourgo, R. 1983. *Proceedings. 19th International Symposium on Combustion*. The Combustion Institute, Pittsburgh, PA, p. 797.

Wachowska, H. 1979. *Fuel*, 58: 99.

Wang, H., Bigham, J.M., and Tuovinen, O.H. 2007. Oxidation of marcasite and pyrite by iron-oxidizing bacteria and archaea. *Hydrometallurgy*, 88(1–4): 127–131.

Wang, H.P., Lo, R., Sommerfeld, D.A., Huai, H., Pugmire, R.J., Shabtai, J., and Eyring, E.M. 1992. *Fuel*, 71: 723.

Weiler, J.F. 1945a. In *Chemistry of Coal Utilization*, H.H. Lowry (Ed.). John Wiley & Sons, Inc., New York, Chapter 8.

Weiler, J.F. 1945b. In *Chemistry of Coal Utilization*, H.H. Lowry (Ed.). John Wiley & Sons, Inc., New York, Chapter 10.

Wender, I., Heredy, L.A., Neuworth, M.B., and Dryden, I.G.C. 1981. In *Chemistry of Coal Utilization*, Second Supplementary Volume, M.A. Elliott (Ed.). John Wiley & Sons, Inc., New York, Chapter 8.

Wilson, B.W., Bean, R.M., Franz. J.A., Thomas, B.L., Cohen, M.S., Aronson, H., and Grey, E.T. 1987. Microbial conversion of low rank coal: Characterization of biodegraded product. *Energy & Fuels*, 1: 80–84.

Yoneyyama, Y., Nakayama, H., Nishi, H., and Kato, T. 1992. *Bulletin of the Chemical Society of Japan*, 65: 1067.

Zarubina, Z.M., Lyalikova, N.N., and Shmuk, Ye.I. 1959. Investigation of the microbiological oxidation of the pyrite of coal. Izvest. Akad. Nauk SSSR, Otdel. Tekh. Nauk Metallurgiya i Toplivo, No. 1: 117–119.

Zher'akova, G. and Kochkan'an, R. 1990. *Fuel*, 69: 898.

13 Thermal Reactivity

13.1 INTRODUCTION

Understanding of the thermal behavior of coal is, just like understanding the chemical behavior of coal (Chapter 12), an essential part of projecting the successful use of coal, say, for conversion and/or utilization processes such as combustion (Chapters 14 and 15), carbonization and briquetting (Chapters 16 and 17), liquefaction (Chapters 18 and 19), gasification (Chapters 20 and 21), or as a source of chemicals (Chapter 24).

Pyrolysis of coal is generally considered to be the thermal decomposition of coal at temperatures in excess of 300°C (570°F) in the absence of oxygen whereby gas, tar, and char are formed. It is the basic process of coking (carbonization) and the starting reaction of combustion, gasification, and hydrogenation. In addition, the application of pyrolysis offers the possibility to convert coal into gases and liquids in the presence of hydrogen at high pressure (hydropyrolysis) by which the yield of volatile matter can be substantially increased.

Briefly and more from a historical perspective, the chemistry of the thermal (pyrolytic) decomposition of organic compounds to carbonaceous materials and volatile products has been the subject of many investigations (for example, see Hurd, 1929; Fitzer et al., 1971; Lewis and Singer, 1981). Since the early days of coal science, and in parallel to investigations on the thermolysis of organic compounds, the thermal decomposition of coal has been the subject of many studies (Howard, 1963, 1981a,b; Jones, 1964). Furthermore, the trend toward the use of coal as a source of liquid and gaseous fuels has caused a marked increase in studies related to the thermal reactions of coal leading to, for example, methods of relating pyrolytic behavior to utilization (Azhakesan et al., 1991).

Before progressing any further into the realm of the thermal decomposition of coal, it is noteworthy (and it should be of little surprise to most readers) that, as a result of the myriad of investigations, several forms of terminology have come into common usage.

The terms *thermal decomposition*, *pyrolysis*, and *carbonization* often may be used interchangeably. However, it is more usual to apply the term *pyrolysis* (a thermochemical decomposition of coal or organic material at elevated temperatures in the absence of oxygen, which typically occurs under pressure and at operating temperatures above 430°C [800°F]) to a process that involves widespread thermal decomposition of coal (with the ensuing production of a char—carbonized residue).

The term *carbonization* is more correctly applied to the process for the production of char or coke when the coal is heated at temperatures in excess of ca. 500°C (ca. 930°F). The ancillary terms volatilization and distillation are also used from time to time but more correctly refer to the formation and removal of volatile products (gases and liquids) during the thermal decomposition process.

13.2 THERMAL DECOMPOSITION PROCESSES

The thermal decomposition process, while written as an orderly process, may be quite disordered. Nevertheless, it is proposed that as the coal particle temperature rises during thermal decomposition (which may also be the initial stages of the combustion process) (Chapters 14 and 15), the bonds between the aromatic clusters in the coal macromolecule break and create lower-molecular-weight fragments that are detached from the macromolecule—the larger fragments of this decomposition process are often (collectively) referred to as the *metaplast*.

During pyrolysis, the metaplast will either vaporize and escape from the coal or be reincorporated into the residual macromolecule by chemical cross-linking. The portion of the metaplast that is vaporized usually consists of the lower-molecular-weight fragments and distills from the hot zone as *tar*. Side chains on the aromatic clusters are released from the coal as light gases. These light gases are generally carbon monoxide (CO), carbon dioxide (CO_2), water (H_2O), and low-molecular-weight hydrocarbons (C_1–C_4 and up to C_8 or C_{10}).

The pyrolysis behavior of a coal is a strong function of coal type or rank. Low-rank coal (such as lignite and subbituminous coal) produces relatively high levels of low-molecular-weight gases and very little tar. Bituminous coal produces significantly more tar than low-rank coal and moderate amounts of light gases. High-rank coal produces relatively low levels of both light gases and tar.

Thus, coal undergoes a large variety of physical and chemical changes when heated to temperatures where thermal decomposition occurs. However, some changes may be noted before the onset of what is often referred to as the *thermal decomposition proper* (i.e., carbon–carbon bond scission and the like) and may manifest themselves as the formation of low-molecular-weight species (Stein, 1981; Hessley et al., 1986).

For example, while the temperature of the onset of thermal decomposition proper is generally recognized to be approximately 350°C (660°F; the so-called cracking temperature), water will appear as a product of heating coal at temperatures below 100°C (120°F). In addition, when a coal is "degassed" at temperatures below 100°C, adsorbed methane and carbon dioxide will appear as "products" of the thermal treatment. On the other hand, a coal such as lignite, which contains many carboxylic functions as part of the coal structure, will evolve carbon dioxide by thermal decarboxylation:

$$RCO_2H \rightarrow RH + CO_2$$

Such changes are usually noted to occur at temperatures just in excess of 100°C (210°F) and more than 50% of the carboxylic acid functions can lose carbon dioxide over the temperature range 100°C–200°C (210°F–390°F).

As the temperature of the thermal treatment is increased to the range 200°C–370°C (390°F–700°F), coals lose a variety of lower-molecular-weight organic species (especially aliphatic compounds), which are believed to arise from moieties that are "loosely bound" to the more thermally stable part of the coal structure. Some of the lower-molecular-weight aromatic species may also be obtained. At higher temperatures (i.e., 370°C; 700°C), methane (indicative of the thermal destruction of the coal structure), polycyclic aromatics, phenols, and nitrogen compounds are produced.

Although nitrogen is typically found in a range of 0.5 to 1.5% w/w and is considered to be a minor component of coal, it is the major source of fuel nitrogen oxide (NOx) during combustion (Pershing and Wendt, 1989). Pyrolysis, the first step in coal conversion processes, is a determinant of the distribution of nitrogen-containing compounds in the final products. Nitrogen exists in coal as pyrroles, pyridines, and quaternary compounds, and has been characterized by x-ray photoelectron spectroscopy (Bruchill and Welch, 1989). In addition, the content of pyrrole-type and pyridine-type nitrogen increases with coal rank (Kelemen et al., 1994). By analogy with nitrogen in petroleum (Speight, 2007 and references cited therein), it is not surprising that more than 90 mol% of the nitrogen atoms in the raw coal remains in the char and tar formed during the primary pyrolysis step—the yield of nitrogen-containing gaseous products (such as hydrogen cyanide, HCN, and ammonia, NH_3) is less than 10 mol% (Tagaki et al., 1999).

The thermal decomposition of coal under nonoxidizing conditions produces a residue (coke) that consists of carbon and mineral matter and during this process the coal passes through several stages. Thus, heated coal generally becomes "plastic" and softens, but as the heating progresses the "plastic" mass may solidify even before the onset of the coking reactions. This behavior has been assumed to be a result of the rupture of chemical bonds in coal, thereby allowing a higher degree of mobility, but continued heating causes the reformation of the bonds or the formation of new bonds with solidification ensuing. In fact, an important physical test, the free swelling index (FSI) (Chapters 8 and 9), is based on the behavior of coal when heated and allows coals to be classified as caking or noncaking as well as presents an indication of the agglomerating properties of the coal.

By way of brief review and at the risk of some repetition, the FSI of coal is a measure of the volume increase of coal when it is heated (without restriction) under prescribed conditions (ASTM, 2011a; ISO, 2011a,b).

The nature of the volume increase is associated with the plastic properties of coal (Loison et al., 1963) and, as might be anticipated, coals that do not exhibit plastic properties when heated do not, therefore, exhibit free swelling. Although this relationship between free swelling and plastic properties may be quite complex, it is presumed that when the coal is in a plastic (or semifluid) condition the gas bubbles formed as a part of the thermal decomposition process within the fluid material cause the swelling phenomenon, which, in turn, is influenced by the thickness of the bubble walls, the fluidity of the coal, and the interfacial tension between the fluid material and the solid particles that are presumed to be present under the test conditions.

The FSI is rank dependent insofar as the high-rank coals (anthracites) do not usually fuse or exhibit a FSI, whereas the FSI of bituminous coals will increase as the rank increases from the high-volatile bituminous coal to the low-volatile bituminous coal. However, and in very general terms, the FSI of bituminous coals generally increases with an increase in rank (Rees, 1966) but the values for individual coals within a rank may vary considerably. The FSI is also subject to other effects such as oxidation (weathering) of the coal.

With respect to the volatile matter produced by the thermal decomposition of coal, rapid thermal decomposition enables the production of lower-molecular-weight hydrocarbons along with the char residue. For bituminous coals, the decomposition increases markedly above 400°C (750°F) and reaches a maximum in the range 700°C–900°C (1290°F–1650°F).

The yield of volatile matter in this process is a function of the coal type and ranges from approximately 20% w/w of the coal for a low-volatile bituminous coal to somewhat more than 55% w/w of the coal for a high-volatile C bituminous coal; subbituminous coals may not show a volatile matter peak with increasing temperature. In addition to tarry products, the rapid pyrolysis of coal produces gases such as hydrogen, methane, and carbon monoxide as well as lesser amounts of hydrocarbons. Pyrolysis of coal is generally defined as the thermal decomposition of coal in the absence of air or other added substances.

13.2.1 Pyrolysis

Pyrolysis is the thermochemical decomposition of coal (and organic material in general) at elevated temperatures in the absence of oxygen. Pyrolysis typically occurs under pressure and at operating temperatures above 430°C (800°F). The pyrolysis process generally produces gas and liquid products and leaves a solid residue richer in carbon content (Solomon and Serio, 1994).

Coal pyrolysis is a complex process involving a large number of chemical reactions and the process occurs with the production of gas, liquor (low-molecular-weight liquids), tar (high-molecular-weight liquids), and char (coke) and there is some variation of the product distribution with the temperature of the decomposition (Figure 13.1) (Owen, 1958).

The heterogeneous nature of coal and the complexity of the pyrolysis process have made it very difficult to perform unambiguous experiments to determine the rates and mechanisms in coal pyrolysis. In tar formation and transport, a consensus is being reached on the central role of the volatility of tar molecules in explaining the variation with operating conditions (pressure, heating rate, particle size, etc.) of the amounts and molecular weight distributions of tars. Progress in the quantitative prediction of tar and char yields is being made through recently developed models for the fragmentation of the macromolecular coal network. These models, which provide quantitative descriptions of the relations between the chemical structure of the coal and the physical and chemical properties

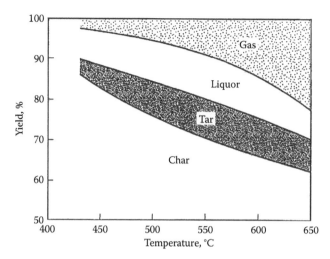

FIGURE 13.1 Distribution of products from coal carbonization at various temperatures.

of the pyrolysis products (gas, tar, soot, and char), are an exciting advance in the understanding of the pyrolysis process. Such models are linking the occurrence of the plastic phase of the coal with the liquid fragments formed during pyrolysis (Solomon et al., 1993).

The role of heating rate on the onset of volatile evolution, volatile yield, product composition, and, to a lesser extent, coal type and particle size were found to be well established. As heating becomes more rapid, the onset of devolatilization shifts to much smaller timescales and to much higher surface temperatures (Maloney et al., 1991; Sunol and Sunol, 1994; Sampath et al., 1996). However, the role of heating rate on coal thermal properties was not found to be well understood. Previous results clearly demonstrated that particle temperature-dependent thermal property assumptions routinely applied in coal combustion models result in large errors (up to 100%) in calculated temperature histories (Maloney et al., 1991; Sampath et al., 1996).

Mechanistically, the pyrolysis of coal has been the subject of many studies (Jamaluddin et al., 1987) and a variety of models have been proposed varying from a sequential scheme of reactions (Chermin and van Krevelen, 1957) to more complex parallel reactions (Pitt, 1962; Anthony et al., 1975; Suuberg et al., 1978; Solomon et al., 1983, 1992; Merrick, 1987), which usually invoke a distribution of activation energies and physical property data but with the same basic fundamental mechanisms (Figure 13.2), realizing that the process may be much more complex and analogous to the mechanism proposed for petroleum asphaltene constituents (Speight, 2007). For example, an important factor in the use of any model is the availability of aliphatic and/or naphthenic (hydroaromatic) hydrogen (Solomon, 1981).

In certain instances, the pyrolysis process may be varied by the addition of a liquid or gas to note the effects of the added material on the product distribution, thereby providing additional information on the nature of the process and/or a more desirable product distribution.

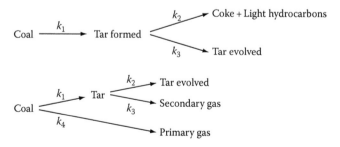

FIGURE 13.2 Hypothetical reaction schemes for coal decomposition.

As part of the products of the process, combustible gases (such as hydrogen and hydrocarbons) are evolved as well as carbon dioxide and water. Other gases, such as hydrogen sulfide, ammonia, and possibly other sulfur and nitrogen compounds, may also appear to varying degrees depending on the composition of the coal and the experimental conditions (Khan, 1989a).

Condensable liquid products and tars will also be produced; these materials are noted for their high aromatic character and high oxygen content (predominantly phenolic oxygen). By analogy with petroleum-derived products, coal products also contain saturated (straight- and branched-chain) material, olefins, aromatics, and naphthenes (saturated hydrocarbon ring systems) with special emphasis on the latter two structural types.

As noted, part of the sulfur can be eliminated as hydrogen sulfide with the remaining sulfur appearing in the char as organically bound sulfur or in the mineral matter ash as metal sulfide or sulfate. This, of course, also depends on temperature and time, and although the amount of sulfur retained in the char/ash tends to vary, it is generally recognized that ca. 50% of the sulfur originally in a coal may be evolved during pyrolysis. However, it should also be recognized that the sulfur that remains in the char/coke will be in a completely different form to that originally present in the coal whether or not the pyrolysis is carried out in the presence of hydrogen (Sugawara et al., 1989, 1991).

In addition to the time and temperature parameters, the rate of removal of the pyrolysis products also has a direct effect on the course of the decomposition and on the nature of the products (Khan, 1989b). In fact, each particular coal will have a definite decomposition temperature (although a general thermal decomposition temperature is 350°C [660°F]), which will be marked by the steady evolution of gaseous products (and an increase in the rate of weight loss) and which can be decreased by pretreatment with organic liquids (i.e., extraction of the lower-molecular-weight species from coal) (Chapters 10 and 11).

Pretreatment of the coal by oxidation increases (as expected) the yields of carbon oxides and water but the coking properties are reduced; conversely, pretreatment with hydrogen improves the coking properties. Finally, the yield of volatile products is also dependent on the particle size of the coal.

The thermal decomposition of coal in the presence of water (or steam) exhibits significant differences over "dry" decomposition. For example, lignite and the lower-rank coals appear to undergo "coalification" by treatment at 315°C–350°C (600°F–660°F) in the presence of water. The effect of water is enhanced if the decomposition is carried out in a "vacuum" and may even be analogous to the stimulus provided by steam in increasing the volatility of heavy crude oils and residua during vacuum distillation.

In addition to water, certain other additives have been noted to have a marked effect on the pyrolysis of coal. Materials such as zinc chloride ($ZnCl_2$), aluminum chloride ($AlCl_3$), sodium hydroxide (NaOH), iron oxide (Fe_2O_3), phosphoric acid (H_3PO_4), clay (aluminosilicate minerals), and active carbon bring about a significant decrease in the amount of coke produced and, with the exception of the phosphoric acid and the carbon, increase the yields of the gases.

Although there is some question as to whether the technique should be included in a section related to the pyrolysis of coal, the thermogravimetric balance has also been employed to investigate the thermal decomposition of coal (Voina and Todor, 1978). The data may be recorded in a variety of ways (Figure 13.3) but virtually always involves a record of the weight percent residue with temperature. The cumulative curve is very similar in shape to the true boiling point (ASTM) curve that is employed as an inspection of various petroleum and petroleum products (ASTM, 2011b,c).

The thermal behavior of coal may also be investigated by use of the differential thermal analysis method. In this technique, the differential change in temperature that occurs when a coal is heated at a given rate is noted to that for a reference material. The derivative is plotted against temperature to produce the thermogram; a positive slope indicates an exothermic reaction whereas a negative slope indicates an endothermic reaction.

For example, peat illustrates exothermic behavior that reaches a maximum at ca. 370°C (700°F) but the maximum is somewhat less pronounced for the higher-rank coals and there is a shift to increasingly higher temperatures. On occasion, an endothermic reaction may be observed in the thermogram at 120°C–150°C (250°F–300°F) and is attributed to the evaporation of the water.

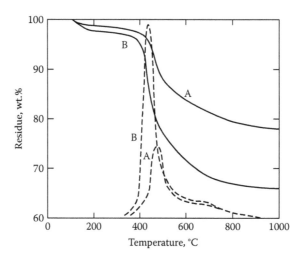

FIGURE 13.3 Thermogravimetric data for two coals (A and B) showing the respective integral curve and the derived curve.

In some coals, it has been observed that initial physical or chemical changes are associated with an endothermic reaction that may start at approximately 250°C (480°F) and continue until approximately 410°C–470°C (770°F–880°F). While other the characteristic endothermic reaction may typically occur at temperatures between 410°C and 610°C (710°F and 1130°F), an endothermic reaction may be observed in the temperature range 750°C–810°C (Bürküt et al., 1994).

On the other hand, thermal treatment of bituminous coal that contains a substantial amount of reactive macerals (such as vitrinite and exinite) shows an exothermic reaction accompanied by softening or plastic behavior followed by solidification (Bürküt et al., 1994).

The mutagenicity of coal pyrolysis products depends on three factors: the chemical structure of the parent coal, pyrolysis temperature, and residence time, as a consequence of the differences in both specific mutagenicity of the pyrolysis products and their yields. Total mutagenicity exhibits the following order with coal type:

High-volatile bituminous > subbituminous > lignite >> anthracite

This order, which is similar to that of volatile matter in the parent coals, reflects the comparative yields of extractable material from the pyrolysis of these coals (Braun et al., 1987).

Total mutagenicity peaks in a temperature range of 1025°C–1225°C (1875°F–2240°F) for all four coals and generally parallels the temperature and residence time trends of aromatic yield. Mutagens form faster and are destroyed faster at higher temperatures. Specific mutagenicity varies with coal type and pyrolysis conditions, but differences exhibit only a secondary effect on total mutagenicity. Specific mutagenic activity reaches peak values at higher temperatures and longer residence times than does organic yield. There is a competition between mutagen formation and mutagen destruction reactions that evinces itself in the residence time/temperature trends of total mutagenicity. Formation reactions prevail at low temperatures and short times; destruction reactions prevail at high temperatures and long times. The resulting products reflect the relative dominance of one set over the other (Braun et al., 1987).

13.2.2 CARBONIZATION

Carbonization is the destructive distillation of organic substances in the absence of air accompanied by the production of carbon and liquid and gaseous products. The coke produced by carbonization of coal is used in the iron and steel industry and as a domestic smokeless fuel.

The thermal decomposition of coal on a commercial scale is often more commonly referred to as carbonization and is more usually achieved by the use of temperatures up to 1500°C (2730°F). The degradation of the coal is quite severe at these temperatures and produces (in addition to the desired coke) substantial amounts of gaseous products.

Low-temperature carbonization, though of somewhat lesser commercial importance, is achieved by the use of temperatures of the order of 500°C–700°C (930°F–1290°F), whereas high-temperature carbonization employs temperatures of the order of 900°C–1500°C (1650°F–2750°F); the temperature range for medium-temperature carbonization processes is self-explanatory.

Four main products are obtained by low-temperature carbonization processes: (1) the final coke or char, (2) an organically complex tar, (3) gases, and (4) aqueous liquor. The proportions are determined, in part, by the rate and the time of heating.

Both fluidized-bed and fixed-bed systems have been used; the former system requires direct contact with another heating medium, whereas fixed-bed systems may be heated directly or indirectly. Carriers can vary and among the carriers for direct heating are steam, air, recycle gases, sand, and metal balls containing a salt to take advantage of the latent heat of fusion. Superheated steam at 540°C–650°C (1000°F–1200°F) has been effective but may actually be economically unfavorable. The commercial aspects of the process usually involve ovens or kilns that are heated externally by a fuel gas, usually by-product gas from the coke ovens themselves.

For any particular coal, the distribution of carbon, hydrogen, and oxygen in the coke is decreased as the temperature of the carbonization is increased. In addition, the yield of gases increases with the carbonization temperature while the yield of the solid (char, coke) product decreases. The yields of tar and low-molecular-weight liquids are to some extent variable but are greatly dependent on the process parameters, especially temperature (Table 13.1 and Figure 13.1), as well as the type of coal employed (Cannon et al., 1944; Davis, 1945; Poutsma, 1987; Ladner, 1988; Wanzl, 1988).

Water has a tendency to reduce the rate of carbonization when certain low-rank coals are employed but often is responsible for increased tar yields. Furthermore, the usual thermal decomposition of coal (i.e., decomposition in the absence of added species) may be markedly influenced by oxidation or by hydrogenation of the coal prior to the thermal process; the presence of mineral matter may also be influential.

There are also indications that finely ground coal has a reduced plasticity (relative to the course goal) and this effect is particularly pronounced for coals that have high proportions of inert material. In addition, high heating rates may greatly increase the fluidity of the coal with the overall effect of an increased plasticity. There is also the need to consider the formation of a mesophase during the

TABLE 13.1
Events That Occur in the Various Temperature Regions of Coal Pyrolysis

Region	Temperature Range (°C)	Reactions	Products
	<350 (<660°F)	Mainly evaporation	Water and volatile organics
Low temperature	400–750 (750°F–1380°F)	Primary degradation	Gas, tar, and liquor
Medium temperature	750–900	Secondary reactions	Gas, tar, liquor plus additional hydrogen
	(1380°F–1650°F)	Secondary reactions	
High temperature	900–1100 (1650°F–2010°F)		
Plasma	>1650 (>3000°F)		Acetylene; carbon black

process since this may/can dictate the progress of the thermal treatment as well as the character of the products (Marsh and Smith, 1978).

Coals that soften and then swell prior to coke formation under atmospheric pressure do not always behave similarly when heated in vacuo. In fact, the reduced pressure actually decreases the degree of softening and swelling. On the other hand, if coal is heated under pressure, the softening and swelling increase and a firmer coke is produced, but to obtain an appreciable effect, higher pressures (on the order of 100 psi and more) may be necessary.

13.2.3 FLASH PYROLYSIS AND HYDROPYROLYSIS

Pyrolysis is the first step of all coal conversion processes and its understanding is essential for effective use of coal. The pyrolysis of coal consists of two series of reactions: first is the primary decomposition that consists of the formation of radical fragments and their stabilization and the other is the secondary gas phase reaction of gaseous components produced by primary decomposition. The product yield during the pyrolysis depends on coal type and operating conditions of pyrolysis (Howard, 1981a,b; Gavalas, 1982; Kawamura et al., 1993).

Among several coal pyrolysis methods, flash pyrolysis is a promising process for producing hydrocarbons and chemicals such as benzene, toluene, and xylene (BTX) (Romovacek and Kubat, 1968; Bonfanti et al., 1977; Giam et al., 1977; Voorhees, 1984; Schobert and Song, 2002). The yield of liquid products from raw coal is limited because of low hydrogen-to-carbon ratio. It is therefore necessary to supply effectively hydrogen from other sources for increasing the tar yield. There are several proposals to supply hydrogen to coal during pyrolysis. Pyrolysis in an atmosphere of hydrogen (hydropyrolysis) is the first choice but it generally requires severe reaction conditions, high hydrogen pressure, high temperature, and long residence time (Kamo et al., 1998; Gu et al., 1999).

The aim of the flash pyrolysis of coal is the production of smaller molecules from it in a shortest possible particle residence time. Therefore, the objective of studying the process chemistry of coal pyrolysis is to investigate the experimental parameters that permit this aim to be achieved and to establish the optimum conditions that produce a favorable product slate. The basic process parameters that influence the product yields during flash pyrolysis of coal are (1) reaction temperature, (2) gas pressure, and (3) residence times of coal particles and ensuing tar vapors. In addition to these major process parameters, product yields can be influenced by other factors such as the nature of the pyrolysis gas and its partial pressure and the gas-to-coal ratio.

Flash pyrolysis in methane, toluene, or methanol atmosphere was proposed but failed to increase the yield of liquid products (Calkins and Bonifaz, 1984; Doolan and Makie, 1985; Hayashi et al., 1996). It is evident that contact at molecular level between hydrogen donor solvents and the coal is essential for increasing the BTX liquid yields (Morgan and Jenkins, 1986; Graff and Brandes, 1987; Kahn, 1989b; Kahn et al., 1989; Miura et al., 1991; Miura, 2000).

Rapid pyrolysis has been briefly mentioned, but in view of the increased importance in this method of the thermal decomposition of coal, more detailed considerations are warranted (Takeuchi and Berkowitz, 1989). The rapid pyrolysis (also often called flash pyrolysis) concept is based on the observation that most coals will release gases, liquids, and tarry materials when heated at certain temperatures, but when the heating is extremely rapid ($102°C/s$ to $106°C/s$) even higher proportions of volatile matter can be released (Weller et al., 1950; Kimber and Gray, 1967; Gibbins and Kandiyoti, 1989; Gibbins et al., 1991).

The premise is that when coal is heated very rapidly to temperatures in excess of those required for decomposition, fragmentation of the coal is very extensive to the point of being "explosive," thereby releasing the liquids from the coal matrix (and reducing the residence time of the liquids within the coal) before secondary reactions to gaseous products can occur. This is especially true in view of the enhanced yields of volatile (liquid and gaseous) products that can be obtained, relative to the standard pyrolytic methods of coal decomposition treatment as well as the standard thermal methods of coal analysis (for example, see ASTM, 2011d; ISO, 2011c) (Chapter 8).

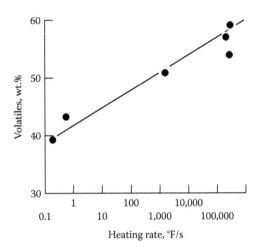

FIGURE 13.4 Relationship of volatile matter yield to heating rate. (From Eddinger, R.T. et al., *Proceedings of the Seventh International Coal Science Conference*, Prague, Czech Republic, 1968.)

Thus, when coal is heated to temperatures in excess of 450°C (840°F) in a fraction of a second rather than over a more prolonged period of time, the chemical constituents of the coal tend to fragment much more extensively. Furthermore, if the resultant transient species are stabilized immediately (and thereby prevented from undergoing recombination), the yields of lower-molecular-weight volatile material may exceed the analytically determined volatile matter content (Chapters 8 and 9) by a substantial amount (Figure 13.4). It is not surprising that this concept has attracted attention as a potentially commercial method for obtaining greater amounts of useful gaseous and liquid hydrocarbons than are usually produced during the more conventional carbonization of coal.

Indeed, both flash pyrolysis and flash hydropyrolysis have come under consideration as possible advanced processes for the conversion of coal to gaseous and liquid fuels on a commercial basis (Bhatt et al., 1981; Bilgesu and Whiteway, 1990). An additional benefit of the rapid heating of heating coal in a hydrogen-containing atmosphere (flash hydropyrolysis; hydrogen costs notwithstanding) is an increase in the overall yield of the volatile products.

Some of the techniques that have been employed to study the flash pyrolysis of coal are (1) laser, (2) microwave, (3) flash tube, (4) plasma, (5) electric arc, (6) shock tube, (7) electric current, and (8) entraining gas (for example, see Hanson and Vanderborgh, 1979; Howard, 1981a,b; Nelson et al., 1988; Mackie et al., 1990; Plotczyk et al., 1990; Maswadeh et al., 1992; Monsef-Mirzai et al., 1992; Pyatenko et al., 1992). Most of the techniques can, at the present time, be classed as research tools that are useful as characterization techniques but the entraining gas technique is under serious consideration for commercialization. In fact, each particular method has some unique aspect that, because of the novelty of the concept, makes a brief discussion here.

Lasers have been employed to produce extremely high temperatures as has microwave heating of coal. Both techniques are reputed to produce considerable quantities of acetylene and the extent of the coal conversion depends on the volatile matter content of the coal, suggesting that the fixed carbon of the coal is not a predominant factor. Flash tubes with short (microsecond or millisecond) irradiation times also produce acetylene as one of the major products and there appears to be little, if any, tar production. Furthermore, the presence of hydrogen appears to reduce the amount of acetylene produced.

Plasmas, which are often produced by exposing a gas stream to an electric arc, are capable of generating temperatures in excess of 10,000°C (18,000°F) and can be employed to study coal reactions, although there is some doubt in regard to the accurate determination of temperature and residence time. Generally, when coal is heated rapidly to temperatures above 1250°C (2280°F), the emitted volatiles are cracked to yield lower hydrocarbons, which, if conditions are favorable, will consist mainly of acetylene. Such conditions involve reaction times of only a few milliseconds coupled with very rapid quenching rates (Nicholson and Littlewood, 1972; Plotczyk et al., 1990). The extent of the reaction depends on the volatile matter content of the coal as well as on the particle size and, again, acetylene is the principal product; the yield of acetylene is increased by the presence of hydrogen.

Electric arc pyrolysis as a means of coal decomposition is accomplished by passing a high-intensity electric arc between two electrodes, one of which is made of coal. The limited data so far available indicate that the products are mainly methane, acetylene, and ethylene with a notable absence of liquid products. Shock tubes produce transient regions of high temperature and high pressure. The products from shock tube experiments have tended to be higher in the proportion of unsaturated hydrocarbons but the overall conversion is usually extremely low.

Techniques using an electric current as the source of heat usually only attain lower temperatures than the previous methods and, as expected, the yields of acetylene are much lower. However, this technique has shown that (1) substantial yields of volatile matter (Chapters 8 and 9) can be achieved at high heating rates, (2) hydrogen is the principal product of coal decomposition at high temperatures, and (3) the yields from the hydropyrolysis of coal are markedly increased if the particle size of the coal is decreased.

The method using an entraining gas may actually be the method that is most amenable to commercial ventures. There have been reports of the near quantitative conversion of the organic material in the coal (bituminous) at 900°C (1650°F) to volatile products (predominantly methane along with other hydrocarbon gases). However, there are still many aspects of this particular technique that are not understood at this time and movement of the concept to commercialization will require assiduous testing followed by pilot work and thence to a demonstration plant.

Thus, while flash pyrolysis and flash hydropyrolysis have not advanced too far beyond the bench-scale studies, there are several potential commercial ventures that employ the pyrolysis of coal as a means of producing liquid products that are becoming more and more viable with the ever-increasing price of petroleum and the more conventional liquid fuels (Chapters 18, 19, and 26).

13.3 PHYSICOCHEMICAL ASPECTS

Coal has been proposed to have a macromolecular network structure to which concepts of cross-linked polymers can be applied. These concepts have been employed to understand and model such properties of coal as (1) the insolubility, (2) the equilibrium swelling and penetration of solvents, (3) the viscoelastic properties, (4) similarities between the parent coal and products of hydroge-nolysis or mild oxidation, (5) cross-linking during char formation, and (6) the formation of coal tar in pyrolysis.

With the success of these concepts in describing coal properties, it appears logical to extend mac-romolecular network concepts to completely describe coal thermal decomposition behavior. This has been done by applying statistical methods to predict how the network behaves when subjected to thermally induced bond breaking, cross-linking, and mass transport processes.

In applying network models to coal thermal decomposition, one considers the coal to consist of aromatic ring clusters linked together. When the coal is heated, the bridges can break and new bridges can form. Various statistical methods can be employed to predict the concentration of single aromatic ring clusters (monomers) and linked clusters up to a totally linked network. By assigning an average or distribution of molecular weights to the monomers, the amounts of tar, extractable products, liquids, or char can then be defined from the distribution of oligomer sizes.

When heated to temperatures where thermal decomposition occurs, coal undergoes a large variety of physical and chemical changes resulting in the evolution of volatile material and the formation of a solid residue that is composed predominantly of carbon. However, some changes may be noted before the onset of what is often referred to as the "thermal decomposition proper" (i.e., carbon–carbon bond scission and the like), and may manifest themselves as the formation of low-molecular-weight species (Stein, 1981; Hessley et al., 1986).

In addition to the chemical changes that occur in various temperature regions (leading to the evolution of a variety of volatile products), the thermal decomposition of coal generally produces a residue (coke) that consists of carbon and mineral matter and during this process the coal passes through several physical stages. Thus, heated coal generally becomes "plastic" and softens, but as the heating progresses the "plastic" mass may solidify even before the onset of the coking reactions. This behavior has been assumed to be a result of the rupture of chemical bonds in coal, thereby allowing a higher degree of mobility, but continued heating causes the reformation of the bonds or the formation of new bonds with solidification ensuing. The FSI test for coal (ASTM, 2011a), an important physical test (Chapters 8 and 9), is based on the behavior of coal when heated and allows coals to be classified as caking or noncaking as well as giving an indication of the agglomerating properties of coal.

13.3.1 THERMAL AND RHEOLOGICAL BEHAVIOR

It is generally assumed that in the temperature range below which the onset of thermal decomposition proper occurs the major changes in coal are due to the evolution of volatile products from thermally susceptible functional groups and very little occurs in the way of rheological changes. However, there may be the occurrence of chemical reactions that lay the foundation for a variety of rheological changes in the higher temperature ranges. Such events have received very little attention and are worthy of some investigation because of the potential effects on the utility of the coal.

For example, preheating coal at ca. 200°C (ca. 390°F) tends to have an adverse effect on the caking properties but may also increase ease of, say, gasification (insofar as caking coals can be difficult to gasify efficiently) (Chapters 20 and 21) and may also increase the ease of dissolution by organic solvents during liquefaction processes (Chapters 18 and 19).

Nevertheless, in the temperature range 350°C–500°C (660°F–930°F), coals (depending on rank) soften, become plastic, and then coalesce to form the solid residue (coke). Two important physicochemical aspects of the change ensue and can be identified as follows: (1) the coal becomes plastic, and (2) at the higher temperatures of this range, the solid material contracts.

Prime quality coking coals have volatile matter contents in the range 20%–32% (dmmf), become plastic before active decomposition occurs and, thus, yield strong coke with good abrasion resistance. Viscosity and rate of devolatilization of the plastic mass are such as to minimize intragranular swelling but enable neighboring coal particles to adhere strongly. The coke thus formed has fairly uniform pores of small diameter surrounded by relatively thick walls, thereby ensuring good resistance to abrasion.

Coal with volatile matter content outside the 20%–32% range undergoes decomposition both before and during the plastic temperature zone (Figure 13.5); usually pronounced swelling occurs with, on occasion, "foaming" giving rise to high porosity and thin-walled pores of large diameter within the coke.

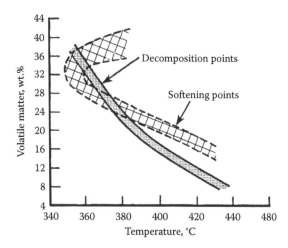

FIGURE 13.5 Relationship of volatile matter yield to softening points and decomposition points of various coals. (From Gibson, J., *Coal and Modern Coal Precessing: An Introduction*. G.J. Pitt and G.R. Millward (Eds.), Academic Press, New York, 1979.)

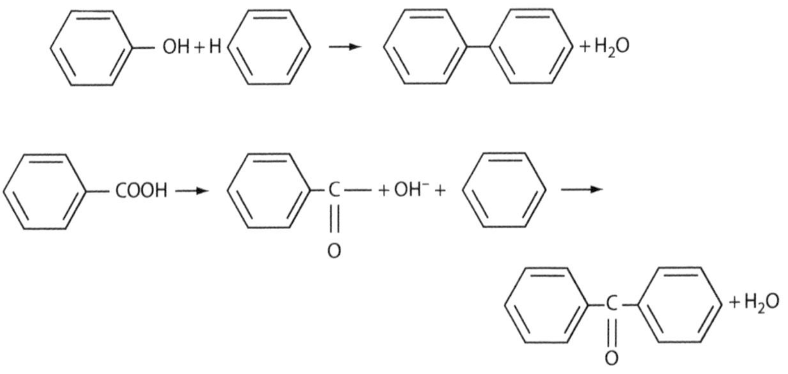

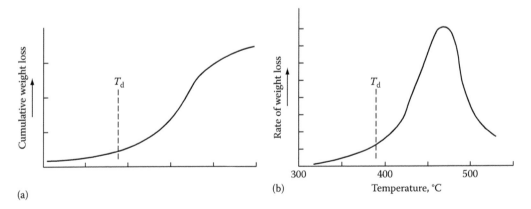

FIGURE 13.6   Possible reactions for the functional groups during coal pyrolysis.

Such coke has a relatively low resistance to abrasion and thus it is apparent that the inherent strength and abrasion resistance are predetermined by the behavior of the original coal in the plastic range.

From a more chemical aspect, in the initial stage of the thermal decomposition of coal (which can commence well below 200°C, or 390°F), the rate is usually slow and the process predominantly involves release of water, oxides of carbon, and hydrogen sulfide. As already noted, these products are considered to be the result of the thermal decomposition of thermally labile substituent groups that occur within the coal or from facile (but usually infrequent) condensation reactions (Figure 13.6).

At temperatures in excess of 200°C (390°F), carbon isomerization appears to commence as evidenced by the evolution of small amounts of alkyl benzenes. The exact mechanism of these low-temperature reactions is unknown and remains open to speculation, but it has been noted that these reactions are significant enough insofar as the original coal structure is sufficiently changed to influence any subsequent thermal behavior of the coal.

As the temperature is raised after the resolidification stage, the solid usually undergoes contraction, but at a rate that is determined by the rate of devolatilization, and will vary from coal to coal. In fact, two temperature ranges of decomposition have been identified; the first occurs shortly after resolidification at ca. 500°C (ca. 930°F) and the second stage occurs at approximately 750°C (approximately 1380°F).

The first stage (often termed "active thermal decomposition") more generally occurs in the range 350°C–550°C (660°F–930°F) and the decomposition temperature is conventionally defined as the temperature at which the weight loss markedly increases (Figure 13.7) and varies significantly with

FIGURE 13.7   Derivation of the decomposition temperature from weight loss data. (a) Cumulative weight loss. (b) Rate of weight loss. (From Berkowitz, N., *An Introduction to Coal Technology*, Academic Press, Inc., New York, 1979.)

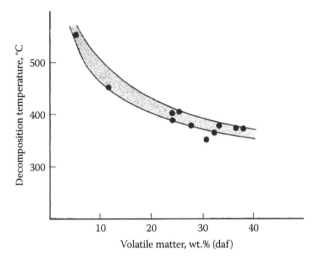

FIGURE 13.8 Variation of volatile matter yield and decomposition temperature. (From Berkowitz, N., *An Introduction to Coal Technology*, Academic Press, Inc., New York, 1979.)

coal rank (Figure 13.8). In general terms, approximately 75% of the volatile matter that is released by the coal (including any tarry material and lighter hydrocarbons) is produced within this temperature. However, the composition of the products (including the composition of the tars) varies considerably with the experimental conditions (heating rate, final temperature, pressure, etc.) (Khan, 1989b). This variation has been attributed to the fact that the so-called active decomposition of the coal involves extensive fragmentation of the constituents and the various reaction alternatives that can occur in the system.

The final stage of the thermal decomposition of coal (which is represented by the relatively flat portion of the weight loss curve) (Figure 13.7) is often referred to as the secondary degasification stage (usually characterized by the elimination of heteroatoms) and is complete when the char is transformed into a graphitic material. However, in general the volatile matter that is released above 800°C–850°C (1470°F–1560°F) has received little attention, the major focus being on the nature of the changes that occur during the graphitization process.

The most important feature that has been noted within the temperature range has been the progressive aromatization of the solid leading to the production of carbon lamellae of increasing diameter and such behavior has been confirmed by the thermal decomposition of polynuclear aromatic compounds to coke or carbon (Lewis and Singer, 1988). The results of such studies have been projected by others to assumptions about the structural nature of coal.

In summary, it has been assumed, on the basis of the behavior of the thermal decomposition of polynuclear aromatic systems, that coal must also consist of large polynuclear aromatic systems (Chapter 10). Be that as it may, such assumptions are highly speculative and, to say the least, somewhat lacking in caution. As an example, similar lines of thought have been applied to structural assumptions about petroleum asphaltene constituents when it is known from other pyrolysis studies that smaller, but polar, systems can produce as much thermal coke as the larger nonpolar highly condensed systems (Speight, 2007).

Nevertheless, to summarize the thermal decomposition of coal, the plastic stage involves the breaking of cross-linkages, comprising either oxygen or nonaromatic carbon bridges between neighboring aromatic groups, leading to mobility of some of the decomposition products. The lower-molecular-weight components can undergo further change to yield (1) the gaseous and (2) the highly complex mixtures found in coal tar. The higher-molecular-weight fractions remain and form semicoke on solidification.

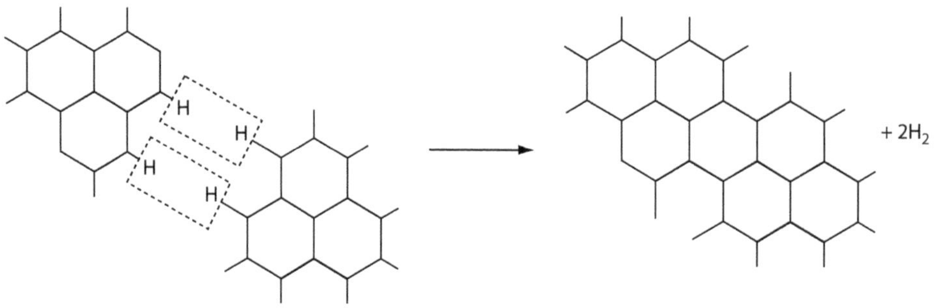

FIGURE 13.9 Condensation reaction leading to coke formation.

The semicoke has viscoelastic properties up to 700°C (1290°F) and transforms to hard coke (a brittle solid) as the temperature rises to approximately 1000°C (approximately 1830°F). The process is accompanied by the elimination of hydrogen, growth and expansion of aromatic lamellae, and resolidification to the semicoke. The predominant gaseous product is hydrogen, which, presumably, arises from reactions involving intermolecular and intramolecular condensation of aromatic systems (Figure 13.9) that ultimately lead to the extensive aromatic (but hydrogen-deficient) structures in the coke.

One particular important aspect of the physicochemical effects involved in thermal treatment of coal is the behavior of coking coal.

Coking coal for steelmaking begins to soften and melt at approximately 400°C (750°F), shows maximum fluidity at approximately 450°C (840°F), and resolidifies at approximately 480°C (895°F). This behavior is considered to be related to the strength and properties of the coke. In the process, the thermal decomposition of aliphatic constituents of coal begins at a temperature around 400°C (750°F)—a temperature region in which the thermal decomposition is operative but does vary according to the type of coal. Furthermore, the rate of thermal decomposition of the aliphatic constituents also varies according to the type of coal. The lower the degree of coalification, the higher is the rate of thermal decomposition in the low-temperature region. On the other hand, the aromatic constituents of coal begin to decrease slightly at a temperature around 500°C (930°F), and this effect is more obvious at 600°C (1110°F) (Fujioka et al., 2006).

13.3.2 KINETIC ASPECTS

The kinetic aspects of the thermal decomposition of coal to coke and volatile matter have received considerable attention (e.g., see Howard, 1981a,b; Lohmann, 1996); as a result of plasticity and other physical measurements, the reactions leading to the formation of coke and volatile matter from (coking) coal have been suggested to involve a stepped sequence (see also Figure 13.2):

Coal → metaplast (k_1)
Metaplast → semicoke + primary volatiles (k_2)
Semicoke → coke + secondary volatiles (k_3)
k_1, k_2, and k_3 are the rate constants.

Thus, in the first reaction sequence, depolymerization occurs and the coal is transformed into the (unstable) intermediate that is responsible for plastic behavior and which is then followed by transformation of the metaplast to semicoke (resolidification). The second reaction is accompanied by swelling (due to gas evolution), and finally the formation of a more compact, coherent, coke occurs by loss of methane and hydrogen from the semicoke.

It is difficult to provide a complete description of the coking process but this kinetic concept does provide a "semiquantitative" explanation of the physical phenomena that occur as part of the

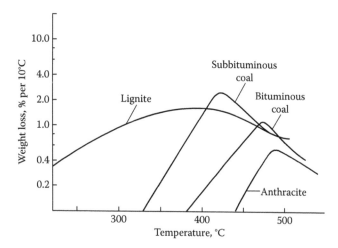

FIGURE 13.10 Variation of weight loss to pyrolysis temperature for various coals. (From Van Krevelen, D.W. et al., *Fuel*, 30, 253, 1951.)

process. There have also been the suggestions that, since the nature and yield of volatile matter depend on the type and rank of the coal (Chapters 2, 8, and 9), there is a direct relationship between the carbon content (or the atomic hydrogen/carbon ratio) of a coal and the extent to which it decomposes at any particular temperature.

For example, as the carbon content of the coal increases, the active thermal decomposition occurs in successively higher temperature ranges and the maximum weight loss decreases quite substantially (Figure 13.10). In addition, different macerals in any one particular coal also generate different amounts of volatile matter, and, thus, similar trends are noted when exinites, vitrinites, and inertinites of the same rank are thermally decomposed (Figure 13.11).

However, the active thermal decomposition of coal can be diminished at any particular stage by terminating the heating program and then maintaining a constant temperature. The residue from the heating will then asymptotically approach an equilibrium value that depends not only on the temperature

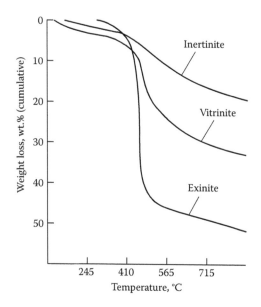

FIGURE 13.11 Variation of weight loss to pyrolysis temperature for various macerals. (From Kroger, C. and Pohl, A., *Brenst. Chem.*, 38, 102, 1957.)

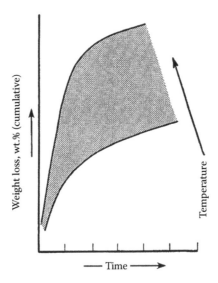

FIGURE 13.12 Representation of variation of weight loss with time at different pyrolysis temperatures.

of the experiment but also on the coal type; any further increase in the release of the volatile matter will only occur after an additional increase in temperature and may require up to 35 h for the equilibrium state to be reached (Figure 13.12). This characteristic behavior pattern consists of a series of near-simultaneous reactions and has been the focus of several kinetic studies (Liguras and Allen, 1992).

Rate data may be obtained by means of thermogravimetric analysis and can be measured while the temperature is being increased at a constant rate. Thus, if the rates of coal decomposition are functions of the undecomposed coal,

$$-df/dt = kf^n$$

where
-df/dt is the rate of the decomposition
k is the velocity constant
n is the order of the reaction
f is the fraction of the coal that is undecomposed

Since

$$k = k_0 k_{-A/T} \quad \text{and} \quad x = dT/dt$$

it follows that

$$-df/f^n = k_0/(xe^{-A/T} dt)$$

A is related to the gas constant and activation energy of the reaction $A = E/R$, and by the process of approximation it can be determined whether the reaction is zero, first, or second order when the integral of $-df/f^n$ is $1 - f$, $\ln f$, or $(1/f) - 1$, respectively.

In more general terms, the thermal decomposition of coal is a complex process (Stein, 1981; Solomon et al., 1992). Activation energies determined by experimental techniques indicate that the decomposition rate(s) is (are) controlled by the scission of carbon–carbon covalent bonds and the like (Poutsma, 1987). In fact, the concepts that bond scission during coal pyrolysis can be induced by other means (McMillen et al., 1989) or can be influenced by cross-linking are sound and deserve consideration in the light of coal complexity and the potential interference of the primary products with one another as well as with the vestiges of the original coal.

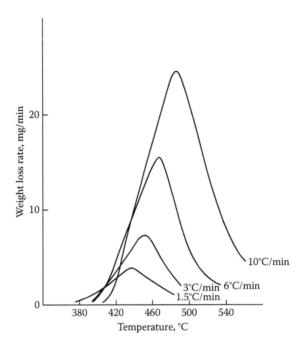

FIGURE 13.13 Relationship of weight loss to heating rate. (From Berkowitz, N., *An Introduction to Coal Technology*, Academic Press, Inc., New York, 1979.)

But it needs to be emphasized that caution must be exercised in drawing too many strict (and/or inflexible) conclusions from activation energy data. This, of course, must also throw some doubt on, and add caution to, the inconsiderate use of model compounds as materials from which to project coal behavior during coal pyrolysis. However, on a positive note, the use of model compounds does offer valuable information about pyrolysis mechanism; it is the means by which the conclusion is drawn with respect to coal that can hurt the effort. Finally, the concept of induced bond scission (McMillen et al., 1989) also opens up the area of coal pyrolysis to the additional concept of selective bond breaking by addition of suitable reagents.

Many factors play a role in producing the observed data and the activation energies are only, at best, a derived "average" for the many processes that are occurring in the system. The fact that the weight loss depends on the rate of heating (Figure 13.13) is perhaps an oversimplified representation of the many factors that influence the data and, therefore, play a major role in the validity of the derived kinetic data.

13.3.3 PETROGRAPHIC COMPOSITION

The influence of the petrographic components on the thermal decomposition of coal has been alluded to briefly in Section 13.3.2 (Figure 13.11), but because of the importance of the effects of the different macerals on coal properties in general, some further notation is warranted here.

The reactive constituents of coal (i.e., vitrinite and exinite) (Table 13.2) soften in the temperature range 400°C–500°C (750°F–930°F) and act as the so-called binder for the "inert" material that does not soften, nor does it undergo subsequent contraction to the same extent as the "reactive" components; it does, however, lose volatile matter and suffer internal structural chemical change. The role of the inert material in, for example, the coking process is to reduce the overall swelling and/or contraction.

The phenolic hydroxyl content of bituminous vitrains drops sharply to low values when the coal is heated to 450°C–500°C (840°F–930°F). It is also plausible to suggest that the hydroaromatic hydrogen content also drops sharply in a similar or somewhat lower temperature range.

Thus, if the desired coke is to be produced, a compositional balance must be achieved between the reactive and inert components of the feed coal. If the coal is a caking coal, there will be a

TABLE 13.2

Relative Reactivity of Macerals during Coking

Coking Characteristic	Petrographic Constituents
Reactive	Vitrinite with reflectance between 0.5% and 2.0%; exinite
Intermediate (partially reactive)	Semifusinite (reflectance less than 2.0%)
Inert	Fusinite, micrinite; vitrinite with reflectance greater than 2.0%; mineral matter (shale, pyrite, etc.)

characteristic transient plastic stage between 350°C (660°F) and 450°C (840°F) and thereby leading to the formation of a predominantly porous coke.

The yields and compositions of the tars obtained from the thermal decomposition of the individual petrographic components (Table 13.3) exhibit the most marked differences in the lower-rank coals, in which the oxygen constituents appear to play a more significant role in the decomposition processes.

Furthermore, if the thermal decomposition of the macerals can be related to the coal precursors (such as lignin) and if, in turn, the thermal decomposition of the precursors can be related to their contributory molecules (e.g., coniferyl alcohol; Chapter 3), several pertinent facts can emerge about coal pyrolysis and decomposition. On this basis, assuming no interference between the decomposition products (McMillen et al., 1989), at 200°C–270°C (320°F–590°F), dimerization and oligomerization reactions appear to predominate whilst side-chain carbon–carbon bond scission, dehydration, rearrangement, and hydrogen transfer reactions appear to be of lesser importance (Masuku, 1992). In addition, the thermolytic reactions of the allylic side chain(s) occur more readily than those of the aryl–alkyl ether linkages. It is also worthy of note at this point that the thermal decomposition of coal in the presence of added hydrogen appears to also involve the catalyst-induced cleavage of heteroatom bonds (Robinson et al., 1991). Although hydrogen is added to this reaction matrix, the same reactions may also occur when thermally induced hydrogen transfer occurs throughout the organic matrix of the coal.

Should this premise be true, it may offer some explanation of the reactions occurring during the early stages of coal pyrolysis as well as present an indication of the chemistry of the onset of carbonization. It has been theorized, from activation energy data, that coal decomposition proceeds, primarily, by way of carbon–carbon bond scission. It would be of interest to determine what changes, if any, are caused by the use of the model compound that is reputed to be a precursor to coal. In addition, it might be proposed that the decomposition of coniferyl alcohol deposited on coal be studied to note what changes are brought about under these conditions.

TABLE 13.3

Yields (% w/w) of Condensable Products from Maceral Decomposition

	Exinite	Vitrinite
Light oils	2.8	1.0
Heavy oils	29.8	2.3
Heavy-oil composition		
Acids	0.3	2.0
Phenols	3.8	21.6
Bases	1.6	6.4
Neutral oils	90.5	70.0

Source: Macrae, J.C., *Fuel*, 22, 117, 1943.

13.4   THERMAL DECOMPOSITION PRODUCTS

The products of the thermal decomposition of coal are solids (coke, char), liquids (oil), and gases. The yields of liquid and gaseous products obtainable from any particular coal can be determined experimentally by quantitative "destructive distillation" of the coal and several such methods have been proven to provide data that can be correlated with the behavior of the coal on an industrial scale. Briefly, these methods are (1) the Fischer assay that was originally developed in Germany by Fischer and Schrader and was, with some modifications, adopted in the United States as well as in a number of European countries and (2) the Gray–King assay that originated in Britain and is now extensively used as a means for classifying coals (Table 13.4).

TABLE 13.4
National Coal Board (United Kingdom) System for Coal Classification (Chapter 2)

Group	Class	Volatile Matter (% dmmf)	Gray–King Coke Type	Description
100	101	6.1[a]	A	Anthracite
	102	6.1–9.0[a]	A	Anthracite
200	201	9.1–13.5	A–G	
	201[a]	9.1–11.5	A–B	Dry-steam coals
	201[b]	11.6–13.5	B–C	
	202	13.6–15.0	B–G	
	203	15.1–17.0	B–G4	Coking steam coals
	204	17.1–19.5	G1–G8	
	206	9.5–19.5	A–B[b]	Heat-altered low-volatile bituminous coals
			A–D[c]	
300	301	19.6–32.0		
	301[a]	19.6–27.5	G4	Prime coking coals
	301[b]	27.6–32.0		
	305	19.6–32.0	G–G3	
	306	19.6–32.0	A–B	Heat-altered medium-volatile bituminous coals
400	401	32.1–36.0		
	402	>36.9	G9	Very strongly caking coals
500	501	32.1–36.0		
	502	>36.0	G5–G8	Strongly caking coals
600	601	32.1–36.0		
	602	>36.0	G1–G4	Medium caking coals
700	701	32.1–36.0		
	702	>36.0	B–G	Weakly caking coals
800	801	32.1–36.0		
	802	>36.0	C–D	Very weakly caking coals
900	901	32.1–36.0		
	902	>36.0	A–B	Noncaking coals

[a] To distinguish between classes 101 and 102, it is sometimes more convenient to use a hydrogen content of 3.35% instead of 6.1% volatile matter.
[b] For volatile matter contents between 9.1% and 15.0%.
[c] For volatile matter contents between 15.1% and 19.5%.

TABLE 13.5

Fischer Assay (500°C [930°F]) of Various Coals[a,b]

Coal	Tar (gal/ton)		Light Oil (gal/ton)		Water (gal/ton)		Gas (scf/ton)	
	Limits[b]	Mean	Limits[b]	Mean	Limits[b]	Mean	Limits[b]	Mean
Bituminous, LV	6.3–12.7	8.6	0.7–1.6	1.0	1.1–6.6	3.2	1600–1960	1760
Bituminous, MV	9.7–25.6	18.9	1.0–2.3	1.7	2.8–6.6	4.1	1390–2240	1940
Bituminous A, HV	22.9–40.7	30.9	1.5–3.3	2.3	3.0–9.2	6.0	1690–2360	1970
Bituminous B, HV	24.3–43.1	30.3	1.6–3.4	2.2	10.2–13.1	11.1	1660–2420	2010
Bituminous C, HV	18.5–38.8	27.0	1.3–2.7	1.9	12.0–19.1	15.9	1560–2070	1800
Subbituminous A	18.4–24.4	20.5	1.4–1.9	1.7				
Subbituminous B	13.2–16.7	15.4	1.1–1.6	1.3	23.3–30.4	27.8	1390–2760	2260
Cannel	53.7–108.3	73.5	3.7–7.4	5.1	2.0–4.8	3.7	1500–2120	1810

Source: Selvig, W.A. and Ode, W.A., Bulletin No. 571, United States Bureau of Mines, Department of the Interior, Washington, DC, 1957.

[a] 1 gal (U.S.) = 3.785 L; 1 ton = 1000 kg; 1 scf = 1 ft³ = 2.83 × 10² m³.

[b] Assay performed on several samples.

The assays are conducted at 500°C or 600°C (930°F or 1110°F), even though they tend at these temperatures to underestimate gas volumes and to show maximum tar plus "light oil" yields, and both are sufficiently precise to yield mass balances of 100 ± 0.3%. The Gray–King assay can also be carried out at 900°C (1650°F) to provide information about the quality of the solid residue.

Typical assay data illustrate how yields of gas, water, tar, and light oil vary with the coal (Table 13.5); anthracites, with less than 4% hydrogen, produce little or no tar, whereas the sapropels (i.e., cannel and boghead coals), which contain substantial amounts of hydrogen, produce up to 400% as much tar as the analogous (equivalent carbon content) humic coals.

The relationship of the ultimate composition of coal-to-tar yield in an assay is emphasized by the development of various formulae that have evolved from the assay data as a means of estimating product yields. For example, thaw yield of water can be expressed as a percentage of the (daf) coal:

$$H_2O = 0.35(\% \ O)^{1.3}$$

It should be remembered that the majority of the products obtained from the thermal decomposition of coal are formed mostly by the random combination of free-radical species that are generated during the process (Gun et al., 1979):

R–R → R• + R•	Thermolysis
R• + H• → RH	Hydrogenation
R• → RH + R¹•	Fragmentation
2R• → RH + R²CH=CH₂	Disproportionation
R• + R³• → RR³	Coupling, condensation

Thus, not only tar yields but also tar composition depends largely on how the coal is heated.

Indeed, of special importance here is the maximum temperature to which the tar vapors are exposed before they are condensed as well as the bonds considered likely to participate in the thermolysis sequence (Benjamin et al., 1978; Miller and Stein, 1979). However, even though bond dissociation energies (Table 13.6) (Szwarc, 1950) will give some indication of the bonds that are most likely to be cleaved during coal pyrolysis, there are other factors involved that may give additional

TABLE 13.6

**Dissociation Energies of the Bond Types
That Occur in Coal**

Compound[a]	Bond Dissociation Energy (kcal/mol)[b]
$C_6H_5—C_6H_5$	103
$RCH_2CH_2—CH_2CH_2R$	83
$C_6H_5—CH_2C_6H_5$	81
$RCH_2—OCH_2R$	80
$C_6H_5CH_2—CH_3$	72
$C_6H_5CH_2—CH_2CH_2C_6H_5$	69
$C_6H_5CH_2—OCH_3$	66
$C_6H_5CH_2—CH_2C_6H_5$	57
$C_6H_5CH_2—OCH_2C_6H_5$	56
$C_6H_5CH_2—OC_6H_5$	51
$C_6H_5CH_2—SCH_3$	51
$CH_2=CHCH_2—CH_2CH=CH_2$	38

[a] C_6H_5 is the phenyl group; R is an alkyl group.
[b] 1 cal = 4.187 J; 1 mol = molecular weight in grams.

strength to an otherwise weak bond and, conversely, strong bonds may be weakened by other factors such as juxtaposition to an activation group or molecular entity.

For example, stabilization of the resulting benzylic free radical, by ring substituents, after the thermolysis of a benzyl ether moiety or the activation effect of oxygen (e.g., phenolic) substituents on an aromatic ring (McMillen et al., 1981) may be cited as examples where the "order" dictated by bond dissociation energies might be modified under the prevalent conditions. Steric factors may, more than likely, also be a consideration.

It should also be noted that the liquids from pyrolysis or carbonization are generally unsatisfactory for immediate use as liquid fuels without some form of further upgrading. While the aromatic and cyclic constituents may indicate good antiknock properties for motor fuel, the presence of oxygen-containing compounds, especially phenols, dictates further treatment.

The distribution of products obtained from coal decomposition is a function of temperature. For example, low-temperature carbonization (500°C–700°C [930°F–1290°F]) does not produce "true" coke; the solid product is a char, or semicoke. Other products include gas, a heavy aromatic hydrocarbon tar, some light hydrocarbon oils, and an aqueous liquor containing soluble products. As noted, product yields depend on the coal but 9–48 US gallons (7.5–40 Imperial gallons; 35–180 L) of tar is usually obtained per ton (1000 kg) of coal. This tar may contain up to 23% of the heating value but by far the majority (68%–87%) of the calorific value of the original coal remains in the char.

In fact, chars from coal thermal decomposition have recently received more attention than in the past and have been recognized as valuable (*added-value*) products (Chapters 16 and 17).

13.4.1 GASEOUS AND LIQUID PRODUCTS

An appreciable fraction of thermally degraded coal may occur as such gases as hydrogen, carbon monoxide, methane, ethane, and ethylene. In fact, gas from a coal carbonization process has a calorific value of ca. 500 Btu/ft³ and, as such, is a medium-Btu gas; upgrading of this gas by methanation is possible.

Until about 1920, coal was the major source of ammonia but the onset of a blossoming petrochemical industry reduced the importance of coal as an ammonia source. However, recent shortages of natural gas and petroleum feedstocks have led to a revival of interest in the production of ammonia from coal. In fact, reliance on coal hydrogenation for the production of synthetic liquid fuels could lead, once again, to an ammonia-from-coal industry

$$N_{coal} + H_2 \rightarrow NH_3 + \text{other products}$$

or the conversion of coal to synthesis gas ($CO + H_2$ mixtures) would enable production of ammonia at the site.

Ammonia can be recovered from coal gas by gas scrubbing to produce a dilute solution of ammonia that can be distilled to produce a concentrated ammonia liquor, but in many instances the ammonia gas can be stripped from scrubbing solutions by addition of base to raise the pH above the pH of ammonia, thereby enabling air stripping of ammonia gas from solutions:

$$NH_4^+ \rightarrow H^+ + NH_3$$

However, ammonia may also be reclaimed from coal gas as ammonium sulfate by reaction with sulfuric acid to produce ammonium sulfate (Figure 13.14):

$$2NH_3 + H_2SO_4 \rightarrow (NH_4)_2SO_4$$

The standard procedure for the manufacture of ammonium sulfate from coal gas involves several steps. The gas is first cooled to ca. 32°C (90°F) in an appropriate condensation system. Most of the tar, which is a very troublesome contaminant of ammonium sulfate (and vice versa), condenses and, in addition, much of the water containing approximately 25% of the ammonia, primarily as ammonium salts, also condenses. This water is rendered basic (lime treatment), thereby converting the ammonium ion to ammonia, which is recovered by being stripped off in a lime still and placed back in the coal gas stream. The coal gas stream is heated to above its dew point (approximately 65°C; 150°F) and the ammonia is adsorbed in 5%–10% sulfuric acid solution contained in a lead-lined saturator at a temperature of 50°C–60°C (120°F–140°F); ammonium sulfate crystals precipitate from the sulfuric acid solution.

In addition to ammonia, coal gas contains small quantities of miscellaneous compounds that may be removed for their own value or to improve the gas product; among the more common of these are pyridine bases and hydrogen cyanide. The former are nitrogen-containing organic materials

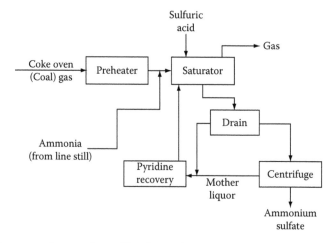

FIGURE 13.14 Recovery of ammonium sulfate from coke oven gas.

that belong to the pyridine series (C_5H_5N) and occur in coal tar. They can be removed from the gas by reaction with sulfuric acid, which converts the bases to nonvolatile sulfates, and the "pyridines" can be recovered as a valuable by-product from the scrubber liquor by neutralizing the liquor with ammonia followed by steam distillation of the uncharged bases. The condensate is saturated with ammonium sulfate, which reduces the solubility of the uncharged bases and causes them to precipitate from solution.

Both hydrogen cyanide and hydrogen sulfide can be removed by alkaline scrubbing of coal gas by conversion to nonvolatile species:

$$HCN + OH^- \rightarrow H_2O + CH^-$$

$$H_2S + OH^- \rightarrow H_2O + HS^-$$

Each process is reversible (by neutralization), which facilitates recovery of hydrogen cyanide and hydrogen sulfide from the scrubber solutions.

Hydrogen sulfide can be removed from coal gas at relatively high temperatures using solid sorbents (Chapter 23). The process (at approximately 540°C, 1005°F) involves the reaction of the hydrogen sulfide with hydrogen in the presence of iron oxide:

$$Fe_2O_3 + 2H_2S + H_2 \rightarrow 2FeS + 3H_2O$$

Ensuing regeneration of the sorbent is accomplished by high-temperature reaction with oxygen to yield a concentrated stream of sulfur dioxide and regenerated iron oxide.

The thermal decomposition of coal produces a light oil that does usually condense with the tar product and may actually occur as a part of the gas stream. There are two major processes for the removal of the light oil from the gas stream: (1) a process that involves countercurrent washing of the gas stream with a gas oil fraction obtained from petroleum and (2) adsorption of the light oil on activated carbon. The light oil product can be recovered from the gas oil or from the adsorbent by steam stripping.

A typical composition of a light oil product from the high temperature (900°C [1650°F]) of coal might contain (Table 13.7) benzene, toluene, and the xylenes, as well as alkanes (e.g., n-hexane), cycloalkanes (e.g., cyclohexane), olefins (e.g., n-hexene), and a wide variety of various aromatic compounds.

The fractional distillation of light oil yields a variety of cuts of which one of the more well-known fractions is benzol (a crude grade of benzene). The light oil fractions can be purified by a

TABLE 13.7

Composition of the Light Oil Product from the High-Temperature Carbonization of Coal

Constituent	Amount (% w/w)
Benzene	61
Toluene	23
o-Xylene	2
m-Xylene	4
p-Xylene	4
Styrene	2
Miscellaneous (aromatic, aliphatic, and olefins)	4

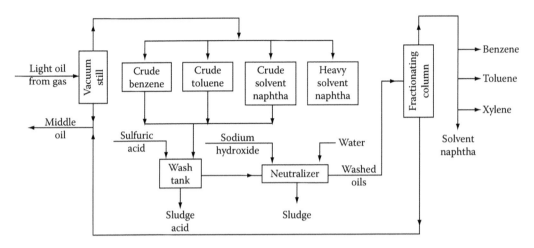

FIGURE 13.15 Refining the light oil product from the high-temperature carbonization of coal.

sulfuric acid wash to remove olefins, neutralized by a sodium hydroxide wash, and finally redistilled to yield products such as benzene, toluene, and xylene, as well as a naphtha solvent (Figure 13.15). Some of the minor components of the light oil, such as indene, benzofuran (coumarone), and dicyclopentadiene (Table 13.8), have been employed as a source of industrial resins.

Sulfur compounds occur to the extent of approximately 1% in the light oil; the most abundant are carbon disulfide and thiophene, followed by methyl- and dimethylthiophenes, ethylthiophenes, propylthiophenes, organic sulfides, and organic disulfides. Carbon disulfide is readily removed by fractional distillation; sulfur compounds such as thiophenes, sulfides, and disulfides can be removed by sulfuric acid washing. Sulfur compounds that escape this treatment can be removed by treatment with solid sodium hydroxide using processes that are standard in the petroleum industry (Speight, 2007).

The tar product from coal carbonization contains some valuable chemicals. Among the most valuable of these are phenols, cresols, pyridine, and naphthalene (Table 13.9). In fact, a high

---

TABLE 13.8
Minor Constituents in the Light Oil
Product from the High-Temperature
Carbonization of Coal

Indene

Coumarone

Dicyclopentadiene

TABLE 13.9

Examples of Chemicals from Coal Tar

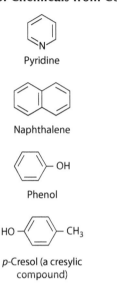

Pyridine

Naphthalene

Phenol

p-Cresol (a cresylic
compound)

---

proportion of the high-temperature tar consists of materials that can satisfy the need for coal tar
fuels, refined tars, pitch, and creosote (Table 13.10).

Thus, raw coal tar represents an excellent source of individual chemicals as well as fractions of specific classes of chemical mixtures. The separation process is essentially a distillation process in which one, or more, of the relatively volatile fractions is removed from the
product. The less-volatile materials consist of creosote oil, and the residue that remains after
the conclusion of the distillation is often referred to as pitch (Nair, 1978). The tar acids (which
will often distill with the various fractions) are extracted from the particular fractions with
aqueous sodium hydroxide whereas the tar bases (the nitrogen-containing organic materials) are
extracted with sulfuric acid in a manner similar to the removal of these products from the gases.

TABLE 13.10

**Representative Composition of Tar
from the High-Temperature
Carbonization of Coal**

Constituent	Amount (% w/w)[a]
Coal tar fuels	29.5
Refined tars	29.5
Pitch	17.7
Creosote	14.2
Naphthalene	4.1
Anthracene	0.6
Phenol	0.6
Cresylic acids (cresols)	3.5
Pyridine bases	0.2

[a] Percent of refined tar.

The extractions leave a neutral naphthalene oil that may be purified further either by fractional distillation or by crystallization.

The creosote fraction of coal tar has a variable composition (Table 13.11) and may contain specific chemical compounds that can be isolated for commercial use. For example, when creosote oil is cooled, a solid referred to as anthracene cake may separate and consists primarily of anthracene, carbazole, and phenanthrene (Table 13.12).

Various processes have been reported as being suitable for the separation of these compounds by distillation and often involve the addition of an organic liquid as a solvent and as a carrier. Fractional crystallization and chemical processes (e.g., carbazole salt formation by treatment with acid) can also be employed.

TABLE 13.11
Representative Composition of Creosote Oil

Chemical Species	Amount (% w/w)
Naphthalene	25
Methyl naphthalenes	5
Phenanthrene	12
Acenaphthene	4
Carbazole	3
Fluorene	4
Fluoranthene	4
Pyrene	3
Anthracene	2
Dibenzofuran	1
Miscellaneous	37

TABLE 13.12
Examples of the Identifiable Constituents from the Anthracene Cake from Creosote Oil

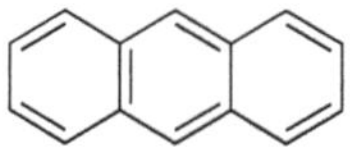

Anthracene

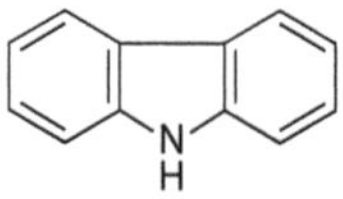

Carbazole

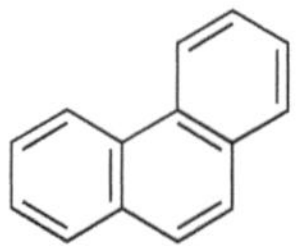

Phenanthrene

TABLE 13.13
Examples of the Oxygen-Containing
Constituents of Coal Tar

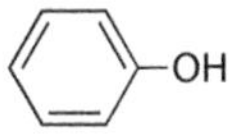

Phenol

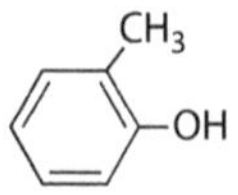

o-Cresol

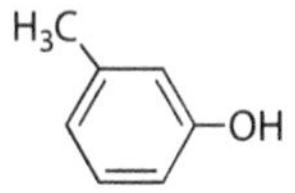

m-Cresol

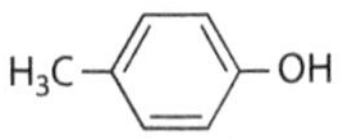

p-Cresol

The tar acids found in coal tar consist of phenol and its derivatives, mixtures of *o*-, *m*-, and *p*-cresol (known collectively as cresylic acids; Table 13.13), and a variety of derivatives of these compounds, all of which contain aromatic hydroxyl functions.

Phenolic compounds are valuable, and various means are used to extract them from coal tar; aqueous caustic soda (NaOH) solutions convert phenols to water-soluble phenolate ions, which can be separated from the organic coal tar, for example:

$$C_6H_5OH + NaOH \rightarrow C_6H_5O^-Na^+ + H_2O$$

Acidification of the aqueous solution reverses the above process and results in the precipitation of the phenolic compounds.

Tar bases are nitrogen-containing compounds that occur in the coal tar and react with hydrogen ions (H) to form positively charged species.

Typical tar bases contain derivatives of quinoline as well as derivatives of isoquinoline and quinaldine (Table 13.14), which react with acid solutions to form water-soluble salts:

$$C_9H_7N + HCl \rightarrow C_9H_7NH^+Cl^-$$

Quinoline   Quinoline
                         hydrochloride

This reaction, as such, can be used for extraction of the tar bases from other neutral materials.

The identification of specific compound types in the products of coal liquids and coal tar (Tables 13.12 through 13.14) has been used many times as indicators of the compound types that occur in coal, albeit bound into the coal structure. However, this may not be the case.

The chemical structure of coal is unknown (Chapter 10), and is likely to remain unknown for some time to come, although it is often speculatively represented as consisting of a large polymeric matrix of aromatic structures, commonly called the coal macromolecule. This macromolecular network consists of clusters of aromatic carbon that are linked to other aromatic structures by bridges,

TABLE 13.14

Examples of Nitrogen-Containing Constituents of Coal Tar

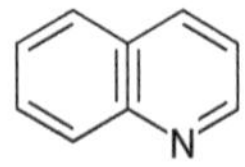

Quinoline

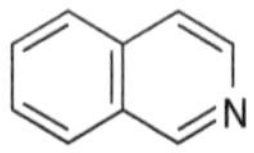

Isoquinoline

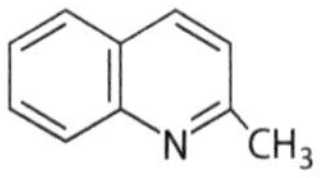

Quinaldine

which are believed to be aliphatic in nature, but may also include other atoms such as oxygen and sulfur. Other attachments to the aromatic clusters are in the form of side chains and are thought to consist mainly of aliphatic and oxygen functional groups.

During coal pyrolysis, covalent bonds throughout the coal macromolecule are broken—bonds in the bridges that connect aromatic clusters are broken along with bonds in the side chains. When these bonds are broken, fragments of the coal molecule are formed. If these fragments are small enough, they will form a volatile product that is released as light gases and tars. The larger fragments remain in the solid phase and can recombine with the coal macromolecule.

However, the various reactions that take place once the reactive intermediate are produced and start to form products include (1) further fragmentation, (2) dehydrogenation, (3) dehydrocyclization, (4) isomerization, and (5) thermal alkylation to mention only the more common reactions that are known to occur during thermal cracking of petroleum constituents (Speight, 2007, 2008).

As a result, relating the structure of the products in the light oil and other products of coal carbonization to coal structure is fraught with speculation and inspired guesswork—which may often be incorrect.

13.4.2 SOLID PRODUCTS

Char from the low-temperature (ca. 650°C [1200°F]) carbonization of coal is not a widely used material at the present time mainly because the production of chemical raw materials and "smokeless fuel" (char) was rendered somewhat uneconomical by the development of the petrochemical industry.

In addition, there is a trend away from the use of coal as a furnace fuel and as a source of raw materials. However, the shortages of petroleum and petroleum products (as well as new considerations of the allowable degree of pollution from fossil fuels) could well lead to a revival of a modified form of low-temperature carbonization of coal.

The two major properties of low-temperature char that make it a very useful fuel are its clean, smokeless combustion and its relatively high reactivity. In fact, current environmental regulations may make the smokeless quality of char even more desirable. Furthermore, carbonization leads to an appreciable loss of sulfur as well as undesirable trace elements, so that the char is actually a "cleaner" fuel than the original coal. Finally, the fact that the low-temperature char is quite reactive means that ignition and burning will be efficient, a desirable property for fuel to be used in open-bed furnaces.

Another potential advantage of low-temperature char involves its production from very low-rank fuels, such as lignite or brown coal. These abundant fuels have very low heat value (Chapters 2 and 9) and high water content, but low-temperature carbonization can convert these fuels to a high-heat (per unit weight) char. Thus, there is the potential for conversion of these coals at the mine site by means of volatile gases (or chars) to generate electricity while the premium quality solid fuel (char) is transported elsewhere for use.

As active thermal decomposition proceeds and is followed by secondary degasification, the solid residues are progressively aromatized and homogenized by growth of graphitic lamellae, which are increased in size by heating to higher temperatures. However, it should be noted here that the properties of the chars that give an indication of lamellar growth and graphitic ordering (as the more familiar "stacks" of lamellae) do not change to any marked degree until temperatures on the order of 500°C–550°C (930°F–1020°F) are reached. In addition, the ultrafine pore structure of coal may be retained until temperatures on the order of 650°C–700°C (1200°F–1290°F) are reached.

REFERENCES

Anthony, D.B., Howard, J.B., Hottel, H.C., and Meissner. 1975. *Proceedings. Fifteenth International Symposium on Combustion*. The Combustion Institute, Naples, Italy, p. 1303.

ASTM. 2011a. *Test Method for Free-Swelling Index of Coal* (ASTM D720). Annual Book of ASTM Standards, Section 05.05. American Society for Testing and Materials, West Conshohocken, PA.

ASTM. 2011b. *Test Method for Distillation of Petroleum Products at Reduced Pressures* (ASTM D1160). Annual Book of ASTM Standards, Section 05.01. American Society for Testing and Materials, West Conshohocken, PA.

ASTM. 2011c. *Test Method for Distillation of Crude Petroleum (15 Theoretical Plate Column)* (ASTM D2892). Annual Book of ASTM Standards, Section 05.02. American Society for Testing and Materials, West Conshohocken, PA.

ASTM. 2011d. *Test Method for Volatile Matter in the Analysis Sample of Coal and Coke* (ASTM D3175). Annual Book of ASTM Standards, Section 05.05. American Society for Testing and Materials, West Conshohocken, PA.

Azhakesan, M., Bartle, K.D., Murdoch, P.L., Taylor, J.M., and Williams, A. 1991. *Fuel*, 70: 322.

Benjamin, B.M., Raaen, F.F., Maupin, P.H., Brown, L.L., and Collins, C.J. 1978. *Fuel*, 57: 269.

Berkowitz, N. 1979. *An Introduction to Coal Technology*. Academic Press, Inc., New York.

Bhatt, B.L., Fallon, P.T., and Steinberg, M. 1981. In *New Approaches in Coal Chemistry*, B.D. Blaustein, B.C. Bockrath, and S. Friedman (Eds.). American Chemical Society, Washington, DC, Chapter 12.

Bilgesu, A. and Whiteway, S.G. 1990. *Fuel Science and Technology International*, 8: 491.

Bonfanti, L.L., Comellas, J.L., Lliberia, R., Vallhonratmatalonga, M., Pich-Santacana, M., and Lopez-Pinol, D.J. 1977. *Journal of Analytical and Applied Pyrolysis*, 44(1): 101.

Braun, A.G., Wornat, M.J., Mitra, A., and Sarofim, A.F. 1987. Organic emissions from coal pyrolysis: Mutagenic effects. *Environmental Health Perspectives*, 73: 215–221.

Burchill, P. and Welch, L.S. 1989. *Fuel*, 68: 100.

Bürküt, Y., Suner, F., and Nakhla, F.M. 1994. Thermal properties of coal. In *Coal: Resources, Properties, Utilization, Pollution*, O. Kural (Ed.). Istanbul Technical University, Istanbul, Turkey, Chapter 6.

Calkins, W.H. and Bonifaz, C. 1984. *Fuel*, 63: 1716.

Cannon, C.G., Griffith, M., and Hirst, W. 1944. *Proceedings. Conference on the Ultrafine Structure of Coals and Cokes*. British Coal Utilization Research Association, Leatherhead, Surrey, p. 131.

Chermin, H.A.G. and van Krevelen, D.W. 1957. *Fuel*, 36: 85.

Davis, J.D. 1945. In *Chemistry of Coal Utilization*, H.H. Lowry (Ed.). John Wiley & Sons, Inc., New York, Chapter 22.

Doolan, K.R. and Makie, J.C. 1985. *Fuel*, 64: 400.

Eddinger, R.T., Jones, J.F., Start, J.F., and Seglin, L. 1968. *Proceedings of the 7th International Coal Science Conference*. Prague, Czech Republic.

Fitzer, E., Mieller, K., and Schaefer, W. 1971. *Chemistry and Physics of Carbon*, 7: 237.

Fujioka, Y., Masayuki Nishifuji, M., Saito, K., and Kato, K. 2006. Analysis of thermal decomposition behavior of coals using high temperature infrared spectrophotometer system. Nippon Steel Technical Report No. 94, July.

Gavalas, G.R. 1982. Coal pyrolysis. *Coal Science and Technology 4, Coal Pyrolysis*. Elsevier, Amsterdam, the Netherlands.

Giam, C.S., Goodwin, T.E., Giam, P.Y., Rion, K.F., and Smith, S.G. 1977. *Analytical Chemistry*, 49(11): 1540.

Gibbins, J.R., Gonenc, Z.S., and Kandiyoti, R. 1991. *Fuel*, 70: 621.

Gibbins, J.R. and Kandiyoti, R. 1989. *Fuel*, 68: 895.

Gibson, J. 1979. In *Coal and Modern Coal Processing: An Introduction*. G.J. Pitt and G.R. Millward (Eds.). Academic Press, New York.

Graff, R.A. and Brandes, S.D. 1987. *Energy & Fuels*, 1: 84.

Gu, H., Tang, L., Zhu, Z., Hu, Z., and Zhang, C. 1999. *Huadong Ligong Daxue Xuebao Bianjibu*, 25(1): 11.

Gun, S.R., Sama, J.K., Chowdhury, P.B., Mukherjee, S.K., and Mukherjee, D.K. 1979. *Fuel*, 58: 176.

Hanson, R.L. and Vanderborgh, N.E. 1979. In *Analytical Methods for Coal and Coal Products*, C. Karr, Jr. (Ed.). Academic Press, Inc., New York, Vol. III, Chapter 40.

Hayashi, J.I., Mori, T., Amamoto, S., Kusakabe, K., and Morooka, S. 1996. *Energy & Fuels*, 10(5): 1099.

Hessley, R.K., Reasoner, J.W., and Riley, J.T. 1986. *Coal Science*, John Wiley & Sons, Inc., New York, Chapter 3.

Howard, H.C. 1963. In *Chemistry of Coal Utilization*, Supplementary Volume, H.H. Lowry (Ed.). John Wiley & Sons, Inc., New York, Chapter 9.

Howard, J.B. 1981a. *Fundamentals of Coal Pyrolysis and Hydropyrolysis*. John Wiley & Sons, Inc., New York, Chapter 12.

Howard, J.B. 1981b. In *Chemistry of Coal Utilization*, Second Supplementary Volume, M.A. Elliott (Ed.). John Wiley & Sons, Inc., New York, Chapter 12.

Hurd, J.D. 1929. *The Pyrolysis of Carbon Compounds*. The Chemical Catalog Co. Inc., New York.

ISO. 2011a. *Hard Coal—Determination of Coking Power, Roga Test* (ISO 335). International Standards Organization, Geneva, Switzerland.

ISO. 2011b. *Coal—Determination of Crucible Swelling Number* (ISO 501). International Standards Organization, Geneva, Switzerland.

ISO. 2011c. *Coal and Coke—Determination of Volatile Matter* (ISO 562). International Standards Organization, Geneva, Switzerland.

Jamaluddin, A.S., Wall, T.F., and Truelove, J.S. 1987. In *Coal Science and Chemistry*, A. Volborth (Ed.). Elsevier, Amsterdam, the Netherlands, p. 61.

Jones, W.I. 1964. *Journal of the Institute of Fuel*, 37: 3.

Khan, M.R. 1989a. *Fuel*, 68: 1439.

Kahn, M.R. 1989b. *Fuel*, 68: 1522.

Kahn, M.R., Chen, W.Y., and Suuberg, E. 1989. *Energy & Fuels*, 3: 223.

Kamo, J.B.T., Kodera, Y., Yamaguchi, H., and Sato, Y. 1998. Preprints. Division of Fuel Chemistry. American Chemical Society, 43(3): 703.

Kawamura, T., Hashimoto, S., Sakawa, M., Kozuru, H., Iida, H., and Okuhara, T. 1993. Nippon Steel Technical Report No. 57. Nippon Steel Corporation, Chiyoda-ku, Tokyo, Japan, April.

Kelemen, S.R., Gorbaty, M.L., and Kwiatek, P.J. 1994. *Energy & Fuels*, 8: 896.

Kimber, G.M. and Gray, M.D. 1967. *Combustion and Flame*, 11: 360.

Ladner, W.R. 1988. *Fuel Processing Technology*, 20: 207.

Lewis, I.C. and Singer, L.S. 1981. *Chemistry and Physics of Carbon*, 17: 1.

Lewis, I.C. and Singer, L.S. 1988. In *Polynuclear Aromatic Compounds*, L.B. Ebert (Ed.). Advances in Chemistry Series No. 217. American Chemical Society, Washington, DC, Chapter 16.

Liguras, D.K. and Allen, D.T. 1992. *Industrial & Engineering Chemistry Research*, 31: 45.

Lohmann, T.W. 1996. Modeling of reaction kinetics in coal pyrolysis. In *Proceedings of the International Workshop*, J. Warnatz and F. Behrendt (Eds.). Modelling of Chemical Reaction Systems.

Loison, R., Peytavy, A., Boyer, A.F., and Grillot, R. 1963. In *Chemistry of Coal Utilization*, Supplementary Volume, H.H. Lowry (Ed.). John Wiley & Sons, Inc., New York, Chapter 4.

Mackie, J.C., Colket, M.B., and Nelson, P.F. 1990. *Journal of Physical Chemistry*, 94: 4099.

Macrae, J.C. 1943. *Fuel*, 22: 117.

Maloney, D.J., Monazam, E.R., Woodruff, S.W., and Lawson, L.O. 1991. Measurement and analysis of temperature histories and size changes for single carbon and coal particles during the early stages of heating and devolatilization. *Combustion and Flame*, 84: 210–220.

Marsh, H. and Smith, J. 1978. In *Analytical Methods for Coal and Coal Products*, C. Karr, Jr. (Ed.). Academic Press, Inc., New York, Vol. II, Chapter 30.

Masuku, C.P. 1992. *Journal of Analytical and Applied Pyrolysis*, 23: 195.

Maswadeh, W.S., Fu, Y., Dubow, J., and Meuzelaar, H.L.C. 1992. Preprints. Division of Fuel Chemistry. American Chemical Society, 37(2): 699.

McMillen, D.F., Malhotra, R., and Nigenda, S.E. 1989. *Fuel*, 68: 380.

McMillen, D.F., Ogier, W.C., and Ross, D.S. 1981. *Journal of Organic Chemistry*, 46: 3322.

Merrick, D. 1987. In *Coal Science and Chemistry*, A. Volborth (Ed.). Elsevier, Amsterdam, the Netherlands, p. 307.

Miller, R.E. and Stein, S.E. 1979. Preprints. Division of Fuel Chemistry. American Chemical Society, 25(1): 271.

Miura, K. 2000. *Fuel Processing Technology*, 62(2–3): 119.

Miura, K., Mae, K., Asaoka, S., Yashimura, T., and Hashimoto, K. 1991. *Energy & Fuels*, 5(2): 340.

Monsef-Mirzai, P., Ravindran, M., McWhinnie, W.R., and Burchill, P. 1992. *Fuel*, 71: 716.

Morgan, M.E. and Jenkins, R.G. 1986. *Fuel*, 65: 764.

Nair, C.S.B. 1978. In *Analytical Methods for Coal and Coal Products*, C. Karr, Jr. (Ed.). Academic Press, Inc., New York, Vol. II, Chapter 33.

Nelson, P.F., Smith, I.W., Tyler, R.J., and Mackie, J.C. 1988. *Energy & Fuels*, 2: 391.

Nicholson, R. and Littlewood, K. 1972. Plasma pyrolysis of coal. *Nature*, 236: 397–400.

Owen, J. 1958. *Proceedings. Residential Conference on Science in the Use of Coal*. Institute of Fuel, London, U.K., p. C34.

Pershing, D.W. and Wendt, J.O.L. 1989. *Industrial & Engineering Chemistry Process Design and Development*, 18: 60.

Pitt, G.J. 1962. *Fuel*, 41: 267.

Plotczyk, W.W., Resztak, A., and Szymanski, A. 1990. Plasma pyrolysis of brown coal. *Journal de Physique Colloques*, 51: C5-43-48.

Poutsma, M.L. 1987. A review of the thermolysis of model compounds relevant to coal processing. Report No. ORNL/TM10637, DE88 003690. Oak Ridge National Laboratory, Oak Ridge, TN.

Pyatenko, A.T., Bukhman, S.V., Lebedinskii, V.S., Nasarov, V.M., and Tolmachev, I.Ya. 1992. *Fuel*, 71: 701.

Rees, O.W. 1966. Chemistry, uses, and limitations of coal analysis. Report of Investigations No. 220. Illinois State Geological Survey, Urbana, IL.

Robinson, N., Eglinton, G., Lafferty, C.J., and Snape, C.E. 1991. *Fuel*, 70: 249.

Romovacek, J.N. and Kubat, J. 1968. *Analytical Chemistry*, 40: 1119.

Sampath, R., Maloney, D.J., Zondlo, J.W., Woodruff, S.D., and Yeboah, Y.D. 1996. Measurements of coal particle shape, mass and temperature histories: Impact of particle irregularity on temperature predictions and measurements. *Proceedings. 26th Symposium (International) on Combustion*. The Combustion Institute, Naples, Italy, July.

Schobert, H.H. and Song, C. 2002. *Fuel*, 81(1): 15.

Selvig, W.A. and Ode, W.A. 1957. Bulletin No. 571, United States Bureau of Mines, Department of the Interior, Washington, DC.

Solomon, P.R. 1981. In *New Approaches in Coal Chemistry*, B.D. Blaustein, B.C. Bockrath, and S. Friedman (Eds.). American Chemical Society, Washington, DC, Chapter 4.

Solomon, P.R., Fletcher, T., and Pugmire, R. 1993. Progress in coal pyrolysis. *Fuel*, 72: 587–597.

Solomon, P.R., Hamblen, D.G., Carangelo, R.M., and Krause, J.L. 1983. *Progress in Energy and Combustion Science*, 9: 323.

Solomon, P.R. and Serio, M. 1994. Progress in coal pyrolysis research. Preprints. Division of Fuel Chemistry. American Chemical Society, 39(1): 69.

Solomon, P.R., Serio, M.A., and Suuberg, E.M. 1992. *Progress in Energy and Combustion Science*, 18(2): 133.

Speight, J.G. 2007. *The Chemistry and Technology of Petroleum*, 4th edn. Taylor & Francis Group, Boca Raton, FL.

Speight, J.G. 2008. *Synthetic Fuels Handbook: Properties, Processes, and Performance*, McGraw-Hill, New York.

Stein, S.E. 1981. In *New Approaches in Coal Chemistry*, B.D. Blaustein, B.C. Bockrath, and S. Friedman (Eds.). American Chemical Society, Washington, DC, Chapter 7.

Sugawara, T., Sugawara, K., Nishiyama, Y., and Sholes, M.A. 1991. *Fuel*, 70: 1091.

Sugawara, T., Sugawara, K., and Ohashi, H. 1989. *Fuel*, 68: 1005.

Sunol, Z.S.G. and Sunol, A.K. 1994. Pyrolysis of coal. In *Coal: Resources, Properties, Utilization, Pollution*, O. Kural (Ed.). Istanbul Technical University, Istanbul, Turkey, Chapter 25.

Suuberg, E.M., Peters, W.A., and Howard, J.B. 1978. *Industrial & Engineering Chemistry Process Design and Development*, 17: 1.

Szwarc, M. 1950. *Chemical Reviews*, 47: 75.

Takagi, H., Isoda, T., Kusakabe, K., and Morooka, S. 1999. Effects of coal structures on denitrogenation during flash pyrolysis. *Energy Fuels*, 13(4): 934–940.

Takeuchi, M. and Berkowitz, N. 1989. *Fuel*, 68: 1311.

Van Krevelen, D.W., van Heerden, C., and Huntjens, F.J. 1951. *Fuel*, 30: 253.

Voina, N.I. and Todor, D.N. 1978. In *Analytical Methods for Coal and Coal Products*, C. Karr, Jr. (Ed.). Academic Press, Inc., New York, Vol. II, Chapter 37.

Voorhees, K.J. 1984. *Analytical Pyrolysis*. Tech. Butterworth, London, U.K.

Wanzl, W. 1988. *Fuel Processing Technology*, 20: 317.

Weller, S., Clark, E.L., and Pelipetz, M.G. 1950. *Industrial & Engineering Chemistry*, 42: 334.

Part II

Technology and Utilization

As global demand for energy continues to rise—especially in rapidly industrializing and developing economies—energy security concerns become ever more important. To provide solid economic growth and to maintain levels of economic performance, energy must be readily available, affordable, and able to provide a reliable source of power without vulnerability to long- or short-term disruptions. Interruption of energy supplies can cause major financial losses and create havoc in economic centers, as well as potential damage to the health and well-being of the population.

Coal is one of the world's most important sources of energy, fueling almost 40% of electricity generation on a worldwide basis. The world currently consumes over 4 billion tons (4.0×10^9 ton) of coal. It is used by a variety of sectors, including power generation, iron and steel production, cement manufacturing, and as a liquid fuel. The majority of coal is either utilized in power generation (steam coal or lignite) or iron and steel production (coking coal).

As a result, coal will continue to play a key role in the world's energy mix, with demand in certain regions set to grow rapidly. Growth in both the steam and coking coal markets will be strongest in developing Asian countries, where demand for electricity and the need for steel in construction, car production, and demands for household appliances will increase as incomes rise.

Over the coming decades, the greatest challenge for many countries will be how to develop and maintain energy security and move away from the thirst for imported oil. While coal will be the answer for many, it is the most polluting fossil fuel. But ensuring energy security for India will not take a silver bullet. Moving away from disproportionate coal usage will require incremental steps on a number of different approaches.

It would be unrealistic to envision a future without coal use in countries such as the United States. Indeed, U.S. coal consumption is set to rise even if sources of energy are diversified. Clean-coal technologies, such as *carbon capture and sequestration* (CCS), are appealing to governments and lawmakers because they require less adaptation than other mitigating routes.

Coal has an important role to play in meeting the demand for a secure energy supply. Coal is abundant and widespread—it is present in almost every country in the world with commercial mining taking place in more than 50 countries. Coal is the most abundant and economical of fossil fuels, and, at current production levels, the known reserves of coal will be available for at least the next 200 years compared to the much more speculative lifetime of oil and gas reserves.

Coal is also readily available from a wide variety of sources in a well-supplied worldwide market. It can be transported to demand centers quickly, safely, and easily by ship and rail. A large number of suppliers are active in the international coal market, ensuring competitive behavior and efficient functioning. Coal can also be easily stored at power stations, and stocks can be drawn on in emergencies. Unlike gaseous, liquid, or intermittent renewable sources, coal can be stockpiled at the power station and stocks drawn on to meet demand.

Coal is also an affordable source of energy, and prices have historically been lower and more stable than the price of oil and gas. Furthermore, coal is likely to remain the most affordable fuel for power generation in many developed and industrializing countries for many decades, perhaps into the twenty-second century.

COMBUSTION AND ELECTRIC POWER GENERATION

Coal-based electricity is well established and highly reliable. Over 41% of global electricity is currently based on coal. The generation technologies are well established, and technical capacity and human expertise is widespread. Ongoing research activities ensure that this capacity is continually being improved and expanded, facilitating innovation in energy efficiency and environment performance.

Unlike gaseous, liquid, or renewable resources, coal can be stockpiled at the power station and stocks drawn on to meet demand without depending on primary supply. Because of the geographical diversity of coal reserves, some power stations can be located at mine-mouth and still be close to the market, thereby minimizing transmission and distribution losses.

CARBONIZATION (BRIQUETTING AND PELLETIZING)

Carbonization is the destructive distillation of coal in the absence of air accompanied by the production of carbon and liquid and gaseous products. The coke produced by the carbonization of coal is used in the iron and steel industry and as a domestic smokeless fuel.

Fundamental research on coal science and the carbonization process found its applications in the development and transfer of various technologies on coke, formed coke, briquettes, and soft coke at different point of time. With the gradual depletion of indigenous good quality coking coal resources, research initiatives have provided the scientific basis and scope of utilizing different type inferior grade coals as blend components for metallurgical coke making. Application of the knowledge base backed by validation tests in pilot plants has helped in optimizing the blend components for coke production.

However, despite considerable effort, there is as yet no practical alternative to traditional coke making. Three different approaches to coal carbonization may be adopted in the future: (1) improvement of coke production by introducing process modifications, (2) the building of a new coking system, and (3) the development of new carbon materials with different specifications and new functions. Improvement in existing technology includes the construction of new heating systems, which reduce NOx formation and the systematic study of blended coal as the feedstock for coke.

Environmental regulations on coke-making technologies, market competitiveness, and demand have activated the development of environment-friendly, improved designs for coke making, which have already entered the market.

LIQUEFACTION

The production of liquid fuels—gasoline and diesel—from coal is not a new process. The first patent was registered in 1913, with the more common Fischer–Tropsch indirect liquefaction process patented in 1925.

Direct coal liquefaction converts coal to a liquid by dissolving coal in a solvent at high temperature and pressure. This process is highly efficient, but the liquid products require further refining ("hydrocracking" or adding hydrogen over a catalyst) to achieve high-grade fuel characteristics.

Indirect coal liquefaction first gasifies the coal with steam to form synthesis gas (syngas—a mixture of hydrogen and carbon monoxide). The sulfur is removed from this gas and the mixture is adjusted according to the desired product. The synthesis gas is then condensed over a catalyst—the Fischer–Tropsch process—to produce high-quality, ultraclean products.

The development of the coal to liquids (CTL) industry can serve to hedge against oil-related energy security risks. Using domestic coal reserves, or accessing the relatively stable international coal market, can allow countries such as the United States to minimize their exposure to oil price volatility while providing the liquid fuels needed for economic stability and growth.

Historically it has been politics rather than economics that have driven development since then— for example, during the Second World War in Germany and as a result of trade embargoes on South Africa from the 1960s. Currently, at the time of writing, China is addressing its oil import dependence by building a commercial scale direct liquefaction plant in Inner Mongolia, which will produce around 50,000 barrels a day of finished gasoline and diesel fuel.

GASIFICATION

Clean coal technology using gasification is a promising alternative to meet the global energy demand. There are indications that the use of coal as a global energy source has caught up with the use of natural gas and will (at current rates of use) surpass natural gas use by 2030.

The most existing coal gasification technologies perform best on high-rank (bituminous) coal and petroleum refinery waste products but are inefficient, less reliable, and expensive to operate when processing low-rank coal. These low-grade coal reserves including low-rank and high-ash coal remain underutilized as global energy sources despite being available in abundance.

Coal gasification for electric power generation enables the use of a technology common in modern natural-gas-fired power plants, the use of *combined cycle* technology to recover more of the energy released by burning the fuel. In a combined cycle plant, the gas is burned in a turbine that generates electricity, and then steam is generated from the hot exhaust and used to power a second generator. This method achieves an efficiency of 45%–50% compared to traditional power plants, which only use one cycle to create electricity at only 35% efficiency.

Coal gasification units offer unique flexibility in that the syngas can also be chemically converted to liquid fuels for use in transportation (coal liquefaction). Alternatively the hydrogen can be refined from the synthesis gas to produce hydrogen fuel. An integrated combined cycle (IGCC) power plant could be used for electric power production during peak demand times and then shifted to liquid fuel production at nonpeak demand times.

The increased efficiency of the *combined cycle* for electrical power generation results in a 50% decrease in carbon dioxide emissions compared to conventional coal power plants. As the technology is required to develop economical methods of *carbon sequestration*, the removal of carbon dioxide from combustion by-products to prevent its release to the atmosphere, coal gasification units could be modified to further reduce their climate change impact because a large part of the carbon dioxide generated can be separated from the syngas *before* combustion.

CLEAN COAL TECHNOLOGIES

On a global macro level, it is clear that coal will continue to form a significant portion of the world's energy supply for the next several decades and that the United States and the world will rely on clean coal technologies to be the bridge leading to future renewable energy solutions.

Clean coal technology is an umbrella term used to describe the technologies being developed that aim to reduce the environmental impact of coal energy generation. These include chemically washing minerals and impurities from the coal, gasification, treating the flue gases with steam to remove sulfur dioxide, and carbon capture and storage (CCS) technologies to capture the carbon dioxide from the flue gas and dewatering lower-rank coals (brown coals) to improve the calorific quality and thus the efficiency of the conversion into electricity.

Clean coal technology usually addresses atmospheric problems resulting from burning coal. Historically, the primary focus was on sulfur dioxide and particulates, due to the fact that it is the most important gas that leads to acid rain.

More recent focus has been on carbon dioxide as well as other pollutants. Concerns exist regarding the economic viability of these technologies and the timeframe of delivery, potentially high hidden economic costs in terms of social and environmental damage, and the costs and viability of disposing of removed carbon and other toxic matter.

The latest in clean coal technologies, CCS, is a means to capture carbon dioxide emissions from coal-fired plants and permanently bury them underground.

Using clean coal technologies for energy will benefit many, and as technology makes coal cleaner to burn and finds ways to protect the environment, it will be a safe alternative. The more options that consumers have for energy needs, the quality and affordability of the energy will improve.

GAS CLEANING

Due to the harmful impact of air pollutants, regulatory agencies have enacted strict regulations to limit their emission. Indeed, the introduction of the strict limiting values imposed by various governments during the twentieth century has seen improvements in gas cleaning technologies of which multistage flue gas cleaning concepts initially tended to predominate. These were characterized by selective separation stages for the individual pollutants. The captured pollutants, such as sulfur dioxide, were usually converted into recyclable products, in this case gypsum.

More recently, a trend has been observed in many countries over the past decade, which moves in the direction of less complex, integrated flue gas cleaning processes. This simplification in the number of flue gas cleaning process elements means, as a rule, that it is no longer feasible or viable to recycle the treated products from flue gas cleaning into the economic trading cycle.

Instead, mixed products are produced that can be disposed in secure landfill areas such as old mines, after suitable physicochemical treatment together with adequate environmental compatibility.

CHEMICALS FROM COAL

An array of products can be made via coal conversion processes. For example, refined coal tar is used in the manufacture of chemicals, such as creosote oil, naphthalene, phenol, and benzene. Ammonia gas recovered from coke ovens is used to manufacture ammonia salts, nitric acid, and agricultural fertilizers. Thousands of different products have coal or coal by-products as components: soap, aspirins, solvents, dyes, plastics, and fibers, such as rayon and nylon.

Coal is also an essential ingredient in the production of specialist products such as (1) activated carbon, which is used in filters for water and air purification and in kidney dialysis machines, (2) carbon fiber, which is an extremely strong but light-weight reinforcement material used in construction, and (3) silicon metal, which is used to produce silicones and silanes that, in turn, are used to make lubricants, water repellents, resins, cosmetics, hair shampoos, and toothpastes.

There are process schemes to convert syngas to ammonia, methanol, substitute natural gas (SNG), and Fischer–Tropsch-derived transportation fuels. These novel and efficient schemes minimize plant energy consumption, water use, and emissions while increasing the output of valuable products.

As petroleum and natural gas supplies decrease, the desirability of producing gas from coal will increase. It is also anticipated that costs of natural gas will increase, allowing coal gasification to compete as an economically viable process.

ENVIRONMENTAL ASPECTS OF COAL USE

The conversion of any feedstock to liquid fuels is an energy-intensive one, and process emissions must be considered. While the CTL process is more carbon dioxide intensive than conventional petroleum refining, there are options for preventing or mitigating emissions. Due to the broad global distribution of coal reserves, emissions may also be avoided through shorter fuel transport distances.

For CTL plants, CCS can be a low-cost method of addressing carbon dioxide concerns and may result in greenhouse gas emissions. Where coprocessing of coal and biomass is undertaken, and combined with CCS, greenhouse gas emissions over the full fuel cycle may be as low as one-fifth of those from fuels provided by conventional oil.

CCS involves the capture of carbon dioxide emissions from the source, followed by transportation to, and storage in, geological formations. Once the carbon has been captured, there are a number of storage options available. The carbon dioxide can be stored in deep saline aquifers or be used to assist in enhanced oil recovery and subsequent carbon dioxide storage.

The problems inherent in coal combustion are well known—from the particulates to the acid rain caused by high-sulfur coal, to mercury pollution, as well as the carbon dioxide and mountaintop removal mining. Burning coal in any fashion is a dirty process, and trying to remove the pollutants from flue gases is challenging.

One of the major environmental advantages of coal gasification is the opportunity to remove impurities such as sulfur and mercury and soot *before* burning the fuel, using readily available chemical engineering processes. In addition, the ash produced is in a vitreous or glasslike state, which can be recycled as concrete aggregate, unlike pulverized coal power plants, which generate ash that must be landfilled, potentially contaminating groundwater.

ENERGY SECURITY

Energy security considerations are potentially even more important in the transport sector. Oil products provide the majority (>90%) of the energy used in transportation. As oil prices have risen dramatically over recent years, the possibility of hedging risks through the use of alternative fuels—coal to liquids (CTL), gas to liquids (GTL), and biomass to liquids (BTL)—is being seriously considered (e.g., see Speight, 2008, 2011a,b, and references cited therein).

Coal, which has a unique role to play in meeting the demand for a secure energy, is well established and is a reliable, secure, and affordable fuel for both power generation and industrial applications. The production and utilization of coal is based on well-proven and widely used technologies and is built on a vast infrastructure and a strong base of expertise worldwide.

Alternative liquid fuels can be made from solid (e.g., coal or biomass) or gaseous (e.g., natural gas) feedstocks and can be used in existing vehicle fleets with no, or little, modification. Coal will have a significant role to play in the provision of these alternative fuels—it is the most affordable of the fossil fuels and is widely distributed around the world.

Coal benefits from a well-established global market, with a large number of suppliers. The production of liquid fuels from coal will not require vast land resources or cause competition with food production. The development of the CTL industry can serve to hedge against oil-related energy security risks. Using domestic (US) coal reserves, or accessing the relatively stable international coal market, can allow the United States (and other coal-rich countries) to minimize their exposure to the price volatility often seen in the petroleum market.

However, it cannot be ignored that coal does face environmental challenges that have implications for energy security and sustainable development. Nevertheless, the coal industry has a proven track record of developing technology pathways, which have successfully addressed (and continue to address) environmental concerns at local and national levels.

In terms of energy security for many countries, a key issue is resource availability—the actual physical amount of the coal resource available within the borders of the country. This leads to the security of a continuous supply of coal as a source of energy, particularly electricity, to meet consumer demand at any given time. In this respect, coal has particular attributes that make a positive contribution to energy security as part of a balanced energy mix.

Many developing and developed countries are able to use their indigenous coal resources to provide the energy needed for economic development. The largest coal-producing countries are not confined to one region—the top five producers are China, the United States, India, Australia, and South Africa. All of these countries use their indigenous coal as the primary fuel for electricity generation, and all except India have a sizeable coal export market.

While coal has furthered social development through electricity generation around the world, it can also be transformed to liquid and gaseous fuels to guard against oil import dependence and price shocks. Coal gasification is a further example where indigenous (or imported) fuels can be transformed to address environmental concerns while enhancing energy security.

While petroleum enjoys a similar global market and can be transported quickly and easily, its market is dominated (in essence) by a single supplier—OPEC—and is affected accordingly.

On the other hand, the gas market is far more regional. Transportation of the fuel (limited by where the pipelines run from and to) has constrained the development of the market to date. This is changing with the advent of liquefied natural gas (LNG) transported by large, expensive vessels, but a truly global market will take many years to develop.

Coal is an affordable source of energy insofar as coal prices have historically been lower and more stable than oil and gas prices, and despite the growth of index and derivative-based sales in recent years, this has typically remained the case. Placing a cost on carbon emissions more directly will, in certain circumstances, put pressure on this interfuel cost relationship. However, coal is likely to remain the most affordable fuel for power generation in many developing and industrialized countries for several decades.

In countries with energy-intensive industries, the impact of fuel and electricity prices will be compounded. High prices can lead to a loss of competitive advantage and in prolonged cases a loss of the industry altogether. Countries with access to indigenous energy supplies, or to affordable fuels from a well-supplied world market, can avoid many of these negative impacts, enabling further economic development and growth.

Overall costs for coal power stations are usually lower than for alternative power generation, and coal will remain one of the key choices for baseload electricity generation. Furthermore, the geographical diversity of coal reserves means that some power stations can be located at mine-mouth and still be close to demand centers, thereby also minimizing transmission and distribution losses.

Renewable energy can reduce dependency on finite energy sources and removes some of the risk around import dependence. Hydropower provides many countries with a substantial amount of their electricity needs and can provide a secure supply—as long as there is enough rainfall to provide sufficient water in the reservoirs. However, when weather conditions deviate from normal, severe problems such as blackouts can occur.

Energy from renewable resources can also cause challenges to those who need to ensure a stable and reliable flow of electricity through the grid. Wind power has been the dominant option for integration to existing energy networks, but it is expensive and suffers from reliability and intermittency problems. It is now generally accepted by power industry engineers and energy analysts that the level of wind power capacity in a grid should not exceed around 10% without the grid operator incurring significant costs to deal with the intermittency issues.

To meet energy security concerns and environmental objectives, a number of measures may be taken that will allow coal to fulfill its vital role in the global energy future.

Policy support for clean and efficient use of coal in power generation can encourage the take-up of existing advanced technologies for low-emission, coal-fired electricity production—providing secure and clean energy. There is also the need for policy support for research and development, and demonstration into new technologies such as CCS can provide a very significant opportunity for major reductions in emissions that are required by our modern societies.

Environmental issues should be addressed in a nondiscriminatory manner, while recognizing the benefits that arise when coal is used as a source of energy. Clear, long-term environmental policies provide certainty, allowing investments to be made in advanced coal technologies that bring enhanced environmental performance. Unlimited pollution in the name of energy generation is not the answer nor is the passage of unrealistic environmental regulations that will cause mankind to freeze in the green darkness.

A balance must be struck, and only then will the full potential of coal as a secure form of energy (for the United States) be realized.

REFERENCES

Speight, J.G. 2008. *Synthetic Fuels Handbook: Properties, Processes, and Performance*. McGraw-Hill, New York.
Speight, J.G. 2011a. *The Refinery of the Future*. Gulf Professional Publishing, Elsevier, Oxford, U.K.
Speight, J.G. (Ed.). 2011b. *The Biofuels Handbook*. Royal Society of Chemistry, London, U.K.

14 Combustion

14.1 INTRODUCTION

There are no documented records of when, or how, man *first discovered* that a certain *black rock* would burn but it can be surmised that coal combustion represents the oldest known use of this fossil fuel. Thus, it is reasonable to presume that coal combustion provided early man with his first practical source of energy giving warmth, light, as well as extending the range of food that could be consumed, and it enabled him to modify metals for a variety of uses.

It is known that coal was employed as a fuel in China about 1100 BC and that Welsh Bronze Age cultures had used coal for funeral pyres. There are also references to coal in the Christian Bible (Cruden, 1930) as well as in the writings of Aristotle, Nicander, and Theophrastus (Hoover and Hoover, 1950) but, all in all, the recorded use of coal during the times classed as *antiquity* is very sketchy.

In more modern times, there are excellent examples of coal mining in Britain from the year 1200 AD, which marked, perhaps, the first documented use of mined coal in England (Galloway, 1882). On one particular note, it is recorded that in 1257 a very singular event occurred that threatened the very existence of coal use and its future as a fuel.

Eleanor, wife of King Henry III and Queen of England, was obliged to leave the town of Nottingham where she had been staying during the absence of the King who was on a military expedition to Wales. The removal of the Queen's royal person from Nottingham was due to the troublesome smoke from the coal being used for heating and cooking. As a result, over the next several decades, a variety of proclamations were issued by Henry and by his son, Edward I, which threatened the population with the loss of various liberties, even life, if the consumption of coal was not seriously decreased and, in some cases, halted (Galloway, 1882).

It is debatable whether or not these worthy kings of England realized the environmental consequences of burning coal or whether they were more interested in the use of wood, from Royal forests of course, and the resulting income therefrom. The Royal positions on the pollution problem, if that really be the issue, have never been resolved and the proclamations did not have any lasting effect; coal burning has continued in England from that time.

There are many other mentions of coal in the historical literature and there is the presumption that the consistent use of coal seems to have evolved in England in the Middle Ages. But the records are somewhat less than complete and it is reasonable (perhaps unwise not) to assume that coal burning in Europe evolved along similar lines to those in England even though coal was exported from the north-eastern England to France and the Low Countries during the fourteenth century (Galloway, 1882). For example, coal is described by one sixteenth century text on ore smelting (Biringuccio, 1540) as stones that occur in many places and have the true nature of charcoal.

Interestingly, in another sixteenth century text that dealt with various natural phenomena, coal is noted to be a variety of bitumen (Agricola, 1546; see also Hoover and Hoover, 1950), thereby bringing to light, inadvertently perhaps, some aspect of the character of coal and the consideration that there may be a relationship between coal and bitumen or heavy petroleum. But, be that as it may, coal use did increase in popularity so that the pyrogenous rock became a major force behind the Industrial Revolution. From that point until the early decades of this century, coal emerged as a major energy source, particularly as a source of combustible energy.

Coal combustion is used in a range of applications, which vary from domestic fires to large industrial furnaces and utility boilers. While, for reasons of economy, the oxidant is usually air, the coal may be in any degree of dispersion. In fact, coal combustion provides the majority of consumable energy to the world and despite the continuing search for alternate sources of energy (whether they are fossil fuels or nonfossil fuels), coal appears to be so firmly entrenched that there is little doubt that coal combustion will remain important into the twenty-first century, particularly where a convenient method of storing energy is required as, for example, in transport applications.

A major concern in the present day combustion of coal is the performance of the process in an environmentally acceptable manner through the use of a variety of environmentally acceptable technologies such as the use of a low-sulfur coal or through the use of post-combustion cleanup of the off-gases (Chapter 23). Thus, there is a marked trend in the modern research to more efficient methods of coal combustion. In fact, the ideal would be a combustion system that is able to accept any coal without a precombustion treatment, or without the need for post-combustion treatment, or without emitting objectionable amounts of sulfur and nitrogen oxides and particulates. *To dream, to dream, perchance to dream.*

14.2   CHEMISTRY AND PHYSICS

Coal is a complex material and, not surprisingly, coal combustion is a complex science because of the variety of physical and chemical properties of coal (Chapters 8 and 9) (Field et al., 1967; Essenhigh, 1981; Morrison, 1986; Gay and Davis, 1987; Brill, 1993; Heitmann, 1993). In addition, it is not only the amount of energy available from coal combustion but also other aspects such as fuel handling, ash removal, emissions, and environmental control techniques that are of extreme importance (Littler, 1981; Reid, 1981; Slack, 1981).

Combustion occurs, chemically, by initiation and propagation of a self-supporting exothermic reaction. The physical processes involved in combustion are principally those which involve the transport of matter and the transport of energy. The conduction of heat, the diffusion of chemical species, and the bulk flow of the gas all follow from the release of chemical energy in an exothermic reaction. Thus, combustion phenomena arise from the interaction of chemical and physical processes.

The first requirement, somewhat difficult with coal because of its molecular complexity, is that the overall stoichiometry of the reaction must always be established. For these purposes, coal is usually represented by carbon, which can react with oxygen in two ways, producing either carbon monoxide or carbon dioxide:

$$C_{coal} + O_2 \rightarrow 2CO$$

$$C_{coal} + O_2 \rightarrow CO_2$$

In direct combustion, coal is burned (i.e., the carbon and hydrogen in the coal are oxidized into carbon dioxide and water) to convert the chemical energy of the coal into thermal energy after which the sensible heat in the products of combustion can then be converted into steam that can be external work or directly into shaft horsepower (e.g., in a gas turbine):

$$C_{coal} + O_2 \rightarrow CO_2$$

$$H_{coal} + O_2 \rightarrow H_2O$$

$$C_{coal} + H_2O \rightarrow CO + H_2$$

In fact, the combustion process actually represents a means of achieving the complete oxidation of coal.

On a more formal basis, the combustion of coal may be simply represented as the staged oxidation of coal carbon to carbon dioxide with any reactions of the hydrogen in the coal being considered to be of secondary importance:

$$C_{coal} + O_2 \rightarrow 2CO$$

$$2CO + O_2 \rightarrow 2CO_2$$

The stoichiometric reaction equations are quite simple but there is a confusing variation of hypotheses about the sequential reaction mechanism, which is caused to a great extent by the heterogeneous nature (solid and gaseous phases) of the reaction. But, for the purposes of this text, the chemistry will remain simple as shown in the earlier equations. Other types of combustion systems may be rate controlled due to the onset of the Boudouard reaction:

$$CO_2 + C \rightarrow 2CO$$

In more general terms, the combustion of carbonaceous materials (which contain hydrogen and oxygen as well as carbon) involves a wide variety of reactions between the many reactants, intermediates, and products (Table 14.1). The reactions occur simultaneously and consecutively (in both forward and reverse directions) and may at times approach a condition of equilibrium. Furthermore, there is a change in the physical and chemical structure of the fuel particle as it burns.

The characteristics and physical properties of coal combustion products (CCP) vary. In general, the size, shape, and chemical composition of these materials determine their beneficial reuse as a component of building materials or as a replacement to other virgin materials such as sand, gravel, or gypsum.

The beneficial use of CCP involves the use of or substitution of CCP for another product based on performance criteria. Using CCP can generate significant environmental, economic benefits. Beneficial use include raw feed for cement clinker, concrete, grout, flowable fill, structural fill, road base/subbase, soil modification, mineral filler, snow and ice traction control, blasting grit and abrasives, roofing granules, mining applications, wallboard, waste stabilization/solidification, soil amendment, and agriculture.

The complex nature of coal as a molecular entity (Berkowitz, 1979; Meyers, 1981; Hessley et al., 1986; Hessley, 1990) has resulted in the chemical explanations of coal combustion being confined to the carbon in the system and, to a much lesser extent with only passing acknowledgement of the hydrogen and other elements, but it must be recognized that the system is extremely complex and that the heteroatoms (nitrogen, oxygen, and sulfur) can exert an influence on the combustion and it is this influence that can bring about the serious environmental concerns.

TABLE 14.1
Simplified Chemistry of Coal Combustion

$C(s) + O_2(g) \rightarrow CO_2(g)$	−169,290 Btu/lb	−94.4 kcal/kg
$2C(s) + O_2(g) \rightarrow 2CO(g)$	−95,100 Btu/lb	−52.8 kcal/kg
$C(s) + CO_2(g) \rightarrow 2CO(g)$	−74,200 Btu/lb	−41.2 kcal/kg
$2CO(g) + O_2(g) \rightarrow 2CO(g)$	−243,490 Btu/lb	−135.3 kcal/kg
$2H_2(g) + O_2(g) \rightarrow 2H_2O(g)$	−208,070 Btu/lb	−115.6 kcal/kg
$C(s) + H_2O(g) \rightarrow CO(g) + H_2(g)$	+56,490 Btu/lb	+31.4 kcal/kg
$C(s) + 2H_2O(g) \rightarrow CO_2 + 2H_2(g)$	+38,780 Btu/lb	+21.5 kcal/kg
$CO(g) + H_2O(g) \rightarrow CO_2(g) + H_2(g)$	−17,710 Btu/lb	−9.8 kcal/kg

Notes: (g) indicates the gaseous state; (s) indicates the solid state; a negative heat of reaction indicates an exothermic reaction; and a positive heat of reaction indicates an endothermic reaction.

For example, the conversion of the coal-bound sulfur and nitrogen (in addition to any reactions or aerial nitrogen with aerial oxygen under the prevailing conditions) to their respective oxides during combustion is a major environmental issue:

$$S_{coal} + O_2 \rightarrow SO_2$$

$$2SO_2 + O_2 \rightarrow 2SO_3$$

$$N_{coal} + O_2 \rightarrow 2NO$$

$$2NO + O_2 \rightarrow 2NO_2$$

Thus,

$$N_{coal} + O_2 \rightarrow NO_2$$

The release of the sulfur and nitrogen from the coal is not as simple as represented here and the equations are simplifications of what are, presumably, complex processes (Crelling et al., 1993; Gavin and Dorrington, 1993).

The sulfur dioxide that escapes into the atmosphere is either deposited locally or is converted to sulfurous and sulfuric acids by reaction with moisture in the atmosphere:

$$SO_2 + H_2O \rightarrow H_2SO_3$$

$$2SO_2 + O_2 \rightarrow 2SO_3$$

$$SO_3 + H_2O \rightarrow H_2SO_4$$

Thus,

$$2SO_2 + O_2 + 2H_2O \rightarrow 2H_2SO_4$$

Nitrogen oxides (Morrison, 1980; Crelling et al., 1993) also contribute to the formation and occurrence of acid rain, in a manner similar to the production of acids from the sulfur oxides, yielding nitrous and nitric acids:

$$NO + H_2O \rightarrow H_2NO_3$$

$$2NO + O_2 \rightarrow 2NO_2$$

$$NO_2 + H_2O \rightarrow HNO_3$$

Thus,

$$4NO_2 + O_2 + 2H_2O \rightarrow 4HNO_3$$

In addition to causing objectionable stack emissions, coal ash (Table 14.2) and volatile inorganic material generated by thermal alteration of mineral matter in coal (Figure 14.1) will adversely affect heat-transfer processes by fouling the heat-absorbing and heat-radiating surfaces and will also influence the performance of the combustion system by causing corrosion, and operating procedures must therefore provide for effective countering of all these hazards (Littler, 1981).

TABLE 14.2

Generalized Composition of Coal Ash

Silica (SiO_2): 40%–90%

Aluminum oxide (Al_2O_3): 20%–60%

Iron (ferric) oxide (Fe_2O_3): 5%–25%

Calcium oxide (CaO): 1%–15%

Magnesium oxide (MgO): 0.5%–4%

Sodium oxide (Na_2O) plus potassium oxide (K_2O): 1%–4%

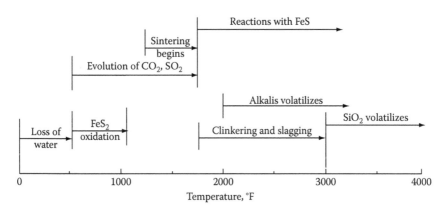

FIGURE 14.1 Changes occurring to the mineral matter during coal combustion. (From Reid, W.Y., *External Corrosion and Deposits: Boilers and Gas Turbines*, Elsevier, New York, 1971.)

14.2.1 INFLUENCE OF COAL QUALITY

Coal quality and/or rank is now generally recognized as having an impact, often significant, on combustion, especially on many areas of power plant operation (Parsons et al., 1987; Rajan and Raghavan, 1989; Bend et al., 1992).

The parameters of rank, mineral matter content (ash content), sulfur content, and moisture content are regarded as determining factors in combustibility as it relates to both heating value and ease of reaction. Thus, lower-rank coals (though having a lower heat content) may be more "reactive" than higher-rank coals, so implying that rank does not influence coal combustibility. At the same time, anthracites (with a low-volatile matter content) are generally more difficult to burn than bituminous coals.

The lower the rank of a coal the greater the wettability with water, but the higher the rank the greater the wettability with tar or pitch. High moisture content is associated with a high unit surface area of the coal (especially for retained moisture after drying) and coals also become harder to grind as the percentage of volatiles decreases.

Lignite usually serves as the more extreme example of low-grade fuel of high moisture content and the problems encountered in lignite combustion are often applicable to other systems (Nowacki, 1980). Lignite gives up moisture more slowly than harder coals but the higher-volatile content tends to offset the effect of high moisture. For the combustion of pulverized material, it appears essential to dry lignite and brown coals to 15%–20% moisture; the lowest possible ash and moisture contents are desired as well as high grindability, high heat content, and high fusion temperature.

Finally, since coal quality can be affected by oxidation or weathering (Joseph and Mahajan, 1991), the question is raised about the effects of oxidation and weathering on combustion and whether oxidized or weathered coal could maintain a self-sustaining flame in an industrial boiler.

The inhibition of volatile matter release due to changes in the char morphology, because of reduced thermoplastic nature of the coal, as a result of the oxidation/weathering suggests that this may not be the case (Bend et al., 1991).

One option for managing coal quality for power generation is to blend one particular coal with others until a satisfactory feedstock is achieved (Jones, 1992). This is similar to current petroleum refinery practice where one refinery actually accepts a blend of various crude oils and operates on the basis of average feedstock composition; the days of one crude oil are no longer with us.

14.2.2 Mechanisms

The exact nature of the coal combustion process is difficult to resolve but can be generally formulated as two processes: (1) the degradation of hydrogen and (2) the degradation of carbon (Barnard and Bradley, 1985). It is also necessary to understand the surface chemistry involved in the burning process and progress is being made in this direction (Brill, 1993). However, coal being a heterogeneous solid adds an increasingly difficult dimension to the combustion chemistry and physics; combustion actually occurs on the surface with the oxidant being adsorbed there prior to reaction. However, the initial reaction at (on) the surface is not necessarily the rate-determining step; the process involves a sequence of reactions, any one of which may control the rate.

The initial step is the transfer of reactant (i.e., oxygen) through the layer of gas adjacent to the surface of the particle. The reactant is then adsorbed and reacts with the solid after which the gaseous products diffuse away from the surface. If the solid is porous, much of the available surface can only be reached by passage of the oxidant along the relatively narrow pores and this may be a rate-controlling step. Rate control may also be exercised by (a) adsorption and chemical reaction, which are considered as chemical reaction control and (b) pore diffusion, by which the products diffuse away from the surface. This latter phenomenon is seldom a rate-controlling step.

In general, rate control will occur if the surface reaction is slow compared with the diffusion processes; whilst diffusion shows a less-marked temperature dependence, reaction control predominates at low temperatures but diffusion control is usually more important at higher temperatures. In fact on a chemical basis, hydrogen degradation outweighs the slower-starting carbon degradation in the early, or initial, stage of combustion. But, at the same time, the carbon monoxide/carbon dioxide ratio is decreased.

After the initial stages of combustion, during which volatile material is evolved (which is also combustible), a nonvolatile carbonaceous residue (coke, char), which can comprise up to 90% of the original mass of the coal, remains. During the combustion of the coke, three different zones (regimes) of combustion can be distinguished (Barnard and Bradley, 1985).

In the first zone (I), the rate of diffusion to and away from the surface is very fast compared with the rate of the surface reaction; such phenomena are observed at low temperatures. At much higher temperatures, the rate at which oxygen molecules are transported from the bulk gas to the external surface is slow enough to be rate controlling (Zone III); the observed rate can be equated to the molar flux of oxygen to unit area of external surface. Finally (Zone II; intermediate between I and III), the oxygen transport to the external surface is rapid but diffusion into the pores before reaction is relatively slow (Figure 14.2) (Mulcahy, 1978).

In practice there are considerable differences between the reactivity of different cokes, some of which can be assigned to variations in pore structure and others to the presence of impurities (e.g., alkali metal salts), which have a pronounced catalytic effect on the surface reaction. Consequently, the temperature ranges corresponding to the three zones differ and are not constant for different cokes/chars (Figure 14.3) (Mulcahy, 1978).

There are alternate ways to consider the mechanism of coal combustion and there is a variety of models proposed for this purpose (Jamaluddin et al., 1987). For example, in a simple model for the

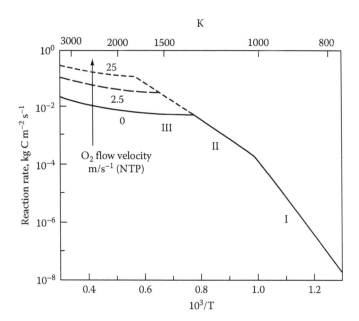

FIGURE 14.2   Illustration of the three combustion zones.

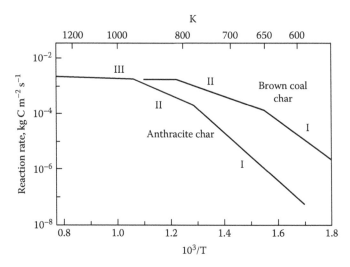

FIGURE 14.3   Illustration of the three combustion zones in still air.

combustion process, the initial step is assumed to be (1) devolatilization, (2) ignition, and (3) rapid burning of the volatiles relative to the char. But the actual mechanism is considered to be somewhat more complex. The process is thought to consist of (1) diffusion of reactive gases to the carbon surface, (2) adsorption, (3) formation of transitory complexes, and (4) desorption of the products, but the overall reaction mechanism for a devolatilized coal char particle is believed to be (1) transport of oxygen to the surface of the particle, (2) reaction with the surface, and (3) transport of the products away from the surface.

As the pressure of the system is increased, the mechanism tends toward diffusion control. This is due to the mass transfer or diffusion rate coefficient being inversely proportional to pressure. The overall rate, however, will increase due to the increased oxygen partial pressure at the higher pressures.

The complexity of coal as a molecular entity (Chapter 10) has resulted in treatments of coal combustion being confined to the carbon in the system and, to a lesser extent, the hydrogen but it must be recognized that the system is extremely complex. Even with this simplification, there are several principal reactions that are considered to be an integral part of the overall combustion of coal (Table 14.1).

In summary, it is more appropriate to consider the combustion of coal (which contains carbon, hydrogen, nitrogen, oxygen, and sulfur) as involving a variety of reactions between (1) the reactants; (2) the intermediate, or transient, species; and (3) the products. The reactions can occur both simultaneously and consecutively (in both forward and reverse directions) and may even approach steady-state (equilibrium) conditions. And, there is a change in the physical and chemical structure of the fuel particle during the process.

14.2.3   IGNITION

The ignition of coal has been described as occurring in just a few hundredths of a second with the onset of burning in less than half a second. The ignition distance has been observed to be on the order of 0.04 in. (1 mm) with the carbon monoxide formed by reaction at the surface burning to carbon dioxide at distances close to the surface (0.5–4 mm). Water is evaporated in the initial stages and the ignition is propagated through a dry bed.

For coals, the ignition temperatures are usually of the order 700°C (1290°F), but may be as low as 600°C (1110°F) or as high as 800°C (1470°F), depending on volatiles evolved. In fact, ignition temperatures depend on rank and generally range from 150°C to 300°C (390°F to 570°F) for lignite to 300°C to 600°C (570°F to 1110°F) for anthracite with some dependence on particle size being noted.

14.2.4   SURFACE EFFECTS

The conditions under which coal ignites and the behavior during ignition will relate to (a) the structure of the volatilized coal and (b) the temperature in the coke-burning state. In addition, some consideration must be given to the manner in which the volatile matter is released. For example, the particle may burn by first releasing all the volatile matter, which may burn simultaneously with the carbon. It does, however, seem unlikely that oxygen would reach the surface in the presence of volatiles and, thus, any oxygen attempting to diffuse through the volatiles layer would react instead.

Generally, it is possible to subdivide the overall reaction sequence into (a) those reactions that could conceivably occur at the surface of the coal char and (b) those reactions that may occur between the gaseous products themselves:

$$C + O_2 \rightarrow CO_2$$

$$2C + O_2 \rightarrow 2CO$$

$$C + CO_2 \rightarrow 2CO$$

$$C + H_2O \rightarrow CO + H_2$$

$$2CO + O_2 \rightarrow 2CO_2$$

$$2H_2 + O_2 \rightarrow 2H_2O$$

$$CO + H_2O \rightarrow CO_2 + H_2$$

During combustion, there are several possibilities for the mode in which the carbon reacts in the particle structure. The carbon may react only from the surface and reaction may proceed uniformly throughout the particle or, alternately, the particle may be regarded as a hollow sphere with burning occurring on both the outer and inner surfaces. In fact, there is evidence that coal particles do form hollow spheres during combustion and such spheres (cenospheres) are believed to be formed during volatilization while the coal is in the plastic stage, such as in the example of coking coals in an inert furnace atmosphere.

Measurements on coal particles of different sizes indicate that the burning times of both the volatiles and residue vary as the square of the initial particle diameter, which is in accord with the surface area proportionally. The porous structure of the char also exerts an effect on the burning operation as does particle temperature up to several hundred degrees above the gas temperature.

There are also indications that the adsorption of small molecules on coal, such as methanol, occurs by a site-specific mechanism (Ramesh et al., 1992). In such cases, it appears that the adsorption occurs first at high-energy sites but with increasing adsorption the (methanol) adsorbate continues to bind to the surface rather than to other (polar) methanol molecules and there is evidence for both physical and chemical adsorption. In addition, at coverages below a monolayer, there appears to be an activation barrier to the adsorption process.

14.2.5 REACTION RATES

In general, residence times may be regarded as ca. 0.5 s and often ca. 0.25 s (Figures 14.4 through 14.7). Particle size significantly affects rate (Figure 14.8). Various combustion systems may be rate controlled due to the occurrence/onset of the Boudouard reaction:

$$CO_2 + C \rightarrow 2CO$$

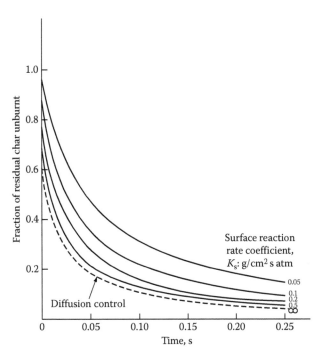

FIGURE 14.4 Effect of the rate of the reaction on burning at the coal surface. (From Field, M.A. et al., *Combustion of Pulverized Coal*, British Coal Utilization Research Association, Leatherhead, Surrey, 1967.)

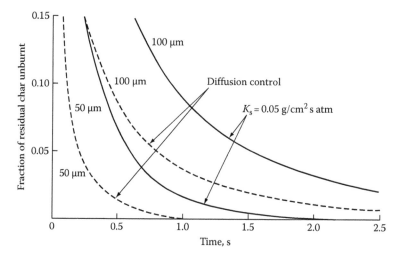

FIGURE 14.5 Burnout times. (From Field, M.A. et al., *Combustion of Pulverized Coal*, British Coal Utilization Research Association, Leatherhead, Surrey, 1967.)

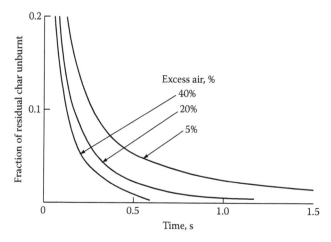

FIGURE 14.6 Influence of excess air on burnout times. (From Field, M.A. et al., *Combustion of Pulverized Coal*, British Coal Utilization Research Association, Leatherhead, Surrey, 1967.)

14.2.6 HEAT BALANCE

The heat balance of the coal combustion process provides a relative weighting of the heat input into the system versus the heat output of the process and can be represented by

$$dH_1 = dH_c + S_H + dH_e$$

where
  dH_1 is the heat input
  S_H is a composite of the sensible and latent heats of air, fuel, and other materials
  dH_e is the heat from exothermic reactions (other than combustion), which may contribute to the overall combustion process

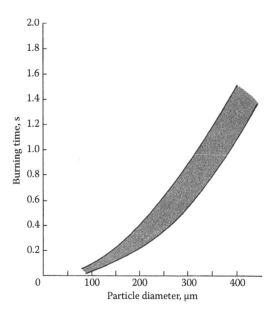

FIGURE 14.7 Burning times of particles in air. (From Field, M.A. et al., *Combustion of Pulverized Coal*, British Coal Utilization Research Association, Leatherhead, Surrey, 1967.)

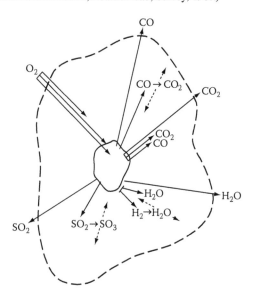

FIGURE 14.8 Reactions that occur within and in the vicinity of a coal particle. (From Berkowitz, N., *An Introduction to Coal Technology*, Academic Press, Inc., New York, 1979.)

and by

$$dH_o = dH_{cu} + S_{HC} + dH_{-E} + dS_{AG} + dH_L$$

where

dH_o is the heat output

dH_{cu} is the heat of combustion of unburned fuel

S_{HC} is the sensible and latent heats in the carbonization products

dH_{-E} is the heat absorbed by endothermic reactions

dS_{AG} is the sensible and latent heats in the combustion products (ash and stack gases)

dH_L is the heat losses to the surroundings by convection, radiation, and combustion

The presence of water vapor in the combustion system appears in the latent heat effects and one consequence of high moisture content in coal combustion is that a part of the heat is lost due to evaporation of the moisture in the coal and is not recouped from the combustion products. It is possible that a small amount (5% w/w) of water in the coal may not exert any marked effect on the overall heat requirements, since the sensible heat of the gases vaporizes the water.

14.2.7 Soot Formation

There is little known about the limiting parameters for soot (small carbonaceous particles) formation during the combustion of coal. It has been noted that conditions favorable to the formation of soot may prevail if contact with oxygen is delayed.

Soot formation is commonly observed in the pyrolysis or combustion process of simple hydrocarbons and coals. Soot does great harm to people's health through carcinogenic effects and the presence of soot in air also leads to visibility reduction, globe temperature decrease, and acid deposition. Soot can suspend in the air for up to 1 month and can be delivered by wind to distant places, which causes wide range pollution. Soot suspended in flames is important to combustion systems because it will significantly enhance radiative heat transfer due to its large surface area (Fletcher et al., 1997).

High temperature is conducive to the formation of soot but soot yields are diminished due to soot oxidation when sufficient oxygen exists in the high-temperature zone. Yields of soot are also diminished with the residence time because of more chances for volatiles reaction with oxygen at the longer residence time. Coal with a high yield of volatile matter and tar is easier to form soot due to the corresponding oxygen scarcity in the same conditions, and release more aromatic hydrocarbons.

14.2.8 Convection and Radiation

In combustion operations, it has been estimated that 20% of the reaction heat is released directly as radiant energy. The remaining heat energy resides in the combustion products, from which about 30% of the energy is then released as radiation. The presence of water vapor in the combustion gases itself may have some appreciable effect upon the gas emissivity and radiation.

14.2.9 Fouling

Coal ash is composed mostly of metal oxides (Table 14.2) (Chapter 7) and the composition affects the softening point. Iron oxides are a particular source of problems and the reducing atmosphere ($CO + H_2$; produced by the water gas reaction) in the fuel bed serves to reduce ferric oxide (Fe_2) to ferrous oxide (Fe) with the production of "clinker," which will contribute to reactor fouling.

Fouling of combustion systems has also been related to the alkali metal content of coal. For example, coal with a total alkali metal content 0.5% w/w (as equivalent sodium oxide, Na_2O) produces deposits that can be removed by the action of a *soot blower* but for coals having more than 0.6% w/w alkali metal (as equivalent sodium oxide) the deposits increase markedly and can be a major problem.

To combat fouling, modern combustion equipment is designed to ensure that particles are cooled to well below their fusion temperatures before they can reach the banks of closely spaced tubes in the upper regions of the boiler. In pulverized fuel systems, provision is usually made for tilting the burners and thereby periodically altering the heat regime. In addition, tube deposits are routinely dislodged by frequent "soot blowing," that is, by inserting perforated lances through which jets of high-pressure air or steam can be sent between boiler tubes.

14.2.10 Additives and Catalysts

The fact that additives may catalyze or otherwise affect combustion processes has been reported many times. For example, salt has long been known to be of some assistance in removing soot deposits from chimneys and coal treated with a more complex mixture of metal oxides (Table 14.3) has been reported to be activated in combustion systems but it is apparently not resolved whether

Combustion

443

TABLE 14.3
Additive Composition to
Mitigate Soot Formation

Component	Percent (w/w)
Barium oxide (BaO)	22.0
Sodium oxide (Na$_2$O)	0.4
Cobalt oxide (CoO)	0.5
Manganese (Mn) salts	97.1

the catalytic effect is with regard to carbon–oxygen reactions or whether it is more indirect since the effect resembles (to a degree) the catalysis of coal–steam systems in which alkali salts serve to catalyze the carbon–steam reaction to produce synthesis gas (carbon monoxide/hydrogen mixtures). This mixture may then, in turn, react to produce hydrocarbons and oxygenated materials in the presence of the multitude of trace metals that can (and do) occur in coal (Chapter 7).

In addition to causing objectionable stack emissions, coal ash and volatile inorganic material generated by thermal alteration of mineral matter in coal (Figure 14.9) will adversely affect heat-transfer processes by fouling heat-absorbing and heat-radiating surfaces and will also influence the performance of the combustion system by causing corrosion, and operating procedures must therefore provide for effective countering of all these hazards (Mitchell, 1989).

Corrosion is mainly caused by oxides of sulfur; but in certain parts of a combustion system, specifically on furnace wall tubes with metal temperature of 290°C–425°C (550°F–800°F) and superheater or reheater tubes with temperatures in the range 600°C–700°C (1110°F–1300°F), corrosion can be induced by tube deposits that destroy protective surface oxide coatings.

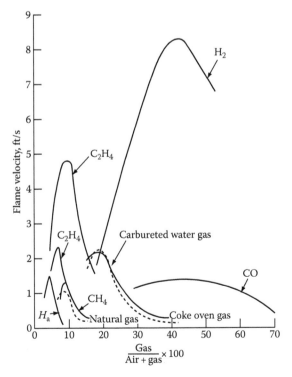

FIGURE 14.9 Relation of flame speed to the air–gas composition. (From Perry, J.H. and Green, D.W., Eds., *Chemical Engineers Handbook*, 8th edn., McGraw-Hill, New York, 2008.)

Corrosion damage that is usually ascribed to sulfur is actually caused by sulfuric acid, which is generated from organic and inorganic sulfur-bearing compounds:

$$2SO_2 + O_2 \rightarrow 2SO_3$$

$$SO_3 + H_2O \rightarrow H_2SO_4$$

Oxidation of sulfur dioxide to sulfur trioxide occurs mostly in flames where (transient) atomic oxygen species are thought to be prevalent by interactions of hydrogen atoms with oxygen and by interactions of carbon monoxide with oxygen and therefore may not occur in the stoichiometric manner shown earlier. The process can, however, be catalyzed by the ferric oxides that form on boiler tube surfaces and show excellent catalytic activity for sulfur dioxide oxidation at approximately 600°C (1110°F), that is, at temperatures that occur in the superheater section of a boiler.

The presence of water has a marked effect on combustion (by participating in various combustion reactions) and there is experimental evidence for the existence of active centers for chain reactions involved in the further combustion of carbon monoxide and hydrogen (which would be reaction intermediates in the combustion of coal). Thus, it is generally assumed that moisture catalyzes the oxidation of coal (Chapter 12).

The endothermic steam–carbon reaction is primarily responsible for cooling effects in furnaces and the presence of moisture is believed to cause heat generation at the surface of the bed and in the combustion by virtue of the (endothermic) formation of carbon monoxide and hydrogen in the bed, which then burn at the surface. On the other hand, the presence of water vapor appears to "assist" in the formation of carbon dioxide:

$$H_2O + C \rightarrow CO + H_2 \text{ (at surface)}$$

$$2CO + O_2 \rightarrow 2CO_2 \text{ (in film)}$$

In fact, moisture appears to play a more integral role in the combustion of hydrogen-deficient carbonaceous fuels (such as coal) than has been generally recognized. The carbon–steam reaction to produce carbon monoxide and hydrogen (which are then oxidized to the final products) is an important stage in the combustion sequence as is the carbon monoxide *shift reaction* to yield carbon dioxide and hydrogen:

$$CO + H_2O \rightarrow CO_2 + H_2$$

Since the whole system involves reactions (and equilibria) between the fuel (i.e., carbon), water, carbon monoxide, hydrogen, and carbon dioxide, the rapid rates of the reactions render it difficult (if not impossible) to determine precisely which of the reactions are the major rate-controlling reactions. In addition, the heterogeneous nature of the system adds a further complication.

While the presence of inert gases would usually be expected to dilute the reactants and therefore diminish the reaction rates, such inert materials may actually, on occasion, accelerate the reaction(s). Indeed, the "addition" of nitrogen to the reaction mixture can be as effective as the addition of oxygen. The nitric oxide formed in the mixture is believed to act as a catalyst:

$$2CO + 2H_2O \rightarrow 2CO_2 + 2H_2$$

$$N_2 + O_2 \rightarrow 2NO$$

$$2NO + 2H_2 \rightarrow N_2 + 2H_2O$$

Thus

$$2CO + O_2 \rightarrow 2CO_2$$

14.2.11 Excess Air

Coal combustion is, on the one hand, a balance of high reaction or flame temperatures that favor carbon monoxide at equilibrium and, on the other hand, the use of excess air that drives the conversion to carbon dioxide. Relative to these two opposing reactions, the rates of reaction are generally controlling and manifested by short residence times with rapid heat transfer such that the system temperature is lowered before equilibrium can occur. Hence, the consideration that (complete) combustion is a nonequilibrium process. This is contrary to gasification by partial combustion, which occurs at lower temperatures and longer residence times without heat transfer (the system remains adiabatic), where equilibrium conditions tend to apply.

Though thermodynamic and rate calculations may be used to maximize flame temperatures and conversions, it is the empirical observation and evidence in each situation that will dictate the optimum air-to-fuel ratio and will depend on fuel analysis including moisture, air humidity and temperature, and other general operating variables. Although 15%–25% excess air is a median range (Tables 14.4 and 14.5), it depends not only on the fuel but also on the type of combustion system and the careful control of that system.

14.2.12 Coal/Air: Transport

The entrained transport of pulverized coal (200 mesh) is accomplished with ratios of about 1.4 lb air per lb coal, but pulverizer performance may sometimes require twice this ratio. If this fuel–air mixture is burned, there is the possibility of flashback unless linear velocities are 55 ft/s.

TABLE 14.4
Excess Air at the Furnace Outlet

Type	Fuel	Excess Air (%)
Solid fuels	Coal	10–40
	Coke	20–40
	Wood	25–50
Liquid fuels	Oil	8–15
Gaseous fuels	Natural gas	5–10
	Refinery gas	8–15
	Blast furnace gas	15–25
	Coke oven gas	5–10

Source: Hoffmann, E.J., *Coal Conversion*, Energon, Laramie, WY, 1978, p. 100.

TABLE 14.5
Excess Air Supplied to Combustion Systems

Fuel	Type of Furnace or Burner	Excess Air (% w/w)
Pulverized coal	Water-cooled furnace for slag-tap or dry ash removal	15–40
Crushed coal	Cyclone furnace	10–15
Coal	Spreader stoker	30–60
	Water-cooled vibrating grate stoker	30–60
	Chain-grate and traveling grate stokers	15–50
Fuel oil	Oil burners, register type	5–10
	Multifuel burners and flat flame	10–20
Wood	Dutch oven (10%–23% through grates) and Hofft type	20–25

Source: Hoffmann, E.J., *Coal Conversion*, Energon, Laramie, WY, 1978, p. 100.

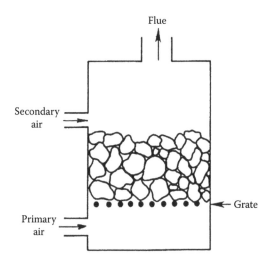

FIGURE 14.10 Simple fireplace.

Furthermore, although gaseous flame speeds are only a few feet per second (Figure 14.10), the coal volatiles mixed with air can form a combustible gaseous boundary layer that is essentially stationary, allowing flame propagation. This propagation cannot occur unless the boundary layer is above a certain minimum, that is, flames will not progress through (for example) a tube if the tube diameter is less than a certain minimum (depending on the combustibles), and a bulk velocity of at least 55 ft/s assures that the effective boundary layer thickness is less than this minimum.

14.2.13 SPONTANEOUS COMBUSTION

The spontaneous combustion of coal (also variously referred to as the spontaneous ignition or autogenous heating of coal) has been recognized as a hazard for some time to the extent that, in the early years of this century, guidelines were laid down for the strict purpose of minimizing the self-heating process (Haslam and Russell, 1926) and have been revised since that time (Allen and Parry, 1954). Indeed, the phenomenon of spontaneous combustion is not limited to coal but has also been observed in other piles of organic debris (Vovelle et al., 1983; Gray et al., 1984; Jones, 1990; Jones et al., 1990).

Almost all types of coal may ignite spontaneously in suitable environmental conditions. This leads to serious safety problems as well as economic losses for coal mines and storage areas. The determination of the liability of a coal type to spontaneous combustion is quite important in dealing with the problem before, during, and after mining.

The spontaneous combustion of coal is believed to center around the basic concept of the oxidation of carbon to carbon dioxide. This particular reaction is exothermic (+94 kcal/mol) and will be self-perpetuating especially since the rates of organic chemical reactions usually double for every 10°C (18°F) rise in temperature. Furthermore, there has also been the suggestion that the heat release that accompanies the wetting of dried (or partially dried) coal may be a significant contributory factor in the onset of burning. Support for such a concept is derived from the observations that stored coal tends to heat up when exposed to rain after a sunny period (during which the coal has been allowed to dry) or when wet coal is placed on a dry pile (Berkowitz and Schein, 1951).

Similar effects have been noted during the storage of hay in the conventional haystacks and ignition has been noted to occur. Thus, any heat generated by climatic changes will also contribute to an increase in the rate of the overall oxidation process. Obviously, if there are no means by which this heat can be dissipated, the continued oxidation will eventually become self-supporting and will ultimately result in the onset of burning.

Spontaneous combustion is usually the culmination of several separate chemical events and although precise knowledge of the phenomenon is still somewhat incomplete but is gradually becoming known, there are means by which the liability of a coal to spontaneously ignite can be tested and evaluated (Brooks et al., 1988).

The relationship between the friability of coal and its rank has a bearing on its tendency to heat or spontaneously ignite (Chakravorty, 1984; Chakravorty and Kar, 1986). The friable, low-volatile coals, because of their high rank, do not oxidize readily despite the excessive fines and the attendant increased surface they produce on handling. Coals of somewhat lower rank, which oxidize more readily, usually are relatively nonfriable; hence, they resist degradation in size with its accompanying increase in the amount of surface exposed to oxidation. But above all, the primary factor in coal stockpile instability is unquestionably oxidation by atmospheric oxygen whilst the role of any secondary factors such as friability seems to exacerbate the primary oxidation effect (Jones and Vais, 1991).

More recent work has presented indications that the tendency for spontaneous combustion is reduced by thermal upgrading and further decreased with increase in treatment temperature (Ogunsola and Mikula, 1992). The decrease in the tendency to spontaneously ignite appears to be due to the loss of the equilibrium moisture as well as the loss of oxygen functional groups. The loss of the equilibrium moisture is an interesting comment because of the previous comment that the presence of indigenous moisture appears to enhance (i.e., increase the rate of) the oxidation reaction.

Coal tends to spontaneously ignite and combust when the moisture within the pore system is removed leaving the pores susceptible to various chemical and physical interactions (Berkowitz and Speight, 1973) that can lead to spontaneous combustion. It is a question of degree and the correct order of reactions being in place. It is obvious that the system is complex and, as noted earlier, spontaneous combustion is the culmination of several interrelated chemical and physical events.

In addition to air and moisture, there are several other factors that influence the oxidation of coal. These include rank, particle size, and volatile matter of the coal being stockpiled. Most of these factors are affected by, or related to, conditions on the surface of the coal. Another important attribute of coal that is largely dependent on surface conditions is its dustiness. In most cases, coals with lower surface moisture, higher friability, and smaller particle size have a higher potential for generating fugitive dust. One of the most effective ways to control dust is to apply treatment that alters the surface of the coal.

Conventional wet- or foam-type dust suppressants add surface moisture and, to a certain extent, agglomerate fines. The control of dusting generally lasts until the added moisture evaporates from the surface. Binding agents, on the other hand, leave a coating on the coal surface. In some cases, a barrier is produced preventing moisture and oxygen from penetrating the surface of the coal. This phenomenon gives rise to the additional benefit of effectively minimizing coal oxidation (Moxon and Richardson, 1987).

Thus, many factors affect heat-producing reactions; the oxidation of carbonaceous matter in coal at ambient temperatures is the major cause for the initiation of spontaneous combustion (Banerjee, 1985; Goodarzi and Gentzis, 1991).

It is generally agreed that the nature of the interaction between coal and oxygen at very low temperatures is fully physical (adsorption) and changes into a chemisorption form starting from an ambient temperature (Banerjee, 1985). The rate of oxygen consumption is extremely high during the first few days (particularly the first few hours) following the exposure of a fresh coal surface to the atmosphere. It then decreases very slowly without causing problems unless generated heat is allowed to accumulate in the environment. Under certain conditions, the accumulation of heat cannot be prevented, and with sufficient oxygen (air) supply, the process may reach higher stages.

The loose coal–oxygen–water complex formed during the initial stage (peroxy complexes) decomposes above 70°C–85°C (160°F–185°F), yielding carbon monoxide, carbon dioxide, and water. The rate of chemical reactions and exothermic change with the rise in temperature, and

TABLE 14.6

Intrinsic Factors (due to the Nature of the Coal) and Other Factors That Affect Coal Combustion

Nature of Coal	Atmospheric, Geological, and Mining Conditions
Pyrites	Temperature
Moisture	Moisture
Particle size and surface area	Barometric pressure
Rank and petrographic constituents	Oxygen concentration
	Bacteria
Chemical constituents	Coal seam and surrounding strata
Mineral matter	Method of working
	Ventilation system and air flow rate
	Timbering

radical changes take place, commencing at approximately about 100°C (212°F), mainly due to loss of moisture (Banerjee, 1985; Handa et al., 1985). This process continues with the rise in temperature, yielding more stable coal–oxygen complexes until the critical temperature is reached. From then on, it is fairly safe to assume that an actual fire incident will result (Chamberlain and Hall, 1973; Feng et al., 1973; Kim, 1977; Banerjee, 1982; Didari, 1988; Goodarzi and Gentzis, 1991).

Thus, the main factors that have significant effects on the spontaneous process are (Table 14.6)

- Pyrite content may accelerate spontaneous combustion.
- Changes in moisture content, that is, the drying or wetting of coal, have apparent effects.
- As the particle size decreases and the exposed surface area increases, the tendency of coal toward spontaneous combustion increases.
- Lower-rank coals are more susceptible to spontaneous combustion than higher-rank coals, which may be attributed to the petrographic constituents of coal.
- Mineral matter content generally decreases the liability of coal to spontaneous heating—certain minerals such as lime, soda, and iron compounds may have an accelerating effect, while others, such as alumina and silica, produce a retarding effect—and oil shale bands adjoining coal seams may play an important role in mine fires.

14.3 COMBUSTION SYSTEMS

A wide range of coal types having either high or low fusion temperatures can be burned using the various combustors. In general, bituminous coal, subbituminous coal, or lignite fit well into the spreader combustion process and these types of coal can be burned in a given unit with the same combustion heat release. However, there may be issues related to the attributes of each coal type as it relates to boiler furnace and gas pass design. There are plants that have substituted lower grades of coal for cost savings as well as for substituting low-sulfur bituminous or subbituminous coal to meet state or local emission requirements.

Coal sizing affects stoker operation. Coals too coarse will not burn at the high rate required for optimum spreader operation and coals too fine can cause operational as well as emission problems without proper design and operating procedures. The theoretical size is equal proportions of 0.75 in. × 0.5 in., 0.5 in. × 0.35 in., and less than 0.25 in. The equal gradation is to allow for the even combustion over the grate surface. This size is not available from a practical standpoint and the spreader feeders have the capabilities to adjust for coal sizing. However, as the amount of fines are smaller than 0.10 in., there is a concomitant increase in fly ash carryover—precipitators or baghouses can readily handle the carryover from spreader stokers just as they do for pulverized coal-fired boilers and circulating fluid bed–fired boilers.

The advantages of a stoker system include (1) the coal does not have to be pulverized, (2) a low level of particulate emissions occurs, simplifying flue gas cleanup, (3) a stoker is easy to operate and can be manually controlled if desired, and (4) the stoke can be built in small sizes. However, some disadvantages are (1) high maintenance is involved due to bulky moving parts, (2) the stoker does not provide efficient gas–solid contact and requires a relatively large furnace volume for a given steam production, due to the low heat release rate per unit area of grate and the fact that the grate takes up furnace volume, and (3) although there may be claims to the contrary, no one stoker can burn all types of coal, mainly due to caking properties of the coal and ash clinker that can form, depending upon the ash fusion temperature—it is often desirable to wash the coal to make it more amenable to stoker firing.

However, a correctly specifically manufactured and configured stoker is an excellent combustor of cellulose waste such as (1) wood—shredded trees to sawdust; (2) garbage—refuse-derived fuel; (3) bagasse—sugarcane residue; (4) industrial residue—paper, plastics, and wood; (5) furfural residue; (6) peanut shells; and (7) shredded tires. Most of these fuels can be burned without auxiliary fuel with proper attention to fuel moisture, design heat release, combustion air system design, and preheated air temperature. Cogeneration and the emphasis on renewable fuels have driven increased use of these fuels. Size distribution of the fuel is important from the standpoint of efficiency, availability, and low emissions.

There are two major methods of coal combustion: fixed-bed combustion and combustion in suspension (Ceely and Daman, 1981).

The first fixed beds (e.g., open fires, fireplaces, and domestic stoves) were simple in principle. Suspension burning of coal began in the early 1900s with the development of pulverized coal-fired systems, and in the 1920s these systems were in widespread use. Spreader stokers, which were developed in the 1930s, combined both principles by providing for the smaller particles of coal to be burned in suspension and larger particles to be burned on a grate.

Coal-burning systems are usually referred to as layer and chambered; the former refers to fixed beds while the latter refers to systems designed for pulverized coal. The feed systems applicable to the combustor are (a) overload, (b) front feed, and (c) underfeed.

As noted in a previous section (Section 14.2), the combustion of coal involves burning in the presence of oxygen to produce heat and carbon dioxide. Over the years, the methods by which this is accomplished have evolved to several types of systems, each with its own particular merits and use.

For example, coal combustion may be achieved using pulverized coal in entrained systems or as sized particles in fixed or slowly moving beds; larger pieces may, in certain instances, also be used. In the case of the fixed- or slowly moving bed combustor, it is usual to employ a mechanical stoker to feed the coal and a grate to support the coal particles as well as to admit air for the combustion process. With regard to the pulverized systems, coal that has been "crushed" to ca. 200 mesh is carried into the system entrained by the air.

Pulverized coal combustion (PCC) involves grinding the feed coal to approximately <70 mm and injecting the powdered coal into the combustor from either wall-mounted burners or corner-mounted (tangential) burners. Combustion takes place within a few seconds at flame temperatures up to 1500°C (2730°F). Supercritical PCC is a variation that seeks to improve thermal efficiency, from the typical values of up to about 40% for PCC to 43%–47% in supercritical systems through higher steam temperatures and pressures.

At this time, some note should also be made of the various size designations and size limits for coal. For example, run-of-mine coal and large-lump coal are variable in size with no top or bottom limits whereas lump coal varies from a minimum size of 1 in. (2.5 cm) to a variable top size. On the other hand, cobble (egg or stove) coal varies in size from 2 in. (5 cm) to 6 in. (15 cm) whereas nut coal falls within the size limits 3/4 in. (2 cm) to 2 in. (5 cm). The coals designated as prepared stoker coals all fall into the size range 1/16 in. (0.2 cm) to 2 in. (5 cm) but they are further subdivided into three classes: large (1/4–2 in. [0.6–5 cm]), intermediate (1/8–1 in. [0.3–2.5 cm]), and small (1/16–3/4 in. [0.2–2 cm]). However, nut (nutty) slack, slack, and fines are only defined in terms of an upper size limit and these are 2 in. (5 cm), 1–1/4 in. (3 cm), and 1/2 in. (1.3 cm), respectively.

14.3.1 FIXED BEDS (SLOWLY MOVING BEDS)

For fuel-bed burning on a grate, a distillation effect occurs and the result is that liquid components that are formed will volatilize before combustion temperatures are reached; cracking may also occur. The ignition of coal in a bed is almost entirely by radiation from hot refractory arches and from the flame burning of volatiles. In fixed beds, the radiant heat above the bed can only penetrate a short distance into the bed. Consequently, convective heat transfer determines the intensity of warming up and ignition. In addition, convective heat transfer also plays an important part in the overall flame-to-surface transmission. The reaction of gases is greatly accelerated by contact with hot surfaces and, while the reaction away from the walls may proceed slowly, reaction at the surface proceeds much more rapidly.

14.3.1.1 Up-Draught Combustion

One of the simplest ways of achieving up-draught combustion involves lumps of coal in a bed supported by a grate (Figure 14.11). Provision is made for the supply of primary air beneath the bed and secondary air above it as there is a connection to a flue in order to provide a draught. This simple principle has provided good service in a number of applications ranging from the domestic fire to the furnace. A fire of this type is ignited at the base, after which the flame front then spreads upward until the whole bed is incandescent; the system is usually termed "up-draught" combustion, but the important technical feature is that the flame front travels in the same direction as the primary air.

The best-known example of coal combustion is provided by the domestic fire and the majority of industrial coal burners operate on the same overfeed principle. Once combustion has commenced, a number of separate reaction zones can be distinguished (Figure 14.12). The fresh, or "green," coal is placed on the upper surface and the heat transferred from the bed below causes evaporation of the volatile material that burns in the secondary air and leaves a residue of fixed carbon or coke.

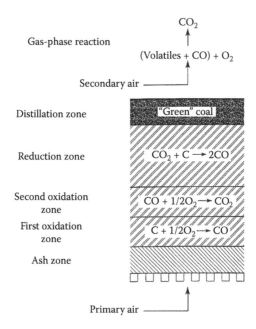

FIGURE 14.11 Combustion process occurring in a solid fuel bed.

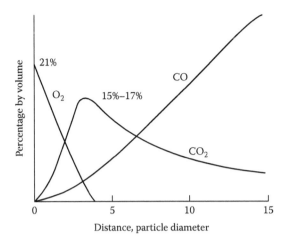

FIGURE 14.12   Variation of gas composition in a solid fuel bed. (From Thring, M.W., *Fuel*, 31, 355, 1952.)

Primary air enters the base of the grate and passes first through the ash zone. The ash performs a useful function in providing an insulation between the high-temperature reaction zone and the grate. In the first oxidation zone, the oxygen reacts at the surface of the carbon to give carbon monoxide:

$$2C_{coal} + O_2 \rightarrow 2CO$$

The carbon monoxide is released from the solid and reacts on mixing with oxygen, in the second oxidation zone, to give carbon dioxide:

$$2CO + O_2 \rightarrow 2CO_2$$

Each of these reactions is strongly exothermic and some of the heat released promotes the initial "carbon–oxygen" reaction at what is usually the hottest part of the fuel bed. At this stage, the oxygen concentration is very much depleted and the carbon dioxide is reduced by the Boudouard reaction in the next layer of fuel:

$$CO_2 + C \rightarrow 2CO$$

This reaction is endothermic and, therefore, the bed temperature will decrease. When the carbon monoxide leaves the fuel, it mixes with secondary air and burns again to carbon dioxide. The concentration changes through the zones (Figure 14.13) (Thring, 1952) and the maximum temperature coincides approximately with the maximum carbon dioxide concentration.

The relative importance of the various factors that control the rate of combustion depends on the temperature of the bed and, in principle, Zone I, II, or III kinetics (as defined earlier) may apply (Mulcahy, 1978).

Many coals contain, or generate, considerable quantities of volatile matter (Chapters 8 and 9) and will also evolve tar at approximately 450°C (840°F). In this simple combustion method, heat is transferred ahead of the flame front by radiation and convection causing the distillation of the volatile tar matter at temperatures below the ignition temperature. The purpose of the secondary air is to burn the volatile matter and in simple appliances it is not difficult to supply the necessary air, but there is rarely sufficient turbulence to mix it with the volatiles and thus the temperature in the zone above the bed can easily fall below the value where ignition is possible. As a result, there may be the emission of a yellow–brown smoke containing pollutants such as hydrocarbons and the oxides of sulfur and nitrogen.

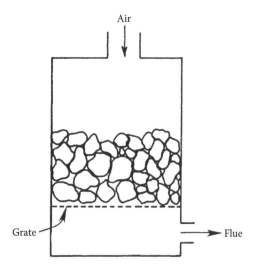

FIGURE 14.13 Principle of downdraft combustion.

14.3.1.2 Down-Draught Combustion

Down-draught combustion has been recognized for many centuries and one of the earliest records reports a method of controlling the emission of smoke, which, in essence, involved inversion of the simple coal fire (Figure 14.14). Thus, air enters at the top of the container and combustion products leave at the bottom. This is generally called a "down-draught" system and an important technical feature is that the flame front travels in the opposite direction of the primary air. In this type of system, volatile matter that was evolved ahead of the flame front would be swept back by the air stream through the flame and into the incandescent part of the bed, where it is combusted thereby reducing its contribution to the pollutants.

However, it is now known that volatile matter will only burn in these circumstances provided it (a) is mixed with oxygen, (b) is maintained at temperatures above 600°C (1110°F), and (c) has sufficient residence time (ca. 0.5 s) for the oxidation to go to completion. Thus, it is essential that the system have sufficient turbulence, temperature, and time, and successful exploitation of the principle has been achieved in the form of a variety of mechanical stokers used in industrial steam raising.

In the *underfeed stoker*, coal is fed by a worm feeder into the bottom of a retort (Figure 14.15). The coal rises vertically in the retort and air enters through tuyeres in the sides. The fire is ignited at the top and the flame front moves downward and its speed is matched by the rising flow of coal, fulfilling the requirement of the flame front traveling in the opposite direction of the primary air.

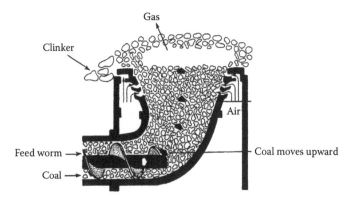

FIGURE 14.14 Underfeed stoker.

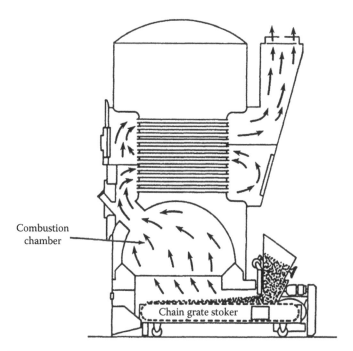

FIGURE 14.15 Chain-grate stoker.

Volatile matter from the coal mixes with the air and ignites as it passes through the incandescent top layer of the bed thereby effectively controlling smoke emission.

Another coal feed system is the chain-grate stoker in which coal is fed from hoppers by gravity to a grate that consists of an endless chain extending into the boiler (Figure 14.16). The horizontal movement of the grate carries the coal into the combustion chamber; the coal is carried forward as a thin layer on the top surface of the grate while air is delivered beneath it. As the coal enters the combustion chamber, the top surface is ignited by radiation from a hot refractory arch and the flame front then travels down through the coal bed while the air comes up through it. At the end of the grate, the coal has been burned, and any residual ash is dropped into a container as the grate turns for the return journey.

Strongly caking coals can cause problems with chain-grate units but not with underfeed or spreader units. Coal size is important with chain-grate stokers since fines lead to excessive caking and high unburned carbon losses. The amount of coal smaller than 0.6 cm (1.4 in.) is usually limited; spreader and underfeed stokers are less sensitive to particle size.

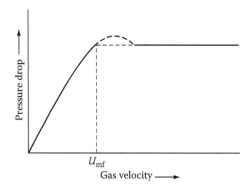

FIGURE 14.16 Pressure drop across a bed of particle as a function of gas velocity—U_{mf} corresponds to the minimum fluidization velocity.

14.3.2 Fluidized Beds

There has been a strong surge of interest and work during the 1970s and 1980s in fluidized-bed combustion (FBC) as a means for providing high heat-transfer rates, controlling sulfur, and reducing nitrogen oxide emissions due to the low temperatures in the combustion zone, which also favor the carbon dioxide equilibrium (Skinner, 1971; Chemical Week, 1981).

FBC offers a technology that can be designed to burn a variety of fuels efficiently and in an environmentally acceptable manner with various forms. In FBC units, coal is combusted in a hot bed of sorbent particles that are suspended by combustion air that is blown in from below through a series of nozzles. Depending on the gas velocity in the bed, fluidized bed can be classified into bubbling fluidized bed (BFB) and circulating fluidized bed (CFB). The fluidized bed can be operated at atmospheric (AFB) and elevated pressure (PFB).

In a fluidized bed, a gas passed slowly upward through a bed of solid particles finds its way through the spaces between the particles and the pressure drop across the bed is directly proportional to the flow rate. If the flow rate is increased, a point is eventually reached at which the frictional drag on the particles becomes equal to their apparent weight (i.e., weight minus any buoyancy force); the bed then expands as the particles adjust their positions to offer less resistance to the flow. The bed is now said to be fluidized and further increase in flow is not accompanied by an increase in pressure drop (Figure 14.17). Above the minimum fluidization velocity, the extra gas over and above that required for fluidization passes through the bed as bubbles and the bed itself is agitated.

In a fluidized-bed combustor, heat and mass transfer between the fluidizing gas and the solid particles are extremely efficient and fluidized-bed reactors are used for carrying out many chemical reactions on an industrial scale (e.g., catalytic cracking of hydrocarbons) (Speight, 2007). An additional advantage of fluidized-bed combustors is that they can use fairly coarse coal particles (ca. 0.04 in., ca. 1 mm diameter) and there is no need for much of the costly crushing equipment associated with the preparation of pulverized fuel.

A fluid bed combustor (Figure 14.17) usually (in the initial stages) consists of sand or some similar inert material that is fluidized by an air stream and raised to the ignition temperature by an external heating source. When the requisite temperature is reached, coal is fed to the vigorously bubbling bed where it becomes thoroughly mixed with the sand. As combustion begins, volatile material is given off and usually burns in the freeboard above the bed. The solid residue, or char, remains in the bed where the temperature is about 900°C (1650°F).

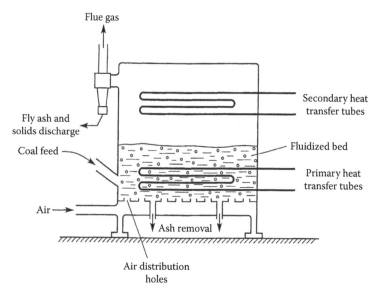

FIGURE 14.17 Fluidized-bed combustor.

The main features of circulating fluidized-bed combustors (CFBC) are

(1) Compatibility with wide range of fuels—conventional boilers for power generation can use only fossil fuels, such as high-grade coal, oil, and gas. The CFBC is also capable of using low-grade coal, biomass, sludge, waste plastics, and waste tires as fuel.
(2) Low polluting NOx and SOx emissions are significantly decreased without special environmental modifications. The operation of circulating fluidized-bed boilers involves a two-stage combustion process: the reducing combustion at the fluidized-bed section and the oxidizing combustion at the freeboard section. Next, the unburned carbon is collected by a high-temperature cyclone located at the boiler exit to recycle to the boiler, thus increasing the efficiency of denitrogenation.
(3) High combustion efficiency—improved combustion efficiency is attained through the use of a circulating fluidization-mode combustion mechanism.

Ash from the coal and the residue from any limestone added to remove sulfur, together with the initial sand, are removed from the bottom of the bed, although organic and inorganic particles eventually become small enough to be carried over (entrained) in the flue gases. This fly ash has to be removed before these gases are discharged to the atmosphere; any unburned material represents a significant loss of efficiency and arrangements are often made for recirculation of the fly ash.

Briefly, *fly ash* is a product of burning finely ground coal in a boiler to produce electricity. It is removed from the plant exhaust gases primarily by electrostatic precipitators or baghouses and secondarily by scrubber systems. Physically, fly ash is a very fine, powdery material composed mostly of silica; nearly all particles are spherical in shape. Fly ash is generally light tan in color and consists mostly of silt-sized and clay-sized glassy spheres. This gives fly ash a consistency somewhat like talcum powder.

Fly ash is a pozzolan, a siliceous material, which, in the presence of water, will react with calcium hydroxide at ordinary temperatures to produce cementitious compounds. Because of its spherical shape and pozzolanic properties, fly ash is useful in cement and concrete applications. The spherical shape and particle size distribution of fly ash also make it a good mineral filler in hot mix asphalt applications and improve the fluidity of flowable fill and grout when it is used for those applications.

Ash-forming elements (such as aluminum, Al; calcium, Ca; iron, Fe; potassium, K; magnesium, Mg; sodium, Na; and silicon, Si) occur in fossil or biofuels as internal or external mineral grains and simple salts such as sodium chloride (NaCl) or potassium chloride (KCl), which arise from the primeval brine, are also associated with the organic matrix of the fuel (Chapters 3 and 7). In PCC approximately 1% (w/w) of the inorganic metals are vaporized, while the rest remains in a condensed form as mineral inclusions (Flagan and Friedlander, 1978).

Depending on the gas/particle temperature and local stoichiometry during coal particle heat-up, devolatilization, and char burnout, the mineral inclusions will undergo phase transformations and approach each other to form a fly ash fraction. The vaporized metal species may undergo several transformations: nucleation, subsequent coagulation, scavenging, heterogeneous condensation, and/or interactions with mineral inclusions in the burning char or residual fly ash particles. The extent of transformation depends on the total specific surface area of the residual fly ash particles, the cooling rate of the flue gas, the local stoichiometry, and the mixing in the gas phase. Local supersaturation with respect to certain chemical species such as sodium sulfate (Na_2SO_4), potassium chloride (KCl), and potassium sulfate (K_2SO_4) may lead to the formation of submicron aerosol particles by homogeneous nucleation (Flagan and Friedlander, 1978).

Among the most significant environmental benefits of using fly ash over conventional cement is that greenhouse gas emissions can be significantly reduced. For every ton of fly ash used for a ton of Portland cement (the most common type of cement in general use around the world), approximately 1 ton of carbon dioxide is prevented from entering the earth's atmosphere. Fly ash does not require the energy-intensive kilning process required by Portland cement.

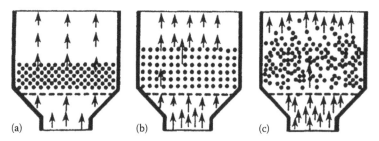

FIGURE 14.18 Illustration of the fluidized-bed concept: (a) gas velocity less than the fluidizing velocity, (b) gas velocity at the minimum fluidizing velocity, and (c) gas velocity greater than the fluidizing velocity.

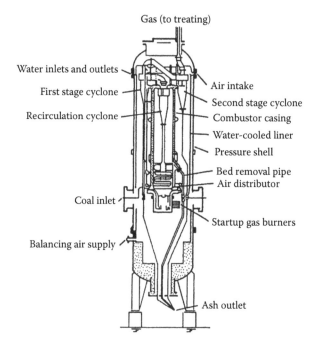

FIGURE 14.19 Pressurized fluidized-bed combustor.

In the simplest terms, fluidized combustion occurs in expanded beds (Figures 14.18 and 14.19). Even though reaction occurs at lower temperatures (900°C [1650°F]), high convective transfer rates exist due to the bed motion. Heat loads higher than in comparably sized radiation furnaces can be effected (i.e., smaller chambers produce the same equivalent heat load) and fluidized systems can operate under substantial pressures (Figure 14.20), thereby allowing more efficient gas cleanup.

Mechanical problems associated with fluidized combustion are encountered with the feeding of coal and particularly with the withdrawal and separation of the ash from the char or unreacted coal for recycling back to the combustion chamber. There are also problems with pollution control. While the sulfur may be removed downstream with suitable ancillary controls, the sulfur may also be captured in the bed, thereby adding to the separations and recycle problems. Capture during combustion, however, is recognized as the ideal and is a source of optimism for fluidized combustion. And ash agglomeration is not guaranteed as a means of ready separation and reducing bed carryover; attrition occurs and particulates occur in the off-gases that require controls. Thus, there remain issues related to sulfur and particulate control along with problems with feeding, withdrawal, separation, and recycle but, in spite of this, fluidized combustion presents an intriguing prospect for direct firing.

The major advantage of FBC technologies is that FBC boilers capitalize on the unique characteristics of fluidization to control the combustion process. The typical combustion temperature is around

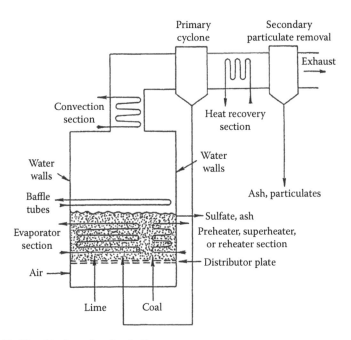

FIGURE 14.20 Fluidized-bed combustion boiler.

850°C (1560°F), which is optimal condition for capturing sulfur dioxide in situ and the nitrogen oxide emission is limited. In addition, FBC boilers are highly flexible to burn different ranks of fuels, including bituminous and subbituminous coals, coal waste, lignite, petroleum coke, with varying sulfur and ash contents, as well as a variety of waste fuels or *opportunity fuels*, such as biomass, which can be converted to synthetic fuels (Johnson et al., 2001; Green, 2003; Tillman et al., 2003; Maciejewska et al., 2006; Speight, 2008) but cannot be accommodated by pulverized coal units.

An important aspect of a fluidized-bed combustor (Figure 14.21) is that only a small concentration of coal (ca. 5%) is necessary to sustain combustion and at typical operating conditions the concentration is less than 1%. The coal feed will, of course, contain a proportion of fine material and inevitably some of these fines will be elutriated before they can be completely burned, but unburned material leaving the furnace is collected by cyclones and, if necessary, can be burned by refiring to the original bed or a separate one.

A fluidized bed is an excellent medium for contacting gases with solids, and this can be exploited in a combustor since sulfur dioxide emissions can be reduced simply by adding limestone ($CaCO_3$) or dolomite ($CaCO_3 \cdot MgCO_3$) to the bed (Abelson et al., 1978). The sulfur oxides react to form calcium sulfate, which leaves the bed as a solid with the ash. The theoretical additive requirement is that limestone amounting to 3% of the coal feed should be added for each 1% sulfur in the fuel. Thus, with efficient fines recycling and a temperature of 800°C–850°C (1470°F–1560°F), the theoretical addition retains about 80% of the sulfur while double the theoretical rate retains about 95%.

Because the combustion temperature is low, the ash fusion characteristics of coals present problems usually only in exceptional circumstances. Since the coal concentration is so low in the bed, it should be possible to burn coals with excessive amounts of mineral matter. In fact, variation in the mineral content of the feed coal does not appoint a major problem providing that the coal feed rate can be varied sufficiently to cope with the changes in calorific value. Furthermore, a wide range of caking properties and of volatile matter contents are acceptable, ranging from very strongly caking coal to noncaking coals of high-volatile matter content. The exception, however, is anthracite because of its very low reactivity and the resulting higher proportion of fine carbon that is elutriated before it can be burnt.

Combustion in a fluidized bed has some particular benefits including suitability for use with low-grade, high-ash coals and the lower bed temperatures compared to those in a conventional furnace.

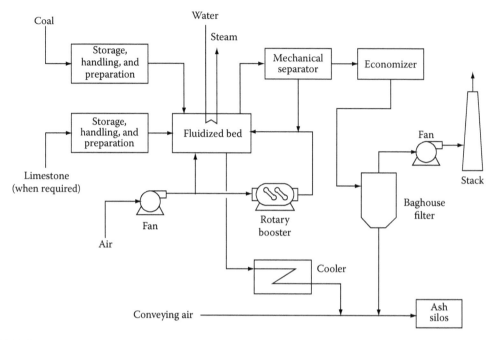

FIGURE 14.21 Atmospheric fluidized-bed process.

As a result of this lower temperature, the nitrogen oxide levels in the flue gas are reduced considerably. In addition, a reduction in sulfur dioxide emissions can also be achieved by mixing the coal with limestone (or dolomite). At the temperature of the bed, the carbonate is converted to the oxide that reacts with any sulfur dioxide to give calcium (or magnesium) sulfate:

$$CaCO_3 \rightarrow CaO + CO_2$$

$$2CaO + 2SO_2 + O_2 \rightarrow 2CaSO_4$$

$$MgCO_3 \rightarrow MgO + CO_2$$

$$2MgO + 2SO_2 + O_2 \rightarrow 2MgSO_4$$

The spent sorbent from FBC may be taken directly to disposal and is much easier than the disposal salts produced by wet limestone scrubbing. These latter species are contained in wet sludge having a high volume and a high content of salt-laden water. The mineral products of FBC, however, are quite dry and in a chemically refractory state and, therefore, disposal is much easier and less likely to result in pollution.

The spent limestone from FBC may be regenerated, thereby reducing the overall requirement for lime and decreasing the disposal problem. Regeneration is accomplished with a synthesis gas (consisting of a mixture of hydrogen and carbon monoxide) to produce a concentrated stream of sulfur dioxide that can be used to synthesize sulfuric acid to produce elemental sulfur:

$$CaSO_4 + H_2 \rightarrow CaO + H_2O + SO_2$$

$$CaSO_4 + CO \rightarrow CaO + CO_2 + SO_2$$

$$CaSO_4 + 4H_2 \rightarrow CaS + 4H_2O$$

$$CaSO_4 + 4CO \rightarrow CaS + 4CO_2$$

The calcium oxide product is supplemented with fresh limestone and returned to the fluidized bed. Two undesirable side reactions in the regeneration of spent lime produce calcium sulfide and result in recirculation of sulfur to the bed.

As new technology is developed, emissions may be reduced by repowering in which aging equipment is replaced by more advanced and efficient substitutes. Such repowering might, for example, involve an exchange in which an aging unit is exchanged for a newer combustion chamber, such as the atmospheric fluidized-bed combustor (AFBC) (Figure 14.21) or the pressurized fluidized-bed combustor (PFBC) (Figures 14.22 and 14.23) (Dainton, 1979; Smith and Hunt, 1979; Thomas, 1986; Argonne, 1990a). In the PFBC, pressure is maintained in the boiler, often an order of magnitude

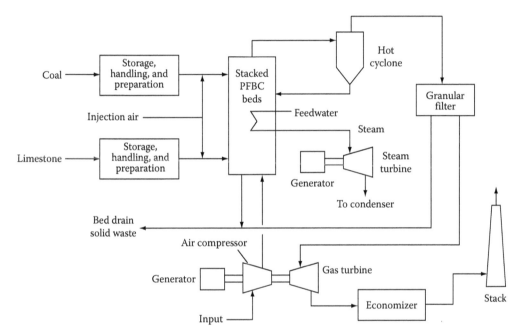

FIGURE 14.22 Pressurized fluidized-bed process.

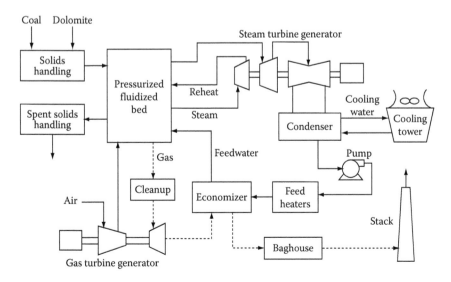

FIGURE 14.23 Steam-cooled pressurized fluidized-bed combined cycle process.

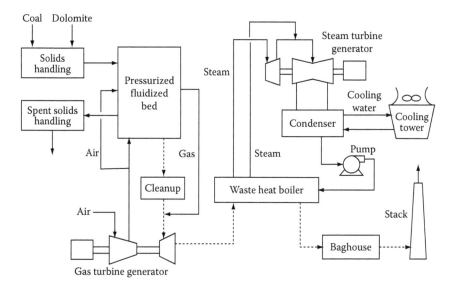

FIGURE 14.24 Air-cooled pressurized fluidized-bed combined cycle process.

greater than in the atmospheric combustor, and additional efficiency is achieved by judicious use of the hot gases in the combustion chamber (combined cycle).

Briefly, in the utility industry, combined cycle technology refers to the combined use of hot combustion gas turbines and steam turbines to generate electricity (Figures 14.22 through 14.24). This process can raise, quite significantly, the overall thermal efficiency of power plants above the thermal efficiency of conventional fossil fuel power plants using either type of turbine alone. Combined cycle plants that incorporate a gasification technology are called integrated gasification combined cycle (IGCC) plants (Figure 14.25) (Argonne, 1990b; Takematsu and Maude, 1991).

Both the atmospheric and pressurized fluid bed combustors burn coal with limestone or dolomite in a fluid bed that, with recent modifications to the system, allows the limestone sorbent to take up about 90% of the sulfur that would normally be emitted as sulfur dioxide. In addition, combustion can achieved at a lower temperature than in a conventional combustor thereby reducing the formation of nitrogen oxide(s).

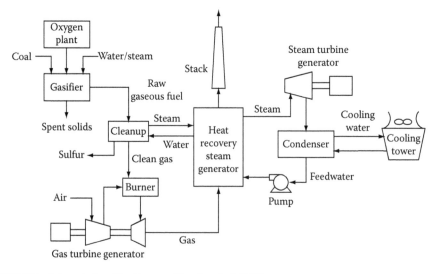

FIGURE 14.25 Integrated gasification combined cycle (IGCC) process.

14.3.3 ENTRAINED SYSTEMS

In entrained systems, fine grinding and increased retention times intensify combustion but the temperature of the carrier and degree of dispersion are also important. In present day practice, the coal is introduced at high velocities that may be greater than 100 ft/s and involve expansion from a jet to the combustion chamber. The pulverized coal is usually suspended in a stream of primary air (ca. 25% of the total air requirement) and the remainder of the air for combustion is introduced as secondary air. The secondary air may be introduced at an inlet region surrounding, or adjacent to, the primary air duet or may even be at some distance away (Figure 14.26). The temperature of the primary air should be regulated within limits; a temperature of at least 60°C (140°F) is required to prevent the condensation of moisture in the lines but at temperatures of 80°C–130°C (175°F–265°F) bituminous coal particles, for instance, will soften and stick.

Types of entrained systems include cyclone furnaces (which have been used for various coals), and other systems have been developed and utilized for the injection of coal–oil slurries into blast furnaces or for the burning of coal–water slurries. The cyclone furnace (developed in the 1940s to burn coal having low ash fusion temperatures) is a horizontally inclined, water-cooled, tubular furnace in which crushed, rather than pulverized, coal is burned with air entering the furnace tangentially (Figure 14.27). Temperatures may be of the order of 1700°C (3100°F) and the ash in the coal is converted to a molten slag that is removed from the base of the unit. Coal fines burn in suspension while the larger pieces are captured by the molten slag and burn rapidly. The heat release rate in a cyclone furnace is more than 50 times greater than in a pulverized coal-fired unit and the size of the unit may vary from 5 to 10 ft diameter.

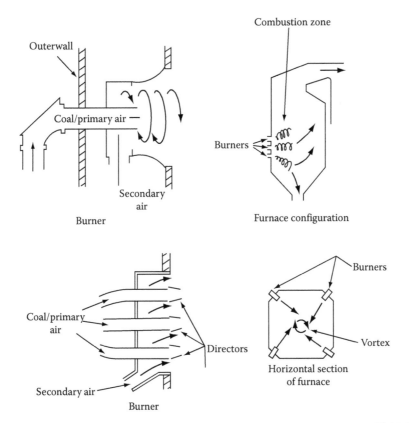

FIGURE 14.26 Pulverized fuel combustion for low-rank and medium-rank coal. (From Field, M.A. et al., *Combustion of Pulverized Coal*, British Coal Utilization Research Association, Leatherhead, Surrey, 1967; Berkowitz, N., *An Introduction to Coal Technology*, Academic Press, Inc., New York, 1979.)

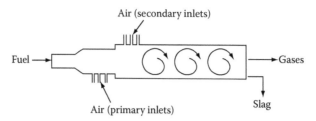

FIGURE 14.27 Cyclone furnace.

Slag, sometimes referred to as *boiler slag*, is the molten bottom ash collected at the base of slag tap and cyclone-type furnaces that is quenched with water. When the molten slag comes in contact with the quenching water, it fractures, crystallizes, and forms pellets. This boiler slag material is made up of hard, black, angular particles that have a smooth, glassy appearance.

Slag is generally a black granular material and the particles that are uniform in size, hard, and durable with a resistance to surface wear. In addition, the permanent black color of this material is desirable for asphalt applications and aids in the melting of snow. Slag is in high demand for beneficial use applications; however, supplies are decreasing because of the removal from service of aging power plants that produce boiler slag.

In the cyclone furnace, the crushed coal is conveyed into the burner by the primary air (20% of the total air), which enters the burner tangentially thereby imparting a whirling motion to the coal. The secondary air is also introduced tangentially into the furnace at a velocity of 300 ft/s. This imparts a whirling action to the coal particles that are thrown to the furnace wall by centrifugal force. These particles are held in the slag layer and burned. Most of the ash is retained in the slag layer, which minimizes the amount of fly ash that is carried over into the boiler.

The slag layer ensures that the total amount of heat absorbed by the water-cooled shell of a cyclone furnace will be relatively small and a secondary furnace is necessary to recover thermal energy. This gas-cooling furnace is similar in construction to a pulverized coal-fired furnace with the possible addition of a slag screen to remove droplets of molten slag carried over from the cyclone unit. Several cyclone furnaces are commonly used with a single secondary furnace.

The cyclone offers the advantage of being able to burn low-ash-fusion coals that create problems when burned in conventional pulverized coal burners. In addition, the cyclone has an inherently low fly ash carryover and minimizes erosion and fouling problems in the boiler and reduces the size of a particulate collector.

An outstanding advantage of the cyclone furnace is its low dust burden in the secondary furnace and, hence, its lesser emission of particulate matter from the stack. Most cyclone furnaces capture about 90% of the ash in the coal and convert it to molten slag. In addition, crushing coal for cyclone furnaces requires less power than pulverization but, on the other hand, the high tangential air velocity for cyclone furnaces requires a wind-box pressure up to 100 cm (40 in.) of water, and the total power requirements of a cyclone furnace may be comparable with those of a pulverized coal-fired unit.

Cyclone furnaces have two major shortcomings: (1) the ash of the coal must be convertible to molten slag at furnace temperatures and (2) the nitrogen oxide emissions are excessive (ca. 1000 ppm) because of the high furnace temperature. Cyclone furnaces have been widely used in areas where the coal contains ash with a low fusion temperature. For successful removal of slag, the slag viscosity cannot exceed 250 poise at 1420°C (2600°F) and many coals do not meet this requirement. However, addition of iron ore, limestone, or dolomite makes it possible to flux the coal ash thereby decreasing the viscosity at furnace temperatures.

A major drawback, however, is that the ash particles in the coal are raised to temperatures near 1400°C (2550°F); some mineral constituents will soften and glaze at these temperatures while others will volatilize. If the ash particles are still soft when they enter the convective heat-transfer part of the boiler, there is a possibility that they will form gluey deposits on the cooling tubes. Corrosive effects may also become more apparent at these high temperatures.

14.3.4 Miscellaneous Systems

The previous sections have given an indication of the combustion systems that are available and in use. There are, however, several other systems that are still in the experimental stage or have seen limited use insofar as they have not yet achieved a high degree of commercial acceptance. Nevertheless, these systems could well be the basis of future coal combustion operations and, as such, are worthy of mention here.

14.3.4.1 Colloidal Fuel-Fired Units

Methods for burning mixtures of pulverized coal in oil (variously called colloidal fuel, coal-in-oil slurry, or coal–oil suspension) have been studied for nearly a century and require the production of coal-in-oil suspensions. Stable short-term suspensions of coal in residual fuel oil are easily attained if the coal is pulverized to 200 mesh (75 μm) and, by adding surfactants, long-term stability can be obtained so that the coal will not settle out of the mixture even over periods as long as a few months. The interaction between the coal and the hydrocarbon allows the apparent viscosity of the coal–oil mixture (COM) to be 10 times greater than the fuel oil base and special precautions may be taken to provide adequate pump capacity, heating systems for the slurry, and properly sized burner nozzles.

The coal–water slurry fuel (CWSF or CWS or CWF) is a fuel that consists of fine coal particles suspended in water. Presence of water in the coal–water slurry reduces harmful emissions into the atmosphere, makes the coal explosion proof, makes use of coal equivalent to use of liquid fuel (e.g., heating oil), and gives other benefits (see the following). A coal–water slurry consists of 55%–70% of fine dispersed coal particles and 30%–45% of water.

The coal–water slurry can be used in place of oil and gas in small, medium, and big heating and power stations. Coal–water slurry is suitable for existing gas, oil, and coal boilers.

The relatively low cost of coal when compared to other energy sources gives coal–water slurries a competitive alternative to heating oil and gas and a relatively environmentally friendly fuel for heat and power generation. One side effect of the coal–water slurry making process is the separation of non-carbon material (such as pyrite and other inorganic mineral matter) mixed with the coal before treatment. This results in a reduction of ash yield to as low as 2% for the treated CWSF.

14.3.4.2 Ignifluid System

The ignifluid fuel-burning concept (Schwarz, 1982) consists essentially of an inclined chain-grate stoker with combustion air supplied at high velocity (approximately 50 ft/s) through the stoker bars at which the crushed coal, at the upper end of the chain-grate stoker, is blown off the grates and burns in suspension; the larger coal particles recirculate in the space above the grates until burning is complete. Ash is agglomerated into spheres approximately 1.0 in. (2.5 cm) in diameter, some of which may remain in the space between the furnace walls and the chain grate to provide a sloping sidewall to redirect recirculating particles of coal into the combustion zone. The ash is eventually conveyed, by the moving stoker gates, to an ash pit.

14.3.4.3 Submerged Combustion Systems

There are several processes that have been proposed for the direct combustion (oxidation or gasification) of coal within a liquid medium, and three of these are described as follows. The Zimpro process is based on the oxidation of crushed coal suspended in an oxygen-saturated hot-pressurized aqueous medium and is an outgrowth of an attempt to produce oxidized chemical products from paper mill wastes and the development of wastewater reclamation systems. The process involves the injection of high-pressure air into a slurry of hot water and coal under high pressure. High rates of oxidation occur at temperatures between 200°C and 350°C (390°F and 660°F); the exact temperature required to complete oxidation with reasonable residence time is dependent on the coal. Since the energy released is used to vaporize water, there is a direct relationship between the liquid temperature and the reactor pressure to which the air used for oxidation must be compressed.

After oxidation, the gases evolved from the slurry contain water vapor, carbon dioxide, nitrogen, and partially oxidized material; unoxidized material is removed from the bottom of the reactor and inevitably is accompanied by some water.

Part of the energy released by oxidation (combustion) can be recovered as work by expansion of the evolved gases. Because these gases contain nitrogen and carbon dioxide as well as some water vapor, a specially designed expansion turbine using corrosion-resistant materials is required. Process heat can be obtained from the gases either directly or by passing them over a heat exchanger before expansion with some loss of the available expansion work. However, if the primary objective is to produce work, the gases need to be heated to a temperature higher than the reactor temperature of 200°C–350°C (390°F–660°F). A small amount of superheat can be achieved by maximizing the amount of partially oxidized gases evolved from the reactor and catalytically oxidizing it; however, to achieve thermodynamically undesirable gas temperatures at the turbine inlet, an additional combustion system beyond the Zimpro process may be needed.

Although it appears that the Zimpro process may not be able to compete with FBC as a means of electricity generation (Table 14.7), there may be circumstances (e.g., cheap, low-grade fuel, and/or process heat) in which the process may be suitable for energy conversion.

Although the Atgas process has been developed mainly to produce substitute natural gas, it has also been proposed as a method for producing sulfur-free low heat content gas for use in power plant boilers. The process depends on the reaction between coal, air, and steam within a bath of molten iron. Sulfur from the coal is retained by the iron from which it is transferred to an overlying high–calcium oxide slag. The proposed gasifier is a cylindrical, refractory line vessel filled with molten iron (1425°C [2600°F]) and a floating layer of slag. Coal and limestone are injected with steam through the slag layer into the molten bath of iron where the coal is devolatilized; the carbon and sulfur are dissolved in the iron. Air introduced just beneath the slag layer oxidizes the carbon to carbon monoxide. The injected steam contributes to the production of hydrogen; the dissolved sulfur is transferred to the slag layer where it forms calcium sulfide and is removed with the slag.

The super-slagging combustor invokes the concept of burning coal (to produce a low heat content gas) within a bed of molten coal-ash slag fluxed with limestone; the sulfur in the coal reacts with the lime to form calcium sulfide. Thus, if such a super-slagging combustor were incorporated into a steam-generating boiler furnace with controlled addition of secondary air, the result might be a direct combustion system that produces flue gas with no sulfur or nitrogen oxides through staged combustion and maintains particulate emissions at a minimum. However, thermochemical

TABLE 14.7
Comparison of the Fluidized-Bed Combustion Process with the Zimpro Process

Item	Fluidized-Bed Combustion	Zimpro
Sulfur in exit gases	10%–20% of input	H₂S approximately 1 ppm Sulfur oxide approximately 1 ppm
NOx in exit gases	50–700 ppm	10–100 ppm
Particulates	Mechanical collectors plus precipitators required	120 mg/m³
Trace metals	Captured by calcium except for mercury and cadmium	Appear in sludge
Heat release	1–2 MW/m³	1–2 MW/m³
Fuel condition	Dry solids	Slurry
Combustion efficiency	87%–97%	95%–98%
Problem areas	Boiler tube corrosion and turbine blade erosion	—

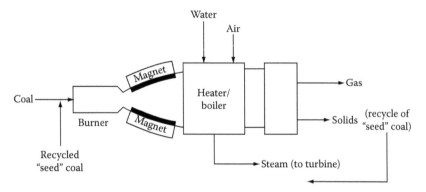

FIGURE 14.28 Magnetohydrodynamic generator.

calculations have shown that the product gas must be in equilibrium with excess carbon if sulfur is to be detained by the calcium oxide fluxed slag at 1540°C (2800°F) and that under even moderately oxidizing conditions sulfur would not be absorbed. This has been demonstrated in preliminary tests where the sulfur in the product gas can be as little as S% of the sulfur in the feed gas.

The magnetohydrodynamic (MHD) process for the generation of electricity invokes the concept that a conducting gas (e.g., a high-temperature combustion gas) *seeded* with an easily ionizable metal (such as potassium or calcium) creates an electric current when flowing at high speed in a magnetic field (Morrison, 1988; Kessler, 1991). The MHD process provides for several alternative generator configurations and operating procedures and can also use media other than a combustion gas. In a closed-cycle system, in which the working fluid would be recycled after reheating upon completion of a passage through the MHD channel, it could be a molten metal or metal vapor or a coal-fired venturi-type generator (Figure 14.28).

14.3.5 FUEL FEEDERS

Fuel feeder/distributors that evenly feed the fuel over the entire grate surface are necessary for even energy rerelease. These feeder/distributors can be mechanical, pneumatic, or a combination of both and must be placed across the width of the front of the stoker in sufficient quantity to achieve even lateral distribution of the fuel and have the means to longitudinally adjust fuel distribution for various types of fuels and sizing. They should be able to bias the feed rate one feeder to another, and to adjust for segregation of fuel sizing from one feeder to another. The performance of the fuel feeder/distributors can adapt to the different characteristics of solid fuels which plays a major part in the ability to operate at lowest possible emissions and highest combustion efficiency (Johnson, 2002).

Coal feeders should have a nonsegregating distributor interfacing between the coal bunker and the stoker feeder. A coal scale is recommended between the nonsegregating spout and the coal bunker. A coal scale provides a method for tracking daily, weekly, or monthly coal usage. All modern coal-scale electronics provide for real-time usage in terms of coal rate per hour, which is useful for tracking efficiency.

For maximum efficiency, best load following characteristics, and lowest emissions, it is recommended that there be a separate metering device for each fuel distributor and that the metering devices be kept full of fuel at all times. It is also important that the metering device be kept in a vertical plane from the front to prevent lateral poor distribution of the feedstock in the furnace (Johnson, 2002).

14.4 COAL–LIQUID MIXTURES

At this point, it is appropriate to consider another form in which coal can be used as a fuel, and this is as COM or coal–water mixture (CWM). A coal–liquid mixture (CLM) consists of finely crushed coal suspended in various liquids and typically small amounts of chemical additives that improve stability and other physical properties. The primary purpose of CLM is to convert coal from a solid to an essentially liquid form, which, with certain equipment modifications, allows the coal to be transported, stored, and burned in a manner similar to fuel oil.

In addition to COM and CWM, CLM technology has been extended to include mixtures of coal–methanol, solvent-refined coal–oil, petroleum-coke oil, and other solid-fuel-liquid mixtures. However, for the purposes of this text, the contents of this section are limited to the more mature technologies involving COM and to CWM (Morrison, 1980; Argonne, 1990c).

14.4.1 COAL–OIL MIXTURES

Renewed interest and significant development efforts in COM have emerged in the past two decades as a result of the search to find a replacement for dwindling oil supplies in the United States and other countries. The possibility of transporting liquefied coal by pipeline as an alternative to transporting solid coal by rail or barge has also stimulated recent development of CWM.

However, initial development work for COM dates back to the previous century, with the earliest known COM patent being issued in the late nineteenth century. In fact, during World War I, COM were evaluated as a fuel for submarines, and during the 1930s, COM were successfully tested as a fuel oil substitute for locomotives and ocean liners (Argonne, 1990c).

More extensive COM research was undertaken in the 1940s due to the wartime constraints on oil supply, and the data collected during this period still serve as a basic source of information. A resumption of readily available oil supplies at prices competitive with coal inhibited the widespread commercialization of the technology at that time. A subsequent constraint on oil supplies, initiated by the 1973 oil embargo, prompted the current era of increased COM research and development. A significant early development was the discovery of low-cost, effective chemical additives for stabilizing the mixture and enhancing other physical and chemical properties. An additional application of COM was as a fuel for blast furnaces.

As a result of these activities, COM have become mature commercial technologies. However, several areas have been identified for further research to improve performance, reliability, and market potential. Principal among these are advanced beneficiation of the coal to further reduce sulfur and ash content, demonstration in a compact boiler designed to burn oil, and increased percentage (by weight) of coal in the mixture.

14.4.2 COAL–WATER MIXTURES

The development of CWM does not have the long history of COM development. In the 1960s, the Germans and Russians also conducted several major CWM combustion tests. However, despite the success of these tests, no further major efforts were undertaken in this country until the major test at 1 million Btu/h input (Argonne, 1990c).

14.4.3 PROCESS TECHNOLOGY

The primary objective for developing CLM fuels is to produce a coal-based fuel that has many of the operational characteristics of oil and can thus serve as a replacement fuel in oil-burning applications, with only minor modifications to fuel storage, handling, and combustion equipment and procedures. To achieve this objective, the principal areas of technology development have been in process for CLM preparation; slurry pumps and other equipment components for storage, handling, and transport; and retrofit modifications or new designs for burners and boilers using CLM (Table 14.8) (Argonne, 1990c).

The basic steps in COM preparation are (1) fine grinding of the coal, (2) mixing the pulverized coal with oil, and (3) stabilizing the mixture by addition of various chemical additives. An additional step of beneficiating the coal to remove sulfur and ash is also typically required. Beneficiation

TABLE 14.8

Technological Aspects of the Use of Coal–Liquid Mixtures

Processes	Characteristics	Representative Technologies
Coal crushing and sizing	Major emphasis on eastern, medium- to high-volatile bituminous coal	Crushing; screening
Grinding	Typically 200 mesh (74 µm or finer); bimodal distributions are under study for CWMs	Dry milling; wet milling; ultrasonic grinding

Beneficiation	Aiming for 15,000 Btu/lb, 0.9% sulfur, 4.0% ash	Physical cleaning
		  Water
		  Heavy media
		  Froth flotation
		  Oil agglomeration
		Chemical cleaning

Coal–liquid mixing	Additives minimize viscosity, increase solids loading capacity, and enhance mixture stability	Proprietary preparation processes involving additives, grinding, and mixing
Transportation and storage	CLMs are highly viscous, non-Newtonian fluids and must be stable with respect to sedimentation and subsidence	Heating; pumping; remixing; agitation; recirculation
Combustion	Fuel parameters: abrasiveness (nozzle and pump wear); atomizability; carbon conversion; flame stability	Burner modification; boiler modifications to prevent fouling, reduce derating, and collect ash

may be performed as part of the preparation of the input coal before it is sent to the COM preparation plant or, in some instances, may be integrated into the COM preparation process. Limits on viscosity and other physical properties affecting handling have limited the coal loading in the product COM to 40%–50% by weight (Table 14.9).

To enable them to serve as a replacement fuel in oil-burning installations, COM are usually prepared from a bituminous coal of medium to high volatility and low ash content, since oil burners have only limited capability for ash removal. To minimize ash deposition and fouling problems, it is also desirable that the feed coal be low in moisture and have a moderate-to-high ash fusion temperature (the temperature of initial deformation) (Table 14.10). A low-sulfur content is also required to maintain the low-sulfur emissions of the oil fuels being replaced.

The coal particle size in a COM is dependent on the application and equipment to be used in the transportation, storage, and combustion. For applications as an oil replacement in utility or industrial boilers or process heaters, where good atomization and stable flames are required, coals are typically ground to 70%–80% −200 mesh (74 µm). Ultrafine grinding to sizes as low as 10 µm can improve

TABLE 14.9

Representative Specifications for a Coal–Oil Mixture

Characteristic	Value
Composition (% w/w) (coal/oil water and additives)	50/43/7
High heating value (million Btu/bbl)	6.2
Density (lb/gal)	10.1
Sulfur (% w/w)	1.0–1.5
Ash (% w/w)	3.0
Viscosity (maximum lb/ft h at 140°F)	30,000

TABLE 14.10

Properties of Bituminous Coals Suitable for Use in Coal–Oil Mixtures

Characteristics	Range of Typical Values
Moisture (% w/w)	4.5–9.0
Ash (% w/w)	4.5–10.0
Volatiles (% w/w)	17–40
Sulfur (% w/w)	0.6–1.5
High heating value (Btu/lb)	12,500–14,500
Ash fusion temperature (°F)	2000–2800+

combustion efficiency, reduce ash deposition, and reduce or eliminate the need for stabilizing additives, but at a higher cost of fuel preparation. On the other hand, in applications such as blast furnaces where flame control is not as critical, COM with coarse coal particles have been successfully used. Pipeline and tanker transport of coal as a COM does not require fine grinding of the coal.

The basic steps for preparation of CWM (Table 14.11) are the same as those for COM, viz., grinding the coal, mixing with a liquid (in this case water) and chemical additives for stabilization, and possibly beneficiating the coal as a separate or integrated step. The grinding and mixing processes for CWM are similar to those for COM. Grinding of the coal particles to −100 mesh (147 μm) is typical, although other size distributions are used in special applications.

TABLE 14.11

Properties of a Representative Coal–Water Mixture

Property	Value	
Particle size	100%, −100 mesh	
Viscosity	Less than 6800 lb/(ft·s) at 113 cycles/s and 77°F (Hakke method)	
Volatile matter	Greater than 30% w/w (dry)	
Coal–water slurry analysis	Total moisture 31.0%	
	Solids content 69.0%	
	As Received	**Moisture-free**
Gross heating value (Btu/lb)	10,170	14,740
Proximate analysis (% w/w)		
Moisture	31.0	—
Volatile matter	27.1	39.3
Fixed carbon	40.1	58.1
Ash	1.8	2.6
Ultimate analysis (% w/w)		
Moisture	31.0	—
Hydrogen	3.8	5.5
Carbon	56.1	81.3
Sulfur	0.6	0.9
Nitrogen	1.1	1.6
Oxygen[a]	5.6	8.1
Ash	1.8	2.6

[a] Oxygen content is measured by subtracting the percentages of all other constituents from 100%.

TABLE 14.12

Coal Properties to be Considered in the Preparation of Coal–Water Mixtures

Coal Characteristic	Impact on CWM	Desired Range[a]
Energy content	High-energy content lowers handling and storage costs	13,000–15,000 Btu/lb
Volatile content	Volatiles improve ignition and combustion	30% or high
Ash content	High ash concentration increases particulate emissions, combustor derating, and coal cleaning costs	5% or lower
Ash chemistry	Corrosive ash with low softening temperatures can cause combustor slagging or fouling and need for derating	Noncorrosive, low in sodium, high softening temperature
Sulfur content	High sulfur content increases SO_2 emissions and sulfur removal costs.	1% or lower
Organic sulfur content	Organic sulfur cannot be removed by physical cleaning	Low enough to ensure 1% total sulfur in CWMs
Surface chemistry	Surface chemistry governs the effectiveness and cost of additives and influences the choice of coal cleaning method	

[a] Range for clean dry coal before mixing with water. CWMs are not limited to coals with characteristics within the ranges shown.

The overall CWM preparation, however, tends to be more sophisticated and controlled than the preparation of a COM so as to achieve a higher coal loading. Coal loadings in CWM of 70% by weight (compared to about 50% in COM) with acceptable viscosity have been achieved, due in large part to the lower viscosity of water compared to oil. The lower water viscosity will, however, allow more rapid settling of the particles, thus presenting a somewhat greater challenge to obtaining mixture stability. CWM with a high solid loading and minimum viscosity also require appropriate dispersants and stabilizers. Characteristics that determine the suitability of stabilizing additives include (1) non-foaming, (2) a structure with both hydrophobic and hydrophilic (water repelling and attracting) portions, (3) water solubility, (4) compatibility with stabilizers, and (5) effectiveness at low concentrations. Gums, salts, clays, and other materials have been used as stabilizers.

The chemical and physical properties of the coal have a major influence on the characteristics of CWM. Experience has shown that CWM with bituminous coals allow higher coal loadings than CWM made of subbituminous coals or lignites. Also, high-volatile bituminous coals are more desirable for CWM because they provide favorable ignition and combustion characteristics (Table 14.12). Since the fuel in CWM is 100% coal, beneficiation to reduce ash and sulfur is relatively more important for CWM compared to COM. As with COM preparation, the elimination of the need for dewatering and other factors are promoting the integration of coal beneficiation with the other CWM preparation process.

14.4.4 Combustion

Tests have demonstrated that heavily loaded CWM can be successfully fired in burners designed for oil. However, the tests also indicated areas needing development, including reducing air preheating requirements, extending burner turndown capability, increasing carbon conversion efficiency, and extending burner lifetime by reducing effects of erosion. These development needs are related to the slow burning and abrasive characteristics of CWM relative to oil.

Developmental goals considered to be necessary for acceptable performance include a turndown ratio of 3:1 or better, minimum burner-tip life of 2000 h, air preheating of less than 150°C (300°F), maximum droplet size of 300 μm, and carbon conversion efficiencies of greater than 99%. Small-scale tests suggest that these coals are achievable, but what is yet required is long-term demonstration in large electric-utility-size boilers in the 100–500 MW range.

A major issue affecting decisions to convert burners from oil to CWM is the *derating* in capacity compared to the original maximum continuous rating with oil firing. In small-scale tests, the capacity with a CWM has been maintained at 90%–100% of maximum continuous rating by modifications, such as adding soot blowers to minimize loss of heat transfer, changing burner-tip design, and enlarging furnace volume. Derating can also be minimized by using high-quality coal and advanced beneficiation in the CWM preparation. Preferred trade-offs between accepting some derating versus maintaining maximum continuous rating (by various facility modifications or use of more refined CWM) are site-specific problems that in most cases will require actual boiler testing.

14.5 COAL COMBUSTION PRODUCTS

CCPs (Table 14.13) are the inorganic residues that remain after pulverized coal is burned. Coarse particles (bottom ash and boiler slag) settle to the bottom of the combustion chamber, and the fine portion (fly ash) is removed from the flue gas by electrostatic precipitators or other gas-scrubbing systems.

The disposal cost of coal combustion by-products has escalated significantly during the last couple of decades due to significant changes in landfill design regulations. Utilization of CCP helps preserve existing licensed landfill capacity and thus reduces the demand for additional landfill sites.

Concerns and public pressure about air quality and acid rain led to the US Congress passing the Clean Air Act Amendments of 1990 (Public Law 101-549), which included stringent restrictions on sulfur oxide emissions. Most electric utilities in the Eastern and Midwestern States use bituminous coal having high sulfur content on the order of 2%–3.5%. Thus, in order to meet the emission standards, many utilities have installed flue-gas-desulfurization (FGD) equipment.

FGD is a chemical process to remove sulfur oxides from the flue gas at coal-burning power plants. Many methods have been developed to varying stages of applicability and the goal of these processes is to chemically combine the sulfur gases released in coal combustion by reacting them with a sorbent, such as limestone (calcium carbonate, $CaCO_3$), lime (calcium oxide, CaO), or ammonia (NH_3) (Chapters 15 and 23).

TABLE 14.13
Types of Coal Combustion Products and Their Characteristics

Product	Characteristics	Texture	Amount Typically Generated (lb per ton)	Major Constituents
Fly ash	Noncombustible particulate matter removed from stack gases	Powdery, silt-like	160	Si, Al, Fe, Ca
Bottom ash	Material collected in dry bottom boilers, heavier than fly ash	Sand-like	40	Si, Al, Fe, Ca
Boiler slag	Material collected in wet bottom boilers or cyclone units	Glassy, angular particles	100	Si, Al, Fe, Ca
FGD material	Solid/semisolid material obtained from flue gas scrubbers	Fine to coarse (dry or wet)	350	Ca, S, Si, Fe, Al[a]

[a] Fixated flue-gas-desulfurization material is a mixture of filter cake (principal constituents: Ca and S), fly ash (principal constituents: Si, Fe, Al), lime, and water.

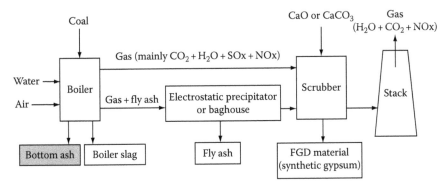

FIGURE 14.29 Flue-gas-desulfurization process based on lime (CaO) or limestone (CaCO$_3$), sorbents used by many flue-gas-desulfurization systems.

In the process, the flue gas emerges from the combustor/boiler and is contacted with the slurry of calcium salts; sulfur dioxide (SO$_2$) reacts with the calcium to form hydrous calcium sulfate (CaSO$_4 \cdot$ 2H$_2$O, gypsum) (Figure 14.29).

Another FGD method uses ammonia (NH$_3$) as the sorbent; the FGD product is ammonium sulfate [(NH$_4$)$_2$SO$_4$)]. Sulfate is the preferred form of sulfur readily assimilated by crops, and ammonium sulfate is the ideal sulfate compound for soil supplements because it also provides nitrogen from ammonium. The use of ammonium sulfate in large-scale fertilizer formulations has been growing gradually. This growth provides a market for FGD products and could make FGD processes based on ammonia—attractive alternatives to the processes based on lime and limestone.

The largest use of CCP (mostly fly ash) is in cement and concrete. The CCP displace Portland cement and significantly reduce emissions of carbon dioxide (CO$_2$). Portland cement manufacture requires the burning of fossil fuels and decomposition of carbonates, which release large amounts of carbon dioxide into the atmosphere.

CCPs have many economic and environmentally safe uses (Table 14.14). For example, in construction, a metric ton of fly ash used in cement and concrete can save the equivalent of a barrel of oil and can reduce carbon dioxide releases that may affect global warming. The use of CCP saves landfill space and can replace clay, sand, limestone, gravel, and natural gypsum.

TABLE 14.14
Uses of Coal Combustion Products

Product	Potential Areas of Major Use
Fly ash	Cement replacement in concrete/grout, structural fill, flowable fill, waste stabilization, surface mine reclamation, soil stabilization, road base, mineral filler
Bottom ash	Concrete block, road subbase, snow and ice control, structural fill, waste stabilization, pipe bedding, cement manufacture
Boiler slag	Blasting grit, roofing granules, snow and ice control, mineral filler, construction backfill, water filtration, drainage media
FGD[a] material	Wallboard, stabilized road base/subbase, structural fill, surface mine reclamation, underground mine injection, livestock pad, low permeability liner, synthetic gypsum raw material, liming substitute, soil conditioning, synthetic aggregate, sludge stabilization

[a] FGD, Flue gas desulfurization.

REFERENCES

Abelson, H.I., Lowenbach, W.A., and Gordon, J.S. 1978. In *Analytical Methods for Coal and Coal Products*, C. Karr, Jr. (Ed.). Academic Press, Inc., New York, Vol. II, Chapter 35.

Agricola, G. 1546. *De Natura Fossilium*. H. Froben and N. Episcopius, Basle.

Allen, R.R. and Parry, V.F. 1954. Report of Investigations No. 5034. United States Bureau of Mines, Washington, DC.

Argonne. 1990a. Environmental consequences of, and control processes for, energy technologies. Pollution Technology Review No. 181. Argonne National Laboratory, Noyes Data Corporation, Park Ridge, NJ, Chapters 3 and 4.

Argonne. 1990b. Environmental consequences of, and control processes for, energy technologies. Pollution Technology Review No. 181. Argonne National Laboratory, Noyes Data Corporation, Park Ridge, NJ, Chapter 8.

Argonne. 1990c. Environmental consequences of, and control processes for, energy technologies. Pollution Technology Review No. 181. Argonne National Laboratory, Noyes Data Corporation, Park Ridge, NJ, Chapter 9.

Banerjee, S.C. 1982. A theoretical design to the determination of risk index of spontaneous fires in coal mines. *Journal of Mines, Metals and Fuels*, 399–406.

Banerjee, S.C. 1985. *Spontaneous Combustion of Coal and Mine Fires*. Oxford and IBH Publishing Company, New Delhi, India.

Barnard, J.A. and Bradley, J.N. 1985. *Flame and Combustion*, 2nd edn. Chapman & Hall, London, U.K.

Bend, S.L., Edwards, I.A.S., and Marsh, H. 1991. In *Coal Science II*, H.H. Schobert, K.D. Bartle, and L.J. Lynch (Eds.). Symposium Series No. 461. American Chemical Society, Washington, DC, Chapter 22.

Bend, S.L., Edwards, I.A.S., and Marsh, H. 1992. *Fuel*, 71: 493.

Berkowitz, N. 1979. *An Introduction to Coal Technology*. Academic Press, Inc., New York.

Berkowitz, N. and Schein, H.G. 1951. *Fuel*, 30: 94.

Berkowitz, N. and Speight, J.G. 1973. *Canadian Institute of Mining and Metallurgical Bulletin*, 66(736): 109; 30: 94.

Biringuccio, V. 1540. *De la Pirotechnia*. V. Roffinello for C. Navo, Venice. Book III, Chapter 10.

Brill, T. 1993. *Chemistry in Britain*, 29(1): 34.

Brooks, K., Svanas, N., and Glasser, D. 1988. *Fuel*, 67: 651.

Ceely, F.J. and Daman, E.L. 1981. In *Chemistry of Coal Utilization*, Second Supplementary Volume, M.A. Elliott (Ed.). John Wiley & Sons, Inc., New York, p. 1313.

Chakravorty, R.N. 1984. Report No. ERP/CRL 84-14. Coal Research Laboratories, Canada Centre for Mineral and Energy Technology, Ottawa, Ontario, Canada.

Chakravorty, R.N. and Kar, K. 1986. Report No. ERP/CRL 86-151. Coal Research Laboratories, Canada Centre for Mineral and Energy Technology, Ottawa, Ontario, Canada.

Chamberlain, E.A.C. and Hall, D.A. 1973. The liability of coals to spontaneous combustion. *Colliery Guardian*, 65–72.

Chemical Week. 1981. Industry takes to burning coal in fluidized beds. September 16, p. 24.

Crelling, J.C., Thomas, K.M., and Marsh, H. 1993. *Fuel*, 72: 349.

Cruden, A. 1930. *Complete Concordance to the Bible*. Butterworth Press, London, U.K.

Dainton, A.D. 1979. In *Coal and Modern Coal Processing: An Introduction*, J.G. Pitt and G.R. Millward (Eds.). Academic Press, Inc., London, U.K., Chapter 6.

Didari, V. 1988. Developing a spontaneous combustion risk index for Turkish coal mines-preliminary studies. *Journal of Mines, Metals and Fuels*, 211–215.

Essenhigh, R.H. 1981. In *Chemistry of Coal Utilization*, Second Supplementary Volume, M.A. Elliott (Ed.). John Wiley & Sons, Inc., New York, p. 1153.

Feng, K.K., Chakravorty, R.N., and Cochrane, T.S. 1973. Spontaneous combustion: A coal mining hazard. *The Canadian Mining and Metallurgical Journal*, 75–84.

Field, M.A., Gill, D.W., Morgan, B.B., and Hawksley, P.G.W. 1967. *Combustion of Pulverized Coal*. British Coal Utilization Research Association, Leatherhead, Surrey.

Flagan, R.C. and Friedlander, S.K. 1978. Particle formation in pulverized coal combustion—A review. In *Recent Developments in Aerosol Science*, D.T. Shaw (Ed.). John Wiley & Sons, Inc., New York.

Fletcher, T.H., Ma, J.L., Rigby, J.R., Brown, A.L., and Webb, B.W. 1997. Soot in coal combustion systems. *Progress in Energy and Combustion Science*, 23: 283–301.

Galloway, R.L. 1882. *A History of Coal Mining in Great Britain*. Macmillan & Co., London, U.K.

Gavin, D.G. and Dorrington, M.A. 1993. *Fuel*, 72: 381.

Gay, A.J. and Davis, P.B. 1987. In *Coal Science and Chemistry*, A. Volborth (Ed.). Elsevier, Amsterdam, the Netherlands, p. 221.

Goodarzi, F. and Gentzis, T. 1991. Geological controls on the self burning of coal seams. In *Geology in Coal Resource Utilization*, D.C. Peters (Ed.). Tech-Books, Fairfax, VA.

Gray, P., Griffiths, J.F., and Hasko, S.M. 1984. *Journal of Chemical Technology and Biotechnology*, 34A: 453.

Green, A. 2003. Overview of co-utilization of domestic fuels. *Proceedings. International Conference on Co-Utilization of Domestic Fuels*. University of Florida, Gainesville, FL.

Handa, T., Nishimoto, T., Morita, M., and Komada, J. 1985. Spontaneous combustion of coal. *Fire Science and Technology*, 5: 21–30.

Haslam, R.T. and Russell, R.P. 1926. *Fuels and their Combustion*. McGraw-Hill, New York, Chapter 4.

Heitmann, H.-G. 1993. *Handbook of Power Plant Chemistry*. CRC Press, Inc., Boca Raton, FL.

Hessley, R.K. 1990. In *Fuel Science and Technology Handbook*, J.G. Speight (Ed.). Marcel Dekker, Inc., New York.

Hessley, R.K., Reasoner, J.W., and Riley, J.T. 1986. *Coal Science*. John Wiley & Sons, Inc., New York.

Hoffman, E.J. 1978. *Coal Conversion*. Energon, Laramie, WY, p. 100.

Hoover, H.C. and Hoover, L.H. 1950. *De Re Metallica*. Dover Publications, Inc., New York, p. 34.

Jamaluddin, A.S., Wall, T.F., and Truelove, J.S. 1987. In *Coal Science and Chemistry*, A. Volborth (Ed.). Elsevier, Amsterdam, the Netherlands, p. 61.

Johnson, D.K., Tillman, D.A., Miller, B.G., Pisupati, S.V., and Clifford, D.J. 2001. Characterizing biomass fuels for cofiring applications. *Proceedings. 2001 Joint International Combustion Symposium*. American Flame Research Committee, Kauai, Hawaii, September.

Johnson, N. 2002. Fundamentals of stoker fired boiler design and operation. *Proceedings. CIBO Emission Controls Technology Conference*. July 15–17, Salt Lake City, UT.

Jones, J.C. 1990. *Fuel*, 69: 399.

Jones, J.C., Rahmah, H., Fowler, O., Vorasurajakart, J., and Bridges, R.G. 1990. *Journal of Life Sciences*, 8: 207.

Jones, J.C. and Vais, M. 1991. *Journal of Hazardous Materials*, 26: 203.

Jones, T.F. 1992. *Mine and Quarry*, 212(6): 9.

Joseph, J.T. and Mahajan, O.P. 1991. In *Coal Science II*, H.H. Schobert, K.D. Bartle, and L.J. Lynch (Eds.). Symposium Series No. 461. American Chemical Society, Washington, DC, Chapter 23.

Kessler, R. 1991. In *Energy and the Environment in the 21st Century*, J.W. Tester, D.O. Wood, and N.A. Ferrari (Eds.). The MIT Press, Cambridge, MA, p. 945.

Kim, A.G. 1977. Laboratory studies on spontaneous heating of coal. Bureau of Mines Information Circular, IC No. 8756. Unites States Department of the Interior, Washington, DC.

Littler, D.J. 1981. In *Energy and Chemistry*, R. Thompson (Ed.). The Royal Society of Chemistry, London, U.K., p. 187.

Maciejewska, A., Veringa, H., Sanders, J, and Peteves, S.D. 2006. Co-firing of biomass with coal: Constraints and role of biomass pre-treatment. Report No. EUR 22461 EN. Institute of Energy, European Commission, Luxembourg.

Meyers, R.A. (Ed.). 1981. *Coal Handbook*. Marcel Dekker, Inc., New York.

Mitchell, R.E. 1989. *Proceedings. Sixth Annual International Pittsburgh Coal Conference*. University of Pittsburgh, Pittsburgh, PA, p. 32.

Morrison, G.F. 1988. Coal-fired MHD. Report No. IEACR/06. IEA Coal Research, International Energy Agency, London, U.K.

Morrison, G.F. 1980. Nitrogen oxides from coal combustion: Abatement and control. Report No. ICTIS/TR11. International Energy Agency, London, U.K.

Morrison, G.F. 1986. Understanding pulverized coal combustion. Report No. ICTIS/TR34. IEA Coal Research, International Energy Agency, London, U.K.

Moxon, N.T. and Richardson, S.B. 1987. The inhibition of coal oxidation and self heating by commercial dust suppressants. *Coal Preparation*, 4: 183–191.

Mulcahy, M.F.R. 1978. *The Metallurgical and Gaseous Fuel Industries*. The Chemical Society, London, U.K.

Nowacki, P. 1980. *Lignite Technology*. Noyes Data Corporation, Park Ridge, NJ.

Ogunsola, O.I. and Mikula, R.J. 1992. *Fuel*, 71: 3.

Parsons, T.H., Higgins, S.T., and Smith, S.R. 1987. *Proceedings. Fourth Annual Pittsburgh Coal Conference*. University of Pittsburgh, Pittsburgh, PA, p. 53.

Perry, J.H. and Green, D.W. (Eds.). 2008. *Chemical Engineers Handbook*, 8th edn. McGraw-Hill, New York.

Rajan, S. and Raghavan, J.K. 1989. *Proceedings. Sixth Annual International Pittsburgh Coal Conference*. University of Pittsburgh, Pittsburgh, PA, p. 979.

Ramesh, R., Francois, M., Somasundaran, P., and Cases, J.M. 1992. *Energy & Fuels*, 6: 239.

Reid, W.Y. 1971. *External Corrosion and Deposits: Boilers and Gas Turbines*. Elsevier, New York.

Reid, W.T. 1981. In *Chemistry of Coal Utilization*, Second Supplementary Volume, M.A. Elliott (Ed.). John Wiley & Sons, Inc., New York, p. 1389.

Schwarz, J. 1982. *Ignifluid Boilers, Reviews in Energy*, 341: 180.

Skinner, D.G. 1971. *Fluidized Combustion of Coal*. Mills and Boon, London, U.K.

Slack, A.V. 1981. In *Chemistry of Coal Utilization*, Second Supplementary Volume, M.A. Elliott (Ed.). John Wiley & Sons, Inc., New York, p. 1447.

Smith, F.B. and Hunt, R.D. 1979. *Philosophical Transactions*, A290: 523.

Speight, J.G. 2008. *Synthetic Fuels Handbook: Properties, Processes, and Performance*. McGraw-Hill, New York.

Speight, J.G. 2007. *The Chemistry and Technology of Petroleum*, 4th edn. Taylor & Francis Group, Boca Raton, FL.

Takematsu, T. and Maude, C. 1991. Coal gasification for IGCC power generation. Report No. IEACR/37. IEA Coal Research, International Energy Agency, London, U.K.

Thomas, J.F. 1986. Atmospheric fluidized bed boilers for industry. Report No. ICTIS/TR35. IEA Coal Research, International Energy Agency, London, U.K.

Thring, M.W. 1952. *Fuel*, 31: 355.

Tillman, D.A., Johnson, D.K., and Miller, B.G. 2003. Analyzing opportunity fuels for firing in coal-fired boilers. *Proceedings of 28th International Technical Conference on Coal Utilization and Fuel Systems*. Coal Technology Association, Gaithersburg, MD, March.

Vovelle, C., Mellottee, H., and Delbourgo, R. 1983. *Proceedings. 19th International Symposium on Combustion*. The Combustion Institute, Pittsburgh, PA, p. 797.

15 Electric Power Generation

15.1 INTRODUCTION

The production of electric power from coal is a mature and well-established technology in the industrialized countries of the world. However, with the advent of stricter environmental controls on effluents from power plants, especially with respect to sulfur oxides, new types of combustion/pollution control technology are emerging. In fact, the increased utilization of coal in place of oil and gas for combustion applications in the United States is motivating near-term development and implementation of alternative technologies for electricity generation from coal.

Coal, with high reserves and a relatively low cost, will continue to be one of the most important energy sources for the United States and elsewhere. In the United States, coal accounts for approximately one-third of all carbon dioxide emissions arising from fossil fuel use (Blasing et al., 2005). Worldwide, coal provides for about one-third of all electricity production, one-quarter of transportation use, and is projected to account approximately one-quarter (or more) of all energy needs by 2030 (Michener and McMullan, 2008). Although alternative sources are being improved, it is likely that alternatives alone will never (in the near term) be sufficient to meet the entire energy demand.

Coal is one of the true measures of the energy strength of the United States. The use of electricity has been an essential part of the US economy since the turn of the century. Coal power, an established electricity source that provides vast quantities of inexpensive, reliable power, has become more important as supplies of oil and natural gas diminish. In addition, known coal reserves are expected to last for centuries at current rates of use.

15.2 ELECTRICITY FROM COAL

Coal has played a major role in electrical production since the first power plants that were built in the United States in the 1880s (Singer, 1981). The earliest power plants used hand fed wood or coal to heat a boiler and produce steam. This steam was used in reciprocating steam engines that turned generators to produce electricity. In 1884, the more efficient high-speed steam turbine was developed by British engineer Charles A. Parsons, which replaced the use of steam engines to generate electricity. In the 1920s, the pulverized coal firing was developed. This process brought advantages that included a higher combustion temperature, improved thermal efficiency, and a lower requirement for excess air for combustion. In the 1940s, the cyclone furnace was developed. This new technology allowed the combustion of poorer grade of coal with less ash production and greater overall efficiency.

Converting coal to electric power appears, on paper, to be a relatively simple process. In most coal-fired power plants, coal is crushed into fine powder and fed into a combustion unit where it is burned. Heat from the burning coal is used to generate steam that is used to spin one or more turbines to generate electricity.

One-quarter of the world's coal reserves are found within the United States, and the energy content of the US coal resources exceeds that of all the world's known recoverable oil. Coal is also the workhorse of the nation's electric power industry, supplying more than half the electricity consumed by Americans.

However, much of the world is becoming increasingly electrified and currently, in the United States, more than half of the electricity generated comes from coal. In the foreseeable future, coal will continue to be the dominant fuel used for electric power production. The low cost and abundance of coal is one of the primary reasons why consumers in the United States benefit from some of the lowest electricity rates of any free-market economy.

Coal combustion technology is a mature technology with potentially increasing the efficiency in progress of advanced materials. In the short term and medium term, it will play an important role in the world energy section. In the long term, the role of coal may be reduced but coal will be difficult to replace.

Power plants for electricity generation are defined by functional type: (1) base load, (2) peak load, and (3) combined cycle—each has advantages and disadvantages.

Base load power plants have the lowest operating cost and generate power most in any given year. There are several different types of base load power plants and the resources available typically determine the type of base load plant used to generate power. Coal and nuclear power plants are the primary types of base load power plants used in the Midwest United States.

Base load power plants are also subdivided into four types: (1) high efficient combined-cycle plants fueled by natural gas, (2) nuclear power plants, (3) steam power plants fueled primarily by coal, and (4) hydropower plants.

Peak load power plants are relatively simple cycle gas turbines that have the highest operating cost but are the cheapest to build. They are operated infrequently and are used to meet peak electricity demands in period of high use and are primarily fueled with natural gas or oil.

The key challenge is to remove the environmental objections to the use of coal in future power plants. New technologies being developed could virtually eliminate the sulfur, nitrogen, and mercury pollutants that are released when coal is burned. It may also be possible to capture greenhouse gases emitted from coal-fired power plants and prevent them from contributing to global warming concerns.

There is also the need to increase the fuel efficiency of coal-fueled power plants. Currently, only one-third of the energy potential of coal is converted to electricity. Higher efficiencies mean even more affordable electricity and fewer greenhouse gases and this is a worthwhile goal for the next decades.

While coal is the major fuel for electric power in the United States, natural gas is the fastest growing fuel (Speight, 2007a,b, 2008) and the importance of coal to electricity generation worldwide is set to continue, with coal fueling 44% of global electricity in 2030. Natural gas is also likely to be a primary fuel for distributed power generators—mini-power plants that would be sited close to where the electricity is needed.

In the context of this chapter and electricity generation (power generation), coal plays a vital role worldwide. Coal-fired power plants currently fuel 41% of global electricity. In some countries, coal fuels a higher percentage of electricity.

The majority of the electricity currently generated in the United States today is produced by facilities that employ steam turbine systems (Figure 15.1). Other power generation systems employ a combination of the aforementioned, such as combined-cycle and cogeneration systems (Figure 15.2). The numbers of these systems being built are increasing as a result of the demands placed on the industry to provide economic and efficient systems.

15.2.1 Coal Properties

Coal type and quality impact generating unit technology choice and design, generating efficiency, capital cost, and performance. Boiler designs usually encompass a broader range of typical coals than initially intended to provide future flexibility. Single coal designs are mostly limited to mine-mouth plants, which today are usually only lignite, subbituminous, or brown coal plants. The energy, carbon, moisture, ash, and sulfur contents, as well as ash characteristics (Chapters 8 and 9),

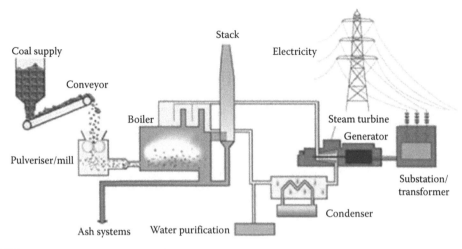

FIGURE 15.1 Electricity generation from coal.

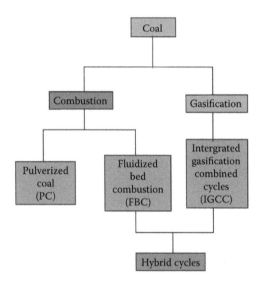

FIGURE 15.2 Options for power generation from coal.

all play an important role in the value and selection of coal, its transportation cost, and the technology choice for power generation.

Each of the coal properties interacts in a significant way with generation technology to affect performance. For example, higher-sulfur content reduces efficiency of pulverized coal combustion (PCC) due to the added energy consumption and operating costs to remove sulfur oxides (SOx) from the flue gas.

Coal having a high mineral matter content (high ash production) requires combustor design changes to manage erosion. Fluidized-bed combustion (FBC) is more suited to coals that produce high yields of ash coals—these include low-carbon coal waste and lignite.

15.2.2 COMBUSTION TECHNOLOGY

The type of system employed at a facility is chosen based on the loads, the availability of fuels, and the energy requirements of the electric power generation facility. At facilities employing these systems, other ancillary processes must be performed to support the generation of electricity.

These ancillary processes may include such supporting operations as coal processing and pollution control, for example.

The process of generating electricity from steam comprises four parts: a heating subsystem (fuel to produce the steam), a steam subsystem (boiler and steam delivery system), a steam turbine, and a condenser (for condensation of used steam). Heat for the system is usually provided by the combustion of coal, natural gas, or oil. The fuel is pumped into the boiler's furnace. The boilers generate steam in the pressurized vessel in small boilers or in the water-wall tube system in modern utility and industrial boilers. Additional elements within or associated with the boiler, such as the superheater, reheater, economizer, and air heaters, improve the boiler's efficiency.

The feeding rate of coal according to the boiler demand and the amount of air available for drying and transporting the pulverized coal fuel is controlled by computers. Pieces of coal are crushed between balls or cylindrical rollers that move between two tracks or races. The raw coal is then fed into the pulverizer along with air heated to about 345°C (650°F) from the boiler. As the coal gets crushed by the rolling action, the hot air dries it and blows the usable fine coal powder out to be used as fuel. The powdered coal from the pulverizer is directly blown to a burner in the boiler. The burner mixes the powdered coal in the air suspension with additional preheated combustion air and forces it out of a nozzle similar in action to fuel being atomized by a fuel injector in modern cars. Under operating conditions, there is enough heat in the combustion zone to ignite all the incoming fuel.

15.2.2.1 Direct Combustion Technology

The essence of the combustion of coal is the presence of air in intimate contact with the coal and complete reaction occurs (Chapters 13 and 14). In such an oxidizing atmosphere, all species leave the combustion system in an oxidized state, namely carbon dioxide (CO_2), water (H_2O), several types of nitrogen oxides (NOx, where x is 1 or 2), and sulfur oxides (mainly SO_2 with some SO_3). If the coal contains little or no sulfur or nitrogen, then no cleanup of the combustion or *stack gases* may be minimal since water and, sometimes (but not always), carbon dioxide are classed as innocuous materials.

There are three main designs of combustion systems in which coal and air can be reacted as the start to power generation and they are (1) the fixed-bed combustor, (2) an entrained-bed combustor, also called a suspended-bed combustor, and (3) a fluidized-bed combustor.

In a *fixed-bed combustor*, the air passes upward through the pulverized coal at a low velocity. The coal is held on a grate, and the bed of hot coal may be several inches thick. The coal remains in a fixed bed since the air velocity is not sufficient to lift the coal particles upward. Ash removal is continuous or semicontinuous by mechanical means. This type of bed, however, does not afford very efficient gas–solid contact. High combustion rates are not possible with this system.

In the *entrained-bed combustor*, the feed coal must be introduced to the combustor as small. The particles (generally <200 mesh) are carried by the gas into the furnace and travel in a suspended state through a hot zone where they are consumed. After combustion, approximately 20% of the ash falls to the bottom of the furnace and is removed there. Combustion gases, which contain about 80% of the ash, pass out of the furnace and are treated to remove the remaining particulates (usually in an electrostatic precipitator) and sulfur compounds (usually in a stack gas scrubber). This type of combustion system (PCC) is in common use for large-scale utility boilers. Higher combustion rates can be attained in pulverized fuel combustors than in fixed-bed combustors.

The third type of combustion system is the *fluidized-bed combustor* in which the velocity of the oxidizing gas is sufficiently high to support or *float* the particles but not carry them out of the top of the reactor. In this combustor, the coal particles are much more concentrated than in a pulverized coal combustor, giving a high surface area of solids with which to react. Heat and mass exchange are very rapid, and the collisions of particles among themselves as well as with heat transfer surfaces submerged in the fluidized bed give much more effective heat transfer to the boiler tubes, compared to pulverized coal combustors. Coals of high mineral content (i.e., high ash-producing coals) can be readily burned in such a system but cannot be easily used in a pulverized coal combustor.

15.2.2.2 Fixed-Bed Combustors

During the nineteenth century, virtually all methods for burning coal used the coal-bed-on-a-grate method. The first technological developments were oriented toward controlling the amount of air supplied to the bed, and developing mechanical devices to transfer coal into the burning zone and remove the ash from the same area to prevent clogging of the chamber.

The first type of furnace was the *spreader stoker*, in which a mechanical shovel moved between a fuel hopper and the fixed grate, spreading the coal onto the burning zone (Chapter 14). Later improvements in ash handling and coal feeding, however, have made this early stoker design obsolete.

The stoker system, however, did have several advantages: (1) the coal does not have to be pulverized, (2) only a low level of particulate emissions occurs, simplifying flue gas cleanup, and (3) the stoker is easy to operate and can be built in sizes varying from small to large.

On the other hand, the stoker also has several disadvantages: (1) high maintenance is involved due to bulky moving parts, (2) there is inefficient gas–solid contact and requires a relatively large furnace volume for a given steam production, due to the low heat release rate per unit area of grate and the fact that the grate takes up furnace volume, and (3) no stoker can burn all types of coal, mainly due to caking properties of the coal and ash clinker that can form, depending upon the ash fusion temperature. It is often desirable to wash the coal to make it more amenable to stoker firing.

As in coal gasification (Chapters 20 and 21), reaction zones can be established in the fixed bed of coal as it sits on the grate. If the coal is ignited at the bottom, air flow helps the combustion front move upward through the bed, but volatiles released above the flame front are entrained by the flue gases without passing through a hot zone.

On the other hand, burning from the top down suffers from incomplete combustion, which sometimes necessitates introduction of secondary air above the bed, if there is not adequate bypassing of air through the bed. Combustion above the bed is promoted by excess air, turbulence, and high temperature. Preheating of the incoming coal also assists complete combustion.

An improvement over the spreader stoker is the *sprinkler stoker*, where the coal is gravity-fed through a drop tube above the fuel bed. Other modifications that have been tried involve pushing burning coke out of the way with fresh coal.

The *underfeed stoker* operates without a grate to support the fuel bed. A screw feeder delivers the raw fuel to the base of the fuel bed and forces it up toward the combustion zone where tuyeres provide air. Some secondary air is introduced above the bed to burn off volatiles and reduce smoking, but most air is under-fire air (50%–60% excess).

However, underfeed stokers tend to exhibit incomplete combustion and can suffer serious screw-feeder erosion problems with abrasive coal (such as high mineral matter coal). These systems also exhibit the tendency to form coke trees in the bed. If the fresh coal becomes plastic and carbonizes on its way up to the combustion front, a central solid mass of coke forms. Since the coke is largely impermeable to air, and remote from the air tuyeres, a coke tree will not burn and has to be broken down manually. Therefore, caking coals are not attractive for the underfeed stoker. Because ash removal at best is only semiautomatic, low-ash fuel is preferred. The combustion temperature should be controlled so that the ash melts and fuses into large chunks rather than a powdery ash that causes grit and dust emission problems when carried over.

The *coking stoker* is designed so that fresh fuel can drop from a hopper in front of a ram; this ram pushes the fuel across a hot grate area where volatiles are released to burn over the fuel bed. The grate may be static or move so that the ash is carried to an ash pit. These stokers lend themselves well to automatic control. As with the underfeed stoker, only fuels that do not cake should be used. The fuel size should be 1–3 in., with a minimum of fines. The ash content of the fuel should be high enough to form a layer on the grate to prevent overheating and damage to the grate.

A subsequent improvement in the spreader system involved moving the grate horizontally, allowing the ash to be carried away from the combustion zone to an ash pit. In the *traveling-grate stoker*, fuel is distributed from a hopper onto the grate, thus forming an even, moving bed with a thickness of 5–7 in. The speed of the grate and the coal feed rates are fixed by the air flow rate. The fresh bed passes beneath an ignition arch that receives radiant heat from the combustion zone to preheat the fresh fuel feed to its ignition temperature. After ignition the bed burns away over a wind box fitted with baffles to distribute the air, providing more flow through the middle combustion areas and less to the ignition and ash zones. Ash discharge is automatic from the end of the moving bed.

This traveling-grate stoker can handle almost all fuels except anthracite, which exhibits unstable burning due to ignition difficulties. Fuel composition changes can be tolerated. To increase burning rates, the grate speed can be increased with a corresponding change in air feed rate. However, a limit is imposed by high air rates blowing the bed off the grate, which causes serious grit and smoke emission and a loss of combustible material. Some air is added from above (20% of the total), and excess air varies from 25% to 50%, depending upon coal rank and particle size. Coal must be sized between 0.25 and 1.25 in. to prevent buildup of fines.

The vibrating-grate stoker represents another variation. Here the ash end of the grate is below a low arch that causes air flow through the bed to move back into the main furnace region. The low arch tends to radiate energy back to the fuel bed, thus helping to keep temperature up and ensure good burnout. Arches of this type are used with low volatile matter coals and are also found in chain or traveling-grate units where such coals are burned.

The *ignifluid* system is a slight departure from the fixed-bed concept in that a certain amount of fluidization of the bed is permitted. As the air flow rate is increased to improve the rate of heat release, the particles can become suspended over the grate, causing great turbulence and of course higher burning rates. This permits use of fuels with low reactivity and high ash content. Since solids carryover may be high, fines must be recycled to the bed to enhance combustion efficiency. This system was the forerunner of FBC units.

15.2.2.3 Entrained Pulverized Coal Firing

The concept of burning coal that has been pulverized into a fine powder arose from the belief that if the coal is made fine enough, it will burn almost as easily and efficiently as a gas. The feeding rate of coal according to the boiler demand and the amount of air available for drying and transporting the pulverized coal fuel is controlled by computers.

Drying of the fuel can be carried out prior to or simultaneous with crushing and pulverization. The fuel is ground immediately before it is needed to avoid storage and the risks of a serious fire and explosion. The coal is crushed to reduce its size to −1 in., followed by pulverization to 70%–80% less than 200 mesh, 99+% less than 50 mesh.

The raw coal is then fed into the pulverizer along with air heated to about 345°C (650°F) from the boiler. As the coal gets crushed by the rolling action, the hot air dries it and blows the usable fine coal powder out to be used as fuel. The powdered coal from the pulverizer is directly blown to a burner in the boiler. The burner mixes the powdered coal in the air suspension with additional preheated combustion air and forces it out of a nozzle similar in action to fuel being atomized by a fuel injector in modern cars. Under operating conditions, there is enough heat in the combustion zone to ignite all the incoming fuel.

After the coal is burned and the combustion gases are formed in the central part of the combustor, they give off heat to radiative superheater surfaces (at temperature exceeding 1595°C [2900°F]) and then at lower temperatures exchange heat convectively with steam tube banks. The gases then pass through the economizer, which is used to heat boiler feed water, exiting at about 300°C (570°F) and the gases then enter the combustion air preheater, where the temperature is about 150°C (300°F).

The particulate removal device, such as an electrostatic precipitator, can be placed either before the air preheater (a hot-side electrostatic precipitator) or after the air preheater (cold-side electrostatic precipitator). A fabric filter can be used in place of the electrostatic precipitator. Alternatively, or in addition, a flue-gas-desulfurization (FGD) unit treats the hot gas before it is released to the atmosphere.

Improvements continue to be made in conventional PCC power station design and new combustion technologies are being developed. These allow more electricity to be produced from less coal—known as improving the thermal efficiency of the power station. Efficiency gains in electricity generation from coal-fired power stations will play a crucial part in reducing CO_2 emissions at a global level.

Efficiency improvements include the most cost-effective and shortest lead time actions for reducing emissions from coal-fired power generation. This is particularly the case in developing countries where existing power plant efficiencies are generally lower and coal use in electricity generation is increasing. Not only do higher efficiency coal-fired power plants emit less carbon dioxide per megawatt (MW), but they are also more suited to retrofitting with carbon dioxide (CO_2) capture systems.

Improving the efficiency of pulverized coal-fired power plants has been the focus of considerable efforts by the coal industry. There is huge scope for achieving significant efficiency improvements as the existing fleet of power plants are replaced over the next 10–20 years with new, higher efficiency supercritical and ultra-supercritical plants and through the wider use of *integrated gasification combined cycle* (IGCC) systems for power generation (Chapters 14 and 26).

Failure to deploy carbon capture and geological storage systems can hamper international efforts to address climate change. The Intergovernmental Panel on Climate Change (IPCC) has identified carbon capture and geological storage as a critical technology to stabilize atmospheric greenhouse gas concentrations in an economically efficient manner. The IPCC believes that carbon capture and geological storage could contribute up to 55% of the cumulative mitigation effort by 2100 while reducing the costs of stabilization to society by 30% or more.

A 1% point improvement in the efficiency of a conventional PCC plant can result in a 2%–3% v/v reduction in carbon dioxide emissions.

15.2.2.4 Cyclone-Fired System

Cyclone-fired furnaces were developed after pulverized coal systems and require less processing of the coal fuel. They can burn poorer-grade coals with higher moisture contents and ash contents to 25%. The crushed coal feed is either stored temporarily in bins or transported directly to the cyclone furnace. The furnace is basically a large cylinder jacketed with water pipes that absorb some of the heat to make steam and protect the burner itself from melting down. A high-powered fan blows the heated air and chunks of coal into one end of the cylinder. At the same time, additional heated combustion air is injected along the curved surface of the cylinder causing the coal and air mixture to swirl in a centrifugal *cyclone motion*. The whirling of the air and coal enhances the burning properties producing high heat densities (about 4700–8300 kW/m²) and high combustion temperatures.

The hot combustion gases leave the other end of the cylinder and enter the boiler to heat the water-filled pipes and produce steam. Like in the pulverized coal burning process, all the fuel that enters the cyclone burns when injected once the furnace is at its operating temperature. Some slag remains on the walls insulating the burner and directing the heat into the boiler, while the rest drains through a trench in the bottom to a collection tank where it is solidified and disposed of. This ability to collect ash is the biggest advantage of the cyclone furnace burning process. Only 40% of the ash leaves with the exhaust gases compared with 80% for pulverized coal burning.

Cyclone furnaces are not without disadvantages. The coal used must have a relatively low-sulfur content in order for most of the ash to melt for collection. In addition, high-powered fans are required to move the larger coal pieces and air forcefully through the furnace, and more nitrogen oxide pollutants are produced compared with PCC. Finally, the actual burner requires yearly replacement of its liners due to the erosion caused by the high velocity of the coal.

Cyclone-fired boilers are used for coals with a low ash fusion temperature, which are difficult to use with a PCC unit. Generally, 80%–90% of the ash leaves the bottom of the boiler as a molten slag, thus reducing the load of fly ash passing through the heat transfer sections to the precipitator or fabric filter to just 10%–20% of that present. As with PCC units, cyclone-fired furnaces operate at close to atmospheric pressure, simplifying the passage of coal and air through the plant. Steam is generated in heat transfer tubes, driving a steam turbine and generator.

A crushed coal feedstock having 95% of <5 μm size is typical. Combustion temperature in the external cyclone furnaces can be anything from 1650°C to over 2000°C (3000°F to 3630°F). In addition, cyclone-fired boilers are suitable for coals with

- Volatile matter yield greater than 15% (dry basis).
- Ash yield between 6% and 25% for bituminous coal, or 4% and 25% for subbituminous coal—the ash must have particular slag viscosity characteristics since ash slag behavior is critical to satisfactory operation, and the maximum temperature at which the slag has a viscosity of 250 cP is 1340°C (2445°F) for bituminous coal and 1260°C (2300°F) for subbituminous coal.
- Moisture content of <20% for bituminous coal and 30% for subbituminous coal—design variations are necessary for firing coals with a high moisture content and some of the moisture may be removed before feeding the coal to the furnace, and the water vapor is routed to bypass the cyclone and sent into the main boiler area.

The cyclone furnace chambers are mounted outside the main boiler shell, which will have a narrow (or tapered) base, together with an arrangement for slag removal.

Primary combustion air carries the particles into the furnace in which the relatively large coal/char particles are retained in the cyclone while the air passes through them, promoting reaction. Secondary air is injected tangentially into the cyclone. This creates a strong swirl, throwing the larger particles toward the furnace walls. Tertiary air enters the center of the burner, along the cyclone axis, and directly into the central vortex. It is used to control the vortex vacuum and hence the position of the main combustion zone, which is the primary source of radiant heat. An increase in tertiary air moves that zone toward the furnace exit and the main boiler.

Larger particles are trapped in the molten and sticky layer that covers the entire surface of the cyclone interior, except for the area in front of the air inlets. The coal is fired under conditions of intense heat input, and while the finest particles may pass through the vortex in the center, the larger ones are thrown toward the walls and are recirculated to achieve adequate burnout.

The combustion gases, residual char, and fly ash pass into a boiler chamber where burnout is completed and there are various stages of heat exchange and heat recovery, producing steam to drive the turbine and generator. As a result of the intense combustion conditions, nitrogen oxide (NOx) formation tends to be considerably higher than in a pulverized coal combustor.

Molten ash flows by gravity from the base of the cyclone furnaces, and is removed from the system at the bottom of the boiler. It drops out into a quench tank, thus losing a substantial amount of heat. Precautions against gas buildup and explosions are essential in and around the slag quench tank.

The principles for flue gas cleaning will be similar to those for pulverized coal combustor boilers, except that in addition to the use of a selective catalytic reduction (SCR) system for NOx reduction, various other combustion methods can be used, such as two-stage combustion, and a reburn concept in which 15%–35% of the coal is pulverized, and added into the main boiler chamber, bypassing the cyclone furnace.

15.2.2.5 Tangential-Fired System

In a tangential-fired furnace, both air and fuel are projected from the corners of the furnace along lines tangent to a vertical cylinder at the center. A rotating motion is created, allowing a high degree of mixing. This system provides great flexibility for multiple fuel firing.

Thus, tangential firing is a method of firing a fuel to heat air in thermal power stations in which the flame envelope rotates ensuring thorough mixing within the furnace, providing complete combustion and uniform heat distribution.

The most effective method for producing intense turbulence is by the impingement of one flame on another. This action is secured through the use of burners located in each of the four corners of the furnace, close to the floor or the water screen. The burner nozzles are so directed that the streams of coal an air are projected along a line tangent to a small circle, lying in a horizontal plane, at the center of the furnace. Intensive mixing occurs where these streams meet. A scrubbing action is present that assures contact between the combustible and oxygen, thus promoting rapid combustion and reducing carbon loss. A rotary motion, similar to that of a cyclone, is imparted to the flame body, which spreads out and fills the furnace area. The ignition at each burner is aided by the flame from the preceding one.

With tangential firing, the furnace is essentially the burner; consequently, air and coal quantities need not be accurately proportional to the individual fuel nozzle assemblies. Turbulence produced in the furnace cavity is sufficient to combine all the fuel and air. This continuously insures uniform and complete combustion so that test performance can be maintained throughout daily operation. With other types of firing, the fuel and air must be accurately proportioned to individual burners making it difficult to always equal test results.

With this type of firing, combustion is extremely rapid, and short flame length results. The mixing is so intense that combustion rates exceeding 35,000 Btu/(ft$^3 \cdot$h) are practical under certain conditions. However, since there is considerable impingement of flame over the furnace walls it is absolutely necessary that they be fully water cooled. This sweeping of the water-cooled surfaces, in the furnace, by the gas increases the evaporation rate.

Thus, in addition to absorption by radiation from the flame envelope, there is transfer by convection, and the resulting furnace temperatures are lower than with other types of burners, even though the heat liberation rates may be somewhat higher. Tangentially fired furnaces are usually clean in the upper zone and, as a result, both the furnace and the boiler are comparatively free from objectionable slag deposits.

15.2.2.6 Fluidized-Bed Combustion

FBC is the newest of the direct coal-burning technologies first attracted worldwide interest in the 1960s and has received increasing attention during the past five decades.

FBC systems use a heated bed of sand-like material suspended (fluidized) within a rising column of air to burn many types and classes of fuel. This technique results in a vast improvement in combustion efficiency of high moisture content fuels, and is adaptable to a variety of biomass fuels or fuels derived from waste materials. The scrubbing action of the bed material on the fuel particle enhances the combustion process by stripping away the carbon dioxide and char layers that normally form around the fuel particle. This allows oxygen to reach the combustible material much more readily and increases the rate and efficiency of the combustion process.

In fluidized-bed combustors, fuel materials are forced by gas into a state of buoyancy. The gas cushion between the solids allows the particles to move freely, thus flowing like a liquid. By using this technology, sulfur dioxide (SO_2) and nitrogen oxides (NOx) emissions are reduced because a sulfur dioxide sorbent, such as limestone, can be used efficiently. Also, because the operating temperature is low, the amount of nitrogen oxide gases formed is lower than those produced using conventional technology.

The FBC system has several advantages over conventional combustors: (1) heat transfer coefficients are higher with a consequent reduction in the required area for heat transfer surface; (2) the coal preparation costs are lower; (3) most units can use coal of high and variable ash content—which allows a higher extraction efficiency in coal mining—as well as other low-grade solid fuels (such as biomass); (4) the high heat generation rates that result in a small boiler size for a given output, with the opportunity of more assembly at the factory and less at the site, particularly for pressurized systems; (5) there is reduced fire-side fouling, corrosion, and erosion because of the low temperature; (6) emission of nitrogen oxides is lower due to the lower operating temperature; and (7) the possibility of passing the cleaned stack gas from a pressurized fluid bed unit through a gas turbine to increase efficiency.

Fluidized-bed combustors are operated at relatively low temperatures, namely 760°C–930°C (1400°F–1700°F). The reason lower temperatures are achieved in these systems compared to the pulverized coal system is that heat transfer surfaces are submerged in the bed and extract heat directly from the burning particles. In a pulverized coal combustor system, heat is not extracted until the coal is essentially burned up, resulting in higher gas temperatures. Because of the great amount of turbulence in a fluidized-bed boiler, the coal particles can be larger, which reduces the amount of grinding required prior to combustion.

Fluidized-bed combustors can operate at atmospheric pressure or in a pressurized chamber. In the pressurized chamber, operating pressures can be 10–20 times the atmospheric pressure (150–300 psi). Pressurized fluidized-bed furnaces provide significant gain in overall thermal efficiency over atmospheric fluidized-bed furnace.

The turbulence in the combustor vapor space combined with the tumultuous scouring effect and thermal inertia of the bed material provide for complete, controlled, and uniform combustion. These factors are essential for maximizing the thermal efficiency, minimizing char, and controlling emissions. The high efficiency of a fluid-bed combustor makes it particularly well suited to problem fuels with low Btu value and high moisture characteristics.

Fluidized-bed combustors can be designed to burn a number of fuels besides coal: (1) low-grade coal, (2) coal-mining wastes such as culm, (3) wood chips and other solid biomass, and (4) shredded tires. With careful design, a combination of fuels can be burned without too much sacrifice in performance.

The feature of the fluidized-bed combustor is the (relatively) low operating temperature, which ensures that (1) the ash does not fuse (melt) and can be easily removed and disposed off in solid form and (2) there is a lower amount of boiler tube fouling and corrosion.

The low temperature also promotes the reaction of sulfur dioxide with metal oxides and carbonates in the ash, most notably calcium and magnesium salts. If limestone (calcium carbonate, $CaCO_3$), is added to the bed of burning coal, the following reaction with sulfur dioxide occurs:

$$2CaCO_3 + 2SO_2 + O_2 \rightarrow 2CaSO_4 + CO_2$$

The production of calcium sulfate ($CaSO_4$) is greatly affected by temperature such that if the temperature exceeds 985°C–1095°C (1800°F–2000°F), the reverse reaction will be dominant, leaving much of the sulfur dioxide unreacted.

However, at lower temperatures, the retention of sulfur dioxide by the calcium carbonate is favored. For temperatures lower than 760°C (1400°F), several difficulties arise: (1) the coal does not burn rapidly, (2) the scavenging or retention of sulfur dioxide does not proceed very rapidly, and (3) heat transfer to the boiler tubes is markedly decreased. Therefore, the operating temperature of a fluidized-bed boiler must be selected with care and may vary according to the specific coal used.

The high combustion efficiency of a fluid bed results in a reduced amount of inorganic material as fine ash. The remaining larger material consists mainly of noncombustibles, such as rocks, and wire brought in with the fuel, and coarse sand-like neutral particles. Low combustion temperatures in the fluidized bed minimize the formation of toxic materials that might go into the ash. Ash samples from FBC systems have consistently tested nontoxic, and in many instances the ash is being sold as input for other products such as cement.

Fluidized-bed combustors are divided into two categories: (1) circulating fluidized beds and (2) bubbling fluidized beds (BFB).

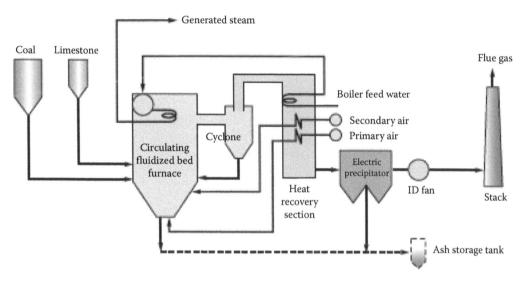

FIGURE 15.3   Circulating fluidized-bed operation.

Circulating fluidized-bed technology uses higher air flows to entrain and move the bed material and recirculate nearly all the bed material with adjacent high-volume, hot cyclone separators.

For example, in a circulating FBC process (Figure 15.3), crushed coal is mixed with limestone and fired in a process resembling a boiling fluid. The limestone removes the sulfur and converts it into an environmentally benign powder that is removed with the ash. The relatively clean flue gas goes on to the heat exchanger. This approach theoretically simplifies feed design, extends the contact between sorbent and flue gas, reduces likelihood of heat-exchanger tube erosion, and improves capture of sulfur dioxide and combustion efficiency.

Whereas pulverized combustion units operate at combustion temperatures on the order of 1400°C–1500°C (2550°F–2730°F), circulating fluidized-bed boilers operate at lower temperatures, ranging from 850°C to 900°C (1560°F to 1650°F), thereby suppressing thermal NOx emissions as the generation of NOx is dependent upon the combustion temperature. In addition, the circulating fluidized bed involves a two-stage combustion process: the reducing combustion at the fluidized-bed section, and the oxidizing combustion at the freeboard section. Next, the unburned carbon is collected by a high-temperature cyclone located at the boiler exit to recycle to the boiler, thus increasing the efficiency of reducing nitrogen oxide production.

The first-generation pressurized fluidized-bed combustor uses *BFB technology*. A relatively stationary fluidized bed is established in the boiler using low air velocities to fluidize the material, and a heat exchanger (boiler tube bundle) immersed in the bed to generate steam. Cyclone separators are used to remove particulate matter from the flue gas prior to entering a gas turbine, which is designed to accept a moderate amount of particulate matter (i.e., *ruggedized*).

BFB units are a preferred choice for wood waste, bark, or sludge—coal is not always the primary choice for such units.

Since the late 1990s, there has been ever increasing interest in BFB combustion. The two fluidized-bed technologies are similar. Both use a bed of inert material (most typically sand) that is fluidized by high-pressure combustion air. The primary differences are that the BFB unit normally operates in a reducing atmosphere (less air than is needed for combustion), does not have as great an ability to absorb sulfur dioxide, and normally is used to burn lower-quality fuels with high-volatile matter. Further, the BFB unit keeps most of the sand in the lower furnace.

Circulating beds fire fuels with high fixed carbon and circulate the hot gases, along with a high-density sand stream, through the entire furnace. By adding materials high in calcium (such as limestone), the BFB will efficiently absorb sulfur dioxide, reducing overall emissions.

If coal is the desired fuel, BFB technology is not the first choice, and a mill should consider circulating fluidized-bed technology. If biomass and other high-volatile/lower-carbon-containing fuels are used, BFB technology should be considered as the preferred combustion process (KEMA, 2009).

15.2.2.7 Arch-Fired Systems

Arch-fired systems (vertical-fired systems) are used to fire solid fuels that are difficult to ignite, such as coals with moisture and ash-free volatile matter yield less than 13% w/w. In this system, the pulverized coal is discharged through a nozzle surrounded by heated combustion air. High-pressure jets are used to prevent fuel–air streams from short circuiting. The firing system produces a looping flame with hot gases discharging at the center.

15.2.2.8 Gas Turbine Systems

A *gas turbine* (also called a *combustion turbine*) is a rotary engine that extracts energy from a flow of combustion gas. It has an upstream compressor coupled to a downstream turbine, and a combustion chamber in-between.

Energy is added to the gas stream in the combustor where fuel is mixed with air and ignited. In the high-pressure environment of the combustor, combustion of the fuel increases the temperature. The products of the combustion are forced into the turbine section. Therefore, the high velocity and volume of the gas flow is directed through a nozzle over the turbine's blades, spinning the turbine that powers the compressor and, for some turbines, drives their mechanical output. The energy given up to the turbine comes from the reduction in the temperature and pressure of the exhaust gas.

Gas turbine systems operate in a manner similar to steam turbine systems except that combustion gases are used to turn the turbine blades instead of steam. In addition to the electric generator, the turbine also drives a rotating compressor to pressurize the air, which is then mixed with either gas or liquid fuel in a combustion chamber. The greater the compression, the higher the temperature and the higher the efficiency that can be achieved in a gas turbine. Exhaust gases are emitted to the atmosphere from the turbine.

Unlike a steam turbine system, gas turbine systems do not have boilers or a steam supply, condensers, or a waste heat disposal system. Therefore, capital costs are much lower for a gas turbine system than for a steam system.

In electrical power applications, gas turbines are typically used for peaking duty, where rapid startup and short runs are needed. Most installed simple gas turbines with no controls have only a 20%–30% efficiency.

15.2.2.9 Combined-Cycle Generation

Combined-cycle generation is a configuration using both gas turbines and steam generators. In a combined-cycle gas turbine (CCGT), the hot exhaust gases of a gas turbine are used to provide all, or a portion of, the heat source for the boiler, which produces steam for the steam generator turbine. This combination increases the thermal efficiency over a coal- or oil-fueled steam generator. The system has an efficiency of about 54%, and the fuel consumption is approximately 25% lower. Combined-cycle systems may have multiple gas turbines driving one steam turbine.

15.2.2.10 Cogeneration

Cogeneration is the simultaneous production of heat and power in a single thermodynamic process. Almost all cogeneration utilizes hot air and steam for the process fluid, although certain types of fuel cells also cogenerate. It is, in essence, the merging of a system designed to produce electric power and a system used for producing industrial heat and steam. Cogeneration accounted for 75% of all nonutility power generation in 1995.

This system is a more efficient way of using energy inputs and allows the recovery of otherwise wasted thermal energy for use in an industrial process. Cogeneration technologies are classified as *topping cycle systems* and *bottoming cycle systems*, depending on whether electrical (topping cycle) or thermal (bottoming cycle) energy is derived first.

15.2.2.11   IGCC Technology

The IGCC approach has been proposed to alleviate problems related to carbon dioxide emissions.

The gasification process converts carbon-containing material into a synthesis gas composed primarily of carbon monoxide and hydrogen for subsequent utilization. The primary product can be used as a fuel to generate electricity, used to synthesize chemicals such as ammonia, oxy-chemicals, liquid fuels, or used to produce hydrogen.

Historically, gasification has been used as industrial processes in the creation of chemicals, with power production as a secondary and subordinate process. In the last decade, the primary application of gasification to power production has become more common due to the demand for high efficiency and low-environmental impact.

There are three types of gasifier classified by the configuration: entrained flow gasifiers, fluidized-bed gasifiers, and moving bed (also called fixed-bed) gasifiers (Chapters 20 and 21).

One of the advantages of the coal gasification technology is that it offers the polygeneration: coproduction of liquid fuels, chemicals, hydrogen, and electricity from the syngas generated from gasification. Chemical gasification plants based on entrained flow and more especially on moving-bed technologies are at present operating all over the world with the biggest plants located in South Africa (Sasol) (Chapters 20 and 21). In addition, gasification is an important step of the indirect liquefaction of coal for production of liquid fuels (Chapters 18 and 19).

IGCC technology has advantages over PCC technology with respect to environmental benefits. After cleanup of the produced syngas, the sulfur in coal can be recovered to elemental form, the nitrogen oxide level will be very low, and carbon dioxide can be more easily removed due to the high concentration, which would be the best selling point for IGCC power production. It is expected that IGCC may have a greater market penetration internationally due to the enforcement of carbon dioxide emission control. In addition, IGCC will consume less water than PCC units.

15.3   STEAM GENERATION

In the steam generation system, heat from combustion causes steam to form in the primary steam generation coils (Stultz and Kitto, 1992). The steam vapor rises into the steam drum, where it is accumulated. The steam then successively passes to the convective and radiant superheaters, which use combustion heat to further heat the steam well above its previous temperature. The steam next flows to the turbine, which has both high- and low-pressure stages.

High-temperature, high-pressure steam is generated in the boiler and then enters the steam turbine. At the other end of the steam turbine is the condenser, which is maintained at a low temperature and pressure. Steam rushing from the high-pressure boiler to the low-pressure condenser drives the turbine blades, which powers the electric generator. Steam expands as it works; hence, the turbine is wider at the exit end of the steam. The theoretical thermal efficiency of the unit is dependent on the high pressure and temperature in the boiler and the low temperature and pressure in condenser.

Steam turbines typically have a thermal efficiency of about 35%, meaning that 35% of the heat of combustion is transformed into electricity. The remaining 65% of the heat either goes up the stack (typically 10%) or is discharged with the condenser cooling water (typically 55%).

Low-pressure steam exiting the turbine enters the condenser shell and is condensed on the condenser tubes. The condenser tubes are maintained at a low temperature by the flow of cooling water. The condenser is necessary for efficient operation by providing a low-pressure sink for the exhausted steam. As the steam is cooled to condensate, the condensate is transported by the boiler feed-water system back to the boiler, where it is used again. Being a low-volume incompressible liquid, the condensate water can be efficiently pumped back into the high-pressure boiler.

A constant flow of low-temperature cooling water in the condenser tubes is required to keep the condenser shell (steam side) at proper pressure and to ensure efficient electricity generation. Through the condensing process, the cooling water is warmed. If the cooling system is an open or a once-through system, this warm water is released back to the source water body. In a closed system, the warm water is cooled by recirculation through cooling towers, lakes, or ponds, where the heat is released into the air through evaporation and/or sensible heat transfer. If a recirculating cooling system is used, only a small amount of makeup water is required to offset the cooling tower blowdown, which must be discharged periodically to control the buildup of solids. Compared to a once-through system, a recirculated system uses about 1/20th of the water (Elliot, 1989).

In a typical process (Figure 15.4), coal is first milled to a fine powder, which increases the surface area and allows it to burn more quickly. In these PCC systems, the powdered coal is blown into the combustion chamber of a boiler where it is burnt at high temperature (see Figure 15.4). The hot gases and heat energy produced converts water—in tubes lining the boiler—into steam.

The returning feed water (condensed steam) requires a great deal of heat for vaporization to produce steam. Thus it is desirable to preheat the feed water before returning it to the boiler by bleeding off a small amount of steam from successive turbine stages. Steam is bled from increasingly hotter stages as the feed water gets hotter. Finally the feed water is fed into the economizer, which is another set of tubes fairly far back in the convection section where the furnace temperature is lower. From here, the feed water returns to the boiler water drum and the primary steam generation coils.

The high-pressure steam is passed into a turbine (Figure 15.5) containing thousands of propeller-like blades. The steam pushes these blades causing the turbine shaft to rotate at high speed.

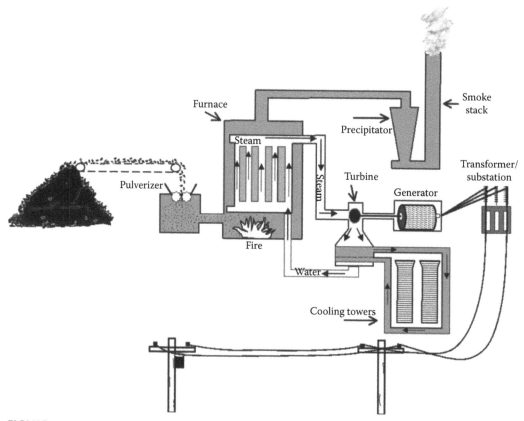

FIGURE 15.4 Pulverized coal combustion system.

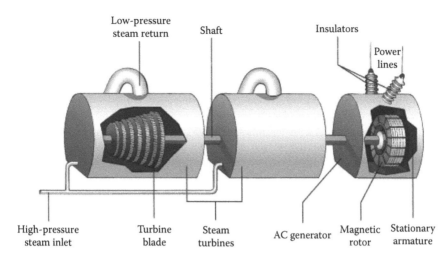

FIGURE 15.5 Steam turbine for electricity generation.

A generator is mounted at one end of the turbine shaft and consists of carefully wound wire coils. Electricity is generated when these are rapidly rotated in a strong magnetic field. After passing through the turbine, the steam is condensed and returned to the boiler to be heated once again.

The electricity generated is transformed into the higher voltages (up to 400,000 V) used for economic, efficient transmission via power line grids. When it nears the point of consumption, such as our homes, the electricity is transformed down to the safer 100–250 voltage systems used in the domestic market.

15.4 POWER PLANT WASTE

The use of coal has a long history of emitting a variety of waste products (typically referred to as *pollutants*) into the environment, which have contributed to several known health episodes. Although regulations on coal usage date back to medieval times (Chapter 14), it was not until major incidents in the 1940s and 1950s in England and the United States, which had severe impacts on human health, that the impetus for legislation in these two countries took hold in order to protect the health of the general public. In the United States, the major development of legislative and regulatory acts occurred from 1955 to 1970.

The regulations are continually changing, as more information on the effect of emissions on health and the environment is obtained, new control technologies are developed, and the demands for a safe living environment are heard and acted upon by various levels of government. Indeed, the use of coal for power generation is a highly regulated industry with regulations being developed and implemented on a regular basis.

In the early days of the power generation industry, coal combustion products (CCPs) were considered to be a waste material. The properties of these materials were not evaluated seriously for other uses and nearly all of the CCP were landfilled. In the course of time, the cementitious and pozzolanic properties of fly ash were recognized. The products were tested to understand their physical properties, chemical properties, and suitability as a construction material. During the last few decades, these "waste" materials have seen a transformation to the status of *by-products* and more recently *products* that are sought for construction and other applications (Chapter 14).

Power plant wastes (or combustion wastes, in the current context) are waste materials that are produced from the burning of coal (Chapter 14). This includes all ash, slag, and particulates removed from flue gas. These wastes are categorized by Environmental Protection Agency (EPA) as a *special waste* and have been exempted from federal hazardous waste regulations under Subtitle C of the

Resource Conservation and Recovery Act (RCRA). In addressing the regulatory status of fossil fuel combustion wastes, EPA divided the wastes into two categories:

1. Large-volume coal combustion wastes (CCW) generated at electric utility and independent power-producing facilities that are managed separately
2. All remaining fossil fuel combustion wastes, including
   a. Large-volume CCW generated at electric utility and independent power-producing facilities that are comanaged with certain other CCWs (referred to as *comanaged wastes*)
   b. CCWs generated at nonutilities
   c. CCWs generated at facilities with FBC technology
   d. Petroleum-coke combustion wastes
   e. Waste from the combustion of mixtures of coal and other fuels
   f. Waste from the combustion of oil
   g. Waste from the combustion of natural gas

In two separate regulatory determinations, EPA determined that neither large-volume wastes nor the remaining fossil fuel combustion wastes warrant regulation as a hazardous waste under Subtitle C of RCRA and therefore remain excluded under 40 CFR §261.4(b)(4). EPA did determine, however, that CCW that are disposed in landfills and surface impoundments should be regulated under Subtitle D of RCRA (i.e., the solid waste regulations), whereas CCW used to fill surface mines or underground mines (*minefill*) should be regulated under authority of Subtitle D of RCRA, the Surface Mining Control and Reclamation Act (SMCRA), or a combination of these authorities.

Wastes from the combustion process include exhaust gases and, when coal or oil is used as the boiler fuel, ash. These wastes are typically controlled to reduce the levels of pollutants exiting the exhaust stack. Bottom ash, another by-product of combustion, also is discharged from the furnace.

15.4.1 COAL ASH

The greatest single problem in operation of coal-fired units is the accumulation of *coal ash* on boiler heat transfer surfaces. Coal ash causes three main problems in large furnaces: (1) buildup of ash on furnace wall tubes, (2) accumulation of small, sticky, molten particles of ash on superheater and reheater tube banks, and (3) corrosion.

Two types of ash are generated during combustion of fossil fuels: (1) bottom ash and (2) fly ash.

It is important to distinguish fly ash, bottom ash, and other CCP from incinerator ash (Chapter 14). CCP result from the burning of coal under controlled conditions and are nonhazardous. Incinerator ash is the ash obtained as a result of burning municipal wastes, medical waste, paper, wood, etc. and is sometimes classified as hazardous waste. The mineralogical composition of fly ash and incinerator ash consequently is very different. The composition of fly ash from a single source is very consistent and uniform, unlike the composition of incinerator ash, which varies tremendously because of the wide variety of waste materials burned.

Ash that collects at the bottom of the boiler is called bottom ash and/or slag. Fly ash is a finer ash material that is borne by the flue gas from the furnace to the end of the boiler. Bottom ashes are collected and discharged from the boiler, economizer, air heaters, electrostatic precipitator, and fabric filters.

Fly ash is collected in the economizer and air heaters or is collected by the particulate control equipment. Coal-fired facilities generate the largest quantity of ash; gas facilities generate so little that separate ash management facilities are not necessary. Fly and bottom ash may be managed separately or together in landfills or in wet surface impoundments.

Ashes differ in characteristics depending upon the content of the fuel burned. For coal, the chemical composition of ash is a function of the type of coal that is burned, the extent to which the coal is prepared before it is burned, and the operating conditions of the boiler. These factors are very plant- and coal-specific.

Generally, however, more than 95% of ash is made up of silicon, aluminum, iron, and calcium in their oxide forms, with magnesium, potassium, sodium, and titanium representing the remaining major constituents. Ash may also contain a wide range of trace constituents in highly variable concentrations. Potential trace constituents include antimony, arsenic, barium, cadmium, chromium, lead, mercury, selenium, strontium, zinc, and other metals.

When ash melts in a pulverized coal flame, the droplets coalesce to form larger drops of *molten slag*. Some of these drops reach the wall tubes of the furnace and adhere to the metal surface, eventually forming a solid layer of slag. Coal-ash slag does not conduct heat readily and thus decreases the amount of heat reaching the wall tubes, lowering the quantity of steam produced at this point. During load changes, there is a differential expansion between the wall tubes and the slag layer, causing huge sheets of slag to peel off the walls.

Slag is controlled more effectively via retractable wall blowers, which use high-velocity jets of air, steam, or water to dislodge the slag and clean the furnace walls—analogous to decoking the delayed coker drums in a petroleum refinery. Each wall blower has an effective range of about 8 ft, so more than 100 blowers may be required in a typical installation. Occasionally, slag accumulations may get so large that manual cleaning is required, because large pieces of slag may fall to the bottom of the furnace. Since these pieces cannot be removed without manual breaking, a furnace shutdown may be required.

Fouling of heat transfer surfaces also occurs when very small particles of ash are carried to the bundles of tubing in the superheaters and reheaters. These fly ash particles collect on the tube surfaces, insulating the metal so that not enough heat is transferred to raise the steam temperature to design levels. These accumulations of ash can plug the normal gas passages. Soot blowers can be employed to remove such ash deposits periodically, but some coal ash may form dense, adherent layers, which are very difficult to remove except by manual cleaning.

Fuel and air nozzles can be adjusted to compensate for changes in heat absorption due to fouling. As furnace walls become coated with ash deposits, the burners are tilted downward and combustion is completed lower in the furnace. This exposes more furnace wall surface to the flame and restores the furnace exit gas temperature and steam temperature to satisfactory levels.

Ash-handling equipment and disposal facilities, if improperly designed, can also limit boiler capacity. The hopper section of the furnace typically slopes at 55° and generally covers a distance of 12–30 ft in a tangentially fired furnace. High fractions of calcium oxide can combine with moisture and other elements to harden the fly ash as it cools in the hoppers. This ash packs and resists flowing and may require prodding.

The ash from most low-rank coals differs from that of higher-rank coals in that the percentage of alkaline earth metals (e.g., calcium and magnesium as the oxides CaO and MgO) exceeds the iron content. In many cases, the sodium content may be in excess of 5%. A higher percentage of basic metals may result in lower ash fusion temperatures. While a lignite-type ash is generally classified as non-slagging, the presence of the alkaline earth metals (sodium and calcium rather than aluminum and silica) increases the tendency for rapid deposit buildup on superheater and reheater tube banks in the form of calcium and/or sodium sulfates. This property requires wider spacing in tube banks, especially for power plants utilizing Western subbituminous coal and lignites.

The measurement of ash fusibility has long been recognized as an index for evaluating performance with regard to slagging and deposit buildup. If the ash is at a temperature above its softening temperature, it probably will settle out as a dust and, as such, is comparatively simple to remove. If, however, the ash is near its softening temperature, the resulting deposit is apt to be porous in structure.

Depending on the strength of the bond, the deposit may fall off due to its own weight or may readily be removed by soot blowing. However, if the deposit is permitted to build up in a zone of high gas temperature, its surface fuses more thoroughly and may exceed the melting temperature with resulting runoff as slag.

The fusibility characteristics of coal ash will vary with its chemical constituents. Most low-rank coals produce an ash high in basic metals and low in iron content. Therefore, they have a higher softening temperature, and consequently, are less susceptible to slagging. The behavior of ash is extremely complex, and while some constituent melt below 1040°C (1900°F), as the calcium and sodium content of the ash increases, the rate of deposition increases on tube surfaces. The sodium oxide content, in particular, can have a catalytic effect on the rate of deposition, and investigations have shown that ash with a sodium content above 5% fouls at an accelerated rate.

However, prediction of fouling behavior based on sodium alone can be misleading. Many factors must be considered when predicting ash deposition, for example, original mineral composition, furnace design, and utility operating practice.

A typical ash deposit structure for a U.S. low-rank coal may consist of three distinct layers that differ in physical character but are quite similar chemically. The first thin *white layer* of very fine powdery ash is deposited all around the tube. This layer, which is usually enriched in sodium sulfate, is always observed during the early period of operation after boiler cleanup.

Next, an inner *sinter layer*, a few millimeters thick, begins to form by initial impaction on the upstream face of the boiler tube. Particles in this deposit are bonded together by surface stickiness. As this layer grows, its outer edge is insulated from the relatively cool boiler tube, thus causing the temperature of the surface of the deposit to increase and approach the temperature of the flue gas. Given a sufficiently high gas temperature and the presence of sufficient sodium (not all sodium in the ash is "active") to flux the remainder of the fly ash material, a melt phase will begin to form at the leading edge of the deposit. This melt material collects particles that impact on the deposit and binds them together into a strong bulk deposit, which is designated the outer sinter layer. Sodium compounds provide the continuous melt phase that binds the deposit, while a deficiency in sodium yields a discontinuous melt and weaker bonding (which can be broken up).

Corrosion is generally caused by oxides of sulfur (which become sulfuric acid in the presence of moisture), but another important corrosion source is buildup of tube deposits that destroy the protective surface oxide coating, due to attack by sulfates of sodium, aluminum, iron, and potassium. Since the dew point can be as high as 120°C–150°C (250°F–300°F), cooler surfaces are subjected to acid attack.

Originally, it was thought that coal-ash corrosion was confined to boilers burning high alkali coals. However, combustors burning medium to low alkali coals also encounter the same problem. In cases where there was no corrosion, either the complex sulfates were absent or the tube wall temperatures were below 595°C (1100°F).

Generally, sufficient sulfur and alkali are present in all bituminous coals to produce corrosive ash deposits on superheater and reheater tubes. In addition, coal containing more than 3.5% w/w sulfur and 0.25% w/w chlorine can be exceptionally difficult to handle and the rate of corrosion is greatly affected by the deposit temperature and metal skin temperature.

15.4.2 Flue-Gas-Desulfurization Waste

If coal or oil is the fuel source, the FGD control technologies result in the generation of solid wastes. Wet lime/limestone scrubbers produce a slurry of ash, unreacted lime, calcium sulfate, and calcium sulfite. Dry scrubber systems produce a mixture of unreacted sorbent (e.g., lime, limestone, sodium carbonates, and calcium carbonates), sulfur salts, and fly ash.

Sludge is typically stabilized with fly ash and sludge produced in a wet scrubber may be disposed of in impoundments or below-grade landfills, or may be stabilized and disposed of in landfills. Dry scrubber sludge may be managed dry or wet.

15.4.3 Waste Heat

All thermal power plants produce waste heat energy as a by-product of the useful electrical energy produced. The amount of waste heat energy equals or exceeds the amount of electrical energy produced. Gas-fired power plants can achieve 50% conversion efficiency while coal and oil plants achieve around 30%–45%. The waste heat produces a temperature rise in the atmosphere, which is small compared to that of greenhouse gas emissions from the same power plant.

Natural draft wet cooling towers at many coal power plants use large hyperbolic chimney-like structures that release the waste heat to the ambient atmosphere by the evaporation of water. However, the mechanical induced-draft or forced-draft wet cooling towers in many large thermal power plants use fans to provide air movement upward through down-flowing water and are not hyperbolic chimney-like structures. The induced or forced-draft cooling towers are typically rectangular, box-like structures filled with a material that enhances the contacting of the up-flowing air and the down-flowing water.

In areas with restricted water use, a dry cooling tower or radiator, directly air cooled, may be necessary, since the cost or environmental consequences of obtaining makeup water for evaporative cooling would be prohibitive. These have lower efficiency and higher-energy consumption in fans than a wet, evaporative cooling tower.

Where economically and environmentally possible, electric companies prefer to use cooling water from the ocean, or a lake or river, or a cooling pond, instead of a cooling tower. This type of cooling can save the cost of a cooling tower and may have lower energy costs for pumping cooling water through heat exchangers. However, the waste heat can cause the temperature of the water to rise detectably. Power plants using natural bodies of water for cooling must be designed to prevent intake of organisms into the cooling cycle. A further environmental impact would be organisms that adapt to the warmer plant water and may be injured if the plant shuts down in cold weather.

15.5   COAL–WATER FUELS

A coal–water fuel (CWF; *coal–water slurry fuel*) is a slurry of coal in water than is typically fed to a combustor for the generation of heat or power. The CWF represents a new type of clean fuel technology that emerged as a way of using coal as a substitute for fuel oil in the petroleum crises of the last century. The basic composition of the fuel is 70% w/w coal, 30% w/w water, and 1% w/w additive to stabilize the slurry.

The CWF can be used in place of oil and gas in small, medium, and large power stations. The CWF is suitable for existing gas, oil, and coal-fired boilers.

The presence of water in the fuel reduces harmful emissions into the atmosphere and makes the coal explosion-proof. By converting the coal into a liquid form, delivery and dispensing of the fuel can be simplified. One side effect of the CWF production process is the separation of non-carbon material mixed in with the coal before treatment. This results in a reduction of ash yield or the treated fuel, making it a viable alternative to diesel fuel #2 for use in large stationary engines or diesel-electric locomotives.

The advantages of CWF include (1) a complete burning fuel with a burn of 96%–99% and (2) a noticeable effect in protecting the environment.

Combustion of the CWF results in hydro-agglomerates of the fly ash constituents, with a concomitant reduction in emissions of particulate matter 80%–90%. In addition, transportation (pipe, tank) of CWF reduces unjustified losses during transportation of coal and improves the ecological environment in areas of its use.

15.6   AIR POLLUTION CONTROL DEVICES

Flue and waste gases from power plants and other industrial operations where coal is used as a feedstock invariably contain constituents that are damaging the climate or environment—these will be constituents such as carbon dioxide (CO_2), nitrogen oxides (NOx), sulfur oxides (SOx), dust, and particles and toxins such as dioxin and mercury.

The processes that have been developed for gas cleaning (Mokhatab et al., 2006; Speight, 2007, 2008) vary from a simple once-through wash operation to complex multistep systems with options for recycle of the gases (Mokhatab et al., 2006). In some cases, process complexities arise because of the need for recovery of the materials used to remove the contaminants or even recovery of the contaminants in the original, or altered, form.

The environmental impact of coal-based power plants has drawn increasing attention, not only for controlling pollutants like sulfur dioxide (SO_2), oxides of nitrogen (NOx), particulates (PM) but also for controlling the emission of carbon dioxide (CO_2), as there is an increasing need to reduce the emissions of carbon dioxide to the atmosphere to alleviate the global warming effect. It induces significant challenges to generate electricity efficiently together with near-zero carbon dioxide emissions.

In the process, carbon dioxide, hydrogen, and other coal by-products are captured so that they can be used for useful purposes. Evolving technologies are also making coal at existing plants cleaner—refined coal technologies remove many of the impurities contained in existing coal. New techniques are helping remove mercury and harmful gases while unlocking more energy potential.

15.6.1 Nitrogen Oxide Emissions

The provisions of the Clean Air Act Amendments have also affected nitrogen oxide (NOx) emissions and their controls for the electric utility industry. The process for reducing nitrogen oxide emissions through combustion control technologies has generally increased the amount of unburned carbon content and the relative coarseness of fly ash at many locations. In particular, post-combustion control technologies for nitrogen oxide emissions such as SCR and selective non-catalytic reduction (SNCR) both utilize ammonia injection into the boiler exhaust gas stream to reduce nitrogen oxide emissions. As a result, the potential for ammonia contamination of the fly ash due to excessive ammonia slip from SCR/SNCR operation is an additional concern.

15.6.2 Sulfur Oxide Emissions

The methods for controlling sulfur oxides from coal, which, in turn, impact the combustion system design can be classified into three types: (1) sulfur removal after combustion, (2) sulfur removal during combustion, and (3) sulfur removal before combustion.

The first approach (i.e., sulfur removal after combustion) is the one that is currently receiving the most attention in many industrialized countries including the United States because it does not represent a significant departure from existing coal-fired power plant technology. Ordinarily, a wet scrubber can be used to remove sulfur oxides from the flue gas. However, currently available sulfur oxide control technology has proven to be expensive, is subject to operational difficulties, and produces a liquid or dry waste product that must be disposed. It is, however, the approach that is now generally accepted by the utility industry.

The second sulfur removal technology (i.e., sulfur removal during combustion) primarily involves the use of fluidized-bed coal combustion. In this type of process, a desulfurization chemical (such as solid limestone) is injected into the burning coal bed where it reacts with sulfur dioxide. The resulting calcium sulfate compound is then removed from the bed in solid form with the coal ash.

The final type of sulfur removal technology (i.e., sulfur removal before or without combustion) is best exemplified by physical coal cleaning (Chapter 6). It is another alternative in this class, although sufficiently high sulfur removal efficiencies are not always possible. Another option that falls into this category is the conversion of coal to low-sulfur products, such as by liquefaction (Chapters 18 and 19) or gasification (Chapters 20 and 21). The greatest obstacle to implementation of this technology is the excessive cost of converting the coal to a clean fuel.

Regulations to reduce sulfur dioxide emissions result in the introduction of wet scrubber FGD systems, which can produce gypsum as a by-product. In 1990, overall annual sulfur dioxide (SO_2) emissions from electric utility companies had fallen 46%. In 1990, the Clean Air Act Amendments were enacted, requiring electric utility companies nationwide to reduce their collective sulfur dioxide emissions.

Many coals of the Western United States (and some eastern US coals) are naturally low in sulfur and can be used to meet the SO_2 compliance requirements. Blending coals of different sulfur contents to achieve a mix that is in compliance with applicable regulation is also common. Wet FGD systems are currently installed on at least one-quarter of the coal-fired utility generating capacity in the United States.

Thus, many governments currently have emissions standards for coal and other power-generating facilities. Each coal facility must meet air-quality standards in order to operate. IGCC generation is one of the most widely discussed clean coal technologies. During this process, coal reacts with steam under high pressure and heat to form a hydrogen gas, which powers a gas turbine. Exhaust from the gas turbine is hot enough to power a conventional steam turbine as well, increasing efficiency.

Air pollution control devices (gas cleaning devices) (Chapter 23) found in fossil fuel-fired systems (particularly steam electric power facilities) include particulate removal equipment, sulfur oxide (SOx) removal equipment, and nitrogen oxide (NOx) removal equipment.

Particulate removal equipment includes electrostatic precipitators, fabric filters, or mechanical particulate collectors, such as cyclones. SOx removal equipment includes sorbent injection technologies and wet and dry scrubbers. Both types of scrubbers result in the formation of calcium sulfate and sulfite as waste products.

The state of the art of the implemented systems for reduction of sulfur dioxide and nitrogen oxides in the current PCC plants are FGD, low NOx burner, and SCR of NOx. A typical modern coal-fired boiler system eliminates up to 97% of the combined sulfur dioxide (SO_2) and nitrogen oxides (NOx) emissions.

A novel process is proposed that replaces the combustion step with solid oxide fuel cells (SOFC) (Adams et al., 2010). A SOFC oxidizes hydrogen electrochemically, producing electricity without the thermodynamic limitations of heat engines. The SOFC is designed to have separate anode (fuel) and cathode (air) sections, so air can be used for oxidation without diluting the fuel stream with nitrogen. This permits easy separation of the anode exhaust (water and carbon dioxide) with a very small energy penalty and no solvent recovery step. The spent air stream, being heated by the SOFC, can provide additional power through the Brayton cycle. Together, these innovations provide higher plant efficiency, significantly reduce the energy penalty of carbon dioxide capture, and facilitate recycle of water in the process.

15.7 CARBON CAPTURE AND STORAGE TECHNOLOGIES

The reduction of carbon dioxide emissions to near-zero level imposes more economic restriction than technical restriction for coal-based power generation, as current state of the art can provide several technologies to capture and store carbon dioxide.

Addressing the challenge of climate change, while meeting the need for affordable energy, will require access to and deployment of the full range of energy-efficient and low-carbon technologies. Capturing carbon dioxide that would otherwise be emitted to the atmosphere and injecting it to be stored in deep geological formations (CCS) is the only technology currently available to make deep cuts in greenhouse gas emissions from fossil fuel use while allowing energy needs to be met securely and affordably.

Carbon capture and storage (Figure 15.6) is not a replacement for taking actions that increase energy efficiency or maximizing the use of renewables or other less carbon-intensive forms of energy. A portfolio approach taking every opportunity to reduce emissions will be required to meet the challenge of climate change.

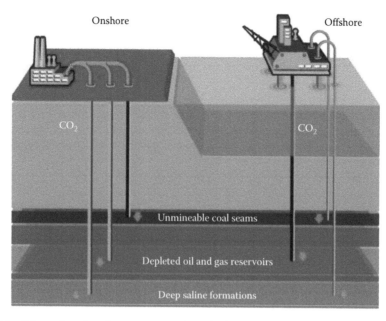

Onshore   Offshore

CO$_2$ CO$_2$

Unmineable coal seams

Depleted oil and gas reservoirs

Deep saline formations

FIGURE 15.6 Schematic of carbon capture and geological storage. (From http://www.worldcoal.org/carbon-capture-storage/ccs-technologies/)

All of the elements of carbon capture and storage have been separately proven and deployed in various fields of commercial activity. In fact, approximately 32 million tons of carbon dioxide is already stored worldwide and this number continues to increase.

The vital next stage is the successful application of a fully integrated, large-scale carbon capture and storage systems fitted to commercial-scale power stations. Failure to deploy carbon capture and storage may hamper international efforts to address climate change. The IPCC has identified carbon capture and storage as a critical technology to stabilize atmospheric greenhouse gas concentrations in an economically efficient manner. The IPCC has concluded that by 2100, carbon capture and storage technologies could contribute up to 55% of the cumulative mitigation effort whilst reducing the costs of stabilization to society by 30% or more.

While carbon dioxide capture technologies are new to the power industry, they have been deployed for the past 60 years by the oil, gas, and chemical industries. They are an integral component of natural gas processing and of many coal gasification processes used for the production of syngas, chemicals, and liquid fuels. There are three main carbon dioxide capture processes for power generation: (1) postcombustion, (2) precombustion, and (3) oxyfuel.

Post-combustion capture involves separating the carbon dioxide from other exhaust gases after combustion of the fossil fuel. Post-combustion capture systems are similar to those that already remove pollutants such as particulates, sulfur oxides, and nitrogen oxides from many power plants.

The most commonly used process for post-combustion carbon dioxide capture involves the use of amine (olamine: for example, ethanolamine H$_2$NCH$_2$CH$_2$OH) washing technology in which a carbon dioxide–rich gas stream, such as a power plant's flue gas, is passed through an amine solution. The carbon dioxide bonds with the amine as it passes through the solution while other gases continue up through the flue. The carbon dioxide in the resulting carbon dioxide–saturated amine solution is then removed from the amines, captured, and is ready for carbon storage. The amine solution can be recycled and reused. The olamine washing technology is in wide commercial use in the petroleum and natural gas industries.

Precombustion capture involves separating the carbon dioxide as part of the combustion or before the coal is burned. The coal is first gasified with a controlled amount of oxygen to produce two gases, hydrogen (H_2) and carbon monoxide (CO). The carbon monoxide is converted to carbon dioxide and removed, leaving pure hydrogen to be burned to produce electricity or used for another purpose. The carbon dioxide is then compressed into a supercritical fluid for transport and geological storage. The hydrogen can be used to generate power in an advanced gas turbine and steam cycle, in fuel cells, or a combination of both.

Oxyfuel combustion (also called *oxyfiring*) involves the combustion of coal in pure oxygen, rather than air, to fuel a conventional steam generator. By avoiding the introduction of nitrogen into the combustion chamber, the amount of carbon dioxide in the power station exhaust stream is greatly concentrated, making it easier to capture and compress.

Each of these capture options has its particular benefits. Post-combustion capture and oxyfuel have the potential to be retrofitted to existing coal-fired power stations and new plants constructed over the next 10–20 years. Pre-combustion capture utilizing IGCC is potentially more flexible, opening up a wider range of possibilities for coal, including a major role in a future hydrogen economy.

All the options for capturing carbon dioxide from power generation have higher capital and operating costs as well as lower efficiencies than conventional power plants without capture. Capture is typically the most expensive part of the carbon capture and storage chain. Costs are higher than for plants without carbon capture and storage because more equipment must be built and operated. Around 10%–40% more energy is required with carbon capture and storage than without. Energy is required mostly to separate carbon dioxide from the other gases and to compress it, but some is also used to transport the carbon dioxide to the injection site and inject it underground.

As carbon capture and storage and power generation technology become more efficient and better integrated, the increased energy use is likely to fall significantly below early levels. Much of the work on capture is focused on lowering costs and improving efficiency as well as improving the integration of the capture and power generation components.

The technology for carbon dioxide transportation and its environmental safety are well established. Carbon dioxide is largely inert and easily handled and is already transported in high-pressure pipelines. In the United States, carbon dioxide is already transported by pipeline for use in *enhanced oil recovery* (EOR) (Speight, 2007a, 2009).

The means of transport depend on the quantity of carbon dioxide to be transported, the terrain, and the distance between the capture plant and storage site. In general, pipelines are used for large volumes over shorter distances. In some situations or locations, transportation of carbon dioxide by ship may be more economic, particularly when the carbon dioxide has to be moved over large distances or overseas.

REFERENCES

Adams, T.A. II, Paul, I., and Barton, P.I. 2010. Clean coal: A new power generation process with high efficiency, carbon capture and zero emissions. In *Proceedings, 20th European Symposium on Computer Aided Process Engineering—ESCAPE 20*, S. Pierucci and G. Buzzi Ferraris (Eds.). Elsevier, Amsterdam, the Netherlands.

Blasing, T.J., Broniak, C.T., and Marland, G. 2005. The annual cycle of fossil-fuel carbon dioxide emissions in the United States. *Tellus, Chemical Physical Meteorology*, 57B: 107–115.

Elliot, C.T. (Ed.). 1989. *Standard Handbook of Power Plant Engineering*. McGraw-Hill, Inc., New York.

KEMA. 2009. *Co-Firing Biomass with Coal: Balancing US Carbon Objectives, Energy Demand and Electricity Affordability*. KEMA Inc., Burlington, MA.

Minchener, A.J. and McMullan, J.T. 2008. Strategy for sustainable power generation from fossil fuels, *Journal of Energy Institute*, 81: 38–44.

Mokhatab, S., Poe, W.A., and Speight, J.G. 2006. *Handbook of Natural Gas Transmission and Processing*. Elsevier, Amsterdam, the Netherlands.

Singer, J.G. (Ed.). 1981. *Combustion: Fossil Power Systems*. Combustion Engineering, Inc., Windsor, CT.

Speight, J.G. 2009. *Enhanced Recovery Methods for Heavy Oil and Tar Sands*. Gulf Publishing Company, Houston, TX.

Speight, J.G. 2007a. *The Chemistry and Technology of Petroleum*, 4th edn. Taylor & Francis Group, Boca Raton, FL.

Speight, J.G. 2007b. *Natural Gas: A Basic Handbook*. GPC Books, Gulf Publishing Company, Houston, TX.

Speight, J.G. 2008. *Synthetic Fuels Handbook: Properties, Processes, and Performance*. McGraw-Hill, New York.

Stultz, S.C. and Kitto, J.B. (Eds.). 1992. *Steam, Its Generation and Use*, 40th edn. The Babcock & Wilcox Company, Barberton, OH.

16 Carbonization

16.1 INTRODUCTION

Carbonization is *the destructive distillation of organic substances in the absence of air accompanied by the production of carbon and liquid and gaseous products* (Porter, 1924).

Next to combustion, carbonization represents one of the most popular, and oldest, uses of coal (Armstrong, 1929; Forbes, 1950). The thermal decomposition of coal on a commercial scale is often more commonly referred to as carbonization and is more usually achieved by the use of temperatures up to 1500°C (2730°F). The degradation of coal is quite severe at these temperatures and produces (in addition to the desired coke) substantial amounts of gaseous products.

There is no doubt that coal has been in use throughout Europe for at least 700 years since the records show that coal was exported from north-eastern England to France and to the Low Countries during the early years of the fourteenth century (Galloway, 1882). The historical records are not complete but it is reasonable (perhaps unwise not) to assume that coal use in Europe evolved along similar lines to those in England where coal carbonization has been documented for at least 200 years.

There are other instances where coal receives acknowledgement in the historical literature, even though the nature of the material (i.e., coal) is not described in any detail or with any degree of modern accuracy (Chapters 1 and 2). For example, coal is described in one sixteenth century text related to ore smelting (Biringuccio, 1540) as stones that occur in many places and have the true nature of charcoal. In another sixteenth century text that dealt with natural phenomena, coal is thought to be a variety of bitumen (Agricola, 1546; see also Hoover and Hoover, 1950). But the use of coal in the smelting industry is acknowledged by these two texts and the conversion of coal to charcoal is but a simple step that, presumably, was known at the time. Thus, it is not surprising that coal use did increase in popularity so that the pyrogenous rock became a major force behind the Industrial Revolution. By this time, the thermal treatment (carbonization) of coal to produce charcoal had become a major industry (Nef, 1957).

Carbonization is essentially a process for the production of a carbonaceous residue (coke) by the thermal decomposition (with simultaneous removal of distillate) of organic substances (Wilson and Wells, 1950; McNeil, 1966; Gibson and Gregory, 1971). The process, which is also referred to as destructive distillation, has been applied to a whole range of organic (carbon-containing) materials, particularly natural products such as wood, sugar, and vegetable matter, to produce charcoal.

Coke is the *solid carbonaceous residue* produced from coal, from which the volatile components are removed in ovens with limited oxygen inlet and temperatures around 1000°C (1832°F). During this process, tar and light oil are produced. Thus, in this present context, the carbonaceous residue from the thermal decomposition of coal will be referred to as *coke* (which is physically dissimilar from charcoal) and has the more familiar honeycomb-type structure.

The term *coke* brings an immediate reference to the blast furnace; the evolution of the blast furnace reduction process occurred over a period of many centuries (ore smelting being known in the sixteenth century and documented in some detail) (Biringuccio, 1540) and most certainly had become well known in almost all areas of the ancient world two millennia ago (Forbes, 1950). By the early Middle Ages, a shaft blast furnace known in Germany as *stuckofen* was well established and iron making was a respected trade and not a sporadic part-time occupation. The fuel for these furnaces was provided by wood charcoal, and, consequently, the vitality of the iron and steel industry was controlled by the availability and cost of wood.

In fact, it was this ready availability of wood that "fueled" the rapid industrial development of Western Europe in the eighteenth century, particularly in England. The supply of wood (in the form of char-grade timber) severely hampered the use of coal. But the timber was consumed at such a drastic rate that the woods of the English Midlands, and presumably the woods and forests in other parts of Europe, were severely decimated. Thus followed the emergence (if not the reemergence) of coal as the fuel of choice for the charcoal and carbonization industries.

The original process of heating coal (in rounded heaps; the hearth process) remained the principal method of coke production for many centuries, although an improved oven in the form of a beehive shape was developed in the Newcastle area of England in about 1759.

The method of coke production was initially the same as for the production of charcoal, that is, stockpiling coal in round heaps (known as milers), igniting the piles, and then covering the sides with a clay-type soil. The smoke generated by the partial combustion of coal tars and gases was soon a major problem near residential areas, and the coking process itself could not be controlled because of climatic elements: rain, wind, and ice. It was this new modification of the charring–coking kiln that laid the foundation for the modern coke blast furnace: the beehive process.

The coke produced by carbonization of coal is used in the iron and steel industry and as a domestic smokeless fuel. However, only a limited range of coals produces acceptable metallurgical cokes. These coals are in the bituminous rank range but not all bituminous coals are caking coals. Prime coking coals are expensive and not always available nationally. It is predicted that remaining indigenous coals available for coke making are poorer in coking quality. In addition, coke ovens need to be rebuilt, and, in most parts of the world, profits from the steel and iron business are insufficient to provide the necessary capital. In future years, technology will have to be extended even further and find new blends or raw materials and optimum operating conditions in order to reduce costs.

This will only be possible with a better understanding of the fundamental aspects of the coking process, the properties of coals, and their functions in the coal-to-coke conversion. This understanding is also necessary because of the gradually increasing stringency of requirements made on coke by modern industrial practices, that is, maximum output with maximum efficiency.

16.2   PHYSICOCHEMICAL ASPECTS

Carbonization is essentially a process for the production of a carbonaceous residue by thermal decomposition (with simultaneous removal of distillate) of organic substances (Davis and Place, 1924; Wilson and Wells, 1950; Gibson and Gregory, 1971):

$$C_{organic} \rightarrow C_{coke/char/carbon} + liquids + gases$$

The process may also be referred to as *destructive distillation* and has been applied to a whole range of organic materials, but more particularly to natural products such as wood, sugar, and vegetable matter to produce charcoal. Coal usually yields "coke," which is physically dissimilar from charcoal and has the more familiar honeycomb-type structure.

However, only a limited range of coals produces acceptable metallurgical cokes. These coals are in the bituminous rank range but not all bituminous coals are caking coals. Prime coking coals are expensive and not always available nationally. It is quite possible that remaining indigenous coals available for coke making are poorer in coking quality. In addition, coke ovens need to be rebuilt, and, in most parts of the world, profits from the steel and iron business may well be insufficient to provide the necessary capital.

In future years, it will be necessary to extend technology even further and find new blends or raw materials and optimum operating conditions in order to reduce costs. This will only be possible with a better understanding of the fundamental aspects of the coking (carbonization) process and properties of coals and their functions in coal-to-coke conversion. This understanding is also necessary because of the gradually increasing stringency of requirements made on coke by modern environmental legislation and modern industrial practices, that is, maximum output with maximum efficiency.

The thermal decomposition of coal is a complex sequence of events (Stein, 1981; Solomon et al., 1992, 1993) (Chapter 13), which can be described in terms of several important physicochemical changes, such as the tendency of the coal to soften and flow when heated (plastic properties; Chapters 8 and 9) or the relationship to carbon-type in the coal (Solomon, 1981). In fact, some coals become quite fluid at temperatures of the order of 400°C–500°C (750°F–930°F) and there is a considerable variation in the degree of maximum plasticity, the temperature of maximum plasticity, as well as the plasticity temperature range for various coals (Kirov and Stephens, 1967; Mochida et al., 1982; Royce et al., 1991). Indeed, significant changes also occur in the structure of the char during the various stages of devolatilization (Fletcher et al., 1992).

The thermal decomposition of coal is a strong function of coal type or rank. Low-rank coals, such as lignite and subbituminous coal, produce relatively high levels of light gases and very little tar. Bituminous coal produces significantly more tar than the low-rank coals and moderate amounts of light gases. Higher-rank coal produces relatively low levels of both light gases and tar.

Several investigators have attempted to isolate and determine the general characteristics of the individual steps that occur during pyrolysis (Suuberg et al., 1978; Hodek et al., 1990). At temperatures of just over 100°C (212°F), the coal residual moisture evolves (Suuberg et al., 1978). The evolution of light gases begins at temperatures of 200°C–500°C (390°F–930°F); these early gases consist mainly of carbon monoxide (CO) and carbon dioxide (CO_2) and low-boiling hydrocarbons. Tar formation was seen in low heating rate experiments to begin at around 330°C (625°F) and increases to temperatures above 530°C (985°F) (Hodek et al., 1990). Cross-linking reactions are thought to occur at different temperatures—depending on the coal and heating rates—due to different kinetics of the competing processes of bond breaking, vaporization, and bond formation. Pyrolysis experiments conducted with a number of coals indicate that early cross-linking begins in the range of 400°C–500°C (750°F–930°F), later cross-linking continues as temperatures increase (Solomon et al., 1988; Ibarra et al., 1991).

The exact temperatures at which these pyrolysis steps occur are dependent on many factors and it is likely that heating rate and coal type have the largest effect on the temperatures at which these steps occur.

For any particular coal, the distribution of carbon, hydrogen, and oxygen in the coke is decreased as the temperature of the carbonization is increased. In addition, the yield of gases increases with the carbonization temperature while the yield of the solid (char, coke) product decreases. The yields of tar and low-molecular-weight liquids are to some extent variable but are greatly dependent on the process parameters, especially temperature (Chapter 13), as well as the type of coal employed (Cannon et al., 1944; Davis, 1945; Poutsma, 1987; Ladner, 1988; Wanzl, 1988).

Thus, as coal is heated to the plastic stage, it is probable that some thermal conversion to lower-molecular-weight species can occur with the release of liquids and gases. These changes (including the liberation of water, carbon dioxide, and traces of hydrocarbons) can occur at temperatures below 300°C (570°F) and are accelerated markedly when a temperature of 350°C (660°F) is reached.

It is worthy of note here that the word *depolymerization* is often used to describe (thermal or chemical) conversion of coal to lower-molecular-weight species. However, the word *depolymerization* as used in this sense is incorrect since coal is not a polymer in the true sense of the word *polymer* but is actually a macromolecular entity made up of different molecular species, some of which bear no relationship to the other molecular species within the coal structure but do give rise to the chemical heterogeneity of coal. Such phenomena also do not support, and speak against, the concept of an *average structure* for coal.

The maximum evolution of tar, hydrocarbons, and combustible gases usually occurs over the temperature range 450°C–500°C (840°F–930°F), whereas products such as hydrogen sulfide and

TABLE 16.1
Terminology of the Various Carbonization Processes

Carbonization Process	Final Temperature		Products	Processes
	°C	°F		
Low temperature	500–700	930–1290	Reactive coke and high tar yield	Rexco (700°C) made in cylindrical vertical retorts. Coalite (650°C) made in vertical tubes
Medium temperature	700–900	1290–1650	Reactive coke with high gas yield, or domestic briquettes	Town gas and gas coke (obsolete). Phurnacite, low-volatile steam coal, pitch-bound briquettes carbonized at 800°C
High temperature	900–1050	1650–1920	Hard, unreactive coal for metallurgical use	Foundry coke (900°C). Blast furnace coke (950°C–1050°C)

Source: Gibson, J., *Coal and Modern Coal Processing: An Introduction*, G.J. Pitt and G.R. Millward, Eds., Academic Press, Inc., New York, 1979.

ammonia are released at temperatures of 250°C–500°C (480°F–930°F) and at temperatures in excess of 500°C (930°F), respectively. At temperatures of the order of 550°C (1020°F), the visible changes to the coal are virtually complete and a semicoke residue has been formed. At temperatures in excess of 550°C (1020°F), the semicoke hardens and shrinks to form coke with the evolution of methane, hydrogen, and traces of carbon monoxide and carbon dioxide (Marsh and Cornford, 1976).

The nature of the physicochemical changes has been a major factor in determining the extent to which coal is carbonized, with the character of the liquid and gaseous by-products being of considerable importance. Thus, coal carbonization processes (Table 16.1) are generally regarded as "low temperature" when the temperature of the process does not exceed 700°C (1290°F) or as "high temperature" if the temperature of the process is at or in excess of 900°C (1650°F). Following from this, the temperature of a "medium-temperature" carbonization process is easily deduced and self-explanatory.

Gases of high calorific value are obtained by low-temperature or medium-temperature carbonization of coal. The gases obtained by the carbonization of any given coal change in a progressive manner with increasing temperature (Table 16.2). The composition of coal gas also changes during

TABLE 16.2
Effect of Carbonization Temperature on Gas Composition

Gas Composition and Yields (% v/v)	Temperature of Carbonization					
	500°C	600°C	700°C	800°C	900°C	1000°C
CO_2	5.7	5.0	4.4	4.0	3.2	2.5
Unsaturates	3.2	4.0	5.2	5.1	4.8	4.5
CO	5.8	6.4	7.5	8.5	9.5	11.0
H_2	20.0	29.0	40.0	47.0	50.0	51.0
CH_4	49.5	47.0	36.0	31.0	29.5	29.0
C_2H_6	14.0	5.3	4.5	3.0	1.0	0.5
Yield (m^3/ton)	62.3	102	176	238	278	312
CV (MJ/m^3)	39.0	29.0	26.5	24.4	22.3	22.3
Yield (MJ/ton)	2118	2960	4660	5810	6200	6960

TABLE 16.3
Effect of Carbonization Reaction Time on Gas Composition

Composition and Yield of Gas (% v/v)	Duration of Carbonization at 950°C				
	0.5 h	0.5–1.5 h	1.5–3.0 h	3.0–4.5 h	4.5–7.5 h
CO	3.2	2.8	2.5	1.8	0.5
Unsaturates	6.5	4.5	2.0	0.3	—
CO	7.5	9.0	7.0	6.5	6.0
H_2	34.0	43.5	53.8	63.5	67.5
CH_4	43.5	37.0	31.5	23.5	22.5
C_2H_6	1.5	0.5	0.2	—	—
CV (MJ/m^3)	26.8	24.3	22.5	18.1	18.1
Gas yield (m^3/ton)	37.7	92.1	63.5	34.8	83.6
MJ/ton	1010	2240	1430	630	1510

the course of carbonization at a given temperature (Table 16.3) and secondary reactions of the volatile products are important in determining gas composition (Xu and Tomita, 1987).

In the early days of the coal carbonization industry, it was noted that the addition of weakly caking or even noncaking coals to a coke oven charge would overcome the difficulties that sometimes occurred when coals having high free swelling indexes (FSIs) and/or Gieseler fluidities (Chapters 8 and 9) were carbonized. This technique then became an "art" and is finally evolving into a "science" but only after much study, investigation, and data correlation.

The concept of blending a charge originates from the fact that coals vary widely in their ability to make coke and, hence, it is customary to blend coals before carbonization in order to produce the desired type of coke (Gruit and Marsh, 1981). For example, foundry coke, a hard and extremely dense product, is usually made from prime coking coals having volatile matter yields of the order of 19%–32% w/w while blast furnace coke, which also should be a hard and strong fuel, is also made from these coals but, because it is used in smaller sizes, higher-volatile coals (32% volatile matter content) are suitable alone or as blends with prime coking coals. For the manufacture of town gas and the softer, reactive cokes, coals of lower ranks have been widely employed; in particular, low-temperature carbonization processes for the production of reactive domestic fuel have favored the use of the less strongly coking coal.

Thus, coal blending is practiced to improve the physical and chemical properties of the coke and to enable a proportion of inferior (or marginal) coking coals to be employed as feeds as well as to improve (and control) the yield of the coke. In addition, blending may also prevent damage to the ovens during, say, the carbonization of coals that develop a comparatively high pressure during the process.

To ensure that a consistent product is obtained, careful segregation, strict proportioning, and effective mixing of the component coals are essential. The cleaned coals are usually passed through 0.5 or 1.0 in. (12.5 or 25 mm) sieves prior to the coke ovens; some stock may be maintained, but the coking properties of coal deteriorate on storage due to oxidation (weathering) (Joseph and Mahajan, 1991) and unless adequate precautions are taken to prevent oxidation, the quantity stored is usually small. If a stock is maintained, it must be consolidated, limited in height, and covered with an impervious coating of tar or a suitable emulsion to prevent the entry of air.

Usually, the different coals are stored in separate bunkers from which the required proportion of each coal is taken and then crushed in a hammer mill to give a product containing approximately 85% below 0.1 in. (3 mm) in size. The hammer mill also acts as a mixer and this may in fact be the only mixing required if the blend consists solely of coking coals, but if finer grinding and more efficient mixing are necessary (as in the case when non-coking coals are also employed as part of the feed), mechanical mixers (Figure 16.1) may be required.

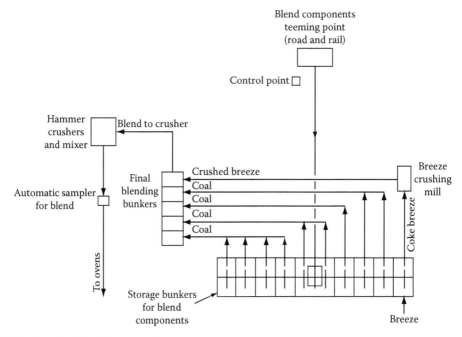

FIGURE 16.1 Coal blending operation.

Before carbonization, the blend may undergo further treatment to enhance the ability of the blend to make better coke or to improve the process economics. For example, blends may contain as much as 10% w/w free moisture that will impede the flow properties and lower the bulk density. Thus, it may be beneficial to preheat (to ca. 200°C [390°F]) the feed before it enters the oven; bulk density will be increased and the carbonization time will be reduced by this simple operation of water removal. The abrasion resistance of the coke may also be improved by this treatment. The preheating operation may be carried out by means of an entrained (dilute phase) system or by means of a fluidized-bed (dense-phase) system, but it is important that oxidation of the coal(s) be prevented to avoid adversely affecting the coking properties.

A significant development in understanding carbonization processes was made with the discovery of mesophase in the plastic stage of carbonization leading to graphitizable carbons, as observed by optical microscopy (Taylor, 1961). The development of spherical mesophase particles from an isotropic mass and their progressive growth and coalescence eventually to form anisotropic structures is well established for pitch-like precursors (Marsh, 1973; Davis et al., 1983; Grew, 1986). Essentially, during the carbonization process dehydrogenative polymerization of aromatic molecules occurs, with a consequential increase in average molecular weight. The final coke structure is related to the properties of mesophase at the time of solidification and these, in turn, are dominantly dependent upon the chemical properties of the parent material. Coke quality improvements are dictated by the quality of the parent feedstock, which predetermines the optical texture of the resultant coke.

16.2.1 EFFECT OF COAL PROPERTIES

The ASTM has defined many testing procedures applicable to coal, as well as to coke (Chapter 8) (Tables 16.4 and 16.5) (Zimmerman, 1979). It is not the purpose here to reproduce all of the tests but to highlight those tests that are more commonly seen as presenting indications of the effect of the coal on the properties of the coke.

TABLE 16.4

Testing Procedures Applicable to Coke

Designation	Title
General	
D121-72	*Standard Definitions of Terms Relating to Coal and Coke*
Sampling	
D346-75	*Standard Method of Collection and Preparation of Coke Samples for Laboratory Analysis*
Chemical	
D1857-68 (reapproved 1974)	*Fusibility of Coal and Coke Ash*
D2361-66 (reapproved 1972)	*Standard Method of Test for Chlorine in Coal*
D2795-69 (reapproved 1974)	*Analysis of Coal and Coke Ash*
D3172-73	*Standard Method for Proximate Analysis of Coal and Coke*
D3173-73	*Standard Method of Test for Moisture in the Analysis Sample of Coal and Coke*
D3174-73	*Ash in the Analysis Sample of Coal and Coke*
D3175-73	*Volatile Matter in the Analysis Sample of Coal and Coke*
D3177-75	*Total Sulfur in the Analysis Sample of Coal and Coke*
D3178-73	*Carbon and Hydrogen in the Analysis Sample of Coal and Coke*
D3179-73	*Nitrogen in the Analysis Sample of Coal and Coke*
Physical	
D167-73	*Specific Gravity and Porosity of Lump Coke*
D292-29 (reapproved 1972)	*Standard Method of Test for Cubic Foot Weight of Coke*
D293-69 (reapproved 1974)	*Standard Method of Sieve Analysis of Coke*
D294-64 (reapproved 1972)	*Standard Method of Tumbler Test for Coke*
D323-70	*Specification for Perforated-Plate Sieves for Testing Purposes*
D2490-70	*Standard Method of Tumbler Test for Small Coke*
D3038-72	*Standard Method of Drop Shatter Test for Coke*
D3402-75	*Standard Methods of Tumbler Test for Coke*
E11-70	*Specification for Wire-Cloth Sieves for Testing Purposes*

Source: ASTM, *Standard Test Methods for Coal and Coke*, American Society for Testing and Materials, West Conshohocken, PA, 2011.

Of the many properties or characteristics of coal (Chapter 9), the following are considered important in coke making: chemistry, petrography, rheology, hardness, moisture, and size. Other coal properties that are useful in estimating the character of the coke include FSI, fluidity, and expansion characteristics.

The chemistry and physics of coal can be expressed in terms of composition, which is defined as ultimate (elemental) analysis, and behavior, which is described in empirical terms by a prescribed method (also defined as proximate analysis) (Chapters 8 and 9). Therefore, it is just as important that coke be defined in terms of analytical data that suits the utility and meets the objective for which the coke was produced (Marsh and Smith, 1978; Patrick and Wilkinson, 1978).

Ultimate analysis is specified when evaluating coals for combustion, gasification, and liquefaction processes, but is less important in coke-making processes. The ultimate analysis includes the determination of carbon, hydrogen, oxygen, sulfur, nitrogen, and ash.

Proximate analysis is often sufficient for meeting coking plant needs and consists of moisture, volatile matter, ash, fixed carbon, and (generally) sulfur. The sulfur content is provided as a part of the proximate analysis and is equivalent to the elemental sulfur determined in the ultimate analysis.

The volatile matter content of a given coal blend will give an indication of the yield of coke to be expected from a blend of coals. Generally, the coking coal blend is made up of varying proportions

TABLE 16.5

Testing Procedures Applicable to Coal

Designation	Title
General	
D121-72	*Standard Definitions of Terms Relating to Coal and Coke*
D 388-66 (reapproved 1972)	*Specifications for Classification of Coals by Rank*
Sampling	
D2013-72	*Preparing Coal Samples for Analysis*
D2234-72	*Collection of a Gross Sample of Coal*
Chemical	
D407-44 (reapproved 1969)	*Definition of the Terms Gross Calorific Value and Net Calorific Value of Solid and Liquid Fuels*
D1412-74	*Test for Equilibrium Moisture of Coal at 96 to 97 Percent Relative Humidity and 30°C*
D1756-62 (reapproved 1974)	*Test for Carbon Dioxide in Coal*
D1757-62 (reapproved 1974)	*Test for Sulfur in Coal Ash*
D1857-68 (reapproved 1974)	*Test for Fusibility of Coal and Coke*
D2015-66 (reapproved 1972)	*Test for Gross Calorific Value of Solid Fuel by the Adiabatic Bomb Calorimeter*
D2361-66 (reapproved 1972)	*Test for Chlorine in Coal*
D2492-68 (reapproved 1974)	*Test for Forms of Sulfur in Coal*
D2795-69 (reapproved 1974)	*Test for Analysis of Coal and Coke Ash*
D2961-74	*Total Moisture in Coal Reduced to Number 8 Top Sieve Size (Limited Purpose Method)*
D3172-73	*Proximate Analysis of Coal and Coke*
D3173-73	*Test for Moisture in the Analysis Sample of Coal and Coke*
D3175-73	*Test for Volatile Matter in the Analysis Sample of Coal and Coke*
D3176-74	*Ultimate Analysis of Coal and Coke*
D3177-75	*Test for Total Sulfur in the Analysis Sample of Coal and Coke*
D3178-73	*Test for Carbon and Hydrogen in the Analysis Sample of Coal and Coke*
D3179-73	*Test for Nitrogen in the Analysis Sample of Coal and Coke*
D3286-73	*Test for Gross Calorific Value of Solid Fuel by the Isothermal-Jacket Bomb Calorimeter*
D3302-74	*Test for Total Moisture in Coal*
Physical	
D197-30 (reapproved 1971)	*Sampling and Fineness Test of Pulverized Coal*
D291-60 (reapproved 1975)	*Test for Cubic Foot Weight of Crushed Bituminous Coal*
D311-30 (reapproved 1969)	*Test for Sieve Analysis of Crushed Bituminous Coal*
D409-71	*Test for Grindability of Coal by the Hardgrove Machine Method*
D410-38 (reapproved 1969)	*Sieve Analysis of Coal*
D431-44 (reapproved 1969)	*Designating the Size of Coal from its Sieve Analysis*
D440-49 (reapproved 1975)	*Drop Shatter Test for Coal*
D720-67 (reapproved 1972)	*Test for Free Swelling Index of Coal*
D1812-69 (reapproved 1974)	*Test for Plastic Properties of Coal by the Gieseler Plastometer*
D2014-71	*Test for Expansion or Contraction of Coal by the Sole Heated Oven*
D2639-74	*Test for Plastic Properties of Coal by the Constant Torque Gieseler Plastomer*
D2796-72	*Definition of Terms Relating to Lithologic Classes and Physical Components of Coal*
D2797-72	*Preparing Coal Samples for Microscopical Analysis by Reflected Light*
D2798-72	*Microscopical Determination of the Reflectance of the Organic Components in a Polished Specimen of Coal*
D2799-72	*Microscopical Determination of Volume Percent of Physical Components of Coal*
E11-70	*Specification for Wire-Cloth Sieves for Testing Purposes*
E323-70	*Specification for Perforated-Plate Sieves for Testing Purposes*

Source: ASTM, *Standard Test Methods for Coal and Coke*, American Society for Testing and Materials, West Conshohocken, PA, 2011.

of high- and medium- or low-volatile coals. All of the mineral matter contained in the coal blend is (usually, entrainment and volatility notwithstanding) retained as mineral ash (mineral oxides) in the coke. A large portion of the organic sulfur is driven off from the coal during carbonization and appears in the by-products.

On the other hand, inorganic sulfur is essentially retained in the coke. The effect that the chemistry of the coal has in the coking process can be summarized by saying that the volatile matter determines to a great extent the yields of coke and by-product, ash in the coal is retained to the coke, and sulfur divided between coke and by-products according to the proportions of inorganic and organic forms.

Coking coal blends are pulverized to permit the intimate mixing of the coking components in the blend. Generally, as the percent passing 3.2 mm increases, coke stability increases for constant moisture in the blend. However, an excessively high degree of pulverization results in a decrease in coke apparent specific gravity, which is also an inverse measure of porosity. Coal pulverization is affected by moisture and the hardness of the coal is measured by the Hardgrove test. Coal with too high moisture content reduces the efficiency of most pulverizing equipment. Different coals usually have different hardness values. Therefore, in preparing a blend for coking, each coal should be pulverized separately according to its needs and then blended. Bulk density of pulverized coal is affected by moisture and particle size. Thus, it can be seen that moisture, hardness, and size are interrelated. Low bulk density can affect productivity by decreasing the amount of coke produced per oven, which in turn requires more time for charging and pushing to meet production demands.

Coal petrography (Chapter 4) has become widely used for predicting coke quality based on coal analysis and has led to a system for predicting coke stability based on petrographic entities and reflectance of coal (Schapiro and Gray, 1960). Thus, an optimum blend of coals could be selected to produce desired coke quality.

When bituminous coals are heated, for the most part, softening occurs causing fluidity (Chapter 9) and, finally, solidifying into a sponge-like coherent solid. It is this flow property that determines whether a particular coal has the potential to make coke. The fluid or swelling property of coal determines whether a coal will make coke. The temperatures at which the softening and solidification occurs are important in that all coals in a blend should have similar softening ranges to permit an interaction of coals in the fluid state during coking. During the coking process, the coal mass softens forming a plastic envelope. This envelope contains water vapor and gas, which exerts pressure on the coke oven walls. The magnitude of this pressure is dependent on the strength of the envelope.

Pressures above 2000 psi are considered unsafe for the structural integrity of the coke oven walls. Another important property is the expansion and subsequent contraction of the coal mass as it is converted to coke. In order for the coke to be pushed from the oven, a certain amount of contraction of the coke is necessary. The consequence of not having contraction at all, or only marginal contraction, is that the coke sticks in the oven or it requires very high power requirements for pushing.

One physical property of significant importance in many coal-processing schemes is particle size. For any given process, there usually exists an optimum particle size range that is desired for the feed material. This desired size range may be a function of the conversion technique employed (i.e., moving bed, fixed bed, fluidized bed, etc.), or it may be determined by the solids handling systems used (i.e., dry lock, hopper, slurry, etc.). Some current coal conversion processes require feed to be pulverized to less than −200 mesh while other processes can handle feed up to 1/4 in. (6 mm).

A related property, which is also an important factor, is the ability of the individual coal particles to retain their initial size during handling and processing. In fluidized-bed processes, attrition and decrepitation of the feed coal may result in unacceptable fines carryover losses. For processes that must withstand high temperatures and blast velocities, only hard and resistant coals are satisfactory feed materials. In addition, the fusibility and caking characteristics are properties of importance to moving bed or fluidized-bed processes.

Another property that is often ignored, since it is not often considered to be a property of the organic matrix of coal, is the behavior of the mineral matter in the coal. Indeed, the mineral matter (or ash) content and the composition of the mineral matter can play a major part in determining the suitability of coal for coke production. For example, mineral matter contents in excess of 12% w/w of the coal may actually "dilute" the coal to such an extent that the caking properties of the coal are markedly diminished. Thus, it is entirely probable that a particular coal may have been rejected as a source of coke but with a cleaning treatment can be upgraded into a satisfactory blend component. There is also evidence that high ash contents have an adverse effect on the quality of the coke. For example, if the ash contains phosphorus there may be limitations on the use of the coke for the production of iron and steel insofar as the phosphorus (which usually occurs in the organically bound state) cannot be easily removed by coal-cleaning techniques.

The FSI is used to indicate the caking property and is measured by comparing the volume changes in a sample after having been heated under controlled conditions. Fluidity is a measure of the plastic behavior of a coal while it is being heated at a constant rate. Significant indices of this measure include softening temperature, temperature of maximum fluidity (i.e., when the stirrer immersed in the heated coal sample is rotating at its maximum rate), maximum fluidity (maximum stirrer rotation), and solidification temperature (the temperature at which the stirrer stops). One way that expansion or contraction of a coal can be measured is by heating only the bottom surface of a sample while a piston applies a constant force to the top surface. The location of the piston upon reaching the conclusion of the heating cycle indicates the degree of expansion or contraction. Size consists are determined by sieving and strength is determined by a drop shatter test.

16.2.2 Determination of Coke Properties

Tests for coke characteristics can be divided into two major classes: chemical and physical. Considering chemical tests first, there are ultimate analysis and proximate analysis for the coke. Another chemical property, reactivity, which refers to the ability of coke to react with carbon dioxide, has also been proposed (Schapiro and Gray, 1963; Miller et al., 1965; Rao and Jalon, 1972; Daly and Budge, 1974); the ASTM does not provide a standard reactivity test procedure.

The physical properties of coke can be categorized into three groups: (1) the static properties considered for a lump of coke (porosity), (2) the static properties of bulk coke (specific gravity, bulk density, and size), and (3) the mechanical properties (shatter, stability and hardness, and high-temperature strength).

Coke pore structure includes pores on the external surface of the lump, internal microscopic pores, and "submicroscopic" pores, whose diameters are only several Angstrom units (10^{-10} m) (Dahme, 1953; Thompson et al., 1971). The calculation of porosity can be made with a knowledge of the apparent specific gravity and the true specific gravity.

$$\% \text{ Porosity} = 100 - 100 \times (\text{apparent specific gravity}) / (\text{true specific gravity})$$

Apparent specific gravity is the average weight per unit volume (relative to water) of individual lumps of coke, and so includes the effect of porosity. True specific gravity is the average weight per unit volume (relative to water) of individual particles of coke, and is measured on a crushed sample to eliminate the effect of porosity.

Bulk density of coke is the weight per unit volume, which is affected by both the apparent specific gravity of the individual lumps and the voids between lumps. Two aspects of coke size are meaningful: size distribution and average size. Size distribution is reported as the series of percentages by weight of coke that passes through one sieve and remains on another sieve with finer apertures. Average size is defined as the weighted average of the products of the average size of each size range and the fraction present in that range.

The usual method for measuring coke strength (resistance to abrasion and impact) is by measuring the size degradation that occurs in a tumbler drum. The different drum tests were devised to maximize the measurement in degradation between various cokes.

As a combustible material, coke is relatively inert and requires relatively high temperatures to initiate a reaction with oxygen that is sufficient to provide the heat necessary for reducing iron oxide and for melting slag and iron. To a good first approximation, this occurs in the immediate vicinity of the blast furnace tuyeres, where heated air (1000°C [1830°F]) reacts with hot coke to produce a temperature sufficient for blast furnace operation (1800°C [3270°F]). Thus, to provide the maximum benefit as a fuel, the amount of contained carbon in the coke should be maximized and hence the ash content minimized. After taking this constraint into account, there are no limitations on coke as a fuel source other than the obvious criterion that the bulk of it survives the passage through the furnace and arrives at the tuyeres.

The principal carbon-bearing gas produced during the combustion of coke at the tuyeres is carbon monoxide, which subsequently reacts with iron oxide to produce iron and carbon dioxide. If this reaction occurs above 1000°C (1830°F), the resulting carbon dioxide can react endothermically with coke (carbon solution) to form additional carbon monoxide. However, below 1000°C (1830°F), coke is relatively inert to carbon dioxide, and hence the carbon dioxide produced below 1000°C (1830°F) reports to the top gas. Of these two possibilities, it is more desirable to encourage the reaction between carbon monoxide and iron oxide below 1000°C (1830°F) since this will reduce the coke rate and hence increase blast furnace productivity. In terms of coke properties, the implication of the desirability to reduce iron oxide below 1000°C (1830°F) is that the coke be relatively unreactive in this temperature range so that the temperature is not reduced by the reaction between coke and carbon dioxide. Also, it is important to note that increased blast temperature increases the production of carbon monoxide at the tuyeres, and hence increases the probability for reaction with iron oxide and consequent generation of carbon dioxide, which subsequently reacts endothermically with coke in the lower portion of the furnace. High blast temperatures are, however, desirable in terms of reducing the heat input that must be supplied by the reaction of carbon and oxygen. There is, therefore, a balance required between coke reactivity and blast temperature in order to simultaneously provide sufficient carbon monoxide for reducing purposes and to minimize coke consumption.

Burden materials are relatively easy to reduce in most blast furnaces. This, in conjunction with high blast temperatures, results in coke rates that are marginal with respect to providing sufficient carbon monoxide during combustion to reduce the burdens below 1000°C (1830°F). Consequently, some carbon solution above 1000°C (1830°F) is required to supplement the generation of carbon monoxide by combustion.

Although the reactivity of coke can be measured by a number of techniques, it is not clear that any of the measurements relate directly to blast furnace performance. Specifically, throughout the world, coke with varying reactivity, as measured by existing reactivity tests, ranging from relatively unreactive to nearly completely reactive, have been used successfully in existing blast furnaces operated under standard accepted operating practices. This is not particularly surprising when one considers the factors that contribute to the reactivity of coke. First, the carbon forms found in coke vary from amorphous carbon to graphite. It is well established that graphite is significantly less reactive than amorphous carbon. Consequently, although the kinetics of graphitization are a function of the carbonization conditions and the nature of the coals employed, the thermal conditions in the blast furnace will increase the degree of graphitization and therefore reduce the reactivity of all cokes; the more reactive coke will tend to be more affected and therefore the difference in reactivity between cokes will be reduced.

There are often questions related to how much graphitization occurs in the range of time/temperature/atmosphere parameters of the blast furnace. It is known that the reactivity of coke is increased significantly by the presence of alkali, which tends to concentrate in the lower portion of the furnace. This effect would be pronounced with the lower-reactive coke and results in reducing the difference between the reactivity of all cokes.

The increase in reactivity due to alkali is so great (5–10 times) that it is most probable that the beneficial effects of graphitization in the furnace are, in large part, negated and all cokes are rendered highly reactive. Consequently, with respect to blast furnace performance, coke reactivity is not likely to be a significant property of coke, although this property has been used successfully to monitor coke plant operations, where essentially every process variable can affect the degree of carbonization and graphitization.

The physical nature of the burden changes as it descends in the blast furnace. Initially, it is charged in alternating layers. This practice increases the permeability in the furnace relative to forming one layer consisting of a uniform mixture of all burden materials. As the burden descends, all the burden materials, with the exception of the coke, soften and melt. Under these conditions, it is imperative that the code retain adequate strength and size in order to ensure a high level of permeability consistent with production requirements. To effect this condition, it is necessary that in the region where molten slag for iron is present, the coke size and size distribution be such that the liquid is properly supported by the gas flow, so that it neither fails to descend with the coke nor fills the voids in the coke layer. In qualitative terms, the proper conditions are achieved with coke that has a large average size and a narrow size distribution.

These conditions suggest the necessity for charging large coke with an initial small size range of sufficient strength, so that size degradation is minimized. It has been demonstrated on operating furnaces that code stability and size are directly related to productivity, which suggests that low-temperature properties are correlated with high-temperature properties and undoubtedly are useful guides. However, the uncertainty in the data, and the effect of specific practices on the sensitivity of the correlations, indicate that additional information, perhaps under actual and for simulated blast furnace conditions, is required before correlations between coke properties and blast furnace performance are sufficiently well defined so that the basic controlling properties of coke can be identified.

16.3 PROCESS CONCEPTS

Both the hearth and beehive processes were deficient in their operation because they (1) lacked the capability for collection of the volatile products (liquids and gases), (2) usually gave a low yield of coke, and (3) caused considerable air pollution. It was not until the mid-nineteenth century, with the introduction of indirectly heated "slot" ovens, that it became possible to collect the liquid and gaseous products for further use.

For many years after the realization of the deficiencies of the beehive oven (in terms of by-product collection), repeated efforts were made to develop a single facility that could produce coke similar in quality to that made in beehive ovens and also provide for recovery of products such as those obtained in gas-producing retorts. Thus, an oven that embodied vertical flues in its walls was developed in France and used both in that country and in the Ruhr district of Germany for producing coke. Known as the Francois oven when built in France and as the Rexroth oven when used in Germany (hence the combination name Francois–Rexroth), this so-called waste-heat oven was 26 ft (8 m) long, 5 ft (1.5 m) high, and 35 in. (0.9 m) wide. The purpose of that design was to exclude air from the coking chamber to the greatest degree possible during the carbonization procedure. The volatile gases were drawn by stack drift through orifices in the top of the coking chamber into vertical flues in the oven walls. Gas from the ovens and air were mixed and burned in a downward direction in these flues and passed to the stack through sole flues under the oven. The development of vertical-flue ovens was pursued later as a means of improving performance and led to two separate and distinct developments.

In 1862, an oven was developed in England that embodied the incorporation of horizontal heating flues in the oven walls. This oven was an improvement over previous models insofar as it provided for a greater yield of coke and required less time for carbonization. In another development, a contribution to the process involved the addition of an exhauster in the chemical-recovery apparatus.

In this system, gas was drawn from the oven by the exhauster, pumped through the chemical-recovery apparatus, and returned to the ovens for use as a fuel, the present method of recovering coal chemicals being based on this same principle. Because of the inclusion of horizontal flues in the oven walls, this system can claim the distinction of being the originator of the horizontal-flue system of oven heating (Armstrong, 1929).

Almost simultaneously, in Belgium, an oven was introduced that used the basic principles of the Francois–Rexroth oven and included the innovative design of reducing the width of the oven and providing additional vertical heating flues in the oven walls; the ovens were, at the time, 30 ft (9 m) long, 3 ft 7 in. (1.1 m) high, and 18 in. (0.5 m) wide. In these ovens, stack draft was used to draw the combustion gases from the vertical downdraft flues through waste-heat boilers, recovering the sensible heat as steam. In addition, one of the chief inclusions in this oven was the control of combustion by dampering the air admitted to the flues. These ovens conformed more nearly to present-day conventional-width ovens and contained a substantial number of vertical flues in the oven walls. Thus, this oven may be considered the predecessor of the modern coke oven.

Evolution of the horizontal-flue oven was also proceeding, but improvements were largely in the direction of by-product recovery rather than in coke oven design. The basic design was improved in Germany and in England. A reduction of coking time to 48 h was the principal contribution of the oven built for a plant in Durham, England. In 1885, the Brussels Gas Works sponsored the construction of twenty-five 4 ton ovens near Mons, Belgium. The design was based on the principle of horizontal.

The by-product coke oven is a long, narrow refractory structure closed by removable doors on each end. Charging ports in the roof are scaled by metal lids and one or more openings are provided to remove the gas generated in the oven during the distillation process. Ovens are arranged in batteries of 30–100 ovens side by side with heating flues in between.

Oven heights increased from 10–11 ft (3.1–3.4 m) in the 1920s to 12–13 ft (3.7–4 m) in the 1940s. Oven lengths were increased from about 37 to 50 ft (11.3–15.2 m) over the same period. Oven widths are expressed as averages since the walls taper, with the widest dimension on the coke side to facilitate pushing the oven. Average oven widths vary from about 14 to 18 in. (0.4–0.5 m). Oven volumes (to the coal line) range from 500 ft (14 m) for a 10 ft (3 m) oven 38 ft (11.6 m) long to 1390 ft^3 (39.4 m) for a 20 ft (6 m) oven 50 ft (15.2 m) long. Since coking time is a function of width, increased volume is directly proportional to increased productivity. Individual coke ovens are assembled into batteries that operate as a unit. Heating flues are in the walls between the ovens and the entire brickwork structure is supported by steel buckstays (upright structural columns), which in turn are held together by adjustable tie rods.

Some early oven doors were of the guillotine type, which could be raised or lowered vertically with a simple hoist. These were replaced when machinery was developed to install and remove doors horizontally with the door in a vertical position. All the early doors were sealed with a clay–water mixture known as luting material. Most oven doors are self-sealing; a thin metal strip on the door is pressed against the metal door jamb by screw or spring pressure exerted against door latches.

Improvement in the control of coke oven gas discharge evolved through the availability of better gas exhausters, coupled with more sophisticated pressure control devices to maintain uniform pressure at the ovens.

The gas evolved from coal carbonization contains a variety of compounds, such as ammonia, hydrogen sulfide, hydrogen cyanide, naphthalene (and other aromatic compounds), and miscellaneous hydrocarbons (Manka, 1979; Tucker, 1979); these latter are usually described collectively as "light oil." As the gas passes out of the oven through a vertical standpipe, around a return bend or gooseneck into a collecting main, it is sprayed and cooled, by means of *flushing liquor*, from approximately 705°C (1300°F) to approximately 65°C (150°F). Some of the tar condenses at this time and flows to the bottom of the main, where it is carried away by the flushing liquor.

A coke oven battery operates on a very delicate pressure balance and gas generation creates higher pressures, which force the gas into the collecting main. Pressure control is extremely

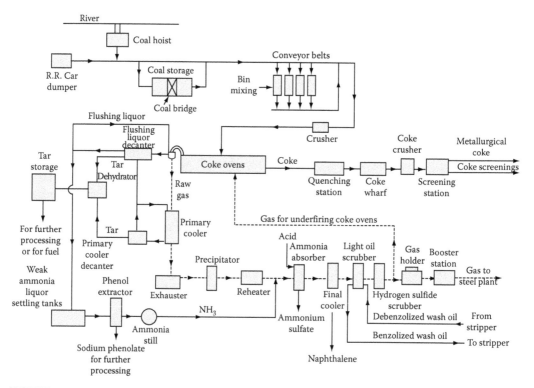

FIGURE 16.2 By-products processing. (From Meyers, R.A., Ed., *Coal Handbook*, Marcel Dekker, Inc., New York, 1981.)

important since excessive pressure in the oven will cause emissions to leak through the door seals, charging ports, and many other openings. On the other hand, negative pressure will pull air through these same areas and cause combustion within the oven resulting in damage to the interior (refractories).

Ammonia, a valuable chemical from coal (Chapter 24), is often recoverable from both the liquid and gas streams and can be readily separated from the liquid, by an ammonia still. The fixed salts must be treated with lime or caustic in a "lime leg." The gaseous ammonia may be absorbed with sulfuric acid to produce ammonium sulfate, processed to anhydrous ammonia, or destroyed by combustion.

Naphthalene must be removed from the gas stream to avoid plugging of lines and valves in subsequent gas usages. It may be scrubbed from the gas with hydrocarbon (usually wash oil) and then separated by distillation; another removal method is to reabsorb it in tar. Naphthalene may be extracted in the crystalline form as pure material.

The light oil (consisting primarily of benzene, toluene, and xylene) may be left in the gas and burned with it or extracted for subsequent distillation. Extraction is accomplished by scrubbing with a wash oil in packed or spray towers. The light oils are then separated from the wash oil by distillation (Figure 16.2).

On another separate, but very important issue, environmental requirements are mandating sulfur removal from the gas before burning and several process options are available (Chapters 23, 25, and 26).

16.3.1 LOW-TEMPERATURE CARBONIZATION

Low-temperature carbonization was mainly developed as a process to supply "town" gas for lighting purposes as well as to provide a "smokeless" (devolatilized) solid fuel for domestic consumption (Wilson and Clendenin, 1963; Seglin and Bresler, 1981). Low-temperature carbonization, though of somewhat lesser commercial importance, is achieved by the use of temperatures of the order of 500°C–700°C (930°F–1290°F) whereas high-temperature carbonization employs temperatures of the order of 900°C–1500°C (1650°F–2750°F); the temperature range for medium-temperature carbonization processes is self-explanatory.

Four main products are obtained by low-temperature carbonization processes: (1) the final coke or char, (2) an organically complex tar, (3) gases, and (4) aqueous liquor. The proportions are determined, in part, by the rate and the time of heating.

Both fluidized-bed and fixed-bed systems have been used; the former system requires direct contact with another heating medium, whereas fixed-bed systems may be heated directly or indirectly. Carriers can vary and among the carriers for direct heating are steam, air, recycle gases, sand, and metal balls containing a salt to take advantage of the latent heat of fusion. Superheated steam at 540°C–650°C (1000°F–1200°F) has been effective but may actually be economically unfavorable. The commercial aspects of the process usually involve ovens or kilns that are heated externally by a fuel gas, usually by-product gas from the coke ovens themselves.

The process by-product (tar) was also considered to be valuable insofar as it was used as a feedstock for an emerging chemical industry and was also converted to gasoline, heating oils, and lubricants. The coals that were preferred for low-temperature carbonization were usually lignites or subbituminous (as well as high-volatile bituminous) coals that yield porous solid products over the temperature range 600°C–700°C (1110°F–1290°F).

The reactivity of the semicoke product was usually equivalent to that of the parent coals. Certain of the higher-rank (caking) coals were less suitable for the process (unless steps were taken to destroy the caking properties) because of the tendency of these higher-rank coals to adhere to the walls of the carbonization chamber.

On a commercial scale, the low-temperature carbonization of coal was employed extensively in the industrialized nations of Europe but suffered a major decline after 1945 as oil and natural gas became more widely available, but the subsequent rapid escalation in oil prices as well as newer and more restrictive environmental regulations have stimulated (and reactivated) interest in the recovery of hydrocarbon liquids from coal by low-temperature thermal processing.

The options for efficient low-temperature carbonization of coal include vertical and horizontal retorts that have been used for batch and continuous processes. In addition, stationary and revolving horizontal retorts have also been operated successfully and there are also several process options employing fluidized or gas-entrained coal. During the last half century, the production of coke from batch-type carbonization of coal has been supported by a variety of continuous retorting processes that allow much greater throughput rates than were previously possible. These processes employ rectangular or cylindrical vessels of sufficient height to carbonize the coal while it travels from the top of the vessel to the bottom and usually employ the principle of heating the coal by means of a countercurrent flow of hot combustion gas. Most notable of these types of carbonizers are the Lurgi-Spulgas retort (Figure 16.3) and the Koppers continuous "steaming" oven (Figure 16.4).

A more recent concept termed "mild gasification" has also been investigated in detail (Cha et al., 1988; Merriam et al., 1992). The purpose of the process is to produce added-value products (such as high heat content char), which will facilitate removal of impurities and low value materials (such as water), thereby making the economics of shipment more favorable. The process conditions are similar to the low-to-medium temperature carbonization processes and do offer an attractive option for coal use.

16.3.2 High-Temperature Carbonization

When heated at temperatures in excess of 700°C (1290°F), low-temperature chars lose their reactivity through devolatilization and also suffer a decrease in porosity. High-temperature carbonization is, therefore, employed for the production of coke (Eisenhut, 1981).

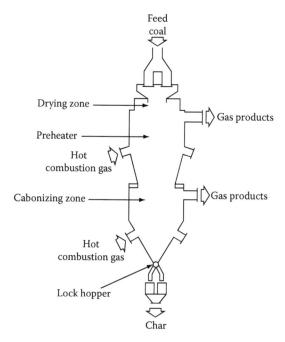

FIGURE 16.3 Lurgi-Spulgas retort.

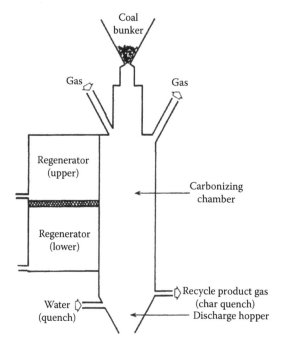

FIGURE 16.4 Koppers continuous steaming coke oven.

Coke making has seen various adaptations of conventional wood-charring methods to the production of coke with the eventual evolution of the beehive oven, which by the mid-nineteenth century had become the most common vessel for the coking of coal.

The beehive oven (Figure 16.5) is actually a simple brick structure into which coal can be introduced through an opening at the apex of the dome. Beehive ovens are charged as soon as practicable

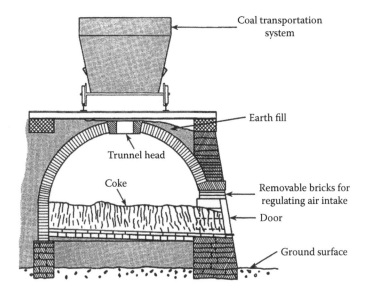

FIGURE 16.5   Beehive coke oven. (From Meyers, R.A., Ed., *Coal Handbook*, Marcel Dekker, Inc., New York, 1981.)

after drawing in order that stored-up heat from the previous charge will be sufficient to start the new coking charge. New ovens must be heated up gradually to the coking temperature by wood and coal fires, after which small charges of coal are coked until the ovens reach normal working conditions. With the oven in readiness for charging, the door is partially bricked up and the charge of 5–12 ton (5–12 × 10³ kg) is dropped through the trunnel head from the car above, leaving the coal in a cone-shaped pile in the oven.

To secure uniform coking of the coal, the pile must be level so that the coal will lie in a bed of uniform depth over the entire bottom of the oven. This leveling may be done by machine or by hand. After leveling the coal, the door opening is then bricked up to about 1 1/2 in. (3.8 cm) of the top. Leveling of the coal can be achieved by means of a side entrance so that a bed (ca. 2 ft, 0.6 m, thick) can be produced.

The heat for the process is supplied by burning the volatile matter that was released from the coal and, consequently, the carbonization would progress from the top of the bed to the base of the bed and the coke was (at process completion) retrieved from the side of the oven. In this manner, approximately 5 ton (5 × 10³ kg) of coal could be charged to the oven to produce the coke over a 72 h period. The coking process begins very soon after leveling is completed, as the ovens retain enough heat in the brick of the walls and the foam backing to start liberation of the volatile matter from the coal. As more heat is absorbed by the coal charge, the temperature of the oven soon reaches the "kindling" (ignition) points of the combustible gases, which, in the presence of the air admitted to the oven, ignite with a slight explosion at first and then continue to burn quietly in the crown of the ovens or as small candle-like flames at the surface of the coking mass, thus supplying heat to continue the process.

Coking proceeds from the top of the coal downward, so that the coking time depends mainly upon the depth of the coal. The generation of gas thus rapidly approaches a maximum, which is maintained for a period, then declines to practically nothing. The burning of volatile matter during this period must be regulated by gradually closing up the opening at the top of the door for admission of air. This regulation is necessary to maintain the temperature at a maximum, and conserve coke, as an excess of air at the beginning of the coking period tends to cool the oven and later consumes the carbon of the coke. The yield is also reduced by improper leveling. If the coal is not of uniform depth in the oven, the thin areas coke burn through before the thick, and some of the coke of the thin sections is consumed by combustion while the coking of the thick portions is being

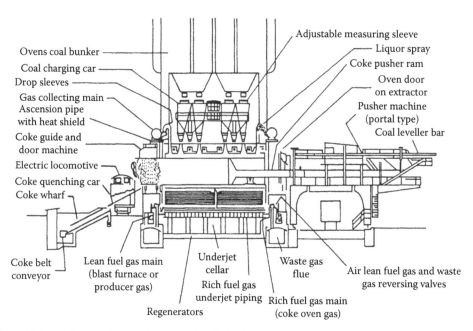

FIGURE 16.6 Coke oven battery (cross-sectional view).

completed. On the other hand, if the process is stopped when the thin areas have coked through, there will be a loss due to un-coked butts in the thick areas.

In the coking of bituminous coals in beehive ovens, coking proceeds downward from the top of the charge in which the coal, at increasing depths, passes through a plastic state as the temperature rises. This produces expansion and contraction of the charge, with the result that the cake is criss-crossed by a great number of irregular vertical fissures, thus giving it a long columnar structure. These very irregular columns extend from the top to the bottom of the cake. This structure affords a means by which beehive coke can be distinguished from by-product coke.

Some beehive ovens (with various improvements and additions of waste-heat boilers, thereby allowing heat recovery from the combustion products) may still be in operation but the beehive oven has generally been replaced by wall-heated horizontal chamber ("slot") ovens in which higher temperatures can be achieved as well as a better control over the quality of the coke. Modern slot-type coke ovens are approximately 50 ft (15 m) long, approximately 20 ft (6 m) high, with the width chosen to suit the carbonization behavior of the coal to be processed. For example, the most common widths are 18 in. (0.5 m) and 20 in. (0.6 m), but some ovens may be as narrow as 12 in. (0.3 m) while others are 22 in. (0.7 m) wide.

Several of these chambers (usually 20 or more, alternating with similar cells that contain heating ducts) are constructed in the form of a battery over a common firing system through which the hot combustion gas is conveyed to the ducts (Figure 16.6). The flat roof of the battery acts as the surface for a mobile (electrically powered) charging car from which the coal (25–40 ton; 25–40×10^3 kg) enters each oven through three openings along the top. The coke product is pushed from the rear of the oven through the opened front section onto a quenching platform or into rail cars that will move the coke through water sprays. The gas and tar by-products of the process are collected for further processing or for on-site use as fuel.

16.4 COAL TAR

Coal tar is the volatile material that is released during the thermal decomposition of coal and which condenses at room temperature. The tar may be composed of solid material (pitch) and liquid or semisolid materials (coal tar).

Thus, the carbonization of coal to produce *coal gas* for street and house lighting in the closing years of the eighteenth century produced substantial quantities of tar, which (during the following 50 years) was mostly discarded as a troublesome and unnecessary by-product (Weiss, 1940).

However, the development of a western European chemical industry (Chapter 24) brought increasing importance to coal tar as a source of the precursors that were to be used for the synthesis of dyes as well as raw materials for the production of solvents, pharmaceuticals, synthetic fibers, and plastics (Karr, 1963; Weiler, 1963; Aristoff et al., 1981; McNeil, 1981). Coal tar can also be upgraded to gasoline and other liquid fuels.

In fact, in the manner of crude petroleum, most high-temperature tars are first fractionated by distillation into (1) light oil, (2) middle (or tar acid) oil, and (3) heavy (or anthracene) oil. This primary separation is carried out by means of batch stills (vertical or horizontal; 3000–8000 US gallon [11–30 × 10^3 L] capacity) or by means of continuous "pipe" stills in which the tar is heated to a predetermined temperature before injection into a fractionating tower.

The light oil fraction (bp 220°C [430°F], c.f. petroleum naphtha bp 205°C [400°F]) consists mostly of benzene (45%–72% w/w), toluene (11%–19% w/w), xylene (3%–8% w/w), styrene (1%–1.5% w/w), and indene (1%–1.5% w/w) and is processed either into gasoline and aviation fuel components or is fractionated further to provide solvents and petrochemical feedstocks. In either case, upgrading involves removal of sulfur compounds, nitrogen compounds, and unsaturated materials. This is usually accomplished by acid-washing in batch agitators or by hydrogenation over a suitable catalyst (e.g., cobalt–molybdenum or nickel–tungsten on a support). Thus, in the acid wash, the crude material is mixed with strong sulfuric acid, neutralized (with ammoniacal liquor or caustic soda), and, after separation of the aqueous phase, steam-distilled or stripped of higher-molecular-weight material by centrifuging. The hydrogenation process conditions vary with the nature of the material to be removed (e.g., sulfur or hydrogenation of olefin material), but could typically be 300°C–400°C (570°F–750°F) and 500–1500 psi hydrogen.

The middle oil usually boils over the range 220°C–375°C (430°F–710°F) and after extraction of the tar acids, tar bases, and naphthalene can be processed to diesel fuel, kerosene, or creosote.

Coal tar creosote consists of aromatic hydrocarbons, anthracene, naphthalene, and phenanthrene derivatives. At least 75% of the coal tar creosote mixture is polycyclic aromatic hydrocarbons (PAHs). Unlike the coal tars and coal tar creosotes, coal tar pitch is a residue produced during the distillation of coal tar. The pitch is a shiny, dark brown to black residue that contains PAHs and their methyl and polymethyl derivatives, as well as heteronuclear compounds.

Coal tar pitch is the tar distillation residue produced during coking operations. The grade of pitch thus produced is dependent on distillation conditions, including time and temperature. The fraction consists primarily of condensed ring aromatics, including 2–6 ring systems, with minor amounts of phenolic compounds and aromatic nitrogen bases. The number of constituents in coal tar pitch is estimated to be in the thousands.

In this context, it should be noted that the tar acids, which are mostly phenol, cresols, and xylenols, can be recovered by mixing the crude middle oils with a dilute solution of caustic soda, separating the aqueous layer, and passing steam through it to remove residual hydrocarbons. The acids are then recovered by treatment of the aqueous extract with carbon dioxide or with dilute sulfuric acid and are then fractionated by distillation in vacuo.

Tar bases are isolated by treating the acid-free oil with dilute sulfuric acid and the bases are regenerated from the acid solution by addition of an excess of alkali (e.g., caustic soda or lime slurry). The mixture is then fractionated to produce pyridine, picoline, lutidine, aniline, quinoline, and isoquinoline.

The temperature to which the distillation of the heavy oil fraction is taken depends on the type of residue pitch (Nair, 1978) that is desired but usually lies within the range 450°C–550°C (840°F–1020°F). In all cases, the distillate is an excellent source of hydrocarbons such as anthracene, phenanthrene, acenaphthene, fluorene, and chrysene. The residual coal tar pitches are complex mixtures that contain several thousand compounds (mostly condensed aromatic compounds) and may, by analogy, be likened to the vacuum residua that are produced in a petroleum refinery and represent the materials in petroleum that have boiling points in excess of 565°C (1050°F).

One important aspect of coal tar chemistry relates to the presence and structure of the *nitrogen species* in the tar. Initial work on tar nitrogen chemistry indicated that the nitrogen structure of tar is similar to the nitrogen structures in the parent coal (Solomon and Colket, 1978). This was due to the belief that nitrogen in coal exists in tightly bound compounds and hence the most thermally stable structures in the coal during devolatilization. Furthermore, during the thermal reactions that led to tar formation these nitrogen compounds were released without rupture as part of the tar.

However, pyridine-type and pyrrole-type nitrogen functionalities were identified in the tar (Nelson et al., 1991; Kelemen et al., 1998). The data also indicated that the variety and complexity of the nitrogen-containing compounds of the tars increased with increasing rank. Additionally, it was found that the pyrrole/pyridine ratio of the nitrogen in coal tar increased with increasing coal rank.

The chemical structure of *coal char* has also been studied by several researchers (Solomon et al., 1990a,b; Ibarra et al., 1991; Pugmire et al., 1991; Fletcher et al., 1992). The results indicated that the char structure did not change significantly until temperatures exceeded approximately 330°C (625°F). At higher temperatures, the most significant changes to the char structure were those of the side chains and bridges of the coal. Low-rank coal, which typically contains relatively large amounts of oxygen-containing functional groups (mainly carboxyl and hydroxyl groups), demonstrated cross-linking reactions at lower temperatures. The decrease in tar formation during pyrolysis for low-rank coals as compared to bituminous coal was attributed to this low-temperature cross-linking behavior.

16.5   COKE

Coke is the solid carbonaceous residue that remains after all of the volatile matter and products have been driven off from coal during the carbonization process. The feedstock is coking coal, which is heated in absence of air in a coke oven without oxygen at temperatures as high as 1000°C (1830°F) so that the fixed carbon and residual ash are fused together.

Coke from coal is gray, hard, and porous and has a heating value of 24.8 million Btu/ton. Some coke-making processes produce valuable by-products that include coal tar, light oil, coal gas, and ammonia. Metallurgical coke is used as a fuel and as a reducing agent in smelting iron ore in a blast furnace. The product is cast iron and is too rich in dissolved carbon, and so must be treated further to make steel.

The coke must be strong enough to resist the weight of overburden in the blast furnace, which is why coking coal is so important in making steel by the conventional route. However, the alternative route to is direct reduced iron, where any carbonaceous fuel can be used to make sponge or pelletized iron.

On the other hand, petroleum coke is the solid residue obtained in petroleum refining, which resembles coal-based coke but often contains too many impurities to be useful in metallurgical applications (Speight, 2007).

Coking coals are hard-coking coals, medium-coking coals, and weak-coking coals. Coking coals are bituminous coals with mineral matter low ash and sulfur content and having good coking properties. The mean maximum reflectance and the crucible swelling number values as well as the Giesler fluidity of the coal decide the coking quality of coal (Chapters 8 and 9).

In the coking process, different varieties of coals are suitably blended and charged in the coke ovens. The blended coking coal is crushed to −3 mm and is charged into the coke oven chamber for carbonization either by gravity from top of the oven or as a stamped cake from the side of the oven. In by-product coke ovens, the coal is indirectly heated at 1200°C–1300°C (2190°F–2370°F) for 16–18 h to form coke that contains about 88% fixed carbon.

The coking process in these ovens also produces such by-products as coke oven gas; ammonia, which is converted to ammonium sulfate; coal tar, which can be distilled into useful secondary products like pitch, anthracene oil, naphthalene, etc.; and benzol for producing chemical products such as benzene, toluene, and xylene.

Coking coals can also be carbonized in heat recovery/energy recovery coke ovens. In these ovens, the volatiles evolved during coal carbonization are not recovered as by-products but are combusted completely in the presence of controlled quantity of air. The heat of the volatiles of evolving gases is utilized for coking the coal mass and thus no external heating is required. The heat consequent to combustion is only partially utilized in the process and the balance heat in the waste gases is gainfully utilized to produce steam and power.

The hot coke pushed from the coke ovens is quenched with either water in wet quenching or nitrogen gas in the case of dry quenching. In case of dry quenching, the heat picked by nitrogen gas during quenching is utilized for production of steam and power.

The coke used in the blast furnace should have a high carbon content, low mineral matter, and sulfur contents and should have an appropriate porosity as well as good strength to ensure that it gives good reactivity and does not pulverize to choke the gas flow in the blast furnace even at high temperatures.

Formed coke is a carbonized coal that has been *briquetted* and is made from weakly caking coals or even from noncaking coals that may have been rejected as coals suitable for coke production. The formed coke may also be used in place of lump coke in certain operations.

Formed coke has a greater mechanical strength than the domestic briquettes that are commonly used in many parts of the world as "smokeless" fuel and is actually made by a very similar process. The crushed coal is heated at temperatures in the range 600°C–800°C (1110°F–1470°F) after which the char is mixed thoroughly with a binder, formed into briquettes, and heated again but at temperatures in the range 900°C–1000°C (1650°F–830°F). The initial charring operation prevents swelling and/or sticking of the briquettes during the high-temperature treatment by virtue of the removal of volatile matter that could conceivably cause such a phenomenon.

Coke briquettes are manufactured from coals normally considered to be low quality for use as a main component in coking blends. They are manufactured for use as a domestic fuel (e.g., Phurnacite) or as metallurgical fuel (e.g., FMC, Bergbau-Lurgi, or Ancit processes), usually called "formed coke" in the latter case. To produce Phurnacite, a coal that is noncaking and has a low volatile matter content is briquetted with a binder and carbonized at 750°C–800°C (1380°F–1470°F) (Figure 16.7). Anthracite may be added and the process can be modified to process coals of higher volatile matter content or caking capacity after modification by oxidation or chemicals (e.g., iron oxide or lime) to prevent swelling and sticking together during carbonization. As a variation, low-volatile char prepared by pyrolysis of high-volatile coal may be briquetted with binder prepared from the pyrolysis products. Treatment at higher temperatures is necessary for the production of blast furnace coke; otherwise, the strength and volatile matter content may render the product only suitable for use as a domestic fuel or as a metallurgical reductant where strength and volatile matter content are of somewhat less importance.

In an alternative method of manufacture, the coal is carbonized at temperatures of the order of 400°C–700°C (750°F–1290°F). The actual temperature depends on the properties of the coal, but, nevertheless, a product will be obtained that can then be hot-briquetted with caking coal without any apparent adverse effects.

16.6 CHARCOAL

Charcoal is the solid carbon residue following the pyrolysis (carbonization or destructive distillation) of carbonaceous raw materials, such as coal and wood. Charcoal manufacture is also used in forest management for disposal of refuse.

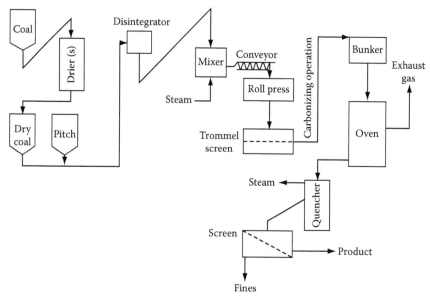

FIGURE 16.7 Phurnacite process. (From Pitt, G.J. and Millward, G.R., Eds., *Coal and Modern Coal Processing: An Introduction*, Academic Press Inc., New York, 1979.)

Charcoal manufacturing kilns generally can be classified as either batch or continuous multiple hearth kilns; continuous multiple hearth kilns are more commonly used than are batch kilns. Batch units such as the Missouri-type charcoal kiln (Figure 16.8) are small manually loaded and manually unloaded kilns producing typically 16–18 of charcoal during a 3 week cycle. Continuous units (Figure 16.9) produce up to 3 ton/h of charcoal.

During the manufacturing process, the feedstock is heated, driving off water and highly volatile organic compounds (VOCs). When the emission of the VOCs ceases, external application of

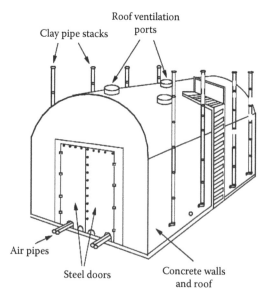

FIGURE 16.8 Missouri-type charcoal kiln.

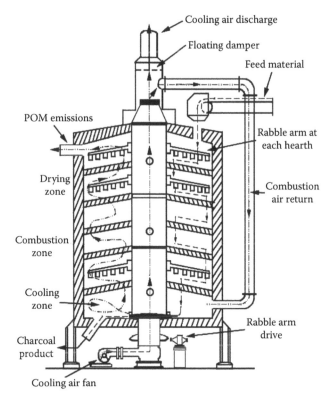

FIGURE 16.9 Continuous multi-heath kiln for charcoal production.

heat is no longer required because the carbonization reactions become exothermic. As the exothermic reactions decrease, heat is again applied to remove the less-volatile tarry materials from the product charcoal.

Fabrication of briquettes from raw material may be either an integral part of a charcoal-producing facility, or an independent operation, with charcoal being received as raw material.

REFERENCES

Agricola, G. 1546. *De Natura Fossilium*. H. Froben and N. Episcopius, Basle, Switzerland.

Aristoff, E., Rieve, R.W., and Shalit, H. 1981. In *Chemistry of Coal Utilization*, Second Supplementary Volume, M.A. Elliott (Ed.). John Wiley & Sons, Inc., New York, Chapter 16.

Armstrong, J. 1929. *Carbonization Technology and Engineering*. Lippincott Publishers, London, U.K.

ASTM. 2011. *Standard Test Methods for Coal and Coke*. American Society for Testing and Materials, West Conshohocken, PA.

Biringuccio, V. 1540. *De la Pirotechnia*. V. Roffinello for C. Navo, Venice, Book III, Chapter 10.

Cannon, C.G., Griffith, M., and Hirst, W. 1944. *Proceedings of Conference on the Ultrafine Structure of Coals and Cokes*. British Coal Utilization Research Association, Leatherhead, Surrey, p. 131.

Cha, C.Y., Merriam, N.W., Jha, M.C., and Breault, R.W. 1988. Report No. DOE/MC/24268-2700 (DE89000967). Morgantown Energy Technology Center, Morgantown, WV.

Dahme, A. 1953. *Ing. Arch.*, 21: 346.

Daly, T.A. and Budge, C.F. 1974. *Fuel*, 53: 8.

Davis, J.D. 1945. In *Chemistry of Coal Utilization*, H.H. Lowry (Ed.). John Wiley & Sons, Inc., New York, Chapter 22.

Davis, A., Hoover, D.S., Wakely, L.D., and Mitchell, G.D. 1983. *Journal of Microscopy*, 132: 315.

Davis, J.D. and Place, P.B. 1924. Thermal reactions of coal during carbonization. *Industrial & Engineering Chemistry*, 16(6): 589–592.

Eisenhut, W. 1981. In *Chemistry of Coal Utilization*, Second Supplementary Volume, M.A. Elliott (Ed.). John Wiley & Sons, Inc., New York, Chapter 14.

Fletcher, T.H., Solum, M.S., Grant, D.M., and Pugmire, R.J. 1992. *Energy & Fuels*, 6: 643–650.

Forbes, R.J. 1950. *Metallurgy in Antiquity*. Heinman Publishers, Leiden, the Netherlands.

Galloway, R.L. 1882. *A History of Coal Mining in Great Britain*. Macmillan & Co. London, U.K.

Gibson, J. 1979. In *Coal and Modern Coal Processing: An Introduction*, G.J. Pitt and G.R. Millward (Eds.). Academic Press, Inc., New York.

Gibson, J. and Gregory, D.H. 1971. *Carbonization of Coal*. Mills & Boon, London, U.K.

Grew, R.A. 1986. *Carbon*, 24: 677.

Gruit, A. and Marsh, H. 1981. *Fuel*, 61: 1105.

Hodek, W., Kraemer, M., and Juntgen, H. 1990. *Fuel*, 70: 424–428.

Hoover, H.C. and Hoover, L.H. 1950. *De Re Metallica*. Dover Publications, Inc., New York, p. 34.

Ibarra, J.V., Moliner, R., and Gavilan, M.P. 1991. *Fuel*, 70: 408–413.

Joseph, J.T. and Mahajan, O.P. 1991. In *Coal Science II*, H.H. Schobert, K.D. Bartle, and L.J. Lynch (Eds.). Symposium Series No. 461. American Chemical Society, Washington, DC, Chapter 23.

Karr, C. Jr. 1963. In *Chemistry of Coal Utilization*, Supplementary Volume, H.H. Lowry (Ed.). John Wiley & Sons, Inc., New York, Chapter 13.

Kelemen, S.R., Gorbaty, M.L., Kwiatek, P.J., Fletcher, T.H., Watt, M., Solum, M.S., and Pugmire, R.J. 1998. *Energy & Fuels*, 12: 159–173.

Kirov, N.Y. and Stephens, J.N. 1967. *Physical Aspects of Coal Carbonization*. National Coal Board Advisory Committee, Australia.

Ladner, W.R. 1988. *Fuel Processing Technology*, 20: 207.

Manka, D.P. 1979. In *Analytical Methods for Coal and Coal Products*, C. Karr, Jr. (Ed.). Academic Press, Inc., New York, Vol. III, Chapter 38.

Marsh, H. 1973. *Fuel*, 52: 205.

Marsh, H. and Cornford, C. 1976. In *Petroleum-Derived Carbons*, M.L. Devine and T.M. O'Grady (Eds.). Symposium Series No. 21. American Chemical Society, Washington, DC, p. 266.

Marsh, H. and Smith, J. 1978. In *Analytical Methods for Coal and Coal Products*, C. Karr, Jr. (Ed.). Academic Press, Inc., New York, Vol. II, Chapter 30.

McNeil, D. 1966. *Coal Carbonization Products*. Pergamon Press, London, U.K.

McNeil, D. 1981. In *Chemistry of Coal Utilization*, Second Supplementary Volume, M.A. Elliott (Ed.). John Wiley & Sons, Inc., New York, Chapter 17.

Merriam, N.W., Sethi, V., Thomas, K., and Grimes, R.W. 1992. *Proceedings of COAL PREP '92*, Cincinnati, OH, May 3–7. Also cited as Report No. DOE/MC/11076-92/C0041 (DE92019176). Morgantown Energy Technology Center, Morgantown, WV.

Meyers, R.A. (Ed.). 1981. *Coal Handbook*. Marcel Dekker, Inc., New York.

Miller, L.D., Bouman, R.W., and Chang, M.C. 1965. *Blast Furnace Steel Plant*, 53(5): 381.

Mochida, I., Korai, Y., H. Fujitsu, H., Takeshita, K., Komatsubara, K., and Koba, K. 1982. *Fuel*, 61: 1083.

Nair, C.S.B. 1978. In *Analytical Methods for Coal and Coal Products*, C. Karr, Jr. (Ed.). Academic Press, Inc., New York, Vol. II, Chapter 33.

Nef, J.U. 1957. In *A History of Technology*, C. Singer, E.J. Holmyard, A.R. Hall, and T.I. Williams (Eds.). Clarendon Press, Oxford, England, Vol. III, Chapter 3.

Nelson, P.F., Kelly, M.D., and Wornat, M.J. 1991. *Fuel*, 70: 403.

Patrick J.W. and Wilkinson, H.C. 1978. In *Analytical Methods for Coal and Coal Products*, C. Karr, Jr. (Ed.). Academic Press, Inc., New York, Vol. II, Chapter 29.

Pitt, G.J. and Millward, G.R. (Eds.). 1979. *Coal and Modern Coal Processing: An Introduction*, Academic Press Inc., New York.

Porter, H.C. 1924. *Chemical & Engineering News*, 2(15): 12.

Poutsma, M.L. 1987. A review of the thermolysis of model compounds relevant to coal processing. Report No. ORNL/TM10637, DE88 003690. Oak Ridge National Laboratory, Oak Ridge, TN.

Pugmire, R.J., Solum, M.S., Grant, D.M., Richfield, S., and Fletcher, T.H. 1991. *Fuel*, 70: 414.

Rao, Y.K. and Jalon, B.P. 1972. *Metallurgical Transactions*, 3: 2465.

Royce, A.J., Readyhough, P.J., and Silveston, P.L. 1991. In *Coal Science II*, H.H. Schobert, K.D. Bartle, and L.J. Lynch (Eds.). Symposium Series No. 461. American Chemical Society, Washington, DC, Chapter 24.

Schapiro, N. and Gray, R.J. 1963. *Blast Furnace Steel Plant*, 51(4): 273.

Seglin, L. and Bresler, S.A. 1981. In *Chemistry of Coal Utilization*, Second Supplementary Volume, M.A. Elliott (Ed.). John Wiley & Sons, Inc., New York, Chapter 13.

Solomon, P.R. 1981. Relation between coal aromatic carbon concentration and proximate analysis fixed carbon. *Fuel*, 60: 3.

Solomon, P.R. and Colket, M.B. 1978. *Fuel*, 57: 749.

Solomon, P.R., Hamblen, D.G., Carangelo, R.M., Serio, M.A., and Deshpande, G.V. 1988. *Energy & Fuels*, 2: 405–422.

Solomon, P.R., Hamblen, D.G., Serio, M.A., Yu, Z.-Z., and Charpenay, S. 1993. *Fuel*, 72: 469.

Solomon, P.R., Serio, M.A., Carangelo, R.M., and Bassilakis, R. 1990b. *Energy & Fuels*, 4: 319–333.

Solomon, P.R., Serio, M.A., Despande, G.V., and Kroo, E. 1990a. *Energy & Fuels*, 4: 42–54.

Solomon, P.R., Serio, M.A., and Suuberg, E.M. 1992. *Progress in Energy and Combustion Science*, 18(2): 133.

Speight, J.G. 2007. *The Chemistry and Technology of Petroleum*, 4th edn. Taylor & Francis Group, Boca Raton, FL.

Stein, S.E. 1981. In *New Approaches in Coal Chemistry*, B.D. Blaustein, B.C. Bockrath, and S. Friedman (Eds.). American Chemical Society, Washington, DC, Chapter 7.

Suuberg, E.M., Peters, W.A., and Howard, J.B. 1978. *Proceedings of the 17th Internatiuonal Symposium on Combustion*. The Combustion Institute, Pittsburgh, PA, pp. 117–130.

Taylor, G.H. 1961. *Fuel*, 40: 465.

Thompson R.R., Mantione, A.F., and Aikman, R.P. 1971. *Blast Furnace Steel Plant*, 59(3): 161.

Tucker, S.P. 1979. In *Analytical Methods for Coal and Coal Products*, C. Karr, Jr. (Ed.). Academic Press, Inc., New York, Vol. III, Chapter 43.

Wanzl, W. 1988. *Fuel Processing Technology*, 20: 317.

Weiler, J.F. 1963. In *Chemistry of Coal Utilization*, Supplementary Volume, H.H. Lowry (Ed.). John Wiley & Sons, Inc., New York, Chapter 14.

Weiss, J.M. 1940. By-products of coal carbonization. *Industrial & Engineering Chemistry*, 32: 1161–1162.

Wilson, P.J. Jr. and Clendenin, J.D. 1963. In *Chemistry of Coal Utilization*, Supplementary Volume, H.H. Lowry (Ed.). John Wiley & Sons, Inc., New York, Chapter 10.

Wilson, P.J. Jr. and Wells, J.H. 1950. *Coal, Coke, and Coal Chemicals*. McGraw-Hill, Inc., New York.

Xu, W.-C. and Tomita, A. 1987. In *Proceedings of the International Conference on Coal Science*, J.A. Moulijn (Ed.). Elsevier, Amsterdam, the Netherlands, p. 629.

Zimmerman, R.E. 1979. *Evaluating and Testing the Coking Properties of Coal*. Miller Freeman, San Francisco, CA, 1979.

17 Briquetting and Pelletizing

17.1 INTRODUCTION

Carbonization is the destructive distillation of coal in the absence of air accompanied by the production of carbon and liquid and gaseous products (Chapter 16). Originally, the concept behind the carbonization was to produce coke and it is from this technology that the industrial processes of briquetting and pelletizing arose.

The objective of briquetting is to convert a low-grade fuel (such as coal fines) into a higher-quality fuel. The briquetting of coal fines and char fines from coal and lignite, coke breeze, charcoal fines, and similar materials is important for producing fuel for a number of uses. Formed coke made from noncaking coal for use in a blast furnace or in a cupola furnace involves the briquetting of treated char fines and further processing of the briquettes by means of carbonizing and curing.

Generally, *briquetting* is a process where some type of material is compressed under high pressure. During the compression of the material, temperatures can rise sufficiently to make the raw material liberate various adhesives. However, to make this process successful, the moisture content of the raw material must be minimum 6% w/w. The high temperature also causes the moisture in the raw material to evaporate. At high moisture content, steam pockets may build up during this process, thus leading to expansion that will demolish the briquette.

On the other hand, *pelletizing* is the process of compressing or molding a material into the shape of a pellet. In addition to coal and carbonaceous fuels, a wide range of different materials are pelletized including chemicals, iron ore, and animal feed. The process of pelletizing combines mixing of the raw material, forming the pellet, and a thermal treatment baking the soft raw pellet to hard spheres. The raw material is rolled into a ball and then fired in a kiln to sinter the particles into a hard sphere.

The production of smokeless fuel briquettes, both for domestic and industrial use, from bituminous coal using a binder is an old art and there are several commercial processes available (Franke, 1930; Haake and Meyer, 1930; Rhys Jones, 1963; Schinzel, 1981; Perlack et al., 1986). But the increasing use of petroleum, gas, and electricity for heating purposes and the reduction in the number of individual heating plants have caused a reduction in the use of coal briquettes. In spite of this, and contradictory as it may seem, the increased use of petroleum has been of some value to the briquetting industry. During the last 20 years the intense competition from petroleum products, as well as an increase in demand by consumers for briquette quality, led to automation of the production of briquettes and a continuous improvement in quality.

However, because of the economic and environmental issues arising from the use of coal, and other fossil fuels, recent trends have been toward the production of added-value products from coal. This is especially true of those coals (e.g., Wyoming coal), which, although containing low amounts of sulfur (usually 0.5% w/w, daf), contain substantial amounts of mineral matter (ca. 10% w/w) and water (ca. 25% w/w) (Merriam and Sethi, 1992; Merriam et al., 1992). Shipping costs for the water and mineral matter (as part of the coal) diminish the value of the mined material. Therefore, there is a strong incentive to produce materials that are low in mineral matter and water, thereby increasing the heat content and decreasing the shipping costs on a unit weight basis (Btu/lb, kJ/kg) of the material shipped.

Whilst the term *added value* might be new, the premise of creating such materials from coal is not. And many of the new initiatives are based on the older briquetting and pelletizing technologies. For example, the introduction of processes that produce a clean coal insofar as the product is a carbonaceous pellet and contains little if any impurities. The production of coal briquettes is closely related to the carbonization process by which fuel-grade coke is produced (e.g., see Reerink, 1956; Rhys Jones, 1963). In fact, briquettes are manufactured from coals that are normally considered to be of insufficient quality to produce coke. Thus, a noncaking coal having a low content of volatile matter might be "briquetted" with a binder at 750°C–800°C (1380°F–1470°F).

The current emphasis on the use of coal is the production of added-value products using a variety of processes by which the total coal can be used with very little wastage of the products. Such is the character of the mild gasification process (Chapter 22) and a variety of other clean coal processes (United States Department of Energy, 1993) (Chapter 22), which have brought about renewed interest in briquetting and pelletization of coal.

By way of a brief summary, the mild gasification process is a modification of the conventional coal gasification (Cha et al., 1988; Merriam and Jha, 1991) (Chapters 20 and 22) and produces gaseous, liquid, and solid products by heating coal in an oxygen-free reactor. The process is less a gasification process and more a pyrolysis process insofar as volatile matter is removed by thermal means to leave a carbonaceous char/residue. The char can be upgraded further to remove both ash and pyritic sulfur, mixed back with coal-derived liquids, and burned in both coal- and oil-fired boilers. The process is of interest since a slurry of the liquid fuel and the upgraded char has the potential of being a very versatile fuel that can be burned in both coal- and oil-fired boilers. If the char is upgraded to a high degree, even feedstock coal with a high sulfur content can be used without alternating heat rates or capacity factors.

17.2 GENERAL CONCEPTS

17.2.1 COAL BRIQUETTES

The production of briquettes (and/or pellets) from coal is a thermal decomposition process, akin to the carbonization process (Chapter 16). Therefore, at the risk of some repetition but for the sake of completeness, a repeat of the general principles of the carbonization process is warranted here.

Carbonization is essentially a thermal treatment process for the production of a carbonaceous residue (with the simultaneous removal of distillate) from a variety of organic substances (Wilson and Wells, 1950; McNeil, 1966; Gibson and Gregory, 1971):

$$C_{organic} \rightarrow C_{coke/char/carbon} + liquids + gases$$

The process may also be referred to as destructive distillation and has been applied to a range of organic materials, but more particularly to naturally occurring materials such as wood, sugar, and vegetable matter. The carbonaceous residue from the thermal decomposition of coal is usually referred to as *coke* (Chapters 13 and 16) (which is physically dissimilar from charcoal) and has the more familiar honeycomb-type structure.

The thermal decomposition of coal is a complex sequence of events (Stein, 1981; Solomon et al., 1992) (Chapters 13 and 16) that can be described in terms of several important physicochemical changes, such as the tendency of the coal to soften and flow when heated (Chapters 8 and 9) or the relationship to carbon type in the coal (Solomon, 1981). In fact, some coals become quite fluid at temperatures of the order of 400°C–500°C (750°F–930°F) and there is a considerable variation in the degree of maximum plasticity, the temperature of maximum plasticity, as well as the plasticity temperature range for various coals (Kirov and Stephens, 1967; Mochida et al., 1982; Royce et al., 1991). Indeed, significant changes also occur in the structure of the char during the various stages of devolatilization (Fletcher et al., 1992).

For a coal, the distribution of carbon, hydrogen, and oxygen in the carbonaceous product is decreased as the temperature of the carbonization is increased. In addition, the yield of gases increases with the carbonization temperature, while the yield of the solid (char, coke) product decreases.

The yields of tar and low-molecular-weight liquids are to some extent variable but are greatly dependent on the process parameters, especially temperature (Chapters 13 and 16), as well as the type of coal employed (Cannon et al., 1944; Davis, 1945; Poutsma, 1987; Ladner, 1988; Wanzl, 1988).

Thus, as coal is heated to the plastic stage, it is probable that some thermal "depolymerization" can occur with the release of liquids and gases. These changes (including the liberation of water, carbon dioxide, and traces of hydrocarbons) can occur at temperatures below 300°C (570°F) and are accelerated markedly when a temperature of 350°C (660°F) is reached. The maximum evolution of tarry material, hydrocarbons, and combustible gases usually occurs over the temperature range 450°C–500°C (840°F–930°F). Products such as hydrogen sulfide and ammonia are released at temperatures of 250°C–500°C (480°F–930°F) and at temperatures in excess of 500°C (930°F), respectively. At temperatures on the order of 550°C (1020°F), the visible changes to the coal are virtually complete and a semicoke residue has been formed. At temperatures in excess of 550°C (1020°F), the semicoke hardens and shrinks to form coke with the evolution of methane, hydrogen, and traces of carbon monoxide and carbon dioxide (Marsh and Cornford, 1976).

The nature of the physicochemical changes has been a major factor in determining the extent to which coal is carbonized, with the character of the liquid and gaseous by-products being of considerable importance. Thus, coal carbonization processes (Table 17.1) are generally regarded as "low temperature" when the temperature of the process does not exceed 700°C (1290°F) or as "high temperature" if the temperature of the process is at or in excess of 900°C (1650°F). Following this, the temperature of a "medium-temperature" carbonization process is easily deduced and is self-explanatory.

The concept of blending a charge originates from the fact that coals vary widely in their ability to make coke, and, hence, it is customary to blend coals before carbonization in order to produce the desired type of carbonaceous product (Gruit and Marsh, 1981). For example, for the manufacture of softer and more reactive cokes, coals of the lower ranks have been widely employed. In particular, the low-temperature carbonization processes have been used for the production of reactive domestic fuel (briquettes) and have favored the use of the less strongly coking coals.

As already noted, briquettes are manufactured from coals that are normally considered to be of insufficient quality to produce coke; a noncaking coal having a low content of volatile matter might be "briquetted" with a binder at 750°C–800°C (1380°F–1470°F). However, the process can be modified to produce briquettes from higher value (i.e., higher volatile matter content coals or caking coals) by oxidative or by chemical (e.g., iron oxide or lime) treatment to prevent swelling and caking during the thermal process. Alternatively, the *poor-quality* coal might undergo partial carbonization at temperatures in the range 400°C–700°C (750°F–1290°F) and then hot briquetted

TABLE 17.1

Classification by Temperature of the Various Carbonization Procedures

Carbonization Process	Final Temperature		Products	Process
	°C	°F		
Low temperature	500–700	930–1290	Reactive coke and high tar yield	Rexco (700°C) made in cylindrical vertical retorts. Coalite (650°C) made in vertical tubes
Medium temperature	700–900	1290–1650	Reactive coke with high gas yield, or domestic briquettes	Town gas and gas coke (obsolete). Phurnacite, low volatile steam coal, pitch-bound briquettes carbonized at 800°C
High temperature	900–1050	1650–1920	Hard, unreactive coal for metallurgical use	Foundry coke (900°C); blast furnace coke (950°C–1050°C)

using a caking coal as the binder (Gregory, 1956). Thus, coal blending is practiced to improve the physical and chemical properties of the carbonaceous product and to enable a proportion of inferior or marginal coking coals to be employed as feeds as well as to improve (and control) the yield of the carbonaceous product.

17.3 BRIQUETTING TECHNOLOGY

The process of making briquettes usually consists of crushing, screening, mixing (with binder or adhesive), and pressing. The details of each process vary widely. For example, the pressure employed in pressing can be from 70 to 5600 kg/cm and the temperature can be from room temperature to around 500°C depending on the property of raw coal and the usage of briquettes.

A briquetting plant (Figure 17.1) typically consists of the mixer-dryer with drive, the fan, and the gas ducts. From the mixing chamber, flue gas at 300°C–500°C (570°F–930°F) passes through the coal–binder mixture, suction being provided by the fan. After cooling, the material enters the roll presses from the mixer-dryer.

In a more detailed perspective (Figure 17.1), briquetting of bituminous coal with a binder requires several consecutive processing steps but more than anything else, water, which occurs in the briquetting process as surface moisture of the coal, as steam that is added in the steam-heated pug mill for heating the coal–binder mixture, and as condensate from the steam, has a major influence on the process. Water, amongst other factors, influences the wetting and adhesion between coal and bituminous binder and on the strength of the resultant briquette. Indeed, interfacial interactions between the coal and the binder are extremely important in determining the strength of the briquette (Clarke and Marsh, 1989b). Systematic investigations has shown that the moisture content of briquetting coal, before the addition of the bituminous binder, should be as low as possible preferably not exceeding 4% w/w. Coal with a higher moisture content than 4% w/w has been used successfully by adding surface-active materials such as tar or a similar bituminous binder.

The means by which the coal is mixed with the binder and the method of heating the mixture largely determines the quality of the final product. The dried coal leaves the dry coal bunker at a temperature of 30°C–80°C (85°F–175°F) depending on size of the bunker and the transportation distance. After the metered addition of the solid or liquid binder, the mixture is conveyed to a pug mill heated by superheated steam, where its temperature is raised to about 20°C–25°C (36°F–45°F) above the softening point of the binders.

After leaving the steam-heated pug mill, the hot briquetting material is transported to the presses, usually by means of screw conveyors, which often have additional built-in shearing devices. In addition to transporting the material, the conveyors also allow the feed material to loose water through

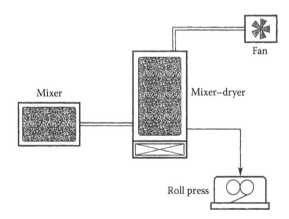

FIGURE 17.1 Fundamental stages in a briquetting operation.

evaporation as well as cooling of the hot material before the pressing. Failure to allow water loss can result in a poor-quality briquette and fracture of the product after exiting the press.

Double-roll presses and rotary table presses are available for briquetting bituminous coal. Different processes for controlling the briquetting process were derived by development of methods for measuring pressing pressure with double-roll presses. Normally double-roll presses have a roll diameter between 24 and 60 in. (600–1500 mm) and the compression ratio does not usually exceed 2.5:1. In general, the compression time for the briquette form is less than 0.05.

17.3.1 Briquetting Techniques

Briquetting requires a binder to be mixed with the charcoal fines and a press to form the mixture into a cake or briquette, which is then passed through a drying oven to cure or it is set by drying out the water so that the briquette is strong enough to be used in the same burning apparatus as normal lump charcoal.

Charcoal is a material totally lacking plasticity and hence needs addition of a sticking or agglomerating material to enable a briquette to be formed. The binder should preferably be combustible, though a noncombustible binder effective at low concentrations can be suitable. Starch is preferred as a binder though it is usually expensive. Highly plastic clays are suitable providing not more than about 15% is used. Tar and pitch from coal distillation or from charcoal retorts have been used for special purpose briquettes but they must be carbonized again before use to form a properly bonded briquette. They are of good quality but costly to produce.

The press for briquetting must be well designed, strongly built, and capable of agglomerating the mixture of charcoal and binder sufficiently for it to be handled through the curing or drying process. The output of briquettes must justify the capital and running costs of the machine. Briquetting machines for charcoal are usually costly precision machines capable of a high output. Brick-making presses have been used but there do not appear to be any commercially effective, really low cost machines for this purpose. Charcoal is quite abrasive so that equipment for screening fines, grinding, mixing with the binder and briquetting must be abrasive resistant and well designed.

The binders that have been tried are many but, as stated, the most common effective binder is starch. About 4%–8% of starch made into paste with hot water is adequate. First, the fines are dried and screened. Undersized fines are rejected and oversized hammer-milled. This powder is blended with the starch paste and fed to the briquetting press. The briquettes are dried in a continuous oven at about 80°C (175°F). The starch sets through loss of water, binding the charcoal into a briquette that can be handled and burned like ordinary lump charcoal in domestic stoves and grates. Generally briquettes are not suitable for use as industrial charcoal in blast furnaces and foundry cupolas, since the bond disintegrates on slight heating. For this, briquettes bonded with tar or pitch and subsequently carbonized in charcoal furnaces to produce a metallurgical charcoal briquette of adequate crushing strength are needed.

It is possible to add material to aid combustion of briquettes, such as waxes, sodium nitrate, and so on, during manufacture to give a more acceptable product. Also clay as a binder, silica, and so on can be mixed with the fines to reduce the cost of the briquette. This, of course, lowers the calorific value and is merely a form of adulteration for which the user pays, though claims may be made that burning is improved. But well-made briquettes are an acceptable, convenient product. The virtual absence of fines and dust and their uniformity are attractive for barbecue purposes. Generally, they sell at around the same cost per kilogram as lump charcoal in high price markets and have more or less the same calorific value as commercial charcoal of 10%–15% moisture content.

Successful briquette operations are found mostly in developed countries. An example is the industry based on carbonization of sawdust and bark in the southern United States using rotary multiple-hearth furnaces that produce perhaps 25–50 tons of fine charcoal per day. When briquetted, this charcoal, intended for barbecues, can be sold in retail outlets. The furnace gases are burned to produce steam of electric power, thus transforming the waste sawdust and bark into two useful products: electric power and charcoal briquettes. Air pollution and waste disposal problems are minimized at the same time.

17.3.2 TECHNOLOGY OVERVIEW

The coal briquette carbonization production process consists of a carbonization stage and a forming stage (Figure 17.2). In the carbonization stage, an internal-heating, low-temperature fluidized-bed carbonization furnace (approximately 450°C [840°F]) produces smokeless semicoke containing approximately 20% volatile matter. The carbonization furnace has a simple structure, with no perforated plates or agitator, making it easy to operate and maintain.

In the forming stage, the smokeless semicoke and auxiliary raw materials, hydrated lime and clay, are thoroughly mixed at a predetermined mixing ratio. After pulverizing, the mixture is blended with a caking additive while water is added to adjust the water content of the mixture. The mixture is kneaded to uniformly distribute the caking additive, and to increase the viscosity in order to make the forming of the briquettes easy. The mixture is then introduced into the molding machine to prepare the briquettes. The briquettes are dried and cooled.

17.3.2.1 Carbonization Stage

The raw coal (10% or lower surface water content, 5–50 mm particle size) is preliminarily dried in a rotary dryer. The gas exhausted from the dryer passes through a multi-cyclone to remove the dust before venting the gas to the atmosphere. The most efficient process for carbonizing semicoke and one that retains approximately 20% of the volatile matter in the semicoke is the internal-heating, low-temperature fluidized-bed carbonization furnace (Chapter 16).

The preliminarily dried raw coal is charged to the middle section of furnace, and is subjected to fluidization carbonization. The semicoke is discharged from the top of the furnace together with the carbonization gas. The semicoke separated from carbonization gas by the primary cyclone and the secondary cyclone.

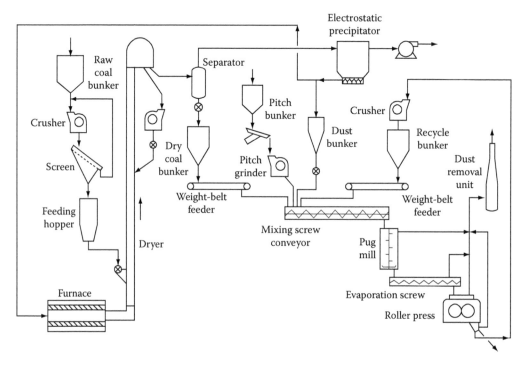

FIGURE 17.2 Details of a briquetting process.

After cooled, the semicoke is transferred to a stockyard, and the carbonization gas is supplied to the refractory-lined combustion furnace, where the carbonization gas is mixed with air to combust. The generated hot gas is injected into the raw coal dryer and to the succeeding briquette dryer to use as the drying heat source for the preliminary heating the raw coal and the drying heat source of the formed oval briquettes.

17.3.2.2   Forming Stage

The semicoke (Coalite) produced in the carbonization stage is the raw material for the briquette, containing adequate amounts of volatile matter, little ash, and sulfur and emitting no smoke or odor. The semicoke, as the primary raw material, is mixed with hydrated lime (sulfur-fixing agent), clay (to assist forming), and a caking additive.

To attain uniform composition and improved formability, the blended raw materials are fully kneaded. The forming of the mixture is carried out using a roll-molding machine at normal temperatures and under approximately 1000 kg/cm (300–500 kg/cm^2) of line pressure.

Since semicoke is used as a raw material for the briquettes, they are highly ignitable and can readily ignite in the furnace if the dryer is not operated at low temperatures. The dryer was designed with this temperature-sensitive concern in mind.

17.4   CHARCOAL BRIQUETTES

Charcoal is a desirable fuel because it produces a hot, long-lasting, virtually smokeless fire (Emrich, 1985). Combined with other materials and formed into uniform chunks called briquettes, it is popularly used for outdoor cooking in the United States.

Basic charcoal is produced by burning a carbon-rich material such as wood in a low-oxygen atmosphere. This process drives off the moisture and volatile gases that were present in the original fuel. The resulting charred material not only burns longer and more steadily than whole wood, but it is much lighter (one-fifth to one-third of its original weight).

Charcoal has been manufactured since prehistoric times. Around 5300 years ago, a hapless traveler perished in the Tyrolean Alps. Recently, when his body was recovered from a glacier, scientists found that he had been carrying a small box containing bits of charred wood wrapped in maple leaves. The man had no fire-starting tools such as flint with him, so it appears that he may have carried smoldering charcoal instead.

As much as 6000 years ago, charcoal was the preferred fuel for smelting copper. After the invention of the blast furnace around 1400 AD, charcoal was used extensively throughout Europe for iron smelting. By the eighteenth century, forest depletion led to a preference for coke (a coal-based form of charcoal) as an alternative fuel.

Plentiful forests in the eastern United States made charcoal a popular fuel, particularly for blacksmithing. It was also used in the western United States through the late 1800s for extracting silver from ore, for railroad fueling, and for residential and commercial heating.

Charcoal's transition from a heating and industrial fuel to a recreational cooking material took place around 1920 when Henry Ford invented the charcoal briquette. Not only did Ford succeed in making profitable use of the sawdust and scrap wood generated in his automobile factory, but his sideline business also encouraged recreational use of cars for picnic outings. Barbecue grills and Ford Charcoal were sold at the company's automobile dealerships, some of which devoted half of their space to the cooking supplies business.

Historically, charcoal was produced by piling wood in a cone-shaped mound and covering it with dirt, turf, or ashes, leaving air intake holes around the bottom of the pile and a chimney port at the top (Moscowitz, 1978). The wood was set afire and allowed to burn slowly; then the air holes were covered so the pile would cool slowly. In more modern times, the single-use charcoal pit was replaced by a stone, brick, or concrete kiln that would hold 25–75 cords of wood (1 cord = 4 ft × 4 ft × 8 ft). A large batch might burn for 3–4 weeks and take 7–10 days to cool.

This method of charcoal production generates a significant amount of smoke. In fact, changes in the color of the smoke signal transitions to different stages of the process. Initially, its whitish hue indicates the presence of steam, as water vapors are driven out of the wood. As other wood components such as resins and sugars burn, the smoke becomes yellowish. Finally, the smoke changes to a wispy blue, indicating that charring is complete; this is the appropriate time to smother the fire and let the kiln's contents cool.

Basic charcoal is produced by burning a carbon-rich material such as wood in a low-oxygen atmosphere. This process drives off the moisture and volatile gases that were present in the original fuel. The resulting charred material not only burns longer and more steadily than whole wood, but it is much lighter (one-fifth to one-third of its original weight).

An alternative method of producing charcoal was developed in the early 1900s by Orin Stafford, who then helped Henry Ford establish his briquette business. Called the retort method, this involves passing wood through a series of hearths or ovens. It is a continuous process wherein wood constantly enters one end of a furnace and charred material leaves the other; in contrast, the traditional kiln process burns wood in discrete batches. Virtually no visible smoke is emitted from a retort, because the constant level of output can effectively be treated with emission control devices such as afterburners.

17.4.1 RAW MATERIALS

Charcoal briquettes are made of two primary ingredients (comprising about 90% of the final product) and several minor ones. One of the primary ingredients, known as char, is basically the traditional charcoal, as described earlier. It is responsible for the briquette's ability to light easily and to produce the desired wood-smoke flavor. The most desirable raw material for this component is hardwood such as beech, birch, hard maple, hickory, and oak. Some manufacturers also use softwoods like pine, or other organic materials like fruit pits and nut shells.

The other primary ingredient, used to produce a high-temperature, long-lasting fire, is coal. Various types of coal may be used, ranging from subbituminous lignite to anthracite.

Minor ingredients include a binding agent (typically starch made from corn, milo, or wheat), an accelerant (such as nitrate), and an ash-whitening agent (such as lime) to let the backyard barbecue *aficionado* know when the briquettes are ready to cook over.

As a final note to this section, charcoal fines, as used for the production of charcoal briquettes, have a much lower purity than lump charcoal. The fines contain, in addition to charcoal, fragments, mineral sand, and clay picked up from the earth and the surface of the fuel wood and its bark. The fine-powdered charcoal produced from bark, twigs, and leaves has a higher ash content than normal wood charcoal. Most of this undesired high ash material can be separated by screening the fines and rejecting undersize material passing, say, a 2–4 mm screen. This fine material may still contain more than 50% charcoal depending on the level of contamination but, nevertheless, it is difficult to find uses for it. Material retained on the screen will be mostly fragments of good charcoal and, after hammer-milling, is suitable for briquetting. Fines cannot be burned by the usual simple charcoal burning methods and hence are more or less unsalable but if fines could be fully used, overall charcoal production would rise by 10%–20%.

17.4.2 MANUFACTURING PROCESS

The first step in the manufacturing process is to char the wood. Some manufacturers use the kiln (batch) method, while others use the retort (continuous) method (Moscowitz, 1978).

In the batch process, it typically takes a day or two to load a typical-size concrete kiln with about 50 cords of wood. When the fire is started, air intake ports and exhaust vents are fully open to draw in enough oxygen to produce a hot fire. During the week-long burning period, ports and vents are adjusted to maintain a temperature between about 450°C–510°C (840°F–950°F). At the end of the desired burning period, air intake ports are closed; exhaust vents are sealed an hour or two later, after smoking has stopped, to avoid pressure buildup within the kiln. Following a 2 week cooling period, the kiln is emptied, and the carbonized wood (char) is pulverized.

In the *continuous process*, wood is sized (broken into pieces of the proper dimension) in a hammer mill. A particle size of about 0.1 in. (3 mm) is common, although the exact size depends on the type of wood being used (e.g., bark, dry sawdust, wet wood). The wood then passes through a large drum dryer that reduces its moisture content by about half (to approximately 25%). Next, it is fed into the top of the multiple-hearth furnace (retort).

Externally, the retort looks like a steel silo, 40–50 ft (12.2–15.2 m) tall and 20–30 ft (6.1–9.14 m) in diameter. Inside, it contains a stack of hearths (3–6, depending on the desired production capacity). The top chamber is the lowest-temperature hearth, on the order of 275°C (525°F), while the bottom chamber burns at about 650°C (1200°F). External heat, from oil- or gas-fired burners, is needed only at the beginning and ending stages of the furnace; at the intermediate levels, the evolving wood gases burn and supply enough heat to maintain desired temperature levels.

Within each chamber, the wood is stirred by rabble arms extending out from a center shaft that runs vertically through the entire retort. This slow stirring process (1–2 rpm) ensures uniform combustion and moves the material through the retort. On alternate levels, the rabble arms push the burning wood either toward a hole around the central shaft or toward openings around the outer edge of the floor so the material can fall to the next lower level. As the smoldering char exits the final chamber, it is quenched with a cold-water spray. It may then be used immediately, or it may be stored in a silo until it is needed.

A typical retort can produce approximately 5500 lb (2.5 metric tons) of char per hour.

17.4.3 FEEDSTOCKS AND BRIQUETTE PRODUCTION

Typically, briquetting using charcoal requires a binder to be mixed with the charcoal fines and a press to form the mixture into a cake or briquette, which is then passed through a drying oven to cure or it is set by drying out the water so that the briquette is strong enough to be used in the same burning apparatus as normal lump charcoal.

Charcoal is a material totally lacking plasticity and hence needs addition of a sticking or agglomerating material to enable a briquette to be formed. The binder should preferably be combustible, though a noncombustible binder effective at low concentrations can be suitable. Starch is preferred as a binder though it is usually expensive. Highly plastic clays are suitable providing not more than about 15% is used. Tar and pitch from coal distillation or from charcoal retorts have been used for special purpose briquettes but they must be carbonized again before use to form a properly bonded briquette.

The press for briquetting must be well designed, strongly built, and capable of agglomerating the mixture of charcoal and binder sufficiently for it to be handled through the curing or drying process. Briquetting machines for charcoal are usually costly precision machines capable of a high output. Brick-making presses have been used but there do not appear to be any commercially effective, really low cost machines for this purpose.

The most common effective binder is starch—approximately 4%–8% w/w starch made into paste with hot water is adequate. First, the fines are dried and screened. Undersized fines are rejected and oversized hammer-milled. This powder is blended with the starch paste and fed to the briquetting press. The briquettes are dried in a continuous oven at about 80°C (175°F). The starch sets through loss of water, binding the charcoal into a briquette that can be handled and burned like ordinary lump charcoal in domestic stoves and grates. Generally briquettes are not suitable for use as industrial charcoal in blast furnaces and foundry cupolas, since the bond disintegrates on slight heating. For this, briquettes bonded with tar or pitch and subsequently carbonized in charcoal furnaces to produce a metallurgical charcoal briquette of adequate crushing strength are needed.

It is possible to add material to aid combustion of briquettes, such as waxes, sodium nitrate, and so on, during manufacture to give a more acceptable product. Also clay as a binder, silica, and so on can be mixed with the fines to reduce the cost of the briquette. This lowers the calorific value and is merely a form of adulteration for which the user pays, though claims may be made that burning is improved.

Lower grades of coal may also be carbonized for use in charcoal. Crushed coal is first dried and then heated to about 590°C (1100°F) to drive off the volatile components. After being air-cooled, it is stored until needed.

Charcoal and minor ingredients such as the starch binder are fed in the proper proportions into a paddle mixer, where they are thoroughly blended. At this point, the material has about a 35% moisture content, giving it a consistency somewhat like damp topsoil.

The blended material is dropped into a press consisting of two opposing rollers containing briquette-sized indentations. Because of the moisture content, the binding agent, the temperature (about 40°C or 105°F), and the pressure from the rollers, the briquettes hold their shape as they drop out the bottom of the press.

The briquettes drop onto a conveyor, which carries them through a single-pass dryer that heats them to about 135°C (275°F) for 3–4 h, reducing their moisture content to around 5%. Briquettes can be produced at a rate of 2,200–20,000 lb (1–9 metric tons) per hour. The briquettes are either bagged immediately or stored in silos to await the next scheduled packaging run.

17.4.4 By-Product Waste

During the late nineteenth and early twentieth centuries, recovery of acetic acid and methanol as by-products of the wood-charring process became so important that the charcoal itself essentially became the by-product. After the development of more-efficient and less-costly techniques for synthesizing acetic acid and methanol, charcoal production declined significantly until it was revitalized by the development of briquettes for recreational cooking.

The batch process for charring wood produces significant amounts of particulate-laden smoke. Fitting the exhaust vents with afterburners can reduce the emissions by as much as 85%, but because of the relatively high cost of the treatment, it is not commonly used.

Not only does the more constant level of operation of retorts make it easier to control their emissions with afterburners, but it allows for productive use of combustible off-gases. For example, these gases can be used to fuel wood dryers and briquette dryers, or to produce steam and electricity.

Charcoal briquette production is environmentally friendly in another way: the largest briquette manufacturer in the United States uses only waste products for its wood supply. Wood shavings, sawdust, and bark from pallet manufacturers, flooring manufacturers, and lumber mills are converted from piles of waste into useful briquettes.

17.5 BIO-BRIQUETTES

Bio-briquettes (biomass briquettes) are a biofuel substitute to coal and charcoal. They are used to heat industrial boilers in order to produce electricity from steam. The most common use of bio-briquettes is in the developing world, where energy sources are not as widely available. There has been a move to the use of briquettes in the developed world through the use of co-firing, when the briquettes are combined with coal in order to create the heat supplied to the boiler. This reduces carbon dioxide emissions by partially replacing coal used in power plants with materials that are already contained in the carbon cycle.

Co-firing relates to the combustion of two different types of materials. The process is primarily used to decrease carbon dioxide emissions despite the resulting lower energy efficiency and higher variable cost. The combination of materials usually contains a high-carbon-emitting substance such as coal and a lesser carbon dioxide–emitting material such as biomass. Even though carbon dioxide will still be emitted through the combustion of biomass, the net carbon emitted is nearly negligible. This is due to the fact that the materials gathered for the composition of the briquettes are still contained in the carbon cycle whereas fossil fuel combustion releases carbon dioxide that has been sequestered for millennia.

Boilers in power plants (Chapter 15) are traditionally heated by the combustion of coal, but if co-firing is implemented, the carbon dioxide emissions decrease while still maintaining the heat inputted to the boiler. Implementing co-firing requires few modifications to the current characteristics to power plants, as only the furl for the boiler is changed.

Co-firing is considered the most cost-efficient means of biomass. A higher combustion rate will occur when co-firing is implemented in a boiler when compared to burning only biomass. The compressed biomass is also much easier to transport since it is more dense, therefore allowing more biomass to be transported per shipment when compared to loose biomass. It is generally agreed that a near-term solution for the greenhouse gas emission problem may lie in co-firing (Prabir et al., 2011)

Manufacturers mainly use three methods to create the briquettes, each depending on the way the biomass is dried out. Although biomass briquettes are usually manufactured, biomass has been used throughout history all over the world from simply starting campfires to the mass generation of electricity.

17.5.1 PRODUCTION

Typically bio-briquettes are prepared by blending coal with 10%–25% w/w biomass, such as wood, bagasse (fibrous reside of processed sugar cane stalks), straw, and corn stalks. A desulfurizing agent, such as slaked lime [$Ca(OH)_2$], is also added in an amount corresponding to the sulfur content of the coal. Owing to the high-pressure briquetting (1–3 tons/cm^2), the coal particles and the fibrous biomass material in the bio-briquette strongly intertwine and adhere to each other. As a result, they do not separate from each other during combustion, and the low ignition temperature biomass simultaneously combusts with the coal. The combined combustion gives favorable ignition and fuel properties, emits little dust and soot, and generates sandy combustion ash, leaving no clinker. Furthermore, since the desulfurizing agent also adheres to the coal particles, the agent effectively reacts with the sulfur in the coal to fix about 60%–80% of the sulfur into the ash.

Many coal ranks can be used, including bituminous coal, subbituminous coal, and brown coal. In particular, the bio-briquettes produced with low-grade coal containing large amounts of ash and having low calorific value combust cleanly; thus the bio-briquette technology is an effective technology to produce clean fuel for household heaters and small industrial boilers.

In the shows the basics of the bio-briquette production process (Figure 17.3), the raw materials, coal and biomass, are pulverized to a size of approximately 3 mm or smaller and then dried. The dried mixture is further blended with a desulfurizing agent, such as slaked lime [$Ca(OH)_2$].

The mixture is formed by compression molding in a high-pressure briquetting machine. Powder coal may be utilized without being pulverized. A small amount of binder may be added to some coal ranks. The production process does not involve high temperatures, and is centered on a dry, high-pressure briquetting machine. The process has a simple flow, which is safe and which does not require skilled operating technique. Owing to the high-pressure briquetting process, the coal particles and the biomass strongly intertwine and adhere to each other; thus the process produces rigid formed coal, which does not separate during combustion.

Bio-briquette combustion decreases the generation of dust and soot to one-fifth to one-tenth that of direct coal combustion. Direct coal combustion increases the generation of dust and soot because the volatile matter released at low temperatures (200°C–400°C [390°F–750°F]) does not completely combust. To the contrary, bio-briquettes simultaneously combust the low ignition point biomass, which permeates the coal particles, assuring the combustion of volatile matter at low combustion temperatures. As a result, the amount of generated dust and soot is significantly reduced.

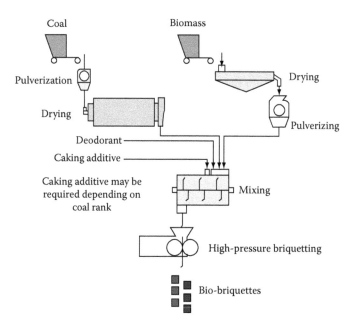

FIGURE 17.3 Bio-briquette production process.

Bio-briquettes prepared by blending biomass with coal have a significantly shorter ignition time. In addition, because of the low expansibility and caking property of bio-briquettes, sufficient air flow is maintained between the briquettes during continuous combustion such as in a fireplace. As a result, the bio-briquettes have superior combustion-sustaining properties and do not die out in a fireplace or other heater even when the air supply is decreased. This makes it easy to adjust the combustion rate.

Since fibrous biomass is intertwined with the coal particles, there is no fear of the fused ash in the coal adhering and forming clinker-lumps during combustion. Instead, the ash falls in a sandy form through the grate. Therefore, aeration is maintained to stabilize the combustion state. Furthermore, since no clinker is formed, the ash contains very small amounts of unburned coal.

The bio-briquettes are formed under high compressive force. Because of this, the desulfurizing agent and the coal particles strongly adhere to each other, and they effectively react during combustion. With the addition of a desulfurizing agent at a ratio of approximately 1.2:2 of Ca/S, 60%–80% of the sulfur in the coal is fixed in the ash.

17.5.2 Bio-Briquettes versus Coal Briquettes

The use of biomass briquettes has been steadily increasing as industries realize the benefits of decreasing pollution through the use of biomass briquettes. Briquettes provide higher calorific value per unit weight than coal when used for firing industrial boilers. Along with higher calorific value, biomass briquettes are claimed to save 30%–40% of boiler fuel cost. However, in the long run, briquettes can only limit the use of coal to a small extent, but it is increasingly being pursued by industries and factories all over the world. Both raw materials can be produced or mined domestically in the United States, creating a fuel source that is free from foreign dependence and less polluting than raw fossil fuel incineration.

Environmentally, the use of biomass briquettes produces much fewer greenhouse gases when compared to coal. Biomass briquettes are also fairly resistant to water degradation, an improvement over the difficulties encountered with the burning of wet coal. However, the briquettes are best used only as a supplement to coal.

The use of co-firing creates an energy that is not as high as pure coal, but emits fewer pollutants and cuts down on the release of previously sequestered carbon (Montross et al., 2010). The continuous release of carbon and other greenhouse gases into the atmosphere leads to an increase in global temperatures. The use of co-firing does not stop this process but decreases the relative emissions of coal power plants.

17.6 BRIQUETTE PROPERTIES

17.6.1 Influence of Coal Type

Coal briquettes can be a very dangerous fuel if they produce carbon monoxide while burning in a closed room but in normal combustion these conditions will not appear. If a chimney is provided to extract the smoke, briquettes are 100% safe even if the stove is installed in a bedroom.

If the coal used for briquette is rich in volatile matter, the coal has to be carbonized first. Carbonized briquettes are superior to uncarbonized briquettes for the same reasons that charcoal is preferred to fuel wood. The sulfur content is also reduced during carbonization as the sulfur is converted to hydrogen sulfide gas.

Coal carbonization can be carried out by a wide range of technologies and coal types, from one-person "village coal piles" to advanced technology plants that recover a series of by-products with few pollutants discharged but rather high investment. The scale of production for such an advanced plant varies from hundreds of tons to thousands of tons per day. The primitive technology plants practice limited or no by-product recovery. Where no by-product recovery is pursued, these plants are heavy polluters, but with merit of saving investment. To build an advanced technology plant may be too expensive for developing countries, especially when the briquette market is limited. So intermediate technology may be appropriate, where major pollutants can be recovered and the environmental damage is tolerable.

Feed materials for briquettes include coals of different properties: pretreated coals, low-temperature coke, high-temperature coke breeze, or mixtures of these. Various binders are used in manufacturing the briquettes. The binder may be obtained partly during the pretreatment of the coal or during the carbonization of the briquette. The thermal posttreatment of briquettes at different temperatures, with or without the influence of oxygen in the air, leads to briquettes that (depending on the feed material, the treatment method, and the treatment temperature) have properties more or less similar to those of coke.

If anthracite or low-volatile coals are used as feed materials, the comminuted coal usually mixed with an additional binder such as pitch or sulfite liquor. The briquettes undergo carbonization in one of several steps. The thermal posttreatment depends on the type of briquettes being treated and the desired properties of the final product. Heating rate and residence time in the hot zone are the deciding factors; however, the composition properties of the raw briquettes also influence the processing steps and the nature of the final product.

It is worth remembering here that, at high heating rate, many bituminous coals undergo expansion or shrinking during heating. These physical phenomena can lead to cracking or bursting of the briquette. In fact, there are "critical temperature intervals" in the process during which the heating rate must be decreased to reduce any stresses in the briquette structure. This will avoid cracking and can increase the strength of the briquette.

If the binder pitch softens, the briquettes may be deformed, or glued together as a result of the exuding of binder from the briquette surface. Such difficulties are avoided by an oxidative hardening pretreatment or by maintaining the briquettes in a quiescent state during the temperature interval used for binder softening.

17.6.2 INFLUENCE OF BINDER

17.6.2.1 Coal Tar

Coke-oven tar (or the solid pitch) has been, and continues to be, widely used as a binder in the production of bituminous coal briquettes. However, it is being replaced to an increasing extent by petroleum-derived products.

The composition of bituminous coal-tar and tar pitches is very complex and consists of a wide variety of components (Lang, 1966). But the coke-forming components of the pitch are of great importance for briquetting the currently preferred low-volatile coals, and especially for briquetting anthracite coals. These coals do not cake or melt while burning in the furnace; thus, the stability of the briquettes depends entirely on the skeleton of pitch coke formed in the briquette structure. Furthermore, the content of coke-forming components in the tar pitch increases as the rate of pitch distillation is decreased.

Pitch viscosity also exerts a strong influence on the process as do the coal–binder interactions (Clarke and Marsh, 1989a). During the briquetting process, the properties of the binder pitch are changed by heat, duration of the process, air, and steam. These factors can change pitch composition in the thin layer present in the heated briquetting matrix.

17.6.2.2 Petroleum Residua

Petroleum residua used in the briquetting of coals have often been given the collective name "bitumen" (asphalt), but this is now recognized as being incorrect terminology (Speight, 1991). Indeed, the term "native asphalt" that is often applied to bitumen is also incorrect. Bitumen is a high-viscosity material that occurs naturally in several areas of the world. Recovery may require thermal stimulation as a viscosity reducer or mining methods (Speight, 1990). Nevertheless, the terminology is still employed and must be recognized by researchers in the field.

Briefly, bitumen is a naturally occurring material having a viscosity and density of a certain order of magnitude whereas asphalt is a product of petroleum refining (Table 17.2) and the name is commonly designated as the products used in road construction, roofing, as well as others

---

TABLE 17.2
**Subdivision of Fossil Fuels and Derivatives
into Various Class Subgroups**

Natural Materials	Manufactured Materials	Derived Materials
Petroleum	Wax	Oils
Heavy oil	Residuum	Resins
Mineral wax	Asphalt[a]	Asphaltenes
Bitumen	Tar[b]	Carbenes
Asphaltite	Pitch[c]	Carboids
Asphaltoid	Coke	
Migrabitumen	Synthetic crude oil	
Bituminous rock		
Bituminous sand		
Kerogen		
Natural gas		
Coal		

[a] Asphalt: a product of petroleum refining; a material for highway construction.
[b] Tar: the "liquid" products from the thermal decomposition of coal.
[c] Pitch: the "solid" products from the thermal decomposition of coal.

(Speight, 1991, 1992). At first, only the materials developed for road building were used as binders but new requirements for binder quality have recently led to materials whose properties are significantly different from road-building bitumen.

The most important data used to characterize bitumen are the softening point (ASTM, 2011b), the penetration (ASTM, 2011a), the Conradson carbon residue (ASTM, 2011d), and the plasticity range (e.g., ASTM, 2011e). These data indicate the thermoplastic behavior of the binder and content of coke-forming components. These two properties are considerably different in bitumen and coal-tar pitch. For example, petroleum-derived materials have a lower content of coke-forming components than coal-tar pitch, and this may be a disadvantage when petroleum-derived materials are used as binders for noncaking coal briquettes, especially anthracites. The use of petroleum-derived materials having a lower-than-desirable propensity for coke formation, as indicated by a low Conradson carbon residue, produces briquettes that have a low stability during firing.

The use of propane asphalt as a briquette binder has also been investigated; propane asphalt is the propane-insoluble portion of petroleum residua (Speight, 1991). The asphalt has a lower penetration value than asphalts obtained by distillation or by oxidation. Propane asphalt has a relatively high temperature sensitivity, which may cause the briquettes to stick together. It is possible to alter the temperature sensitivity of the propane asphalt by the "conventional" methods of treatment that are used to alter asphalt properties for highway use (Speight, 1991, 1992).

17.6.2.3 Sulfite Liquor

Sulfite liquor is a by-product obtained during the treatment of wood with sulfites (calcium, sodium, magnesium, and ammonium) to produce cellulose. The liquor has significant adhesion properties, thereby fulfilling one of the binder requirements.

The sulfite liquor and the sulfite pitch remaining after water removal are water soluble. Briquettes made with sulfite liquor as binder can be made water resistant (referred to as "hardening the briquette") by heating to 220°C–350°C (430°F–660°F).

Slightly sintering or lightly caking coals can be briquetted with calcium sulfite liquor or with magnesium or sodium sulfite liquor as binder. Briquettes that are stable during firing can be made from anthracite coals only when using ammonium sulfite liquor. This low-ash binder produces a favorable coke skeleton in the briquette structure during combustion. The briquettes made from anthracite and an ammonium sulfite binder liquor are marketed under the name Extrazit.

17.6.2.4 Starch Materials

The adhesive poser of starch enables it to serve as a binder for the production of hard coal briquettes of sufficient strength. Starch at 1%–3% w/w of the coal may be added as a solid or suspension. Starch is used in the manufacture of charcoal briquettes, which are used in increasing amounts as barbeque briquettes.

The use of dry starch makes it possible to use briquetting coals with high moisture content, since 8% water is necessary to produce the starch suspension for a better quality briquette. Starch is not practically used as a binder in fuel production because it is very costly and lacks coke-forming components.

17.6.2.5 Lime

Although not always recognized as a typical binder, lime has also been used in briquetting. The production of briquettes with lime involves the conversion of calcium hydroxide to calcium carbonate, with carbon dioxide for hardening the briquettes.

The product is a fuel with low pollution potential since the desulfurization of the coal occurs as a result of the presence of the lime (Kural, 1994).

17.6.3 TYPES AND SIZES

The shape and the size of bituminous coal briquettes depend essentially on their application. Briquettes for industrial use are usually cubes or bricks weighing 1–4 lb (0.5–2.0 kg). The briquettes manufactured for domestic use must be small enough to be loaded conveniently through the appliance filling aperture but not small enough to drop through the grate bars or to give high resistance to air flow, even in deep beds.

17.6.4 MECHANICAL AND BURNING PROPERTIES

There are few, if any, recognized specifications or standard tests available for measuring the mechanical properties of bituminous coal briquettes. The properties currently measured are the breaking strengths and the abrasion or attrition of the briquettes.

The attrition testing of briquettes by means of the tumbler tests (ASTM, 2011c) permits the comparison of briquettes that have the same weight but different shapes (breaking or crushing strengths of briquettes that vary in shape cannot be compared). The test involves placing the briquette between two parallel plates and loading one plate until the briquette breaks. The larger the surface of contact between the testing plate and the briquette, the larger the point compressive strength.

Competition from petroleum derivatives (Speight, 1991) and the desire to produce environmentally acceptable household fuels from bituminous coal require that more and more attention be given to the burning properties of briquettes. Briquettes must have a high heating value: anthracite pitch briquettes average 13,300–13,900 Btu/lb.

17.7 PROCESSES

The processes available for the production of briquettes are many and, often, not too varied. It is not the purpose to give a comprehensive review of the various processes but to present those processes that are of interest because of historical uses or because of continued use on a commercial scale. In addition, processes normally classified as "hot briquetting" processes are also included.

17.7.1 ANTHRACINE PROCESS

In the anthracine process, fines are briquetted with a pitch binder and treated with air-enriched flue gas at 350°C–380°C (660°F–715°F) thereby undergoing an oxidative thermal treatment. The main product is fuel briquettes that evolve little smoke. The Coppee process is similar to the anthracine process but uses higher-volatile coals as feedstock.

17.7.2 THERMAL OXIDATION IN A SAND BED

The sand-bed oven is a closed, long kiln with a bottom consisting of screen plates. These screens support a bed of sand that is fluidized by intermittent blowing of steam, air, recycle gas, or inert gas. The briquettes in the kiln are heated by the hot sand and travel from the feed point to the extraction point where they pass the screen and are separated from the sand.

17.7.3 INIEX PROCESS

In this process, which was developed for use with briquettes made from anthracite and low-volatile coal, the pitch-bound briquettes are thermally treated in two stages (Langhoff, 1974). In the first stage, the green briquettes are carbonized in the sand-bed kiln at 550°C (1020°F) and, in the second stage, the carbonized briquettes are then fed into a Rushing gas kiln and are coked at 1000°C (1830°F).

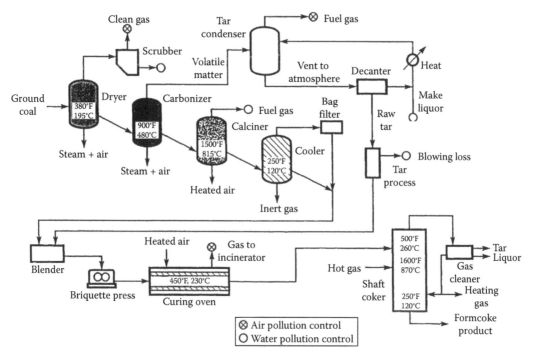

FIGURE 17.4 FMC form coke process.

17.7.4 FOUNDRY FORM COKE PROCESS

This process uses high-temperature coke particles (0.04 in. [1 mm]) as the feedstock. The coke is briquetted with coal-tar pitch and the green briquettes undergo thermal oxidation in a sand-bed kiln.

17.7.5 FMC PROCESS

The FMC process (Figure 17.4) is a multistage process for the manufacture of coke briquettes from high-volatile coals (Coal Age, 1960). In the process, the comminuted coal is oxidized, carbonized at low temperature, and calcined. On cooling, the low-temperature tar or an extraneous binder is used. Weakly caking coals, which are not suitable for the production of coke by the conventional process, can be converted by means of this process into a suitable blast furnace coke.

17.7.6 ANCIT PROCESS

In the Ancit process (Figure 17.5), two components, an inert noncaking component (such as low-volatile coal, anthracite, or coke fines) and a binder component consisting of caking coals, are employed. The main feed coal is comminuted wet by being heated to approximately 600°C (1110°F). The off-gas from this reactor heats the binder coal in another, smaller flow reactor. The two coals, comminuted at different temperatures, are mixed in a vertical, cylindrical mixer (at 460°C–520°C [860°F–970°F]). At this temperature, the binder coal is softened and the mixture is then briquetted on double-roll presses.

17.8 PELLETIZING

Pelletizing is the particle-enlargement process, in which tumbling of moist fines in drums, disks, or in conical devices leads to the formation of pellets. Physical forces, such as interfacial attraction, surface tension, van der Waals interactions, combined with the applied mechanical energy of

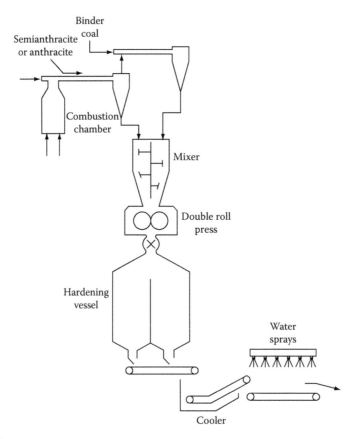

FIGURE 17.5   Ancit process.

tumbling, that bring particles together are responsible for pelletizing (Sastry and Fuerstenau, 1982; Holuszko and Laskowski, 2004).

The materials used for pelletizing fuels generally have a size consistency below 0.004 in. (0.1 mm); and the majority of the particles are smaller than 0.002 in. (0.060 mm). Addition of liquid on rotating disks or drums agglomerates the fuel to spherical shapes. The strength of the pellets depends on the water content of the agglomerated material and is greatest when the optimum action of the capillary forces of the water is obtained.

Coal pellets can be used immediately after drying or they may be carbonized for property enhancement.

17.8.1   Cold Pelletizing

The fines produced during coal preparation can be converted to larger lumps by pelletizing. In the process, some of the properties that are important for sintering such as reactivity and ignition property may change. The strength of coke pellet agglomerated with water only is small and varies with the size of the coke fines.

The carbonization of coke pellets is attracting increased attention and pelletizing of coal–ore mixtures, with an addition of lime if necessary, has received attention. The pellets are heated to 550°C–650°C (1020°F–1200°F) and subsequently carbonized with sand as heat carrier.

Another process for pelletizing caking coal fines and carbonizing the pellets involves the conversion of caking coal fines into pellets. In the second stage, the pellets are coated with hematite ore (Fe_2O_3) and the ore-covered pellets are then coked whereupon soften and sinter together to form a coherent coke. Addition of heavy oil to the coal increases the stability of the pellets in the oven. The amount of ferric oxide (Fe_2O_3) is critical, and a definite level exists that cannot be exceeded if the coke is to retain its polyhedron shape.

17.8.2 HOT PELLETIZING

A hot pelletizing process shapes the coal in the temperature range of the softening of caking coal, possibly in the presence of binder, using pelletizing equipment. An advantage is the possibility of mixing in larger amounts of binder than can be done in briquetting. The binder cokes at the temperatures that are used during pellet shaping and the possibility of desulfurizing the coke pellets during calcining is possible by addition of hydrogen under pressure.

REFERENCES

ASTM. 2011a. *Test Method for Penetration of Bituminous Materials* (ASTM D5). Annual Book of ASTM Standards, Sections 04.03 and 04.04. American Society for Testing and Materials, West Conshohocken, PA.

ASTM. 2011b. *Test Method for Softening Point of Bitumen Ring- and Ball Apparatus* (ASTM D36). Annual Book of ASTM Standards, Section 04.04. American Society for Testing and Materials, West Conshohocken, PA.

ASTM. 2011c. *Method for Tumbler Test for Coal* (ASTM D441). Annual Book of ASTM Standards, Section 05.05. American Society for Testing and Materials, West Conshohocken, PA.

ASTM. 2011d. *Test Method for Coking Value of Tar and Pitch (Modified Conradson)* (ASTM D2416). Annual Book of ASTM Standards, Section 04.04. American Society for Testing and Materials, West Conshohocken, PA.

ASTM. 2011e. *Test Method for the Plastic Properties of Coal by the Constant-Torque Gieseler Plastometer* (ASTM D2639). Annual Book of ASTM Standards, Section 05.05. American Society for Testing and Materials, West Conshohocken, PA.

Cannon, C.G., Griffith, M., and Hirst, W. 1944. *Proceedings of Conference on the Ultrafine Structure of Coals and Cokes*. British Coal Utilization Research Association, Leatherhead, Surrey, p. 131.

Cha, C.Y., Merriam, N.W., Jha, M.C., and Breault, R.W. 1988. Report No. DOE/MC/24268-2700 (DE89000967). Morgantown Energy Technology Center, Morgantown, WV.

Clarke, D.E. and Marsh, H. 1989a. *Fuel*, 68: 1023.

Clarke, D.E. and Marsh, H. 1989b. *Fuel*, 68: 1031.

Coal Age. 1960. 65(2): 26.

Davis, J.D. 1945. In *Chemistry of Coal Utilization*, H.H. Lowry (Ed.). John Wiley & Sons, Inc., New York, Chapter 22.

Emrich, W. 1985. *Handbook of Charcoal Making: The Traditional and Industrial Methods*. Kluwer Academic Publishers, Hingham, MA.

Fletcher, T.H., Solum, M.S., Grant, D.M., and Pugmire, R.J. 1992. *Energy & Fuels*, 6: 643.

Franke, G. 1930. *Handbuch der Brikettierung*. Enke AG, Stuttgart, Germany, Vol. 1, p. 2.

Gibson, J. and Gregory, D.H. 1971. *Carbonization of Coal*. Mills and Boon, London, U.K.

Gregory, D.H. 1956. *Proceedings of International Conference on Chemical Engineering in the Coal Industry*. Pergamon Press, Inc., London, U.K., p. 56.

Gruit, A. and Marsh, H. 1981. *Fuel*, 61: 1105.

Haake, A. and Meyer, A. 1930. *Handbuch der Brikettierung*. Enke AG, Stuttgart, Germany, Vol. 1, p. 57.

Holuszko, M.E. and Laskowski, J.S. 2004. Use of pelletization to assess the effect of particle-particle interactions on coal handleability. *Physicochemical Problems of Mineral Processing*, 38: 23–35.

Kirov, N.Y. and Stephens, J.N. 1967. *Physical Aspects of Coal Carbonization*. National Coal Board Advisory Committee, Australia.

Kural, O. 1994. Briquetting of coal. In *Coal: Resources, Properties, Utilization, Pollution*, O. Kural (Ed.). Istanbul Technical University, Istanbul, Turkey, Chapter 23.

Ladner, W.R. 1988. *Fuel Processing Technology*, 20: 207.

Lang, K.F. 1966. *Fortschr. Chem. Forsch.*, 7(1): 172.

Langhoff, J. 1974. *Aufbereitungstechnok*, (4): 181.

Marsh, H. and Cornford, C. 1976. *Petroleum-Derived Carbons*, M.L. Devine and T.M. O'Grady (Eds.). Symposium Series No. 21. American Chemical Society, Washington, DC, p. 266.

McNeil, D. 1966. *Coal Carbonization Products*. Pergamon Press, London, U.K.

Merriam, N.W. and Jha, M.C. 1991. Development of an advanced continuous mild gasification process for the production of co-products, Final Report. Report No. WRI-91-R068 (DE-AC21-87MC24268). Western Research Institute, Laramie, WY.

Merriam, N.W. and Sethi, V.K. 1992. Status of a process to produce a premium solid fuel from Powder River Basin coal. Report No. WRI 92-R056 (DE-FC21-86MC11076). Western Research Institute, Laramie, WY.

Merriam, N.W., Sethi, V., Thomas, K., and Grimes, R.W. 1992. *Proceedings of COAL PREP '92*. Cincinnati, OH, May 3–7. Also cited as Report No. DOE/MC/11076-92/C0041 (DE92019176). Morgantown Energy Technology Center, Morgantown, WV.

Mochida, I., Korai, Y., Fujitsu, H., Takeshita, K., Komatsubara, K., and Koba, K. 1982. *Fuel*, 61: 1083.

Montross, M., O'Daniel, B., Patil, D., Swoder, N., and Taulbee, D. 2010. Combustion of briquettes and fuel pellets prepared from blends of biomass and fine coal. *Proceedings of International Coal Preparation 2010 Conference*, R.Q. Honaker (Ed.). Society for Mining, Metallurgy, and Exploration (SME), Englewood, CO, pp. 161–170.

Moscowitz, C.M. 1978. *Source Assessment: Charcoal Manufacturing: State of the Art*. Environmental Protection Agency, Office of Research and Development, Industrial Environmental Research Laboratory, Cincinnati, OH.

Perlack, R.D., Stevenson, G.G., and Shelton, R.B. 1986. Prospects for coal briquettes as a substitute fuel for wood and charcoal in US Agency for international development assisted countries. Report No. ORNL/TM9770. Oak Ridge National Laboratory, Oak Ridge, TN.

Poutsma, M.L. 1987. A review of the thermolysis of model compounds relevant to coal processing. Report No. ORNL/TM10637, DE88 003690. Oak Ridge National Laboratory, Oak Ridge, TN.

Prabir, B., Butler, J., and Leon, M. 2011. Biomass co-firing options on the emission reduction and electricity generation costs in coal-fired power plants. *Renewable Energy*, 36: 282–288.

Reerink, W. 1956. *Proceedings of International Conference on Chemical Engineering in the Coal Industry*. Pergamon Press, Inc., London, U.K., p. 40.

Rhys Jones, D.C. 1963. *Chemistry of Coal Utilization*, Supplementary Volume. John Wiley & Sons, Inc., New York, Chapter 16.

Royce, A.J., Readyhough, P.J., and Silveston, P.L. 1991. In *Coal Science II*, H.H. Schobert, K.D. Bartle, and L.J. Lynch (Eds.). Symposium Series No. 461. American Chemical Society, Washington, DC, Chapter 24.

Sastry, K.V.S. and Fuerstenau, D.W. 1982, Pelletization of fine coals. Report No. EPRI CS-2198. Research Project 1030. University of California, Berkeley, California.

Schinzel, W. 1981. *Chemistry of Coal Utilization*, Second Supplementary Volume. John Wiley & Sons, Inc., New York, Chapter 11.

Solomon, P.R. 1981. Relation between coal aromatic carbon concentration and proximate analysis fixed carbon. *Fuel*, 60: 3.

Solomon, P.R., Serio, M.A., and Suuberg, E.M. 1992. *Progress in Energy Combustion and Science*, 18(2): 133.

Speight, J.G. 1990. In *Fuel Science and Technology Handbook*, J.G. Speight (Ed.). Marcel Dekker, Inc., New York, Chapter 12.

Speight, J.G. 1991. *The Chemistry and Technology of Petroleum*, 2nd edn. Marcel Dekker, Inc., New York.

Speight, J.G. 1992. *Kirk-Othmer Encyclopedia of Chemical Technology*, 4th edn., John Wiley & Sons Inc., New York, Vol. 3, p. 689.

Stein, S.E. 1981. In *New Approaches in Coal Chemistry*, B.D. Blaustein, B.C. Bockrath, and S. Friedman (Eds.). American Chemical Society, Washington, DC, Chapter 7.

United States Department of Energy. 1993. Clean coal technology demonstration program. DOE/FE-0272. United States Department of Energy, Washington, DC, February.

Wanzl, W. 1988. *Fuel Processing Technology*, 20: 317.

Wilson, P.J. and Wells, J.H. 1950. *Coal, Coke, and Coal Chemicals*. McGraw-Hill, Inc., New York.

18 Liquefaction

18.1 INTRODUCTION

Coal liquefaction is a process used to convert coal, a solid fuel, into a substitute for liquid fuels such as diesel and gasoline. Coal liquefaction has historically been used in countries without a secure supply of petroleum, such as Germany (during World War II) and South Africa (since the early 1970s). The technology used in coal liquefaction is quite old and was first implemented during the nineteenth century to provide gas for indoor lighting. Coal liquefaction may be used in the future to produce oil for transportation and heating, in case crude oil supplies are ever disrupted.

The production of liquid fuels from coal is not new and has received considerable attention (Berthelot, 1869; Batchelder, 1962; Stranges, 1983, 1987) since the concept does represent alternate pathways to liquid fuels (Figure 18.1) (Donath, 1963; Anderson and Tillman, 1979; Whitehurst et al., 1980; Gorin, 1981; Speight, 2011a). In fact, the concept is often cited as a viable option for alleviating projected shortages of liquid fuels as well as offering some measure of energy independence for those countries with vast resources of coal who are also net importers of crude oil.

The first commercial production of liquids from coal was obtained during carbonization primarily for the production of coke and gas early in the nineteenth century. Successful research on coal liquefaction by direct hydrogenation and indirect synthesis began in the early part of the twentieth century (Bergius, 1912, 1913, 1925; Graham and Skinner, 1929; Donath, 1963; Wu and Storch, 1968). This culminated in the production of approximately 100,000 bbl/day (15.9 × 10⁶ L/day) of liquid fuels for the German war effort in the early to mid-1940s (Extance, 2011).

While the current objectives are in essence the same, there is a marked attempt to reduce the operating conditions (ca. 400°C [ca. 750°F]; ca. 3000 psi hydrogen) that were prevalent in the early work. Furthermore, there are also serious efforts being made to reduce the hydrogen requirements (an ever expensive commodity) for the process by maximizing the use of hydrogen in the coal itself or by the use of a coal-derived solvent that is capable of donating hydrogen (in situ) to the process (Longanbach, 1981; Rudnick and Whitehurst, 1981).

The aim of this chapter is to review the physicochemical aspects of coal liquefaction and to outline the process options that have evolved during the last two decades. Of course, there are a variety of such options that have at one time or another been considered unsuitable as a result of the state, and flux, of this largely untried segment of coal technology. In spite of the interest in coal liquefaction processes that emerged during the 1970s and the 1980s, petroleum prices always remained (by design, of course) lower than the price of the liquid fuels from sources such as coal. Perhaps, it is time to rethink the price that we would be willing to pay for a measure of energy independence. On the assumption (hopefully not a futile assumption) that liquid fuels from coal will, one day, be a commercial reality and in order to present an overall view of coal liquefaction, the majority of the once-common and promising process options are included here (Chapter 19).

18.2 PHYSICOCHEMICAL ASPECTS

The thermal decomposition of coal on a commercial scale is an old art and is often more commonly referred to as carbonization. Coal decomposition to a mix of solid, liquid, and gaseous products is usually achieved by the use of temperatures up to 1500°C (2730°F) (Chapters 13 and 16).

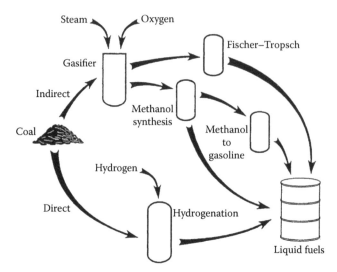

FIGURE 18.1 Routes for the production of liquid products from coal.

However, carbonization is a process for the production of a carbonaceous residue (coke) by the thermal decomposition (with simultaneous removal of distillate) of organic substances (Wilson and Wells, 1950; McNeil, 1966; Gibson and Gregory, 1971). The process, which is also referred to as destructive distillation, has been applied to a whole range of organic (carbon-containing) materials particularly natural products such as wood, sugar, and vegetable matter to produce charcoal. In this present context, the carbonaceous residue from the thermal decomposition of coal is usually referred to as "coke" (which is physically dissimilar from charcoal) and has the more familiar honeycomb-type structure. But, coal carbonization is not a process that has been designed for the production of liquids as the major products.

Carbonization is essentially a process for the production of a carbonaceous residue by thermal decomposition (with simultaneous removal of distillate) of organic substances:

$$C_{organic} \rightarrow C_{coke/char/carbon} + liquids + gases$$

The process may also be referred to as destructive distillation and has been applied to a whole range of organic materials, but more particularly to natural products such as wood, sugar, and vegetable matter to produce charcoal. Coal usually yields "coke," which is physically dissimilar from charcoal and has the more familiar honeycomb-type structure.

The thermal decomposition of coal is a complex sequence of events (Stein, 1981; Solomon et al., 1992, 1993) (Chapters 13 and 16) that can be described in terms of several important physicochemical changes, such as the tendency of the coal to soften and flow when heated (Chapters 8 and 9) or the relationship to carbon type in the coal (Solomon, 1981). In fact, some coals become quite fluid at temperatures of the order of 400°C–500°C (750°F–930°F) and there is a considerable variation in the degree of maximum plasticity, the temperature of maximum plasticity, as well as the plasticity temperature range for various coals (Kirov and Stephens, 1967; Mochida et al., 1982). Indeed, significant changes also occur in the structure of the char during the various stages of devolatilization (Fletcher et al., 1992).

Just as coal carbonization involves a complex sequence of chemical and physical events, the chemistry of coal liquefaction (although often represented by simple process schematics) (Figure 18.2) is also extremely complex, not so much from the model compound perspective but more from the interactions that can occur between the constituents of the coal liquids. Even though many schemes for the chemical sequences, which ultimately result in the production of liquids from coal, have been

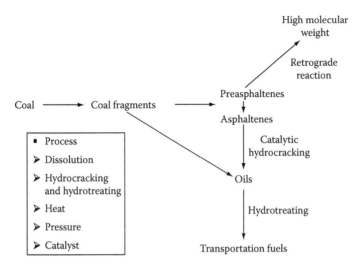

FIGURE 18.2 General chemistry of direct liquefaction.

formulated, the exact chemistry involved is still largely speculative, largely because the interactions of the constituents with each other are generally ignored. Indeed, the so-called structure of coal itself is still only speculative (Chapter 10).

The rates of the fragmentation reactions are very fast and depend on the nature of coals (Chakrabartty, 1981). The microcomponents, as identified by the microscopic examination of coal particles as *vitrinite* and *exinite* (Chapter 4), are the most reactive materials that undergo rapid fragmentation reactions. The other components, known as *semifusinite* and *fusinite* (Chapter 4), produce insignificant amounts of liquid products but may act as nuclei for coke formation, which might be deleterious to ultimate conversion.

When coal is heated in a slurrying vehicle, it is *liquefied* at 400°C to −500°C (750°F–930°F). Though the reaction mechanism involving conversion of coal to oil is very complex, it appears that the interaction of coal with solvent at the initial stage of the reactions plays the vital role to determine the sequential conversion of coal substances—first to a pyridine-soluble solid and thereafter to benzene-soluble liquid hydrocarbons and light oils (Chakrabartty, 1981). Thus, the isolation and identification of the products of coal–solvent interactions to yield pyridine-soluble matter may provide information regarding the suitability of the sample coal for liquefaction.

Therefore, it is essential that the liquefaction schemes postulated heretofore be treated with the utmost caution and looked upon for what they are, the most optimistic outline of what we believe current knowledge indicates to be the pathways involved in the conversion of coal to liquids.

The chemical objectives of coal liquefaction are (1) to reduce the effect of weak (i.e., van der Waals as well as hydrogen) bonds and, thus, separate fairly large units of coal structure into smaller units, (2) to bring about the decomposition of "key" aromatic–aliphatic, aromatic–aromatic, and a variety of carbon–heteroatom (i.e., nitrogen, oxygen, and sulfur) bonds within the coal to form smaller fragments, and, finally, (3) to increase the hydrogen/carbon atomic ratio from approximately 0.8 (equivalent to ca. 6% w/w hydrogen, daf) to produce a low-sulfur, ash-free liquid product (having H/C 1.7) that is comparable to crude oil, gasoline, or even heavy oil and bitumen.

However, it is obvious (Table 18.1) that the hydrogen content of the coal must be virtually doubled in order to achieve the ultimate goal of producing liquid fuels from the coal. Thus, as already noted, hydrogen can represent a major cost item of the liquefaction process and, accordingly, several process options have been designed to limit (or control) the hydrogen consumption or even to increase the hydrogen/carbon atomic ratio without the need for added gas-phase hydrogen.

As already noted, the exact chemistry of coal liquefaction is, at best, difficult to define but there is the suggestion that the first reaction step in the direct liquefaction of coal is the thermal rupture of

TABLE 18.1

Carbon and Hydrogen Content of Various Liquid Fuels and Bituminous Coal

Fuel	%C	%H	H/C
Gasoline	86.0	14.0	1.95
Crude oil	85.8	13.0	1.82
Lloyminster heavy oil (Alberta)	83.7	10.9	1.56
Oil sands bitumen (Athabasca deposit) (Alberta)	84.2	10.3	1.47
High-volatile C bituminous coal (Alberta)	77.7	4.9	0.76

the coal system to smaller units (Whitehurst et al., 1980; Durie, 1982; Coughlin and Davoudzadeh, 1983; Abichandani et al., 1984). In addition, coal liquefaction can also be represented as the addition of hydrogen to the structural entities in coal (either as molecular hydrogen and/or through the agency of a hydrogen donor) (Neavel, 1976; Squires, 1978; Hessley et al., 1986).

Thus, while various structures (e.g., Figure 18.3) have been postulated for the structure of coal—albeit with high degrees of uncertainty and very little certainty (Chapter 10)—the representation of coal as any one of these structures is extremely difficult and, hence, projecting a thermal decomposition route and the accompanying chemistry is even more precarious. However, it is apparent that the means of stabilization of the thermally produced fragments is abstraction of hydrogen for interaction with other species within the coal structure. An example of such stabilization is the abstraction of hydrogen from a hydroaromatic system within the coal.

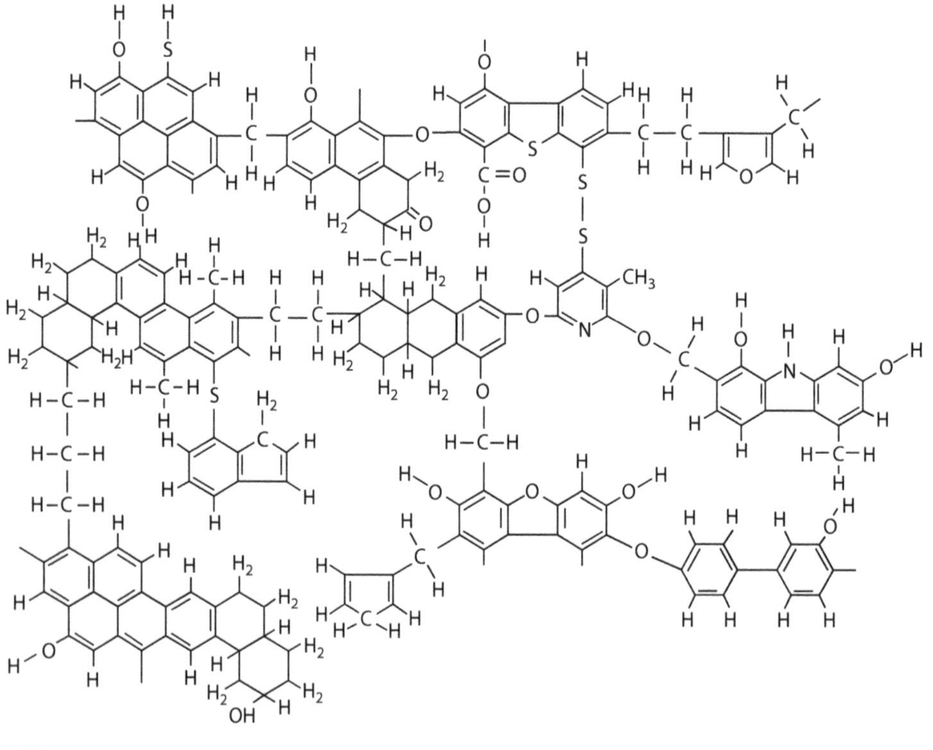

FIGURE 18.3 Hypothetical representation of the functional groups and internuclear bridges in coal. (From Cooper, B.R., Ed., *Scientific Problems of Coal Utilization*, United States Department of Energy, Washington, DC, 1978.)

When the limited hydrogen donor capacity (e.g., hydroaromatic hydrogen) of the coal is consumed, the free radicals (reactive species) continue to form and undergo recombination (often referred to, incorrectly, as "polymerization"), resulting in the formation of char (coke). The hydrogen present in the coal is thus redistributed by means of the hydrogen transfer process to form a volatile fraction that has a hydrogen/carbon atomic ratio of ca. 1.5:1.8 (as compared to a hydrogen/carbon atomic ratio of 0.8:0.9 in the original coal) and a hydrogen-deficient solid (char or coke).

In the presence of a solvent that has hydrogen-donating capabilities, the free radicals may abstract hydrogen from the solvent and, thereby, achieve stabilization (Gun et al., 1979). Thus, processes that produce char or coke are limited because the reactive intermediates lose their reactive capabilities and yield lower-molecular-weight species. Coal pretreatment also appears to have a beneficial effect on the liquefaction process (Rindt et al., 1992; Shams et al., 1992).

In addition, if molecular hydrogen and an appropriate catalyst are added to the system, chemisorption of the hydrogen on the catalyst can also yield a hydrogen species that can stabilize the reactive intermediates (Cusumano et al., 1978; Hessley et al., 1986). And the introduction of the catalyst and hydrogen can promote ring hydrogenation (where the ring may also be part of a condensed ring system) followed by carbon bond scission with ring opening, thereby reducing the size of the larger ring clusters.

It is also possible for molecular hydrogen itself to stabilize free radicals in the absence of a catalyst or donor solvent but at a reaction rate that is much slower and, therefore, which may compete unfavorably with the rate of recombination of the radicals.

As the temperature of the reaction is increased, the extent of bond scission (via thermal rupture of relatively labile bonds) also increases. This continued bond rupture ("depolymerization" of the coal) reduces the viscosity of the system through a reduction in the mean molecular size of the constituents but may also cause an increase in the amount of hydrogen consumed in the process by virtue of the increased production of gases.

As a further note on the chemistry of coal liquefaction, the thermal decomposition of aliphatic hydrocarbons, of the size that are desired in a liquid fuel product, is quite extensive in the range 450°C–550°C (840°F–1020°F) and aliphatic side chains on aromatic ring systems also decompose readily in this range. However, the aromatic ring systems themselves do not decompose (to nonring systems) readily in this temperature range and, in fact, any attempts to accomplish such a goal by the use of long residence times may only result in increased coke formation.

The majority of the coal liquefaction processes involve the addition of a coal-derived solvent to the coal prior to heating the coal to the desired process temperature. This is, essentially, a means of facilitating the transfer of the coal to a high-pressure region (usually the reactor) and also to diminish the sticking that might occur by virtue of the plastic properties of the coal (Chapter 9).

When the coal particles are surrounded by a liquid, several minutes are required to raise the temperature of the particles to that required for the reaction (i.e., 400°C–500°C [750°F–930°F]) and residence times in the preheater/reactor may be of the order of 0.25–1.0 h. The use of dry coal particles with hydrogen gas (in a turbulent flow condition) can shorten the heat-up time to a matter of several seconds (0.01 h).

The inorganic matter within the coal can have a detrimental effect on the process. In addition to poisoning catalysts, there is the potential for leaching harmful elements from residues and ashes after disposal and certain minerals cause abrasive and chemical wear as well as clogging and buildup in the reactors (Harris and Yust, 1977; Russell and Rimmer, 1979). But there is also the suggestion that the mineral matter in coal might have catalytic properties and act as an enhancement to the process (Mukherjee et al., 1972; Bockrath and Schroeder, 1981).

The inorganic constituents of coal may be of the order of 5%–15% w/w of the coal and can be composed of a wide variety of elements (Chapter 7), some of which are known to possess catalytic properties. Thus, while some coals of a particular rank do not appear to undergo any appreciable reaction in a hydrogen atmosphere in the absence of an added catalyst, there are those coals that will exhibit more reactivity, possibly because of the catalytic nature of the mineral matter (Mukherjee et al., 1972; Whitehurst et al., 1980; Bockrath and Schroeder, 1981; Davidson, 1983). In fact, disposable catalysts, such as the mineral constituents of coal, have been investigated with some degree of success (Narain et al., 1987).

If a material is to act as a catalyst for the direct reaction of molecular hydrogen with coal, the catalyst must bring the hydrogen and the reactive intermediate together at the precise moment, otherwise reaction will be difficult and may result in coke formation. However, since a coal particle and a catalyst particle (each with a distinct diameter, possibly of the order of several microns) may have some degree of difficulty in establishing the intimate contact necessary for efficient reaction, the coal must be reduced to a size (several angstroms; angstrom, $\text{Å} = 1 \times 10^{-10}$ m) that will enable it to enter the catalyst pore system; this can be achieved by "dissolving" the coal in a solvent prior to contact with the catalyst.

On the other hand, the catalyst may be reduced to a similar size (several angstroms) and allowed to enter the pore systems of the coal. This can also be achieved by treating the coal with a liquid solution of the catalyst or by allowing the catalyst to diffuse as a vapor onto the coal pore system; such systems are often referred to as dispersed catalyst systems (Hirschon and Wilson, 1991). The molecular hydrogen then diffuses to the reaction site and (in either of the earlier cases) the three components are in the desired intimate contact when the reaction temperature is attained. If such an operation is successful by virtue of the careful preparation of the two principal components (the coal and the catalyst), recombination of the reactive species will be minimized and coke formation will be virtually inhibited whereas the production of the desired smaller molecular species will be maximized.

The removal of the heteroatoms (i.e., nitrogen, oxygen, and sulfur) originally present in the coal represents an important facet of coal liquefaction technology. While the majority of the oxygen and sulfur in the coal are not located in ring systems, this is not the case for the nitrogen. Thus, in the temperature range necessary for the liquefaction process to proceed, the heteroatoms that are external to the ring systems may be removed, catalytically or noncatalytically, to produce water (in the case of oxygen) and hydrogen sulfide (in the case of sulfur).

Little, if any, nitrogen is removed as ammonia and in fact the removal of sulfur (as hydrogen sulfide) and nitrogen (as ammonia) requires catalytic hydrogenation of the ring system in which they may be located followed by rupture of the ring. In this respect, it has been noted that the less mature coals (lignite and subbituminous coals) are easier to hydrogenate than the more mature (bituminous) coals. This is believed to be related to the lesser degree of occurrence of condensed ring systems in the younger coals, which renders them more amenable to decomposition than the more mature coals. However, there are several facets of coal nature and behavior that must be given consideration when attempting to relate coal liquefaction behavior to coal "structure" and the overall process is doubtless much more complex than such a simple correlation would indicate (Given et al., 1975; Abdel-Baset et al., 1978; Dooley et al., 1979; Garr et al., 1979; Gates, 1979; Gray et al., 1980; Whitehurst et al., 1980; Yarzab et al., 1980; Gray and Shah, 1981; Baldwin, 1991; Snape et al., 1991).

Another aspect of coal behavior in relation to liquefaction that has also received (and is still receiving) some attention is the relationship of liquid yield to petrographic composition (Chapter 4) (Gagarin and Krichko, 1992). For example, vitrinite can be converted to liquid products readily as can exinite, but fusinite is quite resistant to liquefaction conditions and, thus, the petrographic composition of coals (whatever the rank) may be an important variable in determining the yields of liquid products.

Indeed, early work (Storch et al., 1951; Wu and Storch, 1968) on the production of liquids from coal indicated that the lower-rank coals (particularly the brown coals, i.e., lignites) were more reactive and required a lower hydrogen pressure for liquid production than the bituminous coals. More recent work (Sturm et al., 1981) has shown that the liquid from the lower-rank coals were easier to upgrade but that coal rank alone could not account for the complexity of the liquid products, as noted elsewhere content of the liquids produced decreased.

In terms of the physicochemical behavior of coal during the liquefaction process, some mention must also be made of the phenomenon of coal plasticity (Chapter 9). Plasticity is particularly evident in coals of the bituminous rank; during the plastic condition (which usually occurs in the temperature range 325°C–350°C [615°F–660°F]), the plastic mass has a tendency to adhere quite strongly to a variety of surfaces. Thus, reactor plugging could be a common result of the tendency of various coals to exhibit plastic behavior and there has been a considerable amount of effort directed to resolving this particular problem.

The actual physics and chemistry of coal plasticity are not well understood and most theories conclude that the plastic state is the result of thermal rupture of chemical bonds that are responsible for the orientation of the coal units with respect to one another as well as the degeneration of a variety of physical bonds. However, neither the anthracites nor the lignites and subbituminous coals display well-defined plastic behavior, but subjecting the lower-rank coals to liquefaction temperatures in the presence of hydrogen may often induce some degree of plasticity with the resultant problems.

Finally, some mention should also be made here of the process parameters insofar as maximizing the yield of liquid products requires a careful balance between temperature, pressure, heat-up time, and residence time of any particular coal particle in the reactor as well as the addition of an appropriate catalyst (Whitehurst et al., 1980). In general, a short residence in the reactor hot zone is more conducive to higher liquid yields, especially in the presence of a hydrogen-enriched solvent (Longanbach, 1981).

Under a specific set of process parameters, anthracite can be classed as "unreactive" while the high-rank bituminous coals require more severe conditions than the lower-rank coals. Thus, the lower-rank subbituminous coals and lignites produce overall lower yields and lower ratios of liquid/gases than the bituminous coals. Obviously, these rules are mere generalities and caution should be exercised in applying such rules on a universal basis.

18.3   PROCESS CLASSIFICATION

The process options for coal liquefaction (Figure 18.4) can generally be divided into four categories: (1) pyrolysis, (2) solvent extraction, (3) catalytic liquefaction, and (4) indirect liquefaction.

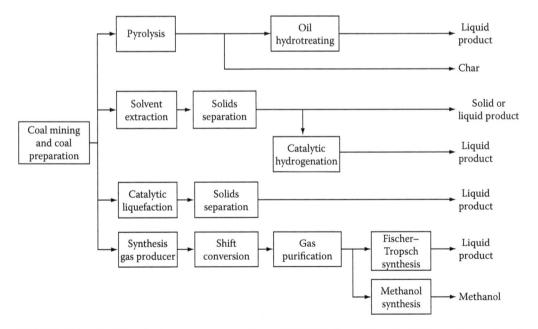

FIGURE 18.4   Coal liquefaction process types. (From US ERDA Report No. 76–67, Energy Research and Development Administration, Washington, DC, 1976.)

It is worthy of note here that a single-stage direct liquefaction process gives distillates through the use of a single primary reactor. Such processes may include an integrated hydrotreating reactor, which is intended to upgrade the primary distillates without directly increasing the overall conversion.

On the other hand, a two-stage direct liquefaction process is designed to give distillate products using two reactors in series. The primary function of the first-stage reactor is coal dissolution and is operated with or without a catalyst or with a low-activity disposable catalyst. The high-boiling products produced in the first stage are hydrotreated in the second stage in the presence of a high-activity catalyst to produce additional distillate.

The advances in refining petroleum residua (Speight, 2007, 2011b) and the ability to exert more control over the process may cause coal refiners to lean toward the second option.

18.3.1 Pyrolysis Processes

The first category of coal liquefaction processes, pyrolysis processes, involves heating coal to temperatures in excess of 400°C (750°F), which results in the conversion of the coal to gases, liquids, and char. The char is hydrogen deficient, thereby enabling inter- or intramolecular hydrogen transfer processes to be operative, resulting in relatively hydrogen-rich gases and liquids. Unfortunately, the char produced often amounts to more than 45% w/w of the feed coal and, therefore, such processes have often been considered to be uneconomical or inefficient use of the carbon in the coal.

Pyrolysis (or carbonization; Chapters 13 and 16) is perhaps the oldest technique for obtaining liquid products directly from coal and involves heating the coal in the absence of air (or oxygen) to produce heavy and light oils, gases, and char. In the presence of hydrogen, the process is called *hydrocarbonization*.

The composition and relative amounts of the products formed are dependent on process parameters such as heating rate, pressure, coal type coal (and product) residence time, coal particle size, and reactor configuration. A major disadvantage of this type of process is the large yields of char (Table 18.2) that markedly reduce the yield of liquid products.

On the other hand, pyrolysis and hydrocarbonization processes are relatively less complex than, say, liquid-phase hydrogenation processes. The operating pressures for pyrolysis processes are usually less than 100 psi (more often between 5 and 25 psi) but the hydrocarbonization processes require hydrogen pressures of the order of 300–1000 psi. In both categories of process, the operating temperature can be as high as 600°C (1110°F).

There are three types of pyrolysis reactors that are of interest: (1) a mechanically agitated reactor, (2) an entrained-flow reactor, and (3) a fluidized-bed reactor.

The agitated reactor (such as the one used for the Toscoal process) may be quite complex but the entrained-flow reactor (such as the one used for the Occidental process) has the advantage of either down-flow or up-flow operation and can provide short residence times. In addition, the coal can be heated rapidly, leading to higher yields of liquid (and gaseous) products that may well exceed the volatile matter content of the coal as determined by the appropriate test (Kimber and Gray, 1967) (Chapters 8 and 13). The short residence time also allows a high throughput of coal and the potential for small reactors.

Fluidized-bed reactors are employed in the COED, Clean Coke, and Coalcon processes and are reported to have been successful for processing noncaking coals but are not usually recommended for processing caking coals. Indeed, there is evidence that even weakly caking coals will agglomerate when heated at rates in excess of 75°C (135°F) in a fluidized-bed reactor that is operating at temperatures in excess of 500°C (930°F). Methods for the prevention of agglomeration have been suggested and include bed recirculation, feed preoxidation, staged preheating, feed dilution, and agitated-bed operations. In fact, the concept of an agitated-bed operation has been included into the Lurgi–Ruhrgas process.

TABLE 18.2

Summary of Selected Pyrolysis and Hydropyrolysis Processes

Process	Developer	Reactor Type	Reaction Temperature °C	Reaction Temperature °F	Reaction Pressure (psi)	Coal Residence Time	Yield (% w/w) Char	Yield (% w/w) Oil	Yield (% w/w) Gas
Lurgi–Ruhrgas	Lurgi–Ruhrgas	Mechanical mixer	450–600	840–1110	15	20 s	55–45	15–25	30
COED	FMC	Multiple fluidized bed	290–815	550–1500	20–25	1–4 h	60.7	20.1	15.1
Occidental coal pyrolysis	Occidental	Entrained flow	580	1075	15	2 s	56.7	35.0	6.6
Toscoal	Tosco	Kiln-type retort vessel	425–540	795–1005	15	5 min	80–90	5–10	5–10
Clean Coke	U.S. Steel	Fluidized bed	650–750	1200–1380	100–150	50 min	66.4	13.9	14.6
Union Carbide Corporation	Union carbide	Fluidized bed	565	1050	1000	5–11 min	38.4	29.0	16.2

Source: Braunstein, H.M. et al., Eds., *Environmental, Health, and Control Aspects of Coal Conversion*, Oak Ridge National Laboratory, Oak Ridge, TN, 1977, Vol. 1.

18.3.2 SOLVENT EXTRACTION PROCESSES

Solvent extraction processes generally utilize liquids (often derived from the feed coal) as "donor" solvents, which are capable of donating hydrogen to the system under the conditions of the reaction. The overall result is an increase (relative to pyrolysis processes) in the amount of coal that is converted to lower-molecular-weight, that is, soluble, products. Reaction temperatures usually have an upper limit of 510°C (950°F); hydrogen may be supplied (under pressure) during the process if it may be introduced by hydrogenation of the solvent prior to the process. Such hydrogen may be produced from any unreacted coal, from feed coal, or from by-product gases.

Solvent extraction processes are those processes in which coal is mixed with a solvent that is capable of effecting the transfer of hydrogen from the solvent to the coal (or even from gaseous hydrogen to the coal) at temperatures up to 500°C (930°F) and pressures up to 5000 psi.

High-temperature solvent extraction processes of coal have been developed in three different process configurations: (1) extraction in the absence of hydrogen but using a recycle solvent that has been hydrogenated in a separate process stage, (2) extraction in the presence of hydrogen with a recycle solvent that has not been previously hydrogenated, and (3) extraction in the presence of hydrogen using a hydrogenated recycle solvent (Table 18.3). In each of these concepts, the distillates of process-derived liquids have been used successfully as the recycle solvent that is recovered continuously in the process.

Solvent extraction can also be achieved under milder conditions but the product may be a high-nitrogen solid or a heavy oil with only low yields of light oils and gases. The more severe conditions are more effective for sulfur and nitrogen removal to produce a "lighter" liquid product that is more amenable to downstream processing.

A more novel aspect of the solvent extraction process type is the use of bitumen and/or heavy oil as process "solvents" and, in fact, the coprocessing of coal with a variety of petroleum-based

TABLE 18.3
Summary of Selected Solvent Extraction Processes

Process	Developer	Reactor	Temperature °C	Temperature °F	Pressure (psi)	Residence Time (h)
Consol synthetic fuel (CSF)	Conoco	Stirred tank	400	750	150–450	<1
Solvent-refined coal (SRC)	Pittsburgh and Midway Mining Co.	Plug flow	~450	840	1000–1500	<1
Solvent-refined lignite (SRL)	University of North Dakota	Plug flow	370–480	700–895	1000–3000	~1.4
Costeam	ERDA	Stirred tank	375–450	705–840	2000–4000	1–2
Exxon donor solvent (EDS)	Exxon Research and Engineering Co.	Plug flow	425–480	795–895	1500–2000	0.25–2.0

Source: Braunstein, H.M. et al., Eds., *Environmental, Health, and Control Aspects of Coal Conversion*, Oak Ridge National Laboratory, Oak Ridge, TN, 1977, Vol. 1.

feedstocks (e.g., heavy oils) has received much attention lately (Moschopedis et al., 1980, 1982; Speight and Moschopedis, 1986; Curtis et al., 1987; Schulman et al., 1988; Curtis and Hwang, 1992; Rosal et al., 1992). The concept is a variation of the solvent extraction process for coal liquefaction. Whether the coprocessing option is a means of producing more liquids or whether the coal should act as a scavenger for the metals and nitrogen species in the petroleum material is dependent upon the process conditions. What is certain is that special efforts should be made to ensure the compatibility of feedstocks and products. Incompatibility can, at any stage of the liquefaction operation, lead to expensive shutdowns as well as to (in respect of the heavier feedstocks) the onset of coke formation (Speight, 1992).

18.3.3 CATALYTIC LIQUEFACTION PROCESSES

The final category of direct liquefaction process employs the concept of catalytic liquefaction in which a suitable catalyst is used to add hydrogen to the coal. These processes usually require a liquid medium with the catalyst dispersed throughout or may even employ a fixed-bed reactor (see p. 394). On the other hand, the catalyst may also be dispersed within the coal whereupon the combined coal–catalyst system can be injected into the reactor.

Catalytic liquefaction processes are those processes (Table 18.4) in which the coal is brought into contact with a catalyst in the presence of hydrogen (Lytle et al., 1979; Buchanan et al., 1992). Many processes of this type have the advantage of eliminating the need for a hydrogen donor solvent (and the subsequent rehydrogenation of the spent solvent) but there is still the need for an adequate supply of hydrogen. The nature of the process also virtually guarantees that the catalyst will be deactivated by the mineral matter in the coal as well as by coke lay-down during the process.

In order to achieve the direct hydrogenation of the coal, the catalyst and the coal must be in intimate contact, but if this is not the case, process inefficiency is the general rule. On the other hand, there has been the tendency of late to achieve coal–catalyst contact by the use of a hydrogen donor solvent. Alternatively, a catalyst with a sufficiently high vapor pressure may be employed so that catalyst deposition on the coal surface) is achieved under process conditions.

Processes in which the coal and the catalyst are in intimate contact in the presence of hydrogen gas are often referred to as solid–gas catalytic hydrocarbonization or dry coal hydrogenation. The major features of these processes are (1) rapid heating to temperatures of the order of 450°C–600°C

TABLE 18.4
Summary of Selected Catalytic Liquefaction Processes

Process	Developer	Reactor	Catalyst	Temperature °C	°F	Pressure (psi)
(a) *Catalytic liquefaction processes*						
H-coal	Hydrocarbon Research, Inc.	Ebullated bed	Co-Mo/A1$_2$O$_3$	450	840	2250–3000
Synthoil	ERDA	Fixed bed	Co-Mo/Al$_2$O$_3$	450	840	2000–4000
CCL	Gulf	Fixed bed	Co-Mo/Al$_2$O$_3$	400	750	2000
Multistage	Lummus	Expanded bed	Co-Mo/Al$_2$O$_3$	400–430	750–805	1000
(b) *Catalytic hydrogenation processes*						
Bergius	Bergius	Plug flow	Iron oxide	480	895	3000–10000
University of Utah	University of Utah	Entrained flow	Zinc chloride, tin chloride	500–550	930–1020	1500–2500
Schroeder	Schroeder	Entrained flow	(NH$_4$)$_2$MoO$_4$	500	930	2000
Zinc chloride	Conoco	Liquid phase	Zinc chloride	360–440	680–825	1500–3500

Source: Braunstein, H.M. et al., Eds., *Environmental, Health, and Control Aspects of Coal Conversion*, Oak Ridge National Laboratory, Oak Ridge, TN, 1977, Vol. 1.

(840°F–1110°F), (2) short residence times, and (3) reactor effluent quenching. Instead of using a slurry-feed system, the coal is entrained in a rapidly moving stream of hydrogen and has a residence time of (usually) less than 20 s in the reactor at 500°C (930°F) and 2000 psi.

18.3.4 INDIRECT LIQUEFACTION PROCESSES

The other category of coal liquefaction processes invokes the concept of the indirect liquefaction of coal; that is, the coal is not converted directly into liquid products (Figure 18.5).

The indirect liquefaction of coal involves a two-stage conversion operation in which coal is first converted (by reaction with steam and oxygen) (Chapters 20 and 21) to produce a gaseous mixture that is composed primarily of carbon monoxide and hydrogen (syngas; synthesis gas).

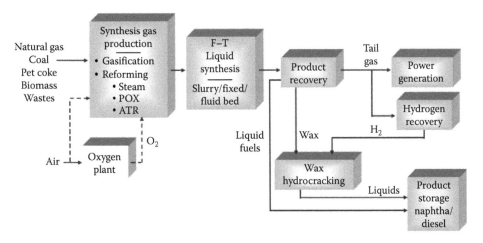

FIGURE 18.5 Indirect coal liquefaction with potential for coprocessing. (From Miller, C.L., Coal conversion—Pathway to alternate fuels, *EIA Energy Outlook Modeling and Data Conference*, Washington, DC, March 28, 2007.)

The gas stream is subsequently purified (to remove sulfur, nitrogen, and any particulate matter) after which it is catalytically converted to a mixture of liquid hydrocarbon products. In addition to the production of fuel gas, coal may be gasified to prepare synthesis gas (carbon monoxide, CO, and hydrogen, H_2), which may be used to produce a variety of products, including ammonia, methanol, and liquid hydrocarbon fuels (Lee, 1990).

The synthesis of hydrocarbons from carbon monoxide and hydrogen (synthesis gas) (the Fischer–Tropsch synthesis) is a procedure for the indirect liquefaction of coal (Storch et al., 1951; Batchelder, 1962; Dry et al., 1966; Dry and Ferrira, 1967, 1968; Dry, 1976; Anderson, 1984; Jones et al., 1992; Speight, 2008). This process is the only coal liquefaction scheme currently in use on a relatively large commercial scale; South Africa is currently using the Fischer–Tropsch process in their Sasol complex (Singh, 1981; Speight, 2008), although Germany produced roughly 156 million barrels of synthetic petroleum annually using the Fischer–Tropsch process during World War II (Stranges, 1983). Thus, the indirect liquefaction of coal involves (as a first stage) the production of a synthesis gas mixture from the coal (Chapters 20 and 21) followed by the catalytic formation of hydrocarbons and oxygenated species from the synthesis gas.

Coal is converted to gaseous products at temperatures in excess of 800°C (1470°F), and at moderate pressures, to produce synthesis gas:

$$C + H_2O \rightarrow CO + H_2$$

The gasification may be attained by means of any one of several processes (Chapters 20 and 21) or even by gasification of the coal in place (underground) or in situ gasification of coal (Chapters 20 and 21). The exothermic nature of the process and the decrease in the total gas volume in going from reactants to products suggest the most suitable experimental conditions to use in order to maximize product yields. The process should be favored by high pressure and relatively low reaction temperature. In practice, the Fischer–Tropsch reaction is carried out at room temperatures of 200°C–350°C (390°F–660°F) and at pressures of 75–4000 psi; the hydrogen–carbon monoxide ratio is usually at ca. 2.2:1 or 2.5:1 (Speight, 2008).

Since up to three volumes of hydrogen may be required to achieve the next stage of the liquids production, the synthesis gas must then be converted by means of the water–gas shift reaction) to the desired level of hydrogen after which the gaseous mix is purified (acid gas removal, etc.) and converted to a wide variety of hydrocarbons:

$$CO + H_2O \rightarrow CO_2 + H_2$$

$$CO + (2n + 1)H_2 \rightarrow C_nH_{2n+2} + nH_2O$$

These reactions result primarily in low- and medium-boiling aliphatic compounds; present commercial objectives are focused on the conditions that result in the production of n-hydrocarbons as well as olefin and oxygenated materials.

Although the efficiency of a two-stage process as compared with a single-stage liquefaction process has often been cited as a disadvantage of the indirect liquefaction of coal, the advantages of the indirect approach are (1) the gasification stage can tolerate a higher degree of impurities (e.g., mineral matter) in the coal; (2) the feed to the liquefaction stage can be maintained at the desired hydrogen level irrespective of the nature and quality of the feed coal; (3) a "clean" product is produced; (4) the method is adaptable to the use of the products from the underground gasification of coal; and (5) there is the added advantage that hydrogen from a variety of other hydrogen production methods may be employed, thereby increasing the overall production of liquids from a given amount of coal.

18.4 REACTORS

Several types of reactor are available for use in liquefaction processes and any particular type of reactor can exhibit a marked influence on process performance. The simplest type of reactor is the non-catalytic reactor (Figure 18.6), which consists, essentially, of a vessel (or even an open tube) through which the reactants pass. The reactants are usually in the fluid state but may often contain solids such as would be the case for a coal slurry. This particular type of reactor is usually employed for coal liquefaction in the presence of a solvent.

The second type of non-catalytic reactor is the continuous-flow, stirred-tank reactor (Figure 18.7), which has the notable feature of encouraging complete mixing of all of the ingredients, and if there is added catalyst (suspended in the fluid phase) the reactor may be referred to as a slurry reactor.

The fixed-bed catalytic reactor contains a bed of catalyst particles through which the reacting fluid flows (Figure 18.8); the catalysis of the desired reactions occurs as the fluid flows through the reactor. The liquid may pass through the reactor in a downward flow or in an upward flow but the problems that often accompany the latter operation (especially with regard to the heavier, less conventional feedstocks) must be recognized. In the downward-flowing mode, the reactor may often be referred to as a trickle-bed reactor.

Another type of reactor is the fluidized-bed reactor, in which the powdered catalyst particles are suspended in a stream of up-flowing liquid or gas (Figure 18.9). Another form of this type of reactor is the ebullating-bed reactor (Figure 18.10). The uniform features of these two types of reactor are the efficient mixing of the solid particles (the catalyst) and the fluid (the reactant) that occurs throughout the whole reactor.

In the entrained-flow reactor (Figure 18.11), the solid particles travel with the reacting fluid through the reactor. Such a reactor has also been described as a dilute or lean-phase fluidized bed with pneumatic transport of solids.

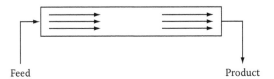

Feed   Product

FIGURE 18.6 Simplified representation of a non-catalytic reactor.

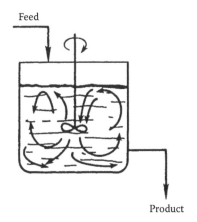

Feed

Product

FIGURE 18.7 Simplified representation of a continuous-flow stirred-tank reactor.

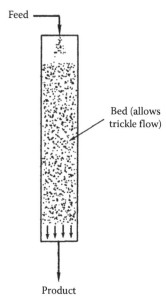

FIGURE 18.8 Simplified representation of a plug-flow (fixed-bed) trickle-bed reactor.

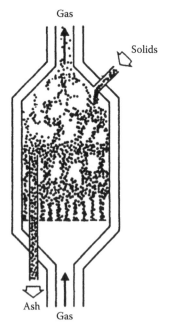

FIGURE 18.9 Fluidized-bed reactor.

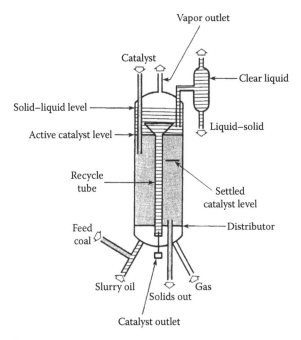

FIGURE 18.10 Ebullating-bed reactor.

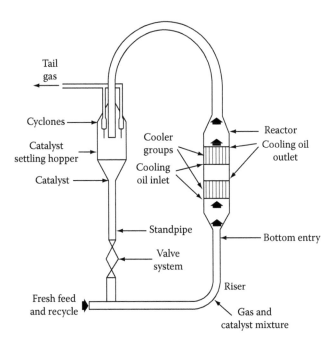

FIGURE 18.11 Entrained-flow reactor.

18.5 PRODUCTS

Coal liquefaction and the upgrading of the coal liquids are being considered as future alternatives of petroleum to produce synthetic liquid fuels due to the declining crude oil reserves and the high dependence on the foreign oil supplies. Since the late 1970s, coal liquefaction processes have been developed into integrated two-stage processes, in which coal is hydroliquefied in the first stage and the coal liquids are upgraded in the second stage. Upgrading of the coal liquids is an important aspect of this approach and may determine whether such liquefaction can be economically feasible.

However, coal liquids have remained largely unacceptable as refinery feedstocks because of their high concentrations of aromatic compounds and high heteroatom and metals content (Speight, 2008). Successful upgrading process will have to achieve significant reductions in the content of the aromatic components. However, the hydrogenation of coal liquids with multi-ring, aromatic hydrocarbons with hydrogen is a difficult process from the technological point of view due to the stable structures of the aromatic compounds and the poor dynamic yields at low pressure and low temperature (Stanislaus and Cooper, 1994). The catalysts commonly used for coal liquid hydrotreating are mixtures of nickel and molybdenum oxides supported on alumina. They deactivate rapidly by active site suppression and often also exhibit undesirable pore choking (Dadyburjor and Raje, 1994). Hence, other pathways to upgrade the coal liquids need to be explored to provide basic information to underpin future technology.

Liquid products from coal are generally different from those produced by petroleum refining, particularly as they can contain substantial amounts of phenols. Therefore, there will always be some question about the place of coal liquids in refining operations. For this reason, there have been some investigations on the "next-step" processing of coal liquids (Steedman, 1989).

The mild hydrogenation of coal under appropriate conditions to produce bitumen-like materials requires some comparison and evaluation of these with petroleum asphalts.

Petroleum-derived asphalts are the products of processing the nonvolatile constituents of petroleum (Parson, 1977; Speight, 2007) and can be generally defined in terms of asphaltene constituents, resin constituents, and oil constituents (Figure 18.12); the asphaltene constituents are considered to impart the plastic properties generally associated with asphalts. These constituents represent various stages in an oxidation or blowing process; for instance, air blowing at elevated temperatures will convert resins to asphaltene constituents and further reaction leads to decomposition and the formation of coke.

Asphalts are also regarded as colloidal suspensions, in which the oily constituents are the dispersant and the asphaltene constituents, principally, the dispersoid. Oily constituents of an aromatic nature lead to sol-like dispersions of lower temperature susceptibility and non-Newtonian behavior; that is, viscosity changes less with temperature but is affected by rate of shear for the velocity gradient.

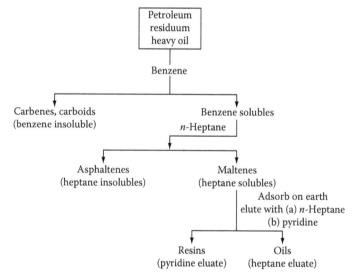

FIGURE 18.12 Feedstock fractionation. (From Speight, J.G., *The Chemistry and Technology of Petroleum*, 4th edn., Taylor & Francis Group, Boca Raton, FL, 2007.)

The chemical nature of asphalt is generally different from that of the synthetic bitumen derived from coal hydrogenation, but at the same time there may be enough similarity to consider that asphaltene-like materials may be dispersed in oil constituents of varying aromaticity, subject to the conditions of performance—and may be further treated to meet specifications. However, a difficulty encountered in evaluating and comparing asphalt properties lies in devising and correlating elemental laboratory tests to actual road performance for these non-Newtonian substances. The range of usual tests, such as shear viscosity and viscosity–temperature susceptibility, has not been entirely adequate.

In general, asphaltene constituents provide the major element of viscosity, and viscosity–temperature susceptibility increases with viscosity. These effects are contradictory as some statements are found to the effect that asphaltene constituents decrease temperature susceptibility and give high fluidity flow viscosity). In fact, it seems to be a predominant opinion that asphaltene constituents lower the temperature susceptibility and that by increasing the asphaltene content (at least up to a point) the temperature susceptibility will be improved. This depends on a lot of "ifs" and such qualifying phrases as "other things being equal."

The presence of free-carbon solids has been noted in coal tars, and undoubtedly, there are unreacted coal particles present in the products from hydrogenation. The presence of solid particles increases viscosity, though there is supposedly little accompanying effect upon the viscosity–temperature susceptibility.

While it is conceded that synthetic bitumen obtained from the hydrogenation of coal is not the same as a petroleum asphalt, still there are similarities that may permit characterization in similar terms. The same element of immiscibility exists between oils and asphaltene constituents or tars (oxygenated materials). It has been noted that these raw synthetic bitumen contain volatile oils and hard, asphaltene-like constituents. The oils are predominantly aromatic, giving stable sol-like dispersions with high temperature susceptibility and exhibiting Newtonian behavior.

Such synthetic bitumen lends itself to a variety of modifications and uses. Steam distillation under vacuum, a universal practice in asphalt processing, would remove volatiles and the nonvolatile residue could be air-blown to asphalt. The residue from distillation, if pitch-like, could be fluted with a nonvolatile oil and excessive asphaltene constituents could be precipitated by solvent refining.

Hydrogenation conditions for the production of synthetic bitumen from coal are of the order of 425°C (800°F) and hydrogen pressures (cold) of 2000 psi. The product obtained will also be related to the residence time and to the degree of contact. An iron catalyst, such as the sulfate, is apparently necessary in order to yield bitumen with asphalt-like properties. These are relatively mild conditions for hydrogenation.

The mild hydrogenation of coal under pressure in the presence of a solvent carrier or vehicle can be used also to produce a solid pitch-like product as well as a semisolid asphalt-like material. The former hydrogenation is for the purposes of producing an ash-free solid fuel, and the latter for purposes of producing a synthetic bitumen or asphalt for use as a road binder. The degree of hydrogenation and severity of conditions determine the nature and range of the product. Hydrocarbon gases and liquids are produced as by-products and occur increasingly with severity of reaction. At the limit of severity, methane is the sole hydrocarbon product (hydrogasification).

18.5.1 Characterization

As a first step in the characterization of coal liquids, it is generally recognized that some degree of fractionation is necessary (Whitehurst et al., 1980) followed by one or more forms of chromatography to identify the constituents (Kershaw, 1989a–c; Philp and de las Heras, 1992). The fractionation of coal liquids is based largely on schemes developed for the characterization of petroleum (Dooley et al., 1978; Speight, 2007), but because of the difference between coal liquids and petroleum, some modification of the basic procedure is usually required to make the procedure applicable to coal liquids (Ruberto et al., 1976; Bartle, 1989) since it must be recognized that coal liquids are

significantly different from petroleum in terms of heteroatom content, heteroatom distribution and type, as well as molecular weight (Callen et al., 1978).

Once fractionation has been achieved, there are a variety of techniques, especially for the analysis of polynuclear aromatic systems (Lee et al., 1981), that might also be applied to coal liquids. Specifically, these techniques are mass spectrometry (Aczel and Lumpkin, 1979; Zakett and Cooks, 1981; Batts and Batts, 1989), infrared spectroscopy (Fredericks, 1989), nuclear magnetic resonance spectroscopy (Hara et al., 1981; Attalla et al., 1989), as well as ultraviolet and luminescence spectroscopy (Kershaw, 1989a; Mille et al., 1990). There has also been the tendency to apply average structure determinations to the products of coal liquefaction (Kershaw, 1989b) and the same degree of caution is urged when the techniques are applied to petroleum constituents (Speight, 2007).

The characterization of heavy fractions of a coal liquid is especially important and complex because of two factors: (1) with increasing boiling point, there is an increase in the structural diversity of hydrocarbons and (2) heteroatomic compounds that contain oxygen, nitrogen, sulfur, or trace elements (Wilson et al., 1983) and are usually concentrated in the higher boiling fractions.

The initial separation of coal liquid fractions is usually achieved by distillation (Table 18.5); the cutoff points for the distillation may be somewhat arbitrarily chosen but usually parallel petroleum cuts; one typical scheme calls for collecting fractions from the initial boiling point to 200°C (390°F), 200°C–300°C (390°F–570°F), and greater than 370°C (700°F).

The highest boiling fraction of coal liquids contains the greatest variety of compounds, which are difficult to separate and characterize, but with appropriate chromatographic techniques it is possible to separate the higher boiling coal liquid material into fractions identified as (1) saturates, (2) monoaromatics, (3) diaromatics, and (4) polyaromatics plus polar compounds (Figure 18.13). The fractions separated in this manner are fractionated largely by virtue of physical–chemical interactions with the columns. These interactions are mostly of an acid–base nature. The aromatic and saturated compounds may be further separated by a number of means. For example, passage of the saturates over 5 Å molecular sieves divides this group into n-paraffins and compounds other than

TABLE 18.5
Distillation Characteristics of Synthoil

Cut Point			
°C	°F	Percent in Fraction	Total Percent of Product
>175[a]	<345	Trace	0.0
200	390	3.4	3.4
225	435	2.0	9.4
250	480	7.6	17.0
275	525	7.4	24.4
200[b]	390	6.7	31.1
225	435	8.3	39.4
250	480	9.3	48.7
275	525	9.0	57.7
300	570	7.4	65.1
Residuum		34.8	99.9

Source: Woodward, P.W. et al., Compositional analysis of synthoil from West Virginia, Report No. BERC/RI-76/2, United States Department of Energy, Washington, DC, 1976.

[a] Subsequent distillation temperatures apply to 747 mmHg pressure up to the point noted.
[b] Subsequent distillation temperatures apply to 40 mmHg pressure.

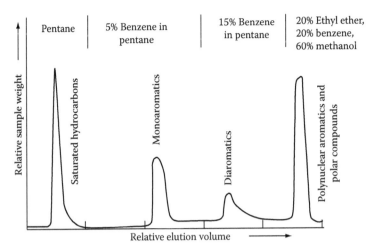

FIGURE 18.13 Separation of high-boiling coal liquids on S-230 alumina/silica by stepwise elution.

n-paraffins. The aromatic compounds may be separated by an alumina column into (1) a monoaromatic fraction, (2) a di- and triaromatic fraction, and (3) a polyaromatic fraction.

Basic compounds containing nitrogen may be precipitated from organic solvents as hydrogen chloride adducts. For example, the asphaltene fraction mentioned is soluble in toluene and bubbling hydrogen chloride through a solution of asphaltene constituents in toluene results in the formation of the adducts that can be separated physically from the constituents remaining in solution:

$$\equiv N + HCl \rightarrow \equiv NH^+Cl^-$$

Such a vast number of individual compounds are contained in higher boiling coal liquids that it is impossible to characterize such a sample totally by individual compounds.

However, separation and characterization of the constituents of coal liquids can be performed by compound type (Figure 18.14). This involves utilization of a separation scheme whereupon the entire sample or individual distillation cuts are separated into seven major classes of compounds: acids, bases, neutral nitrogen compounds, saturates, monoaromatics, diaromatics, and polyaromatics plus polar compounds. To avoid excessive dilution and use of solvents, a recycle column (Figure 18.15) may be employed that enables recycling of carrier solvent and accumulation of product in a flask from which fresh solvent is distilled.

Coal liquids can also be separated by various chromatographic techniques, including size exclusion chromatography, which also provides, in addition to fractionation, an indication of the molecular weight distribution within the sample (Wong, 1987; Bartle, 1989).

18.5.2 COMPOUND TYPES

The distribution of liquids produced from coal depends on the character of the coal and on the process conditions and, particularly, on the degree of "hydrogen addition" to the coal (Aczel et al., 1978; Schiller, 1978; Schwager et al., 1978; Wooton et al., 1978; Whitehurst et al., 1980; Kershaw, 1989c). Nevertheless, it is in this area that mass spectrometry has proved to be extremely useful, providing valuable evidence about the nature of the compound types in coal liquids and their relationship to the original coal (Anderson et al., 1981; Batts and Batts, 1989).

There is evidence for the similarity of the polynuclear aromatic types produced (thermally) from coal under different conditions suggesting that the major compound types are either representative of similar aromatic moieties in the original coal or a result of the processing conditions involving complex reactions that lead to similar final products, that is, heat is an "equalizer" in terms

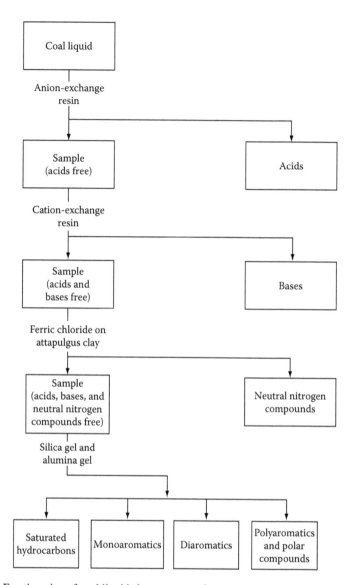

FIGURE 18.14 Fractionation of coal liquids into compound types.

of structural differentiation. It may also be that many of the same complex reactions occur during diagenesis (Nishioka and Lee, 1988). However, other work (Sturm et al., 1981) gives the distinct impression that there are other factors involved in liquids production since coal rank alone could not account for the observed variations in complexity of the coal liquids. This is an interesting observation since there has been a more recent suggestion (Berkowitz, 1988) that there exists a somewhat lesser dependence of the molecular chemistry of coal on rank and that the chemistry of coal is heavily influenced by its source as well as by the early formation history. It is also possible that coals of similar rank may, therefore, be chemically much more diverse than is usually supposed.

In more generic terms, the liquid products may be classified as neutral oils (essentially pure hydrocarbons), tar acids (phenols), and tar bases (basic nitrogen compounds) (Farcasiu, 1977).

The neutral oils making up 80%–85% of hydrogenated coal distillate are approximately half aromatic compounds, including polycyclic aromatic hydrocarbons. Typical components of the neutral oils are benzene, naphthalene, and phenanthrene (Wood, 1987). Hydroaromatic compounds (cycloparaffins, often called *naphthenes*—see Speight, 2007) are another important component of

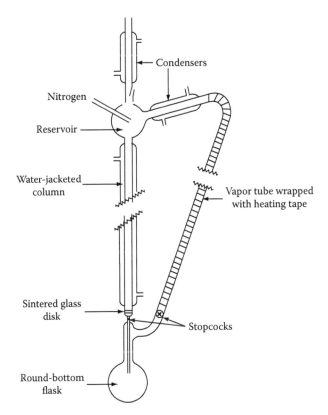

FIGURE 18.15   Chromatographic column equipped for solvent recycle.

neutral oils. Hydroaromatic compounds are formed at high hydrogen pressures in the presence of a catalyst, but in the presence of another species capable of accepting hydrogen (such as unreacted coal), hydroaromatic species lose hydrogen to form the thermodynamically more stable aromatic compounds and are important intermediates in the transfer of hydrogen to unreacted coal during liquid-phase coal hydrogenation and solvent refining of coal.

The next most abundant component of neutral oil consists of liquid olefins. The olefins are reactive and tend to undergo polymerization, oxidation, and other reactions causing changes in the properties of the product with time. On the other hand, olefins are excellent raw materials for the manufacture of synthetic polymers and other chemicals and, thus, can be valuable chemical by-products in coal liquids.

In addition to neutral oil, coal liquids contain tar acids (consisting of phenolic compounds), which may constitute from 5% to 15% w/w of many coal liquids. They constitute one of the major differences between coal liquids and natural petroleum, which has a much lower content of oxygen-containing compounds and, although tar acids are valuable chemical raw materials, they are troublesome to catalysts in refining processes.

Tar bases containing basic nitrogen make up 2%–4% w/w of coal hydrogenation liquids. Tar bases are made of a variety of compounds, such as pyridine, quinoline, aniline, and higher-molecular-weight analogs.

It is particularly pertinent here to compare hydrogenated coal liquids with crude oil. For example, a coal liquid prepared by hydrogenating Illinois No. 6 coal was distilled and only the material boiling below 525°C (975°F) was used for comparison and, thus, the coal liquid differed from conventional crude oil in having no heavy ends. This synthetic crude oil had an API gravity of 25.2, specific gravity of 0.9030, pour point of −19°C (66°F), Reid vapor pressure of 1.0 psi, viscosity at 15.6°C (60°F) of 37.9 SU, and sulfur content of 0.13%. These values are not at all out of line

TABLE 18.6

Distillation Yields from Coal Synthetic Crude Oil and Petroleum

	Yield (% v/v)	
Distillation Fraction	Synthetic Crude[a]	Natural Crude[b]
Off-gas	0.2	0.5–2.0
Naphtha		
Light (IBP–77°C)	3	2–5
Medium (77°C–190°C)	34	12–22
Heavy (190°C–205°C)	7	1–3
Kerosene (205°C–250°C)	20	7–10
Heavy fuel oil (250°C–315°C)	16	11–15
Gas oil		
Light (315°C–345°C)	3	5–7
Heavy (345°C–510°C)	12	18–30
Residuum (>510°C)	4	10–40

[a] Prepared by hydrogenation of Illinois No. 6 coal.
[b] Typical ranges of values expected in natural crude oil.

compared with typical crude oils. It is interesting that the sulfur content is much lower than that of high-sulfur oil such as Wyoming sour crude (2.9%), or even low-sulfur Louisiana crude (0.38%).

The fractional distillation yields of coal-derived synthetic crude oil compared with typical petroleum values (Table 18.6) are useful in determining the utility and ability to refine of the product. However, it should be noted that the nature of the synthetic crude oil depends on the manner of preparation and this particular oil should be considered as only an example.

In addition, since this coal crude was distilled from a coal hydrogenation mixture leaving pitch, coke, and other residue behind, it is "unnaturally" deficient in residue compared with petroleum-based crude oil. The distillation yields show that the coal crude is especially rich in the medium naphtha-kerosene fraction boiling up to 250°C (480°F) and can be attributed to the presence of benzene (bp 80°C [176°F]), toluene (bp 111°C [232°F]), o-xylene (bp 144°C [291°F]), m-xylene (bp 139°C [282°F]), p-xylene (bp 138°C [280°F]), naphthalene (bp 218°C [424°F]), and tetralin (bp 207°C [405°F]), as well as various alkyl (methyl) derivatives of the aforementioned compounds, hydroaromatic analogs, and alkanes.

Synthetic crude oil prepared by the hydrogenation of coal is deficient for diesel fuel constituents since the products in this boiling point range are largely aromatic, which makes the "cut" unsuitable for use in a diesel engine.

The olefins in coal synthetic crude oil are reactive and tend to form gums and undesirable deposits and may be removed for chemical synthesis or may undergo alkylation processes to form highly branched, high-octane gasoline. Alkylation is normally catalyzed by sulfuric or hydrofluoric acid and involves the reaction of an olefin with an alkane:

$$RCH{=}CH_2 + R^1CH_3 \rightarrow RCH_2CH_2CH_2R^1$$

In addition, the high naphthene content makes synthetic crude oil a potentially good raw material for reforming processes in which molecules are thermally or catalytically rearranged to molecules more suitable for gasoline without cracking the molecular structure.

A major difference between natural crude oil and synthetic crude oil is the high content of oxygen compounds (phenols) in the latter. If these compounds are subjected to processes in which hydrogen is present under conditions conducive to reactivity of the oxygen compounds, such as catalytic cracking or reforming, much of the oxygen ends up as water and, thus, constitutes a somewhat wasteful use of expensive hydrogen.

Fischer–Tropsch synthesis is employed in the manufacture of liquid hydrocarbons in the first commercial coal-to-liquids plant constructed since World War II, the coal conversion plant operated in Sasol, South Africa (Speight, 2008). The reactions occur between hydrogen and carbon monoxide at pressure over iron catalysts activated with alkali, and a large number of products may result depending on various conditions. Fischer–Tropsch synthesis is especially useful for the production of aliphatic hydrocarbons; some olefins are produced (Speight, 2008):

$$(2n+1)H_2 + nCO \rightarrow C_nH_{2n+2} + nH_2O$$

$$2nH_2 + nCO \rightarrow C_nH_{2n} + nH_2O$$

Other reactions yielding carbon dioxide also occur to produce saturated hydrocarbons or olefins. In addition to hydrocarbons, Fischer–Tropsch synthesis also produces oxygen-containing compounds (Figure 18.16):

$$(n+1)H_2 + 2nCO \rightarrow C_nH_{2n+2} + nCO_2$$

$$nH_2 + 2nCO \rightarrow C_nH_{2n} + nCO_2$$

$$2nH_2 + nCO \rightarrow C_nH_{2n+1}OH + (n-1)H_2O$$

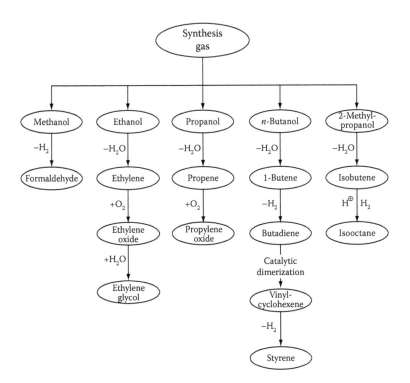

FIGURE 18.16 Alcohols and olefins from synthesis gas.

Some of the products of Fischer–Tropsch synthesis are troublesome to the process, particularly through catalyst poisoning. These include waxes, salts, and organometallic compounds, which form deposits on the catalyst surface, thus poisoning it.

18.5.3 Asphaltene Constituents

There has been significant emphasis on determining the nature of the asphaltene-type constituents. As with petroleum, the asphaltene portion of coal-derived liquids are responsible for many of the problems that arise when the liquids are processed. Therefore, a brief notation about the character of coal-derived asphaltene constituents is necessary, recognizing that more complete reviews are available. Such information aids in understanding of this complex fraction and may also help dispel some prior, but erroneous, notions about the chemical character of coal-derived asphaltene constituents.

Addition of a normal paraffin liquid (e.g., n-pentane, n-hexane, n-heptane, etc.) to either natural petroleum or coal synthetic crude oil results in the formation of a precipitate known as asphaltene (Mima et al., 1978; Steffgen et al., 1979; Steedman, 1985; Speight, 2007) (Figure 18.12). In both natural petroleum and coal crude oil, asphaltene constituents are held in solution by interaction with other organic matter present. When dried, asphaltene constituents are brown-to-black powders resembling finely divided coal. The benzene-insoluble residue contains pyridine-soluble material sometimes designated as preasphaltenes, the analog of which in the petroleum industry might be the carbenes and carboids that are usually isolated (as they are not usually found in natural petroleum) from thermally treated stocks (Speight, 2007).

Asphaltene constituents in coal liquids are important for several reasons: (1) they contribute to many problems in the processing of crude oil; (2) they contribute to instabilities in processing; (3) they lead to excessive viscosity, contributing to pumping expense, clogging, and slow processing; (4) the presence of asphaltene constituents can lead to incompatibility with some solvents used in petroleum processing; and (5) asphaltene constituents lead to coke formation under some circumstances.

The origin of coal asphaltene constituents has been the subject of much speculation insofar as they have been considered to be not only the initial products of coal liquefaction but also the secondary products of coal liquefaction:

$$Coal \rightarrow asphaltene\ constituents \rightarrow oil$$

$$Coal \rightarrow oil \rightarrow asphaltene\ constituents$$

Although current concepts tend to favor the former hypothesis, the matter is still not completely resolved.

Nevertheless, structural studies of the asphaltene constituents from coal liquefaction processes have progressed to the point where various molecular types have been identified. It has also been pointed out that the character of the asphaltene constituents is process dependent and the structural character of the asphaltene may even bear some relationship to the structural types in the parent coal (Snape et al., 1984).

Coal asphaltene constituents are quite different in nature from petroleum asphaltene constituents (Table 18.7). The molecular weight of asphaltene constituents from coal liquids may be some 8–10 times lower than the observed molecular weight of petroleum asphaltene constituents, although this latter can be revised to lower values for a variety of reasons (Steedman, 1985).

Coal asphaltene constituents have frequently been defined in terms of an acid–base complex with the asphaltene existing as a composite of the two systems, acid and base:

$$HNR(OH)OH \quad acid$$

$$N=R(OR)=O \quad base$$

TABLE 18.7
Comparison of the Asphaltene Fraction from Coal Liquids and from Petroleum

Inspection	Coal Asphaltenes	Petroleum Asphaltenes
Carbon (% w/w)	86.93	81.7
Hydrogen (% w/w)	6.83	7.60
Nitrogen (% w/w)	1.36	1.23
Sulfur (% w/w)	1.09	7.72
Oxygen (% w/w)	3.8	1.7
Vanadium (ppm)	9	1200
Nickel (ppm)	3	390
VPO (MW)	726	5400
H/C (atomic ratio)	0.94	1.12

Source: Speight, J.G. and Long, R.B., *Atomic and Nuclear Methods in Fossil Energy Research*, R.H. Filby, Ed., Plenum Publishing Corporation, New York, 1982, p. 295.

However, a more recent report indicates that pentane insolubility of the bulk of the coal asphaltene constituents cannot be ascribed to hydrogen-bonding effects between the acidic and basic components.

Examination of coal asphaltene constituents by a variety of techniques (chromatographic and spectroscopic) has resulted in the positive identification of many compound types. Thus, from mass spectroscopic investigations major oxygen compound types include phenols, dihydroxy-benzenes, and the hydroxy and dihydroxy derivatives of phenanthrenes, pyrenes, benzopyrenes, coronenes, and other condensed aromatics. Nitrogen compounds present include carbazoles, benzocarbazoles, and hydroxycarbazoles. Aromatic constituents and aromatic sulfur compounds found include naphthene–naphthalenes, phenanthrenes, pyrenes, benzopyrenes, benzperylenes, coronenes, and dibenzothiophenes (for example, see Wilson et al., 1983; Boduszynski et al., 1986; Schmidt et al., 1987; Ostman and Colmsjo, 1988; Kershaw, 1989a–c).

Nuclear magnetic resonance spectroscopy indicates that the asphaltene constituents consist predominantly of aromatic molecules carrying short alkyl groups that may be either open chains or saturated rings condensed to the aromatic rings. In addition, data from ultraviolet spectroscopy have provided indications that the aromatic portion of certain coal asphaltene constituents may contain condensed structures composed of more than five aromatic rings, but the ultraviolet data exclude the presence of anthracene and tetracene homologs. Finally, there are also indications from the various techniques that high proportions of the total carbon exist in aromatic locations (Steedman, 1985).

Again, it is necessary to proceed with caution when deriving molecular structures by mathematical manipulation of the data (Chapter 10) because of the complex nature of the liquid products. Nevertheless, these techniques do offer valuable primary information about these particular products, which, in turn, is of some assistance in understanding the means by which coal is degraded in, say, a liquefaction system and may therefore be of assistance in maximizing liquid yields.

In summary, coal-derived asphaltene constituents are the fraction of a coal-derived liquid that is insoluble in low-boiling liquid hydrocarbons such as pentane or heptane. Coal-derived asphaltene constituents are a complex mixture by virtue of their thermal mode of formation from coal. They are a complex state of matter and, although they can be conveniently described using a molecular weight polarity grid, there is still the tendency to construct "average" structures on the basis of spectroscopic investigations. However, there is the need to recognize the deficiencies of the spectroscopic techniques and it is the too-literal interpretation of the data that lead to misconceptions of the structural concepts of coal-derived asphaltene constituents.

Current evidence indicates that coal-derived asphaltene constituents are a collection of predominantly one-to-four ring condensed aromatic systems that contain basic and nonbasic nitrogen constituents as well as oxygen (acidic and etheric) functions. These functionalities play a role in intramolecular relationships within the asphaltene fraction and also with the other constituents of the coal-derived liquid. This latter effect influences the viscosity of the liquid. Thus, coal-derived asphaltene constituents are an extremely complex solubility class by virtue of their thermal derivation from coal.

As with petroleum asphaltene (Speight, 2007), it is difficult (if not impossible) to accurately depict the structure of coal-derived asphaltene constituents. A simple assessment should include the statement "small-ring polynuclear aromatic systems," basic and nonbasic nitrogen as well as phenolic and etheric "oxygen." How these types enjoy an intramolecular existence is another matter which is difficult and whose definition is left to future research.

18.5.4 REFINING

The production of liquid fuels from coal is, in the simplest sense, a means by which hydrogen is added to the coal to improve the atomic hydrogen/carbon ratio and, at the same time, to bring about a reduction in the molecular weight of the product relative to that of the feed resulting in the production of, the at least, catalytic cracking feedstocks and, ultimately, liquid fuels with high atomic hydrogen/carbon ratios (Figure 18.17) (deRosset et al., 1979; Givens et al., 1979; Schneider et al., 1979; Crynes, 1981).

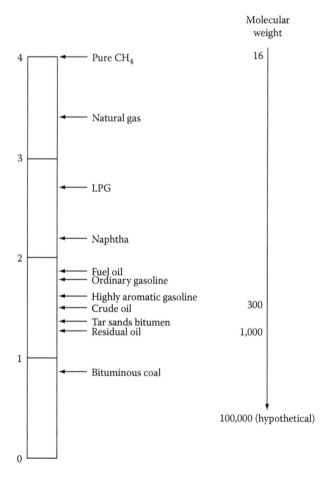

FIGURE 18.17 Atomic hydrogen/carbon ratios for various fuels and fuel sources.

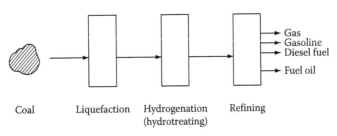

Coal   Liquefaction   Hydrogenation   Refining
                            (hydrotreating)

FIGURE 18.18 Simplified representation of the production of refined liquid fuels from coal liquids.

The manner by which hydrogenation can occur will vary from process to process and may even occur as part of the process by the use of a hydrogen atmosphere and a solvent capable of donating hydrogen to the system (Che et al., 1979) and the type of catalyst employed (Cusumano et al., 1978; Ahmed and Crynes, 1979; Frumkin et al., 1981; O'Rear et al., 1981; Potts et al., 1981). Nevertheless, in the more general sense, at some stage of the operation, the liquid products need to be stabilized (i.e., freed from unsaturated materials as well as nitrogen, oxygen, and sulfur species) by what may be simply referred to as a hydrotreating operation. In the simplest sense, this operation may be viewed as occurring downstream of the liquefaction process (Figure 18.18).

Current concepts for refining the products of coal liquefaction processes rely for the most part on the already existing petroleum refineries, although it must be recognized that the acidity (i.e., phenol content) of the coal liquids and the potential incompatibility of the coal liquids with conventional petroleum (or even heavy oil) feedstocks may pose severe problems within the refinery system (*European Chemical News*, 1981).

The first essential step in refining coal liquids is severe catalytic hydrogenation to remove most of the nitrogen, sulfur, and oxygen and to convert at least part of the high-boiling material to lower-boiling distillates that might be further refined. This is analogous to the hydrodesulfurization of heavy oils using a preliminary cracking technique so that after product separation (by distillation) the most suitable choice of process conditions can be made (Speight, 2000; Ancheyta and Speight, 2007).

However, a major limiting factor in refining coal liquids is due to the high aromatics content and to the condensed nature of many of the aromatic ring systems (Dooley et al., 1979). Thus, to produce liquid fuels of the types currently in demand, each condensed aromatic ring would have to be hydrogenated (saturated) and cracked to produce the lower-boiling distillate material. The hydrogen demand for such conversions and the effect of these polynuclear aromatic systems (especially those that contain nitrogen and other heteroatoms) on catalysts is a very worthy hurdle to overcome. Nevertheless, it is a hurdle that can be surpassed, and by a variety of process conditions.

It is beyond the scope of the present text to present a detailed account of how the liquid (or even solid) products of coal liquefaction may be refined, but it is the intent of this section to present a brief outline of a petroleum refinery to indicate which methods may be available to upgrade to liquefaction products.

While appearing to be a relatively facile system, the petroleum refinery is actually a complex integrated series of operations (Figure 18.19) that ultimately results in the production of high-value, salable materials from low-value feedstocks (Speight, 2007, 2008, 2011b). The operation can vary from the relatively "simple" distillation (or "skimming") process to the much more complex thermal conversion units in which the crude feedstocks are thermally degraded to lighter (lower-molecular-weight) marketable products. Processes involving the use of a variety of complex and expensive catalysts are also a necessary part of any refinery and such processes will play an important role in the processing of the products from coal liquefaction units (Boucher et al., 1981).

In the catalytic cracking process, the object is to produce gasoline, heating oil, and the like from a heavier feedstock such as gas oil (Table 18.8), by means of an aluminosilicate base catalyst.

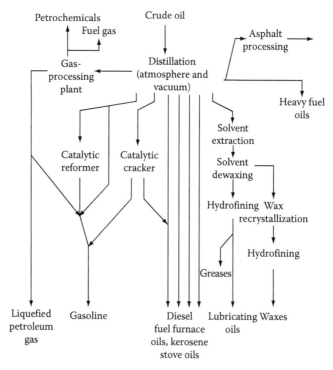

FIGURE 18.19 Petroleum refinery.

TABLE 18.8
Boiling Ranges of Various Petroleum Fractions

Fraction	Boiling Range	
	°C	°F
Fuel gas	−160 to −40	−260 to −40
Propane	−40	−40
Butane(s)	−12 to −1	11–30
Light naphtha	−1 to +150	30–300
Heavy naphtha	150–205	300–400
Gasoline	−1 to 180	30–355
Kerosene	205–260	400–500
Stove oil	205–290	400–550
Light gas oil	260–315	500–600
Heavy gas oil	315–425	600–800
Lubricating oil	<400	>750
Vacuum gas oil	425–600	800–1100
Residuum	>600	>1100

Sources: Speight, J.G., *The Chemistry and Technology of Petroleum*, 4th edn., Taylor & Francis Group, Boca Raton, FL, 2007; Speight, J.G., *Synthetic Fuels Handbook: Properties, Processes, and Performance*, McGraw-Hill, New York, 2008.

However, the reactions that occur are varied and, especially with the heavier or more aromatic feedstocks, there is the inevitable deposition of carbon (coke) on the catalyst and the accompanying decrease in catalyst activity.

Hydrocracking is a process that accomplishes the same goals as catalytic cracking but the presence of hydrogen and more specific catalyst often allows a much better control of the reaction and therefore results in a better distribution of products. The hydrocracker is operated at elevated pressures (several thousand psi in the case of the heavier feedstocks) and employs a bifunctional catalyst that has sites capable of promoting the hydrogenation reactions as well as the cracking reactions.

Thus, while current refinery technology may suffice to a point for the refining of the liquid and solid products from liquefaction processes, there are many aspects of the operation that may need some modification when the products from coal become a major refinery feedstock. Of course, this may dictate the creation and evolution of a completely new refining technology.

REFERENCES

Abdel-Baset, M.B., Yarzab, R.F., and Given, P.H. 1978. *Fuel*, 57: 89.

Abichandani, J.S., Wieland, J.H., Shah, Y.T., and Cronauer, D.C. 1984. *AIChE Journal*, 30: 295.

Aczel, T. and Lumpkin, H.E. 1979. In *Refining of Synthetic Crudes*, M.L. Gorbaty and B.M. Harney (Eds.). Advances in Chemistry Series No. 179. American Chemical Society, Washington, DC, Chapter 2.

Aczel, T., Williams, R.B., Brown, R.A., and Pancirov, R.J. 1978. In *Analytical Methods for Coal and Coal Products*, C. Karr, Jr. (Ed.). Academic Press, Inc., New York, Vol. I, Chapter 17.

Ahmed, M.M. and Crynes, B.L. 1979. In *Refining of Synthetic Crudes*, M.L. Gorbaty and B.M. Harney (Eds.). Advances in Chemistry Series No. 179. American Chemical Society, Washington, DC, Chapter 11.

Ancheyta, J. and Speight, J.G. 2007. *Hydroprocessing Heavy Oils and Residua*. Taylor & Francis Group, Boca Raton, FL.

Anderson, R.B. 1984. In *Catalysis on the Energy Scene*, S. Kaliaguine and A. Mahay (Eds.). Elsevier, Amsterdam, the Netherlands, p. 457.

Anderson, L.L., Chung, K.E., Pugmire, R.J., and Shabtai, J. 1981. In *New Approaches in Coal Chemistry*, B.D. Blaustein, B.C. Bockrath, and S. Friedman (Eds.). American Chemical Society, Washington, DC, Chapter 13.

Anderson, L.L. and Tillman, D.A. 1979. *Synthetic Fuels from Coal: Overview and Assessment*. John Wiley & Sons, Inc., New York.

Attalla, M.I., Vassallo, A.M., and Wilson, M.A. 1989. In *Spectroscopic Analysis of Coal Liquids*, J. Kershaw (Ed.). Elsevier, Amsterdam, the Netherlands, Chapter 7.

Baldwin, R.M. 1991. In *Coal Science II*, H.H. Schobert, K.D. Bartle, and L.J. Lynch (Eds.). Symposium Series No. 461. American Chemical Society, Washington, DC, Chapter 13.

Bartle, K.D. 1989. In *Spectroscopic Analysis of Coal Liquids*, J. Kershaw (Ed.). Elsevier, Amsterdam, the Netherlands, Chapter 2.

Batchelder, H.R. 1962. In *Advances in Petroleum Chemistry and Refining*, J.J. McKetta, Jr. (Ed.). Interscience Publishers, Inc., New York, Vol. V, Chapter 1.

Batts, D.D. and Batts, J.E. 1989. In *Spectroscopic Analysis of Coal Liquids*, J. Kershaw (Ed.). Elsevier, Amsterdam, the Netherlands, Chapter 4.

Bergius, F. 1912. *Z. Angewandte Chemie*, 23: 2540.

Bergius, F. 1913. *Journal of the Indian Chemical Society*, 32: 462.

Bergius, F. 1925. *Fuel*, 4: 458.

Berkowitz, N. 1988. In *Polynuclear Aromatic Compounds*, L.B. Ebert (Ed.). Advances in Chemistry Series No. 217. American Chemical Society, Washington, DC, Chapter 14.

Berthelot, M. 1869. *Bulletin de la Societe Chimique de France*, 11: 278.

Bockrath, B.C. and Schroeder, K.T. 1981. In *New Approaches in Coal Chemistry*, B.D. Blaustein, B.C. Bockrath, and S. Friedman (Eds.). American Chemical Society, Washington, DC, Chapter 11.

Boduszynski, M.M., Hurtubise, R.J., Allen, T.W., and Silver, H.F. 1986. *Fuel*, 65: 223.

Boucher, L.J., Holy, N.L., and Davis, B.H. 1981. In *New Approaches in Coal Chemistry*, B.D. Blaustein, B.C. Bockrath, and S. Friedman (Eds.). American Chemical Society, Washington, DC, Chapter 18.

Braunstein, H.M., Copenhaver, E.D., and Pfuderer, H.A. (Eds.). 1977. *Environmental, Health, and Control Aspects of Coal Conversion*. Oak Ridge National Laboratory, Oak Ridge, TN, Vol. 1.

Buchanan, A.C., Britt, P.F., and Biggs, C.A. 1992. *Energy & Fuels*, 6: 110.

Callen, R.B., Simpson, C.A., and Bendoraitis, J.G. 1978. In *Analytical Chemistry of Liquid Fuel Sources: Tar Sands, Oil Shale, Coal, and Petroleum*, P.C. Uden, S. Siggia, and H.B. Jensen (Eds.). Advances in Chemistry Series No. 170. American Chemical Society, Washington, DC, Chapter 21.

Chakrabartty, S.K. 1981. A simple method to characterize feed-coals for liquefaction. *Canadian Journal of Chemistry*, 59: 1487–1489.

Che, S.C., Qader, S.A., and Knell, E.W. 1979. In *Refining of Synthetic Crudes*, M.L. Gorbaty and B.M. Harney (Eds.). Advances in Chemistry Series No. 179. American Chemical Society, Washington, DC, Chapter 10.

Cooper, B.R. (Ed.). 1978. *Scientific Problems of Coal Utilization*. United States Department of Energy, Washington, DC.

Coughlin, R.W. and Davoudzadeh, F. 1983. *Nature*, 303: 789.

Crynes, B.L. 1981. In *Chemistry of Coal Utilization*, Second Supplementary Volume, M.A. Elliott (Ed.). John Wiley & Sons, Inc., New York, Chapter 29.

Curtis, C.W., Guin, J.A., Pass, M.C., and Tsai, K.J. 1987. *Fuel Science and Technology International*, 5: 245.

Curtis, C.W. and Hwang, J.-S. 1992. *Fuel Processing Technology*, 30: 47.

Cusumano, J.A., Dalla Betta, R.A., and Levy, R.B. 1978. *Catalysis in Coal Conversion*. Academic Press, Inc., New York.

Dadyburjor, D.B. and Raje, A.P. 1994. *Journal of Catalysis*, 145: 16.

Davidson, R.M. 1983. Mineral effects in coal conversion. Report No. ICTIS/TR22. International Energy Agency, London, U.K.

deRosset, A.J., Tan, G., and Gatsis, J.G. 1979. In *Refining of Synthetic Crudes*, M.L. Gorbaty and B.M. Harney (Eds.). Advances in Chemistry Series No. 179. American Chemical Society, Washington, DC, Chapter 7.

Donath, E.E. 1963. In *Chemistry of Coal Utilization*, Supplementary Volume, H.H. Lowry (Ed.). John Wiley & Sons, Inc., New York, Chapter 22.

Dooley, J.E., Lanning, W.C., and Thompson, C.J. 1979. In *Refining of Synthetic Crudes*, M.L. Gorbaty and B.M. Harney (Eds.). Advances in Chemistry Series No. 179. American Chemical Society, Washington, DC, Chapter 1.

Dooley, J.E., Thompson, C.J., and Scheppele, S.E. 1978. In *Analytical Methods for Coal and Coal Products*, C. Karr, Jr. (Ed.). Academic Press, Inc., New York, Vol. I, Chapter 16.

Dry, M.E. 1976. *Industrial & Engineering Chemistry of Product Research and Development*, 15(4): 282.

Dry, M.E., Du Plessis, J.A.K., and Leuteritz, G.M. 1966. *Catalysis*, 6: 194.

Dry, M.E. and Ferrira, L.C. 1967. *Journal of Catalysis*, 7: 352.

Dry, M.E. and Ferrira, L.C. 1968. *Journal of Catalysis*, 11: 18.

Dry, M.E. and Erasmus, H.B. de W., 1987. *Annual Review of Energy*, 12: 1.

Durie, R.A. 1982. *Fuel*, 61: 883.

European Chemical News. 1981. October 5, p. 25.

Extance, A. 2011. Liquid assets. *Chemistry World*, 8(5): 51–53.

Farcasiu, M. 1977. *Fuel*, 56: 9.

Fletcher, T.H., Kerstein, A.R., Pugmire, R.J., Solum, M.S., and Grant, D.M. 1992. Chemical percolation model for devolatilization. 3. Direct use of carbon-13 NMR data to predict effects of coal type. *Energy Fuels*, 6(4): 414–431.

Fredericks, P.M. 1989. In *Spectroscopic Analysis of Coal Liquids*, J. Kershaw (Ed.). Elsevier, Amsterdam, the Netherlands, Chapter 5.

Frumkin, H.A. Jr., Sullivan, R.F., and Stangeland, B.E. 1981. In *Upgrading Coal Liquids*, R.F. Sullivan (Ed.). Symposium Series No. 156. American Chemical Society, Washington, DC, Chapter 3.

Gagarin, S.G. and Krichko, A.A. 1992. *Fuel*, 71: 785.

Garr, G.T., Lytle, J.M., and Wood, R.E. 1979. *Fuel Processing Technology*, 2: 179.

Gates, B.C. 1979. *Chemtech*, 9(2): 97.

Gibson, J. and Gregory, D.H. 1971. *Carbonization of Coal*. Mills & Boon, London, U.K.

Given, P.H., Cronauer, D.C., Spackman, W., Lovell, H.L., Davis, A., and Biswas, B. 1975. *Fuel*, 54: 40.

Givens, E.N., Collura, M.A., Skinner, R.W., and Greskovich E.J. 1979. In *Refining of Synthetic Crudes*, M.L. Gorbaty and B.M. Harney (Eds.). Advances in Chemistry Series No. 179. American Chemical Society, Washington, DC, Chapter 8.

Gorin, E. 1981. In *Chemistry of Coal Utilization*, Second Supplementary Volume, M.A. Elliott (Ed.). John Wiley & Sons, Inc., New York, Chapter 27.

Graham, J.I. and Skinner, D.G. 1929. The action of hydrogen on coal. *Journal of Indian Chemical Society*, 48: 129–136.

Gray, D., Barrass, G., Jezko, J., and Kershaw, J.R. 1980. *Fuel*, 59: 146.

Gray, J.A. and Shah, Y.T. 1981. *Energy Science and Technology*, 3: 24.

Gun, S.R., Sama, J.K., Chowdhury, P.B., Mukherjee, S.K., and Mukherjee, D.K. 1979. *Fuel*, 58: 176; 179.

Hara, T., Tewari, K.C., Li, N.C., and Schweighardt, F.K. 1981. In *New Approaches in Coal Chemistry*, B.D. Blaustein, B.C. Bockrath, and S. Friedman (Eds.). American Chemical Society, Washington, DC, Chapter 17.

Harris, L.A. and Yust, C.S. 1977. *Fuel*, 56: 456.

Hessley, R.K., Reasoner, J.W., and Riley, J.T. 1986. *Coal Science,* John Wiley & Sons, Inc., New York.

Hirschon, A.S. and Wilson, R.B. Jr. 1991. In *Coal Science II*, H.H. Schobert, K.D. Bartle, and L.J. Lynch (Eds.). Symposium Series No. 461. American Chemical Society, Washington, DC, Chapter 21.

Jones, C.J., Jager, B., and Dry, M.D. 1992. *Oil and Gas Journal,* 90(3): 53.

Kershaw, J. 1989a. In *Spectroscopic Analysis of Coal Liquids*, J. Kershaw (Ed.). Elsevier, Amsterdam, the Netherlands, Chapter 6.

Kershaw, J. 1989b. In *Spectroscopic Analysis of Coal Liquids*, J. Kershaw (Ed.). Elsevier, Amsterdam, the Netherlands, Chapter 8.

Kershaw, J. 1989c. In *Spectroscopic Analysis of Coal Liquids*, J. Kershaw (Ed.). Elsevier, Amsterdam, the Netherlands, Chapter 10.

Kimber, G.M. and Gray, M.D. 1967. *Combustion and Flame*, 11: 360.

Kirov, N.Y. and Stephens, J.N. 1967. *Physical Aspects of Coal Carbonization*. National Coal Research Advisory Committee, Australia Department of National Development, Canberra, Australian Capital Territory, Australia.

Lee, S. 1990. *Methanol Synthesis Technology*. CRC Press, Boca Raton, FL.

Lee, M.L., Novotny, M.V., and Bartle, K.D. 1981. *Analytical Chemistry of Polycyclic Aromatic Compounds*. Academic Press, Inc., New York.

Longanbach, J.R. 1981. In *New Approaches in Coal Chemistry*, B.D. Blaustein, B.C. Bockrath, and S. Friedman (Eds.). American Chemical Society, Washington, DC, Chapter 8.

Lytle, J.M., Hsieth, B.C.B., Anderson, L.L., and Wood, R.E. 1979. *Fuel Processing Technology*, 2: 235.

McNeil, D. 1966. *Coal Carbonization Products*. Pergamon Press, London, U.K.

Mille, G., Kister, J., Doumenq, P., and Aune, J.P. 1990. In *Advanced Methodologies in Coal Characterization*, H. Charcosset and B. Nickel-Pepin-Donat (Eds.). Elsevier, Amsterdam, the Netherlands, Chapter 13.

Miller, C.L. 2007. Coal conversion—Pathway to alternate fuels. *EIA Energy Outlook Modeling and Data Conference*, Washington, DC, March 28.

Mima, M.J., Schultz, H., and McKinstry, W.E. 1978. In *Analytical Methods for Coal and Coal Products*, C. Karr, Jr. (Ed.). Academic Press, Inc., New York, Vol. I, Chapter 19.

Mochida, I., Tamaru, K., Korai, Y., Fujitsu, H., and Takeshita, K. 1982. *Carbon*, 20: 231.

Moschopedis, S.E., Hawkins, R.W., Fryer, J.F. and Speight, J.G. 1980. *Fuel*, 59: 647.

Moschopedis, S.E., Hawkins, R.W., and Speight, J.G. 1982. *Fuel Processing Technology*, 5: 213.

Mukherjee, D.K., Sama, J.K., Chowdhury, P.B., and Lahiri, A. 1972. *Proceedings. Symposium on Chemicals from Oil and Coal*. Central Fuel Research Institute, Dhanbad, Jharkhand, India, p. 143.

Narain, N.K., Cillo, D.L., Stiegel, G.J., and Tischer, R.E. 1987. In *Coal Science and Chemistry*, A. Volborth (Ed.). Elsevier, Amsterdam, the Netherlands, p. 111.

Neavel, R.C. 1976. *Fuel*, 55: 237.

Nishioka, M. and Lee, M.L. 1988. In *Polynuclear Aromatic Compounds*, L.B. Ebert (Ed.). Advances in Chemistry Series No. 217. American Chemical Society, Washington, DC, Chapter 14.

O'Rear, D.J., Sullivan, R.F., and Stangeland, B.E. 1981. In *Upgrading Coal Liquids*, R.F. Sullivan (Ed.). Symposium Series No. 156. American Chemical Society, Washington, DC, Chapter 4.

Ostman, C.E. and Colmsjo, A.L. 1988. *Fuel*, 67: 396.

Parson, J.E. 1977. *Asphalts and Road Materials: Modern Technology*. Noyes Data Corporation, Park Ridge, NJ.

Philp, R.P. and de las Heras, F.X. 1992. In *Chromatography. Part B: Applications*, 5th edn., E. Heftmann (Ed.). Elsevier, Amsterdam, the Netherlands, Chapter 21.

Potts, J.D., Hastings, K.E., Chillingworth, R.S., and Unger, H. 1981. In *Upgrading Coal Liquids*, R.F. Sullivan (Ed.). Symposium Series No. 156. American Chemical Society, Washington, DC, Chapter 6.

Rindt, J.R., Hetland, M.D., Sauer, R.S., Sukalski, W.M., and Haug, L.W. 1992. *Fuel*, 71:1033.

Rosal, R., Cabo, L.F., Dietz, F.V., and Sastre, H. 1992. *Fuel Processing Technology*. 31: 209.

Ruberto, R.G., Jewell, D.M., Jensen, R.K., and Cronauer, D.C. 1976. In *Shale Oil, Tar Sands, and Related Fuel Sources*, T.F. Yen (Ed.). Advances in Chemistry Series No. 151. American Chemical Society, Washington, DC, Chapter 3.

Rudnick, L.R. and Whitehurst, D.D. 1981. In *New Approaches in Coal Chemistry*, B.D. Blaustein, B.C. Bockrath, and S. Friedman (Eds.). American Chemical Society, Washington, DC, Chapter 9.

Russell, S.J. and Rimmer, S.M. 1979. In *Analytical Methods for Coal and Coal Products*, C. Karr, Jr. (Ed.). Academic Press, Inc., New York, Vol. III, Chapter 42.

Schiller, J.E. 1978. In *Analytical Chemistry of Liquid Fuel Sources: Tar Sands, Oil Shale, Coal, and Petroleum*, P.C. Uden, S. Siggia, and H.B. Jensen (Eds.). Advances in Chemistry Series No. 170. American Chemical Society, Washington, DC, Chapter 4.

Schmidt, C.E., Sprecher, R.F., and Batts, B.D. 1987. *Analytical Chemistry*, 59: 2027.

Schneider, A., Hollstein, E.J., and Janoski, E.J. 1979. In *Refining of Synthetic Crudes*, M.L. Gorbaty and B.M. Harney (Eds.). Advances in Chemistry Series No. 179. American Chemical Society, Washington, DC, Chapter 6.

Schulman, B.L., Biasca, F.E., Dickenson, R.L., and Simbeck, D.R. 1988. Report No. DOE/FE/60457-H3. Contract No. DE-AC01-84FE60457. United States Department of Energy, Washington, DC.

Schwager, I., Farmanian, P.A., and Yen, T.F. 1978. In *Analytical Chemistry of Liquid Fuel Sources: Tar Sands, Oil Shale, Coal, and Petroleum*, P.C. Uden, S. Siggia, and H.B. Jensen (Eds.). Advances in Chemistry Series No. 170. American Chemical Society, Washington, DC, Chapter 5.

Shams, K., Miller, R.L., and Baldwin, R.M. 1992. *Fuel*, 71: 1015.

Singh, M. 1981. *Hydrocarbon Processing*, 60(6): 138.

Snape, C.E., Derbyshire, F.J., Stephens, H.P., Kottenstett, R.J., and Smith, N.W. 1991. In *Coal Science II*, H.H. Schobert, K.D. Bartle, and L.J. Lynch (Eds.). Symposium Series No. 461. American Chemical Society, Washington, DC, Chapter 14.

Snape, C.E., Ladner, W.R., Petrakis, L., and Gates, B.C. 1984. *Fuel Processing Technology*, 8: 155.

Solomon, P.R. 1981. New approaches in coal chemistry. In *ACS Symposium Series 169*. American Chemical Society, Washington, DC, pp. 61–71.

Solomon, P.R., Hamblen, D.G., Serio, M.A., Yu, Z.-Z., and Charpenay, S. 1993. *Fuel*, 72: 469.

Solomon, P.R., Serio, M.A., and Suuberg, E.M. 1992. *Progress in Energy and Combustion Science*, 18(2): 133.

Speight, J.G. 1992. *Proceedings. 4th International Conference on the Stability and Handling of Liquid Fuels*. United States Department of Energy, Washington, DC, Vol. 1, p. 169.

Speight, J.G. 1994. Fuels, synthetic. In *Kirk-Othmer Encyclopedia of Chemical Technology*, 4th edn., Vol. 12. John Wiley & Sons Inc., New York, p. 126.

Speight, J.G. 2000. *The Desulfurization of Heavy Oils and Residua*, 2nd edn. Marcel Dekker, Inc., New York, Chapter 6.

Speight, J.G. 2011a. *Handbook of Industrial Hydrocarbon Processes*. Gulf Professional Publishing, Elsevier, Oxford, U.K.

Speight, J.G. 2011b. *The Refinery of the Future*. Gulf Professional Publishing, Elsevier, Oxford, U.K.

Speight, J.G. and Long, R.B. 1982. In *Atomic and Nuclear Methods in Fossil Energy Research*, R.H. Filby (Ed.). Plenum Publishing Corporation, New York, p. 295.

Speight, J.G. and Moschopedis, S.E. 1986. *Fuel Processing Technology*, 13: 215.

Speight, J.G. 2007. *The Chemistry and Technology of Petroleum*, 4th edn. Taylor & Francis Group, Boca Raton, FL.

Speight, J.G. 2008. *Synthetic Fuels Handbook: Properties, Processes, and Performance*. McGraw-Hill, New York.

Squires, A.M. 1978. *Applied Energy*, 4: 161.

Stanislaus, A. and Cooper, A.H. 1994. *Catalysis Reviews, Science and Engineering*, 36(1): 75.

Steedman, W. 1985. *Fuel Processing Technology*, 10: 209.

Steedman, W. 1989. In *Spectroscopic Analysis of Coal Liquids*, J. Kershaw (Ed.). Elsevier, Amsterdam, the Netherlands, Chapter 3.

Steffgen, F.W., Schroeder, K.T., and Bockrath, B.C. 1979. *Analytical Chemistry*, 51: 1164.

Stein, S.E. 1981. In *New Approaches in Coal Chemistry*, B.D. Blaustein, B.C. Bockrath, and S. Friedman (Eds.). American Chemical Society, Washington, DC, Chapter 7.

Storch, H.H., Golumbic, N., and Anderson, R.B. 1951. *The Fischer Tropsch and Related Syntheses*. John Wiley & Sons, Inc., New York.

Stranges, A.N. 1983. *Journal of Chemical Education*, 60: 617.

Stranges, A.N. 1987. *Fuel Processing Technology*, 16: 205.

Sturm, G.P. Jr., Dooley, J.E., Thomson, J.S., Woodward, P.W., and Vogh, J.W. 1981. In *Upgrading Coal Liquids*, R.F. Sullivan (Ed.). Symposium Series No. 156. American Chemical Society, Washington, DC, Chapter 1.

Whitehurst, D.D., Mitchell, T.O., and Farcasiu, M. 1980. *Coal Liquefaction: The Chemistry and Technology of Thermal Processes*. Academic Press, Inc., New York.

Wilson, B.W., Pelroy, R.A., Mahlum, D.D., Frazier, M.E., and Wright, C.W. 1983. In *Advanced Techniques in Synthetic Fuels Analysis*, C.W. Wright, W.C. Weimer, and W.D. Felix (Eds.). Technical Information Center, United States Department of Energy, Washington, DC, p. 231.

Wilson, P.J. Jr. and Wells, J.H. 1950. *Coal, Coke, and Coal Chemicals*. McGraw-Hill, Inc., New York.

Wong, J.L. 1987. In *Coal Science and Coal Chemistry*, A. Volborth (Ed.). Elsevier, Amsterdam, the Netherlands, p. 461.

Wood, K.V. 1987. In *Coal Science and Chemistry*, A. Volborth (Ed.). Elsevier, Amsterdam, the Netherlands, p. 207.

Woodward, P.W., Sturm, G.P. Jr., Vogh, J.W., Holmes, S.A., and Dooley, J.E. 1976. Compositional analyses of synthoil from West Virginia, Report No. BERC/RI-76/2. United States Department of Energy, Washington, DC.

Wooton, D.L., Coleman, W.M., Glass, T.E., Dorn, H.C., and Taylor, L.T. 1978. In *Analytical Chemistry of Liquid Fuel Sources: Tar Sands, Oil Shale, Coal, and Petroleum*, P.C. Uden, S. Siggia, and H.B. Jensen (Eds.). Advances in Chemistry Series No. 170. American Chemical Society, Washington, DC, Chapter 3.

Wu, W.R.K. and Storch, H.H. 1968. Bulletin No. 633. United States Bureau of Mines, Washington, DC.

Yarzab, R.F., Given, P.H., Spackman, W., and Davis, A. 1980. *Fuel*, 59: 81.

Zakett, D. and Cooks, R.G. 1981. In *New Approaches in Coal Chemistry*, B.D. Blaustein, B.C. Bockrath, and S. Friedman (Eds.). American Chemical Society, Washington, DC, Chapter 16.

19 Liquefaction Processes

19.1 INTRODUCTION

In any text about coal (but more specifically in any chapter about coal liquefaction), it is appropriate to include a listing of the types of processes available as well as a description of the different processes. However necessary as this might be, it is not the intent to reproduce all of the liquefaction processes here (a virtually impossible task where so many of the processes might be derived from a single concept) but to give selected examples of specific processes.

In terms of process history, processes for coal liquefaction have been varied and have evolved along different paths (Nowacki, 1979; Whitehurst et al., 1980; Alpert and Wolk, 1981). In this particular section, those processes for which an adequate amount of information is available in the literature will be described. However, the status of any particular process may change markedly with time, especially since most of the concepts were first developed during the 1970s. In fact, even at that time, processes that were considered feasible at the beginning of any particular year may have been rejected as being unsuitable by the end of that year. On the other hand, processes are often revived on the basis of advances in technology that were unforeseen at the time of process rejection.

Inclusion of a process in this section is based upon several factors and omission of a process from this section is not to be construed as a negative comment about the suitability of that process. The first factor given consideration is whether or not the concept was close to commercialization, or at least has moved toward commercialization (Figure 19.1) after initial testing had indicated a workable concept. Another factor is that the concept must have shown promise during pre-commercialization testing. A further consideration is concept novelty and the ability of the concept to produce the desired products.

In this way, it is anticipated that a comprehensive overview of the process types available for coal liquefaction will be made available to the reader. Finally, many of the references used here are taken from the original articles of the 1970s and early 1980s because of their relevance to the particular the process. The fact that older references are used is an indication of the relevance of that work to the proposed processes. And the use of dated references should not be construed as meaning that attempts to bring coal liquefaction processes to commercialization are *old hat* and that the production of liquids from coal is *going the way of the dinosaur*. Perhaps the opposite is the case.

Much research continues and the valuable work carried out in the United States and in other countries shows that the efforts to produce liquids from coal are far from dinosaur-type extinction.

19.2 PYROLYSIS PROCESSES

19.2.1 LURGI–RUHRGAS PROCESS

The Lurgi–Ruhrgas process (Figure 19.2) was developed as a low-pressure process for the production of liquids from lower rank coals in Europe. It is based on the premise that when coal is heated very rapidly to temperatures in excess of those required for decomposition, fragmentation is very extensive (Weller et al., 1950) to the point of being "explosive."

In this process (Nowacki, 1979), crushed coal is fed into a mixer where it is heated rapidly to 450°C–600°C (840°F–1110°F) by direct contact with hot, recirculating char particles which have

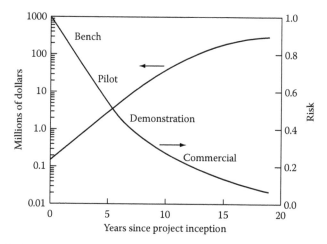

FIGURE 19.1 Relative costs over time for the development of a concept from bench-scale to commercialization.

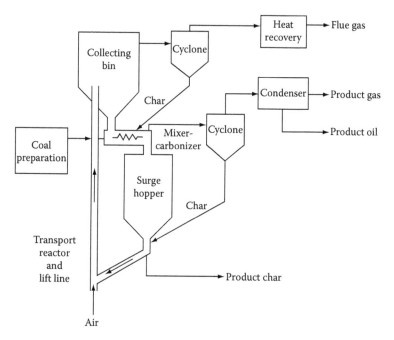

FIGURE 19.2 Lurgi–Ruhrgas process.

been previously heated in a partial oxidation process in an entrained-flow reactor. The gases from the mixer are freed of particulate matter in a cyclone and are then passed through a series of condensers to collect the liquid products; these latter are hydrotreated to yield stable products.

The char forms approximately 50% of the products while the liquid yields are on the order of 18%, and the remainder (approximately 32%) are gases having an approximate calorific value of 700–850 Btu/ft³.

The high gas yield is due to the relatively long residence times of the products in the reactor by virtue of which thermal decomposition occurs to yield gaseous products that are, in actual fact, a secondary product insofar as they are not formed from the coal. The majority of the sulfur originally in the coal occurs in the char.

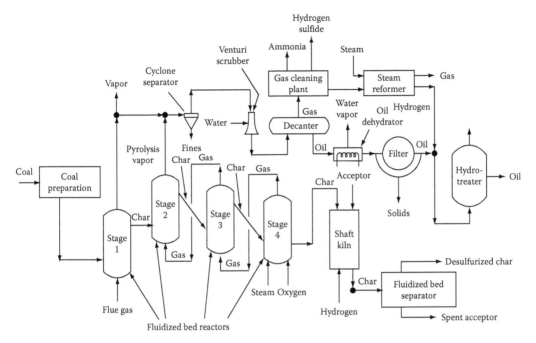

FIGURE 19.3 COED process.

19.2.2 CHAR-OIL ENERGY DEVELOPMENT PROCESS

In the char-oil energy development (COED) process (Figure 19.3), crushed coal (0.125 in. [3 mm]) is dried and heated to successively higher temperatures in a series of reactors, which results in the incremental removal of volatile matter from the coal (Jones, 1973, 1975; McCray et al., 1979). The maximum temperature of each reactor has to be carefully controlled so that it remains appreciably lower than the temperature at which the coal will cake (or agglomerate), thereby reducing the effectiveness of the individual fluidized beds. Thus, reactor temperatures and the number of reactors are dictated by the character of the coal feedstock; in the final reactor, the product char is used to generate heat for the process.

The liquid products are filtered to remove particulate matter and then hydrotreated (370°C–425°C [700°F–800°F]; 1750–2500 psi hydrogen) to produce a synthetic crude oil (Table 19.1) from which the majority (90%) of the nitrogen, oxygen, and sulfur have been removed. As already noted for the Lurgi–Ruhrgas process, the majority of the sulfur originally in the coal occurs in the char and, therefore, combustion of the char may be an environmental problem unless there is some effort made to install stack gas cleaning equipment.

19.2.3 OCCIDENTAL FLASH PYROLYSIS PROCESS

The occidental flash pyrolysis process originally called the Garrett process (Sass, 1974; Howard-Smith and Werner, 1976) (Figure 19.4) involves the pyrolysis of the feed coal particles at a temperature not in excess of 760°C (1400°F) in an entrained system to produce a liquid product, char, and gas.

The process utilizes the concept of short residence time of the feed coal thereby increasing the throughput of coal per unit of cross-sectional area with the subsequent higher production of liquid products.

Milled and screened coal is transported to the reactor by hot nitrogen (approximately 300 psi) where recycled hot char heats the coal rapidly to 510°C–730°C (950°F–1350°F). A portion of the

TABLE 19.1

Product Inspections of Synthetic Crude Oil from Different Coal by the COED Process

	Coal	
Inspection	Illinois No. 6	Utah King
Hydrocarbon-type analysis (% v/v)		
Paraffins	10.4	23.7
Olefins		
Naphthenes	41.4	42.2
Aromatics	48.2	34.1
API gravity	28.6	28.5
ASTM distillation (°F)		
Initial boiling point (IBP)	108	260
50% distilled	465	562
End point (EP)	746	868
Fractionation yields (% w/w)		
IBP–180°F	2.5	
180°F–390°F	30.2	5.0
390°F–525°F	26.7	35.0
390°F–650°F	51.0	65.0
650–EP	16.3	30.0
390–EP	67.3	95.0

Source: Jones, J.F., *Energ. Sources*, 2, 179, 1975.

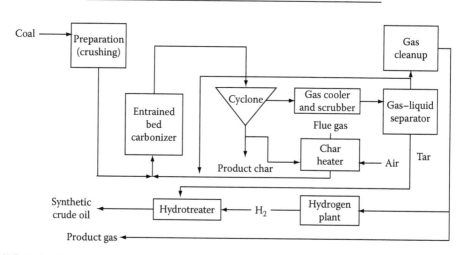

FIGURE 19.4 Occidental flash pyrolysis process.

char product is cooled (as product) but the remainder is heated in a partial oxidation (burning) process to raise its temperature to 650°C–925°C (1200°F–1700°F) for recycle to the reactor.

Cyclones remove fine char particles from the pyrolysis overhead prior to quenching in the collector system. This system consists of two tar recovery stages and a vacuum flash unit. In the first recovery stage, the pyrolysis vapor is condensed by quenching at approximately 99°C (approximately 210°F) to remove the majority of the heavier constituents; cooling to approximately 25°C (80°F) in the second stage causes water and light oils to condense. The vacuum flash unit is employed to separate the higher boiling constituents into fractions of varying volatility.

A related flash pyrolysis process, the Rockwell process, uses hydrogen as the carrier gas in entrained flow reactors which is claimed to improve the yield of products (Oberg and Falk, 1980).

19.2.4 TOSCOAL PROCESS

In the Toscoal process (Carlson et al., 1973, 1975; Cartez and DaDelfa, 1981) (Figure 19.5), crushed coal is charged to a rotating drum which contains preheated ceramic balls whereupon the coal is decomposed to produce the usual liquid product, char, and gas (Table 19.2) The process is analogous to the Tosco process for producing overhead oil from oil shale with the added note that the char replaces the spent shale.

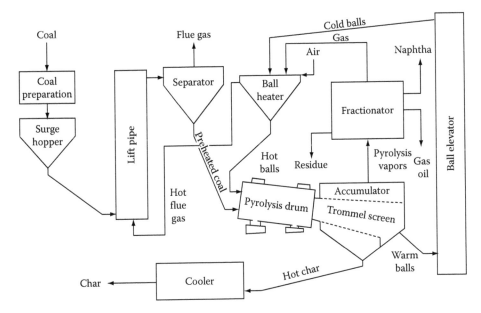

FIGURE 19.5 Toscoal process.

TABLE 19.2

Production of Liquids from As-Mined Wyodak Coal by the Toscoal Process

Temperature			
°C	425	480	520
°F	800	900	970
Yield (% w/w)			
Gas ($\leq C_3$)	6.0	7.8	6.3
Oil ($\geq C_4$)	5.7	7.2	9.3
Char	52.5	50.6	48.4
Water	35.1	35.1	35.1
Recovery (% w/w)	99.3	100.7	99.1

Source: Carlson, F.B. et al., *Preprints. Symposium on Clean Fuels from Coal II,* Institute of Gas Technology, Chicago, IL, 1975, p. 504.

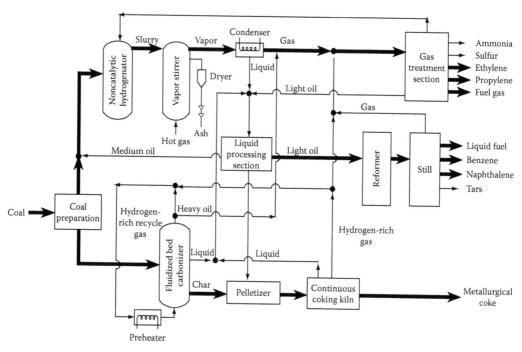

FIGURE 19.6 Clean coke process.

19.2.5 CLEAN COKE PROCESS

The clean coke process (Figure 19.6) involves feeding oxidized clean coal into a fluidized bed reactor at temperatures up to 800°C (1470°F) whereupon the coal reacts to produce tar, gas, and low-sulfur char. Alternatively, the coal may be processed by noncatalytic hydrogenation at 455°C–480°C (850°F–900°F) and pressures of up to 5000 psi hydrogen.

The liquid products from both the carbonization and hydrogenation stages of the process are combined for further processing to yield synthetic liquid fuels and a variety of other coal products. The gases are treated for sulfur and ammonia removal and are also capable of acting as a source of valuable materials other than the usual fuel gas.

19.2.6 COALCON PROCESS

The Coalcon process (Figure 19.7) differs from other processes insofar as it involves the use of a dry, noncatalytic fluidized bed of coal particles suspended in hydrogen gas instead of a fixed-bed or liquid phase system (Nowacki, 1979). The process also converts organic sulfur in the coal to hydrogen sulfide.

In this process, hot, oxygen-free flue gas is employed to heat the coal to approximately 325°C (615°F) and also to carry the coal to a feed hopper where it is pressurized and fed (by gravity) to the reactor. In the reactor, the coal is fluidized (with hydrogen at a pressure some 250 psi above the reactor pressure) whereupon the high temperature and pressure (160°C [1040°F]; 555 psi) brings about conversion of the coal.

A fractionator is employed to subdivide the overhead stream into (1) gases, hydrogen, carbon monoxide, carbon dioxide, and methane; (2) light oil; (3) heavy oil; and (4) water. Most of the char is removed from the base of the reactor, quenched with water, and cooled. The char can then be used as a feed to a Koppers–Totzek gasifier and reacted with oxygen and steam to produce hydrogen for the process.

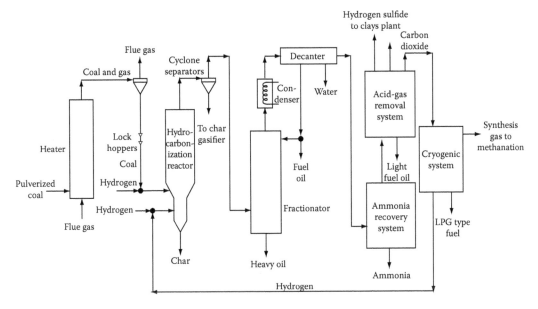

FIGURE 19.7 Coalcon process.

19.3 SOLVENT EXTRACTION PROCESSES

19.3.1 CONSOL SYNTHETIC FUEL PROCESS

This process involves the conversion of (high-sulfur) coal to a solid residue and a synthetic crude oil by extraction of coal with a coal-derived solvent which has been previously hydrogenated (Nowacki, 1979). In this process (Figure 19.8), crushed coal (0.375 in. [10 mm]) is partially dried and heated to 230°C (450°F) before being mixed with the coal-derived solvent. The coal is then extracted in a reactor at 405°C (765°F) and 150–400 psi. The liquid product and the solid residue (which also contains any

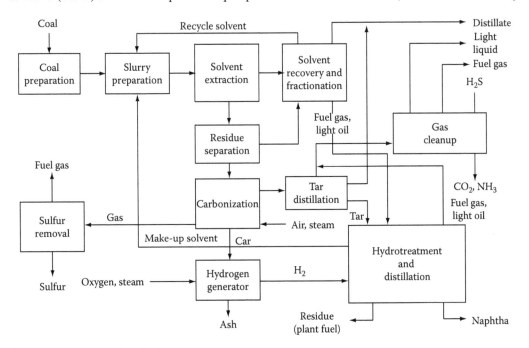

FIGURE 19.8 Consol synthetic fuel process.

TABLE 19.3
Products from Pittsburgh Seam Coal by the CSF Process

Product	Product per Ton of Coal Processed[a]	Product Inspections		
		API Gravity	Btu	% of Total Product
Gas	3.424 mscf		933/scf	
Naphtha	0.52 lb	58°	5.22×10^6/bbl	5.6
Fuel oil	1.52 lb	10.3°	6.4×10^6/bbl	12.8
Ammonia	11.00 lb			
Sulfur	71.00 lb			
Ash	213.60 lb			

[a] Coal contained 14.4% w/w moisture and 10.8% w/w ash.

unreacted coal as well as indigent mineral matter) are then separated in a series of hydrocyclones (Table 19.3). This residue is conveyed as a concentrated slurry to a low-temperature carbonization reactor where it is heated in a fluidized bed of char at 495°C (925°F) and 15 psi to eventually recover more distillate. The overall liquids from the process are fractionated from which the process solvent is derived.

19.3.2 COSTEAM PROCESS

The Costeam process (Appell et al., 1969; Vera and Bell, 1978; Nowacki, 1980) is designed to convert low-rank coals (such as lignite and subbituminous coals) to a low-sulfur product by the noncatalytic reaction of a coal/oil slurry with either carbon monoxide or synthetic gas (carbon monoxide/hydrogen). In the process (Figure 19.9), pulverized coal is introduced into a slurrying tank where it is mixed with process solvent (a process-derived liquid) after which the slurry is fed (with the gas choice) into

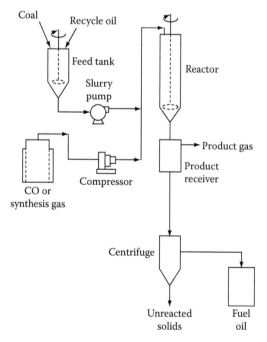

FIGURE 19.9 Costeam process.

a stirred reactor at 425°C (800°F) and 4000 psi. Additional hydrogen for the process can also be provided by the reactor indigent to the coal by means of the water–gas shift reaction.

19.3.3 EXTRACTIVE COKING PROCESS

The extractive coking process is of interest because it is a modification of the delayed coking process (used in the petroleum industry; Speight, 2007) but also involves the use of a hydrogen donor solvent under mild conditions. In this process, crushed coal and hydrogenated recycle solvent are charged to a coke drum whereupon the coal and solvent react at approximately 400°C (750°F) for up to 1 h and at a sufficient pressure of hydrogen to maintain substantially all of the solvent in the liquid phase. At the termination of the reaction, the drum pressure is reduced to allow the solvent and any light ends to vaporize. The contents of the drum are then raised to approximately 450°C (840°F) at which temperature the drum residue cokes to produce a further amount of overhead product and coke; the coke is eventually removed from the drum hydraulically.

The overhead product is fractionated to yield recycle solvent (which is hydrotreated prior to use in the process) and a variety of other products: (1) gases; (2) light oil (C_4: 230°C [C_4: 445°F]); and (3) middle distillates (230°C–400°C [445°F–750°F]).

19.3.4 EXTRACTION BY SUPERCRITICAL FLUIDS

In the extraction by supercritical fluids process, pulverized coal is treated with compressed gases at temperatures on the order of 175°C–200°C (345°F–390°F) whereupon a portion of the coal passes into solution in the compressed gas (Nowacki, 1979; Ceylan and Olcay, 1981). The solution is then transferred to a second vessel and in so doing leaves the inorganic matter and undissolved coal behind. Release of the pressure on the second vessel causes separation of the extract; the gas is recompressed and recycled to the process.

The quoted advantages for this process over the liquid solvent extraction processes are as follows: (1) filtering to remove insoluble residue may not be necessary; (2) recovery of the gaseous solvent appears to be virtually complete; and (3) more mobile liquids of higher hydrogen content are obtained. The disadvantages are as follows: (1) the yield of extract is considerably less than that obtained by conventional (liquid) extraction and (2) operation at the higher pressures may cause a severe escalation of the process economics.

19.3.5 EXXON DONOR SOLVENT PROCESS

The Exxon donor solvent process (Mitchell et al., 1979; Nowacki, 1979) (Figure 19.10) involves the noncatalytic liquefaction of crushed coal in the presence of hydrogen and a hydrogen donor solvent at 425°C–470°C (800°F–880°F) and 1500–2000 psi. The donor solvent may be a mid-distillate boiling range liquid (205°C–455°C [400°F–850°F]) that is also process-derived.

The solvent is first hydrogenated and then mixed with the fresh coal after which the slurry is passed through a preheater and into the reactor. The slurry product is separated by distillation into gas, naphtha, middle distillates, and a vacuum bottoms slurry; this latter product is coked to produce additional liquid products. The process has the capability of producing high yields of low-sulfur liquids from bituminous (or subbituminous) coals and lignites (Table 19.4).

19.3.6 POTT–BROCHE PROCESS

The Pott–Broche process is now mainly of historical interest (Donath, 1963; Dryden, 1963) but several of the processes described previously actually employ the basic Pott–Broche concepts. In the process, bituminous coal is dissolved in a process-derived solvent at 450°C (840°F) and 2200 psi whereupon approximately 75% (daf) of the feed coal can be dissolved and the products are a light oil and a heavy oil.

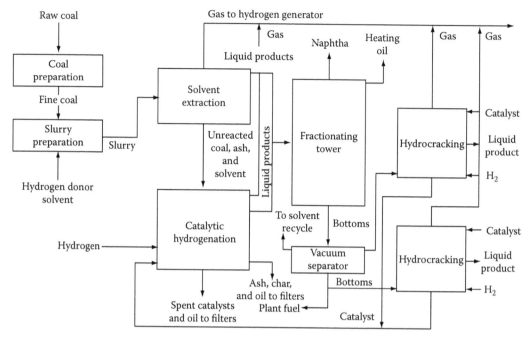

FIGURE 19.10   Exxon donor solvent process.

TABLE 19.4
Yields for Liquefaction of Illinois No. 6 Bituminous Coal from the Exxon Donor Solvent Process

	Liquefaction	Liquefaction Plus Coking
H_2O, CO_2, CO	10	10
H_2S, NH_3	4	4
C_1–C_3	6	9
C_4, C_5	3	4
Naphtha	15	16
Fuel oil	17	25
Liquefaction bottoms	48	—
Coke and ash	—	35
	—	—
Liquid yield	100	100
On dry coal (% w/w)	35	45
H_2 consumption, scf/bbl liquid	5600	4100

19.3.7   Solvent Refined Coal Process

The solvent refined coal (SRC) process has been conveniently described in two forms: the SRC I process and the SRC II process (Schmid, 1975; Baughman, 1978). In the SRC I process, high-sulfur, high-ash coals are converted to a low-ash solid fuel whereas the SRC II process results in a liquid product (rather than a solid product) from a recycle of the product slurry, thereby increasing the conversion of the coal to lower molecular weight species (Figures 19.11 and 19.12).

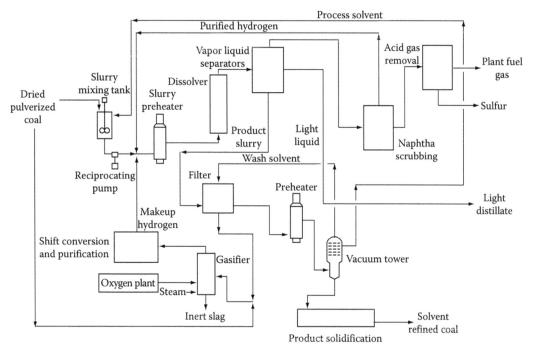

FIGURE 19.11   SRC I process.

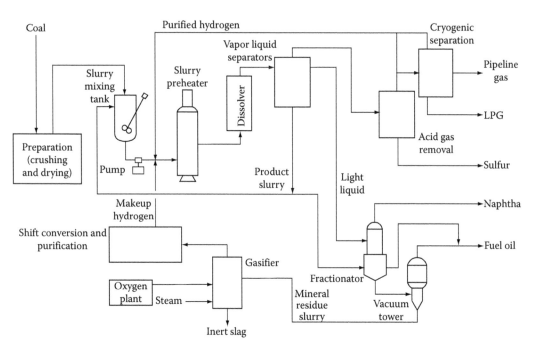

FIGURE 19.12   SRC II process.

TABLE 19.5
Inspection of Products from the
Solvent-Refined Coal Process
(See Also Table 19.6)

	Analysis (% w/w)	
Component	Raw Coal	Product
Carbon	70.7	88.2
Hydrogen	4.7	5.2
Nitrogen	1.1	1.5
Sulfur	3.4	1.2
Oxygen	10.3	3.4
Ash	7.1	0.5
Moisture	2.7	0.0
	100.0	100.0
Volatile matter	38.7	36.5
Fixed carbon	51.5	63.0
Ash	7.1	0.5
Moisture	2.7	0.0
	100.0	100.0
	12,821 Btu/lb	15,768 Btu/lb

Thus, pulverized coal (0.125 in. [3 mm]) is mixed with solvent (approximate solvent/coal ratio = 1.5 for SRC I and 2.0 for SRC II) combined with hydrogen and preheated to 300°C–370°C (700°F–750°F) and charged to a reactor operating at 450°C–465°C (840°F–870°F). The solvent is also thermally decomposed in the reactor with the overall production of substances as light as methane. The hot effluent from the reactor is treated in a series of high-pressure separators in which gases and the light hydrocarbons are removed from the stream. Any undissolved solid material is removed from the resulting slurry which, after removal of the process solvent for recycle, affords the solid solvent-refined coal product (Table 19.5).

In the SRC II process, however, the slurry stream is divided into two streams: one that is recycled to provide liquid for the coal–slurry mixing and one that is fractionated for recovery of the "primary" products of the process. The SRC II process has the option of producing various proportions of solid and liquid products (Table 19.6).

19.3.8 SOLVENT REFINED LIGNITE PROCESS

The solvent refined lignite (SRL) process (Figure 19.13) is a noncatalytic process that is claimed to recover approximately 70% of the lignite feed as a solid product (m.p. 150°C–200°C [300°F–390°F]) as well as quantities of liquids and gases.

Pulverized lignite (undried) is slurried with recycle solvent at temperatures on the order of 480°C (895°F) and pressures up to 3000 psi hydrogen (or synthesis gas). The products from the reactor are separated by a series of pressure differential systems to produce the gases, liquids, and the solid product (which may also contain unconverted lignite and mineral ash) which can be de-ashed using an aromatic naphtha; this solid product can contain as much as 40% w/w of ash and unreacted lignite (Table 19.7).

The major difference between the SRC process and the SRL process is the option (in the latter process) of using synthesis gas in place of hydrogen. This is, in fact, an opportune application since the low-rank high moisture lignites provide the necessary moisture for the water–gas shift reaction.

TABLE 19.6
Inspection of Products from the
Solvent-Refined Coal II (SRC II) Process
(See Also Table 19.5)

	Solid Fuel	Distillate Fuel
Gravity (°API)	−18.3	5.0
Boiling range, °C	425+	
°F	800+	400–800
Fusion point, °C	175	
°F	350	
Flash point, °C	—	
°F	—	168
Viscosity, SUS at 100°F	—	50
Sulfur	0.8	0.3
Nitrogen	2.0	0.9
Heating value (Btu/lb)	16,000	17,300

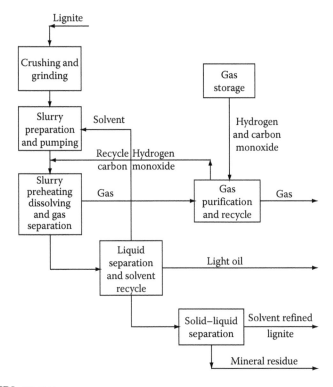

FIGURE 19.13 SRL process.

19.3.9 UOP Process

The UOP process (Nowacki, 1979) involves solvent extraction of the coal in a flow reactor at 400°C–450°C (750°F–840°F) and 1500–3000 psi and, after filtration, the product stream is hydrogenated to afford a low-sulfur liquid product which accounts for the majority (approximately 70%) of the feed coal (daf); gases make up approximately 8% of the feed coal.

TABLE 19.7
Product Yields from the SRL Process

Liquid hourly space velocity	0.90	1.41
Gas hourly space velocity	164	321
Solvent/coal ratio	2.30	1.91
Coal charged ([lb/h]/ft³ reactor)	18.6	32.9
Gas charged (scf/ton coal)	17,700	19,500
H₂ equivalent consumed (% w/w daf coal)	2.35	1.50
Yield (% w/w daf coal)		
Gas	10.4	15.9
Total product yield	69.4	66.8
Light oil	—	(9.5)
Heavy product	(69.4)	(57.3)
H₂O and ash	−6.5	−4.4
Unconverted daf coal	26.7	21.7
Solvent recycle, %	85.9	89.2
Conditions		
Temperature, °C	370	400
°F	700	750
Preheater outlet	395	410
Reactor exit temperature	310	314

Source: Baughman, G.L., *Synthetic Fuels Data Handbook*,
          Cameron Engineers, Denver, CO, 1978.

19.4 CATALYTIC LIQUEFACTION PROCESSES

19.4.1 H-COAL PROCESS

The H-Coal process (Stotler and Schutter, 1979) (Figure 19.14) is an extension of the ebullated-bed technology that is used to convert heavy oils and residua into lighter liquid products. Crushed (−60 mesh) coal is slurried with recycle oil and pumped to a pressure of up to 3000 psi after which the mixture is preheated and introduced into the bottom of the ebullated-bed reactor. The overhead products are fractionated into gases, light distillate, and heavy distillate.

The process conditions (which may be on the order of 345°C–370°C [650°F–700°F]) can be altered appropriately to produce different product slates (Table 19.8). For example, synthetic crude production usually requires higher temperatures and higher hydrogen (partial) pressures than the process conditions for the production of low-sulfur residual oils. This latter operational mode requires that less hydrogen be consumed. In this respect, it must be remembered that the H-Coal process can require as much as 20,000 ft³ hydrogen per ton of coal processed.

19.4.2 SYNTHOIL PROCESS

The Synthoil process (Akhtar et al., 1975) (Figure 19.15) is a hydrodesulfurization process in which crushed, partially dried coal is mixed with process oil to form a slurry (35%–50% w/w coal) which is combined with hydrogen and charged (after preheating) to a fixed-bed reactor to produce a low-sulfur liquid product (Table 19.9).

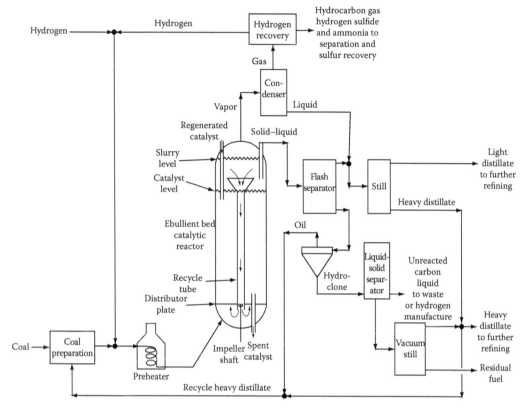

FIGURE 19.14 H-Coal process.

TABLE 19.8

Effect of Processing Options and Coal on the Product Composition from the H-Coal Process

	Illinois		Wyodak Synthetic Crude
	Synthetic Crude	Low-Sulfur Fuel Oil	
Product (% w/w)			
C_1–C_3 hydrocarbons	10.7	5.4	10.2
C_4–200°C distillate	17.2	12.1	26.1
200°C–340°C distillate	28.2	19.3	19.8
340°C–525°C distillate	18.6	17.3	6.5
525°C + residual oil	10.0	29.5	11.1
Unreacted ash-free coal	5.2	6.8	9.8
Gases	15.0	12.8	22.7
Total (100.0 + H_2 reacted	104.9	103.2	106.2
Conversion, %	94.8	93.2	90.2
Hydrogen consumption, scf/ton	18,600	12,200	23,600

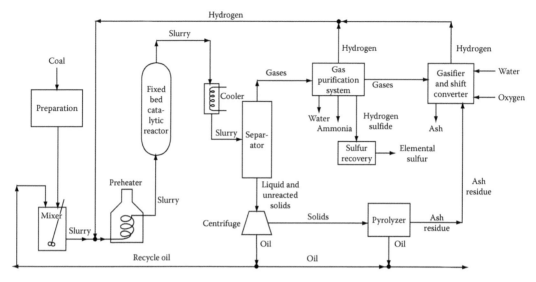

FIGURE 19.15 Synthoil process.

TABLE 19.9

Liquefaction of Kentucky Coal by the Synthoil Process

Process condition

    Liquid feed throughout: 140 (lb/h)/ft^3 reactor volume

    Slurry feed: 45 coal + 55 recycle oil (w/w)

    Hydrogen recycle rate: 125 scf/h

    Pressure: 4000 psi

Temperature: 450°F; 840°F

Sulfur in feed coal % w/w	4.6
Sulfur in recycle oil (product oil) % w/w	0.19
Product inspections	
Solubility characteristics % w/w	
Oil (pentane-soluble)	79.5
Asphaltene (pentane-insoluble)	17.4
Benzene insolubles (organic)	2.1
Ash	1.0
Elemental analysis, % w/w	
Carbon	89.9
Hydrogen	9.2
Nitrogen	0.6
Sulfur	21.30
Viscosity (SSF at 82°C)	21.30
Calorific value (Btu/lb)	17,700

19.4.3 GULF CATALYTIC COAL LIQUIDS PROCESS

In the Gulf process (Figure 19.16), ground coal which has been slurried with a recycle solvent is introduced with hydrogen into a catalytic, fixed-bed reactor at 480°C (900°F) and 2000 psi. The products are flashed to remove gases from the liquids which are then separated into a synthetic crude oil, recycle solvent, and solids. The solids are coked to yield more liquids.

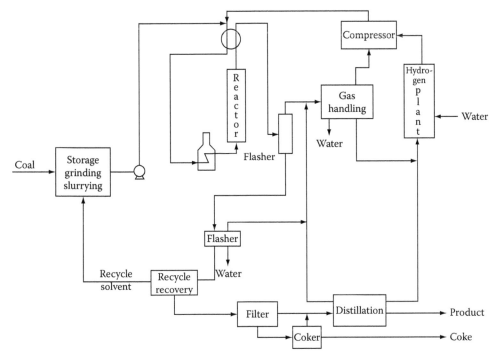

FIGURE 19.16 Gulf catalytic coal liquids (CCL) process.

19.4.4 MULTISTAGE LIQUEFACTION PROCESS

The multistage liquefaction process (Figure 19.17) involves the liquefaction of crushed, dried coal which is slurried with an aromatic recycle solvent and the whole is then charged to a reactor at 415°C–440°C (780°F–825°F) at approximately 2000 psi. An expanded bed reactor system is employed to circumvent the problems that arise through reactor plugging, catalyst deactivation, and inequitable liquid distribution.

The overhead liquid product is fractionated into light and heavy liquids and the residue is de-ashed by treatment with a solvent that promotes the settling of an "under-flow" stream that contains, essentially, all of the ash that occurred in the original coal feed.

19.4.5 BERGIUS PROCESS

The Bergius process (Storch, 1945), like the Pott–Broche process, is more of historical interest than current commercial interest but it was a process that literally paved the way for the development of catalytic liquefaction of coal.

The process involves the conversion of coal (slurried with a heavy oil) in the presence of hydrogen and a catalyst (iron oxide) at 350°F–500°F (830°F–930°F) and 3,000–10,000 psi.

The products were usually separated into light oils, middle distillates, and residuum. The residuum fraction was filtered (or centrifuged) to remove any solid material (unreacted coal, mineral matter, and ash) and the remaining material was used as a recycle oil for the liquefaction stage. The lighter liquid products were generally hydrotreated to produce stable liquid fuels.

19.4.6 UNIVERSITY OF UTAH PROCESS

In the University of Utah process (Figure 19.18), crushed coal is impregnated with zinc chloride (approximately 3% w/w of the coal) and the mix is fed from pressurized lock-hoppers into

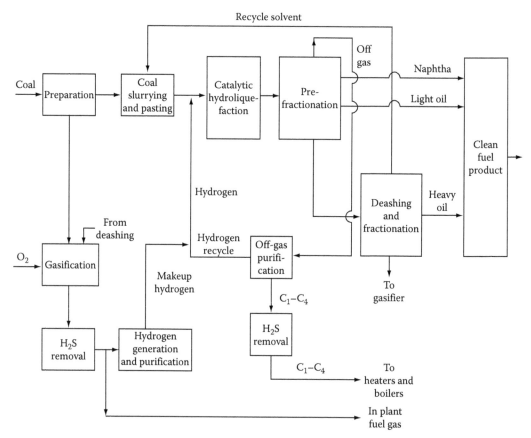

FIGURE 19.17 Multistage liquefaction process.

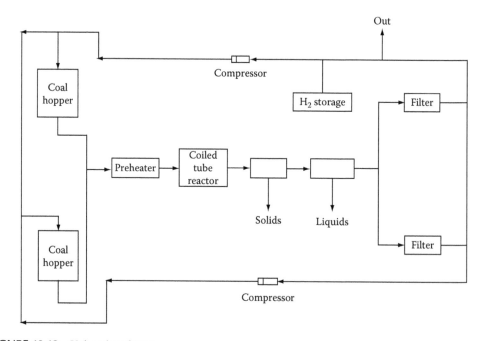

FIGURE 19.18 University of Utah process.

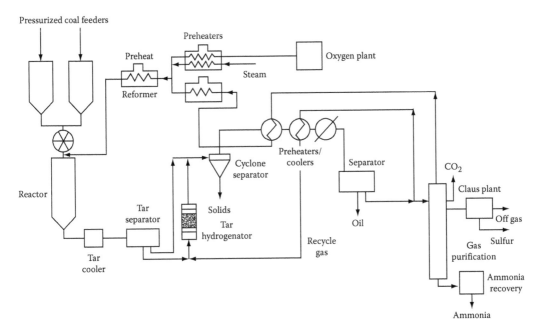

FIGURE 19.19 Schroeder process.

a moving stream of hydrogen leading into a coiled-tube reactor at a temperature of 500°F–550°F (130°F–1020°F) at a pressure of 1650–2100 psi.

The residence time of the solids in the reactor is usually less than 12 s, whereupon approximately 60% of the coal is converted to liquid products. Up to 85% of the zinc can be recovered by water-washing the char; on the other hand, acid washes of the char are reported to raise the recovery of the zinc from 95% to 99%.

The use of zinc chloride as a means of liquefying coal is not surprising since there has been an interest in process catalysis by Lewis acids (e.g., zinc chloride) from the very early days of coal liquefaction technology (Nowacki, 1979). The chemistry involved in the hydrocracking (and other) reactions is not fully understood and therefore any mechanistic speculations are not warranted here.

19.4.7 SCHROEDER PROCESS

In the Schroeder process (Figure 19.19), pulverized, dry coal is hydrogenated at pressures and temperatures of approximately 500°C (930°F) and 2000 psi (13.8) in an entrained system in which the residence time of the feed coal is less than 1 min. The products are separated into light liquids (approximately 30%), heavy liquids (approximately 5%), char (5%), and gases (30%). The heavy liquids can be hydrotreated to produce further quantities of usable (light) liquids. The reaction may be accelerated by the use of ammonium molybdate (1% w/w) that has been impregnated on to the coal.

19.4.8 LIQUID PHASE ZINC CHLORIDE PROCESS

The zinc chloride catalyst process (Figure 19.20) is a means of converting coal into gasoline-type liquids by severe catalytic hydrocracking (Zielke et al., 1969; Nowacki, 1979; Green et al., 1980). Pulverized and dried coal is first slurried with a process-derived recycle oil and then charged to a hydrocracking reactor (355°C–440°C [675°F–825°F]; 1500–3000 psi) where it is mixed with the zinc chloride catalyst in an atmosphere of hydrogen. Liquid products are separated by distillation while the spent catalyst residue (which contains nitrogen compounds, sulfur compounds,

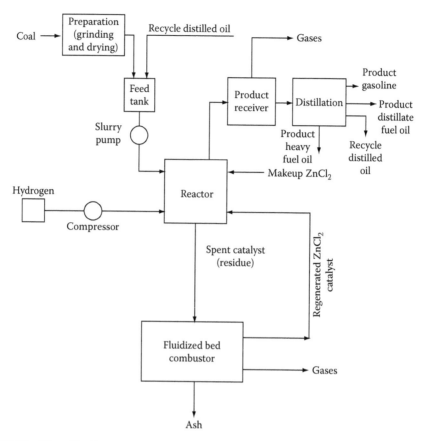

FIGURE 19.20   Zinc chloride process.

and carbonaceous materials) is charged to a fluidized bed reactor for catalyst recovery. The zinc chloride is actually separated as a vapor and, after condensation, is led back to the reactor; fresh makeup zinc chloride is also added to the reactor.

19.4.9   MISCELLANEOUS PROCESSES

Other process concepts similar to those outlined previously have also appeared in the literature from time to time and an example of such a process is the *disposable catalyst process* (Figure 19.21) in which the coal (dried and slurried with a recycle oil) is mixed with an iron-containing (red mud) catalyst after which hydrogen (2000–4000 psi) is added and the whole is introduced into the reactor. The reactions occur in the liquid phase and the liquid product is separated by distillation to produce a liquid fuel oil and a heavier slurrying oil as well as residuum-type oil. This latter product may be processed further to yield further quantities of lighter products.

19.5   INDIRECT LIQUEFACTION PROCESSES

19.5.1   FISCHER–TROPSCH PROCESS

To date, the Fischer–Tropsch is the only commercially operating method for the indirect production of liquids from coal; the Sasol plant in South Africa (Figure 19.22) has been in operation since 1956 (Hoogendorn and Salomon, 1957; Bodle and Vygas, 1974; Singh, 1981; Swart et al., 1981; Dry and Erasmus, 1987; Chadeesingh, 2011).

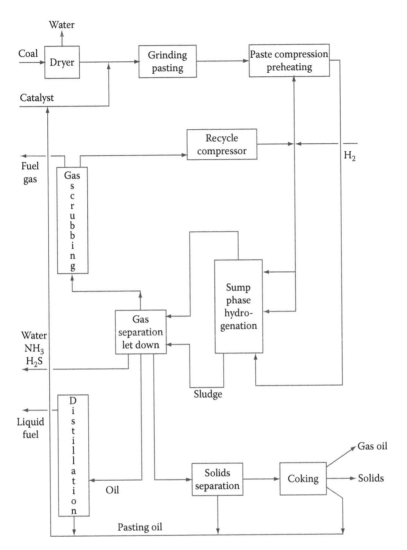

FIGURE 19.21 Disposable catalyst process.

After generation of the synthesis gas, conversion to liquid hydrocarbons, waxes, alcohols, and ketones is achieved using an iron or a cobalt catalyst (Table 19.10) in fixed-bed or entrained-bed reactors. A variety of catalysts, among them magnetite (iron oxide), have been proposed and used for the Fischer–Tropsch synthesis (Kugler and Steffgen, 1979; Cooper et al., 1984; Hindermann et al., 1984; Mirodatos et al., 1984; Moser and Slocum, 1992). Magnesium oxide (MgO) is frequently added as a structural, or surface, promoter, and potassium oxide (or other alkali metal oxide) is often added as a chemical promoter (Dry and Ferreira, 1967, 1968).

The structural promoter functions to provide a stable, high-area catalyst, while the chemical promoter alters the selectivity of the process. The effectiveness of the alkali metal oxide promoter increases with increasing basicity. Increasing the basicity of the catalyst shifts the selectivity of the reaction toward the heavier or longer-chain hydrocarbon products (Dry and Ferreira, 1967). By the proper choice of catalyst basicity and ratio, the product selectivity in the Fischer–Tropsch process can be adjusted to yield from 5% to 75% methane. Likewise, the proportion of hydrocarbons in the gasoline range roughly can be adjusted to produce 0%–40% of the total hydrocarbon yield.

A number of problems are associated with the liquefaction of coal by the Fischer–Tropsch process. First and most obvious is the inherent inefficiency of any process that takes the molecules apart and

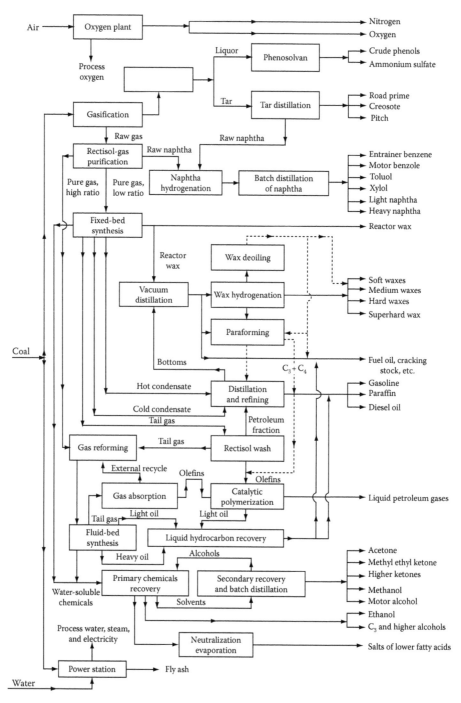

FIGURE 19.22 General flowsheet of the SASOL plant. (From Hoogendorn, J.C. and Salomon, J.M., *Br. Chem. Eng.*, 2, 238, 1957.)

then rebuilds them. Second, the hydrocarbon mixture from the process is largely aliphatic, with a preponderance of straight-chain hydrocarbons. The branched-chain alkanes are primarily mono- and dimethyl-substituted chains. There are few highly branched systems in the product. This results in a product (fuel) with a fairly low octane number (approximately 40), which must be further processed to yield a suitable fuel. Third, there is a need to develop better catalysts that exert even more control

TABLE 19.10
Product Composition from the SASOL Fischer–Tropsch Process

	Composition (% v/v)	
	Fixed-Bed Reactor	Entrained-Bed Reactor
Liquefied petroleum gas (C_3–C_4)	5.6	7.7
Gasoline (C_5–C_{11})	33.4	72.3
Middle oils (diesel, furnace, oil, etc.)	16.6	3.4
Waxy oil	10.3	3.0
Medium wax	11.8	—
Hard wax	18.0	—
Alcohols and ketones	4.3	12.6
Organic acids	Traces	1.0

Source: Bodle, W.W. and Vygas, K.C. *Oil and Gas Journal*, 72, 34, 73, 1974.

over the product distribution. Fourth, the synthesis gas must have a very low sulfur content, since the catalysts are very sensitive to sulfur poisoning (Karn et al., 1963; Madon and Taylor, 1979). Finally, the highly exothermic nature of the reaction causes rather rapid deterioration of the process catalyst and requires effective methods of heat removal from the reactor.

19.5.2 METHANOL SYNTHESIS

The synthesis of methanol from carbon monoxide/hydrogen mixtures is also achieved by a catalytic process according to the following stoichiometric equation (Kasem, 1979; Muetterties and Stein, 1979; Klier, 1984; Lee, 1990; Chadeesingh, 2011) or by way of a carbon dioxide/hydrogen reaction:

$$CO + 2H_2 \rightarrow CH_3OH$$

$$CO + 3H_2 \rightarrow CH_3OH + H_2O$$

Conversion is thermodynamically favored by relatively low temperatures and high pressures and, in the initial attempts to produce methanol from coal, conditions were usually on the order of 300°C–375°C (570°F–705°F) and 4000–5250 psi, although recent improvements in catalyst behavior and performance have led to reductions in the operating pressure to the range 500–1500 psi.

Even though the process is generally regarded as being highly selective, the crude methanol product still requires purification. This is usually accomplished by distillation which removes dimethyl ether and the higher molecular weight alcohols.

Methanol can itself be used as a motor fuel. It has been used as a racing fuel for some time. Its use as a fuel does present some special problems. Methanol is toxic, hygroscopic, and (for a given volume of fuel) contains about half the energy content of gasoline. There are, therefore, good reasons to convert methanol to gasoline.

19.5.3 METHANOL-TO-GASOLINE

Although methanol itself can be used as a transportation fuel, current needs (without engine modifications) dictate the need to convert the methanol to the immediately usable gasoline (Baughman, 1978). The conversion of methanol to gasoline is based on the use of a new type of zeolite (which has a unique channel structure) as well as a series of narrow-pore zeolites.

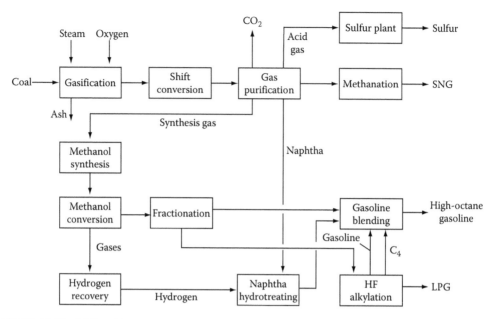

FIGURE 19.23 M-Gasoline process.

The technology of producing gasoline from coal via methanol production (Figure 19.23) involves coal gasification by reaction with oxygen and steam under pressure, the composition of the synthesis gas produced adjusted by a shift reaction (which converts carbon monoxide and steam to hydrogen and carbon dioxide), purified, and then converted to methanol. The crude intermediate methanol, containing about 15% water, is the feed to the M-Gasoline unit, which can be a two-stage fixed-bed conversion unit.

The Mobil M process was developed by Mobil Oil Corporation for the purpose of converting methanol directly into high-octane gasoline. In this process, methanol is first dehydrated to yield dimethyl ether, followed by further dehydration using a ZSM-5 catalyst to yield a series of C_5^+ hydrocarbon products.

The distribution of hydrocarbon products is controlled by the zeolite catalyst. The zeolite is a crystalline aluminosilicate with pores, or cavities. It is the size and geometry of the zeolite pores that control product distribution. Molecules with critical dimensions larger than the pore size cannot be made in the process. Both fixed-bed and fluidized-bed reactors have been employed in developing the Mobil M process.

The crude methanol is vaporized and fed to the first converter stage at approximately 300°C (570°F) and 300 psi, where it is catalytically converted to an equilibrium mixture of dimethyl ether, water, and unchanged methanol:

$$2CH_3OH \rightarrow CH_3OCH_3 + H_2O$$

The first stage product is then combined with a recycle stream and led to the second-stage converter. The inlet temperature is about 345°C (650°F) and the outlet approximately 400°C (750°F). The second-stage converter, filled with catalyst, produces gasoline constituents. The raw product is fractionated, the light ends alkylated, the naphtha recovered from the conversion steps is hydrotreated, and the products blended to produce a gasoline with a 92 research octane-clear (RON) rating.

In the methanol-to-gasoline process (Figure 19.24), methanol is passed in the vapor phase upward through a catalyst bed at 415°C (775°F) and at 25 psi whereupon the methanol is converted to hydrocarbons and water. The catalyst is removed from the products at the top of the reactor, and overhead product then condensed and separated from the water; small quantities of coke may also be produced (Table 19.11). Some degree of catalyst regeneration is necessary from time to time.

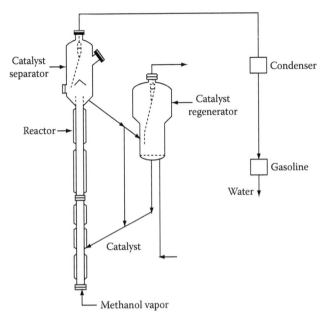

FIGURE 19.24 Mobil methanol-to-gasoline process. (From Baughman, G.L., *Synthetic Fuels Data Handbook*, Cameron Engineers, Denver, CO, 1978.)

TABLE 19.11
Product Yields from the
Methanol-to-Gasoline Process

Bed temperature, °C	410
°F	775
Pressure (psi)	25
Yields, % w/w	
Methanol + ether	0.2
Hydrocarbons	43.5
Water	56.0
CO, CO_2	0.1
Coke	0.2
	100.0
Hydrocarbon products, % w/w	
Light gas	5.6
Propane	5.9
Propylene	5.0
Isobutane	14.5
n-Butane	1.7
Butenes	7.3
Gasoline (C_5^+)	60.0
	100.0

Source: Baughman, G.L., *Synthetic Fuels Data Handbook*, Cameron Engineers, Denver, CO, 1978.

A key characteristic of the M-gasoline process is the ability of the catalysts to selectively produce high-octane gasoline constituents. This selectivity is greater than that obtained from classical Fischer–Tropsch-type catalysts. To illustrate, the gasoline selectivity using Mobil catalysts is stated to be as high as 85%, whereas that from Fischer–Tropsch is on the order of 50%.

19.6 COAL LIQUIDS REFINERY

Refinery feedstocks from coal (*coal liquids*) have not been dealt with elsewhere in this text but descriptions are available from other sources (Speight, 2008, 2011). Once the liquids are produced, the next issue is the means by which these liquids can be refined to produce the necessary fuel products.

The Bergius process was one of the early processes for the production of liquid fuels from coal. In the process, lignite or subbituminous coal is finely ground and mixed with heavy oil recycled from the process. Catalyst is typically added to the mixture and the mixture is pumped into a reactor. The reaction occurs between 400°C and 500°C under a pressure of hydrogen and produces heavy oil, middle oil, gasoline, and gas:

$$nC_{coal} + (n+1)H_2 \rightarrow C_nH_{2n+2}$$

A number of catalysts have been developed over the years, including catalysts containing tungsten, molybdenum, tin, or nickel.

The different fractions can be sent to a refinery for further processing to yield synthetic fuel or a fuel blending stock of the desired quality. It has been reported that as much as 97% of the coal carbon can be converted to synthetic fuel but this very much depends on the coal typo, the reactor configuration, and the process parameters.

However, liquid products from coal are generally different from those produced by petroleum refining, particularly as they can contain substantial amounts of phenols. Therefore, there will always be some question about the place of coal liquids in refining operations. For this reason, there have been some investigations of the characterization and *next-step* processing of coal liquids.

The composition of coal liquids produced from coal depends very much on the character of the coal and on the process conditions and, particularly, on the degree of *hydrogen addition* to the coal. In fact, current concepts for refining the products of coal liquefaction processes have relied, for the most part, on already-existing petroleum refineries, although it must be recognized that the acidity (i.e., phenol content) of the coal liquids and their potential incompatibility with conventional petroleum (including heavy oil) may pose new issues within the refinery system (Speight, 1994, 2007, 2008).

The other category of coal liquefaction processes invokes the concept of the indirect liquefaction of coal. In these processes, the coal is not converted directly into liquid products but involves a two-stage conversion operation in which coal is first converted (by reaction with steam and oxygen) to produce a gaseous mixture that is composed primarily of carbon monoxide and hydrogen (syngas; synthesis gas). The gas stream is subsequently purified (to remove sulfur, nitrogen, and any particulate matter) after which it is catalytically converted to a mixture of liquid hydrocarbon products.

The synthesis of hydrocarbons from carbon monoxide and hydrogen (synthesis gas) (the Fischer–Tropsch synthesis) is a procedure for the indirect liquefaction of coal (Speight, 2008, and references cited therein; Chadeesingh, 2011, and references cited therein).

Thus, coal is converted to gaseous products at temperatures in excess of 800°C (1470°F), and at moderate pressures, to produce synthesis gas:

$$[C]_{coal} + H_2O \rightarrow CO + H_2$$

The gasification may be attained by means of any one of several processes or even by gasification of coal in place (underground, or in situ, gasification of coal).

In practice, the Fischer–Tropsch reaction is carried out at temperatures of 200°C–350°C (390°F–660°F) and at pressures of 75–4000 psi. The hydrogen/carbon monoxide ratio is usually 2.2:1 or 2.5:1. Since up to three volumes of hydrogen may be required to achieve the next stage of the liquids production, the synthesis gas must then be converted by means of the water–gas shift reaction to the desired level of hydrogen:

$$CO + H_2O \rightarrow CO_2 + H_2$$

After this, the gaseous mix is purified and converted to a wide variety of hydrocarbons:

$$nCO + (2n+1)H_2 \rightarrow C_nH_{2n+2} + nH_2O$$

These reactions result primarily in low- and medium-boiling aliphatic compounds suitable for gasoline and diesel fuel. Synthesis gas can also be converted to methanol, which can be used as a fuel, fuel additive, or further processed into gasoline via the Mobil M-gas process.

In terms of liquids from coal that can be integrated into a refinery, this represents the most attractive option and does not threaten to bring on incompatibility problems as can occur when phenols are present in the coal liquids.

A major challenge for refining coal is related to air pollution issues and mining hazards but conversion of coal to liquids is not likely to be abandoned because of the following: (1) abundance; (2) low, relatively nonvolatile prices; (3) gasification that can be the key to environmental acceptance; and (4) synthetic fuels from coal *via gasification* that can be cleaner than the crude oil-derived hydrocarbon fuels derived from petroleum. While such a scheme is not meant to replace other fuel-production systems, it would certainly be a fit into a conventional refinery—gasification is used in many refineries to produce hydrogen and a gasification unit is part of the flexicoking process (Speight, 2011).

REFERENCES

Akhtar, S., Mazzocco, N.J., Weintraub, M., and Yovorsky, P.M. 1975. *Energy Communications*, 1(1): 21.

Alpert, S.B. and Wolk, R.H. 1981. In *Chemistry of Coal Utilization*, Second Supplementary Volume, M.A. Elliott (Ed.). John Wiley & Sons, Inc., New York, Chapter 28.

Appell, H.R., Wender, I., and Miller, R.D. 1969. *Chemical Industry*, 47: 1703.

Baughman, G.L. 1978. *Synthetic Fuels Data Handbook*. Cameron Engineers, Denver, CO.

Bodle, W.W. and Vygas, K.C. 1974. *Oil and Gas Journal*, 72(34): 73.

Carlson, F.B., Atwood, M.T., and Yardumian, L.H. 1973. *Chemical Engineering Progress*, 69(3): 50.

Carlson, F.B., Yardumian, L.H., and Atwood, M.T. 1975. *Preprints. Symposium on Clean Fuels from Coal II*, Institute of Gas Technology, Chicago, IL, p. 504.

Cartez, D.H. and DaDelfa, C.J. 1981. *Hydrocarbon Processing*, 60(2): 111.

Ceylan, R. and Olcay, A. 1981. *Fuel*, 60: 197.

Chadeesingh, R. 2011. The Fischer-Tropsch process. In *The Biofuels Handbook*, J.G. Speight (Ed.). The Royal Society of Chemistry, London, U.K., Part 3, Chapter 5, pp. 476–517.

Cooper, W.C., Cameron, J., and Farrand, B. 1984. In *Catalysis on the Energy Scene*, S. Kaliaguine and A. Mahay (Eds.). Elsevier, Amsterdam, the Netherlands, p. 481.

Donath, E.E. 1963. In *Chemistry of Coal Utilization*, Supplementary Volume, H.H. Lowry (Ed.). John Wiley & Sons, Inc., New York, Chapter 22.

Dry, M.E. and Erasmus, H.B. de W., 1987. *Annual Review of Energy*, 12: 1.

Dry, M.E. and Ferreira, L.C. 1967. *Journal of Catalysis*, 7: 352.

Dry, M.E. and Ferreira, L.C. 1968. *Journal of Catalysis*, 11: 18.

Dryden, I.G.C. 1963. In *Chemistry of Coal Utilization*, Supplementary Volume, H.H. Lowry (Ed.). John Wiley & Sons, Inc., New York, Chapter 6.

Green, C.R., Biasca, F.E., Pell, M., Struck, R.T., and Zielke, C.W. 1980. *Coal Processing Technology*, Vol. 6. American Institute of Chemical Engineers, New York, p. 102.

Hindermann, J.P., Kiennemann, A., and Laurin, M. 1984. In *Catalysis on the Energy Scene*, S. Kaliaguine and A. Mahay (Eds.). Elsevier, Amsterdam, the Netherlands, p. 489.

Hoogendorn, J.C. and Salomon, J.M. 1957. *British Chemical Engineering*, 2: 238.

Howard-Smith, I. and Werner, G.J. 1976. *Coal Conversion Technology*. Noyes Data Corporation, Park Ridge, NJ, p. 71.

Jones, J.F. 1973. *Proceedings of Symposium on Clean Fuels from Coal*, Institute of Gas Technology, Chicago, IL, p. 383.

Jones, J.F. 1975. *Energy Sources*, 2: 179.

Karn, F.S., Shultz, J.F., Kelley, R.E., and Anderson, R.B. 1963. *Industrial and Engineering Chemistry Product Research and Development*, 2: 43.

Kasem, A. 1979. *Three Clean Fuels from Coal: Technology and Economics*. Marcel Dekker, Inc., New York.

Klier, K. 1984. In *Catalysis on the Energy Scene*, S. Kaliaguine and A. Mahay (Eds.). Elsevier, Amsterdam, the Netherlands, p. 439.

Kugler, E.L. and Steffgen, F.W. (Eds.). 1979. *Hydrocarbon Synthesis from Carbon Monoxide and Hydrogen*. Advances in Chemistry Series No. 178. American Chemical Society, Washington, DC, Chapter 8.

Lee, S. 1990. *Methanol Synthesis Technology*. CRC Press, Boca Raton, FL.

Madon, R.J. and Taylor, W.F. 1979. In *Hydrocarbon Synthesis from Carbon Monoxide and Hydrogen*, E.L. Kugler and F.W. Steffgen (Eds.). Advances in Chemistry Series No. 178. American Chemical Society, Washington, DC, Chapter 8.

McCray, F.L., McClintock, N., and Bloom, R. 1979. *Coal Processing Technology*, Vol. 5. American Institute of Chemical Engineers, New York, p. 156.

Mirodatos, C., Dalmon, J.A., and Martic, G.A. 1984. In *Catalysis on the Energy Scene*, S. Kaliaguine and A. Mahay (Eds.). Elsevier, Amsterdam, the Netherlands, p. 505.

Mitchell, W.N., Trachte, K.L., and Zaczepinsky, S. 1979. *Industrial and Engineering Chemistry Product Research and Development*, 18: 311.

Moser, W.R. and Slocum, D.W. (Eds.). 1992. *Homogeneous Transition Metal Catalyzed Reactions*. Advances in Chemistry Series No. 230. American Chemical Society, Washington, DC.

Muetterties, E.L. and Stein, J. 1979. *Chemical Reviews*, 79(6): 479.

Nowacki, P. 1979. *Coal Liquefaction Processes*. Noyes Data Corporation, Park Ridge, NJ.

Nowacki, P. 1980. *Lignite Technology*. Noyes Data Corporation, Park Ridge, NJ.

Oberg, C. and Falk, A. 1980. *Coal Processing Technology*, Vol. 6. American Institute of Chemical Engineers, New York, p. 159.

Sass, A. 1974. *Chemical Engineering Progress*, 70(1): 72.

Schmid, B.K. 1975. *Chemical Engineering Progress*, 75(4): 75.

Singh, M. 1981. *Hydrocarbon Processing*, 60(6): 138.

Speight, J.G. 1994. Fuels, synthetic. In *Kirk-Othmer Encyclopedia of Chemical Technology*, 4th edn., Vol. 12. John Wiley & Sons Inc., New York, p. 126.

Speight, J.G. 2007. *The Chemistry and Technology of Petroleum*, 4th edn., Taylor & Francis Group, Boca Raton, FL.

Speight, J.G. 2008. *Synthetic Fuels Handbook: Properties, Processes, and Performance*. McGraw-Hill, New York.

Speight, J.G. 2011. *The Refinery of the Future*. Gulf Professional Publishing, Elsevier, Oxford, U.K.

Storch, H.H. 1945. In *Chemistry of Coal Utilization*, H.H. Lowry (Ed.). John Wiley & Sons, Inc., New York, Chapter 38.

Stotler, H.H. and Schutter, R.T. 1979. *Coal Processing Technology*, Vol. 5. American Institute of Chemical Engineers, New York, p. 73.

Swart, J.S., Czaikowski, C.J., and Couser, R.E. 1981. *Oil and Gas Journal*, August 31: 62.

Vera, M.J. and Bell, A.T. 1978. *Fuel*, 57: 194.

Weller, S., Clark, E.L., and Pelipetz, M.G. 1950. *Industrial and Engineering Chemistry Research*, 42: 334.

Whitehurst, D.D., Mitchell, T.O., and Farcasiu, M. 1980. *Coal Liquefaction*. Academic Press, Inc., New York.

Zielke, C.W., Struck, R.T., and Gorin, E. 1969. *Industrial and Engineering Chemistry Product Research and Development*, 8(4): 552.

20 Gasification

20.1 INTRODUCTION

The gasification of coal or a derivative (i.e., char produced from coal) is, essentially, the conversion of coal (by any one of a variety of processes) to produce combustible gases (Massey, 1979; Radovic et al., 1983; Radovic and Walker, 1984; Garcia and Radovic, 1986; Calemma and Radovic, 1991; Kristiansen, 1996; Speight, 2008). With the rapid increase in the use of coal from the fifteenth century onward (Nef, 1957; Taylor and Singer, 1957), it is not surprising the concept of using coal to produce a flammable gas, especially the use of the water and hot coal, became commonplace (Elton, 1958). In fact, the production of gas from coal has been a vastly expanding area of coal technology, leading to numerous research and development programs. As a result, the characteristics of rank, mineral matter, particle size, and reaction conditions are all recognized as having a bearing on the outcome of the process, not only in terms of gas yields but also on gas properties (Massey, 1974; Hanson et al., 2002).

Gasification has been considered for many years as an alternative to combustion of solid or liquid fuels. It is easier to clean gaseous mixtures than it is to clean solid or high-viscosity liquid fuels. Clean gas can be used in internal combustion-based power plant that would suffer from severe fouling or corrosion if solid or low-quality liquid fuels were burnt inside them.

Recent developments in gas turbine technology have resulted in combined cycle units with efficiencies close to 60% when generating electricity from natural gas. Gas turbine improvements lead to a number of power plants where "dirty" fuels (usually coal, residual oil or petroleum coke) are gasified, the gas is cleaned and used in a combined cycle gas turbine power plant. Such power plants generally have higher capital cost, higher operating cost, and lower availability than conventional combustion and steam cycle power plants on the same fuel. Efficiencies of the most sophisticated plants have been broadly similar to the best conventional steam plants with losses in gasification and gas cleaning being balanced by the high efficiency of combined cycle power plants. Environmental aspects resulting from the gas cleaning before the main combustion stage have often been excellent, even in plants with exceptionally high levels of contaminants in the feedstock fuels.

Power plants under about 350 MWe cannot use the latest high-efficiency combined cycle technologies. Those below about 250 MWe cannot use particularly high-efficiency steam turbines because of friction losses and leaks in small dimension gas paths. Those below about 100 MWe cannot economically use reheat steam cycles, giving a further efficiency drop. Moving further down in size gives a steady reduction in efficiency of the gas turbine, whichever manufacturer is selected. The scale effect of gas turbine efficiencies is due to flow paths and pressure drops and can only be partly compensated for with additional components such as intercoolers or reheaters.

At smaller sizes, reciprocating engines become relatively more attractive compared with rotating machinery. Their electricity generation efficiency is higher for power generation unit sizes of a few tens of MWe and less. Their major disadvantage is often the frequent and expensive maintenance required.

These technical considerations indicate some of the incentives for large unit size of power plant. Labor requirements per unit of installed capacity provide yet another driver toward large unit size. With the exception of power plants using easily handled fuels (generally natural gas) where there is a significant heat demand as well as a power demand, the trend has been toward larger power plants.

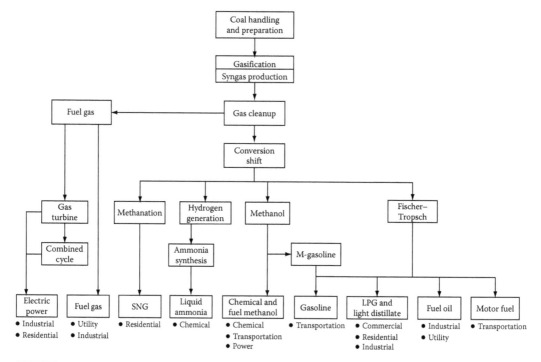

FIGURE 20.1   Uses of the various gaseous products from coal.

Gasification and pyrolysis processes can be classified as entrained gasifiers, fluidized-bed gasifiers (bubbling bed or circulating, atmospheric or pressurized), small industrial-scale gasifiers (fixed-bed or grate, which can be up-draught or down-draught), and hybrid systems.

Gasification agents are normally air, oxygen-enriched air, or oxygen. Steam is sometimes added for temperature control, heating value enhancement or to permit the use of external heat (allothermal gasification). The major chemical reactions break and oxidize hydrocarbons to give a product gas of carbon monoxide, carbon dioxide, hydrogen, and water. Other important components include hydrogen sulfide, various compounds of sulfur and carbon, ammonia, light hydrocarbons, and heavy hydrocarbons (tars).

The products from the gasification of coal may be of low, medium, or high heat (high-Btu) content as dictated by the process as well as by the ultimate use for the gas (Figure 20.1) (Fryer and Speight, 1976; Mahajan and Walker, 1978; Anderson and Tillman, 1979; Cavagnaro, 1980; Bodle and Huebler 1981; Argonne, 1990; Baker and Rodriguez, 1990; Probstein and Hicks, 1990; Lahaye and Ehrburger, 1991; Matsukata et al., 1992).

20.2   GASEOUS PRODUCTS

The products of coal gasification are varied insofar as the gas composition varies with the system employed (Table 20.1). It is emphasized that the gas product must be first freed from any pollutants such as particulate matter and sulfur compounds before further use, particularly when the intended use is a water gas shift or methanation (Cusumano et al., 1978; Probstein and Hicks,1990).

20.2.1   Low-Heat-Content Gas (Low-Btu Gas)

During the production of coal gas by oxidation with air, the oxygen is not separated from the air and, as a result, the gas product invariably has a low heat content (ca. 150–300 Btu/ft [5.6–11.2 MJ/m]). Low- heat-content gas is also the usual product of in situ gasification of coal which is used essentially

TABLE 20.1
Composition of Gas from Various Gasification Processes

Process	Gas Composition (% v/v)						
	CO	H$_2$	CO$_2$	H$_2$O	CH$_4$	H$_2$S	H$_2$/CO
Lurgi (Lurgi Mineraloltechnik GmbH)	9.2	20.1	14.7	50.2	4.7	0.3	2.2
Koppers–Totzek (Heinrich Koppers GmbH)	50.4	33.1	5.6	9.6	0	0.3	0.66
Winkler (Davy Powergas, Inc.)	25.7	32.2	15.8	23.1	2.4	0.3	1.3
Synthane (U.S. Bureau of Mines)	10.5	17.5	18.2	37.1	15.4	0.3	1.7
Bi-Gas (Bituminous Coal Research, Inc.)	22.9	12.7	7.3	48.0	8.1	0.7	0.55
CO$_2$ acceptor (Consolidation Coal Co.)	14.1	44.6	5.5	17.1	17.3	0.03	3.2
HYGAS: steam oxygen (Institute of Gas Technology)	18.0	22.8	18.5	24.4	14.1	0.9	1.3
HYGAS: steam-iron (Institute of Gas Technology)	7.4	22.5	7.1	32.9	26.2	1.5	3.0

TABLE 20.2
Coal Gasification Reactions

$2C + O_2 \rightarrow 2CO$

$C + O_2 \rightarrow CO_2$

$C + CO_2 \rightarrow 2CO$

$CO + H_2O \rightarrow CO_2 + H_2$ (shift reaction)

$C + H_2O \rightarrow CO + H_2$ (water gas reaction)

$C + 2H_2 \rightarrow CH_4$

$2H_2 + O_2 \rightarrow 2H_2O$

$CO + 2H_2 \rightarrow CH_3OH$

$CO + 3H_2 \rightarrow CH_4 + H_2O$ (methanation reaction)

$CO_2 + 4H_2 \rightarrow CH_4 + 2H_2O$

$C + 2H_2O \rightarrow 2H_2 + CO_2$

$2C + H_2 \rightarrow C_2H_2$

$CH_4 + 2H_2O \rightarrow CO_2 + 4H_2$

TABLE 20.3
Coal Gasification Products

Product	Characteristics
Low-Btu gas (150–300 Btu/scf)	Around 50% nitrogen, with smaller quantities of combustible H$_2$ and CO, CO$_2$ and trace gases, such as methane
Medium-Btu gas (300–550 Btu/scf)	Predominantly CO and H$_2$, with some incombustible gases and sometimes methane
High-Btu gas (980–1080 Btu/scf)	Almost pure methane

as a technique for obtaining energy from coal without the necessity of mining the coal. The process is, in essence, a technique for utilization of coal which cannot be mined by other techniques.

Several important chemical reactions (Table 20.2), and a host of side reactions, are involved in the manufacture of low-heat-content gas under the high-temperature conditions employed. Low-heat-content gas contains several components (Table 20.3), four of which are always major components present at levels of at least several percent; a fifth component, methane, is marginally a major component.

The nitrogen content of low-heat-content gas ranges from somewhat less than 33% v/v to slightly more than 50% v/v and cannot be removed by any reasonable means; the presence of nitrogen at these levels makes the product gas "low heat content" by definition. The nitrogen also strongly limits the applicability of the gas to chemical synthesis. Two other noncombustible components (water, H_2O, and carbon dioxide, CO) further lower the heating value of the gas; water can be removed by condensation and carbon dioxide by relatively straightforward chemical means.

The two major combustible components are hydrogen and carbon monoxide; the H_2/CO ratio varies from approximately 2:3 to about 3:2. Methane may also make an appreciable contribution to the heat content of the gas. Of the minor components hydrogen sulfide is the most significant and the amount produced is, in fact, proportional to the sulfur content of the feed coal. Any hydrogen sulfide present must be removed by one, or more, of several procedures (Chapter 23) (Speight, 2007).

Low-heat-content gas is of interest to industry as a fuel gas or even, on occasion, as a raw material from which ammonia, methanol (Chapter 24), and other compounds may be synthesized.

20.2.2 Medium-Heat-Content Gas (Medium-Btu Gas)

Medium-heat-content gas has a heating value in the range of 300–550 Btu/ft (11.2–18.6 MJ/m) and the composition is much like that of low-heat-content gas, except that there is virtually no nitrogen. The primary combustible gases in medium-heat-content gas are hydrogen and carbon monoxide (Kasem, 1979).

Medium-heat-content gas is considerably more versatile than low-heat-content gas; like low-heat-content gas, medium-heat-content gas may be used directly as a fuel to raise steam, or used through a combined power cycle to drive a gas turbine, with the hot exhaust gases employed to raise steam, but medium-heat-content gas is especially amenable to synthesize methane (by methanation), higher hydrocarbons (by Fischer–Tropsch synthesis), methanol, and a variety of synthetic chemicals (Davis and Occelli, 2010; Chadeesingh, 2011). The reactions used to produce medium-heat-content gas are the same as those employed for low-heat-content gas synthesis, the major difference being the application of a nitrogen barrier (such as the use of pure oxygen) to keep diluent nitrogen out of the system.

In medium-heat-content gas, the H_2/CO ratio varies from 2:3 to ca. 3:1 and the increased heating value correlates with higher methane and hydrogen contents as well as with lower carbon dioxide contents. Furthermore, the very nature of the gasification process used to produce the medium-heat-content gas has a marked effect upon the ease of subsequent processing. For example, the CO_2-acceptor product is quite amenable to use for methane production because of the following properties: (1) the desired H_2/CO ratio just exceeding 3:1, (2) an initially high methane content, and (3) relatively low water and carbon dioxide contents. Other gases may require appreciable shift reaction and removal of large quantities of water and carbon dioxide prior to methanation.

20.2.3 High-Heat-Content Gas (High-Btu Gas)

High-heat-content gas is essentially pure methane and often referred to as synthetic natural gas or substitute natural gas (SNG) (Kasem, 1979; c.f. Speight, 1990). However, to qualify as substitute natural gas, a product must contain at least 95% methane; the energy content of SNG is 980–1080 Btu/ft (36.5–40.2 MJ/m). The commonly accepted approach to the synthesis of high-heat-content gas is the catalytic reaction of hydrogen and carbon monoxide:

$$3H_2 + CO \rightarrow CH_4 + H_2O$$

To avoid catalyst poisoning, the feed gases for this reaction must be quite pure, and, therefore, impurities in the product are rare. The large quantities of water produced are removed by condensation and recirculated as very pure water through the gasification system. The hydrogen is usually present in slight excess to ensure that the toxic carbon monoxide is reacted; this small quantity of hydrogen will lower the heat content to a small degree.

The carbon monoxide/hydrogen reaction is somewhat inefficient as a means of producing methane because the reaction liberates large quantities of heat. In addition, the methanation catalyst is troublesome and prone to poisoning by sulfur compounds and the decomposition of metals can destroy the catalyst. Thus, hydrogasification may be employed to minimize the need for methanation:

$$C_{coal} + 2H_2 \rightarrow CH_4$$

The product of hydrogasification is far from pure methane and additional methanation is required after hydrogen sulfide and other impurities are removed.

20.2.4 METHANE

Several exothermic reactions may occur simultaneously within a methanation unit (Table 20.2). A variety of metals have been used as catalysts for the methanation reaction; the most common, and to some extent the most effective methanation catalysts, appear to be nickel and ruthenium, with nickel being the most widely used (Seglin, 1975; Cusumano et al., 1978; Tucci and Thompson, 1979; Watson, 1980). The synthesis gas must be desulfurized before the methanation step since sulfur compounds will rapidly deactivate (poison) the catalysts (Cusumano et al., 1978). A problem may arise when the concentration of carbon monoxide is excessive in the stream to be methanated since large amounts of heat must be removed from the system to prevent high temperatures and deactivation of the catalyst by sintering as well as the deposition of carbon (Cusumano et al., 1978). To eliminate this problem, temperatures should be maintained below 400°C (750°F).

20.2.5 HYDROGEN

Hydrogen is produced from coal-by-coal gasification (Johnson et al., 2007). Although several gasifier types exist, entrained-flow gasifiers are considered most appropriate for producing both hydrogen and electricity from coal since they operate at temperatures high enough (approximately 1500°C [2730°F]) to enable high carbon conversion and prevent downstream fouling from tars and other residuals. Of the three major commercial entrained-flow gasifiers (Shell, GE, and E-Gas), the GE (formerly ChevronTexaco) gasifier is preferred for hydrogen production since the simple vessel design and slurry-feed allow for high operating pressures in the process.

In the process, the coal undergoes three processes in its conversation to synthesis gas (syngas); the first two processes, pyrolysis and combustion, occur very rapidly. In pyrolysis, char is produced as the coal heats up and volatiles are released. In the combustion process, the volatile products and some of the char react with oxygen to produce various products (primarily carbon dioxide and carbon monoxide) and the heat required for subsequent gasification reactions. Finally, in the gasification process, the coal char reacts with steam to produce hydrogen (H_2) and carbon monoxide (CO).

Combustion:

$$2C_{coal} + O_2 \rightarrow 2CO + H_2O$$

Gasification:

$$C_{coal} + H_2O \rightarrow H_2 + CO$$

$$CO + H_2O \rightarrow H_2 + CO_2$$

The resulting syngas is approximately 63% CO, 34% H_2, and 3% CO_2. At the gasifier temperature, the ash and other coal mineral matter liquefies and exits at the bottom of the gasifier as slag, a sand-like inert material that can be sold as a coproduct to other industries (e.g., road building) (Chapter 15). The synthesis gas exits the gasifier at pressure and high temperature and must be cooled prior to the syngas cleaning stage.

Although processes that use the high temperature to raise high-pressure steam are more efficient for electricity production (Chapter 15), full-quench cooling, by which the synthesis gas is cooled by the direct injection of water, is more appropriate for hydrogen production. Full-quench cooling provides the necessary steam to facilitate the water gas shift reaction, in which carbon monoxide is converted to hydrogen and carbon dioxide in the presence of a catalyst.

Water gas shift reaction:

$$CO + H_2O \rightarrow CO_2 + H_2$$

This reaction maximizes the hydrogen content of the synthesis gas, which consists primarily of hydrogen and carbon dioxide at this stage. The synthesis gas is then scrubbed of particulate matter and sulfur is removed via physical absorption (Chapter 23). The carbon dioxide is captured by physical absorption or a membrane and either vented or sequestered.

Unlike pulverized coal combustion plants in which expensive emissions control technologies are required to scrub contaminants from large volumes of flue gas, smaller and less expensive emissions control technologies are appropriate for coal gasification plants since the clean-up occurs in the syngas. The synthesis gas is at high pressure and contains contaminants at high partial pressures, which facilitates clean-up. For this reason, emissions control is both more effective and less expensive in gasification facilities.

Since the synthesis gas is at high pressure and has a high concentration of carbon dioxide, a physical solvent can be used to capture carbon dioxide (Chapter 23), which is desorbed from the solvent by pressure reduction and the solvent is recycled into the system.

At this point, the hydrogen-rich synthesis gas is sufficiently pure for some stationary fuel cell applications and use in hydrogen internal combustion engines. However, for use in vehicles featuring proton exchange membrane fuel cells, the hydrogen must be purified to 99.999% using a pressure swing adsorption (PSA) unit. The high-purity hydrogen exits the PSA unit sufficiently compressed for pipeline transport to refueling stations. The purge gas from the PSA unit is compressed and directed to a combined cycle (gas and steam turbine) for coproduction of electricity.

As with other processes, the characteristics of the coal feedstock (e.g., heating value and ash, moisture, and sulfur content) have a substantial impact on plant efficiency and emissions. As a result, the cost of producing hydrogen from coal gasification can vary substantially depending on the proximity to appropriate coal types.

Coals of the Western United States tend to have lower heating values, lower sulfur contents, and higher moisture contents relative to bituminous coals from the Eastern United States. The efficiency loss associated with high-moisture- and ash-content coals is more significant for slurry-feed gasifiers. Consequently, dry-feed gasifiers, such as the Shell gasifier, may be more appropriate for low-quality coals.

One of the reasons that Powder River Basin (Wyoming) coals are widely used for pulverized coal combustion power plants (despite the relatively low heating value) is the low sulfur content. With the increased restrictions on sulfur dioxide emission regulations, coal combustion power plants looking to avoid expensive and efficiency-reducing flue gas desulfurization retrofits have switched to low-sulfur Powder River Basin coal. There is also the possibility that western coals can be combined with petroleum coke in order to increase the heating value and decrease the moisture content of the gasification feedstock.

Several technologies are being pursued to increase hydrogen conversion efficiency, improve plant reliability, and lower hydrogen costs. These technologies include high-temperature syngas cleaning and carbon dioxide capture, improved hydrogen-rich syngas turbines and air separation units, co-capture of carbon dioxide and hydrogen sulfide (H_2S), solid oxide fuel cell topping cycles for coproduction of electricity, and flexible gasification systems that can operate on a variety of available feedstocks (i.e., various coals, biomass, and waste).

Gasification is one of the critical technologies that enable hydrogen production from solid hydrocarbons such as coal and biomass (Speight, 2008). Gasifiers produce a syngas that has multiple applications and can be used for hydrogen production, electricity generation, and chemical plants. Integrated gasification combined cycle (IGCC) plants utilize the syngas in a combined cycle power plant (gas turbine and steam turbine) to produce electricity (Chapter 15).

With the increasing costs of petroleum, the gasification-based coal refinery is another concept for the production of fuels, electricity, and chemical products (Speight, 2011). Coal gasification has also been used for production of liquid fuels (Fischer–Tropsch diesel and methanol) via a catalytic conversion of synthesis gas into liquid hydrocarbons (Speight, 2008 and references cited therein; Chadeesingh, 2011).

20.2.6 OTHER PRODUCTS

The major products produced by gasification of coal have been described earlier. However, there is a series of products that are called by older (even archaic) names that should also be mentioned here as clarification.

Producer gas is a low-Btu gas obtained from a coal gasifier (fixed-bed) upon introduction of air instead of oxygen into the fuel bed. The composition of the producer gas is approximately 28% v/v carbon monoxide, 55% v/v nitrogen, 12% v/v hydrogen, and 5% v/v methane with some carbon dioxide.

Water gas is a medium-Btu gas which is produced by the introduction of steam into the hot fuel bed of the gasifier. The composition of the gas is approximately 50% v/v hydrogen and 40% v/v carbon monoxide with small amounts of nitrogen and carbon dioxide.

Town gas is a medium-Btu gas that is produced in the coke ovens and has the approximate composition: 55% v/v hydrogen, 27% v/v methane, 6% v/v carbon monoxide, 10% v/v nitrogen, and 2% v/v carbon dioxide. Carbon monoxide can be removed from the gas by catalytic treatment with steam to produce carbon dioxide and hydrogen.

Synthetic natural gas (SNG) is methane obtained from the reaction of carbon monoxide or carbon with hydrogen. Depending on the methane concentration, the heating value can be in the range of high-Btu gases.

20.3 PHYSICOCHEMICAL ASPECTS

Gasification of coal/char in a CO_2 atmosphere can be divided into two stages, the first stage due to pyrolysis (removal of moisture content and devolatilization) which is comparatively at lower temperature and char gasification by different O_2/CO_2 mixtures at high temperature. In N_2 and CO_2 environments from room temperature to 1000°C, the mass loss rate of coal pyrolysis in N_2 is lower than that of CO_2 may be due to the difference in properties of the bulk gases. The gasification process of pulverized coal in O_2/CO_2 environment is almost the same as compared with that in O_2/N_2 at the same oxygen concentration; but this effect is little bit delayed at high temperature. This may be due to the lower rate of diffusion of oxygen through CO_2 and the higher specific heat capacity of CO_2. However, with the increase of O_2 concentration, the mass loss rate of coal also increases and hence it shortens the burnout time of coal. The optimum value of O_2/CO_2 for the reaction of O_2 with the functional group present in the coal sample was found to be about 8%.

The combination of pyrolysis and gasification process can be the unique and fruitful technique as it can save the prior use of gasifying medium and the production of fresh char simultaneously in one process. With the increase of heating rate, coal particles are faster heated in a short period of time and burnt in a higher temperature region, but the increase in heating rate has almost no substantial effect on the combustion mechanism of coal. Also the increase of heating rate causes a decrease in activation energy value. Activation energy values were calculated by different well-known methods at different fractions from 90% to 15% of the original coal within the temperature range of about 400°C–600°C and it was found that Coats–Redfern approach showed the highest value of E and Freeman–Carroll method showed the least value of E at every fraction of converted coal (Irfan, 2009).

20.3.1 REACTIONS

Coal gasification involves the thermal decomposition of coal and the reaction of the coal carbon and other pyrolysis products with oxygen, water, and fuel gases such as methane (Table 20.2).

The presence of oxygen, hydrogen, water vapor, carbon oxides, and other compounds in the reaction atmosphere during pyrolysis may either support or inhibit numerous reactions with coal and with the products evolved. The distribution of weight and chemical composition of the products are also influenced by the prevailing conditions (i.e., temperature, heating rate, pressure, residence time, etc.) and, last but not least, the nature of the feedstock (Wang and Mark, 1992).

If air is used for combustion, the product gas will have a heat content of ca. 150–300 Btu/ft^3 (depending on process design characteristics) and will contain undesirable constituents such as carbon dioxide, hydrogen sulfide, and nitrogen. The use of pure oxygen, although expensive, results in a product gas having a heat content of 300–400 Btu/ft (11.2–14.9 MJ/m) with carbon dioxide and hydrogen sulfide as by-products (both of which can be removed from low- or medium-heat-content [low- or medium-Btu] gas by any of several available processes).

If a high-heat-content (high-Btu) gas (900–1000 Btu/ft [33.5–37.3 MJ/m]) is required, efforts must be made to increase the methane content of the gas. The reactions which generate methane are all exothermic and have negative values (Table 20.4), but the reaction rates are relatively slow and catalysts may, therefore, be necessary to accelerate the reactions to acceptable commercial rates. Indeed, the overall reactivity of coal and char may be subject to catalytic effects. It is also possible that the mineral constituents of coal and char may modify the reactivity by a direct catalytic effect (Cusumano et al., 1978; Wen, 1980; Davidson, 1983; Baker and Rodriguez, 1990; Martinez-Alonso and Tascon, 1991; Mims, 1991).

20.3.2 PROCESS CONCEPTS

While there has been some discussion (Chapters 13 and 16) on the influence of physical process parameters and the effect of coal type on coal conversion, a note is warranted here regarding the influence of these various parameters on the gasification of coal. Most notable effects are those due to coal character, and often due to the maceral content.

In regard to the maceral content, differences have been noted between the different maceral groups with inertinite being the most reactive (Huang et al., 1991). In more general terms of the character of the coal, gasification technologies generally require some initial processing of the coal feedstock with the type and degree of pretreatment, a function of the process and/or the type of coal. For example, the Lurgi process will accept "lump" coal (1 in. [25 mm], to 28 mesh), but it must be noncaking coal (Chapter 2) with the fines removed. The caking, agglomerating coals tend to form a plastic mass in the bottom of a gasifier and subsequently plug up the system thereby markedly reducing process efficiency. Thus, some attempts to reduce caking tendencies are necessary and can involve preliminary partial oxidation of the coal, thereby destroying the caking properties.

TABLE 20.4
Thermodynamic Data for Methanation Reactions

Temperature		Reaction[a]				
K	°C	1	2	3	4	5
A. Heat of reaction, ΔH_f (kcal)						
300	27	−9.838	−49.298	−39.460	−59.136	−41.227
400	127	−9.710	−50.360	−40.650	−60.070	−41.434
500	227	−9.518	−51.297	−41.779	−60.815	−41.499
600	327	−9.292	−52.084	−42.792	−61.376	−41.460
700	427	−9.050	−52.730	−43.680	−61.780	−41.350
800	527	−8.799	−53.248	−44.449	−62.047	−41.190
900	627	−8.549	−53.654	−45.105	−62.203	−40.996
1000	727	−8.304	−53.957	−45.653	−62.261	−40.729
B. Free energy of reaction, ΔF^0 (kcal)						
300	27	−6.827	−33.904	−27.077	−40.731	−28.621
400	127	−5.841	−28.610	−22.769	−34.451	−27.385
500	227	−4.894	−23.062	−18.168	−27.954	−20.111
600	327	−3.991	−17.338	−13.347	−21.329	−15.836
700	427	−3.127	−11.493	−8.366	−14.620	−11.574
800	527	−2.298	−5.567	−3.269	−7.865	−7.332
900	627	−1.500	+0.594	+1.921	−1.079	−3.108
1000	727	−0.729	−6.444	+7.173	+5.715	+1.090
C. Equilibrium constant, log K_p						
300	27	4.973	24.698	19.724	29.670	20.849
400	127	3.191	15.630	12.44	18.822	13.322
500	227	2.139	10.080	7.940	12.219	8.790
600	327	1.453	6.314	4.861	7.768	5.768
700	427	0.976	3.588	2.611	4.564	3.613
800	527	0.628	1.512	0.893	2.148	2.003
900	627	0.364	−0.144	−0.466	0.261	0.755
1000	727	0.159	−1.408	−1.568	−1.248	−0.238

Source: Mills, G.A. and Steffgen, F.W., *Catal. Rev.*, 15, 165, 1975.

[a] Reactions:

1. $CO + H_2O \rightarrow CO_2 + H_2$
2. $CO + 3H_2 \rightarrow CH_4 + H_2O$
3. $2H_2 + 2CO \rightarrow CH_4 + CO_2$
4. $CO_2 + 4H_2 \rightarrow CH_4 + 2H_2O$
5. $2CO \rightarrow CO_2 + C$

Depending on the type of coal being processed and the analysis of the gas product desired, pressure also plays a role in product definition (Figure 20.2). In fact, some (or all) of the following processing steps (Figure 20.3) will be required: (1) pretreatment of the coal (if caking is a problem); (2) primary gasification of the coal; (3) secondary gasification of the carbonaceous residue from the primary gasifier; (4) removal of carbon dioxide, hydrogen sulfide, and other acid gases; (5) shift conversion for adjustment of the carbon monoxide/hydrogen mole ratio to the desired ratio; and (6) catalytic methanation of the carbon monoxide/hydrogen mixture to form methane. If high-heat-content (high-Btu) gas is desired, all of these processing steps are required since coal gasifiers do not yield methane in the concentrations required (Mills, 1969; Graff et al., 1976; Cusumano et al., 1978; Mills, 1982).

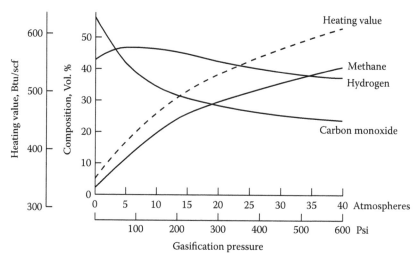

FIGURE 20.2 Variation of gas composition and heating value with pressure. (From Baughman, G.L., *Synthetic Fuels Data Handbook*, Cameron Engineers, Denver, CO, 1978.)

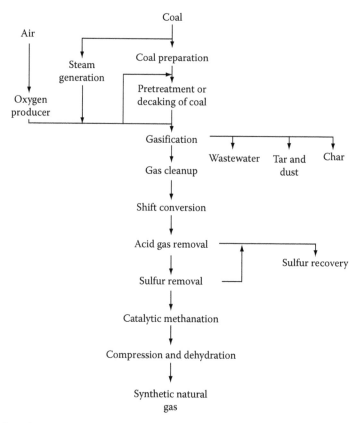

FIGURE 20.3 Steps involved in a coal gasification operation. (From Braunstein, H.M. et al., Environmental health and control aspects of coal conversion: An information overview, Report ORNL-EIS-94, Oak Ridge National Laboratory, Oak Ridge, TN, 1977.)

20.3.2.1 Pretreatment

Some coals display caking, or agglomerating, characteristics when heated (Chapters 8 and 9), and these coals are usually not amenable to treatment by gasification processes employing fluidized-bed or moving-bed reactors; in fact, caked coal is difficult to handle in fixed-bed reactors. The pretreatment involves a mild oxidation treatment which destroys the caking characteristics of coals and usually consists of low-temperature heating of the coal in the presence of air or oxygen.

20.3.2.2 Primary Gasification

Primary gasification involves thermal decomposition of the raw coal via various chemical processes (e.g., see Table 20.2) and many schemes involve pressures ranging from atmospheric to 1000 psi. Air or oxygen may be admitted to support combustion to provide the necessary heat. The product is usually a low-heat-content (low-Btu) gas ranging from a carbon monoxide/hydrogen mixture to mixtures containing varying amounts of carbon monoxide, carbon dioxide, hydrogen, water, methane, hydrogen sulfide, nitrogen, and typical products of thermal decomposition such as tar (themselves being complex mixtures; see Dutcher et al., 1983), hydrocarbon oils, and phenols.

A solid char product may also be produced, and may represent the bulk of the weight of the original coal. This type of coal being processed determines (to a large extent) the amount of char produced and the analysis of the gas product.

20.3.2.3 Secondary Gasification

Secondary gasification usually involves the gasification of char from the primary gasifier. This is usually done by reacting the hot char with water vapor to produce carbon monoxide and hydrogen:

$$C_{char} + H_2O \rightarrow CO + H_2$$

20.3.2.4 Shift Conversion

The gaseous product from a gasifier generally contains large amounts of carbon monoxide and hydrogen, plus lesser amounts of other gases. Carbon monoxide and hydrogen (if they are present in the mole ratio of 1:3) can be reacted in the presence of a catalyst to produce methane (Cusumano et al., 1978).

However, some adjustment to the ideal (1:3) is usually required and, to accomplish this, all or part of the stream is treated according to the waste gas shift (shift conversion) reaction. This involves reacting carbon monoxide with steam to produce carbon dioxide and hydrogen whereby the desired 1:3 mole ratio of carbon monoxide to hydrogen may be obtained:

$$CO + H_2O \rightarrow CO_2 + H_2$$

20.3.2.5 Hydrogasification

Not all high-heat-content (high-Btu) gasification technologies depend entirely on catalytic methanation and, in fact, a number of gasification processes use hydrogasification, that is, the direct addition of hydrogen to coal under pressure to form methane (Anthony and Howard, 1976):

$$C_{char} + 2H_2 \rightarrow CH_4$$

The hydrogen-rich gas for hydrogasification can be manufactured from steam by using the char that leaves the hydrogasifier. Appreciable quantities of methane are formed directly in the primary gasifier and the heat released by methane formation is at a sufficiently high temperature to be used in the steam–carbon reaction to produce hydrogen so that less oxygen is used to produce heat for the steam–carbon reaction. Hence, less heat is lost in the low-temperature methanation step, thereby leading to a higher overall process efficiency.

20.4 PROCESS TYPES AND REACTORS

20.4.1 REACTORS

Some mention must also be made here of the reactor types that have been used for coal gasification processes.

There are three fundamental reactor types: (1) a gasifier reactor, (2) a devolatilizer, and (3) a hydrogasifier (Figure 20.4) with the choice of a particular design depending on the ultimate product gas desired.

Reactors may also be designed to operate either at atmospheric pressure or at high pressure. In the latter type of operation, the hydrogasification process is optimized and the quality of the product gas (in terms of heat, or Btu, content) is improved. In addition, the reactor size may be reduced and the need to pressurize the gas before it is introduced into a pipeline is eliminated (if a high-heat-content gas is to be the ultimate product). However, high-pressure systems may have problems associated with the introduction of the coal into the reactor.

Contacting the solid coal with reactant gases to accomplish the required gasification is a second major mechanical problem. There are three types of contacting methods: a descending bed of solids with up-flowing gas, a fluidized bed of solids, and entrained solids (Figure 20.5).

The descending-bed-of-solids system is often referred to as a moving or fixed bed or, on occasion, a countercurrent descend-bed reactor (Figure 20.6) (Bodle and Huebler, 1981; Probstein and Hicks, 1990; Hobbs et al., 1992). Lumps of coal (ca. 1/8–1 in. [3–25 mm], diameter) are laid down

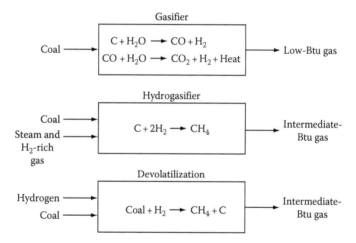

FIGURE 20.4 General chemistry of the various gasification systems.

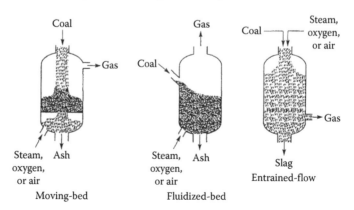

FIGURE 20.5 Various types of reactor systems.

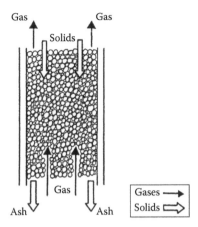

FIGURE 20.6 Countercurrent descending-bed reactor (From Meyers, R.A., Ed., *Coal Handbook*, Marcel Dekker Inc., New York, 1981).

at the top of a vessel while reactant gases are introduced at the bottom of the vessel and flow at relatively low velocity upward through the interstices between the coal lumps.

As the coal descends, it is reacted first by devolatilization using the sensible heat from the rising gas, then hydrogenated by the hydrogen in the reactant gas, and finally burned to an ash. The reactions are, therefore, carried out in a countercurrent fashion.

The fluidized-bed system requires coal to be ground to about mesh or less (Figure 20.7) (see also Arnold et al., 1992). The reactant gases are introduced through a perforated deck near the bottom of the vessel. The volume rate of gas flow is such that its velocity (1–2 ft/s [0.3–0.6 m/s]) is high enough to suspend the solids but not high enough to blow them out of the top of the vessel. The result is a violently "boiling" bed of solids having very intimate contact with the upward-flowing gas. This gives a very uniform temperature distribution. The solid flows rapidly and repeatedly from bottom to top and back again, while the gas flows rather uniformly upward. The reactor is said to be completely back-mixed and no countercurrent flow is possible. If a degree of countercurrent flow is desired, two or more fluid-bed stages are placed one above the other. Reaction rates are faster than in the moving bed because of the intimate contact between gas and solids and the increased solids surface area due to the smaller particle size.

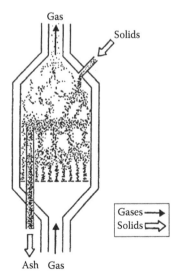

FIGURE 20.7 Fluidized-bed reactor (From Meyers, R.A., Ed., *Coal Handbook*, Marcel Dekker Inc., New York, 1981).

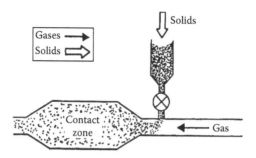

FIGURE 20.8 Entrained-flow reactor (From Meyers, R.A., Ed., *Coal Handbook*, Marcel Dekker Inc., New York, 1981).

The entrained flow reactor (Figure 20.8) uses a still finer grind of coal (80% through 200 mesh)—fine enough that it can be conveyed pneumatically by the reactant gases. Velocity of the mixture must be about 20 ft/s (6.1 m/s) or higher depending upon the fineness of the coal. In this case, there is little or no mixing of the solids and gases, except when the gas initially meets the solids.

In addition, the extensive variety of gasification processes that are being developed are all influenced to a large extent by the following mechanics: (1) feeding solid coal into reactors (often at high pressure), (2) contacting the coal with reactant gases, (3) removing solid (or liquid) ash (slag), and (4) collecting of fine partially reacted dust (carried out with the gaseous products).

20.4.2 Process Types

In any text about coal (but more specifically in any chapter about coal gasification), it is appropriate to include a listing of the types of processes available as well as a description of the different processes. Thus, it is the intent here to give selected examples of specific processes. This is in direct contrast to the inclusion of an appendix to the chapter on coal liquefaction (Chapters 18 and 19); in fact, the situation is quite different with respect to liquefaction processes.

In contrast to liquefaction processes, gasification processes have been evolving since the early days of the nineteenth century when town gas became a common way of bringing illumination and heat not only to factories but also to the domestic consumer. But it must not be assumed that all processes included here have been successful in the move to commercialization. Some still remain conceptual but are included here because of their novelty and/or their promise for future generations of gasifiers.

There has been a general tendency to classify gasification processes by virtue of the heat content of the gas which is produced; it is also possible to classify gasification processes according to the type of reactor vessel and whether or not the system reacts under pressure. However, for the purposes of the present text gasification processes are segregated according to the bed types, which differ in their ability to accept (and use) caking coals.

Thus, gasification processes can be generally divided into four categories based on reactor (bed) configuration: (1) fixed bed, (2) fluidized bed, (3) entrained bed, and (4) molten salt (Table 20.5).

20.4.2.1 Fixed-Bed Processes

In a fixed-bed process the coal is supported by a grate, and combustion gases (steam, air, oxygen, etc.) pass through the supported coal whereupon the hot produced gases exit from the top of the reactor. Heat is supplied internally or from an outside source, but caking coals cannot be used in an unmodified fixed-bed reactor.

20.4.2.2 Fluid-Bed Processes

The fluidized-bed system uses finely sized coal particles and the bed exhibits liquid-like characteristics when a gas flows upward through the bed. Gas flowing through the coal produces turbulent lifting and separation of particles and the result is an expanded bed having greater coal surface area to promote the chemical reaction, but such systems have only a limited ability to handle caking coals.

TABLE 20.5

Categories of Gasification Processes

A. Fixed-bed processes

    1. Lurgi (dry ash and slagging)

    2. Wellman–Galusha

    3. Foster–Wheeler Stoic

    4. Slagging fixed-bed (Grand Forks Energy Research Center)

    5. Woodall–Duckham

B. Fluidized-bed processes

    1. Winkler

    2. Hydrane

    3. Hygas

    4. Synthane

    5. U-Gas

    6. Union Carbide (Coalcon)

    7. Westinghouse pressurized

    8. Carbon dioxide acceptor

    9. Agglomerating burner

    10. COED/COGAS

C. Entrained-bed processes

    1. Koppers–Totzek

    2. Bi-Gas

    3. Combustion engineering entrained bed

    4. Texaco

D. Molten salt

    1. Atgas

    2. Atomics International (Rockgas)

    3. Pullman-Kellogg

Source: Braunstein, H.M. et al., Environmental health and control aspects of coal conversion: An information overview, Report ORNL-EIS-94, Oak Ridge National Laboratory, Oak Ridge, TN, 1977.

20.4.2.3 Entrained-Bed Processes

An entrained-bed system uses finely sized coal particles blown into the gas steam prior to entry into the reactor and combustion occurs with the coal particles suspended in the gas phase; the entrained system is suitable for both caking and noncaking coals. The molten salt system employs a bath of molten salt to convert the charged coal (Cover et al., 1973; Howard-Smith and Werner, 1976; Koh et al., 1978).

20.4.2.4 Molten Salt Processes

Molten salt processes, as the name implies, use a molten medium of an inorganic salt to generate the heat to decompose the coal into products.

In molten bath gasifiers, crushed coal, steam air, and/or oxygen are injected into a bath of molten salt, iron, or coal ash. The coal appears to *dissolve* in the melt where the volatiles crack and are converted into carbon monoxide and hydrogen. The fixed carbon reacts with oxygen and steam to produce carbon monoxide and hydrogen. Unreacted carbon and ash float on the surface from which they are discharged.

High temperatures, around 900°C (1650°F) and above, depending on the nature of the melt, are required to maintain the bath molten. Such temperature levels favor high reaction rates and throughputs and low residence times. Consequently, tars and oils are not produced in any great quantity, if at all.

Gasification may be enhanced by the catalytic properties of the melt used. Molten salts, which are generally less corrosive and have lower melting points than molten metals, can strongly catalyze the steam–coal reaction and lead to very high conversion efficiencies.

20.5   GASIFICATION OF COAL WITH BIOMASS AND WASTE

Pyrolysis and gasification of fossil fuels, biomass materials, and wastes have been used for many years to convert organic solids and liquids into useful gaseous, liquid, and cleaner solid fuels.

Gasification generally has a nonnegative environmental image whereas combustion has a negative public image. The public view is not always based on objective analysis of emissions and other effects but nevertheless it is still a very important factor. In fact, the technology of co-gasification can result in very clean power plants using a range of fuels but there are considerable economic and environmental challenges.

Waste may be municipal solid waste (MSW), which has had minimal presorting, or refuse-derived fuel (RDF), which has had significant pretreatment, usually mechanical screening and shredding. Other more specific wastes, possibly including petroleum coke, may provide niche opportunities for co-utilization.

Currently, neither biomass nor wastes are produced, or naturally gathered at sites in quantities sufficient to fuel a modern large and efficient power plant. The disruption, transport issues, fuel use, and public opinion all act against gathering hundreds of MWe worth of such fuels at a single location. Biomass or waste-fired power plants are therefore inherently limited in size and hence in efficiency, labor costs per unit electricity produced, and in other economies of scale. The production rates of municipal refuse follow reasonably predictable patterns over time periods of a few years. Recent experience with the very limited current *biomass for energy* harvesting has shown unpredictable variations in harvesting capability with long periods of zero production over large areas during wet weather and the foot and mouth outbreak.

The situation is very different for coal. This is generally mined or imported and thus large quantities are available from a single source or a number of closely located sources, and supply has been reliable and predictable. However, the economics of new coal-fired power plants of any technology or size have not encouraged any new coal-fired power plant in the gas generation market.

Combining biomass, refuse, and coal overcomes the potential unreliability of biomass, the potential longer-term changes in refuse, and the size limitation of a power plant using only waste and/or biomass. It also allows benefit from a premium electricity price for electricity from biomass and the gate fee associated with waste. If the power plant is gasification-based, rather than direct combustion, further benefits may be available. These include a premium price for the electricity from waste, the range of technologies available for the gas to electricity part of the process, gas cleaning prior to the main combustion stage instead of after combustion, and public image, which is currently generally better for gasification than for combustion. These considerations lead to the current study of co-gasification of wastes/biomass with coal (Speight, 2008).

For large-scale power generation (>50 MWe), the gasification field is dominated by plant based on the pressurized, oxygen-blown, entrained flow or fixed-bed gasification of fossil fuels. Entrained gasifier operational experience to date has largely been with well-controlled fuel feedstocks with short-term trial work at low co-gasification ratios and with easily handled fuels.

There is less single-fuel experience with the British Gas Lurgi (BGL) than with entrained gasifiers. However, the Lurgi gasifier is better suited to difficult-to-mill feedstocks than are entrained gasifiers and has the most operational experience with fuels of widely differing mechanical properties.

The use of waste materials as co-gasification feedstocks may attract significant disposal credits. Cleaner biomass materials are renewable fuels and may attract premium prices for the electricity generated. Availability of sufficient fuel locally for an economic plant size is often a major issue, as is the reliability of the fuel supply. The use of more predictably available coal alongside these fuels overcomes some of these difficulties and risks. Coal could be regarded as the "flywheel" which keeps the plant running when the fuels producing the better revenue streams are not available in sufficient quantities.

Coal characteristics are very different to younger hydrocarbon fuels such as biomass and wastes. Hydrogen-to-carbon ratios are higher for younger fuels, as is the oxygen content. This means that the reactivity is very different under gasification conditions. Gas cleaning issues can also be very different, with sulfur a major concern for coal gasification but chlorine compounds and tars more important for waste and biomass gasification. There are no current proposals for adjacent gasifiers and gas cleaning systems, one handling biomass or waste and one coal, alongside each other and feeding the same power production equipment. However, there are some advantages to such a design compared with mixing fuels in the same gasifier and gas cleaning system.

Most small- to medium-sized biomass/waste gasifiers are air blown, operating at atmospheric pressure and at temperatures in the range of 800°C–100°C (1470°F–2190°F). They face very different challenges to large gasification plants—the use of small-scale air separation plant should oxygen gasification be preferred. Pressurized operation, which eases gas cleaning, may not be practical.

Biomass fuel producers, coal producers and, to a lesser extent, waste companies are enthusiastic about supplying co-gasification power plant and realize the benefits of co-gasification with, or use on the same site as, "competitor" fuels.

The benefits of a co-gasification technology involving coal and biomass include the use of a reliable coal supply with gate-fee waste and biomass which allows the economies of scale from a larger plant than could be supplied just with waste and biomass.

The United States has wide experience in gasification technologies, components for fuel handling, general engineering, scientific understanding, and project development. In addition, the technology offers a future option for refineries for hydrogen production and fuel development (Speight, 2011).

Oil refineries and petrochemical plants are opportunities for gasifiers when the hydrogen is particularly valuable (Speight, 2011).

Electricity production or combined electricity and heat production remain the most likely area for the application of gasification or co-gasification. The lowest investment cost per unit of electricity generated is the use of the gas in an existing large power station. This has been done in several large utility boilers, often with the gas fired alongside the main fuel. This option allows a comparatively small thermal output of gas to be used with the same efficiency as the main fuel in the boiler as a large, efficient steam turbine can be used. It is anticipated that addition of gas from a biomass or wood gasifier into the natural gas feed to a gas turbine is technically possible, but there will be concerns as to the balance of commercial risks to a large power plant and the benefits of using the gas from the gasifier.

The use of fuel cells with gasifiers is frequently discussed but the current cost of fuel cells is such that their use for mainstream electricity generation is uneconomic.

Furthermore, the disposal of municipal and industrial wastes has become an important problem because the traditional means of disposal, landfill, has become environmentally much less acceptable than previously. New, much stricter regulation of these disposal methods will make the economics of waste processing for resource recovery much more favorable. One method of processing waste streams is to convert the energy value of the combustible waste into a fuel. One type of fuel attainable from wastes is a low heating value gas, usually 100–150 Btu/scf, which can be used to generate process steam or electricity (Gay et al. 1980). Co-processing such waste with coal is also an option (Speight, 2008).

20.6 UNDERGROUND GASIFICATION

This section outlines the methods available for the in situ gasification of coal. In terms of the combustion process itself, both forward and reverse combustion are employed. Forward combustion is defined as movement of the combustion front in the same direction as the injected air whereas reverse combustion involves movement of the combustion front moves in the opposite direction to the injected air (Bodle and Huebler, 1981).

The concept of the gasification of coal in the ground is not new; in 1868, Siemens in Germany proposed that slack and waste be gasified in the mine without any effort being made to bring this waste material to the surface. However, it was left to Mendeleev (in 1888) to actually propose that coal be gasified in the undisturbed (in situ) state.

A further innovative suggestion followed in 1909 in England when Betts proposed that coal be gasified underground by ignition of the coal at the base of one shaft (or borehole) and that a supply of air and steam be made available to the burning coal and the product gases be led to the surface through a different borehole. The utility of such a concept was acknowledged in 1912 when it was suggested that coal be combusted in place to produce a mixture of carbon monoxide and steam which could then be used to drive gas engines at the mine site for the production of electricity.

A considerable volume of investigative work has since been performed on the in situ gasification of coal in the former USSR, but it is only in recent years that the concept has been revived in Russia, Western Europe, and North America as a means of gas (or even liquids) production (Elder, 1963; Gregg and Edgar, 1978; Thompson, 1978; King and Magee, 1979; Olness, 1981; Zvyaghintsev, 1981).

The aim of the underground (or in situ) gasification of coal is to convert the coal into combustible gases by combustion of a coal seam in the presence of air, oxygen, or oxygen and steam. Thus, not only could mining and the ever present dangers (Chapter 5) be partially or fully eliminated, but the usable coal reserves would be increased because seams that were considered to be inaccessible, unworkable, or uneconomical to mine could be put to use. In addition, strip mining and the accompanying environmental impacts, the problems of spoil banks, acid mine drainage, and the problems associated with the use of high-ash coal would be minimized or even eliminated.

The principles of underground gasification are very similar to those involved in the above-ground gasification of coal. The concept involves the drilling and subsequent linking of two boreholes (Figure 20.9) so that gas will pass between the two (King and Magee, 1979). Combustion is then initiated at the bottom of one borehole (injection well) and is maintained by the continuous injection of air. In the initial reaction zone (combustion zone), carbon dioxide is generated by the reaction of

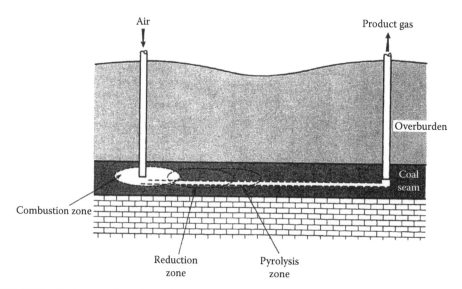

FIGURE 20.9 Underground coal gasification system. (From Berkowitz, N., *An Introduction to Coal Technology*, Academic Press, Inc., New York, 1979.)

oxygen (air) with the coal after which the carbon dioxide reacts with coal (partially devolatilized) further along the seam (reduction zone) to produce carbon monoxide:

$$C_{coal} + O_2 \rightarrow CO_2$$

$$C_{coal} + CO_2 \rightarrow 2CO$$

In addition, at the high temperatures that can frequently occur, moisture injected with oxygen or even moisture inherent in the seam may also react with the coal to produce carbon monoxide and hydrogen:

$$C_{coal} + H_2O \rightarrow CO + H_2$$

The coal itself may also decompose (pyrolysis zone) by the ever-increasing temperature to produce hydrocarbons and tars which contribute to the product gas mix. The gas so produced varies in character and composition but usually falls into the low-heat (low-Btu) category ranging from 4.7 to 6.5 MJ/m (125–175 Btu/ft) (King and Magee, 1979).

Both shaft and shaftless systems (and combinations of these systems) constitute the methods for underground gasification. Selection of the method to be used depends on such parameters as the permeability of the seam, the geology of the deposit, the seam thickness, depth, inclination, and the amount of mining desired. The shaft system involves driving large diameter openings into the seam and may therefore require some underground labor, whereas the shaftless system employs boreholes for gaining access to the coal and therefore does not require mining.

20.7   ENVIRONMENTAL ASPECTS

During the next five decades fossil fuels, of which the use of coal will be prominent, are expected to provide the lowest-cost options for producing liquid and gaseous fuels (Speight, 2007, 2008, 2009, 2011).

It is estimated that global coal resources can last more than 100 years at current consumption rates and considering that the United States has the largest coal reserves in the world (Chapters 1 and 26), producing fuels from coal would improve the energy security of the United States by reducing dependence on imported transportation fuels. Moreover, unlike natural gas or petroleum, the price of coal has remained low and stable over the past decade. If this trend continues, coal gasification will offer one of the lowest cost mechanisms for producing hydrogen in the near term and can help promote a hydrogen economy until renewable technologies are improved.

Despite these benefits, the use of coal also presents several serious environmental challenges, including significant air quality, climate change, and mining impacts. However, coal gasification technologies have been demonstrated that provide order-of-magnitude reductions in criteria pollutant emissions and, when coupled with carbon capture and sequestration (CCS), the potential for significant reductions in carbon dioxide emissions. Therefore, although coal is a finite nonrenewable resource, coal-derived hydrogen with carbon capture and storage can increase domestic energy independence, provide near-term carbon dioxide and criteria pollutant reduction benefits, and facilitate the transition to a more sustainable hydrogen-based transportation system. Carbon capture and storage is one of the critical enabling technologies that could lead to coal-based hydrogen production for use as a transportation fuel. However, there are other risks to the environment that need to be addressed.

For example, the risk of fire or explosion is a significant concern which has to be addressed at an early stage of the design of every gasifier. Co-gasification units are likely to be small and have limited availability of highly qualified staff, or may even run unmanned. Hence, safety systems have to be automated and comprehensive. These may add significantly to the cost of a commercial unit compared with a research prototype.

The relative merits of air and oxygen as the oxidant for co-gasification are closely balanced. However, oxygen and nitrogen both have major safety concerns, as do air separation units. Oxygen will cause combustion of materials at temperatures where they are stable in air. Nitrogen is perhaps an even greater danger as an oxygen displacer leading to nitrogen asphyxiation.

Gasifiers require flares for start-up, shutdown, and fault conditions as the gas, inherently toxic before the combustion stage, cannot be simply vented. Although flares are standard items on oil refineries, they could be a source of difficulties for gasification power plant. They have to be sized for the full gas production rate and hence have a large potential thermal output. In addition, the gas from gasifiers has very different radiative properties to natural gas combustion in flares. Visual effects of flare operation and heat fluxes to surrounding equipment and buildings can be significant. The potential for ignition of stored fuel needs particular assessment. Flares may appear to be a simple component of a gasification power plant but they have caused considerable problems on many gasification power plant designs and have often had to be redesigned or relocated.

The increasing costs of conventional waste management and disposal options, and the desire, in most developed countries, to divert an increasing proportion of mixed organic waste materials from landfill disposal, for environmental reasons, will render the investment in energy from waste projects increasingly attractive. Most new projects involving the recovery of energy from municipal waste materials will involve the installation of new purpose-designed incineration plant with heat recovery and power generation. However, advanced thermal processes for MSW which are based on pyrolysis or gasification processes are also being introduced to the. These processes offer significant environmental and other attractions and these processes will likely have an increasing role to play but the rate of increase of use is difficult to predict.

In addition to the product gas, coal gasification plants produce other effluent streams (Figure 20.10) that must be disposed of in an economically and environmentally acceptable manner (Argonne, 1990).

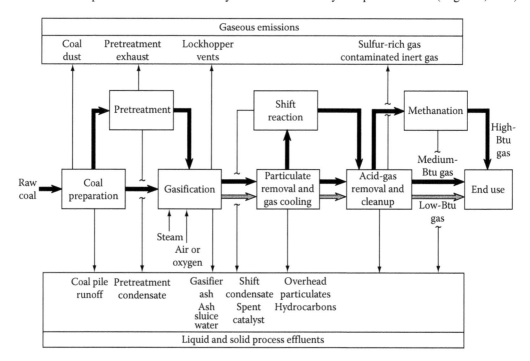

FIGURE 20.10 Process effluents from coal gasification. (From Braunstein, H.M. et al., Environmental health and control aspects of coal conversion: An information overview, Report ORNL-EIS-94, Oak Ridge National Laboratory, Oak Ridge, TN, 1977.)

Some of the streams contain valuable by-products such as hydrocarbon oils, sulfur, ammonia, as well as valuable elements in (and properties of) the ash (Braunstein et al., 1977; Probstein and Hicks, 1990).

The gas from any gasification process is inherently toxic, because of essential components such as carbon monoxide and unwanted components. However, this inherent toxicity is not the reason for gas cleaning because the gas should never be released to the atmosphere directly.

Gas cleaning (Chapter 23) is usually required to prevent damage to the electricity generating system using the gas and to prevent unwanted emissions from the combustion process using this gas. The relatively small volume and the reactive nature of species to be removed is a major advantage in gas cleaning from large pressurized oxygen-blown coal gasifiers when compared with conventional coal combustion plants.

The relatively small volume throughput advantages of gasifiers (compared with combustion processes) are to a great extent lost when gasifiers are used at smaller scales, atmospheric pressure, and are air-blown.

The effect of nitrogen is more than would be expected from simple nitrogen dilution by bringing four volumes of nitrogen into the system with every one volume of oxygen. The effect is not always fully appreciated, so some explanation may be helpful for readers new to the observation. The nitrogen needs sensible heat to raise it to the gasification reaction temperature. It retains some of this heat on reaching the gas cleaning stage. It increases the number of oxygen molecules needed for gasification reactions if air is used instead of oxygen. This further reduces the ratio of chemical energy to sensible heat in the gas and hence increases the volume of gas to be cleaned than what would be expected from the simple dilution calculation.

The major gas emissions are carbon dioxide and water vapor. Wood and refuse inherently emit more carbon dioxide than coal during an electricity generation process, whatever technology is used. Carbon dioxide emission reporting and trading is not always sophisticated enough to distinguish between emission of carbon dioxide for which the carbon was in the atmosphere only a few years ago and that for which the carbon was last in the atmosphere millions of years ago.

Sulfur compounds generally result in sulfur dioxide emissions. Sulfur removal from coal gas is a relatively straightforward process. Co-gasification of coal with a low sulfur substance such as biomass or waste reduces the hydrogen sulfide partial pressure and hence requires a larger gas cleaning system. The tar from wood gasification and the hydrochloric acid from waste gasification further complicate the issue. Objective assessment of potentially toxic emissions probably regards liquid products from gas cleaning as the greatest concern.

Gas cleaning to a high standard, before the combustion stage, is possible, albeit with an energy consumption, capital cost, and operational cost penalty. However, the production of oxides of nitrogen is more dependent on the gas-to-electricity conversion stage than the fuel composition.

All of the major operations associated with low-, medium-, and high-Btu gasification technology (coal pretreatment, gasification, raw gas cleaning, and gas beneficiation) can produce potentially hazardous air emissions. Auxiliary operations, such as sulfur recovery and combustion of fuel for electricity and steam generation, could account for a major portion of the emissions from a gasification plant.

Dust emissions from coal storage, handling, and crushing/sizing can be controlled with available techniques. Controlling air emissions from coal drying, briquetting, and partial oxidation processes is more difficult because of the volatile organics and possible trace metals liberated as the coal is heated.

The coal gasification process itself appears to be the most serious potential source of air emissions. The feeding of coal and the withdrawal of ash release emissions of coal or ash dust and organic and inorganic gases that are potentially toxic and carcinogenic. Because of their reduced production of tars and condensable organics, slagging gasifiers pose less severe emission problems at the coal inlet and ash outlet.

Gasifiers and associated equipment also will be sources of potentially hazardous fugitive leaks. These leaks may be more severe from pressurized gasifiers and/or gasifiers operating at high temperatures.

TABLE 20.6

By-Products from Coal Gasification and Methods for Removal

By-Product	Removal Process
CO_2	Acid gas scrubbing
H_2S	Acid gas scrubbing, Stretford process, amine treatment, Rectisol process
COS, CS_2	Removed with H_2S
NH_3	Scrubbing and ammonia stripping
HF, HCl, HCN	Scrubbing
Ash	Removed from gasifier for landfill (or minefill)
Suspended particles	Cyclone separators, electrostatic precipitators, scrubbing
Tar, oils	Scrubbing

Depending on subsequent processing and final use, various products and by-products must be removed from the low- and medium-heat-content products that come from a gasifier (Table 20.6). In all cases, hydrogen sulfide and other sulfur compounds must be removed because (in addition to the environmental aspects of gas use) they can poison catalysts in subsequent processing. This may be essentially all of the cleanup that is necessary for low-heat-content gas destined for combustion, whereas gas that is to be methanated requires virtually complete removal of essentially all components except hydrogen and carbon monoxide.

Acid gas removal refers to the removal of carbon dioxide, hydrogen sulfide, and other gases with acidic qualities and may be achieved in several ways (Table 20.7) (Chapter 23) (Speight, 2007, 2008). The first of these is absorption by a solvent, including hot carbonate solutions, amines, or methanol.

On the other hand, absorption onto a solid surface may be employed and solids employed for this purpose include activated carbon, charcoal, and iron sponge. Finally, acid gases may be chemically converted by reactants such as ferric oxide, zinc oxide, calcium oxide, and vanadium pentoxide. Some of these processes remove both carbon dioxide and hydrogen sulfide, whereas others selectively remove hydrogen sulfide (Speight, 2007).

TABLE 20.7

Acid Gas Removal Processes

Sorbent or Reactant	Gases Removed	Process
A. Solvent absorption		
20%–30% potassium carbonate in hot water solution + catalyst	H_2S, CO_2	Benfield
15% monoethanolamine in water	H_2S, CO_2	Amine
Cold methanol	H_2S, CO_2	Rectisol
B. Solid surface adsorption		
Carbon	H_2S	Activated carbon
Iron metal	H_2S	Iron sorption
C. Chemical reaction of acid gases		
Ferric oxide	H_2S	Iron sponge
Zinc oxide	H_2S	Zinc oxide

These processes require regeneration of the sorbent by release of hydrogen sulfide which is subsequently reacted with sulfur dioxide to produce elemental sulfur (Claus reaction):

$$2H_2S + SO_2 \rightarrow 2H_2O + 3S$$

The Stretford process for acid gas treatment allows for direct recovery of sulfur without going through a Claus reaction, that is,

$$2H_2S + O_2 \rightarrow 2H_2O + 2S$$

The absorbent is an aqueous solution of sodium carbonate, sodium metavanadate, and anthraquinone disulfonic acid:

$$H_2S + Na_2CO_3 \rightarrow NaHS + NaHCO_3$$

$$HS^- + 2VS^{5+} \rightarrow S + 2V^{4+} + H^+$$

The vanadium (IV) product reacts with the anthraquinone disulfonic acid to regenerate vanadium (V); oxygen is provided by air blowing.

Most processes for ammonia removal from gas streams involve absorption of the ammonia in water, and steam may be used to strip a concentrated stream of ammonia from the scrubber water. The moist ammonia from the steam stripping may be condensed to yield a fairly concentrated ammonia solution. Alternatively, the stripped ammonia may be absorbed in a solvent such as phosphoric acid to yield a solution that is subsequently treated to recover anhydrous ammonia.

Particulate matter may be removed from gas products by a number of means such as (1) water scrubbing or (2) a cyclone separator. Part of the ash from a conversion operation occurs as fly ash and is removed with particulate matter. Some gasifiers produce a melted bottom ash (slag) which is quenched in water while other processes produce a "self-agglomerating" ash consisting of softened particles.

Wastewater streams are generated in many coal gasification processes from the quenching and water washing of the raw gas, which removes water-soluble impurities such as ammonia, hydrogen sulfide, and phenols. The nature of the contaminants depends, to a large extent, on the composition of the feedstock coal as well as on the gasifier operation, and the techniques that are used for quenching and washing the raw gas. Hence, waste-water management has to be an integral part of any gasification plant (Luthy and Walters, 1979).

Liquid/liquid extraction processes are used to remove high concentrations of phenols from industrial waste-waters. For water that contains dissolved ammonia and hydrogen sulfide, most designs place the phenol extraction process before water stripping to prevent phenol-initiated problems in the stripper and ammonia-recovery steps. The extent of phenol removal depends on the types of phenols in the feed as well as other organic compounds.

Water stripping is a form of distillation in which trace amounts of relatively volatile materials, such as dissolved gases, are removed from a large volume of waste-water. The stripping is affected by either indirect reboiling of the feed-water by steam or by direct injection of steam into the bottom of a column. This technique is used extensively for sour-water strippers in almost all refinery installations. For coal gasification processes, dissolved base ionization may complicate the design procedures. For example, acidifying the feed will fix the ammonia in solution to allow selective removal of hydrogen sulfide. Unless cations, such as ammonia and sodium, are present to fix the hydrogen sulfide, removals greater than 98% are commonly reported.

The stripped water can be sent to a biological waste-water treatment plant for further processing. The gases require further treatment, such as ammonia recovery, sulfur removal recovery, or incineration/combustion.

Insoluble entrained organic compounds can be removed from waste-water by physical separators; the principal contaminant removed is free oil. If suspended solids are present in the waste-water, these solids will also settle in the basin.

In addition to the various water-treating systems, a variety of treatments may be considered depending upon the particular circumstances of contaminants and degree of final water purity required. These treatment systems include the following: (1) clarification, (2) filtration, (3) air flotation, (4) carbon adsorption, (5) vapor compression distillation, and (6) biological treatment. In the last case, that is, biological treatment, techniques such as trickle filtration, sludge activation, and digestion processes are used. Such treatments are worthy of detailed consideration as they generally offer attractive options that are acceptable to a variety of environmental regulators.

Additional changes to gas cleaning strategies for co-firing coal with other feedstocks (Figure 20.11) compared with coal alone are required. Biomass and waste gasification results in a higher tar content of the gas and often a higher hydrochloric acid content of the gas and more potentially toxic and carcinogenic organic compounds compared with coal gasification. Waste may require additional removal of toxic compounds, particularly mercury and heavy metals. Sulfur compound removal is much more important in coal gasification gas-cleaning but may be unnecessary in biomass and some waste gasification situations. It is clear that gas cleaning associated with co-gasification is not the same as gas cleaning for other single feedstock gasifiers. The differences may even be enough for separate gasifiers and gas cleaning streams for coal, biomass, and waste to be preferred to a single larger co-gasification and gas cleaning system, even if all the systems feed gas to the same electricity generation unit.

Biomass firing and probably waste firing will require tar removal. Catalytic cracking and thermal cracking, if they prove reliable, are generally regarded as the best processes as they retain much of the chemical energy of tars in the gas phase. However, experience to date on the reliability of tar

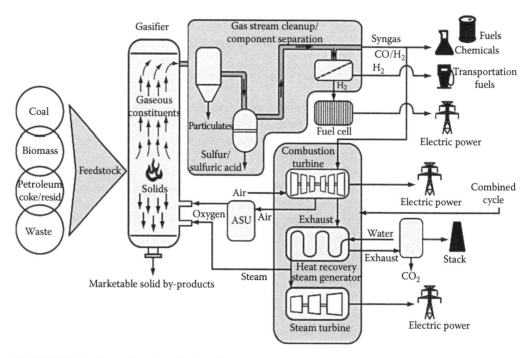

FIGURE 20.11   Co-gasification of coal and other carbonaceous feedstocks.

cracking processes has been at best variable. Condensation and/or wet scrubbing are better proven than tar cracking processes. However, the collected tars are often toxic, carcinogenic, or difficult to break down even in combustion or oxygen gasification processes. Oxygen-blown entrained gasifiers are particularly good at breakdown of the most difficult tars but it is likely to be rare to find such a gasifier conveniently close to a smaller co-gasification unit.

The issue then arises as to what advantages, technical or economic, are offered by the co-utilization of waste materials with coal. There is no simple answer to this question. The partial replacement of coal with waste and biomass materials can be a way of introducing renewable energy. In a number of countries, this is regarded as being of significant environmental benefit, and government subsidies and other inducements are available to encourage these activities. The coal can be regarded as being beneficial because the security of supply of a number of the waste and biomass materials is uncertain, and the quality of the delivered fuel is subject to only limited control. These are significant risk areas in most waste/biomass energy conversion projects, and the co-utilization of coal can be regarded as providing a means of reducing these risks, in that the supply of coal to a prescribed quality specification is assured in most industrial countries.

The co-utilization of RDF materials with coal in gasification and pyrolysis plant will also be of increasing relevance in the medium- to longer-term future, even though there are relatively few plants currently in operation. Apart from the limited extent of demonstration of the technologies and the lack of a turnkey supplier for plant, electricity market economics are a major hurdle to the introduction of co-gasification.

The difficulties of co-gasification of very different fuels in small gasifiers outweigh the benefits of combining fuels for economies of scale, at least at the gasification and gas cleaning stages. For oxygen-blown gasifiers, the situation is more likely to be dictated by considerations of fuel pretreatment and conclusions will be specific to each gasifier technology.

Co-gasification adds additional complexity to the gas cleaning system as the different products from coal, waste, and biomass all have to be addressed. There will be wastes from gas cleaning, including some form of chlorides, sulfur compounds, and potentially very toxic and carcinogenic tars and organic compounds. It may or may not be possible to use these tars by recycling to the gasifier.

Biomass growth specifically for power generation can be an excellent environmental option, particularly if the crop is managed with environmental care. There are many benefits over current set-aside arrangements for arable land.

Waste is an area of considerable public concern. The waste industry has had a bad public image, not always based on logical or scientific evaluation of the facts. However, whatever the facts the public concern is a major issue in the use of waste by any technology. Any link with waste or traffic is an area of particular concern for brand image. Major issues are therefore local planning consents, the environmental permits, and the support of the local population.

The technology of co-gasification of waste/biomass with coal can result in a very clean power plant using a range of fuels. In addition, the use of coal allows economies of scale and reliability of fuel supply. There are also considerable economic and permitting challenges in producing a financially attractive project.

REFERENCES

Anderson, L.L. and Tillman, D.A. 1979. *Synthetic Fuels from Coal: Overview and Assessment*. John Wiley & Sons, Inc., New York.

Anthony, D.B. and Howard, J.B. 1976. *AIChE Journal*, 22: 625.

Argonne. 1990. *Environmental Consequences of, and Control Processes for, Energy Technologies*. Argonne National Laboratory. Pollution Technology Review No. 181. Noyes Data Corporation, Park Ridge, NJ, Chapter 5.

Arnold, M.S.J., Gale, J.J., and Laughlin, M.K. 1992. *Canadian Journal of Chemical Engineering*, 70: 991.

Baker, R.T.K. and Rodriguez, N.M. 1990. In *Fuel Science and Technology Handbook*, J.G. Speight (Ed.). Marcel Dekker, Inc., New York, Chapter 22.

Baughman, G.L. 1978. *Synthetic Fuels Data Handbook*. Cameron Engineers, Denver, CO.

Berkowitz, N. 1979. *An Introduction to Coal Technology*. Academic Press, Inc., New York.

Bodle, W.W. and Huebler, J. 1981. In *Coal Handbook*, R.A. Meyers (Ed.). Marcel Dekker, Inc., New York, Chapter 10.

Braunstein, H.M., Copenhaver, E.D., and Pfuderer, H.A. 1977. Environmental health and control aspects of coal conversion: An information overview. Report ORNL-EIS-94. Oak Ridge National Laboratory, Oak Ridge, TN.

Calemma, V. and Radovic, L.R. 1991. On the gasification reactivity of Italian Sulcis coal. *Fuel*, 70: 1027.

Cavagnaro, D.M. 1980. *Coal Gasification Technology*. National Technical Information Service, Springfield, VA.

Chadeesingh, R. 2011. The Fischer–Tropsch process. In *The Biofuels Handbook*, J.G. Speight (Ed.). The Royal Society of Chemistry, London, U.K., Part 3, Chapter 5, pp. 476–517.

Cover, A.E., Schreiner, W.C., and Skapendas, G.T. 1973. *Chemical Engineering Progress*, 69(3): 31.

Cusumano, J.A., Dalla Betta, R.A., and Levy, R.B. 1978. *Catalysis in Coal Conversion*. Academic Press, Inc., New York.

Davidson, R.M. 1983. Mineral effects in coal conversion. Report No. ICTIS/TR22. International Energy Agency, London, U.K.

Davis, B.H. and Occelli, M.L. 2010. *Advances in Fischer-Tropsch Synthesis, Catalysts, and Catalysis*. Taylor & Francis Group, Boca Raton, FL.

Dutcher, J.S., Royer, R.E., Mitchell, C.E., and Dahl, A.R. 1983. In *Advanced Techniques in Synthetic Fuels Analysis*, C.W. Wright, W.C. Weimer, and W.D. Felic (Eds.). Technical Information Center, United States Department of Energy, Washington, DC, p. 12.

Elder, J.L. 1963. In *Chemistry of Coal Utilization*, Supplementary Volume. H.H. Lowry (Ed.). John Wiley & Sons, Inc., New York, Chapter 21.

Elton, A. 1958. In *A History of Technology*, C. Singer, E.J. Holmyard, A.R. Hall, and T.I. Williams (Eds.). Clarendon Press, Oxford, England, Vol. IV, Chapter 9.

Fryer, J.F. and Speight, J.G. 1976. *Coal Gasification: Selected Abstract and Titles*. Information Series No. 74. Alberta Research Council, Edmonton, Alberta, Canada.

Garcia, X. and Radovic, L.R. 1986. Gasification reactivity of Chilean coals. *Fuel*, 65: 292.

Gay, R.L., Barclay, K.M., Grantham, L.F., and Yosim, S.J. 1980. Fuel production from solid waste. *Symposium on Thermal Conversion of Solid Waste and Biomass*, Symposium Series No. 130, American Chemical Society, Washington, DC, Chapter 17, pp. 227–236.

Graff, R.A.A., Dobner, S., and Squires, A.M. 1976. *Fuel*, 55: 109.

Gregg, D.W. and Edgar, T.F. 1978. *AIChE Journal*, 24: 753.

Hanson, S., Patrick, J.W., and Walker, A. 2002. The effect of coal particle size on pyrolysis and steam gasification. *Fuel*, 81: 531–537.

Hobbs, M.L., Radulovic, P.T., and Smoot, D.L. 1992. *AIChE Journal*, 38: 681.

Howard-Smith, I. and Werner, G.J. 1976. *Coal Conversion Technology*. Noyes Data Corporation, Park Ridge, NJ.

Huang, Y.-H., Yamashita, H., and Tomita, A. 1991. *Fuel Processing Technology*, 29: 75.

Irfan, M.F. 2009. Research Report: Pulverized coal pyrolysis & gasification in $N_2/O_2/CO_2$ mixtures by thermogravimetric analysis. Novel Carbon Resource Sciences Newsletter, Kyushu University, Fukuoka, Japan, Vol. 2, pp. 27–33.

Johnson, N., Yang, C., and Ogden, J. 2007. Hydrogen production via coal gasification. Advanced Energy Pathways Project, Task 4.1 Technology Assessments of Vehicle Fuels and Technologies, Public Interest Energy Research (PIER) Program, California Energy Commission, Sacramento, CA, May.

Kasem, A. 1979. *Three Clean Fuels from Coal: Technology and Economics*. Marcel Dekker, Inc., New York.

King, R.B. and Magee, R.A. 1979. In *Analytical Methods for Coal and Coal Products*, C. Karr Jr. (Ed.). Academic Press, Inc., New York, Vol. III, Chapter 41.

Koh, A.L., Harty, R.B., and Johnson, J.G. 1978. *Chemical Engineering Progress*, 74(8): 73.

Kristiansen, A. 1996. IEA Coal Research report IEACR/86. Understanding coal gasification. International Energy Agency, London, U.K.

Lahaye, J. and Ehrburger, P. (Eds.). 1991. *Fundamental Issues in Control of Carbon Gasification Reactivity*. Kluwer Academic Publishers, Dordrecht, the Netherlands.

Luthy, R.B. and Walters, R.W. 1979. In *Analytical Methods for Coal and Coal Products*, C. Karr Jr. (Ed.). Academic Press, Inc., New York, Vol. III, Chapter 44.

Mahajan, O.P. and Walker, P.L. Jr. 1978. In *Analytical Methods for Coal and Coal Products*. C. Karr Jr. (Ed.). Academic Press, Inc., New York, Vol. II, Chapter 32.

Martinez-Alonso, A. and Tascon, J.M.D. 1991. In *Fundamental Issues in Control of Carbon Gasification Reactivity*, J. Lahaye and P. Ehrburger (Eds.). Kluwer Academic Publishers, Dordrecht, the Netherlands.

Massey, L.G. (Ed.). 1974. *Coal Gasification*. Advances in Chemistry Series No. 131. American Chemical Society, Washington, DC.

Massey, L.G. 1979. In *Coal Conversion Technology*, C.Y. Wen and E.S. Lee (Eds.). Addison-Wesley Publishers, Inc., Reading, MA, p. 313.

Matsukata, M., Kikuchi, E., and Morita, Y. 1992. *Fuel*, 71: 819.

Meyers, R.A. (Ed.). 1981. *Coal Handbook*, Marcel Dekker Inc., New York.

Mills, G.A. 1969. *Industrial and Engineering Chemistry*, 61(7): 6.

Mills, G.A. 1982. *Chemtech*, 12: 294.

Mills, G.A. and Steffgen, F.W. 1975. *Catalysis Review*, 15: 165.

Mims, C.A. 1991. In *Fundamental Issues in Control of Carbon Gasification Reactivity*, J. Lahaye and P. Ehrburger (Eds.). Kluwer Academic Publishers, Dordrecht, the Netherlands, p. 383.

Nef, J.U. 1957. In *A History of Technology*, C. Singer, E.J. Holmyard, A.R. Hall, and T.I. Williams (Eds.). Clarendon Press, Oxford, England, Vol. III, Chapter 3.

Olness, D.U. 1981. The Podmoskovnaya underground coal gasification station. Report UCRL-53144. Lawrence Livermore National Laboratory, Livermore, CA.

Probstein, R.F. and Hicks, R.E. 1990. *Synthetic Fuels*. pH Press, Cambridge, MA, Chapter 4.

Radovic, L.R. and Walker, P.L. Jr. 1984. Reactivities of chars obtained as residues in selected coal conversion processes. *Fuel Processing Technology*, 8: 149.

Radovic, L.R., Walker, P.L. Jr., and Jenkins, R.G. 1983. Importance of carbon active sites in the gasification of coal chars. *Fuel*, 62: 849.

Seglin, L. (Ed.). 1975. *Methanation of Synthesis Gas*. Advances in Chemistry Series No. 146. American Chemical Society, Washington, DC.

Speight, J.G. 1990. In *Fuel Science and Technology Handbook*, J.G. Speight (Ed.). Marcel Dekker, Inc., New York, Chapter 33.

Speight, J.G. 2007. *Natural Gas: A Basic Handbook*. GPC Books, Gulf Publishing Company, Houston, TX.

Speight, J.G. 2008. *Synthetic Fuels Handbook: Properties, Processes, and Performance*. McGraw-Hill, New York.

Speight, J.G. 2009. *Enhanced Recovery Methods for Heavy Oil and Tar Sands*. Gulf Publishing Company, Houston, TX.

Speight, J.G. 2011. *The Refinery of the Future*. Gulf Professional Publishing, Elsevier, Oxford, U.K.

Taylor, F.S. and Singer, C. 1957. In *A History of Technology*, C. Singer, E.J. Holmyard, A.R. Hall, and T.I. Williams (Eds.). Clarendon Press, Oxford, England, Vol. II, Chapter 10.

Thompson, P.N. 1978. *Endeavor*, 2: 93.

Tucci, E.R. and Thompson, W.J. 1979. *Hydrocarbon Processing*, 58(2): 123.

Wang, W. and Mark, T.K. 1992. *Fuel*, 71: 871.

Watson, G.H. 1980. Methanation catalysts. Report ICTIS/TR09 International Energy Agency, London, U.K.

Zvyaghintsev, K.N. 1981. *Trends of Development of Underground Coal Gasification in the U.S.S.R. Natural Resources Forum (USSR)*, 5: 99.

21 Gasification Processes

21.1 INTRODUCTION

In any text about coal (but more specifically in any text that contains a chapter about coal gasification), it is appropriate to include a listing of the types of processes available as well as a description of the different processes. Thus, it is the intent here to give selected examples of specific processes.

In contrast to liquefaction processes, gasification processes have been evolving since the early days of the nineteenth century when town gas became a common way of bringing illumination and heat not only to factories but also to the domestic consumer. But it must not be assumed that all processes included here have been successful in the move to commercialization. Some still remain conceptual but are included here because of their novelty and/or their promise for future generations of gasifiers.

There has been a general tendency to classify gasification processes by virtue of the heat content of the gas which is produced; it is also possible to classify gasification processes according to the type of reactor vessel and whether or not the system reacts under pressure. However, for the purposes of the present text gasification processes are segregated according to the bed types, which differ in their ability to accept (and use) various types of coal (Collot, 2002, 2006).

Thus, gasification processes can be generally divided into four categories based on reactor (bed) configuration: (1) fixed bed, (2) fluidized bed, (3) entrained bed, and (4) molten salt (Table 21.1).

21.2 FIXED-BED PROCESSES

In a fixed-bed process the coal is supported by a grate and combustion gases (steam, air, oxygen, etc.) pass through the supported coal whereupon the hot produced gases exit from the top of the reactor. Heat is supplied internally or from an outside source, but caking coals cannot be used in an unmodified fixed-bed reactor.

21.2.1 LURGI PROCESS

The Lurgi process (Figure 21.1), which was developed in Germany before World War II, is a process that is adequately suited for large-scale commercial production of synthetic natural gas (Verma, 1978).

The older Lurgi process is a dry ash gasification process which differs significantly from the more recently developed slagging process (Baughman, 1978; Massey, 1979). The dry ash Lurgi gasifier is a pressurized vertical kiln which accepts crushed (1/4 × 1 3/4 in. [6 × 44 mm]) noncaking coal and reacts the moving bed of coal with steam and either air or oxygen. The coal is gasified at 350–450 psi and devolatilization takes place in the temperature range 615°C–760°C (1140°F–1400°F); residence time in the reactor is approximately 1 h.

Hydrogen is supplied by injected steam and the necessary heat is supplied by the combustion of a portion of the product char. The revolving grate, located at the bottom of the gasifier, supports the bed of coal, removes the ash, and allows steam and oxygen (or air) to be introduced.

The Lurgi product gas has a high methane content relative to the products from non-pressurized gasifiers. With oxygen injection, the gas has a heat content of ca. 450 Btu/ft (16.8 MJ/m).

The crude gas which leaves the gasifier contains tar, oil, phenols, ammonia, coal fines, and ash particles. The steam is first quenched to remove the tar and oil, and, prior to methanation, part of

---

TABLE 21.1
Categories of Gasification Processes

A. Fixed-bed processes
    1. Lurgi (dry ash and slagging)
    2. Wellman–Galusha
    3. Foster–Wheeler Stoic
    4. Slagging fixed bed (Grand Forks Energy Research Center)
    5. Woodall–Duckham

B. Fluidized-bed processes
    1. Winkler
    2. Hydrane
    3. Hygas
    4. Synthane
    5. U-Gas
    6. Union Carbide (Coalcon)
    7. Westinghouse pressurized
    8. Carbon dioxide acceptor
    9. Agglomerating burner
    10. COED/COGAS

C. Entrained-bed processes
    1. Koppers–Totzek
    2. Bi-Gas
    3. Combustion engineering entrained bed
    4. Texaco

D. Molten salt
    1. Atgas
    2. Atomics International (Rockgas)
    3. Pullman–Kellogg

Source: Braunstein, H.M. et al., Environmental health and control aspects of coal conversion: An information overview, Report ORNL-EIS-94, Oak Ridge National Laboratory, Oak Ridge, TN, 1977.

---

the gas passes through a shift converter and is then washed to remove naphtha and unsaturated hydrocarbons; a subsequent step removes the acid gases. The gas is then methanated to produce a high-heat-content pipeline quality product.

Dry ash Lurgi gasification is a highly advanced technology that, with the addition of a stirrer, can now process caking coals or those with low ash fusion temperatures. The agglomerated ash from such coals cannot be easily removed through the grate at the bottom of the gasifier. In addition, the low operating temperature encourages the production of byproduct tars, oils, and phenols. A modification of the Lurgi process, known as the slagging Lurgi (Figure 21.2), is being developed to make it possible to process caking coals.

Thus, while minimizing waste-water treatment costs in a conventional Lurgi gasifier, the temperature is intentionally kept low (by injecting an excess amount of steam, usually 6–10 mol of steam per mole of oxygen) to minimize ash agglomeration (Marinov et al., 1992). In a slagging Lurgi gasifier, the steam injection rate is reduced to 1–1.5 mol of steam per mole of oxygen and the higher operating temperature causes the coal ash to melt and run off as a slag. As in a conventional Lurgi unit, coal is fed to the gasifier through a lock hopper system and a distributor. As it passes down through the bed, the coal is preheated and devolatilized by the upward flowing steam of hot product gas. The coal is then gasified with steam and oxygen which are injected near the bottom of

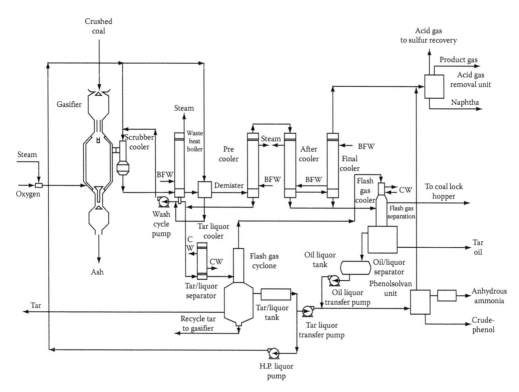

FIGURE 21.1 Lurgi process.

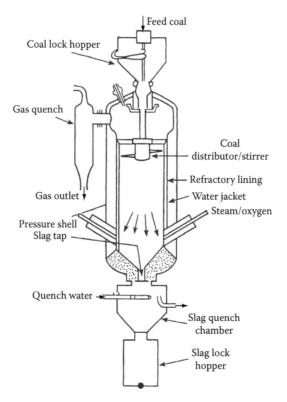

FIGURE 21.2 Slagging Lurgi gasifier.

the vessel. The entire bed rests on a hearth through which the molten ash, or slag, can pass through the slag tap hole. The slag is then quenched with water and is finally removed through a lock hopper.

The amount of unreacted steam passing through the bed is minimized in the slagging Lurgi process and, thus, the product gases can be removed from the unit faster with minimum fines carryover. This aspect of the process, together with the higher operating temperature, leads to higher output rates for slagging units than conventional dry ash units.

21.2.2 Wellman–Galusha Process

The Wellman–Galusha process has been in commercial use for more than 45 years (Howard-Smith and Werner, 1976). There are two types of gasifiers, the standard type (Figure 21.3) and the agitated type (Figure 21.4), and the rated capacity of an agitated unit may be (25% or more) higher than that of a standard gasifier of the same size. In addition, an agitated gasifier is capable of treating volatile caking bituminous coals.

The gasifier is water-jacketed, and, therefore, the inner wall of the vessel does not require a refractory lining. Agitated units include a varying-speed revolving horizontal arm which also spirals vertically below the surface of the coal bed to minimize channeling and to provide a uniform bed for gasification. A rotating grate is located at the bottom of the gasifier to remove the ash from the bed uniformly. Steam and oxygen are injected at the bottom of the bed through tuyeres.

Crushed coal is fed to the gasifier through a lock hopper and vertical feed pipes. The fuel valves are operated so as to maintain a relatively constant flow of coal to the gasifier to assist in maintaining the stability of the bed and, therefore, the quality of the product gas. The air or oxygen which is injected into the gasifier passes over the top of the water jacket and thereby picks up the steam required for the process. The air–steam mix is introduced into the ash bin section underneath the

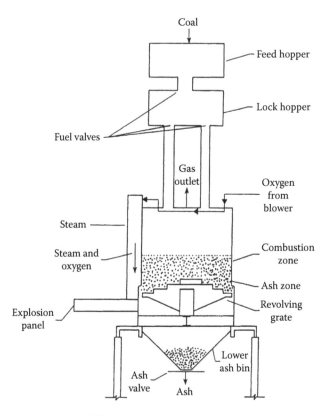

FIGURE 21.3 Wellman–Galusha gasifier.

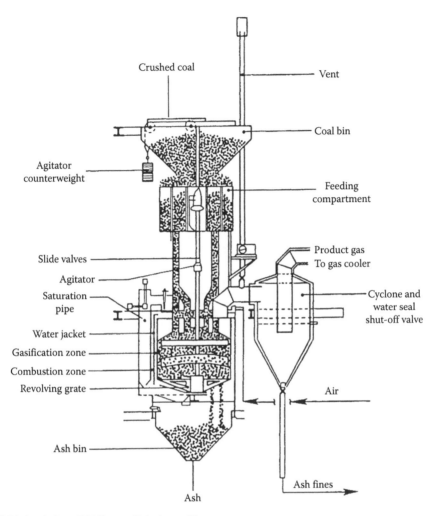

FIGURE 21.4 Agitated Wellman–Galusha gasifier.

grate and is then distributed through the grate into the bed, and passes upward through the ash, combustion, and gasification zones.

The product gas contains primarily carbon monoxide, carbon dioxide, hydrogen, and nitrogen (if air is injected) which, being hot, dries and preheats the incoming coal before leaving the gasifier. The product gas is passed through a cyclone, in which fine ash and char particles are removed, and if the sulfur content of the gas is acceptable, it may be used directly. If, however, the sulfur content is too high, the gas can be scrubbed and cooled in a direct-contact countercurrent water cooler and then treated for sulfur removal. The use of air to support combustion will yield a low-heat-content gas, but the use of oxygen will yield a medium-heat-content gas.

21.2.3 Foster–Wheeler Stoic Process

The Foster–Wheeler process (Figure 21.5) involves the use of a two-stage gasifier in which the upper stage is the distillation zone and the lower stage is the gasification zone (Probstein and Hicks, 1990). The coal is charged to the top of the unit and is reduced to coke by the time it reaches the bottom of the distillation zone. Ash is removed from the base of the gasifier through a water seal, and steam and air are introduced through a grate at the base of the coal bed.

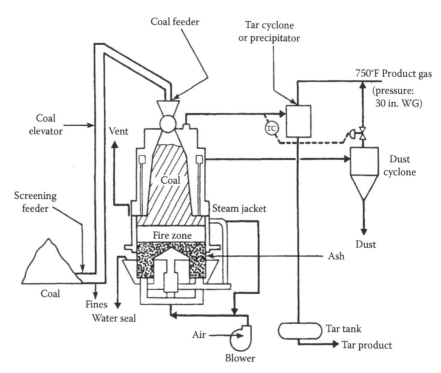

FIGURE 21.5 Foster–Wheeler Stoic process.

The steam–air mixture is preheated by passage through the bed of hot ash and enters the fire zone, a narrow band 4–10 in. (100–250 mm) deep, operating at 980°C (1800°F). A partial oxidation reaction takes place in the fire zone and produces carbon monoxide, some carbon dioxide, and hydrogen as well as heat for the balance of the gasification reactions. The two-stage operation of the gasifier is designed to heat the coal gradually with only part of the hot reducing gas and, thus, the oils and tars formed are sufficiently fluid to be handled easily. At temperatures in excess of 480°C (900°F), the plastic components resolidify and decompose to yield coke and a hydrogen-rich gas, some of which reacts with carbon to form a small additional amount of methane.

The gas from the distillation zone contains almost all of the oil and tar and devolatilized methane as well as the residual water entering with the coal.

The coke is reduced to ash in the fire zone which then moves down onto the grate and out of the gasifier via the water seal. The bed of ash between the fire zone and the grate is cooled by the incoming blast of air and steam. Water jacketing is used in the gasification zone to cool the shell and, at the same time, to generate the steam required for the gasification reaction. The steam also helps to cool the fire zone so that the ash will not agglomerate, fuse, and form large clinkers that would block the flow out of the units.

21.2.4 WOODALL–DUCKHAM PROCESS

The Woodall–Duckham process employs a gasifier (Figure 21.6) which is a vertical cylindrical vessel having a rotating grate in the bottom for ash removal (Howard-Smith and Werner, 1976). There are three functional zones within the reactor: (1) a water-jacketed gasification zone, (2) a refractory-lined distillation zone, and (3) a refractory-lined drying zone.

Upon entering the gasifier, which operates at atmospheric pressure, the coal (0.25–1.5 in. [6–38 mm]) is contacted with upward flowing hot gases from the gasification zone and any moisture present in the

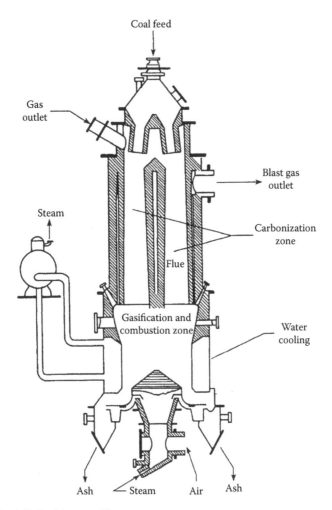

Coal feed

Gas
outlet

Blast gas
outlet

Steam

Carbonization
zone

Flue

Gasification and
combustion zone

Water
cooling

Ash ⟵ Steam   Air   Ash

FIGURE 21.6   Woodall–Duckham gasifier.

coal is driven off by the hot (120°C [250°F]) gases. The coal then falls into the distillation zone where the volatile matter present is driven off by the ascending hot gas, but since a relatively slow heating rate prevails, negligible cracking of the tar and oil occurs.

The devolatilized char (noncaking coals) or semicoke (caking coals) is further heated by the hot gases until the material passes into the gasification zone, where it is contacted counter-currently with steam and oxygen (or air) in a fixed bed whereupon the remaining carbon is mostly gasified. The temperature in the gasification zone may depend on the type of coal being processed but is usually of the order of 1205°C (2200°F).

In the lower portions of this zone, the descending ash is contacted with incoming steam and air, thereby effecting gas preheating and ash cooling. The ash is removed from the gasifier through a rotating grate, which also serves to distribute the air and steam evenly over the entire cross section of the gasifier.

The product gas from the gasifier is withdrawn at two points in the vessel; the gas withdrawn between the distillation and drying zones is known as clear gas whereas the gas withdrawn near the top of the vessel is called top gas. Varying the portion of gas withdrawn through the lower tap affords a means of control of the temperature of the distillation zone. Thus, as the flow of clear gas is reduced, more hot gas is forced through the distillation zone, thus increasing the temperature.

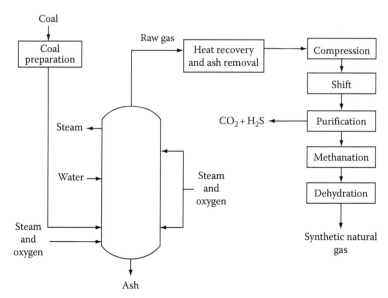

FIGURE 21.7 Winkler process.

21.3 FLUIDIZED-BED PROCESSES

The fluidized-bed system uses finely sized coal particles and the bed exhibits liquid-like character-istics when a gas flows upward through the bed. Gas flowing through the coal produces turbulent lifting and separation of particles, and the result is an expanded bed having greater coal surface area to promote the chemical reaction, but such systems have only a limited ability to handle caking coals.

21.3.1 WINKLER PROCESS

In the Winkler process dried, crushed coal (−8 mesh) is fed to the fluidized-bed gasifier (Figure 21.7) through a variable-speed screw feeder whereupon the coal is contacted with steam and oxy-gen injected near the bottom of the vessel (Howard-Smith and Werner, 1976; Baughman, 1978). The upward flow of steam and oxygen maintains the bed in a fluidized state at a temperature of 815°C–980°C (1600°F–1800°F) with a pressure that is marginally higher than atmospheric. The high operating temperature reduces the amount of tars and other heavy hydrocarbons in the product (Nowacki, 1980).

Approximately 70% of the coal ash is carried over the gas flow and 30% is removed from the bottom of the vessel by screw conveyors. Unreacted carbon contained in the carryover ash is con-sumed by the injection of supplemental steam and oxygen in the space above the bed. To moderate the bed temperature and thereby minimize ash melting, a heat exchange surface is provided in the dilute phase to remove heat and generate steam. The raw gas leaving the gasifier, rich in carbon monoxide and hydrogen, is cooled to approximately 705°C (1300°F) in a boiler and then passes through (1) a heat exchanger to superheat steam, (2) a waste heat boiler, and (3) a cyclone to remove entrained char. Further gas cleanup is effected by wet scrubbers, electrostatic precipitators, and sulfur removal equipment.

21.3.2 HYDRANE PROCESS

In the Hydrane process, more than 90% of the methane produced is generated within the gas-ifier (Figure 21.8), thus minimizing the necessity for subsequent catalytic methanation processes (Howard-Smith and Werner, 1976). Coal enters the gasifier and falls through a counter-currently

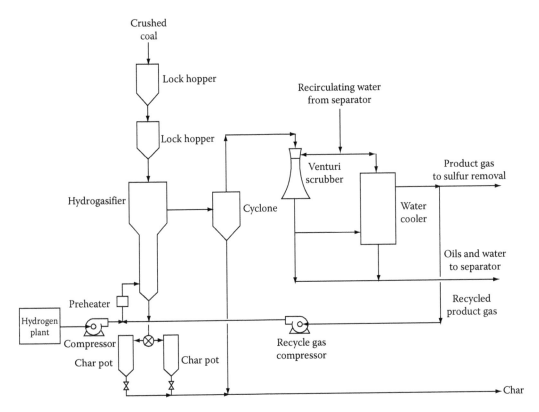

FIGURE 21.8 Hydrane process.

flowing stream of hot gases rising from the second-stage fluidized bed, therefore allowing agglomerating coals to be hydrogasified (any agglomerating properties are usually destroyed during the free-fall preheating step). The preheated solids then enter the second zone (consisting of a bed fluidized by hydrogen) which permits maximum coal conversion without cracking the methane product to carbon.

The product gas requires purification and light methanation so that the final product meets pipeline specifications. The char from the gasifier may be either completely consumed in synthesis gas production or used for power production.

21.3.3 Hygas Process

The Hygas process involves the direct hydrogasification of coal in the presence of pressurized hydrogen and steam in two stages (Bair and Lau, 1980; Nowacki, 1980). Crushed coal (−14 mesh) is slurried with an aromatic recycle oil to form a slurry which is pressurized to approximately 100 psi and injected into the top section of the gasifier (Figure 21.9) which contains a fluidized bed of hot coal particles. If the coal is a caking coal, it is pretreated in a fluidized bed at 400°C–455°C (750°F–810°F) and at atmospheric pressure to destroy the caking tendencies. The slurry oil is vaporized and removed together with the hot gases passing upward from the lower reactor stages; the vaporized oil is subsequently recovered for reuse.

The dry coal particles flow into a lift pipe which serves as the first stage of hydrogasification. In the pipe, the coal is contacted with hot gas (methane, carbon oxides, hydrogen, and steam) from the lower sections of the reactor and the coal reacts with the hydrogen, forming additional methane (ca. 33% of the methane in the final product is produced in this section).

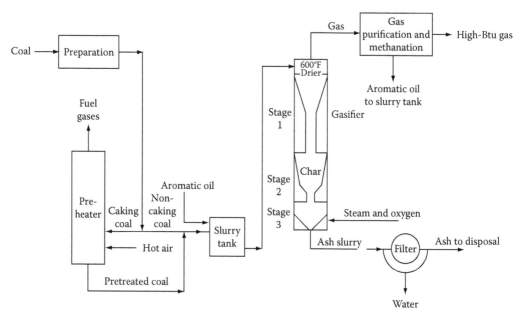

FIGURE 21.9 Hygas process.

In the second stage, partially converted coal is contacted with hydrogen-rich gas at 760°C–925°C (1400°F–1700°F) where a further 33% of the methane in the final product is generated. The hot residual char is then transferred to the third stage where it reacts with steam and oxygen in a fluid-ized bed to yield a mixture of hydrogen-rich gases. The gaseous product from this stage passes upward through the reactor while ash is removed from the bottom of the steam–oxygen zone. The gas leaving the reactor is cooled and rinsed in a water quench before being purified, shifted, and methanated. Following methanation, the product gas has a heat content of ca. 930–950 Btu/ft (34.7–35.4 MJ/ft).

21.3.4 SYNTHANE PROCESS

The Synthane process (Weiss, 1978) (Figure 21.10) was developed as a result of investigations deal-ing with the pretreatment of caking coals when it was noted that the proper combination of (1) oxygen content of the fluidizing gas, (2) temperature, and (3) residence time made possible the pretreatment of caking coals. Additional work later showed that pretreatment, carbonization, and gasification were all possible within a single vessel.

The Synthane gasifier itself is a vertical, cylindrical, fluidized-bed reactor which operates at approximately 100 psi and up to 980°C (1800°F).

Coal (crushed to −20 mesh) is dried and then pressurized to approximately 1000 psi. Caking coals are then fed to a fluidized-bed pre-treater by high-pressure steam and oxygen to provide a mild oxidation of the coal particles to minimize agglomeration in the gasifier. The coal then over-flows from the top of the pre-treater into the top of the fluidized-bed gasifier and falls through the hot gases rising from the fluidized bed and is devolatilized. Steam and oxygen enter the gasifier just below the fluidizing gas distributor and gasification reaction occurs within the fluidized bed. Unreacted char flows downward into a bed fluidized and cooled with steam and is removed through lock hoppers; the char can then be burned to produce process steam.

The product gas, containing methane, hydrogen, carbon monoxide, carbon dioxide, and ethane, is passed through a Venturi scrubber and a water scrubber to remove carryover ash, char, and tars. Acid gas impurities are absorbed in a hot potassium carbonate scrubber and any remaining traces of sulfur in the product gas are removed by passing the gas through activated charcoal.

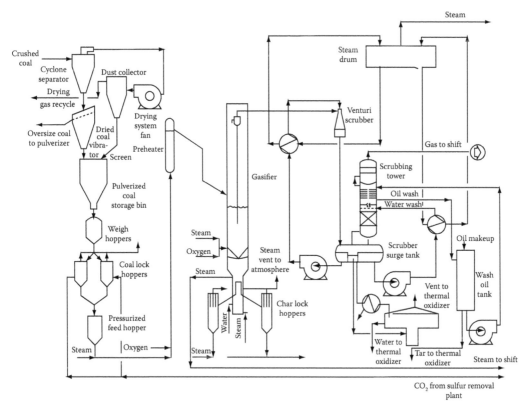

FIGURE 21.10   Synthane process.

21.3.5   U-Gas Process

The U-Gas reactor (Figure 21.11) is a vertical vessel containing an internal cyclone for returning elutriated fines to the bed. A sloped grid at the bottom of the vessel serves as both an ash outlet and a steam/air distributor.

The feed coal is first crushed to −14 mesh and (lignite, or noncaking subbituminous coals) fed to the gasifier. However, caking coals must first be pretreated by contact with air in a fluidized-bed reactor operating at gasifier pressure and approximately 370°C–425°C (700°F–800°F) to prevent agglomeration in the gasifier. The coal is then gasified with a mixture of air and steam in a single fluidized bed at 50–350 psi and 1040°C (1900°F).

The gases leaving the pre-treater and gasifier are passed through heat exchangers to recover the sensible heat and are treated to remove the sulfur compounds. With air-blown operation, the product gas typically has a heat content of ca. 155 Btu/ft whereas if the gasifier is oxygen-blown, the product gas may have a heat content of ca. 300 Btu/ft³.

21.3.6   Coalcon Process (Hydrocarbonization Process)

In the process (Figure 21.12), coal is first crushed, milled, and fed to the coal-preheating unit where the coal is entrained in a hot, oxygen-free flue gas which helps maintain the reactor heat efficiency and also drives off some volatiles and moisture. After heating to approximately 325°C (620°F), the coal is held in the coal feed hopper before it is pressurized to the operating pressure of the system (approximately 550 psi). The coal is then dropped from the lock hoppers into another coal holding vessel from whence it is fed into an injection vessel where it is fluidized with hydrogen at 560°C (1040°F) and 750 psi.

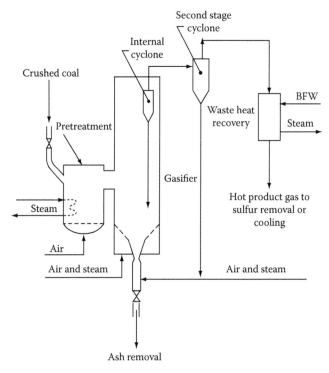

FIGURE 21.11   U-Gas process.

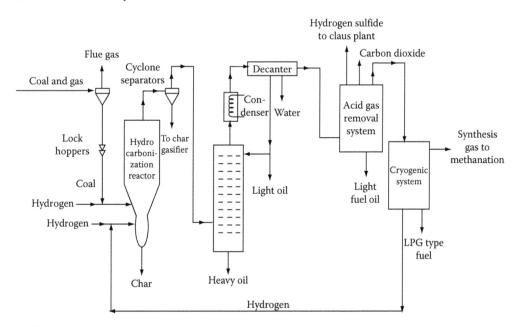

FIGURE 21.12   Coalcon process.

This mixture enters the reactor and the solids which are not gasified are removed from the reactor through the bottom of the reactor vessel. This char product may either be used to generate hydrogen or be burned to generate steam.

The gas passing upward through the reactor "carries" fines which are subsequently removed by two cyclones. The gas from the cyclones is then sent to a fractionator for cooling and separation.

Four streams are produced by the fractionator for cooling and separation: (1) overhead gas (hydrogen, carbon monoxide, carbon dioxide, methane), (2) light liquids, (3) heavy liquids, and (4) waste-water.

The heavy oil product is cooled to about 40°C (105°F) and pumped to storage. The overhead product is condensed and fed to a decanter where the light fuel oil, overhead gas, and waste-water are separated. Some of the light fuel oil is sent to the fractionator as reflux and the remainder is sent to storage as product. The remaining gas is treated in a series of separation and purification systems which include ammonia removal and recovery, acid gas removal, and a cryogenic gas processing system; the latter system manufactures fuel gas, synthesis gas, and a hydrogen-rich gas stream. The hydrogen stream is recycled to the reactor while the synthesis gas is sent to a methanation reactor for upgrading to high-heat-content pipeline gas.

21.3.7 PRESSURIZED FLUID-BED PROCESS

In the process (Figure 21.13), coal is crushed to −6 + 100 mesh, dried, and transported to a reactor vessel for devolatilization desulfurization and partial hydrogasification (Archer et al., 1975).

A central draft tube is used primarily for recirculating solids, and the dense, dry char collects in the fluidized bed at the top of the draft tube and is withdrawn at this point. Dolomite or calcium oxide (sorbent) is added to the fluidized bed to absorb the sulfur present as hydrogen sulfide in the fuel gas and spent sorbent is withdrawn from the bottom of the reactor and regenerated. The heat for devolatilization is supplied primarily by the high-temperature fuel gas produced in the gasifier combustor. After separation of fines and ash, product gas is cooled and scrubbed with water for final purification.

Final gasification occurs in a fluid-bed gasifier-combustor; char from the devolatilizer-desulfurizer is burned with air in the lower section of the gasifier at 1040°C–1095°C (1900°F–2000°F) to provide the heat for gasification. Heat is transported from the combustor to the gasifier by combustion gases flowing upward and by fines that escape upward and are trapped and recycled to the space between the combustor and gasifier. Ash from the combustion of fines agglomerates on the ash from the char and segregates in the lower section for removal.

21.3.8 CARBON DIOXIDE ACCEPTOR PROCESS

In the carbon dioxide acceptor process (Curran et al., 1969; Nowacki, 1980) (Figure 21.14), coal is first crushed to 8 × 100 mesh in hot-gas-swept impact mills in which the moisture content is reduced from ca. 38% w/w to ca. 16% w/w. The coal is further dried to 0.5% w/w moisture in

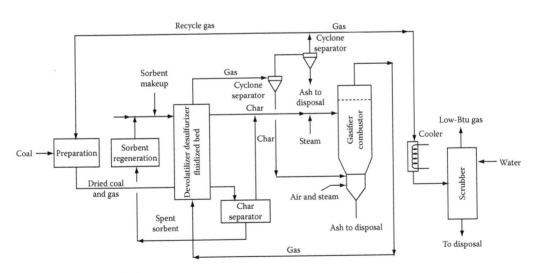

FIGURE 21.13 Westinghouse process.

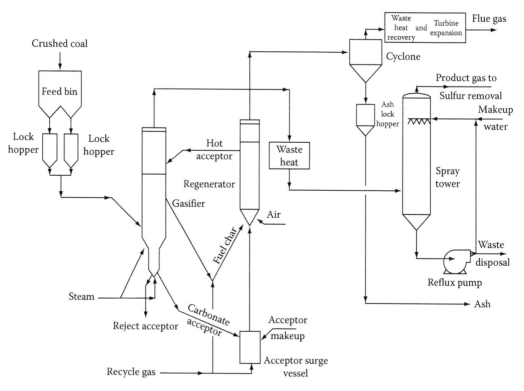

FIGURE 21.14 CO_2-acceptor process.

flash dryers and is then conveyed to fluidized-bed preheaters where the temperature is increased to approximately 260°C (500°F).

The coal is fed to the gasifier, entering near the bottom of a fluidized bed of char whereupon rapid devolatilization occurs, followed by the gasification of the fixed carbon with steam. The temperature in the gasifier ranges from 805°C to 845°C (1480°F to 1550°F). Heat for the gasification reactions is supplied by a circulating stream of lime-bearing material ("acceptor"), usually limestone or dolomite, which supplies the necessary heat through the exothermic carbon-dioxide acceptor reaction:

$$CaO + CO_2 \rightarrow CaCO_3 + heat$$

This acceptor, having been crushed to 6 × 14 mesh, enters the gasifier above the fluidized char bed, showers through the bed, and is collected in the gasifier base. Steam for hydrogasification enters the gasifier through a distributor ring in the base. Product gas from the gasifier passes through a steam-generating heat exchanger and then passes to a gas cleanup section.

Spent acceptor leaving the gasifier is calcined in a regenerator vessel at approximately 1010°C (1850°F), at which temperature the carbon dioxide acceptor reaction is reversed; the calcined acceptor is then returned to the gasifier.

Both the product gas from the gasifier and the flue gas from the regenerator are quenched and purified. The flue gas is either recycled to the regenerator or flared whereas the product gas is sent to the methanation section.

The methanation facilities, which convert the low-heat-content synthesis gas into a high-heat-content product, include a shift converter, carbon dioxide absorber, hydrodesulfurizer, zinc oxide sulfur guard, and a packed-tube methanator.

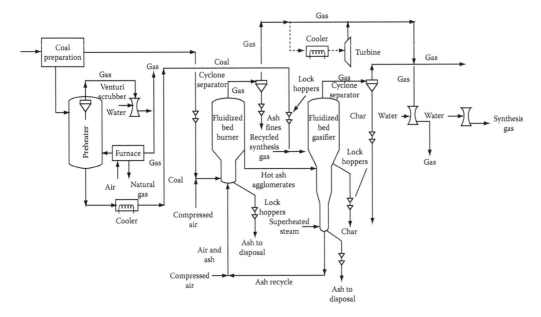

FIGURE 21.15   Agglomerating burner process.

21.3.9   AGGLOMERATING BURNER PROCESS

In the agglomerating burner process (Figure 21.15), coal is crushed and separated into two sizes: −100 and −8 + 100 mesh. Caking coal in the −8 + 100 mesh range is fed to a fluidized-bed pre-treater, where it is mixed with gas and air at atmospheric pressure and 400°C (750°F) after which it is cooled, charged to the gasifier, and reacted with steam in the fluidized bed. The 100-mesh coal is burned with air in a fluidized-bed combustor in a manner allowing agglomeration of the ash at a temperature approaching the ash fusion point (1150°C [2100°F]).

The hot ash agglomerates are transferred continuously from burner to gasifier and coal is fed through lock hoppers and is conveyed by inert gas to the gasifier. Hot agglomerated ash flows downward and transfers a portion of its sensible heat to support the coal gasification reactions. Product gas from the gasifier is sent to the gas cleanup section. Most of the agglomerated ash is recycled to the burner for reheating, but ash equivalent to the ash content of the coal fed to the burner is removed from the system continuously to maintain a constant quantity of ash agglomerates in the cycle.

21.3.10   COED/COGAS PROCESS

The COED/COGAS process (Figure 21.16) involves the flow of coal through four fluidized-bed pyrolysis stages, each operating at a higher temperature than the preceding unit, and the temperatures of the stages are selected to be just below the maximum temperature to which the particular feed coal can be heated without agglomerating.

The optimum stage temperatures (and even the number of stages) vary depending on the properties of the feed coal. Typical operating temperatures are 315°C–345°C (600°F–650°F) in the first stage, 425°C–455°C (800°F–850°F) in the second stage, 540°C (1000°F) in the third stage, and 870°C (1600°F) in the fourth stage.

Heat for the process is provided by burning a portion of the product char with oxygen in the presence of steam in the fourth stage. Hot gases from this stage flow countercurrent to the char and provide the hot fluidizing medium for the intermediate pyrolysis stages. The gases leaving both the

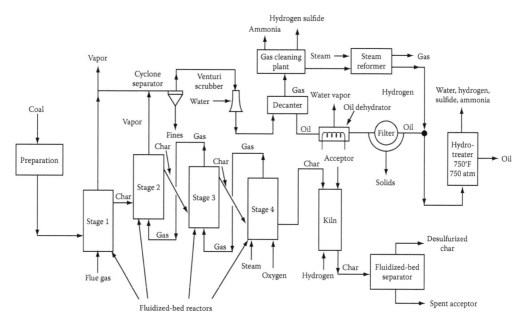

FIGURE 21.16 COED/COGAS process.

first and second stages are passed to cyclones which remove the fines, but the vapors leaving the cyclones need to be quenched in a Venturi scrubber to condense the oil, and the gases and oil are separated in a decanter. The gas is desulfurized and then steam-reformed to yield hydrogen and fuel gas; the hydrogen is returned to the process.

The oil from the decanter is dehydrated, filtered, and hydrotreated to remove nitrogen, sulfur, and oxygen (forming ammonia, hydrogen sulfide, and water, respectively) to form a heavy synthetic crude oil (ca. 25° API). The char produced by the process is desulfurized in a shift kiln, where hydrogen is treated with the char to produce hydrogen sulfide which is then absorbed by an acceptor, such as limestone or dolomite.

The COGAS process involves the gasification of the COED char to produce a synthesis gas composed of carbon monoxide and hydrogen. The heat for the char gasification reaction is provided by the combustion of part of the char.

21.3.11 CATALYTIC GASIFICATION PROCESS

The process is based on the concept that alkali metal salts (such as potassium carbonate, sodium carbonate, potassium sulfide, sodium sulfide, and the like) will catalyze the steam gasification of coal. In addition, tests with potassium carbonate showed that this material also acts as a catalyst for the methanation reaction.

Thus, in this process (Figure 21.17), crushed (−8 mesh) coal is treated with an aqueous solution of the catalyst after which the feed is dried and charged (through a system of lock hoppers) to the fluidized-bed gasifier at 700°C (1300°F) and 500 psi.

The bed is fluidized by a mixture of steam and recycled carbon monoxide-hydrogen. Unreacted steam is condensed and the acid gases (CO_2, H_2S) are removed by conventional acid gas treatment. The product (methane) is separated from the carbon monoxide and hydrogen by a cryogenic process. The solid product (which is actually a mixture of char, coal minerals, and the catalyst) is removed on a continuous basis; the majority of the catalyst is recovered and recycled.

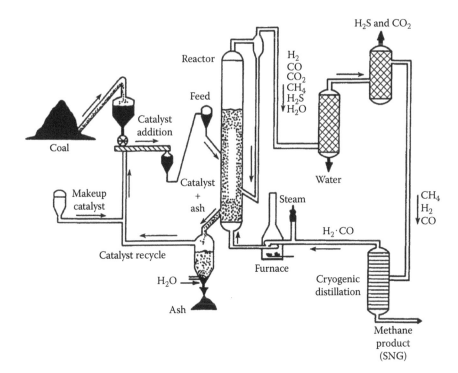

FIGURE 21.17 Exxon catalytic gasification process.

21.4 ENTRAINED-BED PROCESSES

An entrained-bed system uses finely sized coal particles blown into the gas steam prior to entry into the reactor and combustion occurs with the coal particles suspended in the gas phase; the entrained system is suitable for both caking and noncaking coals. The molten salt system employs a bath of molten salt to convert the charged coal (Cover et al., 1973; Howard-Smith and Werner, 1976; Koh et al., 1978).

21.4.1 KOPPERS–TOTZEK PROCESS

The Koppers–Totzek process (Baughman, 1978; Michaels and Leonard, 1978) (Figure 21.18) is an entrained-solids process which operates at atmospheric pressure. The reactor (Figure 21.19) is a relatively small, cylindrical, refractory-lined coal "burner" into which coal, oxygen, and steam are charged through at least two burner heads.

The feed coal for the process is crushed (so that 70% will pass through a 200-mesh screen), mixed with oxygen and low-pressure steam, and injected into the gasifier through a burner head. The heads are spaced 180° or 90° apart (representing two-headed or four-headed opposed burner arrangements) and are designed such that steam envelopes the flame and protects the reactor walls from excessive heat.

The reactor typically operates at an exit temperature of about 1480°C (2700°F) and the pressure is maintained just slightly above atmospheric. Only about 85%–90% of the total carbon may be gasified in a single pass through the gasifier because carbon conversion is a function of the reactivity of the coal and approaches 100% for lignite.

The heat in the reactor causes the formation of slag from mineral ash and this is removed from the bottom of the gasifier through a water seal. Gases and vaporized hydrocarbons produced by the coal at medium temperatures immediately pass through a zone of very high

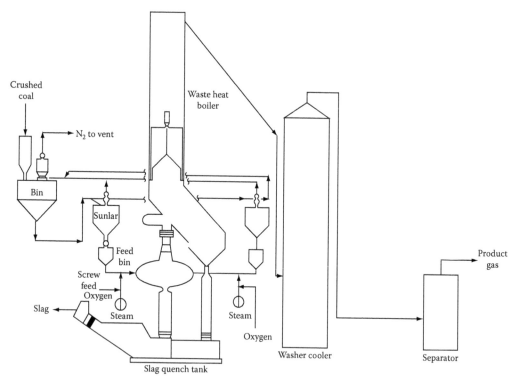

FIGURE 21.18 Koppers–Totzek process.

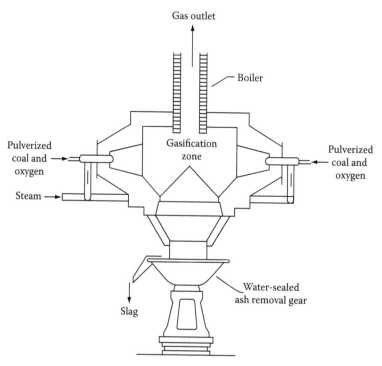

FIGURE 21.19 Koppers–Totzek gasifier.

temperature in which they decompose so rapidly that coal particles in the plastic stage do not agglomerate, and thus, any type of coal can be gasified irrespective of caking tendencies, ash content, or ash fusion temperature.

In addition, the high operating temperature ensures that the gas product contains no ammonia, tars, phenols, or condensable hydrocarbons. The raw gas can be upgraded to synthesis gas by reacting all or part of the carbon monoxide content with steam to produce additional hydrogen plus carbon dioxide.

21.4.2 Bi-Gas Process

The Bi-Gas process (Figure 21.20) is a two-stage, high-pressure, oxygen-blown slagging system using pulverized coal in an entrained flow (Hegarty and Moody, 1973). The coal is pulverized so that 70% passes through 200 mesh, is mixed with water, and is then fed to a cyclone where the solids are concentrated into a slurry. This slurry is concentrated in a thickener and centrifuged, re-pulped, and mixed with flux to produce the desired consistency. The blended slurry is transported at high pressure to a steam preheater where it is contacted with hot recycle gas in a spray drier which nearly instantaneously vaporizes the surface moisture. The coal is then conveyed to a cyclone at the top of the gasifier vessel by a stream of water vapor and inert recycle gas, as well as additional recycled gas from the methanator.

The coal is separated from the recycle gas in the cyclone and the coal flows by gravity into the gasifier through injector nozzles near the throat which separates the stages. Steam is injected through a separate annulus in the injector and the two streams combine at the injector tip and mix with hot synthesis gas (from stage 1). A mixing temperature of approximately 1205°C (2200°F) is rapidly attained and the coal is decomposed to produce methane, synthesis gas, and char. The raw gas and char rise (through stage 2) and leave the vessel (at approximately 925°C [1700°F]) and are quenched to 425°C (800°F) by atomized water prior to separation in a cyclone.

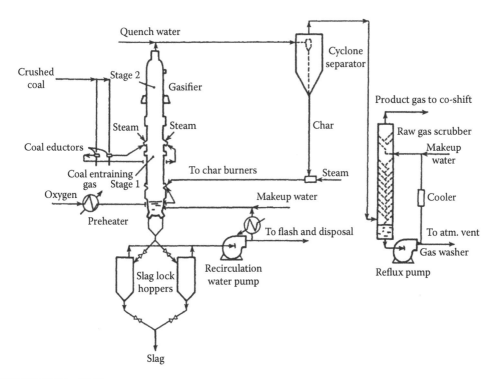

FIGURE 21.20 Bi-Gas process.

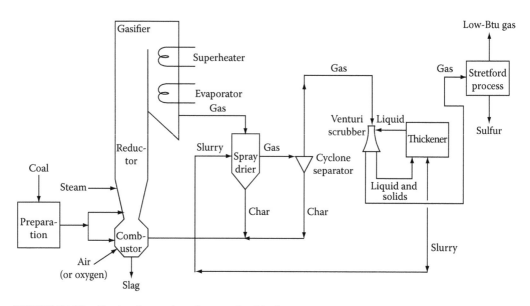

FIGURE 21.21 Combustion engineering entrained-bed process.

The char is recycled to the gasifier and the synthesis gas passes through a scrubber for additional cooling and purification. The clean gas is then sent to a shift converter to adjust the ratio of carbon monoxide to hydrogen. After shift conversion, the gas is improved by (1) hydrogen sulfide removal, (2) carbon dioxide removal, and (3) methanation.

21.4.3 ENTRAINED-BED PROCESS

The process (Figure 21.21) is based on an air-blown, atmospheric pressure, entrained-bed slagging gasifier and in the process part of the coal char is charged to the combustion section of the gasifier to supply the heat necessary for the endothermic gasification reaction (Patterson, 1976).

In the combustion section, substantially all of the ash in the system is converted to molten slag, which is removed via the bottom of the gasifier. The remainder of the coal is fed to the reaction section of the gasifier where it is contacted with hot gases entering the reaction zone from the combustor. The gasification process takes place in the entrainment section of the reactor where the coal is devolatilized and reacts with the hot gases to produce the desired product gas.

The gas contains solid particles and hydrogen sulfide; the former are removed and recycled to the combustor by means of a spray drier, cyclone separators, and Venturi scrubbers and the hydrogen sulfide is removed and elemental sulfur produced by the Stretford process.

21.4.4 TEXACO PROCESS

The Texaco process (Figure 21.22) gasifies coal under high pressure in an entrained bed by the injection of oxygen (or air) and steam with concurrent gas/solid flow (Probstein and Hicks, 1990). The coal is crushed in a two-stage system (the second step performed under an inert atmosphere) and is then mixed with water or oil to form a pumpable slurry which is pumped under pressure into the gasifier vessel (a refractory-lined chamber inside a pressure vessel). In this unit, the slurry reacts with either air or oxygen at high temperature; the product gas from the reactor contains primarily carbon monoxide and hydrogen, but may contain appreciable quantities of nitrogen if the reactor is air-blown. Oils or tars are not usually produced by the process and methane is the only hydrocarbon gas generated.

The product gases and molten slag produced in the reaction zone pass downward through a water spray chamber and a slag quench bath and the cooled gas and slag are then removed for further

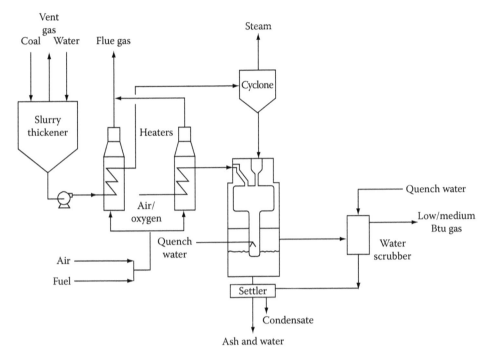

FIGURE 21.22 Texaco process.

treatment. In most cases the gas leaving the quench unit, once it is separated from the slag, is treated to remove the carbon fines and ash. The gas is then subsequently recycled to the slurry preparation system, treated for acid gas removal and elemental sulfur is recovered from the hydrogen sulfide-rich stream. Texaco has also modified the partial oxidation, which is used to gasify crude oil, to gasify coal; the effluent gas stream has little, or no, hydrocarbon content (Cornilis et al., 1981).

21.5 MOLTEN SALT PROCESSES

Molten salt processes, as the name implies, use a molten medium of an inorganic salt to generate the heat to decompose the coal into products and there are a number of applications of the molten bath gasification.

21.5.1 Atgas Process

The Atgas process (Figure 21.23) features the use of a molten iron bath (1550°C [2820°F]) into which coal, steam, and oxygen are injected (Karnavos et al., 1973). Coal and limestone are injected into the molten iron through tubes (lances) using steam as the carrier (La Rosa and McGarvey, 1975).

The coal devolatilizes with some thermal cracking of the volatile constituents leaving the fixed carbon and sulfur to "dissolve" in the iron whereupon carbon is oxidized to carbon monoxide by oxygen introduced through lances placed at a shallow depth in the bath. The sulfur, both organic and pyritic, migrates from the molten iron to the slag layer where it reacts with lime to produce calcium sulfide.

The product gas, which leaves the gasifier at ca. 1425°C (2600°F), is cooled, compressed, and fed to a shift converter where a portion of the carbon monoxide is reacted with steam to attain a carbon monoxide to hydrogen ratio of 1:3. The carbon dioxide so produced is removed and the gas is again cooled and enters a methanator where carbon monoxide and hydrogen react to form methane. Excess water is

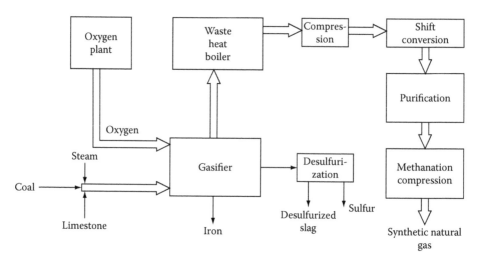

FIGURE 21.23 Atgas process.

removed from the methane-rich product and, depending on the type of coal used and the extent of purification desired, the final gas product may have a heat content of 920 Btu/ft (34.3 MJ/m).

21.5.2 MOLTEN SALT (ROCKGAS) PROCESS

In the Rockgas (Rockwell International) molten salt gasification process (Figure 21.24), coal and sodium carbonate are first transported by compressed air (at 150–300 psi) into the bottom of the molten bed (ca. 980°C [1800°F]; 300 psi) in the gasifier. This melt bed is composed of sodium carbonate along with any sodium sulfide and sulfate formed during the process (Rosemary and Trilling, 1978).

The fuel gas produced has a heat content of ca. 150 Btu/ft and is composed primarily of carbon monoxide, hydrogen, and nitrogen. The melt also contains ash and sulfur residue from the coal and hence part of the melt must be continuously withdrawn from the reactor for purification while additional fresh sodium carbonate is added.

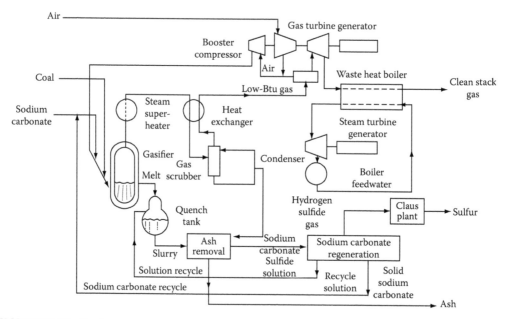

FIGURE 21.24 Rockgas process.

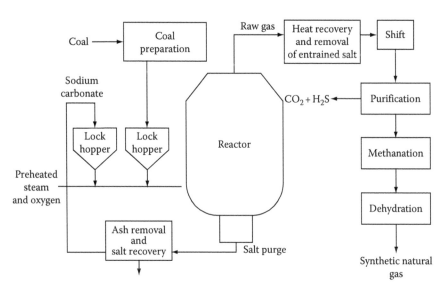

FIGURE 21.25 Pullman–Kellogg process.

21.5.3 PULLMAN–KELLOGG SALT PROCESS

In the process (Figure 21.25), air is bubbled into the bottom of the gasifier through multiple inlet nozzles and coal (sized to 1/4 in. [6 mm]) is fed beneath the surface of the molten salt bath using a central coal feed tube whereupon natural circulation and agitation of the melt disperses the coal. The main gasification reaction is a partial oxidation reaction and any volatile matter in the coal reacts to produce a fuel gas free of oils, tars, as well as ammonia. A water–gas shift equilibrium exists above the melt and, accordingly, in the reducing environment, carbon dioxide and water concentrations are minimal. Sulfur in the coal reacts with the melt to form sodium sulfide.

21.5.4 RUMMEL SINGLE-SHAFT MOLTEN BATH PROCESS

The Rummel single-shaft gasifier is a tall narrow water-cooled unit with the slag bath at the base that operates under atmospheric pressure (Rummel, 1959). Coal particles in suspension with the gasifying reactants are injected into the slag bath through the tangential nozzles. The coal is rapidly converted to gaseous products. The ash is converted to slag and the continuous overflow of the slag is quenched with water. The product gas is cooled at the top of the reactor.

21.6 UNDERGROUND GASIFICATION

The present section outlines the methods available for the in situ gasification of coal. In terms of the combustion process itself, both forward and reverse combustion are employed. Forward combustion is defined as movement of the combustion front in the same direction as the injected air, whereas reverse combustion involves movement of the combustion front moves in the opposite direction to the injected air.

The concept of the gasification of coal in the ground is not new; in 1868, Siemens in Germany proposed that slack and waste be gasified in the mine without any effort being made to bring this waste material to the surface. However, it was left to Mendeleev (in 1888) to actually propose that coal be gasified in the undisturbed (in situ) state.

A further innovative suggestion followed in 1909 in England when Betts proposed that coal be gasified underground by ignition of the coal at the base of one shaft (or borehole) and that a supply of air and steam be made available to the burning coal and the product gases be led to the surface through a different borehole. The utility of such a concept was acknowledged in 1912 when it was suggested that coal be combusted in place to produce a mixture of carbon monoxide and steam which could then be used to drive gas engines at the mine site for the production of electricity.

A considerable volume of investigative work has since been performed on the in situ gasification of coal in the former USSR, but it is only in recent years that the concept has been revived in Russia, Western Europe, and North America as a means of gas (or even liquids) production (Elder, 1963; Gregg and Edgar, 1978; Thompson, 1978; King and Magee, 1979; Olness, 1981; Zvyaghintsev,1981).

The aim of the underground (or in situ) gasification of coal is to convert the coal into combustible gases by combustion of a coal seam in the presence of air, oxygen, or oxygen and steam. Thus, not only could mining and the ever present dangers (Chapter 5) be partially or fully eliminated, but the usable coal reserves would be increased because seams that were considered to be inaccessible, unworkable, or uneconomical to mine could be put to use. In addition, strip mining and the accompanying environmental impacts, the problems of spoil banks, acid mine drainage, and the problems associated with the use of high-ash coal would be minimized or even eliminated.

The principles of underground gasification are very similar to those involved in the aboveground gasification of coal. The concept involves the drilling and subsequent linking of two boreholes (Figure 21.26) so that gas will pass between the two (King and Magee, 1979). Combustion is then initiated at the bottom of one borehole (injection well) and is maintained by the continuous injection of air. In the initial reaction zone (combustion zone), carbon dioxide is generated by the reaction of oxygen (air) with the coal after which the carbon dioxide reacts with coal (partially devolatilized) further along the seam (reduction zone) to produce carbon monoxide:

$$C_{coal} + O_2 \rightarrow CO_2$$

$$C_{coal} + CO_2 \rightarrow 2CO$$

In addition, at high temperatures that can frequently occur, moisture injected with oxygen or even moisture inherent in the seam may also react with the coal to produce carbon monoxide and hydrogen:

$$C_{coal} + H_2O \rightarrow CO + H_2$$

The coal itself may also decompose (pyrolysis zone) by the ever increasing temperature to produce hydrocarbons and tars which contribute to the product gas mix. The gas so produced varies in

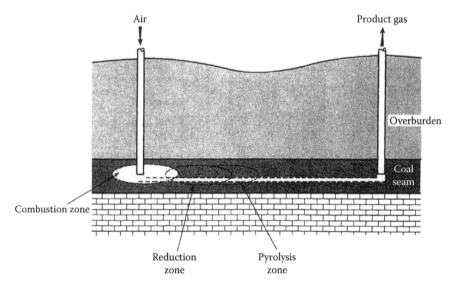

FIGURE 21.26 Underground coal gasification system. (From Berkowitz, N., *An Introduction to Coal Technology*, Academic Press, Inc., New York, 1979.)

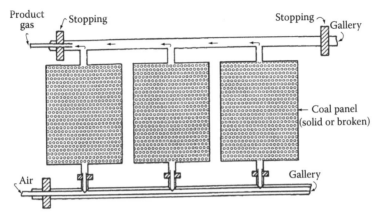

FIGURE 21.27 Chamber method of underground coal gasification. (From Braunstein, H.M. et al., Environmental health and control aspects of coal conversion: An information overview, Report ORNL-EIS-94, Oak Ridge National Laboratory, Oak Ridge, TN, 1977.)

character and composition but usually falls into the low-heat (low-Btu) category ranging from 4.7 to 6.5 MJ/m (125–175 Btu/ft) (King and Magee, 1979).

Both shaft and shaftless systems (and combinations of these systems) constitute the methods for underground gasification. Selection of the method to be used depends on such parameters as the permeability of the seam, the geology of the deposit, the seam thickness, depth, inclination, and the amount of mining desired. The shaft system involves driving large diameter openings into the seam and may therefore require some underground labor, whereas the shaftless system employs boreholes for gaining access to the coal and therefore does not require mining.

21.6.1 CHAMBER METHOD (WAREHOUSE METHOD)

In the chamber method (Figure 21.27), coal panels are isolated with brick-work and underground galleries. This method relies on the natural porosity of coal for flow through the system. The panel is ignited at one side and any gas produced can be removed at the opposite end of the panel.

Gasification and combustion rates are usually low and the product gas may have a variable composition and even contain unconsumed oxygen. Variations of this method consist of breaking the coal by hand or drilling several holes into the seam and charging them with dynamite in the hope that the coal will be crushed in advance of the reaction zone by a series of explosions.

21.6.2 BOREHOLE PRODUCER METHOD

In the borehole producer method (Figure 21.28), parallel galleries (used for air inlet and product gas withdrawal) are constructed in the coal bed. The boreholes are drilled from one gallery to another about 5 ft (1.5 m) apart and are filled with valves at their inlets and with iron seatings at their outlets. Electric ignition of the coal in each borehole is achieved by remote control.

The borehole producer method was designed to gasify generally flat-lying seams and is usually constructed by driving three parallel galleries (ca. 490 ft; ca. 150 m apart) into the coal from an access road and then connecting them by approximately 4 in. (10 cm) diameter holes at 13–16 ft (4–5 m) intervals. Incoming air is directed to the operating panels by control valves placed at the drill hole inlets.

21.6.3 STREAM METHOD

The stream method can be applied generally to steeply pitched coal beds; inclined galleries following the dip of the coal seam are constructed parallel to each other and are connected at the bottom by a horizontal gallery or "fire drift" (Figure 21.29).

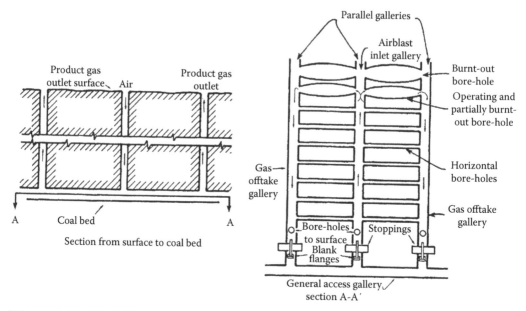

FIGURE 21.28 Borehole producer method of underground coal gasification. (From Braunstein, H.M. et al., Environmental health and control aspects of coal conversion: An information overview, Report ORNL-EIS-94, Oak Ridge National Laboratory, Oak Ridge, TN, 1977.)

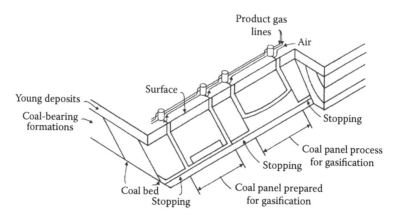

FIGURE 21.29 Stream method of underground coal gasification. (From Braunstein, H.M. et al., Environmental health and control aspects of coal conversion: An information overview, Report ORNL-EIS-94, Oak Ridge National Laboratory, Oak Ridge, TN, 1977.)

A fire in the horizontal gallery starts the gasification, which proceeds upward with air coming down one inclined gallery and gas departing through the other. An advantage of this method is that the fire zone advances upward, and the ash, together with any roof that may fall, collects below the fire zone.

21.6.4 Shaftless Methods

One example of a shaftless system for underground gasification of coal is the percolation or filtration method (Figure 21.30) in which two boreholes are drilled from the surface through the coal seam. The distance between boreholes depends on the seam permeability.

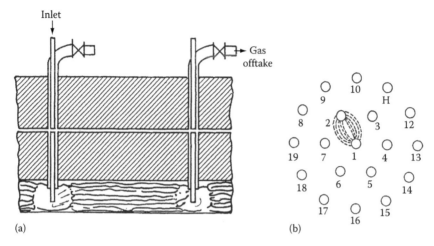

FIGURE 21.30 Shaftless method (percolation method) of underground coal gasification: (a) section through boreholes and (b) plan of boreholes. (From Braunstein, H.M. et al., Environmental health and control aspects of coal conversion: An information overview, Report ORNL-EIS-94, Oak Ridge National Laboratory, Oak Ridge, TN, 1977.)

Air or air and steam are blown through one hole, and gas is removed from the second; reverse combustion is permitted in this method. As the burning progresses, the permeability of the seam increases and compressed air blown through the seam also enlarges cracks in the seam. When combustion of a zone nears completion, the process is transferred to another pair of boreholes.

REFERENCES

Archer, D.W., Keairns, D.L., and Vidt, E.J. 1975. *Energy Communications*, 1: 115.

Bair, W.G. and Lau, F.S. 1980. In *Coal Processing Technology*. American Institute of Chemical Engineers, New York, Vol. VI, p. 166.

Baughman, G.L. 1978. *Synthetic Fuels Data Handbook*. Cameron Engineers, Denver, CO.

Berkowitz, N. 1979. *An Introduction to Coal Technology*. Academic Press, Inc., New York.

Braunstein, H.M., Copenhaver, E.D., and Pfuderer, H.A. 1977. Environmental health and control aspects of coal conversion: An information overview. Report ORNL-EIS-94. Oak Ridge National Laboratory, Oak Ridge, TN.

Collot, A.G. 2002. Matching gasifiers to coals. Report No. CCC/65. Clean Coal Centre, International Energy Agency, London, U.K.

Collot, A.G. 2006. Matching gasification technologies to coal properties. *International Journal of Coal Geology*, 65: 191–212.

Cornilis, B., Hibbel, J., Ruprecht, P., Durrfeld, R., and Langhoff, J. 1981. *Hydrocarbon Processing*, 60(1): 149.

Cover, A.E., Schreiner, W.C., and Skapendas, G.T. 1973. *Chemical Engineering Progress*, 69(3): 31.

Curran, C.P., Fink, C.E., and Gorin, E. 1969. *Industrial and Engineering Chemistry Process Design and Development*, 8(4): 559.

Elder, J.L. 1963. In *Chemistry of Coal Utilization*, Supplementary Vol., H.H. Lowry (Ed.). John Wiley & Sons, Inc., New York, Chapter 21.

Gregg, D.W. and Edgar, T.F. 1978. *AIChE Journal*, 24: 753.

Hegarty, W.P. and Moody, B.E. 1973. *Chemical Engineering Progress*, 69(3): 37.

Howard-Smith, I. and Werner, G.J. 1976. *Coal Conversion Technology*. Noyes Data Corporation, Park Ridge, NJ.

Karnavos, J.A., LaRosa, P.J., and Pelczarski, E.A. 1973. *Chemical Engineering Progress*, 69(3): 54.

King, R.B. and Magee, R.A. 1979. In *Analytical Methods for Coal and Coal Products*, Vol. III, C. Karr, Jr. (Ed.). Academic Press Inc., New York, Chapter 41.

Koh, A.L., Harty, R.B., and Johnson, J.G. 1978. *Chemical Engineering Progress*, 74(8): 73.

La Rosa, P. and McGarvey, R.J. 1975. In *Proceedings. Clean Fuels from Coal. Symposium II*, Institute of Gas Technology, Chicago, IL.

Marinov, V., Marinov, S., Lazarov, L., and Stefanova, M. 1992. *Fuel Processing Technology*, 31: 181.

Massey, L.G. 1979. In *Coal Conversion Technology*, C.Y. Wen and E.S. Lee (Eds.). Addison-Wesley Publishers, Inc., Reading, MA, p. 313.

Michaels, H.J. and Leonard, H.F. 1978. *Chemical Engineering Progress*, 74(8): 85.

Nowacki, P. 1980. *Lignite Technology*. Noyes Data Corporation, Park Ridge, NJ.

Olness, D.U. 1981. The Podmoskovnaya underground coal gasification station. Report UCRL-53144. Lawrence Livermore National Laboratory, Livermore, CA.

Patterson, R.C. 1976. *Combustion*, 47: 28.

Probstein, R.F. and Hicks, R.E. 1990. *Synthetic Fuels*. pH Press, Cambridge, MA, Chapter 4.

Rosemary, J.K. and Trilling, C.A. 1978. SNG production by the Rockgas process. In *Energy Technology V: Challenges to Technology. Proceedings. Fifth Conference*, Washington, DC, February 27–March 1. (A79-15879 04-44) Washington, DC, Government Institutes, Inc., pp. 462–471.

Rummel, R. 1959. Gasification in a slag bath. *Coke Gas*, 21(247): 493–501.

Thompson, P.N. 1978. *Endeavor*, 2: 93.

Verma, A. 1978. *Chemtech*, 8: 372, 626.

Weiss, A.J. 1978. *Hydrocarbon Processing*, 57(6): 125.

Zvyaghintsev, K.N. 1981. *Trends of Development of Underground Coal Gasification in the U.S.S.R. Natural Resources Forum (USSR)*, 5: 99.

22 Clean Coal Technologies

22.1 INTRODUCTION

The use of coal, indeed the use of fossil fuel resources, has increased by several orders of magnitude since the early decades of the twentieth century (Miller, 2005; Speight, 2008). The expansion of the industrialized system which requires the generation and use of vast amounts of electrical energy as well as the increased use of automobiles, to mention only two examples, have been the major driving forces behind the expansion of fossil fuel usage. But, in concert with this increased usage, there has also followed the onset of detrimental side effects. Emissions of nonindigenous chemicals into the environment or the ejection into the environment of chemicals that are indigenous but in quantities that exceed their natural occurrence are major issues. Thus, the expansion and evolution of industry in the service of man is, perhaps, an excellent example of medicine almost killing the patient.

Emissions resulting from the use of the various fossil fuels have had deleterious effects on the environment and promise further detriment unless adequate curbs are taken to control not only the nature but also the amount of gaseous products being released into the atmosphere. In this respect, coal is often considered to be a *dirty* fuel and is usually cited as the most environmentally obnoxious of the fossil fuels. However, coal is not necessarily a dirty fuel; perhaps coal is no dirtier than the so-called clean nuclear fuels which have a habit of being really dirty when nuclear plants go awry. In fact, it is quite possible that nuclear power will only be chosen as a viable energy option if the demand for electricity accelerates, nuclear costs (including the potential for accidents) are contained, and global warming concerns escalate (Ahearne, 1993; Bell, 2011). Thus, it is more appropriate to consider the way in which we use coal as being detrimental to the environment. Such a change in thinking opens new avenues for coal use.

Fossil fuel use is a necessary part of the modern world and the need for stringent controls over the amounts and types of emissions from the use of, in the present context, coal is real. The necessity for the cleanup of process gases is real but to intimate that coal is the major cause of all our environmental concerns is unjust. Gaseous products and by-products are produced from a variety of industries (Austin, 1984; Probstein and Hicks, 1990; Speight, 1990, 1991). These gaseous products all contain quantities of noxious materials that are a severe detriment to the environment.

Coal is an abundant energy source and forms a major part of the earth's fossil fuel resources (Berkowitz, 1979; Hessley et al., 1986; Hessley, 1990), the amount available being subject to the method of estimation and to the definition of the resources (Chapter 1) (EIA, 1988, 1989, 1991a,b, 1992). In general, but in terms of the measurable reserves, coal constitutes ca. 33% of the fossil fuel supplies of the world (Figure 22.1). In terms of the energy content, coal (68%) and natural gas (13%) are the major energy-containing fossil fuels, petroleum (19%) making up the remainder (EIA, 1989, 1991a,b, 1992).

Coal is the most familiar of the fossil fuels not necessarily because of its use throughout the preceding centuries (Galloway, 1882) but more because of its common use during the nineteenth century. Coal was largely responsible not only for the onset but also for the continuation of the industrial revolution. Coal occurs in various forms defined in a variety by rank or type (Chapter 2) and is not only a solid hydrocarbonaceous material with the potential to produce considerable quantities of carbon dioxide as a result of combustion, but many coals also contain considerable quantities of sulfur (Table 22.1). Sulfur content varies (Table 22.2) but, nevertheless, opens up not only the possibility but also the reality of sulfur dioxide production (Manowitz and Lipfert, 1990; Tomas-Alonso, 2005).

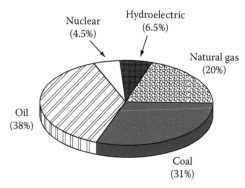

FIGURE 22.1 Distribution of world energy supplies.

TABLE 22.1
Sulfur Content of Various Coals of the United States

Region	No. of Samples	Organic S (%)	Pyritic S (%)	Total S (%)
N. Appalachian	227	1.00	2.07	3.01
S. Appalachian	35	0.67	0.37	1.04
E. Midwest	95	1.63	2.29	3.92
W. Midwest	44	1.67	3.58	5.25
Western	44	0.45	0.23	0.68
Alabama	10	0.64	0.69	1.33

TABLE 22.2
Distribution of Sulfur within the Various Coal Ranks of the United States

	Sulfur Content Range[a] (% w/w)			
Rank	**0–0.7**	**0.8–1.0**	**1.1–3.0**	**3.1+**
Anthracite	9.65	0.6	2.9	
Bituminous	14.3	15.2	26.2	44.3
Subbituminous	66.0	33.6	0.4	
Lignite	77.0	13.7	9.3	
U.S. average	46	19	15	20

[a] Percentage of total in each range.

It is predictable that coal will be the primary source of energy for the next several decades well into the next century and therefore the message is clear: until other energy sources supplant coal, the challenge is to develop technological concepts that will provide the maximum, and environmentally efficient, recovery of energy from coal (Dryden, 1975; Yeager and Baruch, 1987; Fulkerson et al., 1990).

There are a number of processes that are being used in coal-fired power stations that improve the efficiency and environmental acceptability of coal extraction, preparation and use, and many more are under development (Bris et al., 2007). These processes are collectively known as *clean coal technologies*. Designation of a technology as a *clean coal technology* does not imply that it reduces emissions to zero or near zero. For this reason, the term has been criticized as being misleading; it might be more appropriate to refer to *cleaner coal technologies* (Balat, 2008a,b; Nordstrand et al., 2008; Franco and Diaz, 2009).

Nevertheless, *clean coal technologies* reduce emissions of several pollutants, reduce waste, and increase the amount of energy gained from each ton of coal (Nordstrand et al., 2008; Franco and Diaz, 2009). They include various chemical and physical treatments applied pre- or postcombustion and may be broadly divided into processes relating either to (1) combustion efficiency or (2) pollution control.

An example is gasification of coal by burning it in oxygen to produce a cleaner gaseous fuel known as *syngas* or *synthesis gas* (a mixture mainly of hydrogen and carbon monoxide (Chapters 20 and 21), which is comparable in its combustion efficiency to natural gas. This reduces the emissions of sulfur, nitrogen oxides, and mercury, resulting in a much cleaner fuel (Sondreal et al., 2004, 2006; Srivastava et al., 2005; Lee et al., 2006; Yang et al., 2007; Nordstrand et al., 2008; Wang et al., 2008). The resulting hydrogen gas can be used for electricity generation or as a transport fuel. The gasification process also facilitates capture of carbon dioxide emissions from the combustion effluent (see discussion of carbon capture and storage [CCS] in the following).

Integrated gasification combined-cycle (IGCC) systems combine gasification with a heat recovery system that feeds a secondary steam-powered generator, thereby increasing the power generated from a given amount of coal. These systems are currently being employed in many new coal-fired power plants worldwide.

CCS is a technology under development that offers much higher prospects of emissions reductions than other clean coal technologies. CCS involves capture of carbon dioxide either before or after combustion of the fuel; transport of the captured carbon dioxide to the site of storage; and injection of the carbon dioxide in deep underground reservoirs for long-term storage (*geosequestration*). CCS is proposed as a means of reducing to near-zero the greenhouse gas (IEA Coal Research, 1992) emissions of fossil fuel burning in power generation and carbon dioxide production from other industrial processes such as cement manufacturing and purification of natural gas. Many clean coal technologies are being developed with CCS in mind, for example, concentrating carbon dioxide in the combustion exhaust to ease the separation and capture of carbon dioxide. The majority of the CCS effort is being invested in incorporating CCS into new power generation plant designs—current figures indicate that it is cheaper to build a new IGCC plant that produces a pure carbon dioxide exhaust stream than to retrofit an existing plant with postcombustion carbon dioxide capture technology (Clarke, 1991).

Many coal-fired electricity-generating plants are of a conventional design, with typical efficiencies of about 33%–35%.; only approximately 35% of the usable energy in the coal is actually converted into electricity and the rest appears as waste heat. Plants with greater energy conversion efficiency (up to approximately 42%–45%) are possible with combined cycles that recycle heat using very high-temperature steam. Even higher efficiencies are expected from plants that utilize additional heat-capturing cycles, technological developments in turbine efficiencies, and higher process temperatures.

Clean coal technologies seek to reduce harsh environmental effects by using multiple technologies to clean coal and contain its emissions.

When coal burns, it releases carbon dioxide and other emissions in *flue gas*. Some clean coal technologies purify the coal before it burns. One type of coal preparation, *coal washing*, removes unwanted minerals by mixing crushed coal with a liquid and allowing the impurities to separate and settle (Luttrell et al., 2006).

Other systems control the coal burn to minimize emissions of sulfur dioxide, nitrogen oxides, and particulates (Darcovkic et al., 1997; Bhanarkar et al., 2008). *Wet scrubbers*, or flue gas desulfurization (FGD) systems (Table 22.3), remove sulfur dioxide, a major cause of *acid rain*, by spraying flue gas with limestone and water. The mixture reacts with the sulfur dioxide to form synthetic gypsum, a component of drywall.

Low-NOx (nitrogen oxide) burners (Table 22.4) reduce the creation of nitrogen oxides, a cause of ground-level ozone, by restricting oxygen and manipulating the combustion process (Bris et al., 2007). *Electrostatic precipitators* remove particulates that aggravate asthma and cause respiratory ailments by charging particles with an electrical field and then capturing them on collection plates.

TABLE 22.3
FGD Technologies

1. *Wet processes: limestone—gypsum*

   It is the most common FGD process—the flue gas is treated with limestone slurry for the sulfur dioxide removal. Approximately 95% of the SO_2 from the flue gas can be eliminated. The method can be used for medium-to high-sulfur coals. A slurry waste or a saleable slurry by-product is obtained.

2. *Sea-water washing*

   Flue gas is treated with sea water to neutralize the sulfur dioxide——up to 98% of the sulfur dioxide can be removed using this method. The process also removes almost 100% of HCl in the flue gas. Treatment of the waste water with air is required to reduce its chemical oxygen demand and acidity before discharging back to the sea.

3. *Ammonia scrubbing*

   The ammonia/ammonium sulfate is employed as scrubbing agent in this process. About 93% of the sulfur dioxide from the flue gas can be eliminated at the commercial scale. Ammonium sulfate is obtained as a saleable product.

4. *Wellman—Lord process*

   The process uses sodium sulfite solution for scrubbing sulfur dioxide from flue gas—the removal rate of sulfur dioxide is up to 98%. The method includes reagent regeneration stage in the process scheme which reduces sorbent consumption. Depending on the plant design, different saleable by-products (elemental sulfur, sulfuric acid, or liquid sulfur dioxide) can be produced.

5. *Semi-dry processes: CFB*

   Hydrated lime is used to remove the sulfur dioxide, sulfur trioxide, and hydrogen chloride from the flue gas in this method. Water is injected into the bed to obtain an operation close to the adiabatic saturation temperature. More than 95% sulfur dioxide removal efficiency can be achieved by using this process. The final product is a dry powdered mixture of calcium compounds requiring disposal operations.

6. *Spray dry process*

   Lime or calcium oxide is usually employed as sorbent in this process in which the flue gas is given into a reactor vessel and the lime slurry is atomized into the same vessel. The water in the slurry is completely evaporated in the spray dry absorber. It is possible to remove 85%–90% of the sulfur dioxide for moderately high-sulfur fuels. A solid by-product needing disposal operation is produced.

7. *Duct spray dry process*

   This process is very similar to the conventional spray drying. However, the slaked lime slurry is directly fed into the ductwork. A moderate degree of desulfurization is possible by using this method. A dry powdered mixture of calcium compounds is produced as the final product.

8. *Dry processes furnace sorbent injection*

   The process considers the injection of lime into wall and tangentially fired boilers for the sulfur dioxide absorption. This process removes between as little as 30% and as much as 90% of the sulfur in the flue gas—inefficient use of the absorbent and the costs of by-product disposal (a mixture of ash and calcium compounds) add to operation expenses.

9. *Sodium bicarbonate injection process*

   The dry soda sorbents are injected as dry powders into the flue gas duct downstream of the air heater to react with acidic compounds such as sulfur dioxide, sulfur trioxide, and hydrogen chloride. It is possible to remove up to 70% of the sulfur dioxide and approximately 90% of the hydrogen chloride. The final product is a dry powdered mixture of sodium compounds and fly ash.

Sources: Kohl, A.L. and Nielsen, R., *Gas Purification*, Gulf Publishing Company, Houston, TX, 1997; IEA, *Control and Minimization of Coal-Fired Power Plant Emissions*, International Energy Agency, Paris, France, 2003; Miller, B.G., *Coal Energy Systems*, Elsevier, Amsterdam, the Netherlands, 2005; Breeze, P., *Power Generation Technologies*, Elsevier, Amsterdam, the Netherlands, 2005; Mokhatab, S. et al., *Handbook of Natural Gas Transmission and Processing*, Elsevier, Amsterdam, the Netherlands, 2006; Suarez-Ruiz, I. and Ward, C.R., Applied coal petrology: The role of petrology in coal utilization, *Coal Combustion*, I. Suarez-Ruiz and J.C. Ward, Eds., Elsevier, Amsterdam, the Netherlands, 2008, Chapter 4.

TABLE 22.4
NOx Control Technologies in Electricity Generation

1. *Low-NOx burners*

   This technology relies on the principle of staging the combustion air within the burner to reduce NOx formation. Unburned carbon levels are higher in low-NOx firing systems. A 35%–55% NOx reduction is possible.

2. *Furnace air staging*

   This process involves staging the combustion by diverting 5%–20% of combustion air from the burners and injecting above the main combustion zone and can reduce the NOx emissions by 20%–60% depending on the initial nitrogen oxide levels in the boiler, fuel combustion equipment design, and fuel type. The amount of unburned carbon is on the order of 35%–50%.

3. *Fuel staging (reburning)*

   In this technology, the burners in the primary combustion zone are worked with low excess air. Up to 30% of the total fuel heat input is given above the main combustion zone to create a fuel-rich zone during combustion. Reburning of coal takes relatively longer residence time, approximately 50% of NOx reduction can be achieved.

4. *Selective non-catalytic reduction*

   This process involves the injection of a reagent such as ammonia (NH_3) or urea (H_2NCONH_2) into the hot flue gas stream to reduce the NOx to nitrogen and water. Using this technology, 30%–60% NOx reductions can be achieved.

5. *Selective catalytic reduction*

   The process considers ammonia (NH_3) injection into the flue gas which is then passed through the layers of a catalyst made from a base metal, a zeolite, or a precious metal. This technology can achieve NOx reductions about 85–95. It is most suitable for low-sulfur coals (up to 1.5% of sulfur) due to the corrosive effects of sulfuric acid formed during the operation.

Sources: Miller, B.G., *Coal Energy Systems*, Elsevier, Amsterdam, the Netherlands, 2005; Breeze, P., *Power Generation Technologies*, Elsevier, Amsterdam, the Netherlands, 2005; Graus, W. and Worrell, E., *Energy Policy*, 37, 2147, 2009; Suarez-Ruiz, I. and Ward, C.R., Applied coal petrology: The role of petrology in coal utilization, *Coal Combustion*, I. Suarez-Ruiz and J.C. Ward, Eds., Elsevier, Amsterdam, the Netherlands, 2008, Chapter 4; Franco, A. and Diaz, A.R., *Energy*, 34, 348, 2009.

Gasification avoids burning coal altogether. With IGCC systems, steam and hot pressurized air or oxygen combine with coal. The resulting *synthesis gas (syngas)*, a mixture of carbon monoxide and hydrogen, is then cleaned and burned in a gas turbine to make electricity. The heat energy from the gas turbine also powers a steam turbine. Since IGCC power plants create two forms of energy, they have the potential to reach a fuel efficiency of 50%.

The clean coal technology field is moving in the direction of coal gasification with a second stage so as to produce a concentrated and pressurized carbon dioxide stream followed by its separation and geological storage. This technology has the potential to provide what may be called *zero emissions*—but which, in reality, is extremely low emissions of the conventional coal pollutants, and as low-as-engineered carbon dioxide emissions.

This has come about as a result of the realization that efficiency improvements, together with the use of natural gas and renewables such as wind, will not provide the deep cuts in greenhouse gas emissions necessary to meet future national targets (Ember et al., 1986; Omer, 2008).

22.2 HISTORICAL PERSPECTIVES

Coal has been the principal source of fuel and energy for many hundreds of years; perhaps even millennia, even though its use is documented somewhat less completely than the use of petroleum and its derivatives (Speight, 1991). Nevertheless, coal has been the principal source of solid and gaseous fuels during this and the last century. In fact, each town of any size had a plant for the gasification of coal (hence, the use of the term *town gas*). Most of the natural gas produced at the petroleum fields was vented to the air or burned in a flare stack; only a small amount of the natural gas from the petroleum fields was pipelined to industrial areas for commercial use. It was only in the years after World War II that natural gas became a popular fuel commodity leading to the recognition that it has at the present time. Coal has probably been known and used for an equal length of time but the records are somewhat less than complete.

There are frequent references to coal in the Christian Bible (Cruden, 1930) but, all in all, the recorded use of coal in antiquity is very sketchy. However, there are excellent examples of coal mining in Britain from the year 1200 AD which marked, perhaps, the first documented use of mined coal in England (Galloway, 1882). There is another historical note, which is worth mentioning here, that a singular environmentally significant event occurred in England in the year 1257 that threatened the very existence of coal use and its future as a fuel (Chapter 16).

Thus, Eleanor, Queen of Henry III of England, was obliged to leave the town of Nottingham where she had been staying during the absence of the King on an expedition into Wales. The Queen was unable to remain in Nottingham due to the troublesome smoke from the coal being used for heating and cooking. Over the next several decades, a variety of proclamations were issued by Henry and by his son, Edward I, which threatened the population with the loss of various liberties, perhaps even the loss of significant part(s) of the miscreant's anatomy and even loss of life, if the consumption of coal was not seriously decreased and, in some cases, halted (Galloway, 1882).

By the late 1500s, an increasing shortage of wood in Europe resulted in the search for another form of combustible energy and coal became even more popular with the English, French, Germans, and Belgians being very willing to exploit the resource. In the mid-to-late 1700s, the use of coal increased dramatically in Britain with the successful development of coke smelters and the ensuing use of coal to produce steam power. By the 1800s, and the industrial revolution well under way, coal was supplying most of Britain's energy requirements. By this time, the use of gas from coal (*town gas*) for lighting was also established.

In contrast, in the United States, where the population density was much lower than in Europe, wood was more plentiful and many colonial fires were fueled by this resource; any coal (often in limited quantities or for specific uses such as in smelters) required for energy was imported from Britain and/or from Nova Scotia. But after the Revolutionary War, coal entered the picture as an increasingly popular source of energy. As an example, the state of Virginia supplied coal to New York City. However, attempts to open the market to accept coal as a fuel were generally ineffective in the United States, and the progress was slow, if not extremely slow. It was not until the period from 1850 to 1885 that coal use in the United States increased, spurred by the emerging railroad industry as a fuel for the locomotives as well as for the manufacture of steel rails. At last, coal seemed to be undergoing a transformation as a fuel on both sides of the Atlantic Ocean (Frosch and Gallopoulos, 1989).

22.3 MODERN PERSPECTIVES

The increased use and popularity of coal is due, no doubt, to the relative ease of accessibility which has remained virtually unchanged over the centuries. On the other hand, petroleum is now an occasional exception because of a variety of physical and political reasons.

The relatively simple means by which coal can be used has also been a major factor in determining its popularity. In addition, coal can be interchanged to the three fuel types insofar as one form can be readily converted to another:

$$\text{Gas} \rightarrow \text{Liquid} \rightarrow \text{Solid.}$$

Indeed, the conversion of coal to fuel products and to chemicals as evidenced by the birth and evolution of the coal chemicals industry in the nineteenth century served to increase the popularity of coal.

The prognosis for the continued use of coal is good. Projections that the era of fossil fuels (gas, petroleum, and coal) will be almost over when the cumulative production of the fossil resources reaches 85% of their initial total reserves (Hubbert, 1969) may or may not have some merit. In fact, the relative scarcity (compared to a few decades ago) of petroleum is real but it seems that the remaining reserves coal, and perhaps natural gas, make it likely that there will be an adequate supply of energy for several decades (Martin, 1985; MacDonald, 1990; Banks 1992); but energy in what form and at what cost? (Bending et al., 1987; Hertzmark, 1987; Meyers 1987; Sathaye et al., 1987). The environmental issues are very real and must be attended to.

The use of coal in an environmentally detrimental manner is to be deplored. The use of coal in an environmentally acceptable manner is to be applauded. Technologies that ameliorate the effects of coal combustion on acid rain deposition, urban air pollution, and global warming must be pursued vigorously (Vallero, 2008). There is a challenge that must not be ignored and the effects of acid rain in soil and water leave no doubt about the need to control its causes (Mohnen, 1988). Indeed, recognition of the need to address these issues is the driving force behind recent energy strategies as well as a variety of research and development programs (Stigliani and Shaw, 1990; United States Department of Energy, 1990; United States General Accounting Office, 1990).

As new technology is developed, emissions may be reduced by repowering in which aging equipment is replaced by more advanced and efficient substitutes (Hyland, 1991). Such repowering might, for example, involve an exchange in which an aging unit is exchanged for a newer combustion chamber, such as the atmospheric fluidized-bed combustor (AFBC) or the pressurized fluidized-bed combustor (PFBC) (Chapters 14 and 15).

In the pressurized fluid bed combustor, pressure is maintained in the boiler, often an order of magnitude greater than in the atmospheric combustor, and additional efficiency is achieved by judicious use of the hot gases in the combustion chamber (combined cycle). Both the atmospheric and pressurized fluid bed combustors burn coal with limestone or dolomite in a fluid bed which allows, with recent modifications to the system, the limestone sorbent to take up about 90% of the sulfur that would normally be emitted as sulfur dioxide. In addition, boiler reconfiguration can allow combustion to be achieved more efficiently than in a conventional combustor thereby reducing the formation of nitrogen oxide(s) (Baldwin et al., 1992; Coal Voice, 1992).

An important repowering approach attracting great interest is the IGCC system (Notestein, 1990; Takematsu and Maude, 1991) (Chapters 20 and 21).

The major innovation introduced with the IGCC technology is the conversion of coal into synthesis gas, a mixture of mainly hydrogen (H_2) and carbon monoxide (CO) with lesser quantities of methane (CH_4), carbon dioxide (CO_2), and hydrogen sulfide (H_2S). Up to 99% of the hydrogen sulfide can be removed by commercially available processes (Chapter 23) before the gas is burned. The synthesis gas then powers a combined cycle in which the hot gases are burned in a combustion chamber to power a gas turbine and the exhaust gases from the turbine generate steam to drive a steam turbine.

The carbon oxides (carbon monoxide, CO, and carbon dioxide, CO_2) are also of importance insofar as coal can produce either or both of these gases during use; and both gases have the potential to harm the environment. Reduction in the emission of these gases, particularly carbon dioxide which is the final combustion product of coal, can be achieved by trapping the carbon dioxide at the time of coal usage.

However, it is not only the production of carbon dioxide from coal that needs to be decreased. The production of pollutants such as sulfur dioxide (SO_2) and oxides of nitrogen (NOx, where x = 1 or 2) also needs attention. These gases react with the water in the atmosphere and the result is an acid:

$$SO_2 + H_2O \rightarrow H_2SO_3 \text{ (sulfurous acid)}$$

$$2SO_2 + O_2 \rightarrow 2SO_3$$

$$SO_3 + H_2O \rightarrow H_2SO_4 \text{ (sulfuric acid)}$$

$$NO + H_2O \rightarrow HNO_2 \text{ (nitrous acid)}$$

$$2NO + O_2 \rightarrow 2NO_2$$

$$NO_2 + H_2O \rightarrow HNO_3 \text{ (nitric acid)}$$

Indeed, the careless combustion of coal can account for the large majority of the sulfur oxides and nitrogen oxides released to the atmosphere. Whichever technologies succeed in reducing the amounts of these gases in the atmosphere should also succeed in reducing the amounts of urban smog, those notorious brown and gray clouds that are easily recognizable at some considerable distances from urban areas, not only by their appearance but also by their odor:

$$SO_2 + H_2O \rightarrow H_2SO_4 \text{ (sulfurous acid)}$$

$$2SO_2 + O_2 \rightarrow 2SO_3$$

$$SO_3 + H_2O \rightarrow H_2SO_4 \text{ (sulfuric acid)}$$

$$2NO + H_2O \rightarrow HNO_2 + HNO_3 \text{ (nitrous acid + nitric acid)}$$

$$2NO + O_2 \rightarrow 2NO_2$$

$$NO_2 + H_2O \rightarrow HNO_3 \text{ (nitric acid)}$$

Current awareness of these issues by a variety of levels of government has resulted, in the United States, of the institution of the Clean Coal Program to facilitate the development of pollution abatement technologies. And it has led to successful partnerships between government and industry (United States Department of Energy, 1993). In addition, there is the potential that new laws, such as the passage in 1990 of the Clean Air Act Amendments in the United States (United States Congress, 1990; Stensvaag, 1991; Elliott, 1992), will be a positive factor and supportive of the controlled clean use of coal. However, there will be a cost but industry is supportive of the measure and confident that the goals can be met (Locheed, 1993).

Indeed, recognition of the production of these atmospheric pollutants in considerable quantities every year has led to the institution of national emission standards for many pollutants. Using sulfur dioxide as the example, the various standards are not only very specific (Table 22.5) (Kyte, 1991) but will become more stringent with the passage of time (IEA Coal Research, 1991). Atmospheric pollution is being taken very seriously and there is also the threat, or promise, of heavy fines and/or jail terms for any pollution-minded miscreants who seek to flaunt the laws (Vallero, 2008). Be that as it may, a trend to the increased use of coal will require more stringent approaches to environmental protection issues than we have ever known at any time in the past.

TABLE 22.5
Examples of Gaseous Emissions from Selected Processes

Application	Components Removed
Chemical plant off-gases	
From metallurgical roasters	SO_2, SO_3, CO
From Claus plants	H_2S, SO_2, COS
From natural gas cleaning	H_2S, HCN, CS_2, CH_x, COS
From gasifier gas cleaning	H_2S, COS, CHS
From coke oven gas cleaning	H_2S, HCN, NH_3, CHS
From viscose plants	H_2S, CS_2
Flue gases	
From industrial steam generators	SO_2, SO_3, CH, CO, NOx
From power plants	SO_2, SO_3, CH, CO, NOx

The need to protect the environment is strong. One example is the passage of amendments to the Clean Air Act which attests to this fact.

Thus, as the alternatives in energy vacillate from coal to oil to gas and back again to be followed, presumably, by the eras of nuclear fuels and solar energy, there is an even greater need to ensure that emissions from coal use are clean. It is time to move away from the uncontrolled and irresponsible use of coal and to show that coal can be used in an environmentally safe manner.

22.4   CLEAN COAL TECHNOLOGY

The term clean coal technology refers to a new generation of advanced coal utilization technologies (Table 22.6) that are environmentally cleaner and in many cases more efficient and less costly than the older, and more *conventional* coal-using processes (United States Department of Energy, 1992). Clean coal technologies offer the potential for a more clean use of coal thereby having a direct effect on the environment and contribute to the resolution of issues relating to acid rain and global climate change.

There are a number of technological concepts that fall under the umbrella of *clean coal*, including installation of air pollution control equipment, and more innovative ideas such as carbon capture and gasification of coal for use in combined-cycle plants similar in design to natural gas-fired power plants. There is also the potential for producing synthetic liquid fuels from coal as means of

TABLE 22.6
Advanced Coal Combustion Systems

Parameters	Conventional Pulverized Fired	Supercritical Pulverized Fired	PFBC/CFBC	IGCC	Hybrid Cycle (Gasification in Combustion)
Maturity of technology	Proven and commercially available	Proven and commercially available	Proven and commercially available	Proven and commercially available	Proven and becoming commercially available
Range of units available	All commercial sizes available (common unit size in the range 300–1000 MW e)	All commercial sizes available	Up to 350 mw sizes available	250–300 MWe, currently limited by the size of large gas turbine units available	Demonstration plant proposed at around 90 MWe
Fuel flexibility	Burns a wide range of coals	Burns a wide range of coals	Will burn a wide range of coals, as well as low grade coals; best suited for low-mineral-matter coal	Can use a wide range of coals—not designed for low-grade, high-mineral-matter coal	Can use a wide range of coals; designed to utilize low-grade, high-mineral-matter coal efficiently
Thermal efficiency (LHV)	Limited by steam conditions—41% with modern designs	At least 45%	44%, increases possible with supercritical steam cycle	43% but over 50% possible with advanced gas	43% but over 50% possible with advanced gas
Operational flexibility	Can operate at low load, but performance would be limited	Can operate at low load, but performance would be limited	Can operate at low load, but performance would be limited	Can operate at base load	Reasonable performance at low load

offsetting petroleum depletion (Speight, 2008). In each case, there are economic and environmental benefits of each option, as well as the disadvantages and obstacles to their implementation.

While some of these concepts are technically viable, there is no single technology or combination of technologies that is capable of addressing all of the environmental or economic challenges likely to arise from continued dependence on coal as a major source of energy in the coming decades, underscoring the need to assemble and prove viable alternatives to address these challenges over the long term.

Moreover, viable clean coal technologies also promote the continued use of coal, thereby offering some degree of energy security to those countries that are net oil importers but having plentiful supplies of coal. Clean coal combustion technologies can reduce emissions of sulfur oxides (SO), nitrogen oxides (NO), and other pollutants at various points of coal use from a mine to a power plant or factory.

Clean combustion refers to optimizing the process of burning coal, or other fuels, to release more useful heat and generate fewer harmful pollutants from the outset, prior to the effects of any pollution control exhaust posttreatment. While the combustion process itself has little effect on the release of certain pollutants intrinsic to the physical material of coal on a per-unit-combusted basis, such as mercury, arsenic, lead, or antimony present in coal ash, it can have a significant impact on the formation of pollutants that form due to combustion itself, such as smog-forming nitrogen oxides (NOx) as well as carbon monoxide (CO) and other incomplete combustion by-products such as volatile organic compounds (VOCs) and black carbon (soot).

One of the challenges of designing an optimized combustion system is that soot, VOC, and carbon monoxide emissions tend to form due to insufficient oxygen supply or insufficient mixing of fuel and air in the combustion chamber and are primarily eliminated through a more complete combustion, whereas NOx tends to form due to an overabundance of oxygen and forms preferentially at higher temperatures typically associated with more complete combustion.

Combustion system improvements must therefore balance nitrogen oxide control with formation of pollutants resulting from incomplete burning like carbon monoxide or, in the case of waste combustors, dioxins and furans. The preferred method is to promote more complete combustion to avoid the formation of a wide range of organic pollutants, and then to reduce nitrogen oxide emissions through a combination of effective control over combustion temperature and exhaust posttreatment with ammonia or other chemicals to dissociate nitrogen oxide particles into benign atmospheric nitrogen and oxygen.

Thus, there are (1) precombustion cleaning, (2) cleaning during combustion, (3) postcombustion cleaning, and (4) cleaning by conversion.

22.4.1 PRECOMBUSTION CLEANING

Application of coal-cleaning/upgrading methods before combustion process can improve the economic value of coal and makes it more environmentally friendly. There are a number of different applications to be used for this purpose such as physical, chemical, biological cleaning methods, drying, briquetting, blending, etc. (IEA, 2003; Breeze, 2005; Miller, 2005). Physical cleaning methods are most effective for removing the ash content, pyritic sulfur, and trace elements associated with major inorganic elements such as mercury (Xu et al 2003; Lee et al., 2006; Sondreal et al., 2004, 2006; Yang et al., 2007; Wang et al., 2008).

Precombustion cleaning involves the removal of any, or at least of a part, of pollution-generating impurities from coal by physical, chemical, or biological means. A substantial amount of the coal used in utility boilers does receive some form of cleaning before it is burned. The major objective of many of the precombustion cleaning processes is the reduction of the sulfur content (usually pyritic, FeS_2, sulfur). The wider use of conventional coal-cleaning processes will allow the sulfur dioxide emissions to be reduced markedly.

Coal-cleaning processes could achieve trace element rejections of 50%–80% (Luttrell, et al., 2000; Xu et al., 2003). Drying removes the excess moisture and reduces the weight and volume of the coal, rendering it more economical to transport and increases the heating value. Solar drying is the simplest option; it involves leaving the coal in an open storage area before transporting it. Drying coal can also be realized by applying heat to remove the moisture. This is most often carried out at the power station by utilizing surplus energy in the plant flue gases. Blending of coals of different types at power plants is also effective for saving the costs, meeting the quality requirements, and improving the combustion behavior of the fuel (Breeze, 2005). Briquetting of coal using the appropriate binders could provide reduction in sulfur dioxide emissions and fixation of some toxic elements.

Low-quality coals can be upgraded by using physical (coal washing plants), chemical (leaching), and biotechnological (microorganisms) processes before the combustion process. However, the most promising solution for this purpose is the use of coal washing plants. In general, gravity-based separation techniques are used for the removal of mineral matter from coal. Dense media separators such as heavy media separators, heavy media cyclones, and jigs are used worldwide for coal upgrading.

Enhanced gravity separators, selective agglomeration, and froth flotation are the effective alternative coal-cleaning methods for the mineral matter and pyrite removal. The ash content of coal could be reduced by over 50% by using these devices (WCI, 2005). Sometimes the properties of the coal may not suitable for the wet washing or wet washing plants may not be considered as feasible by the power plant operators due to its additional capital and operational costs; under these circumstances dry-cleaning methods such as air-heavy medium devices, air tables, and air jigs can be used for the cleaning of coals (Cicek, 2008).

The removal of mineral matter content including the pyritic sulfur improves the efficiency of power plant; it could provide a reduction of up to 40% in SO_2 emissions and about 5% in carbon dioxide emissions (Breeze, 2005; IEA, 2008).

Traditionally, a research to improve precombustion cleaning has concentrated on two major categories of cleaning technology: physical cleaning and chemical cleaning (Wheelock, 1977). A new category of coal cleaning, biological cleaning, has recently attracted much interest as advances have been made in microbial and enzymatic techniques for liberating sulfur and ash from coal (Couch, 1987, 1991; Dugan et al., 1989; Beier, 1990; Faison, 1991).

22.4.1.1 Physical Cleaning

Generally, precombustion coal cleaning is achieved by the use of physical techniques, some of which have been used for more than a century. Physical cleaning methods typically separate undesirable matter from coal by relying on differences in densities or variations in surface properties.

Physical cleaning can remove only matter that is physically distinct from the coal, such as small dirt particles, rocks, and pyrite. Physical cleaning methods cannot remove sulfur that is chemically combined with the coal (organic sulfur), nor can they remove nitrogen from the coal. Currently, physical cleaning can remove 30%–50% of the pyritic sulfur and about 60% of the ash-forming minerals in coal.

Advanced physical cleaning techniques are expected to be more effective than older techniques (Feibus et al., 1986). And increased efficiency can be achieved by grinding the coal to a much smaller size consistency whereupon the coal will release more of the pyrite and other mineral matter. Thermal treatment can be used to reduce moisture and modify surface characteristics to prevent reabsorption. New coal-cleaning processes can remove more than 90% of the pyritic sulfur and undesirable minerals from the coal.

22.4.1.2 Chemical/Biological Cleaning

Removing organic sulfur that is chemically bound to the coal is a more difficult task than removing pyritic sulfur through physical means (Wheelock, 1977). Currently, chemical and biological processes are being used to remove organic sulfur.

One chemical technique that has shown promise is molten caustic leaching. This technique exposes coal to a hot sodium- or potassium-based chemical which leaches sulfur and mineral matter from the coal. Biological cleaning represents some of the most exotic techniques in coal cleaning insofar as bacteria are *cultured* (*developed*) to consume the organic sulfur in coal. Other approaches involve using fungi and the injection of sulfur-digesting enzymes directly into the coal.

Chemical or biological coal cleaning appears to be capable of removing as much as 90% of the total sulfur (pyritic and organic) in coal. Some chemical techniques also can remove 99% of the ash.

22.4.1.3 Fuel Switching

Fuel switching is the substitution of one type of fuel for another, especially the use of a more environmentally friendly fuel as a source of energy in place of a less environmentally friendly fuel.

Thus, fuel switching is an emission control measure that involves the exchange of a less pure fuel to a cleaner fuel. The environmental impact of the fuel-switching methods is a potential reduction of greenhouse gases or other element that gives negative impact to the environment. The common potential reductions include lower emissions of carbon dioxide, sulfur dioxide, and nitrogen oxides.

The capability to use substitute energy sources means that the combustors (e.g., boilers, furnaces, ovens, and blast furnaces) of a facility had the machinery or equipment either in place or available for installation so that substitutions could actually have been introduced within a specific time period without extensive modifications.

Fuel-switching capability does not depend on the relative prices of energy sources; it depends only on the characteristics of the equipment, environmental issues, and legal constraints.

22.4.2 Cleaning during Combustion

Cleaning during combustion can involve modification of the manner in which coal is burned or, alternately, the use of pollutant-absorbing substances which can be injected into the combustion chamber.

Clean combustion consists of removing the pollutants from coal as it is burned. This can be accomplished by controlling the combustion parameters (fuel, air/oxygen, and temperature) to minimize the formation of pollutants and/or by injecting pollutant-absorbing substances into the combustion chamber to capture the pollutants as they are formed (Martin, 1986).

22.4.2.1 Advanced Combustion

Advanced combustion systems control or remove sulfur dioxide (SO_2), nitrogen oxides (NO), and/or particulate matter from coal combustion gases before they enter a steam generator or heater. Pollutants are controlled by the combustion parameters and/or sorbents used during the combustion process. Nitrogen oxides can be controlled through staged combustion, coal reburning, or by a method of controlling combustion flame temperature. On the other hand, sulfur dioxide is controlled by means of a sorbent injected in the combustion chamber. Ash can be controlled by operating at high temperatures and converting it into molten slag, but such high temperatures may not be conducive to removal of sulfur dioxide and nitrogen oxides.

Some advanced combustion systems are designed to reduce only nitrogen oxide emissions while others are designed to reduce or capture several pollutants (sulfur dioxide, nitrogen oxides, and ash). Depending on the specific technology, these systems are capable of reducing nitrogen oxide emissions by 50%–70%, sulfur dioxide emissions by 50%–95%, and ash by 50%–90%, relative to present conventional technology.

Efficiency improvements of conventional solid coal combustion plants can take a number of forms, and each reduces the environmental footprint of a power plant's output work by using rather than wasting more of the energy contained in coal's chemical bonds and released when it is burned. Technical improvements that fall into this category include improved furnace and boiler design to keep heat inside the power generation cycle rather than releasing it through ash quenching and condensing of steam, reductions in parasitic load demand from pumps, induced draft fans, and other plant components, and improved efficiency of the steam turbines used to transform heat energy into electricity.

The steam temperature can be raised to levels as high as 580°C–600°C (930°F–1110°F) and pressure over 4500 psi. Under these conditions, water enters a *supercritical* phase with properties in between those of liquid and gas. This supercritical water can dissolve a variety of organic compounds and gases, and when hydrogen peroxide and liquid oxygen are added, combustion is triggered. Turbines based on this principle (*supercritical turbines*) offer outputs of over 500 MW.

The supercritical turbines can burn low-grade fossil fuels and can completely stop nitrogen oxide emissions and keep emissions of sulfur dioxide to a minimum. For example, lignite or brown coal has a high water content and is not normally used for power generation but when lignite is added to water that has been heated to 600°C (1110°F) at a pressure of 4500 psi, it will completely burn up in 1 min while emitting no nitrogen oxides and only 1% of its original sulfur content as sulfur dioxide. This also eliminates the need for desulfurization and denitrogenation equipment and soot collectors. Although large amounts of energy are required to create supercritical water, operating costs could be significantly different from existing power generating facilities because there would be no need to control gas emissions. The demand for cooling water is also reduced, almost proportionally to an increase in the efficiency (Figures 22.2 through 22.6).

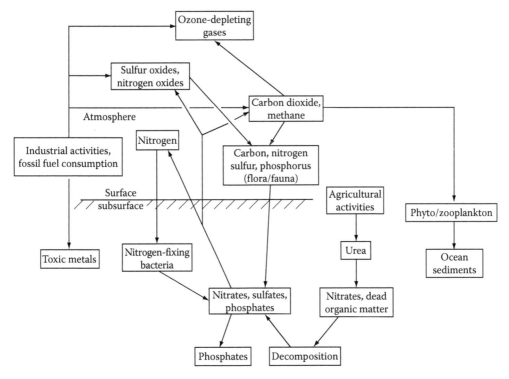

FIGURE 22.2 Various gaseous cycles. (From Clark, W.C., *Sci. Am.*, 261(3), 46, 1989.)

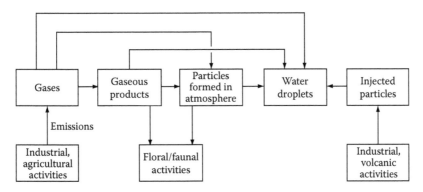

FIGURE 22.3 Fate of gaseous emissions in the atmosphere. (From Graedel, T.E. and Crutzen, P.J., *Sci. Am.*, 261(3), 58, 1989.)

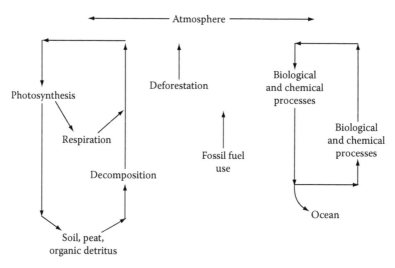

FIGURE 22.4 Carbon cycles. (From Schneider, S.H., *Sci. Am.*, 261(3), 70, 1989.)

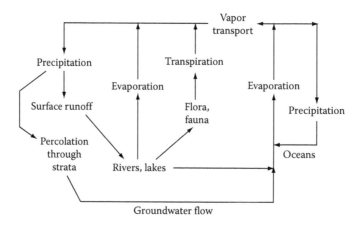

FIGURE 22.5 Water cycle. (From Maurits la Rivière, J.W., *Sci. Am.*, 261(3), 80, 1989.)

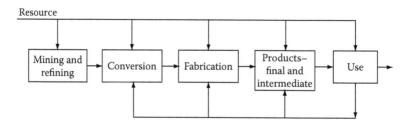

FIGURE 22.6 Industrial ecosystem.

Currently, supercritical power plants reach thermal efficiencies of just over 40%, although a few of the more plants have attained high efficiency up to 45%. A number of steam generator and turbine manufacturers around the world now claim that steam temperatures up to 700°C (1290°F) (*ultra-supercritical conditions*) are possible which might raise plant efficiencies to over 50%, but require the use of expensive nickel-based alloys for reaction equipment and power generators because of the corrosive properties.

The main competition to supercritical system is from new gas turbine combined-cycle plants which are now expedited to achieve an overall efficiency of 60%, making a huge difference in generating and life-cycle costs. However, the new gas turbines will release exhaust into waste heat recovery steam generator at temperatures above 600°C (1110°F), thus necessitating the use of the high-chromium steel and nickel alloys as used in the supercritical coal-fired plants.

The advancements that have been achieved in conventional coal plant performance demonstrate significant promise and room for further improvements; however, they also demonstrate the limitations of existing technology, as coal combustion and turbine design have been continually improving for many years, yet power generation from this source still generates considerable pollution. Potential areas of further improvement are being exhausted. *Conventional clean coal* offers promise for the future, perhaps more so than any other form of coal power, but it is still far from unproblematic. Analogous and in some cases greater improvements are likely to occur in alternative energy sources as well, as has certainly been the case with wind power and other renewable energy sources over the past decade (Omer, 2008), and coal may not remain the winner in pure economic terms that it is today with many of the changes listed previously as fuel prices and capital costs of new plants and retrofits continue to increase.

22.4.2.2 Fluidized-Bed Combustion

Fluidized-bed combustion has the ability to reduce emissions by controlling combustion parameters and by injecting a sorbent, or a pollutant absorbent (such as crushed limestone), into the combustion chamber along with the coal (Martin, 1986; Yeager and Preston 1986).

Pulverized coal mixed with crushed limestone is suspended on jets of air (or fluidized) in the combustion chamber. As the coal burns, sulfur is released, and the limestone captures the sulfur before it can escape from the boiler. The sulfur combines with the limestone to form a mixture of calcium sulfite ($CaSO_3$) and calcium sulfate ($CaSO_4$).

The temperature in the fluidized-bed combustor is on the order of 800°C–900°C (1470°F–1650°F) compared with 1300°C–1500°C (2370°F–2730°F) in *pulverized coal combustion* systems (PCC). Low temperature helps minimize the production of nitrogen oxides and, with the addition of a sorbent (typically limestone) into the fluidized bed, much of the sulfur dioxide formed can be captured. The other advantages of fluidized-bed combustors are compactness, ability to burn low calorific values (as low as 1800 kcal/kg), and production of ash which is less erosive.

Fluidized-bed combustors are essentially of two types: (1) bubbling bed, and (2) circulating bed. While bubbling beds have low fluidization velocities to prevent solids from being elutriated, circulating beds employ high velocities to actually promote elutriation. Both of these technologies operate at atmospheric pressure. The circulating bed can remove 90%–95% of the sulfur content from coal while the bubbling bed can achieve 70%–90% sulfur removal.

The fluidized mixing of fuel and sorbent enhances both the coal-burning and sulfur-capturing processes and allows for reduced combustion temperatures of 760°C–870°C (1400°F–1600°F), or almost half the temperature of a conventional boiler. This temperature range is below the threshold where most of the nitrogen oxides form and, thus, fluidized-bed combustors have the potential to reduce the emission of both sulfur dioxide and nitrogen oxides.

Fluidized-bed combustors can be either atmospheric or pressurized (Yeager and Preston, 1986). The atmospheric type operates at normal atmospheric pressure while the pressurized type operates at pressures 6–16 times higher than normal atmospheric pressure. The pressurized fluid-bed boiler offers a higher efficiency and less waste products than the atmospheric fluid-bed boiler. There is also a circulating (entrained) bed combustor which allows for finer coal feed, better fuel mixing, higher efficiency, as well as an increased sulfur dioxide capture.

Unlike conventional pulverized-coal combustor, the circulating fluidized-bed (CFB) combustor is capable of burning fuel with volatile content as low as 8%–9% w/w (e.g., anthracite coke, petroleum, etc., with minimal carbon loss). Fuels with low ash-melting temperature such as wood and biomass have been proved to be feedstocks in CFB combustor due to the low operating temperature of 850°C–900°C

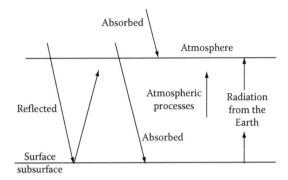

FIGURE 22.7 Representation of heat trapping by the atmosphere of the earth (Greenhouse effect). (From Schneider, S.H., *Sci. Am.*, 261(3), 70, 1989.)

(1560°F–1650°F). The CFB combustor boiler is not bound by the tight restrictions on ash content either and can effectively burn fuels with mineral matter content up to 70% w/w (Figure 22.7).

The CFB combustor can successfully burn agricultural wastes, urban waste, wood, and other form of biomass which are the low melting temperature as fuels. The low furnace temperature precludes the production of *thermal NOx* which appears above a temperature of 1200°C–1300°C (2190°F–2370°F). Besides, in a CFB combustor boiler, the lower bed is operated at near sub-stoichiometric conditions to minimize the oxidation of *fuel-bound nitrogen*. The remainder of the combustion air is added higher up in the furnace to complete the combustion. With the staged combustion, approximately 90% w/w of fuel-bound nitrogen is converted to elemental nitrogen as the main product.

The *pressurized bed* was developed in the late 1980s to further improve the efficiency levels in coal-fired plants. In this concept, the conventional combustion chamber of the gas turbine is replaced by a PFBC. The products of combustion pass through a hot gas cleaning system before entering the turbine. The heat of the exhaust gas from the gas turbine is utilized in the downstream steam turbine. This technology is called PFBC combined cycle (Figure 22.8).

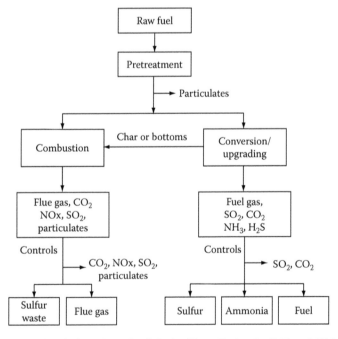

FIGURE 22.8 Sources of emissions from fossil fuels. (From Probstein, R.F. and Hicks, R.E., *Synthetic Fuels*, pH Press, Cambridge, MA, 1990.)

The bed is operated at a pressure between 75 and 300 psi and operating the plant at such low pressures allows some additional energy to be captured by venting the exhaust gases through a gas turbine which is then combined with the normal steam turbine to achieve plant efficiency levels of up to 50%. The steam turbine is the major source of power in PFBC, contributing about 80% of the total power output; the remaining 20% is produced in gas turbines.

The *pulsed atmospheric fluidized-bed combustor* (PAFBC) is a bubbling fluidized-bed coal combustor combined with a pulse combustor (Pence and Beasley, 1995; Muller, 1996). Fluidized-bed combustors allow the use of high-sulfur coal, performing extremely well environmentally with particularly low nitrogen oxide and sulfur dioxide emissions; however, they have the disadvantage of requiring coarse coal with no fines. Adding the pulse combustor to the fluidized bed allows the use of very fine coal. The pulsing stabilizes the hydrodynamic and thermodynamic characteristics while minimizing particulate carry over.

The pulse combustor is a combustion chamber with no moving parts operating on either gas or fine coal. The chamber and exit tube are designed in a manner which results in a self-sustaining, periodic combustion process. The frequency of the resonance varies with chamber size and exit tube length. The exit tube of the pulse combustor is immersed in the fluidized bed. The raw coal is pneumatically separated with the coal fines carried to the pulse combustor and the coarse coal to the fluidized-bed combustor via a screw feeder.

22.4.3 POSTCOMBUSTION CLEANING

Postcombustion cleaning involves the use of processes that remove pollutants from the flue gases exiting the boiler (Kuhr et al., 1988; Frazier et al., 1991). Finally, cleaning by coal conversion (which is a departure from traditional coal-burning methods) involves the conversion of coal into a gas or liquid that can be cleaned and then used as fuel.

The use of coal in electricity production plants inevitably generates some pollutants such as sulfur oxides (SOx), nitrogen oxides (NOx), coal combustion products (CCPs), and trace elements. Several technologies are utilized alone or together throughout the coal combustion to eliminate these pollutant emissions (Kuhr et al., 1988; Frazier et al., 1991; Xu et al., 2003).

22.4.3.1 Sulfur Oxide Emissions

The sulfur presents in both inorganic and organic forms in coal. The inorganic sulfur occurs as sulfide minerals (pyritic sulfur, FeS_2) and/or a range of sulfate compounds (sulfate sulfur). Pyritic and organically bound sulfur constitute the majority of sulfur content, and sulfates are at very low concentrations in coal. Approximately 95% of sulfur content is converted to sulfur dioxide (SO_2) during the combustion process and a small amount of sulfur trioxide (SO_3) is also formed. Sulfur dioxide is a major contributor to acid rain formation and harmful to the plants and soil (IEA, 2003; Breeze, 2005; Miller, 2005; Suarez-Ruiz and Ward, 2008; Franco and Diaz, 2009; Graus and Worrell, 2009).

The sulfur gas produced by burning coal can be partially removed with scrubbers or filters. In conventional coal plants, the most common form of sulfur dioxide control is through the use of scrubbers. To remove the SO_2, the exhaust from a coal-fired power plant is passed through a mixture of lime or limestone and water, which absorbs the SO_2 before the exhaust gas is released through the smokestack. Scrubbers can reduce sulfur emissions by up to 90%, but smaller particulates are less likely to be absorbed by the limestone and can pass out the smokestack into the atmosphere. In addition, scrubbers require more energy to operate, thus increasing the amount of coal that must be burned to power their operation.

Other coal plants use *fluidized-bed combustion* (Chapters 14 and 15) instead of a standard furnace. Fluidized-bed technology was developed in an effort to find a combustion process that could limit emissions without the need for external emission controls such as scrubbers. A fluidized

TABLE 22.7
Stack Gas Scrubbing Chemistry

Process	Chemical Reaction
Lime slurry scrubbing	$Ca(OH)_2 + SO_2 \rightarrow CaSO_3 + H_2O$
Limestone slurry scrubbing	$CaCO_3 + SO_2 \rightarrow CaSO_3 + CO_2(g)$
Magnesium oxide scrubbing	$Mg(OH)_2(slurry) + SO_2 \rightarrow MgSO_3 + 2H_2O$
Sodium-base scrubbing	$Na_2SO_3 + H_2O + SO_2 \rightarrow 2NaHSO_3$
	$2NaHSO_3 + heat \rightarrow Na_2SO_43 + H_2O + SO_2$[a]
Double alkali	$2NaOH + SO_2 \rightarrow H_2O$
	$Ca(OH)_2 + Na_2SO_3 \rightarrow CaSO_3(s) + 2NaOH$[b]

[a] Regeneration of Na_2SO_3
[b] Regeneration of NaOH

bed consists of small particles of ash, limestone, and other nonflammable materials, which are suspended in an upward flow of hot air.

Powdered coal and limestone are blown into the bed at high temperature to create a tumbling action, which spurs more effective chemical reactions and heat transfer. During this burning process, the limestone binds with sulfur released from the coal and prevents it from being released into the atmosphere. Fluidized-bed combustion plants generate lower sulfur emissions than standard coal plants, but they are also more complex and expensive to maintain. According to the Union of Concerned Scientists, sulfur emissions decreased by 33% between 1975 and 1990 through the use of scrubbers and fluidized-bed combustors, as well as switching to low-sulfur coal.

Conventional technology (wet scrubbers) uses limestone or lime (and in some cases other alkaline agents) (Table 22.7) to remove sulfur pollutants from the flue gas before it exits the stack (Slack, 1986; Speight, 1993). Such processes can be plagued by corrosion and plugging and also produce a wet waste product (sludge) which has high disposal costs. However, the reliability of wet scrubbers has improved significantly, and they have demonstrated the ability to remove more than 90% of the sulfur dioxide (Slack, 1986).

Flue gas purification (*air pollution control*, APC) refers to technologies used to condition a power plant's emissions after the combustion of fuel but before the release of gaseous and suspended particulate combustion by-products into the atmosphere. Each device or system corresponds to a given pollutant or category of pollutants to be removed from the flue gas stream. Reducing chemicals such as ammonia or urea, along with catalysts in the case of selective catalytic reduction (SCR) systems, are used to treat the exhaust to remove nitrogen oxides.

Slaked or slurried lime is used to neutralize acid gases such as sulfur dioxide. Packed beds or spray injection of activated carbon, with its high surface area to volume ratio, are used to adsorb heavy metals and other particulate fly ash. Electrostatic precipitators and fabric filters remove adsorbed and residual particulates entrained in the flue gases as well as reagents from other air pollution control processes. Most air pollution control devices are applied on the *cold side* of the heat exchangers once the heat used to do work has been transferred to the boiler fluid. The main exception is in control of nitrogen oxides, wherein the reducing agents ammonia and its precursor urea are typically added on the hot side to eliminate nitrogen oxides in order to meet the temperature range requirements for the reduction reaction.

22.4.3.2 Nitrogen Oxide Emissions

The principal atmospheric oxides of nitrogen include nitric oxide (NO), nitrogen dioxide (NO_2), and nitrous oxide (N_2O). Collectively, nitric oxide and nitrogen dioxide are commonly referred to as NOx, which are generated by the reaction of the nitrogenous compounds of the coal and nitrogen in the air with the oxygen used during the combustion process (Miller, 2005; Suarez-Ruiz and Ward, 2008; Franco and Diaz, 2009). The majority of nitrogen oxide emissions are in the form of nitric oxide (NO). A small fraction of nitrogen dioxide (NO_2) and nitrous oxide (N_2O) can also be formed.

Three primary sources of nitrogen oxide formation are documented: (1) thermal NOx, which is generated by the high-temperature (above 1600°C) reaction of oxygen and nitrogen from the combustion air; (2) prompt NOx, which is the fixation of atmospheric (molecular) nitrogen by hydrocarbon fragments in the reducing atmosphere in the flame zone; and (3) fuel NOx, which originates from the nitrogen in the coal. Nitrogen oxides are being considered responsible for the formation of acid rain (atmospheric NOx eventually forms nitric acid and contributes to acid rain), as well as contributing to the formation of urban smog which is known as ozone pollution (Miller, 2005; Bris et al., 2007; Suarez-Ruiz and Ward, 2008; Franco and Diaz, 2009).

NOx pollutants do not form in significant amounts at temperatures below 2800°F. Initially, the focus of NOx controls was on finding ways to burn fuel in stages. *Low-NOx* burners use a staged combustion process, which uses a lower flame temperature during some phases of combustion to reduce the amount of NOx that forms. These burners also limit the amount of air in the initial stages of combustion, when the nitrogen naturally occurring in the coal is released, so that there is less oxygen present to bond with the nitrogen. These burners can reduce nitrogen oxide emissions by 40% or more. In the case of fluidized-bed technology, combustion occurs at temperatures of 760°C–930°C (1400°F–1700°F), lower than the threshold at which nitrogen oxide pollutants form.

In 2005, the EPA passed the Clean Air Interstate Rule, which requires a 61% cut in nitrogen oxide emissions from power plants by 2015. This level of emissions reduction requires a different technology. Selective catalytic reduction (SCR) and selective non-catalytic reduction (SNCR) both convert NOx into water (H_2O) and nitrogen (N_2). SCR is capable of reducing NOx emissions by approximately 90%. SNCR is a simpler and less expensive technology than SCR, but it also provides a lower level of NOx reduction.

In general, NOx control technologies can be divided into two main groups: The first group of NOx control technologies considers the reduction of NOx produced in the primary combustion zone and the other involves the reduction of NOx existing in the flue gas (Miller, 2005). There are also multi-pollutant control technologies to eliminate NOx emissions together with the other pollutants such as sulfur dioxide (SO_2), mercury (Hg), particulate matter (PM), and/or air toxics. Several NOx control technologies such as low-NOx burners (LNB), furnace air staging (OFA), fuel staging (reburning), selective catalytic reduction (SCR), selective non-catalytic reduction (SNCR), flue gas recirculation, co-firing, and flue gas treatment are applied at the power plants. Sometimes, these NOx control technologies can be used together to increase the control efficiency (LNB + OFA, LNB with SCR, LNB with OFA or SCR). Recently, hybrid NOx control systems incorporating both a redesigned SNCR system and a compact induct SCR systems have been introduced for more effective emission control (Figure 22.9).

22.4.3.3 Fly Ash Emissions (Trace Element Emissions)

Combustion of coal produces various forms of CCPs due to the mineral impurities in coal body. Fly ash is one of the most important CCPs requiring efficient control during coal combustion (Akar et al., 2009). Especially, the fine particulates (PM_{10}) in the fly ash may pass through the dust collection devices and be released to the atmosphere.

The fly ash particles can be a potential source of contamination because of the high concentration and surface associations of some trace elements in their composition. Coal contains various trace elements originating from different minerals and macerals in its body (Pavlish et al., 2003; Xu et al., 2003; Balat, 2008b). These elements show different behavior during coal combustion (Figure 22.10) (Suarez-Ruiz and Ward, 2008).

During the combustion process, the low volatile trace elements tend to stay in the bottom ash or to be distributed between the bottom and fly ash. The more volatile trace elements (volatile and especially highly volatile trace elements) are vaporized in the furnace and they may be incorporated

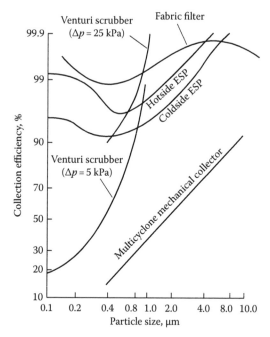

FIGURE 22.9 Efficiency of various gas cleaning systems. (From Probstein, R.F. and Hicks, R.E., *Synthetic Fuels*, pH Press, Cambridge, MA, 1990.)

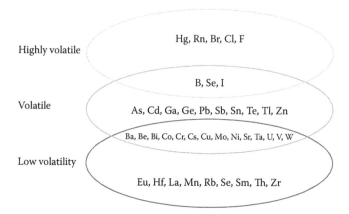

FIGURE 22.10 Behavior of trace elements during combustion. (From Clarke, L.B., Management of by-products from IGCC power generation, IEA Coal Research, IEACR/38, International Energy Agency, London, U.K.)

with any fouling/slagging deposits or mostly condense onto the existing fine fly ash or totally emitted in vapor phase (Xu, et al., 2003; Suarez-Ruiz and Ward, 2008; Vejahati et al., 2010).

There are several methods of removing fine particulate matter before it can be released from the smokestack. Wet scrubbers remove dust pollutants by capturing them in liquid droplets and then collecting the liquid for disposal. Electrostatic precipitators add electrical charges to particles in the flue gas and collect the particles on plates to remove them from the air stream. Wet electrostatic precipitators combine the functions of a standard dry electrostatic precipitator with a wet scrubber and spray moisture to the air flow to help collect extremely fine particulate matter ($PM_{2.5}$), making the process more effective. Fabric filter baghouses are another means of controlling particulate matter emissions. As dust enters the baghouse compartment, larger particles fall out of the system, while smaller dust particles are collected onto cloth filters.

TABLE 22.8
Particulate Control Technologies for Coal Combustion

1. *Electrostatic precipitation*

   Uses electrical forces to capture the particles from the flue gas and collect them onto a grounded plate within an electrical field. The process has a somewhat lower performance for the particle sizes between 0.1 and 1 μm (90%–95%). Overall efficiency of these devices is over 99%. They work with low pressure drops which minimizes the fan costs.

2. *Cyclone/multi-cyclones*

   Inertial collectors use centrifugal force to separate the particles from the flue gas stream. The inlet flue gas is forced to follow a circular or conical path at high velocity in a cyclone. The particles are forced to move against the walls by the centrifugal force and settle down into hoppers. These devices separate the particles in the size range of 1.0–100 μm with 50%–90% efficiency.

3. *Fabric filter (baghouses)*

   Flue gas is forced to pass through a filter and the dust particles are collected on the surface of the permeable fabric. These systems have very high collection efficiencies (about 99.9%) for both coarse and fine particles (0.01–100 μm). The fabric needs replacement at every 2–4 years. Hot gases must be cooled; the system generally operates in the temperature range of 120°C–180°C (250°F–355°F).

4. *Wet scrubbers*

   This technology involves contacting a flue gas stream with a scrubbing liquid by applying different methods. There are different scrubber designs such as spray tower, dynamic scrubber, collision scrubber, and Venturi scrubber which render the removal of gaseous and particle emissions at the same time and neutralize corrosive gases with a removal efficiency from 90%–99.9% for particles in the size range of 0.5–100 μm. The efficiency is somewhat lower for fine particles (less than 1 μm size). There can be corrosion problems and freezing in cold climates.

Source: Miller, B.G., *Coal Energy Systems*, Elsevier, Amsterdam, the Netherlands, 2005; Mokhatab, S. et al., *Handbook of Natural Gas Transmission and Processing*, Elsevier, Amsterdam, the Netherlands, 2006.

Particulate emissions generated in coal combustion are categorized in three main groups, namely, PM_{10}, $PM_{2.5}$, and PM_1.

The ultrafine fractions of these particulates (PM_1 and $PM_{2.5}$) may remain air-suspended for a long time and have deleterious impacts on the environment and human health (Breeze, 2005; Miller, 2005; Bhanarkar et al., 2008; Suárez-Ruiz and Crelling, 2008). The finest fraction of the particulates (<1 μm in size) are mostly originated from the ash-forming species vaporized during combustion. The remaining particles are referred as the residual ash; they are larger than 1 μm and are generally formed by the mineral impurities in coal (Senior et al., 2000; Ohlstrom et al., 2006).

Emissions of primary ash particles can be controlled efficiently (up to 99.99%) by the combination of an efficient electrostatic precipitator or baghouse and wet FGD system (Tables 22.3 and 22.8). Most of trace metals, except for volatile elements such as mercury and selenium, are captured with the primary particles (Sondreal et al., 2004, 2006; Yang et al., 2007; Wang et al., 2008).

The fine particulates (<2.5 μm, <$PM_{2.5}$) have minor impact on the control efficiency due to their small size and small proportion in the total particulate mass; however, they are the most dangerous particulates in terms of human health, considering that it is crucial to capture these particles with high collection efficiencies using improved technologies (Table 22.8) (Ohlstrom et al., 2006).

22.4.3.4   Mercury

Mercury is identified as the toxic of greatest concern in all the air toxics emitted from power plants due to its persistence and bioaccumulation in the foods and environment (Senior et al., 2000). It generally occurs in coal in association with sulfide minerals (FeS_2 and HgS), it may also be organically bound to the coal macerals (Pavlish et al., 2003; Sondreal et al., 2004; Kolker et al., 2006). Mercury concentration in coals is about 0.1–0.15 g/ton and it passes to the flue gas during the combustion process as a mixture of different chemical states or species at varying percentages; namely, elemental mercury (HgO), oxidized mercury (Hg^{2+}), and particulate-bound mercury (Hg_p) (Lee et al., 2006). The total concentration of mercury in the flue gas can be between 1 and 20 $\mu g/m^3$ (Yang et al., 2007; Wang et al., 2008).

Elemental mercury (Hg) is highly volatile and insoluble; therefore, it is hard to capture by emission controlling equipment. It is almost completely released to the atmosphere and can be transported over long distances (Pavlish et al., 2003; Wang et al., 2008). The long life time in the atmosphere and long distance atmospheric transport of elemental mercury make it a global environmental threat (Sondreal et al., 2004). The oxidized form of mercury (generally considered to be mercuric chloride) is soluble and tends to form surface associations with particulate matter. The oxidized mercury and particulate-bound mercury have a short life time (a few days) in the atmosphere. If they released from the power plant, they tend to be deposited near the source of emission (Pavlish et al., 2003). Efficient control of oxidized mercury and particulate-bound mercury is possible by using conventional emission control equipment (Senior et al., 2000; Pavlish et al., 2003; Sondreal et al., 2004; Wang et al., 2008).

In combustion systems, mercury is oxidized by kinetically controlled reactions. Chlorine species promotes homogeneous oxidation of mercury (Pavlish et al., 2003). Calcium likely reacts with chlorine during combustion and its high concentrations may have a reducing effect on the positive influence of chlorine in mercury oxidation (Yang et al., 2007). The presence of fly ash and sorbents promotes heterogeneous oxidation of mercury.

HCl, NO, and NO_2 affect mercury oxidation and capture process positively, whereas the interaction of SO_2 with NO_2 in the flue gas greatly reduces the capture of elemental mercury by the fly ash and sorbents (Senior et al., 2000, Pavlish et al., 2003; Sondreal et al., 2004). Currently, there is no mature single-best technology for mercury reduction. Combination of the existing pollution control devices can provide some degree of mercury removal from the flue gas. The rate of mercury reduction strongly depends on the type of coal, mercury speciation in the flue gas, and the configuration of the existing pollution control devices (Yang et al., 2007).

Electrostatic precipitator, fabric filter bag house, and wet FGD systems can remove some particulate-bound and oxidized forms of mercury from the flue gas. The efficiency of the mercury removal by these devices can range from 0% to 90%. However, elemental mercury cannot be controlled effectively by conventional air pollution control equipment (Pavlish et al., 2003; Yang et al., 2007; Wang et al., 2008; Pavlish et al., 2003). The scrubber systems have moderate–high costs and need additional equipment and installation area (Pavlish et al., 2003). The highest reduction rate can be achieved by the use of fabric filters. Cold-side ESPs are much more effective than hot-side ESPs. The cost of these systems is moderate–high, additional equipment and laydown space may be needed (Pavlish et al., 2003). In general, higher removal efficiencies can be achieved for the combustion bituminous coals by using these air pollution control equipment. They are less effective for subbituminous coal combustion and almost useless in lignite coal combustion (Pavlish et al., 2003; Kolker et al., 2006).

Sorbent injection technology seems to have the highest potential to remove both elemental and oxidized mercury from the flue gas (Yang et al., 2007). Different sorbents such as activated carbon, chemically treated sorbents and coal additives, calcium-based sorbents, petroleum coke, zeolites, fly ash, other chemically treated carbons or carbon substitutes, etc., are injected into the upstream of either an ESP or a fabric filter baghouse to control mercury emissions (Pavlish et al., 2003; Yang et al., 2007). The cost of this process is low-to-moderate and separate injection systems may be required (Pavlish et al., 2003).

22.4.3.5 Advanced Postcombustion Cleaning

Advanced postcombustion cleaning technologies encompass two approaches: (1) using the existing flue gas ductwork to inject a sorbent and (2) inserting one or more separate vessels into the downstream ductwork where pollutant absorbents are added. These advanced technologies offer several advantages over conventional technologies: (1) regeneration of the sulfur-absorbing chemical; (2) increased residence time with the sulfur absorbent; (3) reduced physical size requirements; (4) a dry, environmentally benign, waste product that may have commercial value.

In-duct sorbent cleaning occurs out, as the name indicates, inside the ductwork leading from the boiler to the smokestack. Sulfur dioxide absorbers (e.g., hydrated lime) are sprayed into the center of the duct. By controlling the humidity of the flue gas and the spray pattern of the sorbent, 50%–70% of the sulfur dioxide can be removed and the reaction produces dry particles that can be collected downstream. In-duct sorbent injection is an attractive option for retrofitting smaller, older plants where space requirements might be limited.

When separate vessels are used, one or more process chambers are inserted in the flue gas ductwork, and various sorbents are injected to remove the pollutants. The separate vessels provide a longer residence time for the absorbent to react with the gas, and pollutant capture is greater. This approach, at some increase in cost over the in-duct injection procedure, has the potential of capturing more than 90% of the pollutants. Technologies such as the spray dryer and SCR represent approaches that use separate vessels.

22.4.4   Conversion and Added-Value Products

Techniques that convert coal into another form of fuel bypass the conventional *coal fuel path* of combustion. The most common system is that in which coal is converted into a gaseous fuel. In other techniques, liquid products are the result, while in others, a combination of gases, liquids, and solids is produced.

22.4.4.1   Integrated Gasification Combined-Cycle Systems

Integrated gasification combined cycle (IGCC) is another approach to reducing the environmental footprint of coal power. The IGCC systems are distinct from the CCS systems, although there is the potential to combine the two in order to achieve greater environmental benefits than either method alone, albeit at great expense.

Gasification, or incomplete combustion in an oxygen-poor environment, produces an intermediate gaseous fuel known as synthesis gas, or syngas for short, composed mainly of the combustible gases hydrogen (H_2) and carbon monoxide (CO). Coal gasification was used to produce the gas burned to light the streets of Paris and a number of other cities beginning in the late 1800s, as well as in the first step of the Fischer–Tropsch process used to produce substitute liquid fuels in Nazi Germany when the war effort strained that country's energy supplies to the breaking point (Chapter 20). The usual combustion by-products of water and carbon dioxide ultimately form when syngas is burned as well.

Combined-cycle power generation, as the name suggests, uses a multistage process to generate electricity. The first stage involves the recovery of energy released by a gas as it burns and expands inside a combustion turbine using the Brayton cycle; the second stage involves the transfer of heat from flue gases to a working fluid, typically water in a boiler, used to turn a steam turbine as in a conventional power plant using the Rankine cycle. Because the first cycle takes place at extremely high temperatures needed to rapidly expand gas to do work in the combustion turbine, the flue gas still contains enough heat at the end of the cycle to make additional energy recovery feasible using heat exchangers. The net power output from the two power generating cycles is combined and fed into the grid. The most obvious disadvantage of generating power this way is the far higher capital cost of constructing an IGCC plant compared to conventional generation facilities.

The IGCC process basically has four steps: (1) combustion gases are formed by reacting coal with high-temperature steam and oxygen (or air); (2) the gases are purified; (3) the clean gases are burned and the hot exhaust gas is passed through a gas turbine to generate electricity; and (4) the residual heat in the exhaust is used to boil water for a conventional steam turbine generator to produce additional electricity.

This combination of gas and steam turbines accounts for the name combined cycle. Gasification combined-cycle systems are among the cleanest and most efficient of the emerging clean coal technologies. Sulfur species, nitrogen species, and particulate matter are removed before the fuel is burned in the gas turbine. Thus, there is a much lower volume of gas to be treated than in a post-combustion scrubber.

The gas stream must have extremely low levels of impurities not only to avoid pollution but to protect turbine components from chipping or corroding. As in the case of clean combustion, much of the sulfur-containing gas can be captured by a sorbent injected into the gasifier.

Many coal gasifiers release fuel gas at temperatures well in excess of 1095°C (2000°F). Loss of efficiency often occurred when the gas had to be cooled before cleaning, although perhaps the simplest of all gas cleaning processes, namely, the iron oxide process (Speight, 1993), was often eminently suitable for the task of hot gas cleaning; zinc oxide was also used on occasion in place of the iron oxide. A more efficient chemical is zinc ferrite ($ZnFeO_2$) and passage of the hot gas through a bed of zinc ferrite particles will cause removal of sulfur contaminants at temperatures in excess of 540°C (1000°F). The zinc ferrite can be regenerated and reused with little loss of effectiveness. During the regeneration stage, salable sulfur is produced and the method is capable of removing more than 99% of the sulfur in coal.

High levels of nitrogen removal are also possible. Some of the coal-nitrogen is converted to ammonia which can be almost totally removed by commercially available processes. Nitrogen oxide formation can be held to allowable levels by staging the combustion process at the turbine or by adding moisture to hold down flame temperature.

The theoretical advantage an IGCC plant has over a conventional coal plant is in the higher system efficiency of the *combined cycle*, a concept originally developed for natural gas-fired plants and used in many such plants for meeting intermediate and peak loads.

While the overall combined cycle is more efficient than conventional pulverized coal plants, energy losses do occur in the transformation of coal into a gaseous fuel, mostly due to the heat input needed for gasification. As a result of the added gasification process needed for combined-cycle systems using coal, the IGCC process is less efficient than the same combined power generation cycle run on a fuel that does not require pretreatment such as natural gas or fuel oils.

As a result, most estimates place the efficiency of full-size IGCC plants at around 45% of the total energy released from burning coal converted into usable electric power, an improvement over the 30%–40% efficiencies achievable in conventional plants but still considerably lower than the 60% electric efficiency achievable in today's most advanced combined-cycle plants.

Additionally, the use of a gasification process generates additional environmental problems distinct from those of conventional coal-fired generation. One of the challenges of designing IGCC plants is management of slag, the semi-liquid by-product that forms from trace elements in coal that do not gasify such as silicon, aluminum, and other metals.

Slag, like combustion ash, contains a high proportion of heavy metals and other contaminants, but in the more potentially hazardous form of wastewater rather than relatively inert solids, and the limited field experience with IGCC plants because of the potential for creating water quality problems. When taking into account the added challenges of managing these unique by-products from the gasification reaction, as well as the dramatically increased capital costs of IGCC plants relative to conventional coal plants, the theoretical environmental benefits that could be gained by achieving a higher plant efficiency and thereby conserving a still relatively inexpensive fuel appear less attractive.

22.4.4.2 Mild Gasification

Mild gasification is a modification of the conventional coal gasification (Chapters 20 and 21) and produces gaseous, liquid, and solid products by heating coal in an oxygen-free reactor. In fact, the process is less a gasification process and more a pyrolysis process insofar as the goal is to remove condensable volatile hydrocarbons and leaving a carbonaceous char/residue, in lieu of converting the entire charge of coal. The char can be upgraded further to remove both ash and pyritic sulfur, mixed back with coal-derived liquids, and burned in both coal- and oil-fired boilers.

A slurry of coal-derived fuel and upgraded char has the potential of being a very versatile fuel that can be burned in both coal- and oil-fired boilers. If the char is upgraded to a high degree, even feedstock coal with a high sulfur content can be used without alternating heat rates or capacity factors.

22.4.4.3 Coal Liquefaction

Two primary methods exist for converting coal into liquid fuels: (1) direct liquefaction (Chapters 18 and 19) and (2) indirect liquefaction using the Fischer–Tropsch process (Chapters 20 and 21).

Direct liquefaction is the conversion of coal directly to liquid products. In general chemical terms, coal liquefaction involves addition of hydrogen to the coal by various techniques so that the ratio of hydrogen to carbon in the product is increased to a level comparable to petroleum-based fuels. *Indirect liquefaction* is coal gasification followed by conversion of the synthesis gas (a carbon monoxide, CO, hydrogen, mixture) to liquid fuels.

While the use of coal as a feedstock to produce liquids to replace petroleum-derived fuels does not technically fall under the same *clean coal* umbrella as CCS or IGCC, it is similar enough to these concepts as an alternative use of coal and ties into related concerns of oil and gas supply problems enough that the possibility of doing so merits some discussion here. A number of processes have been proposed to produce liquid fuels from coal, most of which are claimed to become cost-competitive at sustained oil prices over $35 per barrel, and almost all of which are very similar to the gasification stage of IGCC plants (when the product of incomplete combustion is a liquid rather than a gas, the process is called *pyrolysis* rather than *gasification*).

As production of oil and gas, more versatile and energy-dense resources than coal, peaks and then declines, increased dependence on relatively more abundant but lower-quality solid fuels appears likely in the absence of greenhouse gas emission constraints. While coal, oil and gas are all viable fuels for electricity generation, many other energy-using technologies such as internal combustion engines require higher quality liquid or gaseous fuels and cannot run on coal in its native form. Since it is a solid fuel and has a lower energy content than oil or gas, coal cannot serve as a direct replacement for the myriad uses of liquid fuel today without first being converted into a liquid itself, and it is difficult to envision a scenario in which such large-scale substitution could take place without creating a major source of pollution and wasting large quantities of energy, exacerbating regional and possibly even global coal shortages in the future.

Coprocessing (Chapters 18 and 19), a recent development for coal liquefaction technology, involves the production of liquids from a mixture of coal and heavy petroleum residue, with the residual oil providing all or most of the hydrogen needed for the conversion process. Once produced, the coal-derived liquid can be refined by sulfur mineral matter (ash) removal before use.

22.4.4.4 Biomass Co-Firing

One of the most effective ways of reducing pollution associated with burning coal for electricity is to directly replace it with a renewable fuel of similar quality, usually wood.

Within the family of fossil fuels, coal contains the highest ash (inorganic) content and produces the most climate-altering greenhouse gases, nitrogen oxides, carbon monoxide, sulfur dioxide, heavy metal emissions, and waste to be disposed. Coal also produces more of these pollutants than wood, a relatively similar solid fuel whose physical and chemical properties make it a decent replacement for coal in generating base load power, at least up to a point (Omer, 2008). While wood is less energy dense than higher quality coals, it is also renewable, produces lower quantities of most air emissions, avoids waste and damage to the landscape associated with mining coal, and is carbon neutral assuming the sources of biomass are sustainably managed (Beer, 2007). Since wood is physically similar to coal and is comparable to lower-quality coals such as lignite in energy density, the two fuels can burn in the same furnaces at the same time so long as certain constraints are met.

While direct co-firing of biomass with coal can be effective as a way of reducing harmful emissions, there are limitations to this practice as well. The lower energy content of wood compared to most coal used in electricity generation today renders long-distance transport inconvenient. To maintain positive net energy and avoid exorbitant costs, wood-fired plants, including plants where it is co-fired with coal, must be located within a certain radius of sources of harvestable wood, determined by the fuel's growth rate and energy content. This limitation places a practical size limit on direct-fired biomass power plants, typically 50–150 MW of electric power. This size is much smaller than the larger multi-gigawatt size typical of coal-fired power stations.

There is also a limit to how much wood can be practically burned in a coal furnace due to the fuels' differing requirements for emissions control; for instance, wood produces fewer total particulates, but they tend to be of a larger size than coal particulate emissions, resulting in a greater overall mass of particulate emissions. Wood also has different ash-handling requirements, since it primarily generates bottom ash that remains in the furnace, while coal ash is lighter, higher in metal content, and is more likely to be entrained in flue gases exiting the furnace. While wood can make a useful substitute for some of the coal used in power generation, the physical properties of the fuel prevent it from being a fully acceptable replacement for all uses.

While wood and other biomass can substitute for some quantity of coal-fired generation, physical differences in the two fuels as well as insufficient total energy resources in sustainably managed biomass make it an insufficient replacement to match the raw power and infrastructure in place for coal-fired utility generation.

22.5 MANAGING WASTES FROM COAL USE

Burning coal, such as for power generation, gives rise to a variety of wastes which must be controlled or at least accounted for. Thus, coal plants, in addition to gaseous and liquid wastes, also produce solid wastes which either must be removed or serious attempts be made to mitigate the problem (Saroff and Robey, 1992). Some of this solid waste is removed with the bed ash through the bottom of the boiler. Small ash particles, or fly ash, that escape the boiler are captured with dust collectors (cyclones and baghouses). More than 90% of the sulfur released from coal can be captured in this manner.

While it is possible to control some of the toxic emissions released by coal-fired power plants, the resulting waste creates more problems for the environment. The pollution controls used to capture harmful emissions concentrate toxins and heavy metals such as mercury into coal ash and sludge. Toxic substances in ash and sludge include arsenic, mercury, chromium, and cadmium.

The *clean coal technologies* are a variety of evolving responses to late twentieth century environmental concerns, including that of global warming due to carbon dioxide releases to the atmosphere. However, many of the elements have in fact been applied for many years, and they will be only briefly mentioned here:

- Coal cleaning by "washing" has been standard practice in developed countries for some time. It reduces emissions of ash and sulfur dioxide when the coal is burned.
- Electrostatic precipitators and fabric filters can remove 99% of the fly ash from the flue gases; these technologies are in widespread use.
- FGD reduces the output of sulfur dioxide to the atmosphere by up to 97%, the task depending on the level of sulfur in the coal and the extent of the reduction. It is widely used where needed in developed countries.
- Low-NOx burners allow coal-fired plants to reduce nitrogen oxide emissions by up to 40%. Coupled with reburning techniques, NOx can be reduced 70% and SCR can clean up 90% of NOx emissions.
- Increased efficiency of plant—up to 46% thermal efficiency now (and 50% expected in future)—means that newer plants create less emissions per kWh than older ones.

- Advanced technologies such as IGCC and PFBC will enable higher thermal efficiencies still—up to 50% in the future.
- Ultra-clean coal (UCC) from new processing technologies which reduce ash below 0.25% and sulfur to very low levels mean that pulverized coal might be used as fuel for very large marine engines, in place of heavy fuel oil. There are at least two UCC technologies under development. Wastes from UCC are likely to be a problem.
- Gasification, including underground coal gasification (UCG) in situ, uses steam and oxygen to turn the coal into carbon monoxide and hydrogen.
- Sequestration refers to disposal of liquid carbon dioxide, once captured, into deep geological strata.

Some of these impose operating costs without concomitant benefit to the operator, though external costs will almost certainly be increasingly factored in through carbon taxes or similar which will change the economics of burning coal.

However, waste products can be used productively. In 1999 the EU used half of its coal fly ash and bottom ash in building materials (where fly ash can replace cement), and 87% of the gypsum from FGD.

Carbon dioxide from burning coal is the main focus of attention today, since it is implicated in global warming, and the Kyoto Protocol requires that emissions decline, notwithstanding increasing energy demand.

CCS technologies are in the forefront of measures to enjoy *clean coal*. CCS involves two distinct aspects: (1) capture and (2) storage of carbon dioxide.

22.6   CARBON DIOXIDE CAPTURE AND SEQUESTRATION

Carbon capture and sequestration (*carbon capture and storage*, CCS) is the *clean coal* concept being promoted most prominently. If successful, the concept would offer a way to continue burning coal for electricity while avoiding major costs expected under greenhouse gas emission regulations. Such regulations appear likely over the long run, whether they take the form of energy and climate legislation passed by Congress or by the Environmental Protection Agency, which is authorized to regulate the emissions under its Clean Air Act authority. The concept in any plant with *CCS* is to pump the carbon dioxide emissions that are the chief by-product of coal combustion underground or into some other permanent or semi-permanent reservoir rather than directly into the atmosphere.

A number of means exist to capture carbon dioxide from gas streams, but they have not yet been optimized for the scale required in coal-burning power plants. The focus has often been on obtaining pure carbon dioxide for industrial purposes rather than reducing carbon dioxide levels in power plant emissions.

Where there is carbon dioxide mixed with methane from natural gas wells, its separation is well proven. Several processes are used, including hot potassium carbonate which is energy-intensive and requires a large plant, an olamine process which yields high-purity carbon dioxide, amine scrubbing, and membrane processes (Mokhatab et al., 2006; Speight, 2007a,b).

Capture of carbon dioxide from flue gas streams following combustion in air is much more difficult and expensive, as the carbon dioxide concentration is only about 14% at best. The main process (Figure 22.11) treats carbon dioxide like any other pollutant, and as flue gases are passed through an amine solution the CO_2 is absorbed. It can later be released by heating the solution. This amine scrubbing process is also used for taking CO_2 out of natural gas. There is an energy cost involved. For new power plants this is quoted as 20%–25% of plant output, due to both reduced plant efficiency and the energy requirements of the actual process.

No commercial-scale power plants are operating with this process yet. At the new 1300 MWe Mountaineer power plant in West Virginia, less than 2% of the plant's off-gas is being treated for

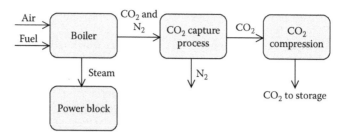

FIGURE 22.11 Primary steps for CCS.

CO_2 recovery, using chilled amine technology. Subject to federal grants, there are plans to capture and sequester 20% of the plant's CO_2—approximate two million tons of CO_2 per year.

Oxyfuel combustion, where coal is burned in oxygen rather than air, means that the flue gas is mostly CO_2 and hence it can more readily be captured by amine scrubbing—at about half the cost of capture from conventional plants.

Oxyfuel combustion is one of the most promising technical options for CO_2 capture from coal-fired power generation (Buhre et al., 2005). The possibility to use advanced steam technology and simplified flue gas processing has brought it into an economically competitive position. Theoretically, a relatively simple flue gas cleaning system is preferable for CO_2 capture, which is essential to achieve a low-cost capture technology. The conceptual development of the flue gas cleaning system focuses on fully understanding the characteristics of flue gas under coal-fired oxyfuel combustion conditions and the differences in the design criteria/requirements in comparison to that of conventional coal-fired power generation (Anheden et al., 2004; Jordal et al., 2004). The relations between the flue gas cleaning and other processes associated with the oxyfuel combustion, such as CO_2 capture, transport and storage, and plant emissions, are also evaluated in order to optimize the flue gas cleaning system within the CO_2 capture chain.

The IGCC plant is a means of using coal and steam to produce hydrogen and carbon monoxide (CO) from the coal and these are then burned in a gas turbine with secondary steam turbine (i.e., combined cycle) to produce electricity. If the IGCC gasifier is fed with oxygen rather than air, the flue gas contains highly concentrated CO_2 which can readily be captured postcombustion as seen earlier.

Further development of this oxygen-fed IGCC process will add a shift reactor to oxidize the CO with water so that the gas stream is basically just hydrogen and carbon dioxide. The CO_2 with some H_2S and Hg impurities is separated before combustion (with about 85% CO_2 recovery) and the hydrogen alone becomes the fuel for electricity generation (or other uses) while the concentrated pressurized carbon dioxide is readily disposed of. (The H_2S is oxidized to water and sulfur, which is saleable.) No commercial scale power plants are operating with this process yet. Currently, IGCC plants typically have a 45% thermal efficiency.

Capture of carbon dioxide from coal gasification is already achieved at low marginal cost in some plants. One (albeit where the high capital cost has been largely written off) is the Great Plains Synfuels Plant in North Dakota, where six million tons of lignite is gasified each year to produce clean synthetic natural gas.

Oxyfuel technology has potential for retrofit to existing pulverized coal plants, which are the backbone of electricity generation in many countries.

Captured carbon dioxide gas can be put to good use, even on a commercial basis, for enhanced oil recovery. This is well demonstrated in West Texas, and today over 5800 km of pipelines connect oilfields to a number of carbon dioxide sources in the United States.

Geological storage is an obvious method for sequestration of carbon dioxide. The geological features being considered for carbon dioxide storage fall into three categories: (1) deep saline formations, (2) depleted oil and gas fields, and (3) unmineable coal seams.

TABLE 22.9
Geological Formations Suitable for Carbon Dioxide Storage

Structural storage	When the carbon dioxide is pumped deep underground, it is initially more buoyant than water and will rise up through the porous rocks until it reaches the top of the formation where it can become trapped by an impermeable layer of cap-rock, such as shale. The wells that were drilled to place the carbon dioxide in storage can be sealed with plugs made of steel and cement.
Residual storage	Reservoir rocks act like a tight, rigid sponge. Air in a sponge is residually trapped and the sponge usually has to be squeezed several times to replace the air with water. When liquid carbon dioxide is pumped into a rock formation, much of it becomes stuck within the pore spaces of the rock and does not move.
Dissolution storage	Carbon dioxide dissolves in salty water, just like sugar dissolves in tea. The water with carbon dioxide dissolved in it is then heavier than the water around it (without carbon dioxide) and so sinks to the bottom of the rock formation.
Mineral storage	Carbon dioxide dissolved in salt water is weakly acidic and can react with the minerals in the surrounding rocks, forming new minerals, as a coating on the rock (much like shellfish use calcium and carbon from seawater to form their shells). This process can be rapid or very slow (depending on the chemistry of the rocks and water) and it effectively binds the carbon dioxide to the rocks.

As carbon dioxide is pumped deep underground, it is compressed by the higher pressures and becomes essentially a liquid. There are a number of different types of geological trapping mechanisms (depending on the physical and chemical characteristics of the rocks and fluids) which can be utilized for carbon dioxide storage and these are as follows: (1) deep saline formations, (2) depleted oil and gas fields, and (3) coal seam storage—providing the formations are of the correct geological character (Table 22.9).

Deep saline formations are underground formations of permeable reservoir rock, such as sandstones, that are saturated with very salty water (which would never be used as drinking water) and covered by a layer of impermeable cap rock (e.g., shale or clay) which acts as a seal. In the case of gas and oilfields, it was this cap rock that trapped the oil and gas underground for millions of years. Carbon dioxide injected into the formation is contained beneath the cap rock and in the groundwater flow and, in time, dissolves into the saline water in the reservoir. Carbon dioxide storage in deep saline formations is expected to take place at depths below 2500 ft (800 m). Saline aquifers have the largest storage potential globally but are the least well-explored and researched of the geological options. However, a number of storage projects are now using saline formations and have proven their viability and potential.

Depleted oil and gas fields are well-explored and geologically well-defined and have a proven ability to store hydrocarbons over geological time spans of millions of years. Carbon dioxide is already widely used in the oil industry for enhanced oil recovery (EOR) from mature oilfields (Speight, 2007a, 2009). When carbon dioxide is injected into an oilfield, it can mix with the crude oil, causing it to swell and thereby reducing its viscosity, helping to maintain or increase the pressure in the reservoir. The combination of these processes allows more of the crude oil to flow to the production wells. In other situations, the carbon dioxide is not soluble in the oil and injection of carbon dioxide raises the pressure in the reservoir, helping to sweep the oil toward the production well (Figure 22.12) (Speight, 2007a, 2009).

Coal seam storage involves another form of trapping in which the injected carbon dioxide is adsorbed onto (accumulates on) the surface of the in situ coal in preference to other gases (such as methane) which are displaced. The effectiveness of the technique depends on the permeability of the coal seam. It is generally accepted that coal seam storage is most likely to be

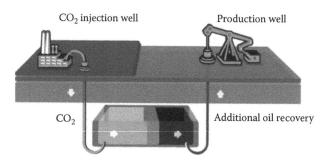

FIGURE 22.12 Schematic diagram of CO_2-enhanced oil recovery.

feasible when undertaken in conjunction with enhanced coalbed methane recovery (ECBM) in which the commercial production of coal seam methane is assisted by the displacement effect of the carbon dioxide.

Such storage projects are carefully tracked through measurement, monitoring and verification procedures both during and after the period when the carbon dioxide is being injected. These procedures address the effectiveness and safety of storage activities and the behavior of the injected carbon dioxide underground.

Measurement, monitoring, and verification procedures are used to measure the amount of carbon dioxide stored at a specific geological storage site, to ensure that the carbon dioxide is behaving as expected. The techniques used for measurement, monitoring, and verification procedure are largely new applications of existing technologies. These technologies now monitor oil and gas fields and waste storage sites. They measure injection rates and pressures, subsurface distributions of carbon dioxide, injection well integrity, and local environmental impacts.

At the Great Plains Synfuels Plant, North Dakota, some 13,000 tons per day of carbon dioxide gas is captured and 5,000 tons of this is piped 320 km into Canada for enhanced oil recovery. This Weyburn oilfield sequesters about 85 m^3 of carbon dioxide per barrel of oil produced, a total of 19 million tons over the project's 20 year life. The first phase of its operation has been judged a success.

Overall in the United States, 32 million tons of CO_2 is used annually for enhanced oil recovery, 10% of this from anthropogenic sources.

The world's first industrial-scale CO_2 storage was at Norway's Sleipner gas field in the North Sea, where about one million tons per year of compressed liquid CO_2 separated from methane is injected into a deep reservoir (saline aquifer) about a kilometer below the sea bed and remains safely in place. The natural gas contains 9% CO_2 which must be reduced before sale or export. The overall Utsira sandstone formation there, about 1 km below the sea bed, is said to be capable of storing 600 billion tons of CO_2.

Another scheme separating carbon dioxide and using it for enhanced oil recovery is at In Salah, Algeria.

West Australia's Gorgon natural gas project will tap natural gas with 14% CO_2. Capture and geosequestration of this is expected to reduce the project's emissions from 13.3 to 5.3 million tons of CO_2 per year, 3.0 million tons of the reduction being injected into geological reservoirs.

Injecting carbon dioxide into deep, unmineable coal seams where it is adsorbed to displace methane (effectively: natural gas) is another potential use or disposal strategy. Currently, the economics of enhanced coal bed methane extraction are not as favorable as enhanced oil recovery, but the potential is large.

While the scale of envisaged need for CO_2 disposal far exceeds today's uses, they do demonstrate the practicality. Safety and permanence of disposition are key considerations in sequestration.

Research on geosequestration is ongoing in several parts of the world. The main potential appears to be deep saline aquifers and depleted oil and gas fields. In both, the CO_2 is expected to remain as a supercritical gas for thousands of years, with some dissolving.

Large-scale storage of CO_2 from power generation will require an extensive pipeline network in densely populated areas. This has safety implications.

Given that rock strata have held CO_2 and methane for millions of years, there seems no reason that carefully chosen ones cannot hold sequestered CO_2. However, the eruption of a million tons of CO_2 from Lake Nyos in Cameroon in 1986 asphyxiated 1700 people, so the consequences of major releases of heavier-than-air gas are potentially serious.

REFERENCES

Ahearne, J.F. 1993. *American Scientist*, 81(1): 24–35.

Akar, G., Arslan, V., Ertem, M.E., and Ipekoglu, U. 2009. Relationship between ash fusion temperatures and coal mineral matter in some Turkish coal ashes. *Asian Journal of Chemistry*, 21: 2105–2109.

Anheden, M., Andersson, A., Bernstone, C., Eriksson, S., Yan, J., Liljemark, S., and Wall, C. 2004. CO_2 quality requirement for a system with CO_2 capture, transport and storage. *Proceedings of Greenhouse Gas Control Technologies Conference*, GHGT-7, Vancouver, British Columbia, Canada, September 5–9.

Austin, G.T. 1984. *Shreve's Chemical Process Industries*. McGraw-Hill, New York.

Balat, M. 2008a. The future of clean coal. *Future Energy: Improved, Sustainable and Clean Options for Our Planet*, T.M. Letcher (Ed.). Elsevier, Amsterdam, the Netherlands, Chapter 2.

Balat, M. 2008b. Coal-fired power generation: Proven technologies and pollution control systems. *Energy Sources, Part A*, 30: 132–140.

Baldwin, A.L., Elia, G., and Corbett, R. 1992. PETC review, Issue No. 6 (Summer). Pittsburgh Energy Technology Center, Office of Fossil Energy, United States Department of Energy, Pittsburgh, PA, p. 8.

Banks, F.E. 1992. *OPEC Bulletin*, XXIII(2): 20.

Beer, J.M. 2007. High efficiency electric power generation: The environmental role. *Progress in Energy and Combustion Science*, 33: 107–134.

Beier, E. 1990. In *Bioprocessing and Biotreatment of Coal*, D.L. Wise (Ed.). Marcel Dekker, Inc., New York, p. 549.

Bell, L. 2011. *Climate of Corruption: Politics and Power behind the Global Warming*. Greenleaf Book Group Press, Austin, TX.

Bending, R.C., Cattell, R.K., and Eden, R.J. 1987. *Annual Review of Energy*, 12: 185.

Berkowitz, N. 1979. *An Introduction to Coal Technology*. Academic Press, Inc., New York.

Bhanarkar, A.D., Gavane, A.G., Tajne, D.S., Tamhane, S.M., and Nema, P. 2008. Composition and size distribution of particulates emissions from a coal-fired power plant in India. *Fuel*, 87: 2095–2101.

Breeze, P. 2005. *Power Generation Technologies*. Elsevier, Amsterdam, the Netherlands.

Bris, T.L., Cadavid, F., Caillat, S., Pietrzyk, S., Blondin, J., and Baudoin, B. 2007. Coal combustion modelling of large power plant, for NOx abatement. *Fuel*, 86: 2213–2220.

Buhre, B.J.P., Elliott, L.K., Sheng, C.D., Gupts, R.P., and Wall, T.F. 2005. Oxy-fuel combustion technology for coal-fired power generation. *Progress in Energy and Combustion Science*, 31: 283–307.

Cicek, T. 2008. Dry cleaning of Turkish coal. *Energy Sources, Part A*, 30: 593–605.

Clark, W.C. 1989. *Scientific American*, 261(3): 46.

Clarke, L.B. 1991. Management of by-products from IGCC power generation. IEA Coal Research, IEACR/38, International Energy Agency, London, U.K.

Coal Voice. 1992. 14(4): 8.

Couch, G.R. 1987. Biotechnology and coal. Report No. ICTRS/TR38. IEA Coal Research, International Energy Agency, London, U.K.

Couch, G.R. 1991. Advanced coal cleaning technology. Report No. IEACR/44. IEA Coal Research, International Energy Agency, London, U.K.

Cruden, A. 1930. *Complete Concordance to the Bible*. Butterworth Press, London, U.K.

Darcovkic, K., Jonasson, K.A., and Capes, C.E. 1997. Developments in the control of fine particulate air emissions. *Advanced Powder Technology*, 8(3): 179–215.

Dryden, I.G.C. 1975. *The Efficient Use of Energy*. IPC Business Press Ltd., Guildford, Surrey, England.

Dugan, P.R., McIlwain, M.E., Quigley, D., and Stoner, D.L. 1989. *Proceedings of the Sixth Annual International Pittsburgh Coal Conference*, University of Pittsburgh, Pittsburgh, PA, Vol. 1, p. 135.

EIA. 1988. International energy outlook: Projections to 2000. Report No. DOE/EIA-0484(87). United States Department of Energy, Energy Information Administration, Washington, DC.

EIA. 1989. Annual energy outlook: Long term projections. Report No. DOE/EIA-0383(89). United States Department of Energy, Energy Information Administration, Washington, DC.

EIA. 1991a. Annual energy review 1990. Report No. DOE/EIA 0384(90). United States Department of Energy, Energy Information Administration, Washington, DC.

EIA. 1991b. Annual outlook for oil and gas 1991. Report No. DOE/EIA-0517(91). United States Department of Energy, Energy Information Administration, Washington, DC.

EIA. 1992. Annual energy outlook: With projections to 2010. Report No. DOE/EIA-0383(92). United States Department of Energy, Energy Information Administration, Washington, DC.

Elliott, T.C. 1992. *Power*, January: 17.

Ember, L.R., Layman, P.L., Lepkowski, W., and Zurer, P.S. 1986. *Chemical and Engineering News*, November 24: 14.

Faison, B.D. 1991. *Critical Reviews in Biotechnology*, 11: 347.

Feibus, H., Voelker, G., and Spadone, S. 1986. In *Acid Rain Control II: The Promise of New Technology*. D.S. Gilleland (Ed.). Southern Illinois University Press, Carbondale, IL, Chapter 3.

Franco, A. and Diaz, A.R. 2009. The future challenges for clean coal technologies: Joining efficiency increase and pollutant emission control. *Energy*, 34: 348–354.

Frazier, W.F., Gaswirth, H., Mongillo, R.J., Shattuck, D.M., and Wedig, C.P. 1991. Flue gas desulfurization systems designed and operated to meet the clean air act amendments of 1990. Presented at the Industrial Gas Cleaning Institute, Inc., IGCI Forum 91, Washington, DC, September 11–13, TP 91-55.

Frosch, R.A. and Gallopoulos, N.E. 1989. Strategies for manufacturing. *Scientific American*, 261(3): 144.

Fulkerson, W., Judkins, R.R., and Sanghvi, M.K. 1990. In *Energy for Planet Earth*. G.R. Davis (Ed.). W.H. Freeman and Co., New York, Chapter 8.

Galloway, R.L. 1882. *A History of Coal Mining in Great Britain*. Macmillan & Co., London, U.K.

Graedel, T.E. and Crutzen, P.J. 1989. The changing atmosphere. *Scientific American*, 261(3): 58.

Graus, W. and Worrell, E. 2009. Trend in efficiency and capacity of fossil power generation in the EU. *Energy Policy*, 37: 2147–2160.

Hertzmark, D.I. 1987. *Annual Review of Energy*, 12: 23.

Hessley, R.K. 1990. *Fuel Science and Technology Handbook*. J.G. Speight (Ed.). Marcel Dekker, Inc., New York.

Hessley, R.K., Reasoner, J.W., and Riley, J.T. 1986. *Coal Science*. John Wiley & Sons, Inc., New York.

Hubbert, M.K. 1969. Energy resources. In *Resources and Man*, P. Cloud (Ed.). Freeman, San Francisco, CA, pp. 157–242.

Hyland, R.P. 1991. *Hydrocarbon Processing*, 70(5): 113.

IEA Coal Research. 1991. *Emissions Standard Data Base*. International Energy Agency Coal Research, London, U.K.

IEA Coal Research. 1992. *Greenhouse Gases Bulletin*. International Energy Agency Coal Research, London, U.K.

IEA. 2003. *Control and Minimization of Coal-Fired Power Plant Emissions*. International Energy Agency, Paris, France.

IEA (International Energy Agency). 2008. *Clean Coal Technologies-Accelerating Commercial and Policy Drivers for Deployment*. OECD/IEA, Paris, France.

Jordal, K., Anheden, M., Yan, J., and Strömberg, L. 2004. Oxyfuel combustion for coal-fired power generation with CO_2 capture—Opportunities and challenges. *Proceedings of Greenhouse Gas Control Technologies Conference*, GHGT-7, Vancouver, British Columbia, Canada, September 5–9.

Kohl, A.L. and Nielsen, R. 1997. *Gas Purification*. Gulf Publishing Company, Houston, TX.

Kolker, A., Senior, C.L., and Quick, J.C. 2006. Mercury in coal and the impact of coal quality on mercury emissions from combustion systems. *Applied Geochemistry*, 21: 1821–1836.

Kuhr, R.W., Wedig, C.P., and Davidson, L.N. 1988. The status of new developments in flue gas and simultaneous cleanup. Presented at the 1988 Joint Power Generation Conference, Philadelphia, PA, TP 88-89.

Kyte, W.S. 1991. Desulphurisation 2: Technologies and strategies for reducing sulphur emissions. Symposium Series No. 123, Institution of Chemical Engineers, Rugby, Warwickshire, England.

Lee, S.H., Rhim, Y.J., Cho, S.P., and Baek, J.I. 2006. Carbon-based novel sorbent for removing gas-phase mercury. *Fuel*, 85: 219–226.

Locheed, T.A. 1993. *Chemical Processing*, 56(1): 41.

Luttrell, D.H. 2006. Development of an advanced de-shaling technology to improve the energy efficiency of coal handling, processing, and utilization operations. Contract No. DE-FC26-05NT42501. Industrial Technologies Program, Mining of the Future, U.S. Department of Energy, Washington, DC.

Luttrell, G.H., Kohmuench, J.N., and Yoon, R.H. 2000. An evaluation of coal preparation technologies for controlling trace element emissions. *Fuel Processing Technology*, 65/66: 407–422.

MacDonald, G.J. 1990. *Annual Reviews of Energy*, 15: 53.

Manowitz, B. and Lipfert, F.W. 1990. In *Geochemistry of Sulfur in Fossil Fuels*, W.L. Orr and C.M. White (Eds.). American Chemical Society, Washington, DC, Chapter 3.

Martin, A.J. 1985. In *Prospects for the World Oil Industry*, T. Niblock and R. Lawless (Eds.). Croom Helm Publishers, Beckenham, Kent, U.K., Chapter 1.

Martin, G.B. 1986. In *Acid Rain Control II: The Promise of New Technology*. Southern Illinois University Press, Carbondale, IL, Chapter 5.

Maurits la Rivière, J.W. 1989. Threats to the world's water. *Scientific American*, 261(3): 80.

Meyers, S. 1987. *Annual Review of Energy*, 12: 81.

Miller, B.G. 2005. *Coal Energy Systems*. Elsevier, Amsterdam, the Netherlands.

Mohnen, V.A. 1988. *Scientific American*, 259(2): 30.

Mokhatab, S., Poe, W.A., and Speight, J.G. 2006. *Handbook of Natural Gas Transmission and Processing*. Elsevier, Amsterdam, the Netherlands.

Muller, B.T. 1996. Update-pulse enhanced fluidized bed combustion. *13th Annual Pittsburgh Coal Conference*, Pittsburgh, PA, September.

Nordstrand, D., Duong, D.N.B., and Miller, B.G. 2008. Combustion engineering issues for solid fuel systems. *Post-Combustion Emissions Control*, B.G. Miller and D. Tillman (Eds.). Elsevier, London, U.K., Chapter 9.

Notestein, J.E. 1990. Commercial gasifier for IGCC applications study report. Report No. DOE/METC-91/6118 (DE91002051). Morgantown Energy Technology Center, Office of Fossil Energy, United States Department of Energy, Morgantown, WV.

Ohlström, M., Jokiniemi, J., Hokkinen, J., Makkonen, P., and Tissari, J. 2006. Combating particulate emissions in energy generation and industry technology, Environment and Health Technology Program, VTT Technical Research Center of Finland, Espoo, Finland.

Omer, A.M. 2008. Energy, environment and sustainable development. *Renewable and Sustainable Energy Reviews*, 12: 2265–2300.

Pavlish, J.H., Sondreal, E.A., Mann, M.D., Olson, E.S., Galbreath, K.C., Laudal, D.L., and Benson, S.A. 2003. Status review of mercury control options for coal-fired power plants. *Fuel Processing Technology*, 82: 89–165.

Pence, D.V. and Beasley, D.E. 1995. Multi-fractal signal simulation of pressure fluctuations in a bubbling gas-fluidized bed. *Proceedings of the Sixth International Symposium on Gas-Solid Flows*, Hilton Head, SC, August.

Probstein, R.F. and Hicks, R.E. 1990. *Synthetic Fuels*. pH Press, Cambridge, MA.

Saroff, L. and Robey, R. 1992. In *Power Generation Technology: 1993*, R. Knox (Ed.). Sterling Publications, Inc., London, U.K., p. 71.

Sathaye, J., Ghirardi, A., and Schipper, L. 1987. *Annual Review of Energy*, 12: 253.

Schneider, S.H. 1989. The changing climate. *Scientific American*, 261(3): 70.

Slack, A.V. 1986. In *Acid Rain Control II: The Promise of New Technology*, Southern Illinois University Press, Carbondale, IL, Chapter 7.

Sondreal, E.A., Benson, S.A., and Pavlish, J.H. 2006. Status of research on air quality: Mercury, trace elements, and particulate matter. *Fuel Processing Technology*, 65/66: 5–22.

Sondreal, E.A., Benson, S.A., Pavlish, J.H., and Ralston, N.V.C. 2004. An overview of air quality III: Mercury, trace elements, and particulate matter. *Fuel Processing Technology*, 85: 425–440.

Senior, C.L. Zeng, T., Che, J., Ames, M.R., Sarofim, A.F., Olmez, I., Huggins, F.E. Shah, N., Huffman, G.P., Kolker, A., Mroczkowski, S., Palmer, C., and Finkelman, R. 2000. Distribution of trace elements in selected pulverized coal as a function of particle size and density. *Fuel Processing Technology*, 63: 215–241.

Speight, J.G. 1990. *Fuel Science and Technology Handbook*. Marcel Dekker, Inc., New York.

Speight, J.G. 1991. *The Chemistry and Technology of Petroleum*, 2nd edn. Marcel Dekker, Inc., New York.

Speight, J.G. 1993. *Gas Processing: Environmental Aspects and Methods*. Butterworth Heinemann, Oxford, England.

Speight, J.G. 2007a. *The Chemistry and Technology of Petroleum*, 4th edn. Taylor & Francis Group, Boca Raton, FL.

Speight, J.G. 2007b. *Natural Gas: A Basic Handbook*. GPC Books, Gulf Publishing Company, Houston, TX.

Speight, J.G. 2008. *Synthetic Fuels Handbook: Properties, Processes, and Performance*. McGraw-Hill, New York.

Speight, J.G. 2009. *Enhanced Recovery Methods for Heavy Oil and Tar Sands*. Gulf Publishing Company, Houston, TX.

Srivastava, R.K., Hall, R.E., Khan, S., Culligan, K., and Lani, B.W. 2005. Nitrogen oxides emission control options for coal-fired electric utility boilers. *Journal of the Air & Waste Management Association*, 55: 1367–1388.

Stensvaag, J.-M. 1991. *Clean Air Act Amendments: Law and Practice.* John Wiley & Sons, Inc., New York.

Stigliani, W.M. and Shaw, R.W. 1990. *Annual Reviews of Energy*, 15: 201.

Suárez-Ruiz, I. and Crelling, J.C. 2008. *Applied Coal Petrology: The Role of Petrology in Coal Utilization.* Academic Press, Inc., New York.

Suarez-Ruiz, I. and Ward, C.R. 2008. Applied coal petrology: The role of petrology in coal utilization. *Coal Combustion*, I. Suarez-Ruiz and J.C. Ward (Eds.). Elsevier, Amsterdam, the Netherlands, Chapter 4.

Takematsu, T. and Maude, C. 1991. (Eds.). Coal gasification for IGCC power generation. Report No. IEACR/37. IEA Coal Research, International Energy Agency, London, U.K.

Tomas-Alonso, F. 2005. A new perspective about recovering SO_2 offgas in coal power plants: Energy saving. Part III. Selection of the best methods. *Energy Sources*, 1051–1060.

United States Congress. 1990. Public law 101-549. An act to amend the clean air act to provide for attainment and maintenance of health protective national ambient air quality standards, and for other purposes, November 15.

United States Department of Energy. 1990. Gas research program implementation plan. DOE/FE-0187P. United States Department of Energy, Washington, DC, April.

United States Department of Energy. 1992. Clean coal technology: The new coal era. DOE/FE-0217P. United States Department of Energy, Washington, DC, March.

United States Department of Energy. 1993. Clean coal technology demonstration program. DOE/FE-0272. United States Department of Energy, Washington, DC, February.

United States General Accounting Office. 1990. Energy policy: Developing strategies for energy policies in the 1990s. Report to Congressional Committees. GAO/RCED-90-85. United States General Accounting Office, Washington, DC, June.

Vallero, D. 2008. *Fundamentals of Air Pollution*, 4th edn. Elsevier, London, U.K.

Vejahati, F., Xu, Z., and Gupta, R., 2010. Trace elements in coal: Associations with coal and minerals and their behavior during coal utilization—A review. *Fuel*, 89: 904–911.

Wang, Y., Duan, Y., Yang, L., Jiang, Y., Wu, C., Wang, Q., and Yang, X. 2008. Comparison of mercury removal characteristic between fabric filter and electrostatic precipitators of coal-fired power plants. *Journal of Fuel Chemistry and Technology*, 36(1): 23–29.

WCI. 2005. *The Coal Resource—A Comprehensive Overview of Coal.* World Coal Institute, London, U.K.

Wheelock, T.D. 1977. Coal desulfurization: Chemical and physical methods. Symposium Series No. 64. American Chemical Society, Washington, DC.

Xu, M., Yan, R., Zheng, C., Qiao, Y., Han, J., and Sheng, C. 2003. Status of trace element emission in a coal combustion process: A review. *Fuel Processing Technology*, 85: 215–237.

Yang, H., Xua, Z., Fan, M., Bland, A.E., and Judkins, R.R. 2007. Adsorbents for capturing mercury in coal-fired boiler flue gas. *Journal of Hazardous Materials*, 146:1–11.

Yeager, K.E. and Baruch, S.B. 1987. *Annual Review of Energy*, 12: 471.

Yeager, K.E. and Preston, G.T. 1986. In *Acid Rain Control II: The Promise of New Technology*. Southern Illinois University Press, Carbondale, IL, Chapter 4.

23 Gas Cleaning

23.1 INTRODUCTION

Flue and waste gases from power plants and other industrial operations where coal is used as a feedstock invariably contain constituents, such as carbon dioxide (CO_2), nitrogen oxides (NOx), sulfur oxides (SOx), dust and particles, and toxins such as dioxin and mercury, that are damaging to the climate or environment.

The processes that have been developed for gas cleaning (Mokhatab et al., 2006; Speight, 1993, 2007, 2008) vary from a simple once-through wash operation to complex multistep systems with options for recycle of the gases (Mokhatab et al., 2006). In some cases, process complexities arise because of the need for recovery of the materials used to remove the contaminants or even recovery of the contaminants in the original, or altered, form.

The purpose of preliminary cleaning of gases that arise from coal utilization is the removal of materials such as mechanically carried solid particles (either process products and/or dust) as well as liquid vapors (i.e., water, tars, and aromatics such as benzenes and/or naphthalenes); in some instances, preliminary cleaning might also include the removal of ammonia gas.

For example, cleaning of town gas is the means by which the crude "tarry" gases from retorts or coke-ovens are (first), in a preliminary step, freed from tarry matter, condensable aromatics (such as naphthalene) and (second) purified by the removal of materials such as hydrogen sulfide, other sulfur compounds, and any other unwanted components that will adversely affect the use of the gas (Figure 23.1).

In more general terms, gas cleaning is divided into removal of particulate impurities and removal of gaseous impurities. For the purpose of this chapter, the latter operation includes the removal of hydrogen sulfide, CO_2, sulfur dioxide, and the like.

There is also need for subdivision of these two categories as dictated by needs and process capabilities: (1) coarse cleaning whereby substantial amounts of unwanted impurities are removed in the simplest, most convenient, manner; (2) fine cleaning for the removal of residual impurities to a degree sufficient for the majority of normal chemical plant operations, such as catalysis or preparation of normal commercial products, or cleaning to a degree sufficient to discharge an effluent gas to atmosphere through a chimney; (3) ultrafine cleaning where the extra step (as well as the extra expense) is justified by the nature of the subsequent operations or the need to produce a particularly pure product.

To make matters even more complicated, a further subdivision of the processes, which applies particularly to the removal of gaseous impurities, is by process character insofar as there are processes that rely upon chemical and physical properties/characteristics of the gas stream to enhance separation of the constituents.

Since coal is a complex, heterogeneous material, there are a wide variety of constituents that are not required in a final product and must be removed during processing. Coal composition and characteristics vary significantly; there are varying amounts of sulfur, nitrogen, and trace-metal species that must be disposed of in the correct manner (Argonne, 1990). Thus, whether the process be gasification to produce a pipeline quality gas (Figure 23.2) or a series of similar steps for gas cleaning before ejection into the atmosphere, the stages required during this processing are numerous and can account for a major portion of a gas cleaning facility.

Generally, the majority of the sulfur that occurs naturally in the coal is driven into the product gas. Thermodynamically, the majority of the sulfur should exist as hydrogen sulfide, with smaller amounts

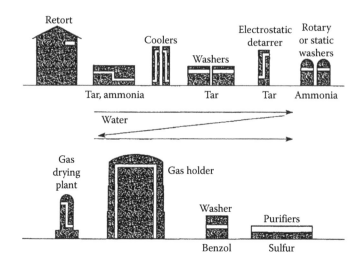

FIGURE 23.1 Layout of a coal gas cleaning plant.

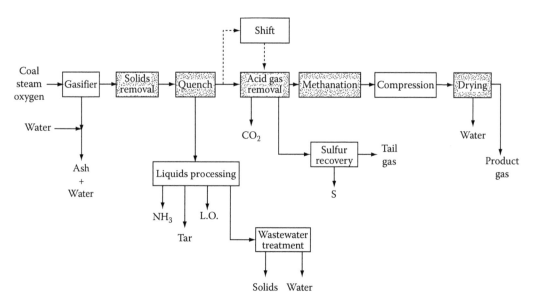

FIGURE 23.2 Production of pipeline-quality gas.

of carbonyl sulfide (COS) and carbon disulfide (CS2). However, data from some operations (coke ovens) show higher than expected (from thermodynamic considerations) concentrations of COS and CS2.

The existence of mercaptans, thiophenes, and other organic sulfur compounds in (gasifier) product gas will probably be a function of the degree of severity of the process, contacting schemes, and heat-up rate. Those processes that tend to produce tars and oils may also tend to drive off high-molecular-weight organic sulfur compounds into the raw product gas.

In general terms, the gaseous emissions from coal conversion facilities may be broadly classed as those originating from four processing steps: pretreatment, conversion, and upgrading, as well as those from ancillary processes (Figure 23.3).

In conventional power plants, pulverized coal is burned in a boiler, where the heat vaporizes water in steam tubes. The resulting steam turns the blades of a turbine, and the mechanical energy of the turbine is converted to electricity by a generator. Waste gases, such as sulfur dioxide, NOx, and CO_2, produced in the boiler during combustion.

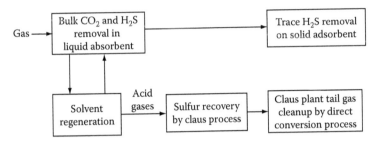

FIGURE 23.3 Process sequence for acid gas removal.

The sulfur dioxide, for example, that is produced during the combustion of coal or petroleum in power plants fuel will react with oxygen and water in the atmosphere to yield the environmentally detrimental sulfuric acid, which is a contributor to acid rain:

$$S_{coal} + O_2 \rightarrow SO_2$$

$$2SO_2 + O_2 \rightarrow 2SO_3$$

$$SO_2 + H_2O \rightarrow H_2SO_3$$

$$2SO_2 + O_2 \rightarrow 2SO_3$$

$$SO_3 + H_2O \rightarrow H_2SO_4$$

Thus,

$$2SO_2 + O_2 + 2H_2O \rightarrow 2H_2SO_4$$

Using coal as an example, the other two fossil fuels, petroleum and natural gas, are not necessarily free from blame even though there is the threat to *outlaw* coal in some countries (Coal and Synfuels Technology, 1992).

The three major types of pollutants emitted by a coal-fired power plant are particulate matter, sulfur dioxide, and NOx. Particulate matter is generally the fine inorganic matter that is produced from mineral matter in the coal, although it may also be finely divided carbon that can also be produced during coal combustion or utilization. However, the material is mostly the finely divided inorganic matter and, for this reason, is referred to as *fly ash* (Chapters 14 and 15), which can be carried out of the stack with the hot exhaust gases. In fact, the practice of burning finely divided coal can contribute to fly ash emissions. Sulfur dioxide is produced by the oxidation of organic sulfur in the coal and is normally cited as the most troublesome of pollutants. 10%–50% of the nitrogen inherent in the organic coal structure is converted to nitric oxide during combustion:

$$N_{coal} + O_2 \rightarrow 2NO$$

$$2NO + O_2 \rightarrow 2NO_2$$

$$N_{coal} + O_2 \rightarrow NO_2$$

$$NO + H_2O \rightarrow H_2NO_3$$

$$NO_2 + H_2O \rightarrow HNO_3$$

Thus,

$$4NO_2 + O_2 + 2H_2O \rightarrow 4HNO_3$$

The hydrogen chloride, although not usually considered to be a major pollutant, is produced from mineral matter and other inorganic contaminants:

$$Cl_{\text{coal minerals}} + H_{\text{coal}} \rightarrow HCl$$

Hydrogen chloride quickly picks up water from the atmosphere to form droplets of hydrochloric acid and, like sulfur dioxide, is a contributor to acid rain. However, hydrogen chloride may exert severe local effects because, unlike sulfur dioxide, it does not need to participate in any further chemical reaction to become an acid. Under atmospheric conditions that favor a buildup of stack emissions in the area of a large power plant, the amount of hydrochloric acid in rain water could be quite high.

The nitrogen in the coal tends to gasify simultaneously with the carbon to form ammonia and cyanides (by reaction of ammonia with coal). High temperature processes, however, do not usually produce significant amounts of ammonia in the effluent gases, presumably because of the thermodynamic potential for ammonia to decompose to molecular nitrogen and hydrogen at high temperatures. On the other hand, in low temperature gasifiers, the occurrence of high-molecular weight nitrogenous compounds is anticipated because of the quantity of tars and oils generated; these compounds include pyridines, pyrroles, azoles, indoles, quinolines, anilines, amines, and similar compounds. In addition, thiocyanates have been reported in the effluents (quench water) from gasifiers to the exclusion of cyanides. However, the reaction

$$H_2S + HCN \rightarrow HCNS + H_2$$

is not favored at the operating conditions. Conceivably, the reaction could occur by the action of hydrogen cyanide with sulfur

$$S_{\text{coal}} + HCN \rightarrow HCNS$$

but the probability of quantitative reaction of hydrogen cyanide within the gasifier is small. There is also the possibility that the majority of the thiocyanates in gasifier effluents are due to the formation of thiocyanates by contact with air, which is highly favored in aqueous media:

$$2H_2S + 2HCN + O_2 \rightarrow 2HCNS + 2H_2O$$

Very little, if anything, can be done during the pretreatment of coal to eliminate nitrogen since the nitrogen is part of the organic coal structure. The situation is less clear in the case of the sources of hydrogen chloride sources; both organic chlorine and inorganic chloride salts contribute to the formation of hydrogen chloride formation during combustion. Coal-cleaning processes can reduce the mineral matter content, but pretreatment processes do not remove organically bound chlorine, which is more likely to be the precursor to hydrogen chloride in a combustion process.

Pretreatment washing processes are also successful methods for removing inorganic sulfur but they do not affect the organic sulfur content. Thus, even before combustion begins, some of the sulfur can be removed from coal. For instance, commercially available processing methods crush the coal and separate the resulting particles on the basis of density, thereby removing up to about 30% of the sulfur. But while pretreatment washing may remove up to 90% of the pyritic sulfur, up to 20% of the combustible coal may also be removed, and a balance must be struck between the value of the sulfur removed and coal lost to the cleaning process.

A wide range of metals occur in the coal in greater abundance than normally found in the earth's crust due to a concentration effect by the vegetation that initially formed the coal deposits. During coal processing (e.g., gasification), some of these elements may be present in the effluent either as vaporized metal or as a volatile compound.

23.2 ENVIRONMENTAL LEGISLATION

The environmental aspects of coal use have been a major factor in the various processes, and the see-sawing movement of the fossil fuel base between petroleum, natural gas, and coal increased the need for pollutant control for large, coal-fired power plants. These power plants emit pollutants that, by atmospheric chemical transformations, may become even more harmful secondary pollutants (Moran et al., 1986).

It has been recognized for some time that gaseous pollutants, especially sulfur dioxide, aggravate existing respiratory disease in humans and contributes to its development (e.g., Houghson, 1966). Sulfur dioxide gas by itself can irritate the upper respiratory tract. It can also be carried deep into the respiratory tract by airborne adsorbents that can cause near-irreversible damage in the lungs. There is also the belief that sulfur dioxide is a contributor to increased respiratory disease death rates. Sulfur dioxide also contributes to the various types of smog that occur in many industrialized areas of the world. Indeed, sulfur dioxide is also harmful to a variety of flora including forage, forest, fiber, and cereal crops as well as many vegetable crops. Vegetation just cannot grow, let alone flourish, in an atmosphere polluted by sulfur dioxide. Indeed, the occurrence of zones of dead vegetation was a common sight at points downwind of sulfide ore smelters.

Thus, it has become very apparent over the last three decades that abatement of air pollution needs to be mandatory now and in the future. Four main avenues of action are open to decrease the amount of sulfur dioxide emitted from stacks of power-generating plants (1) burn low-sulfur fuels, (2) desulfurize available fuels, (3) remove SOx from flue gases, or (4) generate power by nuclear reactors.

Low-sulfur fuels are expensive and not readily available in many areas where population density is the greatest. Desulfurization of fuels is also expensive, and the technology for desulfurizing coal is still in the development stage. A few nuclear power plants are being built today. Safety and health concerns about nuclear facilities make it unlikely that there will be an outburst of building such facilities for some years to come.

The use of coal involves, at some stage by deliberate means or by accidental means, the generation of gaseous mixtures that can be quite obnoxious in terms of environmental contamination. Coal combustion and gasification produce hydrogen sulfide as a by-product of the primary process. Thereafter, the usual practice is to utilize a Claus sulfur recovery unit to convert the hydrogen sulfide to elemental sulfur (Speight, 1990, 2000). While the problem, in the former heavy industrialized centers, may be seemingly less acute than it was decades ago, predominantly because of an increased environmental consciousness, the generation of such noxious materials is still an issue. Industry continues to march forward; but the increased need to maintain a clean, livable, environment is more evident now than at any time in the past.

In the past, a certain amount of pollution was recognized as being almost inevitable, perhaps even fashionable. But now, this is not the case. Any industry found guilty of emitting noxious materials can suffer heavy fines. And there is also the possibility of a jail term for the offending executives! Pollution of the environment will not be tolerated.

Thus, while industry marches on using many of the same processes that were in use in the early days of the century, more stringent methods for cleanup are necessary before any product/by-product can be released to the atmosphere. And this is where gas processing becomes an important aspect of industrial life. Furthermore, gas-cleaning processes are now required to be more efficient than ever before.

23.3 GENERAL ASPECTS

Contrary to the general belief of some scientists and engineers, all gas-cleaning systems are *not* alike, and having a good understanding of the type of gaseous effluents from coal-based processes is necessary to implement the appropriate solution.

The design of a gas-cleaning system must always take into account the operation of the upstream installations since every process will have a specific set of requirements. In some cases, the application of a dry-dusting removal unit may not be possible and thus requires a special process design of the wet gas-cleaning plant.

Thus, the gas-cleaning process must always be of optimal design for both the upstream and downstream processes.

The *average* (a most inaccurate term to say the least since it bears very little relationship to the chemistry of the combustion of the different sulfur forms) sulfur content of coal burned to generate electricity is generally assumed to be on the order of 2.5% wt/wt. Many coals have a much higher sulfur content (Chapters 8 and 9; Berkowitz, 1979; Hessley et al., 1986; Hessley, 1990) and, because of a variety of geographical, economic, as well as political issues, such coals are (or have to be) used for power generation.

Organic sulfur comprises 50%–60% of the total sulfur present in coal; it is an integral part of the coal structure and cannot be removed by mechanical means (Chapter 6). Pyritic sulfur accounts for most of the remaining sulfur in coal. Gravity separation techniques can readily remove pyritic sulfur from coal if the pyrite particles in the coal are fairly large. The coal industry has used these techniques for many years. Many American coals permit the removal of about half of the pyritic sulfur in this way. The pyrite in some coals, however, is too fine to permit separation by these methods.

Gas processing, although generally simple in chemical and/or physical principles, is often confusing because of the frequent changes in terminology and, often, lack of cross-referencing (Mokhatab et al., 2006; Speight, 2007, 2008). Although gas processing employs different process types, there is always overlap between the various concepts. And, with the variety of possible constituents and process operating conditions, a "universal" purification system cannot be specified for economic application in all cases.

Nevertheless, the first step in gas cleaning (Figure 23.2) is usually a device to remove large particles of coal and other solid materials. This is followed by cooling, quenching, or washing, to condense tars and oils and remove dust and water-soluble materials—phenols, chlorides, ammonia, hydrogen cyanide, thiocyanate, and perhaps some sulfur compounds from the gas stream. Water washing is desirable for simplicity in gas cleaning; however, the purification of this water is not simple.

Cleanup steps and their sequence can be affected by the type of gas produced and its end use (Mokhatab et al., 2006; Speight, 2007, 2008). The minimum requirement in this respect would be the application of low heat-value (low-Btu) gas produced from low-sulfur anthracite coal as a fuel gas. The gas may pass directly from the gasifier to the burners, and, in this case, the burners are the cleanup system. Many variations on this theme are possible, and, in addition, the order of the cleanup stages may be varied.

The selection of a particular process-type (Table 23.1) for a gas-cleaning operation is not a simple choice. Many factors have to be considered, not the least of which is the constitution of the gas stream that requires treatment. Indeed, process selectivity indicates the preference with which the process will remove one acid gas component relative to (or in preference to) another. For example, some processes remove both hydrogen sulfide and CO_2 while other processes are designed to remove hydrogen sulfide only (Table 23.2).

Gas cleaning by sorption by a liquid or solid sorbent is one of the most widely applied operations in the chemical and process industries (Table 23.1). Some processes have the potential for sorbent regeneration, but, in a few cases, the process is applied in a nonregenerative manner. The interaction between sorbate and sorbent may either be physical in nature or consist of physical sorption followed by chemical reaction. Other gas stream treatments use the principle of chemical conversion of the contaminants with the production of "harmless" (non-contaminant) products or to substances, which can be removed much more readily than the impurities from which they are derived (Mokhatab et al., 2006; Speight, 2007, 2008).

TABLE 23.1

Summary of Gas Cleaning Processes

Sorbent	Nature of Interaction	Regeneration	Examples
Liquid	Absorption + chemical reaction	Yes	Many processes for the removal of CO_2 and H_2S from various gases, with solvents such as water + MEA, DEA, and DIPA Agents improving physical solubility may be added (Sulfinol process); H_2S may be recovered as such or oxidized to S
Liquid + solid	Absorption + chemical reaction	Varies	Some slurry wash processes for FGD
Liquid	Physical adsorption	Yes	CO_2 and/or H_2S from hydrocarbon gases; solvents: N-methyl pyrrolidone, propylene carbonate, methanol
Solid	Physical adsorption	Yes	Purification of natural gas (H_2S, CO_2) with molecular sieves
		Yes	Gas drying operatives (cyclic regenerative), molecular sieves
		Varies	Odor removal from waste gases (active carbon)
Solid	Chemical reaction	No	H_2S from process gases, with ZnO
		Yes	SO_2 from flue gases, with CuO/Al_2O_3

TABLE 23.2

Processes for Hydrogen Sulfide and CO_2 Removal from Gas Streams

Process	Sorbent	Removes
Amine	Monethanolamine, 15% in water	CO_2, H_2S
Economine	Diglycolamine, 50%–70% in water	CO_2, H_2S
Alkazid	Solution M or DIK (potassium salt of dimethylamine acetic acid), 25% in water	H_2S, small amount of CO_2
Benfield, Catacarb	Hot potassium carbonate, 20%–30% in water (also contains catalyst)	CO_2, H_2S; selective to H_2S
Purisol (Lurgi)	N-Methyl-2-pyrrolidone	H_2S, CO_2
Fluor	Propylene carbonate	H_2S, CO_2
Selexol (Allied)	Dimethyl ether polyethylene glycol	H_2S, CO_2
Rectisol (Lurgi)	Methanol	H_2S, CO_2
Sulfinol (Shell)	Tetrahydrothiophene dioxide (sulfolane) plus diisopropanolamine	H_2S, CO_2; selective to H_2S
Giammarco-Vetrocoke	K_3AsO_3 activated with arsenic	H_2S
Stretford	Water solution of Na_2CO_3 and anthraquinone disulfonic acid with activator of sodium metavanadate	H_2S
Activated carbon	Carbon	H_2S
Iron sponge	Iron oxide	H_2S
Adip	Alkanolamine solution	H_2S; some COS, CO_2 and mercaptans
SNPA-DEA	Diethanolamine solution	H_2S, CO_2
Takahax	Sodium 1,4-naphthoquinone 2-sulfonate	H_2S

There are many variables in gas cleaning, and the precise area of application of a given process is difficult to define although there are several factors that need to be considered: (1) the types and concentrations of contaminants in the gas; (2) the degree of contaminant removal desired; (3) the selectivity of acid gas removal required; (4) the temperature, pressure, volume, and composition of the gas to be processed; (5) the CO_2 to hydrogen sulfide ratio in the gas; (6) the desirability of sulfur recovery due to process economics or environmental issues.

Any gases, such as hydrogen sulfide and/or CO_2, that are the products of coal processing can be removed by the application of an amine-washing procedure:

$$2RNH_2 + H_2S \rightarrow (RNH_3)_2S$$

$$(RNH_3)_2S + H_2S \rightarrow 2RNH_3HS$$

$$2RNH_2 + CO_2 + H_2O \rightarrow (RNH_3)_2CO_3$$

$$(RNH_3)_2CO_3 + H_2O \rightarrow 2RNH_3HCO_3$$

There are also *solvent extraction* methods for producing low-sulfur and low-mineral matter coal, but hydrotreatment of the coal extract is also required. In these methods, the organic material is extracted from the inorganic material in coal. A study has indicated that solvent-refined coal will probably not penetrate the power generation industry on a large scale for several years to come.

In addition to hydrogen sulfide and CO_2, gas streams may contain other contaminants such as sulfur dioxide, mercaptans, and COS.

The presence of these impurities may eliminate some of the sweetening processes since some processes will remove large amounts of acid gas but not to a sufficiently low concentration. On the other hand, there are those processes that are not designed to remove (or are incapable of removing) large amounts of acid gases whereas they are capable of removing the acid gas impurities to very low levels when the acid gases are there in low-to-medium concentrations in the gas stream.

Many different methods have been developed for CO_2 and hydrogen sulfide removal, some of which are briefly discussed later. Concentrates of hydrogen sulfide obtained as by-products of gas desulfurization are often converted by partial oxidation to elemental sulfur (Claus process).

For large gas volumes containing high concentrations of CO_2 mixed with hydrogen sulfide, a likely but by no means unique sequence of treatments is possible (Probstein and Hicks, 1990). Most of the CO_2 and hydrogen sulfide are removed in a regenerable liquid absorbent, which is continuously circulated, and the final traces of hydrogen sulfide, which, for example in a refining scenario, might poison processing catalysts, are removed in a solid adsorbent that can be regenerated or discarded. Off-gas from the solvent-section of the process may be treated in a Claus unit for the recovery of sulfur. The final cleanup of the Claus plant off gas, usually referred to as "tail gas," can be done by a direct conversion process.

23.4 PARTICULATE MATTER REMOVAL

The detrimental effects of particulate matter on the atmosphere have been of some concern for several decades. In fact, the total output of particulate matter into the atmosphere has increased in Europe since medieval times (Brimblecombe, 1976), and, although the sources are various, there is special concern because of the issue of particulate matter from fossil fuel use (Cawse, 1982). Species such as mercury, selenium, and vanadium, which can be ejected into the atmosphere from fossil fuel combustion (Kothny, 1973; Lakin, 1973; Zoller et al., 1973), are particularly harmful to the flora and fauna mercury. Thus, there is the need to remove such materials from gas streams that are generated during fossil fuel processing.

There are many types of particulate collection devices in use, and they involve a number of different principles for the removal of particles from gas streams (Licht, 1988). However, the selection of an appropriate particle removal device must be based upon equipment performance as anticipated/predicted under the process conditions. To enter into a detailed description of the various devices available for particulate removal is well beyond the scope of this text. However, it is essential for the reader to be aware of the equipment available for particulate removal and the means by which this might be accomplished.

23.4.1 CYCLONES

Cyclones are low-cost particle collectors that have many potential applications in coal gasification systems; however, they have low efficiency for collecting particles smaller than 10 μm but above this size collection efficiency can be at least 90%. Conventional applications for cyclones include use as precleaners, entrainment separators, and for controlling dust emissions from coal grinding and pulverizing (Chapter 6).

Cyclone collectors utilize the principle of centrifugal force to separate particulates from a gas stream; vortex flow is induced by the design of the gas inlet duct (Figure 23.4). The main vortex is characterized by axial flow away from the gas inlet and radial flow outward from the edge of the cyclone body. The central core has the same rotary direction, but the axial and radial velocity components are in the opposite direction to that of the main vortex.

The major requirements of the dust discharge are that upward gas flow into the cyclone body should be minimal while continuous discharge of the dust is maintained. This is often accomplished by the inclusion of baffles or straightening vanes at the discharge to suppress the vortex at this point, thus minimizing the up-flow.

23.4.2 ELECTROSTATIC PRECIPITATORS

Electrostatic precipitators (Parker and Calvert, 1977; White, 1977) are efficient collectors of fine particulates matter and are capable of reducing the amount of submicron particles by 90%, or more; they have the capability of collecting liquid mists as well as dust.

The basic components of a precipitator are the discharge (or corona) electrode, the collection electrode (plates), and the precipitator shell. Other components are the dust hopper, high-voltage equipment, and gas distributors. Suspended particles are charged by collisions with the negative

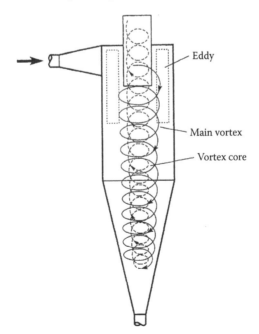

FIGURE 23.4 Vortex and eddy flows in a cyclone.

ions passing through the zone between the corona and the collection electrode. Movement of the particles to the collection electrode is governed by the interaction of the electric field and charged particles in the moving gas stream.

23.4.3 GRANULAR-BED FILTERS

Granular-bed filters comprise a class of filtration equipment that is distinguished by a bed of separate, closely packed granules that serve as the filter medium. These devices are under study because of their potential ability to collect particulates at high temperature and pressure. Although eminently suitable for low-temperature cleanup of gas streams, an attractive feature of granular-bed filters is the potential for high-temperature cleanup of process gas. It is also conceivable that these filters can be used for simultaneous control of particulates and hydrogen sulfide (by adsorption) at high temperature because of their ability to use almost any material as the filter medium.

Granular bed filters are classified according to the method used to remove collected dust from the filter medium continuously moving, intermittently moving, or fixed-bed filters. Cleaning is required to prevent interstitial plugging, which can cause pressure drop. This requirement distinguishes granular bed filters from fluidized-bed systems, which utilize the product gas to keep the granules in motion and prevent plugging.

Continuously moving systems may be arranged in a cross-flow configuration in which the gas passes horizontally through the granular layer while the granules and collected dust move continuously downward and are collected at the bottom. The dust and granules are then separated, and the cleaned granules returned to the bed. In the intermittently moving-bed concept, the bed is stationary during filtration. Accumulated dust and the surface layer of granules are removed from the panel by a backwash pulse and replaced by fresh granules from the overhead hoppers. Fixed-bed systems use either backwash air and/or mechanical agitation to remove collected dust.

23.4.4 WET SCRUBBERS

A wet scrubber is a simple method to clean exhaust air or exhaust gas and remove toxic or smelling compounds. In the flue gas scrubber, the gas gets in close contact with fine water drops in a cocurrent or counter current flow. This method is more effective when the water drop size gets smaller and the total surface between water or washing fluid and the gas gets larger. The water or washing fluid is recirculated normally in order to save water and reduce the amount of waste water.

A wet scrubber can also be combined with other flue gas-cleaning methods. Gases can be first cleaned by a washer, then treated by a thermal reactor, and then treated again by a wet scrubber.

The following components can be removed from coal gas: (1) water-soluble substances will be dissolved, (2) dust will be precipitated, (3) chemicals that can be hydrolyzed are decomposed, and (4) steam will be condensed. The result is decontamination, detoxification, dust removal, and dehumidification of the coal gas, as well as removal of many odiferous constituents of the gas.

By dissolving water-soluble components, the water or washing liquid will be contaminated in many cases. The dissolved components are frequently acid or basic chemicals (such as hydrogen chloride, NOx, and SOx). An optional neutralization unit, installed in the wet scrubber, is often required to keep the pH value of the washing liquid and the waste water at a neutral level. Furthermore, absorption of acid components is improved by using basic washing liquid, and removal of basic chemicals is more effective by using acid washing liquid.

When hot raw gas is to be cooled directly or if it has already been cooled in a waste-heat recovery system, it will normally be treated in a wet scrubber. Wet scrubbers are devices that utilize gas/liquid contacting to cool the gas stream, condense high-boiling hydrocarbons, dissolve some constituents, and separate particles from gas streams. There are many different wet scrubber designs, but all utilize similar mechanisms.

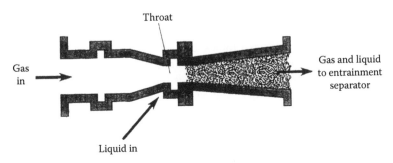

FIGURE 23.5 A Venturi scrubber.

A type of wet scrubber that has been widely applied to coal gas is the Venturi scrubber (Figure 23.5); gas streams are passed through a tube to contact with water, which is added at the throat. The throat promotes intimate mixing of the gas and liquid. The liquid, which forms droplets ranging in size from 100 to 1000 μm, collects particulates mainly by inertial impaction.

Other wet scrubber designs that might be applied to coal gas are plate scrubbers (sieves, bubble caps, and impingement plates), massive packing (rings, saddles), fibrous packing plastic, spun glass, fiber glass, steel wool), centrifugal (cyclone), and directional baffles (louvres, zigzags, disk, and donut; Sundberg, 1974; Strauss, 1975; Semrau, 1977).

23.5 ACID GAS REMOVAL

A variety of processes are commercially available for the removal of acid gas from gas streams (Figure 23.6) (Probstein and Hicks, 1990), and the processes generally fall into one of several categories. But, moreover, several factors control the choice of an acid gas removal process, and these are (1) gas flow rate, (2) concentration of acid gases in the gas stream, and (3) the necessity to remove CO_2 as well as hydrogen sulfide.

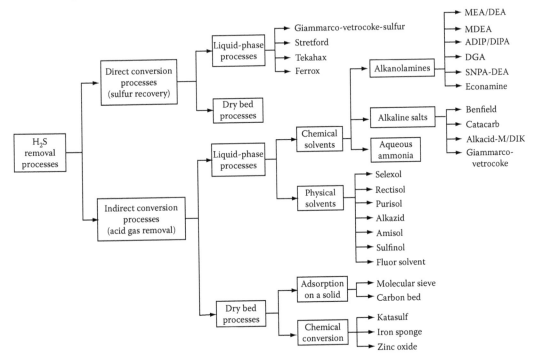

FIGURE 23.6 Processes for hydrogen sulfide removal.

The removal of acid gases from gas streams can be generally classified into two categories: (1) chemical absorption processes and (2) physical absorption processes. There are several such processes that fit into these categories (Tables 23.3 and 23.4); the features of the individual process may vary (Table 23.4; van den Berg and de Jong, 1980; Bodle and Heubler, 1981).

In more general and simple process terms, acid gas removal is considered to be hydrogen sulfide and CO_2 removal; the removal of SOx and NOx is often achieved by contact of the gas with

TABLE 23.3

Simple Classification System for Acid Gas Removal Processes

Chemical Absorption (Chemical Solvent Processes)	Physical Absorption (Physical Solvent Processes)
Alkanolamines	
MEA	Selexol
SNPA:DEA (DEA)	Rectisol
UCAP (TEA)	Sulfinol[a]
Selectamine (MDEA)	
Econamine (DGA)	
ADIP (DIPA)	
Alkaline salt solutions	
Hot potassium carbonate	
Catacarb	
Benfield	
Giammarco-Vetrocoke	
Nonregenerable	
Caustic	

[a] A combined physical/chemical solvent process.

TABLE 23.4

Summary of the Various Acid Gas Removal Processes

Feature	Chemical Adsorption		Physical Absorption
	Amine Processes	Carbonate Processes	
Absorbents	MEA, DEA, DGA, MDEA	K_2CO_3, K_2CO_3 + MEA K_2CO_3 + DEA, K_2CO_3 + arsenic trioxide	Selexol, purisol, rectisol
Operating pressure, psi	Insensitive to pressure	>200	250–1000
Operating temp, °F	100–400	200–250	Ambient temperature
Recovery of absorbents	Reboiled stripping	Stripping	Flashing, reboiled, or steam stripping
Utility cost	High	Medium	Low–medium
Selectivity, H_2S, CO_2	Selective for some amines (MDEA)	May be selective	Selective for H_2S
Effect of O_2 in the feed	Formation of degradation products	None	Sulfur precipitation at low temperature
COS and CS_2 removal	MEA not removed; DEA slightly removed; DGA removed	Converted to CO_2 and H_2S and removed	Removed
Operating problems	Solution degradation; foaming; corrosion	Column instability; erosion; corrosion	Absorption of heavy hydrocarbons

an alkaline solution as used in various processes that offer variations in the relative selectivity for hydrogen sulfide, CO_2, and hydrocarbons (Bodle and Heubler, 1981; Wesch, 1992; Mokhatab et al., 2006; Speight, 2007, 2008).

Most of the treating agents rely upon physical absorption and chemical reaction. When only CO_2 is to be removed in large quantities, or when only partial removal is necessary, a hot carbonate solution or one of the physical solvents is the most economical selection. The sulfinol solvent (a mixture of an aqueous amine, chemical solvent, with sulfolane, physical solvent) is reported to be particularly advantageous (Taylor et al., 1991). Hydrogen sulfide may be removed solely by the use of several processes (Grosick and Kovacic, 1981; Mokhatab et al., 2006; Speight, 2007, 2008).

Most sulfur removal processes concentrate on removing the hydrogen sulfide because, over the years, it has been considered the greatest health hazard and also because it is corrosive, particularly when water is present. With increasing emphasis on eliminating or minimizing sulfur discharge to the atmosphere, attention in the newer and more effective processes is turning to removal of other sulfur compounds from gas. Generally, specifications dictate a very low hydrogen sulfide content to be transmitted by pipeline.

A number of processes are available for the removal of hydrogen sulfide from gas streams. These processes can be categorized as those based on physical absorption, adsorption by a solid, or chemical reaction.

Physical absorption processes suffer from the fact that they frequently encounter difficulty in reaching the low concentrations of hydrogen sulfide required in the sweetened gas stream. However, there are processes that, with proper attention and care to regeneration cycles, can meet this specification.

Solid bed adsorption processes suffer from that fact that they are generally restricted to low concentrations of hydrogen sulfide in the entering sour gas stream. The development of a short-cycle adsorption unit for hydrogen sulfide removal might help remove part of this low-concentration restriction for the solid bed absorption processes.

In general, chemical processes are able to meet the regulated hydrogen sulfide levels with little difficulty. However, they suffer from the fact that, in general, a material that will react satisfactorily with hydrogen sulfide will also react with CO_2.

The most well-known hydrogen sulfide removal process is based on the reaction of hydrogen sulfide with iron oxide (often also called the Iron Sponge process or the Dry Box method) in which the gas is passed through a bed of wood chips impregnated with iron oxide after which the bed is regenerated by passage of air through the bed:

$$Fe_2O_3 + 3H_2S \rightarrow Fe_2S_3 + 3H_2O$$

$$2Fe_2S_3 + 3O_2 \rightarrow 2Fe_2O_3 + 6S$$

The bed is maintained in a moist state by circulation of water or a solution of soda ash.

The method is suitable only for small-to-moderate quantities of hydrogen sulfide. Approximately 90% of the hydrogen sulfide can be removed per bed, but bed clogging by elemental sulfur occurs, and the bed must be discarded and the use of several beds in series is not usually economical.

Removal of larger amounts of hydrogen sulfide from gas streams requires continuous processes, such as the Ferrox process or the Stretford process (Mokhatab et al., 2006; Speight, 2007, 2008). The Ferrox process is based on the same chemistry as the iron oxide process except that it is fluid and continuous. The Stretford process employs a solution containing vanadium salts and anthraquinone disulfonic acid (Maddox, 1974).

Most hydrogen sulfide removal processes involve fairly simple chemistry (Table 23.5) with the potential for regeneration with "return" of the hydrogen sulfide. However, if the quantity involved

TABLE 23.5
Chemistry of Hydrogen Sulfide Removal from Gas Streams

Name	Reaction	Regeneration
Caustic soda	$2NaOH + H_2S \rightarrow NaS + 2H_2O$	None
Lime	$Ca(OH)_2\, H_2S \rightarrow CaS + 2H_2O$	None
Iron oxide	$Fe_2O_3 + 3H_2S \rightarrow Fe_2S_3 + 3H_2O$	Partly by air
Seaboard	$Na_2CO_3 + H_2S \rightleftharpoons NaHCO_3 + NaHS$	Air blowing
Thylox	$Na_4As_2S_5O_2 + H_2S \rightarrow Na_4As_2S_4O + H_2O$	Air blowing
	$Na_4As_2S_4O + 1/2O_2 \rightarrow Na_4As_2S_5O_2 + S$	
Girbotol	$2RNH_2 + H_2S \rightleftharpoons (RNH_3)_2S$	Steaming
Phosphate	$K_3PO_4 + H_2S \rightleftharpoons KHS + K_2HPO_4$	Steaming
Phenolate	$NaOC_6H_5 + H_2S \rightleftharpoons NaHS + C_6H_5OH$	Steaming
Carbonate	$Na_2CO_3 + H_2S \rightleftharpoons NaHCO_3 + NaHS$	Steaming

does not justify installation of a sulfur recovery plant, usually a Claus plant, it is will be necessary to select a process that produces elemental sulfur directly.

$$3H_2S + 3O_2 \rightarrow 2SO_2 + 2H_2O$$

$$2H_2S + SO_2 \rightarrow 3S + 2H_2O$$

The conversion can be achieved by reacting the hydrogen sulfide gas directly with air in a burner reactor if the gas can be burnt with a stable flame.

Other equilibria that should be taken into account are the formation of sulfur dimer, hexamer, and octamer as well as the dissociation of hydrogen sulfide:

$$2H_2S + O_2 \rightarrow 2S + 2H_2O$$

$$H_2S \rightarrow S + H_2$$

COS and CS2 may be formed, especially when the gas is burned with less than the stoichiometric amount of air in the presence of hydrocarbon impurities or large amounts of CO_2.

Equilibrium data on the reaction between hydrogen sulfide and sulfur dioxide indicate that the equilibrium conversion is almost complete (100%) at relatively low temperatures and diminishes at first at higher temperatures, in accordance with the exothermic nature of the reaction. A further rise in temperature causes the equilibrium conversion to increase again. This is a consequence of the dissociation of the polymeric sulfur into monatomic sulfur.

Catalysis by alumina is necessary to obtain good equilibrium conversions: the thermal Claus reaction is fast only above 500°C (930°F) (Dowling et al., 1990; Chou et al., 1991). There is also a lower temperature limit, which is not caused by low rates but by sulfur condensation in the catalyst pores and consequent deactivation of the catalyst. The lower limit at which satisfactory operation is still possible depends on the pore size and size distribution of the catalyst; with alumina-based catalysts having wide pores, the conversion proceeds satisfactorily at ca. 200°C (390°F) (Lagas et al., 1989; Luinstra and d'Haiene, 1989).

In all Claus process configurations, several conversion steps in adiabatic, that is, cheap, reactors are used, with intermittent and final condensation of the sulfur produced. There are three main process forms, depending on the concentration of hydrogen sulfide and other sulfur compounds in the gas to be converted, that is, the straight-through, split-flow oxidation process.

The straight-through process is applicable when the gas stream contains more than 50% v/v hydrogen sulfide. Feed gases of this type can be burnt with the stoichiometric amount of air to give sulfur. The combustion reactor is followed by a combined waste heat boiler and sulfur condenser from which liquid sulfur and steam are obtained. The gases are then reheated by in-line fuel combustion to the temperature of the first catalytic convertor, which is usually kept at about 350°C (660°F) to decompose any COS and any CS2 formed in the combustion step. A second catalytic convertor, operating at as low a temperature as possible, is also employed to obtain high final conversions.

Caution is necessary to avoid condensation of an aqueous phase in the system because of the extreme corrosive nature of the liquid phase. Another operating issue concerns mist formation, a phenomenon which occurs very readily when condensing sulfur, and a series of demisters are necessary to prevent this. Residual sulfur is converted to sulfur dioxide by incineration of the tail gas from the process to prevent emission of other sulfur compounds and to dilute the effluent to reduce ground level sulfur dioxide concentrations.

Molecular sieves and membranes have been undergoing development for the removal of hydrogen sulfide and CO_2 from gas streams, especially when the amount of the acid gas(es) is low (Benson, 1981; Chiu, 1990; Winnick, 1991; Mokhatab et al., 2006; Speight, 2007, 2008). The most appropriate use of the sieves and the membranes would be the use of the sieve to selectively remove hydrogen sulfide, without removing much of the CO_2, and/or the use of membranes permeable to hydrogen sulfide but not to CO_2.

23.6 REMOVAL OF SULFUR-CONTAINING GASES

Since the time when English kings recognized that the burning of coal could produce noxious fumes (Chapters 14 and 15), there has been a series of attempts (not always continuous) to mitigate the amounts of noxious gases entering the atmosphere, the least of which have been attempts to reduce the amount of sulfur oxide(s) (particularly sulfur dioxide) released to the environment.

Historically, the first method for removing sulfur dioxide from flue gases consisted of simple water scrubbing of the flue gas to absorb sulfur dioxide into solution (Plumley, 1971), and the method was first used in London in during the 1930s. Since then, various regulatory organizations in many countries have set standards for sulfur dioxide emissions, which must be met immediately, or in the very near future.

Sulfur dioxide represents a high percentage of the sulfur oxide pollutants generated in combustion. The removal of the sulfur dioxide from the combustion gases before they are released to the stack is essential, and a considerable number of procedures exist for flue gas desulfurization (FGD). These procedures may be classified as wet or dry (Mokhatab et al., 2006; Speight, 2007, 2008) depending on whether a water mixture is used to absorb the sulfur dioxide or whether the acceptor is dry.

There are a variety of processes that are designed for sulfur dioxide removal from stack gas, but scrubbing process utilizing limestone ($CaCO_3$) or lime [$Ca(OH)_2$] slurries has received more attention than other stack gas scrubbing processes. Attempts have been made to use dry limestone or dolomite ($CaCO_3$, $MgCO_3$) within the combustor as an "in situ" method for sulfur dioxide removal, thereby eliminating the wet sludge from wet processes. This involves injection of dry carbonate mineral with the coal followed by recovery of the calcined product along with sulfite and sulfate salts:

$$S_{coal} + O_2 \rightarrow SO_2$$

$$2SO_2 + O_2 \rightarrow 2SO_3$$

$$2CaCO_3 + SO_2 + SO_3 \rightarrow CaSO_3 + CaSO_4 + 2CO_2$$

The majority of the stack gas scrubbing processes are designed to remove sulfur dioxide from the gas streams; some processes show the potential for the removal of NOx. However, there is the current line of thinking that pursues the options that enable SOx and NOx to be controlled, at least as far as possible, by modification of the combustion process. Sulfur (as already noted) can be removed by injecting limestone, with the coal into a boiler while modifications of the combustion chamber, as well as methods of flame temperature regulation and techniques that lower combustion temperatures, such as injection of steam into the combustion region are claimed to reduce emissions of NOx.

The procedures can be classified further as regenerative or nonregenerative depending upon whether the chemical used to remove the sulfur dioxide can be regenerated and used again or whether the chemical passes to disposal. In the wet regenerative processes, either elemental sulfur or sulfuric acid is recovered. In the wet nonregenerative lime and/or limestone procedures, calcium sulfite sludge or calcium sulfate sludge is produced that is disposed. Wet scrubbing processes contact the flue gas with a solution or slurry for sulfur dioxide removal.

In a wet limestone process, the flue gas contacts a limestone slurry in a scrubber:

$$CaCO_3 + SO_2 \rightarrow CaSO_3 + CO_2$$

$$2CaCO_3 + O_2 \rightarrow 2CaSO_4$$

Both the sulfite and the sulfate absorb water of hydration to form the sulfite/sulfate sludge.

In a wet lime process, a gas stream containing the sulfur dioxide is reacted with a wet lime slurry to form the sulfite with subsequent conversion, by oxidation, to the sulfate:

$$Ca(OH)_2 + SO_2 \rightarrow CaSO_3 + H_2O$$

In the double alkali nonregenerative procedure, flue gas is scrubbed with a soluble alkali such as sodium sulfite, which is subsequently regenerated with lime to form insoluble calcium sulfite after which disposal of the calcium sulfite slurry occurs, and the spent absorbent, sodium bisulfite, is regenerated by thermal means:

$$Na_2SO_3 + SO_2 + H_2O \rightarrow 2NaHSO_3$$

$$2NaHSO_3 + Ca(OH)_2 \rightarrow Na_2SO_3 + 2H_2O + CaSO_3$$

$$2NaHSO_3 + \rightarrow Na_2SO_3 + SO_2 + H_2O$$

In all of these procedures, water exits the system as vapor in the flue gas.

In either the lime/limestone or sodium sulfite/lime scrubbing processes, the hydrated calcium sulfite and calcium sulfate can be of some environmental concern when the issue of disposal arises. This has, more than anything else, promoted efforts to develop alternate dry scrubbing procedures for the removal of sulfur dioxide. And the dry systems have the additional advantage of reducing the pumping requirements necessary for the wet systems.

Thus, the tendency is to advocate the use of dry lime scrubbing systems thereby producing a waste stream that can be handled by "conventional" fly ash removal procedures. There are also dry processes, such as the metal oxide processes, in which sulfur dioxide can be removed from gas streams by reaction with a metal oxide. These processes, which are able to operate at high temperatures (approximately 400°C [750°F]), are suitable for hot gas desulfurization without an energy-wasteful cooling step. The metal oxides can usually be regenerated by aerial oxidation to convert any metal sulfide(s) back to the oxide(s) or by the use of a mixture of hydrogen and steam.

In the dry desulfurization process, metal oxides are reduced by coal gasification gas, and the reduced metal oxides remove sulfur compounds from the gas and are then converted to sulfides. This method can be used more than once by letting the sulfides react with oxygen to release the sulfur contents as SO_2 to return them to metal oxides.

Membranes have found increasing industrial use in the past two decades (Porter, 1990; Hsieh, 1991; Ho and Sirkar, 1992) and have also been suggested as being appropriate for the separation of hydrogen sulfide and sulfur dioxide, and they are applicable to higher temperature conditions (McKee et al., 1991; Shaver et al., 1991; Winnick, 1991). The removal of COS from gas streams, especially those that are destined for the manufacture of synthesis gas, has also been investigated using the principle of hydrogenation:

$$COS + H_2 \rightarrow CO + H_2S$$

$$COS + H_2O \rightarrow CO_2 + H_2S$$

23.7   REMOVAL OF NITROGEN-CONTAINING GASES

The occurrence of nitrogen in natural gas can be a major issue if the quantity is sufficient to lower the heating value. Thus, several plants for the removal of nitrogen from the natural gas have been built, but it must be recognized that nitrogen removal requires liquefaction and fractionation of the entire gas stream, which may affect process economics. In many cases, the nitrogen-containing natural gas is blended with a gas having a higher heating value and sold at a reduced price depending upon the thermal value (Btu/ft; kJ/m).

Of equal interest is the occurrence of nitrogen compounds in gases produced by coal combustion. These compounds, the oxides, originate from the organically bound nitrogen in the coal:

$$2N_{coal} + O_2 \rightarrow 2NO$$

$$2NO + O_2 \rightarrow 2NO_2$$

NOx are formed during burning by oxidation, at the high temperatures, of the nitrogen in the fuel and in the air. This has given rise to the terminology "thermal" and "fuel" as a means of distinguishing between the two sources of NOx. But, be that as it may, NOx from whatever the sources are pollutants that must be removed from gas streams.

Indeed, many coal-fired boilers are being built with burners designed to reduce NOx formation by delaying fuel/air mixing or distributed fuel addition, thereby establishing fuel-rich combustion zones within the burner whereby the reduced oxygen level maintains a low level of NOx production (Slack, 1981; Wendt and Mereb, 1990). Other procedures employ ammonia to reduce the NOx by injection of ammonia and oxygen into the postcombustion zone:

$$4NO + 4NH_3 + O_2 \rightarrow 4N_2 + 6H_2O$$

Vanadium oxide/aluminum oxide and iron/chromium, as well as the systems based on iron oxide itself, have also been reported to be successful for the removal of NOx from gas streams.

Nitrogen compounds must be absorbed in several chemical and related processes, the most important of which is the absorption of nitrogen peroxide in water for the manufacture of nitric acid. Absorption of nitrous gases also takes place in the lead-chamber process used for the production of sulfuric acid, in the metallurgical industries where metals are treated with nitric acid, and in the purification of several tail gases.

At this point it is worth considering the production of nitric acid from nitric oxide (using simple chemistry), as might be envisaged in the formation of nitrous and nitric acids in an industrial setting or even in the atmosphere; this latter phenomenon would result in the deposition of acid rain. Thus, nitric oxide is oxidized to nitrogen peroxide, and dimerization of nitrogen peroxide gives nitrogen tetroxide after which combination of nitrogen peroxide and nitric oxide gives nitrogen trioxide. In the presence of water vapor, nitrogen trioxide can be hydrated to nitrous acid. In the liquid phase, the gross reaction equation of the formation of nitric acid is

$$2NO + O_2 \rightarrow 2NO_2$$

$$NO + NO_2 \rightarrow N_2O_3$$

$$N_2O_3 + H_2O \rightarrow HNO_2$$

Thus,

$$3NO_2 + H_2O \rightarrow 2HNO_3 + NO$$

REFERENCES

Argonne. 1990. Environmental Consequences of, and Control Processes for, Energy Technologies. Argonne National Laboratory. Pollution Technology Review No. 181. Noyes Data Corporation, Park Ridge, NJ, Chapter 11.

Benson, H.E. 1981. In *Chemistry of Coal Utilization,* Second Supplementary Volume, M.A. Elliott (Ed.). John Wiley & Sons, Inc., New York, Chapter 25.

Berkowitz, N. 1979. *Introduction to Coal Technology.* Academic Press, Inc., New York.

Bodle W.W. and Huebler, J. 1981. In *Coal Handbook.* R.A. Meyers (Ed.). Marcel Dekker, Inc., New York, Chapter 10.

Brimblecombe, P.J. 1976. *Air Pollution Control Association,* 26: 941.

Cawse, P.A. 1982. In *Environmental Chemistry.* H.M.J. Bowen (Ed.). The Royal Society of Chemistry, London, U.K., Vol. 2, Chapter 1.

Chiu, C.-H. 1990. *Hydrocarbon Processing,* 69(1): 69.

Chou, J.S., Chen, D.H., Walker, R.E., and Maddox, R.N. 1991. *Hydrocarbon Processing,* 70(4): 38.

Coal and Synfuels Technology. 1992. Puerto Rico considers outlawing coal, March 23, p. 5.

Dowling, N.I., Hyne, J.B., and Brown, D.M. 1990. *Industrial and Engineering Chemistry Research,* 29: 2332 and references cited therein.

Grosick, H.A. and Kovacic, J.E. 1981. In *Chemistry of Coal Utilization,* Second Supplementary Volume, M.A. Elliott (Ed.). John Wiley & Sons, Inc., New York, Chapter 18.

Hessley, R.K. 1990. In *Fuel Science and Technology Handbook.* J.G. Speight (Ed.). Marcel Dekker, Inc., New York.

Hessley, R.K., Reasoner, J.W., and Riley, J.T. 1986. *Coal Science.* John Wiley & Sons, Inc., New York.

Ho, W.S.W. and Sirkar, K. 1992. *Membrane Handbook.* Van Nostrand Reinhold, New York.

Houghson, R.V. 1966. *Chemical Engineering,* August 29: 71.

Hsieh, H.P. 1991. *Catalysis Reviews—Science and Engineering,* 33(1&2): 1.

Kothny, E.L. 1973. In *Trace Metals in the Environment.* E.L. Kothny (Ed.). Advances in Chemistry Series No. 123. American Chemical Society, Washington, DC, Chapter 4.

Lagas, J.A., Borboom, J., and Heijkoop, G. 1989. *Hydrocarbon Processing,* 68(4): 40.

Lakin, H.W. 1973. In *Trace Metals in the Environment.* E.L. Kothny (Ed.). Advances in Chemistry Series No. 123. American Chemical Society, Washington, DC, Chapter 6.

Licht, W. 1988. *Air Pollution Control Engineering.* Marcel Dekker, Inc., New York.

Luinstra, E.A. and d'Haene, P.E. 1989. *Hydrocarbon Processing,* 68(7): 53.

Maddox, R.N. 1974. *Gas and Liquid Sweetening.* Campbell Publishing Company, Norman, OK.

McKee, R.L., Changela, M.K., and Reading, G.J. 1991. *Hydrocarbon Processing,* 70(4): 63.

Mokhatab, S., Poe, W.A., and Speight, J.G. 2006. *Handbook of Natural Gas Transmission and Processing.* Elsevier, Amsterdam, the Netherlands.

Moran, J.M., Morgan, M.D., and Wiersma, J.H. 1986. *Introduction to Environmental Science*, 2nd edn. W.H. Freeman and Company, New York.

Parker, R. and Calvert, S. 1977. *High-Temperature and High-Pressure Particulate Control Requirements*. Report No. EPA-600/7–77-071. United States Environmental Protection Agency, Washington, DC.

Plumley, A.L. 1971. Combustion, p. 36 (October).

Porter, M.C. 1990. *Handbook of Industrial Membrane Technology*. Noyes Publications, Park Ridge, NJ.

Probstein, R.F. and Hicks, R.E. 1990. *Synthetic Fuels*. pH Press, Cambridge, MA.

Semrau, K.T. 1977. *Chemical Engineering*, 84(September 26): 87.

Shaver, K.G., Poffenbarger, G.L., and Grotewold, D.R. 1991. *Hydrocarbon Processing*, 70(6): 77.

Slack, A.V. 1981. In *Chemistry of Coal Utilization*. Second Supplementary Volume, M.A. Elliott (Ed.). John Wiley & Sons, Inc., New York, Chapter 22.

Speight, J.G. (Ed.). 1990. *Fuel Science and Technology Handbook*. Marcel Dekker, Inc., New York.

Speight, J.G. 1993. *Gas Processing: Environmental Aspects and Methods*. Butterworth Heinemann, Oxford, U.K.

Speight, J.G. 2000. *The Desulfurization of Heavy Oils and Residua*, 2nd edn. Marcel Dekker, Inc., New York.

Speight, J.G. 2007. *The Chemistry and Technology of Petroleum*, 4th edn. Taylor & Francis Group, Boca Raton, FL.

Speight, J.G. 2008. *Synthetic Fuels Handbook: Properties, Processes, and Performance*. McGraw-Hill, New York.

Strauss, W. 1975. *Industrial Gas Cleaning*, 2nd edn. Pergamon Press, Elmsford, NY.

Sundberg, R.E. 1974. *Journal of the Air Pollution Control Association*, 24(8):48.

Taylor, N.A., Hugill, J.A., van Kessel, M.M., and Verburg, R.P.J. 1991. *Oil and Gas Journal*, 89(33): 57.

Van den Berg, P.J. and de Jong, W.A. 1980. *Introduction to Chemical Process Technology*. Delft University Press, Delft, the Netherlands, Chapter VI.

Wendt, J.O.L. and Mereb, J.B. 1990. *Nitrogen Oxide Abatement by Distributed Fuel Addition*. Report No. DE92005212. Office of Scientific and Technical Information, Oak Ridge, TN.

Wesch, I.M. 1992. *Oil and Gas Journal*, 90(8): 58.

White, H.J. 1977. *Air Pollution Control Association*, 27(2): 114.

Winnick, J. 1991. *High-Temperature Membranes for and Separations*. Report No. DE92003115. Office of Scientific and Technical Information, Oak Ridge, TN.

Zoller, W.H., Gordon, G.E., Gladney, E.S., and Jones, A.G. 1973. In *Trace Metals in the Environment*. E.L. Kothny (Ed.). Advances in Chemistry Series No. 123. American chemical Society, Washington, DC, Chapter 3.

24 Chemicals from Coal

24.1 HISTORICAL ASPECTS

The birth of coal chemical industry first appeared in the late eighteenth century, and in the nineteenth century the complete system of coal chemical industry was set up. After entering the twentieth century, raw materials of organic chemicals were changed into coal from the former agricultural and forestry products, and then coal chemical industry became an important part of chemical industry. After World War II, the petrochemical industry saw rapid development, which weakened the position taken up by coal chemical industry by changing raw materials from coal to petroleum and natural gas. The organic matters and chemical structures of coal with condensed rings as their core units connected by bridged bonds can transform coal into various fuels and chemical products through hot working and catalytic processing.

Coal carbonization is the earliest and most important method. Coal carbonization is mainly used to produce cokes for metallurgy and some secondary products like coal gas, benzene, and methylbenzene. Coal gasification takes up an important position in chemical industry. City gas and varieties of fuel gases can be produced by coal gasification. The common role of low-temperature carbonization, direct coal liquefaction, and indirect coal liquefaction is to produce liquid fuels.

Thus, for many years, chemicals that have been used for the manufacture of such diverse materials as nylon, styrene, fertilizers, activated carbon, drugs, and medicine as well as many others have been made from coal (Gibbs, 1961; Serrurier, 1976; Wender, 1976; Pitt and Millward, 1979; Ayers, 1981; Matsunaga, 1987; Tirodkar and Belgoankar, 1991). These products will expand in the future as petroleum and natural gas resources become strained to supply petrochemical feedstocks, and coal becomes a predominant chemical feedstock once more (Spitz, 1989; Agreda, 1990; Song and Schobert, 1992). Although many traditional markets for coal tar chemicals have been taken over by the petrochemical industry, the position can change suddenly as oil prices fluctuate upward. Therefore, the concept of using coal as a major source of chemicals can be very real indeed.

The ways in which coal may be converted to chemicals include carbonization (Chapters 16 and 17), hydrogenation (Chapters 18 and 19), oxidation (Chapter 12), solvent extraction (Chapters 11, 18, and 19), hydrolysis (Chapter 12), halogenation (Chapter 12), and gasification (followed by conversion of the synthesis gas; Chapters 20 and 21). In some cases, such processing does not produce chemicals in the sense that the products are relatively pure and can be marketed as even industrial grade chemicals.

A complete description of the processes to produce all of the possible chemical products is beyond the scope of this text. In fact, the production of chemicals from coal has been reported in numerous texts; therefore, it is not the purpose of this text to repeat these earlier works. It is, however, the goal of this chapter to present indications of the extent to which chemicals can be produced from coal as well as indications of the variety of chemical types that arise from coal (e.g., see Lowry, 1945; Donath, 1963).

On the basis of the thermal chemistry of coal (Chapters 13 and 16), many primary products of coal reactions are high-molecular weight species, often aromatic in nature, that bear some relation to the carbon skeletal of coal. The secondary products (i.e., products formed by decomposition of the primary products) of the thermal decomposition of coal are lower molecular weight species but are less related to the carbon species in the original coal as the secondary reaction conditions become more severe (higher temperatures and/or longer reaction times) (Xu and Tomita, 1987).

In very general terms, it is these primary and secondary decomposition reactions of coal that are the means to produce chemical from coal. There is some leeway in terms of choice of the reaction conditions, and there is also the option of the complete decomposition of coal (i.e., gasification) and the production of chemicals from the synthesis gas (a mixture of carbon monoxide, CO, and hydrogen, H_2) produced by the gasification process (Chapters 20 and 21).

24.2 COAL TAR CHEMICALS

The coal carbonization industry was established initially as a means of producing coke (Chapter 16), but a secondary industry emerged (in fact, became necessary) to deal with the secondary or by-products (namely, gas, ammonia liquor, crude benzole, and tar) produced during carbonization (Table 24.1).

Coal tar is black or dark brown-colored liquid or a high-viscosity semisolid that is one of the by-products formed when coal is carbonized (Chapter 16). Coal tars are complex and variable mixtures of polycyclic aromatic hydrocarbons (PAHs), phenols, and heterocyclic compounds. Because of its flammable composition coal, tar is often used for fire boilers, in order to create heat. Before any heavy oil flows easily, they must be heated.

Coal tar, coal tar pitch, and coal tar creosote are very similar mixtures obtained from the distillation of coal tars. The physical and chemical properties of each are similar, although limited data are available for coal tar and coal tar pitch.

Coal tars are by-products of the carbonization of coal to produce coke and/or natural gas. Physically, they are usually viscous liquids or semisolids that are black or dark brown with a naphthalene-like odor. The coal tars are complex combinations of PAHs, phenols, heterocyclic oxygen, sulfur, and nitrogen compounds.

By comparison, coal tar creosotes are distillation products of coal tar. They have an oily liquid consistency and range in color from yellowish-dark green to brown. The coal tar creosotes consist of aromatic hydrocarbons, anthracene, naphthalene, and phenanthrene derivatives. At least 75% of the coal tar creosote mixture is PAHs. Unlike the coal tars and coal tar creosotes, coal tar pitch is a residue produced during the distillation of coal tar. The pitch is a shiny, dark brown to black residue that contains PAHs and their methyl and polymethyl derivatives as well as heteronuclear compounds.

As an aside, the nomenclature of the coal tar industry, like that of the petroleum industry (Speight, 2007), needs refinement and clarification. Almost any black, undefined, semisolid-to-liquid material is popularly, and often incorrectly, described as tar or pitch whether it be a manufactured product or a naturally occurring substance (Chapter 16). However, to be correct and to avoid any ambiguity, use of these terms should be applied with caution. The term "tar" is usually applied to the volatile and nonvolatile soluble products that are produced during the carbonization or destructive distillation (thermal decomposition with the simultaneous removal of distillate) of various organic materials. By way of further definition, distillation of the tar yields an oil (volatile organic products often referred to as benzole) and a nonvolatile pitch. In addition, the origin of the

---

TABLE 24.1
Products (% w/w) from Coal Carbonization

Product	Low Temperature	High Temperature
Gas	5.0	20.0
Liquor	15.0	2.0
Light oils	2.0	0.5
Tar	10.0	4.0
Coke	70.0	75.0

tar or pitch should be made clear by the use of an appropriate descriptor, that is, coal tar, wood tar, coal tar pitch, and the like.

Thus, the eventual primary products of the carbonization process (Chapter 16) are coke, coal tar, and crude benzole (which should not be mistaken for benzene although benzene can be isolated from benzole), ammonia liquor, and gas. The benzole fraction contains a variety of compounds, both aromatic and aliphatic in nature, and can be conveniently regarded as an analog of petroleum naphtha (Speight, 2007).

The yield of by-product tar from a coke oven is, on average, 8.5–9.5 U.S. gallons (32–36 L) per ton of coal carbonized, but the yield from a continuous vertical retort is ca. 15.5–19.0 U.S. gallons (60–75 L) per ton of coal carbonized. In low-temperature retorts, the yield of tar varies over the range of 19.0–36.0 U.S. gallons (75–135 L) per ton of coal.

Crude coal tar sometimes referred to as crude coke oven tar or simply coal tar is a byproduct collected during the carbonization of coal to make coke. Coke is used as a fuel and as a reducing agent in smelting iron ore in a blast furnace to manufacture steel and in foundry operations. Crude coal tar consists of a mixture of naturally occurring compounds containing H_2 and carbon that are common to all organic fuels including coal, petroleum, and wood.

Crude coal tar is a raw material, which is further distilled to produce various carbon products, refined tars, and oils used as essential components in the production of aluminum, rubber, concrete, plasticizers, coatings, and specialty chemicals. Crude coal tars have been processed in the United States since Koppers Company completed the first by-product coke ovens around 1912.

During the distillation of crude coal tar, light and medium weight oils are removed from the crude coal tar to produce various refined coal tar products. These light and medium oils represent 20%–50% by weight of the crude coal tar depending upon the refined product that is desired. Coal tar contains hundreds of chemical compounds that will have varying amounts of PAHs depending upon the source.

Refined tar-based coatings have a great advantage over asphalt in that it has better chemical resistance than asphalt coatings. Refined tar-based coatings hold up better under exposures of petroleum oils and inorganic acids. Another outstanding quality of refined tar-based coatings is their extremely low permeability to moisture and high dielectric resistance, both of which contribute to the corrosion resistance (Munger, 1984).

Coal tar is a complex mixture, and the components range from low-boiling, low-molecular-weight species, such as benzene, to high-molecular-weight polynuclear aromatic compounds. Similar classes of chemical compounds occur in the tars, usually with little regard to the method of manufacture, but there are marked variations in the proportions present in the tars due to the type of coal, the type of carbonizing equipment, and the method of recovery (Chapter 16).

Coke-oven tar contains relatively low proportions (ca. 3%) of tar acids (phenols), and vertical retort tars may contain up to 30% phenolic compounds. Moreover, the phenols in coke-oven tars mainly comprise phenol, methyl, and poly-methyl phenols (e.g., cresols and xylenols) and naphthols; those in vertical retort tar are mainly xylenols and higher-boiling phenols.

Coke-oven tars contain only minor quantities of nonaromatic hydrocarbons, while the vertical retort tars may have up to 6% of paraffinic compounds. Low-temperature tars are more paraffinic and phenolic (as might be expected from relative lack of secondary reactions) than are the continuous vertical retort tars. Coke-oven tars are comparatively rich in naphthalene and anthracene, and distillation is often the means by which various chemicals can be recovered from these particular products. On the other hand, another objective of primary distillation is to obtain a pitch or refined-tar residue of the desired softening point. If the main outlet for the pitch is as a briquetting (Chapter 17) or electrode binder, primary distillation is aimed at achieving a medium-soft pitch as product or for the production of road tar/asphalt (Yan, 1986).

In terms of composition, the compounds positively identified as pitch components consist predominantly of condensed polynuclear aromatic hydrocarbons or heterocyclic compounds containing from three to six rings. Some methyl and hydroxyl substituent groups have also been observed, and it is reasonable to assume that vertical retort pitches contain paraffinic constituents in addition (McNeil, 1966). Pitches are often characterized by solvent analysis, and many specifications quote limits for the amounts insoluble in certain solvents (Mallinson, 1956; Hoiberg, 1966).

Primary distillation of crude tar produces pitch (residue) and several distillate fractions, the amounts and boiling ranges of which are influenced by the nature of the crude tar (which depends upon the coal feedstock) and the processing conditions. For example, in the case of the tar from continuous vertical retorts, the objective is to concentrate the tar acids, (phenol, cresols, and xylenols) into carbolic oil fractions. On the other hand, the objective with coke oven tar is to concentrate the naphthalene and anthracene components into naphthalene oil and anthracene oil, respectively.

The products of tar distillation can be divided into refined products, made by the further processing of the fractions, and bulk products which are pitch, creosote, and their blends.

Coal tar light oil, or crude benzole, is similar in chemical composition to the crude benzole recovered from the carbonization gases at gas works and in coke-oven plants. The main components are benzene, toluene, and xylene(s) with minor quantities of aromatic hydrocarbons, paraffins, naphthenes (cyclic aliphatic compounds), and phenols, as well as sulfur and nitrogen compounds.

The first step in refining benzole is steam distillation that is employed to remove compounds boiling below benzene. To obtain pure products, the benzole can be distilled to yield a fraction containing benzene, toluene, and xylene(s) (BTX). Benzene is used in the manufacture of numerous products including nylon, gammexane, polystyrene, phenol, nitrobenzene, and aniline. On the other hand, toluene is a starting material in the preparation of saccharin, trinitrotoluene, and polyurethane foams. The xylenes present in the light oil are not always separated into the individual pure isomers since xylene mixtures can be marketed as specialty solvents.

Higher boiling fractions of the distillate from the tar contain pyridine bases, naphtha, and coumarone resins. Other tar bases occur in the higher boiling range, and these are mainly quinoline, isoquinoline, and quinaldine.

Pyridine has long been used as a solvent in the production of rubber chemicals, textile water-repellant agents, and in the synthesis of drugs. The derivatives 2-benzylpyridine and 2-aminopyridine are used in the preparation of antihistamines. Another market for pyridine is in the manufacture of the nonpersistent herbicides diquat and paraquat.

Alpha-picoline (2-picoline; 2-methylpryridine) is used for the production of 2-vinylpyridine, which, when copolymerized with butadiene and styrene, produces a product that can be used as a latex adhesive which is used in the manufacture of car tires. Other uses are in the preparation of 2-beta-methoxyethyl-pyridine (known as promintic, an anthelmintic for cattle) and in the synthesis of a 2-picoline quaternary compound (amprolium), which is used against coccidiosis in young poultry. Beta-picoline (3-picoline; 3-methylpryridine) can be oxidized to nicotinic acid, which, with the amide form (nicotinamide), belongs to the vitamin B complex; both products are widely used to fortify human and animal diets. Gama-picoline (4-picoline; 4-methylpyridine) is an intermediate in the manufacture of isonicotinic acid hydra-zide (isoniazide), which is a tuberculostatic drug. 2,6-Lutidine (2,6-dimethylpyridine) can be converted to dipicolinic acid, which is used as a stabilizer for hydrogen peroxide and peracetic acid.

Solvent naphtha and heavy naphtha are the mixtures obtained when the 150°C–200°C (300°F–390°F) fraction, after removal of tar acids and tar bases, is fractionated. These naphtha fractions are used as solvents.

The tar-acid-free and tar-base-free coke oven naphtha can be fractionated to give a narrow-boiling fraction (170°C–185°C [340°F–365°F]) containing coumarone and indene. This is treated with strong sulfuric acid to remove unsaturated components and is then washed and redistilled. The concentrate is heated with a catalyst (such as a boron fluoride/phenol complex) to polymerize the indene and part of the coumarone. Unreacted oil is distilled off, and the resins obtained vary from pale amber to dark brown in color. They are used in the production of flooring tiles and in paints and polishes.

Naphthalene and several tar acids are the important products extracted from volatile oils from coal tar. It is necessary to first extract the phenolic compounds from the oils and then to process the phenol-depleted oils for naphthalene recovery.

Tar acids are produced by the extraction of the oils with aqueous caustic soda at a temperature sufficient to prevent naphthalene from crystallizing. The phenols react with the sodium hydroxide to give the corresponding sodium salts as an aqueous extract known variously as crude sodium phenate, sodium phenolate, sodium carbolate, or sodium cresylate. The extract is separated from the phenol-free oils, which are then taken for naphthalene recovery.

Phenol is a key industrial chemical; however, the output of phenol from coal tar is exceeded by that of synthetic phenol. Phenol is used for the production of phenol-formaldehyde resins, while other important uses in the plastics field include the production of polyamides such as nylon, of epoxy resins, and polycarbonates based on bisphenol A and of oil-soluble resins from p-t-butyl and p-octyl phenols. Phenol is used in the manufacture of pentachlorophenol, which is used as a fungicide and in timber preservation. Aspirin and many other pharmaceuticals, certain detergents, and tanning agents are all derived from phenol, and another important use is in the manufacture of 2,4 dichlorophenoxyacetic acid (2,4-D), which is a selective weed killer.

Ortho-cresol is mainly used in making the selective weed killers 4-chloro-2-methyl-phenoxyacetic acid and the corresponding propionic (MCPP) and butyric (MCPB) acids as well as 2,4-dinitro-o-cresol, a general herbicide/insecticide. Para-cresol finds its largest use in the production of BHT (2,6-ditert-butyl-4-hydroxytoluene), an antioxidant. Meta/para-cresol mixtures are used in the production of phenoplasts, tritolyl phosphate plasticizers, and petroleum additives. Other outlets for cresylic acids are as agents for froth flotation, metal degreasing, as solvents for wire-coating resins, antioxidants, cutting oils, nonionic detergents, and disinfectants.

Naphthalene is probably the most abundant component in high-temperature coal tars. The primary fractionation of the crude tar concentrates the naphthalene into oils, which, in the case of coke-oven tar, contain the majority (75%–90%) of the total naphthalene. After separation, naphthalene can be oxidized to produce phthalic anhydride, which is used in the manufacture of alkyd and glyptal resins and plasticizers for polyvinyl chloride and other plastics.

The main chemical extracted on the commercial scale from the higher-boiling oils (bp 250°C, 480°F) is crude anthracene. The majority of the crude anthracene is used in the manufacture of dyes after purification and oxidation to anthraquinone.

Creosote is the residual distillate oils obtained when the valuable components, such as naphthalene, anthracene, tar acids, and tar bases have been removed from the corresponding fractions.

Coal tar creosote is a brownish-black/yellowish-dark green oily liquid with a characteristic sharp odor, obtained by the fractional distillation of crude coal tars. The approximate distillation range is 200°C–400°C (390°F–750°F). The chemical composition of creosotes is influenced by the origin of the coal and also by the nature of the distilling process; as a result, the creosote components are rarely consistent in their type and concentration.

Major uses for creosotes have been as a timber preservative, as fluxing oils for pitch and bitumen, and in the manufacture of lampblack and carbon black. However, the use of creosote as a timber preservative has recently come under close scrutiny, as have many other ill-defined products of coal processing. Issues related to the seepage of such complex chemical mixtures into the surrounding environment have brought an awareness of the potential environmental and health hazards related to the use of such chemicals. Stringent testing is now required before such chemicals can be used.

As a corollary to this section, where the emphasis has been on the production of bulk chemicals from coal, a tendency-to-be-forgotten item must also be included. That is the mineral ash from coal processes. Coal minerals are a very important part of the coal matrix and offer the potential for the recovery of valuable inorganic materials (Chapter 7). However, there is another aspect of the mineral content of coal that must be addressed and that relates to the use of the ash as materials for roadbed stabilization, landfill cover, cementing (due to the content of pozzolanic materials), and wall construction to mention only a few (Golden, 1987).

24.3 FISCHER–TROPSCH CHEMICALS

At the risk of some repetition, it is necessary to deal once again with the production of chemicals from coal by gasification followed by conversion of the synthesis gas mixture (CO and H_2) to higher molecular weight liquid fuels and other chemicals (Chapters 20 and 21) (Penner, 1987).

The synthesis of hydrocarbons from the hydrogenation of CO was discovered in 1902 by Sabatier and Sanderens who produced methane by passing CO and H_2 over nickel, iron, and cobalt catalysts (Hindermann et al., 1993). At about the same time, the first commercial H_2 from syngas produced from steam methane reforming was commercialized. Haber and Bosch discovered the synthesis of ammonia from H_2 and nitrogen in 1910, and the first industrial ammonia synthesis plant was commissioned in 1913. The production of liquid hydrocarbons and oxygenates from syngas conversion over iron catalysts was discovered in 1923 by Fischer and Tropsch. Variations on this synthesis pathway were soon to follow for the selective production of methanol, mixed alcohols, and iso-hydrocarbon products. Another outgrowth of Fischer–Tropsch synthesis (FTS) was the hydroformylation of olefins discovered in 1938.

The production of synthesis gas (syngas) involves reaction of coal with steam and oxygen (Chapters 20 and 21). The gas stream is subsequently purified (to remove sulfur, nitrogen, and any particulate matter) after which it is catalytically converted to a mixture of liquid hydrocarbon products. In addition, synthesis gas may also be used to produce a variety of products, including ammonia and methanol.

In principle, syngas can be produced from any hydrocarbon feedstock. These include natural gas, naphtha, residual oil, petroleum coke, coal, and biomass. The lowest cost routes for syngas production, however, are based on natural gas. The cheapest option is remote or stranded reserves.

The choice of technology for syngas production also depends on the scale of the synthesis operation. Syngas production from solid fuels can require an even greater capital investment with the addition of feedstock handling and more complex syngas purification operations. The greatest impact on improving gas-to-liquids plant economics is to decrease capital costs associated with syngas production and improve thermal efficiency through better heat integration and utilization. Improved thermal efficiency can be obtained by combining the gas-to-liquids plant with a power generation plant to take advantage of the availability of low-pressure steam.

The synthesis of hydrocarbons from CO and H_2 (synthesis gas; the FTS) is a procedure for the indirect liquefaction of coal (Storch et al., 1951; Batchelder, 1962; Dry, 1976; Anderson, 1984; Dry and Erasmus, 1987; Dry, 1988; Jones et al., 1992). This process is the only coal liquefaction scheme currently in use on a relatively large commercial scale; South Africa is currently using the Fischer–Tropsch process on a commercial scale in their SASOL complex, although Germany produced roughly 156 million barrels of synthetic petroleum annually using the Fischer–Tropsch process during the Second World War (Stranges, 1983).

24.3.1 PROCESS

Coal is converted to gaseous products at temperatures in excess of 800°C (1470°F), and at moderate pressures, to produce synthesis gas:

$$C + H_2O \rightarrow CO + H_2$$

The gasification may be attained by means of any one of several processes or even by gasification of the coal in place (underground, or in situ, gasification of coal (Chapters 20 and 21). The exothermic nature of the process and the decrease in the total gas volume in going from reactants to products suggest the most suitable experimental conditions to use in order to maximize product yields. The process should be favored by high pressure and relatively low reaction temperature.

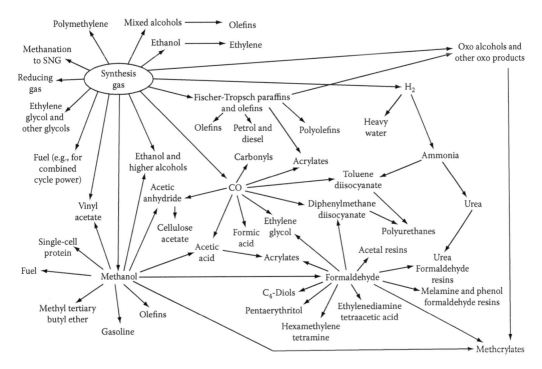

FIGURE 24.1 Routes to chemicals from methane via synthesis gas.

In practice, the Fischer–Tropsch reaction is generally carried out at temperatures in the range of 200°C–350°C (390°F–660°F) and at pressures of 75–4000 psi; the H_2/CO ratio is usually at ca. 2.2:1 or 2.5:1. Since up to three volumes of H_2 may be required to achieve the next stage of the liquids production, the synthesis gas must then be converted by means of the water-gas shift reaction) to the desired level of H_2 after which the gaseous mix is purified (acid gas removal, etc.) and converted to a wide variety of hydrocarbons:

$$CO + H_2O \rightarrow CO_2 + H_2$$

$$CO + (2n+1)H_2 \rightarrow C_nH_{2n+2} + H_2O$$

These reactions result primarily in low- and medium-boiling aliphatic compounds; present commercial objectives are focused on the conditions that result in the production of n-hydrocarbons as well as olefins and oxygenated materials (Figure 24.1).

24.3.2 Catalysts

Catalysts play a major role in syngas conversion reactions. In fact, fuels and chemicals synthesis from syngas does not occur in the absence of appropriate catalysts. The basic concept of a catalytic reaction is that reactants adsorb onto the catalyst surface and rearrange and combine into products that desorb from the surface. One of the fundamental functional differences between syngas synthesis catalysts is whether or not the adsorbed CO molecule dissociates on the catalyst surface.

For hydrocarbon and synthesis of higher molecular weight alcohols, dissociation of CO is a necessary reaction condition. For methanol synthesis, the CO molecule remains intact. H_2 has two roles in catalytic syngas synthesis reactions. In addition to being a reactant needed for hydrogenation of CO, it is commonly used to reduce the metalized synthesis catalysts and activate the metal surface.

A variety of catalysts can be used for the Fischer–Tropsch process, but the most common are the transition metals cobalt, iron, and ruthenium. Nickel can also be used but tends to favor methane formation (*methanation*).

Cobalt-based catalysts are highly active, although iron may be more suitable for low-H_2-content synthesis gases such as those derived from coal due to its promotion of the water-gas-shift reaction. In addition to the active metal, the catalysts typically contain a number of promoters, such as potassium and copper.

Group 1 alkali metals (including potassium) are poisons for cobalt catalysts but are promoters for iron catalysts. Catalysts are supported on high-surface-area binders/supports such as silica, alumina, and zeolites (Spath and Dayton, 2003). Cobalt catalysts are more active for FTS when the feedstock is natural gas. Natural gas has a high H_2 to carbon ratio, so the water-gas-shift is not needed for cobalt catalysts. Iron catalysts are preferred for lower quality feedstocks such as coal or biomass.

Unlike the other metals used for this process (Co, Ni, Ru), which remain in the metallic state during synthesis, iron catalysts tend to form a number of phases, including various oxides and carbides during the reaction. Control of these phase transformations can be important in maintaining catalytic activity and preventing breakdown of the catalyst particles.

Fischer–Tropsch catalysts are sensitive to poisoning by sulfur-containing compounds. The sensitivity of the catalyst to sulfur is greater for cobalt-based catalysts than for their iron counterparts.

Promoters also have an important influence on activity. Alkali metal oxides and copper are common promoters, but the formulation depends on the primary metal, iron versus cobalt (Spath and Dayton, 2003). Alkali oxides on cobalt catalysts generally cause activity to drop severely even with very low alkali loadings. C_{5+} and carbon dioxide selectivity increase while methane and C_2–C_4 selectivity decrease. In addition, the olefin to paraffin ratio increases.

24.3.3 PRODUCT DISTRIBUTION

The product distribution of hydrocarbons formed during the Fischer–Tropsch process follows an Anderson-Schulz-Flory distribution (Spath and Dayton, 2003):

$$W_n/n = (1-\alpha)^2 \alpha^{n-1}$$

where
   W_n is the weight fraction of hydrocarbon molecules containing n carbon atoms
   α is the chain growth probability or the probability that a molecule will continue reacting to form a longer chain

In general, α is largely determined by the catalyst and the specific process conditions.

According to the previous equation, methane will always be the largest single product; however, by increasing α close to one, the total amount of methane formed can be minimized compared to the sum of all of the various long-chained products. Increasing α increases the formation of long-chain hydrocarbons—waxes—which are solid at room temperature. Therefore, for the production of liquid transportation fuels, it may be necessary to crack the Fischer–Tropsch longer chain products.

It has been proposed that zeolites or other catalyst substrates with fixed sized pores can restrict the formation of hydrocarbons longer than some characteristic size (usually $n < 10$). This would tend to drive the reaction to minimum methane formation without producing the waxy products.

24.4 CHEMICALS FROM METHANE

The products of coal gasification are varied insofar as the gas composition varies with the system employed (Chapters 20 and 21). It is emphasized that the gas product must be first freed from any pollutants such as particulate matter and sulfur compounds before further use, particularly when the intended use is a water gas shift or methanation (Cusumano et al., 1978; Probstein and Hicks, 1990).

The commonly accepted approach to the synthesis of methane from coal is the catalytic reaction of H_2 and CO (Chapters 20 and 21):

$$CO + 3H_2 \rightarrow CH_4 + H_2O$$

A variety of metals have been used as catalysts for the methanation reaction; the most common and, to some extent, the most effective methanation catalysts appear to be nickel and ruthenium, with nickel being the most widely used (Seglin, 1975; Cusumano et al., 1978; Tucci and Thompson, 1979; Watson, 1980). The synthesis gas must be desulfurized before the methanation step since sulfur compounds will rapidly deactivate (poison) the catalysts (Cusumano et al., 1978).

The production of methane from coal does not depend entirely on catalytic methanation, and, in fact, a number of gasification processes use hydrogasification, that is, the direct addition of H_2 to coal under pressure to form methane (Anthony and Howard, 1976):

$$C_{coal} + 2H_2 \rightarrow CH_4$$

The H_2-rich gas for the hydrogasification process can be manufactured from steam by using the char that leaves the hydrogasifier. Appreciable quantities of methane are formed directly in the primary gasifier, and the heat released by methane formation is at a sufficiently high temperature to be used in the steam-carbon reaction to produce H_2 so that less oxygen is used to produce heat for the steam-carbon reaction. Hence, less heat is lost in the low-temperature methanation step, thereby leading to higher overall process efficiency.

Methane is a major raw material for many chemical processes and the potential number of chemicals that can be produced from methane is almost limitless (Fox, 1993; Solomons and Fryhle, 2011). Indeed, methane can be converted to a wide variety of chemicals (Figure 24.2; Table 24.2), in addition to being a source of synthesis gas. This leads to a wide variety of chemicals (Figure 24.1)

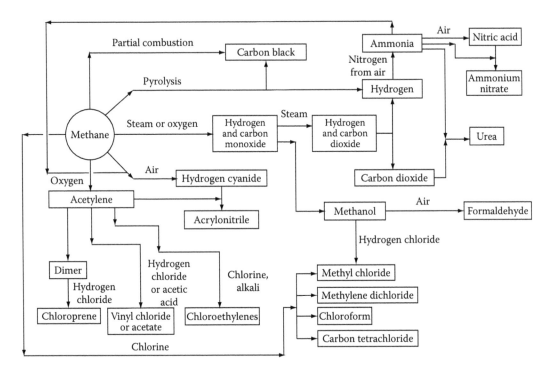

FIGURE 24.2 Production of chemicals from methane.

TABLE 24.2

Examples of Chemicals Produced from Methane

Basic Derivative/Source	Uses
Ammonia	Agricultural chemicals (as ammonia, salts, urea)
	Fibers, plastics
	Industrial explosives
Carbon black	Rubber compounding
	Printing ink, paint
Methanol	Formaldehyde (mainly for resins)
	Methyl esters (polyester fibers), amines, and other chemicals
	Solvents
Chloromethanes	Chlorofluorocarbons for refrigerants, aerosols, solvents, cleaners, grain fumigants
Hydrogen cyanide	Acrylonitrile
	Adiponitrile
	Methyl methacrylate

that involve chemistry (i.e., the chemistry of methane and other one-carbon compounds; Samsa and Hedman, 1984). In this aspect, the use of coal for chemicals is similar to the chemistry employed in the synthesis of chemicals from the gasification products of coal (Speight, 1990).

In the chemical industry, methane is the feedstock of choice for the production of H_2, methanol, acetic acid, and acetic anhydride. To produce any of these chemicals, methane is first made to react with steam in the presence of a nickel catalyst at high temperatures (700°C–1100°C [1290°F–2010°F]):

$$CH_4 + H_2O \rightarrow CO + 3H_2$$

The synthesis gas is then reacted in various ways to produce a wide variety of products.

In addition, acetylene is prepared by passing methane through an electric arc. When methane is made to react with chlorine (gas), various chloromethanes are produced: chloromethane (CH_3Cl), dichloromethane (CH_2Cl_2), chloroform ($CHCl_3$), and carbon tetrachloride (CCl_4). However, the use of these chemicals, however, is declining—acetylene may be replaced by less costly substitutes— and the chloromethanes are used less often because of health and environmental concerns.

It must be recognized that there are many other options for the formation of chemical intermediates and chemicals from methane by indirect routes, that is, where other compounds are prepared from the methane that are then used as further sources of petrochemical products (Figures 24.1 and 24.2).

In summary, methane can be an important source of petrochemical intermediates and solvents (Sasma and Hedman, 1984).

REFERENCES

Agreda, V.H. 1990. *Fuel*, 69: 132.
Anderson, R.B. 1984. In *Catalysis on the Energy Scene*, S. Kaliaguine and A. Mahay (Eds.). Elsevier Science Publishers, Amsterdam, the Netherlands, p. 457.
Anthony, D.B. and Howard, J.B. 1976. *AIChE Journal*, 22: 625.
Ayers, R.F. 1981. Coal utilization. In *Coal Handbook*, R.F. Meyers (Ed.). Marcel Dekker, Inc., New York, Chapter 7.
Batchelder, H.R. 1962. In *Advances in Petroleum Chemistry and Refining*, J.J. McKetta, Jr. (Ed.). Interscience Publishers, Inc., New York, Vol. V, Chapter 1.
Cusumano, J.A., Dalla Betta, R.A., and Levy, R.B. 1978. *Catalysis in Coal Conversion*. Academic Press, Inc., New York.

Donath, E.E. 1963. In *Chemistry of Coal Utilization*. Supplementary Volume. H.H. Lowry (Ed.). John Wiley & Sons, Inc., New York, Chapter 22.

Dry, M.E. 1976. *Industrial and Engineering Chemistry Product Research and Development*, 15(4): 282.

Dry, M.E. 1988. In *Methane Conversion*, B.M. Bibby, C.D. Chang, R.F. Howe, and S. Yurchak (Eds.). Elsevier Science Publishers, Amsterdam, the Netherlands, p. 447.

Dry, M.E. and Erasmus, H.B. de W. 1987. *Annual Review of Energy*, 12: 21.

Fox, J.M. III. 1993. *Catalysis Reviews—Science and Engineering*, 35: 169.

Gibbs, F.W. 1961. *Organic Chemistry Today*. Pergamon Books Ltd., London, U.K.

Golden, D.M. 1987. *Proceedings. Fourth Annual Pittsburgh Coal Conference*. University of Pittsburgh, Pittsburgh, PA, p. 302.

Hindermann, J.P., Hutchings, G.J., and Kiennemann, A. 1993. Mechanistic aspects of the formation of hydrocarbons alcohols from CO hydrogenation. *Catalysis Reviews—Science and Engineering*, 35: 1.

Hoiberg, A.J. 1966. *Bituminous Materials: Asphalts, Tars, and Pitches*. Coal tars and pitches. Interscience Publishers, Inc., New York, Vol. III.

Jones, C.J., Jager, B., and Dry, M.D. 1992. *Oil and Gas Journal*, 90(3): 53.

Lowry, H.H. (Ed.). 1945. In *Chemistry of Coal Utilization*. John Wiley & Sons, Inc., New York, Vol. II.

Mallinson, H. 1956. *40 Jahre Teerforschung*. Chemie und Technik Verlagsgesellschaft m.b.H., Heidelberg, Germany.

Matsunaga, E. 1987. *Proceedings. Fourth Annual Pittsburgh Coal Conference*. University of Pittsburgh, Pittsburgh, PA, p. 482.

McNeil, D. 1966. *Coal Carbonization Products*. Pergamon Press, London, U.K.

Munger, C.G. 1984. *Corrosion Prevention by Protective Coating*. NACE International, Houston, TX, p. 32.

Penner, S.S. 1987. *Proceedings. Fourth Annual Pittsburgh Coal Conference*. University of Pittsburgh, Pittsburgh, PA, p. 493.

Pitt, G.J. and Millward, G.R. (Eds.). 1979. In *Coal and Modern Coal Processing: An Introduction*. Academic Press, Inc., New York.

Probstein, R.F. and Hicks, R.E. 1990. *Synthetic Fuels*. pH Press, Cambridge, MA, Chapter 4.

Samsa, M.E. and Hedman, B.A. 1984. *Proceedings. International Gas Research Conference*. Gas Research Institute, Chicago, IL.

Seglin, L. (Ed.). 1975. In *Methanation of Synthesis Gas*, Advances in Chemistry Series No. 146. American Chemical Society, Washington, DC.

Serrurier, R. 1976. *Hydrocarbon Processing*, 55(9): 253.

Solomons, T.W.G. and Fryhle, C.B. 2011. *Organic Chemistry*, 10th edn. John Wiley & Sons, Inc. Hoboken, NJ.

Song, C. and Schobert, H.H. 1992. Preprints, *Division of Fuel Chemistry—American Chemical Society*, 37(2): 524.

Spath, P.L. and Dayton, D.C. 2003. *Preliminary Screening—Technical and Economic Assessment of Synthesis Gas to Fuels and Chemicals with Emphasis on the Potential for Biomass-Derived Syngas*. Report No. NREL/TP-510-3492. Contract No. DE-AC36-99-GO10337. National Renewable Energy Laboratory, Golden, CO.

Speight, J.G. 1990. In *Fuel Science and Technology Handbook,* J.G. Speight (Ed.). Marcel Dekker, Inc., New York, Chapter 37.

Speight, J.G. 2007. *The Chemistry and Technology of Petroleum*, 4th edn. Marcel Dekker, Inc., New York.

Spitz, P.H. 1989. *Chemtech*, February: 92.

Storch, H.H., Golumbic, N., and Anderson, R.B. 1951. *The Fischer Tropsch and Related Syntheses*. John Wiley & Sons, Inc., New York.

Stranges, A.N. 1983. *Journal of Chemical Education*, 60: 617.

Tirodkar, R.B. and Belgoankar, V.H. 1991. *Paintindia*, 41(2): 30.

Tucci, E.R. and Thompson, W.J. 1979. *Hydrocarbon Processing*, 58(2): 123.

Watson, G.H. 1980. *Methanation Catalysts*, Report ICTIS/TR09. International Energy Agency, London, U.K.

Wender, I. 1976. *Catalysis Reviews—Science and Engineering*, 14(1): 97.

Xu, W.-C. and Tomita, A. 1987. *Proceedings. International Conference on Coal Science*, J.A. Moulijn (Ed.). Elsevier Science Publishers, Amsterdam, the Netherlands, p. 629.

Yan, T.-Y. 1986. *Industrial & Engineering Chemistry Product Research Development*, 25: 637.

25 Environmental Aspects of Coal Use

25.1 INTRODUCTION

Coal is one of the many vital commodities that contributes on a large scale to energy supply and, unfortunately, to environmental pollution, including acid rain, the greenhouse effect, and *allegedly* global warming (global climate change) (Bell, 2011). Whatever the effects, the risks attached to the coal fuel cycle could be minimized by the introduction of new clean coal technologies (Chapter 22), remembering that there is no single substitute for coal fuel in the generation of energy.

Concerns about the impacts of coal on the environment and human health are not new; they may date from the first use of coal as a fuel in China in about 1100 BC. In the thirteenth century the concern about the *sulfurous air* in London attracted the attention of the British when Eleanor, wife of King Henry III and Queen of England, was obliged to leave the town of Nottingham where she had been staying during the absence of the King who was on a military expedition to Wales. The removal of the Queen's royal person from Nottingham was due to the troublesome smoke from the coal being used for heating and cooking. As a result, over the next several decades, a variety of proclamations were issued by Henry and by his son, Edward I, which threatened the population with the loss of various liberties, even life, if the consumption of coal was not seriously decreased and, in some cases, halted (Galloway, 1882).

Until industrialization, the amounts of coal being used were minuscule and the environmental and health problems were local. However, during the past 200 years, increasingly large amounts of coal have been required to satisfy the ever-growing demand for global energy.

In the 1970s, a legacy of abandoned mined areas and red-stained streams (somewhat reminiscent of the River Nile at the beginning of the 10 plagues) from acid mine drainage of mines and preparation areas in the United States spurred public concern about the environmental impacts of mining (Costello, 2003). These concerns led the federal regulations to guide reclamation and limit off-site impacts to the environment. Most industrialized countries regulate modern mining practices, but in those countries with a long mining history, it will take time to mitigate the legacy of past mining. This legacy includes physical disturbances to the landscape, subsidence and settlement above abandoned underground mines, flooding and increased sedimentation, polluted ground and surface waters, unstable slopes, long-burning fires, miners' safety, and public safety and land disturbance issues. In countries where such regulations do not exist, these issues are a continued concern.

As a result of environmental concerns, various technologies have been developed that capture potentially harmful elements and compounds before they can be emitted to the atmosphere.

This chapter focuses on the environmental and human health issues related to coal production (mining) and coal utilization. Emphasis is placed on those issues that are related to the petrographic, chemical, and mineralogical composition of coal.

The examination commences with the environmental issues associated with (1) coal mining (which, by inference includes coal in the seams), (2) coal preparation, (3) coal transportation and storage, and (4) followed by the emission-related environmental issues from commercial use of coal.

Coal itself is harmless and presents no risk when it is in situ where it was coalified and deposited millions of years ago. When involved in coal-related activities, however, its environmental impacts are deleterious if the coal is utilized in the wrong place at the wrong time and in the wrong amounts.

From the earliest days of coal extraction, potential hazardous situations were created by the coal-mining and coal-use technologies of the time. The consequences of past irresponsible practices have been inherited today. Recent industrial expansion and the population explosion worldwide are contributing further to the pollution being brought about by the coal industry.

A balance must be struck between industrial development and the energy required in order to build self-contained national economies. At one time, oil-fuel and then nuclear power were considered to be the answer to the world energy demands. These assumptions were proved to be inadequate because of (1) the unrest and armed hostile conflicts in the Middle East affecting oil supplies and (2) the catastrophic nuclear accidents in various parts of the world, which have justifiably or unjustifiably—it is not the purpose of this text to decide on the viability of energy from nuclear sources—posed serious questions on the viability and safety of the nuclear industry.

By comparison, coal offers substantial opportunities for diversification of energy supply. Coal reserves are abundant and it is well dispersed geographically. This makes it an invaluable source of energy and fundamental raw material for the generation of electrical power. Large quantities of coal, 3300 Mt, are mined and used in more than 60 countries around the world in order to fulfill the demands of industry.

However, the use of coal does pose serious environment questions some of which have been answered with satisfaction and others which have not been answered to the satisfaction of everyone.

Coal mining has always been regarded as a dangerous industry and to a certain degree will always be so, but today's new technologies offer the prospect of safer working conditions and improved productivity while at the same time meeting the new environmental standards required by legislative regulations. Criticism and judgment by the general public and environmental action groups on the coal industry are, in the main, based on emotion rather than factual evidence. In scientific circles, the reasons for and effects of certain kinds of pollution are still a cause of speculation. It is generally accepted that the proper application of modern technologies to mining and coal utilization can substantially limit its adverse environmental impacts, and coal's value and use in industry will continue in the foreseeable future.

25.2 PRODUCTION

Coal in the ground does not generally pose an environmental threat although coal mineralogy (Chapter 7) can influence groundwater properties and in fires in the coal seams can lead to subsidence, contamination of groundwater, and health and safety issues related to heat and emissions from the fires.

Some of the chemical processes that influence groundwater chemistry in coal-bearing strata include carbon dioxide production, silicate hydrolysis, pyrite oxidation, carbonate mineral dissolution, cation exchange, sulfate reduction, and the precipitation and dissolution of secondary minerals leading to contamination of the groundwater (Banaszak, 1980; Groenewold et al., 1981; Powell and Larson, 1985).

However, coal production and use have diverse impacts on the surrounding earth and atmosphere, generating various pollutants from such stages of the coal fuel cycle as (1) coal production, (2) coal preparation, (3) coal transportation and storage, and (4) coal utilization. In this chapter, these phases form the basis of discussion on the nature of coal pollutants and their consequent effects.

Irrespective of how it is extracted in mines and used in industry, coal produces three distinct types of pollutants: (1) gaseous, (2) liquid, and (3) solid substances, which generally demand quite different preventive or ameliorative measures. In this context, other impacts like noise, subsidence, waste disposal should also be classified as pollutants arising from coal use. Numerous methods have been devised to keep environmental standards at threshold limits and thus minimize pollution damage while at the same time improving worker productivity, coal quality, and accident prevention schemes.

Coal mining has always been, and remains, a dangerous occupation. Mine atmospheres are hazardous in terms of the health and safety of miners and, although technological advancements and legislation have led to improvement in the environmental and safety aspects in mines, pollutants are still produced at significant levels from coal excavation in both surface and underground mines.

25.2.1 Dust

The production of airborne dust particles is a major problem in underground mines where dust explosions with or without explosions due to released gases are the main concern (Colinet et al., 2010). In addition, dust is not solely a localized problem since fine particles can be transported to contaminate areas far from their source.

Dust is a by-product of blasting and (in open pit mines) earth-moving operations. The degree of risk is relative to the physical size of particulate matter, the humidity of air, and the velocity and direction of prevailing winds. Particle size and duration of exposure determine how far dust and droplets penetrate the respiratory tract. Inhaled fine dust remains in the alveoli, but all types of larger particles are removed by the filtering capacity of the respiratory system (Suess et al., 1985). The constituents of particulate matter (*particulates*) having a diameter less than 10 μm (particularly those in the range of 0.25–7 μm) may give rise to such respiratory diseases as chronic bronchitis and pneumoconiosis. If the dust contains silica particles, diseases such as silicosis (progressive nodular fibrosis) become a major threat to health.

On the surface, water spraying, site selection, and screening from winds are advantageous, while in underground working sites suppression by water, dilution and dispersion by ventilation and removal by filtration or electrostatic precipitation are advisable.

25.2.2 Fugitive Emissions

The geological processes of coal formation also produce methane (CH_4), and carbon dioxide (CO_2) may also be present in some coal seams. These are known collectively as seam gas, and remain trapped in the coal seam until the coal is exposed and broken during mining. Methane is the major greenhouse gas emitted from coal mining and handling. The major stages for the emission of greenhouse gases for both underground and surface coal mines are (1) mining emissions, (2) post-mining emissions, (3) low temperature oxidation of coal, and (4) spontaneous ignition of coal

Mining emissions result from the liberation of stored gas during the breakage of coal, and the surrounding strata, during mining operations.

Post-mining emissions results during subsequent handling, processing and transportation of coal as the coal normally continues to emit gas *even after it has been mined*, although more slowly than during the coal breakage stage. After mining has ceased, *abandoned coal mines* may also continue to emit methane.

Emissions from the *low temperature oxidation* of coal arise because once coal is exposed to oxygen in air, the coal oxidizes to produce carbon dioxide. However, the rate of formation of CO_2 by this process is low.

On occasion, when the heat produced by low temperature oxidation is trapped, the temperature rises and an active fire may result (*spontaneous combustion of coal*) and is an extreme case of coal oxidation. Spontaneous combustion is characterized by rapid reactions, sometimes visible flames and rapid carbon dioxide formation, and may be natural or anthropogenic.

25.2.3 Mine Waste Disposal

Significant volumes of earth must be displaced to mine coal; coal mines and the resulting rock waste can disrupt the environment. Furthermore, burning coal releases environmentally harmful chemical compounds into the air.

Waste material from deep mines derives from the sinking of shafts, roadways, and ventilation tunnels, and extraction of the coal seam. This is then hauled to the surface and dumped locally. Previously, when coal was dug by hand, much of the waste was stored underground in voids. However, the implementation of modern mechanical techniques, particularly with the introduction of coal cutting machines on the coal faces, has dramatically increased the proportion of waste material.

The enormous volumes of waste material, which are the by-product of both underground and surface excavation operations, are a major pollution issue and disposal of these solid wastes is the most controversial aspect of coal mining. The primary environmental damage of waste piles (mines tips) are not only noise and dust from moving vehicles but also groundwater contamination, leaching of toxic and acid pollutants, and loss of usable land. Reclamation to some degree restores the land, but reduction in the fertility of the soil and diminished ecological habitat are slow to recover. In fact, old mine tips, inherited from unregulated past practices of dumping mine waste, are a danger to local drainage systems and may have toxic effects on human health.

Another issue related to waste disposal arose because of the oxidation of pyrite that produces acidic compounds, which, with other toxic materials, can be leached into the local water supply. Simultaneously, heat produced from such chemical reactions led to spontaneous combustion of coal particles in the waste tips (Guney, 1968; McNay, 1971). The potential hazards from spontaneous combustion in the spoil heaps can be substantially reduced by controlled tipping, site selection, as well as compaction of waste.

Surface mining has a greater adverse effect on the surroundings than underground mining operations.

For example in strip mining operations, the overburden is removed and the volume produced may be on the order of 25 volumes of overburden for every volume of coal (Doyle, 1976). In addition, the mining operations (which use power shovels, draglines, and bucket-chain/bucket-wheel excavators) alter the topography of the surface and destroy all of the original vegetation and often lead to contamination of surface water and groundwater courses. Nevertheless, surface mining is not considered to be a major contributor to air pollution (Cummins and Given, 1973).

Strip mining has resulted in a great deal of damage to the landscape. Many strip mines have removed acres of vegetation and altered topographic features, such as hills and valleys, leaving soil exposed for erosion. Longwall mining, which allows the mine to collapse, results in widespread land subsidence (land sinking). Coal and rock waste, often dumped indiscriminately during surface and underground mining processes, weathers rapidly, producing *acid drainage*. Acid drainage contains sulfur-bearing compounds that combine with oxygen in water vapor to form sulfuric acid. In addition, weathering of coal mine waste can produce alkaline compounds, heavy metals, and sediments. Acid mine drainage, alkaline compounds, heavy metals, and sediment leached from mine waste into groundwater or washed away by rainwater can pollute streams, rivers, and lakes.

In the past, mine overburden or mine tipple was usually dumped into low-lying areas, often filling wetlands or other sources of water. This resulted in dissolution of heavy metals that seeped into both ground and surface water causing disruption to marine habitats and deteriorate drinking water sources. In addition, pyrite (FeS_2) can form sulfuric acid (H_2SO_4) and iron hydroxide [$Fe(OH)_2$] when exposed to air and water. When rainwater washes over these rocks, the runoff can become acidified affecting local soil environments, rivers, and streams (*acid mine drainage*).

Coal mining can still result in a number of adverse effects on the environment. For example, surface mining eliminates existing vegetation, displaces or destroys wildlife and habitat, degrades air quality, alters current land uses, and to some extent permanently changes the general topography of the area mined.

Area mining occurs on level ground, where workers use excavation equipment to dig a series of long parallel strips, or *cuts*, into the earth. The overburden is cleared from each cut, and the material (known as *spoil*) is stacked alongside the long trench. After the exposed coal is shoveled from the cut, workers dump the spoil back into the trench to help reclaim the mined area.

Once the mine is worked out and coal recovery operations cease, mined out areas are to be converted into productive agricultural land or restored to their former natural beauty. Potential recovery work is comprised of top-soil and sub-soil replacement, compaction, regrading, revegetation of the land, and chemical treatment and management of contaminated water resources. Recultivation necessitates the enrichment of soil for seeding and planting. Generally, land restoration reduces the potential for land destruction and pollution hazards.

25.2.4 Mine Water Drainage

Another significant source of pollution due to coal recovery derives from mine drainage in both currently active and abandoned worked-out mines.

Large quantities of fresh water are used underground for dust suppression, coal washing and cooling of equipment, particularly drilling machines. Additionally, during the extraction of coal, water enters the working sites from water-bearing strata or aquifers. This water picks up coal, shale and clay particles from mine roadways, and it also becomes contaminated with the soluble oxidation production of iron pyrites such as ferrous sulfate ($FeSO_4$), ferric sulfate [$Fe_2(SO_4)_3$], and aluminum sulfate [$Al_2(SO_4)_3$], and sulfuric acid (H_2SO_4).

However, mine tailing dumps produce acid mine drainage that can seep into waterways and aquifers, with serious consequences for ecosystems. If underground mine tunnels collapse, this can cause subsidence of land surfaces. During actual mining operations, methane may be released into the air thereby contributing to the *greenhouse effect*.

As a result, mining enterprises in many countries must secure government permits before mining for coal. In the United States, mining companies must submit plans detailing proposed methods for blasting, road construction, land reclamation, and waste disposal. New land reclamation methods, driven by stringent laws and regulations, require coal mining companies to restore strip-mined landscapes to nearly pre-mined conditions.

Disposal of excess water is an integral and essential phase of mining operations because pollutants in the waterways result in the reduction of the oxygen content of water leading to devastation of the aquatic life. With modem mining practices for controlling water flow, sumps and pools are provided where drainage water accumulates and where suspended coal, shale, and clay particles can ultimately be removed. Sedimentation, with or without the use of flocculants, is applied to process water in lagoons. Occasionally, if geological conditions are favorable, an adequate thickness of strata is left as a wall pillar between the aquifer and mine workings to prevent water flow into mines.

Atmospheric oxidation of iron pyrites (FeS_2) in coal and associated rocks and rapid dissolution of the resultant sulfur oxides and iron sulfates in percolating waters produce very acidic effluents (*acid mine drainage*) (Berkowitz, 1985). The extent and toxicity of the waste streams depend on coal characteristics, local rainfall patterns, topography, and site drainage features. Leaching of such waste could lead to an unacceptable level of contamination of surface and groundwater.

Remedies are mainly aimed at restricting water flows seeping through the porous structure, boreholes and fractures in the water-bearing strata. This can be brought about by sealing of boreholes, grouting of fracture zones, restricting free oxygen into the mine, caving-ins to fill the voids and diverting groundwater courses.

25.2.5 Subsidence

Subsidence is a costly economic impact of underground mining because it creates horizontal and vertical displacement of the surface, which generally causes structural damage to buildings, roads, and railroads as well as pipeline rupture.

The factors contributing to the ground movement and ultimate surface damage are thickness, dip and depth of the coal seam, angle of draw, the nature and thickness of the overburden, and the amount of support left in the *goaf* (*goff* or *gob*—part of a mine from which the mineral has been partially or wholly removed). In addition, seepage of methane through the cracks into the houses has resulted in accumulation and gas explosions causing excessive damage to property.

Mining companies are much more conscious of the problem since wide scale repair and/or replacement of surface structures could be disastrous for the economy of a mining company. In many countries, legislation demands payment of compensation to property owners for damage caused by subsidence. A reduced likelihood of subsidence will occur if the dimensions of the working areas are limited and permanent support practices are applied in the goaf. Surface destruction may be scaled down by backfilling the abandoned workings with stowing material or solid mine waste. However, such preventive methods to reduce ground movement are only one aspect of the problem since it must be achieved without excess loss of coal in the pillars and with minimum interference to normal mining operations.

Underground mine planning and design has as its goal an integrated mine systems design, whereby a mineral is extracted and prepared at a desired market specification and at a minimum unit cost within acceptable social, legal, and regulatory constraints. A large number of individual engineering disciplines contribute to the mine planning and design process, such that it is a multidisciplinary activity. Given the complexity of the mining system, planning assures the correct selection and coordinated operation of all subsystems, while design applies to the traditionally held engineering design of subsystems. A mining operation should be more correctly viewed as a system, because of the diversity of technological processes, facilities, and personal skills required, the large capital invested, the mutual relations that exist between subsystems, etc. The planning process necessitates a systems engineering approach suitable for complex design problem.

Planning must account for both environmental protection, beginning as early as the initial exploration, and for reclamation. It is critical that planning alleviate or mitigate potential impacts of mining for two key reasons: (1) the cost of environmental protection is minimized by incorporating it into the initial design, rather than performing remedial measures to compensate for design deficiencies, and (2) negative publicity or poor public relations may have severe economic consequences. From the start of the planning process, adequate consideration must be given to regulatory affairs. The cost of compliance may be significantly reduced when taken into account in the design or planning process, in a proactive manner, rather than being addressed on an ad hoc basis as problems develop or enforcement actions occur.

From the beginning of the mine design planning stage, data gathering and permitting, environmental considerations are important, although benefits from a strictly economic sense may be intangible. From exploration, where core holes must be sealed and the site reclaimed, through plan development, the impacts on the environment must be considered. These impacts include aesthetics, noise, air quality (dust and pollutants), vibration, water discharge and runoff, subsidence, and process wastes; sources include the underground and surface mine infrastructure, mineral processing plant, access or haul roads, remote facilities, etc. If mining will cause quality deterioration of either surface water or groundwater, remedial and treatment measures must be developed to meet discharge standards. The mine plan must include all the technical measures necessary to handle all the environmental problems from initial data gathering to the mine closure and reclamation of the disturbed surface area.

Reclamation plans include many of the following concerns: drainage control, preservation of top soil, segregation of waste material, erosion and sediment control, solid waste disposal, control of fugitive dust, regrading, and restoration of waste and mine areas. The plan must also consider the effects of mine subsidence, vibration (induced by mining, processing, transport, or subsidence), and impact on surface water and groundwater. These environmental items often dictate the economics of a planned mining operation and determine its viability.

The environmental aspects of underground mining are different to those encountered in surface mining operations. They have been considered to be different, even to the point of being considered (erroneously) to be of lesser importance. But it cannot be denied that underground mining can disturb aquifers either in the construction of the shaft or as a result of other influences such as formation disturbance due to subsidence. Indeed, it is only in the last three decades that subsidence has been elevated to the role of a major environmental issue.

In addition, the transportation of spoil or tipple to the surface from an underground mine where it is then deposited in piles or rows offers a new environmental hazard. The potential for leaching materials from spoil/tipple offers a mode of environmental contamination not often recognized that is of at least equal importance to the potential contamination from surface mining operations. Similarly the discharge of gaseous and liquid effluents from underground mining operations should be of at least equal concern to those arising from surface mining operations. Thus the environmental issues related to coal mining, be they underground or surface, operations are multifaceted (Argonne, 1990).

However, because of the general nature of surface mining operations, and not to fail to recognize the nature of the underground operations, the environmental issues related to coal recovery by surface mining operations often pose additional problems, recognizing of course that site preparation can, in both underground and surface operations, raise similar issues in relation to environmental disturbance and restoration.

Thus, as a matter of course, the mining industry must submit as much information as is necessary (and even more) about any mining projects before development may begin. In addition, the applicant may (will) also be requested to submit a "mining and reclamation plan," which will cover the details of the proposed mining sequence, construction of haulage ways, blasting, topsoil handling, placement of overburden, hydrological balance, and mining plans. In the United States, considerable detail is necessary (as declared by state law) and includes maps and as much information about the plans to grade and reclaim the mined lands.

The plans usually commence at the exploration stage of mine development and, as such, will also apply to underground mining operations. For example, reconnaissance explorations using the popular off-road vehicles create trails and primitive roads that can seriously affect the habitat of the flora and fauna. Drill holes may rupture or penetrate several separated water-bearing strata, causing degradation of the water resources through commingling of waters of different quality. Thus, in all aspects of exploration, up-front planning through geological surveys is a necessity.

The construction phase of mine development can also cause considerable environmental disturbance of a faunal nature. An influx of construction workers may severely overtax the local community and its facilities thereby creating many unpredictable socioeconomic problems. Site preparation, including construction of the tipple, the coal preparation plant, and the coal storage silos, erection of large earth-moving equipment, the construction of haulage roads, and grading can cause environmental changes that last for the life of the mine, and well beyond such as when wildlife patterns are altered. Indeed, when the construction of roads and railroads are considered, any changes made to the environment during development of the mine site can, therefore, be permanent and must be planned carefully as an integral part of the overall mine development plan.

Whether the objective is underground mining or surface mining, preparation of the site or stripping of the overburden from the coal seam is a trauma to the environment. Vegetation is removed, flora and microorganisms are disturbed and/or destroyed, soil and subsoil are removed, underlying strata are ruptured and displaced, hydrological systems may also suffer, and the surface is exposed to weathering (i.e., oxidation, which can result in chemical alteration of the mineral components), as well as other general topographic changes.

For example, there is the ever-present danger of waterways contamination during and after mining operations. Materials in the overburden (such as heavy metals and/or minerals) can be leached off, usually after erosion by rain or by runoff water from snowfall, into a nearby waterway and due caution must be used in such enterprises. However, in a more positive aspect, mining can also be a benefit to the local hydrology; spoil piles may act as "sponges" absorbing large quantities of water for future plant growth and use, retarding and diminishing runoff, making good aquifers for supplying flow to nearby streams on a steady basis.

Some forms of vegetation upon which wildlife depend for food and habitat may never be the same as it was prior to mining because of changes in the topography and drainage. If the habitat will not support certain forms of wildlife, animals will not return but will remain in the nearby undisturbed areas. But most mine operators plant vegetation and restore the terrain and make every attempt to ensure that the site becomes a superior habitat for wildlife (including humankind) than it was prior to mining. Indeed, many mined lands have, by design, become forests of valuable tree species in which animals and plants thrive. In addition, lands have also been restored to become pastures for domestic animals, with wildlife living on the periphery.

Mining can and does cause some long-lasting wildlife habitat impairment or changes of habitat. A few wildlife species may be unable to adjust to these changes, and they do not return to the restored lands but live elsewhere in the neighborhood. But it is more than likely that, given sufficient time after restoration of the mined lands, the long-term impact of mining is favorable for wildlife. Long-term impacts of surface mining on wildlife can be minimized by careful consideration of wildlife presence when making plans for mining.

Toxic minerals and substances exposed during removal of the overburden include acidic materials, highly alkaline materials, and dilute concentrations of heavy metals. These materials can have an adverse effect on the indigenous wildlife by creating a hostile environment (often through poisoning the waterways) and, in some cases, destruction of species. Thus, mine design should include plans to accommodate potentially harmful substances that are generated by weathering of spoil piles.

Another hazard that may become obvious by its presence is the spontaneous ignition and ensuing combustion of coal (Chapter 14). This usually occurs when coal is exposed to changing weather conditions and uncontrollable oxidation reaction, with the evolution of heat, ensues.

Artifacts of historic sites will be destroyed unless they have been systematically investigated before being disturbed by mining. The mining plan should include the provision for systematic archaeological studies of the area to be mined. Such studies by the mining industry often benefit the community and create an appreciation for historic values.

Finally, aesthetic rehabilitation of mined lands should, and is, being done so that the land is aesthetically more pleasing after mining than before. The removal of overburden, however, is disruptive to the landscape and aesthetically repugnant for a temporary period until the lands are restored. From the community and regulatory viewpoint, it is beneficial if the mining company includes consideration of aesthetic values in the mining plan.

25.2.6 Noise, Vibration, and Visibility

Noise from underground mining operations does not have a significant environmental impact (Bauer et al., 2006), being no more than is likely to arise from any industrial plant. However, residential communities in the locality of open-pit mines often find the quality of life diminished due to the effects of noise from mining operations, vibration from blasting, and continuous heavy traffic. Furthermore, clouds of dust may reduce visibility and increase haze promoting severe distress and annoyance.

These adverse conditions can be improved to a certain degree limiting excavation to one section of the mining area at any one time, siting of boundaries and screening banks, erection of barriers, and use of low-noise machinery fitted with effective exhaust silencers. The implementation of environmentally friendly techniques has served to establish better public relations between the mining companies and the local populace.

25.2.7 Reclamation

Surface and underground mining of coal can disrupt the pre-mining environment. Vegetative cover must often be removed. Geologic and soil profiles can be overturned so that rich soils are now buried and leachable, lifeless rock is on top. After mining is complete, reclamation of the land is necessary.

Thus, *reclamation*, as it applies to mining activities, means to rehabilitate the land to a condition where it can support *at least* the same land uses as it supported prior to the mining activity. In some countries, reclamation may not involve restoring the land to its exact condition prior to mining but only to a condition where it can support the same or other desired productive land use. This can involve reestablishment of the biological community (flora and fauna), or alternate biological community on the surface following mining. Reclamation can be done periodically or contemporaneously with mining operations.

Periodic reclamation may occur at area strip mines, where reclamation activities may follow the advancing strip pit by a specified distance. Alternatively, as may occur with a large open pit, reclamation may have to be deferred to the end of the mining operations where the pit expands with time or is needed to provide space for production and mining support traffic.

One increasingly popular reclamation technique is to develop the post-mining landscape into wetlands and marshes. Such landscapes become valuable wildlife habitat along the fly-ways of migratory birds and help replace swamps, marshes, and small lakes that are being drained in other places to create large tracts of farmland amenable to large cultivation machinery. Wetlands may be more economical to develop than rehabilitating and achieving specific yield targets for prime farm land.

Another reclamation technique involves planting natural grasses that were displaced by agriculture. While some grasses have taken longer to establish than agricultural stands, these natural species are more robust against insects and draught and require less fertilization and can create a natural habitat for wildlife and/or recreation.

In mountainous areas (such as those in the Eastern United States), mountain top removal sites may be the only flat land, other than river flood plains, suitable for development of homes and town infrastructure. Whatever the intent, reclamation plans must be considered in advance of mining operations. While the real added cost of reclaiming mined land cannot be denied, it is often most cost effective to integrate reclamation into the mining. Where a mining company has not considered such factors as the need for topsoil or burial of salty or acid forming materials, the result has been sites that are difficult to reclaim.

Much of the reclamation work is part of the overburden handling in the mining sequence, leaving little but final grading and planting operations for post-mining activities. With good reclamation planning, it is possible to have competitive, productive mining operations and unspoiled post-mining landscapes at the same time, as well as town infrastructure.

Successful reclamation is generally defined by regulatory standard, private or community agreement, or the goals of the mining company. Surface mining requires handling the entire surface and overburden mass above the coal seam—reclaiming and restructuring this mass offers a greater opportunity to select and structure the land for the needs and purposes of the human and natural community than would be available in almost any other way. Selecting a post-mining land use must be based on cultural factors in the surrounding community, subject to the constraints of the environment that are beyond human control.

The *choice of post-mining land use* (including the type of land ownership) is often a significant factor in reclamation. Land leased from a farmer would often be reclaimed for agricultural use, while government held land in recent years has often been used for parks, recreation, or wildlife habitat. Surrounding land use should be considered since land may often be reclaimed to its former use or to the predominant land use in the immediate area.

Efforts in recent years have been directed toward the use of recycled material in reclamation. Sewage sludge has been used to add organic material to substitute top soil (Bhumbla, 1992; Logan, 1992). The bacterial content of sewerage sludge is usually not a problem, provided that grazing and human exposure at the site is delayed for several weeks or months after treatment.

Use of coal combustion by-products in reclamation is becoming more common. Fly ash (use of which may have the added benefit of containing mineral nutrients lacking in many soils) can be used to break up and add texture to clay rich spoils and produce a more suitable substitute topsoil. Fluidized bed combustion fly ash is being tested as an alternative to lime reagents in mixing with acid producing spoils, treatment of acid mine drainage, and in strengthening refuse embankments.

Scrubber sludge has also been used as a substitute backfill for regrading mine sites that are deficient in spoil for the establishment of the desired grade. Combustion by-products may contain high surface concentrations of arsenic, selenium, and boron. These elements are often quickly depleted from by-products. There is also the suggestion that combustion by-products will clean, rather than contaminate groundwater (Paul et al., 1993a,b).

25.3 PREPARATION

Coal preparation is an industry that has a substantial impact on the environment, and refuse disposal forms a major part of this multi-faceted problem (Argonne, 1990). Run-of-mine coal from underground mines goes directly to the coal washing section or preparation plants where beneficiation and removal of certain impurities associated with coal substance take place.

In itself, the disposal of waste material from coal cleaning operations presents a twofold problem. First, unless return of the coal wastes to the mine is possible, large areas of land are required, and a fixed and mobile materials handling plant must be provided. Second, land sterilization and spoil-heap drainage can present serious problems related to environmental control.

The danger of spontaneous combustion of coal is a constant threat during mining, during storage, and in refuse piles (Chapter 14). The danger of combustion in refuse piles arises when the coal wastes contain sufficient carbonaceous material and where spoil heaps are constructed by methods that allow ingress of sufficient air to support combustion. Refuse disposal requires the handling of wet granular material and sludge. The latter present severe difficulties during handling because of the ultrafine nature of the constituents and if mixed into main spoil heaps can lead to pile stability problems.

On the other hand, if the sludge is disposed of separately, they usually require additional treatment, including chemical dosing, to achieve acceptable stability.

Mechanical processes based on surface characteristics or the difference in specific gravity can remove 95% of the pyritic sulfur. This is normally equivalent to 30%–35% of the total sulfur and over 50% of the ash content of the coal. Preparation, therefore, improves the quality of coal and hence its environmentally friendly characteristics. On the other hand, however, the aquatic ecosystem is disrupted by solid and liquid residue, and particulate and gaseous effluents are of concern to public health, water source and supply quality, and landscape.

Washing and beneficiation processes produce large quantities of liquid and solid wastes. Liquid residues, so-called slurries, contain 95% of water mixed with very fine coal particles. Environmental and health risks arising from the possible leaching of such wastes could lead to the deterioration of surface water and ground water. Tips formed from the dumping of solid waste from preparation plants become unstable after prolonged wetting, particularly if the slopes contain fine coal particles.

Spontaneous combustion may develop due to the heating of coal particles and slow combustion. Toxic material may be released with the oxidation of compounds present in the waste material. Runoffs, with minimal iron pyrites content, from the tip surface, promote hazards and the acid compounds may leach into domestic water supplies. All these effects may be partly or wholly overcome with proper management of the planning and design of waste heaps and good operating practices. Moreover, encouragement of the commercial use of waste material is beneficial.

The effects of dust particulates, noise and vibration from operating machinery, dumping activities and traffic are harmful to the local environment. These can be reduced to an acceptable level with sensitive planning of operations, siting of boundaries, erection of barriers, and use of machinery assisted with silencers.

25.3.1 Water Treatment

Cleaning plants may require up to 2000 gal/ton of feed coal, although volumes of only half this quantity are now required by the latest jig designs, which are the units in a coal cleaning plant that require the greatest quantities of water (Couch, 1991).

If no water treatment and recycling are practiced, this volume represents a net demand for fresh water, which like tar sand processing (Speight, 2007a, 2008) can be a major issue in coal producing regions that lie in semi-arid areas. To combat the need for water, the installation of recycle water systems has become a major aim, one advantage being that the water requirement reduces to that necessary for replacing water lost in the products so that water consumption drops to 18–30 gal/ton of feed coal.

Unless the coal being washed contains appreciable quantities of soluble salts, cleaning processes do not materially alter water composition. However, difficulties may be encountered when iron carbonates and pyrite and pyrite are present, particularly if the coal passes through stockpiles that allow some oxidation of the iron salts. The contamination is usually indicated by a substantial lowering of pH, from a normal range of 6–7.5 to values of 3 or even less, and may necessitate addition of alkali (caustic soda or lime) to restore near-neutral conditions. This type of problem is not often encountered and the usual type of water treatment involves clarification to remove suspended salts (*slime*) after which the water may be recirculated.

Clarification may be partially achieved using batteries of small-diameter hydrocyclones (see above) to remove particles down to about 50 μm (Couch, 1991). Complete water clarification requires settling in cones and static thickeners, in conjunction with chemical dosing to promote flocculation and rapid settling.

25.3.2 Dust Control

The presence of dust can cause serious problems in many dry cleaning plants and in the pretreatment sections of wet cleaning plants. In addition, dust containment is an integral part of all thermal drying units. The need for dust collection arises because of air pollution control, safety- and health-hazard elimination, and reduction of equipment maintenance costs.

In addition to the economic and health hazards, coal dust can present serious explosion hazards. Certain bituminous coals and anthracites also continue to release measurable quantities of methane gas for days, and sometimes weeks, after they have been mined, which increases the possibility of explosion. Provision of efficient ventilation and restriction of electrical or thermal sources of ignition are required by regulation.

The main sources of dust emissions are crushers, breakers, de-dusters, dry screens, and transfer points between the units in coal pretreatment operations. The size of the coal particles that become airborne generally falls into the range 1–100 μm ($1–100 \times 10^{-6}$ m) and particles larger than this usually deposit close to the point of origin. It is the usual practice to extract the dust clouds from the various sources through ducts that exhaust to a central air cleaning unit, although some equipment may include integral dust suppression devices. Dust collection equipment used at cleaning plants includes (1) dry units such as cyclones, dynamic collectors, and baghouses (fabric filters) and (2) wet units such as dynamic, impingement, or centrifugal devices or gravity, disintegrator, or venturi scrubber systems.

Other sources of dust are windblown losses from storage piles and from railcars during loading and transportation. Various chemical treatments are available for stockpile sealing and are being used for spraying the tops of railcars; active stockpiles present an almost insurmountable problem.

25.3.3 Noise Control

Coal preparation plants invariably contain various types of active machinery, in addition to large throughput tonnages of free-falling solids, and have long been recognized as presenting a serious noise problem. Most of the noise control measures may be classified broadly into measures that

(1) suppress noise formation or (2) provide noise containment. Examples of the former include the installation of mufflers on air pumps and of rubber screen cloths and resilient screen decks and chute liners. Noise containment requires erection of barriers, either walls or curtains.

25.3.4 RECENT LEGISLATION

On September 25, 2009, the United States Environmental Protection Agency (USEPA) set more stringent emissions standards for coal preparation and processing facilities. The final rule, according to USEPA, reflects improvements in air emission control technologies that have been developed since the new source performance standards (NSPS) for these sources were first issued in 1976.

The final rule amends the NSPS for coal preparation and processing plants, establishing new emissions standards for particulate matter (PM), opacity, sulfur dioxide (SO_2), nitrogen oxides (NOx), and carbon monoxide (CO). The rule does *not* establish NSPS for emissions of carbon dioxide (CO_2), nitrous oxide (N_2O), black carbon (a component of PM) or other greenhouse gases from these sources. The revised performance standards are intended to reduce emissions at all new, modified and reconstructed coal preparation and processing facilities at coal mines, power plants, cement plants, coke manufacturing facilities, and other industrial sites that process more than 200 tons of coal per day.

This rule takes the following actions:

- It establishes more stringent standards for emissions of particulate matter and opacity included in the existing guidelines for thermal dryers, pneumatic coal-cleaning equipment and coal-handling equipment.
- It expands the applicability of the thermal dryer standards so that they will apply to both direct contact and indirect contact thermal dryers drying all coal ranks, and to pneumatic coal-cleaning equipment cleaning all coal ranks.
- It establishes an emission limit sulfur dioxide and a combined emissions limit for nitrogen oxides and carom monoxide from thermal dryers. The USEPA promotes the combined standard because a decrease in emissions of one of the two pollutants often leads to an increase in emissions of the other.
- Coal processing and conveying equipment, coal storage systems, and transfer system operations and fugitive emissions from those facilities will be subject to new opacity limits.
- Mechanically vented coal handling and coal processing equipment constructed, reconstructed or modified after April 28, 2008, have to meet new emissions standards for particulate matter.

The rule does *not* establish opacity or limits for particulate matter from open storage piles (which include the equipment used in the loading, unloading, and conveying operations at the facility) or roadways. However, the rule requires owners or operators of these facilities to develop and comply with a fugitive coal dust emissions control plan. This plan must require implementation of one or more control measures to minimize fugitive emissions of coal dust that are appropriate for the particular site.

New, modified and reconstructed coal preparation and processing plants will need to install controls or utilize work practices to meet the revised limits.

25.4 TRANSPORTATION AND STORAGE

25.4.1 TRANSPORTATION

Coal, after washing and upgrading, is transported to industrial centers either by rail, trucks, barges, ships or slurry pipelines (Chapter 6). Generally, the manner of coal handling is substantially affected by particle size, particle size distribution, moisture content, and local weather conditions.

Considerable damage to the ecological system, water quality and aesthetics of the land are inevitable. The choice of the mode of transportation may be instrumental in dealing with pollution hazards. Bulk movement of coal by methods other than road is advantageous from both the economic and environmental standpoint. Transportation by road tends to create dust, increase traffic noise and congestion, and raises the risk of accidents and injuries.

Environmental impacts occur during loading, en route, or during unloading and affect natural systems, manmade buildings and installations, and people (e.g., due to injuries or deaths) (Chadwick et al., 1987).

For example, the major adverse environmental impacts resulting from truck coal hauling are coal dust particle releases during coal loading or unloading, and coal dust entrainment during transport. Some coal will escape from the trucks during transport because the loads are normally uncovered. The coal dust tends to wash off roadways during rainstorms, causing aesthetic unsightliness and contamination of runoff waters. The air pollutant emissions from diesel fuel combustion add to the emissions.

All forms of coal transportation have certain common environmental impacts, which include use of land, structural damage to facilities such as buildings or highways, air pollution from engines that power the transportation systems, and injuries and deaths related to accidents involving workers and the general public (e.g., railway crossing accidents). In addition, fugitive dust emissions are experienced with all forms of coal transport, although precautionary measures are increasingly being taken (Chadwick et al., 1987). It is estimated that 0.02% of the coal loaded is lost as fugitive dust with a similar percentage lost when unloading. Coal losses during transit are estimated to range from 0.05% to 1.0%. The amount is dependent upon mode of transportation and length of trip but can be a sizeable amount, especially for unit train coal transit across the United States.

In many instances, conventional railroads are preferred for the handling of coal since the large capacity wagons can be covered, sprayed by water, and wind guards and chemical binders can be used. In the future, slurry pipelines may prove to be the most economic method of moving huge volumes of coal over long distances. However, land requirement, the use of large quantities of water that are difficult to dispose of, and lack of eminent domain rights are the main drawbacks in establishing a nationwide network system.

Coal stockpiled near the pit-head or in close proximity to industrial sites is a potential source of dust and surface water runoffs. Regular inspection is required to monitor the rate of spontaneous combustion, which causes the emission of local scale toxic gases. Moreover, stockpiling on a large scale or over a long period arouses adverse reaction from the general public on its unsightliness and puts constraints on coal handling. These problems can, to a certain degree, be overcome by storing coal in abandoned pits or silos constructed for the purpose, but authorities still need to look into alternative optional proposals to deal with the environmental matters.

25.4.2 Storage

Storage of coal at coal preparation plants is not practiced to the same extent as it was decades ago. The active principles in reducing coal storage at coal preparation plants were (1) fugitive dust and (2) runoff control.

In fact, in some areas, local environmental regulations may rule out open storage, but it should be considered at least initially, because it can be less expensive from a total cost standpoint, even with environmental controls.

On the other hand, utility coal yards at power plants often need to store coal so that in anticipation of disruption in coal delivery for the plant (such as a train derailment) often means that the plants may require several days of backup. An efficient method that has been used and still is used where environmental regulations permit such actions for storing large quantities of coal is ground placement, where stocks of 30–90 days consumption may be required.

However, in such cases where open storage is allowed, rain and snow runoff can become contaminated by chemical and bacteriological action on pyritic materials contained in the coal. Chemically, this occurs by the same series of reactions that are known to produce acid coal mine drainage and the amount of runoff is dependent upon (1) configuration of the stockpile, (2) particle size of the coal, (3) moisture content of the coal, (4) amount of rainfall, and (5) the intensity of the rain (shower or downpour). Studies indicate runoff ranges from 50% to 95% of the rainfall on the pile; the remainder evaporates or is held in the pile. Constructing coal storage piles so as to encourage runoff and inhibit the amount of water percolating through the piles can minimize the energy wasted in driving off moisture prior to combustion (Cox et al., 1977; Ripp, 1984).

In many countries (including the United States), regulations require mine operators to comply with specific limitations relating to effluents. In order to meet these requirements, the following treatments are typically employed: (1) addition of alkaline material, such as lime or sodium hydroxide, (2) natural or mechanical aeration, and (3) settling. When the pH is raised to 7 or 8 and settling has occurred, most drainage waters will meet the standards for iron content and suspended solids.

In drainage waters, a pH value as high as 10 (sternly alkaline water) may be sometimes required to reduce manganese to effluent limits. Thus, the reduction of manganese to the regulated standard may require discharging water with pH higher than the upper limit, which action requires the operator to obtain a variance.

25.5 COMBUSTION

Coal utilization for power generation (Chapter 15) is of growing environmental concern, due mainly to emissions of carbon dioxide associated with the combustion process. Although coal is only one of many sources represented by this anthropogenic carbon dioxide, the coal industry is searching for and developing technological options to mitigate its contribution to the problem.

Previously, coal was used extensively as a fuel for energy production and as a raw material in the chemical, gas, and metallurgical industries. Today, however, its principal uses are in the utility, industrial, and residential/commercial sectors. In countries with planned economies, 70%–90% of the total quantity of coal consumed is burnt in the utility sector (Chadwick et al., 1987).

The environmental aspects of coal combustion have been a major factor in the various processes, and the movement of the fossil fuel base away from petroleum and natural gas to coal has increased the need for effluent/pollutant control for large, fossil-fueled power plants (Argonne, 1990). Very large amounts of coal are consumed in generating electricity and the emissions from power stations and similar industrial sources represent a potential, and considerable, environmental hazard. These power plants and the accompanying flue gas desulfurization processes emit effluents, which often are pollutants, and which by mere contact with the external environment or by (generally) simple atmospheric chemical transformations, may form secondary pollutants that are more harmful than the initial effluent/pollutant.

Coal-based processes involved in combustion and conversion facilities release gaseous and liquid effluents as well as solid effluents deleterious to the environment and human health. The preference can be made from the following alternatives (Clark et al., 1977):

- Removing the pollutant from the process effluent, for example, passing polluted air through a series of dust collectors that filter the fine particulates
- Removing the pollutant from the process input, desulfurization of coal
- Controlling the process, lowering combustion temperature to minimize the generation of nitrogen oxides and their emission
- Replacing the process with one that does not generate or will minimize the pollutant, for example, pressurized fluidized bed combustion instead of pulverized coal burning, and
- Selecting a type of coal-fuel that eliminates the pollutant, for example, use of low-sulfur coal.

Combustion of coal produces environmentally harmful emissions (Chapter 14). Some gases produced from burning coal are known as greenhouse gases because they trap the earth's heat like the roof of a greenhouse and may contribute to possible global warming. Other emissions from coal combustion can lead to air and water pollution.

During the combustion of coal, the mineral matter is transformed into ash, part of which is fly ash discharged to the atmosphere as particles suspended in the flue gases and part of which is bottom ash removed from the base of the furnace. Furthermore, scrubber sludge is formed where sulfur dioxide scrubber facilities are employed. The quantity, particle size distribution and properties of fly ash are directly related to the combustion technique applied and the constitution of the coal.

Particulates in the form of dust, smoke, fly ash and the like are generated by a variety of chemical and physical processes in the combustion chamber. These effluents exhibit a range of particle size and composition. Particles of relatively large size remain in the furnace, but smaller particles are emitted to the atmosphere with the flue gases.

Of the various methods of conventional combustion techniques, pulverized coal-fired furnaces are preferred for power stations and larger industrial facilities. Combustion processes create very fine effluents of which 70%–90% is carried to the air fly ash and 10%–30% remains as bottom ash (Commission on Energy, 1981). Difficulties in disposing fly ash stem from the enormous quantities collected (Scanlon and Dugan, 1979).

25.5.1 Fly Ash

Coal combustion releases fly ash into the atmosphere (Chapter 14) (Fisher and Natusch, 1979; Adriano et al., 1980). Fly ash contains toxic metals such as arsenic and cadmium. In the United States the Clean Air Act requires that fly ash be removed from coal emissions. As a result, antipollution devices such as air baghouses and electrostatic precipitators are used to trap these pollutants.

Baghouses work like giant vacuum cleaners, drawing coal emissions through giant fabric bags that trap the fly ash inside. Electrostatic precipitators use discharge electrodes (electrically charged parts of an electric circuit) to trap ash particles. In an electrostatic precipitator the electrodes are located between long, positively charged collection plates. As the fly ash passes between these collection plates, the discharge electrodes give each particle a negative charge. These negatively charged particles are then attracted to and held by the positively charged collection plates.

The *furnace sorbent injection process* removes acid gas from coal emissions at less cost than expensive scrubbers (Chapter 23). In this process a highly absorbent material (called a sorbent), such as powdered limestone, is injected into the boilers, where the powdered limestone reacts with the acid gases emitted by the burning coal. The used powder is siphoned away through the boiler out-take and is captured (with fly ash) in a baghouse or electrostatic precipitator.

Another process—the *advanced flue-gas desulfurization*—also removes acid gas from burning coal without expensive scrubbers (Chapter 23). Emissions from burning coal are piped into a container called an *absorber*, where the acid gases react with an absorbing solution (such as a mixture of lime, water, and oxygen). This reaction forms gypsum, a soft white mineral valuable as an ingredient in cement.

Fly-ash effluents escaping from power plants and being emitted into the atmosphere are further contaminated with trace elements suspended on the fly-ash particles. The degree of contamination is relative to the variety of elements present in the coal substance derived from transformation processes that occurred before, during, and after coalification (Chapter 7) and during high-temperature processes in the furnace (Chapter 14).

Mass-balance studies based on coal combustion in power plants show consistent and selective separation of trace elements in the various outlets of the combustion furnace in the following groups (Chadwick et al., 1987):

Group I. Elements, such as Al, Ba, Ca, Ce, Cs, Fe, K, Mg, Mn, and Th, remain condensed at the temperature of coal combustion and divide equally between fly ash and bottom ash.

Group II. Elements, such as As, Cd, Cu, Pb, Sb, Se, and Zn, are volatilized and will, therefore, be depleted from the slag and bottom ash and condense out on the smaller fly-ash particles.

Group III. Elements, such as Br, Hg, and I, are very volatile and mostly remain in the gas phase. They are depleted in all ashes. Any remainder will be associated with fly ash.

Elements, such as Cr, Ni, U, and V, display intermediate behavior between Groups I and II.

Uranium existing in coal as silicate mineral coffinite and uraninite (UO_2) poses a potential environmental hazard. Following combustion of coal, the refractory coffinite remains in the bottom ash and slag while the uraninite is vaporized and is later condensed on the fly-ash particles as the flue gases cool (Chadwick et al., 1987). Comparative radiation exposure assessment studies on coal and nuclear-based electricity generation reveal that emissions from both are very low, but dose levels from coal-fired plants are equal to or slightly lower than from a nuclear power plant (UNGA, 1980; Chadwick et al., 1987).

Emission of fly ash into the atmosphere may cause a number of environmental and human health problems. Particles emitted are mainly in the <1 μm diameter range since larger size particulates are effectively removed by control devices. Such small particles can be inhaled by humans and deposited in various regions of the respiratory system. This may lead to health complications. Airborne particles can diminish visibility and the dispersion and deposition of particles could have adverse effects on soil and water. Leaching of particles may lead to unacceptable and even dangerous contamination levels of groundwater.

The leachates are characterized by excessively high levels of trace elements such as As, Be, Cd, Cr, F, Hg, Pb, Se, and Sr from fugitive fly ash (Berkowitz, 1985). These and other trace pollutants are transformed from relatively harmless oxides into potentially highly toxic substances harmful to ecological systems. Humans are exposed to coal-derived radionuclides from suspension of particles in the air and particles inhaled or ingested. This source of radioactivity represents a very small fraction of the collective dose. Nevertheless, any substantial increase in the utilization of uranium-containing coal would increase the radiological exposure to people living in close proximity to coal-fired power stations.

Pollution control techniques are well established in order to efficiently collect effluent particulates. These include mechanical cyclone collectors, electrostatic precipitators and fabric filters or baghouses, and wet scrubbers. Only electrostatic precipitators and fabric filters are able to carry out the collection of particles with an overall efficiency of 99.5% or more, so meeting the strict emission standards economically. Effluent particles are valuable in certain sectors of industry—they can be used as a lightweight engineering fill material to stabilize soil and as a road subbase (International Energy Agency, 1985). When appropriately added to concrete mix, fly ash yields a stronger and more durable concrete than that produced by regular Portland cement.

25.5.2 GASEOUS EFFLUENTS

The products of the complete combustion of coal are carbon dioxide and water. However, in the exhaust gases from any practical device there will be, in addition to these compounds, carbon monoxide, sulfur oxides (SOx), nitrogen oxides (NOx), unburned hydrocarbons and probably solid particulates (soot) with small amounts of hydrogen chloride and polycyclic organic matter. Traces of other gaseous compounds are also to be found. Of these principal pollutants, concern has largely been focused on sulfur dioxide and nitrogen oxides, so-called acid gases because of their apparent importance on long-term environmental effects.

The lifetime of these species in the atmosphere is relatively short and if they were distributed evenly their harmful effects would be minimal. Unfortunately these man-made effluents are usually concentrated in localized areas and their dispersion is limited by both meteorological and topographical factors. Furthermore, synergistic effects mean that the pollutants interact with each other: in the presence of sunlight, carbon monoxide, nitrogen oxide(s), and unburned hydrocarbons lead to photochemical smog, while when sulfur dioxide concentrations become appreciable, sulfur oxide-based smog is formed.

Thus, carbon dioxide (CO_2) is released to the atmosphere from coal utilization processes. It also occurs naturally in the air, and a large amount is stored in the world's oceans (Crane and Liss, 1985). Since coal became the primary source of energy, the increase of carbon dioxide into the atmosphere is very significant, from 290 ppm in the middle of the last century to 335 ppm today (Chadwick et al., 1987). Moreover, it is anticipated that by the year 2000, carbon dioxide concentration will be about 125% of its 1850 level.

It is worth remembering that even the nontoxic, but nonlife supporting and suffocating, carbon dioxide may have an important effect on the environment. The surface of the earth emits infrared radiation with a peak of energy distribution in the region where carbon dioxide is a strong absorber. This results in the situation whereby this infrared radiation is trapped by the atmosphere and the temperature of the earth's surface is raised. As a result of the combustion of fossil fuels, the concentration of carbon dioxide in the atmosphere is increasing from its present level. Although many factors are involved, it does seem that an increase in the carbon dioxide concentration in the atmosphere would result in a temperature increase at the surface of the earth, which could cause an appreciable reduction in the polar ice caps and this in turn would result in further heating.

While it is true that the present rate of increase of carbon dioxide is fairly level, nevertheless, if by the end of the twenty-second century most of the world's known fossil fuel supplies are consumed, it is estimated that the amount of carbon dioxide in the atmosphere will have risen by a considerable factor from the present level. A change of this magnitude must be expected to bring about substantial climatic alterations.

The carbon dioxide discharged into the air increases, so does the temperature of the lower atmosphere. The carbon dioxide absorbs thermal radiation and reflects back a proportion of the infrared radiations. These processes are contributing to the so-called greenhouse effect or global warming of the atmosphere. Scientific opinions differ on the possible consequences, but man may be the cause of further significant global warming, unprecedented climate changes and imbalance in world ecology.

Carbon monoxide (CO) is generally formed by the partial combustion of carbonaceous material in a limited supply of air. Small quantities of carbon monoxide are produced by the combustion of coal and from spontaneous combustion. In the atmosphere the carbon monoxide is eventually converted to carbon dioxide. These emissions are not considered to be a problem for environmental pollution.

Acid gases (SOx and NOx) emitted into the atmosphere provide the essential components in the formation of acid rain. Sulfur is present in coal as both an organic and inorganic compound. On combustion, most of the sulfur is converted to sulfur dioxide with a small proportion remaining in the ash as sulfite:

$$S_{coal} + O_2 \rightarrow SO_2$$

At flame temperatures in the presence of excess air, some sulfur trioxide is also formed:

$$2SO_2 + O_2 \rightarrow 2SO_3$$

$$SO_3 + H_2O \rightarrow H_2SO_4$$

Only a small amount of sulfur trioxide can have an adverse effect as it brings about the condensation of sulfuric acid and causes severe corrosion. Although diminution of the excess air reduces sulfur trioxide formation considerably, other considerations, such as soot formation, dictate that the excess air level cannot be lowered sufficiently to eliminate sulfur trioxide entirely.

From 10% to 50% of the nitrogen inherent in the organic coal structure is converted to nitrogen oxides during combustion, which also produce acidic products, thereby contributing to the acid rain:

$$2N_{coal} + O_2 \rightarrow 2NO$$

$$NO + H_2O \rightarrow HNO_3$$
Nitrous acid

$$2NO + O_2 \rightarrow 2NO_2$$

$$NO_2 + H_2O \rightarrow HNO_3$$
Nitric acid

Nitrogen in the coal itself also contributes significantly to the formation of gases. Of the nitrogen oxides emitted in power station flue gas, 95% is nitric oxide (NO), which oxidizes rapidly in the atmosphere to form nitrogen dioxide (NO_2) (Commission on Energy, 1981).

Fluidized-bed combustion generates little, if any, thermal nitrogen oxides despite the long residence times since the bed temperatures are usually too low; any residual nitrogen oxide is often derived from the fuel nitrogen rather than from the conversion of aerial nitrogen.

Very little, if anything, can be done as part of the coal pretreatment operation(s) (Chapter 6) to eliminate nitrogen because this element is a part of the organic coal structure (Chapter 10). The situation is less clear in the case of hydrogen chloride sources in coal; both organically bound chlorine and inorganic chloride salts (although it is not known that these contribute appreciably to hydrogen chloride formation during combustion) but these processes do not remove organically bound chlorine, which is a more likely precursor to hydrogen chloride in a combustion process.

However, the management of volatile matter evolution patterns also is of significance in controlling nitrogen oxide emissions. Nitrogen evolving rapidly, frequently under reducing conditions, does not oxidize but converts to nitrogen. Nitrogen volatiles that evolve more slowly can exist in an oxidizing environment and formation of nitrogen oxides from these is more likely than from fuel nitrogen sources.

This principle is one of the principles behind staged combustion as well as oxygen-enhanced combustion approaches where the oxygen is injected at the root of the flame, increasing volatile nitrogen evolution; however, the quantity of oxygen used is insufficient to convert the reducing environment at the base of the flame to an oxidizing environment.

Thus, control of nitrogen oxides has traditionally been achieved with staged combustion and with catalytic conversion processes using ammonia or urea. Catalytic processes using ammonia or urea have proven to be quite effective, but they are very expensive to implement and operate. Staged combustion has provided some control of nitrogen oxides formed from the volatiles, but it has had little effect on the formation of nitrogen oxides formed from the char.

Recently, advanced staging processes, known as low-NOx burners, have been developed. These burners are designed on the basis that volatile nitrogen may be converted to nitrogen rather than nitrogen oxides under locally fuel-rich conditions with sufficient residence time at appropriate temperatures. The amount and chemical form of nitrogen released during devolatilization greatly influences the amount of nitrogen oxide reduction achieved using this strategy. Since the nitrogen in the char is released by heterogeneous oxidation, these burner design modifications have no effect on NOx formed from the char nitrogen.

Low-NOx burners alter the near-burner aerodynamics of the combustor. This alteration influences the devolatilization process and therefore influences the amount and chemical form of nitrogen released during devolatilization (Solomon and Fletcher, 1994). Low-NOx burners provide a less expensive emission control strategy and are often the method of choice to limit the amount of nitrogen oxides formed during combustion.

Most stack gas scrubbing processes are designed for sulfur dioxide removal; nitrogen oxides are controlled as far as possible by modification of combustion design and flame temperature regulation. However, processes for the removal of sulfur dioxide usually do remove some nitrogen oxides; particulate matter can be removed efficiently by commercially well-established electrostatic precipitators.

There are several processes for the removal of sulfur dioxide from stack gas (Table 25.1) (Chapter 23) but to date scrubbing process utilizing limestone ($CaCO_3$) or lime [$Ca(OH)_2$] slurries have received more attention than other stack gas scrubbing processes. Attempts have been made to use dry limestone or dolomite ($CaCO_3 \cdot MgCO_3$) within the combustor as an "in situ" method for sulfur dioxide removal, thus eliminating the wet sludge from wet processes. This involves injection of dry carbonate mineral with the coal followed by recovery of the calcined product along with sulfite and sulfate salts.

$$CaCO_3 + SO_2 \rightarrow CaSO_3 + CO_2$$

$$CaCO_3 + SO_3 \rightarrow CaSO_4 + CO_2$$

The various solid products are removed by standard means, including cyclone separators and electrostatic precipitators.

TABLE 25.1

Simplified Chemistry of Flue Gas Desulfurization Processes

Process	Chemical Reactions[a]
Limestone slurry scrubbing	$CaO_3 + SO_2 \rightarrow CaSO_3 + CO_2$
Lime slurry[b] scrubbing	$CaCO_3 + heat \rightarrow CaO + CO_2$
	$CaO + SO_2 + 2H_2O \rightarrow CaSO_3 \cdot 2H_2O$
Magnesia slurry scrubbing with regeneration	$Mg(OH)_2 + SO_2 \rightarrow MgSO_3 + 2H_2O$
	$MgSO_3 \xrightarrow{\Delta} MgO + SO_2$
Sodium base scrubbing	$NA_2SO_3 + H_2O + SO_2 \rightarrow 2NaHSO_3$
	$2NAHSO_3 \xrightarrow{\Delta} Na_2SO_3 + H_2O + SO_2$
Catalytic oxidation ion	$2SO_2 + O_2 \rightarrow 2SO_3$
	$H_2O + SO_3 \rightarrow H_2SO_4$
Sodium citrate scrubbing	$SO_2 + OH^- \rightarrow HSO_3^-$
	$2H_2S + SO_2 \rightarrow 3S + 2H_2O$

[a] All of these processes (except for the catalytic oxidation process) depend on the absorption of sulfur dioxide by an acid–base reaction. The first two are "throwaway" processes insofar as they produce considerable quantities of waste material. The other processes include sulfur product recovery.

[b] May require as much as 200 lb of calcium oxide per ton of coal.

Some of the sulfur and nitrogen oxides will fall to the ground quite quickly as dry deposition. Both are in the form of dry gases or are adsorbed on other aerosols like fly ash or soot (Park, 1987). These particles may not be in the acidic state while in the air but will become so on contact with moisture. The remaining oxides will be carried up into the atmosphere with the help of tall chimney stacks. Sulfur and nitrogen oxides are immediately absorbed by water droplets in clouds. Oxidation converts the gases into acids.

Subsequent dissolution causes sulfuric and nitric acids to dissociate into a solution of positively charged hydrogen cations and negatively charged sulfate and nitrate anions, which eventually return to earth as acid rain. This is termed wet deposition. Furthermore, nitrogen oxide absorbs ultraviolet light from the sun and triggers off photochemical reactions that produce smog, of which ozone is a part (Clark et al., 1977).

The U.S. Clean Air Act, implemented in 1970 (and revised in 1970 and 1990) is the federal legal basis for air pollution control in the United States. This legislation has significantly reduced emissions of sulfur oxides, known as *acid gases*. For example, the Clean Air Act requires facilities such as coal-burning power plants to burn low-sulfur coal.

High-grade coal (coal with a higher heating value) generally contains more sulfur than low-grade coal such as lignite and subbituminous coal. Therefore, certain processes have been developed to remove sulfur-bearing compounds from high-grade coal prior to burning. The Clean Air Act also requires use of pollution-trapping equipment such as *air scrubbers* (devices installed inside plant smokestacks to remove sulfur dioxide from coal emissions).

In addition, 1990 revisions to the Clean Air Act established a system that allows coal-burning power plants to buy and sell sulfur emission permits with one another. Power plants that reduce their sulfur emissions to below the permitted levels can sell permits to companies wishing to exceed federal levels.

Air pollution with dry deposition is mainly a local problem. This is evident from the infamous London smog of 1952 that caused acute respiratory disorders among the population and several thousand deaths in the city (Bach, 1972). This phenomenon was brought about by a buildup of pollutants in the air from coal-burning industries and power stations, but mainly from domestic fires. The sulfur dioxide concentrations rose to almost seven times the normal levels ultimately producing sulfuric acid droplets, floating free in aerosol form or attached to smoke particles. The catastrophe led to a clean air act and a tall-stacks policy where 200–400 m high smokestacks were built to push the pollutants higher into the atmosphere (Sage, 1980; Pearce, 1982). However, this application inadvertently contributed to the geographical distribution of acid rain beyond national borders.

Besides the adverse consequences of acid rain, many green plants are damaged by exposure to low concentrations of sulfur and nitrogen oxides (Clark et al., 1977). These gases can also severely impair respiratory functions and change the metabolic rates of humans. Sulfur oxides, moreover, affect metals and building materials causing corrosion. On the whole, it is apparent that oxides of sulfur and nitrogen pose a direct threat to the balance of nature devastating both terrestrial and aquatic ecosystems alike.

The scale of these environmental problems can be minimized, if not eliminated, by the use of proper processes invented for the purpose. Control of sulfur dioxide emissions could be accomplished by using low-sulfur content coal, blending high-sulfur coal with low, and treating coal before combustion to remove some of the sulfur. Coal washing, although mainly applied for reducing the mineral content of coal, removes up to 70% of inorganic sulfur and 40% of total sulfur. Another alternative would be to switch from coal to oil fuel or natural gas.

Environmental concerns and strict pollution legislation prompted action in the construction of up to date coal-fired boilers and adaptation of existing plants. At the present time, flue gas desulfurization is the only conventional method employed on a commercial scale for reducing sulfur emissions after coal combustion. Over 90% reduction of sulfur dioxide in flue gases can be achieved by this process. Combustion control techniques of the flames will effectively reduce oxides of nitrogen emissions into the atmosphere.

Flue gas denitrification by selective catalytic reduction has limited application with acceptable results in reducing the effluent nitrogen oxides emissions from power stations. Future technology includes fluidized bed combustion where sulfur removal can be brought about during the combustion of coal with a subsequent reduction of sulfur dioxide emissions of 35%–50% (Burdett et al., 1985). Furthermore, coal can be converted to liquid or gaseous products, so-called synthetic fuels, with sulfur and other impurities largely removed. Coal-oil and coal-water mixtures with a low ash and sulfur content are seriously being considered as alternative replacement fuels for oil in utility and industrial boilers.

Although not generally considered to be a major pollutant, hydrogen chloride from inorganic contaminants is gaining increasing recognition as a pollutant arising from the combustion of coal. Hydrogen chloride quickly picks up water from the atmosphere to form droplets of hydrochloric acid and, like sulfur dioxide, is a contributor to acid rain. However, hydrogen chloride may exert severe local effects because it does not have to undergo any chemical change to become an acid and, therefore, under atmospheric conditions involving a buildup of stack emissions in the area of a large power plant, the concentration of hydrochloric acid in rain droplets could be quite high.

25.5.3 WASTE EFFLUENTS

The nonvolatile oxidation products of the inorganic constituents of the fuel (e.g., sodium and potassium) are left as ash and contribute to the formation of corrosive deposits (Argonne, 1990). When pulverized coal is used, a large proportion of the ash is carried out of the combustion chamber with the exhaust gas stream and has to be removed as completely as possible, usually by use of electrostatic precipitators.

On completion of the coal cycle, a number of solid wastes are formed from the combustion and conversion of coal. For every ton of coal burned or processed, 40–300 kg of ash and 200–300 kg of end products from flue gas desulfurization are produced, respectively (International Energy Agency, 1985). It is reported that countries of the Organization for Economic Co-operation and Development currently yield 150 Mt ash annually from industry and thermal electric power stations (International Energy Agency, 1985). The disposal of such massive quantities of solid waste necessitated stringent regulations and legislation.

Solid waste and dewatered sludge are generally tipped onto selected sites above ground, thrown into mine cavities or dumped in landfill sites. The land required for tipping is enormous, and its appearance is an obvious target for adverse public reaction. The prevailing option is to use waste in landfills, which simply entails dumping the waste into a hole in the ground. This method of discharge is a serious cause for environmental pollution.

During the disposal activities, a large quantity of dust is created, which not only spoils the air opacity but pollutes the soil fertility when it is deposited on the earth's surface. Contaminated soil can affect human health through the food chain. In transportation about 10% water is added to the waste material to dampen the surface and so minimize the rate of dusting (International Energy Agency, 1985). If a dry disposal scheme is in use, like landfill, the bottom ash is generally dewatered in bins, whereas in wet disposal the ash is deposited into a pond or lagoon. In order to prevent the leaching of undesirable toxic solutions into groundwater, discharging in safe areas, compaction and sound management design are essential precautionary steps. Use of solid ash in industry is an alternative that is advantageous, both commercially and for the environment.

The removal of fly ash from combustion effluent streams is as important to the environment as the removal of sulfur oxides and nitrogen oxides. If not removed, the inorganic constituents and the organic constituents of the ash can cause serious environmental consequences, in addition to the solid particles acting as condensation nuclei and (catalytic) surfaces for the conversion of sulfur dioxide to the trioxide.

Because of its organic origin and its intimate comixture with mineral formations, coal contains a large number of elements in minor or trace quantities (Chapter 7). Of the known non-transuranic elements, only a minority have not yet been found.

In regard to the behavior of trace elements in coal-fired power plants, the elements have been divided into two groups, those appearing mainly in the bottom ash (elements or oxides having lower volatility) and those appearing mainly in the fly ash (elements or oxides having higher volatility).

Bottom ash is agglomerated ash particles, formed in pulverized coal furnaces, that are too large to be carried in the flue gases and impinge on the furnace walls or fall through open grates to an ash hopper at the bottom of the furnace. Physically, bottom ash is typically gray to black in color, is quite angular, and has a porous surface structure.

Bottom ash is coarse, with grain sizes spanning from fine sand to fine gravel. Bottom ash can be used as a replacement for aggregate and is usually sufficiently well graded in size to avoid the need for blending with other fine aggregates to meet gradation requirements. The porous surface structure of bottom ash particles make this material less durable than conventional aggregates and better suited for use in base course and shoulder mixtures or in cold mix applications, as opposed to wearing surface mixtures. This porous surface structure also makes this material lighter than conventional aggregate and useful in lightweight concrete applications.

For power plants using dry particulate collection devices, such as electrostatic precipitators, it was believed that the most volatile elements (such as mercury and selenium) could actually escape in the elemental state with the flue gas; wet scrubbers, however, were believed capable of removing most of the elements from the gas streams and transferring them to the liquid effluent.

The material from *flue gas desulfurization* units is the product of a process typically used for reducing sulfur dioxide emissions from the exhaust gas system of a coal-fired boiler. The physical nature of these materials varies from a wet sludge to a dry powdered material depending on the process. The wet sludge from a lime-based reagent wet scrubbing process is predominantly calcium sulfite. The wet product from limestone-based reagent wet scrubbing processes is predominantly calcium sulfate. The dry material from dry scrubbers that is captured in a baghouse consists of a mixture of sulfites and sulfates. This powdered material is referred to as dry flue gas desulfurization ash, dry flue gas desulfurization material, or lime spray dryer ash. Flue gas desulfurization gypsum consists of small, fine particles.

Calcium sulfite flue gas desulfurization material can be used as an embankment and road base material. Calcium sulfate flue gas desulfurization material, once it has been dewatered, can be used in wallboard manufacturing and in place of gypsum for the production of cement. The largest single market for flue gas desulfurization material is in wallboard manufacturing.

Essentially no particulates from coal combustion should be ejected into the atmosphere; all particulate streams should be collected and either returned to the combustors, where they melt and are removed as slag, or are removed as fly ash. Any eventual dispersion of the elements present depends on the possibility of leaching. The concern, therefore, is to identify elements that may be occurring in the gaseous state.

Among the trace elements present in coal with recognized toxic properties, high-volatility elements (beryllium, mercury, and lead) do not form gaseous-hydrides, will condense on cooling and very likely will be almost completely removed by the aqueous condensates formed on gas cooling and/or purification. Arsenic, antimony, and selenium have lower volatility but can form gaseous hydrides, (covalent) hydrides: arsine (AsH_3), stibine (SbH_3), and hydrogen selenide (H_2Se). These, however, have stability characteristics that preclude their formation at the temperature and pressure prevailing in the oil/gas plant gasifiers.

25.6 CARBONIZATION

Coal carbonization is the process for producing metallurgical coke for use in iron-making blast furnaces and other metal smelting processes. Carbonization of coal (Chapter 16) entails heating coal to temperatures as high as 1100°C (2010°F) in the absence of oxygen in order to distill out tars and light oils (Chapter 16). A gaseous by-product referred to as coke oven gas (COG) along with ammonia, water, and sulfur compounds are also thermally removed from the coal.

The coke that remains after this distillation largely consists of carbon, in various crystallographic forms, but also contains the thermally modified remains of various minerals that were in the original coal. These mineral remains, commonly referred to as coke ash, do not combust and are left as a residue after the coke is burned. The coke product also contains a portion of the sulfur that was originally present in the coal and can affect the combustion characteristics of the coke (leading to the emission of sulfur dioxide).

Environmental control typically commences with the arrival of the coal, which is not usually stored at coke plants for lengthy time periods. Besides the costs to maintain such inventories, coal undergoes low temperature oxidation that can adversely impact its coking behavior. This reduces the environmental consequences of runoff from the coal during period of heavy rain or melting snow.

When the coal arrives at the plant (whether it arrives by ship, rail, conveyor, or truck) each type of coal is unloaded into a separate stockpile in the coal field. Reclaiming coal from the stockpile can be accomplished using mobile equipment or bridge-mounted hoists. Coal of each type is moved to its coal bunker. In some plants, it is possible to crush each coal independently prior to it reaching the coal bunker. This crushing is often done in two stages. The first stage may be to simply ensure that no large, that is, usually no larger than about 25 mm, coal lumps remain and to ensure that large pieces of rock or foreign material are removed. The second stage of crushing takes place in a hammer mill or similar facility that is equipped to pulverize the coal.

It is common practice to pulverize coal for coke making to more than 80%, being less than 3 mm in size. In other plants, crushing of the coal takes place after the various coals have been blended. This blending is accomplished by discharging measured, either mass or volumetric, amounts of each coal into mixing bins. This may be accomplished through use of intermediate conveyor belt systems as well. The coal blend is withdrawn from the mix bins on an as-needed basis and is transported to the coal blend silo at the coke battery.

At this stage, the environmental precautions are essentially those outlined for a coal preparation plant (Section 25.3) since this is essentially a coal-cleaning operation.

However, like any thermal cracking processes, coal carbonization tends to produce a relatively small amount of fugitive emissions.

The direct environmental impact of a coal carbonization plant is caused by the emission of gaseous products (such as carbon dioxide, carbon monoxide, hydrogen sulfide, nitrogen oxides, and sulfur dioxide) from carbonization plant. During the process, numerous gas analyses need to be carried out both by analyzing the gases emitted by inserting analytical tube directly into the chimney and at some distance downwind from it. Particulate emissions from carbonization processes can also be considerable.

The combustion and reduction functions of coke are enhanced by maximization of the carbon in the coke—other chemical constituents of coke derived from the coal sulfur and coal minerals should be as low as possible as the presence of these impurities dilutes the amount of carbon in coke. Additionally, these impurities must be melted and prevented from being retained in the molten product iron. They can be removed from the blast furnace by maintaining a molten slag layer that floats on the molten iron and periodically *tapping* this slag from the furnace.

The indirect environmental impact is carried over into the blast furnace where steel is produced. The slag chemistry and properties must be continually adjusted to drive the coke impurities, as well as impurities from other raw materials, into the slag rather than into the pig iron. The materials added to the blast furnace in order to affect this slag volume, chemistry, and properties must also be melted. Slag components arising from coke impurities create a thermal drain on the blast furnace and occupy furnace volume that could have otherwise contributed to productivity.

25.7 LIQUEFACTION

The potential exists for generation of significant levels of atmospheric pollutants from every major operation in a coal liquefaction facility. These pollutants include coal dust, combustion products, fugitive organics, and fugitive gases. The fugitive organics and gases could include carcinogenic polynuclear organics, and toxic gases such as metal carbonyls, hydrogen sulfides, ammonia, sulfurous gases, and cyanides. Many studies are currently underway to characterize these emissions and to establish effective control methods.

Emissions from coal preparation include coal dust from the many handling operations and combustion products from the drying operation. The most significant pollutant from these operations is the coal dust from crushing, screening, and drying activities. Wetting down the surface of the coal, enclosing the operations, and venting effluents to a scrubber or fabric filter are effective means of particulate control.

A major source of emissions from the coal dissolution and liquefaction operation is the atmospheric vent on the slurry mix tank. The slurry mix tank is used for mixing feed coal and recycle solvent. Gases dissolved in the recycle solvent stream under pressure will flash from the solvent as it enters the unpressurized slurry mix tank. These gases can contain hazardous volatile organics and acid gases. Control techniques proposed for this source include scrubbing, incineration, or venting to the combustion air supply for either a power plant or a process heater.

Emissions from process heaters fired with waste process gas or waste liquids will consist of standard combustion products.

The major emission source in the product separation and purification operations is the sulfur recovery plant tail gas. This can contain significant levels of acid or sulfurous gases.

Emissions from the residue gasifier used to supply hydrogen to the system are very similar to those for coal gasifiers. Emissions from auxiliary processes include combustion products from onsite steam/electric power plant and volatile emissions from the waste water system, cooling towers, and fugitive emission sources. Volatile emissions from cooling towers, waste water systems, and fugitive emission sources possibly can include every chemical compound present in the plant. These sources will be the most significant and most difficult to control in a coal liquefaction facility. Compounds that can be present include hazardous organics, metal carbonyls, trace elements such as mercury, and toxic gases such as carbon dioxide, hydrogen sulfide, hydrogen cyanide, ammonia, carbonyl sulfide, and carbon disulfide.

Emission controls for waste water systems involve minimizing the contamination of water with hazardous compounds, enclosing the waste water systems, and venting the waste water systems to a scrubbing or incinerating system. Cooling tower controls focus on good heat exchanger maintenance, to prevent chemical leaks into the system, and on surveillance of cooling water quality. Fugitive emissions from various valves, seals, flanges, and sampling ports are individually small but collectively very significant. Diligent housekeeping and frequent maintenance, combined with a monitoring program, are the best controls for fugitive sources. The selection of durable low leakage components, such as double mechanical seals, is also effective.

Environmental factors, including the potential quantity and the composition of the pollutants, in addition to those compounds in the fuel stream recognized as mutagens that are initially present in internal process and utility streams, are a necessary consideration in the operation of fossil fuel conversion plants (Table 25.2).

Laboratory and bench-scale operations may not provide a true picture of the generation of pollutants. But it is essential that the methods for treating effluent streams to remove pollutants and the types of equipment used for environmental control be a very necessary part of the design and operation of any potential coal liquefaction procedures. Therefore, it is essential that some consideration be given to this aspect of coal liquefaction (Nowacki, 1979, 1980; Kimball and Munro, 1981; Speight, 2011).

TABLE 25.2

Health Effects of the Products and B-Products of Coal Liquefaction Processes

Constituent	Source	Effect
Inorganic		
Ammonia	Gas liquor	Respiratory edema, asphyxia, death
Carbon disulfide	Concentrated acid gas	Nausea, vomiting, convulsions
Carbon monoxide	Coal lock hopper vent gas, gasifier gas	Headache, dizziness, weakness, vomiting, collapse, death
Carbonyl sulfide	Concentrated acid gas	Few data on human toxicity
Hydrogen sulfide	Coal lock hopper vent gas, gasifier gas, concentrated acid gas, catalyst regeneration off-gas	Collapse, coma, and death may occur within a few seconds; may not be detected by smell
Hydrogen cyanide	Concentrated acid gas, coal lock hopper gas	Headache, vertigo, nausea, paralysis, coma, convulsions, death
Mineral dust and ash	Ash or slag	None
Nickel carbonyl	Catalyst regeneration off-gas	Highly toxic, irritation, lung edema
Trace elements[a]	Bottom and fly ash, gasifier gas, solid waste disposal	Element-specific
Sulfur oxides	Combustion flue gases	Intense irritation of respiratory tract
Organic		
Aliphatic hydrocarbons	Evaporative emissions from product storage	Most are not toxic
Aromatic amines	Coal lock hopper vent gas	Cyanosis, methemoglobinemia, vertigo, headache, confusion
Single-ring aromatics	Gas liquor, coal lock hopper vent gas	Irritation, vomiting, convulsions
Aromatic nitrogen heterocyclics	Gas liquor, coal lock hopper vent gas	Skin and lung irritants
Phenols	Gas liquor	
Polycyclic aromatic hydrocarbons	Gas liquor, coal lock hopper vent gas, gasifier gas	

[a] Including arsenic, beryllium, cadmium, lead, manganese, mercury, selenium, and vanadium.

25.7.1 GASEOUS EMISSIONS

The major type of air pollution control in coal liquefaction processes is aimed at desulfurizing the gases generated during the coal conversion process to make the fuels environmentally acceptable (Tables 25.1 and 25.3). The removal of acid gases is achieved by use of a physical solvent process that removes these gases from the main stream after which a selective regeneration process can be used to release a stream of hydrogen sulfide containing part of the carbon dioxide and a stream of nearly pure carbon dioxide. The hydrogen sulfide stream is sent to the sulfur recovery plant where it is oxidized to elemental sulfur (Speight, 2007a,b).

25.7.2 AQUEOUS EFFLUENTS

The generation and treatment of aqueous contaminants is another aspect of coal liquefaction that must not be ignored. Major contaminants are hydrogen sulfide, ammonium sulfide, phenols, cresols, xylenols, and thiocyanates; cyanides and solids (ash and char particles) may also be present. The gaseous contaminants (hydrogen sulfide and ammonia) are removed by steam-stripping.

Waste-water treatment is a combination of processes for the treatment, recycling, and discharge of aqueous effluents (Table 25.4).

TABLE 25.3

Gaseous Emissions from Coal Liquefaction Processes

Source	Emissions
Fugitive emissions	
Vent gases	
Coal storage and pretreatment	Dust, particulates, trace elements
Coal lock hopper	Carbon monoxide, hydrogen sulfide, tars, oils, naphtha, cyanides, carbon disulfide
Ash lock hopper[a]	Particulates, trace elements
Valves, fittings leaks	Acid gases, other
Product storage	Hydrocarbon vapors, ammonia
Process effluents gas purification[a]	Hydrogen sulfide, carbonyl sulfide, carbon disulfide, hydrogen cyanide, carbon monoxide, carbon dioxide, light hydrocarbons, mercaptans, thiophenes
Process emissions	
Preheat, liquefaction preheater,[b] hydrogen generation, hydrotreating,[b] and solids–liquids separation	Particulates, sulfur and nitrogen oxides
Produce fractionation	Sulfur and nitrogen oxides, hydrocarbon vapors
Catalyst regeneration	Nickel and metal carbonyls, carbon monoxide, sulfur compounds, organics
Water cooling drift and evaporation	Ammonia, sodium, calcium sulfides and sulfates, chlorine, phenols, fluorine, water treatment chemicals, trace elements

[a] Indirect liquefaction
[b] Direct liquefaction

TABLE 25.4

Liquid Waste Streams from Coal Liquefaction Processes

Source	Waste Stream
Coal pile runoff	Particulates, trace metals
Cooling tower blowdown	Dissolved and suspended solids
Boiler blowdown	Calcium, sulfates
Hydrogen generation	Sour or foul wastewater, spantamine scrubbing solution
Acid-gas removal[a]	Hydrogen sulfide, cyanides, phenols
Ammonia recovery[b]	Dissolved ammonia
Phenol recovery[b]	Dissolved phenols, cresylics
Spent reagents and sorbents	Sulfides, sulfates, trace metals, dissolved and suspended solids, ammonia, phenols, tars, oils, hydrogen sulfide, carbon oxides
Leachates from gasifier ash, desulfurization sludge, biosludge, and spent catalysts	Trace elements, organics

[a] Indirect liquefaction
[b] Direct liquefaction

TABLE 25.5
Solid Waste Streams from Coal Liquefaction Processes

Source	Waste Stream
Coal pretreatment	Slag, trace minerals
Gasification[a]	Ash, sulfides, thiocyanate, ammonia, organics, minerals
Steam and power generation	Ash, minerals, trace elements
Filter cake, excess residues from solids–liquids separation[b]	Ash, minerals, trace elements, absorbed heavy hydrocarbons
Spent catalyst from hydrotreating[b]	Metals, absorbed organics, sulfur compounds
Spent catalysts from shift conversion, synthesis, and sulfur recovery[a]	Metals, absorbed organics, sulfur compounds
Sludges from waste treatment and product purification	Trace elements, polycyclic aromatic hydrocarbons
Slag from hydrogen generation[b]	Trace metals, sulfides, ammonia, organics, phenols, minerals

[a] Indirect liquefaction only
[b] Direct liquefaction only

25.7.3 SOLID WASTES

A coal liquefaction plant generates three main types of solid-waste materials: ash and slag from the reactors, sludge from various waste-water treatment units, and spent catalysts from the various catalytic units (Table 25.5).

The possibility of leaching of trace metals from the ash into ground or surface waters is also a possibility. The difference in physical properties between slag and fly ash would suggest a different leaching behavior. In fact, leaching behavior may vary within different types of fly ash especially where there is a size dependence of the physical and chemical properties (Fisher and Natusch, 1979).

25.8 GASIFICATION

Coal gasification (Chapters 20 and 21) offers one of the most versatile methods to convert coal into electricity, hydrogen, and other valuable energy-related products. In the gasifier, coal is typically exposed to steam and carefully controlled amounts of air or oxygen under high temperatures and pressures. Under these conditions, chemical reactions are initiated that typically produce a mixture of carbon monoxide, hydrogen, and other gaseous compounds.

Gasification of coal can produce synthesis gas (syngas) not only from coals having a wide range of heat values but also from low-value carbon feedstocks such as petroleum coke, high-sulfur fuel oil, municipal wastes, and biomass. This flexibility increases the economic value of these resources and lowers costs by providing industry with a broader range of feedstock options.

The gas produced by gasification can contain one or more of contaminants such as ash, char, alkali metals, nitrogen compounds, polynuclear aromatic compounds, tar, sulfur-containing compounds (including hydrogen sulfide), and (on occasion) chlorine-containing compounds. The identity and amount of these contaminants depend on the gasification process and the type of coal.

Conventional gas-cleaning systems are generally the technology of choice for tar removal from the product gas. However, scrubbing cools the gas and produces an unwanted waste stream. Removal of the tar by catalytically cracking the larger hydrocarbons reduces or eliminates this waste stream, eliminates the cooling inefficiency of scrubbing, and enhances the product gas quality and quantity.

Thus, once the raw gas leaves the gasifier, treatment facilities for gas cleaning, conditioning, and separation refine the gas stream using proven commercial technologies that are integral to the gasification plant. Indeed, gasification systems, when using appropriate gas clearing methods (Chapter 23), can meet the strictest environmental regulations pertaining to emissions of sulfur dioxide (SO_2), particulate matter, and toxic compounds other than coal contaminates such as mercury, arsenic, selenium, and cadmium. Gasification provides an effective means of capturing and storing or sequestering carbon dioxide (CO_2), which is present at much higher concentrations and at higher pressures than in streams produced from conventional combustion (Chapters 14 and 15), making them easier to capture.

However, emissions are highly variable and depend on gasifier type, feedstock, process conditions (temperature and pressure), and gas conditioning systems. For example, indirect gasification systems generate flue gas emissions from the combustion of additional fuel, char, a portion of the feedstock.

Cogasification of coal with biomass results (depending on the type of biomass) in a change in the composition of the emissions as the ash may contain higher proportions of heavy metals. A major concern with these feedstocks is the potential for heavy metals to leach into the environment following ash disposal.

Incompletely converted coal and ash particulate removal is accomplished with cyclones, wet scrubbing, or high-temperature filters. A cyclone can provide primary particulate control, but is not adequate to meet gas turbine specifications. A high-temperature filter system can be used to remove particulates to acceptable levels for gas turbine applications.

Water scrubbing can remove up to 50% of the tar in the product gas, and when followed by a venturi scrubber, the potential to remove the remaining tars increases to 97%. The wastewater from scrubbing can be cleaned using a combination of a settling chamber, sand filter, and charcoal filter. This method is claimed to clean the wastewater discharge to within drinking water standards.

Gasification can be integrated with other technologies for advanced power generation (Chapters 20 through 22) and the resulting systems are highly efficient. Systems using advances in gasification and related components can achieve efficiencies of up to 60%, compared with an efficiency limit of 40% for conventional plants.

Thus, the environmental benefits of gasification stem from the capability to achieve extremely low sulfur oxide (SOx), nitrogen oxide (NOx) and particulate emissions from burning coal-derived gases. In an *integrated gasification combined-cycle* (IGCC) plant, the synthesis gas produced is virtually free of fuel-bound nitrogen. Selective catalytic reduction (SCR) can be used to reach levels comparable to firing with natural gas if required to meet more stringent emission levels. Other advanced emission control processes are being developed that could reduce nitrogen oxide emissions from hydrogen fired turbines to as low as two parts per million.

Coal gasification may offer a further environmental advantage in addressing concerns over the atmospheric buildup of greenhouse gases, such as carbon dioxide. If oxygen is used in a coal gasifier instead of air, carbon dioxide is emitted as a concentrated gas stream in syngas at high pressure. In this form, it can be captured and sequestered more easily and at lower costs. By contrast, when coal burns or is reacted in air, 79% of which is nitrogen, the resulting carbon dioxide is diluted and more costly to separate.

25.9 CLEAN COAL TECHNOLOGIES

Coal, the fuel that helped initiate and maintain the Industrial Revolution, remains the primary fuel for electricity generation worldwide as well as in the United States. However, in recent years, issues related to global climate change has attained political prominence thereby potentially limiting the role of coal can play under various scenarios of greenhouse gas regulations. There are, however, serious and meaningful questions about the facts relating to climate change and whether or not the evidence on which these so-called facts are based is believable (Bell, 2011).

Nevertheless, coal use can give rise to pollution and the polluting effects of coal mining and combustion on the air, water and soil remain as significant a challenge today, with coal providing such a large fraction of global primary energy, as they did during the early ages of coal use. With both U.S. and worldwide supplies of coal in relative abundance compared to oil and gas, a number of concepts have been proposed to continue taking advantage of this inexpensive and comparatively widespread resource while minimizing the environmental impacts associated with its use.

While there is no one technology that will serve as a panacea to halt pollution, the idea of *clean coal* may be better understood as a collection of different technologies, each with its own benefits and drawbacks that are typically limited by the character of the coal being employed for the generation of energy.

The focus of the United States Clean Coal Program is to seek methods by which coal can be used cleanly and efficiently in a variety of industrial operations (Chapter 22). The capacity of the environment to absorb the effluents and the impacts of energy technologies is not unlimited as some would have us believe. The environment should be considered to be an extremely limited resource, and discharge of chemicals into it should be subject to severe constraints. Indeed, the declining quality of raw materials, especially fossil fuels that produce the energy for industrial and domestic use give rise to a variety of emissions (gaseous, liquid, and solid) and dictates that more material must be processed to provide the needed energy. And the growing magnitude of the effluents from fossil fuel processes has moved above the line where the environment has the capability to absorb such (gaseous) effluents without disruption.

In a more general perspective, coal and its companion fossil fuels are not the only feedstocks that produce emissions that are of harm to the (Table 19.5). In fact, process operators must make very serious attempts to ensure that natural resources such as air, land, and water remain as unpolluted as possible and the environmental aspects of such an operation are carefully addressed. In addition, there must always be reason in the minds of the regulators. It may not always be physically and economically possible or necessary to clean up every last molecule of pollutant. But such rationale must not be used as a free license to pollute nor should the law be so restrictive that industry cannot survive. Rational thought must prevail.

Although the focus of this chapter is on the cleanup of potentially harmful gaseous emissions from coal plants, there is also the need to recognize that gas processing operations, as a consequence of the chemicals employed (Speight, 2007a,b) can also cause other environmental damage. For example, the spill of an acid solution or any solution that might be used in gas washing operations can cause severe damage to the flora and fauna as well as to the aquatic life in the region. Thus, caution is advised from all environmental aspects and not just that most closely related to the process operation.

Indeed, any of the products from a gas cleaning plant can contain contaminants. The very nature of the gas cleaning plant dictates that this be so but there are many options available to assist in the cleanup of the plant products. However, it is very necessary, in view of the scale of such plants, that controls be placed on the release of materials to the environment to minimize the potential damage to the immediate environment; damage to the environment must be avoided.

Coal combustion releases substantial quantities of carbon dioxide (CO_2) to the atmosphere that can participate in one or more of the various cycles that exist in the biosphere and atmosphere (Clark, 1989; Graedel and Crutzen, 1989; Frosch and Gallopoulos, 1989; Maurits la Rivière, 1989; Schneider, 1989; Speight and Lee, 2000) have dominated the global environmental system for millennia. In addition, other coal plant emissions such as sulfur dioxide (SO_2) and nitrogen oxides (NO) are major contributors to acid rain (Manowitz and Lipfert, 1990).

The major waste streams leaving many coal plants might be water vapor and carbon dioxide. The former has little effect on the environment while the effect of the latter on the environment, while being considered proven, is still open to considerable debate, speculation, and often violent scientific and emotional disagreement. Nevertheless, as common sense alone must tell us, it must be assumed that the effects of releasing of unlimited amounts of carbon dioxide to the atmosphere will, and most probably does, cause adverse, if not severe, perturbations to the environment. However, it is all a matter of degree. The discharge of liquid water, because the potential for dissolved contaminants is real, is a different issue. Water may not be water.

If it can be assumed that water used for cooling purposes in a coal utility plant is predominantly the type that passes through cooling towers, the gaseous emissions from such a plant will be carbon dioxide (CO_2), sulfur oxides (SOx, where x = 2 or 3), nitrogen oxides (NOx, where x = 1 or 2), as well as sundry other sulfur compounds of which hydrogen sulfide (H_2S), carbonyl sulfide (COS), carbon disulfide (CS_2), and mercaptans (RSH) are examples.

For the last several years, there has been much concern about global warming (the *greenhouse effect* as the phenomenon is often unaffectionately called) (Ember et al., 1986; Schlesinger and Mitchell, 1987; Smith, 1988; Douglas, 1990; White, 1990) whereby increased concentrations of carbon dioxide in the atmosphere, produced by the combustion of fossil fuels are believed, with much debate and conjecture from both proponents and opponents of the theory, to lead to global long-term climatic changes (Stobaugh and Yergin, 1983). But to maintain the perspective, the possibility that changes in the amount of carbon dioxide in the atmosphere could lead to changes in world climate is not new and has been suggested since at least the end of the last century.

In the greenhouse effect (Schneider, 1989; Speight and Lee., 2000) carbon dioxide and water vapor in the atmosphere absorb (or trap) part of the long wave (infrared) radiation from the earth's surface, while, at the same time, the atmosphere allows passage of the shortwave (visible) radiation from the sun. An increase in the concentration of *greenhouse gases* (such as carbon dioxide) in the atmosphere causes an increase in the absorption of the radiation from the earth, which, ultimately, causes an increase in the surface temperature. As a very short summary, the differential in the behavior of the atmosphere toward outgoing/incoming radiation plays a similar role to that of the glass roof in a small horticultural ecosystem, that is, a greenhouse. Hence the name *greenhouse effect.*

However, the term *greenhouse effect* is deemed by some to be inappropriate because the actual events are much more complex than this simple example absorption/radiation example would indicate since many other physical processes and climatic effects must be included in the model. And there are those who consider that the extent of the coupling of the effects is sufficiently in doubt and that it cannot be stated, or estimated, with any degree of certainty whether or not the surface temperature of the earth will increase, is increasing, or may even be decreasing.

In fact, it can only be stated with any degree of certainty that opinions will differ and the debate will continue. But instincts alone tell us that the discharge of foreign species, or even the discharge of a large surplus of indigenous species, must alter the delicate systems in an adverse fashion. Perhaps it can be likened to the injection of excessive amounts of steroid materials into the human body, without any short-term beneficial effect. Perhaps the atmosphere is unaffected in the short term but then the long-term effects take over. In the United States, the Clean Air Act of 1970 and the ensuing amendments to it (United States Congress, 1990; Stensvaag, 1991), as well as the emissions standards for other countries, provide the basis for the regulatory constraints imposed on air emissions. The specific regulated pollutants are particulates, sulfur dioxide, photochemical oxidants, hydrocarbons, carbon monoxide, nitrogen oxides, and lead (Mintzer, 1990).

There are several areas where pollutants can be produced during coal conversion. Indeed, a major generic source of air polluting emissions from any plant is in the preparation of the raw feedstock, specifically the crushing, screening, and storage operations associated with a coal conversion (e.g., utility) plant and the subsequent feeding to the reactors. Another major source of particulate matter arises from the ancillary combustion operations where fly ash is produced when coal, or ash-containing carbonaceous materials, is burned to generate process heat, raise steam, or produce electric power.

Gases leaving a reactor whether they are product gases from a gasifier or by-product gases from a carbonization reactor contain components that may be categorized as desirable, neutral, or undesirable. A *desirable component* is only defined here as desirable or undesirable primarily from a process view. A desirable component should be present in the end product, an undesirable one be absent. A component may also be undesirable because its presence in a processing stage could be detrimental.

For example, hydrogen sulfide is undesirable in gas streams because its level specified for pipeline gas contracts is often orders of magnitude less than what is present in off-gases. In addition to hydrogen sulfide, undesirable components include ammonia, hydrogen chloride, carbon dioxide, and particulate matter depending upon the ultimate use of the gas stream (Kohl and Nielsen, 1997; Bhanarkar et al., 2010).

Another effect that must also be considered in gas cleaning operations is that the gases exiting reactors such as those used for coal gasification and carbonization are hot with temperatures ranging from 400°C to 1500°C (750°F to 2730°F). Thus, a simple gas cleaning procedure involves cooling the gas, generally by indirect heat exchange, to a temperature at which the water and hydrocarbons condense; tars (liquids) as are often evident as products of coal processing may also be condense.

Gas cooling by scrubbing with water is normally employed as one of several means of gas cleaning and purification. The water will remove from the gas stream any soluble gases and particulate matter. Furthermore, even though the acid gases (carbon dioxide and hydrogen sulfide) by themselves are not very soluble in water, the presence of ammonia (a weak alkali and soluble in water) in the gas stream or in the water will affect the solubilization of the carbon dioxide and hydrogen sulfide. Particulates removal is essential although there is still some debate about the requirements, which are considered by many to be too generic insofar as many standards do not relate, specifically, either to chemical composition or to particulate size. But increasing emphasis is being placed on inhalable particulates less than 15 μm in size, especially on fine particulates less than 2–3 μm.

Among the conventional control technologies are cyclones, scrubbers, electrostatic precipitators, and fabric filters. Electrostatic precipitators, also suitable for fine particulate removal, operate on the principle of charging the particles with ions and then collecting the ionized particles on a surface from which they are subsequently removed. Electrostatic precipitators are in common use for fly-ash removal from power plant flue gases. All of the conventional devices have certain advantages and disadvantages and efficiency varies (Figure 19.9).

Electrostatic precipitators that are suitable for removal of fine particulate matter, such as fly ash, from power plant flue gases operate on the principle of charging the particles with ions followed by collection of the ionized particles on a surface from which they are subsequently removed. Fabric filtration, which is used for removal of extremely fine particulate matter, is accomplished in a *baghouse* in which are hung a number of filter bags through which the gas stream flows.

Amendments to the Unites States Clean Air Act of 1967 were made in 1970 and again in 1990 (United States Congress, 1990) provided for the establishment of national ambient air quality standards for, as an example, sulfur dioxide.

REFERENCES

Adriano, D.C., Page, A.L., Elseewi, A.A., Chang, A.C., and Straughan, I. 1980. Utilization and disposal of fly ash and other coal residues in terrestrial ecosystems: A review. *Journal of Environmental Quality*, 3(9): 333–344.

Argonne. 1990. *Environmental Consequences of, and Control Processes for, Energy Technologies*. Argonne National Laboratory. Pollution Technology Review No. 181. Noyes Data Corporation, Park Ridge, NJ.

Bach, W. 1972. *Atmospheric Pollution*. McGraw-Hill, New York.

Banaszak, K.J. 1980. Coal as aquifers in the United States. In *Surface Mining Hydrology, Sedimentology, and Reclamation*, D.H. Graves and R.W. DeVore (Eds.). Proceedings. Symposium on Surface Mining Hydrology, Sedimentology, and Reclamation, University of Kentucky, Lexington, KY, December 1–5, pp. 235–241.

Bauer, E.R., Babich, D.R., and Vipperman, J.R. 2006. Equipment noise and worker exposure in the coal mining industry. Information Circular 9492. CDC/NIOSH Office of Mine Safety and Health Research. US Department of Health and Human Services, Washington, DC.

Bell, L. 2011. *Climate of Corruption: Politics and Power Behind the Global Warming*. Greenleaf book Group Press, Austin, TX.

Berkowitz, N. 1985. *The Chemistry of Coal*. Elsevier, Amsterdam, the Netherlands.

Bhanarkar, A.D., Majumdar, D., Nema, P., and George, K.V. 2010. Emissions of SO_2, NOx and particulates from a pipe manufacturing plant and prediction of impact on air quality. *Environmental Monitoring and Assessment*, 169(1): 677–685.

Bhumbla, D. 1992. Nitrogen transformations on soils amended with fly ash and sewage sludge mixtures. In *Proceedings. Association of Abandoned Mine Land Programs, 14th Annual Conference*, Chicago, IL.

Burdett, N.A., Cooper, J.R.P., Dearnly, S., Kyte, W.S., and Turnicliffe, M.F. 1985. The application of direct limestone injection to UK power stations, *Journal of the Institute of Engineering*, 58: 64–69, No. B, London, U.K.

Chadwick, M.J., Highton, N.H., and Lindman, N. 1987. *Environmental Impacts of Coal Mining and Utilization*. Pergamon Press, Oxford, U.K.

Clark, W.C. 1989. Managing planet earth. *Scientific American*, 261(3): 46054.

Clark, J.W., Viesmann, W., and Hammer, M. 1977. *Water Supply and Pollution Control*. Harper and Row Publishers, New York.

Colinet, J.F., Rider, J.P., Listak, J.M., Organiscak, J.A., and Wolfe, A.L. 2010. Best practices for dust control in coal mining. Information Circular 9517. CDC/NIOSH Office of Mine Safety and Health Research. US Department of Health and Human Services, Washington, DC.

Commission on Energy and the Environment. 1981. Coal and the environment. H.M. Stationary Office, London, U.K.

Costello, C. 2003. Acid mine drainage: Innovative treatment technologies. U.S. Environmental Protection Agency, Office of Solid Waste and Emergency Response Technology Innovation Office, Washington, DC.

Couch, G.R. 1991. Advanced coal cleaning technology. Report No. IEACR/44. IEA Coal Research, International Energy Agency, London, U.K.

Cox, D.B., Chu, T.-Y., and Ruane, R.J. 1977. Quality and treatment of coal pile runoff. *Proceedings. 3rd Symposium on Coal Preparation. NCA/BCR Coal Conference and Expo IV*, Washington, DC, October, pp. 252–277.

Crane, A. and Liss, P. 1985. Carbon dioxide, climate and the sea. *New Scientist*, 108: 50.

Cummins, A.B. and Given, I.A. 1973. *Mining Engineering Handbook*. Society of Mining Engineers of AIME, New York.

Douglas, J. 1990. *EPRI Journal*. Electric Power Research Institute, June: 4.

Doyle, W.S. 1976. *Strip Mining of Coal: Environmental Solutions*. Noyes Data Corporation, Park Ridge, NJ.

Ember, L.R., Layman, P.L., Lepkowski, W., and Zurer, P.S. 1986. *Chemical and Engineering News*, November 24: 14.

Fisher, G.L. and Natusch, D.F.S. 1979. Size-dependance of the physical and chemical properties of coal fly ash. In *Analytical Methods for Coal and Coal Products*, C.E. Karr Jr. (Ed.). Academic Press, Inc., New York, Vol. III, Chapter 54.

Frosch, R.A. and Gallopoulos, N.E. 1989. Strategies for manufacturing. *Scientific American*, 261(3): 144.

Galloway, R.L. 1882. *A History of Coal Mining in Great Britain*. Macmillan and Co., London, U.K.

Graedel, T.E. and Crutzen, P.J. 1989. The changing atmosphere. *Scientific American*, 261(3): 58–68.

Groenewold, G.H., Rehm, B.W., and Cherry, J.C. 1981. Depositional setting and groundwater quality in coal-bearing sediments and spoils in western North Dakota. In *Recent and Ancient Non-Marine Depositional Environments*, F.G. Etheridge and R.M. Flores (Eds.). Special Publication No. 31. Society for Economic Paleontologists and Mineralogists, Tulsa, OK, pp. 157–167.

Guney, M. 1968. Oxidation and spontaneous combustion of coal. *Colliery Guardian*, 216: 105, January–February, London, U.K.

International Energy Agency. 1985. The clean use of coal. Organization for Economic Co-operation and Development, Paris, France.

Kimball, R.F. and Munro, N.B. 1981. A critical review of the mutagenic and other genotoxic effects of direct coal liquefaction. Report ORNL-5721. Oak Ridge National Laboratory, Oak Ridge, TN.

Kohl, A.L. and Nielsen, R.B. 1997. *Gas Purification*, 5th edn., Gulf Professional Publishers, Houston, TX.

Logan, T. 1992. Mine spoil reclamation with sewage sludge stabilized with cement kiln dust and FGD byproduct. In *Proceedings. Association of Abandoned Mine Land Programs, 14th Annual Conference*, Chicago, IL, August.

Manowitz, B. and Lipfert, F.W. 1990. In *Geochemistry of Sulfur in Fossil Fuels*. W.L. Orr and C.M. White (Eds.). American Chemical Society, Washington, DC, Chapter 3.

Maurits la Rivière, J.W. 1989. Threats to the world's water. *Scientific American*, 261(3): 80–84.

McNay, L.M. 1971. Coal refuse fires: An environmental hazard. Bureau of Mines, US Department of Interior, Washington, DC.

Mintzer, I.M. 1990. Energy, greenhouse gases, and climate change. *Annual Reviews of Energy*, 15: 513–550.

Nowacki, P. 1979. *Coal Liquefaction Processes*. Noyes Data Corporation, Park Ridge, NJ, p. 71.

Nowacki, P. 1980. *Lignite Technology*. Noyes Data Corporation, Park Ridge, NJ.

Park, C.C. 1987. *Acid Rain: Rhetoric and Reality*. Methuen, London, U.K.

Paul, B., Chaturvedula, S., Chatterjee, S., and Paudel, H. 1993b. Return of fly ash to the mine site. In *Proceedings. Pittsburgh Coal Conference*, Pittsburgh, PA, September.

Paul, B., Esling, S., Chaturvedula, S., and Chatterjee, S. 1993a. Use of coal combustion byproducts as structural fill in reclamation: Impacts on groundwater. In *Proceedings. National Mined Land Center Reclamation Conference*, Collinsville, IL, April.

Pearce, F. 1982. The menace of acid rain. *New Scientist*, 95: 419–423.

Powell, J.D. and Larson, J.D. 1985. Relation between ground-water quality and mineralogy in the coal-producing Norton Formation of Buchanan County, Virginia. Paper 2271. Water-Supply Paper 2274, United States Geological Survey, Alexandria, VA.

Ripp, J. 1984. Smart coal-pile management can cut heating-value losses. *Power*, February: 53–54.

Sage, B. 1980. Acid drops from fossil fuels. *New Scientist*, 85: 743–745.

Scanlon, D.H. and Dugan, J.C. 1979. Growth and element uptake of woody plants on fly ash. *Environmental Science and Technology*, 13: 311–315.

Schlesinger, M.E. and Mitchell, J.F.B. 1987. Climate model simulations of the equilibrium climatic response to increased carbon dioxide. *Reviews of Geophysics*, 25: 760–798.

Schneider, S.H. 1989. The changing climate. *Scientific American*, 261(3): 70–79.

Smith, I.M. 1988. CO_2 and climatic change. Report No. IEACR/07. International Energy Agency, Coal Research, London, U.K.

Solomon, P.R. and Fletcher, T.H. 1994. Impact of coal pyrolysis on combustion. *Proceedings of the Combustion Institute*, 25: 463–474.

Speight, J.G. 2007a. *The Chemistry and Technology of Petroleum*, 4th edn. Taylor & Francis Group, Boca Raton, FL.

Speight, J.G. 2007b. *Natural Gas: A Basic Handbook*. GPC Books, Gulf Publishing Company, Houston, TX.

Speight, J.G. 2008. *Synthetic Fuels Handbook: Properties, Processes, and Performance*. McGraw-Hill, New York.

Speight, J.G. 2011. *Handbook of Industrial Hydrocarbon Processes*. Gulf Professional Publishing, Elsevier, Oxford, U.K.

Speight, J.G. and Lee, S. 2000. *Environmental Technology Handbook*, 2nd edn. Taylor & Francis, New York.

Stensvaag, J.-M. 1991. *Clean Air Act Amendments: Law and Practice*, John Wiley & Sons Inc., New York.

Stobaugh, R. and Yergin, D. 1983. *Energy Future*, Vintage Books/Random House, New York.

Suess, M.J., Grefen, K., and Reinisch, D.W. 1985. *Ambient Air Pollutant from Industrial Sources*. WHO, Regional Office for Europe, Elsevier, Amsterdam, the Netherlands.

UNGA (United Nations General Assembly). 1980. Technologically modified exposures to natural radiation. UN Scientific Committee on the Effects of Atomic Radiation, UN Publication No. A/AC.82IR.37I, Vienna, Austria.

United States Congress. 1990. Public law 101-549. An act to amend the clean air act to provide for attainment and maintenance of health protective national ambient air quality standards, and for other purposes. November 15.

White, R.M. 1990. The great climate debate. *Scientific American*, 263(1): 36–43.

26 Coal and Energy Security

26.1 INTRODUCTION

Throughout this book, frequent reference has been made to the use of coal as a source of energy and its role in determining whether or not the United States can develop energy independence (also often referred to as *energy security*) insofar as energy independence means a reduced reliance on foreign source of energy, particularly foreign oil.

The distribution of coal reserves around the world varies significantly from the distribution of petroleum and natural gas reserves. For example, substantial reserves of coal occur in the United States and Russia but not in the petroleum exporting countries of the Middle East. Thus, there is a strong case to be made for energy independence (at a level to be determined). On the other hand, environmental issues may be the basis for equally strong arguments energy independence (at any level) based on coal. However, the environment and coal mining can coexist in harmony when mined land is restored to its pre-mining condition and the lakes, rivers, and streams are protected from environmental damage.

Energy independence has been variously played up and then ignored by the members of the Congress of the United States (as politicians determine which path will garner them the most votes for reelection) since the first Arab oil embargo in 1973. In spite of the calls for energy independence, the United States is still dependent upon imports of foreign petroleum with no end in sight (Speight, 2011a,b).

The dependency on foreign petroleum by the United States has increased steadily since the mid-1980s when the daily imports of petroleum and petroleum products (as percentage of consumption) increased from approximately 50% thence to approximately 60% in the early 1990s to the almost unbelievable current (2011a, 2011b) levels of 65%–70% of the daily petroleum and petroleum products.

Briefly, in 2010 the United States imported 11.8 million barrels per day of petroleum crude oil and refined petroleum products and consumed 19.1 million barrels per day of petroleum products during 2010. While the United States was the largest petroleum consumer in the world, the United States was third in crude oil production at 5.5 million barrels per day. Petroleum products imported by the United States during 2010 included gasoline, diesel fuel, heating oil, jet fuel, chemical feedstocks, asphalt, and other products, and the net imports of petroleum other than crude oil were 2% of the petroleum consumed in the United States during 2010 (U.S. Energy Information Administration, 2011).

It is not *rocket science* to know that when imports of a necessary commodity exceed domestic production of this same commodity a dangerous situation exists. Especially when the commodity is supplied by countries where the governments of those countries cannot, by any stretch of the imagination, be classified as governments that are stable.

Events such as civil war, *coup d'état*, and the occasional (some would say *frequent*) labor strike will occur with sufficient frequency to severely restrict imports into the United States. It is even more worrying that the countries where highest levels of disruption can occur contain the majority of the proven worldwide petroleum reserves and provide the majority of the imports into the United States.

For example, the United States currently requires approximately 18,000,000 barrels of petroleum per day and, hypothetically, through events that are not within the control sphere of the United States, a shortfall of imported petroleum on the order of 2,000,000 barrels per day would leave a large gap in the domestic energy availability.

Furthermore, the petroleum industry itself cannot be held immune from petroleum shortages. Periods when there has been overproduction, when low prices and profits led petroleum producers to devise ways to restrict output and raise prices, and periods of underproduction have been known. Supply and demand is one thing, but holding a country to ransom is another! In addition, the petroleum supply pessimists would have everyone believe that the *era of petroleum* is over, while realistically there is sufficient petroleum to last for another 100 years, provided recovery methods and refining technology advance with time.

In the years to come, technology will focus on the search for new sources of petroleum (including heavy oil), while refining will focus on higher rates of conversion of heavy oil to salable products (Speight, 2009). There is no recognition here of the so-called undiscovered petroleum, which is difficult to define (and may even be mythological). The focus is on petroleum left in the ground when wells have been shut in.

It might be argued that the degree of dependence has no impact on energy security as long as foreign petroleum is imported from secure sources. However, if the degree of dependence on nonsecure sources increases, energy security would be in jeopardy. In this case, vulnerability would increase and economic and national security of individual petroleum-importing countries would be compromised.

Dependency and vulnerability to petroleum imports in the United States and, for that matter, in other petroleum-importing countries can be reduced not only by diversification of suppliers (and this is largely controlled by the petroleum holding countries) but also by diversification of energy sources (Table 26.1). There would be a cost, but this must be measured against the cost of future disruptions due to geo-political issues that cannot be controlled by the United States.

The dependence on foreign petroleum by the United States is a threat to national security and to the economy. Growing demand and shrinking domestic production means that the United States is importing more and more petroleum each year—much of it from the countries controlled by unfriendly and/or unstable governments.

TABLE 26.1

Future Sources of Fuels

Nonconventional oil:

Tar sand—Athabasca Canada (in situ, mining)

Synthetic crude (from coal and/or natural gas)

Nonconventional gas:

Coalbed methane (CBM)

Gas shale

Tight gas sands

Gas in geo-pressured aquifers

Conventional hydrocarbons in nonconventional locations:

Deep water (>500 or <1000 m)

Arctic

Antarctic

Uneconomic conventional hydrocarbons:

Small fields: 1–10 million barrels

Improved EOR technologies

Surrogate nonconventional hydrocarbons:

Oil shale

Gas hydrates

As each year passes, the Congress fails to reduce this dependency and continues to rely on factors outside of the country's control, thereby exposing the United States to greater security risks in the name of unrestrained consumption without any thought of the consequences of such actions (or *inactions*).

Currently, petroleum use outpaces new petroleum discoveries, with the world using about 12 billion more barrels (12×10^9 barrels) per year than are discovered. The growing imbalance between supply and demand means higher petroleum prices and the ever-present threat of supply disruptions.

The only real solution is for the United States to reduce the demand for petroleum and therefore the economic and security risks of dependence on petroleum imports.

While increasing the efficiency of vehicles, it is necessary to develop alternate sources of liquid fuels.

As petroleum supplies decrease, the desirability of producing gas from other carbonaceous feedstocks will increase, especially in those areas where natural gas is in short supply. It is also anticipated that costs of natural gas will increase, allowing coal gasification to compete as an economically viable process. Research in progress on a laboratory and pilot-plant scale should lead to the invention of new process technology by the end of the century, thus accelerating the industrial use of coal gasification.

The conversion of the gaseous products of gasification processes to synthesis gas, a mixture of hydrogen (H_2) and carbon monoxide (CO), in a ratio appropriate to the application, needs additional steps, after purification. The product gases—carbon monoxide, carbon dioxide, hydrogen, methane, and nitrogen—can be used as fuels or as raw materials for chemical or fertilizer manufacture.

Thus, through the careful and responsible use of coal, the United States can become more energy secure by putting coal to work in ever increasing amounts while still paying attention to the environmental consequences. Instead of investing hundreds of billions of dollars expanding petroleum production in the Persian Gulf and other unstable regions, the United States needs to defray some of this money to developing domestic resources.

To immediately ease the addiction to petroleum, the United States should make a national commitment to reduce petroleum dependency by at least 2.5 million barrels per day within a decade and to set longer term goals that secure our energy future without depending on unstable and hostile areas of the world.

Since a reduction in dependency cannot happen overnight. The so-called leaders in Washington (the Members of Congress) must make it happen by use of common-sense policies. Instead, the energy policy leaves the United States too dependent on petroleum and the unstable regimes that supply it.

26.2   ENERGY SECURITY

Energy security is the continuous and uninterrupted availability of energy, to a specific country or region. The security of energy supply conducts a crucial role in decisions that are related to the formulation of energy policy strategies. The economies of many countries are depended by the energy imports in the notion that their balance of payments is affected by the magnitude of the vulnerability that the countries have through imported petroleum.

Energy security has become uncertain over the past four decades because of (1) the political instability of several energy producing countries; (2) the manipulation of energy supplies; (3) the competition over energy sources now that China and India are providing large additional markets for petroleum; (4) attacks on supply infrastructure; as well as (5) accidents, natural disasters, rising terrorism, and dominant countries' reliance to the foreign oil supply.

The limited supplies, uneven distribution, and rising costs of petroleum and natural gas, create a need to change to more sustainable energy sources. With as much dependence that the United States currently has for petroleum and with the peaking limits of oil production (Hubbert's peak), various countries are already feeling (Hubbert's peak) and societies will begin to feel the decline in the resource that we have become dependent upon (Speight, 2011b). Energy security has become one of the leading issues in the world today as oil and other resources have become vital to the world's people.

Briefly, the Hubbert theory of peak oil assumes that petroleum reserves will not be replenished (i.e., that abiogenic replenishment is negligible) and predicts that future world petroleum production must inevitably reach a peak and then decline as these reserves are exhausted (Speight, 2011a,b and references cited therein). Controversy surrounds the theory since predictions for the time of the global peak is dependent on the past production and discovery data used in the calculation.

For the United States, the prediction of petroleum being a depletable resource turned out to be correct (as it would with any naturally occurring resource), and, after the United States peaked in 1971 and thus lost its excess production capacity, the OPEC consortium was (literally) given a free hand at the manipulation of petroleum prices. Since then petroleum production in several other countries has also peaked. However, for a variety of reasons, it is difficult to predict the oil peak in any given region. Based on available production data, proponents have previously (and incorrectly) predicted the peak for the world to be in years 1989, 1995, or in the 1995–2000 period. Other predictions chose 2007 and beyond for the peak of petroleum production.

But more important, several trends that should have been established in the wake of the decreasing crude prices have never been put into practice, for example, and most important, the failure of politicians to recognize the need for a measure of energy independence through the development of alternate resources as well as the development of technologies that would assist in maximizing petroleum recovery.

The fact that petroleum producing countries in the Middle East provide more than 50% of the world's consumption (Speight, 2011b and references cited therein) is indicative for the low diversification of energy sources and the accompanying risks on smooth energy supply. The diversification that is offered by the alternative supplies from Russia and Africa cannot provide a sound solution for a supply disruption that may occur in the Middle East region. An overview of the petroleum market and the related risks and incidents clearly indicate that the risks associated to energy supply are many. War and civil conflicts might have been replaced, to some extent, by weather conditions and monopolistic practices, but they are still playing a crucial factor in energy supply. Therefore, the high dependency that most countries have on energy imports made essential for the policy makers to focus on the concept of security of energy supply. In this context, the need to assess the current energy system and the risks of energy disruptions is essential in order to better design and adopt the required policies is obvious.

Uncertainty about future demand for petroleum—which will influence how quickly the remaining petroleum is used—contributes to the uncertainty about the timing of peak petroleum production. It is very likely that crude petroleum will continue to be a major source of energy well into the future, and world consumption of petroleum products may even grow during the next four decades.

Future world petroleum demand will depend on such uncertain factors as world economic growth, future government policy, and consumer choices.

Environmental concerns about gasoline's emissions of carbon dioxide, which is a greenhouse gas, may encourage future reductions in petroleum demand if these concerns are translated into policies that promote biofuels, although the uncertainty of the extent to which biofuels will be a major facet of energy production in the foreseeable future is still debatable (Lee et al., 2007; Speight, 2008, 2011a,c; Giampietro and Mayumi, 2009; Langeveld et al., 2010; Nersesian, 2010; Seifried and Witzel, 2010 and references cited therein).

Consumer choices about conservation also can affect petroleum demand and thereby influence the timing of a peak. For example, if consumers in the United States were to purchase more fuel-efficient vehicles in greater numbers, this could reduce future petroleum demand in the United States, potentially delaying a time at which petroleum supply is unable to keep pace with petroleum demand. Such uncertainties that lead to changes in future petroleum demand ultimately make estimates of the timing of a peak uncertain. Specifically, using future annual increases in world petroleum consumption, ranging from 0%, to represent no increase, to 3%, to represent a large increase (from the various scenarios examined), it may take up to 75 years for the peak to occur.

Factors that affect petroleum exploration and production also create uncertainty about the rate of production decline and the timing of the peak. The rate of decline after a peak is an important consideration because a decline that is more abrupt will likely have more adverse economic consequences than a decline that is less abrupt.

Consumer actions could help mitigate the consequences of a near-term peak and decline in petroleum production through demand-reducing behaviors such as carpooling; teleworking; and eco-driving measures, such as proper tire inflation and slower driving speeds. Clearly these energy savings come at some cost of convenience and productivity, and limited research has been done to estimate potential fuel savings associated with such efforts. However, estimates by the United States Department of Energy indicate that teleworking could reduce total fuel consumption by 1%–4%, depending on whether teleworking is undertaken for two days per week or the full 5-day week, respectively.

If the peak occurs in the more distant future or the decline following a peak is less severe, alternative technologies have a greater potential to mitigate the consequences. The United States Department of Energy projects that the alternative technologies have the potential to displace up to the equivalent of 34% of annual U.S. consumption of petroleum products in the 2025 through 2030 time frame. However, the United States Department of Energy also considers these projections optimistic—since the assumption is that sufficient time and effort are dedicated to the development of these technologies to overcome the challenges they face.

The prospect of a peak in petroleum production presents problems of global proportion whose consequences will depend critically on our preparedness. The consequences would be most dire if a peak occurred soon, without warning, and were followed by a sharp decline in petroleum production because alternative energy sources, particularly for transportation, are not yet available in large quantities. Such a peak would require sharp reductions in petroleum consumption, and the competition for increasingly scarce energy would drive up prices, possibly to unprecedented levels, causing severe economic damage. While these consequences would be felt globally, the United States, being the largest consumer of petroleum and one of the nation's most heavily dependent on petroleum for transportation, is especially vulnerable.

The subject of energy security has been for many years an important concern among energy policy makers. The devastating short- and long-term effects of the petroleum crisis of 1973 on the global economy made clear since then that the need to guarantee the availability of energy resource supply in a sustainable and timely manner with the energy price being at a level that will not adversely affect the economic performance of the European continent is of utmost importance.

The United States will find that the recent (2011) actions (even though support was voiced for these actions) that gave rise to the name *Arab Spring* are not the answer to stable supplies of petroleum. The continuous destabilization of the Middle East, growing fears about further military intervention in this fragile geopolitical area, environmental catastrophes, the advent of organized terrorist operations across the globe, political risks, and legal reforms have profoundly increased the possibility of potential energy disruptions that will have detrimental effects, considering the dependence of Europe to external energy suppliers.

The popularity of the *energy risk-premium* concept has led to the formulation of a vast pool of knowledge encompassing an abundance of derivatives models. Traders have the ability to hedge against various risks and create risk neutral portfolios using a diversified mix of energy derivative securities. However, up to now the risk-premium concept has not been used in the energy domain to quantify an energy security indicator. The main reason for this is that current techniques used in the energy domain do not incorporate the necessary probabilistic models that reflect on risk parameters associated with rare *catastrophic* events that cause adverse movements on the spot price of the underlying instrument.

A catastrophic event in general (or catastrophe) is an event that *has severe losses, injury, or property damage; affects large population of exposures; and is caused by natural or handmade events.* Examples of catastrophic events include natural disasters (hurricanes, earthquakes, floods) and terrorist attacks. The last 20 years, natural catastrophes have been happening with increasing intensity.

Catastrophic events in the energy context have a slightly different meaning than in many other contexts. They can be events with low frequency of occurrence that cause the spot price of the energy commodity to soar (*price volatility*). Usually a price increase due to the catastrophic event does not have a lasting effect and the spot price tends to return to (or close to) its initial value. To combat such events, it will be necessary for the nonpetroleum producing nations to commence development of sources of energy other than petroleum.

Once all risk indicators associated with catastrophic events have been identified and properly estimated in terms of frequency of occurrence and impact in the underlying spot price, the respective premium can be calculated under the common assumptions of derivatives pricing.

In the longer term, there are many possible alternatives to using petroleum, including using biofuels and improving automotive fuel efficiency, but these alternatives will require large investments, and in some cases, major changes in infrastructure or break-through technological advances. In the past, the private sector has responded to higher petroleum prices by investing in alternatives, and it is doing so now. Investment, however, is determined largely by price expectations, so unless high petroleum prices are sustained, we cannot expect private investment in alternatives to continue at current levels. If a peak were anticipated, petroleum prices would rise, signaling industry to increase efforts to develop alternatives and consumers of energy to conserve and look for more energy-efficient products.

Finally, with the onset of the twenty-first century, petroleum technology is driven by the increasing supply of heavy petroleum with decreasing quality and the fast increases in the demand for clean and ultraclean vehicle fuels and petrochemical raw materials. As feedstocks to refineries change, there must be an accompanying change in refinery technology. This means a movement from conventional means of refining heavy feedstocks using (typically) coking technologies to more innovative processes (including hydrogen management) that will produce the ultimate amounts liquid fuels from the feedstock and maintain emissions within environmental compliance (Penning, 2001; Lerner, 2002; Davis and Patel, 2004; Speight, 2008, 2011a,b).

To meet the challenges from changing restructured over the years from simple crude trends in product slate and the stringent distillation operations into increasingly specifications imposed by environmental complex chemical operations involving legislation, the refining industry in the near transformation of crude petroleum into a variety of future will become increasingly flexible and refined products with specifications that meet innovative with new processing schemes, users requirements.

During the next 20–30 years, the evolution future of petroleum refining and the current refinery layout will be primarily on process modification with some new innovations coming on-stream. The industry will move predictably on to (1) deep conversion of heavy feedstocks, (2) higher hydrocracking and hydrotreating capacity, and (3) more efficient processes.

Unlike most energy technologies, gasification processes can use almost any feedstock provided the correct choice of gasifier is made (Speight, 2008, 2011c). Once in a gaseous form, scrubbers and distillation columns are used to separate the gases and remove impurities. Furthermore, gasification of coal is a proven, mature technology (that still offers/holds significant potential for improvement and growth) that is capable of meeting the future energy needs of the United States in the future. Indeed, gasification of coal blended with other feedstocks such as petroleum residua and biomass offers additional opportunities.

Hence it would not be surprising (it may even be expected) that high conversion refineries will move to gasification of feedstocks for the development of alternative fuels and to enhance equipment usage. A major trend in the refining industry market demand for refined products will be in synthesizing fuels from simple basic reactants (e.g., synthesis gas) when it becomes uneconomical to produce super clean transportation fuels through conventional refining processes. Fischer–Tropsch plants together with *integrated gasification combined cycle* (IGCC) systems will be integrated with, or even into refineries, which will offer the advantage of high-quality products (Davis and Occelli, 2010; Chadeesingh, 2011; Speight, 2011c).

In summary, the petroleum industry is indeed at the verge of a major decision period with the onset of processing high volumes of heavy crude petroleum and residua. Several technology breakthroughs have made this possible but many technical challenges remain and some are being met, including the production of fuels derived from sources other than petroleum (Speight, 2008, 2011a; Høygaard Michaelsen et al., 2009; Luce, 2009).

Furthermore, coal use need not be incompatible with sustainable development. Coal already contributes in a major way to social and economic development. It can also be used in a way that is compatible with environmental protection. With a favorable policy environment to facilitate the continued deployment of existing clean coal technologies and the development of the next generation of technologies, the vision of an ultralow emission energy production system for the twenty-first century can be realized, and the coal industry is committed to working with others in achieving this goal.

26.3 NATIONAL ENERGY PLAN AND COAL UTILIZATION

The National Energy Plan announced by President Carter on April 29, 1977, proposed a significant increase in the utilization of the vast domestic deposits of coal to replace the dwindling supplies of petroleum and natural gas, and increasingly expensive petroleum from foreign sources, to meet national energy needs. And, yet 34 years later, the country has not decreased the dependency on foreign petroleum—in fact the dependency has increased!

As follow-ups to the National Energy Plan, the 2001 *Report of the National Energy Policy Development Group* and the *Energy Independence and Security Act of 2007* sought (1) to move the United States toward greater energy independence and security, (2) to increase the production of clean renewable fuels; to protect consumers, (3) to increase the efficiency of products, buildings, and vehicles, (4) to promote research on and deploy greenhouse gas capture and storage options, and (5) to improve the energy performance of the Federal Government.

Deciding whether or not the Act has been given serious consideration and follow up is difficult. The dependency on foreign petroleum continued to increase and the U.S. Congress has failed to act in a manner conducive to the best interests of the people of the United States.

While the use of fuels from biological sources seems to be on the rise, the increase in use of such fuels is marginal and will require longer times for such source to become a major part of the United States economy and fuel picture.

In the meantime, coal offers an answer through technology that has been known for decades and is already established.

However, there are many factors that influence (even act as a deterrent to) coal use in the United States (and in many other countries) but the most prominent factor is the environmental impact of coal use—especially carbon dioxide emissions associated with global climate change—which is speculated to pose the greatest potential constraint on future coal utilization (Pittock, 2009; Bell, 2011). Furthermore, the uncertainty about future requirements to control these environmental impacts can itself act as a constraint on future coal utilization.

As a result it is not surprising that a number of projects use technologies based on coal gasification (Chapters 20 and 21) rather than coal combustion (Chapters 14 and 15).

IGCC systems generate electricity by first converting coal or other feedstock into a clean gaseous fuel that is then combusted in a gas turbine to generate electricity. The exhaust from the hot gas turbine is used to generate steam that drives a second turbine to generate additional electricity. Like pulverized coal plants, the overall thermal efficiency of an IGCC system depends on many factors, including the gasifier type, coal type, oxidant type, and level of plant integration.

The range of efficiencies for current IGCC systems is comparable to that for current pulverized coal plants, with significant future improvements expected as gas turbine technology develops. Another advantage cited for the IGCC process is its ability to achieve lower levels of air pollutant emissions than PC systems, although modern technology permits emissions from both types of plants to be controlled to very low levels for all regulated air pollutants. Both pulverized coal and IGCC plants also can achieve high levels of carbon dioxide removal.

The gasification process is an established commercial technology (Chapters 20 and 21) that is widely used in the petroleum and petrochemical industries to convert carbon-containing feedstocks (such as coal, petroleum coke, residual petroleum, and biomass) (Speight, 2008, 2011a) to a mixture of carbon monoxide and hydrogen, referred to as synthesis gas (or syngas). For IGCC applications, the syngas is burned in a combined cycle power plant to generate electricity.

A key attraction of IGCC technology is that the incremental cost of capturing carbon dioxide emissions (in addition to regulated air pollutants) is lower than that for a comparably sized pulverized coal plant based on current commercial carbon dioxide capture technology.

Thus, while an IGCC plant without carbon dioxide capture is currently more costly than a pulverized coal plant, an IGCC plant is typically less costly if carbon dioxide capture is added to coal-based power plants using bituminous coal. However, many other factors—especially coal type and, hence, properties—also affect the relative costs of IGCC and pulverized coal power plants. For low-rank coals (subbituminous coal and lignite), preliminary studies indicate that the current cost of an IGCC plant with carbon dioxide capture increases to levels comparable to or higher than the cost of a pulverized coal plant with carbon dioxide capture using the same coals.

The current constraints and the potential for future constraints on greenhouse gas emissions is an increasingly important consideration when comparing the merits of investments in different power generation technologies. Although the properties, availability, and cost of different coal types is an important factor, the potential long-term impact of carbon dioxide capture and storage on the demand for different types of coal (in conjunction with other environmental control requirements) is still speculative at this time.

Recent fluctuations in world petroleum prices, as well as domestic natural gas prices, have stimulated renewed interest in the production of gaseous and liquid fuels from coal. Coal liquefaction technology (Chapters 18 and 19) has long been used to produce high-quality transportation fuels, most notably by SASOL in South Africa, which is the largest commercial facility in the world. Substitute natural gas (SNG) also can be produced from coal through a methanation step using synthesis gas as the starting material (Chapters 20 and 21).

The Great Plains Synfuels Plant near Beulah, North Dakota, operated by the Dakota Gasification Company, provides strong impetus for the use of coal in the United States—the production of synthetic liquid fuel and gaseous fuels. The plant also produces carbon dioxide, which is captured and piped 200 miles north to Saskatchewan, Canada, where it is used for enhanced oil recovery (EOR) in the Weyburn Field. An additional amount of carbon dioxide is discharged to the atmosphere. The plant also produces anhydrous ammonia and ammonium sulfate for agricultural use, as well as a variety of other minor products.

In both the SASOL case and the Great Plains Synfuels Plant case, coal gasification is a key technology and by adjusting the ratio of carbon monoxide and hydrogen in the syngas product, either gaseous or liquid products can be manufactured with the proper choice of catalysts and operating conditions.

However, the overall thermal efficiency of these processes is relatively low, on the order of 50%. Thus, total carbon dioxide emissions for a coal-to-liquid plant, including carbon dioxide from the conversion process and from combustion of the liquid fuel, are roughly twice that of diesel fuel produced from petroleum.

26.4   ELECTRIC POWER GENERATION

The use of electricity has been an essential part of the U.S. economy since the turn of the century. Coal as an established electricity source that provides vast quantities of reliable power has become more important as supplies of oil and natural gas diminish.

The generation of electricity from coal involves combustion of the coal to generate heat, which is used to generate steam that is used to spin one or more turbines to generate electricity.

Coal has played a major role in electrical production since the first power plants that were built in the United States in the 1880s. The earliest power plants used hand fed wood or coal to heat a boiler and produce steam. This steam was used in reciprocating steam engines that turned generators to produce electricity. In 1884, the more efficient high-speed steam turbine was developed by British engineer Charles A. Parsons, which replaced the use of steam engines to generate electricity.

In the 1920s, the pulverized coal firing was developed. This process brought advantages that included a higher combustion temperature, improved thermal efficiency, and a lower requirement for excess air for combustion. In the 1940s, the cyclone furnace was developed. This new technology allowed the combustion of poorer grade of coal with less ash production and greater overall efficiency.

In the modern world, electricity generation from coal is still based on the same methods started over 100 years ago, but improvements in all areas have brought coal power to be the inexpensive power source.

The importance of coal to electricity generation worldwide is set to continue, with coal fuelling 44% of global electricity in 2030. *Steam coal*, also known as *thermal coal*, is used in power stations to generate electricity.

The concept of burning coal, which has been pulverized into a fine powder, stems from the belief that if the coal is made fine enough, it will burn almost as easily and efficiently as a gas. The feeding rate of coal according to the boiler demand and the amount of air available for drying and transporting the pulverized coal fuel is controlled by computers. Pieces of coal are crushed between balls or cylindrical rollers that move between two tracks (*races*). The raw coal is then fed into the pulverizer along with air heated to about 340°C (650°F) from the boiler.

As the coal is crushed by the rolling action, the hot air dries it and blows the usable fine coal powder out to be used as fuel. The powdered coal from the pulverizer is directly blown to a burner in the boiler in which the powdered coal is mixed in the air suspension with additional pre-heated combustion air and forces it out of a nozzle similar in action to fuel being atomized by a fuel injector in modern cars. Under operating conditions, there is enough heat in the combustion zone to ignite all the incoming fuel.

Cyclone furnaces were developed after pulverized coal systems and require less processing of the coal fuel. They can burn poorer grade coals with higher moisture contents and ash contents to 25%. The crushed coal feed is either stored temporarily in bins or transported directly to the cyclone furnace. The furnace is basically a large cylinder jacketed with water pipes that absorb the some of the heat to make steam and protect the burner itself from melting down. A high powered fan blows the heated air and chunks of coal into one end of the cylinder. At the same time additional heated combustion air is injected along the curved surface of the cylinder causing the coal and air mixture to swirl in a centrifugal "cyclone" motion. The whirling of the air and coal enhances the burning properties producing high heat densities (about 4700–8300 kW/m²) and high combustion temperatures.

The hot combustion gases leave the other end of the cylinder and enter the boiler to heat the water-filled pipes and produce steam. As in the pulverized coal burning process, if all the fuel that enters the cyclone burns when injected, the furnace is at its operating temperature. Some slag remains on the walls insulating the burner and directing the heat into the boiler while the rest drains through a trench in the bottom to a collection tank where it is solidified and sent for disposal.

This ability to collect ash is the biggest advantage of the cyclone furnace burning process. Only 40% of the ash leaves with the exhaust gases compared with 80% for pulverized coal burning. However, cyclone furnaces do have some disadvantages: (1) the coal used must have a relatively low sulfur content in order for most of the ash to melt for collection, (2) high-power fans are required to move the larger coal pieces and air forcefully through the furnace, (3) more nitrogen oxide pollutants are produced compared with pulverized coal combustion, and (4) the actual burner may require regular replacement of its liners due to the erosion caused by the high velocity of the coal.

However, improvements continue to be made in conventional pulverized coal-fired power plant design and new combustion technologies are being developed. These allow more electricity to be produced from less coal—known as improving the thermal efficiency of the power station. Efficiency gains in electricity generation from coal-fired power stations will play a crucial part in reducing emissions of carbon dioxide at a global level.

Improving the efficiency of pulverized coal-fired power plants has been the focus of considerable efforts by the coal industry. There is huge scope for achieving significant efficiency improvements as the existing fleets of power plants are replaced over the next two-to-three decades with new, higher efficiency supercritical and ultra-supercritical plants and through the wider use of IGCC systems for power generation.

As a general rule of thumb, 1% point improvement in the efficiency of a conventional pulverized coal combustion plant results in a 2%–3% reduction in emissions of carbon dioxide.

26.5 HYDROGEN FROM COAL

Large quantities of hydrogen are currently used worldwide in the petroleum refining industry to desulfurize and upgrade crude oil and in the manufacture of ammonia for fertilizers. Hydrogen for these applications is produced predominantly by steam reforming of natural gas and as a byproduct from naphtha reforming. Some hydrogen is also produced from coal gasification, coke oven gas, and electrolysis of water.

The gasification process combines the coal with steam in a hot environment to produce a syngas (synthetic gas) composed mostly of carbon monoxide (CO) and hydrogen. In the process, coal is first gasified with oxygen and steam to produce a synthesis gas consisting mainly of carbon monoxide (CO) and hydrogen (H), with some carbon dioxide (CO_2), sulfur, particulates, and trace elements.

Oxygen is added in less than stoichiometric quantities so that complete combustion does not occur. This process is highly exothermic, with temperatures controlled by the addition of steam. Increasing the temperature in the gasifier initiates devolatilization and breaking of weaker chemical bonds to yield tar, oils, phenol derivatives, and hydrocarbon gases. These products generally further react to form hydrogen, carbon monoxide, and carbon dioxide:

$$C + H_2O \rightarrow CO + CO_2 + H_2$$

The coke or char carbon that remains after devolatilization is gasified through reactions with oxygen, steam, and carbon dioxide to form additional amounts of hydrogen and carbon monoxide.

Once the syngas is produced, it can be burned directly in a turbine to produce power or further reacted with more steam to shift the remaining carbon monoxide to carbon dioxide and produce more hydrogen:

$$CO + H_2O \rightarrow CO_2 + H_2$$

The carbon dioxide can be stored in oil and gas fields and the hydrogen can be used for the many applications that make up the hydrogen economy—such as to power a car in an engine or a fuel cell, to power a turbine to produce electricity, or to power a stationary fuel cell to make electricity.

Minerals in the feedstock separate as ash and leave the bottom of the gasifier as an inert slag (or bottom ash), a potentially marketable solid product. The fraction of the ash entrained with the synthesis gas, which is dependent upon the type of gasifier employed, requires removal downstream in particulate control equipment, such as filtration and water scrubbers.

The temperature of the synthesis gas as it leaves the gasifier is generally slightly below 1040°C (1900°F). With current technology, the gas has to be cooled to ambient temperatures to remove contaminants, although with some designs, steam is generated as the synthesis gas is cooled. Depending on the system design, a scrubbing process is used to remove hydrogen cyanide (HCN), ammonia (NH_3), hydrogen chloride (HCl), hydrogen sulfide (H_2S), and particulate matter. The scrubber system usually operates at low temperatures with synthesis gas leaving the process at about 23°C (72°F). Hydrogen sulfide and carbonyl sulfide (COS), once hydrolyzed, are removed by dissolution in, or reaction with, an organic solvent and converted to valuable by-products, such as elemental sulfur or sulfuric acid. The recovery of sulfur is usually near quantitative (99%+ v/v). The residual gas from this separation can be combusted to satisfy process-heating requirements.

This raw clean synthesis gas must be reheated to 315°C–370°C (600°F–700°F) for the first of two water gas shift reactors that produce additional hydrogen through the catalytically assisted equilibrium reaction of carbon monoxide with steam to form carbon dioxide and hydrogen. The exothermic reaction in the water gas shift reactor increases the temperature to approximately 430°C (800°F), which must be cooled to the required inlet temperature for the second water gas shift reactor in the range of 120°C–345°C (250°F–650°F), depending on design. The water gas shift reaction alters the H_2/CO ratio in the final mixture.

Typically, approximately 70% of the heating value of the feedstock fuel is associated with the carbon monoxide and hydrogen components of the gas, but can be higher depending upon the gasifier type. Hydrogen must be separated from the gas product stream (which also contains carbon dioxide and carbon monoxide as well as other trace contaminants) and *polished* to remove remaining sulfur, carbon monoxide, and other contaminants to meet the requirements for various end uses.

Concerns over global climate change and eventual resource depletion of fossil fuel resources have revived the concept of the hydrogen economy where hydrogen is used as an energy carrier. This concept would use hydrogen to provide energy to all sectors including central generating electric power, distributed power, industrial, residential, and transportation. Eventually the hydrogen would be produced from water using energy derived from sustainable resources, for example, nuclear fusion technology and photovoltaic technology.

Combustion of the hydrogen or electrochemical conversion via fuel cell technology would produce water, thus completing the cycle. In the shorter term, the hydrogen could be produced from fossil resources including natural gas, coal, petroleum coke, etc. The use of fossil carbon as a reductant and the conversion inefficiencies associated with hydrogen production from these resources would result in the production of large quantities of carbon dioxide. With the continued concern over climate change this carbon dioxide would have to be sequestered.

Hydrogen currently produced from the gasification of coal is essentially used as an intermediate for the synthesis of chemicals. However, with the increasing awareness of the necessity to control greenhouse gas emissions there is an incentive to move toward the production of hydrogen for power generation with carbon dioxide capture/sequestration.

New concepts for the production of hydrogen from coal are under development and include concepts based on either the steam gasification or the hydrogasification of coal with carbon dioxide capture/separation, membrane reactors directly used for the production of hydrogen from coal, molten bath processes originally used for metal smelting processes and adapted to the production of hydrogen, and plasma melting gasification processes.

26.6 ENERGY SECURITY AND SUSTAINABLE DEVELOPMENT

26.6.1 Continuing Use of Coal

It might be argued that coal, being a finite resource, should have no part in sustainable development. This somewhat biased view overlooks the benefits of coal as well as the ability (or need) to substitute one form of energy source for another. To the extent that substitution is possible, depletion of one type of energy source capital is consistent with sustainability if offset by an increase in other types of energy sources, with any accompanying disadvantages that arise from energy sources that have not been fully developed.

Therefore, the use of coal is consistent with sustainable development if, while meeting our present needs, it produces new capital and options for future generations—such as infrastructure, new technologies, and new knowledge. An associated risk is that coal's use may degrade natural capital, such as the environment to an unacceptable or irreversible extent, leading to unsustainable development.

However, apart from their inherent practical limitations at that time, supplies of biomass, wind, and water were limited. Coal was abundant and available and new technology allowed it to be used for steam raising and iron making. The environmental consequences of rapidly growing and uncontrolled coal use were, of course, unacceptable. Continual technology development over time allowed coal to be used with much greater efficiency and with greatly reduced environmental impact.

For example, the oxygen-fired pulverized coal combustion process (Oxy-Fuel process) offers a low risk step with the development of existing of power generation technology to enable carbon dioxide capture and storage. Oxy-firing of pulverized coal in boilers involves the combustion of pulverized coal in a mixture of oxygen and recirculated flue gas in order to reduce the net volume of flue gases from the process and to substantially increase the concentration of carbon dioxide (CO_2) in the flue gases—compared to the normal pulverized coal combustion in air.

Many developed countries rely on coal to support living standards and industrial development. In fact coal plays a significant economic role in coal producing and consuming countries alike and remains the main fuel of choice for electricity generation worldwide and is an essential input to two-thirds of the world's steel production. The challenge for coal, as for other energy sources, is to ensure that it meets all the objectives of sustainable development and, in particular, ensuring an acceptable environmental performance that is in keeping with modern regulations.

Energy sources, particularly coal, will also become more and more important in power generation in certain parts of the world over the coming decades (China and India in particular) as a result of the significant rise in demand for energy. It is therefore essential for the right framework to be established for the development and distribution of sustainable coal technologies, and thus limit emissions of carbon dioxide from the use of coal for electricity generation.

The improvements already made in coal technologies (increase in energy efficiency and a reduction in acid rain and local atmospheric pollution due to SOx, NOx, and particulate emissions) show that significant technological progress is possible, in particular by applying the principle of carbon capture and storage (CCS).

Technologies for the sustainable use of coal will be based on an optimum combination of "clean coal" technologies (improving yield and reducing atmospheric emissions) and CCS technologies. Continued development of these technologies and demonstrating their commercial viability will lead to their large-scale use.

The sustainable use of fossil fuels, in particular carbon capture and storage, will make it possible to eliminate the majority of the carbon emissions from fossil-fuel power plants. The use of appropriate technologies will also enable the atmospheric pollutants traditionally associated with coal combustion, including NOx and SOx, to be reduced, thus resulting in lower local environmental and health costs.

Coal is an extremely important fuel and will remain so as it is the world's most abundant and widely distributed fossil fuel source. Development of new *clean coal* technologies is necessary so that the coal resources can be utilized for future generations without contributing to serious environmental effects. Much of the challenge is in commercializing the technology so that coal use remains economically competitive despite the cost of achieving low and eventually near *zero emissions*.

As many coal-fired power stations approach retirement, their replacement gives much scope for the introduction of more modern facilities and cleaner production of electricity.

26.6.2 Management of Coal Wastes

Another important aspect of the continuing coal use is the management of wastes, especially waste from coal combustion as might be produced at an electricity generating plant. This does not mean that other uses of coal and the waste generated should be ignored. It is merely a notation of the importance of the management of waste materials from coal use (Chapter 25).

Burning coal, such as for power generation, gives rise to a variety of wastes, which must be controlled or at least accounted for. The *clean coal technologies* (Chapter 22) are a variety of evolving responses to late twentieth century environmental concerns, including that of global climate change due to carbon dioxide releases to the atmosphere. However, many of the elements have in fact been applied for many years.

For example, coal cleaning by washing (Chapter 6) has been standard practice in developed countries for some time. It reduces emissions of ash and sulfur dioxide when the coal is burned. Furthermore, electrostatic precipitators and fabric filters can remove 99% of the fly ash from the flue gases and such technologies are in widespread use (Chapters 14, 15, and 23).

In addition, flue gas desulfurization reduces the output of sulfur dioxide to the atmosphere and low-NOx burners allow coal-fired plants to reduce nitrogen oxide emissions. Both technologies are in wide use.

Other technologies such as IGCC and *pressurized fluidized bed combustion* (PFBC) enable higher thermal efficiencies for the future. Ultraclean coal (UCC) from new processing technologies that reduce ash below 0.25% and sulfur to very low levels mean that pulverized coal might be used as fuel for very large marine engines, in place of heavy fuel oil.

Carbon capture and storage or *carbon capture and sequestration* (CCS) technologies are in the forefront of measures for use of coal as a clean fuel. A number of means exist to capture carbon dioxide from gas streams (Chapter 23), and the focus in the past has often been on obtaining pure carbon dioxide for industrial purposes rather than reducing carbon dioxide levels in power plant emissions.

However, capture of carbon dioxide from flue gas streams following combustion of coal in air is reputed to be more difficult and expensive than from natural gas streams, as the carbon dioxide concentration is only about 14% at best, with nitrogen most of the rest, and the flue gas is hot. The main process treats carbon dioxide like any other pollutant, and as flue gases are passed through an amine solution in which the carbon dioxide is absorbed (Chapter 23). It can later be released by heating the solution.

26.6.3 SUSTAINABLE DEVELOPMENT

The clean coal technology field is moving in the direction of coal gasification with a second stage so as to produce a concentrated and pressurized carbon dioxide stream followed by its separation and geological storage. This technology has the potential to provide extremely low emissions of the conventional coal pollutants, and as low-as-engineered carbon dioxide emissions. Zero emissions from coal use is an element of the future foe coal.

Sustainable development (meeting the needs of the present without compromising the ability of future generations to meet their own needs) has been an important part of public policy debate for the last decade. It has evolved into a widely subscribed ideal for how business and society should interact and function.

Coal has a crucial role in energy production since a supply of affordable and reliable energy is essential for economic development and is a significant contributor to the alleviation of poverty, improved health, and better quality of life.

Technological advances have diminished the traditional disadvantages of coal use, although local and regional environmental impacts remain an issue. The use of state of the art technology can make a contribution to coal meeting stringent environmental standards. However, state of the art technologies are not universally deployed and this remains a high priority for governments, coal users, and suppliers. Improvement in environmental performance is technologically feasible and should be a priority of industry and government to enhance the reputation of coal and its contribution to environmental sustainability.

Coal stands out as an affordable resource that is relatively straightforward to convert to electrical power. It is also abundant and reliable and will inevitably form a significant part of the future energy mix in many countries. Therefore adapting clean coal technologies to coal use is of worldwide particular importance. This presents a challenge to the coal industry in giving practical effect to the notion of sustainable development by helping to facilitate the transfer of environmentally friendly coal technologies to developing economies.

Many of the potential new markets for coal may necessitate that it be processed in ways other than traditional combustion. For example, if coal is to be used as a source of liquid fuels, there will be the need to refine such products, as many of the product constituents are not (in the produced state) compatible with petroleum-based fuels—the liquids from coal would need to be refined further. Such a refinery could well be a system consisting of one or more individual processes integrated in such a way as to allow coal to be processed into two or more products supplying two or more markets.

While there are differing points of view as to the configuration of such a refinery, eventually it would be advisable if there was a close relationship a petroleum refinery in that a full slate of liquid products must be possible and the system must be capable of changing product yields with temporal changes in market conditions.

Indeed, another approach would be to include all processes at a coal refinery and at a petroleum refinery, and to class the facility as an *integrated fuel refinery* or as an *integrated energy facility*.

Such a term would fit the concept of a *gasification refinery,* which would have, as the center piece, gasification technology to produce synthesis gas from which liquid fuels would be manufactured using the Fischer–Tropsch synthesis technology (Speight, 2008, 2011a; Chadeesingh, 2011).

In fact, gasification to produce synthesis gas can proceed from any carbonaceous material, including biomass, as well as coal (Speight, 2008, 2011c). Inorganic components of the feedstock, such as metals and minerals, are trapped in an inert and environmentally safe form as char, which may have use as a fertilizer. Coal-biomass gasification is therefore one of the most technically and economically convincing energy possibilities for future use of coal.

The manufacture of gas mixtures of carbon monoxide and hydrogen has been an important part of chemical technology for about a century. Originally, such mixtures were obtained by the reaction of steam with incandescent coke and were known as *water gas.* Eventually, steam reforming processes, in which steam is reacted with natural gas (methane) or petroleum naphtha over a nickel catalyst, found wide application for the production of synthesis gas.

A modified version of steam reforming known as autothermal reforming, which is a combination of partial oxidation near the reactor inlet with conventional steam reforming further along the reactor, improves the overall reactor efficiency and increases the flexibility of the process. Partial oxidation processes using oxygen instead of steam also found wide application for synthesis gas manufacture, with the special feature that they could utilize low-value feedstocks such as heavy petroleum residua. Furthermore, catalytic partial oxidation employing very short reaction times (milliseconds) at high temperatures (850°C–1000°C [1560°–1830°F]) is providing still another approach to synthesis gas manufacture.

In a gasifier, the carbonaceous material undergoes several different processes: (1) pyrolysis of carbonaceous fuels, (2) combustion, and (3) gasification of the remaining char. The process is very dependent on the properties of the carbonaceous material and determines the structure and composition of the char, which will then undergo gasification reactions.

Obviously, coal will play an important role in energy systems that support sustainable development for the foreseeable future. This is because of coal's unique combination of advantages: (1) it is affordable, (2) it is safe to transport and store, and (3) it is available from a wide range of sources. Coal therefore remains essential in achieving a diverse balanced and secure energy mix in developed countries; it can also meet the growing energy needs of many developing countries.

Concerns about climate change add a most complex challenge to the long-term use of coal in a sustainable development context. In disregarding the great underlying uncertainties of future climate, emissions, and the efficacy of response options, climate change is commonly presented simply as an environmental issue requiring urgent intervention.

Further improvement in the environmental performance of coal will not only be required but will be a necessity. While improved coal technologies have provided very substantial efficiency and emission improvements to date, accelerated technological effort is required to reduce greenhouse gas emissions. Deployment of cleaner and higher efficiency technologies will be important in both developed and developing countries.

Climate change considerations must always be considered. Any response must be affordable and provide the basis for sustainable development, by addressing ongoing economic and social requirements as well as the environmental challenge.

Thus, the challenge is to extract energy from coal in more efficient and cleaner ways. This meets the necessity of moving to a sustainable future by replacing resource depleting technologies with new options of at least equivalent value.

In summary, if coal will be a basic energy source for future sustainable development (and scenario that is highly likely), it is necessary to build new plants that work with this type of clean coal technology, in order to reduce emissions of greenhouse gases and help to achieve the targets. The provision of such technologies could bring benefits for both the environment and the same companies that are currently before an uncertain future.

REFERENCES

Bell, L. 2011. *Climate of Corruption: Politics and Power Behind the Global Warming.* Greenleaf book Group Press, Austin, TX.

Chadeesingh, R. 2011. The Fischer–Tropsch process. In *The Biofuels Handbook*, J.G. Speight (Ed.). Royal Society of Chemistry, London, U.K., Part 3, Chapter 5.

Davis, B.H. and Occelli, M.L. 2010. *Advances in Fischer-Tropsch Synthesis, Catalysts, and Catalysis.* Taylor & Francis Group, Boca Raton, FL.

Davis, R.A. and Patel, N.M. 2004. Refinery hydrogen management. *Petroleum Technology Quarterly*, Spring: 29–35.

Giampietro, M. and Mayumi, K. 2009. *The Biofuel Delusion: The Fallacy of Large-Scale Agro-Biofuel Production.* Earthscan, London, U.K.

Høygaard Michaelsen, N., Egeberg, R., and Nyström, S. 2009. Consider new technology to produce renewable diesel. *Hydrocarbon Processing*, 88(2): 41–44.

Langeveld, H., Sanders, J., and Meeusen, M. (Eds.). 2010. *The Bio-based Economy: Biofuels, Materials, and Chemicals in the Post-Oil Era.* Earthscan, London, U.K.

Lee, S., Speight, J.G., and Loyalka, S.K. 2007. *Handbook of Alternative Fuel Technologies.* CRC Press, Taylor & Francis Group, Boca Raton, FL.

Lerner, B. 2002. The future of refining. *Hydrocarbon Engineering*, 7: 51.

Luce, G.W. 2009. Renewable energy solutions for and energy hungry world. *Hydrocarbon Processing*, 88(2): 19–22.

Nersesian, R.L. 2010. *Energy for the 21st Century: A Comprehensive Guide to Conventional and Alternative Energy Sources.* M.E. Sharpe Inc., Armonk, NY.

Penning, R.T. 2001. Petroleum refining: A look at the future. *Hydrocarbon Processing*, 80(2): 45–46.

Pittock, A.B. 2009. *Climate Change: The Science, Impacts, and Solutions*, 2nd edn. Earthscan, CSIRO Publishing, Collingwood, Victoria, Australia.

Seifried, D. and Witzel, W. 2010. *Renewable Energy: The Facts.* Earthscan, London, U.K.

Speight, J.G. 2008. *Synthetic Fuels Handbook: Properties, Processes, and Performance.* McGraw-Hill, New York, 2008.

Speight, J.G. 2009. *Enhanced Recovery Methods for Heavy Oil and Tar Sands.* Gulf Publishing Company, Houston, TX.

Speight, J.G. 2011a. *The Refinery of the Future.* Gulf Professional Publishing, Elsevier, Oxford, U.K.

Speight, J.G. 2011b. *An Introduction to Petroleum Technology, Economics, and Politics.* Scrivener Publishing, Salem, MA.

Speight, J.G. (Ed.). 2011c. *The Biofuels Handbook.* Royal Society of Chemistry, London, U.K.

U.S. Energy Information Administration. 2011. This week in petroleum, May 25. United States Department of Energy, Washington, DC. http://www.eia.gov/energy_in_brief/foreign_oil_dependence.cfm

Glossary

Abandoned workings: Sections, panels, and other areas that are not ventilated and examined in the manner required for work places.

Abutment: (1) The weight of the rocks above a narrow roadway is transferred to the solid coal along the sides, which act as abutments of the arch of strata spanning the roadway; (2) the weight of the rocks over a longwall face is transferred to the front abutment, that is, the solid coal ahead of the face and the back abutment, that is, the settled packs behind the face.

Accessed: Coal deposits that have been prepared for mining by construction of portals, shafts, slopes, drifts, and haulage ways; by removal of overburden; or by partial mining.

Acid deposition or acid rain: A mixture of wet and dry *deposition* (deposited material) from the atmosphere containing higher than typical amount of nitric and sulfuric acids.

Acid drainage: The runoff of acidic liquids from coal production waste piles. Such runoff can contaminate ground and surface waters.

Acid gas: Hydrogen sulfide (H_2S) or carbon dioxide (CO_2).

Acid mine drainage: Any acid water draining or flowing on, or having drained or flowed off, any area of land affected by mining.

Acid mine water: Mine water that contains free sulfuric acid, mainly due to the weathering of iron pyrites.

Acid rain: A solution of acidic compounds formed when sulfur and nitrogen oxides react with water droplets and airborne particles.

Acre-foot (acre-ft): The volume of coal that covers 1 acre at a thickness of 1 ft ($43,560 \, ft^3/1,613.333 \, yd^3/1,233.482 \, m^3$). The weight of coal in this volume varies according to rank.

Acre-inch (acre-in.): The volume of coal that covers 1 acre at a thickness of 1 in. ($3630 \, ft^3/134.44 \, yd^3/102.7903 \, m^3$). The weight of coal in this volume varies according to rank.

Active workings: Any place in a mine where miners are normally required to work or travel and which are ventilated and inspected regularly.

Adit: A nearly horizontal passage from the surface by which a mine is entered and dewatered; a blind horizontal opening into a mountain, with only one entrance.

Advance: Mining in the same direction, or order of sequence; first mining as distinguished from retreat.

Afterdamp: Gases remaining after an explosion in a mine consisting of carbon dioxide, carbon monoxide, nitrogen, and hydrogen sulfide; the toxic mixture of gases left in a mine following an explosion caused by firedamp (methane), which itself can initiate a much larger explosion of coal dust.

Agglomerating: Coal that, during volatile matter determinations, produces either an agglomerate button capable of supporting a 500 g weight without pulverizing or a button showing swelling or cell structure.

Agglomeration: Formation of larger coal or ash particles by smaller particles sticking together.

Airshaft: A vertical shaft in which air is blown down through the various sections of the underground mine. The air is generated by a large fan on the surface providing oxygen for the miners below.

Air split: The division of a current of air into two or more parts.

Airway: Any passage through which air is carried. Also known as an air course.

Anemometer: Instrument for measuring air velocity.

Angle of dip: The angle at which strata or mineral deposits are inclined to the horizontal plane.

Angle of draw: This angle is assumed to bisect the angle between the vertical and the angle of repose of the material and is 20° for flat seams; for dipping seams, the angle of break increases, being 35.8° from the vertical for a 40° dip; the main break occurs over the seam at an angle from the vertical equal to half the dip.

Angle of repose: The maximum angle from the horizontal plane at which a given material will rest on a given surface without sliding or rolling.

Anthracene oil: The heaviest distillable coal tar fraction, with the distillation range of 270°C–400°C (520°F–750°F), containing creosote oil, anthracene, phenanthrene, carbazole, and so on.

Anthracite (hard coal): A hard, black, shiny coal very high in fixed carbon and low in volatile matter, hydrogen, and oxygen; a rank class of non-agglomerating coals as defined by the American Society for Testing and Materials (ASTM) having more than 86% fixed carbon and less than 14% volatile matter on a dry, mineral-matter-free basis; this class of coal is divisible into the semi-anthracite, anthracite, and meta-anthracite groups on the basis of increasing fixed carbon and decreasing volatile matter.

Anthracosis: *see* Black lung.

Anthraxylon: U.S. Bureau of Mines' term for vitrinite viewed by transmitted light.

Anticline: An upward fold or arch of rock strata.

Aquifer: A water-bearing formation through which water moves more readily than in adjacent formations with lower permeability.

Arching: Fracture processes around a mine opening, leading to stabilization by an arching effect.

Area (of an airway): Average width multiplied by average height of airway, expressed in square feet.

Ash: The noncombustible residue remaining after complete coal combustion; the final form of the mineral matter present in coal.

Ash analysis: Percentages of inorganic oxides present in an ash sample. Ash analyses are used for the evaluation of the corrosion, slagging, and fouling potential of coal ash. The ash constituents of interest are silica (SiO_2), alumina (Al_2O_3), titania (TiO_2), ferric oxide (Fe_2O_3), lime (CaO), magnesia (MgO), potassium oxide (K_2O), sodium oxide (Na_2O), and sulfur trioxide (SO_3). An indication of ash behavior can be estimated from the relative percentages of each constituent.

Ash free: A theoretical analysis calculated from basic analytical data expressed as if the total ash had been removed.

Ash-fusion temperatures: A set of temperatures that characterize the behavior of ash as it is heated. These temperatures are determined by heating cones of ground, pressed ash in both oxidizing and reducing atmospheres.

As-received basis: It represents an analysis of a sample as received at a laboratory.

As-received moisture: The moisture present in a coal sample when delivered.

Attritus: A microscopic coal constituent composed of macerated plant debris intimately mixed with mineral matter and coalified. U.S. Bureau of Mines' usage, viewed by transmitted light.

Auger: A rotary drill that uses a screw device to penetrate, break, and then transport the drilled material (coal).

Auger mining: Mining generally practiced but not restricted to hilly coal-bearing regions of the country that uses a machine designed on the principle of the auger, which bores into an exposed coal seam, conveying the coal to a storage pile or bin for loading and transporting. May be used alone or in combination with conventional surface mining. When used alone, a single cut is made sufficient to expose the coal seam and provide operating space for the machine. When used in combination with surface mining, the last cut pit provides the operating space.

Auxiliary operations: All activities supportive of but not contributing directly to mining.

Auxiliary ventilation: Portion of main ventilating current directed to face of dead end entry by means of an auxiliary fan and tubing.

Azimuth: A surveying term that references the angle measured clockwise from any meridian (the established line of reference); the bearing is used to designate direction; the bearing of a line is the acute horizontal angle between the meridian and the line.

Back: The roof or upper part in any underground mining cavity.

Backfill: The operation of refilling an excavation. Also, the material placed in an excavation in the process of backfilling.

Baghouse: An air pollution control device that removes particulate matter from flue gas, usually achieving a removal rate above 99.9%.

Barrel: Liquid volume measure equal to 42 U.S. gal, commonly used in measuring petroleum or petroleum products.

Barren: Said of rock or vein material containing no minerals of value, and of strata without coal, or containing coal in seams too thin to be workable.

Barricading: Enclosing part of a mine to prevent inflow of noxious gases from a mine fire or an explosion.

Barrier: Something that bars or keeps out. Barrier pillars are solid blocks of coal left between two mines or sections of a mine to prevent accidents due to inrushes of water, gas, or from explosions or a mine fire.

Beam: A bar or straight girder used to support a span of roof between two support props or walls.

Beam building: The creation of a strong, inflexible beam by bolting or otherwise fastening together several weaker layers. In coal mining this is the intended basis for roof bolting.

Bearing: A surveying term used to designate direction. The bearing of a line is the acute horizontal angle between the meridian and the line. The meridian is an established line of reference. Azimuths are angles measured clockwise from any meridian.

Bearing plate: A plate used to distribute a given load. In roof bolting, the plate used between the bolt head and the roof.

Bed: A stratum of coal or other sedimentary deposit.

Bedrock: The rock material directly above and below the coal seam.

Beehive oven: A dome-shaped oven not equipped to recover the by-product gas and liquids evolved during the coking process.

Belt conveyor: A looped belt on which coal or other materials can be carried and which is generally constructed of flame-resistant material or of reinforced rubber or rubber-like substance.

Belt feeder (feeder breaker): A crawler-mounted surge bin often equipped with a crusher or breaker and used in room-and-pillar sections positioned at the end of the section conveyor belt. It allows a quick discharge of the shuttle car. It sizes the coal, and a built-in conveyor feeds it at an appropriate rate onto the conveyor belt.

Belt idler: A roller, usually of cylindrical shape, which is supported on a frame and which, in turn, supports or guides a conveyor belt. Idlers are not powered but turn by contact with the moving belt.

Belt take-up: A belt pulley, generally under a conveyor belt and in the drive pulley, kept under strong tension parallel to the belt line. Its purpose is to automatically compensate for any slack in the belting created by start-up.

Bench: The surface of an excavated area at some point between the material being mined and the original surface of the ground, on which equipment can sit, move, or operate. A working road or base below a high wall, as in contour stripping for coal.

Beneficiation: The treatment of mined material, making it more concentrated or richer; *see* Physical coal cleaning.

Berm: A pile or mound of material capable of restraining a vehicle.

Binder: A streak of impurity in a coal seam.

Bit: The hardened and strengthened device at the end of a drill rod that transmits the energy of breakage to the rock—the size of the bit determines the size of the hole; a bit may be either detachable from or integral with its supporting drill rod.

Bituminous (soft) coal: A relatively soft dark brown to black coal, lower in fixed carbon than anthracite but higher in volatile matter, hydrogen, and oxygen; a rank class of coals as defined by the ASTM high in carbonaceous matter, having less than 86% fixed carbon, and more than 14% volatile matter on a dry, mineral-matter-free basis and more than 10,500 Btu on a moist, mineral-matter-free basis. This class may be either agglomerating or non-agglomerating and is divisible into the high-volatile C, B, A; medium; and low-volatile bituminous coal groups on the basis of increasing heat content and fixed carbon and decreasing volatile matter.

Blackdamp: A deadly gas that is caused from coal burning in an atmosphere that lacks oxygen; mostly a mixture of carbon dioxide and nitrogen found in mines after fires and explosions.

Black lung (anthracosis): A respiratory disease caused by prolonged inhalation of coal dust.

Blasting agent: Any material consisting of a mixture of a fuel and an oxidizer.

Blasting cap: A detonator containing a charge of detonating compound, which is ignited by electric current or the spark of a fuse. Used for detonating explosives.

Blasting circuit: Electric circuits used to fire electric detonators or to ignite an igniter cord by means of an electric starter.

Bleeder or bleeder entries: Special air courses developed and maintained as part of the mine ventilation system and designed to continuously move air–methane mixtures emitted by the gob or at the active face away from the active workings and into mine-return air courses.

Blue gas: A mixture consisting chiefly of carbon monoxide and hydrogen formed by action of steam on hot coal or coke.

Boghead coal: Same as cannel coal except that algal remains can be seen under the microscope.

Boiler slag: A molten ash collected at the base of slag tap and cyclone boilers that is quenched with water and shatters into black, angular particles having a smooth glassy appearance.

Bolt torque: The turning force in foot-pounds applied to a roof bolt to achieve an installed tension.

Bone coal or bone: Impure coal that contains much clay or other fine-grained detrital mineral matter; the term "bone coal" has been erroneously used for cannel coal, canneloid coal, and well-cemented to metamorphosed coaly mudstone and (or) claystone. Bone coal has also been applied to carbonaceous partings; the term *impure coal* accompanied by adjective modifiers such as *silty, shaly,* or *sandy* is the preferred usage because the definition of bone coal does not specify the type or weight percentages of impurities.

Borehole: Any deep or long drill hole, usually associated with a diamond drill.

Boss: Any member of the managerial ranks who is directly in charge of miners (e.g., shift boss, face boss, fire boss).

Bottom: Floor or underlying surface of an underground excavation.

Bottom ash: It consists of agglomerated ash particles formed in pulverized coal boilers that are too large to be carried in the flue gases and impinge on the boiler walls or fall through open grates to an ash hopper at the bottom of the boiler. Bottom ash is typically gray to black in color, is quite angular, and has a porous surface structure.

Box-type magazine: A small, portable magazine used to store limited quantities of explosives or detonators for short periods of time at locations in the mine that are convenient to the blasting sites at which they will be used.

Brattice or brattice cloth: Fire-resistant fabric or plastic partition used in a mine passage to confine the air and force it into the working place; also termed *line brattice, line canvas,* or *line curtain.*

Break line: The line that roughly follows the rear edges of coal pillars that are being mined; the line along which the roof of a coal mine is expected to break.

Breakthrough: A passage for ventilation that is cut through the pillars between rooms.

Bridge carrier: A rubber-tire-mounted mobile conveyor, about 10 m long, used as an intermediate unit to create a system of articulated conveyors between a mining machine and a room or entry conveyor.

Bridge conveyor: A short conveyor hung from the boom of mining or lading machine or haulage system with the other end attached to a receiving bin that dollies along a frame supported by the room or entry conveyor, tailpiece—as the machine boom moves, the bridge conveyor keeps it in constant connection with the tailpiece.

Bright coal: U.S. Bureau of Mines' term for a combination of clarain and vitrain with small amounts of fusain.

Briquetting: A process of applying pressure to coal fines, with or without a binder, to form a compact or agglomerate.

British thermal unit (Btu): The quantity of heat required to raise the temperature of 1 lb of water 1°F at, or near, its point of maximum density of 39.1°F (equivalent to 251.995 g cal/1054.35 J/1.05435 kJ/0.25199 kcal).

Brow: A low place in the roof of a mine, giving insufficient headroom.

Brushing: Digging up the bottom or taking down the top to give more headroom in roadways.

Btu: British thermal unit; a measure of the energy required to raise the temperature of 1 lb of water 1°F.

Bug dust: The fine particles of coal or other material resulting from the boring or cutting of the coal face by drill or machine.

Bump (or burst): A violent dislocation of the mine workings that is attributed to severe stresses in the rock surrounding the workings.

Burn line: The contact between burned and unburned coal in the subsurface. In the absence of definitive information, the subsurface position of a burn line is assumed to be vertically below the surface contact between unaltered and altered rocks.

Butt cleat: A short, poorly defined vertical cleavage plane in a coal seam, usually at right angles to the long face cleat.

Butt entry: A coal mining term that has different meanings in different locations—it can be synonymous with panel entry, sub-main entry, or in its older sense it refers to an entry that is *butt* onto the coal cleavage (i.e., at right angles to the face).

Cage: A rectangular transporting device used to haul mine cars (pit cars) loaded with coal or dirt and rock from the earth below. The cage is also used to transport miners, mules, and supplies to and from the workplace below.

Cage person (cage man or cage woman): A person who works at the cages loading and unloading the mine cars, etc., on to the cages.

Calorie: The quantity of heat required to raise 1 g of water from 15°C to 16°C; a calorie is also termed gram calorie or small calorie (equivalent to 0.00396832 Btu/4.184 J/0.001 kg cal).

Calorific value: The quantity of heat that can be liberated from 1 lb of coal or oil measured in Btu/lb.

Cannel coal: Predominately durain with lesser amounts of vitrain than splint coal and small quantities of fusain. Spores can be seen under the microscope.

Canopy: A protective covering of a cab on a mining machine.

Cap: A miner's safety helmet; a highly sensitive, encapsulated explosive that is used to detonate larger but less-sensitive explosives.

Cap block: A flat piece of wood inserted between the top of the prop and the roof to provide bearing support.

Car: A railway wagon, especially any of the wagons adapted to carrying coal, ore, and waste underground.

Car dump: The mechanism for unloading a loaded car.

Carbide bit: A cutting or drilling bit for rock or coal, made by fusing an insert of molded tungsten carbide to the cutting edge of a steel bit shank.

Carbonization: A process whereby coal is converted to coke by devolatilization.

Carbon monoxide: A colorless, odorless, very toxic gas formed by incomplete combustion of carbon, as in water gas or producer gas production.

Carbureted blue gas: *see* Water gas.

Cast: A directed throw; in strip mining, the overburden is cast from the coal to the previously mined area.

Certified: It describes a person who has passed an examination to do a required job.

Chain conveyor: A conveyor on which the material is moved along solid pans (troughs) by the action of scraper crossbars attached to powered chains.

Chain pillar: The pillar of coal left to protect the gangway or entry and the parallel airways.

Check curtain: Sheet of brattice cloth hung across an airway to control the passage of the air current.

Chock: Large hydraulic jacks used to support roof in longwall and shortwall mining systems.

Clarain: A macroscopic coal constituent (lithotype) known as bright-banded coal, composed of alternating bands of vitrain and durain.

Clastic dike: A vertical or near-vertical seam of sedimentary material that fills a crack in and cuts across sedimentary strata—the dikes are found in sedimentary basin deposits worldwide; dike thickness varies from millimeters to meters and the length is usually many times the width.

Clastic rocks: Rocks composed of fragments (clasts) of preexisting rock; may include sedimentary rocks as well as transported particles whether in suspension or in deposits of sediment.

Clay vein: A fissure that has been infilled as a result of gravity, downward-percolating ground waters, or compactional pressures, which cause unconsolidated clays or thixotropic sand to flow into the fissure.

Clean Air Act Amendments of 1990: A comprehensive set of amendments to the federal law governing U.S. air quality; the Clean Air Act was originally passed in 1970 to address significant air pollution problems in US cities, and the 1990 amendments broadened and strengthened the original law to address specific problems such as acid deposition, urban smog, hazardous air pollutants, and stratospheric ozone depletion.

Clean coal technologies: A number of innovative new technologies designed to use coal in a more efficient and cost-effective manner while enhancing environmental protection; technologies include fluidized-bed combustion (FBC), integrated gasification combined cycle, limestone injection multistage burner, enhanced flue gas desulfurization (FGD) (or *scrubbing*), coal liquefaction, and coal gasification.

Cleats: Natural opening-mode fractures in coal beds, which account for most of the permeability and much of the porosity of coalbed gas reservoirs.

Coal: An organic rock; a stratified combustible carbonaceous rock, formed by partial to complete decomposition of vegetation; varies in color from dark brown to black; not fusible without decomposition and very insoluble.

Coal bed methane (coalbed methane): Methane adsorbed to the surface of coal; often considered to be a part of the coal seam.

Coal dust: Particles of coal that can pass a No. 20 sieve.

Coal gas: The mixture of volatile products (mainly hydrogen, methane, carbon monoxide, and nitrogen) remaining after removal of water and tar, obtained from carbonization of coal, having a heat content of 400–600 Btu/ft^3.

Coal gasification: Production of synthetic gas from coal.

Coalification: The processes involved in the genetic and metamorphic history of the formation of coal deposits from vegetable matter.

Coal liquefaction: Conversion of coal to a liquid.

Coal measures: Strata containing one or more coal beds.

Coal province: An area containing two or more coal regions.

Coal mine: An area of land and all structures, facilities, machinery, tools, equipment, shafts, slopes, tunnels, excavations, and other property, real or personal, placed upon, under, or above the surface of such land by any person, used in extracting coal from its natural deposits in the earth by any means or method, and the work of preparing the coal so extracted, including coal preparation facilities; the British term is *colliery* or (in northern areas) *pit*.

Coal preparation/washing: The treatment of coal to reject waste. In its broadest sense, preparation is any processing of mined coal to prepare it for market, including crushing and screening or sieving the coal to reach a uniform size, which normally results in removal of some non-coal material. The term "coal preparation" most commonly refers to processing, including crushing and screening, passing the material through one or more processes to remove impurities, sizing the product, and loading for shipment. Many of the processes separate rock, clay, and other minerals from coal in a liquid medium; hence, the term "washing" is widely used. In some cases coal passes through a drying step before loading.

Coal rank: It indicates the degree of coalification that has occurred for a particular coal. Coal is formed by the decomposition of plant matter without free access to air and under the influence of moisture, pressure, and temperature. Over the course of the geologic process that forms coal—coalification—the chemical composition of the coal gradually changes to compounds of lower hydrogen content and higher carbon content in aromatic ring structures. As the degree of coalification increases, the percentage of volatile matter decreases and the calorific value increases. The common ranks of coal are anthracite, bituminous, subbituminous, and brown coal/lignite.

Coal region: An area containing one or more coal fields.

Coal reserves: Measured tonnages of coal that have been calculated to occur in a coal seam within a particular property.

Coal sampling: The collection and proper storage and handling of a relatively small quantity of coal for laboratory analysis. Sampling may be done for a wide range of purposes, such as coal resource exploration and assessment, characterization of the reserves or production of a mine, to characterize the results of coal cleaning processes, to monitor coal shipments or receipts for adherence to coal quality contract specifications, or to subject a coal to specific combustion or reactivity tests related to the customer's intended use. During predevelopment phases, such as exploration and resource assessment, sampling typically is from natural outcrops, test pits, old or existing mines in the region, drill cuttings, or drilled cores. Characterization of a mine's reserves or production may use sample collection in the mine, representative cuts from coal conveyors or from handling and loading equipment, or directly from stockpiles or shipments (coal rail cars or barges). Contract specifications rely on sampling from the production flow at the mining or coal handling facility or at the load-out, or from the incoming shipments at the receiver's facility. In all cases, the value of a sample taken depends on its being representative of the coal under consideration, which in turn requires that appropriate sampling procedures be carefully followed.

For coal resource and estimated reserve characterization, appropriate types of samples include the following (alphabetically):

Bench sample: A face or channel sample taken of just that contiguous portion of a coalbed that is considered practical to mine, also known as a *bench*; For example, bench samples may be taken of minable coal where impure coal that makes up part of the geologic coalbed is likely to be left in the mine, where thick partings split the coal into two or more distinct minable seams, or where extremely thick coal beds cannot be recovered by normal mining equipment so that the coal is mined in multiple passes, or benches, usually defined along natural bedding planes.

Column sample: A channel or drill core sample taken to represent the entire geologic coalbed; it includes all partings and impurities that may exist in the coalbed.

Composite sample: A recombined coalbed sample produced by averaging together thickness-weighted coal analyses from partial samples of the coalbed, such as from one or more bench samples, from one or more mine exposures or outcrops where the entire bed could not be accessed in one sample, or from multiple drill cores that were required to retrieve all local sections of a coal seam.

Face channel or channel sample: A sample taken at the exposed coal in a mine by cutting away any loose or weathered coal and then collecting on a clean surface a sample of the coal seam by chopping out a channel of uniform width and depth; a face channel or face sample is taken at or near the working face, the most freshly exposed coal where actual removal and loading of mined coal is taking place. Any partings greater than 3/8 in. and/or mineral concretions greater than 1/2 in. thick and 2 in. in maximum diameter are normally discarded from a channel sample so as better to represent coal that has been mined, crushed, and screened to remove at least gross non-coal materials.

Coal seam: A layer, vein, or deposit of coal. A stratigraphic part of the earth's surface containing coal.

Coal sizes: In the coal industry, the term *5–3/4 in.* means all coal pieces between 5 and 3/4 in. at their widest point. *Plus 5 in.* means coal pieces over 5 in. in size; *1 1/2 in. to 0* or *–1 1/2 in.* means coal pieces 1 1/2 in. and under.

Coal tar: The condensable distillate containing light, middle, and heavy oils obtained by carbonization of coal. About 8 gal of tar is obtained from each ton of bituminous coal.

Coal upgrading: It generally refers to upgrading technology that removes moisture and certain pollutants from lower-rank coals such as subbituminous coal and lignite by raising the calorific value; upgrading technologies are typically precombustion treatments and/or processes that alter the characteristics of a coal before it is burned; the product is often referred to as *refined coal*; may also refer to gasification and liquefaction processes in which the coal is upgraded to a gaseous or liquid product.

Coal washing: The process of separating undesirable materials from coal based on differences in densities; pyrite (FeS_2) is heavier and sinks in water—coal is lighter and floats.

Coal zone: A series of laterally extensive and (or) lenticular coal beds and associated strata that arbitrarily can be viewed as a unit; generally, the coal beds in a coal zone are assigned to the same geologic member or formation.

Coarse coal: Coal pieces larger than 1/2 mm in size.

Cogeneration: A process by which electricity and steam, for space heating or industrial-process heating, are produced simultaneously from the same fuel.

Coke: A gray, hard, porous, and coherent cellular-structured combustible solid, primarily composed of amorphous carbon; produced by destructive distillation or thermal decomposition of certain bituminous coal that passes through a plastic state in the absence of air.

Coke-oven gas: A medium-Btu gas, typically 550 Btu/ft^3, produced as a by-product in the manufacture of coke by heating coal at moderate temperatures.

Colliery: British name for coal mine.

Column flotation: A precombustion coal cleaning technology in which coal particles attach to air bubbles rising in a vertical column—the coal is then removed at the top of the column.

Comminution: Breaking, crushing, or grinding of coal, ore, or rock.

Competent rock: Rock that, because of its physical and geological characteristics, is capable of sustaining openings without any structural support except pillars and walls left during mining (stalls, light props, and roof bolts are not considered structural support).

Compliance coal: A coal or a blend of coals that meets sulfur dioxide emission standards for air quality without the need for FGD.

Concretion: A volume of sedimentary rock in which mineral cement fills the spaces between the sediment grains; often ovoid or spherical in shape, although irregular shapes also occur.

Contact: The place or surface where two different kinds of rocks meet; applies to sedimentary rocks, as the contact between a limestone and a sandstone, for example, and to metamorphic rocks; and it is especially applicable between igneous intrusions and their walls.

Continuous miner: A mechanical mining machine consisting of a cutting head, a coal-gathering device, a chain conveyor with flexible loading boom, and a crawler-equipped chassis. Its function is to excavate the mineral and to load it onto shuttle cars or continuous-haulage systems. It is electrically powered, with a hydraulic subsystem for auxiliary functions. Power is supplied through a trailing cable.

Contour: A line on a map that connects all points on a surface having the same elevation.

Contour mining (contour stripping): The removal of overburden and mining from a coal seam that outcrops or approaches the surface at approximately the same elevation in steep or mountainous areas.

Core drilling: The process by which a cylindrical sample of rock and other strata is obtained through the use of a hollow drilling bit that cuts and retains a section of the rock or other strata penetrated.

Core sample: A cylinder sample generally 1–5 in. diameter drilled out of an area to determine the geologic and chemical analysis of the overburden and coal.

Cover: The overburden of any deposit.

Creep: The forcing of pillars into soft bottom by the weight of a strong roof; in surface mining, a very slow movement of slopes downhill.

Crib: A roof support of prop timbers or ties, laid in alternate cross-layers, log-cabin style; it may or may not be filled with debris and is also maybe called a chock or cog.

Cribbing: The construction of cribs or timbers laid at right angles to each other, sometimes filled with earth, as a roof support or as a support for machinery.

Crop coal: Coal at the outcrop of the seam; usually considered to be of inferior quality due to partial oxidation, but this is not always the case.

Crossbar: The horizontal member of a roof timber set supported by props located either on roadways or at the face.

Crosscut: A passageway driven between the entry and its parallel air course or air courses for ventilation purposes; also, a tunnel driven from one seam to another through or across the intervening measures; sometimes called *crosscut tunnel* or *breakthrough*; in vein mining, an entry perpendicular to the vein.

Cross entry: An entry running at an angle with the main entry.

Crusher: A machine for crushing rock or other materials—among the various types of crushers are the ball mill, gyratory crusher, Handsel mill, hammer mill, jaw crusher, rod mill, rolls, stamp mill, and tube mill.

Cutter; cutting machine: A machine, usually used in coal, that will cut a 10–15 cm slot, which allows room for expansion of the broken coal; also applies to the man who operates the machine and to workers engaged in the cutting of coal by prick or drill.

Cycle mining: A system of mining in more than one working place at a time, that is, a miner takes a lift from the face and moves to another face while permanent roof support is established in the previous working face.

Cyclone: A cone-shaped air-cleaning apparatus that operates by centrifugal separation that is used in particle-collecting and fine-grinding operations.

Cyclone collectors: Equipment in which centrifugal force is used to separate particulates from a gas stream.

Cyclone firing: It refers to slagging combustion of coarsely pulverized coal in a cylindrical (cyclone) burner. Some wet-bottom boilers are not cyclone-fired. The primary by-product is a glassy slag referred to as boiler slag.

Demonstrated reserves: A collective term for the sum of coal in both measured and indicated resources and reserves.

Dense media (heavy media): Liquids, solutions, or suspensions having densities greater than that of water.

Dense-media separation: A coal-cleaning method based on density separation, using a heavy-media suspension of fine particles of magnetite, sand, or clay.

Dense medium: A dense slurry formed by the suspension of heavy particles in water; used to clean coal.

Depleted resources: Resources that have been mined; includes coal recovered, coal lost in mining, and coal reclassified as subeconomic because of mining.

Deposit: Mineral deposit or ore deposit is used to designate a natural occurrence of a useful mineral, or an ore, to a sufficient extent and degree of concentration to invite exploitation.

Depth: The vertical depth below the surface; in the case of incline shafts and boreholes it may mean the distance reached from the beginning of the shaft or hole, the borehole depth, or the inclined depth.

Descending-bed system: Gravity downflow of packed solids contacted with upwardly flowing gases—sometimes referred to as *fixed-bed* or *moving-bed* system.

Detectors: Specialized chemical or electronic instruments used to detect mine gases.

Detonator: A device containing a small detonating charge that is used for detonating an explosive, including, but not limited to, blasting caps, exploders, electric detonators, and delay electric blasting caps.

Development mining: Work undertaken to open up coal reserves as distinguished from the work of actual coal extraction.

Devolatilization: The removal of vaporizable material by the action of heat.

Dewatering: The removal of water from coal by mechanical equipment such as a vibrating screen, filter, or centrifuge.

Diffusion: Blending of a gas and an air, resulting in a homogeneous mixture; blending of two or more gases.

Diffuser fan: A fan mounted on a continuous miner to assist and direct air delivery from the machine to the face.

Dilute: To lower the concentration of a mixture; in this case the concentration of any hazardous gas in mine air by addition of fresh intake air.

Dilution: The contamination of ore with barren wall rock in stopping.

Dip: The inclination of a geologic structure (bed, vein, fault, etc.) from the horizontal surface; dip is always measured downward at right angles to the strike.

Dipping strata: Strata that have a pronounced downdip; the strata that form either side of the typical petroleum anticline are often referred to as dipping strata.

Direct hydrogenation: Hydrogenation of coal without the use of a separate donor solvent hydrogenation step.

Downcast: Air forced down into the mine below, by way of the airshaft that is adjacent to the escape shaft.

Dragline: An excavating machine that uses a bucket operated and suspended by means of lines or cables, one of which hoists or lowers the bucket from a boom; the other, from which the name is derived, allows the bucket to swing out from the machine or to be dragged toward the machine for loading. Mobility of draglines is by crawler mounting or by a walking device for propelling, featuring pontoon-like feet and a circular base or tub. The swing of the machine is based on rollers and rail. The machine usually operates from the highwall.

Drainage: The process of removing surplus ground or surface water either by artificial means or by gravity flow.

Draw slate: A soft slate, shale, or rock from approximately 1–10 cm thick and located immediately above certain coal seams, which falls quite easily when the coal support is withdrawn.

Drift: A horizontal passage underground. A drift follows the vein, as distinguished from a crosscut that intersects it, or a level or gallery, which may do either.

Drift mine: An underground coal mine in which the entry or access is above water level and generally on the slope of a hill, driven horizontally into a coal seam.

Drill: A machine utilizing rotation, percussion (hammering), or a combination of both to make holes; if the hole is much over 0.4 m in diameter, the machine is called a borer.

Drilling: The use of such a machine to create holes for exploration or for loading with explosives.

Dry, ash-free (daf) basis: A coal analysis basis calculated as if moisture and ash were removed.

Drying: The removal of water from coal by thermal drying, screening, or centrifuging.

Dull coal: Coal that absorbs rather than reflects light, containing mostly durain and fusain lithotypes.

Dummy: A bag filled with sand, clay, etc., used for stemming a charged hole.

Dump: To unload; specifically, a load of coal or waste; the mechanism for unloading, for example, a car dump (sometimes called tipple); or the pile created by such unloading, for example, a waste dump (also called heap, pile, tip, spoil pike).

Durain: A macroscopic coal constituent (lithotype) that is hard and dull gray in color.

Ebullating-bed reactor: A system similar to a fluidized bed but operated at higher gas velocities such that a portion of the solids is carried out with the up-flowing gas.

Electrostatic precipitation: Separation of liquid or solid particles from a gas stream by the action of electrically charged wires and plates.

Electrostatic precipitator (ESP): Collection of coal combustion fly ash requires the application of an electrostatic charge to the fly ash, which then is collected on grouped plates in a series of hoppers. Fly ash collected in different hoppers may have differing particle size and chemical composition, depending on the distance of the hopper from the combustor. The ESP ash may also be collected as a composite.

Endothermic reaction: A process in which heat is absorbed.

Entrained flow system: Solids suspended in a moving gas stream and flowing with it.

Entry: An entrance into a series of dugout tunnels and/or passageways in the mine below.

Equilibrium moisture: The moisture capacity of coal at 30°C (86°F) in an atmosphere of 95% relative humidity.

Escape shaft: A stairway reaching from the bottom of the mine to the top of the mine used in case of an emergency.

Excluded minerals: Minerals that may be mined with the coal but are not an intrinsic part of it.

Exinite: A microscopic coal constituent (maceral) or maceral group containing spores and cuticles. Appears dark gray in reflected light.

Exothermic reaction: A process in which heat is evolved.

Exploration: The search for mineral deposits and the work done to prove or establish the extent of a mineral deposit. Alt: prospecting and subsequent evaluation.

Explosive: Any rapidly combustive or expanding substance; the energy released during this rapid combustion or expansion can be used to break rock.

Extraction: The process of mining and removal of cal or ore from a mine.

Face: The solid unbroken surface of a coal bed that is at the advancing end of the mine workplace.

Face cleat: The principal cleavage plane or joint at right angles to the stratification of the coal seam.

Face conveyor: A conveyor used on longwall mining faces and consisting of a metal trough with an integrated return channel. Steel scrapers attached to an endless round link or roller-type chain force through the trough any material deposited inside the trough by the mining machine. Spill plates and guides for mining equipment are attached. For flexibility and ease of installation the conveyor is made up of 5 ft sections. Commonly, two electrically powered drives (one on each end) move chain, scrapers (flights), and material along.

Face supports: Hydraulically powered units used to support the roof along a longwall face. They consist of plates at the roof and floor and two to six hydraulic cylinders that press these plates against the respective surfaces with forces of 200–800 tons.

Factor of safety: The ratio of the ultimate breaking strength of the material to the force exerted against it—if a rope will break under a load of 6000 lb, and it is carrying a load of 2000 lb, its factor of safety is 3 (6000/2000).

Fall: A mass of roof rock or coal that has fallen in any part of a mine.

Fan, auxiliary: A small, portable fan used to supplement the ventilation of an individual working place.

Fan, booster: A large fan installed in the main air current and thus in tandem with the main fan.

Fan signal: Automation device designed to give alarm if the main fan slows down or stops.

Faults: Fractures in the rock sequence along which strata on each side of the fracture appear to have moved, but in different directions; a slip surface between two portions of the earth's surface that have moved relative to each other; a failure surface and is evidence of severe earth stresses.

Fault zone: A fault, instead of being a single clean fracture, may be a zone hundreds or thousands of feet wide; the fault zone consists of numerous interlacing small faults or a confused zone of gouge, breccia, or mylonite.

Fluidized-bed combustion materials: Unburned coal, ash, and spent bed material used for sulfur control. The spent bed material (removed as bottom ash) contains reaction products from the absorption of gaseous sulfur oxides (SO_2 and SO_3).

FGD materials: Derived from a variety of processes used to control sulfur emissions from boiler stacks. These systems include wet scrubbers, spray dry scrubbers, sorbent injectors, and a combined sulfur oxide (SOx) and nitrogen oxide (NOx) process. Sorbents include lime, limestone, sodium-based compounds, and high-calcium coal fly ash.

Fill: Typically any material that is put back in place of the extracted coal—sometimes to provide ground support.

Fine coal: Coal pieces less than 1/2 mm in size.

Fines: The content of fine particles, usually less than 1/8 in., in a coal sample.

Firedamp: An explosive mixture of carbonaceous gases, mainly methane, formed in coal mines by the decomposition of coal.

Fissure: An extensive crack, break, or fracture in the rocks.

Fixed-bed system: *see* Descending-bed system.

Fixed carbon: The combustible residue left after the volatile matter is driven off. In general, the fixed carbon represents that portion of the fuel that must be burned in the solid state.

Float-and-sink analysis: Separation of crushed coal into density fractions using a series of heavy liquids. Each fraction is weighed and analyzed for ash and often for sulfur content. Washability curves are prepared from these data.

Flocculants: Water-soluble or colloidal chemical reagents that when added to finely dispersed suspensions of solids in water promote the formation of flocs of the particles and their rapid settlement.

Floor: The layer of rock directly below a coal seam or the floor of a mine opening.

Flue gas desulfurization (FGD): Removal of the sulfur gases from the flue gases (stack gases) of a coal-fired boiler—typically using a high-calcium sorbent such as lime or limestone; three primary types of flue gas desulfurization processes commonly used by utilities are wet scrubbers, dry scrubbers, and sorbent injection.

Fluidity: The degree of plasticity exhibited by a sample of coal heated in the absence of air under controlled conditions, as described in ASTM Standard Test Methods D1812 and D2639.

Fluidization: *see* Fluidized-bed system.

Fluidized-bed combustion: It accomplishes coal combustion by mixing the coal with a sorbent such as limestone or other bed material. The fuel and bed material mixture is fluidized during the combustion process to allow complete combustion and removal of sulfur gases. Atmospheric FBC (AFBC) systems may be bubbling (BFBC) or circulating (CFBC). Pressurized FBC (PFBC) is an emerging coal combustion technology.

Fluidized-bed system: Solids suspended in space by an upwardly moving gas stream.

Fluid temperature (ash fluid temperature): The temperature at which the coal ash becomes fluid and flows in streams.

Fly ash: Airborne bits of unburnable ash that are carried into the atmosphere by stack gases; coal ash that exits a combustion chamber in the flue gas and is captured by air pollution control equipment such as electrostatic precipitators, baghouses, and wet scrubbers.

Folded strata: Strata that are bent or curved as a result of permanent deformation.

Formation: Any assemblage of rocks that have some character in common, whether of origin, age, or composition. Often, the word is loosely used to indicate anything that has been formed or brought into its present shape.

Fouling: The accumulation of small, sticky molten particles of coal ash on a boiler surface.

Fracture: A general term to include any kind of discontinuity in a body of rock if produced by mechanical failure, whether by shear stress or tensile stress. Fractures include faults, shears, joints, and planes of fracture cleavage.

Free moisture (surface moisture): The part of coal moisture that is removed by air-drying under standard conditions approximating atmospheric equilibrium.

Free swelling index: A measure of the agglomerating tendency of coal heated to 800°C (1470°F) in a crucible. Coals with a high index are referred to as coking coals; those with a low index are referred to as free-burning coal.

Friability: The tendency of coal particles to break down in size during storage, transportation, or handling; quantitatively expressed as the ratio of average particle size after test to average particle size before test, ×100.

Friable: Easy to break, or crumbling naturally. Descriptive of certain rocks and minerals.

Froth flotation: A process for cleaning coal fines in which separation from mineral matter is achieved by attachment of the coal to air bubbles in a water medium, allowing the coal to gather in the froth while the mineral matter sinks.

Fusain: A black macroscopic coal constituent (lithotype) that resembles wood charcoal; extremely soft and friable. Also, U.S. Bureau of Mines' term for mineral charcoal seen by transmitted light microscopy.

Fuse: A cord-like substance used in the ignition of explosives—black powder is entrained in the cord and, when lit, burns along the cord at a set rate; a fuse can be safely used to ignite a cap, which is the primer for an explosive.

Fusinite: A microscopic coal constituent (maceral) with well-preserved cell structure and cell cavities empty or occupied by mineral matter.

Gallery: A horizontal or nearly horizontal underground passage, either natural or artificial.

Gasification: Conversion of coal to gas.

Gasification, underground (in situ gasification): A method of utilizing coal by burning in place and extracting the released gases, tars, and heat.

Gas purification: Gas treatment to remove contaminants such as fly ash, tars, oils, ammonia, and hydrogen sulfide.

Gob: The term applied to that part of the mine from which the coal has been removed and the space more or less filled up with waste; also, the loose waste in a mine, which is also called *goaf*.

Global climate change: It usually refers to the gradual warming of the earth caused by the greenhouse effect; believed to be the result of man-made emissions of greenhouse gases such as carbon dioxide, chlorofluorocarbons (CFC) and methane, although there is no agreement among the scientific community on this controversial issue.

Grade: A term indicating the nature of coal as mainly determined by the sulfur content and the amount and type of ash; not recommended for use in coal resource estimations; definitive statements as to the contents and types of sulfur and ash are preferable—statements indicating high, medium, or low grade are inappropriate without documentation; *see* Quality.

Grain: In petrology, that factor of the texture of a rock composed of distinct particles or crystals which depends upon their absolute size.

Granular bed filters: Equipment that uses a bed of separate, closely packed solids as the separation medium.

Gravity separation: Treatment of coal particles that depends mainly on differences in specific gravity of particles for separation.

Green energy: Energy that can be extracted, generated, and/or consumed without any significant negative impact on the environment.

Grindability index: A number that indicates the ease of pulverizing a coal in comparison to a reference coal. This index is helpful in estimating mill capacity. The two most common methods for determining this index are the Hardgrove Grindability Method and Ball Mill Grindability Method. Coals with a low index are more difficult to pulverize.

Grizzly: Course screening or scalping device that prevents oversized bulk material from entering a material transfer system; constructed of rails, bars, and beams.

Ground control: The regulation and final arresting of the closure of the walls of a mined area; the term generally refers to measures taken to prevent roof falls or coal bursts.

Ground pressure: The pressure to which a rock formation is subjected by the weight of the superimposed rock and rock material or by diastrophic forces created by movements in the rocks forming the earth's crust; such pressures may be great enough to cause rocks having a low compressional strength to deform and be squeezed into and close a borehole or other underground opening not adequately strengthened by an artificial support, such as casing or timber.

Gunite: A cement applied by spraying to the roof and sides of a mine passage.

Hard coal: Coal with a heat content greater than 10,260 Btu/lb on a moist ash-free basis. It includes anthracite, bituminous, and the higher-rank subbituminous coals.

Hardgrove Grindability Index: The weight percent of coal retained on a No. 200 sieve after treatment as specified in ASTM Standard Test Method D409.

Haulage: The horizontal transport of ore, coal, supplies, and waste; the vertical transport of the same is called hoisting.

Haulageway: Any underground entry or passageway that is designed for transport of mined material, personnel, or equipment, usually by the installation of track or belt conveyor.

Headframe: The structure surmounting the shaft that supports the hoist rope pulley, and often the hoist itself.

Heading: A vein above a drift; an interior level or airway driven in a mine; in longwall workings, a narrow passage driven upward from a gangway in starting a working in order to give a loose end.

Head section: A term used in both belt and chain conveyor work to designate that portion of the conveyor used for discharging material.

Heat of combustion, heat value: The amount of heat obtainable from coal expressed in Btu's per pound, joules per kilogram, kilojoules or kilocalories per kilogram, or calories per gram: to convert Btu/lb to kcal/kg, divide by 1.8. To convert kcal/kg to Btu/lb, multiply by 1.8.

Heaving: Applied to the rising of the bottom after removal of the coal; a sharp rise in the floor is called a *hogsback*.

Heavy media: *see* Dense media.

Heavy oil: A heavy coal tar fraction with distillation range usually 250°C–300°C (480°F–570°F), containing naphthalene and coal tar bases.

High-temperature tar: The heavy distillate from the pyrolysis of coal at a temperature of about 800°C (1470°F).

High-volatile bituminous coal: Three related rank groups of bituminous coal as defined by the ASTM, which collectively contain less than 69% fixed carbon on a dry, mineral-matter-free basis; more than 31% volatile matter on a dry, mineral-matter-free basis; and a heat value of more than 10,500 Btu/lb on a moist, mineral-matter-free basis.

Highwall: The unexcavated face of exposed overburden and coal in a surface mine or the face or bank on the uphill side of a contour strip mine excavation.

Highwall mining: A highwall mining system consists of a remotely controlled continuous miner, which extracts coal and conveys it via augers, belt, or chain conveyors to the outside; the cut is typically a rectangular horizontal cut from a highwall bench, reaching depths of several hundred feet or deeper.

Hogsback: A sharp rise in the floor of a seam.

Hoist: A drum on which hoisting rope is wound in the engine house, as the cage or skip is raised in the hoisting shaft.

Hoisting: The vertical transport coal or material.

Horizon: In geology, any given definite position or interval in the stratigraphic column or the scheme of stratigraphic classification; generally used in a relative sense.

Horseback: A mass of material with a slippery surface in the roof; shaped like a horse's back.

Hydraulic: Of or pertaining to fluids in motion. Hydraulic cement has a composition that permits it to set quickly under water. Hydraulic jacks lift through the force transmitted to the movable part of the jack by a liquid. Hydraulic control refers to the mechanical control of various parts of machines, such as coal cutters, loaders, etc., through the operation or action of hydraulic cylinders.

Hydrocyclone: Hydraulic device for separating suspended solid particles from liquids by centrifugal action. Cyclone action splits the inlet flow, a small part of which exits via the lower cone, the remainder overflowing the top of the cylindrical section. Particles are separated according to their densities so that the denser particles exit via the cone underflow and less dense particles exit with the overflow.

Hydrogasification: Reaction of carbonaceous material such as coal with hydrogen to produce methane.

Hydrogenation: Chemical reaction in which hydrogen is added to a substance.

Hydrology: The science that relates to the water systems of the earth.

Hydrophilic: Possessing polar surfaces that are readily wetted by water; literally, water loving.

Hydrophobic: Possessing nonpolar surfaces that are not wetted by water; literally, water hating.

Igneous intrusions: An intrusion into another geologic formation that occurs when magma cools and solidifies before it reaches the surface; coal and associated strata may have been intruded by once molten igneous rocks forcibly injected into the sedimentary sequence from below.

Impure coal: Coal having 25 weight percent or more, but less than 50 weight percent ash on the dry basis (ASTM D2796); impure coal having more than 33 weight percent ash is excluded from resource and reserve estimates unless the coal is cleanable to less than 33 weight percent ash; *see* Bone coal.

Inby: In the direction of the working face.

Incline: Any entry to a mine that is not vertical (shaft) or horizontal (adit). Often incline is reserved for those entries that are too steep for a belt conveyor (+17° to 18°), in which case hoist and guide rails are employed. A belt conveyor incline is termed a slope. Alt: Secondary inclined opening, driven upward to connect levels, sometimes on the dip of a deposit; also called *inclined shaft*.

Included minerals: Minerals that are part of the coal particle and matrix.

Incompetent: Applied to strata, a formation, a rock, or a rock structure not combining sufficient firmness and flexibility to transmit a thrust and to lift a load by bending.

Indicated coal resources: Coal for which estimates of the rank, quality, and quantity have been computed partly from sample analyses and measurements and partly from reasonable geologic projections; the points of observation are 1/2–1 1/2 miles apart. Indicated coal is projected to extend as an 1/2 mile wide belt that lies more than 1/4 mile from the outcrop or points of observation or measurement.

Indirect hydrogenation: Coal is first gasified to make a synthesis gas. The gas is then passed over a catalyst to produce methanol or paraffinic hydrocarbons.

Inferred coal resources: Coal in unexplored extensions of the demonstrated resources for which estimates of the quality and size are based on geologic evidence and projection; quantitative estimates are based largely on broad knowledge of the geologic character of the deposit and for which there are few, if any, samples or measurements—the estimates are based on an assumed continuity or repletion of which there is geologic evidence; this evidence may include comparison with deposits of similar type; bodies that are completely concealed may be included if there is specific geologic evidence of their presence; the points of observation are 1 1/2–6 miles apart.

Inferred reserves (unproved reserves): The term *inferred reserves* is commonly used in addition to, or in place of, *potential reserves.*

Initial deformation temperature (ash initial deformation temperature): The temperature at which coal begins to fuse and become soft.

In situ: In the natural or original position. Applied to a rock, soil, or fossil when occurring in the situation in which it was originally formed or deposited.

Intake: The passage through which fresh air is drawn or forced into a mine or to a section of a mine.

Intermediate section: A term used in belt and chain conveyor network to designate a section of the conveyor frame occupying a position between the head and foot sections.

Immediate roof: The roof strata immediately above the coalbed, requiring support during the excavation of coal.

Isopach: A line, on a map, drawn through points of equal thickness of a designated unit. Synonym for isopachous line; isopachyte.

Jackleg: A percussion drill used for drifting or stopping that is mounted on a telescopic leg, which has an extension of about 2.5 m. The leg and machine are hinged so that the drill need not be in the same direction as the leg.

Jackrock: A caltrop or other object manufactured with one or more rounded or sharpened points, which when placed or thrown present at least one point at such an angle that it is peculiar to and designed for use in puncturing or damaging vehicle tires; jackrocks are commonly used during labor disputes.

Jigs: Machines that produce stratification of the particles in a bed or particles of differing densities by repeated differential agitation of the bed, the heaviest particles migrating to the lowest layer. The jigging action may be carried out in air or with the bed immersed in water or other liquids.

Job safety analysis (JSA): A job breakdown that gives a safe, efficient job procedure.

Joint: A divisional plane or surface that divides a rock and along which there has been no visible movement parallel to the plane or surface.

Kettle bottom: A smooth, rounded piece of rock, cylindrical in shape, which may drop out of the roof of a mine without warning. The origin of this feature is thought to be the remains of the stump of a tree that has been replaced by sediments so that the original form has been rather well preserved.

Kerf: The undercut of a coal face.

Lamp: The electric cap lamp worn for visibility; also, the flame safety lamp used in coal mines to detect methane gas concentrations and oxygen deficiency.

Layout: The design or pattern of the main roadways and workings—the proper layout of mine workings is the responsibility of the manager aided by the planning department.

Lift: The amount of coal obtained from a continuous miner in one mining cycle.

Light oil: A coal tar and coal gas fraction with distillation range between 80°C and 210°C (175°F–410°F) containing mainly benzene with smaller amounts of toluene and xylene.

Lignite: A brownish-black woody-structured coal, lower in fixed carbon and higher in volatile matter and oxygen than either anthracite or bituminous coal similar to the *brown coal* of Europe and Australia; a class of brownish-black, low-rank coal defined by the ASTM as having less than 8300 Btu on a moist, mineral-matter-free basis; in the United States, lignite is separated into two groups: Lignite A (6300–8300 Btu) and lignite B (<6300 Btu).

Liquefaction: The conversion of coal into nearly mineral-free hydrocarbon liquids or low-melting solids by a process of direct or indirect hydrogenation at elevated temperatures and pressures and separation of liquid products from residue by either filtration or distillation or both.

Liquefied petroleum gas (LPG): A mixture of propane and butane.

Lithology: The character of a rock described in terms of its structure, color, mineral composition, grain size, and arrangement of its component parts; all those visible features that in the aggregate impart individuality of the rock; lithology is the basis of correlation in coal mines and commonly is reliable over a distance of a few miles.

Lithotypes: Coal lithotypes represent the macrostructure of coal and are, in fact, descriptive of the coal.

Load: To place explosives in a drill hole. Also, to transfer broken material into a haulage device.

Loader: A crawler-mounted unit equipped with a coal-gathering device mounted to an inclined feed plate at the front side. A chain conveyor with an articulated loading boom discharges the coal at the opposite end into shuttle cars or any other conveying systems. It is used primarily on conventionally mined room-and-pillar sections and resembles a continuous miner without the cutting head.

Loading machine: Any device for transferring excavated coal into the haulage equipment.

Loading pocket: Transfer point at a shaft where bulk material is loaded by bin, hopper, and chute into a skip.

Long ton: A unit of weight in the U.S. Customary System and in the United Kingdom equal to 2240 lb (1.0160469 metric tons/1.1200 short tons/1016.0469 kg).

Longwall mining: A mining method in which a large rectangular section of coal is removed in one continuous operation. Equipment is installed along one side of the section (the longwall face), and the coal is removed in slices 2–4 ft thick. The excavated area behind the equipment is allowed to cave.

Loose coal: Coal fragments larger in size than coal dust.

Low-Btu gas: A nitrogen-rich gas with a heat content of 100–200 Btu/ft^3 produced in gasification processes using air as the oxygen source. The air-blown form of producer gas.

Low voltage: Up to and including 660 V by federal standards.

LPG: *see* Liquefied petroleum gas.

Maceral: Microscopic petrographic units of coal.

Main entry: A main haulage road. Where the coal has cleats, main entries are driven at right angles to the face cleats.

Main fan: A mechanical ventilator installed at the surface; operates by either exhausting or blowing to induce airflow through the mine roadways and workings.

Manhole: A safety hole constructed in the side of a gangway, tunnel, or slope in which miner can be safe from passing locomotives and car. Also called a refuge hole.

Man trip: A carrier of mine personnel, by rail or rubber tire, to and from the work area.

Manway: An entry used exclusively for personnel to travel form the shaft bottom or drift mouth to the working section; it is always on the intake air side in gassy mines. Also, a small passage at one side or both sides of a breast, used as a traveling way for the miner, and sometimes, as an airway, or chute, or both.

Measured coal resources: Coal for which estimates of the rank, quality, and quantity have been computed from sample analyses and measurements from closely spaced and geologically well-known sample sites, such as outcrops, trenches, mine workings, and drill holes. The points of observation and measurement are so closely spaced and the thickness and extent of coals are so well defined that the tonnage is judged to be accurate within 20% of true tonnage. Although the spacing of the points of observation necessary to demonstrate continuity of the coal differs from region to region according to the character of the coal beds, the points of observation are no greater than 1/2 mile apart. Measured coal is projected to extend as a 1/4 mile wide belt from the outcrop or points of observation or measurement.

Meridian: A surveying term that establishes a line of reference. The bearing is used to designate direction. The bearing of a line is the acute horizontal angle between the meridian and the line. Azimuths are angles measured clockwise from any meridian.

Metallurgical coal: Coal used in the steelmaking process to manufacture coke; metallurgical coal; an informally recognized name for bituminous coal that is suitable for making coke by industries that refine, smelt, and work with iron—other uses are space heating, blacksmithing, smelting of base metals, and power generation; generally, metallurgical coal has less than 1% sulfur and less than 8% ash on an as-received basis—most premium metallurgical coal is low- to medium-volatile bituminous coal.

Methane: A potentially explosive gas formed naturally from the decay of vegetative matter, similar to that which formed coal. Methane, which is the principal component of natural gas, is frequently encountered in underground coal mining operations and is kept within safe limits through the use of extensive mine ventilation systems.

Methane monitor: An electronic instrument often mounted on a piece of mining equipment, that detects and measures the methane content of mine air.

Microlithotypes: The microscopic analogs of the coal lithotypes and, hence, represent a part of the fine microstructure of coal; associations of coal macerals with the proviso that the *associations* should occur within an arbitrary minimum bandwidth (50 μm, 50 × 10 mm).

Middlings: Coal of an intermediate specific gravity and quality.

Methanation: A process for catalytic conversion of 1 mol of carbon monoxide and 3 mol of hydrogen to 1 mol of methane and 1 mol of water.

Middle (carbolic or creosote) oil: A coal tar fraction with a distillation range of 200°C–270°C (390°F–520°F), containing mainly naphthalene, phenol, and cresols.

Mine development: The term employed to designate the operations involved in preparing a mine for ore extraction. These operations include tunneling, sinking, crosscutting, drifting, and raising.

Mined land: Land with new surface characteristics due to the removal of minable commodities by surface-mining methods and subsequent surface reclamation.

Mine mouth electric plant: A coal-burning electric-generating plant built near a coal mine.

Miner: A person who is engaged in the business or occupation of extracting ore, coal, precious substances, or other natural materials from the earth's crust.

Mineral: An inorganic compound occurring naturally in the earth's crust, with a distinctive set of physical properties, and a definite chemical composition.

Mineral matter: The solid inorganic material in coal.

Mineral-matter-free basis: A theoretical analysis calculated from basic analytical data expressed as if the total mineral matter had been removed; used in determining the rank of a coal.

Misfire: The complete or partial failure of a blasting charge to explode as planned.

Moisture: The total moisture content of a sample customarily determined by adding the moisture loss obtained when air-drying the sample and the measured moisture content of the dried sample. Moisture does not represent all of the water present in coal, as water of decomposition (combined water) and hydration are not given off under standardized test conditions.

Molten bath gasifier: A reaction system in which coal and air or oxygen with steam are contacted underneath a pool of liquid iron, ash, or salt.

Mountain top removal (mountaintop removal): A form of surface mining in which the summit or summit ridge of a mountain is removed in order to permit easier access to coal seams; after the coal is extracted, the overburden is either put back onto the ridge to approximate the original contours of the mountain or dumped elsewhere, often in neighboring valleys; generally associated with coal mining in the Appalachian Mountain areas.

Moving-bed system: *see* Descending-bed system.

Mud cap: A charge of high explosive fired in contact with the surface of a rock after being covered with a quantity of wet mud, wet earth, or sand, without any borehole being used. Also termed adobe, dobie, and sandblast (illegal in coal mining).

Multiple-seam mining: Mining in areas where several seams are recovered from the same area.

Natural gas: A naturally occurring gas with a heat content over 1000 Btu/ft³, consisting mainly of methane but also containing smaller amounts of the C_2–C_4 hydrocarbons as well as nitrogen, carbon dioxide, and hydrogen sulfide.

Natural ventilation: Ventilation of a mine without the aid of fans or furnaces.

Nip: A device at the end of the trailing cable of a mining machine used for connecting the trailing cable to the trolley wire and ground.

Oil agglomeration: Treatment of a suspension of fine coal particles suspended in water with a light hydrocarbon oil so that the particles are preferentially collected by the oil, which separates as a floating pasty agglomerate and can be removed by skimming. First developed as a method for recovering fine coal particles by Trent in 1914.

Open end pillaring: A method of mining pillars in which no stump is left; the pockets driven are open on the gob side and the roof is supported by timber.

Open-pit mining: Surface mining, a type of mining in which the overburden is removed from the product being mined and is dumped back after mining; may refer specifically to an area from which overburden has been removed, which has not been filled.

Outby; outbye: Nearer to the shaft and hence farther from the working face. Toward the mine entrance. The opposite of inby.

Outcrop: An area at which a coal seam is naturally exposed at the surface.

Overburden: The earth, rock, and other materials that lie above the coal seam.

Overcast (undercast): Enclosed airway, which permits one air current to pass over (under) another without interruption.

Oxidized coal: Bituminous coal, the properties of which have been fundamentally modified as a result of chemisorption of oxygen in the air or oxygen dissolved in groundwater. The chemisorption is a surface phenomenon rarely detectable by chemical analysis but usually detectable by petrographic examination. It reduces the affinity of coal surfaces for oil and seriously impairs coking, caking, and agglutinating properties.

Panel: A coal mining block that generally comprises one operating unit.

Panic bar: A switch, in the shape of a bar, used to cut off power at the machine in case of an emergency.

Parting: A layer or stratum of non-coal material in a coal bed that does not exceed the thickness of coal in either the directly underlying or overlying benches.

Peat: Partially carbonized plant matter, formed by slow decay in water.

Percentage extraction: The proportion of a coal seam that is removed from the mine. The remainder may represent coal in pillars or coal that is too thin or inferior to mine or lost in mining. Shallow coal mines working under townships, reservoirs, etc., may extract 50%, or less, of the entire seam, the remainder being left as pillars to protect the surface. Under favorable conditions, longwall mining may extract from 80% to 95% of the entire seam. With pillar methods of working, the extraction ranges from 50% to 90% depending on local conditions.

Percussion drill: A drill, usually air powered, that delivers its energy through a pounding or hammering action.

Permissible: That which is allowable or permitted. It is most widely applied to mine equipment and explosives of all kinds that are similar in all respects to samples that have passed certain tests of the MSHA and can be used with safety in accordance with specified conditions where hazards from explosive gas or coal dust exist.

Permit: As it pertains to mining, a document issued by a regulatory agency that gives approval for mining operations to take place.

Petrography: A branch of coal petrology specifically deals with the analysis of the maceral composition and rank of coal and therefore plays an essential role in predicting coal behavior.

Petrology: The study of the organic and inorganic constituents of coal and their transformation via metamorphism.

Physical coal cleaning: Processes that employ a number of different operations, including crushing, sizing, dewatering and clarifying, and drying, which improve the quality of the fuel by regulating its size and reducing the quantities of ash, sulfur, and other impurities. In this text the term coal cleaning is synonymous with the terms coal preparation, beneficiation, and washing.

Pick: A tool for loosening or breaking up coal or dirt consisting of a slightly curved bar at both ends and fitted onto a long handle.

Piggy-back: A bridge conveyor.

Pillar: An area of coal left to support the overlying strata in a mine; sometimes left permanently to support surface structures.

Pillar robbing: The systematic removal of the coal pillars between rooms or chambers to regulate the subsidence of the roof; also termed *bridging back the pillar*, *drawing the pillar*, or *pulling the pillar*.

Pinch: A compression of the walls of a vein or the roof and floor of a coal seam so as to *squeeze out* the coal.

Pinning: Roof bolting.

Pipeline gas: A methane-rich gas with a heat content of 950–1050 Btu/ft^3 compressed to 1000 psi.

Pit: Used in reference to a specifically describable area of open-cut mining. May be used to refer to only that part of the open-cut mining area from which coal is being actively removed or may refer to the entire contiguous mined area; also used in Britain to refer to a mine.

Pit car: A small railroad-type car approximately 6 × 3 in size, used to haul coal, dirt, and rock.

Pitch: The inclination of a seam; the rise of a seam; also, the nonvolatile portion of coal tar.

Plan: A map showing features such as mine workings or geological structures on a horizontal plane.

Plasticity: A property of certain coals when heated in the absence of air. For a relative and a semi-quantitative method for determining the relative plastic behavior of coal, refer to ASTM Standard Test Methods D2639 and D1812, respectively.

Pneumoconiosis: A chronic disease of the lung arising from breathing coal dust.

Portal: The structure surrounding the immediate entrance to a mine; the mouth of an adit or tunnel.

Portal bus: Track-mounted, self-propelled personnel carrier that holds 8–12 people.

Post: The vertical member of a timber set.

Potential reserves: Reserves of coal that are believed to exist in the earth.

Power shovel: A large machine for digging, the digging part of which is a bucket as the terminal member of an articulated boom. Power to the bucket is supplied through hydraulic cylinders or cables.

Precision: A measure of the maximum random error or deviation of a single observation. It may be expressed as the standard error or a multiple thereof, depending on the probability level desired.

Preheating (coke making): The heating of coal in a preheating column to 180°C–300°C (355°F–570°F) to dry off all the moisture and leave a hot, dry fluid coal that can be charged by gravity or pipeline.

Preparation: The process of upgrading run-of-mine coal to meet market specifications by washing and sizing.

Preparation (coke making): Fine grinding of coal to ensure adequate fusion of the particles. Usual practice is to grind the coal so that 65%–85% will pass through a 1/8 in. screen opening.

Preparation plant: A place where coal is cleaned, sized, and prepared for market.

Pretreatment: Mild oxidation of coal to eliminate caking (agglomeration) tendencies.

Primary roof: The main roof above the immediate top. Its thickness may vary from a few to several thousand feet.

Primer (booster): A package or cartridge of explosive that is designed specifically to transmit detonation to other explosives and which does not contain a detonator.

Producer gas: Mainly carbon monoxide with smaller amounts of hydrogen, methane, and variable nitrogen, obtained from partial combustion of coal or coke in air or oxygen, having a heat content of 110–160 Btu/ft^3 (air combustion) or 400–500 Btu/ft^3 (oxygen combustion).

Prop: Coal mining term for any single post used as roof support. Props may be timber or steel; if steel, screwed, yieldable, or hydraulic.

Proven reserves: Coal reserves that are actually found (proven), usually by drilling and coring.

Proximate analysis: The determination by prescribed methods of moisture, volatile matter, fixed carbon (by difference), and ash; unless specified, proximate analyses do not include determinations of sulfur or phosphorous or any determinations other than those named; proximate analyses are reported by percent and on as-received, moisture-free, and moisture- and ash-free bases.

Pulverized coal combustion: It refers to any combustion process that uses very finely ground (pulverized) coal in the process.

Pyrite: A hard, heavy, shiny, yellow mineral, iron disulfide (FeS_2), generally in cubic crystals; also called iron pyrites, fool's gold, and sulfur balls—the most common sulfide found in coal mines.

Quality: An informal classification of coal relating to its suitability for use for a particular purpose. It refers to individual measurements such as heat value, fixed carbon, moisture, ash, sulfur, phosphorus, major, minor, and trace elements, coking properties, petrologic properties, and particular organic constituents. The individual quality elements may be aggregated in various ways to classify coal for such special purposes as metallurgical, gas, petrochemical, and blending use; *see* Grade.

Raise: A secondary or tertiary inclined opening, vertical or near-vertical opening driven upward form a level to connect with the level above, or to explore the ground for a limited distance above one level.

Ramp: A secondary or tertiary inclined opening, driven to connect levels, usually driven in a downward direction, and used for haulage.

Rank: A complex property of coals that is descriptive of their degree of coalification (i.e., the stage of metamorphosis of the original vegetal material in the increasing sequence peat, lignite, subbituminous, bituminous, and anthracite).

Ranks of coal: The classification of coal by degree of hardness, moisture, and heat content. *Anthracite* is hard coal, almost pure carbon, used mainly for heating homes. *Bituminous coal* is soft coal. It is the most common coal found in the United States and is used to generate electricity and to make coke for the steel industry. *Subbituminous coal* is a coal with a heating value between bituminous and lignite. It has low fixed carbon and high percentages of volatile matter and moisture. *Lignite* is the softest coal and has the highest moisture content. It is used for generating electricity and for conversion into synthetic gas. In terms of Btu or heat content, anthracite has the highest value, followed by bituminous, subbituminous and lignite.

Raw coal: Run-of-mine coal that has been treated by the removal of tramp material, screening, or crushing.

Reclamation: The process of reconverting mined land to its former or other productive uses.

Recoverable reserves (coal): Unmined coal deposits that can be removed by current technology, taking into account economic, legal, political, and social variables.

Recovery: The proportion or percentage of coal or ore mined from the original seam or deposit.

Red dog: A nonvolatile combustion product of the oxidation of coal or coal refuse. Most commonly applied to material resulting from in situ, uncontrolled burning of coal or coal refuse piles. It is similar to coal ash.

Reduction ratio: The ratio between the feed top size and the product top size; the ratio between the feed and product sizes.

Regulator: A device (wall, door) used to control the volume of air in an air split.

Refined coal: *see* Coal upgrading.

Repeatability: The closeness of agreement between test results carried out by one person with one instrument in one laboratory.

Replicate: A measurement or observation that is part of a series performed on the same sample.

Reproducibility: The measure of agreement between test results carried out by more than one person with more than one instrument in more than one laboratory.

Reserve: The portion of the identified coal resource that can be economically mined at the time of determination. The reserve is derived by applying a recovery factor to that component of the identified coal resource designated as the reserve base.

Resin bolting: A method of permanent roof support in which steel rods are grouted with resin.

Resources: Concentrations of coal in such forms that economic extraction is currently or may become feasible. Coal resources broken down by identified and undiscovered resources. Identified coal resources are classified as demonstrated and inferred. Demonstrated resources are further broken down as measured and indicated. Undiscovered resources are broken down as hypothetical and speculative.

Respirable dust: Dust particles 5 μm or less in size.

Respirable dust sample: A sample collected with an approved coal mine dust sampler unit attached to a miner, or so positioned as to measure the concentration of respirable dust to which the miner is exposed, and operated continuously over an entire work shift of such miner.

Retreat mining: A system of robbing pillars in which the robbing line, or line through the faces of the pillars being extracted, retreats from the boundary toward the shaft or mine mouth.

Return: The air or ventilation that has passed through all the working faces of a split.

Return idler: The idler or roller underneath the cover or cover plates on which the conveyor belt rides after the load that it was carrying has been dumped at the head section and starts the return trip toward the foot section.

Rib: The side of a pillar or the wall of an entry. The solid coal on the side of any underground passage. Same as rib pillar.

Rider: A thin seam of coal overlying a thicker one.

Ripper: A coal extraction machine that works by tearing the coal from the face.

Rob: To extract pillars of coal previously left for support.

Robbed out area: The part of a mine from which the pillars have been removed.

Roll: (1) A high place in the bottom or a low place in the top of a mine passage and (2) a local thickening of roof or floor strata, causing thinning of a coal seam.

Roll protection: A framework, safety canopy, or similar protection for the operator when equipment overturns.

Roof: The stratum of rock or other material above a coal seam; the overhead surface of a coal working place; *see* Back or top.

Roof bolt: A long steel bolt driven into the roof of underground excavations to support the roof, preventing and limiting the extent of roof falls. The unit consists of the bolt (up to 4 ft long), steel plate, expansion shell, and pal nut. The use of roof bolts eliminates the need for timbering by fastening together, or laminating, several weaker layers of roof strata to build a beam.

Roof fall: A coal mine cave-in especially in permanent areas such as entries.

Roof jack: A screw- or pump-type hydraulic extension post made of steel and used as temporary roof support.

Roof sag: The sinking, bending, or curving of the roof, especially in the middle, from weight or pressure.

Roof stress: Unbalanced internal forces in the roof or sides, created when coal is extracted.

Roof support: Posts, jacks, roof bolts, and beams used to support the rock overlying a coal seam in an underground mine. A good roof support plan is part of mine safety and coal extraction.

Roof trusses: A combination of steel rods anchored into the roof to create zones of compression and tension forces and provide better support for weak roof and roof over wide areas.

Room-and-pillar mining: A mining method in which a designated area is divided into regular-shaped coal pillars through the parallel development of entries and crosscuts. After the area is so developed, the remaining pillars are mined by slicing them into smaller pillars.

Room neck: The short passage from the entry into a room.

Round: Planned pattern of drill holes fired in sequence in tunneling, shaft sinking, or stopping. First the cut holes are fired, followed by relief, lifter, and rib holes.

Royalty: The payment of a certain stipulated sum on the mineral produced.

Rubbing surface: The total area (top, bottom, and sides) of an airway.

Run-of-mine coal: Raw coal as it leaves the mine prior to any type of crushing or preparation.

Safety fuse: A train of powder enclosed in cotton, jute yarn, or waterproofing compounds, which burns at a uniform rate; used for firing a cap containing the detonation compound, which in turn sets off the explosive charge.

Safety lamp: A lamp with steel wire gauze covering every opening from the inside to the outside so as to prevent the passage of flame should explosive gas be encountered.

Safety pillar: A large unmined area left between a mining section and mine openings designated for long-term use. It is laid out to absorb the abutment loads created by the mining activity and to prevent any adverse effects on the openings designated for long-term use.

Sampling: Cutting a representative part of an ore (or coal) deposit, which should truly represent its average value.

Sandstone: A sedimentary rock consisting of quartz sand united by some cementing material, such as iron oxide or calcium carbonate.

Scaling: Removal of loose rock from the roof or walls. This work is dangerous and a long bar (called a scaling bar) is often used.

Scoop: A rubber tired, battery- or diesel-powered piece of equipment designed for cleaning runways and hauling supplies.

Scrubbers: Any of several forms of chemical/physical devices that remove sulfur compounds formed during coal combustion. These devices, technically known as FGD systems, combine the sulfur in gaseous emissions with another chemical medium to form inert sludge, which must then be removed for disposal; *see* Flue gas desulfurization.

Seam: Underground layer of coal or other mineral of any thickness.

Secondary roof: The roof strata immediately above the coalbed, requiring support during the excavation of coal.

Section: A portion of the working area of a mine.

Selective mining: The object of selective mining is to obtain a relatively high-grade mine product; this usually entails the use of a much more expensive stopping system and high exploration and development costs in searching for and developing the separate bunches, stringers, lenses, and bands of ore.

Self-contained breathing apparatus: A self-contained supply of oxygen used during rescue work from coal mine fires and explosions; same as SCSR (self-contained self-rescuer).

Self-rescuer: A small filtering device carried by a coal miner underground, either on his belt or in his pocket, to provide him with immediate protection against carbon monoxide and smoke in case of a mine fire or explosion. It is a small canister with a mouthpiece directly attached to it. The wearer breathes through the mouth, the nose being closed by a clip. The canister contains a layer of fused calcium chloride that absorbs water vapor from the mine air. The device is used for escape purposes only because it does not sustain life in atmospheres containing deficient oxygen. The length of time a self-rescuer can be used is governed mainly by the humidity in the mine air, usually between 30 min and 1 h.

Severance: The separation of a mineral interest from other interests in the land by grant or reservation. A mineral dead or grant of the land reserving a mineral interest, by the landowner before leasing, accomplishes a severance as does his execution of a mineral lease.

Shaft: A deep vertical passage used to enter the mine below; a shaft has to be *sunk* or dug out until the vein of coal is reached. When a coal vein or layer is found, the digging begins in a mainly horizontal direction to follow the vein.

Shaft mine: An underground mine in which the main entry or access is by means of a vertical shaft.

Shale: A rock formed by consolidation of clay, mud, or silt, having a laminated structure and composed of minerals essentially unaltered since deposition.

Shale parting (shale break): Typically a layer of shale in a coal seam that runs parallel to the bedding plane of the seam.

Sheave: A large pulley used to guide a cable. Sheaves at the mine were placed at the highest point of the tipple called the headframe. These sheaves guided the cables that raised and lowered the cages.

Shortwall mining: A mining method with a panel layout similar to longwall mining but at a panel width reduced by approximately 50%. It uses continuous-mining equipment for coal cutting and haulage and a specially designed hydraulic roof support.

Shovel: An excavating or coal-loading machine that uses a bucket mounted on and operated by means of a handle or dipper stick that moves longitudinally on gears and is lifted or lowered by cable. The entire machine is mounted on crawlers for mobility, and the upper structure is mounted on rollers and rail for swing or turn.

Shuttle car: A rubber-tired vehicle used to haul coal from the continuous miner or loader to a belt feeder or conveyor belt. It is usually electrically powered, with the power supplied through a trailing cable. Some new models are equipped with diesel engines.

Side slope: The slope of the sides of a canal, dam, or embankment. It is customary to name the horizontal distance first as 1.5–1.0 or frequently 1 1/2:1, meaning a horizontal distance of 1.5–1 ft vertical.

Silicosis: A respiratory disease (fibrosis of the lung) caused by the prolonged inhalation of silica dust.

Sinking: The process by which a shaft is driven.

Skid: A track-mounted vehicle used to hold trips or cars from running out of control. Also it is a flat-bottom personnel or equipment carrier used in low coal.

Skip: A car being hoisted from a slope or shaft.

Skipjack: A triggering mechanism that causes mine cars (pit cars) to dump its load of coal or rock to a designated area at the mine.

Slack: Small coal; the finest-sized soft coal, usually less than 1 in. in diameter.

Slag: The nonmetallic product resulting from the interaction of flux and impurities in the smelting and refining of metals.

Slag cyclone: The primary combustion chamber for a cyclone-fired boiler. Ash from the coal melts in the cyclone and is removed as a slag.

Slagging: The accumulation of coal ash on the wall tubes of a coal-fired boiler furnace, forming a solid layer of ash residue and interfering with heat transfer.

Slag pile: A significant amount of dirt and rock excavated from the earth below that is dumped into a pile.

Slate: A miner's term for any shale or slate accompanying coal. Geologically, it is a dense, fine-textured, metamorphic rock, which has excellent parallel cleavage so that it breaks into thin plates or pencil-like shapes.

Slate bar: The proper long-handled tool used to pry down loose and hazardous material from roof, face, and ribs.

Slickenside: A smooth, striated, polished surface produced on rock by friction.

Slip: A fault. A smooth joint or crack where the strata have moved on each other.

Slope: Primary inclined opening, connecting the surface with the underground workings.

Slope mine: An underground mine with an opening that slopes upward or downward to the coal seam.

Slot oven: A long, narrow refractory chamber charged with coal heated in the absence of air by adjacent heating flues. Ovens are arranged in batteries with heating flues between each pair, like books on a shelf. Typical ovens are 457 mm wide, 12–15 m long, and 4–6 m high.

Sloughing: The slow crumbling and falling away of material from roof, rib, and face.

Slurry: A mixture of pulverized insoluble material and water.

Slurry pipeline: A pipeline that can transport a coal–water mixture for long distances.

Solid: Mineral that has not been undermined, sheared out, or otherwise prepared for blasting.

Sounding: Knocking on a roof to see whether it is sound and safe to work under.

Spad: A flat spike hammered into a wooden plug anchored in a hole drilled into the mine ceiling from which is threaded a plumbline. The spad is an underground survey station similar to the use of stakes in marking survey points on the surface. A pointer spad, or sight spad, is a station that allows a mine foreman to visually align entries or breaks from the main spad.

Span: The horizontal distance between the side supports or solid abutments along sides of a roadway.

Sparging: Bubbling a gas into the bottom of a pool of liquid.

Specific energy: The energy per unit of throughput required to reduce feed material to a desired product size.

Specific gravity: The ratio of weight per unit volume of a substance to the weight of the same unit volume of water.

Splint coal: U.S. Bureau of Mines' term for durain with some vitrain and clarain and small amount of fusain.

Split: Any division or branch of the ventilating current; also, the workings ventilated by one branch. Also, to divide a pillar by driving one or more roads through it.

Split coal: Coal that is disturbed by layers of other geologic material, usually layers of shale.

Spoil: The overburden or non-coal material removed in gaining access to the coal or mineral material in surface mining.

Spontaneous combustion: The self-ignition of coal through oxidation under very specific conditions. Different types of coal vary in their tendency toward self-ignition.

Spragger: A person whose occupation is to apply brake to the mine cars below by means of placing wood blocks or wedges underneath the wheels to prevent them from rolling down slight grades.

Square mile-foot: The volume of coal (27,878,400 ft^3/789,428.38 m^3/1,032,533.33 yd^3) that covers 1 mile^2 to a thickness of 1 ft; the weight of coal varies according to the rank.

Squeeze: The settling, without breaking, of the roof and the gradual upheaval of the floor of a mine due to the weight of the overlying strata.

Stack gas: The product gas evolved during complete combustion of a fuel.

Stage loader: A chain conveyor of a design similar to the face conveyor. It provides a connection between the face conveyor and the section conveyor belt.

Steeply inclined: Said of deposits and coal seams with a dip of 0.7–1 rad (40°–60°).

Stemming: The noncombustible material used on top or in front of a charge or explosive.

Stinkdamp: A mine gas containing a high proportion of hydrogen sulfide.

Stoker firing: The combustion of coal on a grate, which may be stationary or moving.

Strike: The direction of the line of intersection of a bed or vein with the horizontal plane; the strike of a bed is the direction of a straight line that connects two points of equal elevation on the bed; also the withdrawal of labor by miners and their associates.

Strip mine: It refers to a procedure of mining that entails the complete removal of all material from over the product to be mined in a series of rows or strips; also referred to as *open cut, open pit,* or *surface mine.*

Stripping: The removal of earth or non-ore rock materials as required to gain access to the ore or mineral materials wanted. The process of removing overburden or waste material in a surface mining operation.

Stripping ratio: The unit amount of spoil or waste (overburden) that must be removed to gain access to a similar unit amount of ore or mineral material.

Stump: Any small pillar.

Subbituminous coal: A glossy-black-weathering and nonagglomerating coal that is lower in fixed carbon than bituminous coal, with more volatile matter and oxygen; a rank class of nonagglomerating coals having a heat value content of more than 8,300 Btus and less than 11,500 Btus on a moist, mineral-matter-free basis—this class of coal is divisible on the basis of increasing heat value into the subbituminous C, B, and A coal groups.

Subsidence: The gradual lowering of the surface area over an extended period of time as a result of an underground excavation.

Substitute natural gas: *see* Synthetic natural gas.

Subsurface water: Water that occurs beneath the surface of the earth in a liquid, solid, or gaseous state; consists of suspended water and groundwater.

Sulfur (total sulfur): Sulfur found in coal as iron pyrites, sulfates, and organic compounds. It is undesirable because the sulfur oxides formed when it burns contribute to air pollution, and sulfur compounds contribute to combustion-system corrosion and deposits.

Sulfur forms: The analytical percentage by weight of coal sulfate, pyritic, and organic sulfur.

Sump: The bottom of a shaft, or any other place in a mine, that is used as a collecting point for drainage water.

Sumping: To force the cutter bar of a machine into or under the coal. Also called a sumping cut or sumping in.

Support: The all-important function of keeping the mine workings open. As a verb, it refers to this function; as a noun it refers to all the equipment and materials—timber, roof bolts, concrete, steel, etc.—that are used to carry out this function.

Surface mine: A mine in which the coal lies near the surface and can be extracted by removing the covering layers of rock and soil.

Surface mining: A mining method whereby the overlying materials are removed to expose the mineral for extraction; see also Strip mine.

Suspension: Weaker strata hanging from stronger, overlying strata by means of roof bolts.

Sustainable energy: The provision of energy that meets the needs of the present without compromising the ability of future generations to meet their needs; sources include renewable energy sources and, in the near term because of the wealth of reserves, coal and oil shale.

Sweetened gas: Gas from which acid (sour) gases such as H_2S and CO_2 have been removed.

Syncline: A fold in rock in which the strata dip inward from both sides toward the axis; the opposite of anticline.

Syncrude: Synthetic crude oil produced by pyrolysis or hydrogenation of coal or coal extracts.

Syngas: *see* Synthesis gas.

Synthesis gas (syngas): Approximately 2:1 molar mixture of hydrogen and carbon monoxide with varying amounts of carbon dioxide.

Synthetic (substitute) natural gas: Pipeline-quality gas that is interchangeable with natural gas (mainly methane).

Tail gas: Residual gas leaving a process; gas produced in a refinery and not usually required for further processing.

Tailgate: A subsidiary gate road to a conveyor face as opposed to a main gate. The tailgate commonly acts as the return airway and supplies road to the face.

Tailpiece: Also known as foot section pulley. The pulley or roller in the tail or foot section of a belt conveyor around which the belt runs.

Tail section: A term used in both belt and chain conveyor work to designate that portion of the conveyor at the extreme opposite end from the delivery point. In either type of conveyor it consists of a frame and either a sprocket or a drum on which the chain or belt travels, plus such other devices as may be required for adjusting belt or chain tension.

Tension: The act of stretching.

Ten-wheeler: An old time train locomotive consisting of ten wheels, also referred to as a 4–6–0, four small wheels up front near the pilot (cow catcher) and six larger wheels under the middle of the loco and zero wheels at the rear or under the cab where the engineer sat.

Tertiary: Lateral or panel openings (e.g., ramp, crosscut).

Through-steel: A system of dust collection from rock or roof drilling. The drill steel is hollow, and a vacuum is applied at the base, pulling the dust through the steel and into a receptacle on the machine.

Timber: A collective term for underground wooden supports.

Timbering: The setting of timber supports in mine workings or shafts for protection against falls from roof, face, or rib.

Timber set: A timber frame to support the roof, sides, and sometimes the floor of mine roadways or shafts.

Tipple: Originally the place where the mine cars were tipped and emptied of their coal; still used in that sense, although now more generally applied to the surface structures of a mine, including the preparation plant and loading tracks. The tracks, trestles, screens, and so on at the entrance to a colliery where coal is screened and loaded.

Ton: A short or net ton is equal to 2000 lb; a long or British ton is 2240 lb; a metric ton is approximately 2205 lb.

Top: A mine roof; *see* Back.

Top size: The smallest sieve size upon which is retained a total of less than 5% w/w of a total sample.

Torque wrench: A wrench that indicates, as on a dial, the amount of torque (in units of foot-pounds) exerted in tightening a roof bolt.

Town gas: A gaseous mixture of coal gas and carbureted water gas manufactured from coal with a heat content of 600 Btu/ft^3.

Toxic spoil: Acid spoil with pH below 4.0; also spoil having amounts of minerals, such as aluminum, manganese, and iron, that adversely affect plant growth.

Trace element: Any element present in minute quantities, such as lead and mercury.

Trackman: A person whose duty is to lay railroad track to selected areas wherein miners could conveniently load the pit cars with coal.

Tractor: A battery-operated piece of equipment that pulls trailers, skids, or personnel carriers. Also used for supplies.

Tram: Used in connection with moving self-propelled mining equipment. A tramming motor may refer to an electric locomotive used for hauling loaded trips or it may refer to the motor in a cutting machine that supplies the power for moving or tramming the machine.

Transfer: A vertical or inclined connection between two or more levels and used as an ore pass.

Transfer point: Location in the materials handling system, either haulage or hoisting, where bulk material is transferred between conveyances.

Trapper: A person, usually of a young age, who opens and closes doors down below in the mine for drivers. The doors down below are used to guide the downcast or airflow to desired areas of the mine.

Tumbling-bed gasifier: An apparatus in which coal is lifted vertically in a revolving cylinder and dropped through an axially flowing stream of oxygen and steam.

Two-stage gasification: Partial gasification or pyrolysis in a first step followed by complete gasification of the resultant char in a second step.

Trip: A train of mine cars.

Troughing idlers: The idlers, located on the upper framework of a belt conveyor, which support the loaded belt. They are so mounted that the loaded belt forms a trough in the direction of travel, which reduces spillage and increases the carrying capacity of a belt for a given width.

Tunnel: A horizontal, or near-horizontal, underground passage, entry, or haulageway, that is open to the surface at both ends. A tunnel (as opposed to an adit) must pass completely through a hill or mountain.

Ultimate analysis: The analytical percentage by weight of coal carbon, hydrogen, nitrogen, sulfur, oxygen, and ash; the determination by prescribed methods of the ash, carbon, hydrogen, nitrogen, oxygen (by difference), and sulfur contents—quantities of each analyzed substance are reported by percentage for the following conditions: as-received, dried at 105°C (221°F), and moisture- and ash-free.

Undercut: To cut below or undermine the coal face by chipping away the coal by pick or mining machine; in some localities the terms *undermine* or *underhole* are used.

Underground mine: Also known as a *deep mine*, usually located several hundred feet below the earth's surface; an underground mine's coal is removed mechanically and transferred by shuttle car or conveyor to the surface.

Underground mining: The extraction of coal or its products from between enclosing rock strata by underground mining methods, such as room and pillar, longwall, and shortwall, or through in situ gasification.

Underground station: An enlargement of an entry, drift, or level at a shaft at which cages stop to receive and discharge cars, personnel, and material. An underground station is any location where stationary electrical equipment is installed. This includes pump rooms, compressor rooms, hoist rooms, battery-charging rooms.

Undiscovered reserves: Reserves that are yet to be discovered; the term and the associated *speculative data* are often used in reserve estimation.

Unit train: A long train of between 60 and 150 or more hopper cars, carrying only coal between a single mine and destination; a railway train designated to achieve economies of scale by transporting a single commodity (such as coal), loading fully and operating nonstop.

Universal coal cutter: A type of coal cutting machine, which is designed to make horizontal cuts in a coal face at any point between the bottom and top or to make shearing cuts at any point between the two ribs of the place. The cutter bar can be twisted to make cuts at any angle to the horizontal or vertical.

Unproved reserves: *see* Inferred reserves.

Upcast shaft: A shaft through which air leaves the mine.

Valuation: The act or process of valuing or of estimating the value or worth; appraisal.

Vein (coal): A layer of coal found in the earth; the deeper the vein, the older and better the quality of coal.

Velocity: Rate of airflow in linear feet per minute.

Ventilation: The provision of a directed flow of fresh and return air along all underground roadways, traveling roads, workings, and service parts.

Violation: The breaking of any state or federal mining law.

Virgin: Unworked; untouched; often said of areas where there has been no coal mining.

Vitrain: A macroscopic coal constituent (lithotype) that appears as brilliant black bands of uniform appearance and greater than 10^{-2} m thickness.

Vitrinite: A microscopic coal constituent (maceral) that appears translucent by transmitted light and gray in reflected light; termed anthraxylon when viewed by transmitted light.

Void: A general term for pore space or other openings in rock. In addition to pore space, the term includes vesicles, solution cavities, or any openings either primary or secondary.

Volatile matter: Hydrogen, carbon monoxide, methane, tar, other hydrocarbons, carbon dioxide, and water obtained on coal pyrolysis; those products, exclusive of moisture, given off as gas and vapor, determined by definite prescribed methods (ASTM D2361, ASTM D3761, ASTM D3175, ASTM D3176, ASTM D3178, and ASTM D3179).

Vortex flow: The whirling motion of a gas stream in a round vessel that causes separation by downward flow of solid or liquid particulates contained in the gas.

Washability analysis: A procedure used in a laboratory before preparation plant design to determine the cleaning processes to be employed and used during normal operation to evaluate the performance of the cleaning equipment and the amenability of the raw coal feed to the cleaning processes chosen.

Washout: The sudden erosion of soft soil or other support surfaces by a gush of water; if a washout occurs in a crater-like formation, it is a sinkhole.

Waste: Rock or mineral that must be removed from a mine to keep the mining scheme practical, but which has no value.

Water (carbureted blue) gas: A mixture of carbon monoxide and hydrogen formed by the action of air and then steam on hot coal or coke and enriched with hydrocarbon gases from the pyrolysis of oils.

Water gauge (standard U-tube): Instrument that measures differential pressures in inches of water.

Weathering: The action of air and water on coal in surface stockpiles, causing size reduction, oxidation, and decreases of any caking or coking properties.

Wedge: A piece of wood tapering to a thin edge and used for tightening in conventional timbering.

Weight: Fracturing and lowering of the roof strata at the face as a result of mining operations, as in *taking weight.*

White damp: Carbon monoxide, CO. A gas that may be present in the afterdamp of a gas- or coal-dust explosion, or in the gases given off by a mine fire; also one of the constituents of the gases produced by blasting. Rarely found in mines under other circumstances. It is absorbed by the hemoglobin of the blood to the exclusion of oxygen. One-tenth of 1% (.001) may be fatal in 10 min.

Width: The thickness of a lode measured at right angles to the dip.

Winning: The excavation, loading, and removal of coal or ore from the ground; winning follows development.

Winze: Secondary or tertiary vertical or near-vertical opening sunk from a point inside a mine for the purpose of connecting with a lower level or of exploring the ground for a limited depth below a level.

Wire rope: A steel wire rope used for winding in shafts and underground haulages. Wire ropes are made from medium carbon steels. Various constructions of wire rope are designated by the number of strands in the rope and the number of wires in each strand. The following are some common terms encountered: airplane strand, cable-laid rope, cane rope, elevator rope, extra-flexible hoisting rope, flat rope, flattened-strand rope, guy rope, guy strand, hand rope, haulage rope, hawser, hoisting rope, lang lay rope, lay, left lay rope, left twist, non-spinning rope, regular lay, reverse-laid rope, rheostat rope, right lay, right twist, running rope, special flexible hoisting rope, standing rope, towing hawser, and transmission rope.

Working: When a coal seam is being squeezed by pressure from roof and floor, it emits creaking noises and is said to be *working*; this often serves as a warning to the miners that additional support is needed.

Working face: Any place in a mine where material is extracted during a mining cycle.

Working place: From the outby side of the last open crosscut to the face.

Workings: The entire system of openings in a mine for the purpose of exploitation.

Working section: From the faces to the point where coal is loaded onto belts or rail cars to begin its trip to the outside.

Index

Milton Keynes UK
Ingram Content Group UK Ltd.
UKHW051927141024
449569UK00027B/1385